Magnetic Resonance Imaging of the Knee

Second Edition

Magnetic Resonance Imaging of the Knee

Second Edition

Jerrold H. Mink, M.D.
Director
Tower Musculoskeletal Imaging Center
Los Angeles, California; and
Associate Clinical Professor of Radiology
University of California
San Francisco, California

Murray A. Reicher, M.D.
President
Radiology Medical Group
San Diego, California

John V. Crues III, M.D.
Medical Director
Medical Imaging Group
Santa Barbara, California; and
Assistant Clinical Professor of Radiology
University of California
Los Angeles, California

Andrew L. Deutsch, M.D.
Associate Director
Tower Musculoskeletal Imaging Center
Los Angeles, California; and
Associate Clinical Professor of Radiology
University of California
San Diego, California

With Contributions by

Gary M. Blum, M.D.
Shirley McCarthy, M.D., PH.D.
Patrick A. Ruwe, M.D.
Frank G. Shellock, PH.D.
David W. Stoller, M.D.
Phillip F. J. Tirman, M.D.

Raven Press New York

Raven Press, Ltd., 1185 Avenue of the Americas, New York, New York 10036

Made in the United States of America

Library of Congress Cataloging-in-Publication Data

MRI of the knee / Jerrold H. Mink . . . [et al.], with contributions by
Frank G. Shellock . . . [et al].—2nd ed.
p. cm.
Rev. ed. of: Magnetic resonance imaging of the knee / Jerrold H.
Mink, Murray A. Reicher, John V. Crues III. c1987.
Includes bibliographical references and index.
ISBN (invalid) 08816709364
1. Knee—Magnetic resonance imaging. I. Mink, Jerrold H.
II. Mink, Jerrold H. Magnetic resonance imaging of the knee.
[DNLM: 1. Knee—pathology. 2. Magnetic Resonance Imaging. WE
870 M9385]
RC951.56 1992
617.5′82—dc20
DNLM/DLC
for Library of Congress 92-17023
CIP

The material contained in this volume was submitted as previously unpublished material, except in the instances in which some of the illustrative material was derived.*

Great care has been taken to maintain the accuracy of the information contained in the volume. However, neither Raven Press nor the authors can be held responsible for errors or for any consequences arising from the use of the information contained herein.

Materials appearing in this book prepared by individuals as part of their official duties as U.S. Government employees are not covered by the above-mentioned copyright.

* Chapter 10 entitled Muscle Injuries by Jerrold H. Mink is being reprinted from *MRI of the Foot and Ankle* (A. L. Deutsch, J. H. Mink, and R. Kerr, eds, Raven Press: New York, 1992; 281–311.)

9 8 7 6 5 4 3 2 1

To Barbara, Justin and Samantha for their support, love and encouragement; they are the treasures of my life.

I am also deeply indebted to three individuals who have profoundly influenced both my professional and personal life: John Plotke, M.D., my true and dearest friend of 35 (and counting) years; Marshall Bein, M.D., my co-resident and now partner; Leo G. Rigler, M.D., a giant among men and physicians, who remains an inspiration and friend.

J.H.M.

To my family and friends

M.A.R.

To Jane, Drew, Ryan, and Melissa

J.V.C.

To Jeanne, Peter, Phillip, and Dana

A.L.D.

Contents

Contributing Authors

Gary M. Blum, M.D. *Resident in Diagnostic Radiology, Department of Radiology, Santa Barbara Cottage Hospital, P.O. Box 689, Santa Barbara, California 93102*

John V. Crues III, M.D. *Medical Director, Medical Imaging Group, 401 Pueblo Street, Santa Barbara, California 93105; Assistant Clinical Professor of Radiology, University of California, Los Angeles, California 90024-1721*

Andrew L. Deutsch, M.D. *Associate Director, Tower Musculoskeletal Imaging Center, 444 S. San Vicente Boulevard, Los Angeles, California 90048; Associate Clinical Professor of Radiology, University of California, San Diego, California 90033*

Shirley McCarthy, M.D., Ph.D. *Associate Professor of Radiology; Director, MRI, Yale University School of Medicine, 333 Cedar Street, New Haven, Connecticut 06510*

Jerrold H. Mink, M.D. *Director, Tower Musculoskeletal Imaging Center, 444 S. San Vicente Boulevard, Los Angeles, California 90048; Associate Clinical Professor of Radiology, University of California, San Francisco, California 94143*

Murray A. Reicher, M.D. *President, Radiology Medical Group, 3366 5th Avenue, San Diego, California 92103*

Patrick A. Ruwe, M.D. *Instructor, Department of Orthopaedics and Rehabilitation, Yale University School of Medicine, 333 Cedar Street, New Haven, Connecticut 06510*

Frank G. Shellock, Ph.D., F.A.C.S.M., F.A.C.C. *Tower Imaging, Director of Research, Development, and Advanced Applications; Assistant Professor of Radiological Sciences, UCLA School of Medicine, 444 S. San Vicente Boulevard, Los Angeles, California 90048*

David W. Stoller, M.D. *Medical Director, California Advanced Imaging, 3440 California Street, San Francisco, California 94118; Assistant Clinical Professor of Radiology, University of California, San Francisco, California 94143*

Phillip F. J. Tirman, M.D. *Fellow, Magnetic Resonance Imaging, Department of Radiology, Cottage Community MRI Center, 401 West Pueblo Street, Santa Barbara, California 93105*

Preface

In June 1984, I was reviewing radiographs with one of the UCLA radiology residents, Dr. Murray A. Reicher. Knowing that I had a particular interest in arthrography, he told me that he had performed magnetic resonance (MR) studies on the knees of volunteers and that he could consistently image the menisci using a 0.3-tesla system. He went so far as to say that he believed that he could reliably detect meniscal *tears*. I chuckled, hesitantly, for several reasons: First, although I knew virtually nothing about this new imaging modality, I wondered how far it could have come given that researchers in the field could not even agree on its name (MR?, NMR?, MRI?); second, I was accustomed to spending 30% of my work week performing arthrograms. If Murray was correct, what would I do with my free time?

Not inclined to take chances, I restudied multiplanar anatomy of the knee and reviewed Murray's pathologic and clinical material. Dr. John Crues and I began to duplicate and expand on Dr. Reicher's early success using our own 1.5-tesla unit.

We are now a mere two years since that day at UCLA, and Drs. Reicher, Crues, and myself, having reviewed over 2,300 MRI studies of the knee, find ourselves coauthoring a book on MRI of the knee. Old habits are broken slowly and reluctantly, but we have become convinced that for nearly every indication, MRI of the knee should replace conventional arthrography. This book is therefore directed toward radiologists who are involved in MRI and orthopedic radiology as well as the orthopedic surgeon with an interest in the knee. In our experience, MRI is as accurate as arthrography for evaluation of the menisci and more so for the extrasynovial cruciate and collateral ligaments and extensor mechanism. The fact that it is noninvasive (without the risks of arthrography or arthroscopy) and painless is important to both the arthrographer (especially in the acute situation) and to the patient (in *every* situation). A fundamental advantage of MRI is that it can, in many cases, obviate the need for multiple examinations in evaluating knee pain of uncertain etiology. MRI, for example, can detect: occult fractures without the need for both nuclear scans and tomograms or computed tomography; spontaneous osteonecrosis and the frequently associated meniscal tear without the need for both bone scans and arthrograms; marrow disorders without both sulfur colloid scans and computed tomography. Growing evidence indicates that when MRI is used to evaluate the knee preoperatively, it should decrease examination under anesthesia, reduce unnecessary surgery, aide preoperative planning and diagnosis, and improve surgical yield by reducing arthroscopic oversights. (*Taken from the preface to the first edition.*)

Although the first edition is now 5 years old, it is still perfectly adequate to be included in the second edition of this book with only two minor additions: (a) I have found something to do with the other 30% of my time; (b) the authors have probably reviewed over 20,000 knee MR examinations by now.

The second edition is based on still valid concepts developed in 1987 but every chapter has been markedly expanded, and not a single paragraph is left unchanged. The first edition was 168 pages; this volume is 496 pages. The chapter on "Physical Principles" has been rewritten to accommodate the principles of gradient-echo, inversion recovery, fast spin-echo, and fat suppression techniques and protocols, all of which are in widespread clinical use. Chapter 2 on "Bioeffects and Safety" expands on the ever growing list of those extraneous and implanted materials judged compatible/incompatible with a safe MR examination. Dr. Shellock has refined the position of the Safety Committee for the Society of Magnetic Resonance Imaging on the pressing issue of safety of MR in pregnancy.

The "core chapters" on meniscal injuries and cruciate and collateral ligament injuries have both been expanded to include concepts of meniscal healing and repair, anterior cruciate ligament-associated fractures, partial ligament tears, and longitudinal follow-up of the cruciate and collateral ligament injuries. A wealth of scientific studies validating the accuracy of MR of the knee is now readily available, whereas rather little scientific data was present at the time of the first edition. The extensor mechanism was previously discussed in a chapter on "Ligaments of the Knee"; it now occupies a separate chapter which is dominated by discussion of the biology and anatomy of articular cartilage as it relates to chondromalacia of the patella. The latter topic was addressed in but a single paragraph in the first edition of this book.

The ability of radiologists to identify meniscal and ligamentous injury, improvements in equipment design, and advances in surgical skills and techniques have made arthroscopic meniscectomy a rather routine surgical procedure. Highly specialized surgeons also perform meniscal repair and anterior cruciate ligament reconstruction through an arthroscope. It is essential that the radiologist who reads postoperative knee MR examinations be familiar with the normal appearance of partially resected menisci, the concepts of postmeniscectomy articular damage, anterior cruciate ligament graft impingement, and graft isometry. These ideas are addressed in Chapter 4.

In 1987, it became apparent that radiographically occult fractures of bone, including bone bruises, stress responses, and osteochondritis dissecans, could be diagnosed through MR with equal sensitivity and greater specificity than through radionuclide bone scans. We now appreciate the magnitude (such bone injuries are found on as many as 30% of MR examinations of the knee) and the significance of such "occult" bone injuries. The concept of medullary bone injury, its effect on articular cartilage, and its possible long-term relationship to degenerative joint disease are extensively explored in Chapter 8.

While muscle strains are the most common injury to the lower extremity, we avoided publishing even a single word about such trauma in the first edition of this work. MR now has found utility in diagnosis and assessment of other muscle disorders, such as inflammatory myopathies, muscle tumors, and muscular dystrophies. Certainly, spectroscopy will occupy an even greater portion of the third edition of this text.

Clinicians and radiologists agree that MR has become an essential tool in the armamentarium of the physician treating knee disorders. In today's world of changing reimbursements, it is essential that the value of MR be documented and quantified. The final chapter, Cost-Effectiveness of MRI, written by Ruwe and McCarthy at Yale University, must certainly convince skeptics, even those at Medicare, of the real dollar value of MR studies of the knee.

To paraphrase from the first edition, "we do not intend for this book to be a picture book of interesting MR cases that we have seen, but rather we have written a reference book of knee disorders with emphasis on diagnosis." It is no longer sufficient for the radiologist to simply report findings; one must be aware of the implications of those findings for the surgeon, and the prognostic value of the findings and their associated disabilities. Radiologists can no longer be "imagers"—we must return to being complete doctors, both anatomists and physiologists. We hope that this book helps bring the reader closer to that ideal.

Jerrold H. Mink, M.D.

Acknowledgments

People rarely read this page, but they should. Much of the real work is done by the people below, but somehow the editors get all the credit (and royalties!).

Our technical staff is an incredibly skilled, courteous, responsive, and dedicated group of people, without whom we would have little *quality* case material. We are deeply indebted to Joe Fraire, Mark Tournie, Kim Fernandez, Gina Slimp, Lisa Edwards, Brenda Tompkins, Kim Thibodeaux, Ernest Fontelera, Claire Alonzo, Art Aguilar, Andy Andrade, and Angel Soriano.

The high quality of the reproductions can be credited to the skill and devotion of Francisco Chanes; he is never satisfied with mediocrity.

Julie Sassoon has helped to keep our spirits high and our madness to a minimum during those tough times in the production of this book. Her organizational skills, dedication, and wonderful temperament have kept this ship on keel.

My colleagues at Tower Imaging in Los Angeles have been wonderfully encouraging; their support for this project was invaluable.

General Electric Medical Systems, at both the corporate and local levels provided advanced technical support as early as 1985; they continue to help push forward the frontier of MR applications to the musculoskeletal system.

CHAPTER 1

Technical Considerations

John V. Crues III

Magnetic resonance imaging (MRI) represents one more milestone in the prodigious growth in application of magnetic resonance (MR) since the first successful nuclear magnetic resonance (NMR) experiments were performed in 1946 by Purcell, Pound, and Torrey at Harvard (1) and by Block, Hansen, and Packard at Stanford (2). Purcell and Block received the Nobel Prize in physics in 1952 for their landmark experiments. Throughout the 1950s and 1960s, MR became a key component in the armamentarium of chemists and physicists for evaluation of molecular structures. However, even though Gabillard noted the ability to obtain spatial localization in NMR in 1951 (3), it was not until the 1970s that medical applications of MR were seriously considered. In 1971, Damadian et al. published the first article on medical applications of NMR, and in 1972 he applied for the first patent for the field-focusing nuclear magnetic resonance (FONAR) technique for MRI (4). The following year, Lauterbur published an article describing MRI using zeugmatography (5). Other MRI techniques rapidly followed (6–11). Early clinical trials with prototype imaging devices began in the early 1980s.

OVERVIEW OF NMR PHYSICS

A complete discussion of the physics of MRI is beyond the scope of this text. Fortunately for the interested reader, many excellent articles and books have been written on NMR physics and its medical applications (9,12–28). The present discussion will be limited to a minimum of basic concepts that we believe to be most relevant for the MRI medical practitioner. The principles behind resonance and spatial localization necessary for image acquisition and for understanding tissue contrast will be presented. Two-dimensional Fourier transform acquisition will be stressed, including frequency encoding, phase encoding, and the concept of K-space. The importance of MRI spatial resolution and contrast resolution, as well as the signal-to-noise ratio, will be addressed. Surface coils will be discussed in general terms, and their importance for MRI of the knee will be emphasized. The chapter will end with a brief discussion of MRI protocols for knee imaging.

PHYSICAL CONCEPTS

Much like the earth, though on a much smaller scale, a proton consists of a ball of electrically charged matter that spins about its own axis, generating a magnetic field. Like a compass placed into a magnetic field, it tends to align itself in the direction of a magnetic field. Unlike a compass, however, the proton is constrained by the laws of quantum mechanics to point either roughly in the direction of the magnetic field, the parallel state, or in the opposite direction to the magnetic field, the anti-parallel state. A description of the exact motion of the proton is beyond the scope of this chapter and is not required for a practical understanding of MRI (9,12,20,29).

The overall picture of the interactions of the countless protons within the body and the magnetic field can be markedly simplified by using the bulk magnetization vector **M** which is the vector sum of the magnetic moments of all of the individual protons in the imaging volume (2,9,25). This quantity can be easily measured and is the essential parameter that generates the NMR signal, used in MRI. When the body is initially placed within the magnetic field, the body's **M** is stationary

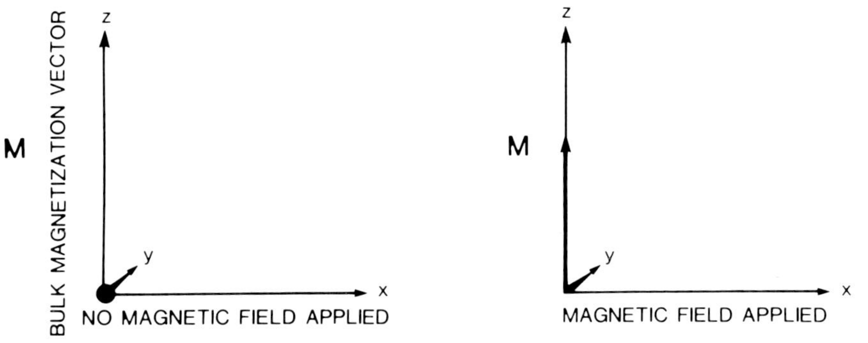

FIG. 1.1. The bulk magnetization vector. With no external magnetic field, the bulk magnetization vector, **M**, is zero. In the presence of an applied magnetic field, **M** is nonzero and points in the direction of the applied field, the *z* direction.

and points in the direction of the main magnetic field (Fig. 1.1). This direction, by convention, is called the *z* direction. By use of radiofrequency energy at a specific frequency given by the Larmor equation:

$$F = \gamma \text{ Bo}, \qquad \text{[Eq. 1.1]}$$

where F is frequency, γ is the gyromagnetic ratio (42 megahertz, MHz per tesla), and Bo is the strength of the main magnetic field, **M** can be rotated away from the *z* axis. If the proper magnitude and length of the radiofrequency pulse are used, then the bulk magnetization vector can be rotated 90° into the *x-y* plane (Fig. 1.2). In this position, the *z* component of **M** is now zero, and the *x-y* component is maximum. Once in the *x-y* plane, the bulk magnetization vector rotates around the *z* axis with the same frequency as given by the Larmor equation. This is the resonant state referred to in the term NMR. This is also referred to as being "in phase," which in physics vernacular means that most of the countless precessing nuclei at any given time are pointing in the same direction in the *x-y* plane (13,25).

A moment's digression to describe electricity generation at Niagara Falls is instructive (Fig. 1.3). An understanding of the production of electric current within wires is essential to an understanding of MRI. At Niagara Falls, water flows over the falls into a turbine, causing the turbine to spin. The turbine is connected via mechanical gears to a magnet that is rotated by the rotating turbines. Current is induced within wires placed in the vicinity of the magnet by the rotating magnetic field. The wires are connected through transformers (which vary the voltage) to household electric outlets.

What does this have to do with MRI? Once the bulk magnetization is rotated into the *x-y* plane, it spins in space with a frequency given by the Larmor equation. Therefore, as with the magnet connected to the turbine at Niagara Falls, if wires are placed in the vicinity of the patient, electric current will be induced in those wires. The induced current is then fed to a computer for analysis. Fortunately, though the nuclei are spinning at 64 MHz, the patient is not!

So far we have described the 90° pulse; however, pulses at any angle can be and are applied to the nuclei to produce almost an endless variety of images and image contrast. These pulse sequences have been named by some of the most imaginative physicists of the twentieth century and include techniques such as GRASS (gradient recalled acquisition in the steady state), FLASH (fast, low angle shot) (18,30–32), STIR (short tau inversion recovery) (33,34), MAVIN (multiple-angle, variable-interval, nonorthogonal) (35), STEAM (stimulated-echo methods) (31,36), CHESS (chemical-shift selective) (37), inversion recovery, RARE (38), and spin echo. Most of the images in this

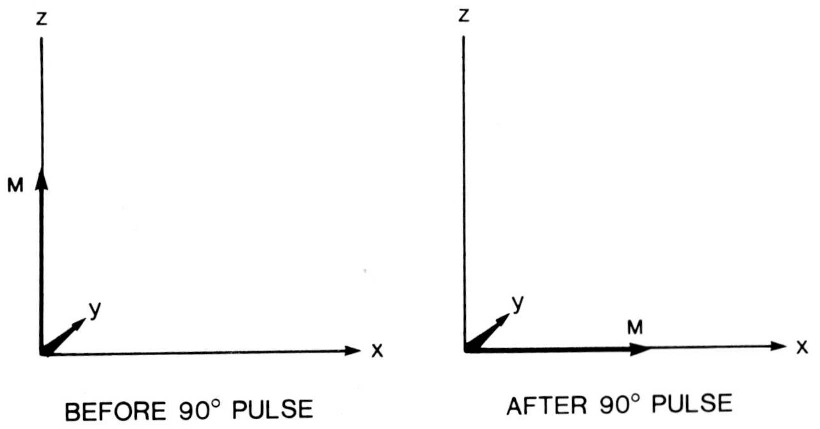

FIG. 1.2. RF pulses. A 90° RF pulse rotates the bulk magnetization vector from pointing in the *z* direction to pointing in the *x-y* plane.

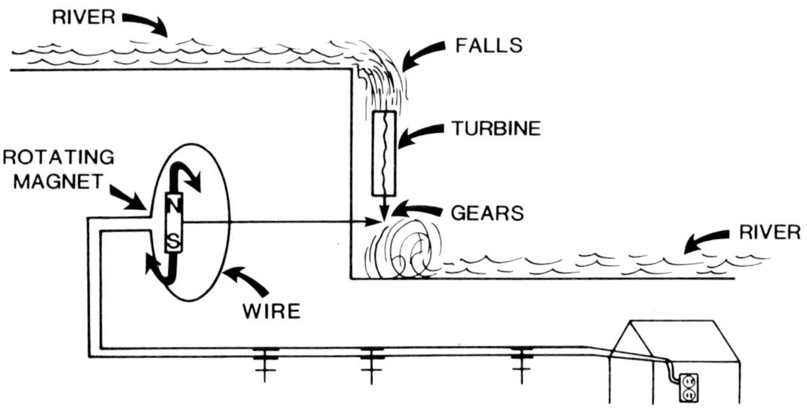

FIG. 1.3. Electricity generation. Current is induced in wires by a rotating magnetic field. In MRI the rotating magnetic field is the bulk magnetization vector during resonance.

book were produced by the Carr-Purcell spin-echo technique (39); however, the STIR and gradient-echo techniques are also important in musculoskeletal imaging.

The spin-echo technique begins with a 90° pulse and ignores the signal that is immediately created, called the free induction decay (FID) (Fig. 1.4). After a time delay, called tau (τ), a 180° pulse is applied, and at a time τ after the 180° pulse, a signal is generated called the spin echo (12,13,16,19,29,39–41). It is important to note, however, that the initial signal of the FID decays because the individual protons in different regions of a voxel precess at slightly different frequencies because of local magnetic field inhomogeneities. Thus, the nuclei rapidly spin out of phase with one another. This causes the decay in the strength of the bulk magnetization vector, **M**. The 180° radiofrequency (RF) pulse rephases that component of the dephasing of the nuclei due to static, time-independent magnetic field inhomogeneities. The time between the 90° pulse and the generation of the spin echo is called TE, the echo delay time. This is an important parameter determining image contrast in musculoskeletal imaging and is determined by the radiologist before scanning a patient. The signal from the echo is fed into the computer and is used to generate the image.

The standard spin-echo technique can be modified in many ways to enhance tissue contrast. Its acquisition time can be markedly shortened by using the "fast spin-echo technique" (see "Spatial Localization" below). One technique frequently used in musculoskeletal imaging is the inversion recovery technique. Inversion recovery pulses begin with a 180° inversion pulse (Fig. 1.5) which is followed after a delay of inversion time (TI) by a standard 90°–180° spin-echo sequence (33,34). The effect of this technique on tissue contrast is discussed below with primary reference to the STIR (short TI inversion recovery) technique under relaxation mechanisms.

Gradient-echo pulse sequences encompass a large family of pulsing techniques which may display a large spectrum of contrast differences. These sequences have in common replacement of the 180° refocusing RF pulse with a rapid reversal of a magnetic field gradient (Fig. 1.6). The sequence begins by a RF pulse which is typically less than the 90° pulse which initiates spin-echo imaging. The amount the initial pulse rotates the bulk magnetization vector in the direction of the *x-y* plane in called the flip angle. Varying the flip angle in gradient echoes is important for two reasons. Small flip angles less than 90° do not completely saturate the protons. If a second pulse sequence is begun in a TR less than the T1 of the protons, significant signal cannot be induced if the flip angle is close to 90° because sufficient time will not have relapsed to allow **M** to regain magnetization along the *z* direction. When small

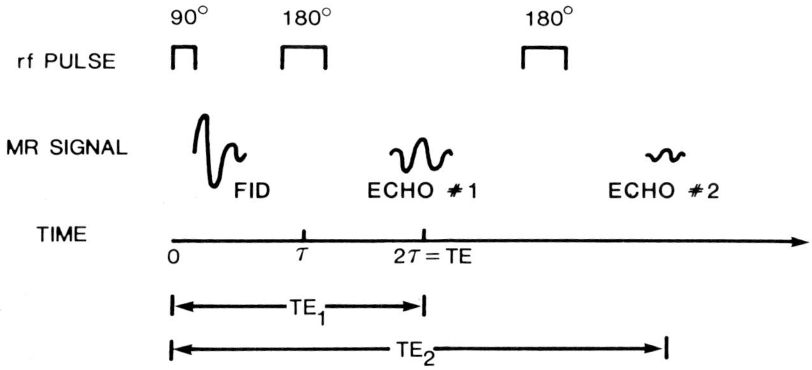

FIG. 1.4. The spin echo. The most frequently used MRI technique involves the Carr-Purcell spin echo. A 90° pulse is followed after a time delay of τ ms by a 180° pulse which causes rephasing of the nuclei leading to a spin echo. For technical reasons the free-induction decay (FID) is ignored and the signal from the echo is used in calculating the image. The timing parameters τ and TE are shown and discussed in the text.

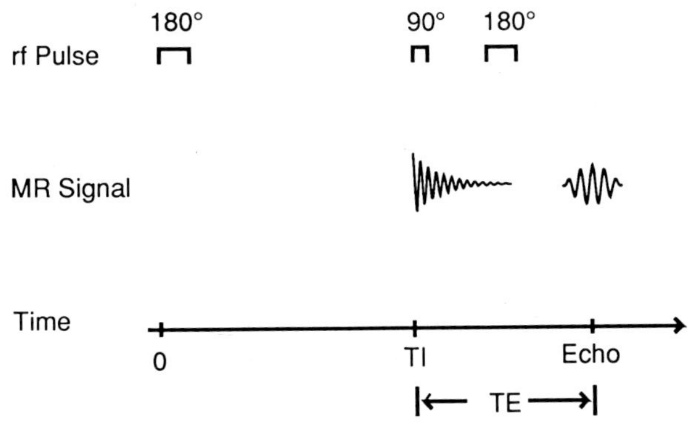

FIG. 1.5. The Inversion Recovery technique. The IR sequence begins with a 180° pulse which inverts **M**. At a time TI (inversion time) later a standard spin echo 90°–180° sequence is performed. The strength of the resultant signal is proportional to the magnitude of **M** at the time of the 90° pulse.

flip angles are used, **M** is not rotated completely into the *x-y* plane and sufficient magnetization along *z* is maintained to allow use of short TRs, which may translate into shorter imaging times (*vide infra*). Secondly, varying the flip angle changes the inherent contrast in the resultant images. As a rough approximation, the smaller the flip angle, the more T2-like is the image contrast.

The 180° refocusing pulse can be replaced by a gradient reversal because the initial gradient causes dephasing of the nuclei along the direction of the gradient by placing nuclei along the gradient in different magnetic fields. Nuclei in different magnetic fields resonate at different frequencies, thus dephasing the nuclei. Phase-encoding gradients are then placed on the object for spatial localization (*vide infra*). The dephasing is reversed by reversing the magnetic field gradient, which brings the nuclei back into phase producing an echo. This process is usually performed rapidly with short TEs because static magnetic field inhomogeneities within tissues (magnetic susceptibility effects) can degrade gradient-echo images, and these artifacts are minimized with short TE acquisitions. The dephasing due to time-independent magnetic field variations within the tissues are better compensated for using 180° RF pulses (*vide infra*). The signal used to generate the image using gradient-echo pulse sequences is primarily from the initial free-induction-decay, not from an RF-induced spin echo (Fig. 1.6). The decay rate of the free-induction-decay is called T2* and is primarily dependent on fixed (non-time dependent) magnetic field inhomogeneities within the tissues as are produced by imperfection in the main magnet and fixed magnetic field gradients within tissues due to differing magnetic properties of adjacent body tissues (the magnetic susceptibility effects). Thus, gradient-echo images are often called T2*-weighted images (see contrast below).

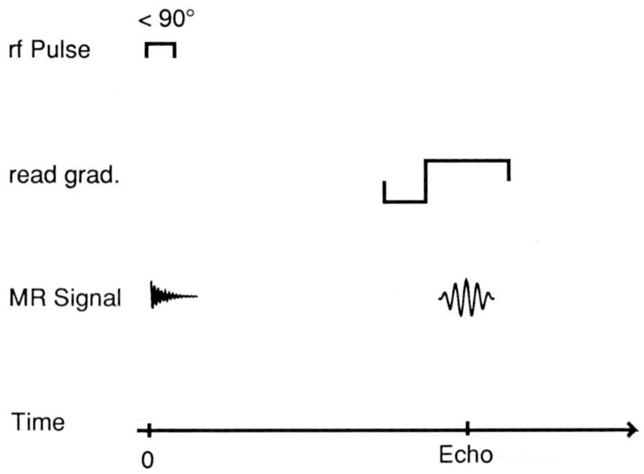

FIG. 1.6. Gradient-echo techniques. Gradient-echo (GRE) techniques are a family of echo techniques which usually begin the sequence with a RF pulse of less than 90° and generate an echo by reversing a magnetic field gradient, rather than using a 180° RF refocusing pulse.

Regular spin echo (90°–180°), inversion recovery (180°–90°–180°), and typical gradient echo (<90°-gradient reversal) are three frequently used pulse sequences from an almost endless list of possible techniques. Clearly, the future will continue to unveil newer techniques and enhancement of present techniques which will further improve musculoskeletal imaging in the years ahead. The burden of continued research and development of new techniques and proof of their clinical utility will be a challenge to students of musculoskeletal imaging which will be richly rewarded by advances in diagnostic sensitivity and accuracy.

SPATIAL LOCALIZATION

Generating an NMR signal is only the beginning of the process of creating a diagnostic study. In order to generate an image, the brightness of each point in the displayed image (called a pixel) must be proportional to the intensity of the MR signal emanating from the corresponding small volume of tissue (called a voxel). A mechanism to spatially localize the origin of the signal must be used. Many techniques have been attempted, but the technique that is most frequently used in medical MRI is the two-dimensional Fourier-transform technique (7,8). This technique uses small linear magnetic field gradients along the x' and y' directions to determine spatial localization in the plane of the image. (In this chapter, z refers to direction of the main magnetic field, and x and y refer to directions orthogonal to z; z' refers to the direction perpendicular to the image slice plane of a two-dimensional Fourier-transform image and is the same as z only for an axial image;

x' and y' are orthogonal directions to z'.) One direction, x', uses frequency encoding for spatial localization. If a small linear magnetic field in the z direction that varies in magnitude along the z' axis is superimposed on the large uniform static magnetic field in the z direction, then nuclei along the x' axis will vary their frequencies of precession directly proportional to the magnetic field gradient according to Eq. 1.1. Thus, nuclei on one side of the body will precess more rapidly than nuclei on the opposite side. The frequencies of the MR signal can therefore be used to localize its site of origin.

Fullerton masterfully illustrated this technique using the analogy of strings in a harp (16). Consider a harpist in one room and a trained musician in an adjoining room. If the harpist plucks a string, the musician can determine the distance from the harpist to the plucked string by listening to the pitch of the note played. If the pitch is low, then the string is far from the harpist; if high the string is near to the harpist. Therefore, the frequency of the note played encodes the location of the string. The training and talent of the musician allows her to decode the frequency information into the desired position. Similarly, the frequency of the emitted RF from nuclei in a magnetic field gradient encodes spatial location within the body which can be decoded by a computer using the mathematics of Fourier transforms. It is the mathematical technique of Fourier transforms which gives the computer "perfect pitch" to provide accurate spatial localization. The interested reader is encouraged to read Fullerton's original article for a more detailed discussion of the analogy between frequency encoding and a wonderful musical instrument (16).

Unfortunately, frequency encoding can only be used in one direction. Spatial encoding in the y' direction requires a different technique called phase encoding. Though simple for the mathematical genius, a detailed description of this technique is beyond the scope of this chapter (13,20). However, the application of the phase-encoding technique to a simplified example is helpful in developing a qualitative understanding of the technique and in explaining important limitations and artifacts of this process (15).

Consider imaging a square beaker of water (Fig. 1.7). As described below, slice location in one direction, z', can be determined by a slice selection gradient at the time of the initial 90° pulse. Spatial localization in another direction, the x', can be determined by frequency encoding. The final dimension, y', requires phase encoding. For the beaker of water the correct plot of signal intensity versus distance along the y' direction is depicted in Fig. 1.8. It is 0 outside of the beaker and a constant value within the beaker because the water gives a uniform signal. According to Fourier's theorem, any arbitrary, well-behaved mathematical function can be exactly reproduced by an infinite series of sine and cosine functions:

$$F(y') = \ldots + K(n)\sin(n\,t) + \ldots + K(m)\cos(m\,t) + . \qquad \text{[Eq. 1.2]}$$

For the beaker of water this series is calculated as the following:

$$SI(y') = 1\sin(t) + 1/3\sin(3\,t) + 1/5\sin(5\,t) + \ldots . \qquad \text{[Eq. 1.3]}$$

This series is schematically diagrammed in Fig. 1.9. Notice that the first sine term (Fig. 1.9A) is a very rough approximation of the desired signal intensity function without sharp margins. Higher order (frequency) sine functions [i.e., the sin(3 t) term] by themselves do not reproduce the desired signal intensity well (Fig. 1.9B). However, when higher ordered terms are added to the lower order terms, a much closer approximation to the desired function is obtained (Fig.

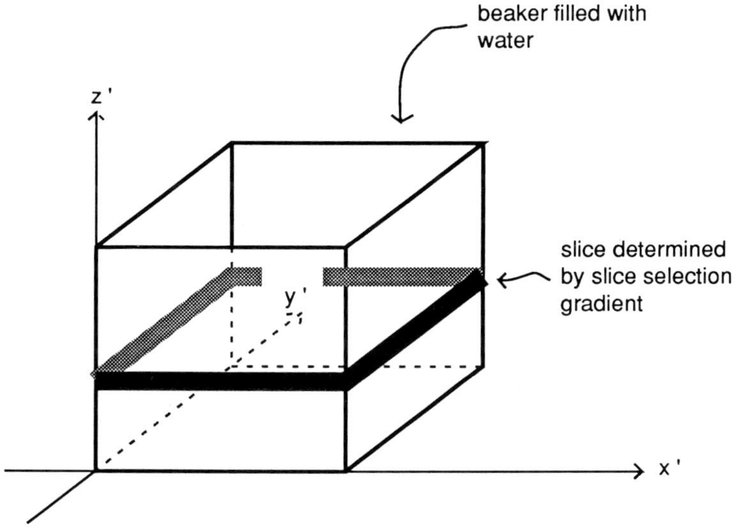

FIG. 1.7. Square beaker of water. Phase encoding as used to generate spatial information in two-dimensional Fourier-transform (2-DFT) MRI may not be intuitively understood initially. A simple square beaker of water may be helpful in conceptualizing 2-DFT imaging. The beaker is illustrated with the frequency encoding direction along the x' axis. The phase encoding direction is along the y' axis, and the plane of imaging is perpendicular to the z' axis.

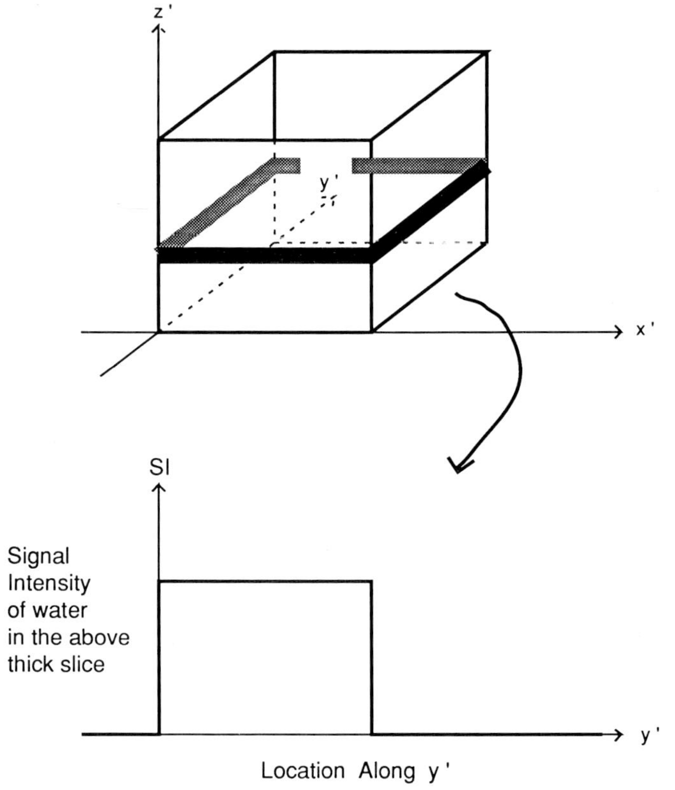

FIG. 1.8. Plot of signal intensity vs. distance along the phase encoding axis for a square beaker of water. An image of the water in a square beaker should show a 0 signal outside of the beaker and a constant high signal within the region in the beaker containing the water. This is plotted above.

1.9C–F). The lower order (lower frequency) terms are necessary for accurate depiction of large anatomical structures. Adding additional higher order terms significantly improves the image, especially edge sharpness which is necessary for high resolution images (Fig. 9D–F).

Fourier series are well and good, but how are the critical coefficients, K, obtained in clinical MR imaging? In a two-dimensional image, a two-dimensional array of K values must be measured in order to completely calculate the image (Fig. 1.10). Each line of these coefficients with a specified Ky' location must be measured by separate pulse sequences. Remember that with each phase encoded pulse, multiple data points along x' are obtained simultaneously by frequency encoding. For instance, when no phase-encoded gradient along the y' direction is applied, the intensity of returned signal from the strip of nuclei in the y' direction located at a specified x' location (located by frequency encoding) is proportional to the Ky''(0) coefficient. The returned signal when a small phase encoded gradient is applied is proportional to K(1). Higher order coefficients are proportional to returned signal when stronger gradients are applied. In clinical practice, phase-encoding gradients often begin negatively, increase to zero, and finish with positive values. A plot of K-space values for our square beaker of water is given in Fig. 1.11.

Why is K-space so important? Though we are interested in seeing an image of the inside of the human body, the MR scanner actually measures numbers in K-space. The Fourier transform (i.e., multiplying values in K-space by their appropriate sine or cosine function, then adding them together), which is easily performed by the computer, is the interpreter between the MR scanner and the radiologist. K-space is the foreign language of the scanner and the actual image of the body part is the language of the radiologist. MR designers, engineers, and radiologists involved in advance applications in MRI need to understand the machine's language to be maximally effective, just as a person who moves to a new country using a different language needs to learn the language of the new country to be most successful. It is of note that values near 0 in K-space are most often important for accurate depiction of large areas of low contrast, whereas values away from 0 in K-space are important for sharp edge display and high spatial resolution.

Remember, only one phase-encoded gradient can be applied during any one TR interval when standard imaging techniques are used. Consequently, imaging time is directly proportional to the number of phase encodings used to calculate the image. Since the more phase encodings used, the more higher order terms are included in the calculation, higher spatial resolution requires longer imaging times. Imaging time is given by the following formula:

$$\text{time} = \{(\text{NEX}) \times (\#\ \text{phase encodings}) \times (\text{TR interval})\}/\text{NP}, \qquad \text{[Eq. 1.4]}$$

where NEX (number of excitations) is the number of times a given phase encoding is repeated to improve signal-to-noise, # is the number of different phase encodings used to calculate the image, TR is the time between 90° (or smaller) pulses, and NP is the number of phase encodings per pulse (usually 1). Thus, a 256 matrix in the y' phase-encoded direction requires twice as much imaging time as a 128 matrix. An entire multislice, multiecho spin-echo imaging sequence is illustrated in Fig. 1.12.

Although a simple square beaker with water in it was used in the above example, Fourier's theorem is applicable to any well-behaved mathematical function. This means that no matter how complicated signal intensity may be in the y' direction, phase encoding can be used to evaluate it.

Next to image quality, the time required to acquire

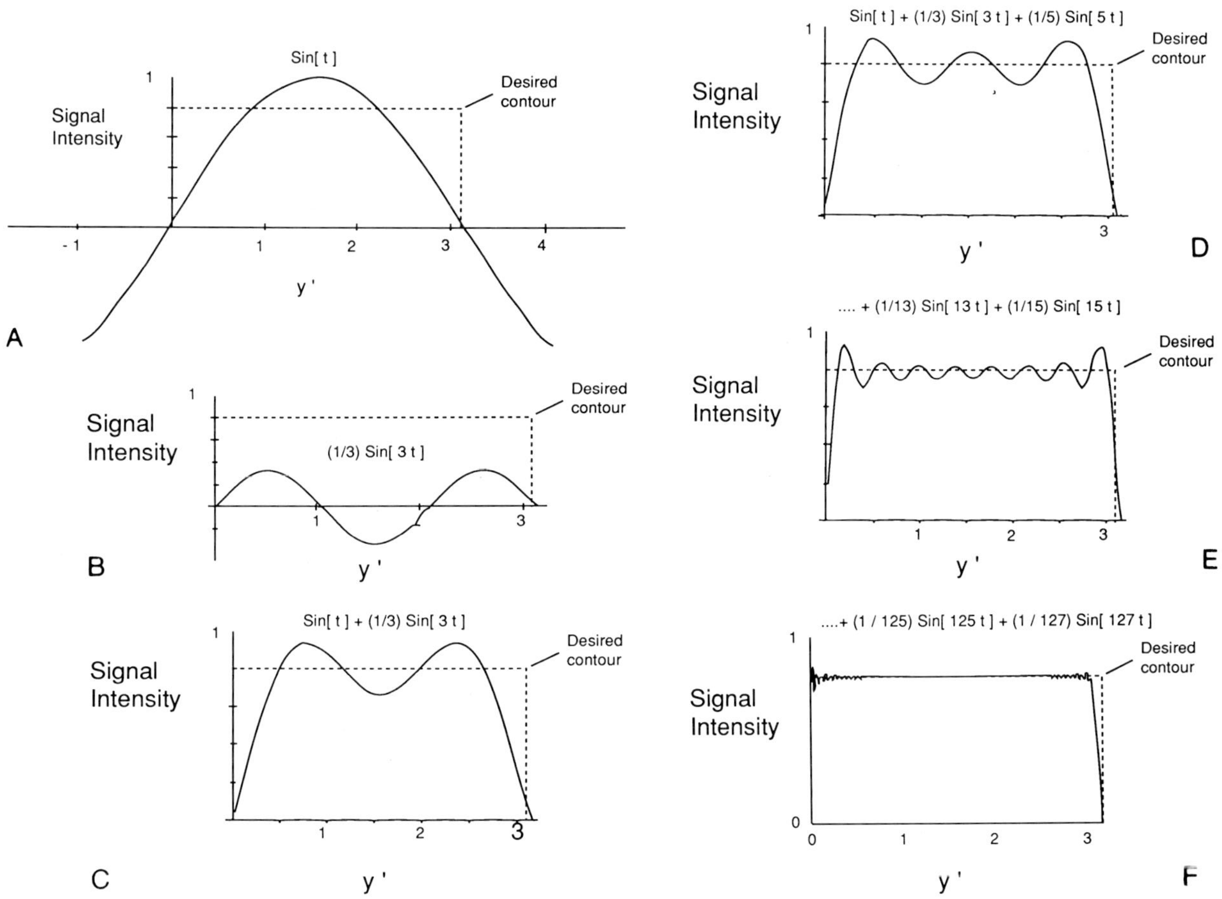

FIG. 1.9. Plot of terms in the Fourier transform of a square beaker of water. **A**: A graph of the first sine term in the Fourier transform depicts a low frequency component of the signal intensity which is important in displaying large features of the imaged object. **B**: The second term displayed by itself is a higher frequency component of the signal intensity, but does not by itself display a close resemblance to the signal intensity of the beaker of water. **C**: When the higher and lower frequency terms are added together, the resultant graph much more closely resembles the desired signal intensity. **D–F**: As more terms in the Fourier series are added, the resultant signal distribution more closely approaches that shown in Fig. 8. Notice, however, that as more terms are added, the "wiggles" become higher in frequency and cluster near the edges of the signal distribution. This effect produces the truncation artifact often noted on MR scans near the boundaries between bright and dark objects.

an image significantly impacts the clinical use of MRI. Consequently, many investigators have attempted to use the information in Eq. 1.4 to modify the length of acquisition time. According to Eq. 1.4, the length of time can be shortened by decreasing several factors. The number of excitations (NEX) is the number of times the data is repeated and averaged together. Decreasing NEX directly reduces the signal-to-noise of the image. Usually the smallest NEX is chosen which routinely will give diagnostic quality images for the body part, coil, and field strength used. The number of phase encodings performed for a given acquisition FOV (field of view) determines the spatial resolution in the phase-encoding direction. Though decreasing the number of phase encodings decreases imaging time, one pays for the decreased time with poorer spatial resolution and increased truncation artifacts (see artifacts below). The TR is often shortened to decrease imaging time; however, for spin-echo imaging, long TRs are necessary for optimal signal-to-noise and, more importantly, image contrast (*vide infra*). Long TR intervals are less important for image contrast using gradient-echo techniques, thus, these techniques usually use short TR intervals with low flip angle initial pulses to shorten imaging time (*vida supra*) (18).

According to Eq. 1.4, another way to shorten imaging times is to increase NP (the number of phase encodings per pulse). Mansfield's echo-planar techniques have performed subsecond imaging times with good image contrast using this technique (42–46).

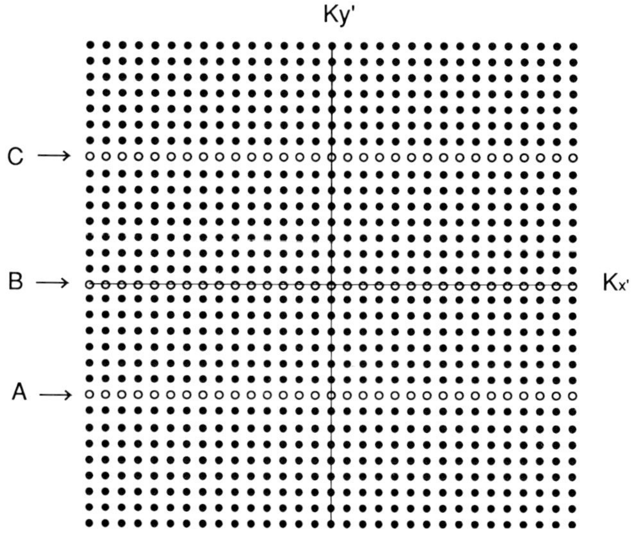

FIG. 1.10. K-space. Discreet values in K-space can be depicted as a two-dimensional plot. In standard MR imaging a line of values with a given K*y'* value is obtained simultaneously with each phase encoding pulse. The values with different K*x'* locations are obtained by frequency encoding. The line labeled *A* would be data obtained with a negative phase encoding gradient applied. Lines *B* and *C* would be data obtained with 0 and positive phase encoding gradients.

These techniques allow multiple rapidly acquired phase-encoded echoes to be obtained from a single excitation pulse. "Fast spin echo" is a similar technique which allows rapid spin-echo T2-weighted images in 1/16 or 1/32 of the usual time by acquiring multiple phase encodings per 90° pulse. With this technique, instead of acquiring one echo at a given TE per 90° pulse, multiple closely spaced 180° pulses are applied and with separated phase-encoding gradients. These induce multiple echoes with different phase encodings centered around a given TE. Thus, with only one 90° pulse, multiple phase encodings can be acquired. Using this technique, NPs of up to 32 can be acquired with spin-echo–type image contrast. This translates into shortening image acquisitions times up to a factor of 32. We have factored this technique into virtually all of our imaging protocols.

The major drawback of this technique in our experience is that, since the TEs are a spectrum and not a single value, there is a mild difference in contrast when compared to standard spin-echo imaging with a single TE. Because of this we still use standard spin-echo imaging in the sagittal plane for more consistent visualization of changes in the anterior cruciate ligament on the T2-weighted images.

We have yet to describe how the slice location is selected. Notice in Fig. 1.12 that a gradient is located along the *z'* direction during the initial RF pulse. If a narrow spectrum of RFs is used in the pulse, then only a narrow band of nuclei at just the right magnetic field strength (Fig. 1.13) will be resonated. All other nuclei will not see the initial pulse and therefore will not be involved in the full spin-echo. The thickness of the slice is determined by the width of the spectrum of the initial RF pulse frequencies as well as by the slope of the gradient. The slice location is determined by the relationship between the exact Larmor frequency of nuclei along the *z'* axis and the center frequency of the initial pulse spectrum. Thus, by decreasing the fre-

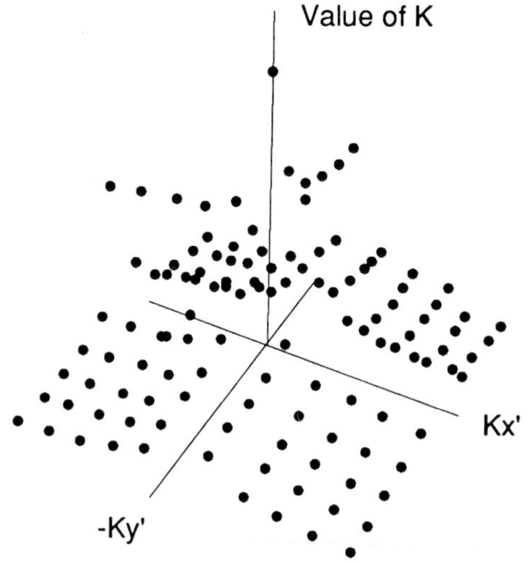

A

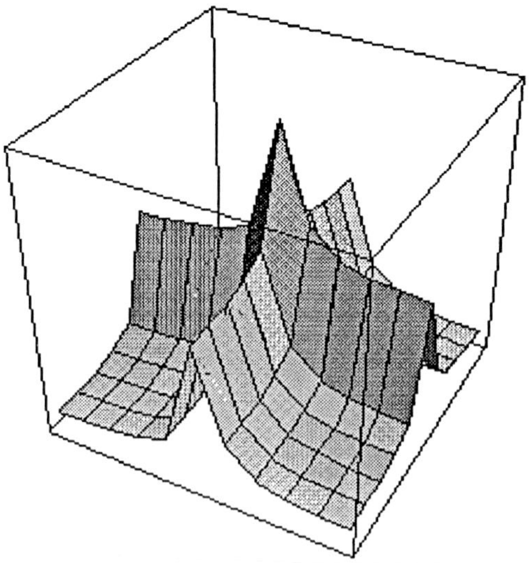

B

FIG. 1.11. K-space for a square beaker of water. **A**: K-space values for a square beaker of water would be maximum at 0 and drop off as 1 divided by the distance from the origin. **B**: Though K-space numbers are discrete numbers in this example, their distribution can be more easily visualized if they are connected and displayed as a rendered surface.

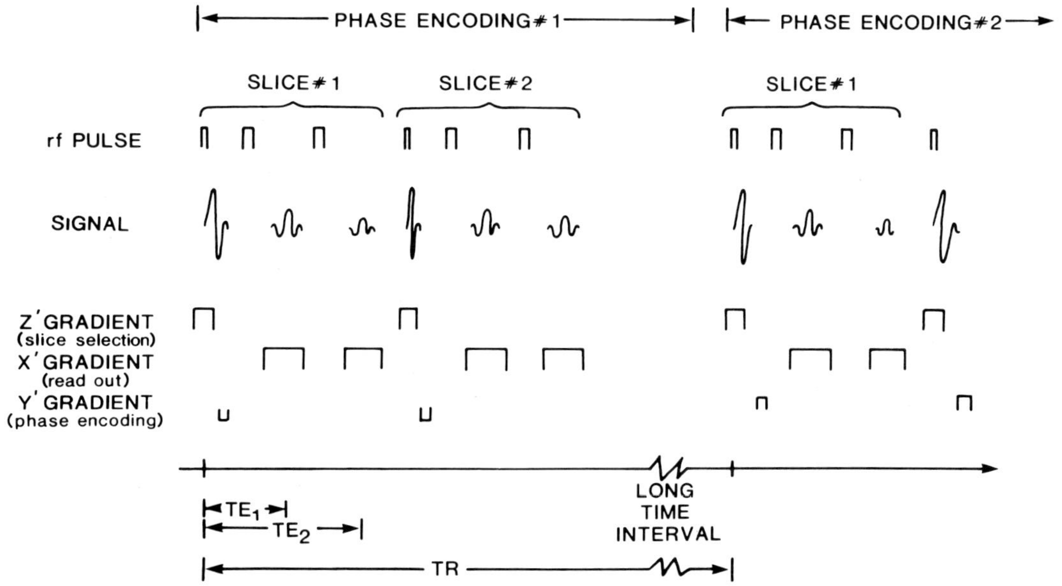

FIG. 1.12. A simplified pulse timing diagram for a multislice, multiecho spin-echo sequence. The slice-selection gradient is on only during the 90° pulse. The readout or frequency-encoding gradient is on during the spin echoes, and the phase-encoding gradient is turned on after the 90° pulse, but before the spin echo. The phase encoding gradient changes for each phase-encoded step.

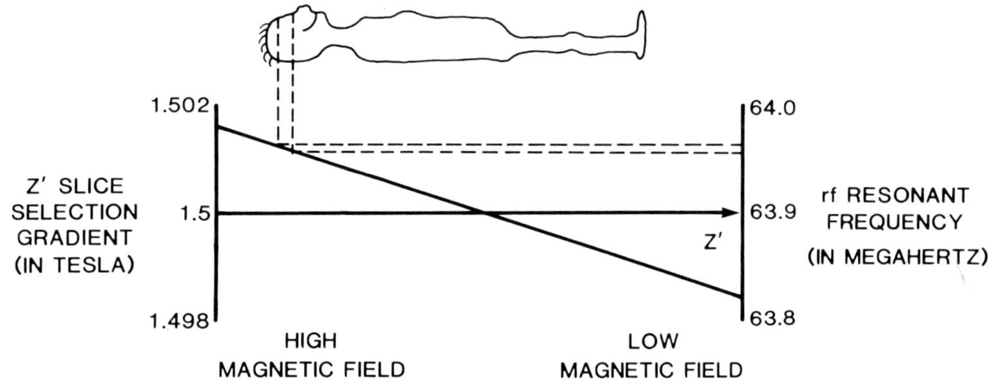

FIG. 1.13. Slice-selection gradient. If a narrow spectrum of radiofrequencies (RF) is used to resonate the nuclei in the presence of a magnetic gradient varying along the *z'* direction, then only nuclei in a slice through the head will be resonated. Lowering the RF will select a more caudal slice. A steeper gradient will create a thinner slice. A narrower RF spectrum will also result in a thinner slice.

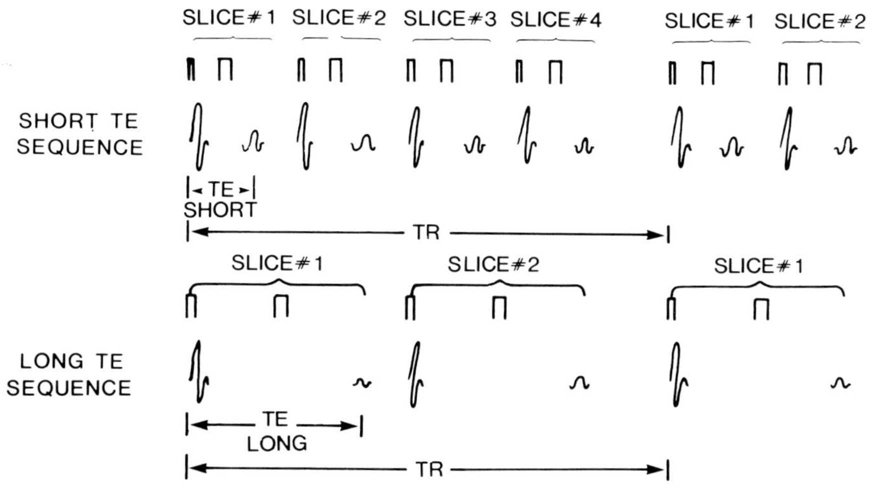

FIG. 1.14. The relationships between TR, TE, and number of obtainable slices. With a given TR, more slices can be obtained with a short TE than with a long TE. Assuming a given TE, more slices can be obtained by lengthening the TR.

quency of the initial pulse, nuclei lower along the z' axis will be excited, and a slice located more caudally can be obtained (Fig. 1.13).

Notice that the TE is typically much shorter than the repetition time (TR) for usual spin-echo sequences. This means that in the interval between the initial RF pulse (90°) for one slice location and the next initial RF pulse (90°) for that same slice location, initial pulses (90°) at other slice locations can be interposed. This is schematically diagrammed in Figs. 1.12 and 1.14 and is the basis for the most frequently used imaging sequence, called the multislice, multiecho, two-dimensional Fourier-transform spin-echo pulse sequence. Notice that the longer the delay in TE, the fewer slice locations that can be placed in a given TR interval (Fig. 1.14). Also notice that with longer TRs, more slice locations can be simultaneously acquired in a given data acquisition. Frequently in MRI, the choices for TR are as much dependent on the number of slices desired as on image contrast (*vide infra*).

RELAXATION MECHANISMS

Of extreme importance in understanding the causes of contrast differences between tissues with MRI is an understanding of the T1, T2, and T2* relaxation processes (47–57). Immediately after the 90° pulse, the bulk magnetization vector begins returning back to its baseline value of magnitude M0 pointing in the z direction. The process is dependent on two basic principles called T1 relaxation and T2 relaxation. T1 relaxation returns the z component of **M** from zero to M0, and T1 is roughly the time it takes for the z component to go from zero after the 90° pulse up to two-thirds of its maximum value, M0 (Fig. 1.15). T2 relaxation causes the x-y component of **M** to return to zero and is defined as roughly the time it takes for the x-y component of the bulk magnetization vector to decrease from its maximum value right after the 90° pulse to roughly one-third of its maximum value (Fig. 1.16).

A simple understanding of the causes of these two processes is helpful in characterizing tissues by MRI (21,58,59). The z component of the bulk magnetization vector, i.e., the component along the direction of the main magnetic field, is strictly determined by the difference in the number of nuclei pointed in the parallel direction versus the number of nuclei pointed in the anti-parallel direction. After the 90° pulse, the number of nuclei pointing parallel and anti-parallel are equal. However, soon after the 90° pulse, more nuclei will lodge in the parallel, low-energy state than in the anti-parallel state. Thus, more will point in the direction of the magnetic field, and the z component of the bulk magnetization vector will go from zero to a positive quantity along the z axis. Any process within the tissues that speeds up the flipping of nuclei will shorten the time it takes for the z component to rise back up to its baseline value. This occurs when hydrogen nuclei are closely associated with large macromolecules, because large macromolecules rotate, and part of their rotation is near the Larmor frequency (21,59,60). The rotations will create local electromagnetic fields which vary at frequencies at or around the Larmor frequency. This will cause rapid switching of nuclei between the parallel and anti-parallel states and will shorten T1. A similar process exists in the arms of triglycerides and is responsible for the short T1 of fat. This process can be markedly accelerated if paramagnetic materials are attached to the macromolecules, since they greatly accentuate the intensity of the locally fluctuating electromagnetic fields (61). Therefore, tissues with short T1 values include proteinaceous cysts, fat, hemorrhage, and tissues containing infused gadolinium, a paramagnetic agent that can be given intravenously.

T2 relaxation processes are somewhat different from T1 processes (13,25,62). Substances with short T2 values contain hydrogen nuclei that are fixed in position.

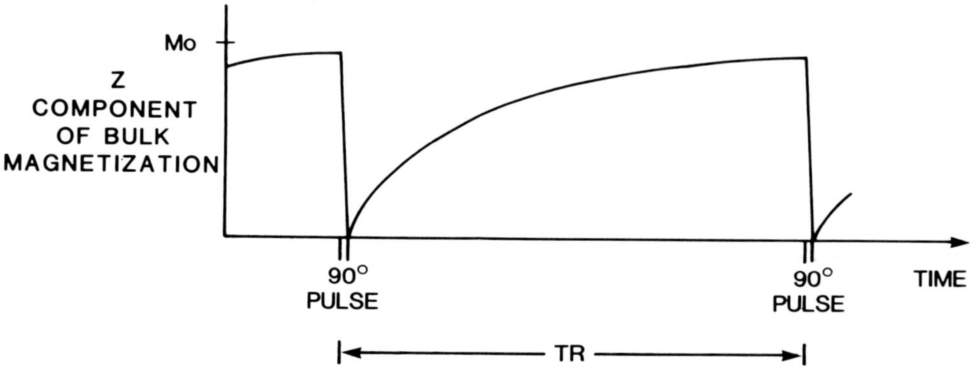

FIG. 1.15. T1 relaxation. After the 90° pulse (or any other pulse), the bulk magnetization vector returns to the baseline value, M0, in a logarithmic fashion. The time it takes for the z component of **M** to return approximately two-thirds of the way back to M0 is called T1.

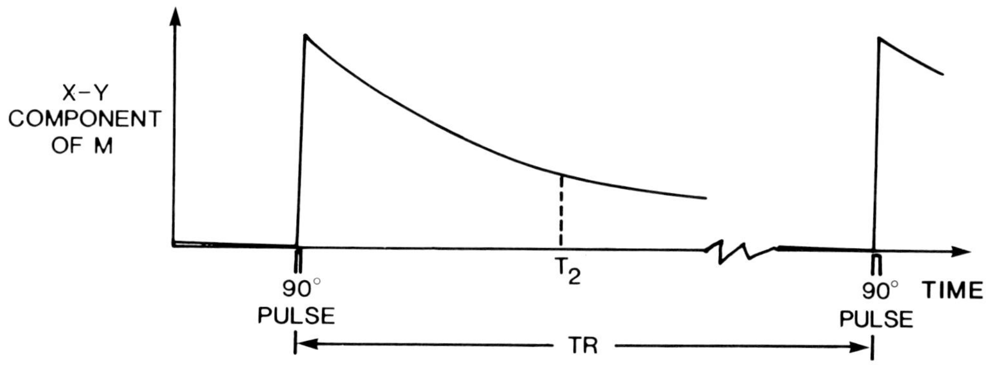

FIG. 1.16. T2 relaxation. After the 90° pulse (or any other RF pulse), the *x-y* component of the bulk magnetization vector returns toward 0 exponentially. The time required to reach approximately one-third of the maximum value is called T2.

Tremendous variations in the strength of magnetic fields exist at the atomic level. If fixed in location, nuclei in different magnetic fields will precess at different rates (as described by Eq. 1); therefore, nuclei in a given region that begin resonating in phase with each other will soon get out of phase. Substances that are fluid, such as urine or water, have long T2 values because the nuclei move rapidly through the substance. Thus, they spend part of their time in regions of low magnetic field strength and part of their time in regions of high magnetic field strength. The net effect over a pulse sequence is that the nuclei effectively experience a single average magnetic field. The nuclei stay in phase longer, and the tissue has a long T2. This is a process called motional narrowing. Liquids tend to have long T2 values, as do tissues that contain excess amount of free water secondary to edema or neoplastic disease (63,64). Because abnormal states such as inflammation or neoplasm tend to be associated with larger amounts of free water than are found in normal tissues, T2-weighted images are in general more sensitive in detecting many disease processes than sequences which are not T2-weighted.

T2* processes are those magnetic effects which lead to dephasing of nuclei after the initial RF pulse (i.e., the decay of the FID). T2* is defined as roughly the time required for the initial FID to decay to one-third of its peak value (Fig. 1.17). The loss of signal is due to dephasing of the nuclei by all magnetic field inhomogeneities within imaged tissues. In practice for the FID, this process is dominated by fixed (time independent) magnetic field inhomogeneities within the tissues. Such fixed tissue inhomogeneities are usually independent of primary physico-chemical properties of the tissues, but are mostly dependent on equipment imperfections and bulk tissue magnetic susceptibility effects. Thus, soft tissue contrast based on the T2* effects is often poor. However, prominent magnetic field gradients may exist at the interfaces between tissues due to differences between the magnetic susceptibility of two different tissues. This effect may lead to accentuated edges between tissues on T2*-weighted images. These effects may or may not be considered advantageous, depending on the interpreter. These constant field inhomogeneities are mostly reversed by 180° RF induced spin echoes, but magnetic inhomogeneities which change between the initial RF pulse and the spin echo are not corrected by the 180° pulse, leading to irreversible signal loss (the T2 effects). Magnetic field inhomogeneities which change with time are usually less important in the FID decay, but are strongly dependent on the basic physico-chemical properties of tissues. Thus, soft tissue contrast is usually superior on 180° RF spin-echo images than gradient-echo images.

Because of the importance of T1 and T2 in tissue characterization, a thorough understanding of the relationship between the TE and TR selected and the resultant T1 and T2 weightings of the subsequent images is critical in MRI (13,41,49–57,62,65,66). Figure 1.18 shows the height of the *z* component of the bulk magnetization vector as a function of time after the 90° pulse for two tissues with differing T1 values. Notice that a tissue with a short T1 may have a higher bulk magnetization vector at the time of the next 90° pulse. Because the magnitude of the *x-y* component of the

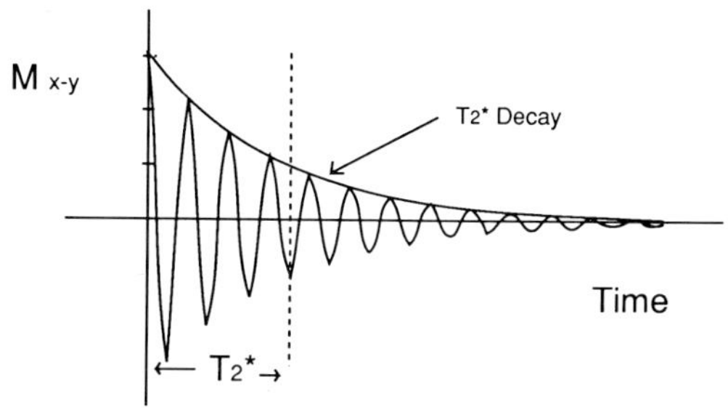

FIG. 1.17. T2* and the free-induction decay (FID). T2* is defined as the time it takes for the FID to decay roughly one-third of its initial peak value.

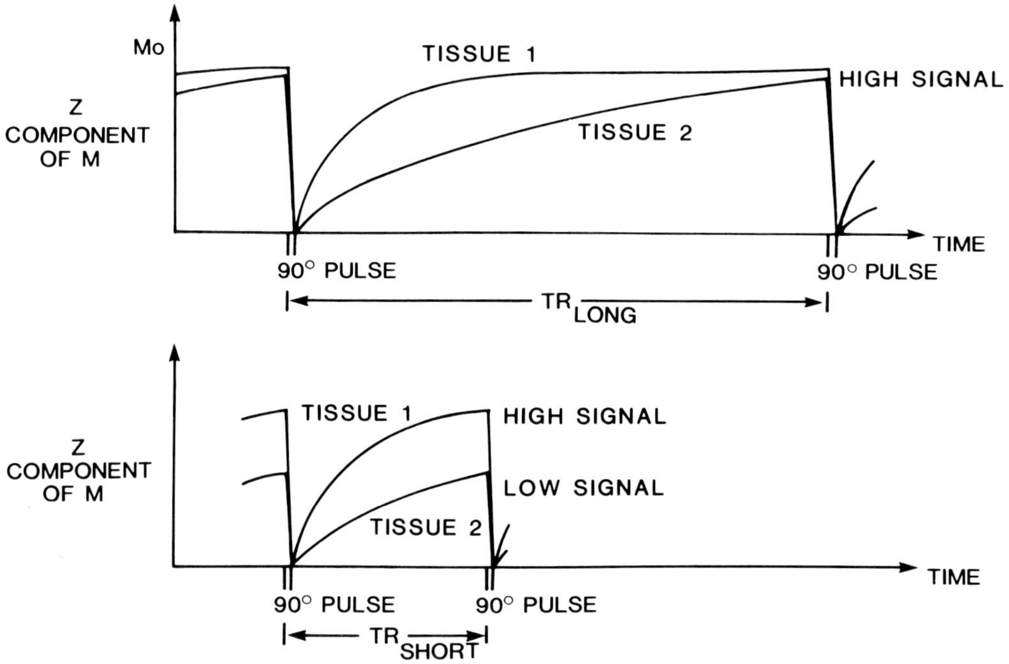

FIG. 1.18. T1 weighting. If two tissues differ in T1 values, then they can best be differentiated by signal intensities with short TR sequences. With a long TR, both tissue 1 (short T1) and tissue 2 (long T1) give high signals. With a short TR, tissue 2 has not regained magnetization and will give a low signal, whereas tissue 1 will have regained magnetization and give a strong signal. The large signal differential will allow easy MR differentiation of the neighboring tissues. Thus, images acquired with short TR values are more T1-weighted than those with long TR values.

bulk magnetization vector after the 90° pulse is proportional to the value of the *z* component right before the 90° pulse, a tissue with a short T1 will produce a brighter signal than will a tissue with a long T1 if all other factors are equal. Notice also that if the TR is short, there is a large difference between the height of the bulk magnetization vector at the time of the next 90° pulse between long- and short-T1 tissues (Fig. 1.18). Therefore, in general with spin-echo imaging, the shorter the TR, the more T1-weighted is the image. Conversely, the longer the TR, the less T1-weighted is the image.

Figure 1.19 shows the time course of T2 decay for two tissues, one with a long T2 and one with a short T2. Notice that the *x-y* component of the bulk magnetization vector, and therefore the intensity of the signal picked up by the receiving wires, is larger for long-T2 substances than for short-T2 substances. Also note that the longer the TE, i.e., the point at which we sample the signal, the more separation there is between

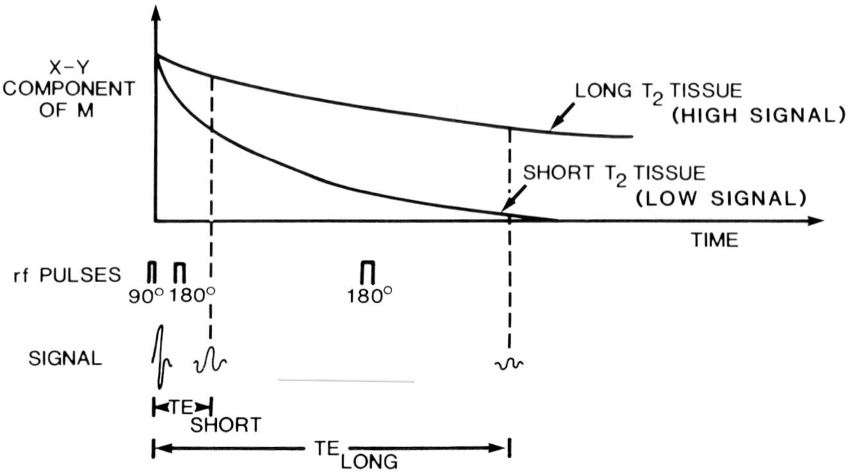

FIG. 1.19. T2 weighting. If other parameters are equal, a long-T2 tissue will emit higher signal than a short-T2 tissue. The difference in signal intensity will be maximal for long TE values. Therefore, long TE acquisitions yield T2-weighted images.

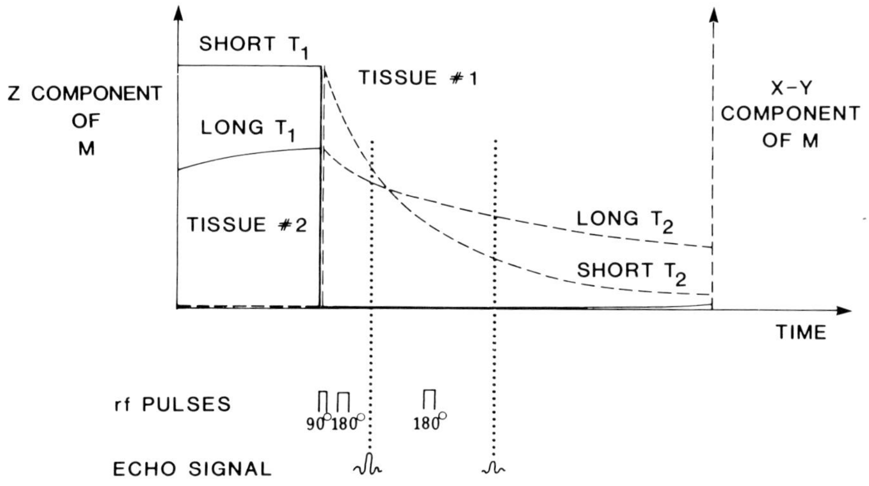

FIG. 1.20. Contrast reversal in MRI. Short T1 and T2 tissues such as fat can be brighter than long T1 and T2 tissues such as joint effusions on first-echo, intermediate weighted images, but darker than effusions on second echo, T2-weighted images. This can be confusing to radiologists familiar with x-ray–based imaging.

tissues with differing T2 values. As a general rule, sequences with longer TE values are more T2-weighted sequences and conversely, sequences with shorter TE values are less T2-weighted sequences.

A frequent dilemma noted by imagers versed in x-ray techniques is that some tissues, such as joint fluid, may appear darker than surrounding fatty tissues on some images (i.e., on T1- and intermediate-weighted images), but brighter than surrounding tissues on other images (i.e., T2-weighted images). This occurs when tissues with long T1 and T2 values (fluid) are adjacent to structures with short T1 and T2 values (fat). On T1- and intermediate-weighted images, such as the first echo in Fig. 1.20, fat will be bright because of its short T1. On the second echo, the contrast reverses because of the long T2 of fluid. Therefore, an understanding of the causes of signal intensities in MRI can eliminate potential confusion and be helpful for tissue characterization.

The STIR technique is a form of inversion recovery sequence. If the T1 is very short (150–200 ms), the 90° pulse of the spin-echo sequence will occur at a time when **M** of fat is near 0, but tissues with longer T1 values will be non-zero (Fig. 1.21) (33,34). The resultant spin-echo signal will be 0 for regions of the body containing tissue with T1 values similar to fat, but higher for regions of the body containing water-based tissues. The contrast in the resultant images depicts low signal in predominantly fatty regions. STIR images are especially useful in evaluating infiltrative disease or edema in high lipid regions, such as bone marrow (67–70). The signal from disease processes may be masked by the high signal from the lipid in standard spin-echo sequences, but the suppression of lipid signal in the STIR images allows improved visualization of underlying abnormal processes. STIR images may also be useful in evaluating articular cartilage (***see chapter 6***) because joint fluid tends to be dark because of the T1-weighting of the images and the water-rich articular cartilage bright (*vida infra*). One drawback of STIR images is that the signal suppression is not specific for fat, but will suppress any tissue which has a short T1 relaxation time. When contrast agents which shorten tissue T1 times, such as gadolinium, are used, STIR images may suppress desired contrast enhancement.

More specific fat suppression using RF pulses just before the spin-echo sequences can specifically suppress signal from fat (71,72). RF fat suppression techniques exploit the small differences in resonant frequency of the proton nuclei in lipid versus the higher resonant frequency of protons in water. The narrow RF pulse centered at the lipid resonant frequency "saturates" fat protons, i.e., rotates their **M** into the *x-y* plane, so that subsequent imaging pulses rotate **M** out of the *x-y* plane, inhibiting resonance of these selected spins during the subsequent imaging pulse sequence. The water protons are unaffected because the saturation pulse does not contain energy at the resonant frequency of water. Unlike the STIR technique, this technique does not use T1 to determine suppression; therefore, nonlipid short T1 tissues are unaffected and continue to display signal. RF saturation is an excellent technique to use with gadolinium to suppress high signal intensity from fat which may mask signal from contrast-enhanced tissues.

Hematomas can be confusing on MRI (73–84). Acute hematomas may appear isointense compared with surrounding soft tissues on T1- and intermediate-weighted images. On T2-weighted images it may be isointense (because of oxyhemoglobin in intact red cells) or of low intensity because of the magnetic-susceptibility effect of deoxyhemoglobin contained within intact red cells (78). Subacute hemorrhage appears bright on T1-weighted images because of the paramagnetic effect of iron in methemoglobin (73) and on T2-weighted images because of motional narrowing of free water collected during the degradation process

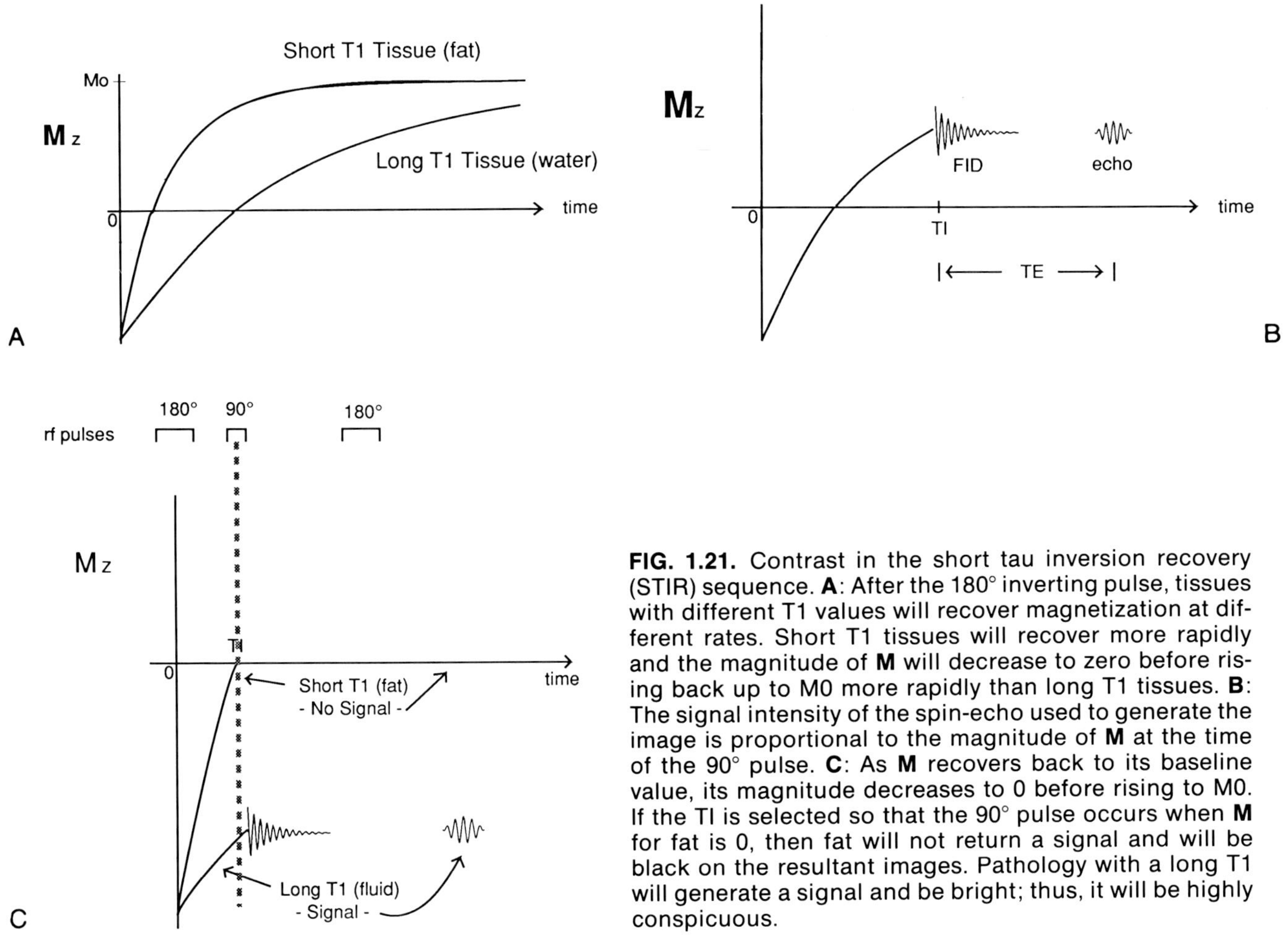

FIG. 1.21. Contrast in the short tau inversion recovery (STIR) sequence. **A**: After the 180° inverting pulse, tissues with different T1 values will recover magnetization at different rates. Short T1 tissues will recover more rapidly and the magnitude of **M** will decrease to zero before rising back up to M0 more rapidly than long T1 tissues. **B**: The signal intensity of the spin-echo used to generate the image is proportional to the magnitude of **M** at the time of the 90° pulse. **C**: As **M** recovers back to its baseline value, its magnitude decreases to 0 before rising to M0. If the TI is selected so that the 90° pulse occurs when **M** for fat is 0, then fat will not return a signal and will be black on the resultant images. Pathology with a long T1 will generate a signal and be bright; thus, it will be highly conspicuous.

(78). Older hematomas are characterized by a black periphery that is secondary to hemosiderin deposition (78,79,85,86). Knowledge of these characteristic changes is helpful in obviating misinterpretation of findings in acute hematomas that may be isointense with surrounding structures and in dating the ages of old hematomas.

In contrast to hematomas, interstitial hemorrhage into medullary bone and muscle is associated with long T2 relaxation times immediately. This is probably due to separation of the red cells from the serum within the interstitium which does not occur in a pool of whole blood. The water molecules are then separated from the magnetic susceptibility effects of the red cells and display long T2 characteristics typical of free water. Thus, medullary hemorrhage in trabecular fractures of acute bony injuries and interstitial hemorrhage into acute muscle and tendon tears are typically bright on T2-weighted images immediately after the injury (*see **Chapter 10***).

Table 1.1 summarizes the foregoing findings concerning T1 and T2 weighting of images. In general, the best T1-weighted image is one with a short TR and a

TABLE 1.1. *Association between pulse parameters and T1 and T2 weighting*

TR \ TE	SHORT (<30 msec)	LONG (>60 msec)
SHORT (≤800 msec)	T_1 WEIGHTED	UGLY
LONG (≥2000 msec)	INTERMEDIATE WEIGHTED	T_2 WEIGHTED

short TE, so that there is good T1 weighting and limited T2 weighting. Conversely, the best T2-weighted image is one with a long TR (so that there is less T1 weighting) and a long TE (so that there is best T2 weighting). Intermediate weighting can be obtained with a long TR and a short TE. This is occasionally referred to as a balanced image, a spin-density, or a proton-density image; however, the latter terms can be confusing because the TR values are not long enough and the TE values are not short enough in routine imaging to give a truly representative image of proton density. Therefore, I prefer to call them intermediate-weighted images. In general, there is little clinical usefulness for short-TR long-TE images, as the signal-to-noise is generally too poor for good tissue contrast.

SIGNAL-TO-NOISE RATIO

A critical determinant of image quality is the ratio between the strength of the signal coming back from the nuclei and the intensity of the "noise" that comes both from the patient and from the electronics (87,88). A low signal-to-noise ratio (S/N) means that the useful signal is not markedly stronger than the noise. The image appears grainy, to use a common radiologic term based on standard film screen radiography. Several determinants of the MR system combine to determine the S/N for a given image (89). The patient is the source for the signal to be evaluated and for a significant amount of noise that is received. This component of the S/N is fixed and cannot be readily changed by the operator. The antenna, electronics, and computer also generate significant noise if proper engineering and design of the system are not employed, including the use of surface coils when indicated (*vide infra*) (90,91). This component of the S/N is manufacture-dependent, and great progress has been made by manufacturers in limiting the noise from this segment of the system. Another determinant of S/N is the magnetic field strength employed for imaging. At present, it is somewhat controversial exactly how S/N varies with field strength; however, an approximate guess is that S/N increases roughly in direct proportion to the field strength.

Of significance to users of MRI are the relationships between parameters of the imaging protocols employed and the S/N. Technical factors that can be selected at the time of imaging which affect S/N include the following: (a) the NEX employed, (b) the voxel size, (c) the TR value, (d) the TE value, and (e) the interslice spacing.

The NEX refers to the number of times a set of phase-encoded gradients is averaged together to calculate the image (see Eq. 1.4). In other words, an image could be obtained with one excitation, and if a 128 matrix were used in the y' direction, 128 90° pulses would be performed. This would require 128 times the TR to produce the image. If two excitations were used, then twice as many 90° pulses would be required to produce the image. Therefore, the data acquisition time would be doubled. The S/N improves proportional to the square root of the number of excitations employed (87). Because examination time is an important factor in determining cost, a heavy price must be paid for using increased excitations to decrease the "graininess" of an image.

The S/N is also directly proportional to the square root of the volume of tissue emitting the signal, i.e., the voxel. As the size of the voxel increases, more tissue is available to return a signal. However, there is also more tissue to generate noise. Because the noise is random, it partly cancels itself and the overall S/N improves proportionally to the square root of the voxel size (87). In turn, the voxel size is dependent on the field of view used and the number of pixels into which the field of view is divided. A pixel is the area of the smallest picture unit that is used to generate the image, i.e., the cross-sectional area of a voxel in the plane of the image. Typical imagers use a 256 matrix along the frequency-encoded direction and either a 128, 192, or 256 matrix along the phase-encoded direction. The depth of the voxel is determined by the slice thickness.

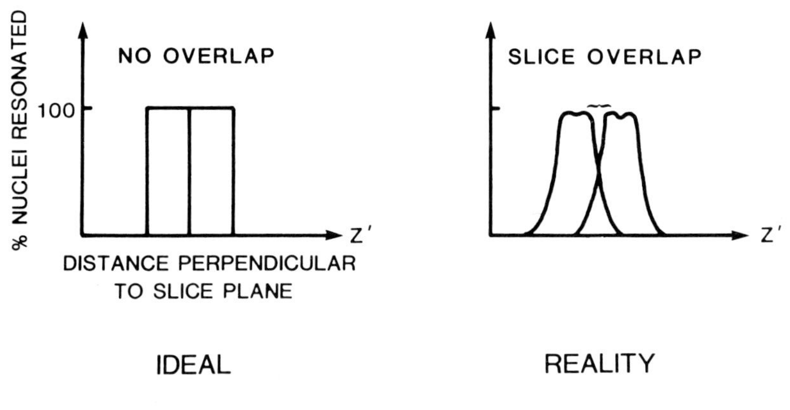

FIG. 1.22. Slice profile. The ideal slice profile would be rectangular. Unfortunately, real profiles have rounded, sloping edges that can overlap and degrade image quality.

TABLE 1.2. *Factors affecting S/N, imaging time, and spatial resolution; N represents number of excitations, and V represents volume of the voxel*

FACTORS AFFECTING S/N	↑ FIELD STRENGTH	NUMBER OF EXCITATIONS	↑ VOXEL SIZE	↑ TR	↑ TE	SLICE SPACING	SURFACE COIL
MAJOR EFFECT ON S/N	↑	↑ $\sqrt{N}$	↑ $\sqrt{V}$	↑	↓	↑	↑
MAJOR EFFECT ON IMAGING TIME	0	↑ N	↓ (WITH ↓ NUMBER OF PHASE ENCODINGS)	↑	0	0	0
MAJOR EFFECT ON SPATIAL RESOLUTION	0	0	↓	0	0	↓ (DECREASE IN PLANE PERPENDICULAR TO SLICE)	0

↑ DIRECTLY INCREASES
↓ DIRECTLY DECREASES
0 INDIRECT EFFECTS ONLY

Therefore, the voxel size can be doubled by going from a 256 matrix along the y' direction to a 128 matrix, thus improving the S/N by the square root of 2. Alternatively, the voxel size can be doubled by increasing the slice thickness from a 5 mm to 10 mm thick section, with the matrix size being unchanged, and again a factor of the square root of 2 in S/N will be gained. Therefore, there is a strong interrelationship involving the time of imaging, the in-plane resolution, the slice thickness, and the size of the field of view.

The TR and TE values markedly affect S/N. As discussed in the previous section, a long TR allows the z component of **M** to relax back to near M0, allowing the maximum signal to be generated after the next 90° pulse (Fig. 1.18). Because the noise is unchanged, long TR values result in better S/N. Conversely, TE is the time delay after the 90° pulse during which the signal rapidly decays before being sampled (Fig. 1.19). The longer the TE, the smaller the signal and the smaller the S/N.

The effects of interslice spacing on S/N and image contrast are somewhat more subtle (92–96). The slice selection technique discussed earlier (Fig. 1.13) does not give perfectly sharp edges to the slice. Instead, the edges of the slice are rounded (Fig. 1.22). If the slices abut one another, then there is a volume of overlap between slices. In other words, some of the nuclei that resonated with the recent 90° pulse of the adjacent slice will again be resonated by the very next 90° pulse. This markedly shortens the effective TR for these nuclei, leading to less signal and decreased tissue contrast. Optimal image quality often requires a 20% separation between adjacent slices. In some MR imagers, interleaving the acquisition by pulsing every other slice in one sequence and then using a second acquisition immediately following the first to fill in the spaces also eliminates image degradation from slice overlaps. Table 1.2 summarizes the effects which various MRI parameters have on S/N.

In the final image, the most important determinant of diagnostic quality is the contrast resolution (87). Contrast resolution is, in essence, a combination of spatial resolution and contrast differences between adjacent anatomic structures or abnormalities. One of the major advantages of MRI over other cross-sectional imaging techniques is the increased inherent contrast between different soft tissues. Maximizing contrast resolution in MRI requires proper selection of imaging parameters and knowledge of the anatomy and abnormality to be evaluated.

ACQUISITION FIELD-OF-VIEW

The importance of the acquisition field-of-view in joint MR imaging cannot be overemphasized. High spatial resolution is essential in evaluating the many small anatomic structures which are critical in joint function,

receive signal from the nuclei and transmit the RF pulses needed in data acquisition. Both maximize efficiency by being locally placed next to the knee. We have had excellent results using a large coil which both receives signal and transmits RF pulses (Fig. 1.25) (GE Medical Systems, Milwaukee, WI).

EXAMINATION PROTOCOL

In designing imaging protocols for the knee, several points must be considered. A technique must be devised to produce images of good spatial and contrast resolution of the bones, tendons, ligaments, and other relevant structures around the knee. The image acquisition time must be minimized to improve patient acceptability, decrease patient motion artifact, and limit the cost of the examination. Our goal was to design protocols that could be performed reliably within 20 to 45 min (Table 1.3).

REFERENCES

1. Purcell EM, Torrey HC, Pound RV. Resonance absorption by nuclear magnetic moments in a solid. *Physiol Rev* 1946;69:37–38.
2. Block F. Nuclear induction. *Physiol Rev* 1946;70:460–473.
3. Gabillard R. Resonance nucleaire. *Compt Rend* 1951; 232:1551–1553.
4. Damadian R, Minkoff L, Goldsmith M, Stanford M, Kout Cher J. Field focus nuclear magnetic resonance (FONAR): Visualization of a tumor in a live animal. *Science* 1976;194:1430–1432.
5. Lauterbur PC. Image formation by induced local interactions: Examples employing nuclear magnetic resonance. *Nature* 1973;242:190–191.
6. Andrew ER. NMR imaging of intact biological systems. *Philos Trans R Soc Lond [Biol]* 1980;289:471–481.
7. Crooks LE. *Overview of NMR imaging techniques*. IN: Kaufman L, Crooks LE, Margulis AR, eds. Nuclear magnetic resonance imaging in medicine. Tokyo: Igaku-Shoin; 1981:42–50.
8. Edelstein WA, Hutchison J, Johnson G. Spin warp NMR imaging and applications to human whole body imaging. *Physiol Med Biol* 1980;25:751–756.
9. Hinshaw WS, Lent AH. An introduction to NMR imaging: From the Block equation to the imaging equation. *Proc. IEEE* 1983;71:338–350.
10. Hoult DI. Rotating frame zeugmatography. *Philos Trans R Soc Lond [Biol]* 1980;289:543–547.
11. Mallard J, Hutchison JMS, Edelstein WA, Ling CR, Foster MA, Johnson G. In vitro NMR imaging in medicine: The Aberdeen approach, both physical and biological. *Philos Trans R Soc Lond [Biol]* 1980;289:519–533.
12. Abraham A. *The principles of nuclear magnetism*. London: Oxford University Press; 1983.
13. Bradley WG, Newton TH, Crooks L. *Physical principles of nuclear magnetic resonance*. IN: Advanced imaging techniques. 1983;15–61.
14. Budinger TF, Lauterbur PC. Nuclear magnetic resonance technology for medical studies. *Science* 1884;226:288–298.
15. Crues JV, Morgan FW. *Physical and technical considerations in magnetic resonance imaging*. IN: Stoller DW, ed. Magnetic resonance imaging in orthopaedics and rheumatology. Philadelphia: Lippencott; 1989.
16. Fullerton GD. Basic concepts for nuclear magnetic resonance imaging. *Magn Reson Imaging* 1982;1:39–55.
17. Gore JC, Emery EW, Orr JS, Doyle FH. Medical nuclear magnetic resonance imaging: I. Physical principles. *Invest Radiol* 1981;16:269–274.
18. Haacke EM, Tkach JA. Fast MR imaging: Techniques and clinical applications. *AJR* 1990;155:951–964.
19. Hahn EL. Spin echoes. *Phys Rev* 1950;50:580–594.
20. Jones JP, Partain CL, Mitchell MR, Patton JA, Stephens WH, Price RR, Kulkarni MV, James AEJ. *Principles of magnetic resonance*. IN: Kressel HY, ed. Magnetic Resonance Annual, 1985. New York: Raven Press; 1985:71–111.
21. Koenig SH, Brown RD. Determinants of proton relaxation rates in tissue. *Magn Reson Imaging* 1984;1:437–449.
22. Pykett IL. NMR imaging in medicine. *Sci Am* 1982;246:78–88.
23. Pykett IL, Newhouse JH, Buonanno FS, Brady TJ, Goldman MR, Kistler JP, Pohost GM. Principles of nuclear magnetic resonance imaging. *Radiology* 1982;143:157–168.
24. Young SW. *Magnetic resonance imaging: Basic principles*. New York: Raven Press; 1988.
25. Wolf GL, Popp C. *NMR–A primer for medical imaging: Clear explanations of basic principles for the radiologist, orthopedic surgeon, internist, technician*. Thorofare, NJ: Slack; 1984.
26. Budinger TF, Margulis AR, eds. *Medical magnetic resonance: A primer—1988*. Berkeley, CA: The Society of Magnetic Resonance in Medicine; 1988:385.
27. Lufkin RB. *The MRI manual*. Littleton, MA: Year Book Medical; 1990:313.
28. Mettler FA, Jr, Muroff LR, Kulkarni MV, eds. *Magnetic resonance imaging and spectroscopy*. New York: Churchill Livingstone; 1986:319.
29. Slichter CP. *Principles of magnetic resonance*. New York: Harper & Row; 1963.
30. Frahm J, Haase A, Matthaei D. Rapid three-dimensional MR imaging using the FLASH technique. *J Comput Assist Tomogr* 1986;10:363–368.
31. Haase A, Frahm J, Matthaei D, Hanicke W, Bomsdorf H, Kunz D, Tischler R. MR imaging using stimulated echoes (STEAM). *Radiology* 1986;160:787–790.
32. Haase A, Matthaei D, Hanicke W, Frahm J. Dynamic digital substraction imaging using fast low-angle shot MR movie sequences. *Radiology* 1986;160:537–541.
33. Bydder GM, Steiner RE, Blumgart LH, Khenia S, Young IR. MR imaging of the liver using short T1 inversion recovery sequences. *J Comput Assist Tomogr* 1985;9(6):1084–1089.
34. Bydder GM, Pennock JM, Steiner RE, Khenia S, Payne JA, Young IR. The short T1 inversion recovery sequence—An approach to MR imaging of the abdomen. *Magn Reson Imaging* 1985;3(3):251–254.
35. Reicher MA, Lufkin RB, Smith S, Flannigan B, Olsen R, Wolf R, Hertz D, Winter J, Hanafee W. Multiple-angle, variable-interval, nonorthogonal MRI. *AJR* 1986;147:363–366.
36. Matthaei D, Frahm J, Haase A, Merboldt KD, Hanicke W. Multipurpose NMR imaging using stimulated echoes. *Magn Reson Med* 1986;3:554–561.
37. Matthaei D, Haase A, Frahm J, Bomsdorf H, Vollman W. Multiple chemical shift selective (CHESS) MR imaging using stimulated echoes. *Radiology* 1986;160:791–794.
38. Friedburg H, Henning J, Schumacher M. [RARE-MR myelography in routine clinical practice. Experience with 175 cases]. *Rofo* 1987;146(5):584–590.
39. Carr HY, Purcell EM. Effect of diffusion on free precession in nuclear magnetic resonance experiments. *Physiol Rev* 1954;94: 630–638.
40. Feinberg DA, Mills CM, Posin JP, Ortendahl DA, Hylton NM, Crooks LE, Watts JC, Kaufman L, Arakawa M, Hoenninger JC, Brant-Zawadzki M. Multiple spin-echo magnetic resonance imaging. *Radiology* 1985;155:437–442.
41. Kneeland JB, Knowles RJR, Cahil PT. Multi-section multi-echo pulse magnetic resonance techniques: Optimization in a clinical setting. 1985;155:159–162.
42. Chrispin A, Small P, Rutter N, Coupland RE, Doyl M, Chapman B, Coxon R, Guilfoyle D, Cawley M, Mansfield P. Transectional echo planar imaging of the heart in cyanotic congenital heart disease. *Pediatr Radiol* 1986;16(4):293–297.
43. Mansfield P. Real-time echo-planar imaging by NMR. *Br Med Bull* 1984;40(2):187–190.
44. Rzedzian R, Chapman B, Mansfield P, Coupland RE, Doyle

M, Chrispin A, Guilfoyle D, Small P. Real time nuclear magnetic resonance clinical imaging in paediatrics. *Lancet* 1983;2(8362):1281–1282.
45. Saini S, Stark DD, Rzedzian RR, Pykett IL, Rummeny E, Hahn PF, Wittenberg J, Ferrucci JT. Forty-millisecond MR imaging of the abdomen at 2.0 T. *Radiology* 1989;173(1):111–116.
46. Stehling MK, Charnley RM, Blamire AM, Ordidge RJ, Coxon R, Gibbs P, Hardcastle JD, Mansfield P. Ultrafast magnetic resonance scanning of the liver with echo-planar imaging. *Br J Radiol* 1990;63(750):430–437.
47. Bradley WG. Effect of relaxation times on magnetic resonance interpretation. *Noninv Med Imaging* 1984;1:193–204.
48. Bydder GM, Young IR. Clinical use of the partial saturation and saturation recovery sequences in MR imaging. *J Comput Assist Tomogr* 1985;9:1020–1032.
49. Foster MA, Hutchison JMS, Mallard JR, Fuller M. Nuclear magnetic resonance pulse sequence and discrimination of high- and low-fat tissues. *Magn Reson Imaging* 1984;2:187–192.
50. Hendrick RE, Nelson TR, Hendee WR. Optimizing tissue contrast in magnetic resonance imaging. *Magn Reson Imaging* 1984;2:193–204.
51. Mitchell MR, Tarr RW, Conturo TE, Partain CL, James AE, Jr. Spin echo technique selection: Basic principles for choosing MRI pulse sequence timing intervals. *Radiographics* 1986;6:245–247.
52. Moran PR. A general approach to T1, T2, and spin-density discrimination sensitivities in NMR imaging sequences. *Magn Reson Imaging* 1984;2:17–22.
53. Perman WH, Hilal SK, Simon HE, Maudsley AA. Contrast manipulation in NMR imaging. *Magn Reson Imaging* 1984;2:23–32.
54. Posin JP, Ortendahl DA, Hylton NM, Kaufman L, Watts JC, Crooks L, Mills CM. Variable magnetic resonance imaging parameters: Effect on detection and characterization of lesions. *Radiology* 1985;155:719–725.
55. Wehrli FW, MacFall JR, Shutts D, Breger R, Herfkins RJ. Mechanisms of contrast in NMR imaging. *J Comput Assist Tomogr* 1984;8:369–380.
56. Wehrli FW, MacFall JR, Glover GH, Grigsby N. The dependence of nuclear magnetic resonance (NMR) image contrast on intrinsic and pulse sequence timing parameters. *Magn Reson Imaging* 1984;2:3–16.
57. Ziedses des Plantes BG, Falke THM, den Boer JA. Pulse sequences and contrast in magnetic resonance imaging. *Radiographics* 1984;4:869–883.
58. Amtey SR. NMR and water proton relaxation in tissues. *Magn Reson Imaging* 1982;1:63–67.
59. Fullerton GD, Potter JL, Dornbluth NC. NMR relaxation of protons in tissues and other macromolecular water solutions. *Magn Reson Imaging* 1982;1:209–228.
60. Soloman I. Relaxation processes in a system of two spins. *Physiol Rev* 1955;99:559–565.
61. Bloemberg N. Proton relaxation times in paramagnetic solutions. *J Chem Phys* 1957;27:572–573.
62. Ehman RL, Gray JE, Bryant RG, Kennedy SD, Earnest F, IV. Spin-spin relaxation time (T2) dependence in MR saturation- and inversion-recovery images. *AJR* 1984;143:903–906.
63. Davis PL, Kaufman L, Crooks LE, Margulis AR. *NMR characteristics of normal and abnormal rat tissues*. IN: Kaufman L, Crooks LE, Margulis AR, eds. Nuclear magnetic resonance imaging in medicine. Tokyo: Igaku-Shoin; 1981:71–100.
64. Shah SS, Ranade SS, Phadke RS, Kasturi SR. Significance of water proton spin-lattice relaxation times in normal and malignant tissues and their subcellular fractions. II. *Magn Reson Imaging* 1982;1:155–164.
65. Bydder GM, Curati WL, Gadian DG, Hall AS, Harman RR, Butsen PR, Gilderdale DJ, Young IR. Use of closely coupled receiver coils in MR imaging: Practical aspects. *J Comput Assist Tomogr* 1985;9:987–996.
66. Wehrli FW. *Principles of magnetic resonance*. IN: Stark DD, Bradley WG, eds. Magnetic resonance imaging. St. Louis: Mosby; 1987.
67. Baker LL, Goodman SB, Perkash I, Lane B, Enzmann DR. Benign versus pathologic compression fractures of vertebral bodies: Assessment with conventional spin-echo, chemical-shift, and STIR MR imaging. *Radiology* 1990;174(2):495–502.
68. Dwyer AJ, Frank JA, Sank VJ, Reinig JW, Hickey AM, Doppman JL. Short-Ti inversion-recovery pulse sequence: Analysis and initial experience in cancer imaging. *Radiology* 1988;168(3):827–836.
69. Golfieri R, Baddeley H, Pringle JS, Souhami R. The role of the STIR sequence in magnetic resonance imaging examination of bone tumors. *Br J Radiol* 1990;63(748):251–256.
70. Shuman WP, Baron RL, Peters MJ, Tazioli PK. Comparison of STIR and spin-echo MR imaging at 1.5 T in 90 lesions of the chest, liver, and pelvis [see comments]. *AJR* 1989;152(4):853–859.
71. Harned EM, Mitchell DG, Burk DJ, Vinitski S, Rifkin MD. Bone marrow findings on magnetic resonance images of the knee: Accentuation by fat suppression. *Magn Reson Imaging* 1990;8(1):27–31.
72. Totterman S, Weiss SL, Szumowski J, Katzberg RW, Hornak JP, Proskin HM, Eisen J. MR fat suppression technique in the evaluation of normal structures of the knee. *J Comput Assist Tomogr* 1989;13(3):473–479.
73. Bradley WG, Schmidt PG. Effect of methemoglobin formation on the MR appearance of subarachnoid hemorrhage. *Radiology* 1985;156:99–103.
74. Cohen MD, Mcguire W, Cory DA, Smith JA. MR appearance of blood and blood products: An in vitro study. *AJR* 1986;146:1293–1297.
75. De La Paz RL, New PFJ, Buonanno FS, Kistler JP, Oot RF, Rosen BR, Taveras JM, Brady TJ. NMR imaging of intracranial hemorrhage. *J Comput Assist Tomogr* 1984;8:599–607.
76. Di Chiro G, Brooks RA, Girton ME, Caporale T, Wright DC, Dwyer AJ, Horne MK, III. Sequential MR studies of intracerebral hematomas in monkeys. *AJNR* 1986;7:193–199.
77. Dooms GC, Fisher MR, Hricak H, Higgins CB. MR imaging of intramuscular hemorrhage. *J Comput Assist Tomogr* 1985;9:908–913.
78. Gomori JM, Grossman RI, Goldberg HI, Zimmerman RA, Bilaniuk LT. Intracranial hematomas: Imaging by high-field MR. *Radiology* 1985;157:87–93.
79. Gomori JM, Grossman RI, Goldberg HI, Hackney DB, Zimmerman RA, Bilaniuk LT. Occult cerebral vascular malformations: High-field MR imaging. *Radiology* 1986;158:707–713.
80. Moon KL, Brant-Zawadzki M, Pitts LH, Mills CM. Nuclear magnetic resonance imaging of CT-isodense subdural hematomas. *AJNR* 1984;5:319–322.
81. Sipponen JT, Sepponen RE, Sivula A. Nuclear magnetic resonance (NMR) imaging of intracerebral hemorrhage in the acute and resolving phases. *J Comput Assist Tomogr* 1983;7:954–959.
82. Sipponen JT, Sepponen RE, Tanttu JI, Sivula A. Intracranial hematomas studied by MR imaging at 0.17 and 0.02 T. *J Comput Assist Tomogr* 1985;9:698–704.
83. Swensen SJ, Keller PL, Berquist TH, McLeod RA, Stephens DH. Magnetic resonance of hemorrhage. *AJR* 1985;145:921–927.
84. Zimmerman RD, Deck MDF. Intracranial hematomas: Imaging by high-field MR. *Radiology* 1985;159:565.
85. Brasch RC, Wesbey GE, Gooding CA, Koerper MA. Magnetic resonance imaging of transfusional hemosiderosis complicating thalassemia major. *Radiology* 1984;150:767–771.
86. Drayer B, Burger P, Darwin R, Riederer S, Herfkins R, Johnson GA. Magnetic resonance imaging of brain iron. *AJNR* 1986;7:373–380.
87. Bradley WG, Kortman KE, Crues JV. Central nervous system high-resolution magnetic resonance imaging: Effect of increasing spatial resolution on resolving power. *Radiology* 1985;156:93–98.
88. Edelstein WA, Glover GH, Hardy CJ, Redington RW. The intrinsic signal-to-noise ratio in NMR imaging. *Magn Reson Imaging* 1986;3:604–618.
89. Bellon EM, Haacke EM, Coleman PE, Sacco DC, Steiger DA, Gangarosa RE. MR artifacts: A review. *AJR* 1986;147:1271–1281.
90. Beyer D, Steinbrich W, Friedman G, Ermers J. Use of surface

coils in magnetic resonance imaging of orbit and knee. *Diagn Imag Clin Med* 1986;55:84–91.

91. Harms SE, Muschler G. Three-dimensional MR imaging of the knee using surface coils. *J Comput Assist Tomogr* 1986; 10:773–777.
92. Chui M, Blakesley D, Mohapatra S. Test method for MR image slice profile. *J Comput Assist Tomogr* 1985;9:1150–1152.
93. Crooks LE, Watts J, Hoenninger J, Arakawa M, Kaufman L, Guenther H, Feinberg D. Thin section definition in magnetic resonance imaging. *Radiology* 1985;154:463–467.
94. Feinberg DA, Crooks LE, Hoenninger JC, Watts JC, Arakawa M, Chang H, Kaufman L. Contiguous thin multisection MR imaging by two-dimensional Fourier transform techniques. *Radiology* 1986;158:811–817.
95. Fitzsimmons JR, Googe RE. Multisection-multiecho MR imaging: Effect on image quality. *Radiology* 1985;157:813–814.
96. Kneeland JB, Shimakawa A, Wehrli FW. Effect of intersection spacing on MR image contrast and study time. *Radiology* 1986;158:819–822.
97. Babcock EE, Brateman L, Weinreb JC, Horner SD, Nunnally RL. Edge artifacts in MR images: Chemical shift effect. *J Comput Assist Tomogr* 1985;9:252–257.
98. Pusey E, Lufkin RB, Brown RKJ, Solomon MA, Stark DD, Tarr RW, Hanafee WN. Magnetic resonance imaging artifacts: Mechanism and clinical significance. *Radiographics* 1986; 6:891–911.
99. Weinreb JC, Brateman L, Babcock EE, Maravilla KR, Cohen JM, Horner SD. Chemical shift artifact in clinical magnetic resonance images at 0.35 T. *AJR* 1985;145:183–185.
100. Purcell EM. Electricity and magnetism. Berkeley Physics Course, vol. 2. New York: McGraw-Hill; 1963.
101. Ehman RL, McNamara MT, Brasch RC, Felmlee JP, Gray JE, Higgins CB. Influence of physiologic motion on the appearance of tissue in MR images. *Radiology* 1986;159:777–782.
102. Schultz CL, Alfidi RJ, Nelson AD, Kopiwoda SY, Clampitt ME. The effect of motion on two-dimensional Fourier transformation magnetic resonance images. *Radiology* 1984; 152:117–121.
103. Wendt RE, III, Murphy PH, Ford JJ, Bryan RN, Burdine JA. Phase alterations of spin echoes by motion along magnetic field gradients. *Magn Reson Med* 1985;2:527–533.
104. Wood ML, Henkelman RM. MR imaging artifacts in magnetic resonance imaging. *Med. Phys* 1985;12:143–151.
105. Lufkin RB, Pusey E, Stark DD, Brown R, Leikind B, Hanafee WN. Boundary artifacts due to truncation errors in MR imaging. *AJR* 1986;147:1283–1287.
106. Wood ML, Henkelman RM. Truncation artifacts in magnetic resonance imaging. *Magn Reson Med* 1985;2:517–526.
107. Gibbs JW. XIX. *Fourier's series*. IN: The Scientific Papers of J. Willard Gibbs, PhD., LL.D. London: Longmans, Green; 1906:258–260.
108. Schenck JF. Willard Gibbs and the Gibbs phenomenon. *AJR* 1990:1127.
109. Carslaw HC. *The approximation curves and the Gibbs phenomenon in Fourier's series*. IN: Introduction to the Theory of Fourier's Series and Integrals. York: Dover; 1930:289–311.
110. Turner DA, Rapoport MI, Erwin WD, McGould M, Silvers RI. Truncation artifact: A potential pitfall in MR imaging of the menisci of the knee. *Radiology* 1991;179(3):629–633.
111. Crues JV, Mink J, Levy TL, Lotysch M, Stoller DW. Meniscal tears of the knee: Accuracy of magnetic resonance imaging. *Radiology* 1987;164:445–448.
112. Lufkin RB, Votruba J, Reicher M, Bassett L, Smith SD, Hanafee WN. Solenoid surface coils in magnetic resonance imaging. *Radiology* 1986;146:409–412.
113. Axel L. Surface coil magnetic resonance imaging. *J Comput Assist Tomogr* 1984;8:381–384.
114. Ehman RL. MR imaging with surface coils. *Radiology* 1985;157:549–550.
115. Fisher MR, Barker B, Amparo BC, Brandt R, Brant-Zawadzki M, Hricak H, Higgins CB. MR imaging using specialized coils. *Radiology* 1985;157:443–447.
116. Hyde JS, Kneeland JB. *High resolution methods using local coils*. IN: Wehrli FW, Kneeland JB, Shaw D, eds. Biomedical magnetic resonance imaging. New York: VCH; 1987.
117. Kneeland JB, Jesmanowicz W, Grist TM, Hyde JS. High-resolution MR imaging using loop-gap resonators: Work in progress. *Radiology* 1986;158:247–250.
118. Kulkarni MV, Patton JA, Price RR. Technical considerations for the use of surface coils in MRI. *AJR* 1986;147:373–378.

CHAPTER 2

MRI Biological Effects, Safety, and Patient Management

Frank G. Shellock

During clinical magnetic resonance imaging (MRI), the patient is exposed to three different types of electromagnetic radiation: a static magnetic field, gradient or time-varying magnetic fields, and radiofrequency (RF) electromagnetic fields. Each of these forms of electromagnetic radiation can cause significant biologic effects if applied at sufficiently high exposure levels. While numerous studies have been performed to identify potential biological effects of MRI and none of these have determined the presence of any significant hazard (1–71), the data are still insufficient to assume absolute safety.

This chapter will describe the current recommended safety guidelines for MRI, review the literature on biological effects associated with exposure to electromagnetic fields with respect to the studies that either indirectly or directly apply to MRI, and also provide an overview of safety and patient management aspects of this imaging technique.

CURRENT SAFETY RECOMMENDATIONS FOR MRI

In 1988, the U.S. Food and Drug Administration (FDA) reclassified MRI scanners from a Class III to a Class II device (74) and revised safety recommendations that were originally issued in 1982 (75). The present recommended FDA guidelines for exposures to the electromagnetic radiation used during MRI are as follows:

1. *Static magnetic field.* Static magnetic field strengths not exceeding 2.0 T are below the level of concern for the static magnetic field. Should the static magnetic field strength exceed 2.0 T, additional evidence of safety must be provided by the sponsor manufacturer.
2. *Gradient magnetic field.* The acceptable exposures to gradient magnetic fields are those that are insufficient to induce peripheral neuromuscular stimulation by at least a factor of three. Alternatively, other safety criteria may be implemented by those involved in MRI, as follows: (a) Demonstrate that the maximum dB/dt of the MRI system is 6 T/sec or less. (b) Demonstrate that for axial gradients, dB/dt < 20 T/sec for $r \geq 120$ μsec, or dB/dt $< (2{,}400\ /\ r)$ T/sec for 12 μsec $< r < 120$ μsec, or dB/dt < 200 T/sec for $r \leq 12$ μsec (r equals the width in microsec of a rectangular pulse or the half period of a sinusoidal dB/dt pulse). For transverse gradients, dB/dt is considered to be below the level of concern when it is less that three times the above limits for axial gradients.
3. *RF electromagnetic fields.* The acceptable exposures are a whole body averaged specific absorption rate (SAR) of $\leq$ 0.04 W/kg, spatial peak in 1 gram of tissue of $\leq$ 8.0 W/kg, and 3.2 W/kg averaged over the head or, alternatively, those that do not cause a body core temperature greater than 1°C except for patients with impaired systemic blood flow and/or perspiration (i.e., patients with compromised thermoregulatory systems), local exposures of less than 38°C in the head, 39°C within the trunk, and 40°C in the extremities (74). The above exposure levels apply to normal clinical environments where the individual is resting and lightly dressed (74).

Other safety considerations are specified in the present labeling of MRI systems, which require the inclusion of statements indicating contraindications for patients with electrically, magnetically, or mechanically activated implants as well as warnings addressing risks for ferromagnetic implants, potential risks of acoustic noise, and the use of MRI during pregnancy and in infants (74).

BIOLOGIC EFFECTS OF STATIC MAGNETIC FIELDS

The literature on static magnetic field related biological effects is contradictory and often confusing. Unfortunately, there is a paucity of data concerning the effects of high intensity static magnetic fields on man. Some of the original investigations on human subjects exposed to static magnetic fields were performed by Vyalov et al. (76,77) who studied workers involved in the permanent magnet industry. These subjects were exposed to static magnetic fields ranging from 0.015 T to 0.35 T and reported feelings of headache, chest pain, fatigue, vertigo, loss of appetite, insomnia, itching, and other nonspecific ailments (76,77). Exposure to additional potentially hazardous environmental working conditions (i.e., elevated room temperature, airborne metallic dust, and chemicals, etc.) may have been partially responsible for the reported symptoms in these study subjects. Since this investigation lacked an appropriate control group, it is difficult to ascertain if there was a definite correlation between the exposure to the static magnetic fields and the reported abnormalities. Subsequent studies performed with more scientific rigor have not substantiated many of the above findings (78–84).

Temperature Effects

There are conflicting statements in the literature regarding the effect of static magnetic fields on body and skin temperatures of mammals. Reports have indicated that static magnetic fields either increase or both increase and decrease temperature depending on the orientation of the organism in the static magnetic fields (17,60). Other articles state that static magnetic fields have no effect on skin and body temperatures of mammals (49,55,80,82).

None of the investigators who identified a static magnetic field effect on temperatures proposed a plausible mechanism for this response. In addition, studies reporting static magnetic field-induced skin and/or body temperature changes used laboratory animals that are known to have labile temperatures or used instrumentation that may have been perturbed by the static magnetic fields (17,60).

A recent investigation indicated that exposure to a 1.5 T static magnetic field does not alter skin and body temperatures in man (55). This study was performed using a special fluoroptic thermometry system demonstrated to be unperturbed by high intensity static magnetic fields (55). Therefore, skin and body temperatures of human subjects are not believed to be effected by exposure to static magnetic fields up to 1.5 T (49,55).

Magnetohydrodynamic Effects

A magnetohydrodynamic effect (i.e., an additional biopotential superimposed on the normal biopotential) may be observed during exposure to static magnetic fields and is caused by blood, a conductive fluid, flowing through a magnetic field. The result is an induced biopotential. This induced biopotential is exhibited by an augmentation of T wave amplitude as well as other nonspecific waveform changes that are apparent on the electrocardiogram (EKG) (Fig. 2.1) and have been observed at static magnetic field strengths of as low as 0.1 T (80,83,84). Alternative lead positions can be used to attenuate the static magnetic field-induced EKG changes in order to facilitate cardiac gated studies (85). The increase in T wave amplitude is directly related to the intensity of the static magnetic field, such that at low static magnetic field strengths, the effects are not as predominant compared with higher ones.

Because no other circulatory alterations appear to coincide with these electrocardiographic changes, no short-term biologic risks are believed to be associated with the magnetohydrodynamic effect that occurs in conjunction with static magnetic field strengths up to 2 T (79,80,83). However, since an elevation in T wave amplitude is indicative of myocardial infarction, ischemia, or potassium toxicity, it may be necessary to monitor the electrocardiogram immediately before and after MRI (especially if performed with high field strength systems) without the influence of the static magnetic field in order to assess these conditions in certain high-risk cardiac patients.

Neurological Effects

Theoretically, electrical impulse conduction in nerve tissue may be affected by exposure to static magnetic fields. However, this is another area in the biological effects literature that contains contradictory information. Some researchers have reported remarkable effects on the function as well as the structures of the central nervous system that were associated with exposure to static magnetic fields (14,22,35, 44,58,61,68,69,70,86,87), while others have failed to

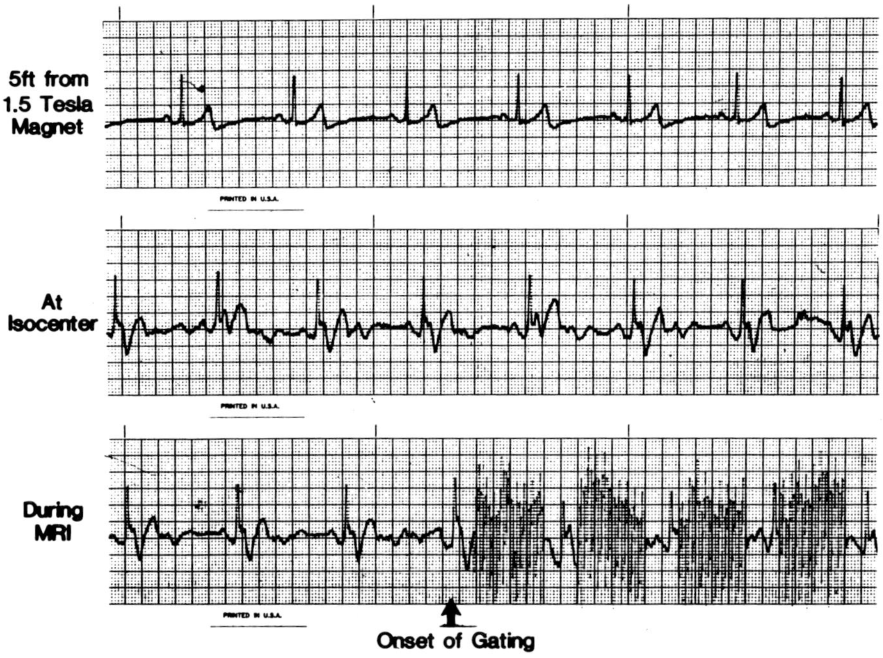

FIG. 2.1. Electrocardiogram recorded outside, inside and during the operation of a 1.5 T MRI scanner. Note the augmented T-wave due to the magnetohydrodynamic effect as well as the other nonspecific changes caused by the static magnetic field. Also, during MRI (onset of gating) there is severe distortion of the ECG waveform.

show any significant changes (18,20,62). Further investigations of potential unwanted biological effects are needed because of the relative lack of clinical studies in this field which are directly applicable to MRI. At the present time, exposure to static magnetic fields up to 2.0 T do not appear to significantly influence bioelectric properties of neurons in man (79,88–90).

In summary, there is no conclusive evidence for irreversible or hazardous biological effects related to acute, short-term exposures of man to static magnetic fields up to field strengths of 2.0 T. However, as of 1991, there are five 4.0 T MR scanners operating at various research sites around the world. A preliminary study has indicated that workers and volunteer subjects exposed to a 4.0 T MRI system have experienced vertigo, nausea, headaches, a metallic taste in their mouths, and magnetophosphenes (i.e., visual light flashes believed to be caused by gradient magnetic field-induced excitation of the optic nerve or retina) (44). Further research is required to study the mechanisms responsible for these undesirable biologic effects and to determine possible means, if any, to counterbalance them.

BIOLOGIC EFFECTS OF GRADIENT MAGNETIC FIELDS

Gradient or time-varying magnetic fields can induce electrical fields and currents in conductive media, including biologic tissue, according to Faraday's law of induction. The potential for interaction between gradient magnetic fields and biologic tissue is inherently dependent on the fundamental field frequency, maximum flux density, average flux density, the presence of harmonic frequencies, the waveform characteristics of the signal, the polarity of the signal, and the current distribution in the body, as well as on the electrical properties and sensitivity of the particular cell membrane (79,88–94).

In general, for animal and human subjects, the induced current is proportional to the conductivity of the biologic tissue and the rate of change of the magnetic flux density (79,96,97). The current density will be enhanced at higher frequencies and magnetic flux densities and further augmented by a larger tissue radius with a greater tissue conductivity (79,96,97). Current paths are affected by tissue types such that tissues with low conductivity (e.g., adipose and bone) will change the pattern of the induced current (79,96,97).

In addition to the gradient magnetic fields operating during MRI, an electrical current can also be induced in an object moving in a static magnetic field, as is the case of a patient introduced into the MRI scanner or when an individual moves in the immediate vicinity of the static magnetic field (44).

The induction of significant electrical currents in conductive tissues is a potential source of harmful biologic effects associated with MRI. For example, intense currents could directly stimulate electrically excitable cells (i.e., nerve cells, skeletal muscle fibers, blood vessels, cardiac muscle fibers, etc.) to depolarize and alter normal physiologic function. The physiologic reactions to induced electrical currents range from mild cutaneous sensations to involuntary muscle contractions, burns, cardiac arrhythmias, and cardiac fibrillation (79–81,96,97,99).

The production of magnetophosphenes, which are visual light flashes or flickers, is considered to be one of the most sensitive physiologic responses to gradient magnetic fields (79,80,98). Magnetophosphenes are supposedly caused by electrical stimulation of the retina and are completely reversible with no associated health effects (79,88,98). There has been no report of the occurrence of magnetophosphenes related to the use of clinical MRI scanners up to field strengths of 1.5 T. However, magnetophosphenes have been observed by normal subjects with a 4.0 T MRI scanner (44).

The current recommendation issued by the U.S. FDA for exposure to gradient magnetic fields during MRI is that the rate of change of the magnetic fields should not be sufficient to cause peripheral nerve stimulation by an adequate margin of safety (at least a factor of three) (74). In the event that significant electrical currents are induced during MR, it is predicted that cutaneous nerves or peripheral skeletal muscle will be stimulated and, thus, give ample warning before the occurrence of a more deleterious response, such as cardiac arrhythmia or fibrillation (79,88,90,91,96).

While the above guidelines are believed to provide a wide margin of safety with respect to exposure to gradient magnetic fields during MRI, it is imperative to realize that the newer echo planar or "snap shot" imaging techniques that require more rapid and complex applications of gradient magnetic fields may easily exceed the recommended levels and must be thoroughly evaluated for potential health hazards. Preliminary studies performed in human subjects have demonstrated that induced eddy currents have resulted in peripheral nerve stimulation producing muscle twitches or contractions in synchrony with field pulses (10,13).

BIOLOGIC EFFECTS OF RADIOFREQUENCY ELECTROMAGNETIC FIELDS

Radiofrequency (RF) radiation is capable of generating heat in tissues as a result of resistive losses. Therefore, the main biologic effects associated with exposure to RF radiation are related to the thermogenic qualities of this electromagnetic field (79,88–92,100–103).

Exposure to RF radiation may also cause athermal, field-specific alterations in biologic systems that are produced without a significant increase in temperature (92,95,98,104). This topic is somewhat controversial due to assertions concerning the role of electromagnetic fields in producing cancer and developmental abnormalities, along with the concomitant ramifications of such effects (92,95,98,104). A report from the U.S. Environmental Protection Agency claimed that the existing evidence on this issue is sufficient to demonstrate a relationship between low level electromagnetic field exposures and the development of cancer (95). To date, there have been no specific studies performed to study potential athermal biological effects of MRI. However, those interested in a thorough review of this topic, particularly as it pertains to MRI, are referred to the extensive article written by Beers (104).

Since the primary biologic responses of exposure to RF radiation are related to tissue heating that results from energy absorption, investigators have typically quantified exposure by means of determining the SAR (81,98,101). SAR is the mass normalized rate at which RF power is coupled to biologic tissue and is indicated in units of watts per kilogram (W/kg). Measurements or estimates of SAR are not trivial, particularly in human subjects, and there are several methods of determining this parameter for RF energy dosimetry (81,98,101–107).

The SAR that is produced during MRI is a complex function of numerous variables including the frequency (which, in turn, is determined by the strength of the static magnetic field), type of RF pulse (i.e., 90° or 180°), repetition time, pulse width, type of RF coil used, volume of tissue within the coil, resistivity of the tissue, configuration of the anatomical region imaged, and other factors (79,88–90). The actual increase in tissue temperature caused by exposure to RF radiation depends on the exposure duration, the ambient conditions, and the specific qualities of the subject's thermoregulatory system (i.e., skin blood flow, sweat rate, etc.).

The efficiency and absorption pattern of RF energy are mainly determined by the physical dimensions of the tissue in relation to the incident wavelength (79,81, 98,101). Therefore, if the tissue size is large relative to the wavelength, energy is predominantly absorbed on the surface; if it is small relative to the wavelength, there is little absorption of RF power (81,98,101). The most efficient absorption of RF energy is when the tissue is approximately 50% the size of the wavelength, and this frequency of peak absorption is known as the "resonant frequency." Exposure to a resonant frequency of RF power is considered to be the most hazardous because of deep and often uneven absorption patterns (81,98,101).

Because of the above relationship between RF energy and physical dimensions, studies that are designed to investigate the effects of exposure to RF radiation during MRI that are intended to be applicable to the clinical setting require tissue volumes and anatomical shapes comparable to that of human subjects. It is notable that no laboratory animal sufficiently mimics or stimulates the thermoregulatory system or re-

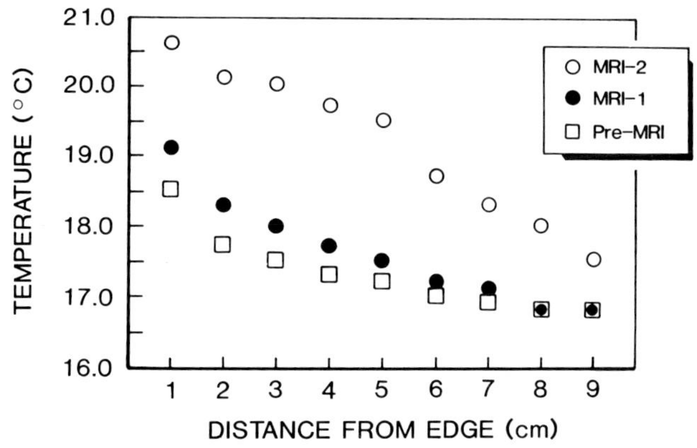

FIG. 2.2. MRI was performed on a spherical, seedless watermelon (diameter, 24 cm) to study the depth of heating. Thermistor needles were inserted into the watermelon before (*opened squares*) and after MRI performed with a 1.5T/64 MHz MRI scanner at whole body averaged SARs of 1.0 W/kg (*closed circles*) and 2.5 W/kg (*opened circles*). Room temperature was 20.5°C; relative humidity was 45%. The temperature plot shows that there is predominantly peripheral heating with small temperature changes in the central portions of the watermelon.

sponses of man. For these reasons, results obtained in laboratory animal experiments cannot simply be "scaled" or extrapolated to human subjects (81,98,101–103).

Little quantitative data have been previously available on thermoregulatory responses of humans exposed to RF radiation prior to the studies performed with MRI. The few studies that existed did not directly apply to MRI because these investigations examined either thermal sensations or therapeutic applications of diathermy, usually involving only localized regions of the body (81,98,101,105). Several studies of RF power absorption during MRI have been performed recently and have yielded useful information about tissue heating results primarily from magnetic induction with a negligible contribution from the electric fields, so that ohmic heating occurs greatest at the surface of the body and approaches zero at the center of the body (Fig. 2.2). Predictive calculations and measurements obtained in phantoms and human subjects exposed to MRI supports this pattern of temperature distribution (50–54,56,106,107).

Although a recent paper reported significant temperature rises in internal organs produced by MRI (59), this study was conducted on anesthetized dogs and is unlikely to be applicable to conscious adult human subjects because of factors related to the physical dimensions and dissimilar thermoregulatory systems of these two species. However, these data may have important implications for the use of MRI in pediatric patients since this patient population is typically sedated or anesthetized for MRI examinations (59).

An investigation during fluoroptic thermometry probes that are unperturbed by electromagnetic fields (108) demonstrated that human subjects exposed to MRI at SAR levels up to 4.0 W/kg (i.e., 10 times higher than the level currently recommended by the U.S. FDA) have no statistically significant increases in body temperatures and elevations in skin temperatures that are not believed to be clinically hazardous (Fig. 2.3)

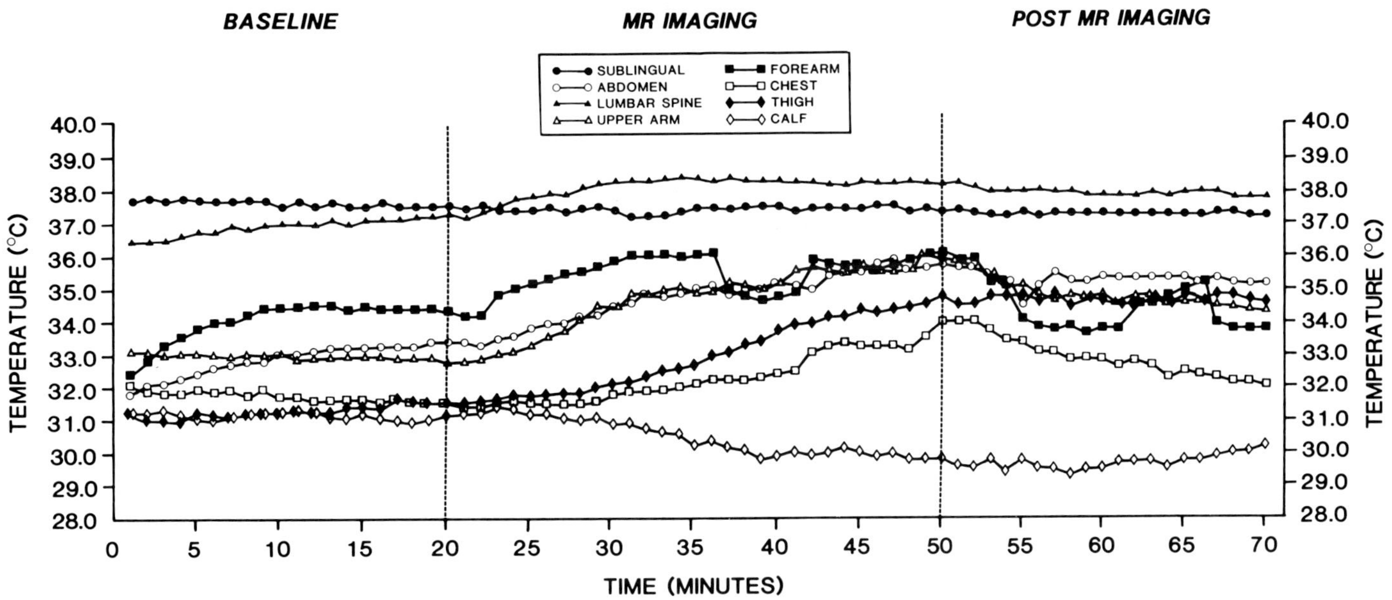

FIG. 2.3. Temperature changes associated with MRI of the abdomen performed at a whole body averaged SAR of 2.8 W/kg. Note that there was little or no change in body temperature, while there were slight to moderate changes in skin temperatures depending on the site of measurement (see ref. 56).

(56). These results imply that the suggested exposure level of 4.0 W/kg for RF radiation during MRI is too conservative for individuals with normal thermoregulatory function (56). However, additional studies are needed to assess physiologic responses of patients with conditions that may impair thermoregulatory function (e.g., elderly patients; patients with underlying health conditions such as fever, diabetes, cardiovascular disease, or obesity; and patients taking mediations that affect thermoregulation such as calcium blockers, beta blockers, diuretics, vasodilators, etc.) before subjecting them to MRI procedures that require high SARs.

Certain human organs that have reduced capabilities for heat dissipation, such as the testis and eye, are particularly sensitive to elevated temperatures. Therefore, these are primary sites of potential harmful effects if RF radiation exposures during MRI are excessive. Laboratory investigations have demonstrated detrimental effects on testicular function (i.e., a reduction or cessation of spermatogenesis, impaired sperm motility, degeneration of seminiferous tubules, etc.) caused by RF radiation-induced heating from exposures sufficient to raise scrotal and/or testicular tissue temperatures between 38 to 42°C (109).

Scrotal skin temperatures (i.e., an index of intratesticular temperature) were measured in volunteer subjects undergoing MRI at a whole body averaged SAR of 1.1 W/kg (57). The highest scrotal skin temperature recorded was 34.2°C (57). These temperature changes were below the threshold known to impair testicular function. However, excessively heating the scrotum during MRI could exacerbate certain preexisting disorders associated with increased scrotal/testicular temperatures (e.g., acute febrile illnesses, varicocele, etc.) in patients who are already oligospermic and may lead to temporary or permanent sterility (57). Therefore, additional studies designed to investigate these issues are needed, particularly if patients are scanned at whole body averaged SARs higher than those previously evaluated.

Dissipation of heat from the eye is a slow and inefficient process due to its lack of vascularization. Acute near-field exposures of RF radiation to the eyes or heads of laboratory animals have been demonstrated to be cataractogenic as a result of the thermal disruption of ocular tissues if the exposure is of sufficient intensity and duration (81,98). An investigation conducted by Sacks et al. (47) revealed that there were no detectable effects produced by MRI on the eyes of rats at exposures that far exceeded typical clinical imaging levels. However, it may not be acceptable to extrapolate these data to human subjects considering the coupling of RF radiation to the anatomy and tissue volume of laboratory rat eyes compared with those of man.

Corneal temperatures have been measured in patients undergoing MRI of the brain using a transmit/receive head coil at local SARs up to 3.1 W/kg (53). The largest corneal temperature change was 1.8°C and the highest temperature measured was 34.4°C. Since the temperature threshold for RF radiation-induced cataractogenesis in animal models has been demonstrated to be between 41° and 55°C for acute, near-field exposures, it does not appear that clinical MRI using a head coil has the potential to cause thermal damage in ocular tissue (53). The effect of MRI at higher SARs and the long-term effects of MRI on ocular tissues remain to be determined.

Theoretically, an RF radiation "hot spot" caused by an uneven distribution of RF power may arise whenever current concentrations are produced in association with restrictive conductive patterns (79). There has been the suggestion that RF radiation "hot spots" may generate thermal "hot spots" under certain conditions during MRI (79). Since RF radiation is mainly absorbed by peripheral tissues, thermography has been used to study the heating pattern associated with MRI at high whole body SARs (51). This study demonstrated no evidence of surface thermal "hot spots" related to MRI of human subjects. The thermoregulatory system apparently responds to the heat challenge by distributing the thermal load, producing a "smearing" effect of the surface temperatures. However, there is a possibility that internal thermal "hot spots" may develop from MRI (59).

STUDIES OF MRI BIOLOGIC EFFECTS

A variety of studies have been performed to specifically investigate the potential biologic effects of MRI (1–73). The results of these MRI-related investigations have been predominantly negative, supporting the widely held view that no significant health risks are associated with the use of this modality. Experiments yielding positive results either identified possible, nonspecific biologic responses, determined short-term biological changes that were not considered to be deleterious, or found bioeffects that require further substantiation.

When perusing these studies, the reader should note that the dosimetry aspects of the exposure(s) to static, gradient, and/or radiofrequency electromagnetic fields have been extremely variable and include those that exceeded clinical exposures, simulated clinical exposures, or involved low-level, chronic exposures. In certain cases, the effects of only one of the electromagnetic fields used for MRI was evaluated. Theoretically, there is a possibility that the combination of static, gradient, and RF electromagnetic fields may produce some unusual and/or unpredictable bioeffects that are unique to MRI (90,104).

"Window" effects are often present with respect to biologic changes that occur in response to electromagnetic radiation. "Window" effects are those biologic alterations associated with a specific spectrum of electromagnetic radiation that are not observed at levels below or above this range (81,98). Virtually all of the experiments conducted to date on MRI biologic effects have been performed at specific "windows," and the results cannot be assumed to apply to all of the various field strengths or frequencies used for clinical MRI.

The selection of biologic systems used for these experiments has also been quite varied. As previously mentioned, since the coupling of electromagnetic radiation to biologic tissues is highly dependent on organism/subject size, anatomical factors, duration of exposure, the sensitivity of the involved tissues, and a myriad of other variables (78,80,81,92,98,104), studies performed on laboratory preparations may not be extrapolated or directly applicable to human subjects or the clinical use of MRI. Therefore, a cautionary approach to the interpretation of the results of some of these studies is advisable.

MRI AND PATIENTS WITH ELECTRICALLY, MAGNETICALLY, OR MECHANICALLY ACTIVATED IMPLANTS AND DEVICES

The U.S. FDA requires labeling of MRI scanners to indicate that the device is contraindicated for patients who have electrically, magnetically, or mechanically activated implants, because the magnetic and electromagnetic fields produced by the MR device may interfere with the operation of these devices (74,75). Therefore, patients with internal cardiac pacemakers, implantable cardiac defibrillators, cochlear implants, neurostimulators, bone-growth stimulators, implantable electronic drug infusion pumps, and other similar devices which could be adversely effected by the electromagnetic fields used for MRI should not be examined with this imaging modality (79,88–90,110–116). Prior ex vivo testing of certain of these implants and devices may indicate that they are, in fact, MRI compatible.

The risks associated with scanning patients with cardiac pacemakers are related to the possibility of movement, reed switch closure or damage, programming changes, inhibition or reversion to an asynchronous mode of operation, electromagnetic interference, and induced currents in lead wires (79,115,116). At least one patient with a pacemaker has been scanned by MRI without incident (117). A letter-to-the-editor recently indicated that a patient who was not pacemaker-dependent underwent MRI knowingly having his pacemaker "disabled" during the procedure (117). Although this patient sustained no apparent discomfort and the pacemaker was not damaged, it is inadvisable to routinely perform this type of maneuver on patients with pacemakers because of the aforementioned hazards.

Of particular concern is the possibility that the pacemaker lead wire(s) or other similar intracardiac wire configuration could act as an antenna in which the gradient and/or RF electromagnetic fields may induce sufficient current to cause fibrillation, a burn, or some other potentially dangerous event (79,112,118). Because of this theoretically deleterious and unpredicted effect, patients referred to MRI with residual external pacing wires, temporary pacing wires, Swan-Ganz catheters, and/or any other type of internally or externally positioned conductive wire or similar device should not undergo MRI because of the associated risks (112,119,120).

Some types of cochlear implants employ a relatively high field strength cobalt samarium magnet used in conjunction with an external magnet to align and retain a radiofrequency transmitter coil on the patient's head, while other types of cochlear implants are electronically activated (110). Consequently, MRI is strictly contraindicated in patients with these implants because of the high likelihood of injuring the patient and/or the possibility of damaging or altering the operation of the cochlear implant.

Because there is a potential for demagnetizing implants that involve magnets (i.e., certain dental implants, sphincters, stoma plugs, ocular implants, and other similar devices) which may necessitate surgery to replace the damaged device, these should be removed from the patient prior to MRI, if possible (111,113). Otherwise, MRI should not be performed on a patient with a magnetically activated implant. A patient with any other similar electrically, magnetically, or mechanically activated implant or device should be excluded from examination with MRI unless the particular implant or device has been previously demonstrated to be unaffected by the electromagnetic fields used for MRI (114).

POTENTIAL HAZARDS AND SPECIAL CONCERNS ASSOCIATED WITH PERFORMING MRI IN PATIENTS WITH METALLIC IMPLANTS, MATERIALS, OR DEVICES

The potential risk associated with performing MRI in patients with ferromagnetic implants, materials, or devices are related to the induction of electric currents, heating, the misinterpretation of an artifact as an abnormality, and the possibility of movement or dislodgement (79,88–90). Electrical currents can be generated in conductive materials by gradient magnetic fields or by movement of conductive materials through the static magnetic field. Conductive implants, materials, or devices may be heated by the induction of electromotive forces subjected to gradient magnetic fields and, under certain conditions, by arcing effects if the object is positioned too close to another conductor and sufficient voltage is generated (79). To date, several incidents of MRI inducing electrical currents in metallic devices have resulted in injuries to patients (119,120). The majority of these incidents were related to the misuse of monitoring devices, whereby current was induced in the leads or cables of the monitors during MRI, resulting in superficial heating and, in some cases, serious burns. No temperature elevations of metallic implants, materials, or devices produced by MRI appear to be clinically significant (115–125).

The various types of artifacts caused by metal, including unusual artifacts arising from tattooed eyeliner and certain types of eye make-up, have been described and are generally well-recognized (125–129). The distortion of the image by metallic implants or materials is caused by disruption of the local magnetic field that perturbs the relationship between position and frequency that is crucial for accurate image reconstruction. Ferromagnetic implants, materials, or devices which incorporate small magnets can produce especially significant artifacts (111,113).

The degree of image distortion depends on the mag-

A

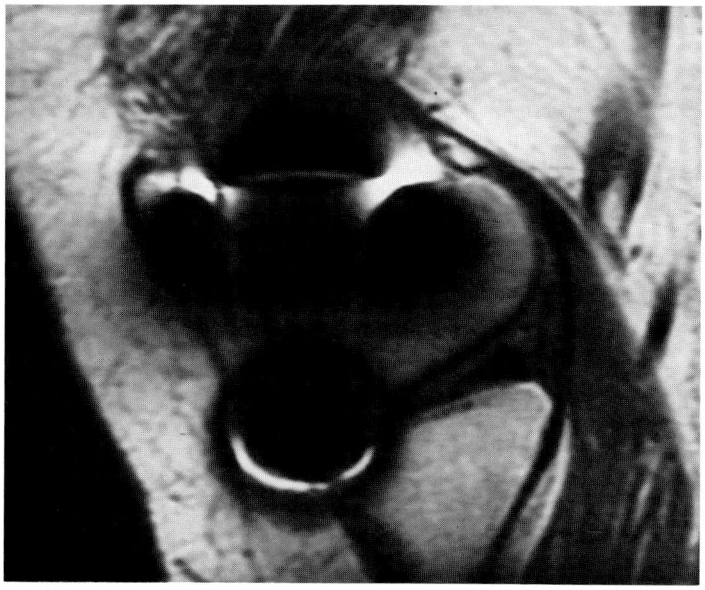

B

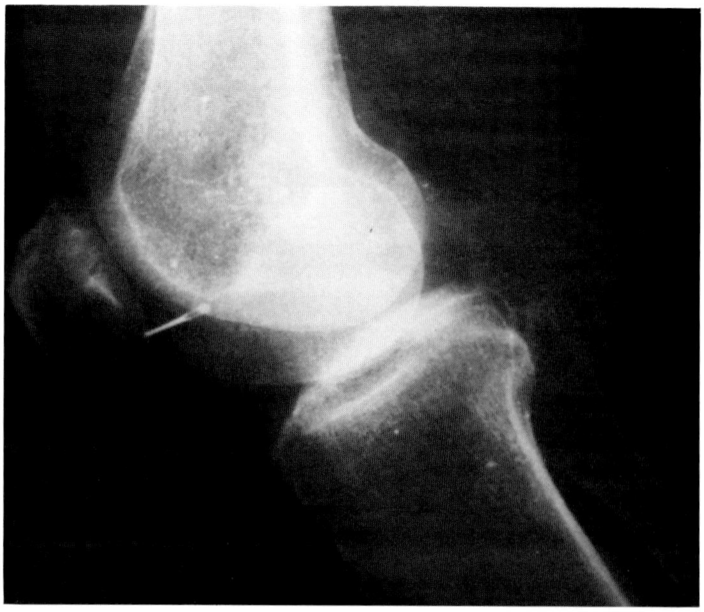

FIG. 2.4. T1-weighted sagittal plane image of the knee obtained from a patient with a sewing needle located beneath the patella **(A)**. Note the unusual image distortion caused by this ferromagnetic foreign body. The lateral plain film radiograph shows the position of the sewing needle **(B)**.

TABLE 2.1. *Metallic implants, materials, and devices tested for movement/deflection forces during exposure to static magnetic fields*

Metallic implant, material, or device	Movement/ deflection	Highest field strength (T)
Aneurysm and hemostatic clips		
Downs multipositional (17-7PH); Sims Surgical, Keene, NH	Yes	1.44
Drake (DR14, DR24); Edward Weck, Triangle Park, NJ	Yes	1.44
Drake (DR16); Edward Weck	Yes	0.147
Drake (301 SS); Edward Weck	Yes	1.5
Gastrointestinal anastomosis clip, Auto Suture SGIA (SS); United States Surgical, Norwalk, Conn	No	1.5
Heifetz (17-7PH); Edward Weck	Yes	1.89
Heifetz (Elgiloy); Edward Weck	No	1.89
Hemoclip no. 10 (316L SS); Edward Weck	No	1.5
Hemoclip (Tantalum); Edward Weck	No	1.5
Housepian implant	Yes	0.147
Kapp (405 SS); V. Mueller	Yes	1.89
Kapp curved (404 SS); V. Mueller	Yes	1.44
Kapp straight (404 SS); V. Mueller	Yes	1.44
Ligaclip, no. 6 (316L SS); Ethicon, Somerille, NJ	No	1.5
Ligaclip (tantalum); Ethicon	No	1.5
Mayfield (301 SS); Codman, Randolph, Mass	Yes	1.5
Mayfield (304 SS); Codman	Yes	1.89
McFadden (301 SS); Codman	Yes	1.5
Olivecrona	No	1.44
Pivot (17-7PH); V. Mueller	Yes	1.89
Scoville (EN58J); Downs Surgical, Decatur, Ga	Yes	1.89
Stevens (50-4190, silver alloy)	No	0.15
Sugita (Elgiloy); Downs Surgical	No	1.89
Sundt-Kees (301 SS); Downs Surgical	Yes	1.5
Sundt-Kees Multi-Angle (17-7PH); Downs Surgical	Yes	1.89
Surgiclip, Auto Suture M-9.5 (SS); United States Surgical	No	1.5
Vari-angle (17-7PH); Codman	Yes	1.89
Vari-angle McFadden (MP35N); Codman	No	1.89
Vari-angle Micro (17-7PM SS); Codman	Yes	0.15
Vari-angle Spring (17-7PM SS); Codman	Yes	0.15
Yasargil (316 SS); Aesculap, Burlingame, Calif	No	1.89
Yasargil (Phynox); Aesculap	No	1.89
Carotid artery vascular clamps		
Crutchfield (SS); Codman	Yes*	1.5
Kindt (SS); V. Mueller	Yes*	1.5
Poppen-Blaylock (SS); Codman	Yes	1.5
Salibi (SS); Codman	Yes*	1.5
Selverstone (SS); Codman	Yes*	1.5
Dental devices and materials		
Brace band (SS); American Dental, Missoula, Mont	No	1.5
Brace wire (chrome alloy); Ormco, San Marcos, Calif	Yes*	1.5
Castable allow; Golden Dental Products, Golden, Colo	Yes*	1.5
Cement-in keeper; Solid State Innovations, Mt Airy, NC	Yes*	1.5
Dental amalgam	No	1.44
GDP Direct Keeper, preformed post; Golden Dental Products	Yes*	1.5
Keeper, preformed post; Parkell Products, Farmingdale, NY	Yes*	1.5
Magna-Dent, large indirect keeper; Dental Ventures of America, Yorba Linda, Calif	Yes*	1.5
Palladium clad magnet; Parkell Products	Yes	1.5
Palladium/palladium keeper; Parkell Products	Yes*	1.5
Palladium/platinum casting alloy; Parkell Products	Yes*	1.5
Permanent crown (amalgam); Ormco	No	1.5
SS clad magnet; Parkell Products	Yes	1.5
SS keeper; Parkell Products	Yes*	1.5
Silver point; Union Broach, New York	No	1.5
Titanium clad magnet; Parkell Products	Yes	1.5
Heart valve prostheses		
Beall; Coratomic, Indiana, Penn	Yes*	2.35

(*continued*)

TABLE 2.1. *(Continued)*

Metallic implant, material, or device	Movement/ deflection	Highest field strength (T)
Bjork-Shiley (convexo/concave); Shiley, Irvine, Calif	No	1.5
Bjork-Shiley (universal/spherical); Shiley	Yes*	1.5
Bjork-Shiley, model MBC; Shiley	Yes*	2.35
Bjork-Shiley, model 25 MBRC 11030; Shiley	Yes*	2.35
Carpentier-Edwards, model 2650; American Edwards Laboratories, Santa Ana, Calif	Yes*	2.35
Carpentier-Edwards (porcine); American Edwards Laboratories	Yes*	2.35
Hall-Kaster, model A7700; Medtronic, Minneapolis	Yes*	1.5
Hancock I (porcine); Johnson and Johnson, Anaheim, Calif	Yes*	1.5
Hancock II (porcine); Johnson and Johnson	Yes*	1.5
Hancock extracorporeal, model 242R; Johnson and Johnson	Yes*	2.35
Hancock extracorporeal; model M 4365-33; Johnson and Johnson	Yes*	2.35
Hancock Vascor, model 505; Johnson and Johnson	No	2.35
Ionescu-Shiley; Universal ISM	Yes*	2.35
Lillehi-Kaster, model 300S; Medical, Inver Grove Heights, Minn	Yes*	2.35
Lillehi-Kaster, model 5009; Medical	Yes*	2.35
Medtronic Hall; Medtronic	Yes*	2.35
Medtronic Hall, model A7700-D-16; Medtronic	Yes*	2.35
Omnicarbon model 3523T029; Medical	Yes*	2.35
Omniscience, model 6522; Medical	Yes*	2.35
Smeloff-Cutter; Cutter Laboratories, Berkeley, Calif	Yes*	2.35
Starr-Edwards, model 1260; American Edwards Laboratories	Yes*	2.35
Starr-Edwards, model 2320; American Edwards Laboratories	Yes*	2.35
Starr-Edwards, model 2400; American Edwards Laboratories	No	1.5
Starr-Edwards, model Pre 6000; American Edwards Laboratories	Yes	2.35
Starr-Edwards, model 6520; American Edwards Laboratories	Yes*	2.35
St. Jude; St. Jude Medical, St. Paul	No	1.5
St. Jude, model A 101; St. Jude Medical	Yes*	2.35
St. Jude, model M 101; St. Jude Medical	Yes*	2.35
Intravascular coils, filters, and stents		
Amplatz IVC filter; Cook, Bloomington, Ind	No	4.7
Cragg nitinol spiral filter	No	4.7
Gianturco Bird's Nest IVC filter; Cook	Yes*†	1.5
Gianturco embolization coil; Cook	Yes*†	1.5
Gianturco zig-zag stent; Cook	Yes*†	1.5
Greenfield vena cava filter (SS); Medi-tech/Boston Scientific, Watertown, Mass	Yes*†	1.5
Greenfield vena cava filter (titanium alloy); Medi-tech/Boston Scientific	No	1.5
Gunther IVC filter; William Cook, Europe	Yes*†	1.5
Maas helical endovascular stent; Medinvent	No	4.7
Maas helical IVC filter; Medinvent, Lausanne, Switzerland	No	4.7
Mobin-Uddin IVC/umbrella filter; American Edwards	No	4.7
Palmaz endovascular stent; Ethicon	Yes*†	1.5
Retrievable IVC filter; Thomas Jefferson University, Philadelphia	Yes*†	1.5
Strecker stent (tantalum); Medi-tech/Boston Scientific	No	1.5
Ocular implants		
Fatio eyelid spring/wire	Yes	1.5
Intraocular lens implant; Binkhorst, iridocapsular lens; platinum-iridium loop	No	1.5
Intraocular lens implant; Binkhorst, iridocapsular lens; platinum loop	No	1.0
Intraocular lens implant; Binkhorst, iridocapsular lens; titanium loop	No	1.0
Intraocular lens implant; Worst, platinum clip lens	No	1.0
Retinal tack (303 SS); Bascom Palmer Eye Institute	No	1.5
Retinal tack (titanium alloy); Coopervision, Irvine, Calif	No	1.5

TABLE 2.1. *(Continued)*

Metallic implant, material, or device	Movement/ deflection	Highest field strength (T)
Retinal tack (303 SS); Duke University	No	1.5
Retinal tack (cobalt/nickel); Grieshaber, Fallsington, Penn	No	1.5
Retinal tack, Norton staple (platinum/rhodium); Norton	No	1.5
Retinal tack (aluminum textraoxide); Ruby	No	1.5
Retinal tack (SS-martensitic); Western European	Yes	1.5
Orthopedic implants, materials, and devices		
AML femoral component bipolar hip prosthesis; Zimmer, Warsaw, Ind	No	1.5
Cervical wire, 20 gauge (316L SS)	No	0.3
Cotrel rods with hooks (316L SS)	No	0.3
Drummond wire (316L SS)	No	0.3
DTT, device for transverse traction (316L SS)	No	0.3
Harrington compression rod with hooks and nuts (316L SS)	No	0.3
Harrington distraction rod with hooks (316L SS)	No	0.3
Harris hip prosthesis; Zimmer	No	1.5
Jewett nail; Zimmer	No	1.5
Kirschner intermedullary rod; Kirschner Medical, Timonium, Md	No	1.5
Perfix interference screw	Yes	1.5
SS mesh; Zimmer	No	1.5
SS plate; Zimmer	No	1.5
SS screw; Zimmer	No	1.5
SS wire; Zimmer	No	1.5
Zielke rod with screw, washer, and nut (316L SS)	No	0.3
Otologic implants		
Austin Tytan piston (titanium); Treace Medical, Nashville	No	1.5
Berger "V" bobbin ventilation tube (titanium); Richards Medical, Memphis	No	1.5
Cochlear implant; 3M/House	Yes	0.6
Cochlear implant; 3M/Vienna	Yes	0.6
Cochlear implant, Nucleus Mini 22-channel; Cochlear, Engelwood, Colo	Yes	1.5
Cody tack	No	0.6
Ehmke hook stapes prosthesis (platinum); Richards Medical	No	1.5
Fisch piston (Teflon, SS); Richards Medical	No	1.5
House single loop (ASTM-318-76 grade 2 SS); Storz, St. Louis	No	1.5
House single loop (tantalum); Storz	No	1.5
House double loop (tantalum); Storz	No	1.5
House double loop (ASTM-318-76 grade 2 SS); Storz	No	1.5
House-type wire loop stapes prosthesis (316L SS); Richards Medical	No	1.5
House-type SS piston and wire (ASTM-318-76 grade 2 SS); Xomed-Treace (Bristol-Meyers, Squibb)	No	1.5
House wire (tantalum); Otomed	No	0.5
House wire (SS); Otomed	No	0.5
McGee piston stapes prosthesis (316L SS); Richards Medical	No	1.5
McGee piston stapes prosthesis (platinum/316L SS); Richards Medical	No	1.5
McGee piston stapes prosthesis (platinum/$Cr_{17}Ni_4$ SS); Richards Medical (recalled by manufacturer)	Yes	1.5
McGee shepherd's crook stapes prosthesis (316L SS); Richards Medical	No	1.5
Plasti-pore piston (316L SS/plasti-pore material); Richards Medical	No	1.5
Platinum ribbon loop stapes prosthesis (platinum); Richards Medical	No	1.5
Reuter bobbin ventilation tube (316L SS); Richards Medical	No	1.5
Richards bucket handle stapes prosthesis (316L SS); Richards Medical	No	1.5
Richards platinum Teflon piston, 0.6 mm (Teflon, platinum); Richards Medical	No	1.5
Richards platinum Teflon piston, 0.8 mm (Teflon, platinum); Richards Medical	No	1.5
Richards piston stapes prosthesis	No	1.5

(continued)

TABLE 2.1. *(Continued)*

Metallic implant, material, or device	Movement/ deflection	Highest field strength (T)	Metallic implant, material, or device	Movement/ deflection	Highest field strength (T)
(platinum/fluoroplastic); Richards Medical			Schuknecht Teflon wire piston, 0.8 mm (Teflon, 316L SS); Richards Medical	No	1.5
Richards shepherd's crook (platinum); Richards Medical	No	0.5	Sheehy incus strut replacement (ASTM-318-76 grade 2 SS); Storz	No	1.5
Richards Teflon piston (Teflon); Richards Medical	No	1.5	Sheehy incus strut (316L SS); Richards Medical	No	1.5
Robinson-Moon-Lippy offset stapes prosthesis (ASTM-318-76 grade 2 SS); Storz	No	1.5	Sheehy-type incus replacement strut (Teflon, 316L SS); Richards Medical	No	1.5
Robinson-Moon offset stapes prosthesis (ASTM-318-76 grade 2 SS); Storz	No	1.5	Silverstein malleus clip ventilation tube (Teflon, 316L SS); Richards Medical	No	1.5
Robinson incus replacement prosthesis (ASTM-318-76 grade 2 SS); Storz	No	1.5	Spoon bobbin ventilation tube (316L SS); Richards Medical	No	1.5
Robinson stapes prosthesis (ASTM-318-76 grade 2 SS); Storz	No	1.5	Tantalum wire loop stapes prosthesis (tantalum); Richards Medical	No	1.5
Ronis piston stapes prosthesis (316L SS/fluoroplastic); Richards Medical	No	1.5	Tef-platinum piston (platinum); Xomed-Treace	No	1.5
Schea cup piston stapes prosthesis (platinum/fluoroplastic); Richards Medical	No	1.5	Total ossicular replacement prosthesis (TORP) (316L SS); Richards Medical	No	1.5
Schea malleus attachment piston (Teflon); Richards Medical	No	1.5	Trapeze ribbon loop stapes prosthesis (platinum); Richards Medical	No	1.5
Schea SS and Teflon wire prosthesis (Teflon, 316L SS); Richards Medical	No	1.5	Williams microclip (316L SS); Richards Medical	No	1.5
Scheer piston stapes prosthesis (316L SS/fluoroplastic); Richards Medical	No	1.5	Xomed stapes (ASTM-318-76 grade 2 SS; Xomed-Treace	No	1.5
Scheer piston (Teflon, 316L SS); Richards Medical	No	1.5	**Pellets and bullets**		
			BB's; Daisy	Yes	1.5
			BB's; Crosman	Yes	1.5
Schuknecht Gelfoam and wire prosthesis, Armstrong style (316L SS); Richards Medical	No	1.5	Bullet, 0.380 inch (copper, plastic, lead); Glaser	No	1.5
Schuknecht piston stapes prosthesis (316L SS/fluoroplastic); Richards Medical	No	1.5	Bullet, 0.44 inch (Teflon, bronze); North American Ordnance	No	1.5
			Bullet, 7.62 × 39 mm (copper, steel); Norinco	Yes	1.5
Schuknecht Tef-wire incus attachment (ASTM-318-76 grade 2 SS); Storz	No	1.5	Bullet, 0.357 inch (copper, lead); Cascade	No	1.5
			Bullet, 0.357 inch (lead); Remington	No	1.5
Schuknecht Tef-wire malleus attachment (ASTM-318-76 grade 2 SS); Storz	No	1.5	Bullet, 0.357 inch (aluminum, lead); Winchester	No	1.5
			Bullet, 9 mm (copper, lead); Remington	No	1.5
Schuknecht Teflon wire piston, 0.6 mm (Teflon, 316L SS); Richards Medical	No	1.5	Bullet, 0.380 inch (copper, nickel, lead); Geco	Yes	1.5

TABLE 2.1. *(Continued)*

Metallic implant, material, or device	Movement/ deflection	Highest field strength (T)	Metallic implant, material, or device	Movement/ deflection	Highest field strength (T)
Bullet, 0.357 inch (nickel, copper, lead); Winchester	No	1.5	Hickman (316L SS); Davol/Subsidiary of C R Bard	Yes*	1.5
Bullet, 0.357 inch (nylon, lead); Smith and Wesson	No	1.5	Hickman, pediatric (titanium); Davol/Subsidiary of C R Bard	No	1.5
Bullet, 0.45 inch (steel, lead); Evansville Ordnance	Yes	1.5	Implantofix II (polysulfone); Burron Medical, Bethlehem, Penn	No	1.5
Bullet, 0.357 inch (copper, lead); Fiocchi	No	1.5	Infusaid, model 400 (titanium); Infusaid	No	1.5
Bullet, 0.357 inch (copper, lead); Hornady	No	1.5	Infusaid, model 600 (titanium); Infusaid	No	1.5
Bullet, 9 mm (copper, lead); Norma	Yes	1.5	Lifeport, model 6013 (Delrin plastic); Strato Medical, Beverly, Mass	No	1.5
Bullet, 0.357 inch (bronze, plastic); Patton-Morgan	No	1.5	Lifeport, model 1013 (titanium); Strato Medical	No	1.5
Bullet, 0.357 inch (copper, lead); Patton-Morgan	No	1.5	Macroport (polysulfone polymer, silicone); Infusaid	No	1.5
Bullet, 0.45 inch (copper, lead); Samson	No	1.5	Microport (polysulfone polymer, silicone); Infusaid	No	1.5
Shot, 12-gauge, size: 00 (copper, lead); Federal	No	1.5	MRI (Delrin plastic, silicone); Davol/Subsidiary of C R Bard	No	1.5
Shot, 7½ (lead)	No	1.5	Norport-AC (titanium); Norfolk Medical, Skokie, Ill	No	1.5
Shot, 4 (lead)	No	1.5	Norport-DL (316L SS); Norfolk Medical	No	1.5
Shot, 00 buckshot (lead)	No	1.5	Norport-LS (titanium); Norfolk Medical	No	1.5
Penile implants			Norport-LS (316L SS); Norfolk Medical	No	1.5
AMS Malleable 600; American Medical Systems, Minnetonka, Minn	No	1.5	Norport-LS (polysulfone); Norfolk Medical	No	1.5
AMS 700 CX Inflatable; American Medical Systems	No	1.5	Norport-PT (titanium); Norfolk Medical	No	1.5
Flexi-Flate Surgitek; Medical Engineering, Racine, Wis	No	1.5	Norport-SP (polysulfone, silicone rubber, Dacron); Norfolk Medical	No	1.5
Flexi-Rod (standard) Surgitek; Medical Engineering	No	1.5	PeriPort (polysulfone, titanium); Infusaid	No	1.5
Flexi-Rod II (firm) Surgitek; Medical Engineering	No	1.5	Port-a-cath P.A.S. port (titanium); Pharmacia Deltec, St Paul	No	1.5
Jonas; Dacomed, Minneapolis	No	1.5	Port-a-cath, titanium dual lumen (titanium); Pharmacia Deltec	No	1.5
Mentor Flexible; Mentor, Minneapolis	No	1.5	Port-a-cath, titanium peritoneal (titanium); Pharmacia Deltec	No	1.5
Mentor Inflatable; Mentor	No	1.5	Port-a-cath, titanium venous low-profile (titanium); Pharmacia Deltec	No	1.5
Omniphase; Dacomed	Yes	1.5	Port-a-cath, titanium venous (titanium); Pharmacia Deltec	No	1.5
Vascular access ports					
Button (polysulfone polymer, silicone); Infusaid, Norwood, Mass	No	1.5			
Dome (titanium); Davol/Subsidiary of C R Bard, Cranston, RI	No	1.5			
Dual Microport (polysulfone polymer, silicone); Infusaid	No	1.5			
Dual Macroport (polysulfone polymer, silicone); Infusaid	No	1.5			

(continued)

TABLE 2.1. *(Continued)*

Metallic implant, material, or device	Movement/ deflection	Highest field strength (T)	Metallic implant, material, or device	Movement/ deflection	Highest field strength (T)
Port-a-cath, venous (316L SS); Pharmacia Deltec	No	1.5	Contraceptive diaphragm, All Flex; Ortho Pharmaceutical, Raritan, NJ	Yes*	1.5
Q-port (316L SS); Quinton Instrument, Seattle	Yes*	1.5	Contraceptive diaphragm, flat spring; Ortho Pharmaceutical	Yes*	1.5
S.E.A. (titanium); Harbor Medical Devices, Boston	No	1.5	Contraceptive diaphragm, Koroflex; Young Drug Products, Piscataway, NJ	Yes*	1.5
Snap-lock (titanium, polysulfone polymer, silicone); Infusaid	No	1.5	Forceps (titanium)	No	1.44
Synchromed, model 8500-1 (titanium, thermoplastic, silicone); Medtronic	No	1.5	Hakim valve and pump	No	1.44
Vasport (titanium/fluoropolymer); Gish Biomedical, Santa Ana, Calif	No	1.5	Intrauterine contraceptive device, Copper T; Searle Pharmaceuticals, Chicago	No	1.5
Miscellaneous			Shunt valve, Holter-type; Holter, Bridgeport, Penn	Yes*	1.5
Artificial urinary sphincter, AMS 800; American Medical Systems	No	1.5	Shunt valve, Holter-Hausner type; Holter-Hausner, Bridgeport, Penn	No	1.5
Cerebral ventricular shunt tube connector, Accu-flow, straight; Codman	No	1.5	Swan-Ganz thermodilution catheter; American-Edwards Laboratories, Irvine, Calif	No‡	1.5
Cerebral ventricular shunt tube connector, Accu-flow right angle; Codman	No	1.5	Tantalum powder	No	1.44
Cerebral ventricular shunt tube connector, Accu-flow, T-connector; Codman	No	1.5	Tissue expander with magnetic port; McGhan Medical, Santa Barbara, Calif	Yes	1.5
Cerebral ventricular shunt tube connector (type unknown)	Yes	0.147	Vascular marker, O-ring washer (302 SS); PIC Design, Middlebury, Conn	Yes*	1.5

Note.—"Highest field strength" refers to the highest intensity of static magnetic field in Tesla that was used for the evaluation of movement or deflection forces of the various implants, materials, and devices tested. IVC = inferior vena cava, SS = stainless steel. Material or materials used to construct the object are indicated, if known. Manufacturer information is provided, if known.

* These metallic implants, materials, or devices were considered to be safe for MR imaging despite being moved/deflected by the static magnetic fields. For example, certain heart-valve prostheses were moved/deflected by the static magnetic fields but the forces were considered to be less than that exerted on the valves by the beating heart.

† Ferromagnetic coils, filters, and stents typically become firmly incorporated into the vessel wall several weeks following placement, and therefore, it is highly unlikely that they will be moved or dislodged by magnetic forces.

‡ While there is no magnetic deflection associated with the triple-lumen thermodilution Swan-Ganz catheter, there has been a report of a catheter "melting" in a patient. Therefore, this catheter would be considered a relative contraindication for MR imaging.

TABLE 2.2. *Metallic implants and foreign bodies contraindicated for MRI*

Aneurysm clips:
- Drake (DR14, DR24), Edward Weck
- Drake (DR16), Edward Weck
- Drake (301 SS), Edward Weck
- Downs multi-positional (17-7PH)
- Heifetz (17-7PH), Edward Weck
- Housepian
- Kapp (405 SS), V. Mueller
- Kapp curved (404 SS), V. Mueller
- Kapp straight (404 SS), V. Mueller
- Mayfield (301 SS), Codman
- Mayfield (304 SS), Codman
- McFadden (301 SS), Codman
- Pivot (17-7PH), V. Mueller
- Scoville (EN58J), Downs Surgical
- Sundt-Kees (301 SS), Downs Surgical
- Sundt-Kees Multi-Angle (17-7PH), Downs Surgical
- Vari-Angle (17-7PH), Codman
- Vari-Angle Micro (17-7PM SS), Codman
- Vari-Angle Spring (17-7PM SS), Codman

Carotid artery vascular clamps:
- Poppen-Blaylock (SS), Codman

Dental devices and materials:
- [a] Palladium clad magnet, Parkell Products
- [a] Titanium clad magnet, Parkell Products
- [a] Stainless steel clad magnet, Parkell Products

Heart valves:
- Starr-Edwards, Model Pre 6000, American Edwards

Intravascular coils, stents, and filters:
- [b] Gianturco embolization coil, Cook
- [b] Gianturco bird nest IVC filter, Cook
- [b] Gianturco zig-zag Stent, Cook
- [b] Gunther IVC filter, Cook
- [b] New retrievable IVC filter, Thomas Jefferson University
- [b] Palmaz endovascular stent, Ethicon

Ocular implants:
- [a] Fatio eyelid spring/wire
- Retinal tack (SS-martensitic), Western European

Otologic implants:
- Cochlear implant (3M/House)
- Cochlear implant (3M/Vienna)
- Cochlear implant, Nucleus Mini 22-channel, Cohlear
- McGee piston stapes prosthesis, (platinum/17Cr-4Ni SS), Richards Medical

Pellets, bullets, shrapnel, etc.:
- BB's, Daisy
- BB's, Crosman
- Bullet, 7.62 × 39 mm (copper, steel), Norinco
- Bullet, 0.380 inch, (copper, nickel, lead), Geco
- Bullet, 0.45 inch, (steel, lead), North American Ordnance
- Bullet, 9 mm, (copper, lead), Norma

Penile implants:
- [b] Penile Implant, OmniPhase, Dacomed Corp.

Miscellaneous:
- Cerebral ventricular shunt tube, connector (type unknown)
- [c] Swan-Ganz Catheter, Thermodilution, American Edwards
- [a] Tissue expander with magnetic port, McGhan Medical

[a] The potential for these metallic implants or devices to produce significant injury to the patient is minimal. However, performing MRI in a patient with one of these devices may be uncomfortable for the individual and/or may result in damage to the implant.

[b] Ferromagnetic coils, filters, and stents typically become firmly incorporated into the vessel wall several weeks following placement and, therefore, it is highly unlikely that they will become dislodged by magnetic forces after a suitable period of time has passed.

[c] While no magnetic deflection is associated with the Swan-Ganz thermodilution catheter, there has been a report of a catheter "melting" in a patient undergoing MRI. Therefore, this catheter is considered contraindicated for MRI.

NOTE: The relative risks of performing MRI in patients with pellets, bullets, or shrapnel are related to whether they are positioned near a vital structure. Manufacturer information is provided if indicated in previously published reference, or if otherwise known.

SS, stainless steel.

netic susceptibility, quantity, shape, orientation, and position of the object in the body as well as the technique used for imaging and image processing (i.e., two-dimensional) Fourier transform reconstruction, back projection, etc.) (129). Artifacts caused by MRI of patients with metallic implants, materials, or devices are typically seen as local or regional distortions of the image and/or signal void which is frequently unpredictable (Fig. 2.4). In some instances, a marked hyperintensity may be observed. Certain nonferromagnetic implants, materials, or devices may likewise cause image artifacts because eddy currents can be generated in these objects by gradient magnetic fields that, in turn, disrupt the local magnetic field (79,118, 129).

Several important factors influence the relative risk of using MRI in patients with ferromagnetic implants or materials, including the strength of the static and gradient magnetic fields; the degree of ferromagnetism of the implant, material, or device; the mass and geometry of the implant, material, or device; its location and orientation in situ; and the length of time it has been in place (79,88–90,118). These factors should be carefully considered before subjecting a patient with a ferromagnetic implant, material, or device to MRI, particularly if it is located in a potentially dangerous area of the body where movement or dislodgement could injure the patient.

Numerous studies have assessed the ferromagnetic qualities of various metallic implants, materials, or devices by measuring the deflection forces or movement associated with static magnetic fields used for MRI (111,124,125,130–143). The results have indicated that patients with certain metallic implants, materials, or devices that are nonferromagnetic or ferromagnetic ones that are minimally affected by the magnetic fields relative to their in vivo applications (i.e., deflection force was considered to be insufficient to move or dislodge the implant or material in situ) can safely undergo MRI. Because prior knowledge of the interaction between the magnetic fields and the specific type of metallic implant, material, or device is essential for safe screening of patients, the literature on this topic has been compiled and is presented in Tables 2.1 and 2.2.

Various types of orthopedic implants, materials, and devices (i.e., hip prostheses, stainless steel mesh, nails, screws, plates, wires, etc.) have been evaluated to determine the presence of ferromagnetic qualities. Of these, only the Perfix interference screw used for reconstruction of the anterior cruciate ligament has significant ferromagnetism. Although the potential for movement or dislodgement of this implant is minimal once it is placed in vivo, the substantial artifact created by the presence of the Perfix interference screw prevents an evaluation of the knee by MRI (Fig. 2.5).

Patients may present for MRI with externally applied devices that must also be assessed for safety before undergoing examination by this imaging technique. For example, a halo vest used for fixation of the cervical spine is an example of a metallic device that may or may not be compatible with MRI. Most halo vests are composed of metallic materials that present unwanted risks and produce mild-to-severe imaging artifacts (144,145). Therefore, patients with halo

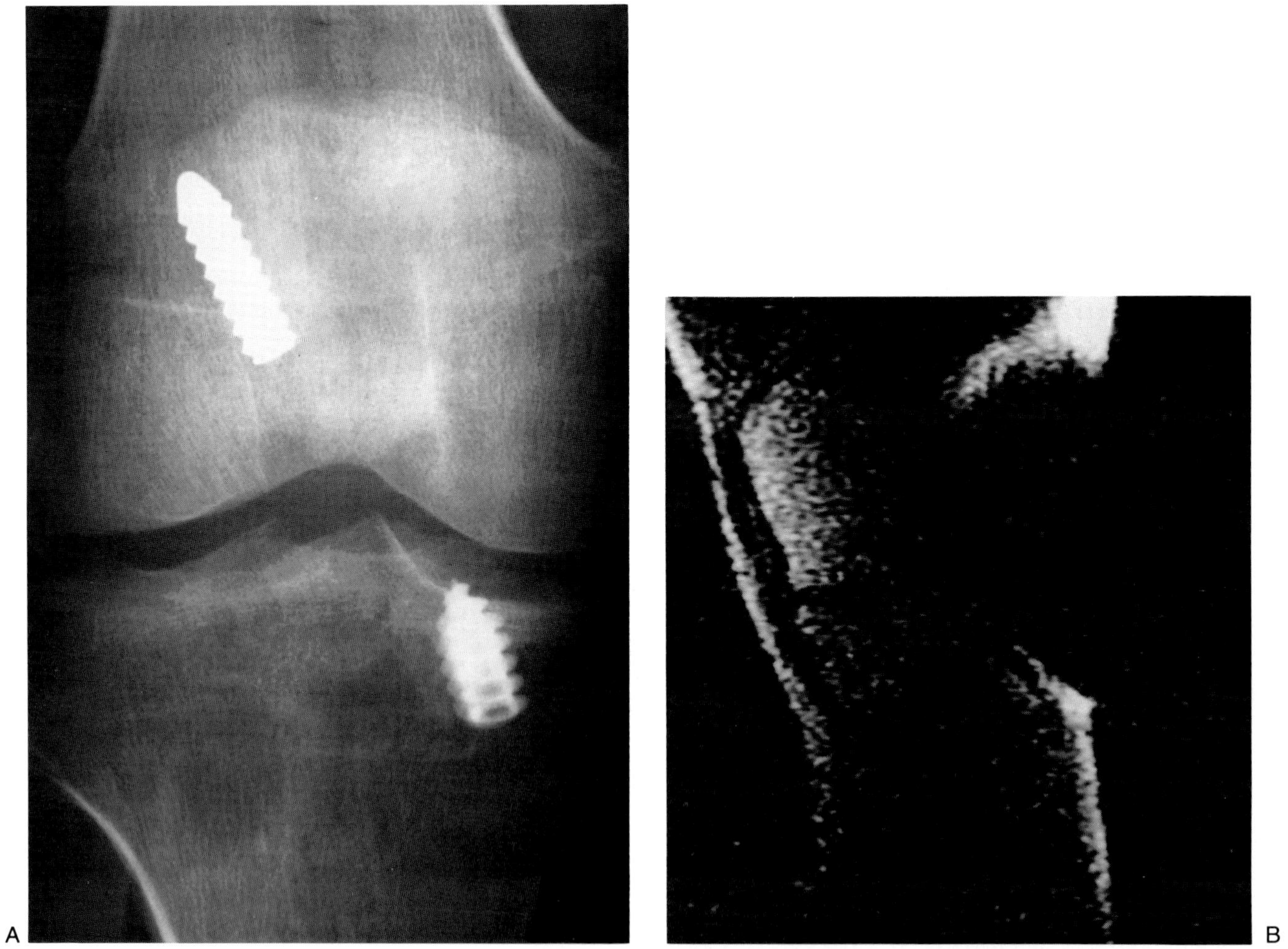

FIG. 2.5. A: Conventional radiograph of a patient that had an anterior cruciate ligament reconstruction. Note the presence of the two Perfix interference screws. **B:** T1-weighted, coronal plane, spin echo image of the same patient. The presence of the two Perfix interference screws prevents an evaluation of the knee by MRI.

vests are not typically regarded as suitable candidates for MRI. This is unfortunate because MRI often provides a useful diagnostic evaluation of the spinal cord and associated soft tissue structures that is unobtainable by other imaging techniques.

A study detailing the effect of various halo vests on image quality has indicated that these devices cause image distortion that is primarily dependent on the type of material used to construct the halo rings and support uprights (144). Halo vests made from stainless steel produced the poorest image quality, followed by those made from aluminum, titanium, and graphite (144). This prior study did not assess deflection forces, heating, or induced currents, each of which may present a potential hazard to the patient undergoing MRI if the halo vest is composed mainly of ferromagnetic and/or conductive materials.

While some commercially available halo devices are composed of nonferromagnetic materials, there is still a theoretical hazard of inducing electric current in the ring portion of any halo device made from conductive materials (145). The current within such a closed-loop is of additional concern for eddy current induction and potential image degradation effects (145,146). Furthermore, there is a potential for the patient to be involved in part of this closed-loop, so that there could be a possible burn or electric injury to the patient.

Specially designed, commercially available halo vests have now been developed which are made from both nonferromagnetic and nonconductive materials that have little or no interaction with the electromagnetic fields used for MRI (Fig. 2.6). Therefore, there are no associated safety-related problems or imaging artifacts related to the use of these devices during MRI of the cervical spine.

Another possible risk related to MRI and ferromagnetic materials is the "missile effect." The missile effect refers to the capability of the static magnetic field to attract ferromagnetic objects (i.e., oxygen tanks, tools, etc.) within the fringe field, which may be subsequently drawn into the magnet by considerable force. This represents one of the greatest potential hazards of the static magnetic field and the occurrence of the missile effect can pose a significant risk to the patient inside the scanner and/or anyone who is in the path of the ferromagnetic missile. In extreme cases, the magnet may need to be "quenched" in order to extract sizable ferromagnetic objects from MRI systems, particularly those with high-field strengths, resulting in considerable financial loss due to replacing cryogens, down-time, etc.

In order to guard against a catastrophe caused by the missile effect, the immediate area around the MRI scanner should be clearly demarcated, labeled with ap-

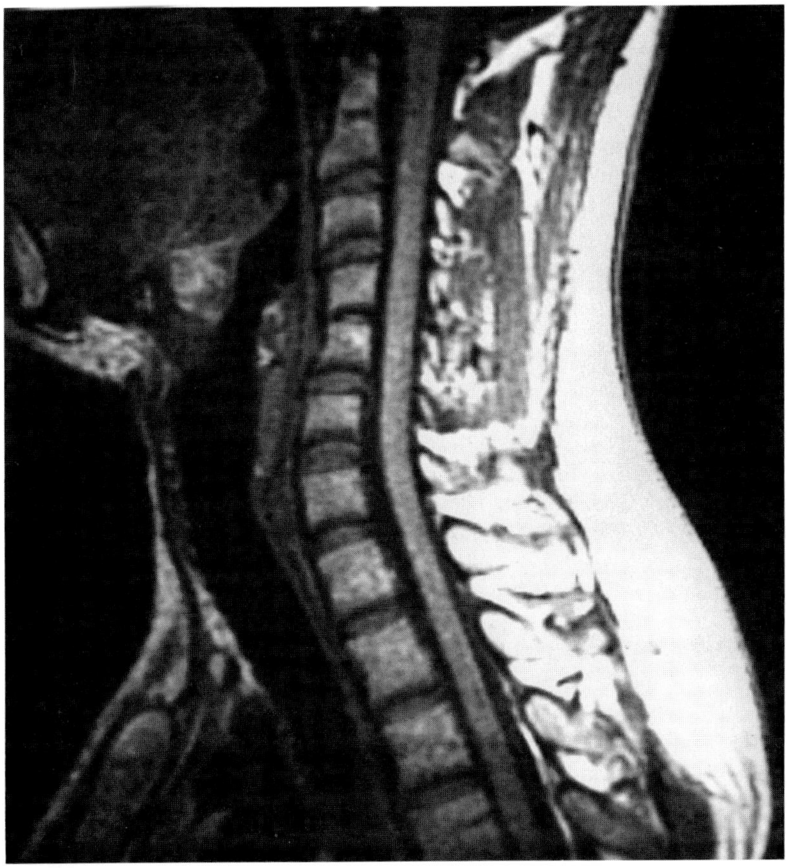

FIG. 2.6. T1-weighted sagittal plane image of the cervical spine obtained from a patient wearing an MRI-compatible halo vest.

propriate warning signs, and secured by trained staff cognizant of MRI safety. In addition, all patients, visitors, and other individuals entering the magnet area should be screened and their metallic personal belongings removed, and they should be educated about the potential hazards of the magnetic fringe field of the MRI scanner.

SCREENING PATIENTS FOR METALLIC FOREIGN BODIES BEFORE MRI

Patients may present to MRI with a history of metallic foreign bodies such as bullets, shrapnel, or other types of metallic fragments. The relative risk of scanning these patients is dependent on the possibility of movement or dislodgement of the metallic foreign body as well as whether it is positioned in a potentially hazardous site within the body (114).

A patient who encounters the static magnetic field of an MRI scanner with an intraocular metallic foreign body is at a particular risk for significant eye injury. The single reported case of a patient who experienced eye damage resulting in blindness underwent MRI on a 0.35 Tesla scanner and had an intraocular metal fragment, 2.0 × 3.5 mm in size, dislodged during the procedure (147). This incident emphasizes the importance of adequately screening patients with suspected intraocular metallic foreign bodies prior to MRI.

Research has demonstrated that small intraocular metallic fragments as little as 0.1 × 0.1 × 0.1 mm in size are detected using standard conventional radiographs (148). In addition, metallic fragments of various sizes and dimensions ranging from 0.1 mm × 0.1 mm × 0.1 mm to 3.0 mm × 1.0 mm × 1.0 mm in size have been examined to determine if they were moved or dislodged from the eyes of laboratory animals during exposure to a 2.0 T MRI system (148). Only the 3.0 mm × 1.0 mm × 1.0 mm fragment moved (i.e., it rotated, but did not cause any discernible clinical damage) (148). Therefore, the use of conventional radiography may be an acceptable technique for identifying or excluding an intraocular metallic foreign body which represents a potential hazard to the patient about to undergo MRI.

Patients with a high suspicion of having an intraocular metallic foreign body (for example, a metal worker exposed to metallic slivers with a history of eye injury) should have conventional radiographs of the orbits to rule out the presence of a metallic fragment prior to exposure to the static magnetic field. If a patient with a suspected ferromagnetic intraocular foreign body has no symptoms and conventional radiographs of the orbits does not demonstrate a radiopaque foreign body, MRI may be performed safely (148). If a patient has no history of injury to the eye but was a metal worker or subjected to metallic splinters or fragments, it is highly unlikely that this patient has a metallic foreign body that would cause an injury during MRI. Therefore, MRI may be performed in this instance without obtaining radiographs of the orbits. However, if the patient presents with a history of an eye injury caused by a metal foreign body, radiographs of the orbits should be obtained whether or not the foreign body was known to be moved. The above is the recommendation of the Safety Committee of the Society for Magnetic Resonance Imaging (114).

Patients referred for examination with MRI may also have foreign bodies from pellets, bullets, or shrapnel. Most of the pellets and bullets that have been previously tested for ferromagnetism are composed of nonferromagnetic materials (135,136). Ferromagnetic ammunition typically comes from foreign countries and/or is used by the military (135,136). Shrapnel usually contains various amounts of steel and, therefore, is ferromagnetic and presents a potential hazard for MRI (135,136). Furthermore, since pellets, bullets, and shrapnel may be contaminated with ferromagnetic materials, these objects represent relative contraindications for MRI (135,136). Patients with these foreign bodies should be regarded on an individual basis with respect to whether the object is positioned near a vital neural, vascular, or soft tissue structure. Using conventional radiography to search for metallic foreign bodies is a sensitive and relatively inexpensive means of identifying patients unsuitable for MRI, and it can also be utilized to screen patients who may have metal fragments in potentially hazardous sites of the body.

Each imaging center should establish a standardized policy for screening patients with suspected foreign bodies. The policy should include guidelines as to which patients require workup by radiographic procedures, the specific procedure to be performed (i.e., number and type of views, position of the anatomy, etc.), and each case should be considered on an individual basis (114). These precautions should be taken in any MRI system regardless of the field strength, magnet type, and the presence or absence of magnetic shielding (114).

POTENTIAL HAZARDS OF ACOUSTIC NOISE ASSOCIATED WITH MRI

A potential risk associated with the use of MRI is related to the acoustic noise produced during the operation of the scanner. Acoustic noise is caused by the activation and deactivation of the gradient magnetic fields. This noise is enhanced with higher static magnetic field strengths, higher gradient magnetic fields, higher gradient duty cycles, and sharper pulse transitions (71,114,121,122). Acoustic noise generated during MRI has been demonstrated to cause patient annoyance, interference with oral communication, and reversible hearing loss in patients who did not wear ear protection (8).

Although gradient magnetic field-related noise levels measured on several commercial MRI devices were considered to be within the recommended safety guidelines (121), these safe levels do not take into consideration that MRI is used for a patient population who may be more sensitive to the effects of acoustic noise. In addition, there has not been any research on the effect of the more significant gradient noises associated with the newer echo planar or ultrafast imaging techniques, which are significantly louder than conventional pulse sequences. Furthermore, the possibility exists that significant gradient coil-induced noise may produce permanent hearing impairment in certain patients who are particularly susceptible to the damaging effects of relatively loud noises.

The safest and least expensive means of preventing problems associated with acoustic noise during MRI is to require the routine use of disposable earplugs or another similar noise reduction device (90,114,118). An acceptable alternative strategy for reducing sound levels during MRI is to use an "antinoise" or destructive interference technique that not only effectively reduces noise but also permits better patient communication (122). While these techniques have not yet found widespread application on clinical MRI systems, they have considerable potential for minimizing acoustic noise and its associated problems (122).

THE USE OF MRI IN PREGNANT PATIENTS

The current guidance of the U.S. FDA requires labeling of MRI devices to indicate that the safety of MRI when used to image the fetus and the infant "has not been established" (74,75). In Great Britain, the acceptable limits of exposure for clinical MRI recommended by the National Radiological Protection Board in 1983 specify that "it might be prudent to exclude pregnant women during the first three months of pregnancy" (147).

Although MRI is not presently considered to be hazardous to the fetus, only a few investigations have examined the teratogenic potential of this imaging modality. By comparison, literally thousands of studies have been performed to examine the possible hazards of ultrasound during pregnancy and controversy still exists concerning the safe use of this diagnostic imaging modality.

A number of mechanisms exist that could conceivably cause adverse biologic effects with respect to the electromagnetic fields used for MRI and the development fetus (78,80,81,92,98,104). Furthermore, it is well known that cells undergoing division, as in the case of the developing fetus during the first trimester, are more susceptible to damage from a variety of physical agents. Therefore, in view of the limited data available, a cautionary approach to the use of MRI during pregnancy is recommended until additional investigations provide more information regarding safety in this population.

MRI is indicated for use in pregnant women if other nonionizing forms of diagnostic imaging are inadequate or if the examination provides important information that would otherwise require exposure to ionizing radiation (i.e., x-ray, computed tomography, etc.). It is recommended that pregnant patients be informed that, to date, there has been no indication that the use of clinical MRI during pregnancy produces deleterious effects. However, as noted by the FDA, the safety of MRI during pregnancy has not been proven (114). The above information is detailed in the Policies, Guidelines, and Recommendations for MRI Imaging Safety and Patient Management developed by the Safety Committee of the Society for Magnetic Resonance Imaging (114).

Patients who are pregnant or suspect they are pregnant must be identified prior to undergoing MRI in order to assess the risks versus the benefits of the examination. Since there is a high spontaneous abortion rate in the general population during the first trimester of pregnancy (i.e., > 30%), particular care should be exercised with the use of MRI during the first trimester because of associated potential medico-legal implications relative to spontaneous abortions.

In consideration of the above recommendations for the use of MRI in pregnant women, it is also prudent to limit the exposure to MRI-related electromagnetic fields for pregnant health workers (i.e., technologists, physicians, nurses, etc.) throughout the duration of pregnancy. At our institution, the policy for pregnant health workers (who are normally subjected to low dose, chronic exposures to electromagnetic fields during the operation of MRI devices) is to limit their exposure to static magnetic fields to a maximum 10 Gauss or less (91). This is equal to or less than the acceptable recommended occupational exposure limits (79) and is normally the upper limit of exposure encountered at the operator's console of most shielded 1.5 T MRI scanners (91). In addition, pregnant health workers should not be exposed to any gradient or RF electromagnetic fields. In most instances, the above recommendations would permit pregnant health workers to scan patients at the console of a 1.5 T MRI system, but would prohibit them from entering the scan room during the operation of the MRI scanner for the duration of their pregnancy (i.e., they should not be involved in positioning the patient prior to imaging, they should not remain in the scan room during the procedure with claustrophobic patients, etc.) (91).

MRI AND PATIENTS WITH CLAUSTROPHOBIA, ANXIETY, OR EMOTIONAL DISTRESS

Claustrophobia and a variety of other psychological reactions including anxiety, panic disorders, and emotional distress may occur in patients before or during the MRI examination (114,150–154). These sensations originate from several factors including the restrictive dimensions of the interior of the MRI scanner, the gradient noises, the duration of the examination, the ambient conditions within the bore of the magnet, etc. (114,150–157). Certain types of MRI system architectures that employ a vertical field magnet offer an unrestricted design that may reduce the instances of psychological problems associated with MRI procedures, but these are typically low-field strength MRI systems and are not widely used in clinical settings.

Fortunately, adverse psychological responses to MRI are usually transient. However, there has been a report of two patients with no prior history of claustrophobia who tolerated MRI with great difficulty and had persistent claustrophobia that required long-term psychiatric treatment (152). Because deleterious psychological responses to MRI usually delay or require cancellation of the examination, several different procedures have been developed and are recommended to avert these problems (114,150–154):

1. Educate the patient concerning the specific aspects of the MRI examination, including the internal dimensions of the scanner, the level of gradient noise to expect, etc.
2. Allow a relative or friend who has been properly screened to remain with the patient during the procedure.
3. Maintain physical or verbal contact with the patient.
4. Place the patient in a prone position to alleviate the "closed-in" feelings, or introduce the patient feet first instead of head first into the scanner.
5. Use a fan in the magnet bore to provide additional air movement.
6. Provide MRI-compatible television or headphones to transmit music to the patient.
7. Use prisms or mirrored glasses or mirrors within the scanner to allow the patient to see out of the scanner.
8. Use a blindfold so that the patient is not aware of the close surroundings.
9. Use relaxation strategies such as controlled breathing or mental imagery.
10. Use hypnotism or psychological "desensitization" techniques.

These procedures used individually or in combination will have varying degrees of success. If all else fails, short-acting sedation or anesthesia may be necessary to accomplish the MRI examination in patients experiencing claustrophobia, anxiety, panic disorders, or emotional distress.

Chloral hydrate is often used as the primary agent of sedation for MRI performed on children less than 2 years old at a dosage of 80 mg/kg (not to exceed 2,000 mg) (166). Supplemental Nembutal may be used during the MRI examination at a dosage of up to 2 mg/kg and is administered intravenously (166). Sleep deprivation is recommended for a higher success rate when sedating children (166). For adults, intravenous diazepam is typically effective for adequate sedation during MRI. In more difficult cases, general anesthesia may be required and should be administered by an anesthesiologist experienced at working in the MRI environment.

PHYSIOLOGICAL MONITORING DURING MRI

Physiologic monitoring is essential for the safe use of MRI in patients that are sedated, anesthetized, comatose, critically ill, or otherwise unable to communicate with the operator. All of these types of patients should be routinely monitored during MRI. In view of the variety of currently available MRI-compatible monitoring equipment, there is no reason to exclude these patients from examination with this imaging technique. Virtually every physiologic parameter that may be obtained under normal circumstances in the critical care unit or operating room is capable of being monitored during MRI. These parameters include heart rate, systemic blood pressure, intracardiac pressure, end-tidal carbon dioxide, oxygen saturation, respiratory rate, skin blood flow, and temperature (158–165).

Table 2.3 lists several examples of MRI-compatible monitors that have been tested and operated successfully at static magnetic field strengths up to 1.5 T. MRI-compatible respirators have also been developed for patients that require ventilatory support and are commercially available from several vendors (Table 2.3) (158–165).

Since the typical MRI scanner is constructed so that patients are positioned inside a cylindrical structure, routine observations and vital signs monitoring are not trivial tasks. Conventional monitoring equipment was not designed to operate in the MRI environment where static, gradient, and RF electromagnetic fields can adversely effect the operation of these devices.

Monitors that contain ferromagnetic components (i.e., transformers, outer casings, etc.) can be strongly attracted by mid- and high-field-strength MRI scanners, posing a serious hazard to the patient and pos-

TABLE 2.3. *Examples of MRI-compatible monitors and respirators*[a]

Device and manufacturer	Function
Omega 1400 In Vivo Laboratories, Inc. Winter Park, FL	blood pressure heart rate
MRI Fiber-Optic Pulse Oximeter Nonin Medical Plymouth, MI	oxygen saturation heart rate
525 Respiratory Rate Monitor Biochem International Waukesha, WI	respiratory rate
MicroSpan Capnometer 8800 Biochem International Waukesha, WI	respiratory rate end-tidal carbon dioxide
Aneuroid Chest Bellows Coulbourn Instruments Allentown, PA	respiratory rate
Laserflow Blood Perfusion Monitor Vasomedics, Inc. St. Paul, MN	skin blood flow
Medpacific LD 5000 Laser-Doppler Perfusion Monitor Medpacific Corporation Seattle, WA	skin blood flow
Fluoroptic Thermometry System Model 3000 Luxtron Mountain View, CA	temperature
Omni-Vent, Series D Columbia Medical Marketing Topeka, KS	respirator
Ventilator, Model 225 Monaghan Medical Corporation Plattsburgh, PA	respirator

[a] Note that these devices may require modifications to make them MRI-compatible, and none of them should be positioned closer than 8 feet from the entrance of the bore of a 1.5 T MRI scanner. Also, monitors with metallic cables, leads, or probes will cause mild-to-moderate imaging artifacts if placed near the imaging area of interest. Consult manufacturers to determine compatibility with specific MRI scanners and for additional safety information.

sible damage to the scanner. Since the intensity of the static magnetic field falls off in proportion to the distance from the magnet, simply placing the monitor at a sufficient distance (e.g., approximately 8 feet from the opening of the bore of a 1.5 T MRI scanner) from the MRI scanner usually is adequate to protect the operation of the device and prevent it from becoming a possible missile. If monitoring equipment is not placed in a permanently fixed position, instructions should be given to all appropriate personnel regarding the hazards of moving monitoring equipment too close to the MRI scanner, and the equipment should be clearly labeled.

Besides being influenced by the static magnetic field, certain monitors may be adversely affected by electromagnetic interference from the gradient and RF magnetic fields during the operation of the MRI scanner (161,163). In these instances, increasing the length of the patient-monitor interface and positioning the equipment outside the RF shielded room will enable the monitor to operate properly (161,163). It is usually necessary to position all monitors with cathode ray tubes at an appropriate location in the magnetic fringe field so that the display is not "bent" or distorted.

Certain monitors emit spurious electromagnetic noise that can result in moderate-to-severe imaging artifacts (Fig. 2.7). These monitors may be modified to work during MRI by adding RF shielded cables or outer casings. Also, special filters may be added to the monitor to inhibit electromagnetic noise.

Of further concern is the fact that some monitoring equipment is potentially harmful to the patient if special precautions are not followed (119,120). For example, there has been a report of an unfortunate accident involving an anesthetized patient who sustained a third-degree burn of the finger associated with using a pulse oximeter during MRI (119). Investigation of this incident revealed that the cable leading from the pulse oximeter to the finger probe may have been looped during MRI and the gradient or RF magnetic fields induced sufficient current to exorbitantly heat the finger probe, resulting in the finger burn (119). This problem may also occur with the use of electrocardiographic lead wires or any other cable that may be looped or form a conductive loop in contact with the patient's tissue (114,119,120). The following procedures are recommended to prevent possible monitor-related accidents from occurring during MRI:

1. Only trained personnel should utilize monitor equipment during MRI.
2. All cables, lead wires, or other conductive materials from monitoring equipment that come into contact with the patient should be positioned so that no conductive loops are formed.
3. Monitoring devices that do not appear to operate properly during MRI should be immediately removed from the patient.

Monitoring the electrocardiogram during MRI is typically required for cardiac imaging, gating to reduce imaging artifacts from the physiologic motion in the brain and spine, and determining the patient's pulse rate. Artifacts caused by the static, gradient, and RF

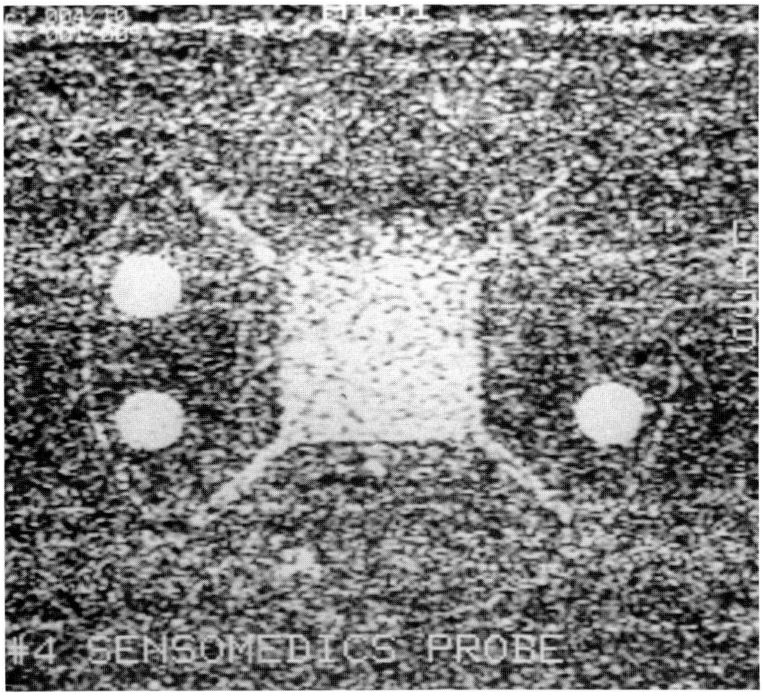

FIG. 2.7. Example of severe image distortion caused by spurious RF noise from a monitor operating within the MRI scanner room. A fluid-filled Plexiglas phantom was scanned with a T1-weighted pulse sequence.

electromagnetic fields may severely distort the morphology of the EKG, making determination of cardiac rhythm during MRI extremely difficult and unreliable (Fig. 2.1). Although sophisticated filtering techniques can be used to attenuate the artifacts from the gradient and RF fields, the static magnetic field produces an augmentation of the T wave, as previously mentioned, and other nonspecific waveform changes that are in direct proportion to the strength of the field. These EKG waveform changes are not easily counterbalanced. In some instances, static magnetic field-induced augmented T waves have a higher amplitude than do the R waves, resulting in false triggering and an inaccurate determination of the heart beats per minute. EKG waveform artifacts can be minimized during MRI by using special filters, using electrodes with a minimal amount of metal, selecting lead wires with a minimal amount of metal, twisting or braiding the lead wires, and using special lead placement techniques (85,165,166).

REFERENCES

1. Adzamli IK, Jolesz FA, Blau M. An assessment of blood-brain barrier integrity under MRI conditions: Brain uptake of radiolabeled Gd-DTPA and In-DTPA-IgG. *J Nucl Med* 1989; 30:839.
2. Barber BJ, Schaefer DJ, Gordon CJ, Zawieja DC, Hecker J. Thermal effects of MR imaging: Worst-case studies in sheep. *AJR* 1990;155:1105–1110.
3. Bartels Von M, Mann K, Matejcek P, Puttkammer M, Schroth G. Magnetresonanztomographie und Sicherheit: Elektroenzephalographische und neuropsychologische Befunde vor und nach MR-Untersuchungen des Gehirns. Fortschr. *Rontgenstr* 1986;145(4):383–385.
4. Besson J, Foreman EI, Eastwood LM, Smith FW, Ashcroft GW. Cognitive evaluation following NMR imaging of the brain. *J Neurol Neurosurg Psychiatry* 1984;47:314–316.
5. Bore PJ, Galloway GJ, Styles P, Radda GK, Flynn G, Pitts PR. Are quenches dangerous? *Magn Reson Imaging* 1986; 3:112–117.
6. Brody AS, Sorette MP, Gooding CA, Listerud J, Clark MR, Mentzer WC, Brasch RC, James TL. Induced alignment of flowing sickle erythrocytes in a magnetic field. A preliminary report. *Invest Radiol* 1985;20:560–566.
7. Brody AS, Embury SH, Mentzer WC, Winkler ML, Gooding CA. Preservation of sickle cell blood flow patterns during MR imaging. An in vivo study. *AJR* 1988;151:139–141.
8. Brummett RE, Talbot JM, Charuhas P. Potential hearing loss resulting from MR imaging. *Radiology* 1988;169:539–540.
9. Carson JJL, Prato FS, Diesbourg LD, Drost DJ, Dixon SJ. Exposure to MR imaging fields increases intracellular free calcium concentration in HL 60 promyelocytic cells. Society of Magnetic Resonance in Medicine: *Book of Abstracts*; 1989:293.
10. Cohen MS, Weisskoff R, Rzedzian R, Kantor H. Sensory stimulation by time-varying magnetic fields. *Magn Reson Med* 1990;14:409–414.
11. Cooke P, Morris PG. The effects of NMR exposure on living organisms. II. A genetic study of human lymphocytes. *Br J Radiol* 1981;54:622–625.
12. Doherty JU, Whitman GJR, Robinson MD, Harken AH, Simson MB, Spear JF, Josephson ME. Changes in cardiac excitability and vulnerability in NMR fields. *Invest Radiol* 1985; 20:129–135.
13. Fischer H. Physiological effects of fast oscillating magnetic field gradient. *Radiology* 1989;173:(P)382.
14. Garber JH, Oldendorf WH, Braun LD, Lufkin RB. MRI gradient fields increase brain mannitol space. *Magn Reson Imaging* 1989;7:605–610.
15. Geard CR, Osmak RS, Hall EJ, Simon HE, Maudsley AA, Hilal SK. Magnetic resonance and ionizing radiation: A comparative evaluation in vitro of oncogenic and genotoxic potential. *Radiology* 1984;152:199–202.
16. Gore JC, McDonnell MJ, Pennock JM, Stanbrook HS. An assessment of the safety of rapidly changing magnetic fields in the rabbit: Implications for NMR imaging. *Magn Reson Imaging* 1982;1:191–195.
17. Gremmel H, Wendhausen H, Wunsch F. Biologische Effekte statischer Magnetifelder bei NMR-Tomographie am Menschen.

Wiss. Radiologische, Klinik, Christian-Albrechts-Universitat zu Kiel, 1983.
18. Gulch RW, Lutz O. Influence of strong static magnetic fields on heart muscle contraction. *Phys Med Biol* 1986;31:763–769.
19. Heinrichs WL, Fong P, Flannery M, Heinrichs SC, Crooks LE, Spindle A, Pederson RA. Midgestational exposure of pregnant balb/c mice to magnetic resonance imaging. *Magn Reson Imaging* 1988;6:305–313.
20. Hong C-Z, Shellock FG. Short-term exposure to a 1.5 Tesla static magnetic field does not effect somato-sensory evoked potentials in man. *Magn Reson Imaging* 1990;8:65–69.
21. Innis NK, Ossenkopp KP, Prato FS, Sestini E. Behavioral effects of exposure to nuclear magnetic resonance imaging: II. Spatial memory tests. *Magn Reson Imaging* 1986;4:281–284.
22. Jehenson P, Duboc D, Lavergne T, Guize L, Guerin F, Degeroges M, Syrota A. Change in human cardiac rhythm by a 2 Tesla static magnetic field. *Radiology* 1988;166:227–230.
23. Kay HH, Herfkens RJ, Kay BK. Effect of magnetic resonance imaging on Xenopus Laevis embryogenesis. *Magn Reson Imaging* 1988;6:501–506.
24. Keltner JR, Roos MS, Brakeman PR, Budinger TF. Magnetohydrodhynamics of blood flow. *Magn Reson Med* 1990; 16:139–149.
25. Kido D, Morris TW, Erickson JL, Plewes DB, Simon JH. Physiologic changes during high field strength MR imaging. *Am J Neuroradiol* 1987;8:263–266.
26. LaPorte R, Kus L, Wisniewski RA, Prechel MM, Azar-Kia B, McNulty JA. Magnetic resonance imaging (MRI) effects on rat pineal neuroendocrine function. *Brain Res* 1990;8:506:294–296.
27. McRobbie D, Foster MA. Cardiac response to pulsed magnetic fields with regard to safety in NMR imaging. *Phys Med Biol* 1985;30:695–702.
28. McRobbie D, Foster MA. Pulsed magnetic field exposure during pregnancy and implications for NMR foetal imaging: A study with mice. *Magn Reson Imaging* 1985;3:231–234.
29. Messmer JM, Porter JH, Fatouros P, Prasad U, Weisberg M. Exposure to magnetic resonance imaging does not produce taste aversion in rats. *Physiol Behav* 1987;40:259–261.
30. Montour JL, Fatouros PP, Prasad UR. Effect of MR imaging on spleen colony formation following gamma radiation. *Radiology* 1988;168:259–260.
31. Niemann G, Schroth G, Klose U, Buettner UW. Influence of magnetic resonance imaging on somatosensory potential in man. *J Neurol* 1988;235:462–465.
32. Ngo FQH, Blue JW, Roberts WK. The effects of a static magnetic field on DNA synthesis and survival of mammalian cells irradiated with fast neutrons. *Magn Reson Med* 1987;5:307–317.
33. Osbakken M, Griffith J, Taczanowsky P. A gross morphologic, histologic, hematologic, and blood chemistry study of adult and neonatal mice chronically exposed to high magnetic fields. *Magn Reson Med* 1986;3:502–517.
34. Ossenkopp KP, Kavaliers M, Prato FS, Teskey GC, Sestini E, Hirst M. Exposure to nuclear magnetic imaging procedure attenuates morphine-induced analgesia in mice. *Life Sci* 1985; 37:1507–1514.
35. Ossenkopp KP, Innis NK, Prato FS, Sestini E. Behavioral effects of exposure to nuclear magnetic resonance imaging: I. Open-field behavior and passive avoidance learning in rats. *Magn Reson Imaging* 1986;4:275–280.
36. Papatheofanis FJ, Paptheofanis BJ. Short-term effect of exposure to intense magnetic fields on hematologic indices of bone metabolism. *Invest Radiol* 1989;24:221–223.
37. Peeling J, Lewis JS, Samoiloff MR, Bock E. Biological effects of magnetic fields on the nematode Panagrullus redivivus. *Magn Reson Imaging* 1988;6:655–660.
38. Prasad N, Bushong SC, Thornby JI, Bryan RN, Hazelwood CF, Harrell JE. Effect of nuclear magnetic resonance on chromosomes of mouse bone marrow cells. *Magn Reson Imaging* 1984;2:37–39.
39. Prasad N, Lotzova E, Thornby JI, Madewell JE, Ford JJ, Bushong SC. Effects of MR imaging on murine natural killer cell cytotoxicity. *AJR* 1987;148:415–417.
40. Prasad N, Lotzova E, Thornby JI, Taber KH. The effect of 2.35-T MR imaging on natural killer cell cytotoxicity with and without interleukin-2. *Radiology* 1990;175:261–263.
41. Prasad N, Wright DA, Ford JJ, Thornby JI. Safety of 4-T MR imaging: A study of effects of developing frog embryos. *Radiology* 1990;174:251–253.
42. Prasad N, Wright DA, Forster JD. Effect of nuclear magnetic resonance on early stages of amphibian development. *Magn Reson Imaging* 1982;1:35–38.
43. Prato FS, Ossenkopp KP, Kavaliers M, Sestini E, Teskey GC. Attenuation of morphine-induced analgesia in mice by exposure to magnetic resonance imaging: Separate effects of the static, radiofrequency and time-varying magnetic fields. *Magn Reson Imaging* 1987;5:9–14.
44. Redington RW, Dumoulin CL, Schneck JF, Roemer PB, Souza SP, Mueller OM, Eisner DR, Piel JE, Edelstein WA. MR imaging and bio-effects in a whole body 4.0 Tesla imaging system. Society of Magn Reson Imaging: *Book of Abstracts*; 1988:20.
45. Reid A, Smith FW, Hutchison JMS. Nuclear magnetic resonance imaging and its safety implications: Follow-up of 181 patients. *Br J Radiol* 1982;55:784–786.
46. Roschamnn P, Wieland J. Response of the human auditory system to RF-pulses in RF-coils for magnetic resonance at 8 MHz-170 MHz. Society of Magnetic Resonance in Medicine: *Book of Abstracts*; 1990:292.
47. Sacks E, Worgul BV, Merriam GR, Hilal S. The effects of nuclear magnetic resonance imaging on ocular tissues. *Arch Ophthalmol* 1986;104:890–893.
48. Schwartz JL, Crooks LE. NMR imaging produces no observable mutations or cytotoxicity in mammalian cells. *AJR* 1982; 139:583–585.
49. Shellock FG, Schaefer DJ, Gordon CJ. Effect of a 1.5 T static magnetic field on body temperature of man. *Magn Reson Med* 1986;3:644–647.
50. Shellock FG, Crues JV. Temperature, heart rate, and blood pressure changes associated with clinical MR imaging at 1.5 T. *Radiology* 1987;163:259–262.
51. Shellock FG, Schaefer DJ, Grundfest W, Crues JV. Thermal effects of high-field (1.5 Tesla) magnetic resonance imaging of the spine: Clinical experience above a specific absorption rate of 0.4 W/kg. *Acta Radiol* 1986;[Suppl 369]:514–516.
52. Shellock FG, Gordon CJ, Schaefer DJ. Thermoregulatory responses to clinical magnetic resonance imaging of the head at 1.5 Tesla: Lack of evidence for direct effects on the hypothalamus. *Acta Radiol* 1986;[Suppl 369]:512–513.
53. Shellock FG, Crues JV. Corneal temperature changes associated with high-field MR imaging using a head coil. *Radiology* 1988;167:809–811.
54. Shellock FG, Crues JV. Temperature changes caused by clinical MR imaging of the brain at 1.5 Tesla using a head coil. *AJNR* 1988;9:287–291.
55. Shellock FG, Schaefer DJ, Crues JV. Effect of a 1.5 Tesla static magnetic field on body and skin temperature in man. *Magn Reson Med* 1989;11:371–375.
56. Shellock FG, Schaefer DJ, Crues JV. Alterations in body and skin temperatures caused by MR imaging: Is the recommended exposure for radiofrequency radiation too conservative? *Br J Radiol* 1989;62:904–909.
57. Shellock FG, Rothman B, Sarti D. Heating of the scrotum by high-field-strength MR imaging. *AJR* 1990;154:1229–1232.
58. Shivers RR, Kavaliers M, Tesky CG, Prato FS, Pelletier RM. Magnetic resonance imaging temporarily alters blood-brain barrier permeability in the rat. *Neurosci Lett* 1987;76:25–31.
59. Shuman WP, Haynor DR, Guy AW, Wesbey GE, Schaefer DJ, Moss AA. Superficial and deep-tissue increases in anesthetized dogs during exposure to high specific absorption rates in a 1.5-T MR imager. *Radiology* 1988;167:551–554.
60. Sperber D, Oldenbourg R, Dransfeld K. Magnetic field induced temperature change in mice. *Naturwissenschaften* 1984; 71:100–101.
61. Stojan L, Sperber D, Dransfeld K. Magnetic-field-induced changes in the human auditory evoked potentials. *Naturwissenschaften* 1988;75:622–623.
62. Sweetland J, Kertesz A, Prato FS, Nantau K. The effect of magnetic resonance imaging on human cognition. *Magn Reson Imaging* 1987;5:129–135.
63. Teskey GC, Prato FS, Ossenkopp KP, Kavaliers M. Exposure to time varying magnetic fields associated with magnetic res-

onance imaging reduces fentanyl-induced analgesia in mice. *Bioelectromagnetics* 1988;9:167–174.
64. Tesky GC, Ossenkopp KP, Prato FS, Sestini E. Survivability and long-term stress reactivity levels following repeated exposure to nuclear magnetic resonance imaging procedures in rats. *Physiol Chemi Phys Med NMR* 1987;19:43–49.
65. Thomas A, Morris PG. The effects of NMR exposure on living organisms. I. A micro assay. *Br J Radiol* 1981;54:615–621.
66. Tyndall DA. MRI effects on the teratogenicity of X-irradiation in the C57BL/6J mouse. *Magn Reson Imaging* 1990;8:423–433.
67. Vogl T, Krimmel K, Fuchs A, Lissner J. Influence of magnetic resonance imaging on human body core and intravascular temperature. *Med Phys* 1988;15:562–566.
68. Von Klitzing L. Do static magnetic fields of NMR influence biological signals. *Clin Phys Physiol Meas* 1986;7(2):157–160.
69. Von Klitzing L. Static magnetic fields increase the power intensity of EEG of man. *Brain Res* 1989;483:201–203.
70. Weiss J, Herrick RC, Taber KH, Plisher GA. Bio-effects of high magnetic fields: A study using a simple animal model. *Magn Reson Imaging* 1990;8[S1]:166.
71. Willis RJ, Brooks WM. Potential hazards of NMR imaging. No evidence of the possible effects of static and changing magnetic fields on cardiac function of the rat and guinea pig. *Magn Reson Imaging* 1984;2:89–95.
72. Withers HR, Mason KA, Davis CA. MR effects on murine spermatogenesis. *Radiology* 1985;156:741–742.
73. Wolff S, Crooks LE, Brown P, Howard R, Painter RB. Tests for DNA and chromosomal damage induced by nuclear magnetic resonance imaging. *Radiology* 1980;136:707–710.
74. FDA, Magnetic resonance diagnostic device. Panel recommendation and report on petitions for MR reclassification. *Federal Register*. 1988;53:7575–7579.
75. FDA, Radiological Health Bureau of U.S.A. Guidelines for evaluating electromagnetic exposure for trials of clinical NMR systems. Washington D.C., 1982.
76. Vyalov AM. Magnetic fields as a factor in the industrial environment. *Vestn Akad Med Nauk SSSR* 1967;8:72–79.
77. Vyalov AM. Clinico-hygenic and experimental data on the effect of magnetic fields under industrial conditions. In: Kholodov Y, ed. Influence of magnetic fields on biological objects. Moscow: Joint Publications Research Service, trans. JPRS-63-38, 1974:20–35.
78. Barnothy MF. *Biological effects of magnetic fields*, vols. 1, 2. New York: Plenum Press; 1964, 1969.
79. Persson BRR, Stahlberg F. *Health and safety of clinical NMR examinations*. Boca Raton, FL: CRC Press, 1989.
80. Tenforde TS. *Magnetic field effects on biological systems*. New York: Plenum Press, 1979.
81. Michaelson SM, Lin JV. Biological effects and health implications of radiofrequency radiation. New York: Plenum Press, 1987.
82. Tenforde TS. Thermoregulation in rodents exposed to high-intensity stationary magnetic fields. *Bioelectromagnetics* 1986;7:341–346.
83. Beischer DE, Knepton JC. Influence of strong magnetic fields on the electrocardiogram of squirrel monkey (Saimiri sciures). *Aerospace Med* 1964;35:939–944.
84. Tenforde TS, Gaffey CT, Moyer BR, Budinger TF. Cardiovascular alterations in Macaca monkeys exposed to stationary magnetic fields. Experimental observations and theoretical analysis. *Bioelectromagnetics* 1983;4:1–9.
85. Dimick RN, Hedlund LW, Herfkens RJ, Fram EK, Utz J. Optimizing electrocardiographic electrode placement for cardiac-gated magnetic resonance imaging. *Invest Radiol* 1987;22:17–22.
86. Abdullakhozhaeva MS, Razykov SR. Structural changes in central nervous system under the influence of a permanent magnetic field. *Bull Exper Biol Med* 1986;102:1585–1587.
87. Hong C-Z. Static magnetic field influence on human nerve function. *Arch Phys Med Rehabil* 1987;68:162–164.
88. Budinger TF. Nuclear magnetic resonance (NMR) in vivo studies: Known thresholds for health effects. *JCAT* 1981;5:800–811.
89. Shellock FG, Crues JV. MRI: Safety considerations in magnetic resonance imaging. *MRI Decis* 1988;2:25–30.
90. Shellock FG. Biological effects and safety aspects of magnetic resonance imaging. *Magn Reson Q* 1989;5:243–261.
91. Kanal E, Shellock FG. Safety considerations in MR imaging. *Radiology* 1990;176:593–606.
92. Adey WR. Tissue interactions with nonionizing electromagnetic fields. *Physiol Rev* 1981;61:435–514.
93. Modan B. Exposure to electromagnetic fields and brain malignancy: A newly discovered menace? *Am J Ind Med* 1988;13:625–627.
94. Brown HD, Chattopadhyay SK. Electromagnetic-field exposure and cancer. *Cancer Biochem Biophys* 1988;9:295–342.
95. Pool R. Electromagnetic fields: The biological evidence. *Science* 1990;249:1378–1381.
96. Reilly JP. Peripheral nerve stimulation by induced electric currents: Exposure to time-varying magnetic fields. *Med Biol Eng Comput* 1989;27:101–112.
97. Bernhardt J. The direct influence of electromagnetic fields on nerve and muscle cells of man within the frequency range of 1 Hz to 30 MHz. *Radiat Environ Phys* 1979;16:309–323.
98. NCRP Report No. 86. Biological effects and exposure criteria for radiofrequency electromagnetic fields. Bethesda: National Council on Radiation Protection and Measurements; 1986.
99. Watson AB, Wright JS, Loughman J. Electrical thresholds for ventricular fibrillation in man. *Med J Aust* 1973;1:1179–1182.
100. Erwin DN. Mechanisms of biological effects of radiofrequency electromagnetic fields: An overview. *Aviat Space Environ Med* 1988;59[11, Suppl.]:A21–A3111.
101. Gordon CJ. Thermal physiology. In: *Biological effects of radiofrequency radiation*. Washington, DC: EA-600/8-83-026A, 1981:4-1–4-28.
102. Gordon CJ. Normalizing the thermal effects of radiofrequency radiation: Body mass versus total body surface area. *Bioelectromagnetics* 1987;8:111–118.
103. Gordon CJ. Effect of radiofrequency radiation exposure on thermoregulation. *ISI Atlas of Science, Plants and Animals*, 1988;1:245–250.
104. Beers J. Biological effects of weak electromagnetic fields from 0 Hz to 200 MHz: A survey of the literature with special emphasis on possible magnetic resonance effects. *Magn Reson Imaging* 1989;7:309–331.
105. Coulter JS, Osbourne SL. Short wave diathermy in heating of human tissues. *Arch Phys Ther* 1936;17:679–687.
106. Bottomly PA, Edelstein WA. Power deposition in whole body NMR imaging. *Med Phys* 1981;8:510–512.
107. Bottomly PA, Redington RW, Edelstein WA, Schenck JF. Estimating radiofrequency power deposition in body NMR imaging. *Magn Reson Med* 1985;2:336–349.
108. Wickersheim KA, Sun MH. Fluoroptic thermometry. *Med Electronics* 1987;Feb:84–91.
109. Berman E. Reproductive effects. In: *Biological effects of radiofrequency radiation*. EPA-600/8-83-026A, 1984.
110. Dormer KJ, Richard GL, Hough JVD, Norquist RE. The use of rare-earth magnet couplers in cochlear implants. *Laryngoscope* 1981;91:1812–1820.
111. Shellock FG. Ex vivo assessment of deflection forces and artifacts associated with high-field MRI of "mini-magnet" dental prostheses. *Magn Reson Imaging* 1989;[Suppl 1]:IT–03.
112. ECRI, Health Devices Alert. A new MRI complication? 1988:May 27;1.
113. Liang MD, Narayanan K, Kanal E. Magnetic port in tissue expanders: A caution for MRI. *Magn Reson Imaging* 1989;7:541–542.
114. Shellock FG, Kanal E. Policies, guidelines, and recommendations for MR imaging safety and patient management. *J Magn Reson Imaging* 1991;1:97–101.
115. Gangarosa RE, Minnis JE, Nobbe J, Praschan D, Genberg RW. Operational safety issues in MRI. *Magn Reson Imaging* 1987;5:287–292.
116. Hayes DL, Holmes DR, Gray JE. Effect of a 1.5 Tesla nuclear magnetic resonance imaging scanner on implanted permanent pacemakers. *JACC* 1987;10:782–786.
117. Alagona P, Toole JC, Maniscalco BS, Glover MU, Abernathy GT, Prida XE. Nuclear magnetic resonance imaging in a patient with a DDD pacemaker. [Letter] *PACE* 1989;12:619.
118. Edelman RR, Shellock FG, Ahladis J. Practical MRI for the

technologist and imaging specialist. In: Edelman RR, Hesselink J, eds. *Clinical magnetic resonance*. Philadelphia: WB Saunders; 1990.
119. Shellock FG, Slimp G. Severe burn of the finger caused by using a pulse oximeter during MRI. *AJR* 1989;153:1105.
120. Kanal E, Shellock FG. Burns associated with clinical MR examinations. *Radiology* 1990;175:585.
121. Hurwitz R, Lane SR, Bell RA, Brant-Zawadzki MN. Acoustic analysis of gradient-coil noise in MR imaging. *Radiology* 1989; 173:545–548.
122. Goldman AM, Grossman WE, Friedlander PC. Reduction of sound levels with antinoise in MR imaging. *Radiology* 1989; 173:549–550.
123. Buchli R, Boesiger P, Meier D. Heating effects of metallic implants by MRI examinations. *Magn Reson Med* 1988;7:255–261.
124. Davis PL, Crooks L, Arakawa M, McRee R, Kaufman L, Margulis AR. Potential hazards in NMR imaging: Heating effects of changing magnetic fields and RF fields on small metallic implants. *AJR* 1981;137:857–860.
125. Shellock FG, Crues JV. High-field MR imaging of metallic biomedical implants: An in vitro evaluation of deflection forces and temperature changes induced in large prostheses. *Radiology* 1987;165:150.
126. Lund G, Nelson JD, Wirtschafter JD, Williams PA. Tatooing of eyelids: Magnetic resonance imaging artifacts. *Ophthalmic Surg* 1986;17:550–553.
127. Sacco R, Steiger DA, Bellon EM, Coleman PE, Haacke EM. Artifacts caused by cosmetics in MR imaging of the head. *AJR* 1987;148:1001–1004.
128. Jackson JG, Acker JD. Permanent eyeliner and MR imaging. *AJR* 1987;149:1080.
129. Pusey E, Lufkin RB, Brown RKJ, et al. Magnetic resonance imaging artifacts: Mechanism and clinical significance. *Radiographics* 1986;6:891–911.
130. Shellock FG, Crues JV. High-field MR imaging of metallic biomedical implants: An ex vivo evaluation of deflection forces. *AJR* 1988;151:389–392.
131. Dujovny M, Kossovsky N, Kossowsky R, et al. Aneurysm clip motion during magnetic resonance imaging: In vivo experimental study with metallurgical factor analysis. *Neurosurgery* 1985;17:543–548.
132. Shellock RG, Schatz CJ, Shelton C, Brown B. Ex vivo evaluation of 9 different ocular and middle-ear implants exposed to a 1.5 Tesla MR scanner. *Radiology* 1990;177(P):271.
133. Shellock FG, Schatz CJ. High-field strength MRI and otologic implants. *AJNR* 1991;12:279–281.
134. Shellock FG, Meeks T. Ex vivo evaluation of ferromagnetism and artifacts for implantable vascular access ports exposed to a 1.5 T MR scanner. *J Magn Reson Imaging* 1991;1:243.
135. Teitelbaum GP, Yee CA, Van Horn DD, Kim HS, Colletti PM. Metallic ballistic fragments: MR imaging safety and artifacts. *Radiology* 1990;175:855–859.
136. Shellock FG, Curtis JS. MR imaging and biomedical implants, materials, and devices. An updated review. *Radiology* 1991; 180:541–550.
137. Holtas S, Olsson M, Romner B, Larsson EM, Saveland H, Brandt L. Comparison of MR imaging and CT in patients with intracranial aneurysm clips. *AJNR* 1988;9:891–897.
138. Huttenbrink KB, Grobe-Nobis W. Experimentells Untersuchungen und theoretische Betrachtungen uber das Verhalten von Stapes-Metall-Prothesen im Magnetfeld eines Kernspintomographen. (Experiments and theoretical considerations on behaviour of metallic stapedectomy-prostheses in nuclear magnetic resonance imaging.) *Laryng Rhinol Otol* 1987;66:127–130.
139. Becker R, Norfray JF, Teitelbaum GP, Bradley WG, Jacobs JB, Wacaser L, Rieman RL. MR imaging in patients with intracranial aneurysm clips. *AJNR* 1988;9:885–889.
140. Randall PA, Kohman LJ, Scalzetti EM, Szeverenyi NM, Panicek DM. Magnetic resonance imaging of prosthetic cardiac valves in vitro and in vivo. *AJC* 1988;62:973–976.
141. Romner B, Olsson M, Ljunggen B, Holtas S, Saveland H, Brandt L, Persson B. Magnetic resonance imaging and aneurysm clips. *J Neurosurg* 1989;70:426–431.
142. Augustiny N, von Schulthess GK, Meier D, Bosiger P. MR imaging of large nonferromagnetic metallic implants at 1.5 T. *JCAT* 1987;11:678–683.
143. Teitelbaum GP, Bradley WG, Klein BD. MR imaging artifacts ferromagnetism, and magnetic torque of intravascular filters, stents, and coils. *Radiology* 1988;166:657–664.
144. Ballock RT, Hajek PC, Byrne TP, Garfin SR. The quality of magnetic resonance imaging, as affected by the composition of the halo arthrosis. *J Bone Joint Surg* 1989;71-A:431–434.
145. Malko JA, Hoffman JC, Jarrett PJ. Eddy-current-induced artifacts caused by an "MR-compatible" halo device. *Radiology* 1989;173:563–564.
146. Shellock FG, Slimp G. MRI-compatible halo vest for cervical spine fixation. *AJR* 1990;1;154:631–632.
147. Kelly WM, Pagle PG, Pearson A, San Diego AG, Soloman MA. Ferromagnetism of intraocular foreign body causes unilateral blindness after MR study. *AJNR* 1986;7:243–245.
148. Williams S, Char DH, Dillon W, Lincoff N, Moseley M. Ferrous intraocular foreign bodies and magnetic resonance imaging. *Am J Ophthalolmol* 1988;105:398–401.
149. National Radiological Protection Board. Revised guidelines on acceptable limits of exposure during nuclear magnetic clinical imaging. *Br J Radiol* 1983;56:974–977.
150. Granet RB, Gelber LJ. Claustrophobia during MR imaging. *N J Med* 1990;87:479–482.
151. Flaherty JA, Hoskinson K. Emotional distress during magnetic resonance imaging. *N Engl J Med* 1989;320:467–468.
152. Fishbain DA, Goldberg M, Labbe E, Zacher D, Steele-Rosomoff R, Rosomoff H. Long-term claustrophobia following magnetic resonance imaging. *Am J Psychiatry* 1988;145:1038–1039.
153. Quirk ME, Letendre AJ, Ciottone RA, Lingley JF. Anxiety in patients undergoing MR imaging. *Radiology* 1989;170:463–466.
154. Quirk ME, Letendre AJ, Ciottone RA, Lingley JF. Evaluation of three psychological interventions to reduce anxiety during MR imaging. *Radiology* 1989;173:759–762.
155. Hricak H, Ampara EG. Body MRI: Alleviation of claustrophobia by prone positioning. *Radiology* 1984;152:819.
156. Weinreb JC, Maravilla KR, Peshock R, Payne J. Magnetic resonance imaging: Improving patient tolerance and safety. *AJR* 1984;143:1285–1287.
157. Klonoff EA, Janata JW, Kaufman B. The use of systematic desensitization to overcome resistance to magnetic resonance imaging (MRI) scanning. *J Behav Ther Exp Psychiatry* 1986; 17:189–192.
158. Karlik SJ, Heatherly T, Pavan F, et al. Patient anesthesia and monitoring at a 1.5 T MRI installation. *Magn Res Med* 1988; 7:210–221.
159. Barnett GH, Ropper AH, Johnson KA. Physiological support and monitoring of critically ill patients during magnetic resonance imaging. *J Neurosurg* 1988;68:246–250.
160. Dunn V, Coffman CE, McGowan JE, Ehrhardt JC. Mechanical ventilation during magnetic resonance imaging. *Magn Reson Imaging* 1985;3:169–172.
161. McArdle CB, Nicholas DA, Richardson CJ, Amparo EG. Monitoring of the neonate undergoing MR imaging: Technical considerations. *Radiology* 1986;159:223–226.
162. Roth JL, Nugent M, Gray JE, Julsrud PR, Berquist TH, Sill JC, Kispert DB. Patient monitoring during magnetic resonance imaging. *Anesthesiology* 1985;62:80–83.
163. Shellock FG. Monitoring during MRI. An evaluation of the effect of high-field MRI on various patient monitors. *Med Electron* 1986;Sept:93–97.
164. Shellock FG. Monitoring sedated patients during MRI (Letter). *Radiology* 1990;177:586.
165. Wendt RE, Rokey R, Vick GW, Johnston DL. Electrocardiographic gating and monitoring during NMR imaging. *Magn Reson Imaging* 1988;6:89–95.
166. Holshoulser B, Hinshaw DB, Shellock FG. Sedation, anesthesia, and physiologic monitoring during MRI. ARRS Syllabus, AN Hasso and DD Stark, eds. 1991;9–15.

CHAPTER 3

An Atlas of Normal Multiplanar Anatomy of the Knee Joint

Murray A. Reicher

Interpretation of highly detailed multiplanar MR images of the knee requires a thorough knowledge of the normal anatomy. Numerous reports now provide compelling evidence that MRI can depict highly detailed knee joint anatomy. The atlas herein presented is based on our previous study of the normal anatomy as seen on MR images of the knee (28) and is corroborated by the reports of other investigators (4,7,8,16,18,19–21,23,31,33,34).

ANATOMIC CORRELATION

The depiction of normal knee joint anatomy in MR images has been evaluated and validated utilizing three basic techniques (28). First, MR images obtained from both cadaver specimens and normal adult patients were compared with corresponding color photographs of three cadaver knees that were cryoplaned in the axial, coronal, and sagittal planes utilizing a technique initially proposed and developed by Rauschning et al. (24). Prior to sectioning, the fresh cadaver specimens were deep frozen and subsequently transferred to a heavy duty cryomicrotome stage. Freezing the specimens guaranteed fixation of the soft tissues in their true natural positions. The calcified frozen specimens were then cryoplaned at intervals as thin as 1 micron.

A specifically designed stable camera stand allowed sequential photography of the cryoplaned surface without variation of magnification or obliquity (Figs. 3.1A–3.12A).

Second, as originally reported in 1985, MR images and cryoplaned sections of the same cadaver specimens were precisely correlated (28). This precise correlation was achieved by first obtaining MR images of each fresh cadaver specimen, carefully marking the plane of the images obtained, then freezing the specimen, and finally cryoplaning the specimen in the same plane as used for imaging. MR images were thus obtained in the precise plane and at the exact interval as the corresponding cryoplaned sections of the same specimen. This technique enabled point-by-point correlation of MR-depicted anatomy and photographed cadaver sections, providing for over 60,000 pixels of anatomic correlation per MR image slice.

Third, the normal anatomy was also evaluated by cross referencing the axial and coronal or sagittal MR images obtained in each cadaver or normal living knee. In a given knee, a series of axial images was obtained first because the axial anatomy was historically best understood. Then a series of coronal or sagittal images was obtained, using computer software capable of selecting these images by placing cursors on the previously obtained axial slices. The knee was not moved between the acquisition of the axial and other images. Using this technique, each point on a sagittal or coronal image could be correlated with a point having the same spatial coordinates on the proper axial slice.

Through these three methods of anatomic study, the normal multiplanar anatomy of the knee joint depicted on MR images was delineated with a high degree of confidence. Improvement in imaging techniques since the first edition of this text has further improved the resolution and reproducibility of knee MR imaging.

In Figs. 3.1 through 3.12, series of coronal, axial, and sagittal MR images are correlated with color photographs of cadaver cryoplaned sections. The MR im-

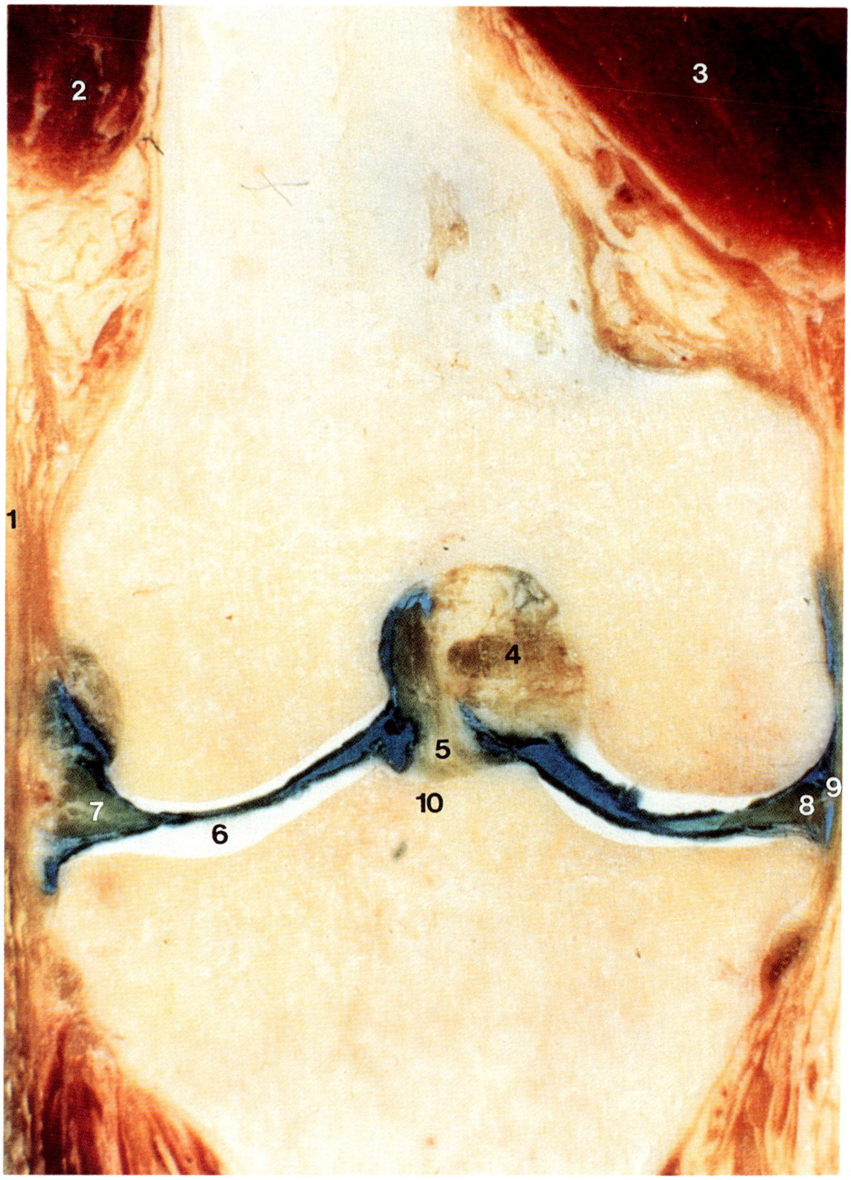

A

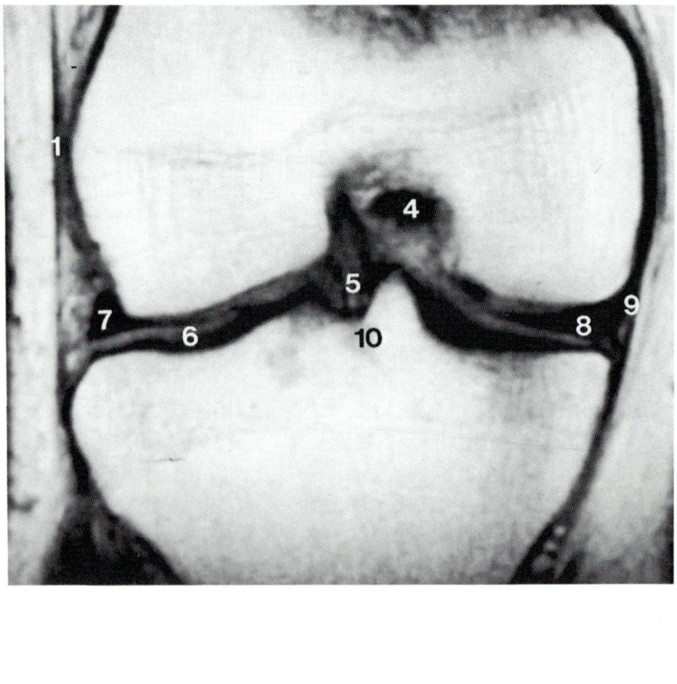

B

FIG. 3.1A,B. Coronal cryosection **(A)** and MRI **(B)** at the level of insertion of the anterior cruciate ligament. 1, Iliotibial band; 2, vastus lateralis muscle; 3, vastus medialis muscle; 4, posterior cruciate ligament; 5, insertion of the anterior cruciate ligament; 6, articular cartilage; 7, lateral meniscus; 8, anterior horn of the medial meniscus; 9, medial capsule; 10, tibial spine bone marrow.

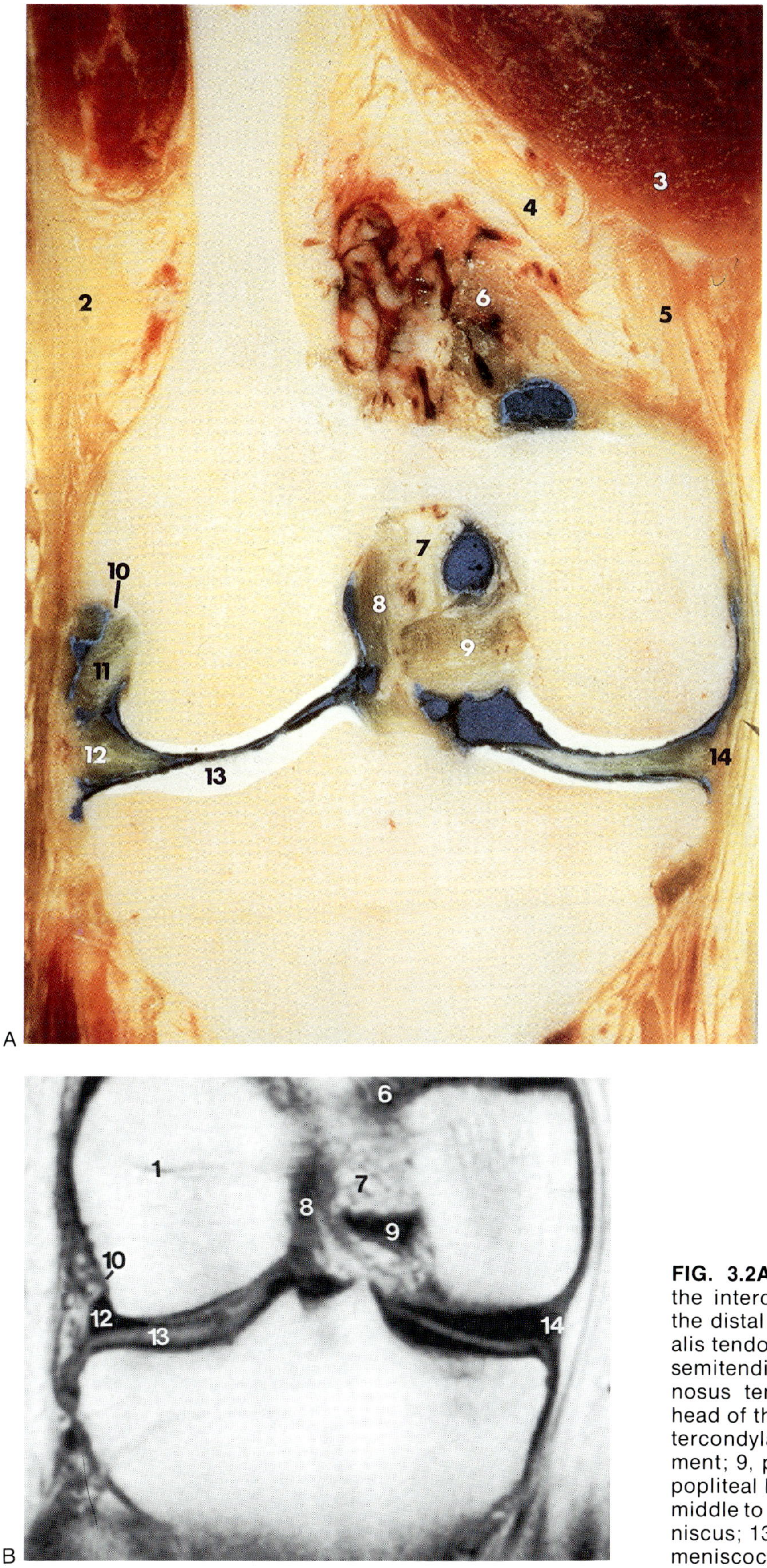

FIG. 3.2A,B. Coronal sections through the intercondylar notch. 1, Remnant of the distal femoral physis; 2, vastus lateralis tendon; 3, vastus medialis muscle; 4, semitendinosus tendon; 5, semimembranosus tendon; 6, origin of the medial head of the gastrocnemius muscle; 7, intercondylar fat; 8, anterior cruciate ligament; 9, posterior cruciate ligament; 10, popliteal hiatus; 11, popliteal tendon; 12, middle to posterior horn of the lateral meniscus; 13, articular cartilage; 14, medial meniscocapsular junction.

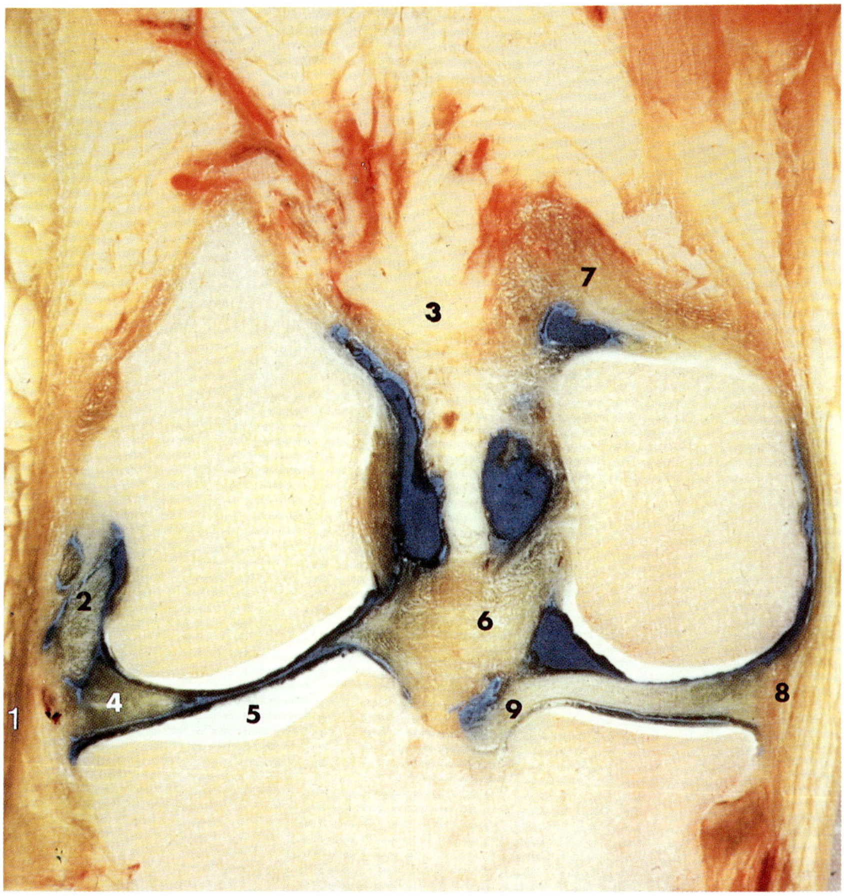

A

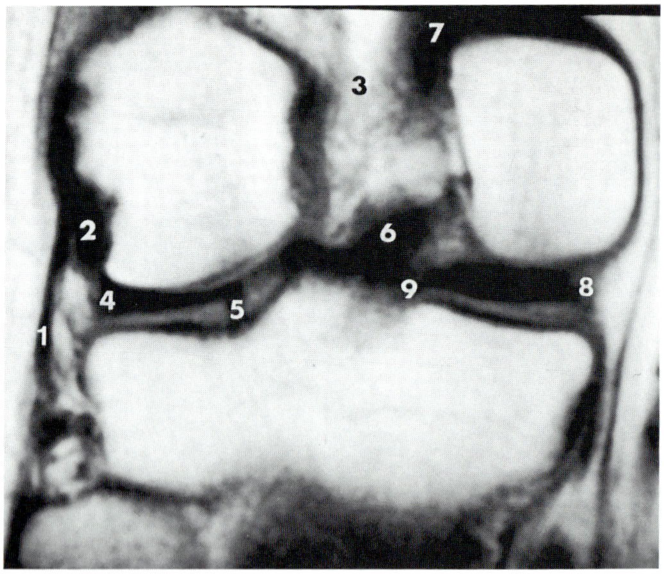

B

FIG. 3.3A,B. Posterior intercondylar coronal section. 1, Fibular collateral ligament; 2, superior portion of the popliteal tendon; 3, intercondylar fat; 4, lateral meniscus; 5, articular cartilage; 6, posterior cruciate ligament; 7, medial head of the gastrocnemius muscle; 8, posteromedial meniscocapsular junction; 9, posteriomesial meniscal attachment.

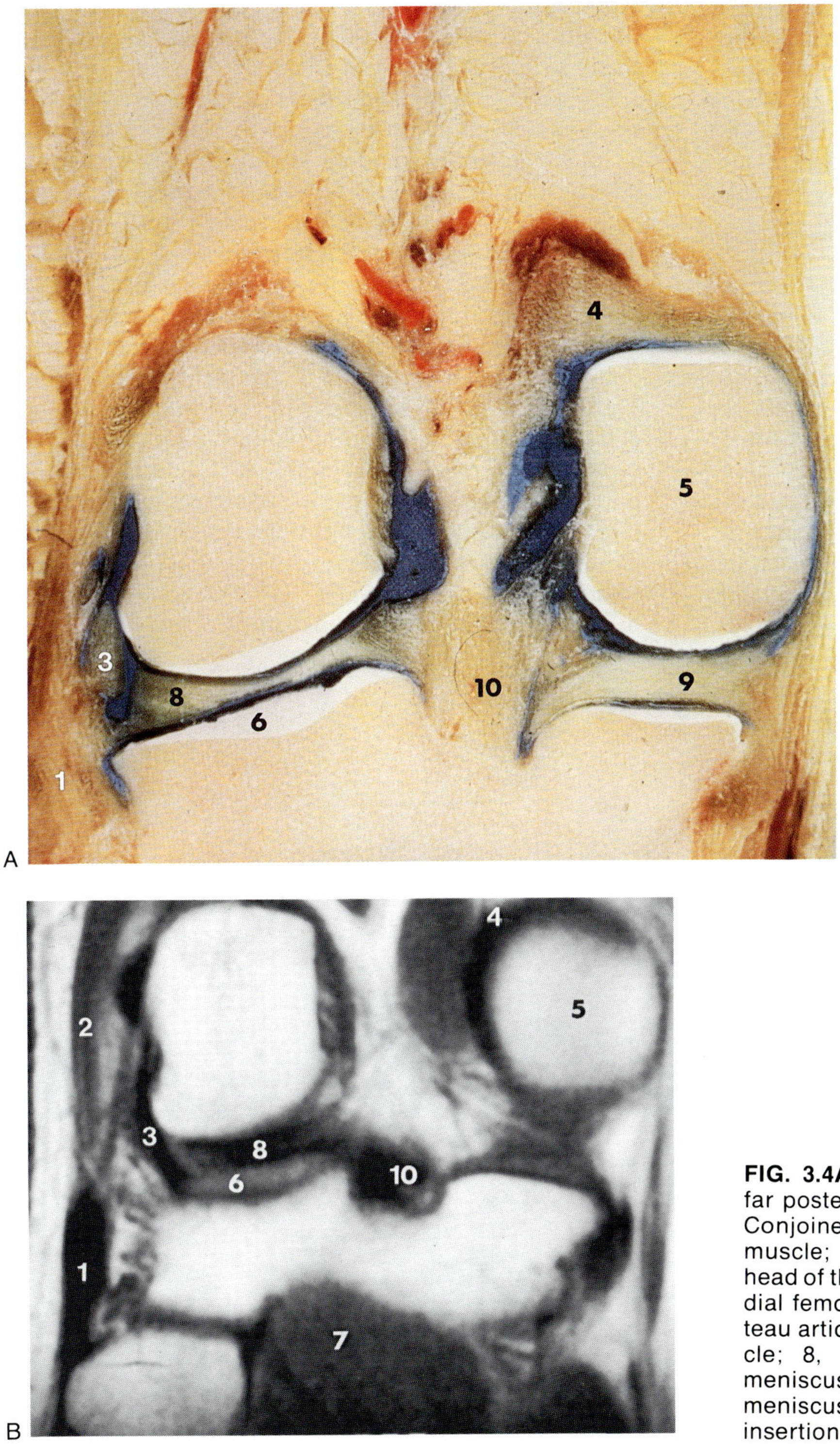

FIG. 3.4A,B. Coronal section along the far posterior margin of the knee joint. 1, Conjoined tendon; 2, biceps femoris muscle; 3, popliteal tendon; 4, medial head of the gastrocnemius muscle; 5, medial femoral condyle; 6, lateral tibial plateau articular cartilage; 7, popliteus muscle; 8, posterior horn of the lateral meniscus; 9, posterior horn of the medial meniscus; 10, posterior cruciate ligament insertion.

ages were obtained on a 1.5 T imaging system, utilizing a specifically designed knee surface coil. Images were obtained with a 256 matrix with pixels of approximately 0.6 mm and slice thickness of 4 mm. In images 3.1B through 3.12B, relatively T1-weighted spin-echo images are presented (TE = 30 msec, TR = 600 msec).

Coronal Cross-Sectional Anatomy

Four pairs of coronal MR images and photographed cryoplaned cadaver sections are presented in Figs. 3.1 through 3.4. The bone marrow, menisci, articular cartilage, and medial and lateral collateral ligaments are clearly delineated. The menisci are homogeneously black structures, but the hyaline articular cartilage yields moderate signal intensity and is easily seen in contrast to the neighboring low-intensity cortical bone. On closer inspection, the hyaline articular cartilage signal intensity can be subdivided into zones correlating with histologic differences in the cartilage layers (22). The depiction of articular cartilage will be more specifically addressed subsequently in this chapter (see ahead to Fig. 3.55) and in Chapter 9.

Both the anterior and posterior cruciate ligaments are depicted on coronal images, the anterior cruciate ligament generally yielding slightly higher signal intensity than the posterior cruciate ligament. Interestingly, ex vivo imaging of the cruciate ligaments has demonstrated no difference in intrinsic signal characteristics between the anterior cruciate ligament (ACL) and the posterior cruciate ligament (PCL) (36). Furthermore, the ACL and PCL are histologically essentially identical with regard to ligament organization. The PCL is, however, larger in cross-sectional diameter, supporting the hypothesis that the typical signal intensity difference between the ACL and PCL noted on clinical images is due to volume-averaging artifact (36). Tendon orientation has been demonstrated to affect MR imaging signal intensity due to a phenomenon referred to as "magic angle" (10). One may logically hypothesize that the difference in ACL and PCL signal

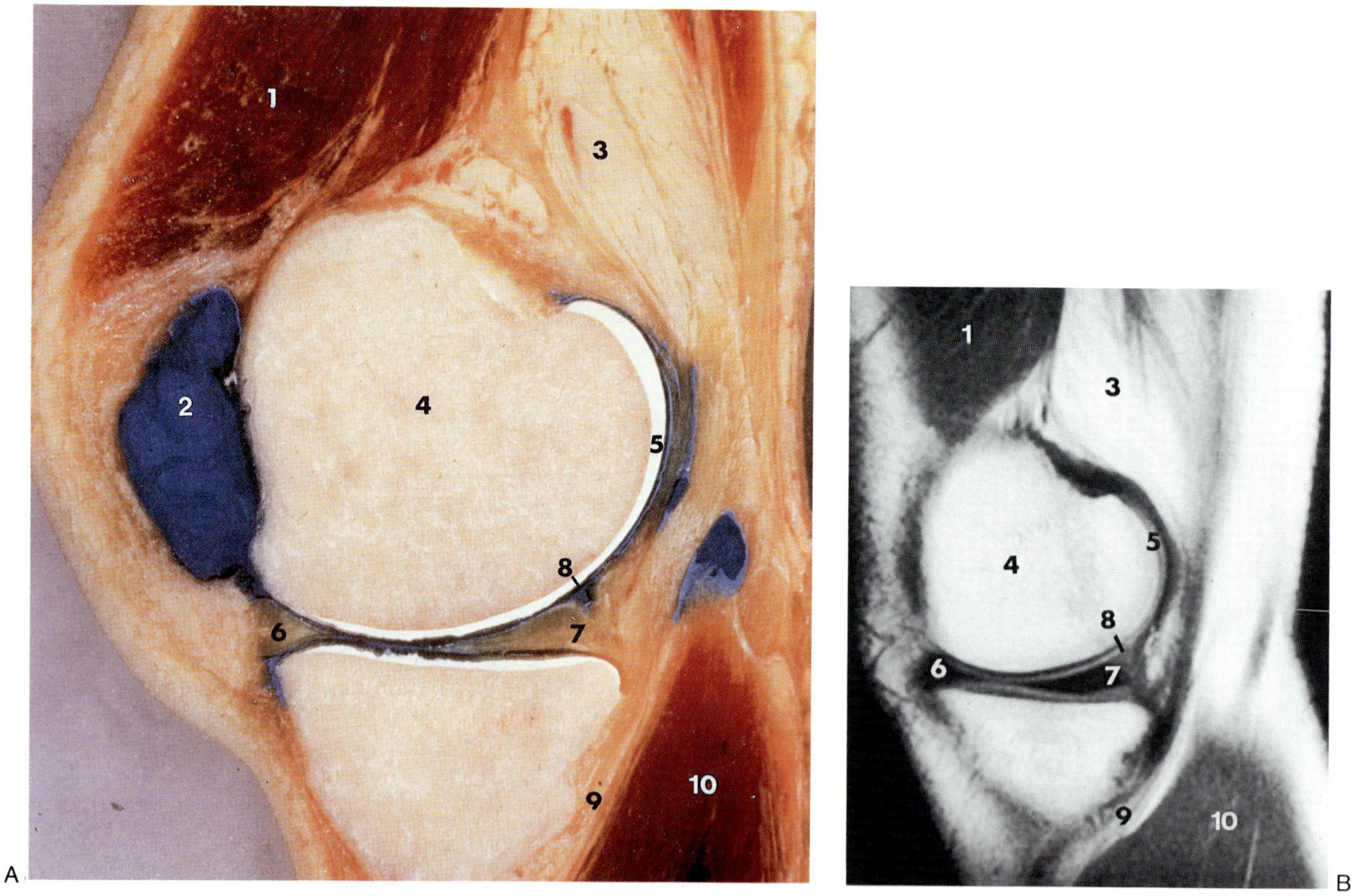

FIG.3.5A,B. Sagittal section through the medial femoral condyle. 1, Vastus medialis muscle; 2, medial recess of the suprapatellar bursa; 3, fat within the popliteal space; 4, medial femoral condyle bone marrow; 5, articular cartilage; 6, anterior horn of the medial meniscus; 7, posterior horn of the medial meniscus; 8, superior recess of the posterior horn of the medial meniscus; 9, pes anserinus; 10, medial gastrocnemius muscle.

intensity on MR images could be in part related to this "magic angle" phenomenon.

In practice, the coronal imaging plane is ideal for initial assessment of the continuity of the medial and lateral collateral ligaments and their relationships to the adjacent menisci. On anterior coronal sections, fat may be seen interposed between the medial patellar retinaculum and the medial meniscus as well as between the superficial and deep layers of the medial capsule. These normal fat stripes should not be mistaken for a menisco-capsular tear. On more posterior sections, the medial collateral ligament and posterior capsule of the knee are seen intimately attached to the medial meniscus, and therefore no interposed fat is normally seen.

Coronal images are also often quite useful in assessing the precise position of osteochondral loose bodies, in assessing displacement of torn menisci, and in defining the precise position of osteochondral defects (3,5,11,15,17,25–27).

Sagittal Cross-Sectional Anatomy

Four matched pairs of sagittal images and cadaver sections are presented in Figs. 3.5 to 3.8. The menisci are clearly depicted as structures of homogeneously low signal intensity. The posterior horn of the medial meniscus is larger than the anterior horn. The posterior and anterior horns of the lateral meniscus are nearly equal in size. Injection of the joint cavity in the cadaver specimens delineates the sites where synovial fluid may be seen in patients with knee joint effusions, including the medial and lateral recesses of the suprapatellar bursa, the semimembranosus-gastrocnemius bursa, the superior recess of the posterior horn of the medial meniscus, and the popliteal tendon sheath bursa. Joint fluid, which produces moderate signal intensity on T1-weighted images and high signal intensity on T2-weighted images, may therefore be seen in the perimeniscal recesses and popliteal tendon sheath bursa in patients with joint effusions (2,34). Fluid in

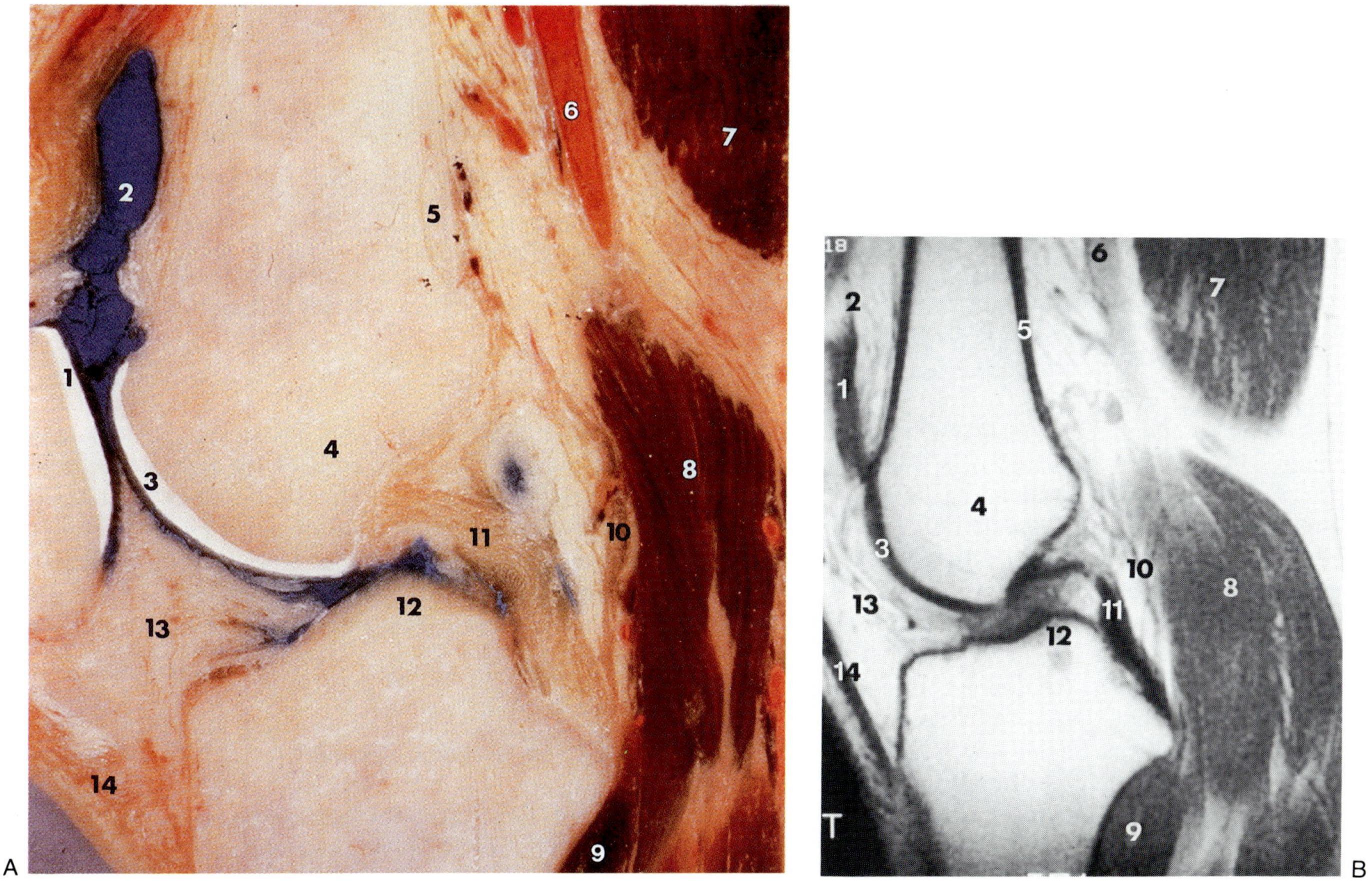

FIG. 3.6A,B. Intercondylar sagittal section at the level of the posterior cruciate ligament. 1, Patellar articular cartilage; 2, suprapatellar bursa; 3, articular cartilage of the femoral trochlear groove; 4, distal femoral bone marrow; 5, cortical bone; 6, popliteal artery; 7, semimembranosus muscle; 8, medial head of the gastrocnemius muscle; 9, popliteus muscle; 10, posterior joint capsule; 11, posterior cruciate ligament; 12, tibial spine; 13, infrapatellar fat pad; 14, patellar tendon.

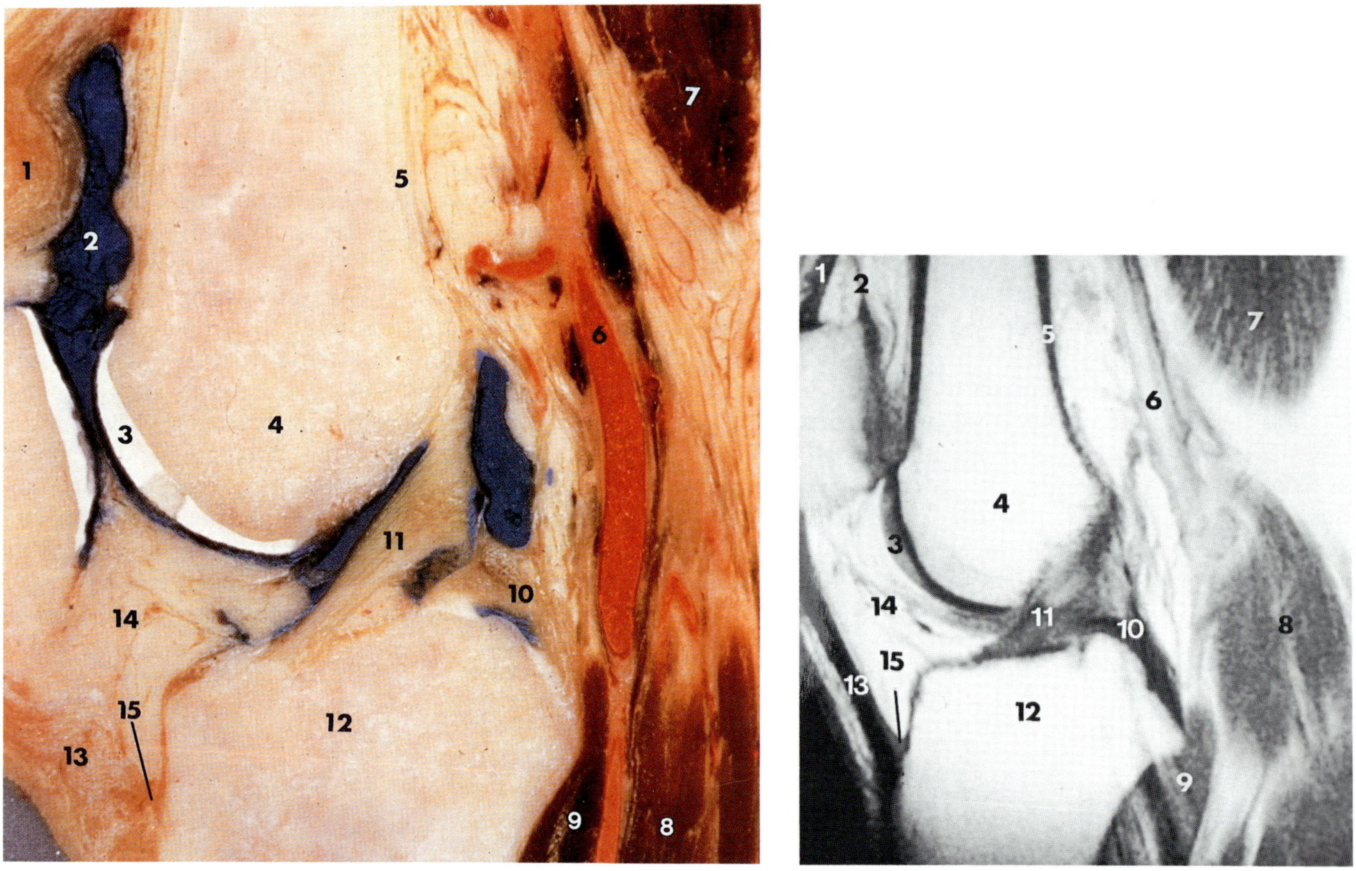

A B

FIG. 3.7A,B. Intercondylar sagittal section at the level of the anterior cruciate ligament. 1, Quadriceps tendon; 2, suprapatellar bursa; 3, articular cartilage along the lateral aspect of the femoral trochlear groove; 4, distal femoral bone marrow; 5, cortical bone; 6, popliteal vasculature; 7, semimembranosus muscle; 8, medial head of the gastrocnemius muscle; 9, popliteal muscle; 10, insertion of the posterior cruciate ligament; 11, anterior cruciate ligament; 12, bone marrow within the proximal tibia; 13, patellar tendon; 14, infrapatellar fat pad; 15, infrapatellar bursa.

these sites should not be mistaken for meniscal pathology.

The continuity of the cruciate ligaments and of the patellar tendon is easily demonstrated on sagittal images. With the knee extended, the posterior cruciate ligament is relatively lax and is seen as a black arc-like band spanning from the distal femur to the posterior tibial plateau. In extension, the anterior cruciate ligament is taut and is seen as a continuous band of moderately low signal intensity tissue outlined by high signal intensity intraarticular fat. The remnant of the epiphyseal growth plate is often depicted on MR images, although this remnant is difficult to appreciate on the cadaver photographs.

In Fig. 3.8, the relationship between the anterior horn of the lateral meniscus and the transverse geniculate ligament is illustrated. Understanding this anatomy is essential to avoid the erroneous misdiagnosis of a meniscal tear, which may be simulated by the normal interface between the transverse geniculate ligament and the anterior horn of the lateral meniscus. Similarly, Fig. 3.8 illustrates the intimate relationship between the popliteal tendon and the posterior horn of the lateral meniscus. Normal synovial fluid collections within the popliteal tendon sheath bursa may also simulate a meniscal tear to those unfamiliar with this precise anatomy.

Axial Cross-Sectional Anatomy

Four matched pairs of axial T1-weighted MR images and cadaver sections are presented in Figs. 3.9 to 3.12. Although there are slight differences in the positions of the patellae in the MR images and color photographs, the correlation is otherwise quite precise. The inherent soft tissue contrast in MR images allows the muscles, tendons, nerves, subcutaneous fat, cortical

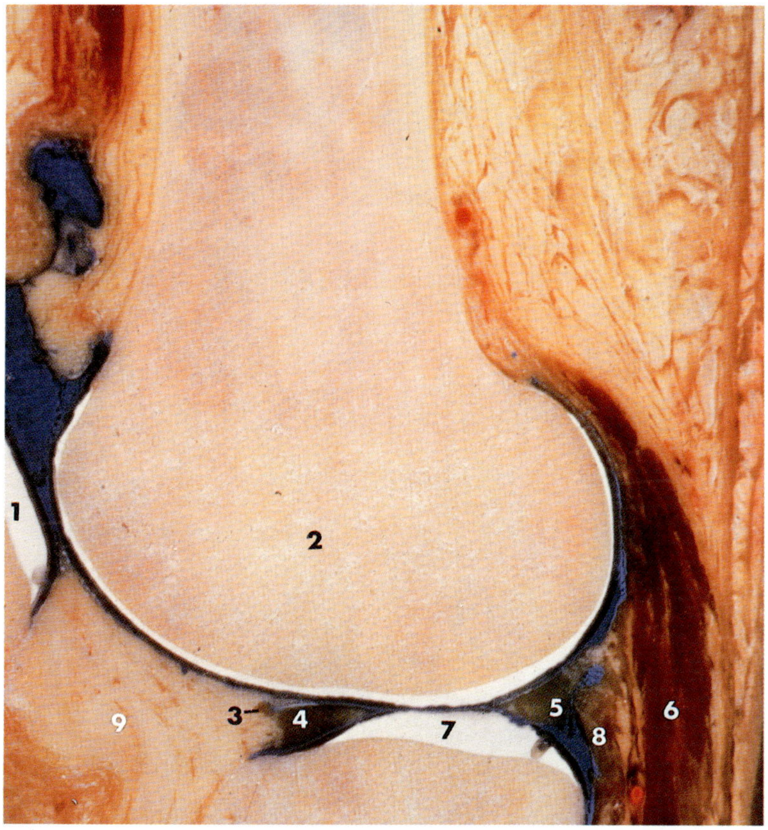

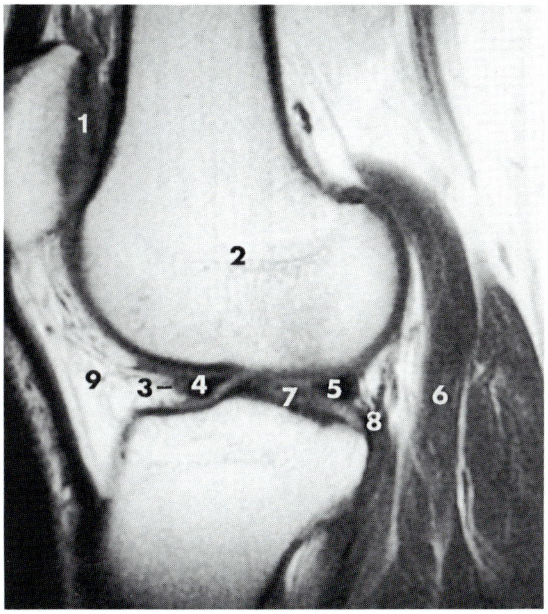

FIG. 3.8A,B. Sagittal section through the lateral femoral condyle. 1, Patellar articular cartilage; 2, lateral femoral condyle; 3, transverse geniculate ligament; 4, anterior horn of the lateral meniscus; 5, posterior horn of the lateral meniscus; 6, lateral head of the gastrocnemius muscle; 7, tibial articular cartilage; 8, popliteal tendon; 9, infrapatellar fat pad.

bone, medullary bone, menisci, and articular cartilage to be easily distinguished. The effect of volume averaging limits resolution of the menisci in the axial plane. Precise identification of the neurovascular bundle and each of the muscle bundles facilitates differentiation of various pathologic entities based on their anatomic positions. For example, a popliteal cyst can be easily distinguished from a popliteal artery aneurysm because the former usually arises in the semimembranosus-gastrocnemius bursa, whereas the latter originates in the popliteal artery.

Homogeneously low signal intensity is yielded by the patellar tendon, cruciate ligaments, collateral ligaments, and muscular tendons. The anterior cruciate ligament courses anteromedially, typically extending approximately 20° in obliquity relative to the sagittal plane (6,28).

Axial images are extremely useful for evaluating congruency of the patella and trochlear groove, for assessing the patellar retinacula, and for precisely quantifying the severity of tears of the patellar tendon, quadriceps tendon, cruciate ligaments, and collateral ligaments.

THE NORMAL EXAMINATION

In the past decade, MR imaging of the knee has grown to play a role in the diagnosis of virtually all knee abnormalities. Just as the indications for knee MRI have expanded, the variety of available imaging techniques has also multiplied. Although spin-echo imaging is emphasized in this text, a variety of other techniques is available, including two-dimensional gradient-echo imaging, three-dimensional gradient-echo imaging, inversion recovery imaging, and a variety of fat suppression methods (1,13,14,29,30,32,35). Furthermore, MRI may be enhanced by intraarticular contrast material or intravascular contrast material. Three-dimensional data acquisition may facilitate reconstruction of images in an infinite array of obliquities utilizing powerful work stations.

Despite the increased variety of available imaging techniques, most MR examinations are still performed with the spin-echo technique. Therefore, the normal examination presented in this chapter continues to emphasize spin-echo imaging.

For most indications, the author's preferred method

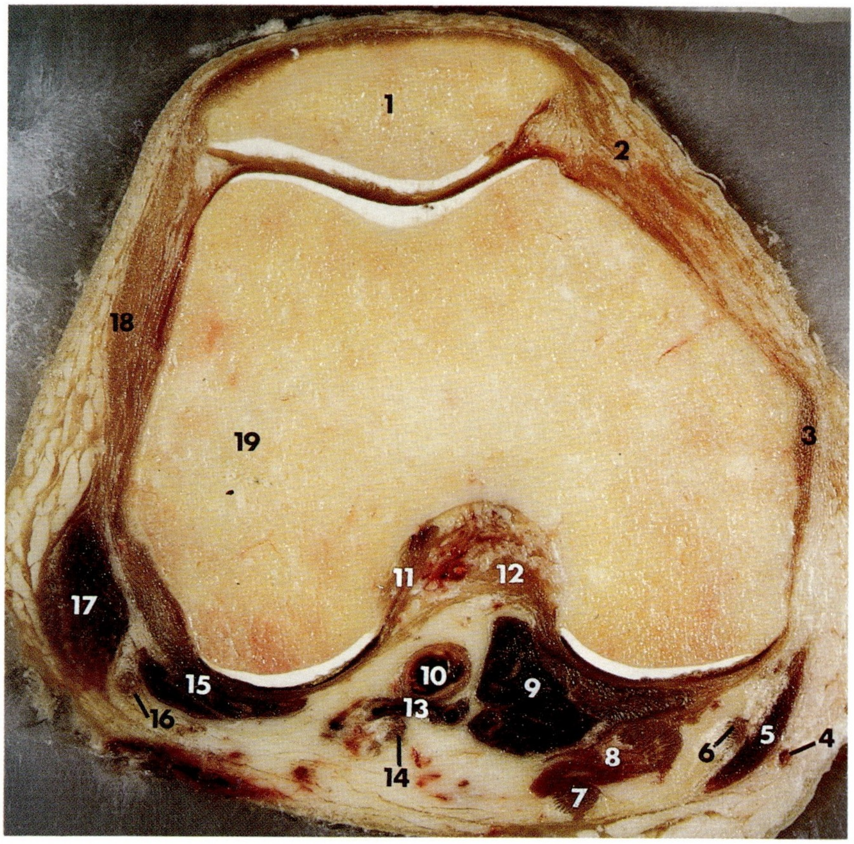

A

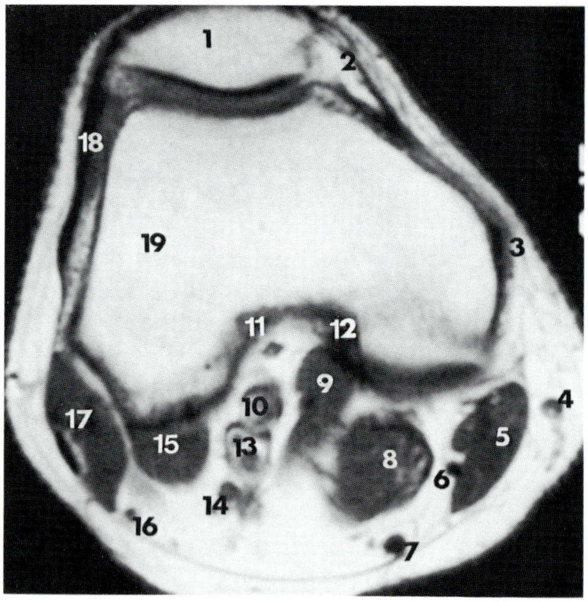

B

FIG. 3.9A,B. Axial section 3.5 cm superior to the tibial plateau. 1, Patella; 2, medial patellar retinaculum; 3, origin of the medial collateral ligament; 4, greater saphenous vein; 5, sartorius muscle; 6, gracilis tendon; 7, semitendinosus tendon; 8, semimembranosus muscle and tendon; 9, medial head of the gastrocnemius muscle; 10, popliteal artery; 11, origin of the anterior cruciate ligament; 12, origin of the posterior cruciate ligament; 13, popliteal vein; 14, tibial nerve; 15, lateral head of the gastrocnemius muscle; 16, common peroneal nerve; 17, biceps femoris muscle and tendon; 18, iliotibial band; 19, lateral femoral condyle bone marrow.

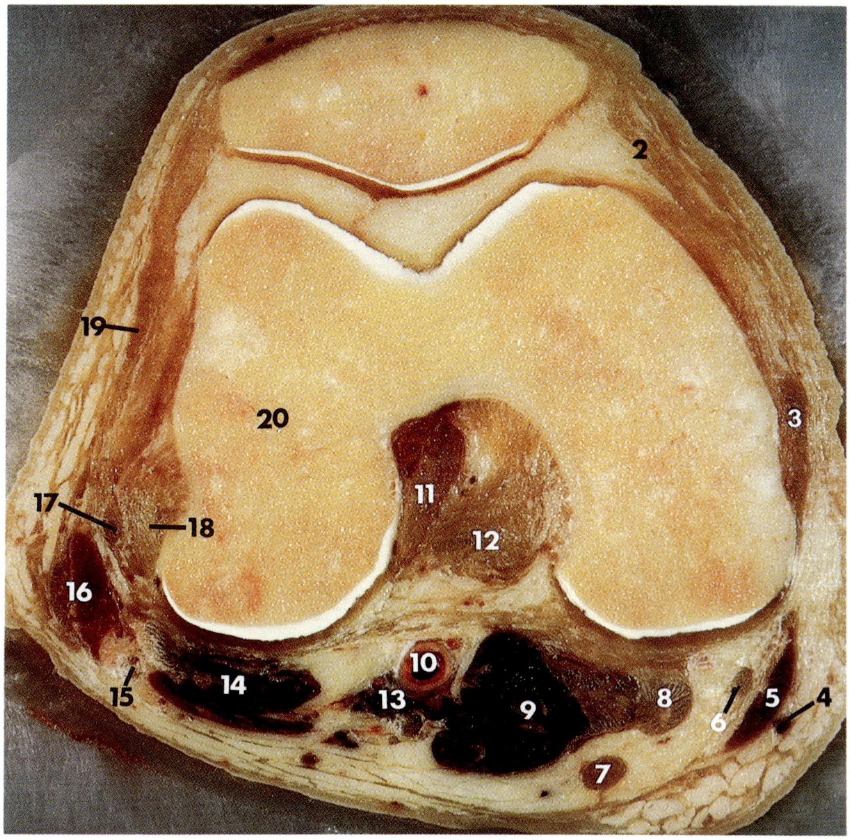

A

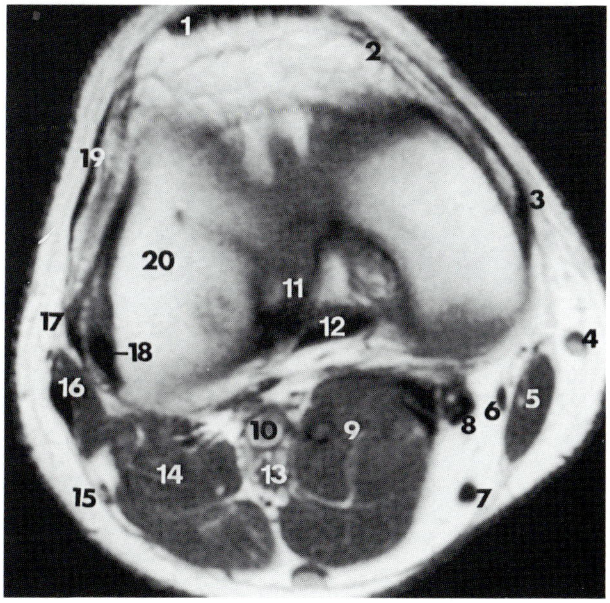

B

FIG. 3.10A,B. Axial section 2.0 cm superior to the tibial plateau. 1, Patellar tendon; 2, medial patellar retinaculum; 3, medial collateral ligament; 4, greater saphenous vein; 5, sartorius muscle; 6, gracilis tendon; 7, semitendinosus tendon; 8, semimembranosus tendon; 9, medial head of the gastrocnemius muscle; 10, popliteal artery; 11, anterior cruciate ligament; 12, posterior cruciate ligament; 13, popliteal vein; 14, lateral head of the gastrocnemius muscle; 15, common peroneal nerve; 16, biceps femoris muscle; 17, fibular collateral ligament; 18, insertion of the popliteal tendon; 19, iliotibial band; 20, lateral femoral condyle.

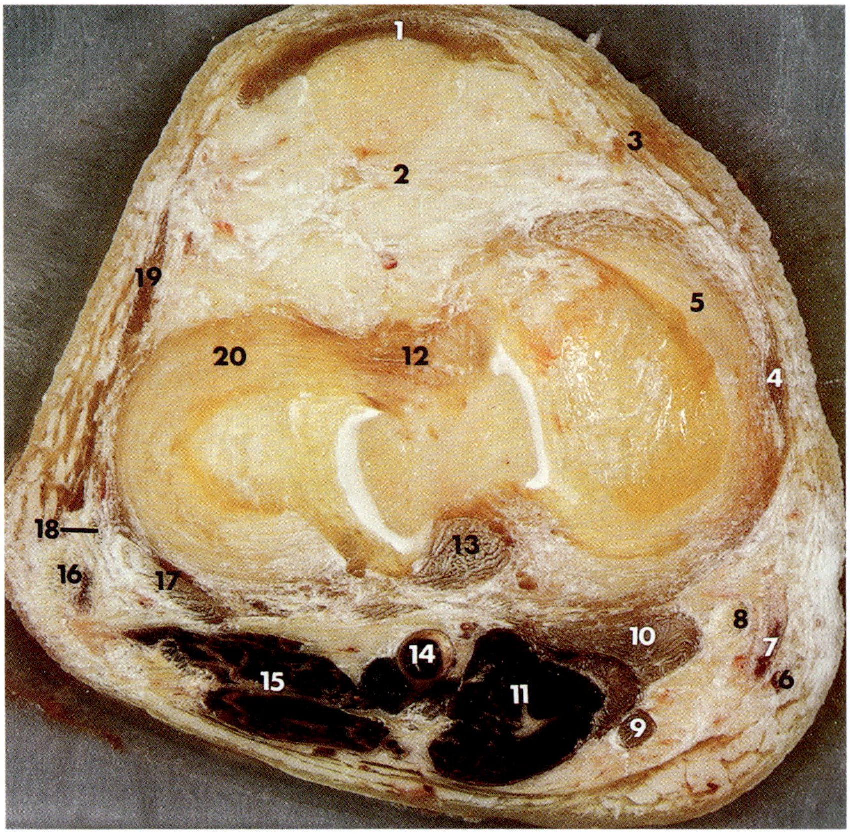

A

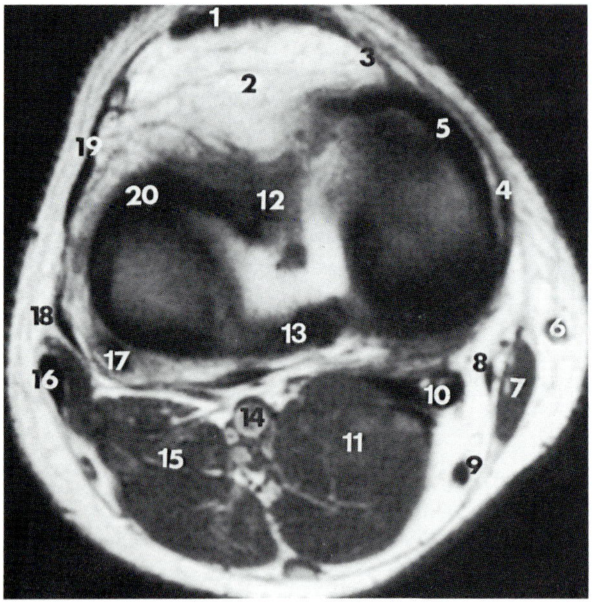

B

FIG. 3.11A,B. Axial section at the level of the menisci. 1, Patellar tendon; 2, intrapatellar fat pad or Hoffa's fat pad; 3, medial patellar retinaculum; 4, medial collateral ligament; 5, medial meniscus; 6, greater saphenous vein; 7, sartorius muscle; 8, gracilis tendon; 9, semitendinosus tendon; 10, semimembranosus tendon; 11, medial gastrocnemius muscle; 12, insertion of the anterior cruciate ligament; 13, insertion of the posterior cruciate ligament; 14, popliteal artery; 15, lateral gastrocnemius muscle; 16, biceps femoris tendon; 17, popliteal tendon; 18, fibular collateral ligament; 19, iliotibial band; 20, anterior horn of the lateral meniscus.

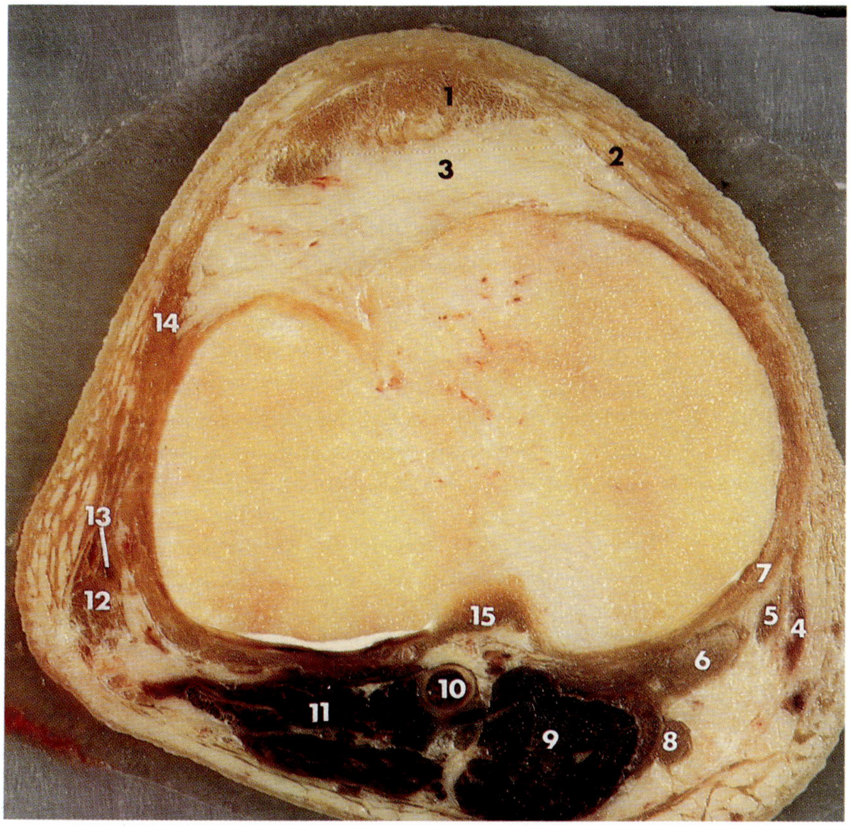

A

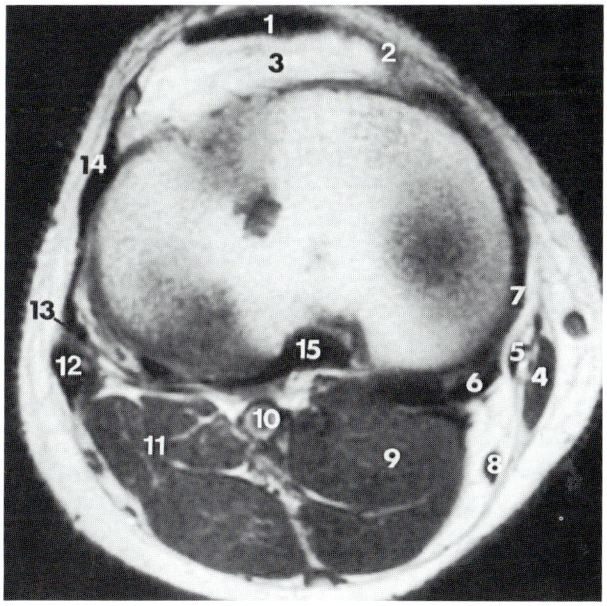

B

FIG. 3.12A,B. Axial section at the level of the tibial plateau. 1, Patellar tendon; 2, medial patellar retinaculum; 3, infrapatellar fat pad; 4, sartorius muscle; 5, gracilis tendon; 6, semimembranosus tendon; 7, posteromedial capsule; 8, semitendinosus tendon; 9, medial gastrocnemius muscle; 10, popliteal artery; 11, lateral gastrocnemius muscle; 12, biceps femoris tendon; 13, fibular collateral ligament; 14, iliotibial band insertion.

for knee MR imaging includes coronal T1-weighted spin-echo images, sagittal intermediate- and T2-weighted spin-echo images, and axial intermediate- and T2-weighted spin-echo images. Images are typically obtained with a 14 to 16 cm field-of-view, and 192 to 256 phase encoding levels. The sagittal and axial images are obtained with a multiecho sequence (TE = 20 msec, 70 msec; TR = 2,000 msec). The coronal T1-weighted spin-echo images are typically obtained with a TE of 20 msec and TR of 500 msec. The coronal imaging sequence, therefore, typically requires 2 to 3 min, and the sagittal and axial series typically each require 6 to 7 min. The image slices are 4 mm thick and one signal excitation is employed. The entire examination can almost always be completed in 20 to 30 min, and yields an array of approximately 90 images of the knee joint.

As a general rule, structures having short T1 values yield high signal intensity on T1-weighted images, and structures having long T2 values yield high signal intensity on T2-weighted images. Substances that lack mobile protons and have very short T2 values yield low signal on both T1- and T2-weighted images. These substances include cortical bone, ligaments, tendons, and menisci. The signal of the medullary canal is primarily determined by its fatty marrow, which has a short T1 and short T2. Within a broad range of T1-weighted to moderately T2-weighted images, the relative signal intensity of most knee structures remains constant. In order of decreasing signal intensity, these structures are subcutaneous fat, medullary bone, articular cartilage, muscles, ligaments, tendons, menisci, and cortical bone. In general, techniques which selectively suppress signal detection from fat tend to increase the sensitivity of detecting disease within medullary bone.

Synovial fluid usually has a moderate T1 and long T2, and therefore yields intermediate signal on T1-weighted sequences, but becomes quite bright with T2 weighting (2). By proper selection of pulse sequences, the radiologist can therefore employ synovial fluid as an inherent intraarticular contrast agent.

In Figs. 3.13 through 3.48, the normal sagittal and axial multiecho pulse sequences are presented. Normal coronal T1-weighted images were depicted in Figs. 3.1 through 3.4.

Sagittal Multiecho Series

The most medial sagittal sections reveal the dark gracilis, sartorius, and semitendinosus tendons sweeping anteriorly to insert on the anteromedial tibia (pes anserinus) (Figs. 3.14 and 3.15). The tibial and femoral hyaline articular cartilage is typically well seen as a thin line of moderate signal intensity interposed between the low-signal meniscus and cortical bone of the tibial plateau. A small amount of joint fluid is also seen in the superior recess of the posterior horn of the medial meniscus (Fig. 3.16). Fibrofatty tissue is often also present in this recess, which is interposed between the posterosuperior portion of the medial meniscus and the posterior capsule (see Figs. 3.16 and 3.17). The posterior margin of the medial meniscus itself extends to within a few millimeters of the posterior margin of the tibial plateau.

The medial third of the medial meniscus is seen as a homogeneously black disk between the convex femur and flat tibial plateau (Fig. 3.14). Because the average width of the mid-zone of the medial meniscus is 10 mm, and 4 mm sections are generally used in scanning, this disk-like or "bow-tie" appearance of the medial meniscus is usually seen on two to three successive images (Figs. 3.14 and 3.15). As the sections progress toward the central free edge of the medial meniscus, the middle third of the medial meniscus becomes thinner until it finally disappears (Fig. 3.16). At this point, the medial meniscus changes in shape from a "bow-tie" to two triangles which respectively represent the anterior and posterior horns. The triangles of the smaller anterior and larger posterior horns of the medial meniscus are typically depicted on three subsequent sections (Figs. 3.16–3.18). As the sagittal sections approach the intercondylar notch, the anterior ligamentous connection of the medial meniscus to the tibia and lateral meniscus can be seen as well as the

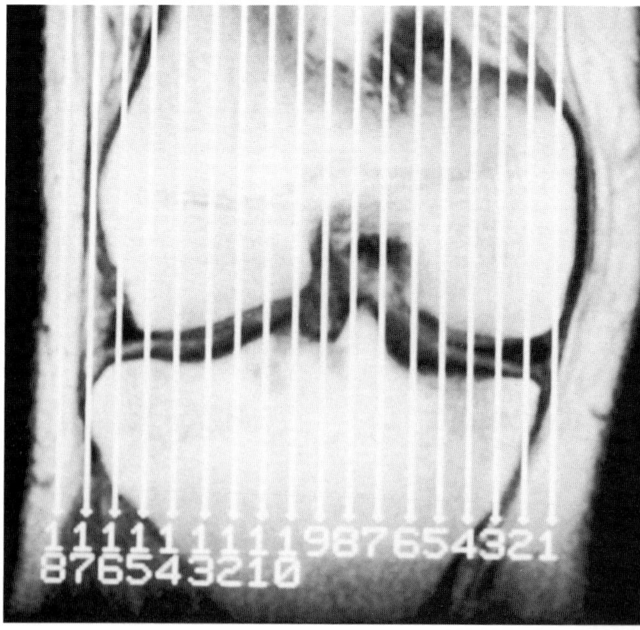

FIG. 3.13. Coronal T1-weighted scan demonstrates the position of the 18 following sagittal sections which course from the medial to the lateral aspect of the knee joint (Figs. 3.14–3.31).

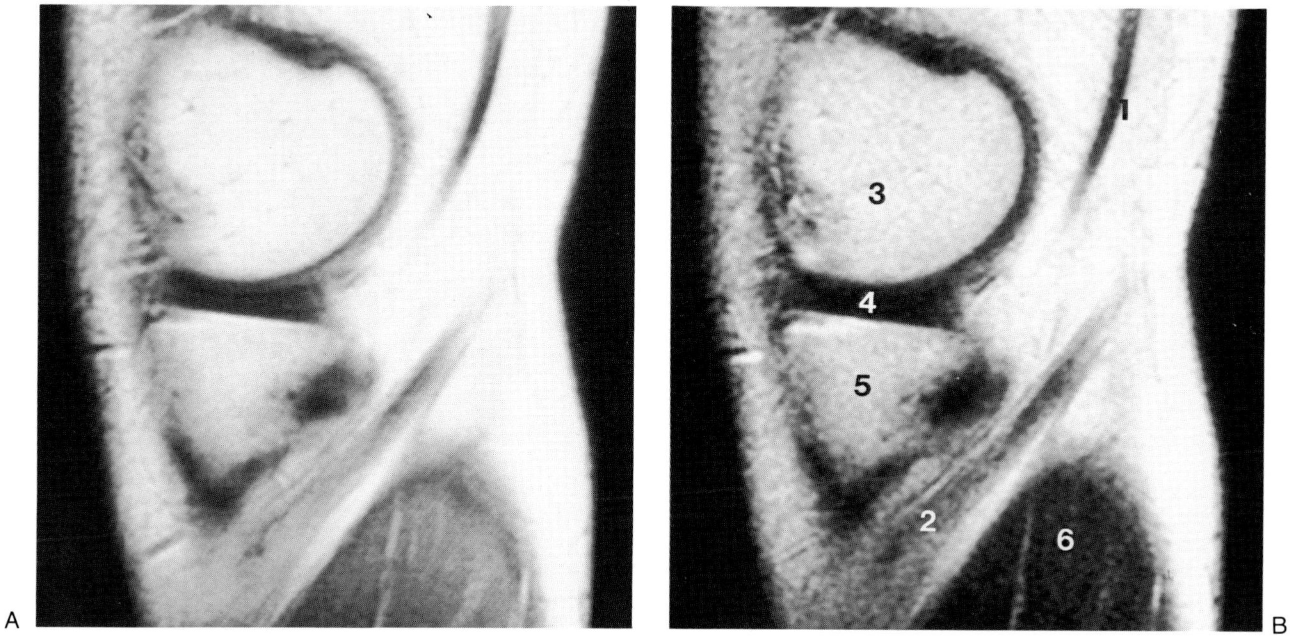

FIG. 3.14. Intermediate-weighted **(A)** and T2-weighted **(B)** spin-echo images at the level of section 1 (see Fig. 3.13). 1, Gracilis tendon; 2, pes anserinus insertion; 3, medial femoral condyle bone marrow; 4, middle third of the medial meniscus; 5, medial proximal tibia.

posterior connection to the posterior cruciate ligament and tibia (Fig. 3.19).

The sagittal series along the medial aspect of the knee joint also typically depicts the normal medial plica (Figs. 3.18 and 3.19), the suprapatellar bursa (Figs. 3.19 and 3.20), the posteromedial joint capsule (Figs. 3.16–3.19), and the gastrocnemius-semimembranosus bursa (Fig. 3.18).

As sections approach the medial aspect of the intercondylar notch, the posterior cruciate ligament is always seen as an arc-like black band (Figs. 3.20 and 3.21). Originating from the anterolateral aspect of the

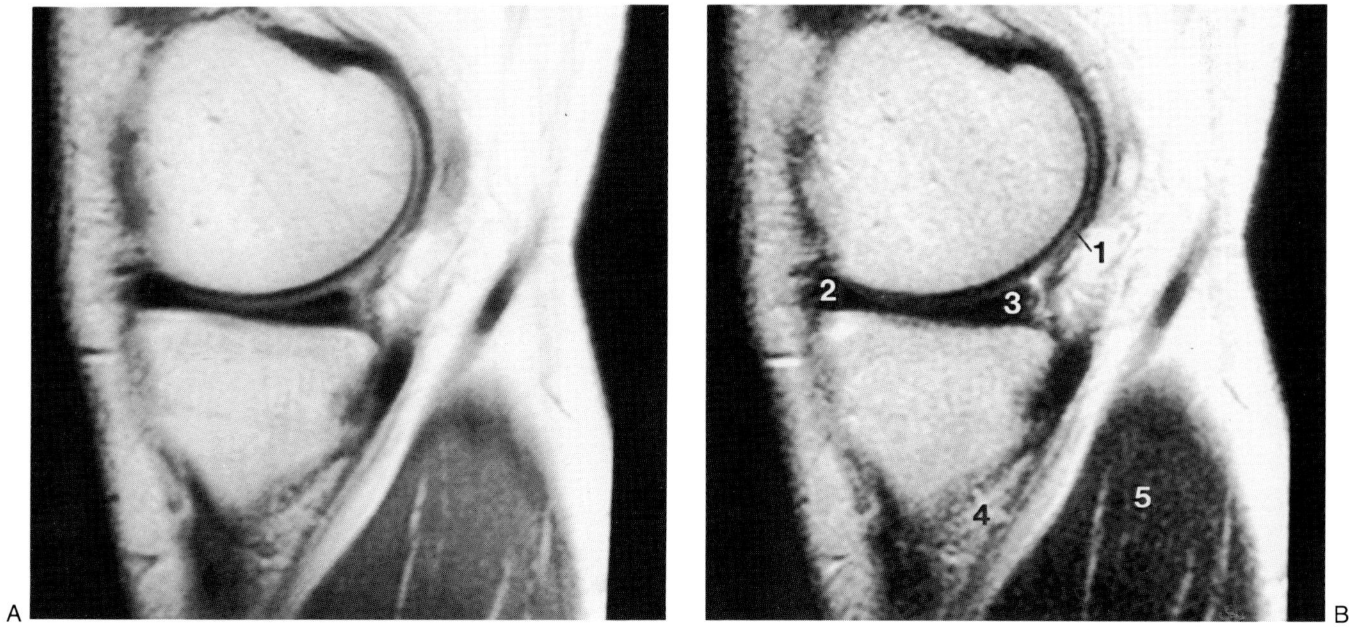

FIG. 3.15A,B. Intermediate- and T2-weighted sagittal images at the level of section 2 (see Fig. 3.13). 1, Hyaline articular cartilage of the medial femoral condyle; 2, anterior horn of the medial meniscus; 3, posterior horn of the medial meniscus; 4, pes anserinus insertion; 5, medial gastrocnemius muscle.

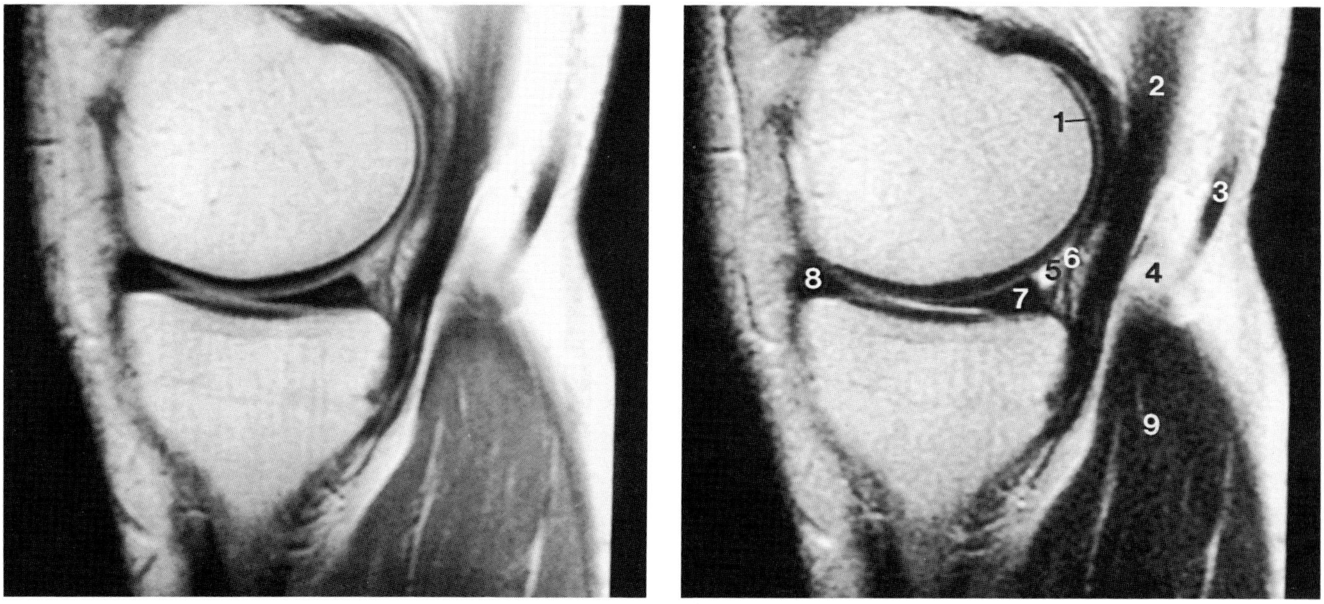

A B

FIG. 3.16A,B. Intermediate- and T2-weighted sagittal images at the level of section 3 (see Fig. 3.13). 1, Black cortical bone; 2, semimembranosus tendon; 3, semitendinosus tendon; 4, semimembranosus-gastrocnemius bursa; 5, superior recess of the posterior horn of the medial meniscus; 6, posterior medial capsule; 7, posterior horn of the medial meniscus; 8, anterior horn of the medial meniscus; 9, medial gastrocnemius muscle.

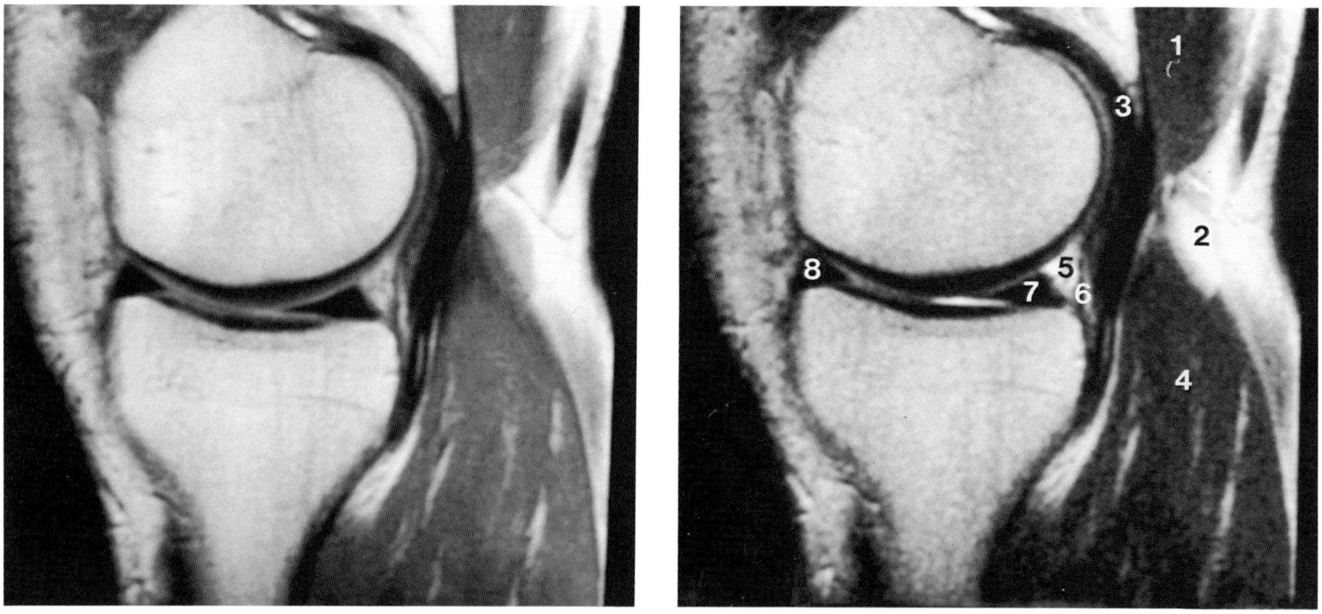

A B

FIG. 3.17A,B. Sagittal intermediate- and T2-weighted images at the level of section 4 (see Fig. 3.13). 1, Semimembranosus muscle; 2, semimembranosus-gastrocnemius bursa; 3, origin of the medial head of the gastrocnemius muscle; 4, medial gastrocnemius muscle; 5, superior recess of the posterior horn of the medial meniscus; 6, posteromedial capsule; 7, posterior horn of the medial meniscus; 8, anterior horn of the medial meniscus.

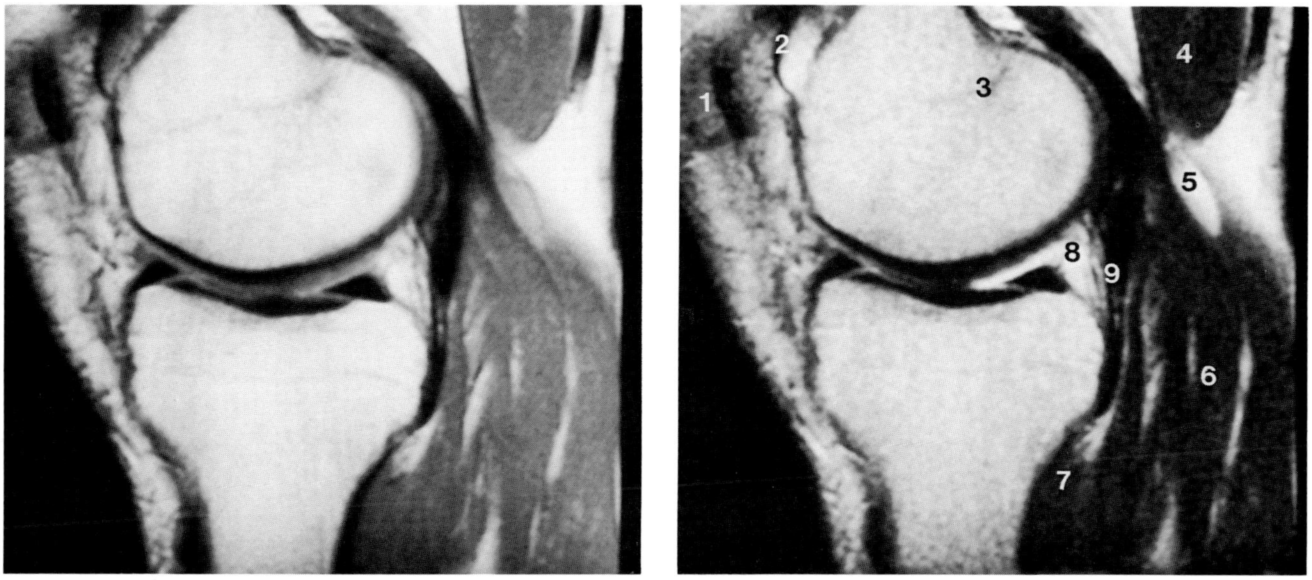

FIG. 3.18A,B. Intermediate- and T2-weighted images at the level of section 5 (see Fig. 3.13). 1, Medial margin of the patella; 2, medial synovial plica; 3, remnant of the femoral physis; 4, semimembranosus muscle; 5, semimembranosus-gastrocnemius bursa; 6, medial gastrocnemius muscle; 7, popliteal muscle; 8, perimeniscal intraarticular fat; 9, posterior joint capsule.

medial femoral condyle, it attaches to the extreme posterior aspect of the tibial plateau. It is slightly posteriorly convex or buckled with the leg in extension, but it becomes taut and thinner during knee flexion. The sagittal sections through the medial aspect of the intercondylar notch also depict the patellar articular cartilage, which normally has a posteriorly convex surface contour (Figs. 3.20 and 3.21), as well as the articular cartilage along the trochlear groove of the femur (Figs. 3.20 and 3.21). Often the accessory meniscofemoral ligaments (the anterior ligament of Humphrey and the posterior ligament of Wrisberg) are seen on these sagittal sections (Figs. 3.20 and 3.21) (12). As sections approach the lateral aspect of the intercondylar notch,

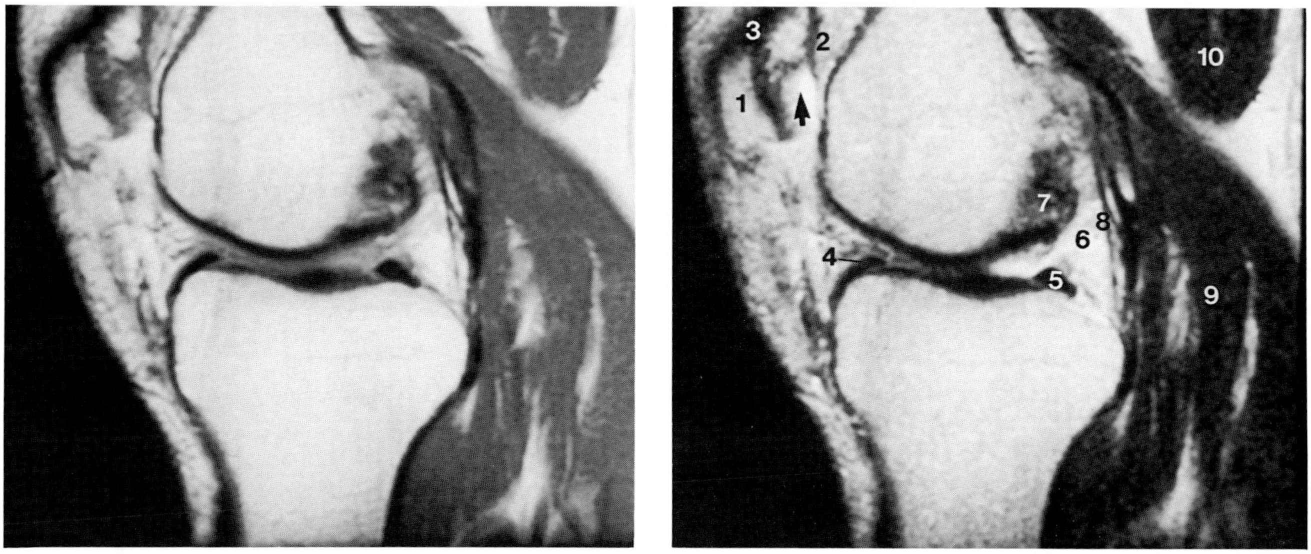

FIG. 3.19A,B. Sagittal intermediate- and T2-weighted images at the level of sagittal section 6 (see Fig. 3.13). 1, Medial aspect of the patella; 2, medial synovial plica; 3, quadriceps tendon; 4, anteromesial tibial attachment of the medial meniscus; 5, posteromesial tibial attachment of the medial meniscus; 6, intercondylar fat; 7, cortical bone along the medial border of the intercondylar notch; 8, posterior joint capsule; 9, medial gastrocnemius muscle; 10, semimembranosus muscle; *arrow*, suprapatellar bursa.

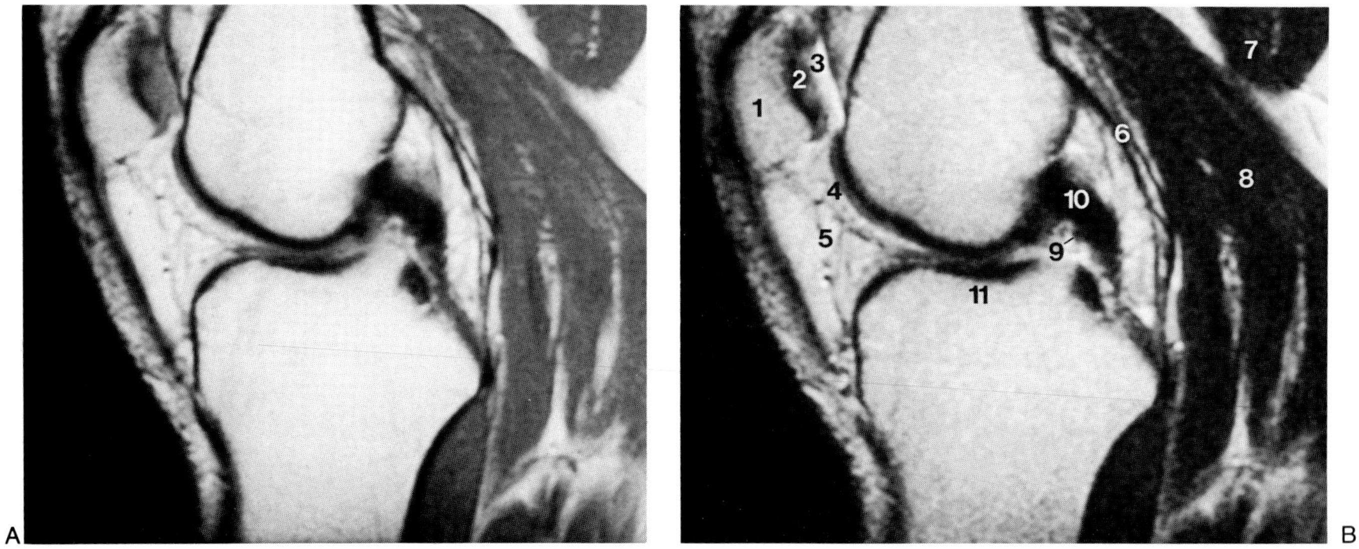

FIG. 3.20A,B. Intermediate- and T2-weighted images at the level of sagittal section 7 (see Fig. 3.13). 1, Central portion of the patella; 2, patellar articular cartilage; 3, suprapatellar bursa; 4, articular cartilage along the femoral trochlear groove; 5, infrapatellar fat pad; 6, posterior joint capsule; 7, semimembranosus muscle; 8, medial gastrocnemius muscle; 9, anterior menisco-femoral ligament of Humphrey; 10, posterior cruciate ligament; 11, tibial spine bone marrow.

the anterior cruciate ligament is seen, originating from the posteromedial aspect of the lateral condyle and inserting 8 mm behind the anterior edge of the proximal tibia. Although precise positioning is necessary to image the normal anterior cruciate ligament on a single section, this can now essentially be universally achieved using one of a variety of techniques (with options including thin sections, oblique imaging, three-dimensional image acquisition, or retrospective image reconstruction). Fibrous septa and blood vessels are often seen in the fat pad anterior to the insertion of the ACL, and occasionally the transverse geniculate ligament can be followed from the medial to the lateral meniscus (Figs. 3.22, 3.24, and 3.25).

Images obtained through the far lateral aspect of the intercondylar notch generally depict the popliteal ar-

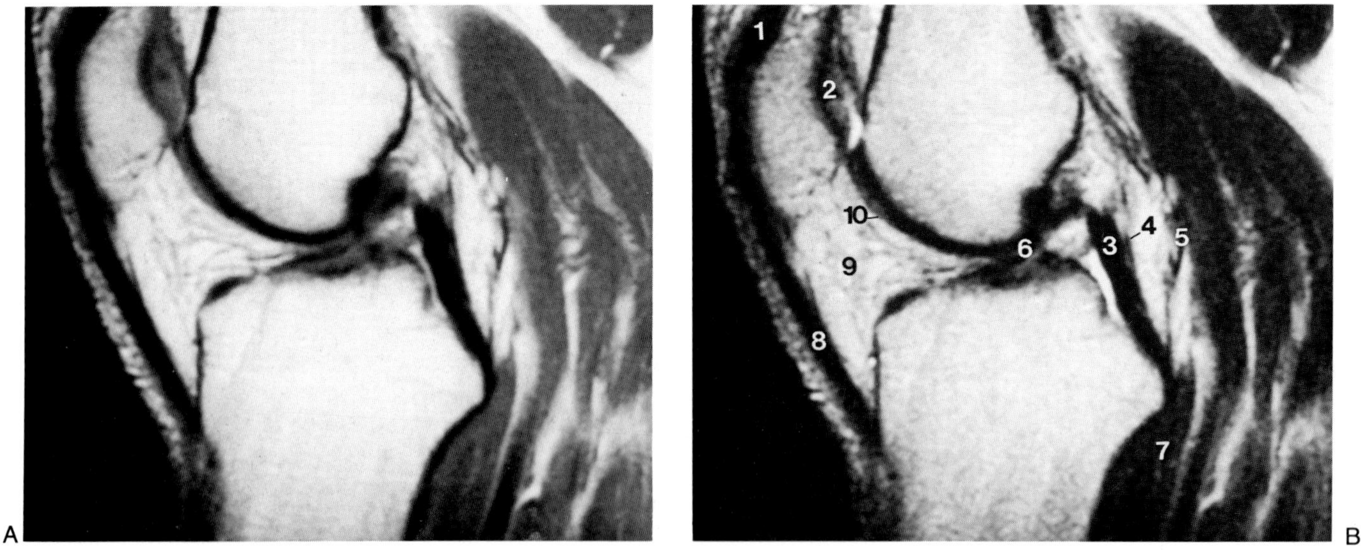

FIG. 3.21A,B. Sagittal intermediate- and T2-weighted images at the level of sagittal section 8 (see Fig. 3.13). 1, Quadriceps tendon; 2, patellar articular cartilage; 3, posterior cruciate ligament; 4, posterior meniscofemoral ligament of Wrisberg; 5, posterior capsule; 6, anterior cruciate ligament; 7, popliteus muscle; 8, patellar tendon; 9, infrapatellar fat pad; 10, articular cartilage along the femoral trochlear groove.

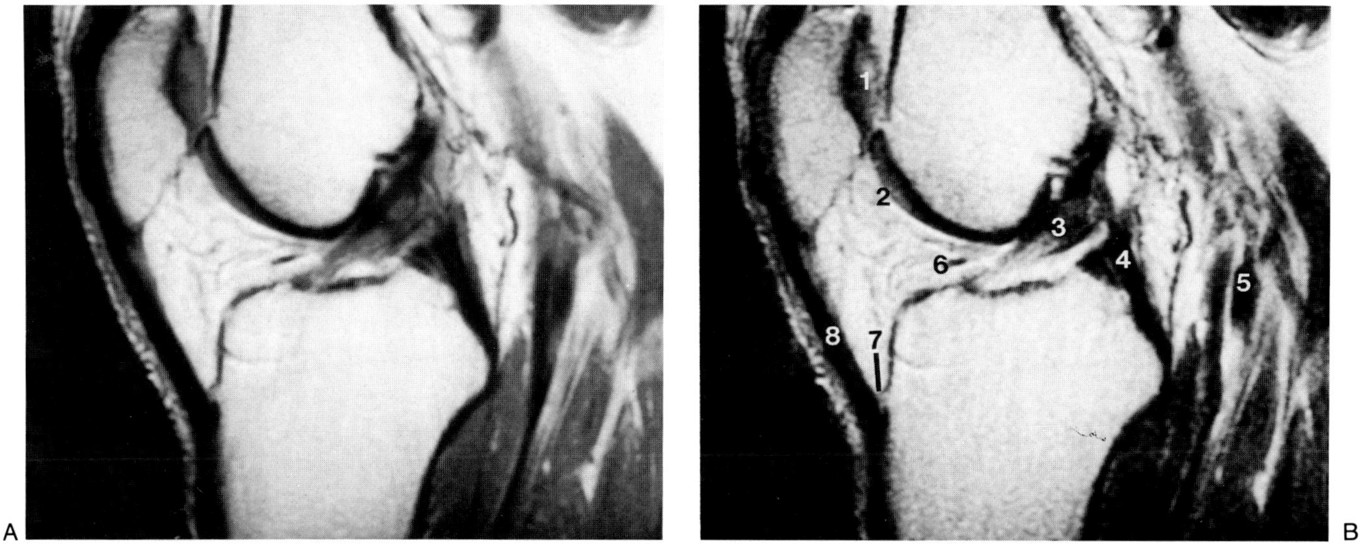

FIG. 3.22A,B. Sagittal intermediate- and T2-weighted images at the level of sagittal section 9 (see Fig. 3.13). 1, Patellar articular cartilage; 2, articular cartilage along the femoral trochlear groove; 3, anterior cruciate ligament; 4, posterior cruciate ligament; 5, popliteal vessels; 6, transverse geniculate ligament; 7, infrapatellar bursa; 8, patellar tendon.

tery and vein. Frequently, phase-encoding artifacts (see Chapter 1) slightly degrade the sagittal images at the level of these vascular structures (Figs. 3.23 and 3.24). A number of techniques can be employed to avoid these artifacts, including flow-compensating pulse sequences and cardiac gating. In practice these artifacts are diagnostically unimportant, and the additional time often required to avoid them is not merited. Sections through the lateral aspect of the intercondylar notch also depict the posteriorly convex surface of the articular cartilage along the lateral patellar facet as well as the articular cartilage along the lateral aspect of the femoral trochlear groove (Fig. 3.24). The patellar tendon is well seen on all of the sagittal sections obtained through the trochlear groove. The infrapatellar bursa is often also subtly identified (Fig. 3.26).

The sections through the lateral aspect of the knee

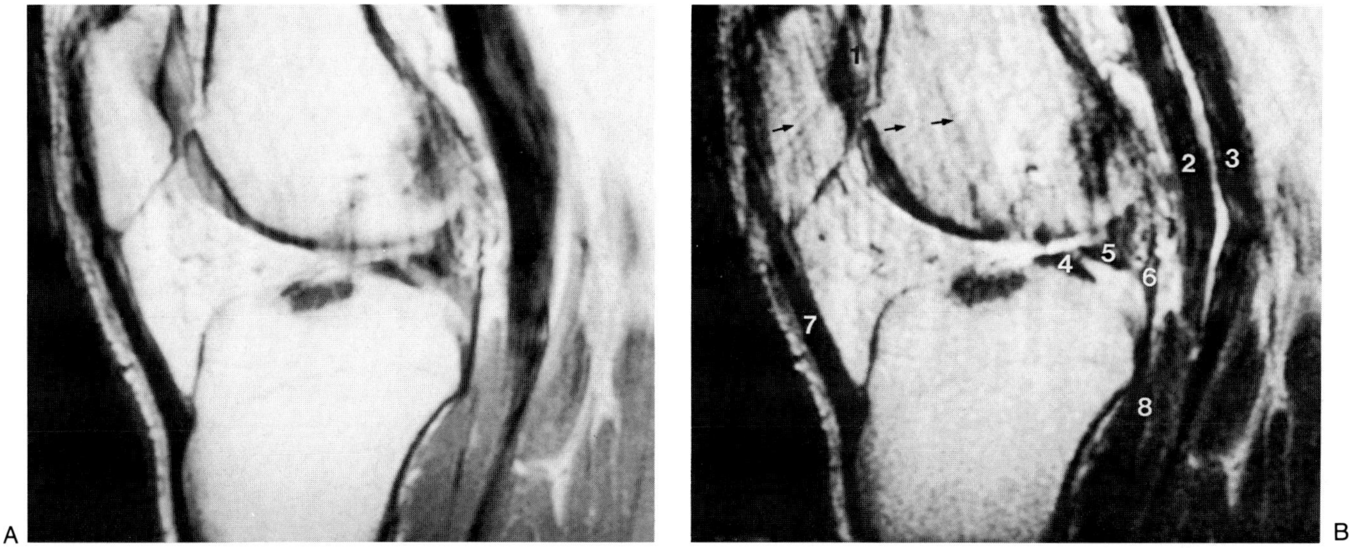

FIG. 3.23A,B. Sagittal intermediate- and T2-weighted images at the level of sagittal section 10 (see Fig. 3.13). 1, Patellar articular cartilage along the lateral patellar facet; 2, popliteal artery; 3, popliteal vein; 4, posteromesial attachment of the lateral meniscus; 5, partial volume imaging of the posterior cruciate ligament; 6, posterior capsule; 7, patellar tendon; 8, popliteal muscle; *small arrows*, phase-encoding artifact due to popliteal vascular flow.

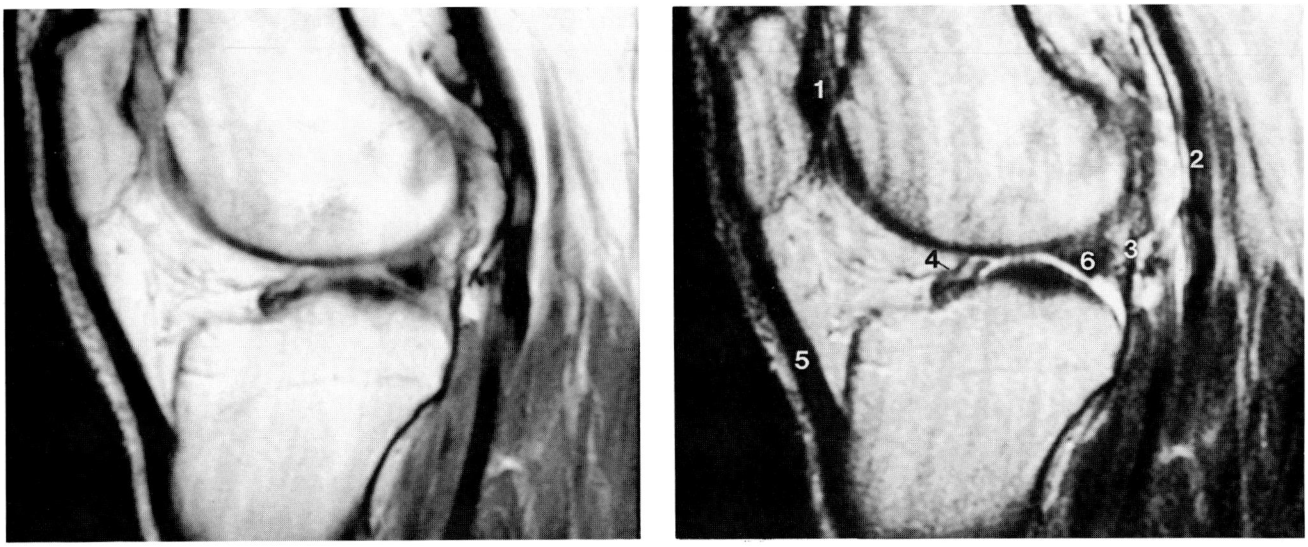

FIG. 3.24A,B. Sagittal intermediate- and T2-weighted images at the level of section 11 (see Fig. 3.13). 1, Articular cartilage along the lateral patellar facet; 2, popliteal vessel; 3, posterior capsule; 4, transverse geniculate ligament; 5, patellar tendon; 6, posterior horn of the lateral meniscus.

joint depict the smaller C-shaped lateral meniscus. Because the knee joint is generally slightly externally rotated, the posterior horn projects more medially than does the anterior horn, and thus appears first as the sections proceed from medial to lateral across the knee. As one proceeds laterally across the knee, the first sections through what appear to be meniscus are, in fact, through the central ligamentous attachments of the menisci (Figs. 3.23 and 3.24). On the next series of sections, the lateral meniscus is depicted as two triangles with sharply pointed apices (Figs. 3.26 and 3.27). On far lateral sections of the lateral meniscus, a disk-like or bow-tie appearance is seen (Figs. 3.28 and 3.29). The bow-tie appearance of the lateral meniscus is more symmetrical than that of the medial meniscus, because the meniscus is nearly uniform in size from front to back.

The popliteal tendon (Figs. 3.27–3.30) originates just below the origin of the fibular collateral ligament. The tendon courses obliquely downward and posteriorly,

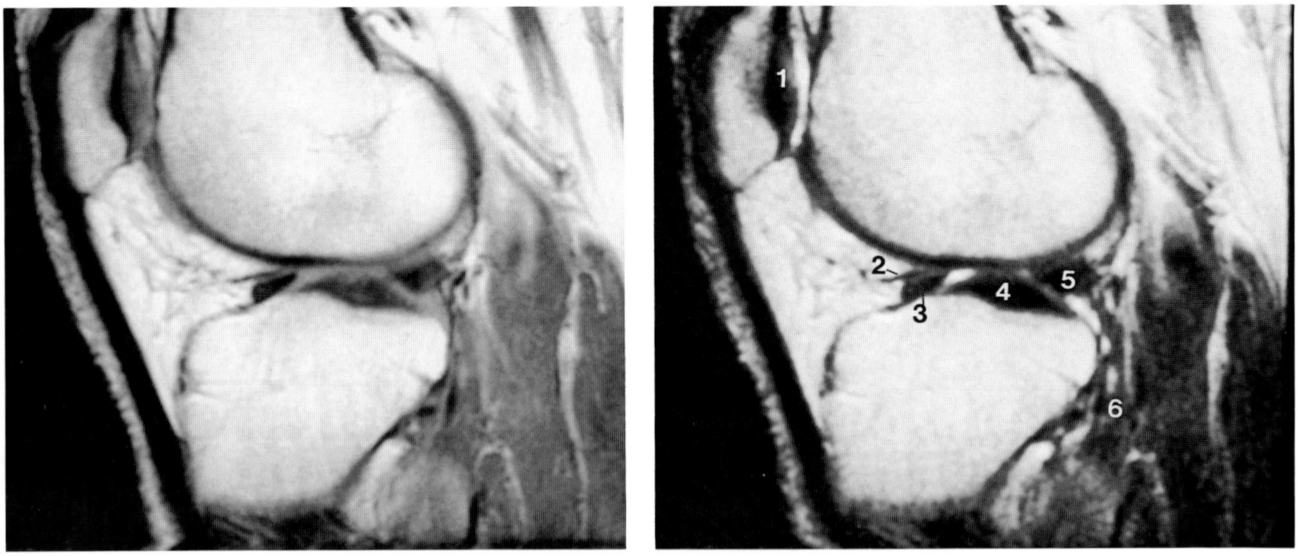

FIG. 3.25A,B. Sagittal intermediate- and T2-weighted images at the level of sagittal section 12 (see Fig. 3.13). 1, Patellar articular cartilage; 2, transverse geniculate ligament; 3, anterior horn of the lateral meniscus; 4, tibial articular cartilage; 5, posterior horn of the lateral meniscus; 6, popliteus muscle.

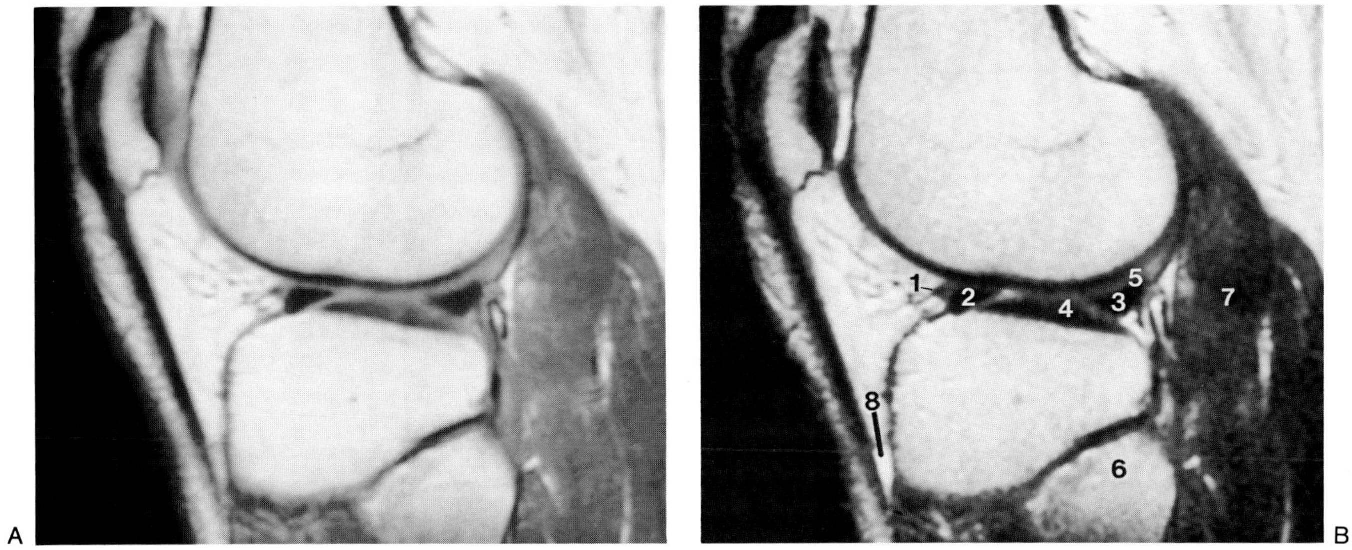

FIG. 3.26A,B. Sagittal intermediate- and T2-weighted images at the level of section 13 (see Fig. 3.13). 1, Transverse geniculate ligament; 2, anterior horn of the lateral meniscus; 3, posterior horn of the lateral meniscus; 4, tibial articular cartilage; 5, femoral articular cartilage; 6, fibular head; 7, lateral gastrocnemius muscle; infrapatellar bursa.

then turns medially toward its junction with the popliteal muscle belly. The popliteal tendon is extracapsular, but is bordered anteromedially by a synovial-lined bursa that is an extension of the knee joint (Figs. 3.27 and 3.28). This bursa is positioned between the posterior horn of the meniscus and the tendon. The superior and inferior meniscal attachments (termed struts or fascicles) of the lateral meniscus form the roof and floor, respectively, of the popliteal bursa (Figs. 3.27 and 3.28). The extreme lateral sections typically depict the union of the fibular collateral ligament and the biceps femoris tendon to form the conjoined tendon which inserts upon the fibular head (Fig. 3.31).

Axial Multiecho Series

Most (but not all) of the axial sections obtained in a typical knee MR examination are presented in Figs.

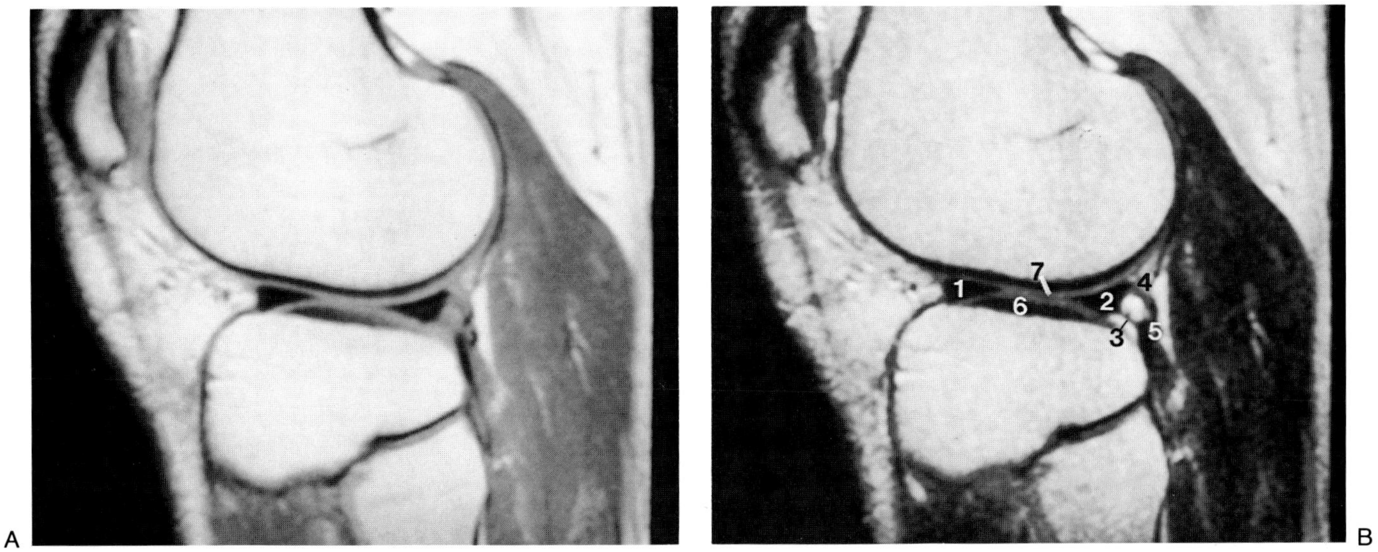

FIG. 3.27A,B. Sagittal intermediate- and T2-weighted scans at the level of section 14 (see Fig. 3.13). 1, Anterior horn of the lateral meniscus; 2, posterior horn of the lateral meniscus; 3, inferior fascicle; 4, superior fascicle; 5, popliteal tendon; 6, tibial plateau cortical bone and overlying articular cartilage; 7, lateral condyle articular cartilage.

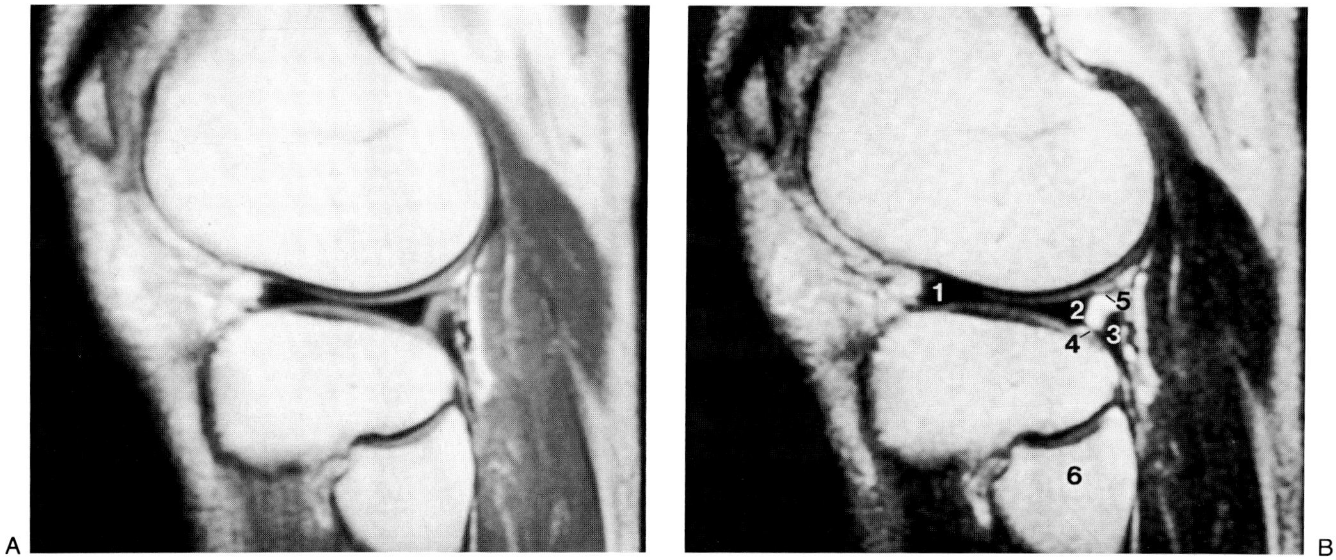

FIG. 3.28A,B. Sagittal intermediate- and T2-weighted scans at the level of section 15 (see Fig. 3.13). 1, Anterior horn of the lateral meniscus; 2, posterior horn of the lateral meniscus; 3, popliteal tendon; 4, inferior fascicle; 5, superior fascicle; 6, fibular head.

3.32 through 3.48. The superior sections routinely include the distal aspect of the quadriceps tendon. This tendon has an elliptical shape and is generally low in signal intensity. There are often small foci of high signal intensity within the quadriceps tendon, particularly adjacent to the musculotendinous junction. The superior axial sections also depict the normal superior extension of the medial synovial plica (Fig. 3.32). The muscular, neural, vascular, and fatty anatomy of the superior popliteal space are also depicted (Figs. 3.32 and 3.33). The popliteal artery is usually relatively low in signal intensity, although a variety of artifacts may alter the signal intensity arising from the popliteal artery. On the typical multi-echo series, the popliteal vein yields intermediate to high signal intensity. When clinically indicated, techniques are now available which can be employed to accurately assess flow within the popliteal arteries and vein (see Chapter 9). Sections obtained through the superior aspect of the patella depict the patellar articular cartilage (Figs. 3.34–3.37).

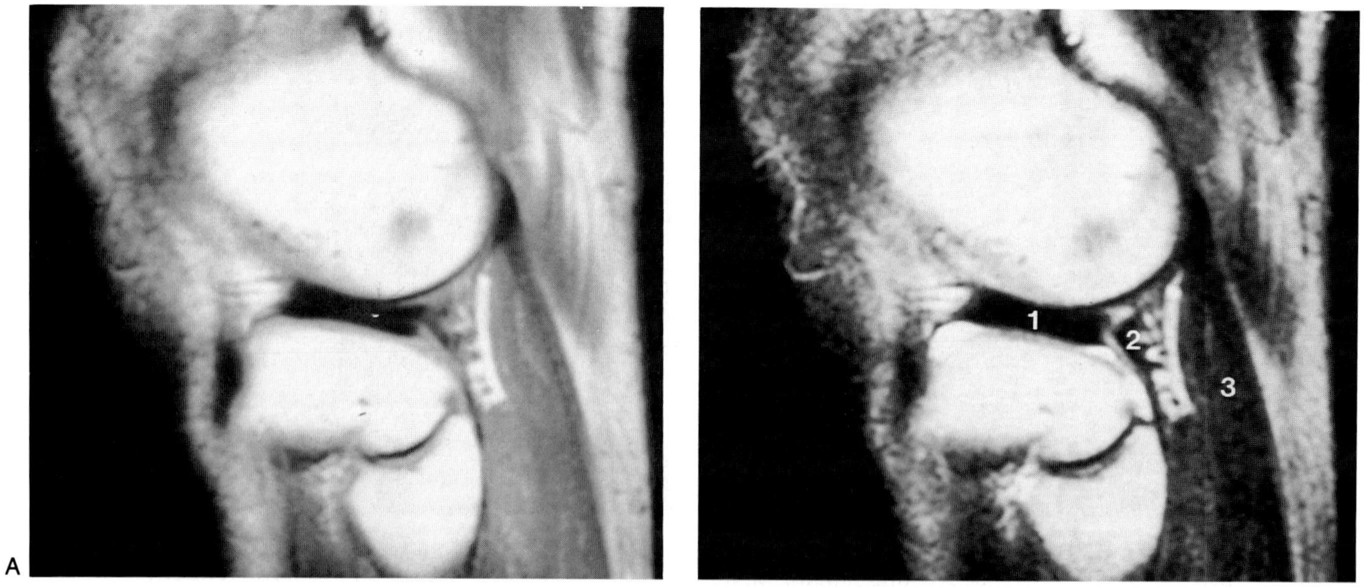

FIG. 3.29A,B. Sagittal intermediate- and T2-weighted images at the level of section 16 (see Fig. 3.13). 1, Middle third of the lateral meniscus; 2, popliteal tendon; 3, lateral gastrocnemius muscle.

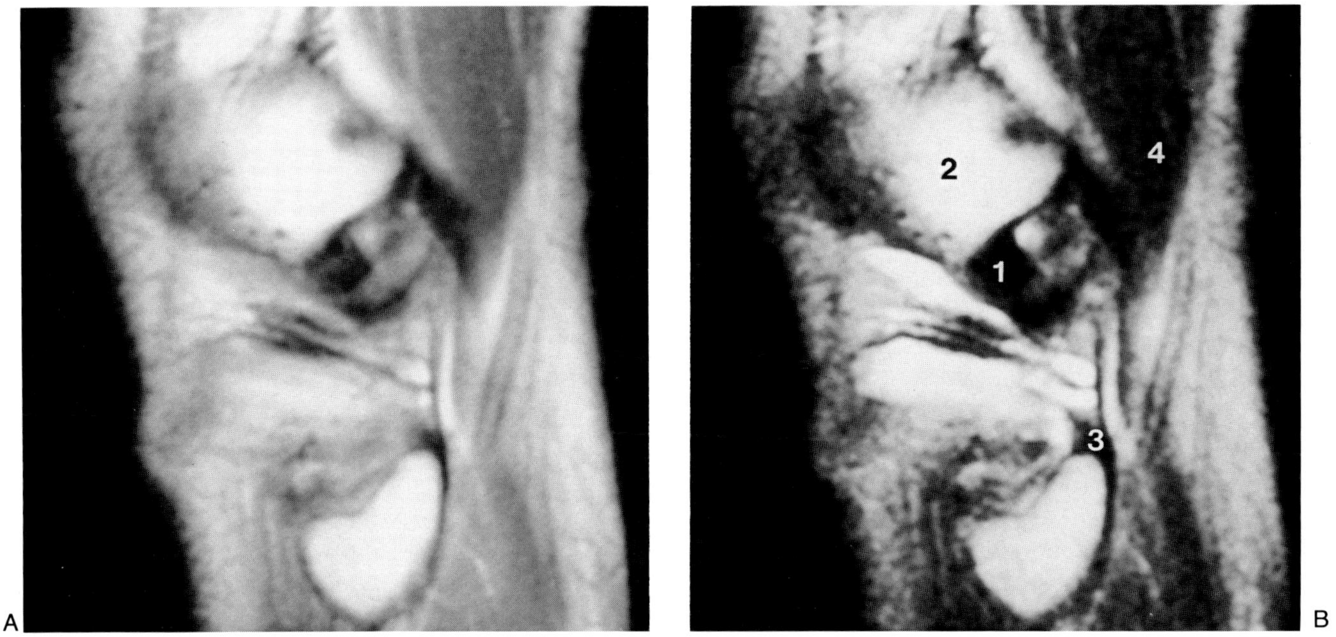

FIG. 3.30A,B. Sagittal intermediate- and T2-weighted scans at the level of section 17 (see Fig. 3.13). 1, Superior portion of the popliteal tendon; 2, lateral femoral condyle; 3, insertion of the conjoined tendon upon the fibular head; 4, biceps femoris muscle.

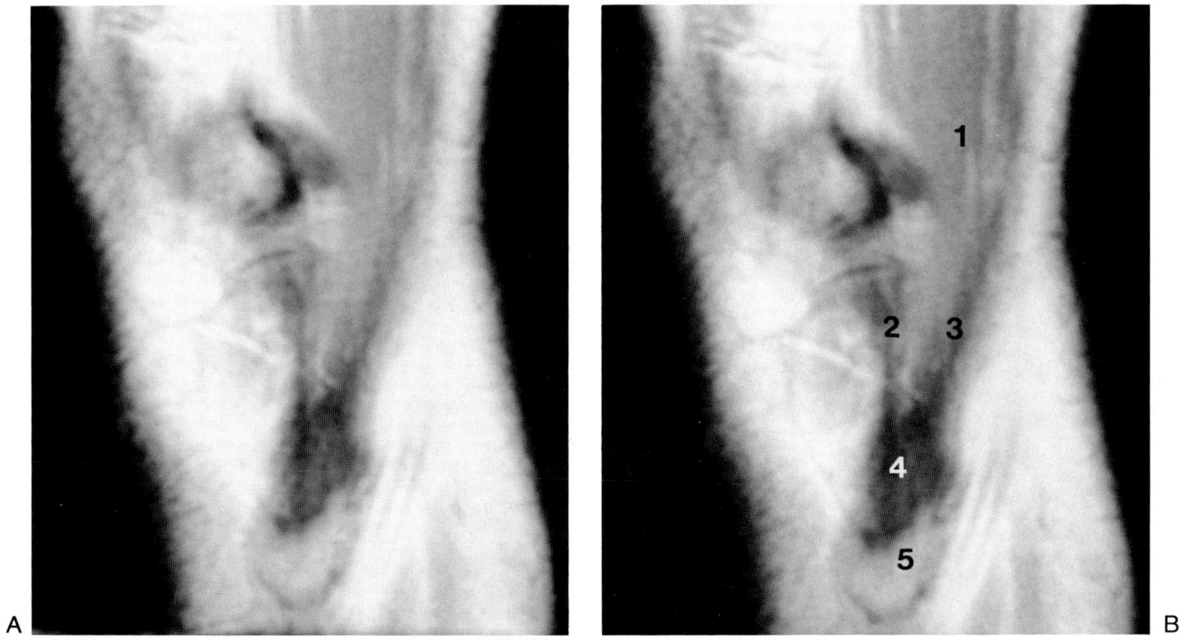

FIG. 3.31A,B. Sagittal intermediate- and T2-weighted images at the level of sagittal section 18 (see Fig. 3.13). 1, Biceps femoris muscle; 2, fibular collateral ligament; 3, biceps femoris tendon; 4, conjoined tendon; 5, fibular head.

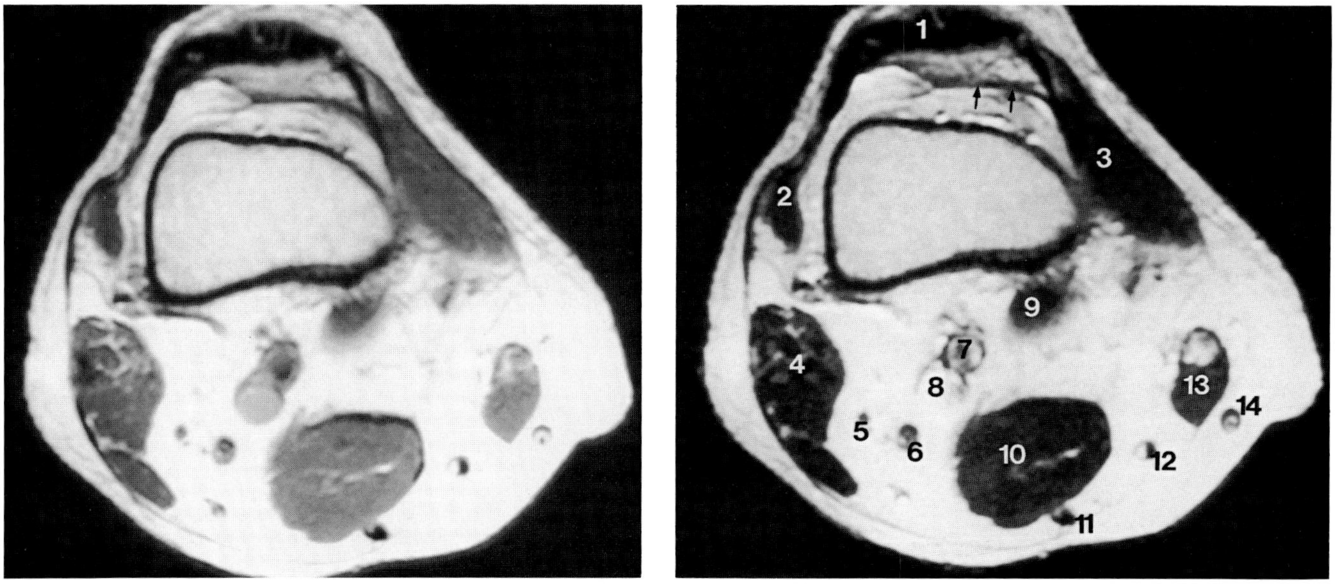

A B

FIG. 3.32A,B. 1, Quadriceps tendon; 2, vastus lateralis; 3, vastus medialis; 4, biceps femoris muscle; 5, common peroneal nerve; 6, tibial nerve; 7, popliteal artery; 8, popliteal vein; 9, medial head of the gastrocnemius tendon; 10, semimembranosus muscle; 11, semitendinosus tendon; 12, gracilis tendon; 13, sartorius muscle; 14, greater saphenous vein; *small arrows*, medial synovial plica.

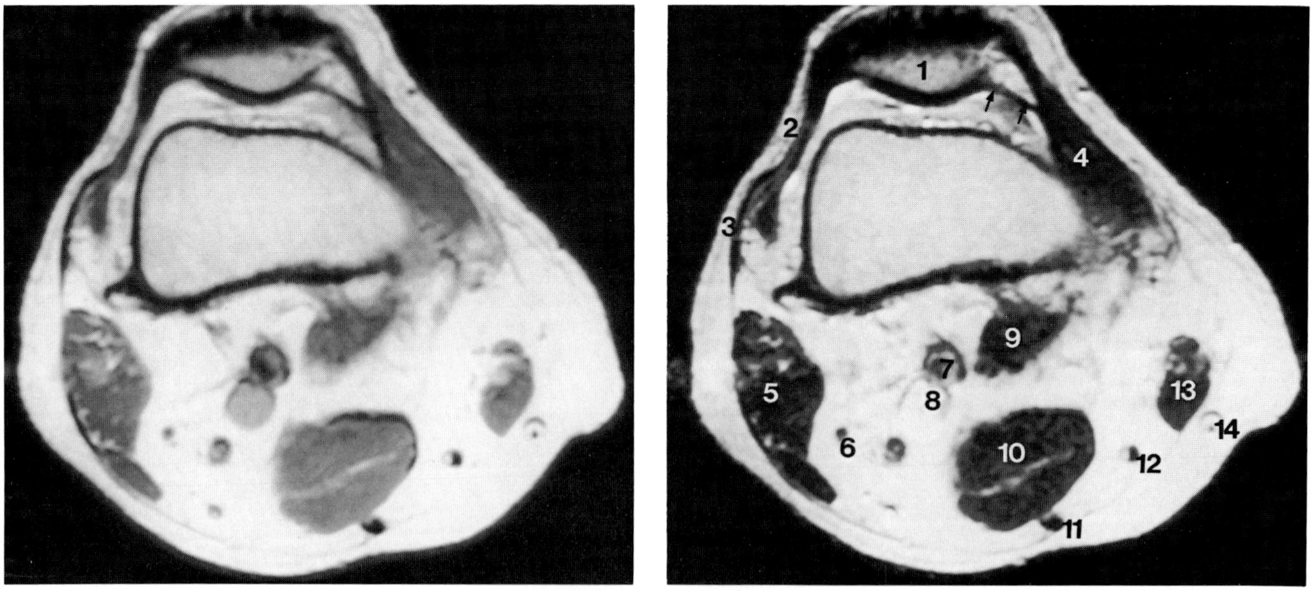

A B

FIG. 3.33A,B. 1, Superior pole of the patella; 2, vastus lateralis tendon; 3, iliotibial band; 4, vastus medialis muscle; 5, biceps femoris muscle; 6, common peroneal nerve; 7, popliteal artery; 8, popliteal vein; 9, medial head of the gastrocnemius muscle; 10, semimembranosus muscle; 11, semitendinosus tendon; 12, gracilis tendon; 13, sartorius muscle; 14, greater saphenous vein; *arrows*, medial synovial plica.

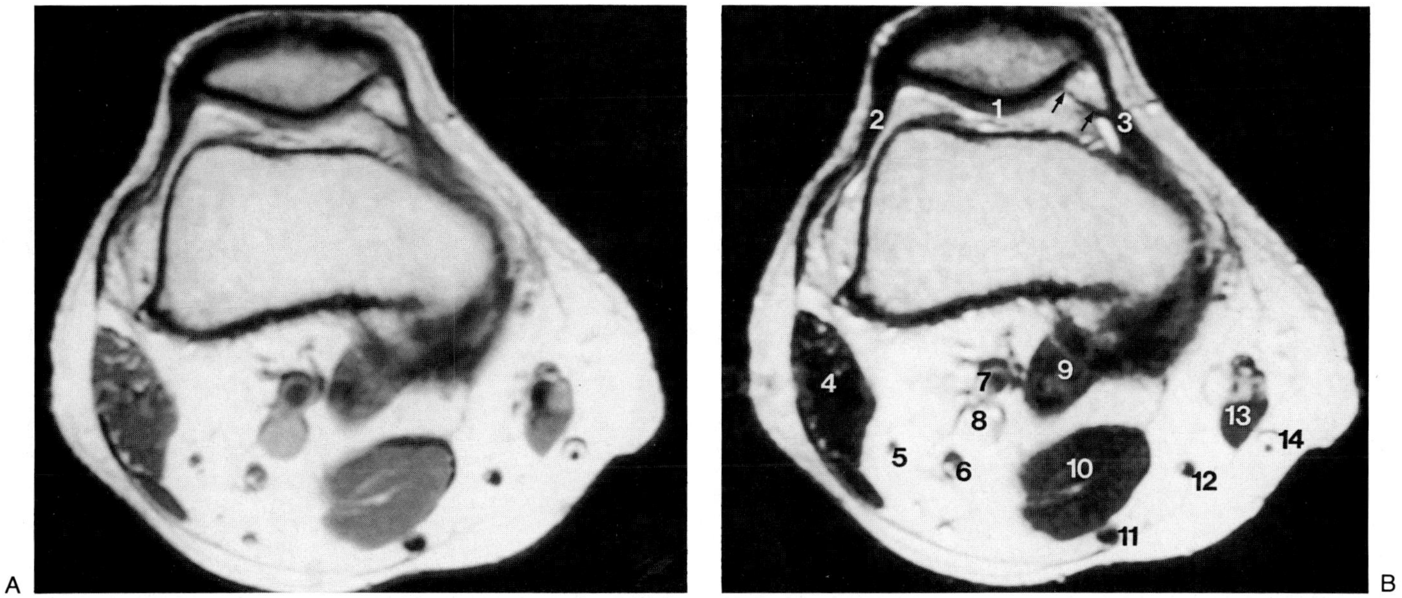

FIG. 3.34A,B. 1, Patellar articular cartilage; 2, vastus lateralis tendon; 3, vastus medialis tendon; 4, biceps femoris muscle; 5, common peroneal nerve; 6, tibial nerve; 7, popliteal artery; 8, popliteal vein; 9, medial head of the gastrocnemius muscle; 10, semimembranosus muscle; 11, semitendinosus tendon; 12, gracilis tendon; 13, sartorius muscle; 14, greater saphenous vein; *arrows*, medial synovial plica.

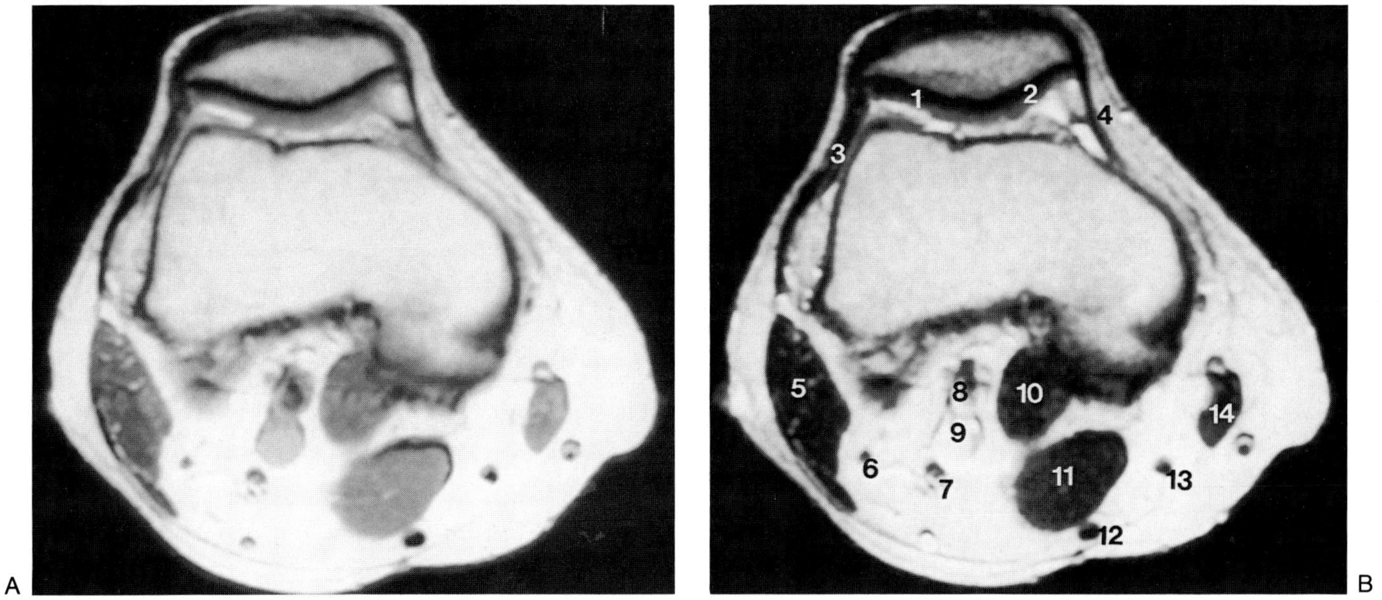

FIG. 3.35A,B. 1, Lateral patellar facet; 2, medial patellar facet; 3, lateral patellar retinaculum; 4, medial patellar retinaculum; 5, biceps femoris muscle; 6, common peroneal nerve; 7, tibial nerve; 8, popliteal artery; 9, popliteal vein; 10, medial head of the gastrocnemius muscle; 11, semimembranosus muscle; 12, semitendinosus tendon; 13, gracilis tendon; 14, sartorius muscle.

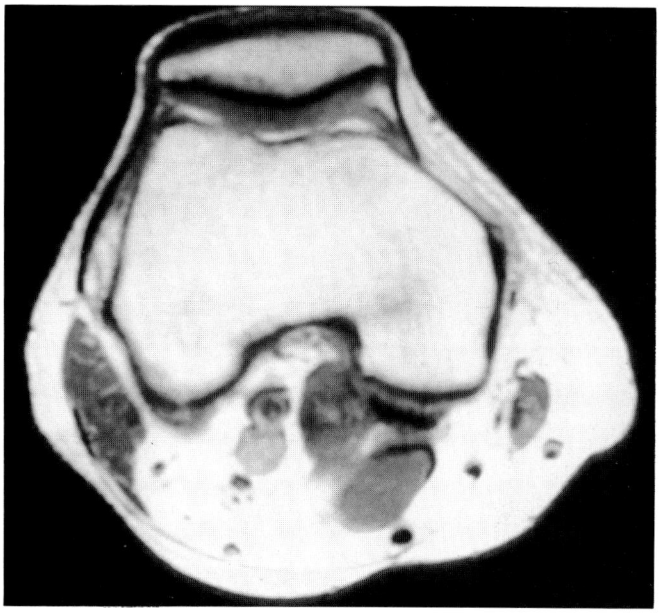

A

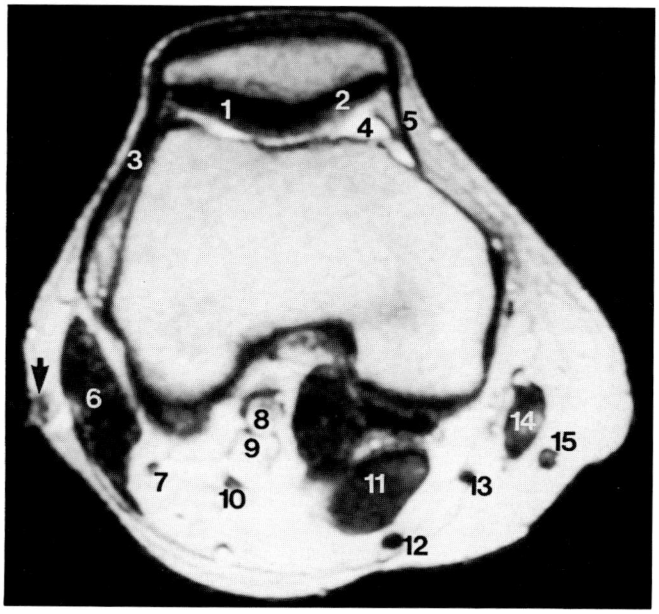

B

FIG. 3.36A,B. 1, Lateral patellar facet articular cartilage; 2, medial patellar facet articular cartilage; 3, lateral retinaculum; 4, medial recess of the suprapatellar bursa; 5, medial retinaculum; 6, biceps femoris muscle; 7, common peroneal nerve; 8, popliteal artery; 9, popliteal vein; 10, tibial nerve; 11, semimembranosus muscle; 12, semitendinosus tendon; 13, gracilis tendon; 14, sartorius muscle; 15, greater saphenous vein; *arrow*, phase-encoding artifact due to popliteal artery flow.

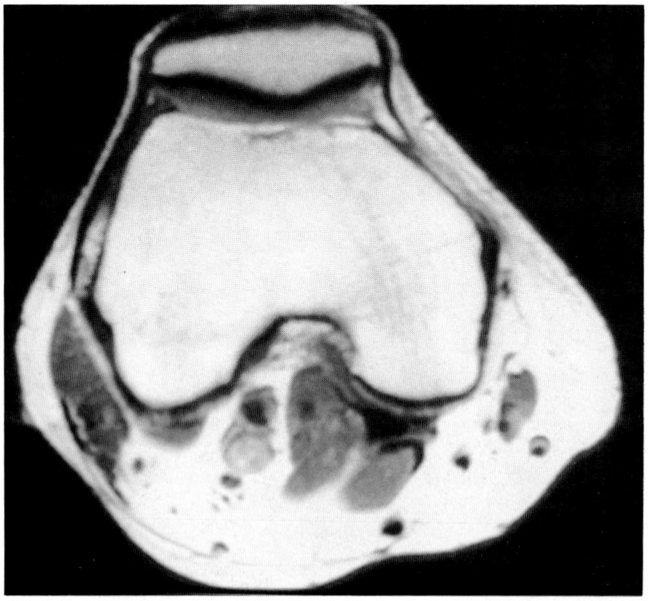

A

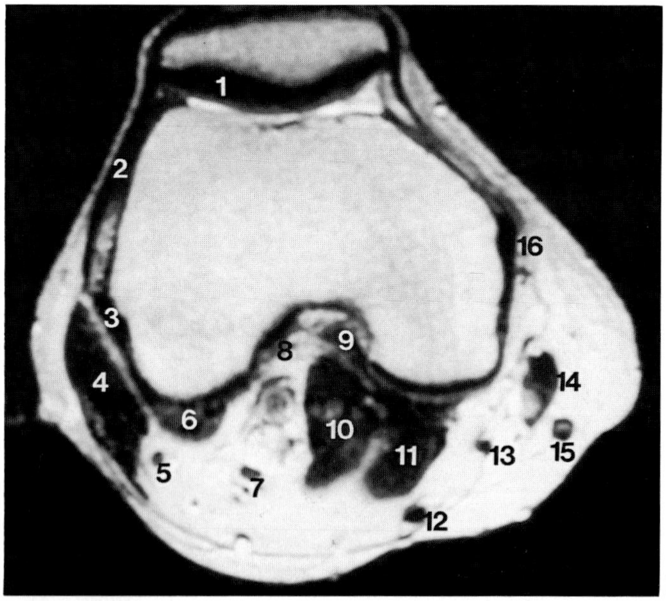

B

FIG. 3.37A,B. 1, Articular cartilage along the lateral patellar facet; 2, lateral patellar retinaculum/iliotibial band; 3, origin of the fibular collateral ligament; 4, biceps femoris muscle; 5, common peroneal nerve; 6, lateral head of the gastrocnemius muscle; 7, tibial nerve; 8, origin of the anterior cruciate ligament; 9, origin of the posterior cruciate ligament; 10, medial head of the gastrocnemius muscle; 11, semimembranosus muscle; 12, semitendinosus tendon; 13, gracilis tendon; 14, sartorius muscle; 15, greater saphenous vein; 16, origin of the medial collateral ligament.

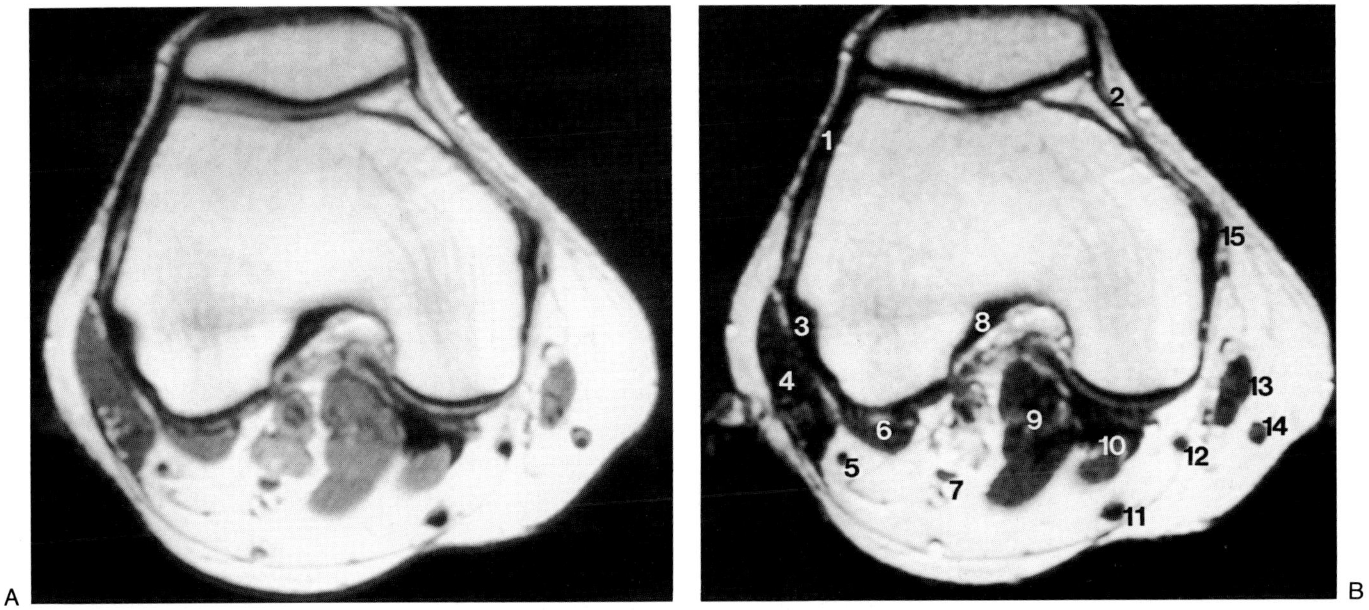

FIG. 3.38A,B. 1, Lateral patellar retinaculum/iliotibial band; 2, medial patellar retinaculum; 3, fibular collateral ligament; 4, biceps femoris muscle; 5, common peroneal nerve; 6, lateral head of the gastrocnemius muscle; 7, tibial nerve; 8, anterior cruciate ligament; 9, medial head of the gastrocnemius muscle; 10, semimembranosus muscle and tendon; 11, semitendinosus tendon; 12, gracilis tendon; 13, sartorius muscle; 14, greater saphenous vein; 15, medial collateral ligament.

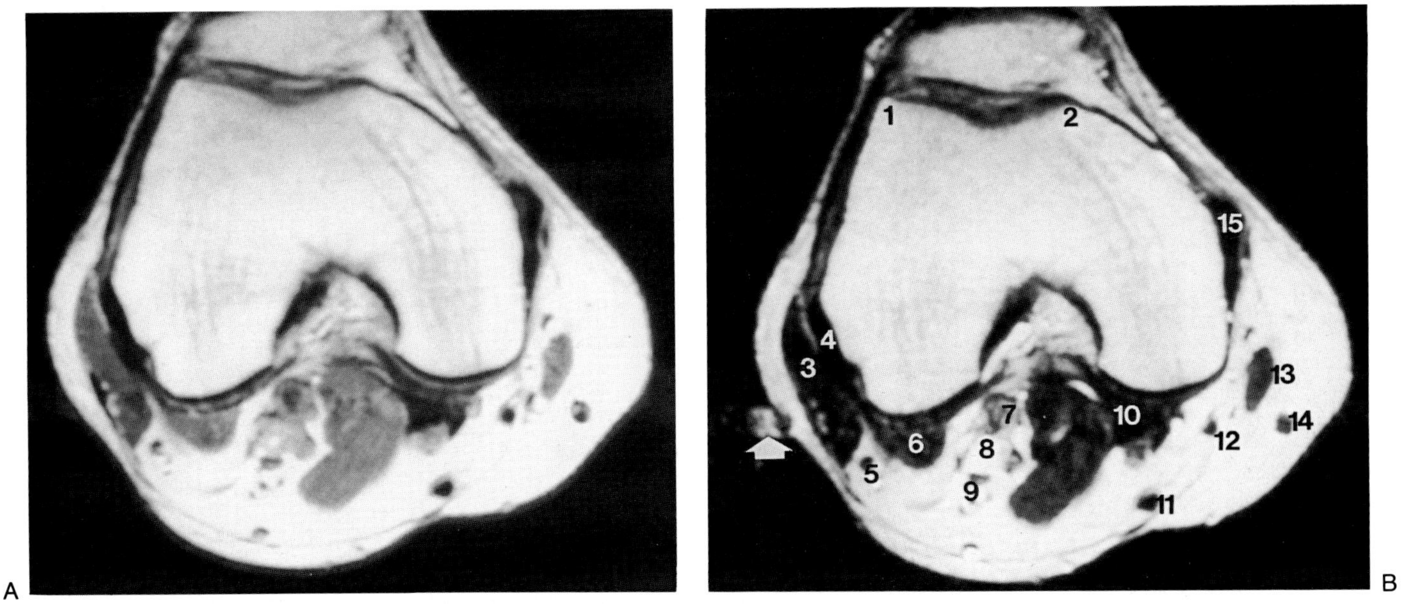

FIG. 3.39A,B. 1, Lateral eminence of the femoral trochlear groove; 2, medial eminence of the femoral trochlear groove; 3, biceps femoris muscle; 4, fibular collateral ligament; 5, common peroneal nerve; 6, lateral gastrocnemius muscle; 7, popliteal artery; 8, popliteal vein; 9, tibial nerve; 10, semimembranosus tendon; 11, semitendinosus tendon; 12, gracilis tendon; 13, sartorius muscle; 14, greater saphenous vein; 15, medial collateral ligament; *white arrow*, phase-encoding artifact due to popliteal artery flow.

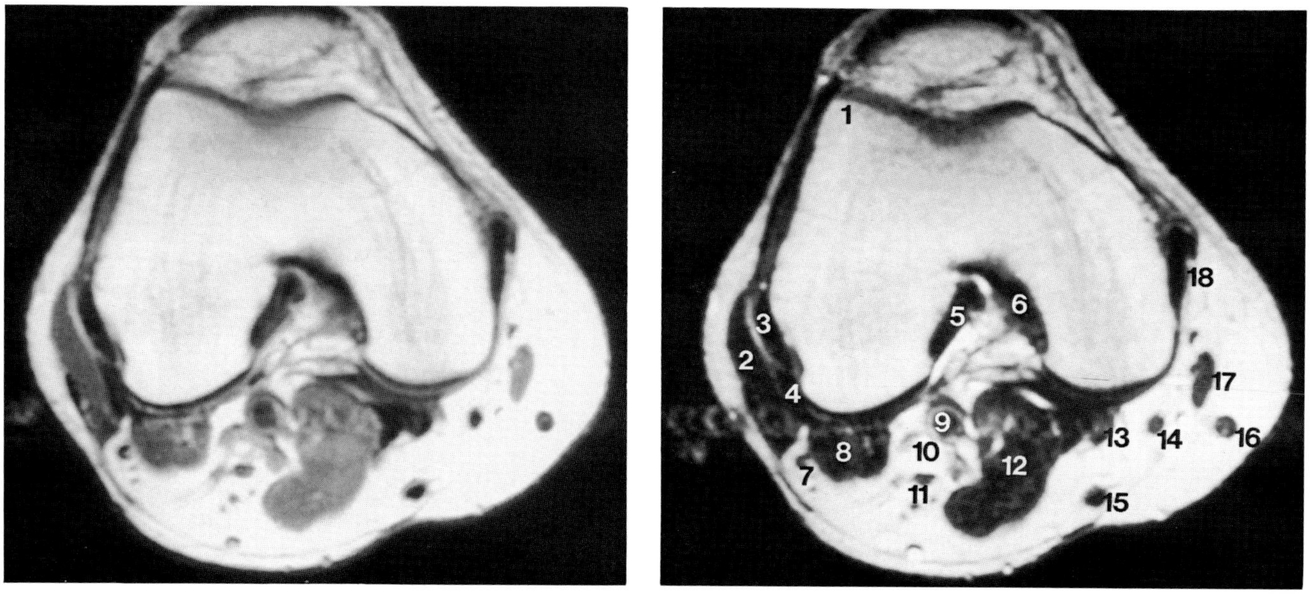

A B

FIG. 3.40A,B. 1, Lateral eminence of the femoral trochlear groove; 2, biceps femoris muscle; 3, fibular collateral ligament; 4, origin of the popliteal tendon; 5, anterior cruciate ligament; 6, posterior cruciate ligament; 7, common peroneal nerve; 8, lateral gastrocnemius muscle; 9, popliteal artery; 10, popliteal vein; 11, tibial nerve; 12, medial gastrocnemius muscle; 13, semimembranosus tendon; 14, gracilis tendon; 15, semitendinosus tendon; 16, greater saphenous vein; 17, sartorius muscle; 18, medial collateral ligament.

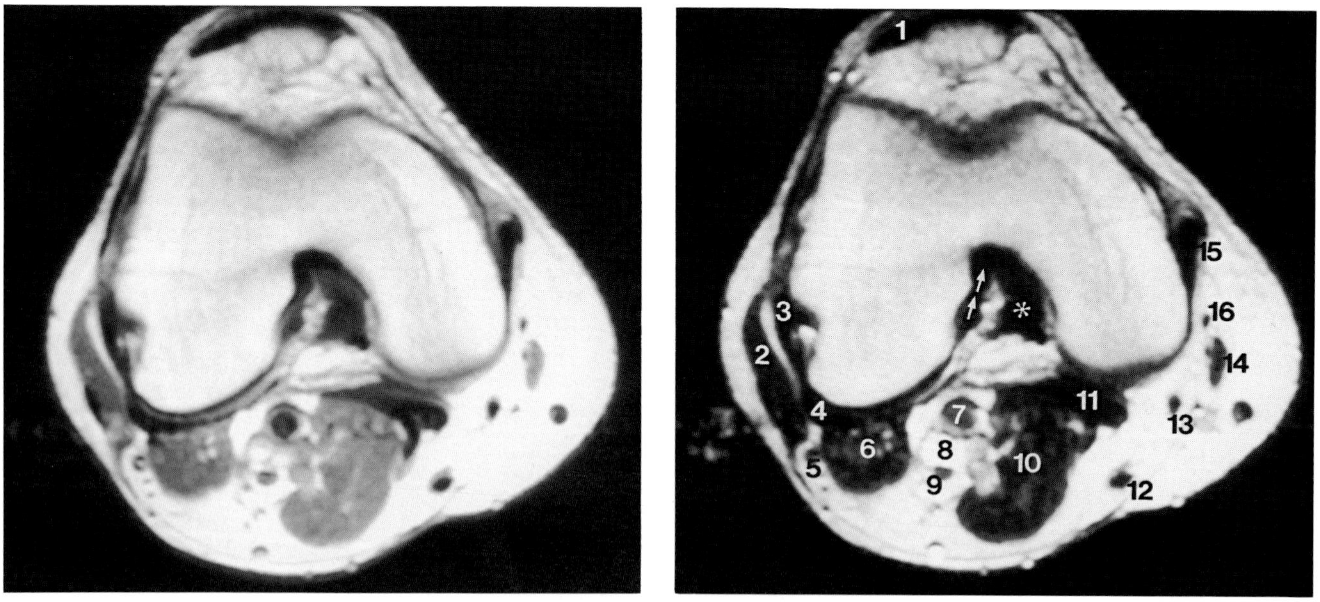

A B

FIG. 3.41A,B. 1, Patellar tendon; biceps femoris muscle; 3, fibular collateral ligament; 4, popliteal tendon; 5, common peroneal nerve; 6, lateral gastrocnemius muscle; 7, popliteal artery; 8, popliteal vein; 9, tibial nerve; 10, medial gastrocnemius muscle; 11, semimembranosus tendon; 12, semitendinosus tendon; 13, gracilis tendon; 14, sartorius tendon; 15, medial collateral ligament; 16, saphenous cutaneous nerve.

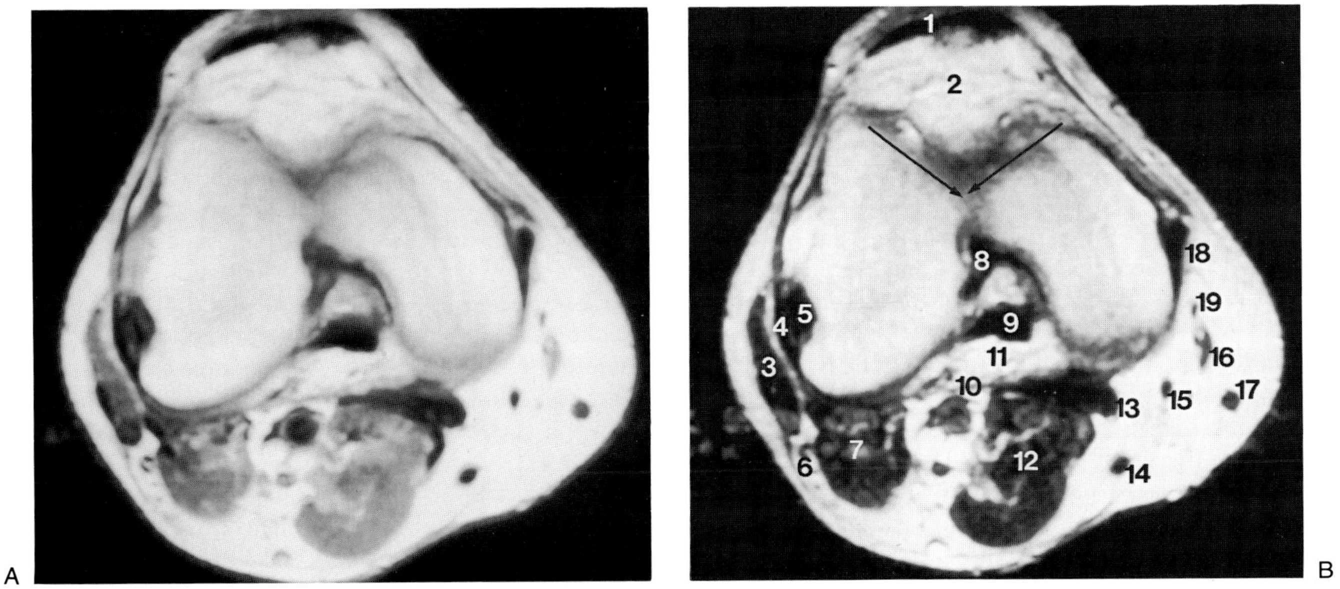

FIG. 3.42A,B. 1, Patellar tendon; 2, infrapatellar fat pad; 3, biceps femoris muscle; 4, fibular collateral ligament; 5, popliteal tendon coursing through the popliteal hiatus; 6, common peroneal nerve; 7, lateral gastrocnemius muscle; 8, anterior cruciate ligament; 9, posterior cruciate ligament; 10, posterior capsule; 11, intraarticular fat; 12, medial gastrocnemius muscle; 13, semimembranosus tendon; 14, semitendinosus tendon; 15, gracilis tendon; 16, sartorius muscle; 17, greater saphenous vein; 18, medial collateral ligament; 19, saphenous cutaneous nerve; *arrows*, the far inferior aspect of the femoral trochlear groove forms a more acute angle compared to the more superior aspect of this groove.

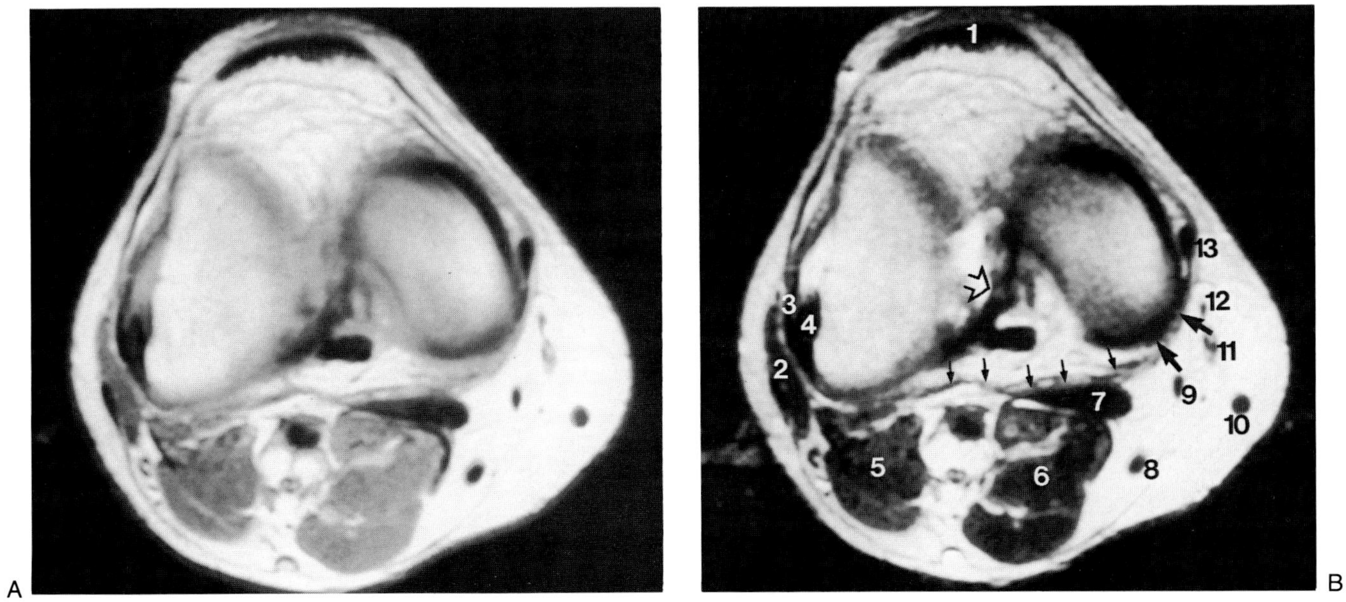

FIG. 3.43A,B. 1, Patellar tendon; 2, biceps femoris muscle; 3, fibular collateral ligament; 4, popliteal tendon; 5, lateral gastrocnemius muscle; 6, medial gastrocnemius muscle; 7, semimembranosus tendon; 8, semitendinosus tendon; 9, gracilis tendon; 10, greater saphenous vein; 11, sartorius tendon; 12, saphenous cutaneous nerve; 13, medial collateral ligament; *small arrows*, posterior capsule; *large arrows*, superior section through the medial meniscus; *open arrow*, anterior cruciate ligament.

On intermediate-weighted scans, synovial fluid is usually isointense to hyaline articular cartilage. T2-weighted spin-echo images allow for better delineation between the high signal intensity synovial fluid and adjacent hyaline cartilage surfaces (Figs. 3.34–3.36). The far superior aspect of the trochlear groove is normally extremely shallow (Figs. 3.34 and 3.35). The lateral retinaculum of the patella is generally thicker than the medial retinaculum (Figs. 3.34 and 3.35).

Axial sections obtained through the midpole of the patella again depict the patellar articular cartilage in outstanding detail. The patellar cartilage is the thickest in the human body, measuring up to 5 mm in thickness. Once again, phase-encoding artifact is often caused by flow within the popliteal vessels (Fig. 3.36).These artifacts may project from medial to lateral or ventral to dorsal, depending on the selection of the phase-encoding axis when the scan is prescribed.

As the axial sections approach the intercondylar notch, the origins of the anterior and posterior cruciate ligaments are seen (Figs. 3.38–3.40). The tendons which more distally comprise the pes anserinus begin to sweep medially and anteriorly. The fibular collateral ligament and tibial collateral ligament are well seen in cross-section as they respectively arise from the lateral and medial femoral condyles. Axial T2-weighted images are therefore often extremely useful in precisely delineating the severity and level of collateral ligament injuries.

As axial sections proceed inferiorly through the intercondylar notch, the oblique course of the anterior cruciate ligament is again seen. Axial images are extremely useful for evaluating the anterior cruciate ligament (Figs. 3.39–3.44) and corroborating findings suspected on the basis of sagittal images. As the axial images approach the popliteal hiatus, a groove within the lateral margin of the lateral femoral condyle, the origin and course of the popliteal tendon may be followed (Figs. 3.40–3.47). Axial images allow precise discrimination of the popliteal tendon, the fibular collateral ligament, and the biceps femoris tendon (Figs. 3.40–3.47).

As the axial sections progress distally through the far inferior aspect of the trochlear groove (Figs. 3.41–3.43), the groove forms a more acute angle. These sections through the inferior aspect of the intercondylar notch also allow the inferior portion of the posterior cruciate ligament to be seen in cross-section, yielding homogeneously low signal intensity (Figs. 3.42 and 3.43). The posterior capsule of the knee joint is consistently delineated as a low signal intensity structure (Figs. 3.43 and 3.44). Fat or fluid yielding moderate to high signal intensity is often seen interposed between the superficial and deep layers of the medial capsule

A

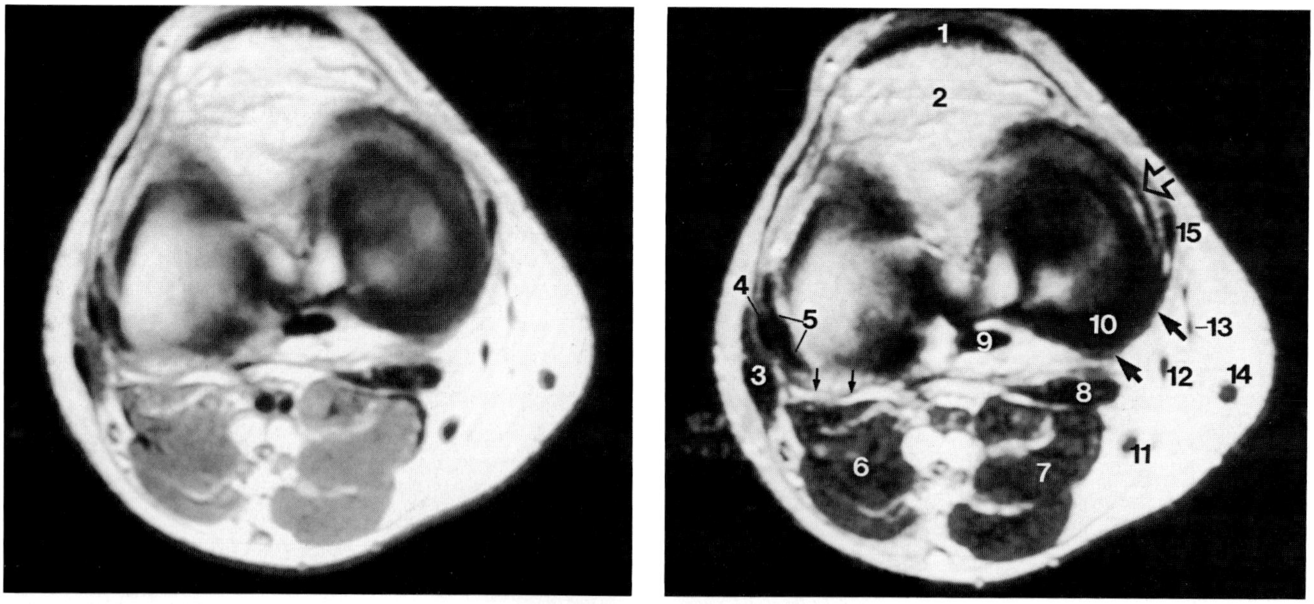

B

FIG. 3.44A,B. 1, Patellar tendon; 2, infrapatellar fat pad; 3, biceps femoris muscle and tendon; 4, fibular collateral ligament; 5, oblique course of the popliteal tendon; 6, lateral gastrocnemius muscle; 7, medial gastrocnemius muscle; 8, semimembranosus tendon; 9, posterior cruciate ligament; 10, posterior horn of the medial meniscus; 11, semitendinosus tendon; 12, gracilis tendon; 13, sartorius tendon; 14, greater saphenous vein; 15, superficial portion of the medial collateral ligament; *small arrows*, posterolateral capsule; *large arrows*, intimate attachment of the posteromedial capsule to the posterior horn of the medial meniscus; *open arrow*, deep portion of the medial collateral ligament.

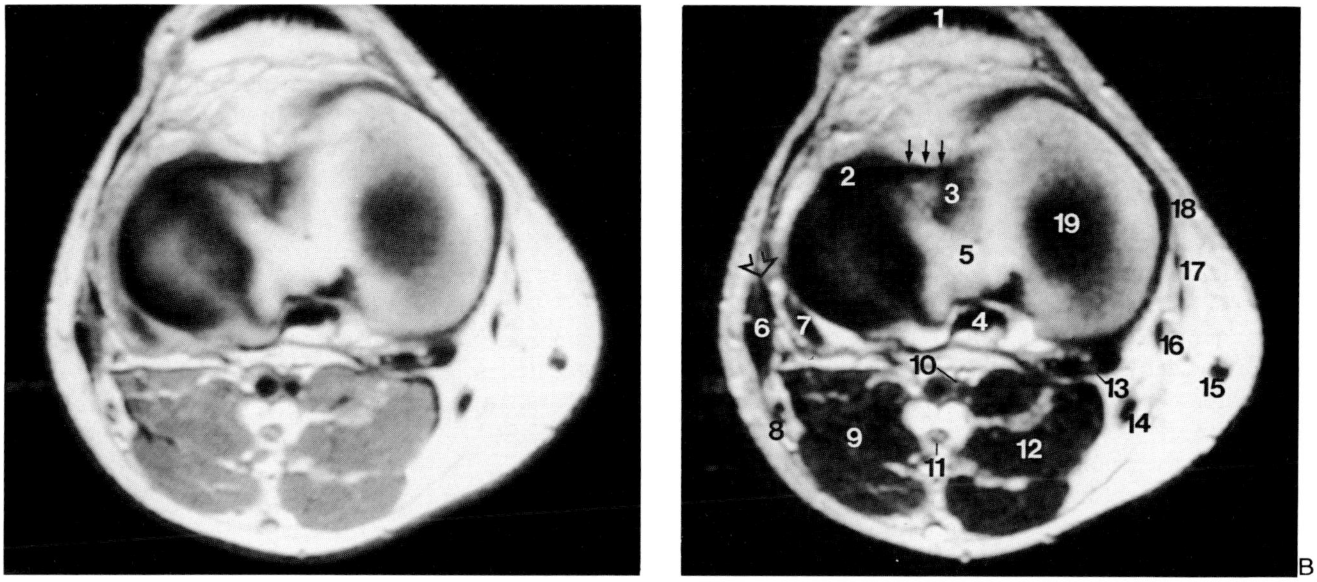

A B

FIG. 3.45A,B. 1, Patellar tendon; 2, anterior horn of the lateral meniscus; 3, anterior cruciate ligament insertion; 4, posterior cruciate ligament; 5, tibial spine; 6, biceps femoris muscle and tendon; 7, oblique course of the popliteal tendon; 8, common peroneal nerve; 9, lateral gastrocnemius muscle; 10, bifurcation of popliteal artery; 11, tibial nerve; 12, medial gastrocnemius muscle; 13, semimembranosus tendon; 14, semitendinosus tendon; 15, greater saphenous vein; 16, gracilis tendon; 17, sartorius tendon; 18, medial collateral ligament insertion; *open arrow*, fibular collateral ligament; *small arrows*, transverse geniculate ligament.

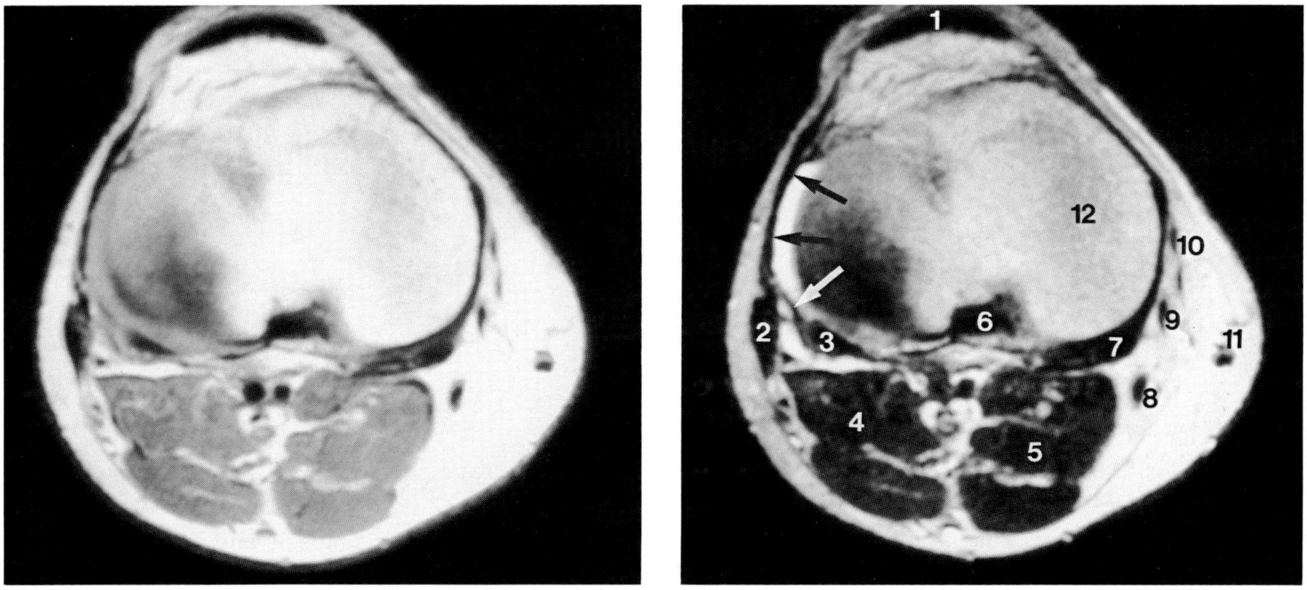

A B

FIG. 3.46A,B. 1, Patellar tendon; 2, conjoined tendon; 3, popliteal tendon; 4, lateral gastrocnemius muscle; 5, medial gastrocnemius muscle; 6, posterior cruciate ligament insertion; 7, semimembranosus tendon insertion; 8, semitendinosus tendon; 9, gracilis tendon; 10, sartorius tendon; 11, greater saphenous vein; 12, medial tibial plateau; *arrows*, lateral joint capsule with underlying joint fluid.

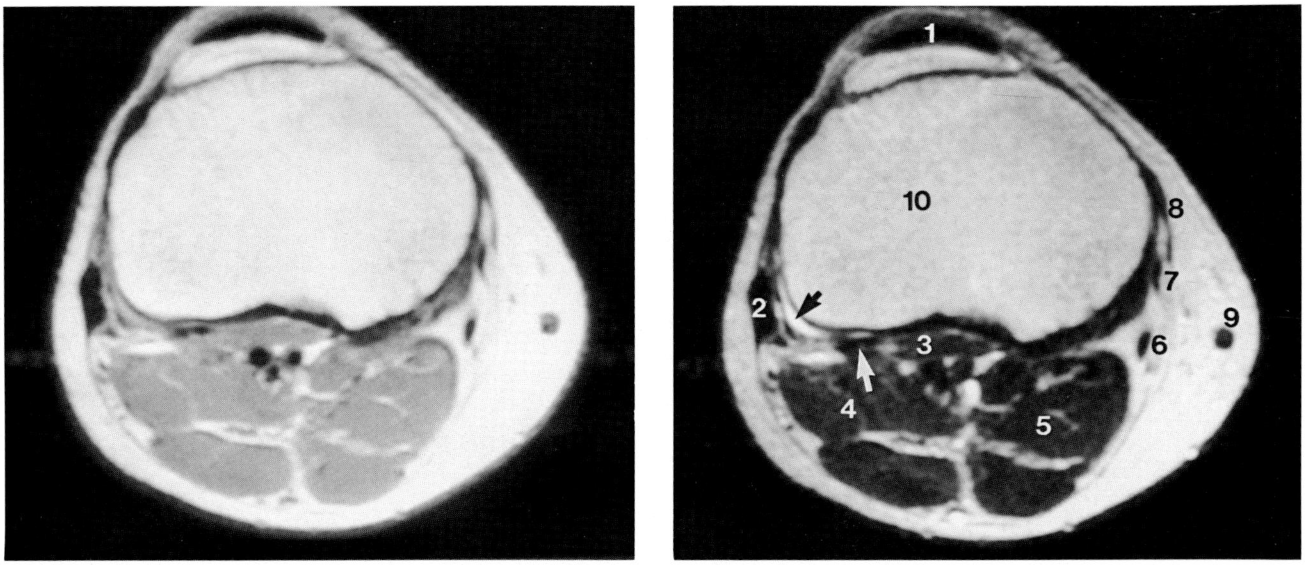

A B

FIG. 3.47A,B. 1, Patellar tendon; 2, conjoined tendon; 3, popliteal muscle; 4, lateral gastrocnemius muscle; 5, medial gastrocnemius muscle; 6, semitendinosus tendon; 7, gracilis tendon; 8, sartorius tendon insertion; 9, greater saphenous vein; 10, proximal tibia; *white arrow*, popliteal musculotendinous junction; *black arrow*, joint fluid deep to posterolateral capsular insertion.

as well as between the medial capsule and the middle third of the medial meniscus (Fig. 3.44). Very importantly, when axial sections reach the inferior margin of the medial meniscus (Fig. 3.44), there is no potential space between the posteromedial capsule and the posteromedial third of the medial meniscus. Careful evaluation of axial images at this level is most important in diagnosing meniscocapsular separations.

Sections obtained at the level of the menisci also depict the infrapatellar fat pad. Scars are almost always present within this fat pad when a patient has undergone prior arthroscopy. This finding may be quite useful in evaluating knee MR examinations when clinical history is limited. These sections also depict the normal flattened arc-like cross-sectional shape of the patellar tendon. The normal patellar tendon is a homogeneously black structure.

Axial sections at the level of the tibial plateau (Figs.

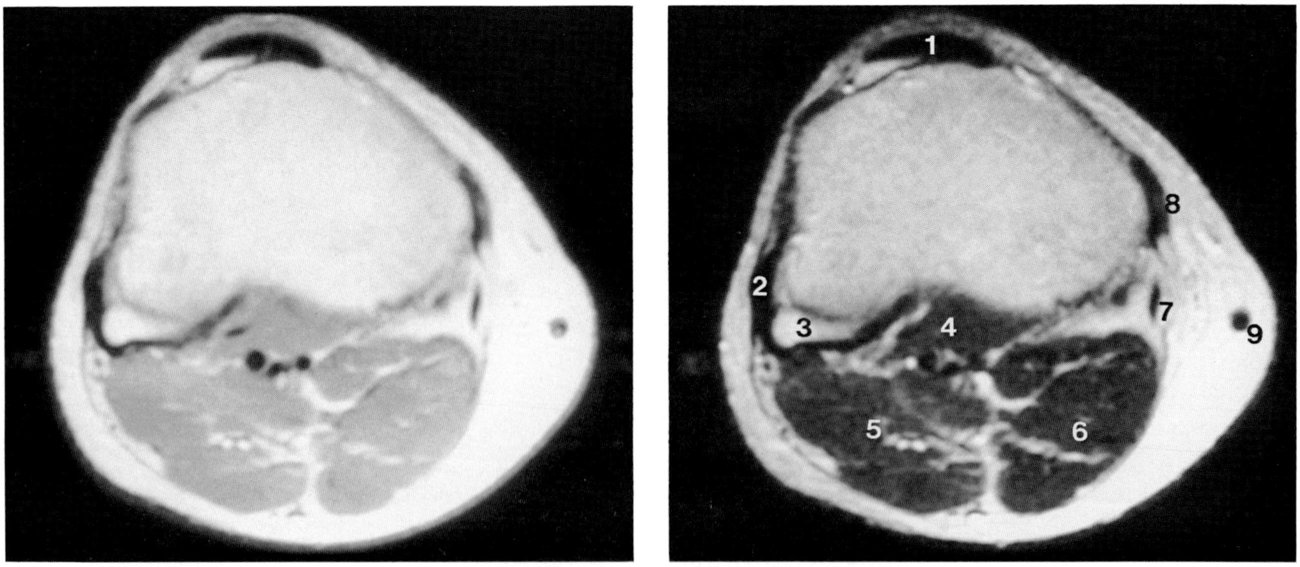

A B

FIG. 3.48A,B. 1, Patellar tendon; 2, conjoined tendon insertion upon fibular head; 3, fibular head; 4, popliteal muscle; 5, lateral gastrocnemius muscle; 6, medial gastrocnemius muscle; 7, semitendinosus tendon; 8, insertion of sartorius and gracilis tendons; 9, greater saphenous vein.

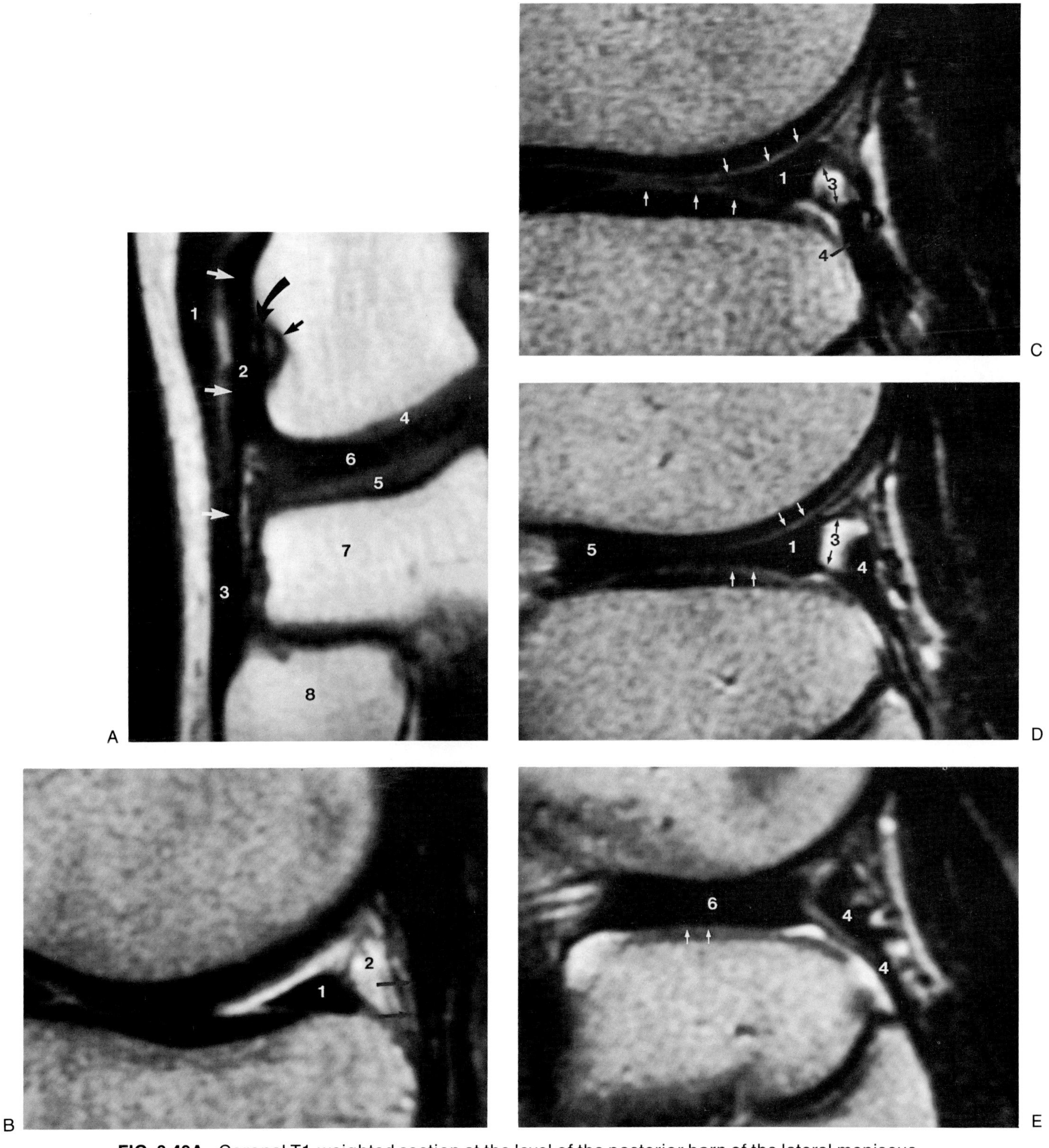

FIG. 3.49A. Coronal T1-weighted section at the level of the posterior horn of the lateral meniscus. 1, Biceps femoris muscle; 2 (and *white arrows*), fibular collateral ligament; 3, conjoined tendon; 4, articular cartilage of the lateral femoral condyle; 5, articular cartilage of the lateral tibial plateau; 6, posterior horn of the lateral meniscus; 7, lateral tibial plateau bone marrow; 8, fibular head; *curved arrow*, popliteal tendon; *small black arrow*, popliteal hiatus. **B–E:** Sagittal T2-weighted spin-echo images coursing through the lateral meniscus from medial to lateral. 1, Posterior horn of the lateral meniscus; 2, intraarticular fibrofatty tissue; 3, superior and inferior fascicles; 4, popliteal tendon; 5, anterior horn of the lateral meniscus; 6, middle third of the lateral meniscus; *small white arrows*, articular cartilage of the lateral femoral condyle and lateral tibial plateau; *black arrows*, posterolateral capsule.

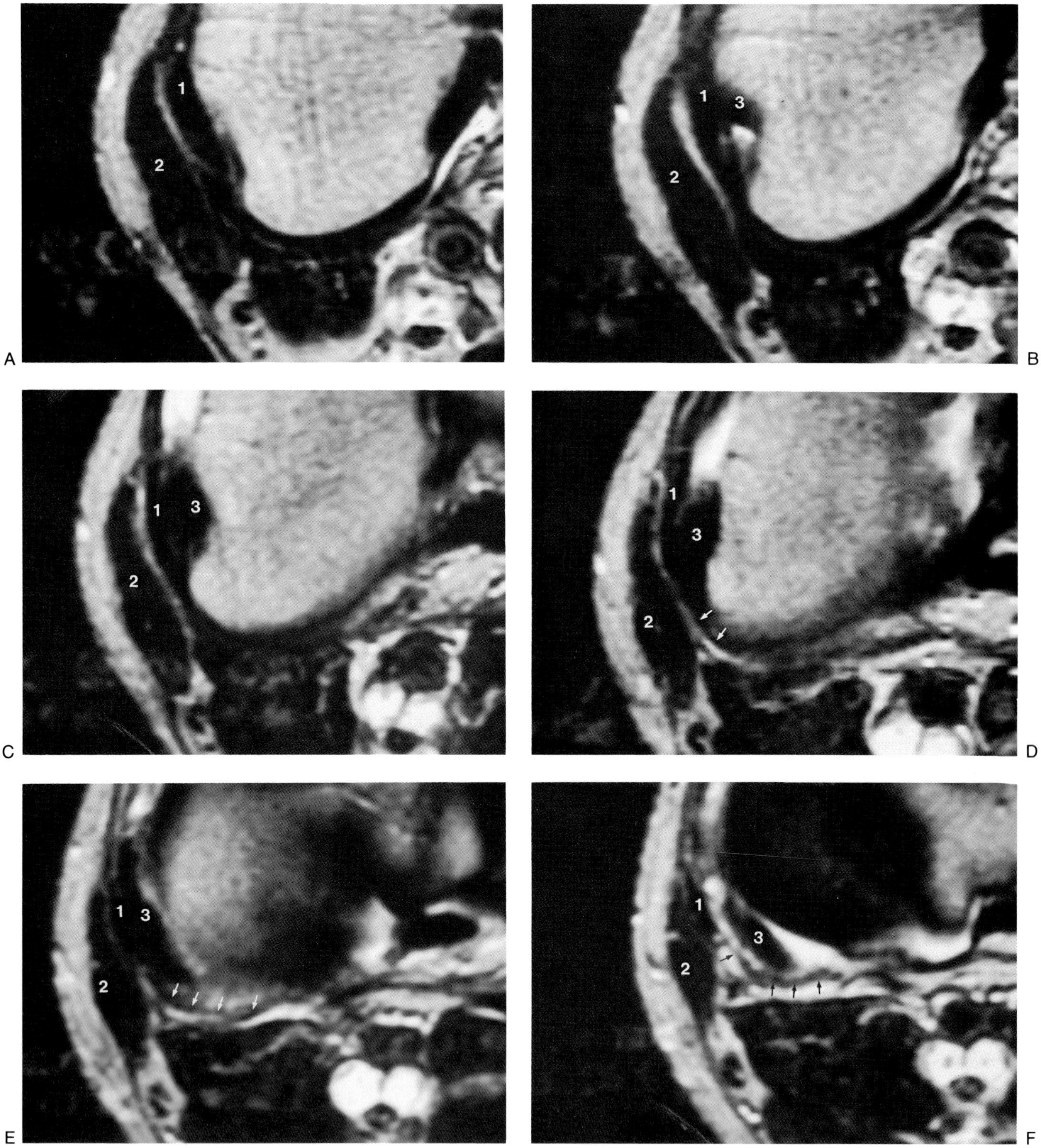

FIG. 3.50A–I. Axial T2-weighted spin-echo images coursing from superior to inferior, emphasizing the posterolateral capsule, tendons, and ligaments. 1, Fibular collateral ligament; 2, biceps femoris muscle and tendon; 3, popliteal tendon; 4, popliteus muscle; *small white and black arrows*, posterolateral capsule encompassing a small amount of high signal intensity joint fluid.

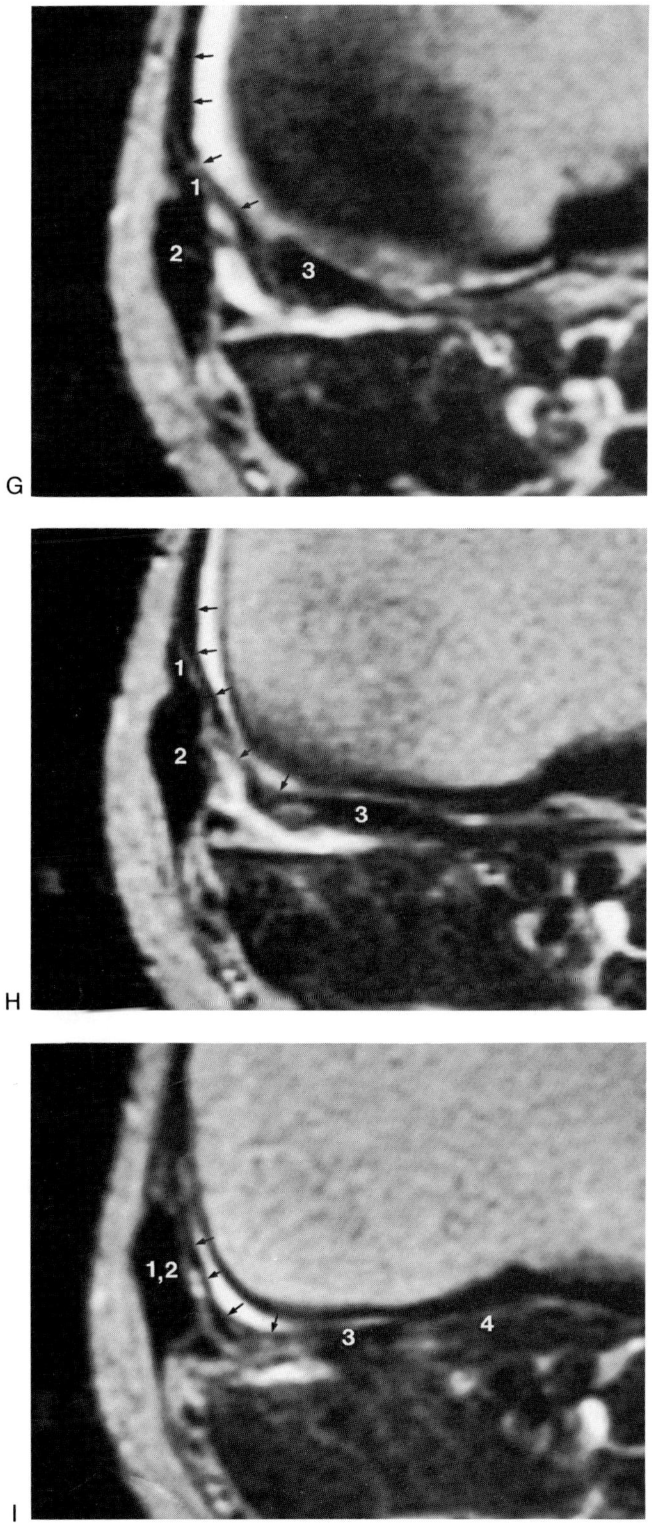

FIG. 3.50. *(Continued)*

3.45 and 3.46) frequently result in low signal intensity within the central portions of both the medial and lateral tibial plateaus. This is due to cross-sectional imaging of the tibial cortical bone. The transverse geniculate ligament is often seen, at least in part (Fig. 3.45). The insertions of the anterior and posterior cruciate ligaments are seen (Fig. 3.45). The popliteal tendon can be seen coursing obliquely downward, neighboring the posterolateral third of the lateral meniscus (Fig. 3.45).

Slightly more inferior sections extending through the proximal tibia depict the insertions of the tibial and fibular collateral ligaments (Figs. 3.46–3.48). The fibular collateral ligament and biceps femoris tendon unite to form a conjoined tendon (Fig. 3.47) which then inserts upon the lateral margin of the proximal fibula (Fig. 3.48). The patellar tendon inserts upon the tibial tubercle (Fig. 3.48). The tendons of the pes anserinus continue to course anteriorly to insert upon the anteromedial tibia (Figs. 3.47 and 3.48). The semimembranous tendon inserts upon the posteromedial margin of the tibia (Fig. 3.46). The popliteus muscle can be seen originating from the posterior tibia, ventral to the tibial and peroneal arteries.

Additional Anatomic Observations

As knee MR images have improved in quality, an increasing abundance of highly detailed anatomic information has become available on routine examinations. Figures 3.49 through 3.55 are photographically magnified images obtained from routine MR examinations utilizing techniques which have been previously described in this chapter. These images are presented here to emphasize the extremely exquisite anatomic detail depicted on routine MR examinations and to stimulate the reader to scrutinize all scans with the utmost diligence. Hopefully the sophistication of knee joint diagnostic evaluation thereby can be raised to a higher level.

Figure 3.49 depicts the meniscal, capsular, and tendinous relationships in the posterolateral corner of the knee joint. In Fig. 3.49A, the popliteal tendon, fibular collateral ligament, biceps femoris tendon, and lateral meniscus each can be easily discriminated. The superior portion of the popliteal tendon courses through a groove in the lateral femoral condyle as it passes upward to its site of insertion. Sagittal images (Figs. 3.49B–3.49E) depict the relationship of the posterior horn of the lateral meniscus to the popliteal tendon. As previously described, the posterior horn of the lateral meniscus connects to the popliteal tendon sheath via the superior and inferior fascicles or struts. The courses of the popliteal tendon, fibular collateral ligament, and biceps femoris tendon are also routinely seen in superb detail on axial images (Figs. 3.50A–3.50I). Axial scans routinely depict the posterolateral

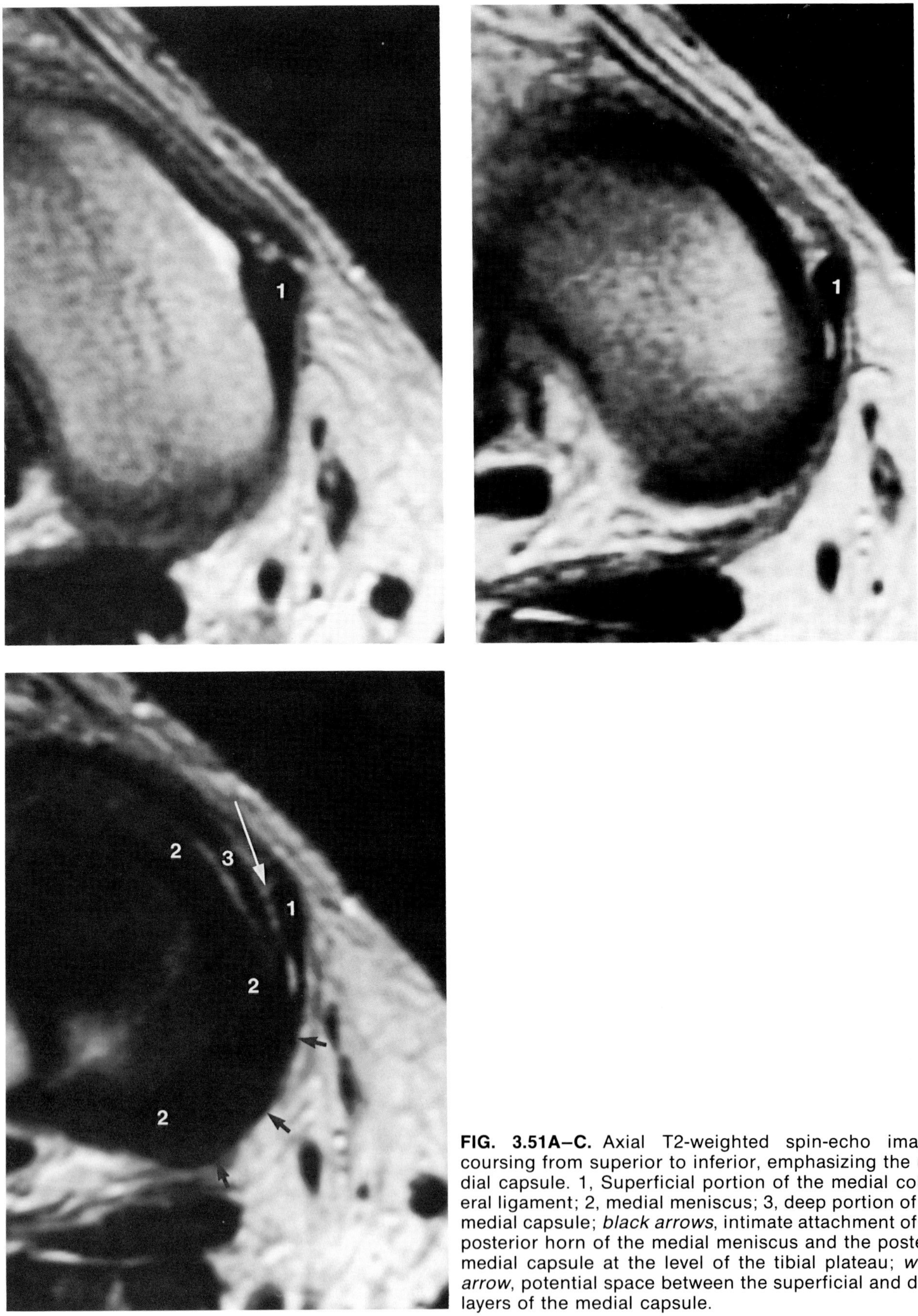

FIG. 3.51A–C. Axial T2-weighted spin-echo images coursing from superior to inferior, emphasizing the medial capsule. 1, Superficial portion of the medial collateral ligament; 2, medial meniscus; 3, deep portion of the medial capsule; *black arrows*, intimate attachment of the posterior horn of the medial meniscus and the posteromedial capsule at the level of the tibial plateau; *white arrow*, potential space between the superficial and deep layers of the medial capsule.

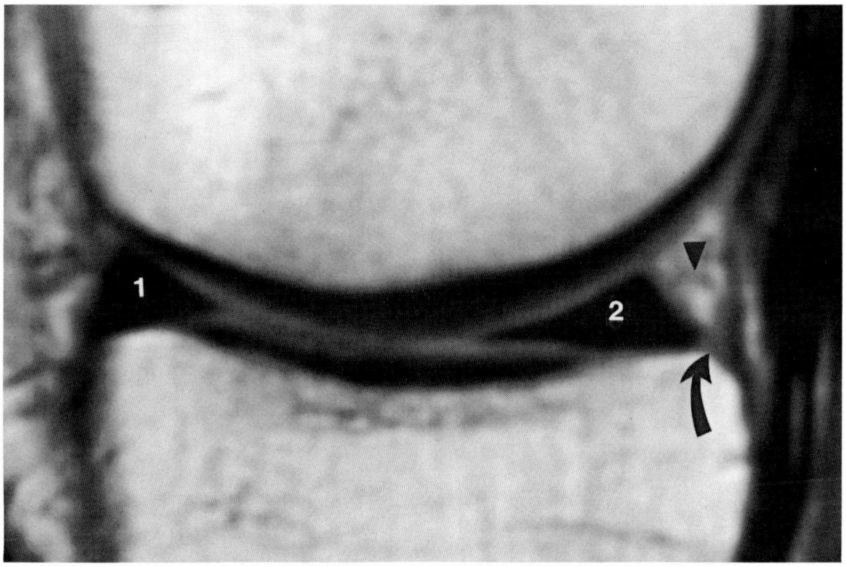

FIG. 3.52. T2-weighted sagittal image of the medial meniscus. 1, Anterior horn of the medial meniscus; 2, posterior horn of the medial meniscus; *arrowhead*, superior recess of the posterior horn of the medial meniscus, broader superiorly and narrower inferiorly; *curved black arrow*, normal position of the inferior aspect of the posterior horn of the medial meniscus relative to the tibial plateau and posteromedial capsule.

capsule, which is particularly well seen with T2-weighted spin-echo images. The popliteal tendon is clearly intracapsular. Injuries of the posterolateral capsule often occur in association with anterior cruciate ligament tears and may play an important role in determining the development of knee joint instability secondary to trauma.

Figures 3.51 and 3.52 emphasize anatomy in the posteromedial corner of the knee joint. The deep and superficial layers of the medial collateral ligament complex can be discriminated on axial scans. Although there is often a small space filled with fibrofatty tissue interposed between the medial collateral ligament and the medial meniscus at the level of the middle third of the medial meniscus, this space disappears along the posteromedial third of the medial meniscus where the inferior portion of the meniscus is directly opposed to the capsule. Understanding this axial anatomy is extremely helpful in differentiating meniscocapsular tears from normal anatomic variations. The superior recess of the posterior horn of the medial meniscus (Fig. 3.52) is generally filled with either fibrofatty tissue or fluid, and should not be mistaken for meniscocapsular detachment.

Figures 3.53 and 3.54 highlight the cruciate ligaments and posterior capsule. Individual fascicles of the anterior cruciate ligament are often discriminated on routine MR images upon close inspection. Partial voluming of synovial fluid or fibrofatty tissue adjacent to the ACL fascicles may account for the apparent increased signal intensity of the anterior cruciate ligament relative to the posterior cruciate ligament on

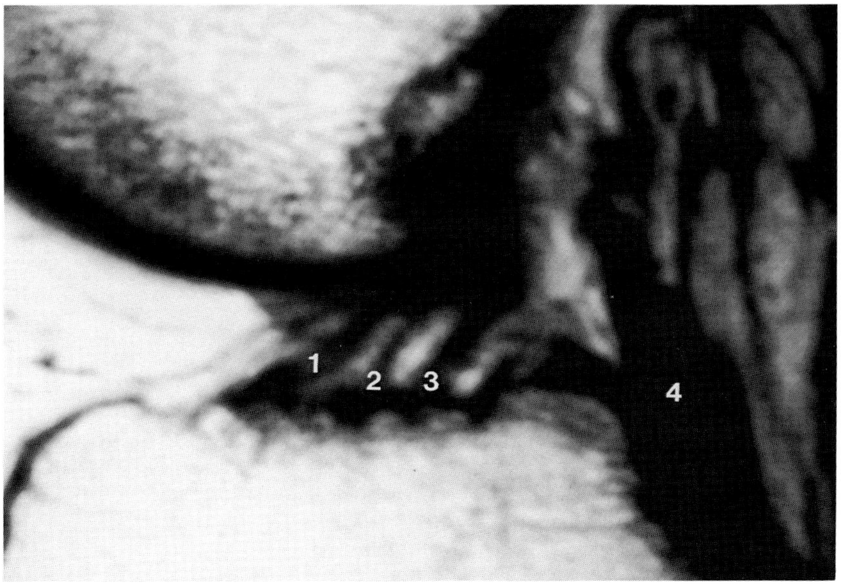

FIG. 3.53. Magnified T2-weighted sagittal section through the cruciate ligaments. 1,2,3, Individual fascicles of the anterior cruciate ligament; 4, posterior cruciate ligament.

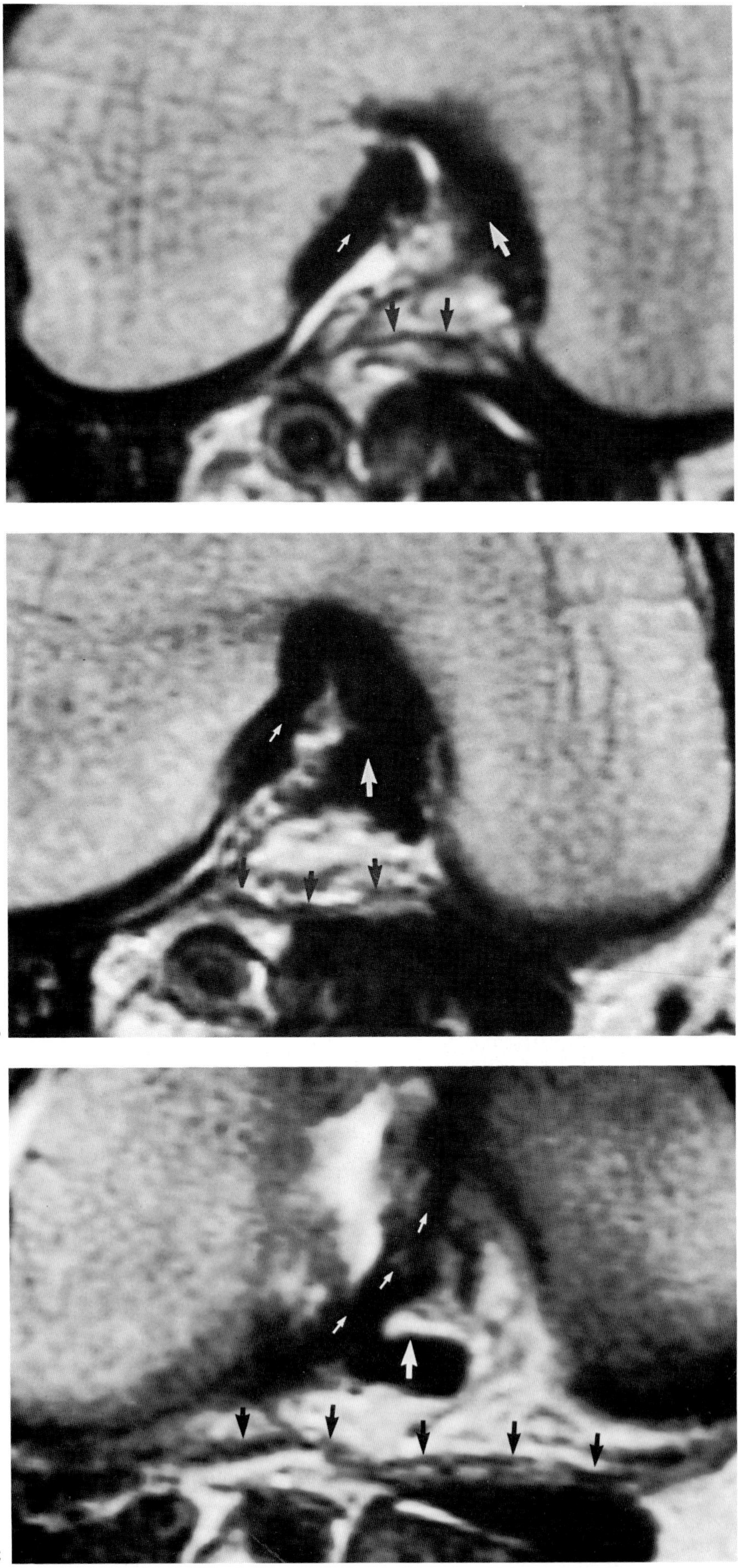

FIG. 3.54A–C. Axial T2-weighted sections coursing from superior to inferior through the intercondylar notch. *Small white arrows*, anterior cruciate ligament; *large white arrows*, posterior cruciate ligament; *black arrows*, posterior capsule.

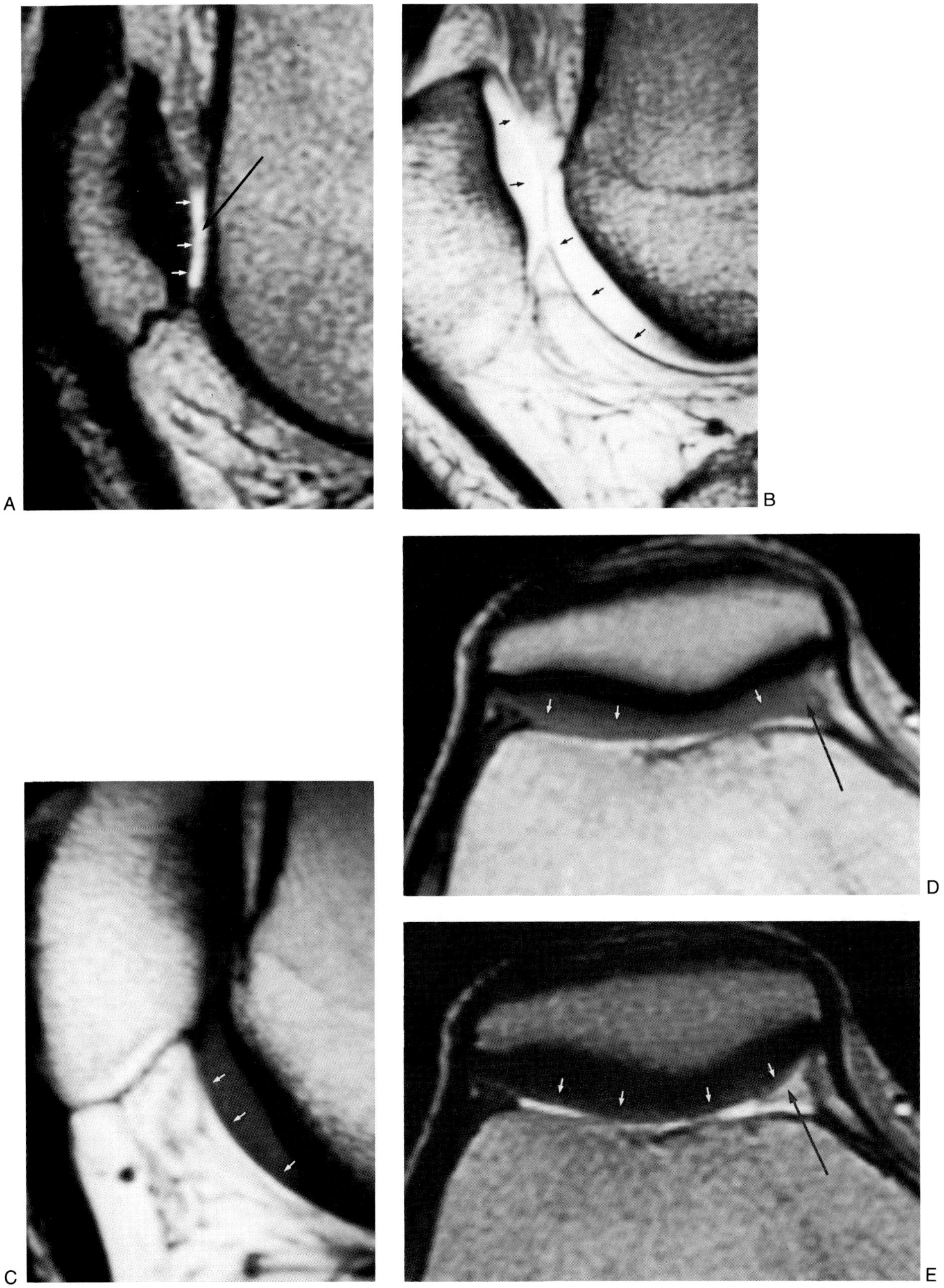

FIG. 3.55A. Sagittal T2-weighted spin-echo image depicts relatively low signal intensity patellar articular cartilage (*white arrows*) demarcated by high signal intensity fluid (*long black arrow*) within the suprapatellar bursa. **B:** T2*-weighted gradient-echo image obtained on mid-field-strength system yields relatively high signal intensity articular cartilage along the patella and femoral trochlear groove (*small black arrows*). **C:** Sagittal T1-weighted spin-echo image yields moderate signal intensity articular cartilage along the femoral trochlear groove (*small white arrows*). **D:** Intermediate-weighted axial spin-echo image yields moderate signal intensity articular cartilage (*small white arrows*), indistinguishable from adjacent joint fluid within the suprapatellar bursa (*long black arrow*). **E:** T2-weighted axial spin-echo image at the same level as in Fig. 2.55D better displays the interface between low signal intensity articular cartilage (*white arrows*) and adjacent high signal intensity joint fluid (*long black arrow*).

most MR pulse sequences. In Fig. 3.54, the course of the cruciate ligaments is depicted in detail in the axial plane. The cruciate ligaments are generally most easily appreciated on T2-weighted images because of the high contrast with synovial fluid and neighboring intercondylar fat. ACL tears suspected on sagittal images routinely should be confirmed and quantified on such axial scans. The posterior capsule is also delineated on detailed axial images.

Figure 3.55 emphasizes the patellar articular cartilage. Many different techniques can be employed to evaluate articular cartilage, including spin-echo and gradient-echo methods (9). It is important to select pulse sequences in which joint fluid yields significantly different signal intensity than articular cartilage (Fig. 3.55D vs. 3.55E). Interestingly, highly detailed anatomic evaluation can be achieved at mid-field strength (Fig. 3.55B) or high field strength (Figs. 3.55A, 3.55C, 3.55D, and 3.55E).

SUMMARY

Exquisite anatomic detail is now routinely available on knee MR examinations. Appreciation of all of the anatomic detail available on these images is essential to accurately and specifically evaluate knee joint disorders.

REFERENCES

1. Adam G, Bohndorf K, Drobnitzky M, Guenther RW. MR imaging of the knee: Three-dimensional volume imaging combined with fast processing. *J Comput Assist Tomogr* 1989;13(6):984–988.
2. Beltran J, Noto AM, Herman LJ, Mosure JC, Burk JM, Christoforidis AJ. Joint effusions: MR imaging. *Radiology* 1986; 158:133–137.
3. Beltran J, Noto AM, Mosure JC, Bools JC, Zuelzer W, Christoforidis AJ. Meniscal tears: MR demonstration of experimentally produced injuries. *Radiology* 1986;158:691–693.
4. Beltran J, Noto AM, Mosure JC, Weiss KL, Zuelzer W, Christoforidis AJ. The knee: Surface-coil MR imaging at 1.5 T. *Radiology* 1986;159:747–751.
5. Buckwalter KA, Braunstein EM, Wilson MR, Wojtys EM, Martel W. Evaluation of hyaline cartilage of the knee with MR imaging. *Radiology* 1986;161(P):239.
6. Buckwalter KA, Pennes DR. Anterior cruciate ligament: Oblique sagittal MR imaging. *Radiology* 1990;175(1):276–277.
7. Burk DL Jr, Dalinka MK, Kanal E, Brunberg JA. High resolution MR imaging of the knee. *Magn Reson Ann* 1988;1–36.
8. Burk DL, Kanal E, Brunberg JA, Johnsone GF, Swensen HE, Wolf GL. 1.5 T surface-coil MRI of the knee. *AJR* 1986;147:293–300.
9. Conway WF, Hayes CW, Loughran T, et al. Cross-sectional imaging of the patellofemoral joint and surrounding structures. *Radiographics* 1991;11:195–217.
10. Erickson SJ, Cox IH, Hyde JS, Carrera MD, Strandt JA, Estkowski LD. Effect of tendon orientation on MR imaging signal intensity: A manifestation of the "magic angle" phenomenon. *Radiology* 1991;181:389–392.
11. Gallimore GW, Harms SE. Knee injuries: High resolution MR imaging. *Radiology* 1986;160:457–461.
12. Grover JS, Bassett LW, Gross ML, Seeger LL, Finerman GAM. Posterior cruciate ligament: MR imaging. *Radiology* 1990; 174:527–530.
13. Haggar AM, Froelich JW, Hearshen DO, Sadasivan K. Meniscal abnormalities of the knee: 3DFT fast-scan GRASS MR imaging. *AJR* 1988;150(6):1341–1344.
14. Harms SE, Flamig DP, Fisher CF, Fulmer JM. New method for fast MR imaging of the knee. *Radiology* 1989;173(3):743–750.
15. Hartzman S, Reicher MA, Bassett LW, Duckwiler GR, Mandelbaum B, Gold RH. Magnetic resonance of the knee joint. Clinical update II: Chronic disorders. *Radiology* 1987;162:553–559.
16. Kean DM, Worthington BS, Preston BJ, et al. Nuclear magnetic resonance imagery of the knee: Examples of normal anatomy and pathology. *Br J Radiol* 1983;56:355–364.
17. Li DKB, Adams ME, McConkey JP. Magnetic resonance imaging of the ligaments and menisci of the knee. *Radiol Clin North Am* 1986;24:209.
18. Li KC, Henkelman M, Poon PY, Rubenstein R. MR imaging of the normal knee. *J Comput Assist Tomogr* 1984;8:1147–1154.
19. Mesgarzadeh M, Schneck CD, Bonakdarpour A. Magnetic resonance imaging of the knee and correlation with normal anatomy. *Radiographics* 1988;8(4):707–733.
20. Middleton WD, Lawson TL, eds. *Anatomy and MRI of the joints*. New York: Raven Press, 1989.
21. Mink JH, Reicher MA, Crues JV III. *Magnetic resonance imaging of the knee*. New York: Raven Press, 1987.
22. Modl JM, Lowell AS, Haughton VM, Kneeland JB. Articular cartilage: Correlation of histologic zones with signal intensity at MR imaging. *Radiology* 1991;181:853–855.
23. Moon KL, Genant HK, Helms CA, Chafetz NI, Crooks LE, Kaufman L. Musculoskeletal applications of nuclear magnetic resonance. *Radiology* 1983;161:161–171.
24. Rauschning W. Computed tomography and cryomicrotomy of lumbar spine specimens. A new technique for multiplanar anatomic correlation. *Spine* 1983;8:170–180.
25. Reicher MA, Bassett LW, Gold RH. High-resolution magnetic resonance imaging of the knee joint: Pathologic correlations. *AJR* 1985;145:903–909.
26. Reicher MA, Hartzmann S, Bassett LW, Mandelbaum B, Duckwiler GR, Gold RH. Magnetic resonance imaging of the knee joint. Clinical update I: Traumatic disorders. *Radiology* 1987; 162:545–552.
27. Reicher MA, Hartzman S, Duckwiler GR, Bassett LW, Anderson LJ, Gold RH. Meniscal injuries: Detection using MR imaging. *Radiology* 1986;159:753–757.
28. Reicher MA, Rauschning W, Gold RH, Bassett LW, Lufkin RB, Glenn W. High-resolution magnetic resonance imaging of the knee joint: Normal anatomy. *AJR* 1985;145:895–902.
29. Reiser MF, Bongartz G, Erlemann R, et al. Magnetic resonance in cartilaginous lesions of the knee joint with three-dimensional gradient-echo imaging. *Skeletal Radiol* 1988;17(7):465–471.
30. Solomon SL, Totty WG, Lee JK. MR imaging of the knee: Comparison of three-dimensional FISP and two-dimensional spin-echo pulse sequences. *Radiology* 1989;173(3):739–742.
31. Soudry M, Lanir A, Angel D, Roffman M, Kaplan N, Mendes DG. Anatomy of the normal knee as seen by magnetic resonance imaging. *J Bone Joint Surg* 1986;68:117–120.
32. Spritzer CE, Vogler JB, Martinez S, et al. MR imaging of the knee: Preliminary results with 3DFT GRASS pulse sequence. *AJR* 1988;150:597–603.
33. Totterman S, Weiss SL, Szumowski J, et al. MR fat suppression technique in the evaluation of normal structures of the knee. *J Comput Assist Tomogr* 1989;13(3):473–479.
34. Watanabe AT, Carter BC, Teitelbaum GP, Seeger LL, Bradley WG. Normal variations in MR imaging of the knee: Appearance and frequency. *AJR* 1989;153:341–344.
35. Wolff SD, Chesnick S, Frank JA, Lim KO, Balaban RS. Magnetization transfer contrast: MR imaging of the knee. *Radiology* 1991;179:623–628.
36. Yao L, Sherwood JW. Ex-vivo MR imaging of the cruciate ligaments. *Computerized medical imaging and graphics* 1991; 15(1):27–30.

CHAPTER 4

The Menisci

John V. Crues III and David W. Stoller

The purpose of this chapter is to describe the use of MRI for diagnosis of meniscal abnormalities. MRI technique, relevant normal knee functional anatomy, and a description of major disease processes that affect the menisci are discussed. The multiple roles of MRI in the clinical evaluation of meniscal diseases and comparisons of MRI with other current diagnostic modalities for evaluating meniscal disease are presented.

MRI TECHNIQUE

The MRI techniques described in this chapter (Table 4.1) evolved over approximately 7 years of clinical experience. However, techniques continue to evolve as new pulse sequences are developed (1,2), and we learn more about the manifold ways in which knee abnormalities and normalcy can present on MRI. A single optimal technique for evaluating all potential knee problems has not been developed. We use short TE images (T1, intermediate weighting, or T2*-weighted) to evaluate meniscal tears because they are more sensitive for disease detection than long TE (T2-weighted) spin-echo images (3). Gradient-echo images are sensitive in the detection of meniscal tears (4–6). Two-dimensional and three-dimensional Fourier transform (3DFT) gradient-echo protocols in some reports have accurately identified the spectrum of degeneration in grades 1 and 2 meniscal signal and meniscal tears in grade 3 meniscal signal (4–6). Gradient-echo contrast (heavily T2*-weighted) may be less sensitive in the detection of subchondral bone pathology. T2-weighted images can be useful for evaluating acute ligamentous tears, characterizing fluid collections, and evaluating lateral meniscal fascicles. Adequate knee imaging requires dedicated knee coils which are available from all MRI manufacturers as well as a growing number of independent MRI surface coil companies.

Special techniques may be helpful in the evaluation of selected knee pathologies. The majority of diagnostic MR images of the knee are acquired in the sagittal, coronal, or axial planes. Some MR interpreters prefer using a radial image acquisition technique whereby gradient-echo images are acquired in multiple planes which are rotated around the center of curvature of the meniscus being evaluated (Table 4.1) (Fig. 4.1). Images acquired in this manner more closely correlate with arthrographic images and the extent of disease may be readily displayed. Drawbacks of this technique include the presence of gradient-echo type contrast in the images rather than short TE spin-echo contrast necessitating additional sequences for evaluation of nonmeniscal disease, increased complexity of anatomy displayed in multiple planes, and increased imaging time. Radial images, however, may be useful in demonstrating the meniscocapsular junction and peripheral meniscal tears. If 3DFT acquisitions are obtained throughout the knee, then reconstructions can be displayed in radial planes without requiring a dedicated two-dimensional radial protocol.

Though contrast agents have been highly beneficial in MRI practice in the brain and spine, the routine uses of intravenous gadolinium-DTPA in the knee examination have not been clearly established (7–11). Gadolinium-enhanced images do provide more selective enhancement for mapping of pannus tissue in rheumatoid arthritis (9). On standard images pannus tissue often has signal characteristics identical to joint fluid from which it cannot be distinguished. After intravenous gadolinium administration pannus enhances, but joint fluid does not (7,9). Intraarticular contrast agents,

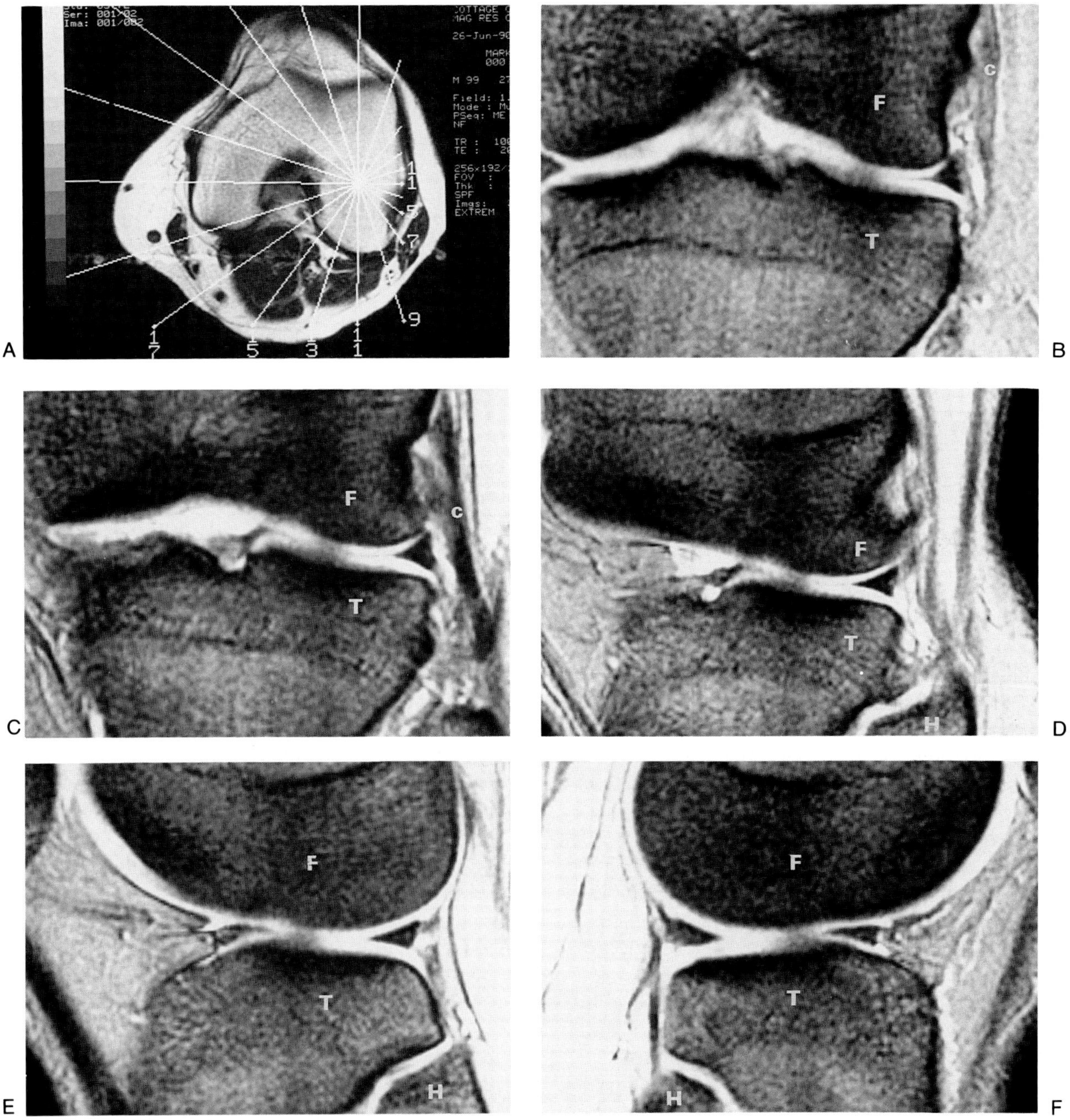

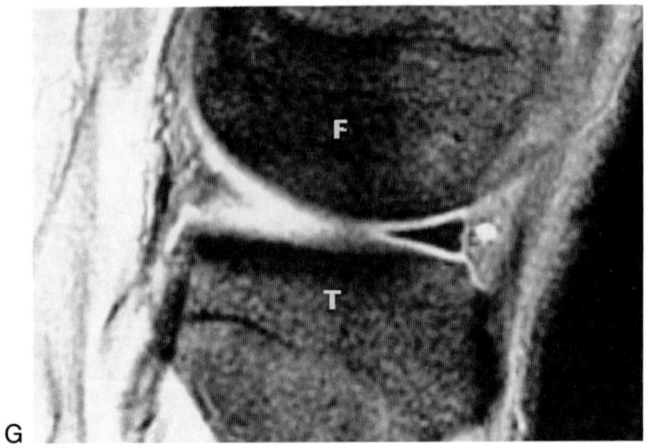

FIG. 4.1. Radial images of the lateral meniscus. **A:** The annotated axial scout image shows the orientations of the radial acquisition planes. **B–G:** Radial images (TR/TE 700/14, flip angle 25°) corresponding to radial plane numbers 3, 5, 7, 9, 11, and 15, respectively. c, fibular (lateral) collateral ligament; F, lateral femoral condyle; H, fibular head; T, lateral tibial plateau.

TABLE 4.1. *Protocols used in evaluating the knee. Either the routine knee protocol or the acute knee protocol are selected. The acute protocol is used if the patient has suffered trauma within the prior 3 weeks or the study is performed to evaluate a mass. A third add-on sequence is selected depending on the clinical history. For suspected patellar chondromalacia an axial 1500 TR double echo spin-echo sequence is usually selected, though the STIR sequence with a TI of 145 may be used. The STIR sequence is also helpful in evaluating subtle marrow disorders. In concert with an appropriate positioning device, kinematic images using a rapid spin-echo technique may be helpful in evaluating patellar tracking disorders.*

Study	Seq.	Plane	FOV	TR	TE	TI	Nex	Sl.Th.	Skip	Φ	A.Mtx	Phase Dir.	Phase O.	Sat.
routine	1	C	14	600	20	–	1	4	1	90	192	S-I	+	Sup
	2	S	14	2000	20/80	–	1	4	1	90	192	S-I	+	Sup
acute	1	C	14	2000	20/80	–	1	4	1	90	192	S-I	+	Sup
	2	S	14	2000	20/80	–	1	4	1	90	192	S-I	+	Sup
add on routines														
Patella	3	A	12	1500	20/80	–	1	4	1	90	192	R-L	–	–
	or													
	3	A	12	1500	25	145	1	4	1	90	192	R-L	+	–
radial menisci														
	3		16	500	12/31	–	2	5	–	25	192	–	+	–
	10 images in radial planes 18° apart													
kinematic														
	3	A	32	400	20	–	.5	5	.5	90	128	A-P	–	–
	multiple images in varying degrees of flexion played in cine loop													

Seq.: sequence, Plane: plane of image, FOV: field of view, TR: repetition time, TE: echo delay time, TI: inversion time for inversion recovery images, Nex: number of excitations, Sl.Th.: slice thickness, Skip: separation between slices, Φ: flip angle, A. Mtx: number of phase encodings, Phase Dir.: direction of phase encoding, Phase O.: over sampling in the phase direction, Sat.: location of presaturation pulses, Sup: superior.

including saline and gadolinium preparations, also appear to have limited utility in the routine knee examination (12,13). Possible indications include the detection of intraarticular loose bodies when the standard examination is negative and the evaluation for recurrent meniscal tear after primary meniscal repair. In this setting persistent meniscal signal can be confusing on the standard MR images (*vida infra*) (see Chapter 7). The ability to see fluid or contrast in the meniscus after intraarticular administration can confirm the presence of recurrent tear, rather than persistent signal from a healed tear (13).

MENISCAL ANATOMY

The knee joint is an extremely efficient structure allowing remarkable motion limited primarily to a single plane. However, it accomplishes this by using an intrinsically unstable design, with a curved surface (the femoral condyles) resting on a flat surface (the tibial plateaus). Stabilization requires intricately designed support structures. Two C-shaped fibrocartilaginous menisci, which are relatively firmly attached to the tibia, cradle the femoral condyles on top of the tibial plateaus (Fig. 4.2). The menisci also distribute torsional and compressive forces during mechanical loading, distribute synovial fluid over the articular cartilages for proper nutrition and lubrication, act as shock absorbers, facilitate complex movements, prevent synovial impingement, and limit abnormal motion (14–22). Flexion and extension of the knee occur primarily between the menisci and the femoral condyles, with motion occurring between the menisci and the tibial plateaus during rotation (18,21,23).

The medial meniscus has a larger radius of curvature than the lateral meniscus (Fig. 4.3) and is tightly attached to the intercondylar tibia by strong ligamentous attachments at the central tips of the anterior and posterior horns (Fig. 4.2). The periphery of the medial meniscus is also relatively firmly attached to the medial tibial plateau by the coronary ligaments and the

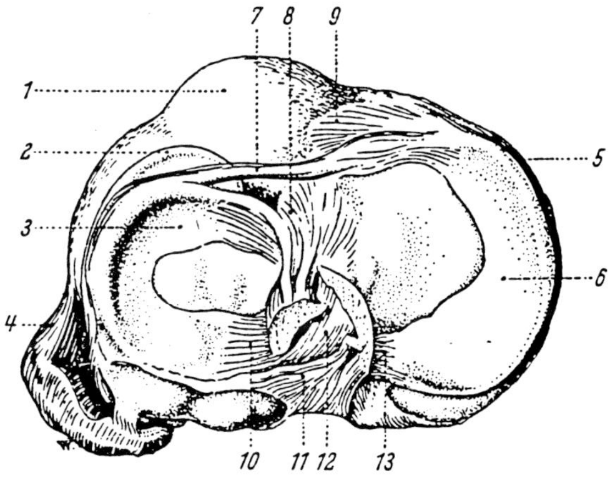

FIG. 4.2. Articular surface of the left tibia, with menisci seen from above: 1, tuberositas tibiae; 2, anterior aspect of articular portion of lateral tibial plateau; 3, lateral meniscus, anterior horn; 4, tibiofibular ligament; 5, medial tibial plateau; 6, medial meniscus, middle portion; 7, transverse ligament; 8, attachment of anterior cruciate ligament; 9, ligamentous attachment of anterior horn of medial meniscus to tibia; 10, posterior attachment of lateral meniscus; 11, ligamentum menisci fibularis; 12, posterior cruciate ligament; 13, posterior attachment of medial meniscus. (From ref. 18.)

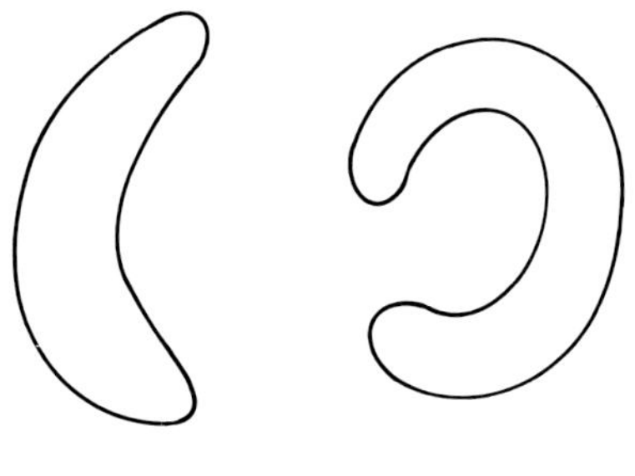

FIG. 4.3. Meniscal shapes. The normal medial meniscus **(A)** is semilunar in shape, whereas the lateral meniscus **(B)** is more circular in shape. (Notice that the medial and lateral are reversed compared with Fig. 4.2.) (Adapted from ref. 53.)

deep fibers of the medial collateral ligament (the medial capsular ligament). The lateral meniscus is more loosely bound to the lateral tibial spine at the central tips of the anterior and posterior horns (Fig. 4.2) (23,24). As in most animals, the posterior horn of the lateral meniscus is also attached to the lateral surface of the medial femoral condyle by the meniscofemoral ligament (ligament of Wrisberg or Robert's ligament) (Fig. 4.4) (25). Wrisberg's ligament may have two branches. The main branch lies posterior to the posterior cruciate ligament and inserts on the medial femoral condyle. The anterior branch (Humphrey's ligament) courses anterior to the posterior cruciate ligament, into which it blends (23). Recently, this terminology has been shortened. The posterior component is often referred to as Wrisberg's ligament and the anterior component is referred to as Humphrey's ligament, rather than the anterior component of Wrisberg's ligament (24). These ligaments are most important to meniscal MRI because of their ability to simulate a meniscal tear of the posterior horn of the lateral meniscus if not properly recognized as normal anatomy (24,26–28). Unique to humans, however, is the existence of a posterior meniscotibial ligament in the lateral compartment. This ligament is thought to stabilize the lateral meniscus on weight bearing in extension, and its absence may be responsible for an acquired form of discoid meniscus (*vide infra*) (23). However, the periphery of the lateral meniscus is more loosely attached to the tibia than is the medial meniscus, allowing more motion (Fig. 4.5). In addition, the popliteal tendon and its sheath penetrate the meniscocapsular junction in such a manner as to produce two separate attachments of the posterior horn of the lateral meniscus to the capsule, the superior and inferior fascicles (see Fig. 4.6) (29–32). The popliteal tendon functions analogously to the pronator teres at the elbow, helping to flex the knee by unlocking the "screw-home mechanism" (*vide infra*) by rotating the tibia with respect to the femur when the knee is in full extension. It also directs anterior and posterior movements of the lateral meniscus during knee extension and flexion (23).

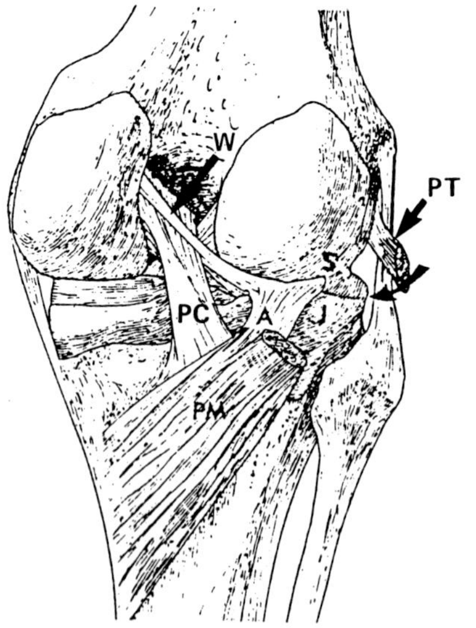

FIG. 4.4. Ligament of Wrisberg (W) extends from the posterior horn of the lateral meniscus near the aponeurosis of the popliteus muscle (A) to the inner aspect of the medial femoral condyle. The popliteal tendon has been severed to improve anatomic visualization. The superior (S) and inferior (I) popliteal recesses communicate (*curved arrow*); (PM) popliteal muscle; (PT) popliteal tendon; (PC) posterior cruciate ligament. (From ref. 31.)

On a sagittal cross-section, the medial meniscus is rectangular in shape peripherally, but separates into two triangular horns more centrally as the slice plane crosses the free edge (see Chapter 3). An average dimension of 10 to 11 mm is a good ballpark number for the breadth of the normal body and posterior horn of the medial meniscus (33). The anterior horn averages 7 to 8 mm. Therefore, the posterior horn of the medial meniscus should be 30% to 50% larger in breadths and

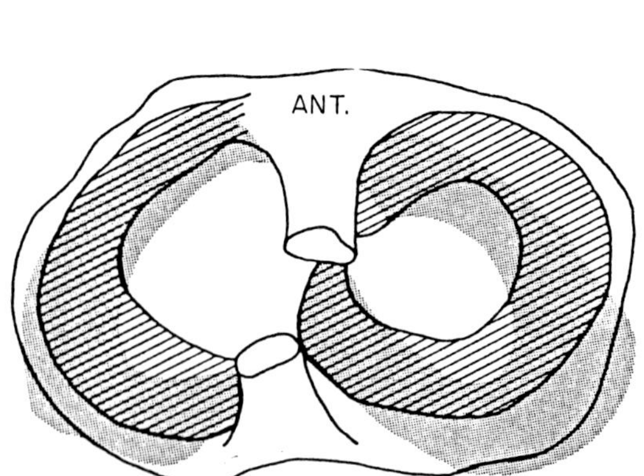

FIG. 4.5. Meniscal motion: schematic representation of the positions of the menisci viewed from above with the knee in full extension (*diagonal lines*) and in flexion (*stippled*). (From ref. 29.)

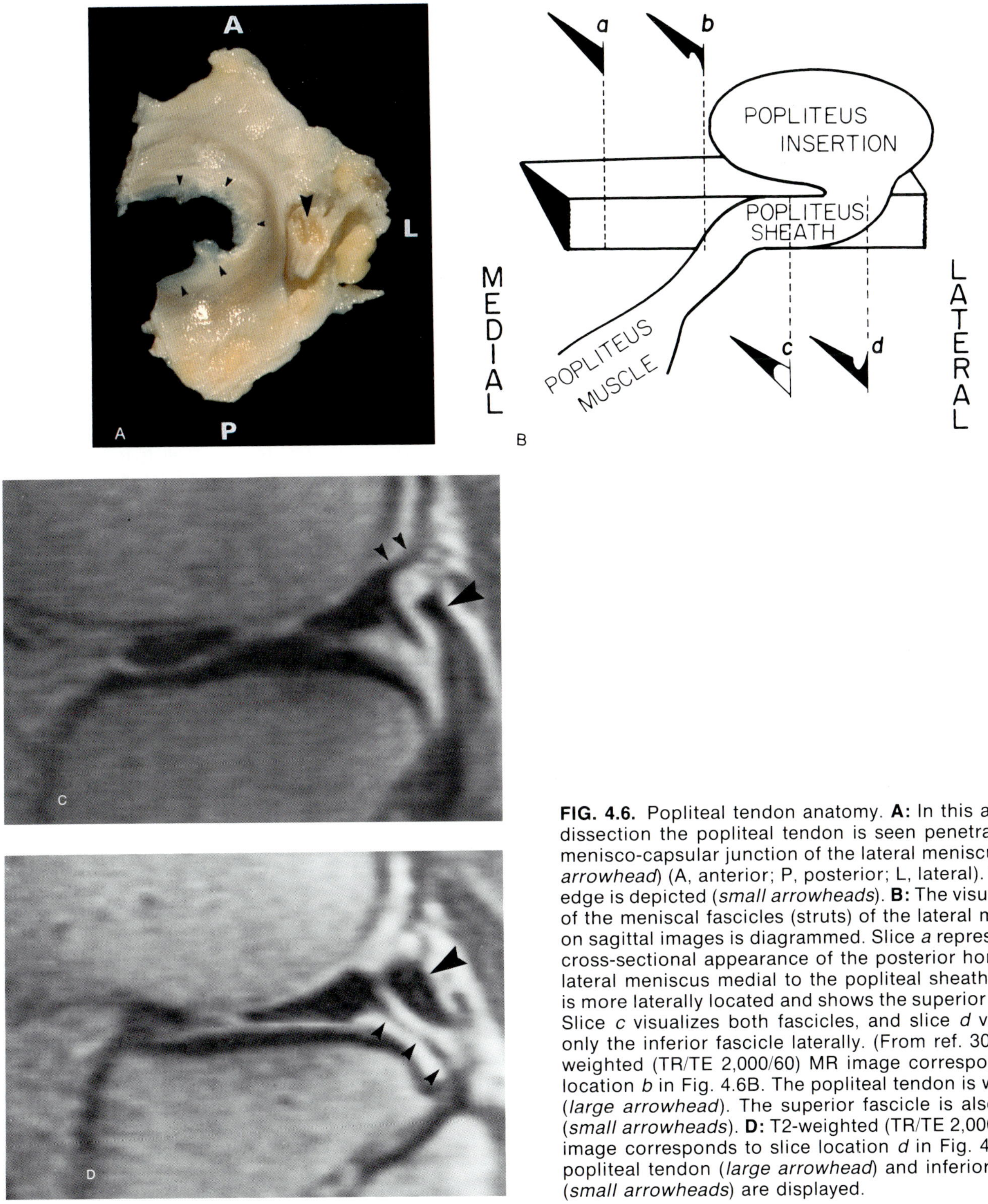

FIG. 4.6. Popliteal tendon anatomy. **A:** In this anatomic dissection the popliteal tendon is seen penetrating the menisco-capsular junction of the lateral meniscus (*large arrowhead*) (A, anterior; P, posterior; L, lateral). The free edge is depicted (*small arrowheads*). **B:** The visualization of the meniscal fascicles (struts) of the lateral meniscus on sagittal images is diagrammed. Slice *a* represents the cross-sectional appearance of the posterior horn of the lateral meniscus medial to the popliteal sheath. Slice *b* is more laterally located and shows the superior fascicle. Slice *c* visualizes both fascicles, and slice *d* visualizes only the inferior fascicle laterally. (From ref. 30.) **C:** T2-weighted (TR/TE 2,000/60) MR image corresponding to location *b* in Fig. 4.6B. The popliteal tendon is well seen (*large arrowhead*). The superior fascicle is also shown (*small arrowheads*). **D:** T2-weighted (TR/TE 2,000/60) MR image corresponds to slice location *d* in Fig. 4.6B. The popliteal tendon (*large arrowhead*) and inferior fascicle (*small arrowheads*) are displayed.

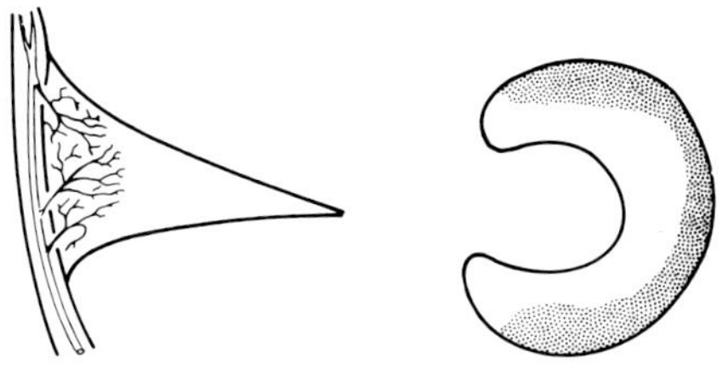

FIG. 4.7. Meniscal vascularity. The meniscal vessels are fed by the parameniscal plexus **(A)** and extend into the peripheral one-third of the meniscus **(B)**. Nerve fibers accompany the meniscal vessels (37). (Adapted from ref. 47.)

approximately 10% to 30% higher in superior-inferior dimension than the anterior horn. The lateral meniscus is "bow-tie" in shape peripherally, but the anterior and posterior horns become triangular in shape when the plane of section crosses the free edge (see Chapter 3). In contradistinction to the medial meniscus, the anterior and posterior horns of the lateral meniscus are symmetric in size and shape, averaging about 10 to 11 mm in breadth (33).

The vascular supply of the meniscus arises from a perimeniscal vascular ring supplied primarily by the inferior and middle geniculate arteries (34–39). In adults, vessels are mostly limited to the peripheral one-third of the meniscus, although the anterior and posterior horns are more richly supplied than the body (Fig. 4.7). Day et al. have recently shown that the meniscus is innervated with nerve fibers accompanying meniscal vessels (37).

Bony ligaments are necessary to maintain the femoral condyles within their meniscal cradles. Intact cruciate ligaments prohibit the femoral condyles from subluxing in the anteroposterior direction and limit external rotation of the femur on tibia (18). Intact collateral ligaments prohibit axial separation of the femur and tibia and constrain excessive varus and valgus angulation. Thus, the knee is a complicated functional unit. In an attempt to simplify imaging knee abnormalities, we have chosen to discuss abnormalities of each of the major components of the knee joint separately. But the reader should always keep in mind that the knee functions as a single unit, and abnormality in one subunit frequently coexists with, influences, or creates abnormal function in the other subunits (14,40–45).

MENISCAL TEARS

The most important clinical pathologic lesions of the menisci are meniscal tears. Tears are a common cause of knee pain and disability. Previously, menisci were thought to be aneural and therefore not themselves a direct source of pain. Pain was thought to result from changes in the rheologic properties of menisci secondary to tears. The tear would cause abnormal buckling under tension and compression, with secondary tugging of peripheral innervated structures (46,47). However, Day et al. have shown that nerves follow the vascular supply into the outer one-third of the meniscus (37). Consequently, it is now reasonable to assume that a tear of peripheral fibrocartilage can be a direct cause of pain. In older tears, innervated synovium can invaginate into the tear and cause pain when compressed during mechanical loading (33,48). Therefore, it is now recognized that multiple mechanisms are responsible for pain in the presence of meniscal tears.

In addition to causing pain, bucket-handle tears can cause locking if the central fragment becomes interposed between the femoral condyle and tibial plateau anteriorly (Fig. 4.8) (18,46). Though it has been said that classic locking against a rubbery substance is almost pathognomonic of a bucket-handle tear with extension to the anterior horn, other internal derangements of the knee, especially anterior cruciate tears, can present similarly (18,46,49). "Giving way" of the knee is also a common complaint secondary to pain caused by meniscal tears, especially horizontal cleavage lesions (46).

Many authors believe that meniscal degeneration and tears may predispose to early articular cartilage degeneration because of incongruous joint motion (15,21,50–52). However, others disagree (33,48,53–55). In some circumstances, early detection and therapy may decrease such secondary degenerative joint changes (56). In the presence of anterior cruciate ligament deficiency, even patients with limited partial meniscectomies may have worse prognoses than if stable tears are left in place (57,58). Present therapeutic trends are favoring meniscal preservation techniques, and the status of the residual meniscus may be important for prognosis. New evidence suggests that x-ray occult bony injuries are a remarkably common, previously unrecognized cause of knee pain clinically indistinguishable from meniscal tears. Therefore an accurate, noninvasive study to evaluate meniscal degenerative changes and tears as well as ligamentous and bony disease has an important clinical role in the evaluation and proper therapeutic selection in patients with suspected meniscal disease.

Pathogenesis of Meniscal Tears

Meniscal tears are generally divided into two major types. The first is the acute traumatic tear which predominantly occurs in young athletes and is thought to be secondary to crush and twisting injuries of the me-

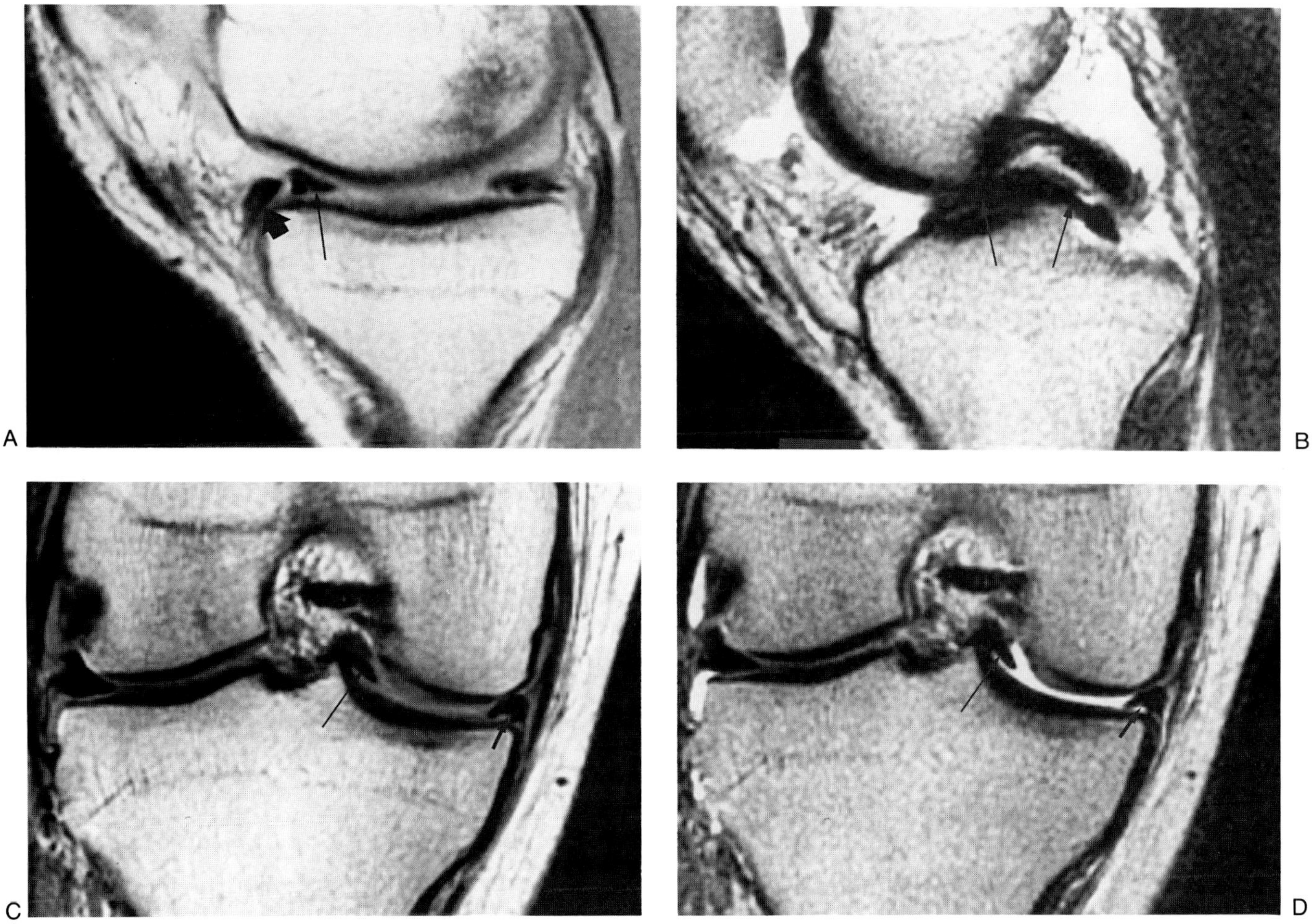

FIG. 4.8. Knee locking from an acute "bucket-handle" meniscal tear. **A:** A sagittal (TR/TE 2,000/20) image reveals two black structures in the anterior portion of the medial compartment (*arrows*). The anterior structure is the anterior horn of the medial meniscus (*fat arrow*). The second is a torn fragment of meniscus which has migrated from the free edge of the body to lie adjacent to the anterior horn (bucket-handle) (*thin arrow*). This added anterior mass impinges on the femoral condyle in extension, impeding the last 15° of extension. The morphology of the torn posterior horn is also abnormal. **B:** The remainder of the bucket-handle fragment can be seen sitting in the notch of the knee inferior to the posterior cruciate ligament on this sagittal (TR/TE 2,000/20) image (*arrows*). **C,D:** The central (*handle*) fragment (*long arrow*) and peripheral rim (*bucket*) (*short arrow*) can be well seen on the coronal intermediate and T2-weighted images (TR/TE 2,000/20,70).

nisci as they are compressed between the femoral condyles and the tibial plateaus. The second occurs in older individuals secondary to degenerative disease (46,48). Degeneration is thought most likely to be secondary to multiple subacute traumatic episodes that lead to chondrocyte death and increased mucinous ground substance, with secondary loss of mechanical integrity of the fibrocartilaginous meniscus (18,33,55). Chronic recurrent trauma produces microscopic tears of the collagen fibers in the mid-plane of the meniscus. Repair of innumerable microscopic injuries is not perfect and may be important in the pathogenesis of degenerative disease. The role of abnormal nutrition in this disease is somewhat unclear (21).

An understanding of the mechanics of knee joint motion is critical to the understanding of meniscal injury. The human is the only animal who maintains constant and complete extension of the knee during activities and in normal stance (23). To assure stability of the extended knee, the joint undergoes a complex maneuver, the "screw-home" mechanism, in the last 15° of extension (18,21). At the end of the knee extension, the tibia rotates with respect to the femur. During rotation, the menisci follow the femoral condyles, and because the center of rotation is about the medial femoral condyle, the lateral meniscus moves posteriorly on the lateral tibial plateau. The anterior cruciate ligament wraps around the posterior cruciate ligament, forcing the femoral condyles into closer apposition of the tibial plateaus, thus "screwing home" the knee. Meniscal tears commonly occur because of abnormal forces exerted across the knee in full extension (21), but because of the stability of the extended knee, meniscal tears rarely occur in this position without associated ligamentous injuries or tibial plateau fractures (18).

As the knee flexes, initially the tibia and femur rotate to unlock the "screw-home mechanism," the ligamentous structures loosen, and the menisci move posteriorly on the tibial plateaus (Fig. 4.5). Because of its tight attachments to the capsule, the medial meniscus moves less than the lateral meniscus and can become trapped between the femoral condyle and posterior lip of the medial tibial plateau in extreme flexion. Rotation may exacerbate this situation (18).

When valgus forces are exerted on the knee, the medial compartment may be slightly opened, allowing the medial meniscus to insinuate more centrally between the femoral condyle and tibial plateau. Compressive forces can then trap the meniscus, causing direct compression and tearing of the centrally fixed meniscus. Subsequent rotation of the femur can displace the periphery of the meniscus, creating shearing forces between the periphery and the trapped central segment. When the elasticity of the meniscus is overcome, it tears. This is probably the mechanism producing longitudinal tears, such as bucket-handle tears, in athletic injuries. Alternatively, forced extension of the femur can markedly increase compressive forces, tearing the trapped meniscus. The medial meniscus is thought to be at greater risk for these mechanisms because of its greater radius of curvature and because the concave surface of the medial tibial plateau favors central displacement (18). For the meniscus to be centrally displaced, the peripheral meniscal attachments must be stretched or ruptured either prior to or simultaneously with a longitudinal tear of the substance of the fibrocartilage (59). Such "meniscocapsular separations" are most significant because of their role in the pathogenesis of acute traumatic tears. Occasionally they may persist and lead to an unstable meniscus. Such patients with persistent pain may benefit from arthroscopic repair (*vida infra*).

The foregoing mechanisms partially account for the acute longitudinal tears most frequently found in the medial meniscus. Tears of the lateral meniscus, however, more typically are transverse or radial tears (59). The latter tears most likely are produced secondary to the small radius of curvature of the lateral meniscus. If the middle third of the lateral meniscus is fixed in the joint by compression between the femoral condyles and tibial plateaus, continued knee motion will result in displacements of the anterior and posterior horns. These displacements lead to distractive forces acting on the free edge which produce radially oriented tears (Fig. 4.9) (21).

Older reports described longitudinal tears to be the most common (18). More recent reports and our experience have found the majority of tears to be horizontal cleavage tears of the posterior horn of the meniscus (33,46,48). Horizontal cleavage tears are usually identified as components of an oblique tear pattern (*vida infra*). These tears are due to weakening of the fibrocartilaginous substance from degenerative changes (21) and are more frequent in patients whose occupations require frequent kneeling and squatting (18) and whose ages are greater than 30 years (46,48). Degenerative changes may begin in the subsurface layers of the under surface of the meniscus. This is probably because of wear and tear as the meniscus slides over the tibial plateau during the rotation of the screw-home mechanism as the knee is axially loaded in extension. The majority of the compressive forces are transmitted across the medial compartment in this maneuver, possibly accounting for the increased incidence of degenerative tears in the posterior horn of the medial meniscus in the series of Smillie (59) and Noble and Hamblen (48).

In other series, horizontal cleavage tears are more common in the lateral meniscus, presumably because in the lateral compartment a higher percentage of the compressive force is exerted through the meniscus than in the medial compartment (33,55). These changes probably are manifest in older age groups secondary to the thinness and rigidity of adult articular cartilage, decreased mobility of older menisci, poorer nutrition,

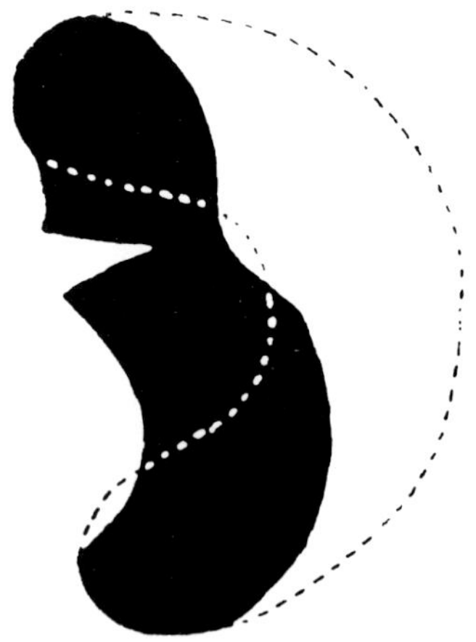

FIG. 4.9. Mechanism of producing transverse tears in the lateral meniscus. If the body of the lateral meniscus becomes trapped between the femoral condyle and tibial plateau, then distractive forces may produce a radial tear because of the small radius of curvature of the lateral meniscus. (*Dotted lines* represent the normal configuration of the lateral meniscus; the *black region* represents the trapped meniscus in which distractive forces have produced a radial, parrot-beak tear.) (From ref. 21.)

less compliant bones as shock absorbers (21), and the added years of living allowing the accumulation of recurrent micro trauma and microcartilaginous tears in the substance of the meniscus. Smillie also stated that degeneration occurs secondary to abnormal fixation of the periphery of the meniscus, creating increased shear forces through the fibrocartilage (59). However, Ferrer-Roca and Vilalta (55) believe that because of difference in the coefficients of friction between the meniscotibial and meniscofemoral surfaces, shearing forces normally occur within the meniscus and may be initiating factors in degenerative disease without abnormal peripheral fixation (55). The shearing forces so produced probably cause recurrent microscopic collagen fiber tears which in concert with reparative processes lead to degenerative changes which weaken the mechanical properties of the mid-plane of the meniscus. This mechanism is probably responsible for the most common meniscal tear, the horizontal cleavage tear. One can also speculate that during the long process of injury leading to degenerative horizontal cleavage tears in older adults, nerves within the periphery may become dysfunctional, leading to the high prevalence of asymptomatic horizontal cleavage over the age of 60 (60). The multiple factors thought to predispose to meniscal tears are summarized in Table 4.2.

MRI-Pathologic Correlation in Meniscal Tears

The normal internal substance of knee menisci produces no significant signal on either T1- or T2-weighted images, presumably secondary to lack of mobile protons within the substance of the meniscus. On the other hand, degenerative changes and meniscal tears produce signal intensity within the meniscus that is most conspicuous on short TE [T1-weighted (SE 800/20), intermediate-weighted (SE 2,000/20), and gradient-echo images]. In contradistinction to the experience of Li et al. (61), in our experience most degenerative changes and meniscal tears are poorly visualized on T2-weighted (SE 2,000/80) images (Fig. 4.10).

TABLE 4.2. *Factors predisposing to meniscal tears*

Constitutional (18,21,33,55)
Primary soft-tissue weakness
Lax ligaments
Discoid menisci
Disruption of the screw-home mechanism (21,288)
Vastus medialis weakness
Hip disease
Medial torsion of the femoral or tibial shaft
Knee fractures
Predisposition to degenerative changes (18,21)
Articular cartilage disease
Prolonged kneeling
Prolonged squatting
Increased age
Sports
Trauma (288)
Sports
Fractures
Direct meniscal trauma
Other
Arthritis
Infection

Although the normal meniscus is without signal on all pulse sequences, this is not because the normal meniscus is without water. Water molecules within meniscal substance are absorbed within large macromolecules, such as collagen. Trapped water molecules cannot translate or rotate. Because of their inability to move, their hydrogen nuclei experience the markedly inhomogeneous local magnetic fields within those macromolecules and are rapidly dephased. Thus, their T2 times are on the order of a few milliseconds or less and no pulse sequences can detect signal from nuclei with such a short T2 (Gary Fullerton, *personal communication*). Microscopic collagen tears and chronic degenerative changes open up macromolecules, allowing increased degrees of freedom to trapped water molecules. This increases the T2 times to tens of milliseconds and allows any short TE sequences (whether spin-echo or gradient-echo) to receive a signal.

On the surface it appears as a paradox: lengthening the T2 from degeneration allows T1-weighted, gradient-echo, and other short TE sequences to detect a signal, but the T2 is still too short in the presence of mild degeneration for true T2-weighted images with TEs around 80 millisec to detect a signal! The intrameniscal signal may also in part be secondary to damaged macromolecules absorbing water and increasing the local spin density. The rotation rates of the absorbed water molecules are slowed by interaction with macromolecules, decreasing the T1 from that of free water which gives a bright signal on short TE images (62,63). The decrease in the translational motion leads to less motional narrowing and a shorter T2 of the adsorbed water so that its signal characteristics are more like those of the exposed water from within the macromolecules (bright on short TE images, but not on long TE images), than like the free water of the joint space (dark on short TE images, but bright on long TE images). Thus, the hallmark of a meniscal tear on MRI is the presence of intrameniscal signal on short TE images.

In an attempt to define the pathologic correlations to the signal intensities produced by menisci on MRI, Stoller et al. (64) compared the grading system proposed by Crues (3,65,66) to pathologic findings in 12 menisci from fresh cadavers and above-the-knee amputations. The MRI grading system is based on the morphologic appearance of intrameniscal signal and its relationship to the articular meniscal surfaces. In con-

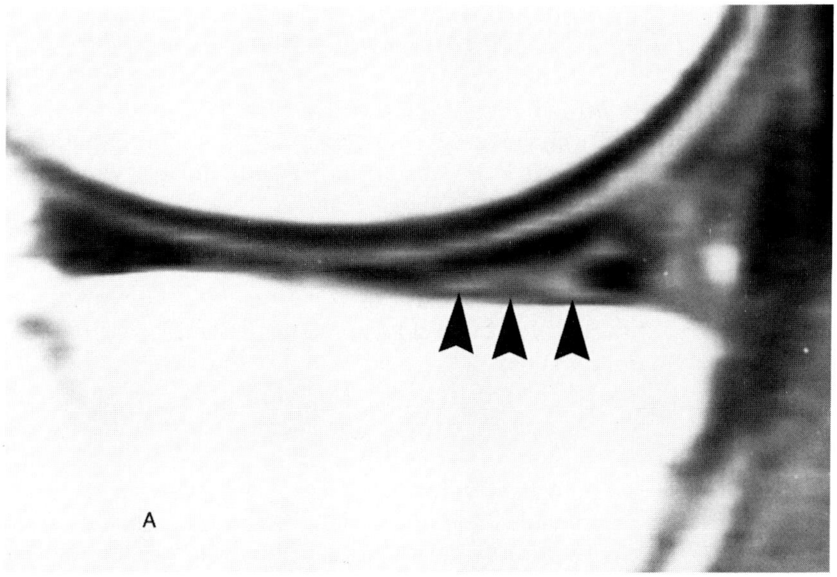

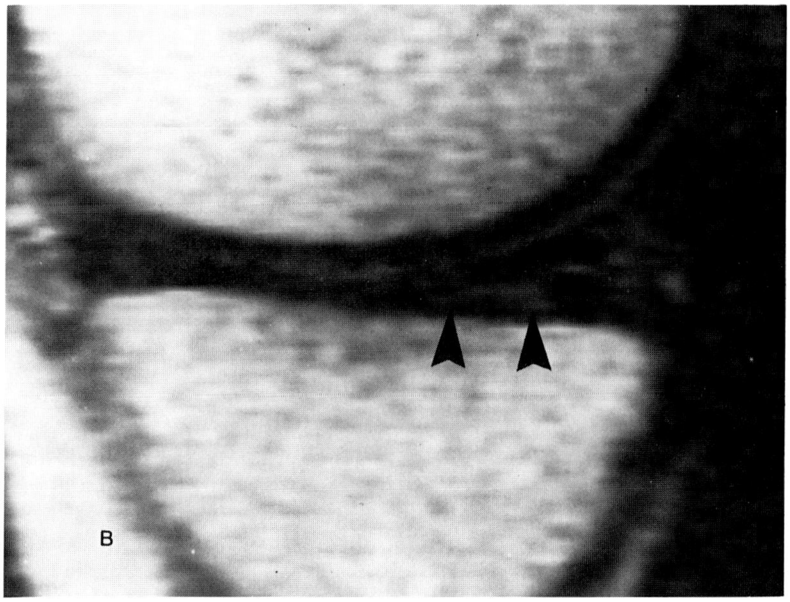

FIG. 4.10. Intermediate- and T2-weighted images of a meniscal tear. **A:** Intermediate-weighted sagittal image (TR/TE 2,000/20) reveals a large tear of the posterior horn of the medial meniscus (*arrowheads*). **B:** T2-weighted image (TR/TE 2,000/80) at the same slice location shows vague increased signal in the region of the tear (*arrowheads*). Smaller tears may not be visible on T2-weighted images.

tradistinction to the findings of other authors (67,68), the intensity of intrameniscal signal in our experience is less useful in discriminating meniscal tears from degenerative disease (3). The articular surfaces are the superior and inferior meniscal border that face the articular cartilage of the femur and tibia, respectively. The histologic staging was determined using established criteria of degeneration on hematoxylin-and-eosin preparations (64). These findings have been confirmed by others (13).

Grade 1. MR signal intensity is globular and is not adjacent to either the superior or inferior articular meniscal surface (Fig. 4.11). Grade 1 meniscal signal corresponds to stage 1 histologic classification of degeneration, in which menisci show foci of mucinous, hyaline, or myxoid degeneration in areas of hypocellularity (chondrocyte-deficient) in hematoxylin-and-eosin preparations (Fig. 4.12). These changes are frequently seen in asymptomatic individuals and represent a response to mechanical stress and loading with increased production of mucopolysaccharide ground substance (33,60).

Grade 2. MR signal is primarily linear signal within a meniscus that does not extend to an articular surface. Typically, however, grade 2 signal does abut the capsular margin of the meniscus (Figs. 4.13 and 4.14). These linear signals correspond to stage 2 histology characterized by more extensive bands of mucinous degeneration bordering hypocellular regions of the meniscus (Fig. 4.14). Although no distinct fibrocartilaginous separation was observed in this stage, microscopic areas of collagen fragmentation were recorded.

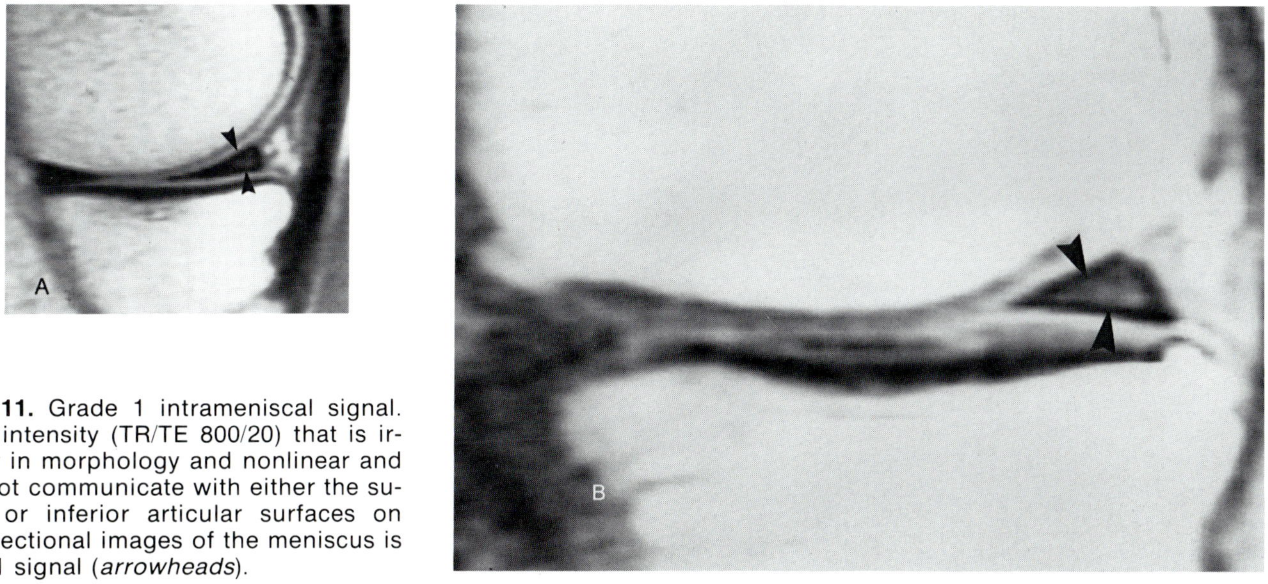

FIG. 4.11. Grade 1 intrameniscal signal. Signal intensity (TR/TE 800/20) that is irregular in morphology and nonlinear and does not communicate with either the superior or inferior articular surfaces on cross-sectional images of the meniscus is grade 1 signal (*arrowheads*).

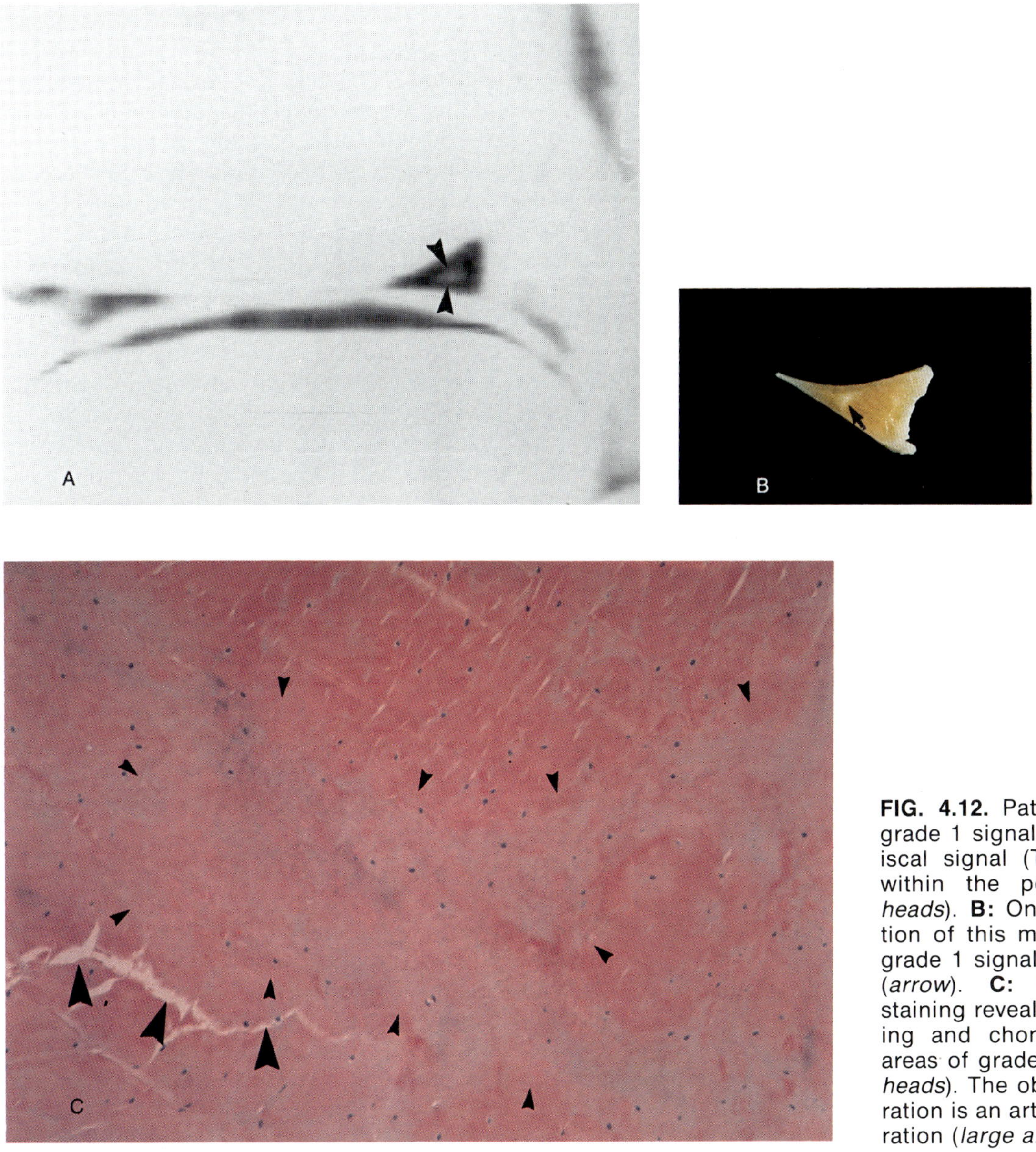

FIG. 4.12. Pathologic correlation of grade 1 signal. **A:** Grade 1 intrameniscal signal (TR/TE 800/20) is seen within the posterior horn (*arrowheads*). **B:** On pathologic cross-section of this meniscus, the region of grade 1 signal is pale in appearance (*arrow*). **C:** Hematoxylin-and-eosin staining reveals regions of pale staining and chondrocyte deficiency in areas of grade 1 signal (*small arrowheads*). The observed meniscal separation is an artifact of the slide preparation (*large arrowheads*).

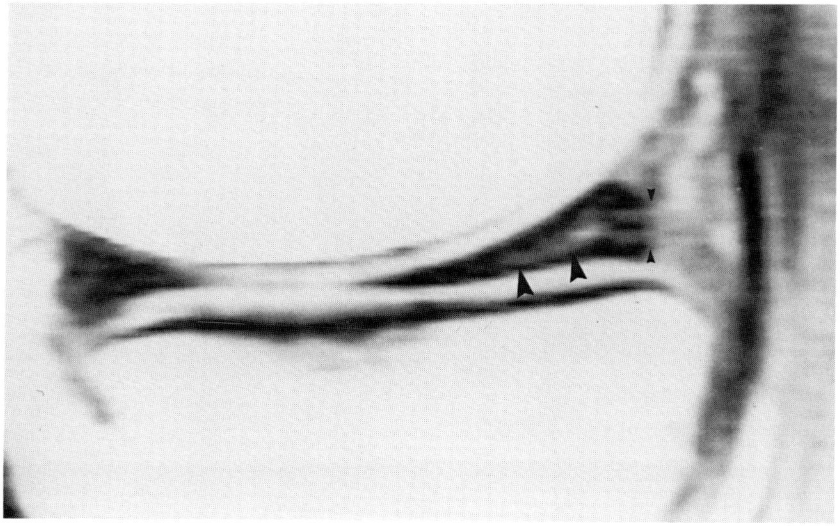

FIG. 4.13. Grade 2 intrameniscal signal (TR/TE 800/20) is linear in morphology (*arrowheads*) and frequently communicates with the menisculo-capsular junction (*small arrowheads*). Grade 2 signal, however, does not communicate with an articular meniscal surface.

The middle perforating collagen bundles are fibers horizontally dividing the meniscus into superior and inferior portions and are not visualized by MRI in the normal meniscus. They are the preferred sites for increased accumulation of mucopolysaccharide ground substance, probably as response to chronic recurrent trauma as described above. This is thought to buffer femoral and tibial frictional forces and when present, to image as linear grade 2 signal intensity. Stage 2 abnormality represents a continuum of progressive degeneration from stage 1 and is not a discrete isolated histologic event. However, the decrease in collagen content with nutritional impairment seen in stages 2 and 3 makes the meniscus more susceptible to injury. Clinically, patients with grade 2 MR signal may not be symptomatic; however, pathologic stage 2 disease has been described as a precursor to frank tears (33,46). Therefore, patients with grade 2 meniscal signal may constitute a high-risk group for the development of symptomatic meniscal tears (69–73). Grade 2 signal has been shown in one study not to progress over a 3 year follow-up in stable knees (74). Though controversial, grade 2 signal intensity in discoid menisci may represent cystic areas or cavities of mucinous degeneration which can be symptomatic and may require treatment with partial menisectomy.

Grade 3. MR intrameniscal signal reaches at least one meniscal articular margin. Grade 3 signal has been further subdivided into grade 3A, which is linear intrameniscal signal abutting an articular margin (Fig. 4.15), and grade 3B, which has irregular morphology and may encompass the entire meniscus (Figs. 4.16 and 4.17). Usually the communication of the grade 3 signal with the articular surface is seen on more than two 4 mm contiguous images (71,73); however, this need not be the case for small radial tears of the lateral meniscus (*vida infra*) (Fig. 4.18). Both grades 3A and 3B are associated with stage 3 abnormality showing a tear within the fibrocartilaginous meniscus with or without gross visible surface extension and extensive stage 2 histology (Fig. 4.19). Grade 3B lesions are associated with more extensive degenerative change in the surrounding meniscus. In clinical applications grade 3 signal intensity is understood to represent either linear or irregular morphologic patterns as described above.

In one study, two menisci containing grade 3A intrameniscal signal were visually normal at gross inspection. However, microscopic sections through these two menisci revealed unequivocal fibrocartilaginous separations within the menisci that did not extend through the surface (Fig. 4.20) (64). Presumably these intrasubstance tears would not have been evident via arthrography or on visual inspection at arthroscopy. Smillie emphasized that such "closed" lesions are common in the early symptomatic horizontal cleavage abnormality and that "it should be recalled that in the earliest stages the horizontal cleavage tear is a 'closed' lesion; there is no outward evidence of injury" (46). He also emphasized that these lesions cannot be detected by contrast arthrography (46). Such closed lesions may be one cause of "false-positive" MRI findings. However, these intrasubstance tears may be detected at arthroscopy and arthrotomy by probing the meniscus, by recognizing loss of mechanical integrity at meniscectomy, or upon saucerization of a discoid meniscus (Fig. 4.16) (75–79). Because these menisci are not rheologically normal, loading forces can cause buckling, with tugging of peripheral structures leading to pain. Also, an intrasubstance tear may be a primary cause of pain, as it is now believed that the meniscus is an innervated structure (37,75,76).

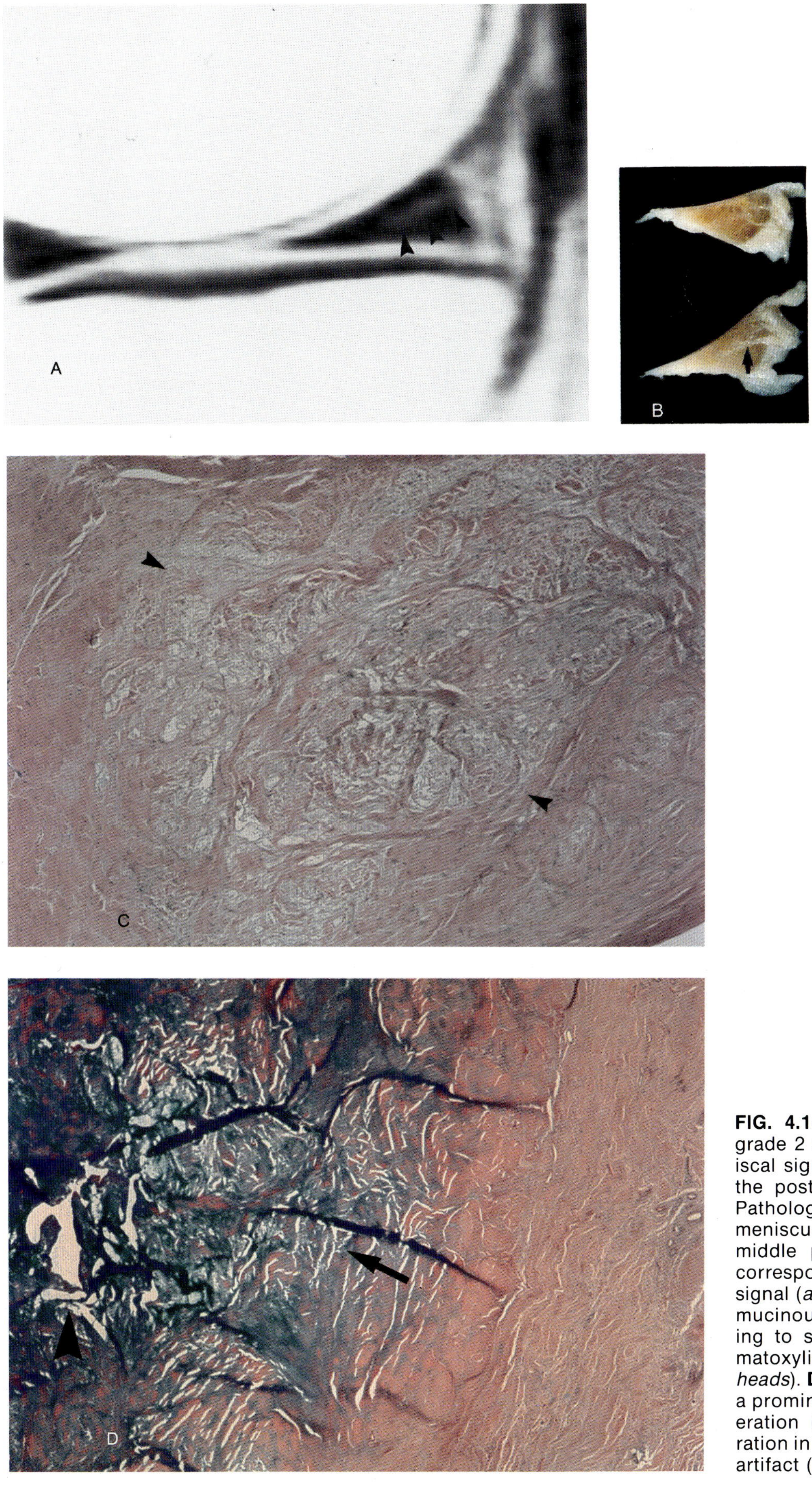

FIG. 4.14. Pathologic correlation of grade 2 signal. **A:** Grade 2 intrameniscal signal (TR/TE 800/20) is seen in the posterior horn (*arrowheads*). **B:** Pathologic cross-section of the same meniscus shows a pale region in the middle perforating collagen bundle corresponding to the intrameniscal signal (*arrow*). **C:** Region of extensive mucinous degeneration corresponding to stage 2 degeneration on hematoxylin-and-eosin staining (*arrowheads*). **D:** An alcian blue stain reveals a prominent area of mucinous degeneration (*arrow*). The apparent separation in the meniscus is a preparation artifact (*large arrowhead*).

A

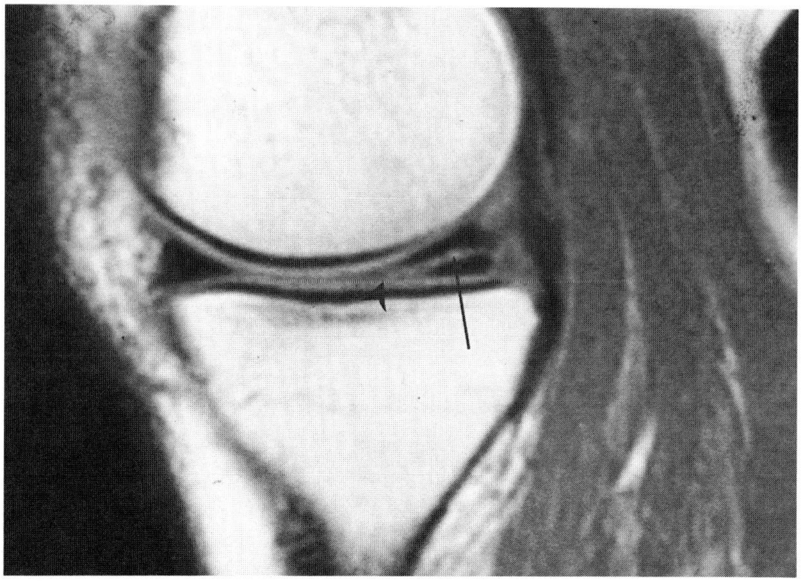

B

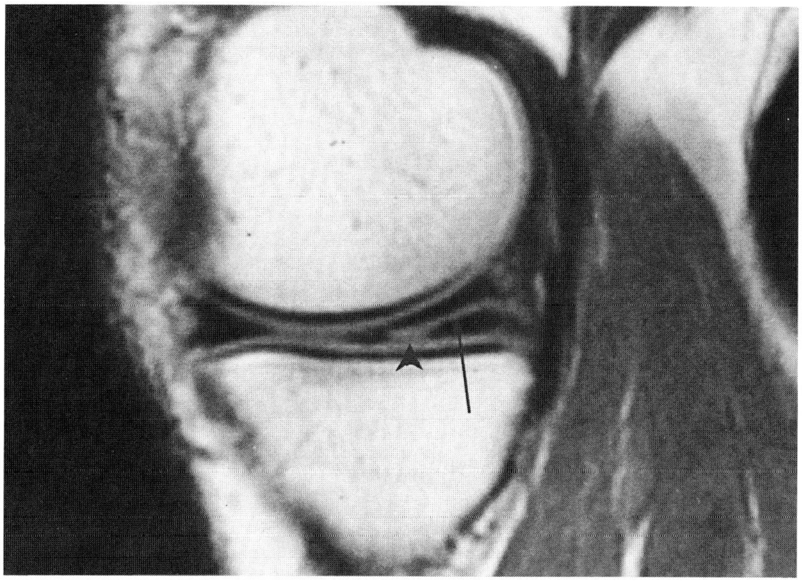

FIG. 4.15. Grade 3 meniscal signal. **A,B:** A and B are adjacent sagittal images (TR/TE 2,000/20) of the medial compartment. Linear signal intensity is seen primarily in the horizontal plane (*arrows*). This is the typical MR appearance the common horizontal cleavage tears of the posterior horn of the medial meniscus. The abnormal signal is primarily located in the central plane of the meniscus, but communicates with the inferior articular surface and free edge at slightly different locations from image to image (*arrowheads*). The surface appearance of this tear is, therefore, oblique in nature.

MRI Appearance of Meniscal Tears

Meniscal tears can be divided into several types, based on the morphology of the tear (18,47,80,81). If one visualizes the triangular cross-section of the meniscus, then tears can be divided into *vertical* tears and *horizontal* tears (Fig. 4.21). The vertical tear is the classic tear in acute traumatic lesions (Fig. 4.22), whereas the horizontal tear is the more common tear in degenerative disease (Figs. 4.23 and 4.24) (33,48, 55,59). Occasionally, both morphologies may coexist in a given meniscus (Fig. 4.24).

Tears can also be described based on circumferential or surface morphology into *longitudinal* tears, in which the tear extends along the meniscus parallel to its margins, and *transverse* or *radial* tears, which extend perpendicular to margins of the meniscus in the radial direction (Fig. 4.25) (81). Tears running at an intermediate angle with respect to the free edge are called *oblique* tears.

A *flap* tear occurs when an attached portion of the meniscus becomes unstable (Fig. 4.26). A flap tear of the posterior horn of the medial meniscus often insinuates into the weightbearing portion of the joint between the femoral condyles and the tibial plateaus, leading to the complaints of intermittent severe pain on weightbearing (Figs. 4.27). Flap tears of the posterior horn of the lateral meniscus often flip superiorly

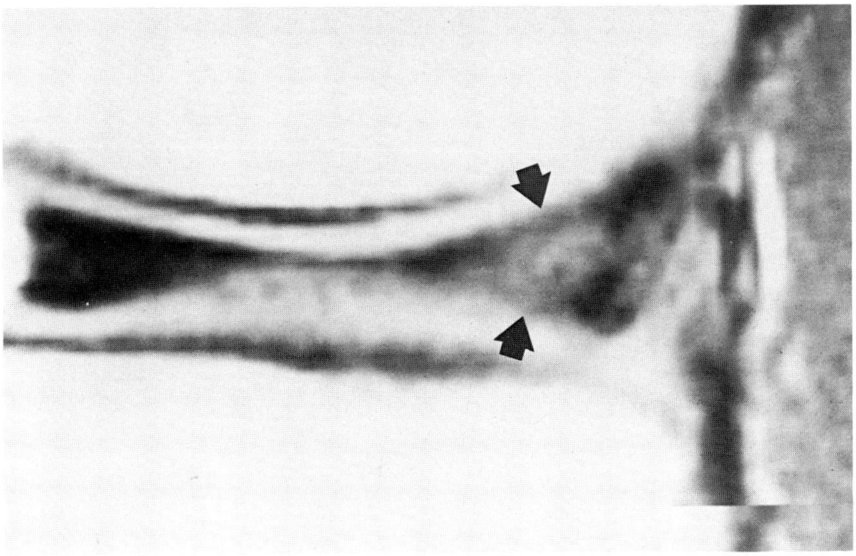

A

B

FIG. 4.16. Grade 3B meniscal signal. **A:** This MR scan was obtained on a high school football player with a 6-week history of severe pain on weightbearing requiring crutches. The MRI revealed a large (discoid) lateral meniscus and ill-defined increased meniscal signal extending to both articular surfaces (TR/TE 2,000/20) (*arrows*). **B:** The signal pattern contained both horizontal and vertical components on the coronal images (TR/TE 2,000/20) (*arrow*). At arthroscopy a discoid meniscus was found, but no tear was visualized or detected with probing. Because of the patients' symptoms and MR findings of grade 3 signal, the meniscus was saucerized and a large mid-substance tear was found. The patient did well after partial meniscectomy. (From refs. 75 and 76.)

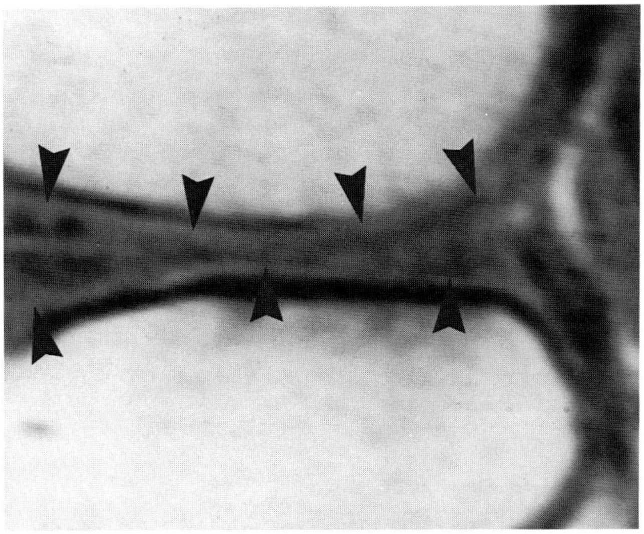

FIG. 4.17. Grade 3B signal can extend throughout the meniscus (*large arrowheads*). This meniscus was described as having the appearance of "shredded wheat" at arthroscopy.

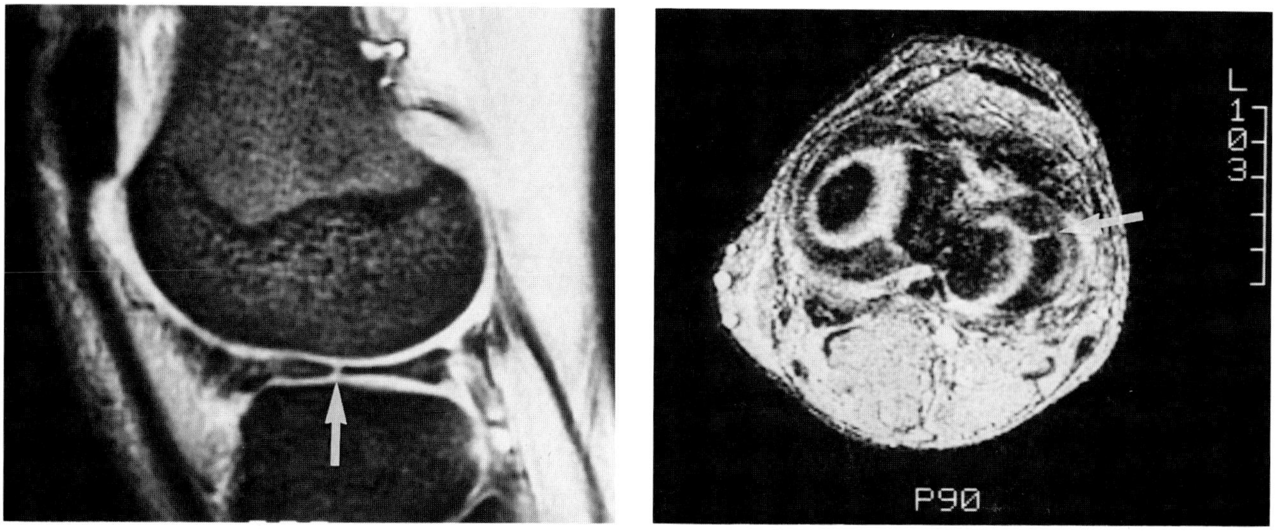

FIG. 4.18. Radial meniscal tear. **A:** Transverse or radial tear (*arrow*) with grade 3 signal intensity involving the anterior horn and body junction of the lateral meniscus on T2*-weighted gradient-echo sagittal image. **B:** An axial display through the meniscus from a 0.7 mm 3DFT (TR/TE 55/15, flip angle 11°) acquisition identifies the radial tear (*arrow*) of the free edge of the lateral meniscus.

on themselves where they also lead to intermittent complaints of severe pain on weightbearing (Fig. 4.28). Flap tears as visualized by these MR findings are usually part of a degenerative horizontal cleavage tear. Complex tears have segments of two or more of these types of tears (Figs. 4.22 and 4.24B).

In addition to location and orientation, tears can be described as being *partial* tears or *full-thickness* tears. In full-thickness tears, the arthroscopist can place a probe through the tear from one side of the meniscus to the other. Tears can also be described as *superficial*, where only the surface is violated. Superficial tears may not be resected or shaved at arthroscopy (3,58). In addition, *intrasubstance* or *closed* tears extend through the substance of the meniscus but do not violate the surface, as described earlier (Figs. 4.16 and 4.20). These tears may be impossible to see at arthrography, arthroscopy, or arthrotomy, but may be symptomatic (75,76).

Because of the newly recognized high prevalence of asymptomatic meniscal tears in some patient populations (see below) and because of the desire to preoperatively evaluate patients which might benefit from primary meniscal repair, increased interest has been placed on MR characterization of meniscal tears. It may be clinically useful to divide meniscal tears into seven common clinically described types (Table 4.3). A more objective system to describe tear types based on surface or cross-sectional morphology of the tear patterns may also be useful (Table 4.4) (81).

The *horizontal cleavage tear* is the most commonly diagnosed meniscal tear. Because it develops within degeneration of the mid-horizontal plane of the meniscus (*vida supra*), MR images typically reveal increased signal horizontally located within the meniscus which extends to an articular surface, usually the inferior surface (Figs. 4.15 and 4.23). These tears may occur anywhere, but most commonly involve the posterior horns, with medial meniscal involvement more frequent than lateral. These tears frequently communicate with the margin of the meniscus at the free edge. As such, the free edge may have the appearance at arthroscopy of the tips of fish scales, and the tear may have an appearance of a fish's mouth, leading to the common designation of *fish-mouth* tear (47). These are

TABLE 4.3. *Types of meniscal tears*

Horizontal cleavage
Parrot-beak
Bucket-handle
Flap
Peripheral
Peripheral meniscocapsular separation
Intrasubstance

TABLE 4.4. *Tear types based on surface and cross-sectional morphology* [81].

Circumferential or surface tear pattern
Longitudinal
Transverse
Oblique
Cross-sectional patterns
Horizontal
Vertical

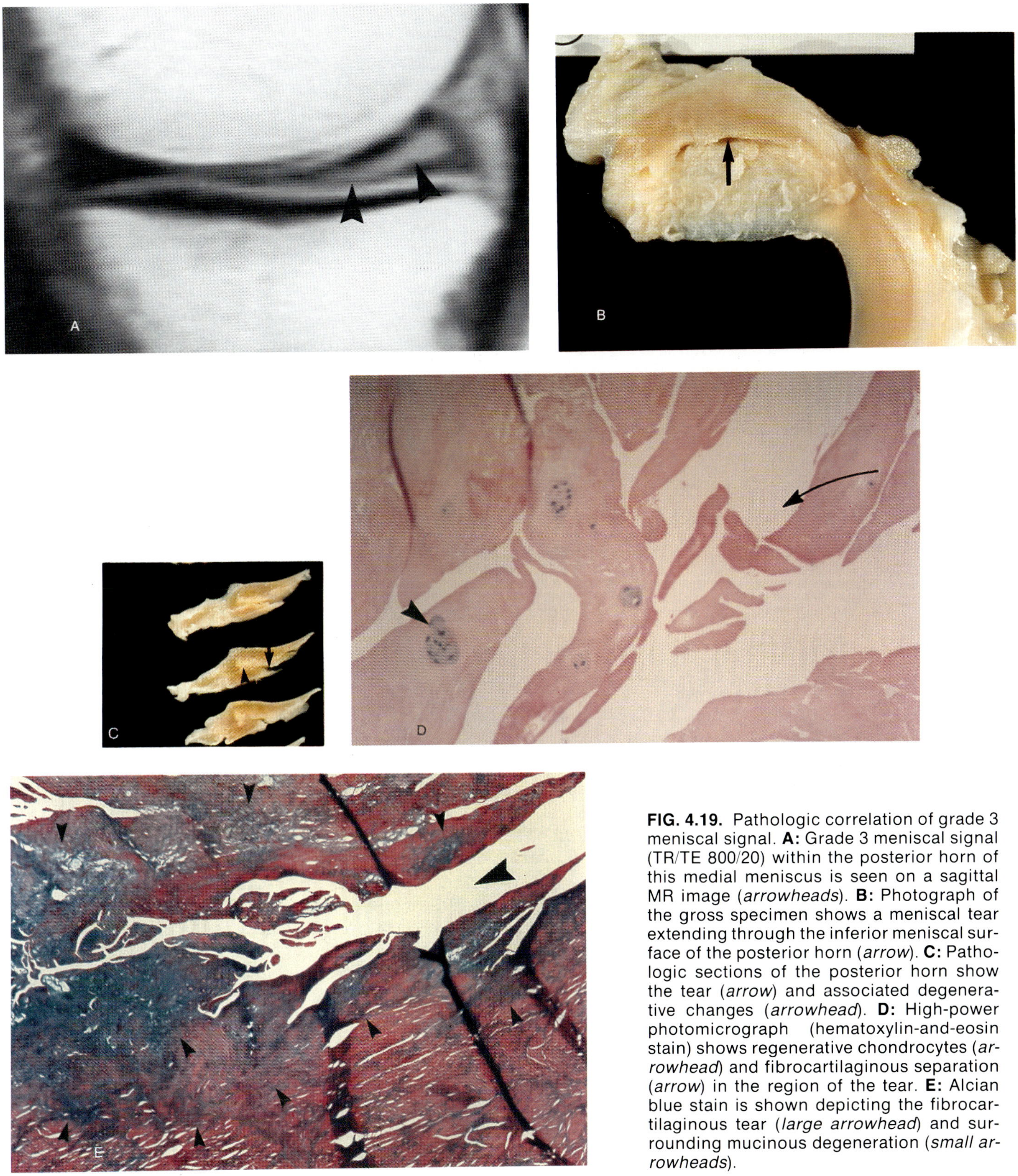

FIG. 4.19. Pathologic correlation of grade 3 meniscal signal. **A:** Grade 3 meniscal signal (TR/TE 800/20) within the posterior horn of this medial meniscus is seen on a sagittal MR image (*arrowheads*). **B:** Photograph of the gross specimen shows a meniscal tear extending through the inferior meniscal surface of the posterior horn (*arrow*). **C:** Pathologic sections of the posterior horn show the tear (*arrow*) and associated degenerative changes (*arrowhead*). **D:** High-power photomicrograph (hematoxylin-and-eosin stain) shows regenerative chondrocytes (*arrowhead*) and fibrocartilaginous separation (*arrow*) in the region of the tear. **E:** Alcian blue stain is shown depicting the fibrocartilaginous tear (*large arrowhead*) and surrounding mucinous degeneration (*small arrowheads*).

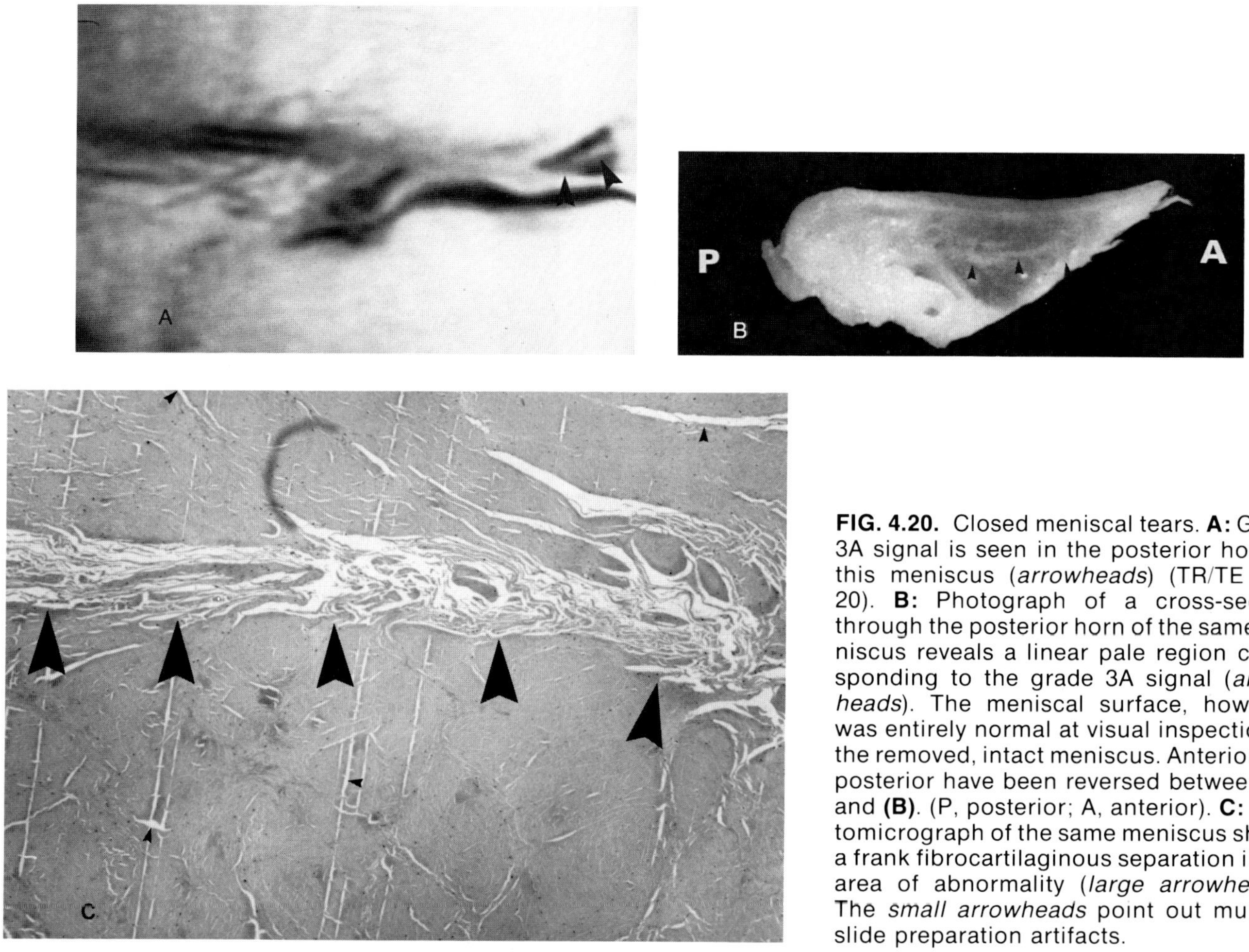

FIG. 4.20. Closed meniscal tears. **A:** Grade 3A signal is seen in the posterior horn of this meniscus (*arrowheads*) (TR/TE 800/20). **B:** Photograph of a cross-section through the posterior horn of the same meniscus reveals a linear pale region corresponding to the grade 3A signal (*arrowheads*). The meniscal surface, however, was entirely normal at visual inspection of the removed, intact meniscus. Anterior and posterior have been reversed between **(A)** and **(B)**. (P, posterior; A, anterior). **C:** Photomicrograph of the same meniscus shows a frank fibrocartilaginous separation in the area of abnormality (*large arrowheads*). The *small arrowheads* point out multiple slide preparation artifacts.

the most commonly visualized asymptomatic tears in the older population (60). In attempting to correlate MR and arthroscopic findings, it must be recognized that a primary horizontal tear pattern based on the cross-section or the sagittal imaging plane may often be oblique at its surface presentation.

Special notice should be taken of small areas of grade 3 signal localized to the tip of the free edge. These are associated with visual fraying of the free edge at arthroscopy and fibrillation of the margin pathologically (Fig. 4.29). It may be difficult or impossible to differentiate between fraying of the free edge and small parrot-beak tears. A *parrot-beak tear* has both

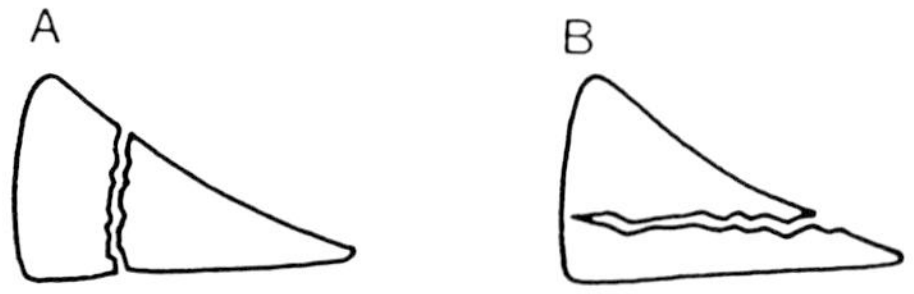

FIG. 4.21. Vertical and horizontal tears. On cross-section images, vertical **(A)** and horizontal **(B)** tears have characteristic orientations. (From ref. 47.)

vertical and horizontal components in an oblique plane at the free edge of the meniscus body (Fig. 4.30) (33,47). If a frank defect is visualized by MR at the free edge of the meniscus body, then the terms transverse or radial tears may more accurately describe the tear morphology (Fig. 4.18). These tears can be small and therefore somewhat subtle in their MRI appearance. They typically occur at the junction of the posterior horn and body of the lateral meniscus and may be seen on only one contiguous 4 mm MR image. High-resolution coronal images through the meniscus may allow better evaluation of the location and extent of these tears than sagittal images alone (Fig. 4.30). They can easily go undetected if not specifically sought on sagittal images and confirmed on coronal images. The clinical significance of these degenerative lesions is controversial (*vida infra*).

A *bucket-handle tear* is a vertical longitudinal tear in which the inner fragment is displaced toward the intercondylar notch of the knee (Figs. 4.8, and 4.31) (82–87). This is the classic type of tear secondary to acute trauma in the young athlete and usually occurs in the medial meniscus. The name derives from the

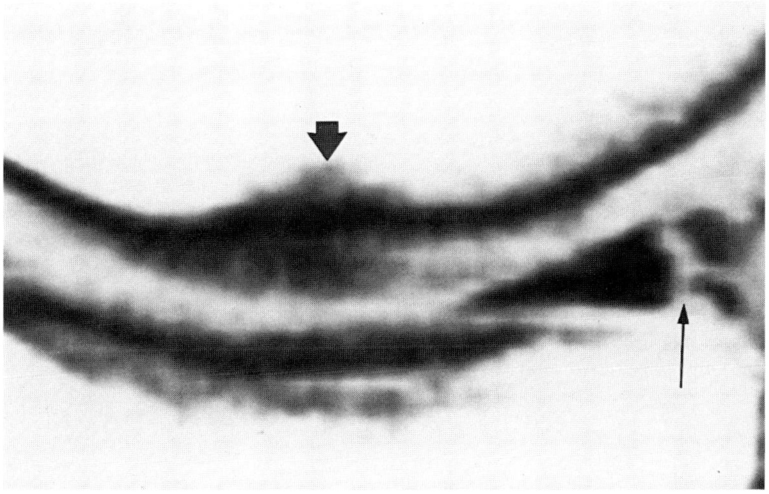
A

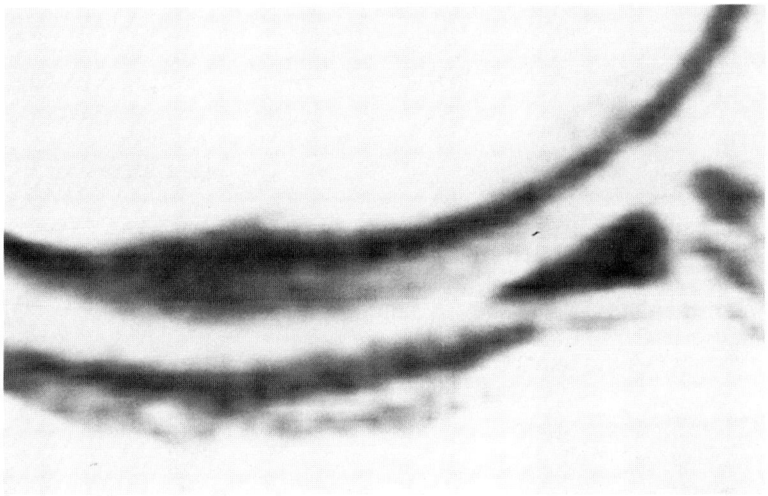
B

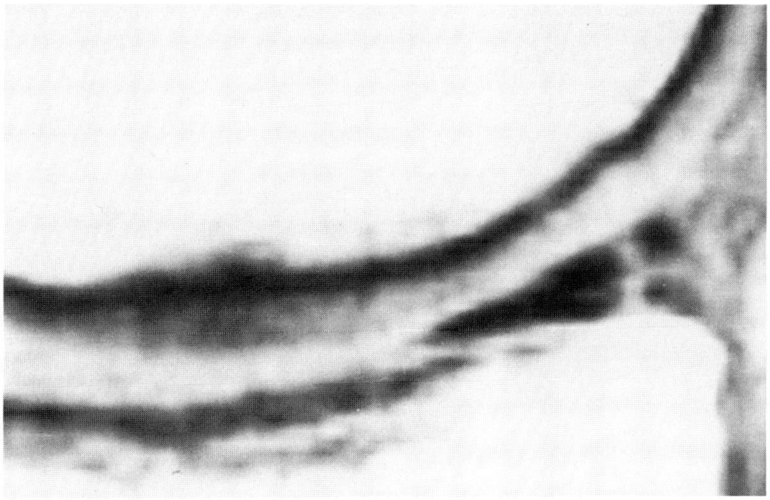
C

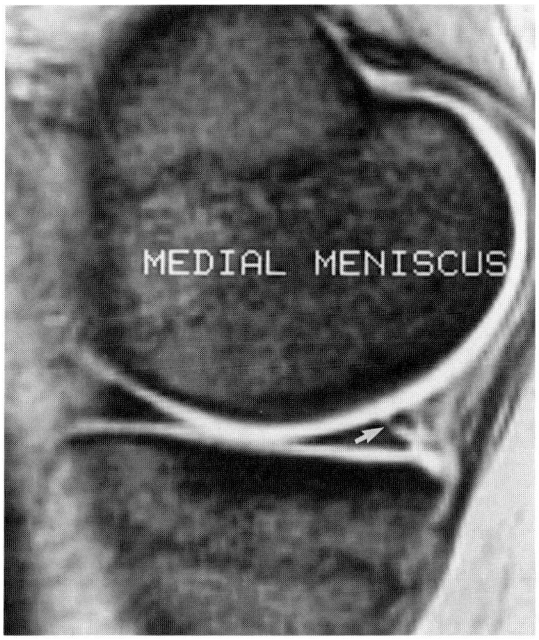

FIG. 4.22. Vertical tear. **A:** This patient suffered an acute sports-related injury to his knee. The MR (TR/TE 2,000/20) revealed a complex, mostly vertical tear of the posterior horn of the medical meniscus (*long arrow*). An associated osteochondral injury was also noted (*fat arrow*). **B:** Ten days later the patient underwent a second MR scan after arthroscopic primary meniscal repair (TR/TE 2,000/20). The signal abnormality is slightly more prominent, but the patient was minimally symptomatic after surgery. **C:** Three months later the patient was asymptomatic, but the grade 3 meniscal signal persisted (TR/TE 2,000/20). This may be an example of a meniscal tear healing with hydrated cartilage which maintains high signal intensity on short TE MR images after healing (150, 151). **D:** A primarily vertical tear in a different patient is seen on this T2*-weighted gradient-echo image.

A 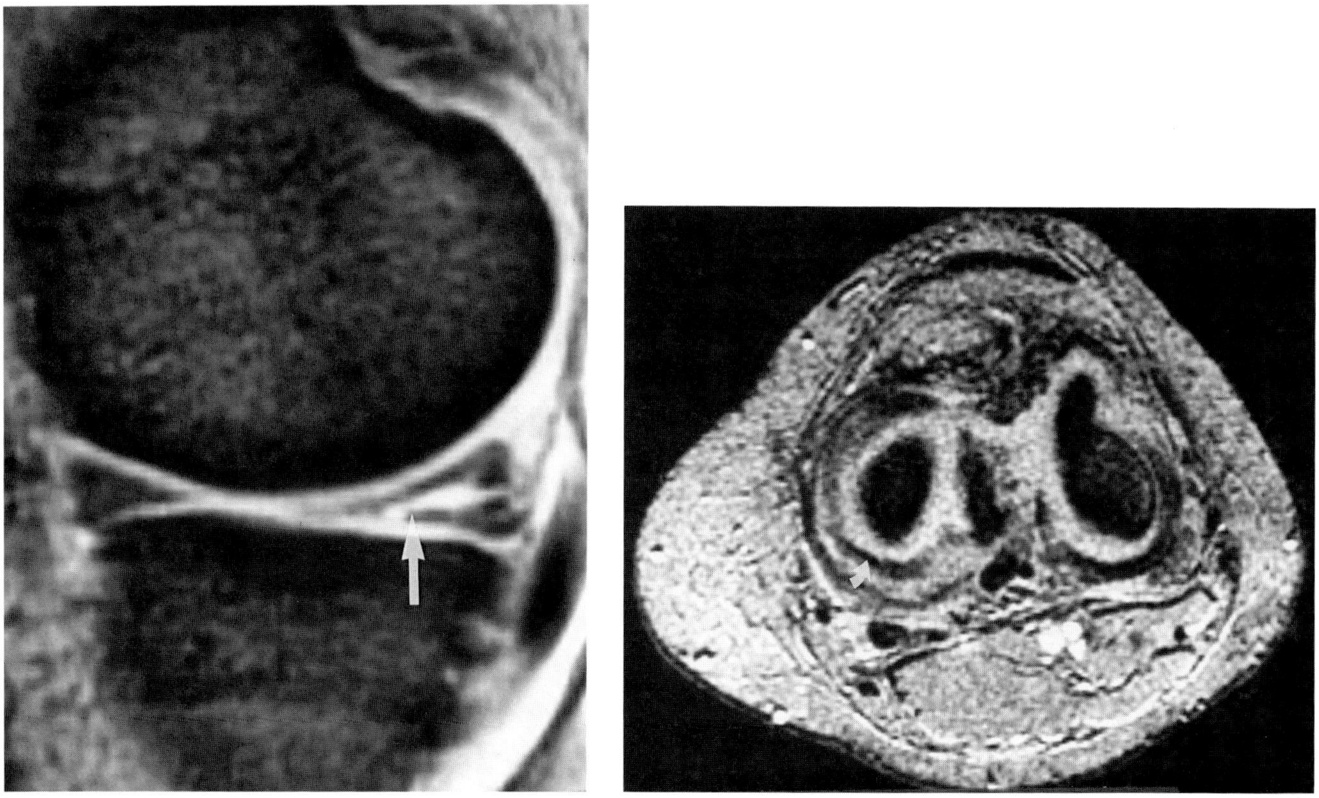 B

FIG. 4.23. Oblique horizontal tear. **A:** The sagittal gradient-echo image shows a tear (*arrow*) in the horizontal plane of the posterior horn of the medial meniscus. This is a typical degenerative-type tear. Without visualizing a change in the direction of grade 3 signal intensity from the inferior to the superior surface of the meniscus, it is difficult to differentiate an oblique from a longitudinal surface tear pattern without a corresponding 3DFT volume series displayed in the axial plane. **B:** The corresponding 3DFT axial MR image shows the tear's morphology in this plane (*arrow*).

A
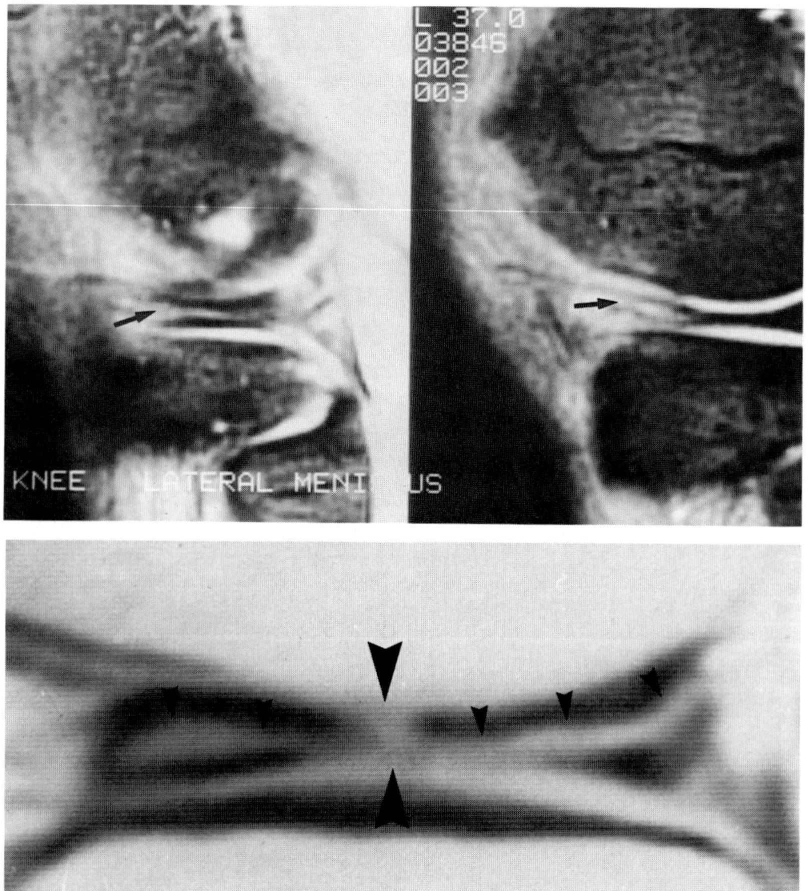

B

FIG. 4.24. Horizontal tears. **A:** The horizontal orientation of a tear is depicted on these gradient-echo T2*-weighted images. **B:** Vertical and horizontal tears. MRI image (TR/TE 800/20) showing horizontal tears of the anterior and posterior horns (*small arrowheads*) and a radial, parrot-beak vertical tear (*large arrowheads*) in the body of this lateral meniscus.

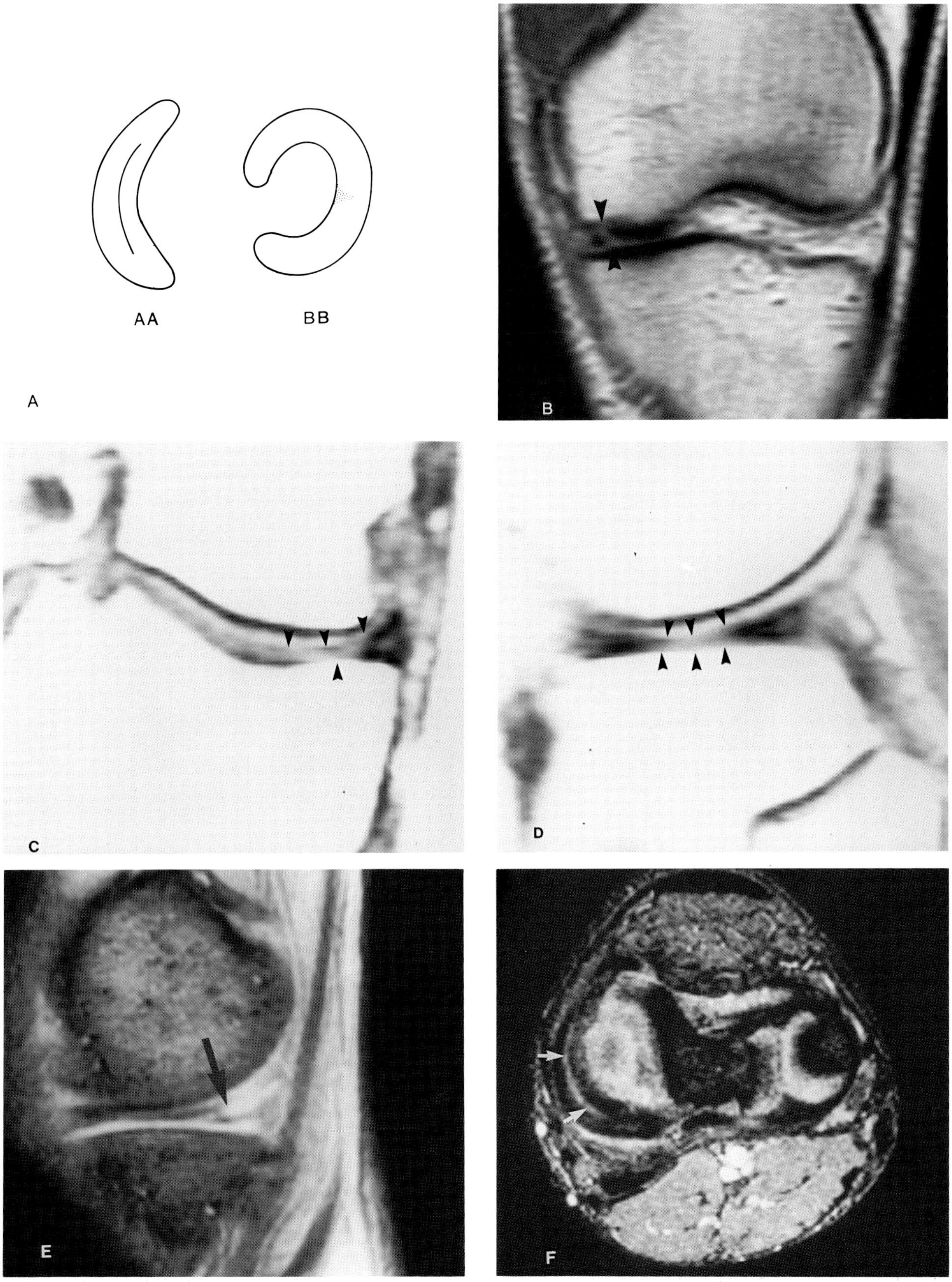

FIG. 4.25. A: Schematic diagram demonstrates the course of a longitudinal tear of the medial meniscus (AA) and transverse tear of the lateral meniscus (BB). **B:** Coronal MR image (TR = 800 msec., TE = 20 msec) reveals a complete vertical longitudinal tear of the meniscus (bucket-handle tear) (*arrowheads*) (same meniscus as in Fig. 4.31). Coronal (**C**) and sagittal (**D**) MR images showing a radial, parrot-beak tear of the lateral meniscus (*arrowheads*). On sagittal images, radial tears cause blunting of the free edge of the meniscus. This may be the only finding of a radial tear on sagittal MR images. **E:** Horizontal cleavage tears usually present with longitudinal extensions to an articular surface as shown in this sagittal gradient-echo image (*arrow*). **F:** The longitudinal tear pattern may be better displayed on axial thin (0.7 mm) 3DFT T2*-weighted displays (*arrows*). This may better display the tear as it might be viewed by the arthroscopist.

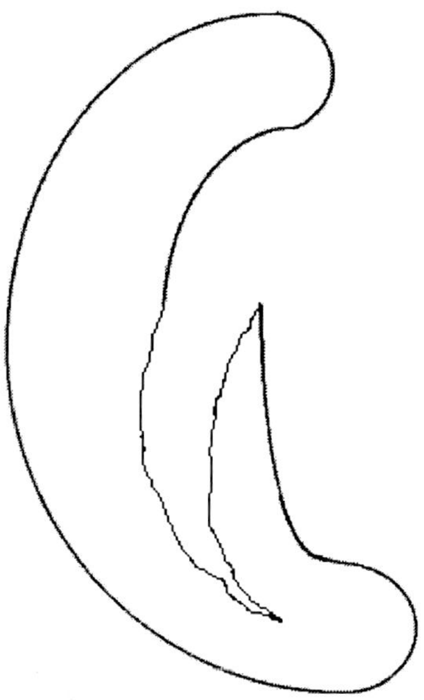

FIG. 4.26. Oblique flap tear. A flap tear is one where a portion of the meniscus becomes unstable in the presence of a meniscal tear. When the tear is oblique in orientation, then an oblique flap tear results. The unstable portion may insinuate into the weightbearing portion of the joint between the femoral condyle and tibial plateau (common on the medial side) or flip upward upon itself (common on the lateral side). Either way the meniscal instability leads to intermittent severe pain on weightbearing.

appearance of the tear, in which the inner displaced fragment resembles the handle and the larger portion of the periphery of the meniscus has the bucket appearance (Fig. 4.31A) (47). This particular type of tear can present with rather subtle findings on MRI, as it can at arthrography (82–84,86–91). The peripheral portion of the meniscus can remodel and have the appearance of a truncated triangle or small meniscus. The normal bow-tie appearance of the body of the meniscus may be attenuated or absent in bucket-handle tears. Patients with bucket-handle tears often present with mechanical symptoms including knee locking. If the meniscal components are black without evidence of extensive degeneration, the patient may benefit from primary repair of the meniscus (*vida infra*). Bucket-handle tears may also be associated with more complex tear patterns with both vertical and horizontal tear planes.

Flap tears are oblique tears which create an attached fragment which is unstable, often causing mechanical symptoms because the fragment occasionally gets trapped between the femoral condyle and tibia (Figs. 4.26) (*vida supra*). Flap tears usually involve the posterior horns where they often present with characteristic abnormal morphology. On the medial side the normal triangular appearance of the posterior horn is lacking. Instead the flap extends into the joint, abnormally prolonging the free edge (Fig. 4.27). Laterally the unstable fragment more commonly flips upward, producing a typical buckled morphology of the posterior horn (Fig. 4.28). These tears usually are associated with pain and apprehension. If large and nondegenerated they may be treated with repair, but in our experience most are treated with partial menisectomy, especially when associated with a more extensive horizontal tear component (Fig. 4.27).

Peripheral tears are nondisplaced tears involving the peripheral one-third of the meniscus (Fig. 4.22). These vertical-peripheral tears occur in the vascularized portion of the meniscus and are the classic indication for primary meniscal repair.

Peripheral meniscocapsular separations are tears in the coronary ligaments which are the capsular attachments of the meniscal rims to the tibia and femur. Typically, the capsular margin of the posterior horn of the medial meniscus will extend within 3 mm of the peripheral edge of the tibial articular cartilage. If this is not the case, then one must consider the possibility of a displaced meniscal capsular separation (Fig. 4.32). However, in practice we have found this sign to be insensitive, nonspecific, and of little diagnostic value. The lateral meniscus, because of increased mobility, bears a less constant relationship to the tibial plateau; however, with marked meniscal displacement, a meniscocapsular separation can be confidently diagnosed (Fig. 4.33). Less sensitive but more specific is the "corner tear" sign. Occasionally a component of the peripheral meniscocapsular separation will extend into an adjacent corner of the peripheral edge of the meniscus. This can occur in both the medial (Fig. 4.34) or the lateral (Fig. 4.35) menisci. Meniscocapsular separation of the medial meniscus can present with fluid extravasating deep to the medial collateral ligament (see "Meniscal Cysts," this chapter). At arthroscopy the meniscal component of the tear is rarely visualized, but the meniscocapsular separation is readily detected. Initially it was suspected that hemorrhage might accompany an acute peripheral-capsular separation and localized high signal adjacent to the meniscal edge on T2-weighted images might be a marker for separation. This is occasionally seen, but is neither sensitive nor specific for meniscocapsular separations in our experience. The use of T2-weighted images, however, permits evaluation of the capsular attachments of the posterior horn of the lateral meniscus, the fascicles (Figs. 4.6). Tears of the fascicles can be visualized, especially on the T2-weighted images (Fig. 4.36).

Intrasubstance tears are cleavages within the fibrocartilage of the meniscus which do not communicate with an articular surface of the meniscus (Fig. 4.16)

A

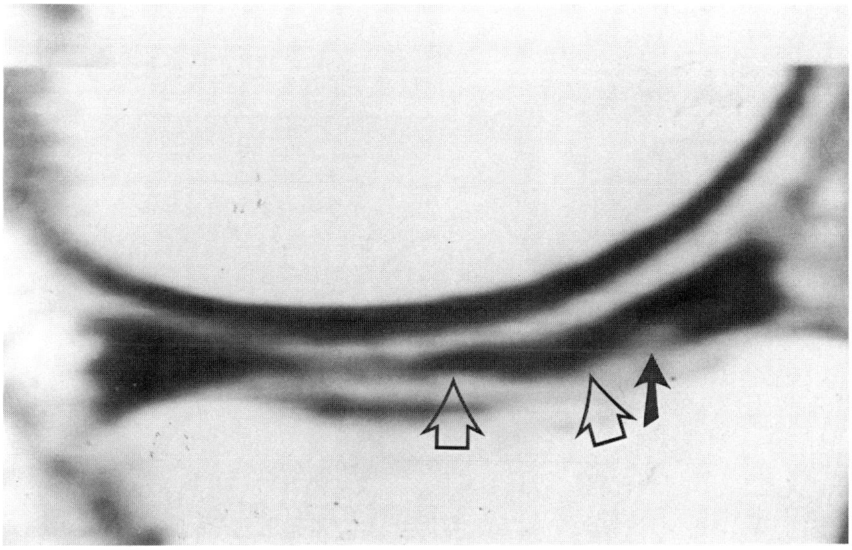

B

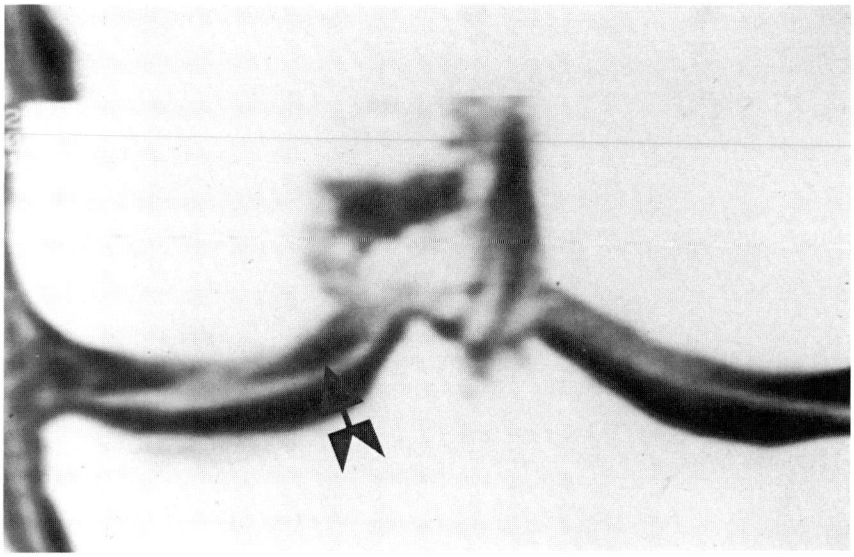

FIG. 4.27. Medial flap tear. **A:** The sagittal images (TR/TE 2,000/20) reveal an abnormally elongated, wavy appearance to the free edge of the posterior horn (*open arrows*). This configuration is compatible with an attached, unstable meniscal fragment. Grade 3 signal is also seen extending to the inferior articular surface (*arrow*). **B:** The coronal images (TR/TE 600/20) better illustrate the displaced position of the attached, unstable meniscal fragment which has insinuated itself between the weight-bearing portions of the femoral condyle and tibial plateau (*arrow*).

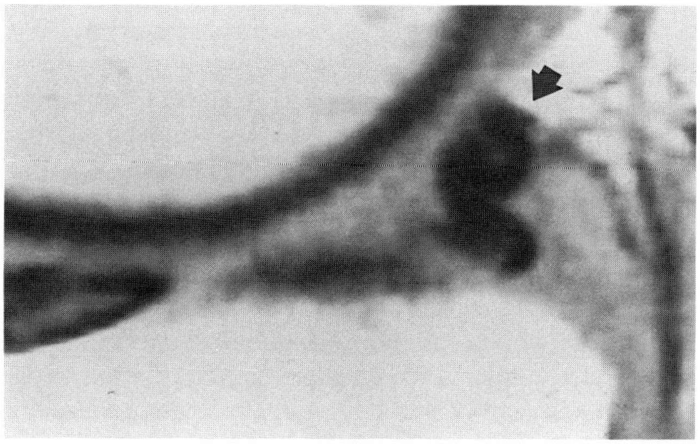

FIG. 4.28. Flap tear of the posterior horn of the lateral meniscus. Flap tears of the posterior horn of the lateral meniscus often present with an attached unstable fragment which flips superiorly on itself. The MR presentation is bizarre, but characteristic (*arrow*).

(*vida supra*). These were initially described as "closed" tears by Smillie who believed that they represented the first symptomatic stage in the development of degenerative type tears (46,59). Symptoms probably result from both secondary changes of the rheologic properties of the meniscus with buckling of the meniscus with stress, and tugging on parameniscal nerves and irritation of intrameniscal nerves (37,46). Pathologic studies have confirmed the existence of intrasubstance tears and correlated them with grade 3 MR signal (Fig. 4.20) (13,64). Symptomatic intrasubstance tears have also been described clinically and correlated with grade 3 meniscal signal (Fig. 4.16) (75,76). Though early meniscal tears may present with grade 2 signal, we are not aware of published surgical or pathological documentation of this association. As described above, grade 2 signal has been correlated with intrasubstance degeneration. We do not believe

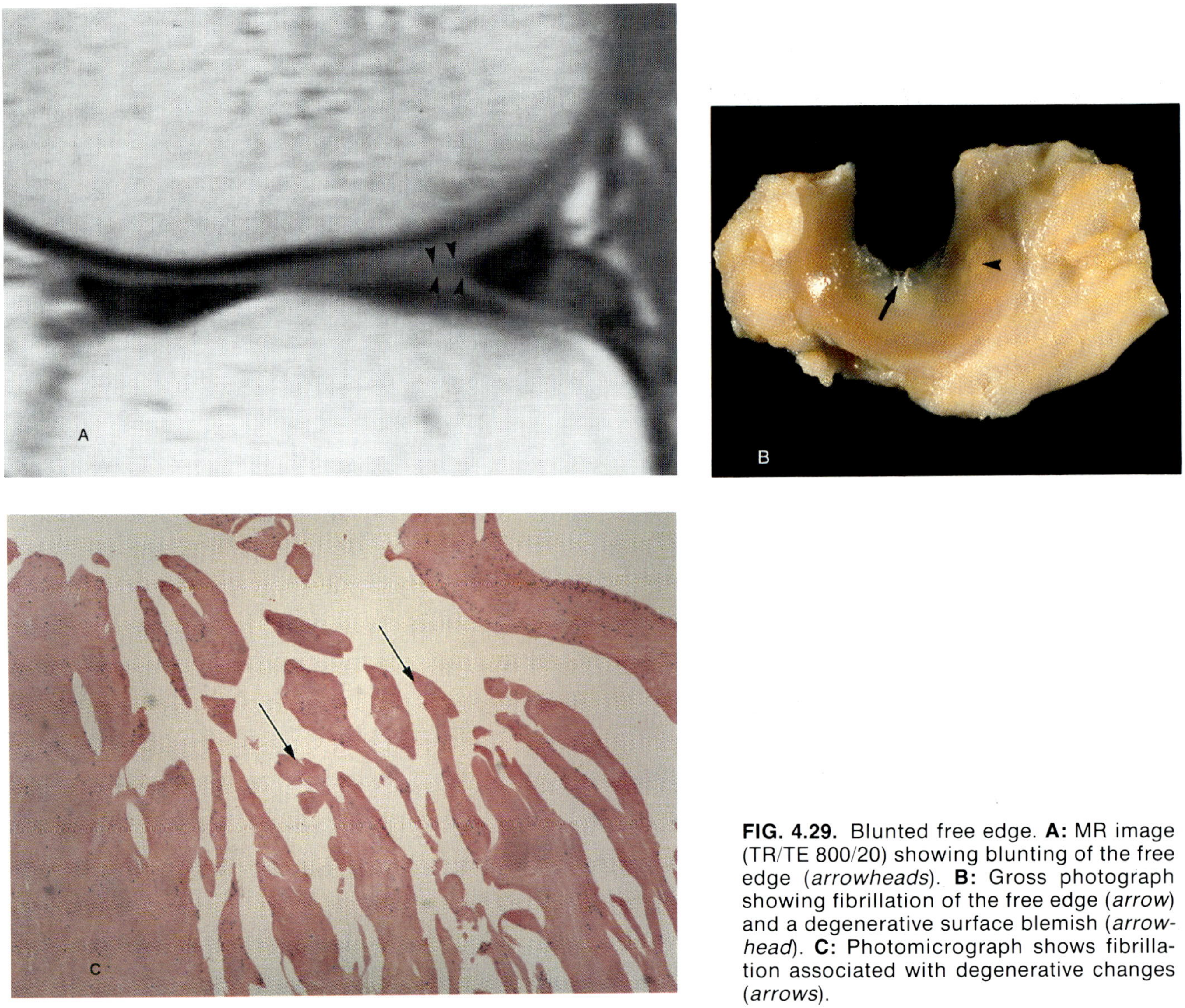

FIG. 4.29. Blunted free edge. **A:** MR image (TR/TE 800/20) showing blunting of the free edge (*arrowheads*). **B:** Gross photograph showing fibrillation of the free edge (*arrow*) and a degenerative surface blemish (*arrowhead*). **C:** Photomicrograph shows fibrillation associated with degenerative changes (*arrows*).

that grade 2 signal is often associated with intrasubstance tears. Grade 3 signal intensity that weakens as it approaches the articular surface is frequently associated with closed or intrasubstance tears (Fig. 4.20). *Only grade 3 signal appears to be highly associated with both intrasubstance and completed meniscal tears.*

Meniscal morphology must also be considered in evaluating meniscal tears. Occasional bucket-handle tears will present with foreshortening of the anteroposterior diameter of the posterior horn of the meniscus and a steeper-than-normal slope of the superior articular margin. There may be no signal from within the remaining rim of the meniscus (Fig. 4.31). This occurs because the cleavage plane within the meniscus may extend obliquely with little associated degeneration within the meniscus. The central fragment then displaces into the notch, leaving a small meniscal rim with normal signal characteristics (82,83). Intrasubstance tears of the posterior horn may present with a decrease in height, so that it is equal to or less than that of the anterior horn. In the absence of a bucket-handle tear, however, menisci of decreased height typically present with abnormal intrameniscal signal (Fig. 4.37). The sharpness of the free edge tip must be evaluated. A blunted free edge can be seen as the only finding of a meniscal tear. An abrupt indentation of the meniscal contour can occasionally be seen at the point where a meniscal tear extends to the surface (92,93). This has been referred to as the "notch sign." Congenital hypoplasia is rare, but should be considered along with juvenile rheumatoid arthritis and bucket-handle tears when small menisci are visualized (94,95). Other congenital anomalies may be seen, but are rare (96).

Clinical Evaluation of Meniscal Tears

Meniscal tears are common causes of knee pain and disability in both young athletes and older individuals. The clinical diagnosis of meniscal disease is based on the history and physical examination. Though knowledge of the mechanism of injury can help determine the location, type, and extent of tear, it generally cannot be extracted reliably and often is not helpful in clinical decision making (18,46).

Initial symptoms generally are not reliable in helping to differentiate between a meniscal tear and an injury to ligamentous, bony, or capsular structures (97). However, knowledge of the rapidity of accumulation of joint fluid is helpful, because injuries limited to menisci rarely are associated with an acute massive effusion. One of the most useful physical signs is acute knee locking, with the patient unable to fully extend the knee because of resistance against a "rubbery substance." This finding is relatively specific for a bucket-handle tear that has extended into the anterior horn, with anterior displacement of the central fragment (Fig. 4.8) (46), but occasionally is can be mimicked by anterior cruciate tears (49).

Various physical signs have been described to help in determining the presence and location of meniscal tears (18). Several are dependent on tenderness at the meniscal margin of a torn meniscus. These include the second Steinmann sign and the Bragard sign. Pain can

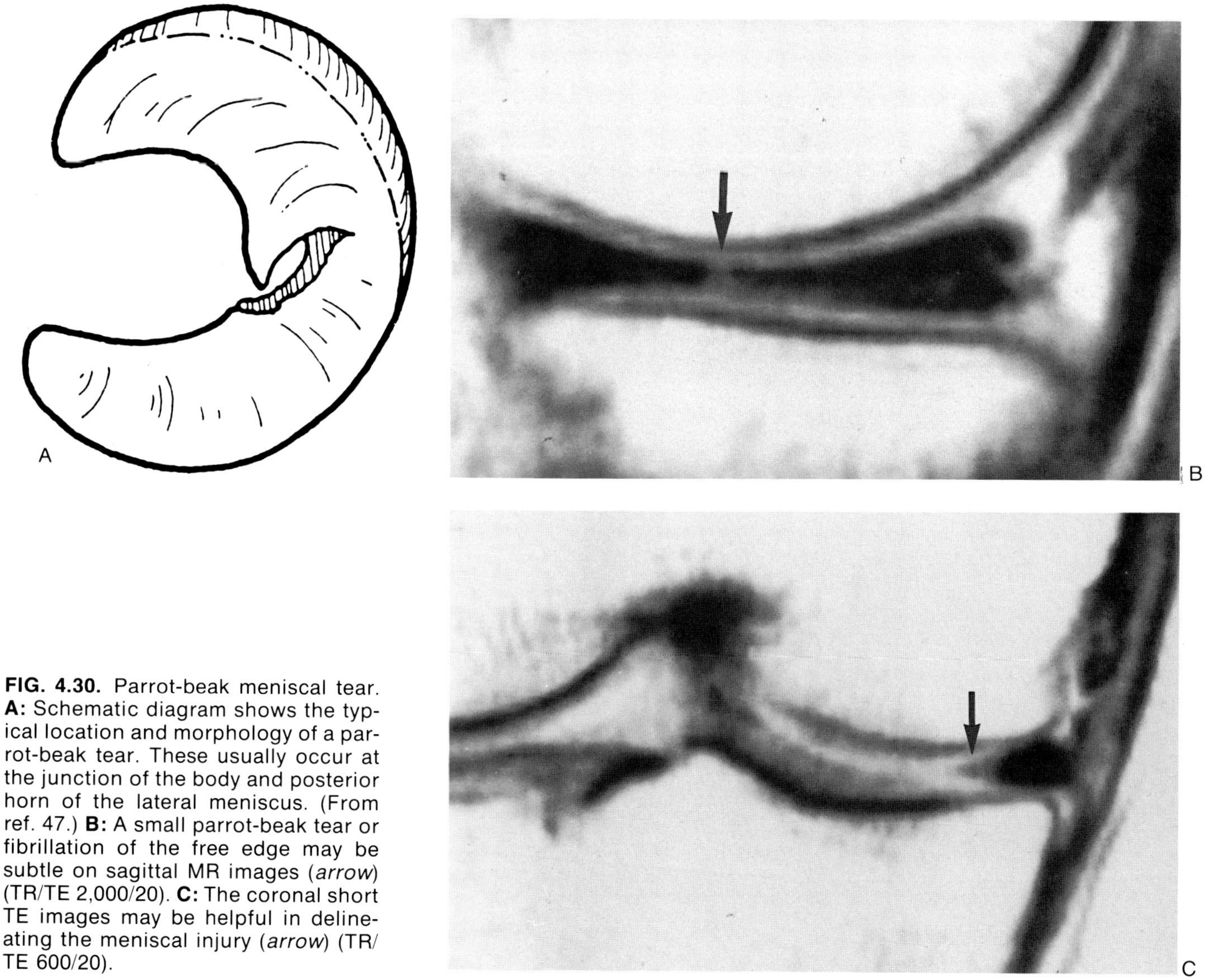

FIG. 4.30. Parrot-beak meniscal tear. **A:** Schematic diagram shows the typical location and morphology of a parrot-beak tear. These usually occur at the junction of the body and posterior horn of the lateral meniscus. (From ref. 47.) **B:** A small parrot-beak tear or fibrillation of the free edge may be subtle on sagittal MR images (*arrow*) (TR/TE 2,000/20). **C:** The coronal short TE images may be helpful in delineating the meniscal injury (*arrow*) (TR/TE 600/20).

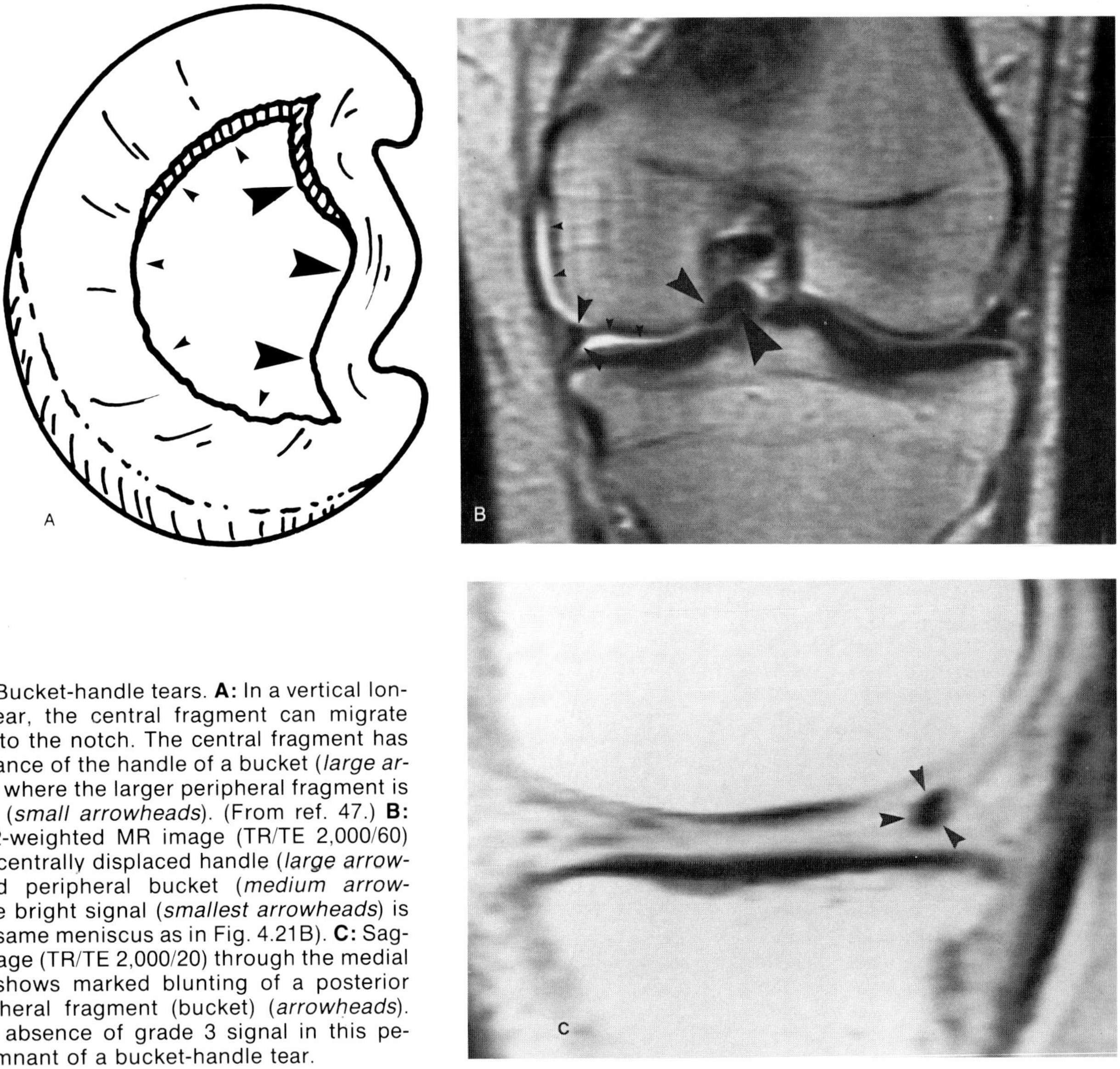

FIG. 4.31. Bucket-handle tears. **A:** In a vertical longitudinal tear, the central fragment can migrate centrally into the notch. The central fragment has the appearance of the handle of a bucket (*large arrowheads*), where the larger peripheral fragment is the bucket (*small arrowheads*). (From ref. 47.) **B:** Coronal T2-weighted MR image (TR/TE 2,000/60) shows the centrally displaced handle (*large arrowheads*) and peripheral bucket (*medium arrowheads*). The bright signal (*smallest arrowheads*) is joint fluid (same meniscus as in Fig. 4.21B). **C:** Sagittal MR image (TR/TE 2,000/20) through the medial meniscus shows marked blunting of a posterior horn peripheral fragment (bucket) (*arrowheads*). Notice the absence of grade 3 signal in this peripheral remnant of a bucket-handle tear.

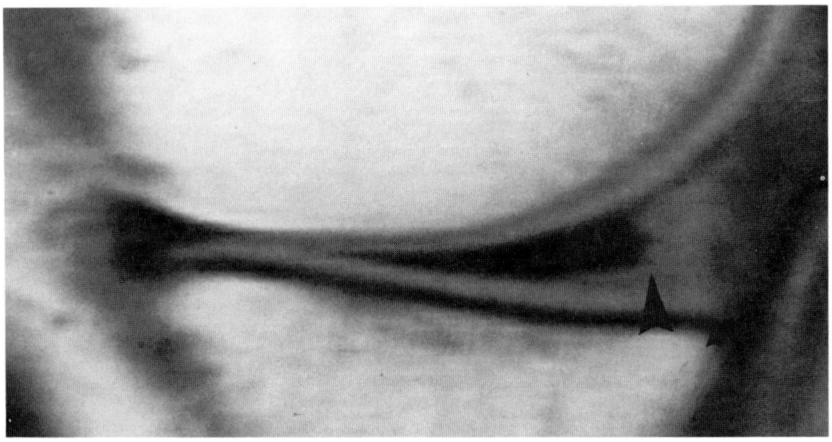

FIG. 4.32. Menisco-capsular separation. The meniscus should almost completely cover the tibial articular cartilage (TR/TE 800/20). If the periphery of the meniscus (*large arrowheads*) does not cover the periphery of the tibial articular cartilage (*small arrowhead*), then a meniscocapsular separation could be suspected.

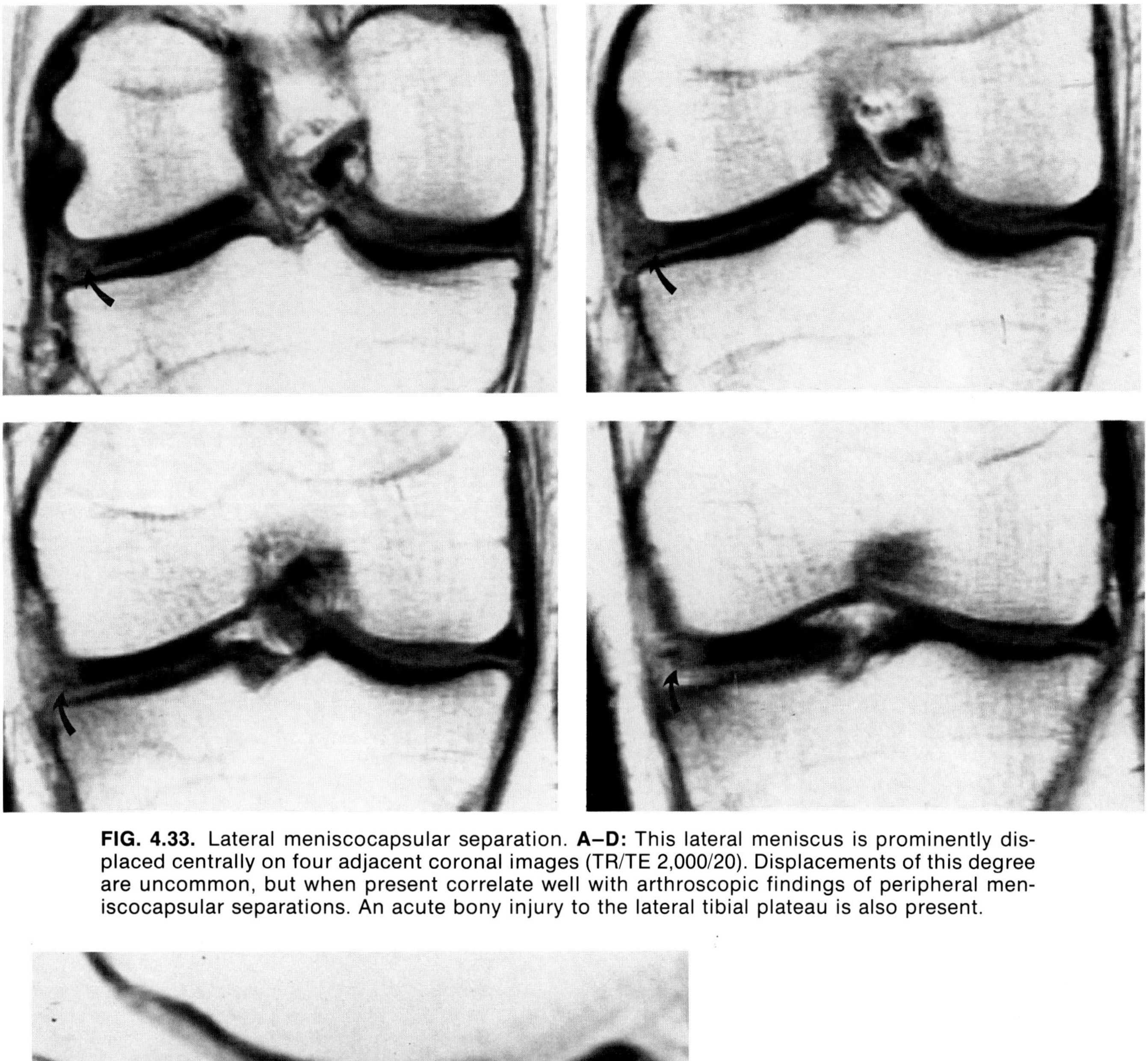

A B C D

FIG. 4.33. Lateral meniscocapsular separation. **A–D:** This lateral meniscus is prominently displaced centrally on four adjacent coronal images (TR/TE 2,000/20). Displacements of this degree are uncommon, but when present correlate well with arthroscopic findings of peripheral meniscocapsular separations. An acute bony injury to the lateral tibial plateau is also present.

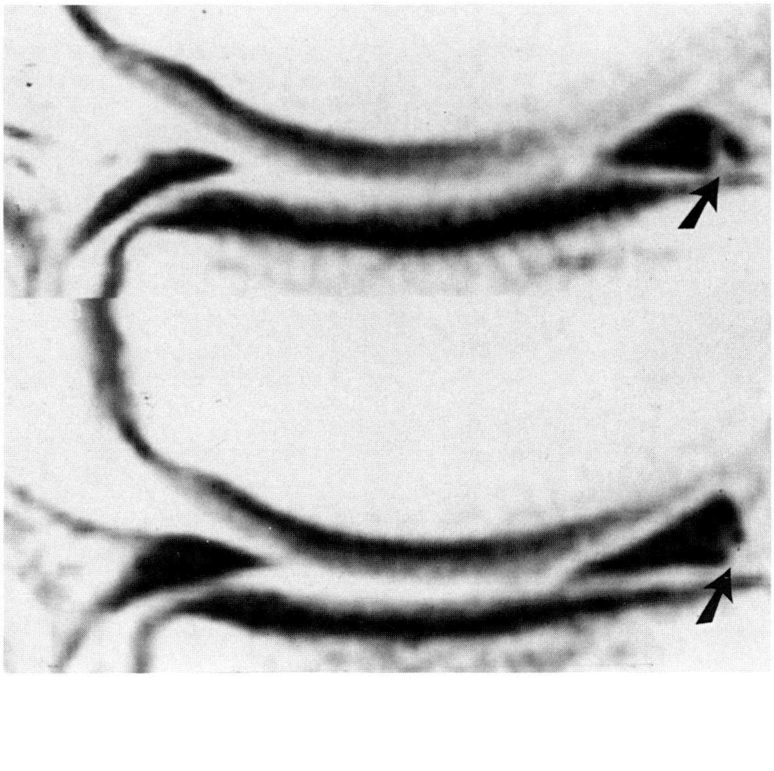

FIG. 4.34. Peripheral meniscocapsular separation. A "corner tear" can be seen at the inferior capsular margin of the posterior horn of the medial meniscus (*arrows*). A peripheral meniscocapsular separation was found at arthroscopy. This sign signifies extension of the capsular attachment tear into a corner of the meniscus, and is a specific but not sensitive sign of a peripheral menisco-capsular separation.

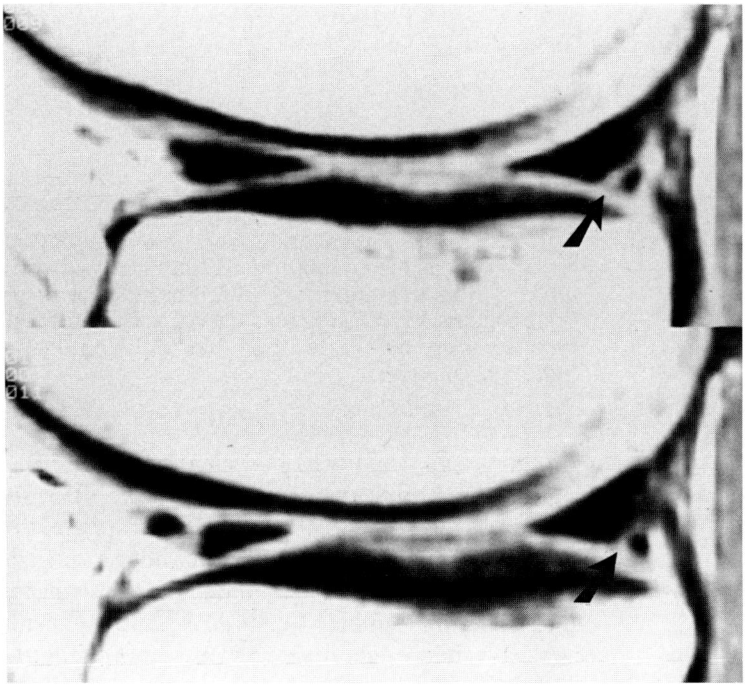

FIG. 4.35. Peripheral meniscocapsular separation. A "corner tear" can be seen at the inferior capsular margin of the posterior horn of the lateral meniscus (*arrows*). As on the medial side, this is a specific but insensitive sign of a peripheral meniscocapsular separation.

also be produced with compression of the torn region of the menisci between the tibial plateau and femoral condyle. This is the basis of the Bohler, Payr, first Steinmann, and Merke signs. The McMurray test is not dependent on production of pain. Instead, the abnormal deformability or shape of a torn meniscus is used to produce an audible or palpable snap as the meniscus moves over bony prominences. A positive meniscal sign may be useful in increasing the examiner's confidence that a meniscal tear is present; however, a negative sign does not exclude a tear.

The accuracy of clinical diagnosis of meniscal le-

A

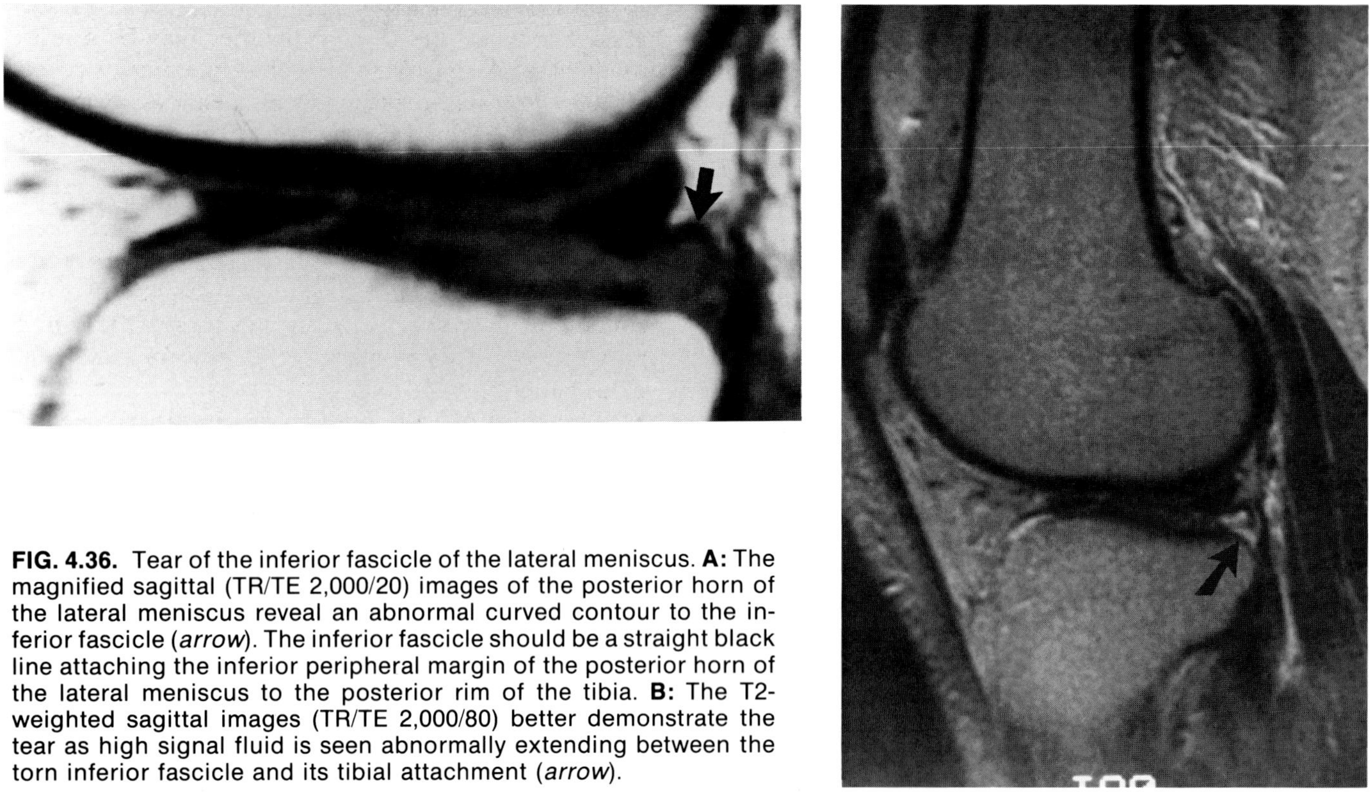

B

FIG. 4.36. Tear of the inferior fascicle of the lateral meniscus. **A:** The magnified sagittal (TR/TE 2,000/20) images of the posterior horn of the lateral meniscus reveal an abnormal curved contour to the inferior fascicle (*arrow*). The inferior fascicle should be a straight black line attaching the inferior peripheral margin of the posterior horn of the lateral meniscus to the posterior rim of the tibia. **B:** The T2-weighted sagittal images (TR/TE 2,000/80) better demonstrate the tear as high signal fluid is seen abnormally extending between the torn inferior fascicle and its tibial attachment (*arrow*).

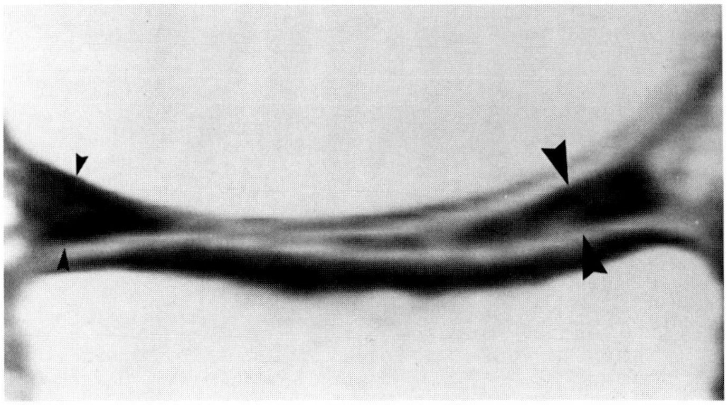

FIG. 4.37. Meniscal collapse. A complex tear of the posterior horn of the medial meniscus is shown with decreased meniscal height (TR/TE 800/20). The height of the posterior horn (*large arrowheads*) of the medial meniscus should be greater than that of the anterior horn (*small arrowheads*).

sions depends markedly on the experience, interest, and expectations of the examiner. It is therefore not surprising that reported diagnostic accuracy rates for clinical examination vary markedly in the literature. Levinsohn and Baker stated that clinical examination accurately diagnosed medial meniscal abnormalities in 91.4% of cases, but only 56.4% laterally (98). This discrepancy in clinical accuracy between the medial and lateral menisci confirms earlier findings by Smillie (46). However, Grenier and du Tremblay (99) found clinical assessment to be 97% accurate in evaluating the lateral meniscus. As meniscal surgery becomes more sophisticated in an attempt to preserve as much meniscus as possible, mere knowledge of the presence of a tear may be insufficient information for optimal surgical planning. The extent, type, and location of the tear, as well as the integrity of the surrounding meniscus, may be important information in planning partial meniscectomies or meniscal repair procedures (*vida supra*) (58,100). Therefore, in practice, clinical examination is essential in the initial evaluation of knee pain. However, further diagnostic studies should be considered important adjuncts if symptoms suggest the need for invasive therapeutic measures (101).

A further complication to the clinical evaluation of meniscal tears is the growing evidence that many, if not most, meniscal tears may be asymptomatic. At least four types of information suggest the common presence of minimally symptomatic or asymptomatic meniscal tears (Table 4.5). First, the initial report of a surgical procedure as treatment for meniscal tears occurred in the mid-1800s (102,103). It was not until a century later that Smillie first described the most common tear of all, the horizontal cleavage tear (59). This length of time may in part be due to the difficulty in evaluating tears at arthrotomy, but it suggests that many horizontal cleavage tears may not be sufficiently symptomatic to have warranted surgery during the century before Smillie. Second, at autopsy meniscal tears have been documented in patients without premorbid clinical complaints of knee pain (64). Third, many patients undergoing arthroscopy and partial meniscectomy for meniscal tears become asymptomatic soon after surgery. Another recognized cohort complained of persistent symptoms resolving in 4 to 6 weeks. We have followed patients with acute sports-related trauma and meniscal injuries who have refused surgery but have become asymptomatic 4 to 6 weeks after the initial event. The symptoms resolved with healing of x-ray occult bony injuries in the face of persistent meniscal disease (Fig. 4.38) (75,76). This suggests that x-ray occult bony injuries may be a more frequent cause of symptoms than previously recognized, even in the presence of meniscal tears (see Chapter 8). Finally, MR scans on large series of asymptomatic individuals have shown a surprisingly high prevalence of grade 3 meniscal abnormalities in people without symptoms, especially when the subjects were over 60 years of age (60) (David O. Davis, *personal communication*). Both the autopsy studies and the MR studies in asymptomatic individuals suggest that a high prevalence of asymptomatic meniscal tears exist in the older patient population.

TABLE 4.5. *Findings suggestive of a high prevalence of asymptomatic meniscal tears*

Delay in recognition of common horizontal cleavage tears
Autopsy findings
Delay in symptom resolution after arthroscopic partial meniscectomies in patients with bony injuries
MR findings in asymptomatic patients

The ability of MRI both to detect other knee pathology which frequently masquerades clinically as meniscal tears (see Chapter 8) and to characterize tears is essential in optimizing treatment protocols in many patients. Neither arthrography nor arthroscopy can sensitively detect x-ray occult bony injuries (104). Optimal treatment usually requires an accurate diagnosis. Since a high percentage of traumatized knees have symptomatic bony injuries, not all meniscal tears are symptomatic, and partial meniscectomy is indicated only for relief of mechanical symptoms and/or pain, then some patients with bony injuries should be fol-

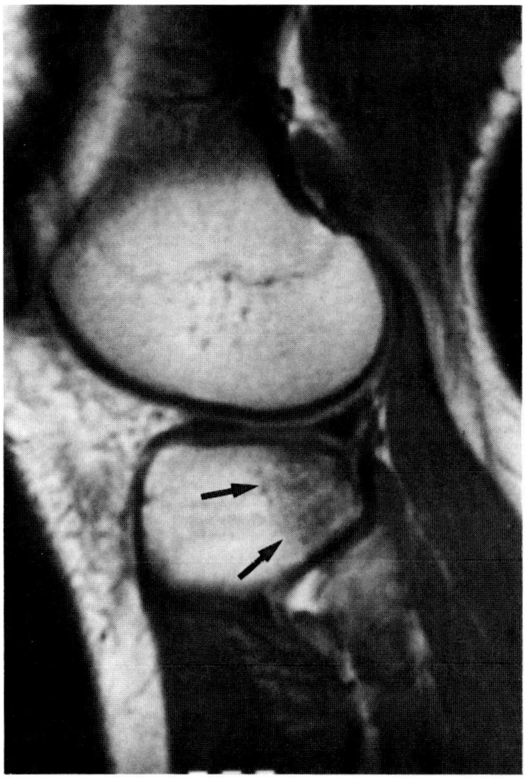

A

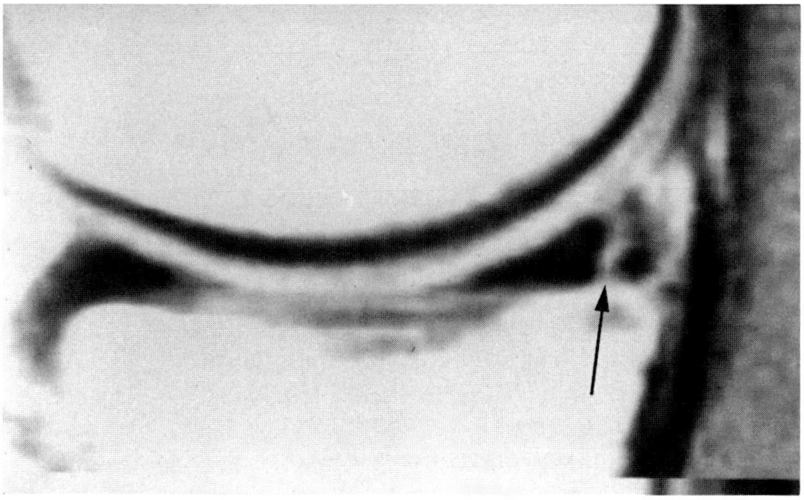

B

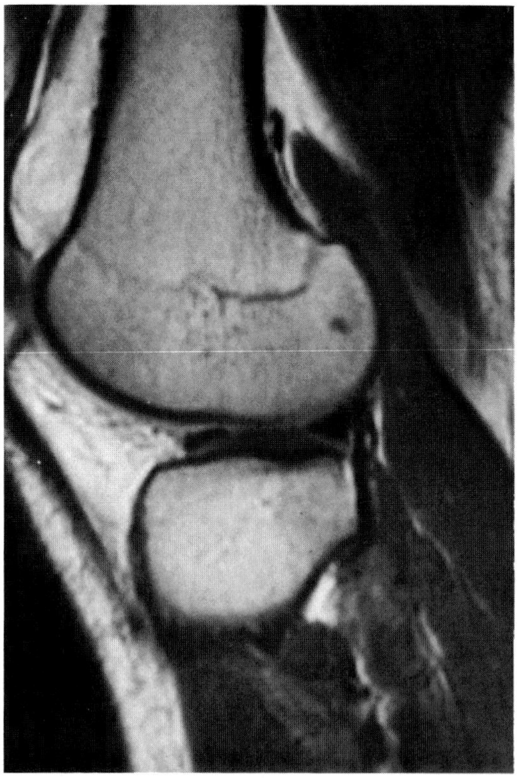

C

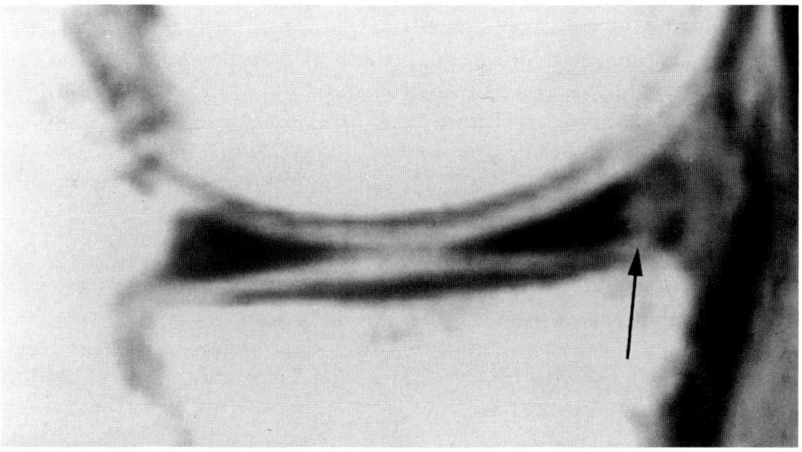

D

FIG. 4.38. Resolution of knee pain with bony injury healing. **A:** The intermediately weighted sagittal image (TR/TE 2,000/20) reveals abnormal decreased signal within the medullary portion of the posterior aspect of the lateral tibial plateau (*arrows*). This scan was obtained approximately 36 hr after a volleyball injury. **B:** The meniscal windows (TR/TE 2,000/20) reveal the presence of a peripheral, vertical tear of the medial meniscus (*arrow*). **C:** The patient refused arthroscopy and was treated conservatively. Six weeks after the injury the patient returned for follow-up without symptoms. The MRI at this time (TR/TE 2,000/20) showed resolution of the medullary bone signal changes. **D:** The meniscal windows (TR/TE 2,000/20) revealed persistent abnormal grade 3 meniscal signal at the tear site (*arrow*).

lowed conservatively for 4 to 6 weeks until the trabecular injuries heal. They can then be reassessed clinically and if symptoms persist, surgical treatment may be indicated.

Accuracy of Diagnostic Imaging Modalities

Abnormal kinematic motion of the knee joint may be diagnostically useful (105), but imaging studies are most frequently used to supplement the physical examination. Arthrography is an established, accurate technique for detecting meniscal tears [29,30–32,56, 77,78,80,87,89–91,98,103,106–121,127,129,130,261], especially in confusing clinical presentations of knee pain (58,122,123). However, in spite of previous predictions of increasing utilization of arthrography (124), many radiologists did not experience a resurgence in interest in spite of a dramatic growth in the use of arthroscopy. The decline of arthrography probably is due to several factors. Although several reports have stressed the complementary nature of arthrography and diagnostic arthroscopy (78,98,115,125), the overall accuracies of the two techniques in expert hands are similar, and the incremental improvement in diagnostic ability is small when both studies are performed. However, in clinical practice, the accuracy of each technique is dependent on the experience and interest of the examiners (78,98,115,125–128). It is well known that arthroscopists have difficulty visualizing the inferior surfaces of the posterior horns of both menisci, where the majority of tears occur (77,78,98,115,125, 129,130). However, modern arthroscopic techniques have decreased error rates in these locations. The most important cause of decreased arthrographic utilization is the reluctance on the part of primary physicians to subject patients to multiple invasive procedures when arthroscopy can be used for both diagnostic and therapeutic interventions. Consequently, the recent trends have been toward increased use of arthroscopy in spite of the high cost, significant complication rates (131), and transient postoperative disabilities (122,132). Though we are not aware of published large series comparing the accuracies of MRI versus arthrography, MR can detect tears not visualized by arthrography (Fig. 4.39).

Because of the invasive nature of traditional diagnostic techniques in meniscal evaluation, much interest has recently been shown in the uses of noninvasive imaging techniques in evaluating meniscal disease. Ultrasound has been shown to be capable of visualizing the meniscus (133). Using the more advanced technology that has rapidly evolved in ultrasound, Selby et al. were able to evaluate normal and abnormal posterior meniscal horns as well as capsular ligaments and articular cartilage. Unfortunately, the entire meniscus and weightbearing areas of the femoral articular cartilages are difficult to evaluate with this technique (134). Bone scintigraphy has also been studied as a diagnostic technique for meniscal tears and may be comparable to arthrography in terms of sensitivity (135). In a study using single photon emission computed tomography (CT) as well as standard planar bone scintigraphy, nuclear techniques were found to be very sensitive for detecting meniscal tears and were recommended as prearthroscopic screening examinations (136). Though nuclear techniques are sensitive indicators of the presence of meniscal disease, their specificities are less than ideal. It is also unclear if the association of a positive bone scan with meniscal tears

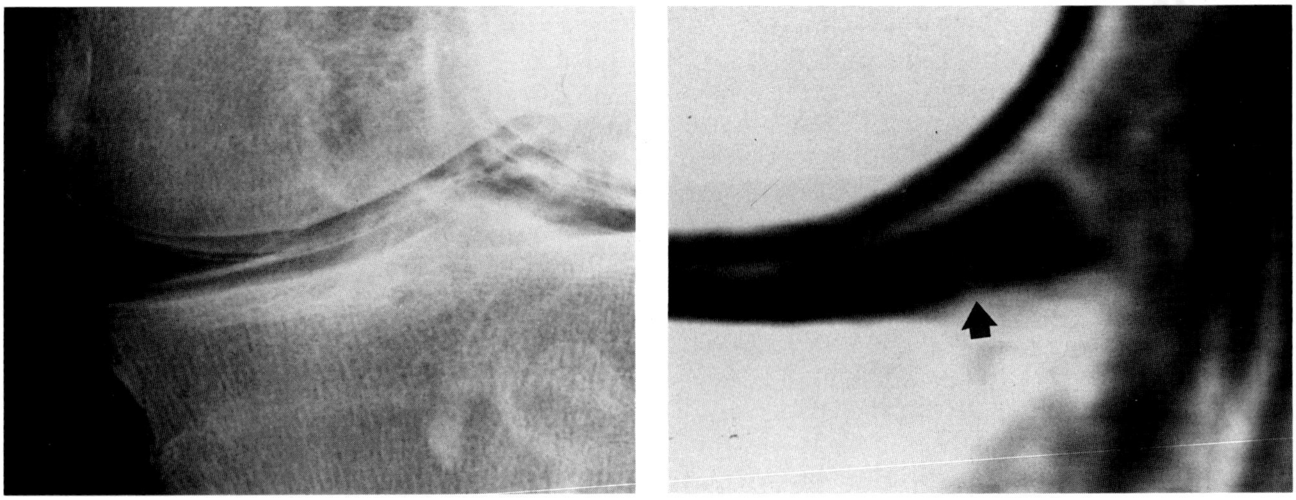

FIG. 4.39. Meniscal tear not visualized at arthrography. **A:** Arthrography was performed by an experienced knee arthrographer and was interpreted as negative for meniscal tear. **B:** Grade 3 signal was seen on MR (TR/TE 2,000/20) extending to the inferior articular surface (*arrow*). A posterior horn tear was seen at arthroscopy.

is due to the scan detecting the meniscal tear or the high association of x-ray occult bony injuries with meniscal tears in traumatized knees. Also, the resolution is poor, and the tear type and extent cannot be determined. Thus, interest in nuclear medicine studies in the evaluation of meniscal tears has waned.

X-ray CT has also been used for evaluating meniscal abnormalities. Passariello et al. (137,138) reported diagnostic accuracies of 89.2% and 96.1% for medial and lateral meniscal lesions, respectively, without the use of intraarticular administration of contrast material. They suggested that if meniscal lesions are clinically suspected, then CT may be an accurate method and less invasive method of evaluating the menisci than arthrography. The accuracy of noncontrast-enhanced CT of the knee for evaluating other causes of knee pain is less clear. CT is also useful for augmenting arthrography, especially for evaluating meniscal cysts and meniscal anatomy in the region of the popliteal tendon sheath (139).

With the advent of MRI, numerous investigators have reported the ability of this technique to detect meniscal tear (1–6, 12,13,24,26–28,60,61,64–73,75,76, 82,83,92,93,140–176). This interest was initially generated by the high contrast resolution of MRI in evaluating soft tissues, thus possibly obviating the need for intraarticular contrast material in evaluating menisci. The findings in early reports were promising (61,156, 157,159,170,177,178), and later reports have substantiated the high accuracy of MRI for evaluating meniscal tears (3,5,60,69,71,73,140,148,150,164,166–169). The major advantages of MRI over arthrography and arthroscopy are its noninvasive nature, its low cost compared with arthroscopy, its lack of complications, and the fact that no recovery period is necessary. The diagnostic accuracies of arthrography (125,127,128) and arthroscopy (126,179,180) are dependent on the skills of the examiner. Using standard techniques and equipment, MRI is able to obtain diagnostic-quality images without variability between examiners as long as adequate protocols are observed. However, as in other areas of medical imaging, interpretation of images is dependent on the experience of the interpreter (170) (J. Fox, *unpublished data*).

MRI is excellent for defining the presence and location of meniscal tears (3,76,140,148,149,164). The sensitivity of MRI for detecting meniscal tears visualized at arthroscopy has been reported to range between 75% and 100% (3,76,148–151,160,164). However, potential difficulties in determining the accuracy of MRI, given possible limitations in the accuracy of arthroscopy, have been discussed by Crues et al. (3) and Reicher et al. (67).

Initial reports commented on the high sensitivity of internal meniscal signal, but the specificity was low (67). In an attempt to improve the specificity, both Reicher et al. (67) and Lotysch et al. (65) defined grading systems for meniscal signal. The latter grading system was compared to histologic staging of meniscal disease by Stoller et al. (64) and Hajek et al. (13), as described earlier. It was prospectively evaluated in 144 surgically proven cases of knee abnormalities (3). This grading system yielded a specificity of 92% with a sensitivity of 89% when compared with arthroscopy. More recent studies have confirmed the high accuracy of this grading system (148,164). The accuracy of MRI is not altered by the presence of an effusion (67).

Reicher et al. (67,68) found that the negative predictive value of MRI was nearly 100%. In their series, 30% of knees arthroscoped for clinically suspected meniscal tears revealed normal menisci. This number might have been higher without preoperative MRI, because surgery often was deferred in patients with negative findings on MRI. Therefore, a negative knee MRI examination is useful in justifying the use of conservative observation when the need for acute surgery is clinically unclear. This is especially important after acute knee trauma because x-ray occult bony injuries are common clinical mimics of meniscal tears (*vida supra* (Fig. 4.41), see Chapter 8).

Intrameniscal tears may be responsible for some of the reported false-positive MRI findings (3,64,75,76). An additional example may be a case reported by Burk et al. in which initial arthroscopy did not report a tear (145). However, subsequent arthroscopy several months later, performed because of continued pain, detected a tear in the location of the previous MRI abnormality. We also have anecdotal reports from surgeons in which areas of grade 3 MR signal looked normal initially at arthroscopy, but subsequent probing directed to the area of grade 3 abnormality on the MRI revealed a tear. Intrasubstance tears may be highly symptomatic (Fig. 4.16) (75,76). As described above, most meniscal tears begin as degeneration in the midplane of the posterior horns which leads to internal cleavages. In normal menisci these probably complete themselves by extending through an articular surface soon after the rheologic properties of the meniscus are changed by this internal tear. This probably happens within months of the initial symptomatic lesion. Symptomatic closed tears appear to be more commonly detected by MRI in discoid lateral menisci (Fig. 4.16) (75,76). This suggests that symptomatic intrasubstance tears may exist for a more extended time period before completing themselves through the surface in abnormally shaped menisci.

MRI may be helpful to the surgeon in locating sites of abnormality for better preoperative planning. This may be especially helpful in evaluating the under surfaces of the posterior horns, which can be difficult to evaluate at routine arthroscopy because of obstruction of vision by the arch of the femoral condyle (68,78,

125,160). In addition, identification of a tear of the peripheral third of the meniscus may be an indication to consider meniscal repair rather than resection (*vide infra*).

There are two situations in which clinical decision making can be particularly difficult in the presence of knee pain. The first concerns knee pain in children, in whom the poor specificity of the clinical history and examination is well recognized (181–184). This is because both the most common pathologic conditions and the modes of presentation of meniscal lesions in children differ from those for adults (15,185,186). These difficulties have been compounded by reports that meniscal lesions in children are detected in only 20% of clinically suspected cases (181), that arthrotomy is unreliable in meniscal diagnosis in children (185), and that children suffer from poor long-term results following meniscectomy (14,187–189). These findings have led some authors to plead for the use of arthroscopy for diagnosis in all children presenting with knee symptoms (184). However, Casscells (190) has stated that because of the low incidence of surgically correctable causes of knee pain in the young, arthroscopy is superfluous in the presence of negative radiographic and physical findings. Though large series are not available to document the accuracy of MR in diagnosing meniscal tears in children, it is likely that grade 2 signal which communicates with the capsular meniscal margin may represent normal vascularity in children under 12 years of age (168,191). Grade 3 signal is probably a reliable indicator of tear (Fig. 4.40). The young person with knee pain may therefore be an important candidate for noninvasive diagnostic evaluation by MRI, possibly obviating the need for surgery in selected cases.

The second difficult clinical situation involves the acutely injured knee. Clinical examination often is impossible without general anesthesia, and arthrography is painful and difficult. Because there is also a high likelihood of surgically correctable lesions in acutely injured knees with hemorrhagic effusion, many authors recommend early arthroscopy for accurate diagnosis and therapy (14,185,192–201). However, if the joint capsule is torn, arthroscopy may be difficult because irrigation fluid may extravasate into tissues, making repair more difficult and occasionally causing neurovascular compression syndromes (194). MRI is an excellent preoperative technique for evaluating the integrity of the menisci and trabecular bone in the acutely injured knee. The examination is rapid and readily accepted by the patient. More importantly, significant traumatic knee injuries not diagnosable by arthrography and arthroscopy are also common in this setting. The most common of these are occult trabecular bone injuries (*vida supra*, see Chapter 8) (75,104, 202–204). We have seen many patients with persistent postoperative knee pain whose postoperative MR has shown trabecular bone injuries not detectable with x-ray or arthroscopic techniques (Fig. 4.41) (104). Other injuries can also be well evaluated by MRI including collateral ligament tears (see Chapter 5) and other extraarticular injuries. As described above, in some patient populations with a high prevalence of asymptomatic meniscal tears, the presence of other symptomatic lesions may affect patient management. Therefore, we believe that MRI is the initial diagnostic study of choice for evaluating acute knee trauma.

Of importance in the interpretation of MR images in athletes is knowledge of common MR findings in asymptomatic athletes. As many as 50% of asymptomatic high level athletes may present with MR findings which could complicate interpretation in the presence

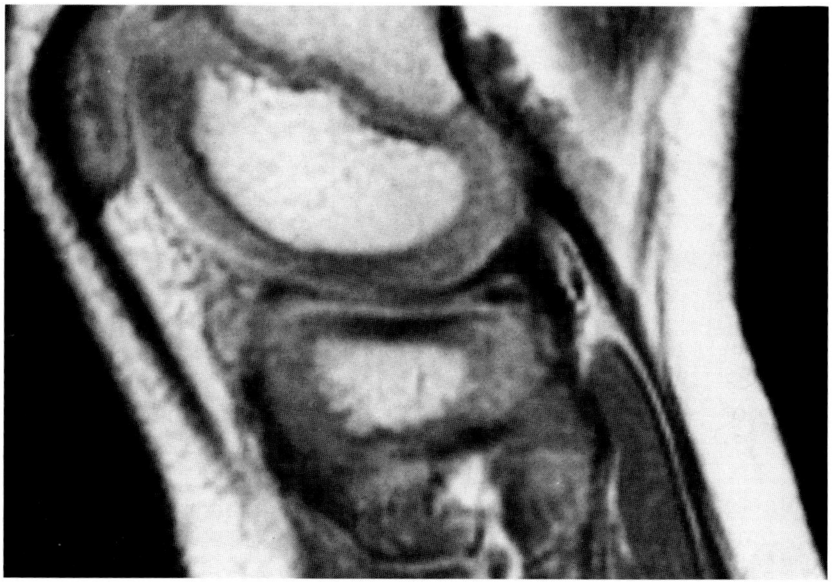

FIG. 4.40. Meniscal tear in a child. Grade 3 signal intensity is visualized in the posterior horn of the lateral meniscus in this child with a meniscal tear (*arrow*).

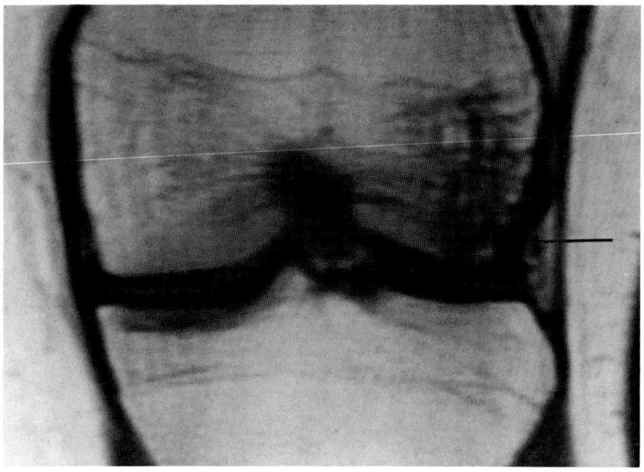

FIG. 4.41. Occult trabecular injury. Bony injury is a common cause of persistent postoperative pain clinically mimicking a meniscal tear. This young woman was thrown from her horse and presented with classic signs and symptoms of a lateral meniscal tear. The MR reveals a normal meniscus, but does display an x-ray occult bony injury to the lateral femoral condyle (*arrow*). Because of the MR findings, arthroscopy was not performed and the patient was followed conservatively.

of acute injury (144). Diffuse meniscal signal changes are occasionally seen in long distance runners and are thought to represent transient edema (158,171,172). Diffuse ill-defined signal can also be seen after acute trauma, and it resolves in 3 to 6 weeks. Such transient signal is probably due to contusion.

Treatment of Meniscal Tear

In recent years, treatment of meniscal tears has undergone considerable evolution and has been the subject of considerable controversy. The controversy has stemmed from three areas of investigation: meniscal function, meniscal repair, and arthroscopic surgery. Much of the impetus toward meniscal repair and partial meniscectomy as substitutes for traditional total meniscectomy is due to improved understanding of meniscal function (15,58,205,206). Whereas tradition held that the semilunar cartilages of the knee were vestigial appendages of little use, more recent investigations have emphasized the importance of the menisci in stabilizing the knee joint, in proper load transmission, in shock absorption, in joint nutrition, and in reducing joint friction (14,16,17,19–22,207–212). Resection of a meniscus has been associated with degenerative changes in the corresponding compartment and impaired function (14,40–44,58,84,182,207, 209,213–221), especially in the presence of anterior cruciate ligament deficiency (57,195,208,212,216, 220,222). Therefore, the present trend in surgical treatment of meniscal tears is to preserve as much meniscus as possible in an attempt to improve outcome. This has lead to the use of partial meniscectomy in lieu of total meniscectomy whenever possible (14,20,43,44,58,84, 85,205,207,217,223–227), especially in the lateral compartment (14,228,229). Because of recent evidence suggesting that small, stable meniscal tears do not predispose to degenerative changes, recommendations have been made not to resect asymptomatic tears when they are small and stable, especially in the degenerated knee and in the presence of anterior cruciate ligament deficiency (53,55,58,206,209). The push toward meniscal preservation has also led to renewed interest in meniscal healing and repair.

Even though Annondale performed the first meniscal repair in the mid-1850s, inconsistent results caused the procedure to fall into disrepute (58,205). Later, King (230) demonstrated that healing requires an intact blood supply. However, traditionally it had been thought that the meniscus was avascular and therefore nonrepairable. Studies have since demonstrated an intact blood supply to the peripheral one-third of the adult meniscus (18,36,37,39,59,231), whereas in children the entire meniscus may be repairable (208). In addition, techniques to augment vascularity in relatively avascular areas show promise (34,208,209, 223,232). It has been stated that up to 90% of meniscal tears may be repairable (208); however, 20% is a more representative figure (14,19,58,205,207,214,224,233–236). The rationale for repair is the belief that the best long-term result will be obtained by returning the meniscus to a shape and function as nearly normal as possible. This may be especially important in children, in whom long-term results of meniscectomies are poor (15,186–189) and meniscal vascularity may be sufficient for most tears to heal without meniscectomy (59,186). The drawbacks of meniscal repair are the technical challenges of the procedure, prolonged rehabilitation, and increased susceptibility to subsequent tears (58,205). We have found that MRI's ability to differentiate between types of tears and to characterize the integrity of residual meniscal components to be helpful in preoperative planning (*vida supra*).

The most potent influence on current trends in the treatment of meniscal disease is the arthroscope. Since Professor Takagi first used a cystoscope to inspect cadaver knees in 1918, many investigators have used and improved arthroscopic techniques (109,110,179,198, 215,237–241). Numerous authors have subsequently emphasized the improved diagnostic accuracy and decreased length of rehabilitation using arthroscopy versus arthrotomy (43,100,126,205,221,225,226,228,242–251). Arthroscopy is, however, a technically demanding subspecialty, with results dependent on the skill of the arthroscopist (126,179,180). This is especially true in therapeutic arthroscopy and arthroscopic meniscal repair (58,205,208,252). The arthroscope, however, has made partial meniscectomy and primary meniscal repair possible. Arthrotomy techniques do not allow proper nondestructive approaches to the injured meniscus to facilitate these meniscal-sparing procedures.

An important area of current interest is the prognosis following meniscal surgery (see Chapter 7). Poor prognosis after meniscal surgery and repair has been shown to correlate with other abnormalities within the knee. These include anterior cruciate ligament deficiency, joint laxity, chondromalacia, degenerative changes, work-related injuries, prior surgery, lateral meniscectomies (40,57,58,99,180,196,205,206,213, 253–257), and the status of the peripheral meniscal rim (252). Horizontal cleavage lesions have a poorer outcome, probably because of the difficulty in determining the true extent of disease at arthroscopy (180). Eastwood (253) also stated that inadequate initial evaluation and treatment of disease lead to a poorer prognosis. Most of these prognostic indicators can be well evaluated by MRI. Some of them, including the true extent of disease and status of the peripheral rim, may be better evaluated by MRI than by other existing techniques, including arthroscopy. Internal degenerative changes may only be nondestructively evaluated by MRI. Clearly, the applications of MRI in directing surgery and predicting prognosis are interesting, but further clinical studies are needed to clarify its role.

MRI OF POSTOPERATIVE MENISCI

The postoperative knee is more completely discussed in Chapter 7. Partial meniscectomy is now a common operative procedure. Asymptomatic postoperative menisci following partial meniscectomy present with altered morphology and an array of meniscal signals (148,150,151,173). Approximately one-third of menisci following meniscal repair show resolution of grade 3 signal (Fig. 4.42) (148). We have also followed patients with grade 3 intrameniscal signal who revealed a spontaneously healed meniscal tear at arthroscopy. Consequently, the presence of grade 3 signal on short TE sequences is not specific for a complete tear in postoperative or post meniscal tear patients. The presence of high signal fluid extending into the meniscus on long TE T2-weighted spin-echo images is probably much more specific for clinically relevant tears in this patient population (Fig. 4.43). Classic x-ray or MR arthrography may be helpful in this setting if adequate liquid is not present in the joint on the routine MRI. Grade 3 signal intensity in the postoperative knee should be correlated with the preoperative MR or arthroscopic findings to differentiate a retear from residual grade 3 signal intensity in a stable meniscal rim.

Unfortunately, it is not uncommon for patients to continue to have or develop knee pain after surgery. Pain developing after surgery may be secondary to several causes. The original tear may not have been totally excised, therefore leading to continued pain from a persistent tear (Fig. 4.43). A second tear can develop in the residual rim left behind after initially successful surgery, or at the site of the surgical repair there may be a recurrent tear. The onset of pain in the latter case develops at variable periods of time after surgery, depending on when the new tear develops. Occasionally meniscal fragments may be left behind after surgery, producing mechanical symptoms and pain secondary to the abnormal mechanics of a foreign body within the knee joint.

Evaluation of a postoperative knee can be difficult (114,145). In our initial series of 144 knees examined with MRI and surgical correlation, 11 cases of post-meniscectomy menisci were evaluated. MRI detected 8 of 10 meniscal remnant tears (80% sensitivity). One residual meniscus yielding grade 3 signal was found to be untorn at surgery. In our subsequent experience, postoperative menisci without grade 3 signal usually have not been associated with tears. Also several occult residual meniscal fragments were seen following "total meniscectomies" that subsequently were removed at surgery, with symptomatic improvement. In a subsequent study of 7 menisci evaluated after primary meniscal repair, MRI finding could be divided into three categories: (a) approximately one-third of asymptomatic postoperative menisci did not reveal grade 3 signal (Fig. 4.42); (b) approximately two-thirds of postoperative menisci displayed persistent grade 3 signal at the operative site despite lack of symptoms (Fig. 4.22). One symptomatic knee revealed a clearly unstable meniscal fragment in the weightbearing portion of the joint between the femoral condyle and tibial plateau (Fig. 4.27). In a larger patient population Deutsch et al. have shown marked variability in MR findings in the postoperative knee (see Chapter 7) (150,151). Therefore, we believe that MRI is useful for evaluation of postmeniscectomy patients who are not symptomatically improved after surgery or who develop pain at some later time. If the menisci do not reveal grade 3 signal, then repeat surgery is unlikely to benefit the patient. If an unstable meniscal fragment is found, then repeat surgery is likely to benefit the patient. In the presence of persistent grade 3 signal, a T2-weighted image may be of benefit. Fluid extending into the meniscus represents a tear, otherwise the surgical decision must be based on clinical criteria. Standard or MR arthrography may be helpful in this setting

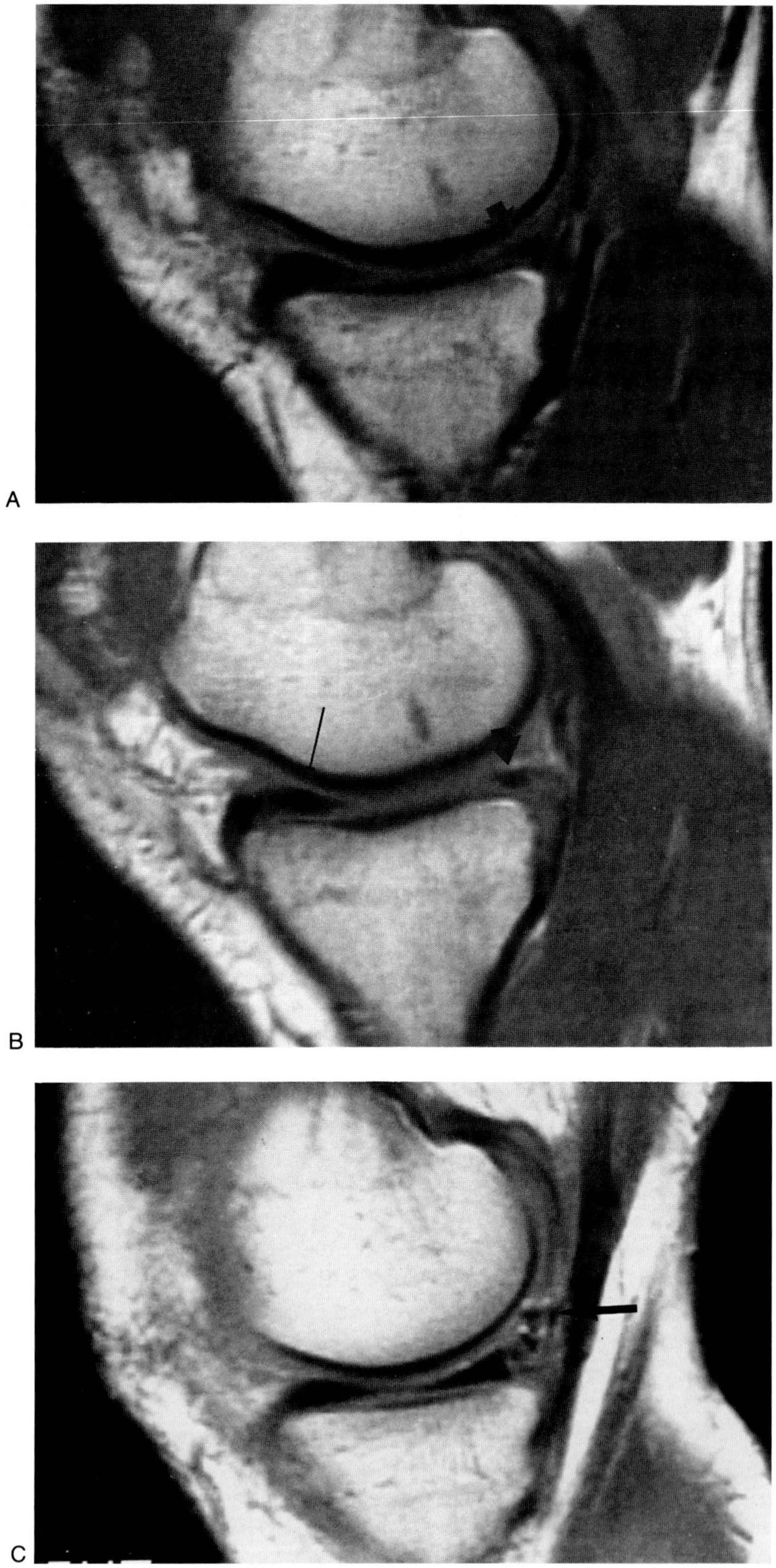

FIG. 4.42. Primary meniscal repair. **A:** The posterior horn of this medial meniscus is too small (*fat arrow*). This is a frequent finding in bucket-handle tears (TR/TE 2,000/20). **B:** The adjacent image shows the small posterior horn (*fat arrow*) as well as the anterior portion of the bucket handle (*long arrow*). **C:** After surgery the signal intensity and morphology of the meniscus is almost normal. Postoperative artifact is seen posteriorly (*arrow*).

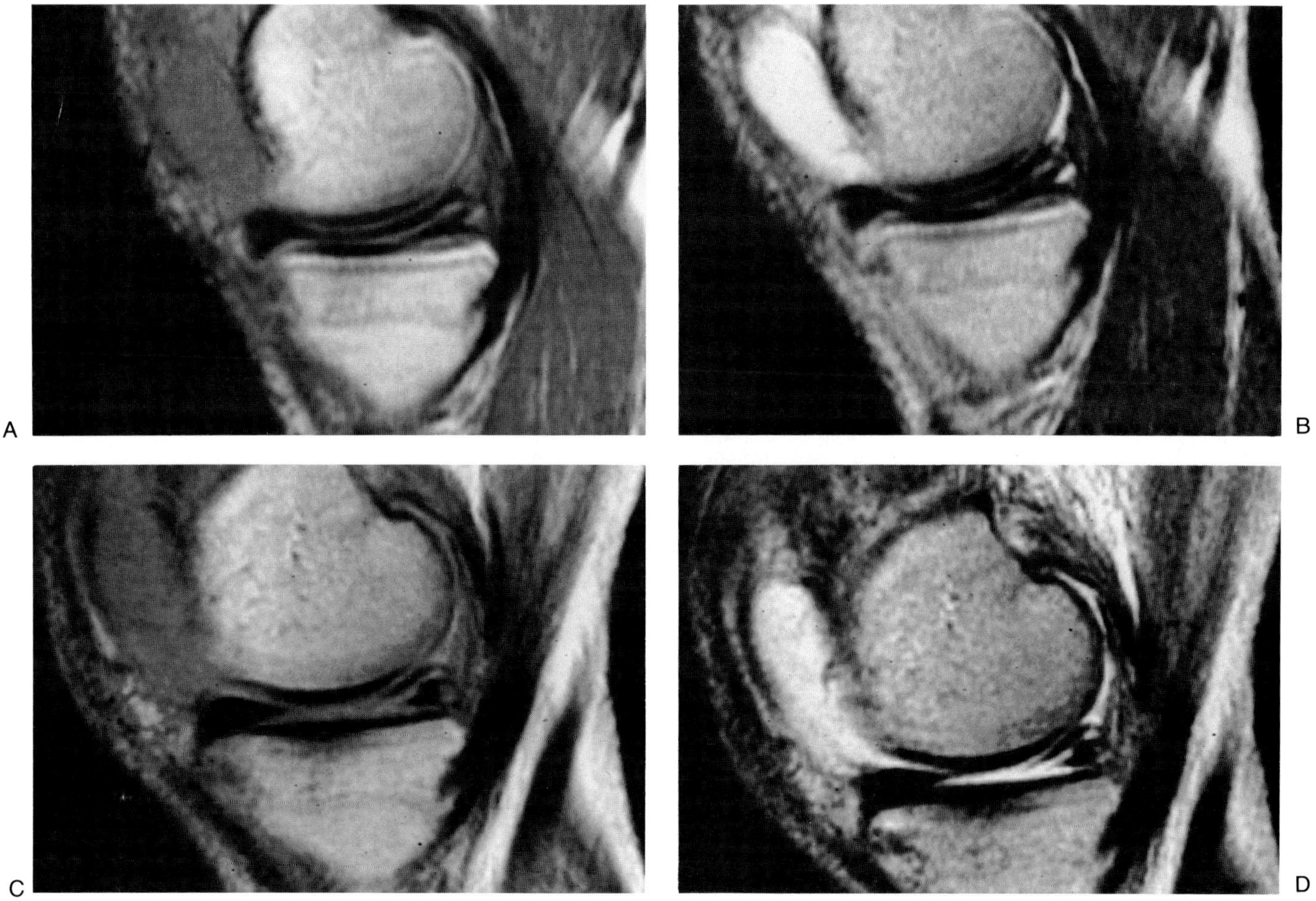

FIG. 4.43. Persistent meniscal tear. **A:** A horizontal cleavage tear is visualized involving the posterior horn and body of this medial meniscus (TR/TE 2,000/20). **B:** The T2-weighted image (TR/TE 2,000/80) also shows high signal intensity within the tear. **C:** Persistent pain after surgery led to a repeat MR scan which revealed grade 3A signal intensity within the posterior horn (TR/TE 2,000/20). **D:** This region is also bright on the T2-weighted image (TR/TE 2,000/80) compatible with a tear.

if insufficient fluid is present in the joint on the routine MRI.

MENISCAL CYSTS

Cysts associated with the menisci are relatively uncommon, being seen in only 1% of meniscectomies (258). These can be divided into four types (Table 4.6). Intrameniscal cysts are expanded fluid collections within a meniscus. They represent an uncommon presentation of meniscal tear. Parameniscal cysts are extravasations of joint fluid through a horizontal cleavage tear into the surrounding parameniscal tissues (55,259) and are much more common than intrameniscal cysts (Fig. 4.44). Thus, intrameniscal and parameniscal cysts are manifestations of meniscal tears and are treated by addressing the meniscal tear (247). If directly resected without addressing the meniscal tear, they usually recur (46,56,97,182). Synovial cysts are cystic outpouchings of the joint capsule and are rare (Fig. 4.45). When symptomatic, synovial cysts can be directly resected. Since the meniscus is usually normal adjacent to synovial cysts, meniscal surgery is not warranted. Rarely peripheral meniscocapsular separations

TABLE 4.6. *Types of cysts associated with knee menisci*

Intrameniscal
Parameniscal
Synovial
Menoiscocapsular separation

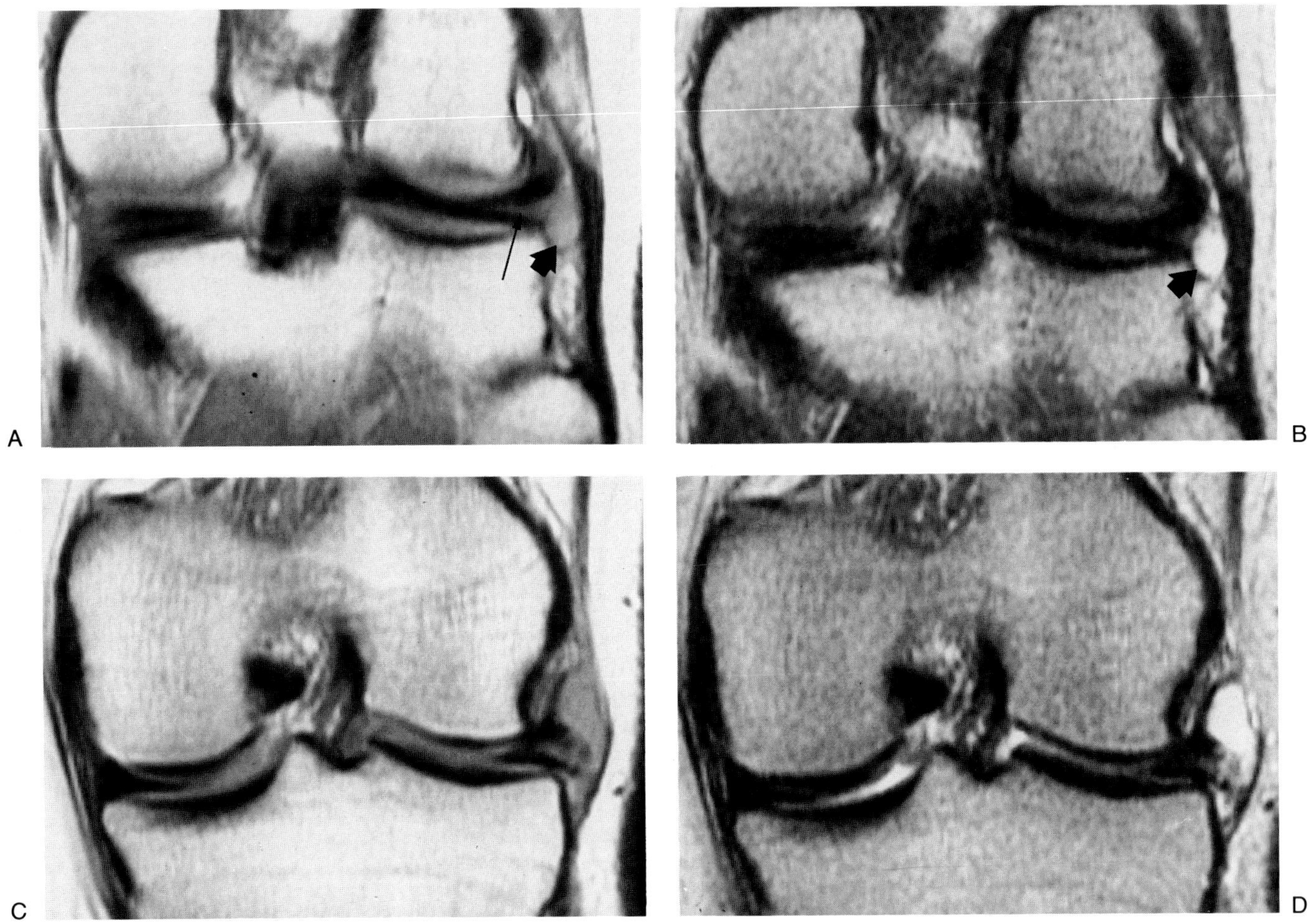

FIG. 4.44. Lateral parameniscal cyst. **A:** A liquid collection is seen adjacent to the lateral meniscus (*fat arrow*) emanating from a horizontal cleavage tear (*long arrow*) (TR/TE 2,000/20). **B:** The meniscal tear is poorly seen on the T2-weighted image, but the parameniscal cyst is clearly visualized (*fat arrow*). **C,D:** The adjacent images also reveal the tear and associated cyst in the parameniscal tissues.

of the medial meniscus can allow joint fluid to extravasate deep to the tibial collateral ligament (*vida supra*) (Fig. 4.46). This finding has been ascribed to bursitis (260). The differentiation between meniscocapsular separation and bursitis is important since therapeutic options may differ between the two diagnoses. Treatment with steroid injections or other antiinflammatory medications may not directly treat the underlying etiology. If a tear is the etiology, then arthroscopic repair may be warranted.

MRI is extremely useful clinically because it can visualize the meniscal tear as the etiology of the first 2 tear types, whereas the meniscus is normal in association with synovial cysts and most meniscocapsular separations (161). Previously the term "ganglion cyst" has been used synonomously with the first three cystic collections. Because of the different etiologies and treatments of these collections, terminology more specific than "ganglion cyst" is currently recommended.

Parameniscal cysts have been reported to be three to four times more common in association with the lateral meniscus than with the medial meniscus (55,56,97). With MRI more medial parameniscal cysts are detected and the predominance of lateral parameniscal cysts over medial cysts appears to be less than in the pre-MRI era. Lateral cysts typically present with lateral joint line swelling and tenderness increasing in size with extension (46,56). Medial cysts are more likely to be asymptomatic, though they tend to be larger than on the lateral side (Fig. 4.47) (55). This is because medial cysts tend to dissect through rather than abut soft tissue planes. Medial cysts may present as soft tissue swellings at some distance from the joint line, connecting to the meniscus via a thin stalk. They may clinically mimic a popliteal cyst (56,97) and have been shown to erode bone (46). Differentiation from

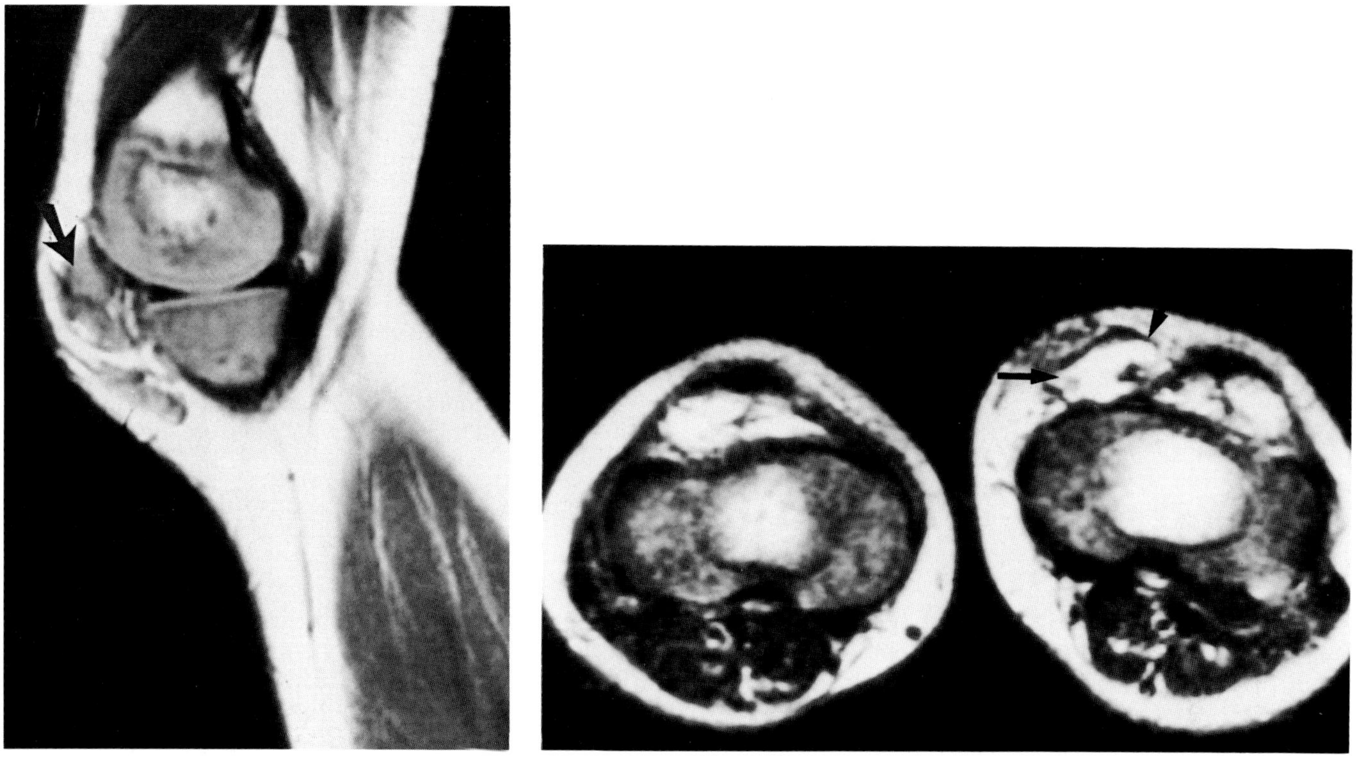

FIG. 4.45. Synovial cyst. **A:** The sagittal images (TR/TE 2,000/20) show an inhomogeneous mass adjacent to the anterior horn of the medial meniscus (*arrow*). **B:** The T2-weighted axial images show the liquid portion of the masses (*arrows*) to be adjacent to the meniscus (*arrowheads*). All images showed the meniscus to be normal. Because of patient symptoms the masses were removed without entering the joint space. A synovial cyst was confirmed on histologic sections.

meniscocapsular separation as described above may be academic since arthroscopic repair may be indicated in both.

Many causes for meniscal cysts have been proposed. Smillie (46) believed meniscal cysts were of traumatic origin, which helps to explain the 3-to-1 male predominance and peak incidence in the third to fourth decades of life. A developmental defect, with synovial cells being enclosed within the meniscus and active secretion of mucin causing progressive enlargement, has been proposed (56,113,258,261). Other possible causes include chronic infection, hemorrhage, and ac-

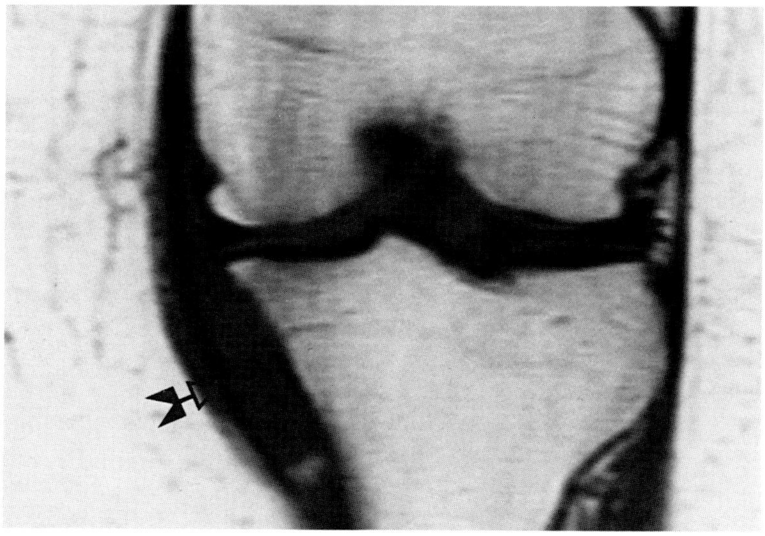

FIG. 4.46. Liquid deep to the medial collateral ligament. A liquid collection deep to the medial collateral is not a rare finding (*arrow*). It probably represents a collection of joint fluid extravasating through a peripheral meniscocapsular separation.

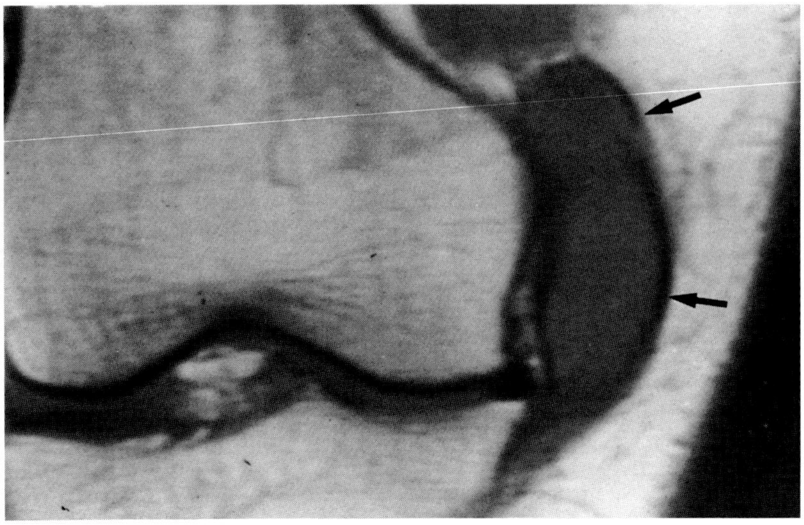

FIG. 4.47. Medial parameniscal cyst. Medial parameniscal cysts typically are larger than lateral cysts and may dissect through soft tissue planes (*arrows*). A horizontal cleavage of the posterior horn was seen on other sections.

cumulation of mucopolysaccharides secreted by mesenchymal cells (55). However, in a detailed histologic study of 70 menisci, Ferrer-Roca and Vilalta (55) found meniscal cysts to be associated with type A myxoid degeneration, i.e., horizontal cleavage tears extending into the parameniscal region, the cyst being secondary to ballooning of the tear in the parameniscal space. The rare intrameniscal cysts are collections of liquid within a meniscal tear and most parameniscal cysts are collections of joint fluid extravasated into the parameniscal tissues through horizontal cleavage tears of the menisci. The MRI characteristics of cyst fluid, in our experience, are similar to joint fluid. Occasionally fluid within the cyst may not be uniformly bright on T2-weighted images because water within the cyst may be resorbed by parameniscal tissues, leaving a more desiccated, paste-like material within the cyst. It is likely that all of these proposed causes, except for hemorrhage and infection, play roles in producing most meniscal cysts (55). This is also consistent with the observation that cysts can communicate with the joint space (56,80,97). Other conditions that may mimic a meniscal cyst clinically are listed in Table 4.7.

Two arthrographic features are characteristic of meniscal cysts. In the medial aspect of the knee, a cyst may present as a mass imprinting the synovium near the meniscus (56). More commonly, contrast is seen filling a horizontal tear, with puddling of contrast material ending in a club-shaped termination in the parameniscal region (56,97). This finding clearly differentiates meniscal cysts from other joint line masses.

TABLE 4.7. *Conditions clinically mimicking meniscal cysts*

Synovial herniation
Tumor
Postoperative abscess
Pes anserine bursa
Chronic capsulitis
Redundant synovium

Adapted from Schuldt and Wolfe (97).

DISCOID MENISCI

Discoid menisci are morphologically enlarged menisci. They may display mild symmetrical enlargement, asymmetric hypertrophy of segments, or massive slabs of cartilage completely interposed between the femoral condyle and the tibial plateau. First described in 1889 by Young (262), it is now recognized that discoid lateral menisci are not rare, with reported incidences of 1.5% to 15.5% (46,119,263–269). The first undisputed discoid medial meniscus was reported in 1941 (263), and it is much less common than its lateral counterpart, being reported in only 0.3% of meniscectomies (183,231,264,266,268–277).

The cause of discoid menisci has been a matter of controversy since Smillie (268) first proposed in 1948 that they were secondary to a premature arrest in the normal developmental sequence of the menisci. Smillie (183) did not recognize a discoid meniscus as a normal stage in meniscal development and cast doubt on his own initial hypothesis. This was subsequently confirmed by other authors (23,25,277,278). In 1957, Kaplan (23) performed a detailed interspecies correlation of the lateral meniscal attachments and found that only humans possess a posterior menisco-tibial ligamentous attachment. He further recognized that lateral discoid menisci in his series lacked this ligamentous attachment and were anchored posteriorly to the femur only via Wrisberg's ligament, as in all lower species. He then hypothesized that the increased motion of the posterior horn lacking the meniscotibial ligament led to increased trauma during mechanical loading in extension. The trauma led to subsequent thickening. Therefore, the predisposition to a discoid lateral meniscus is congenital, but the lesion itself is acquired. Unfortunately, this mechanism cannot account for the development of medial discoid menisci or discoid lateral menisci with intact posterior meniscotibial ligaments (15,272,279,280). Muscoskeletal deformities associated with discoid lateral menisci include hypoplasia of the lateral femoral condyle (269), high fibular head, fibular muscular defects (274), and a markedly enlarged inferior lateral geniculate artery with risk of significant hemorrhage if unrecognized at surgery (183). It is now generally recognized that discoid menisci are best categorized as congenital lesions (183,269,271).

As with the etiology of discoid menisci, classifications of discoid menisci have been debated. Smillie (268) initially classified discoid menisci as primitive, intermediate, and infantile, based on the degree of enlargement. He amended this classification to massive, intermediate, and analogous when he rejected his initial etiologic hypothesis (183). Watanabe combined elements of Smillie's classification and Kaplan's anatomic descriptions into a useful orthopedic classification of complete, incomplete, and Wrisberg-ligament type, where "complete" and "incomplete" refer to the degree of interposition between the femoral condyle and tibial plateau in the presence of an intact posterolateral meniscotibial ligament (272,279,280). The Wrisberg-ligament type refers to discoid lateral menisci lacking the posterior menisco-tibial ligament, as described by Kaplan (23). An arthrographic classification based on detailed morphologic characteristics has also been proposed by Hall (119). The choice of which, if any, classification should be used is a matter of individual preference. Each classification may have useful therapeutic implications.

Classically, the discoid lateral meniscus presents clinically as the "snapping-knee syndrome" (23,46, 102,119,265,271). As the name implies, a snapping sound is heard during flexion and extension, and it is variably associated with pain or lateral joint line tenderness. Multiple causes for the snapping sound have been proposed by Smillie (183), Kaplan (23), Hall (119), and Dickaut and DeLee (272). It seems likely that the snapping-knee syndrome is predominantly seen only in Wrisberg ligament-type discoid menisci and is due to the femoral condyle overriding a deformed anterior horn of the discoid lateral meniscus in the presence of abnormal meniscal displacement. Unfortunately, though the snapping-knee syndrome may be a classic clinical presentation of lateral discoid menisci, it is seen in only a minority of such cases (119,272). Also, no specific clinical findings have been described in medial discoid menisci, which typically present clinically as routine medial meniscal tears.

As with meniscal injuries in general, treatment for discoid menisci is evolving. Initially it was believed that discoid menisci were usually symptomatic, almost always seen in children and adolescents, and required total meniscectomy (48,281). It is now recognized that many discoid menisci are asymptomatic and do not require surgery (119,271,272). Increased understanding of the function of menisci has lead many surgeons to preserve the meniscus whenever possible (15,58,208). If the discoid meniscus is torn and symptomatic, partial meniscectomy and shaping of the meniscus to near normal size and configuration, when feasible, may be advantageous (15,86).

It appears that visualization of a discoid meniscus by MRI was first reported by Reicher et al. (68). Severe findings are helpful in detecting discoid lateral menisci by MRI. The transverse diameter of the body of the lateral meniscus averages about 10 to 11 mm (33). Therefore, using a 5 mm slice thickness, only two sagittal slice locations through the body of the normal lateral meniscus should display the black "bow-tie" (see Chapter 3). If three or more slice locations show a bridge of meniscus between the anterior and posterior horns, then a lateral discoid meniscus is likely (Fig. 4.48). Conversely, normally there are at least two slice locations displaying triangular-shaped anterior and posterior horns. If only one or two sagittal slices display distinct anterior and posterior horns, then a discoid lateral meniscus should be suspect. Discoid menisci may have significantly increased inferior-superior dimensions, and an abnormally thickened bow-tie may be indicative of a lateral discoid meniscus. The height of the normal meniscus rapidly tapers as one moves from the periphery toward the center. If the two adjacent peripheral cuts show similar meniscal heights, then a discoid meniscus should be suspected. An asymmetric discoid meniscus may present with an abnormally large anterior or posterior horn that may be detected by violation of the normal symmetry between the anterior and posterior horns (*vide supra*).

In practice the coronal images are often most helpful. In Smillie's massive or Watanabe's complete discoids, meniscus is seen interposed between the femoral condyle and tibial plateau in all slice locations through the knee (Fig. 4.48E). The more common incomplete discoid meniscus presents with the meniscus occupying more than one-third of the femoral-tibial articulation on the central coronal image which displays the smallest radial cross-section of the meniscal body. In adults the radial diameter of this cross-section should be 12 mm or less (33). In practice using a 13 mm measured distance on coronal MR images through the body of the meniscus from the tip of the free edge to the capsular margin is accurate in discriminating

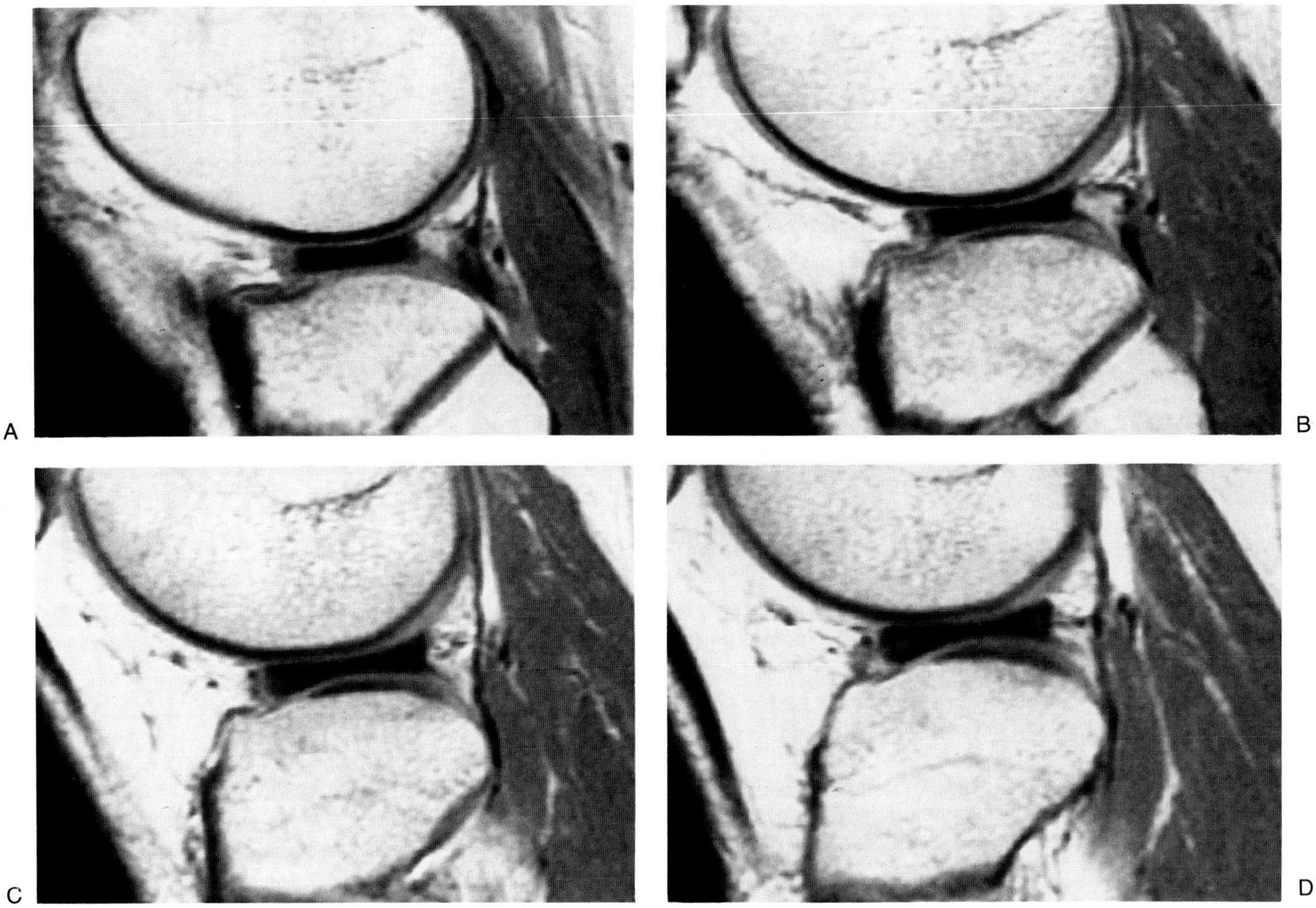

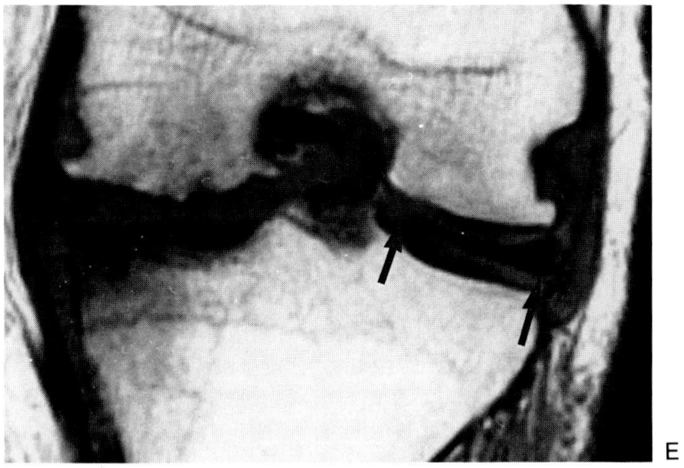

FIG. 4.48. Discoid lateral meniscus. **A–D:** Multiple adjacent sagittal images (TR/TE 2,000/20) through the lateral meniscus fail to reveal the normal sharp free edge of the normal triangularly shaped meniscal cross-section. This is characteristic of a complete slab type discoid meniscus. **E:** The coronal images are very helpful in evaluating discoid menisci, especially those which present as large menisci, but are not a complete slab type. The coronal image through the body which displays the *smallest* radial diameter should reveal a radial diameter of less than 13 mm in normal menisci (*arrows*) (33).

most discoid from normal menisci (Fig. 4.48E). If greater, then an enlarged discoid meniscus should be suspected.

Both arthrotomy and arthroscopy may be more difficult in the presence of discoid menisci (15,23,86,182, 270,272,282–285); therefore, knowledge of the existence of a discoid meniscus can be helpful in presurgical planning. More important, however, is the ability of MRI to detect and localize meniscal tears within discoid menisci (Figs. 4.16 and 4.49). The discoid lateral meniscus is difficult to evaluate arthrographically (31,89–91,113,119,261), yet the presence of a tear within the meniscus is the major determinant of the need for surgical intervention (Fig. 4.16). Also, the location and extent of the tear determine the surgical procedure of choice. It is at present unclear whether or not MRI can reliably differentiate between Wrisberg-ligament and non-Wrisberg discoid lateral me-

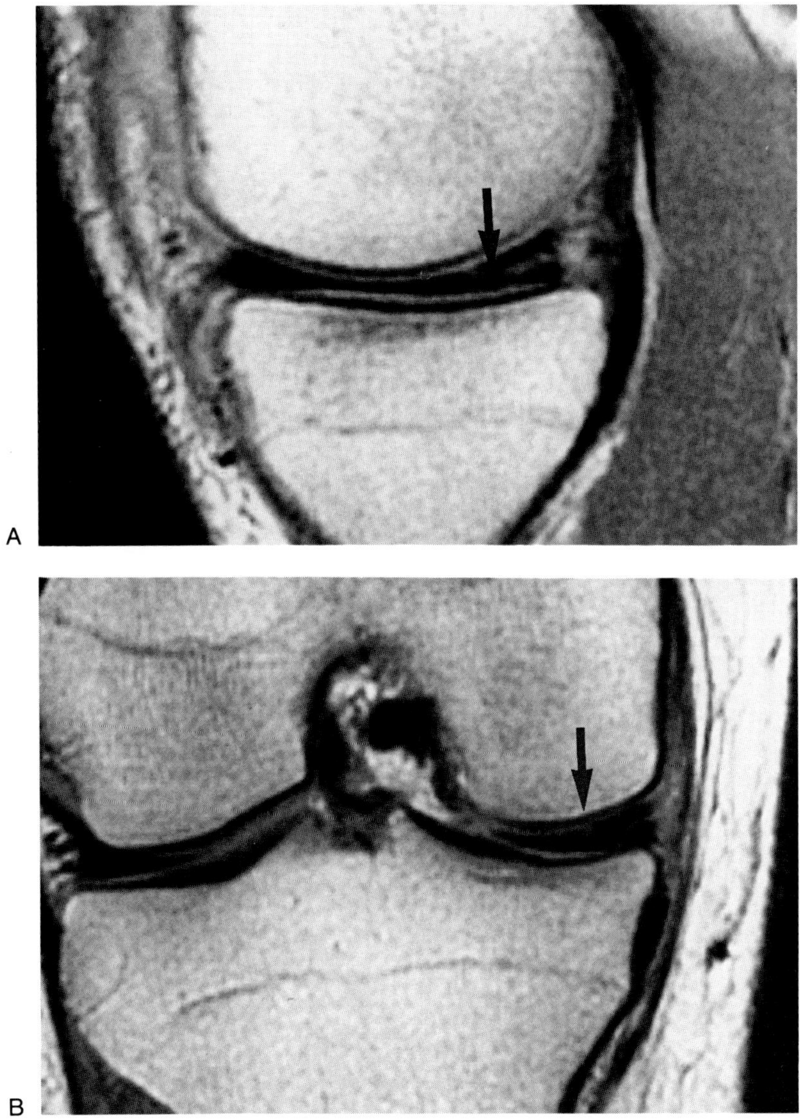

FIG. 4.49. Tear in a discoid medial meniscus. **A:** A sagittal image (TR/TE 2,000/20) reveals a horizontal cleavage tear extending to the superior articular surface of the posterior horn of this discoid medial meniscus (*arrow*). **B:** The discoid nature of the meniscus is better visualized on the coronal image through the body showing the smallest radial diameter. The tear is again visualized (*arrow*). This patient also has an enlarged discoid lateral meniscus as well.

nisci. Because Wrisberg ligament-type discoid menisci usually are symptomatic (272) and may not be candidates for partial meniscectomy (15), this distinction may be important clinically. Symptomatic intrasubstance tears may have a higher incidence in discoid menisci than in normal menisci (*vida supra*) (Fig. 4.16). MRI may be essential in determining the etiology of knee pain in these patients and lead to appropriate therapy (75,76).

In summary, the discoid meniscus is a common congenital meniscal lesion. The clinical presentation of lateral discoid menisci is the snapping-knee syndrome, but this finding is infrequent. Noninvasive evaluation of discoid menisci is helpful in presurgical planning.

CALCIFICATIONS

Calcification within the menisci is frequently seen by x-ray techniques. The causes can be multiple, including calcium pyrophosphate disease (CPPD), hydroxyapatite deposition disease, hyperparathyroidism, hemochromatosis, Wilson's disease, ochronosis, gout, collagen vascular diseases, and idiopathic. Although conventional radiography is more sensitive to chondrocalcinosis, we have used high contrast settings to identify focal, low signal intensity calcifications separate from the adjacent low signal intensity meniscus. Calcium itself does not yield MR signal and therefore appears black on MR images. We have used T2*-weighted gradient-echo images to identify local susceptibility artifact around foci of calcium pyrophosphate in both the meniscus and articular cartilage. Localized magnetic susceptibility associated with CPPD crystals may attenuate intrameniscal signal intensity resulting in a false-negative interpretation in the presence of a meniscal tear.

Bony ossicles can also occasionally be found within menisci. Ossicles may contain marrow fat which gives

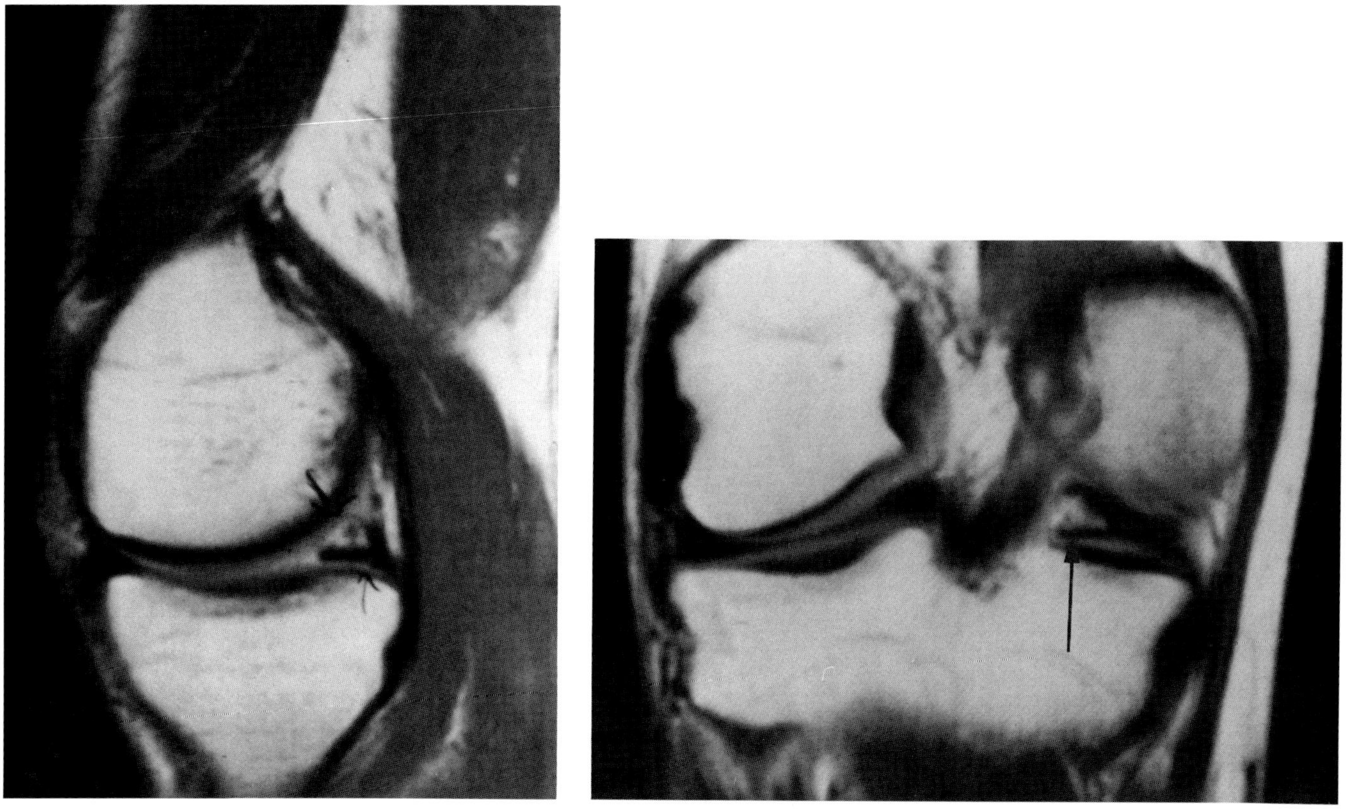

A B

FIG. 4.50. Meniscal ossicle. **A,B:** Meniscal ossicles are an uncommon etiology of increased signal within a meniscus. Medullary fat within the ossicle presents as high signal on T1-weighted images (TR/TE 600/20) (*arrows*). Most ossicles are located in the lateral portion of the posterior horn of the medial meniscus adjacent to the superior articular surface. Often the cortex of the ossicle is lower in signal intensity than the surrounding meniscus.

a characteristic MRI appearance of high signal intensity within the confines of dark, calcified cortices (Fig. 4.50) (286). The etiology of meniscal ossicles is debated. A congenital origin with the ossicle being formed from rests of primitive mesenchyme is the preferred theory of origin at the present time, but recent evidence supports a traumatic etiology (287). The two meniscal ossicles we have seen have both been located in the superior half of the posterior horn of the medial meniscus laterally near the posterior meniscotibial attachment.

THE ROLE OF MRI

The use of MRI for evaluating meniscal abnormalities is as an established diagnostic technique with accuracy comparable to arthrography and arthroscopy. MRI can accurately evaluate the presence and location of meniscal tears and, as such, has been recommended as the diagnostic study of choice in detecting the etiology of acute and chronic knee pain (148,160). MRI also appears to be excellent for excluding the presence of meniscal tears, and patients with negative MRI findings may be candidates for conservative therapy (67,68). Recent evidence that many, if not most, meniscal tears are asymptomatic, especially in the older age groups, has led to increased emphasis on MRI's ability to type meniscal tears in an attempt to better determine the clinical significance of the tear. With the advent of more sophisticated arthroscopic meniscal-sparing treatments, the ability of MRI to type tears is important in surgical planning and patient counseling when primary meniscal repair is a possible option. The ability to use MRI to accurately determine the degree, location, and extent of degenerative meniscal changes that cannot be detected by arthrography or arthroscopy may also be important in clinical decision making. MRI may be helpful in directing the arthroscopist.

We have heard numerous anecdotal reports from surgeons that tears which otherwise might not have been detected were found at arthroscopy because the arthroscopist carefully probed a region of grade 3 MRI abnormality (75,76). A tear initially detected by MRI but visualized only on a second arthroscopy performed for continuing symptoms has been reported (145). In the presence of acute trauma, MRI can be used to evaluate not only the menisci, but also abnormalities which clinically mimic symptoms of a meniscal tear but are inaccessible to the arthrographer or arthroscopist. Examples include occult fractures, ligament diseases, and osteonecrosis. Less common diseases such as cysts and discoid menisci are well evaluated by MRI.

Because of the low sensitivity of clinical findings, arthrography, and arthroscopy for diagnosing the spectrum of common joint abnormalities which clinically masquerade as meniscal tears, MRI is the diagnostic study of choice in evaluating the etiology of knee pain which clinically presents as possible meniscal disease.

REFERENCES

1. Harms S, Flamig D, Fisher C, Fulmer J. New method for fast MR imaging of the knee. *Radiology* 1989;173:743–750.
2. Jacobs B, Applegate GR, Recht M, Thaete FL, Fu F. Comparison of frequency-selective fat saturation with spin-echo MR images in the detection of knee pathology. *JMRI* 1991;1(2):176.
3. Crues JV, Mink J, Levy TL, Lotysch M, Stoller DW. Meniscal tears of the knee: Accuracy of magnetic resonance imaging. *Radiology* 1987;164:445–448.
4. Adam G, Bohndorf K, Drobnitzky M, et al. MR imaging of the knee: Three-dimensional volume imaging combined with fast processing. *JCAT* 1989;13:984.
5. Reeder JD, Matz SO, Becker L, Andelman SM, MR imaging of the knee in the sagittal projection: Comparison of three-dimensional gradient-echo and spin-echo sequences. *AJR* 1989;153(3):537–540.
6. Tyrrell RL, Gluckert K, Pathria M, Modic MT. Fast three-dimensional MR imaging of the knee: Comparison with arthroscopy. *Radiology* 1988;166:865–872.
7. Bjorkengren AG, Geborek P, Rydholm U, Holtas S, Petterson H. MR imaging of the knee in acute rheumatoid arthritis: Synovial uptake of gadolinium-DOTA. *AJR* 1990;155:329–332.
8. Erlemann R, Reiser MF, Peters PE, Vasallo P, Nommensen B, Kusnierz GC, Ritter J, Roessner A. Musculoskeletal neoplasms: Static and dynamic Gd-DTPA–enhanced MR imaging. *Radiology* 1989;171(3):767–773.
9. Konig H, Sieper J, Wolf K. Rheumatoid arthritis: Evaluation of hypervascular and fibrous pannus with dynamic MR imaging enhanced with Gd-DTPA. *Radiology* 1990;176:473–477.
10. Erlemann R, Vassallo P, Bongartz G, Muller-Miny H, Rummeny E, Stober U, Peters PE. Musculoskeletal neoplasms: Fast low-angle shot MR imaging with and without Gd-DTPA. *Radiology* 1990;176:489–495.
11. Chandnani VP, Ho C, Chu P, Trudell D, Resnick D, Knee hualine cartilage evaluated with MR imaging: A cadaveric study involving multiple imaging sequences and intraarticular injection of gadolinium and saline solution. *Radiology* 1991;178:557–561.
12. Hajek PC, Baker LL, Sartoris DJ, Neumann CH, Resnick D. MR arthrography: Anatomic-pathologic investigation. *Radiology* 1987;163:141–147.
13. Hajek PC, Gylys MV, Baker LL, Sartoris DJ, Haghighi P, Resnick D. The high signal intensity meniscus of the knee. Magnetic resonance evaluation and in vivo correlation. *Invest Radiol* 1987;22(11):883–890.
14. DiStefano VJ. Function, post-traumatic sequelae and current concepts of management of knee meniscus injuries: A review article. *Clin Orthop* 1980;151:143–146.
15. Fujikawa K, Iseli R, Mikura Y. Partial resection of the discoid meniscus in the child's knee. *J Bone Joint Surg* 1981;63:391–395.
16. Kettelkamp DB, Jacobs AW. Tibiofemoral contact area—Determination and implications. *J Bone Joint Surg* 1972;54:349–356.
17. Maquet PG, Van de Berg AJ, Simonet JC. Femorotibial weight-bearing areas: Experimental determination. *J Bone Joint Surg* 1975;57:766–771.
18. Ricklin P, Ruttimann A, Del Buono MS. *Meniscal lesions: Practical problems in clinical diagnosis, arthrography and therapy*. New York: Grune & Stratton; 1971.
19. Seedholm BB. Transmission of the load in the knee joint with special reference to the role of the menisci. Part I. Anatomy, analysis, and apparatus. *Eng Med* 1979;8:207–219.
20. Seedholm BB, Hargreaves DJ. Transmission of the load in the knee with special reference to the role of the menisci. Part II. Experimental results, discussion and conclusion. *Eng Med* 1979;8:220–228.
21. Smillie IS. *Biomechanics: Form of the joint menisci and ligaments: Mechanism of injury and sequelae*. IN: Injuries of the knee joint, 4th ed. London: E. & S. Livingstone;1970:23–28.
22. Walker PS, Erkman J. The role of the meniscus in force transmission across the knee. *Clin Orthop* 1975;109:184–192.
23. Kaplan EB. Discoid lateral meniscus of the knee joint. *J Bone Joint Surg* 1957;39:77–87.
24. Watanabe AT, Carter BC, Teitelbaum GP, Bradley WJ. Common pitfalls in magnetic resonance imaging of the knee. *J Bone Joint Surg [Am]* 1989;71(6):857–862.
25. Kaplan EB. The embryology of the menisci of the knee joint. *Bull Hosp Joint Dis* 1955;16:111–124.
26. Carpenter WA. Meniscofemoral ligament simulating tear of the lateral meniscus: MR features. *JCAT* 1990;14(6):1033–1034.
27. Vahey TN, Bennett HT, Arrington LE, Shelbourne KD, Ng J. MR imaging of the knee: Pseudotear of the lateral meniscus caused by the meniscofemoral ligament. *AJR* 1990;154:1237–1239.
28. Watanabe AT, Carter BC, Teitelbaum GP, Seeger LL, Bradley WJ. Normal variations in MR imaging of the knee: Appearance and frequency. *AJR* 1989;153(2):341–344.
29. Harley JD. An anatomic-arthrographic study of the relationships of the lateral meniscus and the popliteus tendon. *AJR* 1977;128:181–187.
30. Jelaso DV. The fascicles of the lateral meniscus: An anatomic-arthrographic correlation. *Radiology* 1975;114:335–339.
31. Pavlov H, Goldman AB. The popliteus bursa: An indicator of subtle pathology. *AJR* 1980;134:313–321.
32. Wickstrom KT, Spitzer RM, Olsson HE. Roentgen anatomy of the posterior horn of the lateral meniscus. *Radiology* 1975;116:617–619.
33. Ferrer-Roca O, Vilalta C. Lesions of the meniscus. Part I: Macroscopic and histologic findings. *Clin Orthop* 1980;146:301–307.
34. Arnoczky SP, Warren RF. Microvasculature of the human meniscus. *Am J Sports Med* 1982;10:90–95.
35. Arnoczky SP, Warren RF. The microvasculature of the meniscus and its response to injury. *Am J Sports Med* 1983;11:131–141.
36. Danzig L, Resnick D, Gonsalues M, Akeson WH. Blood supply to the normal and abnormal menisci of the human knee. *Clin Orthop* 1983;172:271–276.
37. Day B, Mackenzie WG, Shim SS, Leung G. The vascular and nerve supply of the human meniscus. *Arthroscopy* 1985;1:58–62.
38. Fanton GS, Andrish JT. Meniscofluoresis. *Orthop Trans* 1983;7:575.
39. Scapinelli R. Studies of the vasculature of the human knee. *Acta Anat* 1968;70:305–331.
40. Blazina ME, Fox JM, Del Pizzo W, Friedman MJ, Ivey FM. Thomas B. Quigley, M.D., and the meniscus. *Clin Orthop* 1982;164:84–92.
41. Cox JS, Nye CE, Schaeffer WW, Woodstein IJ. The degenerative effects of partial and total resection of the medial meniscus in dogs' knees. *Clin Orthop* 1975;109:178–183.
42. Fairbank TJ. Knee joint changes after meniscectomy. *J Bone Joint Surg* 1948;30:664–670.
43. Kempf FK. Treatment of meniscus defects with reference to our operative method. Subtotal meniscectomy, resection of Hoffa's corpus adiposum, inner drainage. *Arch Orthop Trauma Surg* 1980;96:95–104.

44. McGinty JB, Guess LF, Marvin RA. Partial or total meniscectomy: A comparative analysis. *J Bone Joint Surg* 1977;59:763–769.
45. Smillie IS. *Diseases of the knee joint*. New York: Churchill Livingstone; 1980.
46. Smillie IS. *Clinical features of internal derangements relative to the menisci*. In: Injuries of the knee joint, 4th ed. London: E. & S. Livingstone; 1970;70–97.
47. Zarins B, McInerney VK. *Lesions of the meniscus*. IN: Arthroscopy: Diagnostic and surgical practice. Casscells, SW, ed. Philadelphia: Lea & Febiger; 1984;37–51.
48. Noble J, Hamblen DL. The pathology of the degenerate meniscus lesion. *J Bone Joint Surg [Br]* 1975;57:180–186.
49. Monaco BR, Noble HB, Bachman DC. Incomplete tears of the anterior cruciate and knee locking. *JAMA* 1982;247:1582–1584.
50. Fahmy NRM, Noble J, Williams EA. Relationship between meniscal tears and osteoarthritis of the knee. *J Bone Joint Surg* 1981;63:629.
51. Frankel VH, Burstein AH, Brooks DB. Biomechanics as determined by analysis of the instant centers of motion. *J Bone Joint Surg* 1971;53:945–962.
52. Norman A, Baker ND. Spontaneous osteonecrosis of the knee and medial meniscal tears. *Radiology* 1978;129:653–656.
53. Casscells SW. The torn meniscus, the torn anterior cruciate ligament, and their relation to degenerative joint disease. *Arthroscopy* 1985;1:28–32.
54. Casscells SW. The torn meniscus and degenerative joint disease. *Am Acad Orthop Surgeons Trans* 1981;5:400.
55. Ferrer-Roca O, Vilalta C. Lesions of the meniscus. Part II: Horizontal cleavages and lateral cysts. *Clin Orthop* 1980;146:301–307.
56. Burgan DW. Arthrographic findings in meniscal cysts. *Radiology* 1971;101:579–581.
57. DeHaven KE. Arthroscopy in the diagnosis and management of the anterior cruciate ligament deficient knee. *Clin Orthop* 1983;172:52–56.
58. DeHaven KE. Rationale for meniscus repair or excision. *Clin Sports Med* 1985;4:267–274.
59. Smillie IS. *Surgical pathology of the menisci*. IN: Injuries of the knee joint, 4th ed. London: E. & S. Livingstone; 1970;39–69.
60. Kornick J, Trefelner E, McCarthy S, Lange R, Lynch K, Jokl P. Meniscal abnormalities in the asymptomatic population at MR imaging. *Radiology* 1990;177:463–465.
61. Li DKB, Adams ME, McConkey JP. Magnetic resonance imaging of the ligaments and menisci of the knee. *Radiol Clin North Am* 1986;24:209–227.
62. Fullerton GD, Potter JL, Dornbluth NC. NMR relaxation of protons in tissues and other macromolecular water solutions. *Magn Reson Imaging* 1982;1:209–228.
63. Koenig SH, Brown RD. The importance of motion of water for magnetic resonance imaging. *Invest Radiol* 1985;20:297–305.
64. Stoller DW, Martin C, Crues JV, Kaplan L, Mink JH. Meniscal tears: Pathologic correlation with MR imaging. *Radiology* 1987;163:731–735.
65. Lotysch M, Mink J, Crues JV, Schwartz A. Magnetic resonance imaging in the detection of meniscal injuries. *Magn Reson Imaging* 1986;4:94.
66. Lotysch M, Mink J, Levy T, Schwartz A, Crues JV. Detection of meniscal injuries using magnetic resonance imaging. *Radiology* 1986;161(P):238.
67. Reicher MA, Hartzman S, Duckweiler GR, Bassett LW, Anderson LJ, Gold RH. Meniscal injuries: Detection using MR imaging. *Radiology* 1986;159:753–757.
68. Reicher MA, Hartzman S, Bassett LW, Mandelbaum B, Duckwiler G, Gold RH. MR imaging of the knee. Part II. Traumatic disorders. *Radiology* 1987;162(2):547–551.
69. Dillon EH, Pope CF, Jokl P, Lynch K. The clinical significance of stage 2 meniscal abnormalities on magnetic resonance knee images. *Magn Res Imaging* 1990;8:411–415.
70. Frija G, Schouman-Claeys E, d'Anthouard F, Feron JM, Paraire F. Grossly normal knee menisci: Correlations with pathology and magnetic resonance imaging. *Diagn Interv Radiol* 1989;1:29–34.
71. Kaplan PA, Nelson NL, Garvin KL, Brown DE. MR of the knee: The significance of high signal in the meniscus that does not clearly extend to the surface. *AJR* 1991;156:333–336.
72. Negendank WG, Fernandez-Madrid FR, Heilbrun LK, Teitge RA. Magnetic resonance imaging of meniscal degeneration in asymptomatic knees. *J Orthop Res* 1990;8:311–320.
73. Reinig JW, McDevitt ER, Ove PN. Progression of meniscal degenerative changes in college football players: Evaluation with MR imaging. *Radiology* 1991;181:255–257.
74. Dillon EH, Pope CF, Jokl P, Lynch JK. Follow-up of grade 2 meniscal abnormalities in the stable knee. *Radiology* 1991;181:849–852.
75. Crues JV, Ryu R. MRI of the knee: Part II. *Surg Rounds Orthop* 1989;(May):27–33.
76. Crues JV, Ryu RR. MRI of the knee: Part I. *Surg Rounds Orthop* 1989;(April).
77. Gillies H, Seligson D. Precision in the diagnosis of meniscal lesions: A comparison of clinical evaluation, arthrography, and arthroscopy. *J Bone Joint Surg* 1979;61:343–346.
78. Ireland J, Trickey EL, Stoker DJ. Arthroscopy and arthrography of the knee: A critical review. *J Bone Joint Surg* 1980;62:3–6.
79. Watt I, Tasker T. Pitfalls in double contrast knee arthrography. *Br J Radiol* 1980;53:754–759.
80. Kaye JJ, Nance EP. Meniscal abnormalities in knee arthrography. *Radiol Clin North Am* 1981;19:277–286.
81. Voto S. A nomenclature system for meniscal lesions of the knee. *Surg Rounds Orthop* 1989;Oct:34–36.
82. Singson RD, Feldman F, Staron R, Kiernan H. MR imaging of displaced bucket-handle tear of the medial meniscus. *AJR* 1991;156:121–124.
83. Weiss KL, Morehouse HT, Levy IM. Sagittal MR images of the knee: A low-signal band parallel to the posterior cruciate ligament caused by a displaced bucket-handle tear. *AJR* 1991;156:117–119.
84. Cargill AOR, Jackson JB. Bucket handle tear of the medial meniscus. *J Bone Joint Surg* 1976;58:248–251.
85. Hargreaves DJ, Seedholm BB. On the "bucket-handle" tear: Partial or total meniscectomy? A quantitative study. *J Bone Joint Surg* 1979;61:381.
86. Patel D, Dimakopoulos P, Denoncourt P. Bucket handle tear of a discoid medial meniscus. Arthroscopic diagnosis—Partial excision. A case report. *Orthopedics* 1986;9:607–608.
87. Shakespeare DT, Rigby HS. The bucket-handle tear of the meniscus: A clinical and arthrographic study. *J Bone Joint Surg* 1983;65:383–387.
88. DeSmet AA. Meniscal tears on knee arthrography: Patterns of arthrographic abnormalities. *Skeletal Radiol* 1985;14:280–285.
89. Hall FM. Pitfalls in knee arthrography. *Radiology* 1976; 118:55–62.
90. Hall FM. Further pitfalls in knee arthrography. *J Can Assoc Radiol* 1978;29:179–184.
91. Hall FM. Pitfalls in the assessment of the menisci by knee arthrography. *Radiol Clin North Am* 1981;19:305–328.
92. Burke JW, Podrasky AE, Bradley WG. The "notch" sign: Useful indicator of meniscal tears in the knee. *Radiology* 1989;173(P):232.
93. Davis SJ, Teresi LM, Bradley WG, Burke JW. The "notch" sign: Meniscal contour deformities as indicators of tear in MR imaging of the knee. *JCAT* 1990;14(6):975–980.
94. Senac MJ, Deutsch D, Bernstein BH, Stanley P, Crues J, Stoller DW, Mink J. MR imaging in juvenile rheumatoid arthritis. *AJR* 1988;150(4):873–878.
95. Twyman RS, Ferris BD. Congenital hypoplasia of the medial meniscus: A report of two cases. *Arthroscopy* 1991;7(2):148–150.
96. Suzuki S, Mita F, Ogishima H. Double-layered lateral meniscus: A newly found anomaly. *Arthroscopy* 1991;7(3):276–271.
97. Schuldt DR, Wolfe RD. Clinical and arthrographic findings in meniscal cysts. *Radiology* 1980;134:250–251.
98. Levinsohn EM, Baker BE. Prearthrotomy diagnostic evaluation of the knee: Review of 100 cases diagnosed by arthrography and arthroscopy. *AJR* 1980;134:107–111.

99. Grenier R, du Tremblay P. Clinical judgment versus arthrography for diagnosing knee lesions. *Can J Surg* 1980;23:186–188.
100. DeHaven KE. Meniscus repair—open vs. arthroscopic. *Arthroscopy* 1985;1:173–174.
101. Drez D. Arthroscopic evaluation of the injured athlete's knee. *Clin Sports Med* 1985;4:275–278.
102. Kroiss F. Die Verletzungen der Kniegelenkoszwischenknorpel und ihrer Verbindungen. *Beitr Klin Chir* 1910;66:598–801.
103. Oberholzer J. *Archiv und Atlas der normalen und pathologischen Anatomie in typischen Rontgenbildern, in Rontgendiagnostik der Gelenke mittels Doppelkontrastmethode.* Leipzig: Thieme; 1938.
104. Lynch TCP, Crues JV, Morgan FW, Sheehan WE, Harter LP, Ryu R. Bone abnormalities of the knee. *Radiology* 1989; 171:761–766.
105. Karadimas JE, Rentis GR, Syrmalis L. Kinematics in the diagnosis of meniscal lesions. *Injury* 1980;12:119–122.
106. Andren L, Wehlin L. Double-contrast arthrography of the knee with horizontal roentgen ray beam. *Acta Orthop Scand* 1960;29:307–314.
107. Arndt RD, Horns JW, Gold RH. *Clinical arthrography*, 2nd ed. Baltimore: Williams & Wilkins; 1985.
108. Aye RC, Dorr TW, Drewry GR. Arthrography of the knee in office practice. *Radiology* 1963;80:829–836.
109. Bircher E. Die arthloendoscopie. *Zentralbl Chir* 1922;48:1460–1461.
110. Bircher E. Diagnosis of knee disease. *Beitr Klin Chir* 1922;127:239–250.
111. Bohr H. On the diagnosis of meniscus lesions in the knee with special regard to the value of arthrography. *Acta Orthop Scand* 1959;29:146–149.
112. Butt WP, McIntyre MB. Double-contrast arthrography of the knee. *Radiology* 1969;92:487–499.
113. Dalinka MK, Lally JF, Gohel VK. Arthrography of the lateral meniscus. *AJR* 1974;121:79–85.
114. Debnam JW, Staple TW. Arthrography of the knee after menisectomy. *Radiology* 1974;113:67–71.
115. Dumas JM, Edde DJ. Meniscal abnormalities: Prospective correlation of double-contrast arthrography and arthroscopy. *Radiology* 1986;453–456.
116. Edgren W, Heikel-Laurent U, Risk EB. Arthrography of the knee in the diagnosis of torn semilunar cartilage. *Acta Orthop Scand* 1967;38:235–246.
117. Freiberger RH, Killoran PJ, Cordonna G. Arthrography of the knee by double contrast method. *AJR* 1966;97:736–747.
118. Freiberger RH, Kaye JJ. *Arthrography*. New York: Appleton-Century-Crofts, 1979.
119. Hall FM. Arthrography of the discoid lateral meniscus. *AJR* 1977;128:993–1002.
120. Heiser S, LaBriola JH, Meyers MH. Arthrography of the knee. *Radiology* 1962;79:822–828.
121. Kessler I, Silberman Z, Nissim F. Arthrography of the knee: A critical study of errors and their sources. *AJR* 1961;86:359–365.
122. Crabtree SD, Bedford AF, Edgar MA. The value of arthrography and arthroscopy in association with a sports injuries clinic: A prospective and comparative study of 182 patients. *Injury* 1981;13:220–226.
123. Korn MW, Spitzer RM, Robinson KE. Is there still a role for knee arthrography? *Arthroscopy* 1985;1:40–43.
124. Kaye JJ. Knee arthrography today. *Radiology* 1985;157:265–266.
125. DeHaven KE, Collins HR. Diagnosis of internal derangement of the knee—The role of arthroscopy. *J Bone Joint Surg* 1975;57:802–810.
126. Orbon RJ, Poehling GG. Arthroscopic meniscectomy. *South Med J* 1981;74:1238–1242.
127. Sommer J. Arthrography of the knee joint for lesions of the meniscus. *J Bone Joint Surg* 1966;48:601.
128. St. Pierre RK, Sones PJ, Fleming LL. Arthroscopy and arthrography of the knee: A comparison study. *South Med J* 1981;74:1322–1328.
129. Stoker DJ, Renton P, Fulton A. The value of arthrography in the management of internal derangements of the knee: The first 1000 are the worst. *Clin Radiol* 1981;32:557–566.
130. Thijn CJP. Accuracy of double-contrast arthrography and arthroscopy of the knee joint. *Skeletal Radiol* 1982;8:187–192.
131. Sherman OH, Fox JM, Snyder SJ, Del Pizzo W, Friedman MJ, Ferkel RD, Lawley MJ. Arthroscopy—"No problem surgery"—An analysis of complications in 2640 cases. *J Bone Joint Surg* 1986;68:256–265.
132. Selesnick FH, Noble HB, Bachman DC, Steinberg FL, Internal derangement of the knee: Diagnosis by arthrography, arthroscopy, and arthrotomy. *Clin Orthop* 1985;198:26–30.
133. Dragnot P, Claussen C. Sonographic demonstration of the meniscus. *ROFO* 1980;133:185–187.
134. Selby B, Richardson ML, Montana MA, Teitz CC, Larson RV, Mack LA. High resolution sonography of the menisci of the knee. *Invest Radiol* 1986;21:332–335.
135. Marymont JV, Lynch MA, Henning CE. Evaluation of meniscus tears of the knee by radionuclide imaging. *Am J Sports Med* 1983;11:432–435.
136. Collier BD, Johnson RP, Carrera GF, Isitman AT, Veluvolu P, Knobel J, Hellman RS, Barthelemy CR. Chronic knee pain assessed by SPECT: Comparison with other modalities. *Radiology* 1985;157:795–802.
137. Passariello R, Trecco F, de Paulis F, Masciocchi C, Bonanni G, Zobel BB. Meniscal lesions of the knee joint: CT diagnosis. *Radiology* 1985;157:29–34.
138. Sanchis-Alfonso V, Martinez-Sanjuan V, Sanchis-Cabanilles M, Guinot-Tormo JM, Torres-Hurtado JI, Rubio-Viguera V. The value of high resolution CT in the diagnosis of meniscal lesions of the knee. *Arthroscopy* 1991;7(4):375–380.
139. Ghelman B. Meniscal tears of the knee: Evaluation by high-resolution CT combined with arthrography. *Radiology* 1985;157:23–27.
140. Barronian AD, Zoltan JD, Bucon KA. Magnetic resonance imaging of the knee: Correlation with arthroscopy. *Arthroscopy* 1989;5(3):187–191.
141. Bellon EM, Keith MW, Coleman PE, Shah ZR. Magnetic resonance imaging of internal derangements of the knee. *Radiographics* 1988;8(1):95–118.
142. Beltran J, Noto AM, Mosure JC, Bools JC, Zuelzer W, Christoforidis AJ. Meniscal tears: MR demonstration of experimentally produced injuries. *Radiology* 1986;158:691–693.
143. Beltran J, Noto AM, Mosure JC, Weiss KL, Zuelzer W, Chrostoforidis AJ. The knee: Surface-coil MR imaging at 1.5T. *Radiology* 1986;159:747–751.
144. Brunner MC, Flower SP, Evancho AM, Allman FL, Apple DF, Fajman WA. MRI of the athletic knee. Findings in asymptomatic professional basketball and collegiate football players. *Invest Radiol* 1989;24(1):72–75.
145. Burk DL, Kanal E, Brunberg JA, Johnstone GF, Swensen HE, Wolf GL. 1.5T surface-coil MRI of the knee. *AJR* 1986;147:293–300.
146. Burk DL, Dalinka MK, Kanal E, Schiebler ML, Cohen EK, Prorok RJ, Gefter WB, Kressel HY. Meniscal and ganglion cysts of the knee: MR evaluation. *AJR* 1988;150(2):331–336.
147. Crues JV, Stoller DW. *The menisci*. In: Magnetic resonance imaging of the knee, Mink JH, Reicher MA, Crues JV, eds. Raven Press: New York; 1987:55–92.
148. Crues JV, Ryu R, Morgan FW. Meniscal pathology: The expanding role of magnetic resonance imaging. *Clin Orthop* 1990;252:80–87.
149. Crues JV, Ryu R. *The knee*. IN: Magnetic resonance imaging, Stark D, Bradley WG, eds. St. Louis: Mosby; 1991; [in press].
150. Deutsch AL, Mink JH, Fox JM, Arnoczky SP, Rothman BJ, Stoller DW, Cannon WD. Peripheral meniscal tears: MR findings after conservative treatment or arthroscopic repair. *Radiology* 1990;176:485–488.
151. Deutsch AL, Mink JH, Rothman BJ. Longitudinal follow-up of patients with conservatively treated and arthroscopically repaired peripheral tears: MR findings and implications. *Radiology* 1989;173(P):232–233.
152. Ehman RL. MR imaging of the knee. *Radiology* 1989; 173(P):132–133.

153. Fitzgerald RH, Berquist TH. Magnetic resonance imaging. *J Bone Joint Surg* 1986;68:799–801.
154. Haggar AM, Froelich JW, Hearshen DO, Sadasivan K. Meniscal abnormalities of the knee: 3DFT fast-scan GRASS MR imaging. *AJR* 1988;150(6):1341–1344.
155. Herman LJ, Beltran J. Pitfalls in MR imaging of the knee. *Radiology* 1988;167(3):775–781.
156. Kean DM, Worthington BS, Preston BJ, et al. NMR imaging of the knee: Examples of normal anatomy and pathology. *Br J Radiol* 1983;56:355–364.
157. King CL, Henkelman M, Poon PY, Rubenstein J. MR imaging of the normal knee. *J Comput Assist Tomogr* 1984;8:1147–1154.
158. Kursunoglu-Brahme S, Schwaighofer B, Gundry C, Ho C, Resnick D. Jogging causes acute changes in the knee joint: an MR study in normal volunteers. *AJR* 1990;154:1233–1253.
159. Li KC, Henkelman RM, Poon PY, Rubenstein J, MR imaging of the normal knee. *J Comput Assist Tomogr* 1984;8:1147–1154.
160. Mandelbaum BR, Finerman GM, Reicher MA, Hartzman S, Bassett LW, Gold RH, Rauschning W, Dorey F. Magnetic resonance imaging as a tool for evaluation of traumatic knee injuries: Anatomical and pathoanatomical correlations. *Am J Sports Med* 1986;14:361–370.
161. Matthews LS, Martin DF. Meniscal cysts: MRI with arthroscopic correlation. *Am J Knee Surg* 1989;2(2):87–92.
162. Mink JH, Stoller DW, Martin C, Crues JV. MR imaging of the knee: Pitfalls in interpretation. *Radiology* 1986;161(P):239.
163. Mink JH, Reicher MA, Crues JV. Magnetic resonance imaging of the knee. New York: Raven Press, 1987:178.
164. Mink JH, Levy T, Crues JV. Tears of the anterior cruciate ligament and menisci of the knee: MR imaging evaluation. *Radiology* 1988;167:769–774.
165. Mink JH, Deutsch AL, The Knee. IN: MRI of the musculoskeletal system: A teaching file, Mink JH, ed. New York: *Raven Press*; 1990:298.
166. Niitsu M, Anno I, Fukubayashi T, Shimojo H, Kuno S, Akisada M. Tears of cruciate ligaments and menisci: Evaluation with cine MR imaging. *Radiology* 1991;178:859–864.
167. Quinn S, Brown T. Evaluation of the menisci: How reliable and relevant is arthroscopic correlation? *Radiology* 1990;177(P):225–226.
168. Quinn SF, Brown TF. Meniscal tears diagnosed with MR imaging versus arthroscopy: How reliable a standard is arthroscopy? *Radiology* 1991;181(3):843–847.
169. Raunest J, Oberle K, Loehnert J, Hoetzinger H. The clinical value of magnetic resonance imaging in the evaluation of meniscal disorders. *J Bone Joint Surg* 1991;73-A(1):11–16.
170. Reicher MA, Rauschning W, Gold RH, Bassett LW, Lufkin RB, Glenn W. High-resolution magnetic resonance imaging of the knee joint: Normal anatomy. *AJR* 1985;145:895–902.
171. Shellock FG, Mink JH. Knees of trained long-distance runners: MR imaging before and after competition. *Radiology* 1991;179(3):635–637.
172. Shellock FG, Deutsch AL, Mink JH, Kerr R. Do asymptomatic marathon runners have an increased prevalence of meniscal abnormalities? An MR study of the knee in 23 volunteers. *AJR* 1991;157:1239–1241.
173. Smith DK, Totty WG. The knee after partial meniscectomy: MR imaging features. *Radiology* 1990;176:141–144.
174. Spritzer CE, Vogler JB, Martinez S, Garrett WJ, Johnson GA, McNamara MJ, Lohnes J, Herfkens RJ. MR imaging of the knee: preliminary results with a 3DFT GRASS pulse sequence. *AJR* 1988;150(3):597–603.
175. Stoller DW, *The knee* IN: Magnetic resonance imaging in orthopaedics and rheumatology. Stoller DW, ed. Philadelphia: Lippincott; 1989;152.
176. Turner DA, Rapoport MI, Erwin WD, McGould M, Silvers RI. Truncation artifact: A potential pitfall in MR imaging of the menisci of the knee. *Radiology* 1991;179(3):629–633.
177. Moon KJ, Genant HK, Helms CA, Chafetz NI, Crooks LE, Kaufman L. Musculoskeletal applications of nuclear magnetic resonance. *Radiology* 1983;147(1):161–171.
178. Reicher MA, Bassett LW, Gold RH. High-resolution magnetic resonance imaging of the knee joint: Pathologic correlations. *AJR* 1985;145:903–909.
179. Casscells SW. Arthroscopy of the knee joint. *J Bone Joint Surg* 1971;53:287–298.
180. Grana WA, Conner S, Hollingsworth S. Partial arthroscopic meniscectomy: A preliminary report. *Clin Orthop* 1982;164:78–83.
181. Bergstrom R, Gillquist J, Lysholm J, Hamberg P. Arthroscopy of the knee in children. *J Pediatr Orthop* 1984;4:542–545.
182. Smillie IS. *Treatment, after treatment and complications of lesions of the menisci.* IN: Injuries of the knee joint. London: E. & S. Livingstone; 1970;98–129.
183. Smillie IS. *Injuries of the knee joint,* 5th ed. New York: Churchill Livingstone, 1978.
184. Suman RK, Stother IG, Illingworth G. Diagnostic arthroscopy of the knee in children. *J Bone Joint Surg* 1984;66:535–537.
185. King AG. Meniscal lesions in children and adolescents: A review of the pathology and clinical presentation. *Injury* 1983;15:105–108.
186. Singer KM, Henry J. Knee problems in children and adolescents. *Clin Sports Med* 1985;4:385–398.
187. Manzione M, Pizzutillo PD, Peoples AB, Schweizer PA. Meniscectomy in children: A long term follow-up study. *Am J Sports Med* 1983;11:111–115.
188. Medlar RC, Mandiberg JJ, Lyne ED. Meniscectomies in children. *Am J Sports Med* 1980;8:87–92.
189. Zaman M, Leonard MA. Meniscectomies in children: Results in 59 knees. *Injury* 1981;12:425–428.
190. Casscells SW. The place of arthroscopy in the diagnosis and treatment of internal derangement of the knee: An analysis of 1000 cases. *Clin Orthop* 1980;151:135–142.
191. Quinn SF, Muus C, Sara A, Estrada J, Walling A. Meniscal tears: Correlation with MR imaging. *Radiology* 1988; 166(2):580.
192. Critchley IJ, Bracey DJ. The acutely locked knee—Is a manipulation worthwhile? *Injury* 1985;16:281–283.
193. DeHaven KE. Diagnosis of acute knee injuries with hemarthrosis. *Am J Sports Med* 1980;8:9–14.
194. DeHaven KE. *Evaluation of the acutely injured knee.* IN: Arthroscopy: Diagnostic and surgical practice. Casscells SW, ed. Philadelphia: Lea & Febiger; 1984;64–71.
195. Glinz W, Segantini P, Kagi P. Arthroscopy in acute trauma of the knee joint. *Endoscopy* 1980;12:269–274.
196. Jensen JE, Conn RR, Hazelrigg G, Hewett JE, Systematic evaluation of acute knee injuries. *Clin Sports Med* 1985;4: 295–312.
197. Kohn D. Arthroscopy in acute injuries of anterior cruciate-deficient knees: Fresh and old intraarticular lesions. *Arthroscopy* 1986;1:40–43.
198. Lysholm J, Gillquist J. Arthroscopy in the early diagnosis of injuries to the knee joint. *Acta Orthop Scand* 1981;52:111–118.
199. Noyes FR, Bassett RW, Grood ES, Butler DL. Arthroscopy in acute traumatic hemarthrosis of the knee. Incidence of anterior cruciate tears and other injuries. *J Bone Joint Surg* 1980;62:687–695.
200. Rand JA. The role of arthroscopy in the management of knee injuries in the athlete. *Mayo Clin Proc* 1984;59:77–82.
201. Warren RF, Levy IM. Meniscal lesions associated with anterior cruciate ligament injury. *Clin Orthop* 1983;172:32–37.
202. Deutsch AL, Mink JH, Shellock FG. Magnetic resonance imaging of injuries to bone and articular cartilage. *Orthop Rev* 1989;19(1):66–75.
203. Mink JH, Deutsch AL. Occult cartilage and bone injuries of the knee: Detection, classification, and assessment with MR imaging. *Radiology* 1989;170:823–829.
204. Lee JK, Yao L. Occult intraosseous fracture: Magnetic resonance appearance versus age of injury. *Am J Sports Med* 1989;17(5):620–623.
205. Colville JM, Jackson DW. Reasonable expectations following arthroscopic knee surgery. *Clin Sports Med* 1985;4:279–294.
206. Larson RL. Overview and philosophy of knee injuries. *Clin Sports Med* 1985;4:209–215.
207. Baratz ME, Fu FH, Mengato R. Meniscal tears: The effect of meniscectomy and of repair on intraarticular contact areas and stress in the human knee. A preliminary report. *Am J Sports Med* 1986;14:270–275.

208. Henning C. Arthroscopic repair of meniscal tears. *Orthopedics* 1983;6:1130–1132.
209. Henning CE, Lynch MA. Current concepts of meniscal function and pathology. *Clin Sports Med* 1985;4:259–265.
210. Levy IM, Torzilli PA, Warren RF. The effect of medial meniscectomy on anterior posterior motion of the knee. *J Bone Joint Surg* 1982;64:883–888.
211. Markolf KL, Mensch JS, Amstutz HC. Stiffness and laxity of the knee—The contributions of the support structures. *J Bone Joint Surg* 1976;58:583–593.
212. Shoemaker SC, Markolf KL. The role of the meniscus in the anterior-posterior stability of the loaded anterior cruciate-deficient knee. Effects of partial versus total excision. *J Bone Joint Surg* 1986;68:71–79.
213. Burr DB, Radin EL. Meniscal function and the importance of meniscal regeneration in preventing late medial compartment osteoarthrosis. *Clin Orthop* 1982;171:121–126.
214. Cassidy RE, Shaffer AJ. Repair of peripheral meniscal tears. *Am J Sports Med* 1981;9:209–214.
215. Jackson JP. Degenerative changes in the knee after meniscectomy. *Br Med J* 1968;2:525–527.
216. Johnson RJ, Kettelkamp DB, Clark W. Factors affecting late results after meniscectomy. *J Bone Joint Surg* 1974;56:719–729.
217. Jones RE, Smith EC, Reisch JS. Effects of medial meniscectomy in patients older than 40. *J Bone Joint Surg* 1978;60:783–786.
218. Kobayashi A. Discoid meniscus of the knee joint. *Clin Orthop Surg [Jpn]* 1975;10:10–24.
219. Krause WR, Pope MH, Johnson RJ. Mechanical changes in the knee after meniscectomy. *J Bone Joint Surg* 1976;58:599–604.
220. Lynch MA, Henning CE, Glick KR. Knee joint surface changes. Long-term follow-up meniscus tear treatment in stable anterior cruciate ligament reconstructions. *Clin Orthop* 1983;172:148–153.
221. Whipple TL, Caspari RB, Meyers JF. Arthroscopic meniscectomy. An interim report at three to four years after operation. *Clin Orthop* 1984;183:105–114.
222. McDaniel WJ, Dameron TB. Untreated ruptures of the anterior cruciate ligament. *J Bone Joint Surg* 1980;62:696–705.
223. Aarstrand T. Treatment of meniscal rupture of the knee joint. *ACTA Chir Scand* 1954;107:146–157.
224. Hamberg P, Gillquist J, Lysholm J. Suture of new and old peripheral meniscal tears. *J Bone Joint Surg* 1983;65:193–197.
225. Martens MA, Bachaert M, Heyman E, Mulier JC. Partial arthroscopic meniscectomy versus total open meniscectomy. *Arch Orthop Trauma Surg* 1986;105:31–35.
226. Northmore-Ball MD, Dandy DJ. Long-term results of arthroscopic partial meniscectomy. *Clin Orthop* 1982;167:34–42.
227. Tapper EM, Hoover NW. Late results after meniscectomy. *J Bone Joint Surg* 1969;51:517–526.
228. Simpson DA, Thomas NP, Aichroth PM. Open and closed meniscectomy. A comparative analysis. *J Bone Joint Surg* 1986;68:301–304.
229. Yocum LA, Kerlan RK, Jobe FW, Carter VS, Shields CL, Lombardo SJ, Collins HR. Isolated lateral meniscectomy: A study of twenty-six patients with isolated tears. *J Bone Joint Surg* 1979;61:338–342.
230. King D. The healing of semi-lunar cartilages. *J Bone Joint Surg* 1936;18:333–342.
231. Dwyer FC, Taylor C. Congenital discoid internal cartilage. *Br Med J* 1945;2:287.
232. Cabaud HE, Rodkey WG, Fitzwater JE. Medial meniscus repairs. An experimental and morphologic study. *Am J Sports Med* 1981;9:129–134.
233. DeHaven KE. Meniscus repair in the athlete. *Clin Orthop* 1985;198:31–35.
234. El-Khoury GY, Usta HY, Berger RA. Meniscotibial (coronary) ligament tears. *Skeletal Radiol* 1984;11:191–196.
235. Stone RG, Miller GA. A technique of arthroscopic suture of torn menisci. *Arthroscopy* 1985;1:226–232.
236. Warren RF. Arthroscopic meniscus repair. *Arthroscopy* 1985;1:170–172.
237. Jayson MIV, Dixon ASJ. Arthroscopy of the knee in rheumatic diseases. *Ann Rheum Dis* 1968;27:503–511.
238. Kawashima W. Arthroscopy of the tuberculous knee joint in its early stage. *J Jpn Orthop Assoc* 1943;18:651–653.
239. Kreuscher PH. Semilunar cartilage disease: A plea for early recognition by means of the arthroscope and the early treatment of this condition. *Ill Med J* 1925;47:290–292.
240. Robles Gil J, Katona G, Barriso MR. Arthroscopy as an aid to diagnosis and investigation. *Excerpta Med Int Series* 1967;143:16–17.
241. Takagi K. Practical experiences using Takagi's arthroscope. *J Jpn Orthop Assoc* 1933;8:132.
242. Chana GS, Tubbs N. Early results of arthroscopic surgery of the knee. *Injury* 1981;13:227–229.
243. Gillquist J. Operative arthroscopy. *Endoscopy* 1980;12:281–287.
244. Gillquist J, Oretorp N. Arthroscopic partial meniscectomy. Technique and long-term results. *Clin Orthop* 1982;167:29–33.
245. Henard DC. Medial meniscectomy: A prospective clinical study. *South Med J* 1983;76:1145–1146.
246. Lysholm J, Gillquist J. Arthroscopic meniscectomy in athletes. *Am J Sports Med* 1983;11:436–438.
247. Parisien JS. Arthroscopic treatment of cysts of the menisci: A preliminary report. *Clin Orthop* 1990;257:154–158.
248. Pettrone FA. Meniscectomy: Arthrotomy versus arthroscopy. *Am J Sports Med* 1982;10:355–359.
249. Tregonning RJ. Diagnostic arthroscopy of the knee joint: Meniscal findings in a prospective study of 200 examinations. *N Z Med J* 1981;94:81–83.
250. Tregonning RJ. Closed and open partial meniscectomy for bucket handle tears: A comparative evaluation. *N Z Med J* 1983;96:375–377.
251. Zarins B, Boyle J, Harris BA. Knee rehabilitation following arthroscopic meniscectomy. *Clin Orthop* 1985;198:36–42.
252. Scott GA, Jolly BL, Henning CE. Combined posterior incision and arthroscopic intra-articular repair of the meniscus: An examination of factors affecting healing. *J Bone Joint Surg* 1986;68:847–861.
253. Eastwood DM. The failures of arthroscopic partial meniscectomy. *Injury* 1985;16:587–590.
254. Ferkel RD, Davis JR, Friedman MJ, Fox JM, Del Pizzo W, Snyder SJ, Berasi CC. Arthroscopic partial medial meniscectomy: An analysis of unsatisfactory results. *Arthroscopy* 1985;1:44–52.
255. Hershman EB, Nisonson B. Arthroscopic meniscectomy: A follow-up report. *Am J Sports Med* 1983;11:253–257.
256. O'Beirne J, O'Dwyer T, O'Rourke JS, Quinlan W. The diagnosis of knee injuries in casualty—A prospective study. *Injury* 1984;15:232–235.
257. Rand JA. Arthroscopic management of degenerative meniscus tears in patients with degenerative arthritis. *Arthroscopy* 1985;1:253–258.
258. Becton JL, Young HH. Cysts of the semilunar cartilage of the knee. *Arch Surg* 1985;90:708–712.
259. Bresler M, Olson M. Lateral meniscal cyst. *Orthopedics*, 1990;13(6):672–674.
260. Lee JK, Yao L. Tibial collateral ligament bursae: MR imaging. *Radiology* 1991;178:855–857.
261. McIntyre JL. Arthrography of the lateral meniscus. *Radiology* 1972;105:531–536.
262. Young RB. The external semilunar cartilage as a complete disc. IN: Memoirs and memoranda in anatomy. Cleland JY, Young RB, eds. London: Williams & Norgate; 1889.
263. Cave EF, Staples OS. Congenital discoid meniscus. A cause of internal derangement of the knee. *Am J Surg* 1941;54:371–376.
264. Jeannopoulos CL. Observations on discoid menisci. *J Bone Joint Surg* 1950;32:649–652.
265. Middleton DS. Congenital disc-shaped lateral meniscus with snapping knee. *Br J Surg* 1936;24:246–255.
266. Nathan PA, Cole SC. Discoid meniscus. A clinical and pathologic study. *Clin Orthop* 1969;64:107–113.
267. Nemoto HN. Study on discoid meniscus of the knee. *Nigata Med J* 1950;64:404–410.

268. Smillie IS. The congenital discoid meniscus. *J Bone Joint Surg* 1948;30:671–682.
269. Weiner B, Rosenberg N. Discoid medial meniscus: Associations with bone changes in the tibia. *J Bone Joint Surg* 1974;56:171.
270. Berson BL, Hermann G. Torn discoid menisci of the knee in adults: Four case reports. *J Bone Joint Surg* 1979;61:303–304.
271. Dickason JM, Del PW, Blazina ME, Fox JM, Friedman MJ, Snyder SJ. A series of ten discoid medial menisci. *Clin Orthop* 1982.
272. Dickaut SC, DeLee JC. The discoid lateral-meniscus syndrome. *J Bone Joint Surg* 1982;64:1068–1073.
273. Murdoch G. Congenital discoid medial semilunar cartilage. *J Bone Joint Surg* 1956;38:564–566.
274. Resnick D, Goergen TG, Kay JJ, Ghelman B, Woody PR. Discoid medial meniscus. *Radiology* 1976;121:575–576.
275. Riachi E, Phares A. An unusual deformity of the medial semilunar cartilage. *J Bone Joint Surg* 1963;45:145–147.
276. Richmond PA. Two cases of discoid medial cartilage. *J Bone Joint Surg* 1958;40:268–269.
277. Ross JA, Tough ICK, English TA. Congenital discoid cartilage. Report of a case of discoid medial cartilage with an embryological note. *J Bone Joint Surg* 1958;40:262–267.
278. Amako T. On the injuries of the menisci in the knee joint of Japanese. *J Jpn Orthop Surg Soc* 1960;33:12.
279. Watanabe M. *Arthroscopy of the knee joint*. IN: Disorders of the knee. Helfet AJ, ed. Philadelphia: Lippincott; 1974:145.
280. Watanabe M, Takeda S, Ikeuchi H. *Atlas of arthroscopy*, 2nd ed. Tokyo: Igaku Shoin; 1969.
281. Ikeuchi H. Arthroscopic treatment of the discoid lateral meniscus. Technique and long term results. *Clin Orthop* 1982;167:19–28.
282. Albertsson M, Gillquist J. Discoid lateral menisci: A report of 29 cases. *Arthroscopy* 1988;4(3):211–214.
283. Husson JL, Meadeb J, Cameau J, Blouet JM, Masse A, Duval JM. [Medial and lateral discoid menisci in the same knee. Apropos of a case]. *Bull Assoc Anat (Nancy)*, 1985;69(206):201–208.
284. Sugawara O, Miyatsu M, Yamashita I, Takemitsu Y, Onozawa T. Problems with repeated arthroscopic surgery in the discoid meniscus. *Arthroscopy* 1991;7(1):68–71.
285. Vandermeer RD, Cunningham FK. Arthroscopic treatment of the discoid lateral meniscus: Results of long-term follow-up. *Arthroscopy* 1989;5(2):101–109.
286. Lynch CP. Case 10 IN: MRI of the musculoskeletal system. Crues JV, ed. New York: Raven Press; 1991:22–23.
287. Berg EE. The meniscal ossicle: The consequence of a meniscal avulsion. *Arthroscopy* 1991;7(2):241–243.
288. Perry CR, Evans LG, Rice S, Fogarty J, Burdge RE. A new surgical approach to fractures of the lateral tibial plateau. *J Bone Joint Surg* 1984;66:1236–40.

ADDITIONAL READINGS

Covall DJ, Wasilewski SA. Roentgenographic changes after arthroscopic meniscectomy: Five-year follow-up in patients more than 45 years old. *Arthroscopy* 1992;8:242–246.

Holder J, Haghighi P, Pathria MN, et al. Meniscal changes in the elderly: correlation of MR imaging and histologic findings. *Radiology* 1992;184:221–225.

CHAPTER 5

The Cruciate and Collateral Ligaments

Jerrold H. Mink

THE ANTERIOR CRUCIATE LIGAMENT

Anatomic Considerations

The anterior cruciate ligament (ACL) arises from a semicircular fossa on the posterior, medial, nonarticular aspect of the lateral femoral condyle and inserts in front of and lateral to the anterior tibial spine, 23 mm from the anterior tibial cortex (1). The ACL has a mean length of 35 mm (+/− 10 mm) and a mean mid-portion width of 11 mm (1,2). Just proximal to its insertion on the tibia, the ACL passes underneath the transverse meniscal ligament, and fibers of the ACL blend with the anterior central attachment of the lateral meniscus. The anterior, as well as the posterior cruciate ligament, is completely enveloped by a mesentery-like fold of synovium that originates from the posterior intercondylar area and inserts on the anterior tibial insertion of the ACL itself where it joins with the tissue of the joint capsule distal to the infrapatellar fat pad. The cruciate ligaments are therefore intraarticular but extrasynovial. It is this synovial membrane, richly supplied with vessels, that is the primary vascular supply to the ACL via the branches of the middle genicular artery, as well as some branches of the medial and lateral inferior genicular arteries (3). The vessels within the sheath arborize to form a network of periligamentous channels that engulf the entire ACL, although its central core has a rather limited blood supply. Importantly, the ligament/osseous junctions do not contribute significantly to the vascular supply of the ACL, a fact that explains why a mid-substance ACL tear most often results in ischemia and the subsequent inability of the ACL to properly repair itself.

The ACL is not a single mass of tissue but rather a collection of individual fascicles. Each fascicle is composed of nonparallel, interlacing networks of collagen fibrils that are grouped into fibers, which in turn are arranged into fiber bundles. These fiber bundles are oriented either in a linear fashion from femur to tibia, or more commonly in an outward, spiral fashion about the long axis of the ACL. It is this spiral arrangement of the fibers that allows winding and unwinding of the ligament along its long axis and permits the ACL to function as a constraint to abnormal joint motion. The fiber bundles have been divided into two groups, named for their insertion sites on the tibia: the larger and stronger anteromedial band (AMB), and the smaller posterolateral band (PLB). During knee extension, the PLB is taut. With progressive flexion, the femoral attachment of the ACL assumes a more horizontal orientation, causing the AMB to tighten and the PLB to become quite relaxed. Therefore, some portion of the ACL is taut throughout the range of motion (1). When taken as a unit, the ACL is maximally taut when the knee is in full extension and in 5° and 20° of flexion, is most lax in 40° to 50° degrees of flexion, and then becomes increasingly taut as flexion increases from 70° to 90° degrees (2). Internal rotation of the knee increases tension of the ACL at any degree of flexion.

Clinical Assessment

Acute rupture of the ACL is frequently a dramatic event. Many patients report the hearing of a "pop" at the time of injury, a sensation which should very strongly suggest the diagnosis of an ACL tear (1). The unique helicoid arrangement of the ACL allows it to store considerable energy before its elastic limit is reached, at which time the ligament bursts convulsively. Following an ACL tear, the patient experiences sudden, severe pain, although he/she may be able to walk off the field of play or down the ski slope. Marked swelling does not occur immediately, but is delayed for 6 to 24 hr after the injury since the artery to the ACL is a rather diminutive structure, and the hemorrhage from it has been described as a leaky faucet. Rapid accumulation of joint fluid in the patient with a suspected ACL tear should suggest the alternative/concomitant diagnosis of an osteochondral fracture or an avulsion of the intercondylar eminence, especially if gross fat is present in the effusion. A hemarthrosis indicates the presence of a partial tear in 15% to 28% of cases, and a complete tear in 44% to 66% of cases (4,5). However, subsynovial tears of the ACL have been found in patients who do not have hemarthrosis (6).

The classic mechanism of injury of the ACL is a change of direction during deceleration with the foot fixed in position and subsequent contact with another person/object, usually from the lateral side of the knee. This is the classic American football injury, one which frequently results in O'Donoghue's unhappy triad, consisting of a tear of the ACL, disruption of the medial collateral ligament, and a tear of the medial meniscus. Recently however, ACL injuries have been occurring more frequently in noncontact sports. These injuries occur when the body and femur rotate externally about a fixed tibia with the leg in extension (2,7). Alpine skiing has produced a virtual epidemic of ACL injuries as a result of changes in equipment, technique, and terrain. Some estimates suggest that over 100,000 ACL tears/year occur as a result of ski-related injuries alone (8). The Monday morning orthopedic waiting room filled with patients suffering malleolar or distal tibial fractures in the 1970s has been replaced by individuals with ACL ruptures; this change has occurred coincident with the use of high, stiff retention boots coupled to firm cambered skis, with the skier undergoing rapid directional changes by means of avalement (sitting back and using the tail of the skis to enhance the stiff, high boot) (1). Ski-related injuries tend to occur at the end of the day when the skier is tired; he/she is unable to remain centered over the skis; and resorts to a hip-flexed, knee-extended position to relieve the burning of fatigued quadriceps. This is the skier at risk, since the ACL is stretched over the lateral femoral condyle (8). Injuries often occur at a relatively slow speed at which the bindings are not able to recognize abnormal forces quickly enough to protect the ACL, and the skiers' muscles respond too slowly to prevent injury.

The mechanism of rupture of the ACL is identical to those forces that may produce patellar dislocation. Up to 10% of knees that are clinically suspected of suffering an ACL tear also have patellar dislocations with osteochondral fracture (1,9).

A computer graphics study has recently supported the concept of primary notch stenosis as being etiologic in the cause of ACL tears (10–13). The notch may be congenitally small, or may decrease in size as a result of hypertrophic changes of the intercondylar eminence/spines.

In a recent study of patients who had suffered an ACL tear, the lesion was proximal in 18%, mid-substance in 72%, and distal in 4% (2). In skiers however, the distribution of rupture is different: proximal in 80%, midsubstance in 7%, and distal in 10% (1). It should be noted, however, that it is quite difficult to precisely localize the site and/or the extent of the tear at arthroscopy (14–16).

Partial rupture of the ACL, representing 10% to 28% of all ACL injuries, usually presents with signs and symptoms of mechanical interference with joint function, but the clinical diagnosis is often incorrect (17). Partial tears occur because of the slightly different functions of the AMB and the PLB, and it is usually the AMB that is torn (5,18,19). Such tears may heal without residue but frequently become complete tears within 1 year of initial injury (6,17,19,20). There is little correlation between the percent ligament tear and the subsequent clinical result.

While acute tears may be isolated, meniscal tears, osteochondral fractures, medial collateral ligament ruptures, and capsular disruptions are frequently found in the knee that has suffered an acute ACL tear (1). ACL ruptures are accompanied by medial meniscal tears in 43%, lateral tears in 66%, with an overall incidence of meniscal tears of 65% to 80% (4,21,22).

Avulsion of the attachment of the ACL occurs almost exclusively at the tibia since the tibial end is stronger than the proximal end. Less than 5% of ACL ruptures occur in adults as a result of avulsion, but the relative strength of the ACL compared to that of the adjacent bone and physeal plate explain why most ACL injuries in children are avulsive or result in a physeal fracture (23). It is essential to rapidly diagnose ACL avulsion; the fragment may block extension and preclude normal knee function; reattachment of the bony fragment has a high rate of success in restoring ACL function. Avulsive injuries are occasionally associated with elevation of "wings" of articular cartilage from either the medial or lateral aspects of the tibial plateau, but are uncommonly associated with other intraarticular injury (1). Congenital absence of the ACL is an uncommon condition that presents in children with signs and symptoms of ACL insufficiency who do not have a history of trauma. These children often have other anomalies of the involved limb, such as congenital dislocation of the knee, proximal femoral focal deficiency, absence of the fibular rays of the foot, and valgus knee and leg length discrepancy (1). Aplasia of the intercondylar eminence should suggest the diagnosis of isolated, congenital absence of the ACL.

The primary function of the ACL is to limit anterior translation of the tibia with respect to the femur and to serve as a rotational guide for the femur and tibia (24). The AMB further serves to prevent hyperextension of the joint. Accurate diagnosis, and ultimately treatment of knee ligament injuries, depends on the ability of clinicians to consistently and reproducibly perform reliable tests of ligamentous integrity (25,26). In the acute phase, soft tissue edema, hemarthrosis, hamstring spasm, and pain often make assessment of abnormal motion quite difficult. Absence of the classic anterior drawer sign (which is performed at 90° of flexion) is not a dependable sign of ACL integrity; its sensitivity in detection of a complete ACL tear is 78% (25,27–34). The posterior horn of the medial meniscus and the normal contour of the knee stabilize the ACL-deficient knee in an anteroposterior direction and serve to limit the accuracy of the anterior drawer sign. Additionally, the rather small forces applied to an ACL-deficient knee during physical examination may be insufficient to demonstrate instability because of the integrity of secondary restraints (e.g., the capsule and the medial collateral ligament), which generally do not become stretched and insufficient until 3 to 6 months following injury (2,27,35). False-positive anterior drawer signs occur in those patients with torn posterior cruciate ligaments who have, in fact, true positive posterior drawer maneuvers (7). The Lachman test, which is the anterior drawer sign performed in 20° of flexion, is only 89% sensitive for the detection of ACL tears in patients who have had arthroscopic confirmation of the tear (25,32,36). Up to 66% of clinical stable knees that present with a hemarthrosis prove to have complete tears of the ACL (4,5).

The pivot shift phenomenon is a momentary anterior subluxation of the tibia in relationship to the femur at 25° of knee flexion (37,38). It is thought to be an examination which preferentially tests for the integrity of the PMB (which is nearly always torn if the AMB is torn). A transient subluxation or reduction of subluxation is diagnostic of a complete ACL tear with anterolateral rotary instability (33). Nonetheless, several cases have been reported in which the test was negative in anesthetized patients who had arthroscopic evidence of an ACL tear (33). Multiple variations of the pivot shift test have been devised to test the same phenomenon of subluxation, with each test named for those who originally described them (e.g., MacIntosh, Slocum, Galway, Losee) (38). Hyperextension is a less reliable indicator of the integrity of the ACL, but a normal knee/abnormal knee difference of 5° to 10° is usually due to an ACL rupture (32).

Clinical quantitation of abnormal knee motion may be invalid because of the wide variations that occur, both in how the tests are performed and the amount of displacement induced (37). Only 6 of 11 examiners were able to reliably estimate the extent of pathologic motion of the knee by performing a variety of clinical tests (39). Since clinical tests may vary in their accuracy and may vary with the skill of the examiner, instrumented testing is frequently utilized (14,40). The KT-1000 arthrometer is commonly used to measure the anteroposterior displacement of the tibia to determine the integrity of the ACL following trauma, to document the amount of pathologic laxity, and to assess the success of reconstructive ACL procedures. A right-left difference of 3 mm is classified as pathologic (16).

While general systemic factors such as the patient's age, general physical condition, and activity level certainly come into play, several unique features of the ACL also play a role in its inability to heal itself: the blood supply; the peculiar anatomy of the ACL; the local biomechanical forces at the knee; the local biological environment; and the intrinsic capacity of ACL cells to effect repair (16). Within 2 weeks following a tear, arthroscopy reveals a contracted fibrotic ACL that is covered by a synovial cap (16,41). These irreversible changes do not allow for late surgical mobilization and reconstitution. While interruption of the vascular supply perhaps plays a role in rapid atrophy, release of collagenase from the synovium and the ACL itself, and the effect of blood in the joint, contribute to this observed effect.

As a result of the inherently poor capacity for healing of the torn ACL, patients frequently experience long-term disability (see Chapter 7). Chronic incompetence of the ACL leads to progressive deterioration of knee function. This process occurs through attritional stretching of the secondary restraints, destruction of the articular cartilage, and ultimate tearing of one or both of the menisci (up to 91%) (42). In a group of high school and college athletes who had suffered an ACL tear, less than half were able to return to their sports activities by 6 months because of pain and swelling, and only 11% were able to play strenuous sports 5 years later. In general, Noyes' "rule of thirds" applies to the patient with a torn ACL; one-third of patients will do well, one-third will have limitation of activity, and one-third will require ACL reconstruction.

Arthroscopy of the ACL

Advances in instrumentation and surgical techniques have so improved arthroscopy that it is 95% accurate for the diagnosis of internal derangement of the knee, and today has become the gold standard against which other imaging modalities are judged (23,48). Nonetheless, diagnostic arthroscopy is "an invasive, high cost examination that exposes the patient to surgical risk factors and postoperative discomfort"

(28,43). Sherman found that contrary to public and orthopedic opinion, arthroscopy is definitely not "no problem" surgery (44). They found a 4.8% rate of several complications and an 8.2% rate of overall complications as a result of this simple procedure. Complications include sepsis, nerve injury, instrument breakage, and reflex sympathetic dystrophy. Arthroscopy of the acutely injured knee is considerably more technically challenging than is elective arthroscopy. Blood frequently obscures the vision of the surgeon; hemorrhage and edema of the synovium and the fat pad may produce sufficient mass to further limit vision (45). Finally, arthroscopy demands a high degree of manual dexterity and three-dimensional perception. "No other orthopedic procedure requires as much experience for the operator to achieve proficiency," and it is a skill "which is learned poorly, or not at all," by some orthopedic surgeons (28).

It would appear that direct observation of the ACL with an arthroscope would be 100% accurate in assessment of a structure as large and significant as the ACL. In fact, arthroscopists misdiagnose even very severe ACL lesions (6,28,46). Casual arthroscopic assessment of the suspect ACL may be normal, but with closer inspection and probing, a complete tear is often identified (6,7). This potential error of overlooking a complete ACL tear is most likely to occur at or near either the femoral or tibial attachment or in those tears which remain in continuity. Two sources of arthroscopic error may lead to "false-positive MR" in the diagnosis of complete ACL rupture (46). The normal ligamentum mucosum, the vestige of the embryonic vertical septum of the knee, may be mistaken by the arthroscopist for a normal ACL. It lies just anterior to the ACL, and if the ACL is torn, the ligamentum mucosum will thicken and even more closely simulate a normal ACL. Chronic disruption of the ACL results in scarring and reattachment of the ACL to the proximal lateral synovium of the PCL (1). The retracted ligament assumes a more vertical orientation than the normal ACL (the vertical strut) (46). The inexperienced arthroscopist who does not critically assess the orientation of the ACL may misinterpret this vertical strut as a normal ligament. This error would be most grievous in those patients who are candidates for repair of a meniscal tear since such repairs are unlikely to succeed in patients with ACL insufficiency (47).

Even when the correct diagnosis of a partial ACL tear is made at arthroscopy, assessment of the extent of damage to the injured ligament may be impossible (14–16,18). The macroscopic appearance of the injured ACL may underestimate the exact extent of ligamentous disruption since damage to the blood supply cannot be appropriately assessed at arthroscopy; additionally, the degree of disruption of the ACL at the time of arthroscopy does not correlate with the eventual outcome (6,7,18–20). Kennedy found that the ACL could be macroscopically intact, while at microscopy one would see widespread disruption of collagen fibrils (48). He suggested that microscopic failure (a) may be a significant cause of later clinical instability or of secondary tears, and (b) may explain why arthroscopists are unable to predict from surgical evaluation which knees would later develop instability.

It should be apparent that a number of potential sources of very significant arthroscopic errors exist, any one of which may lead to an incorrect assessment of the status of the ACL and account for discrepancies between arthroscopic and MR findings.

Imaging Assessment

Conventional radiographs may demonstrate findings suggestive of chronic ACL disruption (49). The intercondylar region reacts to rotatory laxity through early peaking of the tibial eminences and later through stenosis of the femoral intercondylar notch. The presence of any one or more of the following conventional radiographic signs is highly indicative of failure of ACL insufficiency: (a) peaking of the intercondylar tubercles; (b) spurring of the intercondylar eminence; (c) osteophytosis of the inferior pole of the patella; (d) stenosis of the notch; and (e) osteoarthritic changes of the knee joint itself.

Arthrography is a relatively accurate examination for the assessment of the ACL, although the test is technically demanding and occasionally uncomfortable for the patient. Early reports of evaluation of the ACL by arthrography indicated an accuracy of 50%, but the addition of planar and computed tomography, cross table lateral upright, and vertical beam decubitus radiographs have produced 87% to 95% accuracy rates (1,31,41,50,51). An abnormal ACL is diagnosed by arthrography when the ligament is absent or has a wavy or irregular appearance on a lateral stress (anterior drawer) radiograph. Errors occur when the examiner mistakes the normal ligamentum mucosum for an intact ACL, or when the injury to the ACL has occurred in the presence of an intact synovial sheath.

MR Imaging

The standard imaging techniques and protocols utilized in assessment of the knee, including the cruciate and collateral ligaments, have been thoroughly discussed in Chapter 2. Several comments regarding specialized situations are worthy of comment.

Standard contiguous 5 mm thick sections in both the sagittal and coronal planes will permit accurate assessment of the ACL in nearly every case. The normal ACL presents one of two appearances (Fig. 5.1). Most

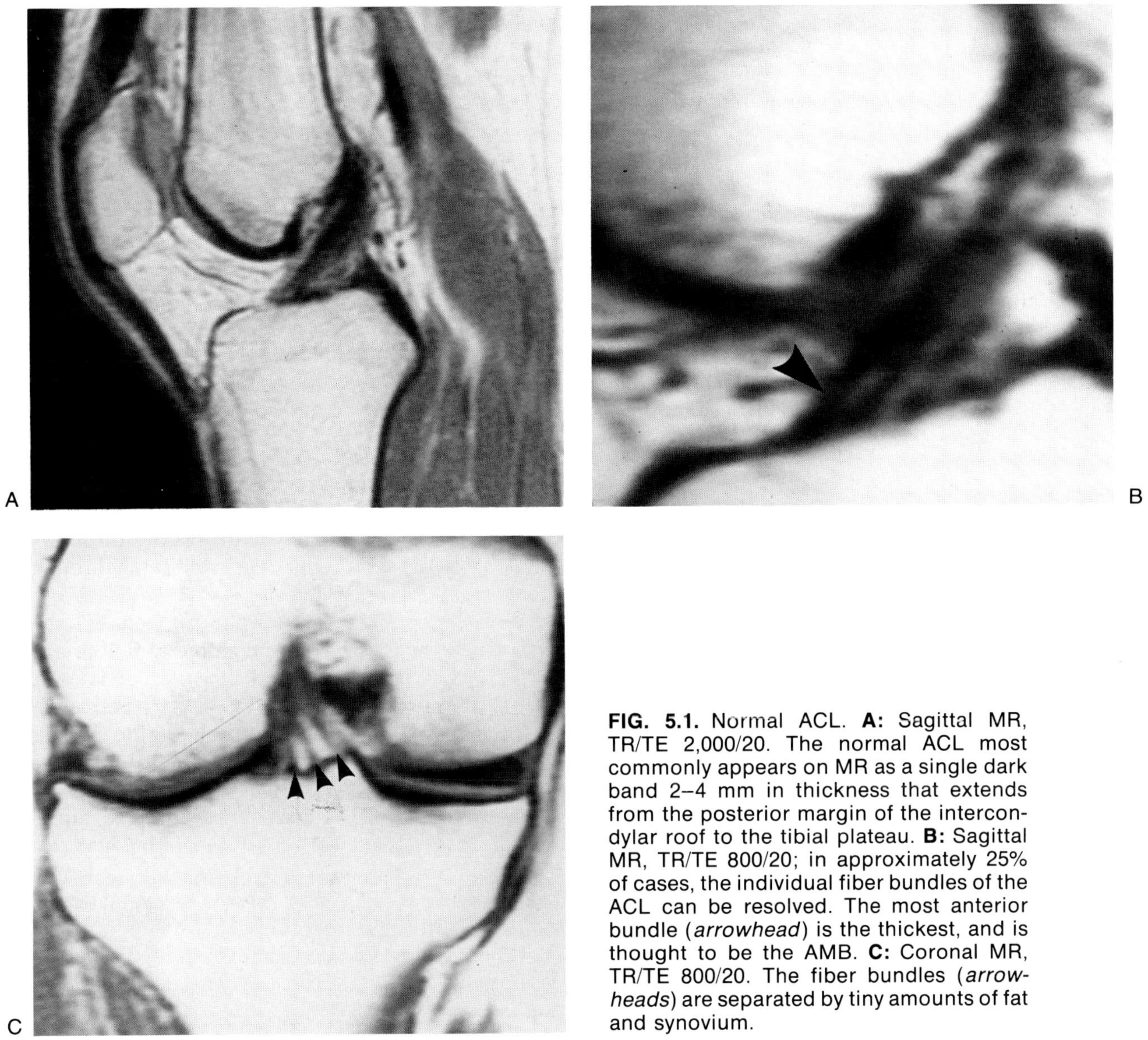

FIG. 5.1. Normal ACL. **A:** Sagittal MR, TR/TE 2,000/20. The normal ACL most commonly appears on MR as a single dark band 2–4 mm in thickness that extends from the posterior margin of the intercondylar roof to the tibial plateau. **B:** Sagittal MR, TR/TE 800/20; in approximately 25% of cases, the individual fiber bundles of the ACL can be resolved. The most anterior bundle (*arrowhead*) is the thickest, and is thought to be the AMB. **C:** Coronal MR, TR/TE 800/20. The fiber bundles (*arrowheads*) are separated by tiny amounts of fat and synovium.

commonly, the ACL is represented by a single, dark band, 3 to 4 mm in thickness, which, in the fully extended knee, is "ruler straight" along its anterior margin. Thickness, as determined by MR, is variable and has not proved to be a reliable criterion of normalcy. As noted above, however, the degree of tension in the ligament is dependent on the degree of flexion of the knee at the time of the examination (52). If the knee is in extension, the ACL is taut; with increasing degrees of flexion, the ACL becomes more lax. Therefore, as in arthrographic examinations, the configuration of the anterior margin of the ACL must be interpreted with full knowledge of the position of the leg.

In 30% to 40% of MR examinations, the normal, distal ACL is represented by two or three linear dark bands that are slightly separated from one another by stripes of intermediate to bright signal (T1WI) intensity thought to be synovium and fat (see Fig. 5.1) (see Chapter 11, Figs. 11.11–20). It is believed that these dark structures represent the individual bundles (AMB and PLB) that compose the ACL (5,52). At the tibia, the AMB appears as a compact, sharply marginated band which forms the anterior border of the ACL (5). The PMB is a less compact, somewhat triangular structure that has a slightly intermediate signal intensity. At the femoral attachment, the two fiber bundles are not separable from one another.

The ACL will be seen in its entirety on a single sagittal image in 90% of patients undergoing MR (52). In the other 10% of patients, portions of the ACL are seen on sequential images, and the examiner must mentally "compose" the integrity of the ligament. Some examiners find the inability to visualize the ACL on a single image confusing and disconcerting. In selected instances, especially in those patients with a high likelihood of an ACL tear (elite athletes, patients with synovitis, etc.), we elect to perform an additional sequence in an oblique sagittal plane in an effort to more optimally image the ligament (Fig. 5.2). Following performance of the high resolution coronal sequence, an oblique sagittal plane parallel to the superolateral to inferomedial trajectory of the normal ACL is selected. The image section thickness is reduced to either 3 or 4 mm; the TR/TE is changed to a "pure" proton density image (1,200/40). While we do not routinely utilize such sequences, others have found them to be accurate and potentially more useful than standard sagittal images (5). Oblique images, however, may result in some degree of distortion of the other intraarticular structures of the knee, and therefore we recommend that if oblique scanning of the ACL is utilized, it should be performed in addition to, rather than instead of, the orthogonal sagittal sequence.

Occasionally there is considerable partial volume effect of the proximal attachment of the ACL with the medial aspect of the lateral femoral condyle, especially when the knee is excessively internally rotated, in slight valgus, or in children or small adults. This degree of partial volume effect may be so significant as to result in the appearance of a "mass" at the proximal end of the ACL (Fig. 5.3). Two MR findings will suggest to the examiner that he/she is dealing with an ACL pseudomass rather than the true hematoma/mass that

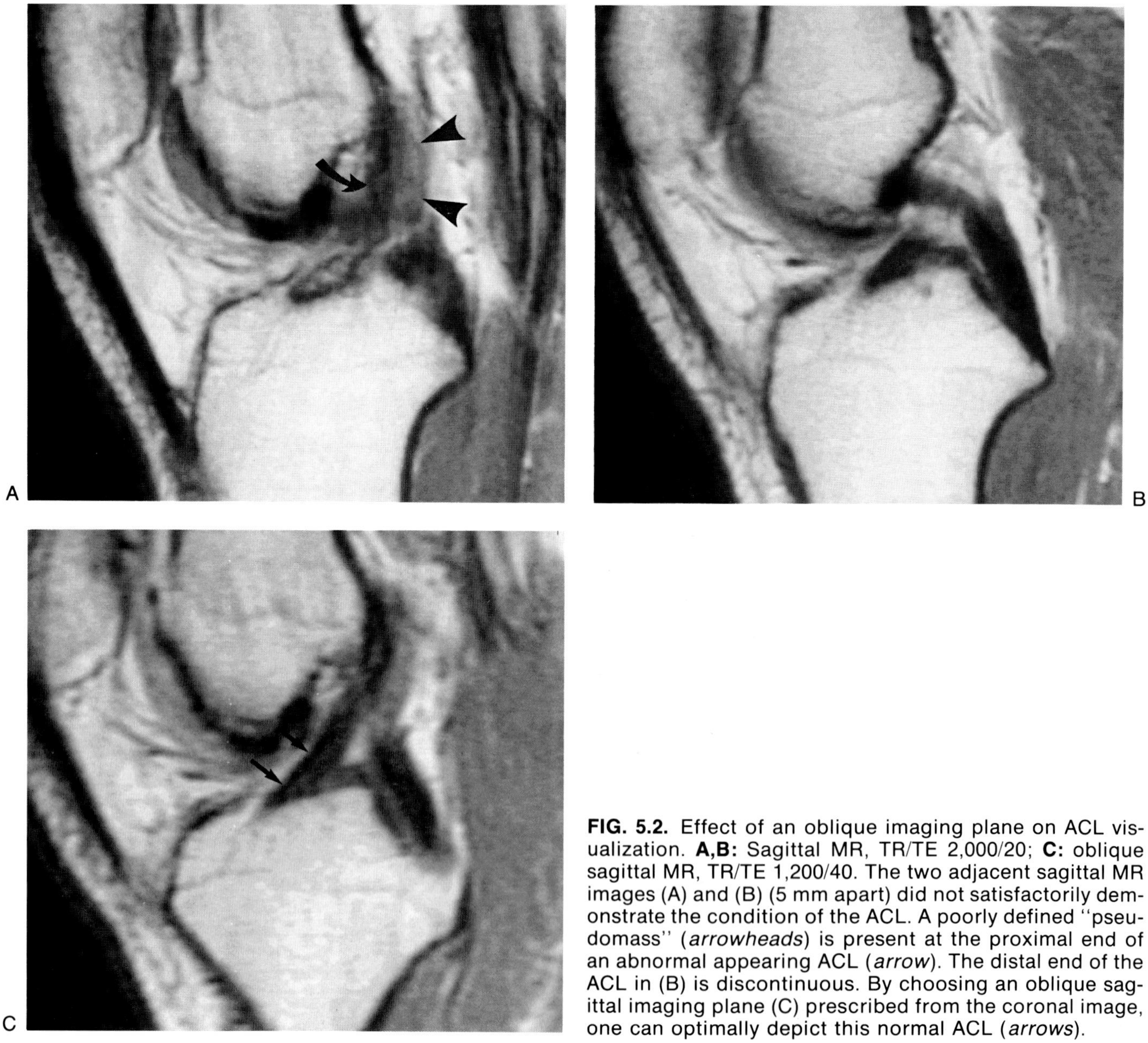

FIG. 5.2. Effect of an oblique imaging plane on ACL visualization. **A,B:** Sagittal MR, TR/TE 2,000/20; **C:** oblique sagittal MR, TR/TE 1,200/40. The two adjacent sagittal MR images (A) and (B) (5 mm apart) did not satisfactorily demonstrate the condition of the ACL. A poorly defined "pseudomass" (*arrowheads*) is present at the proximal end of an abnormal appearing ACL (*arrow*). The distal end of the ACL in (B) is discontinuous. By choosing an oblique sagittal imaging plane (C) prescribed from the coronal image, one can optimally depict this normal ACL (*arrows*).

accompanies a rupture. In cases in which a pseudomass is present, the high resolution coronal images will demonstrate a normal ACL; a true mass would clearly be evident in any plane. To exclude the possibility that one is not mistaking a partial volume pseudomass effect for a true mass, one should confirm that the plane of section intersects the intercondylar roof. On sections through the intercondylar roof, the entire superoposterior to inferoanterior course of the roof, manifest on MR by a dark cortical line, is identified. In our experience, a mass identified on an image in which the entire intercondylar roof is incompletely visualized from front to back is frequently a pseudotear.

Acute Ruptures

For the purposes of this treatise, acute ruptures of the ACL are those that come to MR examination within 1 to 2 weeks of the injury. Several different MR patterns of acute ruptures share a combination of characteristics including loss of continuity, mass effect, retraction of the torn ends, waviness or concavity of the anterior margin of the ACL, and increased signal within the ligament (47,52–55).

An interstitial tear is defined as one in which the injury pattern on an MR examination is over virtually the entire length of the ACL (Fig. 5.4). The ligament

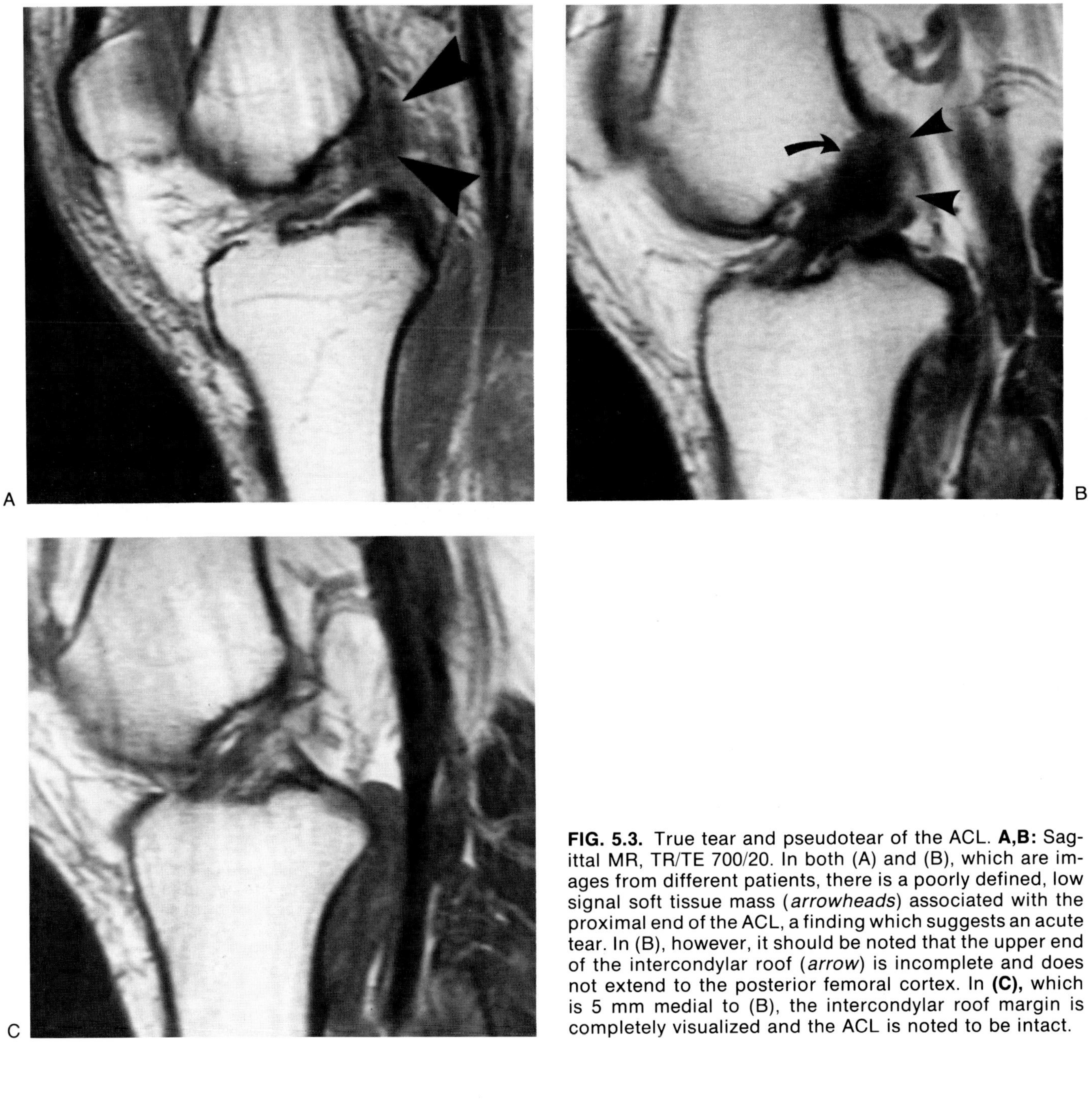

FIG. 5.3. True tear and pseudotear of the ACL. **A,B:** Sagittal MR, TR/TE 700/20. In both (A) and (B), which are images from different patients, there is a poorly defined, low signal soft tissue mass (*arrowheads*) associated with the proximal end of the ACL, a finding which suggests an acute tear. In (B), however, it should be noted that the upper end of the intercondylar roof (*arrow*) is incomplete and does not extend to the posterior femoral cortex. In **(C),** which is 5 mm medial to (B), the intercondylar roof margin is completely visualized and the ACL is noted to be intact.

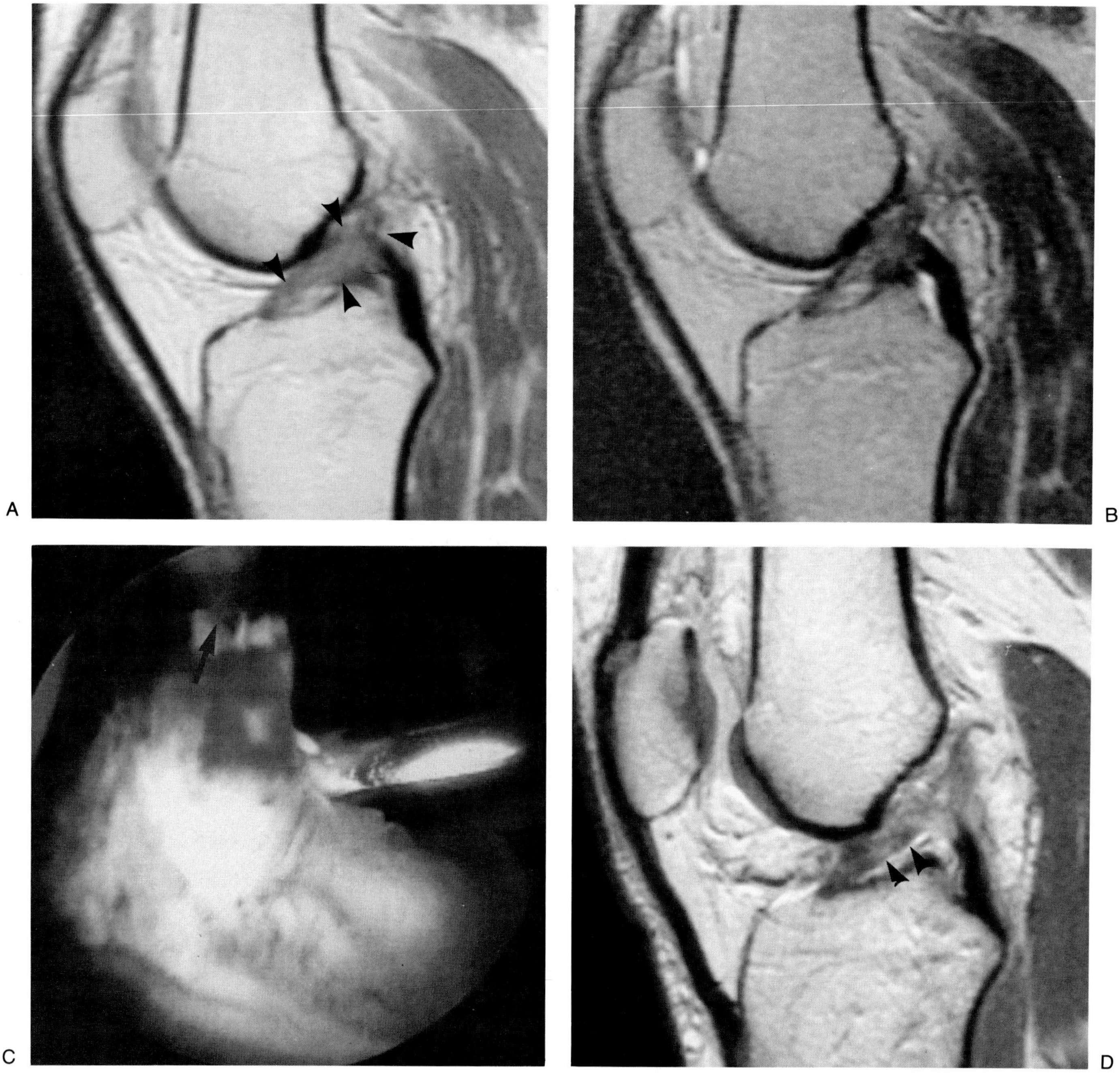

FIG. 5.4. Acute interstitial ruptures of the ACL. **A:** Sagittal MR, TR/TE 2,000/20; **B:** sagittal MR, TR/TE 2,000/80. A poorly defined mass (*arrowheads*) replaces virtually the entire ACL, which has a sausage-like character. Only tiny foci become bright on the long TR/TE sequence. **C:** This surgical photograph demonstrates that in spite of the fact that the MR suggested some element of continuity, the proximal end of the ligament (*large arrow*) is completely torn. **D:** Sagittal MR, TR/TE 2,000/20. In another patient, only the distal ACL has a ligament-like character (*arrowheads*). The axis of the ACL is not parallel to the intercondylar roof. At surgery, the ACL was completely torn.

is widened throughout its length, assuming a sausage-like appearance. On T1 or proton density sequences, the ACL has a diffusely increased signal intensity, only small segments of which become brighter on long TR/TE sequences; some segments of the torn ligament may actually decrease in signal intensity on the T2WI. At surgery, ligaments demonstrating this pattern of injury on MR are nearly always completely torn, and even the interstitial tears that appear to be only partially torn at arthroscopy eventually become functionally complete. This pattern of injury occurs in nearly 50% of all ACL tears (56).

Focal mass-like ruptures most typically affect the proximal end of the ACL (55). The tear may be manifest on MR by a normal or slightly serpiginous distal ACL terminating in a proximal mass. The appearance would suggest that the ACL is in continuity due to the fact that the hematoma and synovium hold the torn ligament edges together anatomically, yet the ligament is usually incompetent (Fig. 5.5). The proximal mass

A

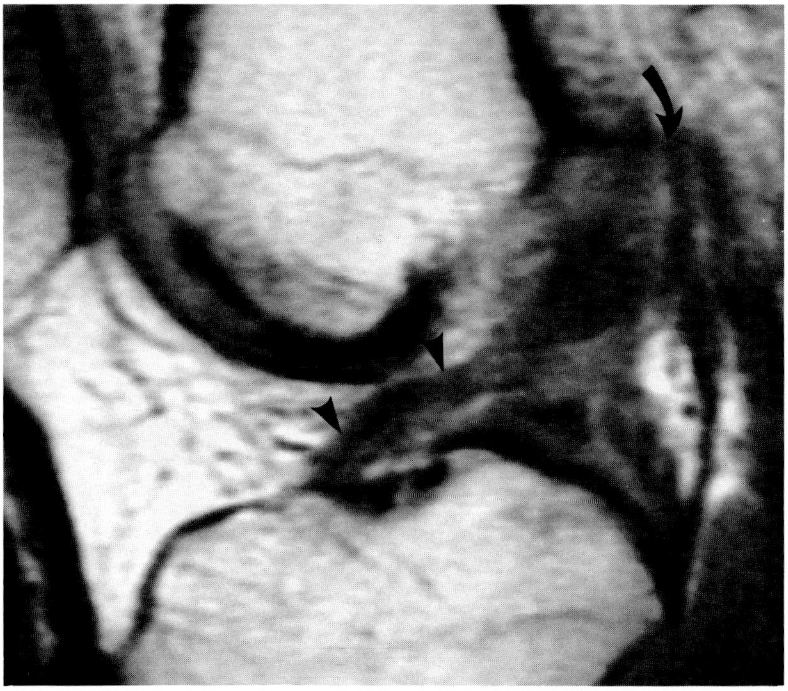

B

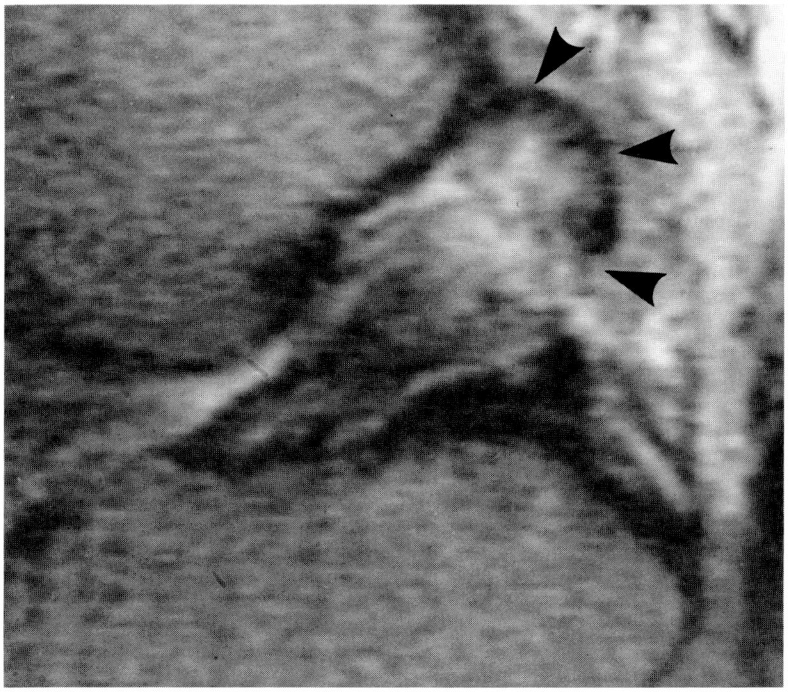

FIG 5.5. Acute proximal focal ruptures of the ACL. **A:** Sagittal MR, TR/TE 2,000/20. Nearly the entire course of the ACL can be identified (*arrowheads*), but the ligament is serpiginous and lies low in the joint. A mass is present proximally (*arrow*). **B:** Sagittal MR, TR/TE 2,000/80. A large soft tissue mass (*arrowheads*), slightly bright on the long TR/TE sequence, represents a hematoma, torn synovium, and ligamentous remnants.

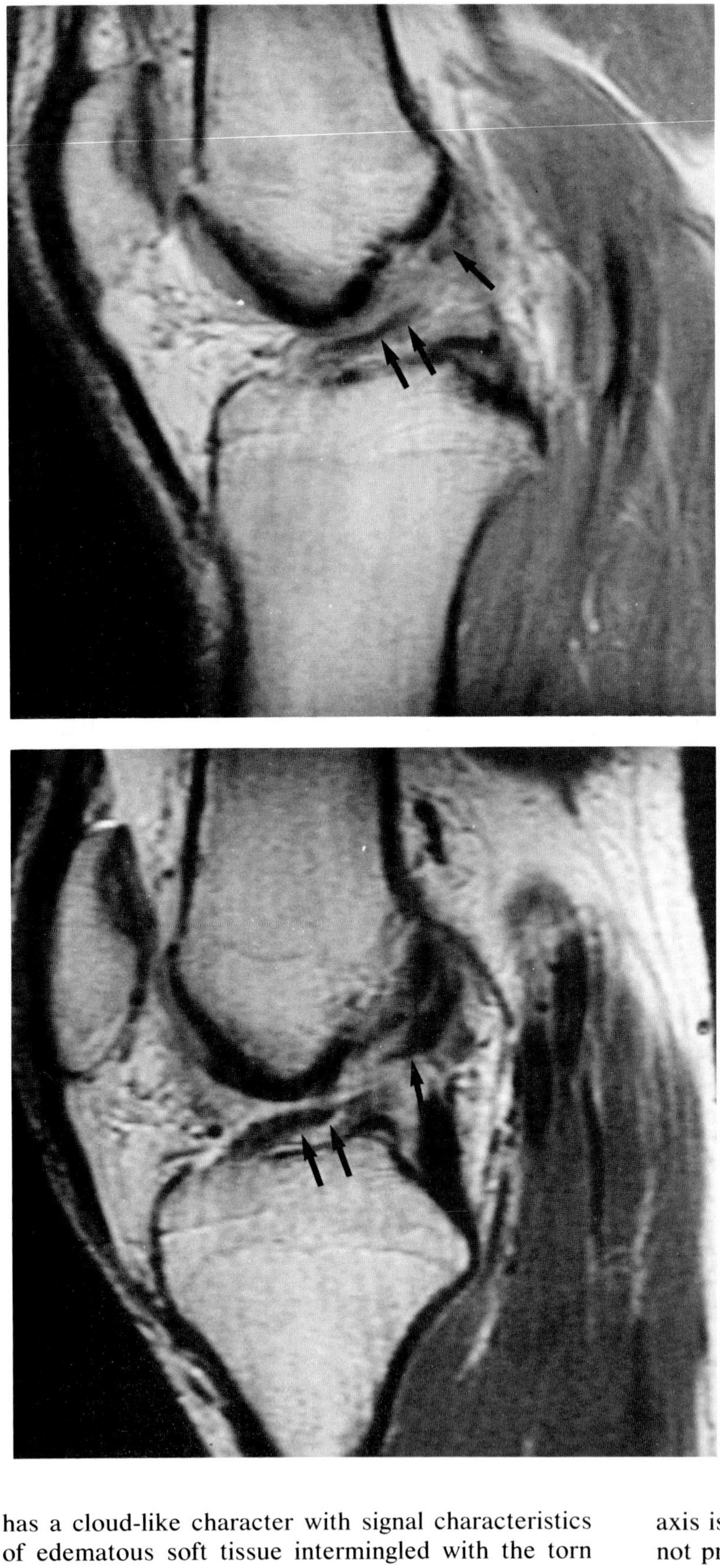

A

B

FIG 5.6. Acute mid-substance tears of the ACL. **A,B:** Sagittal MR, TR/TE 2,000/20, in two different patients. Both patients experienced a pop in the knee and had MRI the morning following the injury. The proximal (*arrow*), and distal (*double arrow*) segments of both ACL are identified, but the mid-portion is sharply angulated and discontinuous. The hematoma and synovium frequently hold the ends of the torn ligament together for 1–3 weeks.

has a cloud-like character with signal characteristics of edematous soft tissue intermingled with the torn dark ligament fibers (57). The completeness of the rupture of this tear or the interstitial-type tear can be determined by examination of the axis of the ACL. If the axis is not parallel to the intercondylar roof and does not project superiorly to the femoral attachment site, the tear is always complete.

Another common MR pattern of an acute ACL tear is a focal angular deformity of the mid-segment of the

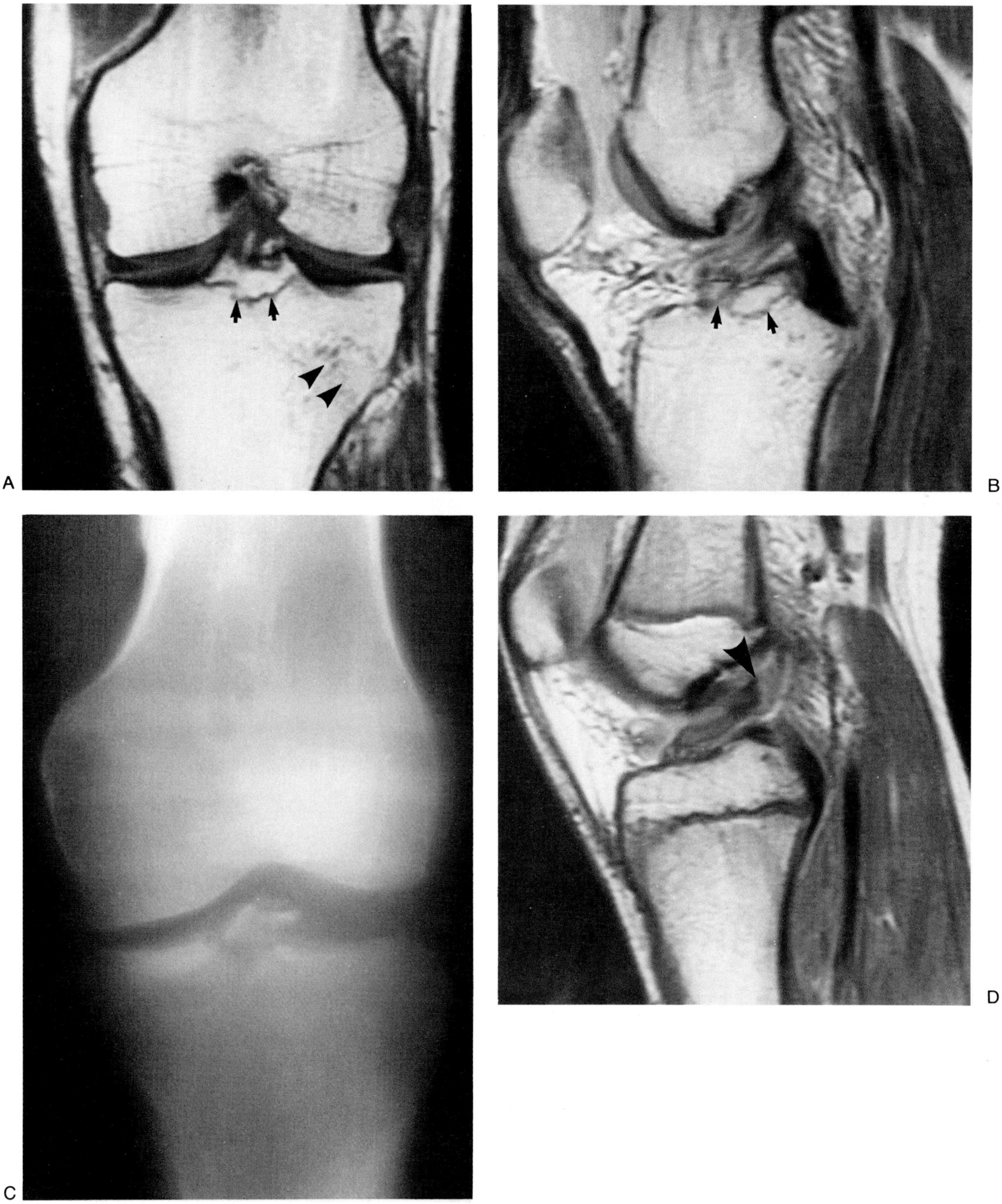

FIG. 5.7. Avulsion ruptures of the ACL. **A:** Coronal MR, TR/TE 2,000/20; **B:** sagittal MR, TR/TE 2,000/20. Both images demonstrate a minimally displaced avulsion fracture of the insertion of the ACL (*arrows*). There is surprisingly little bone marrow edema associated with the fracture. A medullary contusion (*arrowheads*) is present in the lateral proximal tibia. **C:** This anteroposterior tomogram demonstrates the avulsion of the intercondylar eminence which is displaced more than on the MR which was performed 1 week earlier. **D:** Detachment of the proximal ACL; sagittal MR, TR/TE 2,000/20. This 15-year-old suffered an ACL tear during a soccer game. The femoral end of the ACL (*arrowhead*) has become detached from the femur with a small flake of bone which was not identified on the MR. The fragment was much too small to be reattached at the time of surgery.

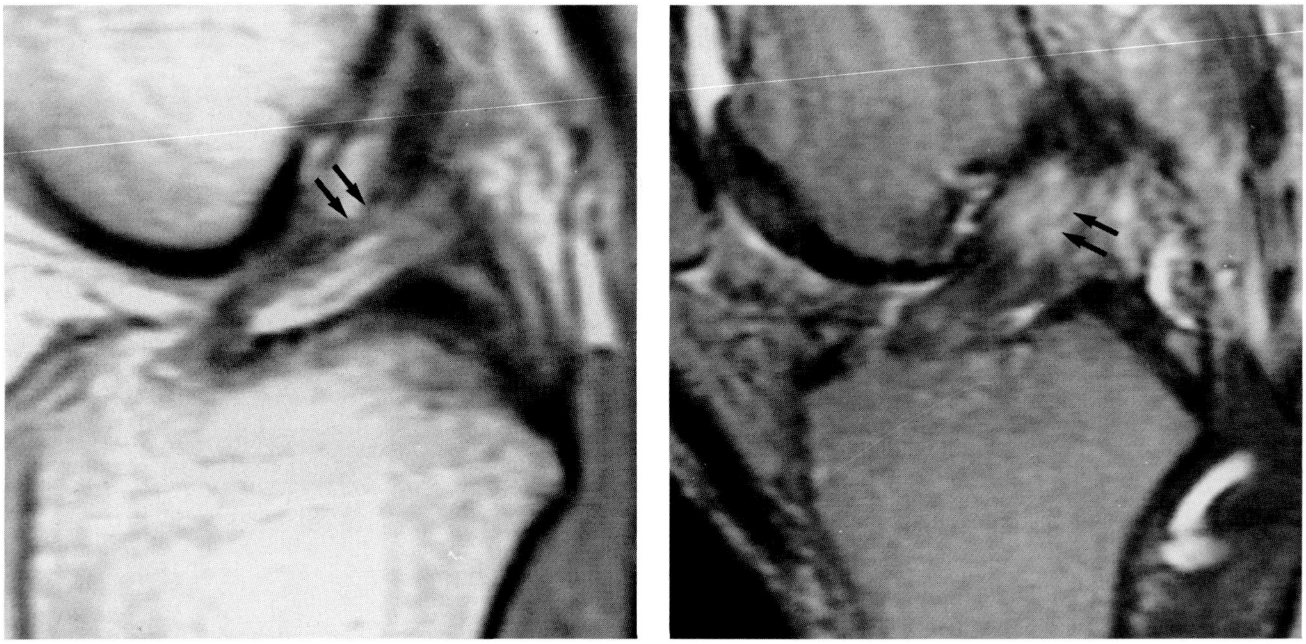
A B

FIG. 5.8. Partial tears of the ACL. **A:** Sagittal MR, TR/TE 2,000/20. The mid-portion of the ACL demonstrates a slight angulation (*arrows*) at the site of a surgically demonstrated partial ACL tear. **B:** Sagittal MR, TR/TE 2,000/80. In another patient, there is edema/hemorrhage (*arrows*) within the substance of the ACL yet ligament fibers are seen traversing this high signal region. There is no clinical or imaging follow-up in either of these two patients.

ACL in which there is rather little mass effect (Fig. 5.6). The segment distal to the tear may fall onto the tibial plateau, but this feature is more characteristic of subacute tears.

Avulsive injuries of either attachment of the ACL are uncommon. We have never encountered a proximal detachment that involved a substantial fragment of bone (Fig. 5.7). At the tibia, however, avulsions of the intercondylar eminence occur in 5% of cases (2,23). Remarkably, these serious but easily repairable injuries may be somewhat difficult to fully appreciate on MR due to several factors. Of the seven avulsive injuries that we have encountered, none have been displaced more than 2 mm. The lack of displacement results in rather little "laxity" of the ACL. The avulsed fragment, composed of medullary bone, has signal characteristics identical to that of fat in the intercondylar notch and may therefore be difficult to identify. Finally, we have not identified significant bone marrow edema in the bed of the avulsed fragment.

The distinction between a partial tear and a complete tear of the ACL by MR may be extremely difficult. As stated above, the problem is compounded by: the fact that partial tears diagnosed by arthroscopy may rapidly progress to completeness (17); the lack of anatomic/pathologic correlation; the surgeon's inability to reliably determine the status of the ligament; and the tendency of arthroscopy to underestimate the degree of damage to the ACL (5,7,18,53,58). If a partial tear is truly present, it is the AMB that is torn and the tear is most fully appreciated at the distal end of the ligament (5), but on most occasions it is not possible to

FIG. 5.9. Longitudinal studies of partially torn ACLs. **A:** Sagittal MR, TR/TE 2,000/20. There is a poorly defined soft tissue mass (*arrows*) at the proximal end of the ACL, with a suggestion of some degree of ligamentous continuity. **B:** Sagittal MR, TR/TE 2,000/80. Five mm medial to (A), a portion of the ACL (*arrowhead*) is seen lying on the tibial plateau. At arthroscopy, the surgeon described the ACL as partially torn, and the knee was stable. **C:** Sagittal MR, TR/TE 2,000/20. This examination, performed 4 months after (A) and (B), demonstrates a vague mass (*arrowheads*) in the expected position of the ACL, but there is no clearly defined ligament. The knee was unstable on reexamination. **D,E:** Sagittal MR, TR/TE 2,000/20 and 2,000/80. In another patient, the ACL has a minimal angulation (*arrows*) within its mid-substance, as well as some increase in signal. The lesion was diagnosed as a partial ACL tear at surgery. **F:** Sagittal MR, TR/TE 1,200/40. Eight weeks later her knee remains clinically stable, and she has resumed full activities. The ACL has a slightly concave contour, but the ligament is intact (*arrowheads*).

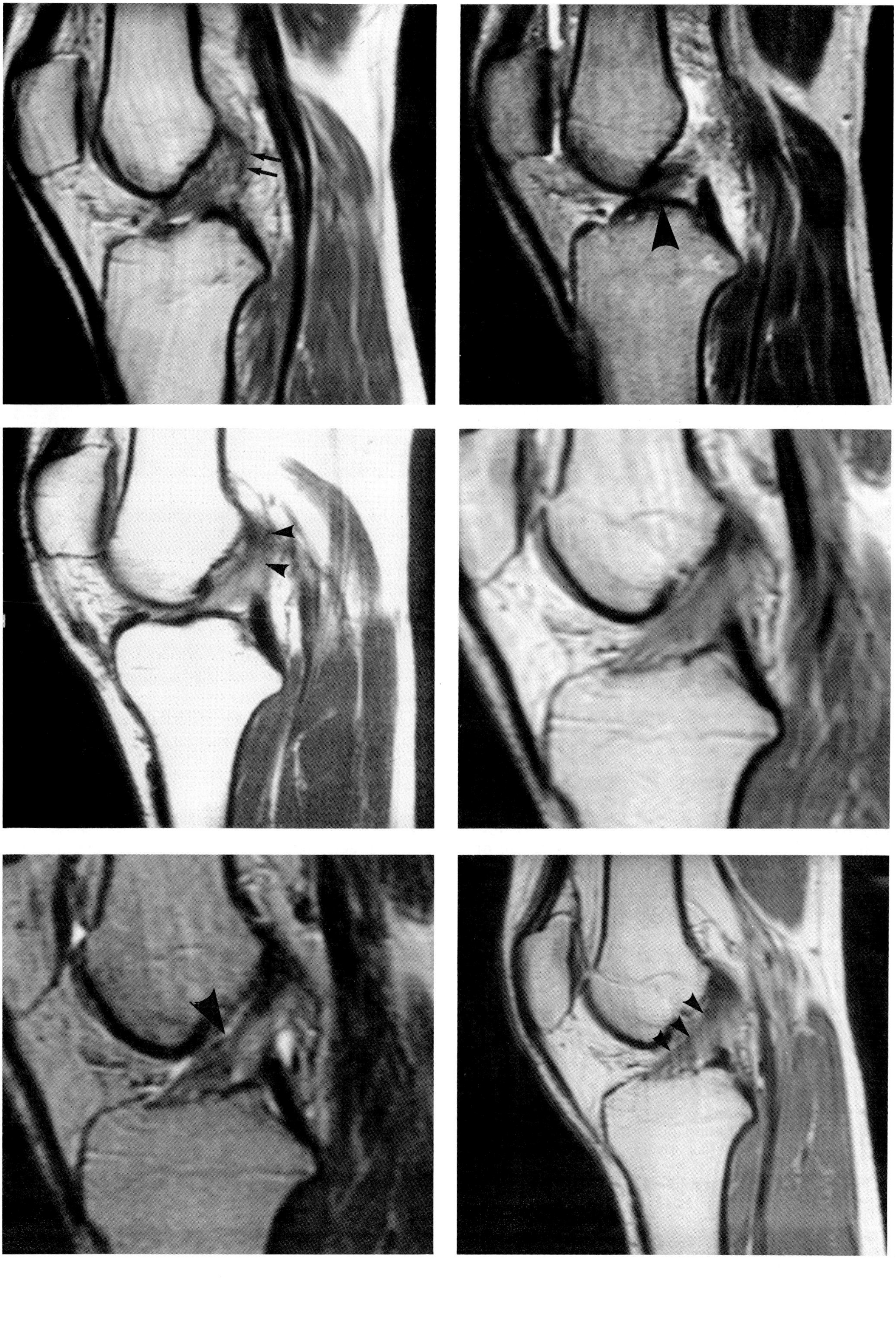

A B C D E F

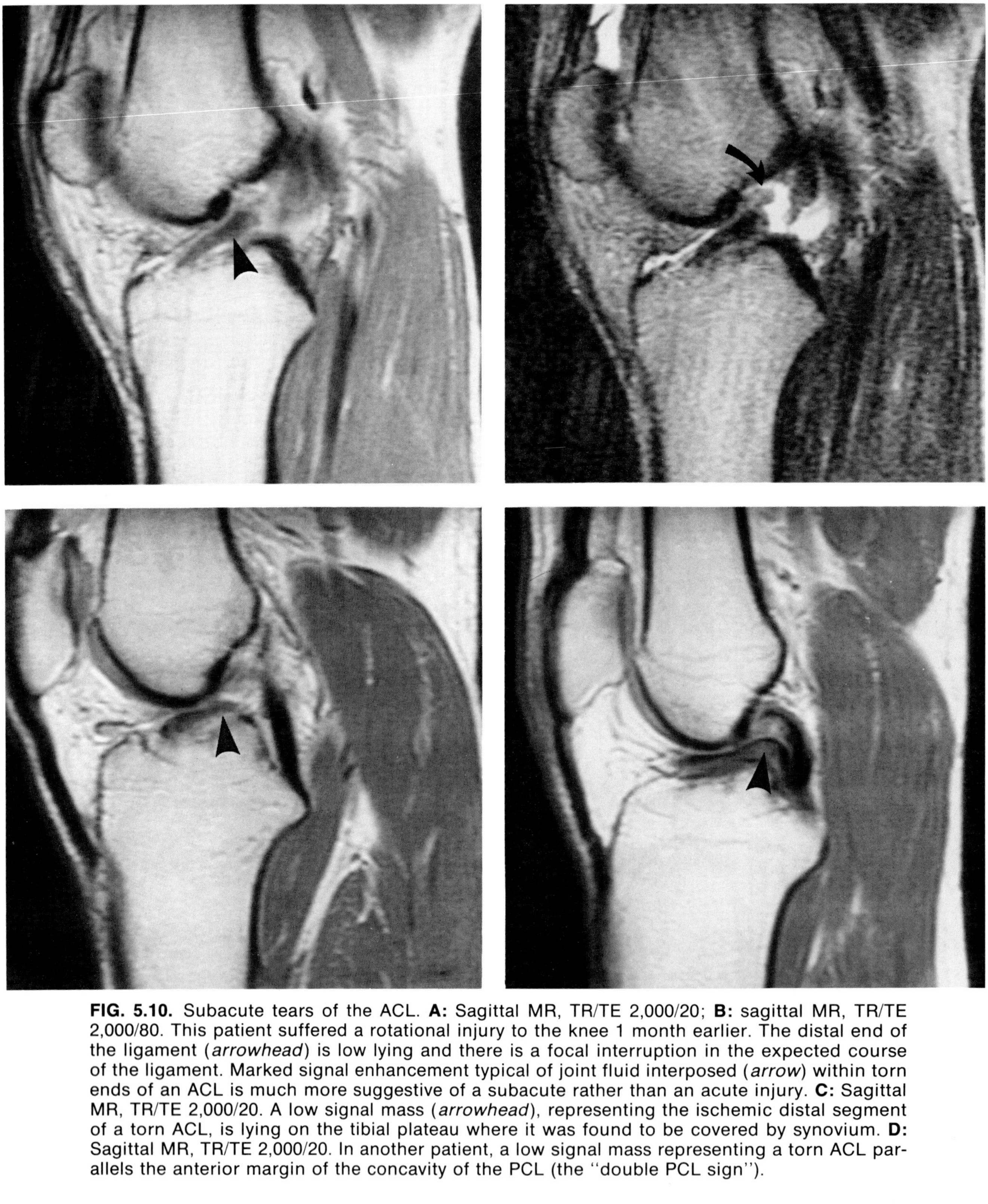

A B C D

FIG. 5.10. Subacute tears of the ACL. **A:** Sagittal MR, TR/TE 2,000/20; **B:** sagittal MR, TR/TE 2,000/80. This patient suffered a rotational injury to the knee 1 month earlier. The distal end of the ligament (*arrowhead*) is low lying and there is a focal interruption in the expected course of the ligament. Marked signal enhancement typical of joint fluid interposed (*arrow*) within torn ends of an ACL is much more suggestive of a subacute rather than an acute injury. **C:** Sagittal MR, TR/TE 2,000/20. A low signal mass (*arrowhead*), representing the ischemic distal segment of a torn ACL, is lying on the tibial plateau where it was found to be covered by synovium. **D:** Sagittal MR, TR/TE 2,000/20. In another patient, a low signal mass representing a torn ACL parallels the anterior margin of the concavity of the PCL (the "double PCL sign").

distinguish which band is torn. The diagnosis of a partial tear may be made when a discrete focus of high signal is present within the substance of an ACL in which the majority of the fibers are clearly seen to be intact from femur to tibia. A minimal degree of midsubstance angulation may be identified (Figs. 5.8 and 5.9). The knee suffering a partial tear and one which is stable at the time of initial examination may become unstable over time because of injury progression. On the other hand, the edema, hemorrhage, and synovial reaction that surround the ACL injury may completely obscure normal, intact ACL fibers and lead the radiologist to suggest that a complete tear (interstitial type) is present. Because of the difficulties noted above, we currently describe ACL tears in the following manner. A diffuse interstitial-type tear with a normal or near normal axis may be either completely or only partially torn, and we therefore describe the lesion as "at least a partial, but probably a complete rupture"; if the axis is low, the tear is certainly complete. If there is a proximal mass-like tear and the axis of the ligament is low, the tear is unequivocally complete; if the axis is normal, the tear may be only partial. After 4 to 6 weeks, the synovitis accompanying the rupture resolves somewhat, and the devascularized segments of the ACL will retract and be found on the tibial plateau. If a partial tear is initially diagnosed and if it is critical to know the precise extent of the anatomic abnormality suffered by the ACL, it is best to repeat the examination in 2 months following the injury.

Subacute Tears

The precise site and degree of an ACL tear is more readily identified in the subacute (2–8 weeks) than in the acute phase. The proximal and distal ends will have retracted slightly, the hematoma will have partially resolved, and bright joint fluid will intravasate into the defect. This finding is quite unlike the edematous soft tissue mass which dominates the intercondylar notch in acute ruptures (57). Subacute tears of the ACL are often manifest by the presence of an intermediate to dark mass within the joint due to retraction of the torn ligament and the development of a synovial cap (Fig. 5.10). During the subacute period, the mass-like appearance of the torn ACL may masquerade as one of several lesions. The "double PCL" sign describes a curvilinear dark structure anterior and parallel to the normal posterior cruciate ligament that represents the displaced fragment of a bucket-handle tear. This appearance can also be simulated by the distal segment of an ACL that has torn proximally. Alternatively, the proximal end of an ACL that is completely torn distally may become displaced behind the posterior cruciate

A

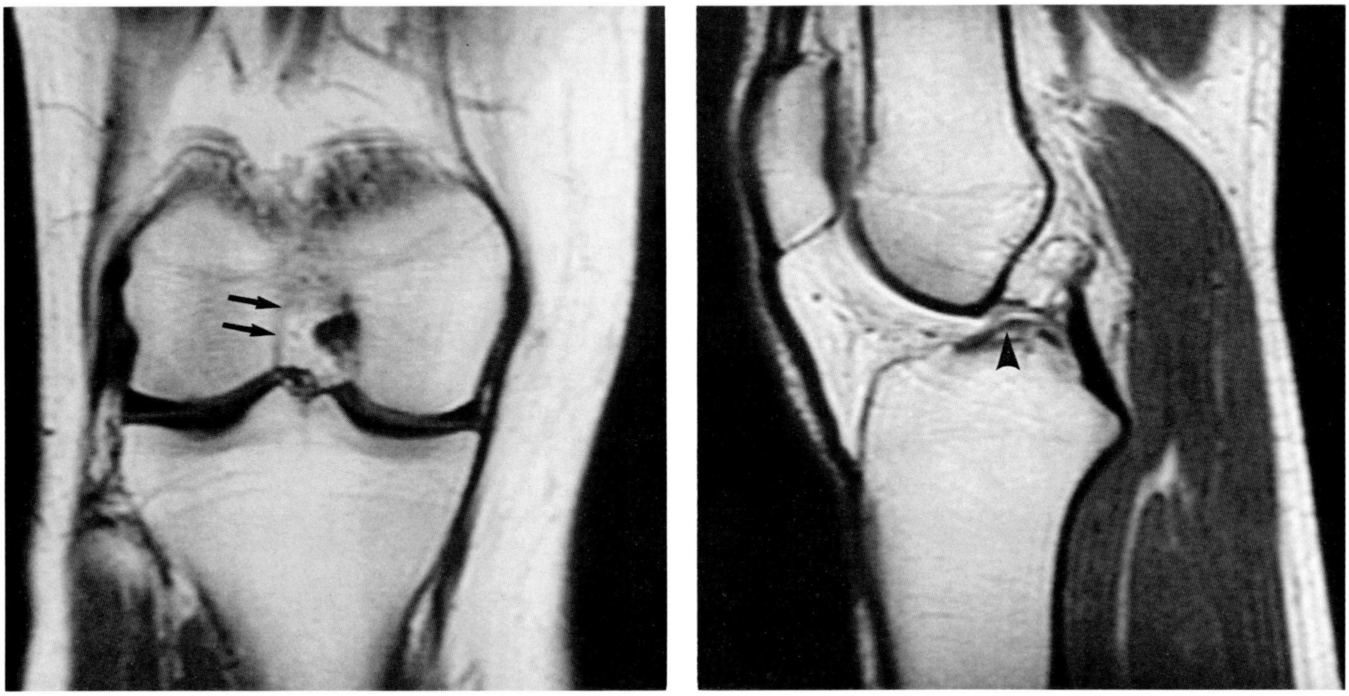

B

FIG. 5.11. Chronic tear of the ACL. **A:** Coronal MR, TR/TE 2,000/20; **B:** sagittal MR, TR/TE 800/20. There is an "empty lateral wall" (*arrow*) seen on the coronal image; the ACL was not identified on any other section. This patient had suffered a ACL rupture many years earlier. The only residual is a small nubbin of tissue (*arrowhead*) lying on the tibial plateau.

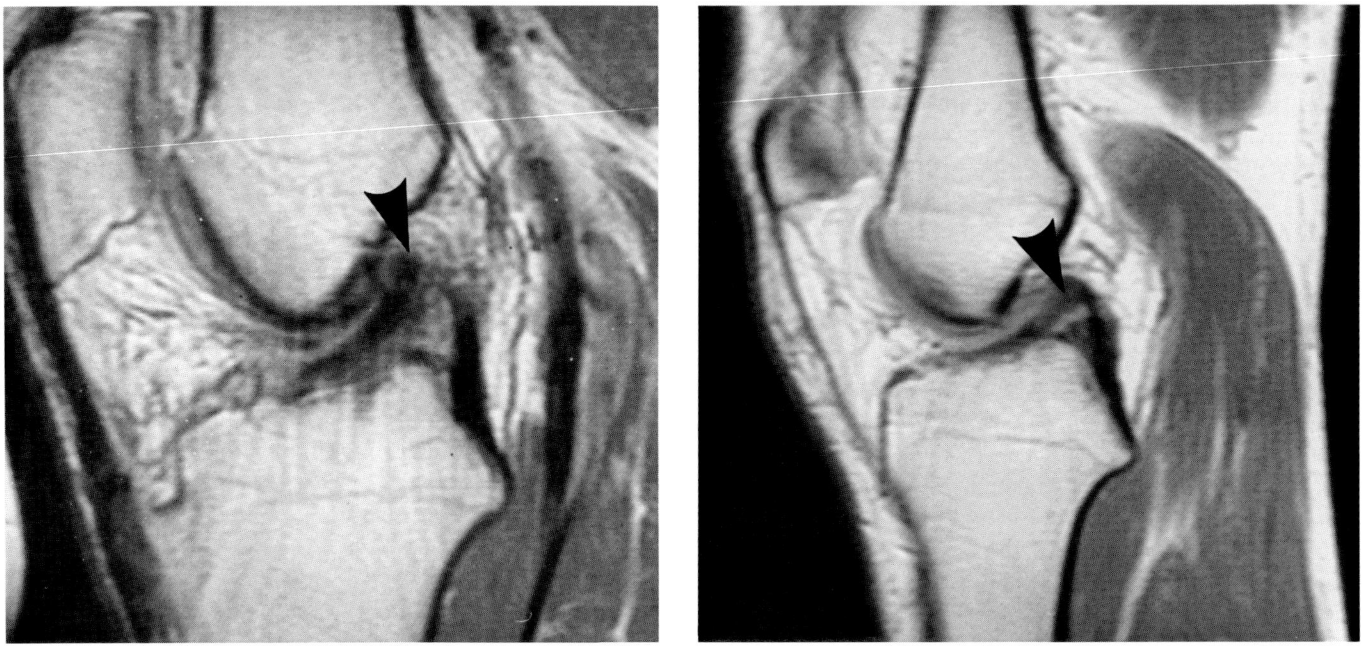
A B

FIG. 5.12. Chronic tear of the ACL; reattachment to the PCL. **A:** Sagittal MR, TR/TE 2,000/20. This professional football player had suffered an arthroscopically confirmed tear of the ACL 5 years earlier. The distal one-half of the ACL has a relatively normal axis. It extends proximally only to the level of the PCL along whose lateral side it has reattached itself (*arrowhead*). **B:** Sagittal MR, TR/TE 2,000/20. In another patient, a small low lying segment of the ACL is attached (*arrowhead*) to the anterior margin of the PCL.

ligament, where it mimics a large, loose body situated in the posterior capsule.

Chronic Tears

The MR image of the chronically torn ACL is dependent on the degree of atrophy of the ligament and the degree of instability that supervenes. In many cases the ACL itself is not visualized; the lateral intercondylar notch is empty, or a tiny nubbin of soft tissue representing a retracted, ischemic ACL remnant may be identified on the tibial plateau. In such cases, the "ACL" may be difficult to find at surgery (41) (Fig. 5.11). On the coronal sequence, the "bald" medial face of the lateral femoral condyle, corresponding to the arthroscopically described "empty lateral wall," is evident (46).

The chronically torn ACL may represent as an intact band of low signal in the lateral aspect of the intercondylar notch, simulating a virgin ligament (57). The structure, however, has either a low lying axis or a more vertical orientation, and the proximal-most segment of the ACL (that portion cephalad to the PCL) is never visualized. In fact, the proximal end of a torn ACL may secondarily reattach itself to the lateral aspect of the posterior cruciate ligament (Fig. 5.12). The ACL, residing in the same synovial envelope with the PCL, falls slightly following complete rupture. It finds

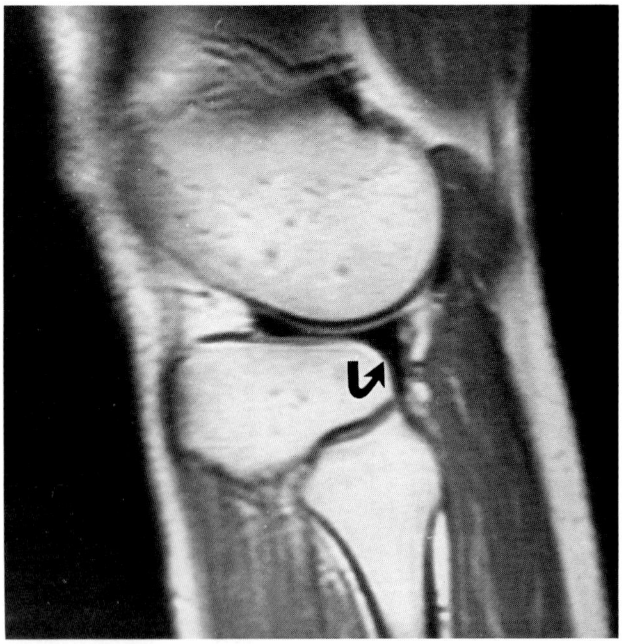

FIG. 5.13. ACL deficiency with instability. Sagittal MR, TR/TE 2,000/20. There is gross femoral tibial instability which is manifest by a forward shift of the tibia, and "uncovering" of the undersurface of the posterior horn of the lateral meniscus (*arrow*).

a secondary blood supply from the PCL to which it adheres, and the scarring which invariably occurs has a low signal intensity identical to that of the normal ACL. Reattached ACLs are commonly mistaken for intact ligaments both on MR images and, remarkably, even at arthroscopy (46,57). Interestingly, the surgeon may actually elect to anastomose the ACL stump to the PCL, a procedure (Wittek procedure) which provides some element of restabilization but which is not intended to be permanent (1).

Instability of the ACL-deficient knee is manifest on MR by an abnormal relationship of the tibia to the femur (47,59) (Fig. 5.13). With chronic dysfunction of the ACL, the tibia is free to sublux anteriorly; the lack of an anterior restraint forms the basis of the anterior drawer and pivot shift test. The fully relaxed patient lying in the MR scanner may allow his tibia to shift. The MR anterior drawer sign appears to be highly sensitive and specific for the diagnosis of an ACL tear (59). A line is drawn along the posterior femoral cortex parallel to the long axis of the tibia; if the posterior margin of the tibial plateau is anterior to this line, the sign is said to be positive. In such patients, the posterior cruciate ligament becomes foreshortened, and assumes a serpiginous (S-shaped) or proximally buckled (sea-horse) configuration (47,52) (Fig. 5.14). One must be cautious in interpretation of this sign since some variability in the normal length of the PCL is present, and quantification of the normal versus abnormal degree of PCL buckling is impossible. Forward

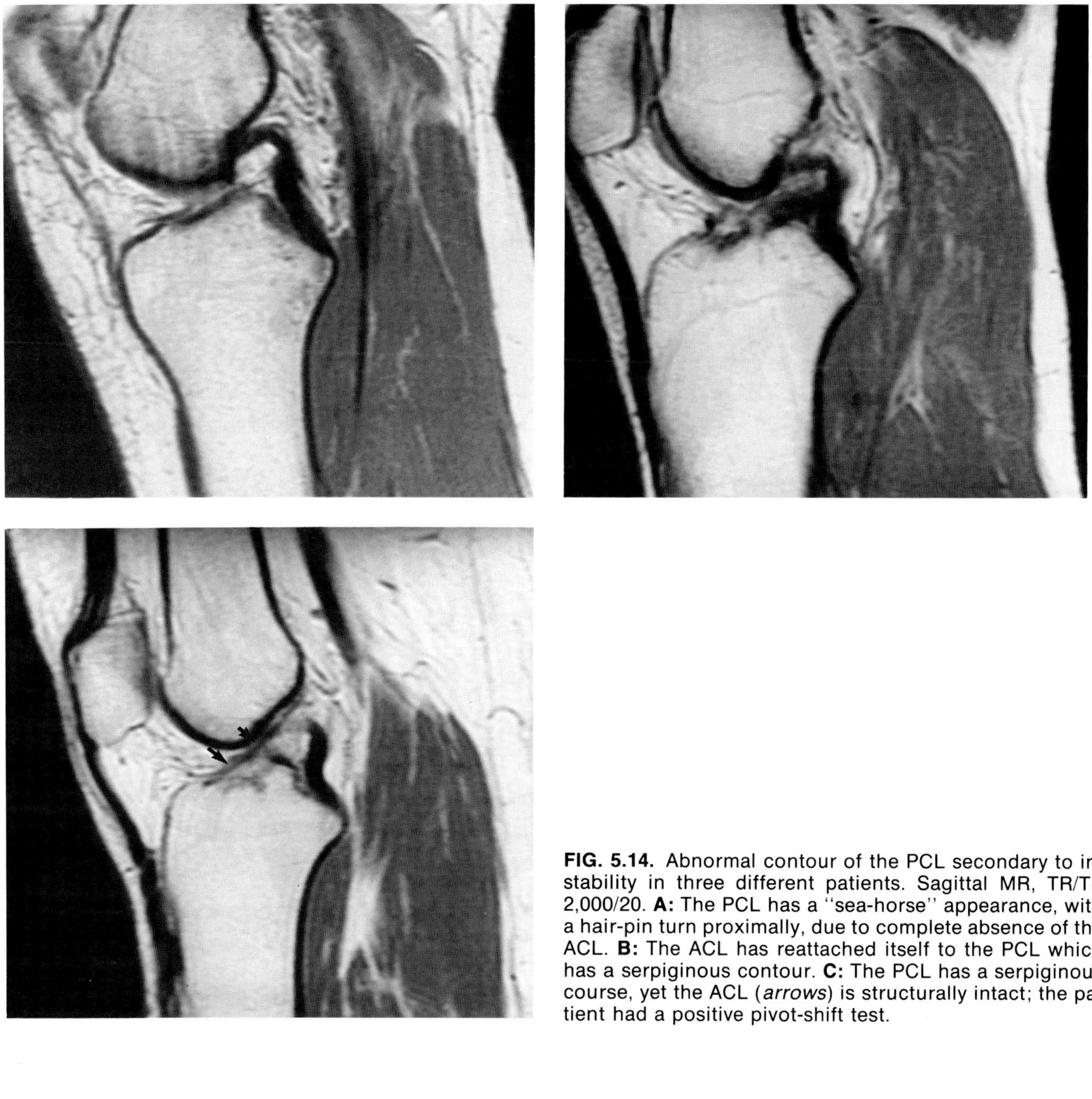

A B C

FIG. 5.14. Abnormal contour of the PCL secondary to instability in three different patients. Sagittal MR, TR/TE 2,000/20. **A:** The PCL has a "sea-horse" appearance, with a hair-pin turn proximally, due to complete absence of the ACL. **B:** The ACL has reattached itself to the PCL which has a serpiginous contour. **C:** The PCL has a serpiginous course, yet the ACL (*arrows*) is structurally intact; the patient had a positive pivot-shift test.

subluxation of the tibia may occur in patients with a structurally intact but obviously functionally deficient ACL.

The cruciate ligament cyst, also known as an intraarticular ganglion, may be associated with a remote ACL injury (60–62) (Fig. 5.15). Just like its extraarticular counterpart, the lesion is a mass which is filled with a jelly-like viscous fluid (63,64). *Extraarticular* ganglia become manifest by virtue of their mass effect, or are found incidentally on imaging studies; they occur most frequently in the vastus lateralis or semimembranosus muscles, and may or may not demonstrate a connection to the adjacent joint. *Intraarticular* ganglia, located in a rather small space, may exert sufficient mass effect to interfere with the full range of joint motion (61). Intraarticular ganglia are associated with the proximal or mid-portions of either cruciate ligament. Two potential etiologies for these cysts have been proposed: congenital and traumatic. In the latter case, the cyst may form following injury to the ACL, with secondary fluid accumulation within an internally herniated synovial sac, or rupture of the synovial membrane covering the ACL may allow fluid to form a cyst within the boundaries of the intercondylar notch. On MR, the cyst presents as a localized collection of fluid inseparable from the cruciate ligaments. In our experience, the cyst has most commonly been found near the proximal attachment of the ACL, and has been an incidental finding. The major differential diagnosis rests with a fortuitous finding of joint fluid adjacent to the cruciate synovial envelope. The rounded or ovoid configuration, and the frequent finding of rather little

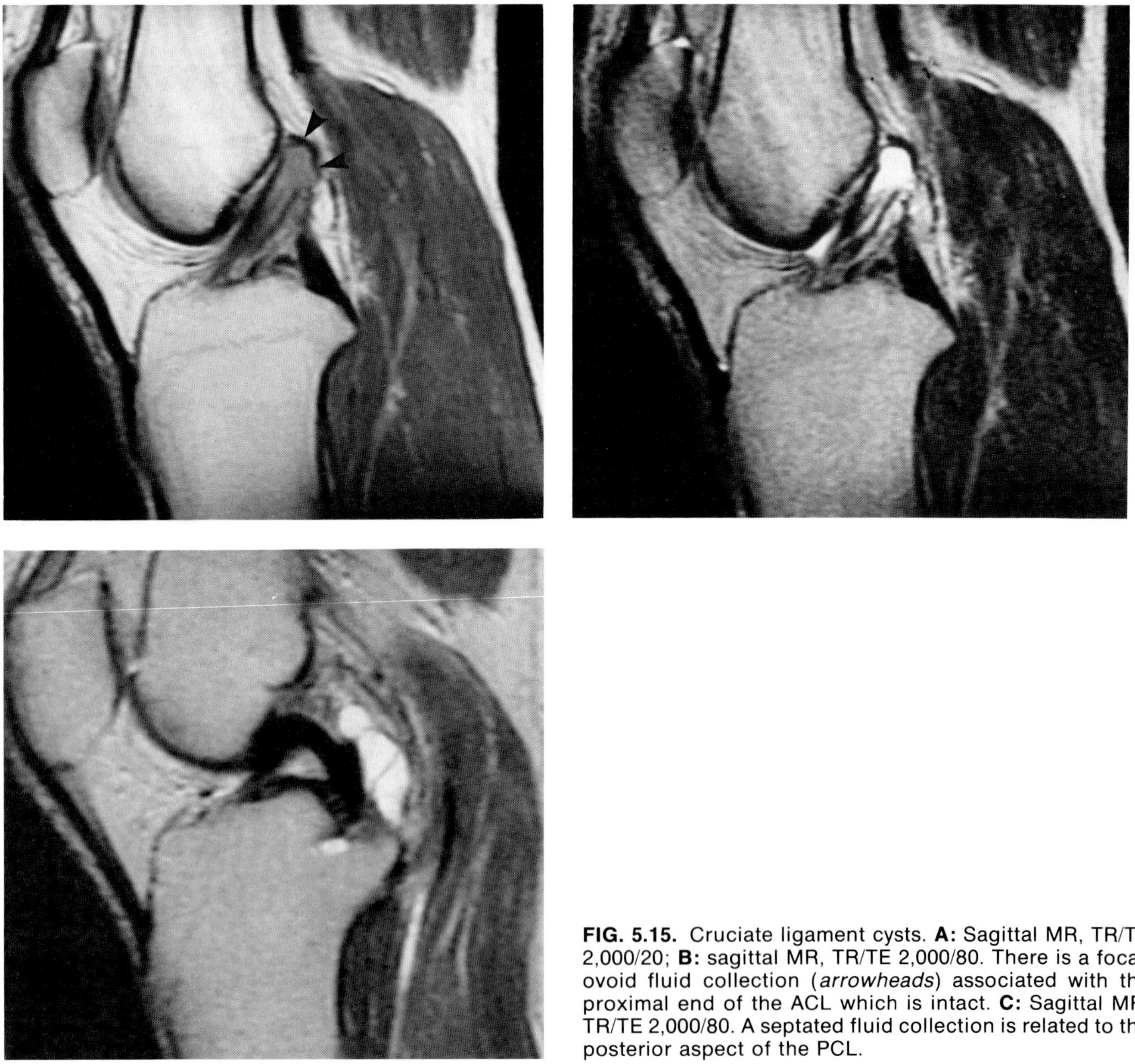

A B C

FIG. 5.15. Cruciate ligament cysts. **A:** Sagittal MR, TR/TE 2,000/20; **B:** sagittal MR, TR/TE 2,000/80. There is a focal, ovoid fluid collection (*arrowheads*) associated with the proximal end of the ACL which is intact. **C:** Sagittal MR, TR/TE 2,000/80. A septated fluid collection is related to the posterior aspect of the PCL.

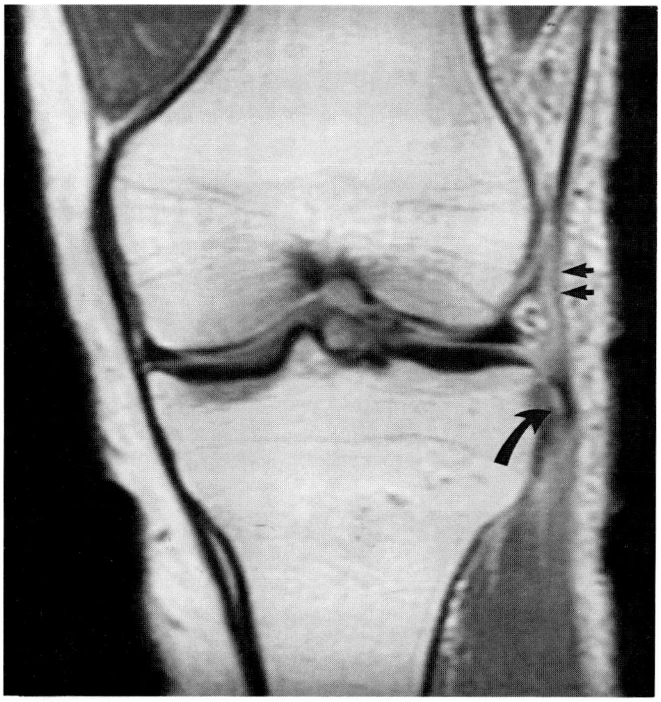
A

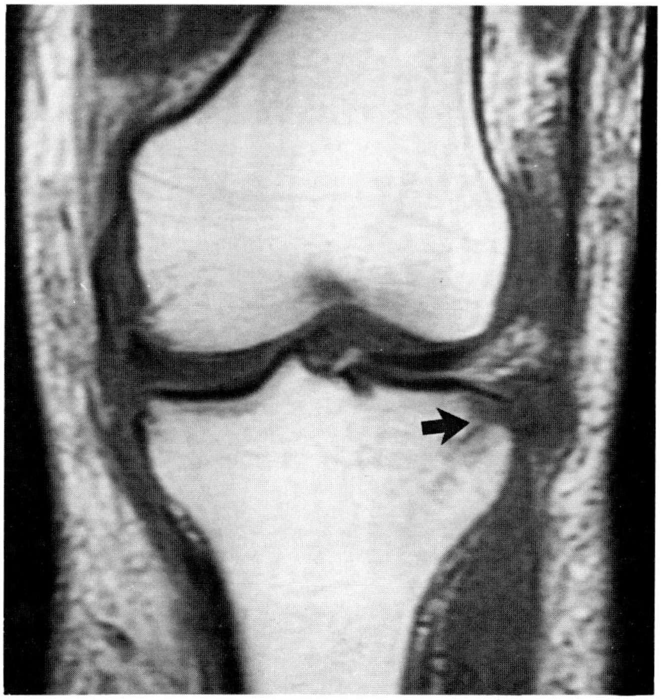
B

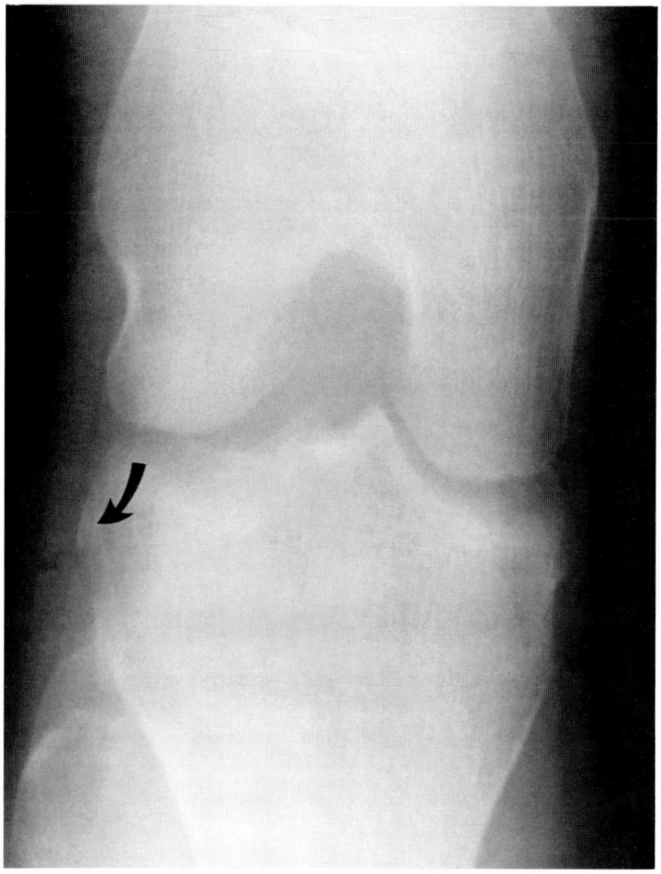
C

FIG. 5.16. Segond fracture. **A:** Coronal MR, TR/TE 2,000/20. The middle third of the lateral capsular ligament has avulsed a small fragment of bone (*arrow*) from the tibial epiphysis. The distal iliotibial band (*double arrows*) is poorly defined because this image section is posterior to its attachment. **B:** Coronal MR, TR/TE 2,000/20. Edema of the medullary bone (*arrow*) is seen at the site of avulsion of a very small Segond fracture; the fragment was visible on conventional radiographs but, because of its small size, it cannot be identified on MR. The patient had suffered a tear of both the ACL and the TCL. **C:** The conventional radiograph demonstrates the class Segond lesion (*arrow*).

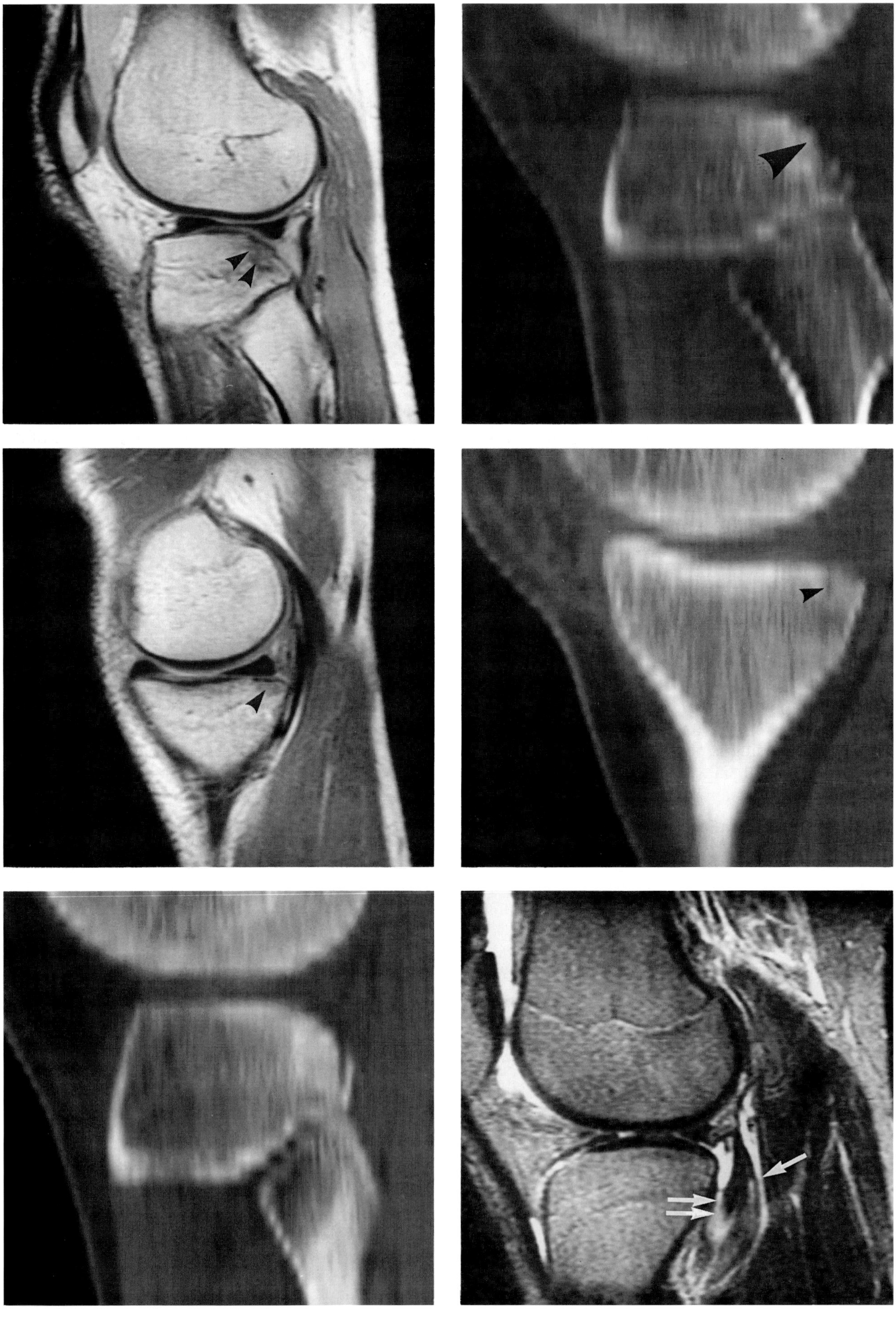
A
B
C
D
E
F

joint fluid in the posterior capsule and suprapatellar bursa, are helpful secondary signs in suggesting the diagnosis of a cyst.

ACL Associated Bone Injuries

A variety of bony injuries commonly accompany ruptures of the ACL (65–74) (Figs. 5.16 and 5.17). (See Chapter 8.) Eighty-five percent of patients with acute ACL tears have an associated bony lesion, with the vast majority occurring in the lateral compartment (65,69,72). This tendency to the lateral compartment probably reflects the action of the most common forces involved in production of an ACL tear, that is, axial loading, rotation, and valgus stress.

The Segond fracture, an avulsion of the proximal lateral tibia occurring at the level of the tibial epiphysis, has a high association with ACL (75–100%) and meniscal tears (67%) (34,75–78). The injury results from avulsion of the meniscotibial portion of the middle third of the lateral capsular ligament occurring during internal rotation and varus stress (76,78). The avulsion fragment is remarkably constant in location and configuration. The typical fragment is elliptical in shape, measuring approximately 10 × 3 mm, with its proximal border 4 mm distal to the subarticular cortex of the lateral tibial plateau (77,78). The diagnosis of a Segond fracture can be made at the time of MR imaging, and thereby alert the examiner to closely search for associated meniscal and ligamentous damage. On MR, the Segond fracture is manifest by (a) the presence of a small, linear, flake-like fracture of lateral tibial rim (this finding is seen in only 25% of patients because of the small size of the fragment and the fact that its signal is the same as that of the periarticular fat) and (b) focal bone marrow edema of the tibial rim in the expected insertion site of the middle third of the capsular ligament. The fracture fragment itself is typically best seen on coronal MR images and anteroposterior radiographs, while the bone marrow edema is best visualized on either coronal or axial MR sequences (Fig. 5.16) (76).

A bone bruise is a poorly marginated zone of decreased signal (T1WI) in the tibial and/or femoral metaepiphysis with normal overlying articular cartilage and cortical bone (52,66,79). Other authors have termed this lesion a type 1 bony injury, or a reticular occult subcortical fracture (67,72). It is the most common ACL-associated bone injury. The lesion is thought to represent trabecular microfracture, a supposition supported by the relationship to acute trauma, positive nuclear bone scan, rapid clinical resolution, and a small amount of pathologic material (80). Trabecular fractures are generally thought to be painful but self-limited abnormalities. Affected bones may be at increased risk for the subsequent development of insufficiency fractures if the bone is not adequately protected during trabecular healing (67).

Transient forward subluxation of the lateral tibial plateau during ACL rupture, especially if accompanied by valgus forces, produces an impaction fracture of the mid-portion of the weightbearing surface of the lateral femoral condyle; the site of impaction is invariably directly over the anterior horn of the lateral meniscus (1,47,52,66,68,72,74,80). Distinction between the normal femoral concavity over the anterior horn of the lateral meniscus and an impaction injury can be difficult to identify on conventional radiographs. (See Chapter 11, Figs. 11.11–21.) The "lateral notch sign" is an exaggeration of the normal lateral femoral condylar indentation (greater than 2 mm) (1). On MR, the subcortical bone demonstrates decreased signal (edema and impacted medullary bone) on T1WI, and mixed signal on long TR/TE sequences. The overlying cartilaginous contour may be normal or minimally depressed, but often its signal is inhomogeneous. The experience of arthroscopists in detecting these lesions is variable, probably reflecting the varying degrees of actual impaction, and the fact that young cartilage is quite elastic and may rebound into normal position following impaction (65,66,72,80). This lateral femoral condylar lesion is highly specific for an ACL tear, although it is only found in 30% of cases (68). If the ACL ruptures because of a hyperextension injury, the lateral notch sign may be positive and accompanied by a "kissing" impaction fracture of the posterior lateral tibial plateau (74).

FIG. 5.17. ACL rupture-associated fractures. **A:** Sagittal MR, TR/TE 2,000/20. This patient suffered a ruptured ACL. The image through the lateral tibial plateau demonstrates flattening of the posterolateral tibial plateau, with adjacent edema (*small arrowheads*). **B:** The lesion (*arrowhead*) is also well visualized on a sagittal reformatted high resolution CT examination. **C:** Sagittal MR, TR/TE 2,000/20. In the same patient, a minimally displaced fracture of the posterior medial tibial plateau (*arrowhead*) represents an avulsion of the central tendon of the semimembranous tendon. **D:** The fracture is identified on a sagittal reformatted CT. **E:** In another patient, this reformatted CT scan demonstrates that there is an avulsive injury of the posterior rim of the lateral tibial plateau that has occurred as a result of the pull of the arcuate ligament. **F:** Capsular rupture at the popliteus; sagittal MR, TR/TE 2,000/80. During ACL rupture, the posterolateral capsule may tear, permitting joint fluid (*double arrows*) to extravasate along the popliteus muscle/tendon (*single arrow*). Such extravasation implies a tear of the attachment of the arcuate complex.

Several different mechanisms of injury at the time of ACL rupture may lead to fractures of the posterior tibia either medially or laterally. During external rotation and abduction with the knee flexed, an avulsion of the posterior tibial attachment of the central tendon of the semimembranosus muscle will result in a small flake-like fracture of the posterior *medial* tibial plateau (Fig. 5.17) (71). This fracture occurs at the insertion of the central tendon on the infraglenoid tubercle. On MR, this fracture may be represented by an obvious bony fragment pulled from a bed on the posterior tibia or, if the fragment is small, only the bone marrow edema of the tibia at the site of avulsion can be identified.

Avulsive or impaction-type fractures may also be found along the posterior aspect of the *lateral* tibial plateau. The arcuate ligament, a portion of the deep lamina of the posterior capsule, has two points of insertion, the posterolateral tibia and the fibular head. Either insertion may be avulsed during injury, resulting in either an avulsive fracture of the fibular head or the posterior tibial rim (1). The rupture of the capsule allows joint fluid to extravasate along the tendon/muscle belly of the popliteus. Transient anterior subluxation of the knee occurs during ACL rupture. The posterior lateral tibia impacts the anterior, lateral femoral condyle, resulting in a fracture of the femur directly over the anterior horn of the lateral meniscus, and also/instead an impacted fracture of the posterior lateral tibial plateau.

Accuracy of MR

Numerous recent articles have addressed the role of MR in the management of acute knee injuries, and in particular, its value and accuracy in assessing the integrity of the ACL (25,26,43,55,57,58,81,82,83–87). These studies have been performed using a variety of MR imagers and imaging protocols and have been conducted by examiners of different skill levels. Eleven recent studies, utilizing *high field strength* imagers, have examined a total of 1,661 patients. In these reports, the sensitivity of MR for the detection of a complete ACL tear has varied from 92% to 100% (six of the 11 studies reported 100% sensitivity), the specificity from 82% to 100%, and the accuracy from 92% to 100% (25,43,55,82–88) Since arthroscopy is not a true gold standard, that is, it is not 100% accurate, it is best to refer to the values given above as "agreement rates of MR and arthroscopy" (6,28,46). The negative predictive value of MR, which is the percentage of normal MR that do not have a tear at arthroscopy, ranged between 89% and 100% in those studies for which the data were available (four studies reported 100%). It must be concluded that MR of the knee, performed on a high field strength MR scanner, is an extremely accurate method for detecting complete ACL tears.

Three recent studies reviewed the accuracy of assessment of the ACL on MR examinations which were performed on *low or mid field strength* imagers (26,58,82). These studies were composed of 301 patients. The sensitivities for MR to detect ACL tears ranged from 61% to 97%, the specificity from 71% to 89%, and the accuracy from 79% to 91%. The data herein presented is not known to be statistically valid and is presented only to summarize the most recent radiologic literature. It must be recognized that *there are no well controlled studies of identical cohorts of patients studied on high and low-mid field strength MR imagers,* with results critically compared to the findings at arthroscopy.

Two other components regarding accuracy of assessment of the ACL by MR are worthy of mention. Several authors have advocated the use of dynamic, or kinematic, MR for assessment of the patellofemoral joint (89); Niitsu and his colleagues have described a method of examining the ACL by means of cine MR imaging (90). Both the static and dynamic examinations were performed using gradient techniques. Cine MR imaging more readily detected ACL tears than did static imaging, and partial tears were more readily detected with the dynamic technique than with standard static imaging. The use of nonorthogonal imaging has been shown to be a highly accurate means of assessing the ACL. Oblique imaging may have superior sensitivity, specificity, and accuracy when compared with standard orthogonal sequences (5).

POSTERIOR CRUCIATE LIGAMENT

Anatomic Considerations

The posterior cruciate ligament (PCL) originates from a fan-shaped attachment on the posterior aspect of the lateral surface of the medial femoral condyle and courses posteriorly and distally across the joint to attach to the extreme posterior intercondylar region of the tibia, 1 cm below the tibial articular surface (91). The PCL lies in the same synovial envelope as the ACL, and like the ACL it is intraarticular but extrasynovial. The PCL is 38 mm long and 13 mm wide, but it is most narrow at its mid-portion, fanning out proximally and to a lesser degree distally on the tibia (91). Quite like the ACL, the PCL is composed of individual fascicles that are grouped into two major fiber bundles. The anterolateral group, which comprises most of the PCL, and the smaller posteromedial group are named for their attachment locations on both the femur (anterior and posterior), and the tibia (medial and lateral). The fibers twist on their longitudinal axis in a clockwise direction during flexion and unwind during extension; this action is analogous, but directly opposite that of the ACL. The anterolateral bundle tightens on flexion and relaxes in extension; the posterior bundle is reversed in action.

The meniscofemoral femoral ligament (MFL), a fibrous cord that connects the posterior horn of the lateral meniscus to the medial femoral condyle, has been called "the third cruciate ligament" because of its relationship to the origin of the PCL from the femur (91). The MFL is a fibrous cord that connects the posterior horn of the lateral meniscus to the medial condyle of the femur. Anatomically, two distinct branches can be identified. The more anterior is known as the ligament of Humphrey; it is directed in an oblique craniomedial orientation from the posterior lateral meniscus, anterior to the PCL toward its condylar insertion site. The posterior branch, also known as the ligament of Wrisberg, or Roberts ligament, has a similar orientation but runs immediately posterior to the PCL. It is the larger of the two branches, and may be as large as one-half the diameter of the PCL. The MFL functions to pull the posterior horn of the lateral meniscus anteriorly and medially during flexion, balancing the action of the popliteus muscle whose fibers attach to the posterolateral meniscus. While both branches of the MFL may exist in the same knee, there are considerable variations in their presence and size. Various anatomic dissections have found that the branches of the meniscofemoral ligaments may be found in 80% to 100% of knees; the anterior is found in 34% and the posterior in 60% of the specimens.

Clinical Considerations

The PCL provides 90% of the total restraint to posterior displacement of the tibia on the femur in the flexed position, and is the primary stabilizer of the knee in flexion, extension, and internal rotation (91,92). The PCL is twice as strong as the ACL or the medial collateral ligament, a fact accounting for its lower rate of injury (48). Ruptures of the PCL are thought to comprise 3% to 20% of knee ligament injuries, but the true rate is probably higher given the fact that acute tears often go undiagnosed (see below). Since the PCL is a large, strong structure, it is ruptured only following forces and injuries of great magnitude, and therefore it is common that other intraarticular injuries occur simultaneously (1,92). Ten of the 11 knees suffering PCL ruptures had other significant internal knee derangements (93). Most commonly, the medial collateral ligament (50%), the medial meniscus (30%), the ACL (65%), the posterior capsule, and the fibular collateral ligament are simultaneously torn, and the peroneal nerve and popliteal artery may also be damaged (92,94–96). The latter is most likely to occur if the PCL is torn at the time of a dislocation of the knee; in such cases, an arteriogram for assessment of the popliteal artery is necessary (1). Bilateral PCL rupture is exceedingly rare (97).

Approximately one-half of PCL tears occur during athletic endeavors, and the remainder occur during traffic and industrial mishaps. The PCL is injured by one of three mechanisms: (a) hyperextension of the knee, as occurs in football, (b) hyperflexion of the knee; and (c) posterior displacement of the tibia on the femur with the knee in flexion (92,98). The latter injury, resulting from high velocity forces, is the most common mode of injury and is associated with ecchymosis or abrasion to the pretibial region. Patellar fractures, hip dislocations, or diaphyseal fractures of the femur occur concurrently. If a knee strikes a dashboard, or as in football, if a planted flexed knee is struck from the front, isolated injury to the PCL may occur (99). The PCL injury resulting from a "dashboard" injury occurs distally, occasionally via tibial avulsion. In a series of 59 acute PCL tears, the injury site was femoral in 21, mid-substance in 13, and tibial in 25; in the latter, avulsion of the insertion site may occur (93,96,100).

The most commonly performed tests for function of the PCL are the posterior drawer, the 90° quadriceps active test, the posterior sag sign, and varus or valgus instability tests with the knee in complete extension (98). Physical examination for suspected posterior instability of the knee is only 50% to 60% accurate. Four of eleven patients with proven tears of the PCL had normal physical examinations (93). The drawer sign is not positive in all cases of acute PCL disruption because of the presence of hamstring spasm and a tense hemarthrosis. The posterior drawer sign may be equivocal or normal if the arcuate ligament complex is intact (101,102). Under anesthesia, false-negative posterior drawer signs have been reported even at the time the surgeon was viewing the torn PCL arthroscopically (4,28,93). Pain, swelling, and protective muscle spasm and hemarthrosis will serve to delay the diagnosis of a PCL tear; 42 of 102 PCL tears were not clinically diagnosed until 3 to 4 months after injury (100).

Many individuals who suffer PCL tears do well if treated conservatively (101,103). This is especially true for those individuals who demonstrate unidirectional instability. Those with a complex of ligamentous injuries are more likely to demonstrate a higher incidence of degenerative changes on radiographs, poorer functional outcomes, and an increased incidence of patellofemoral disease (101,104). Repair of an acute bony avulsion provides excellent static and functional results, and should be performed within 2 weeks (102). Primary repairs may succeed if performed acutely, but late attempts at repair are notoriously variable. A variety of reconstructive procedures have been designed for the chronically torn PCL. Patellar tendon reconstructions (similar to that performed for the ACL), patellar tendon–quadriceps tendon flaps, use of the medial tendinous insertion of the gastrocnemius, and the use of synthetics have all been proposed with varying degrees of success, but at this time, satisfactory results following reconstruction cannot be routinely expected (99,105,106); good results occur in 54% to 80% and poor results in 5% to 32% (100). Nonoperative treatment of the isolated mid-substance tear may be a viable alternative to the technically challenging reconstruction (102).

Following chronic rupture of the PCL, the posterior control of the joint falls upon the patella and the patellar tendon; large forces are applied to the articular surface of the patella and its femoral groove. The patellofemoral joint is therefore placed at great mechanical disadvantage in controlling extension of the knee joint and anteroposterior motion of the joint. The end result of PCL injury may well be tricompartmental osteoarthritis (103).

Arthroscopy of the PCL

While the PCL is a very large structure, it is not easily assessed at the time of arthroscopy (4,28, 91,92,95,100,103,107,108). The inferior two-thirds of the PCL are poorly seen from an anterior approach because of the presence of large amounts of synovium and subsynovial fat, the presence of an intact ACL, and the fact that the PCL attachment is well below the articular surface of the tibia (1). Ideally, the PCL should be viewed by means of posteromedial and posterolateral arthroscopic portals. Occasionally, the synovial envelope of the PCL needs to be opened to find the tear (95). The patient with a PCL tear may have a taut, intact ligament of Humphrey and the arthroscopist may mistake it for an intact PCL (1). The arthroscopic appearance of the PCL is altered in the knee in which a chronically torn ACL has reattached itself to the PCL. Finally, rupture of the posterior capsule may result in inability of the surgeon to maintain distention of the joint and optimally assess the intraarticular structures of the knee (100). Extravasation of joint fluid through a capsular rent into the popliteal space can result in catastrophic neural or vascular compromise.

Imaging Assessment

The PCL, by virtue of its size and low signal intensity on all spin-echo imaging sequences, is perhaps the most easily identified structure within the knee on MR examinations. On sagittal images made through the more medial aspect of the intercondylar notch, the PCL is seen to arise from the inferior portion of the intercondylar roof and to insert on the posterior aspect of the tibial plateau below the joint line. The entire course of the PCL is readily visualized on a single sagittal image on 95% of properly performed knee MR examinations (Fig. 5.18) (52,93). The mid-portion of the PCL appears slightly thickened, often due to close approximation of the ligaments of Humphrey and/or Wrisberg. Either of the two branches of the meniscofemoral ligament may be present in up to 58% of MR examinations; both branches are uncommonly present in the same knee (11% of MR examination) (91,93).

Conventional spin-echo sequences have been the standard protocol used in assessment of suspected internal derangement disorders of the knee. Some centers, however, now utilize a variety of gradient-recalled sequences as their primary imaging mode (GRE). Areas of high signal intensity on GRE sequences have been identified in the normal PCL on images in which the ligament lies at an angle of 45° to 85° to the axis of the main magnetic field (109). It is at this angle that the effective T2 of the PCL fibers is the longest. In patients with high signal foci in the PCL on GRE studies, spin-echo images and arthroscopic assessment are often normal.

The MR morphology of the PCL is critically dependent on the degree of flexion of the knee and the integrity of the ACL and the secondary restraints. When the knee is in extension or a minimal degree of flexion, the PCL has a gently convex posterior margin; if the knee is flexed more than 10°, the ligament becomes taut and perhaps slightly thinner than in the extended position (47,52,93). Minimal degrees of buckling or a slight degree of accentuated convexity of the proximal end of the PCL are normal. Disruption of either the PCL (or the ACL) must be suspected when the PCL assumes a kinked, sharply buckled, hair-pin turn, S-shaped, or so called "sea-horse" appearance (Figs. 5.14 and 5.19) (52–54). Such descriptions describe a PCL which is either: (a) foreshortened, as a result of a distal complete tear of the PCL, distal avulsion of the PCL, or complete unstable tear of the ACL; or (b) elongated and stretched as a result of an intrasubstance PCL tear (47). In the coronal plane, the PCL is usually not seen on a single image; such single-image visualization implies that the PCL has a more vertical course than normal, which is invariably due to buckling and foreshortening as detailed above.

Acute Ruptures

Acute tears of the PCL can be divided into two broad groups: those occurring within the substance of the PCL, and those occurring less commonly through the mechanism of avulsion. Nearly all substance tears of

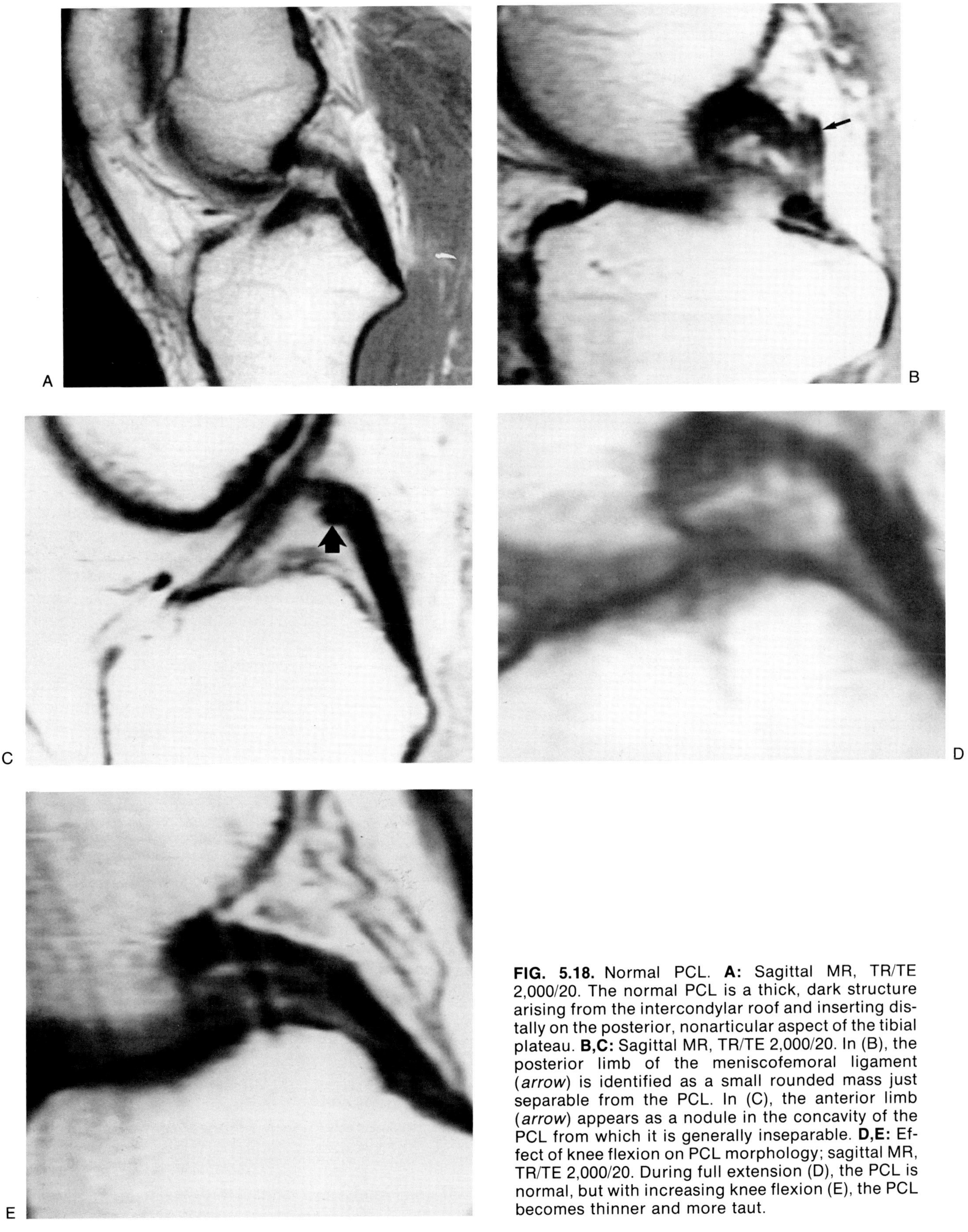

FIG. 5.18. Normal PCL. **A:** Sagittal MR, TR/TE 2,000/20. The normal PCL is a thick, dark structure arising from the intercondylar roof and inserting distally on the posterior, nonarticular aspect of the tibial plateau. **B,C:** Sagittal MR, TR/TE 2,000/20. In (B), the posterior limb of the meniscofemoral ligament (*arrow*) is identified as a small rounded mass just separable from the PCL. In (C), the anterior limb (*arrow*) appears as a nodule in the concavity of the PCL from which it is generally inseparable. **D,E:** Effect of knee flexion on PCL morphology; sagittal MR, TR/TE 2,000/20. During full extension (D), the PCL is normal, but with increasing knee flexion (E), the PCL becomes thinner and more taut.

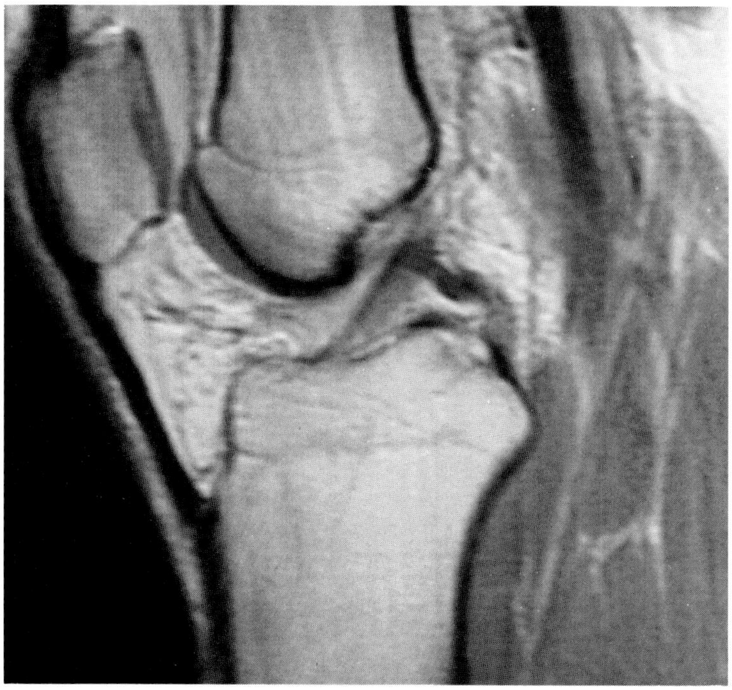

A

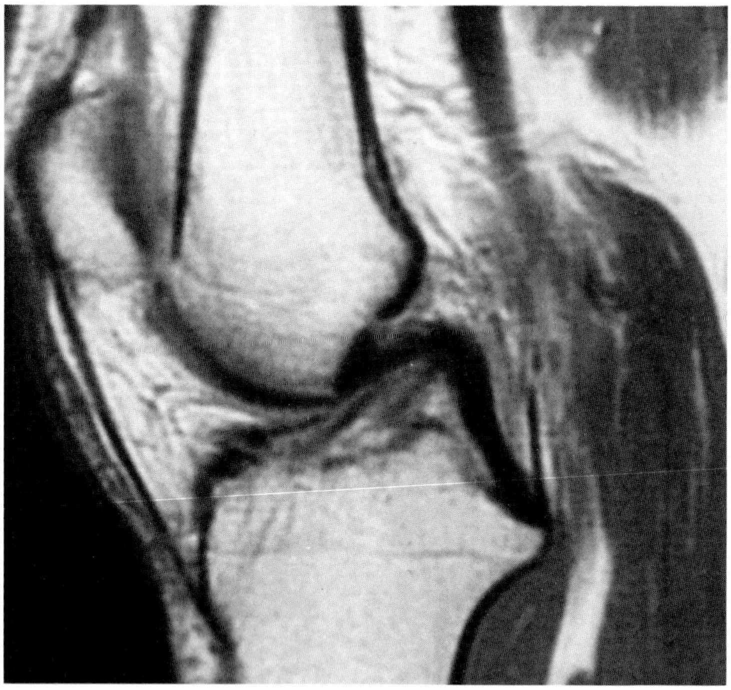

B

FIG. 5.19. Abnormal morphology of the PCL. **A,B:** Sagittal MR, TR/TE 2,000/20, from two different patients. On clinical examination, both had positive posterior drawer signs. At arthroscopy, both PCL were diffusely replaced by scar which was physiologically insufficient to restrain abnormal knee motion. In (B), the signal of the PCL is normal, but in (A) the signal intensity is increased. (See Fig. 5.14.)

FIG. 5.20. Ruptures of the PCL. **A:** Sagittal MR, TR/TE 2,000/20; **B:** sagittal MR, TR/TE 2,000/80; **C:** axial MR, TR/TE 2,200/35/160. All three sequences demonstrate diffuse widening and increased signal intensity of the PCL (*arrowheads*). Quite like interstitial ACL tears, the signal within the torn PCL does not become hyperintense on long TR/TE sequences. The ligament remains in continuity. **D:** Sagittal MR, TR/TE 2,000/20. In another patient, an interstitial-type tear affects only the distal half of the ligament which remains in continuity. **E:** Sagittal MR, TR/TE 2,000/20. Detachments of the PCL are unusual except for avulsions of the tibial insertion. In this patient, the proximal PCL has torn from the femur and has fallen on the tibial plateau.

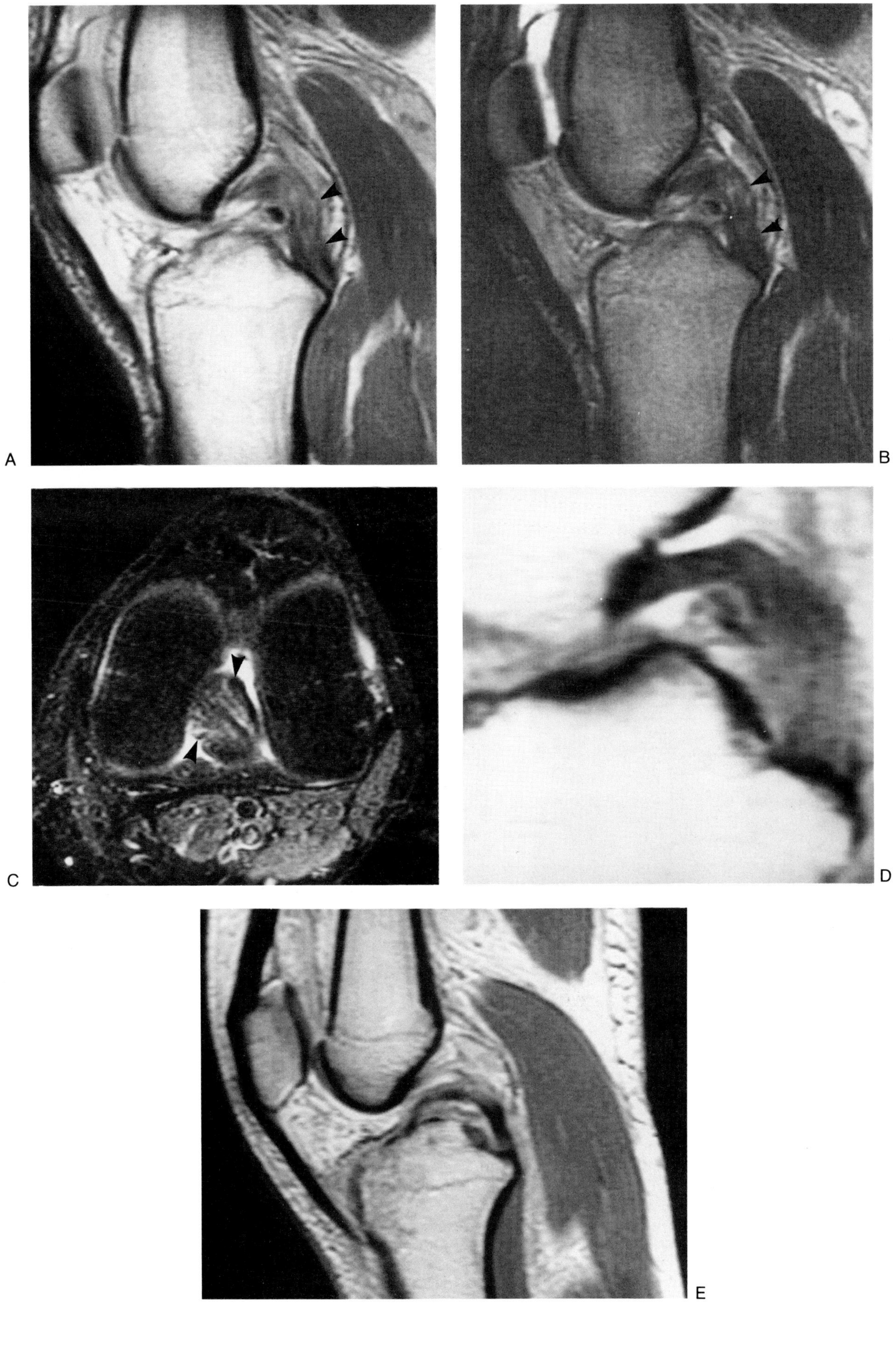
A
B
C
D
E

the PCL demonstrate morphologic characteristics described for an interstitial tear of the ACL (Fig. 5.20) (47,52). A long segment, or perhaps the entire length of the PCL, is diffusely widened, but the torn PCL has less of a mass-like character than does the torn ACL (52). The signal on T1WI is inhomogeneously increased; on long TR/TE sequences, there are small areas within the ligament that increase in signal intensity due to the presence of intraligamentous edema, but it is unusual to identify coalescent areas of joint fluid within the PCL. On rare occasions, the PCL may be detached and lie free in the joint. We have never identified a discrete gap within the course of a torn PCL.

Avulsive ruptures affect the tibial attachment of the PCL (Fig. 5.21). In our experience, however, the fragment is displaced further away from the bone than occurs with ACL avulsions, and there is often extensive bone marrow edema at the avulsion site.

Chronic Ruptures

Since most PCL tears occur interstitially, the "continuity" of the ligament is maintained. Spontaneous repair of the torn PCL occurs over the torn ligament and the interposed hematoma, and a dense fibrous scar may replace the normal ligament, unlike the ACL in which atrophy rapidly occurs (Figs. 5.19 and 5.22). Both fibrous scar and the normal PCL have low signal character at all imaging sequences, and the anatomically "intact" but functionally torn PCL may be a source of interpretive error (47,53,102). An accentuated serpiginous course, ligamentous redundancy,

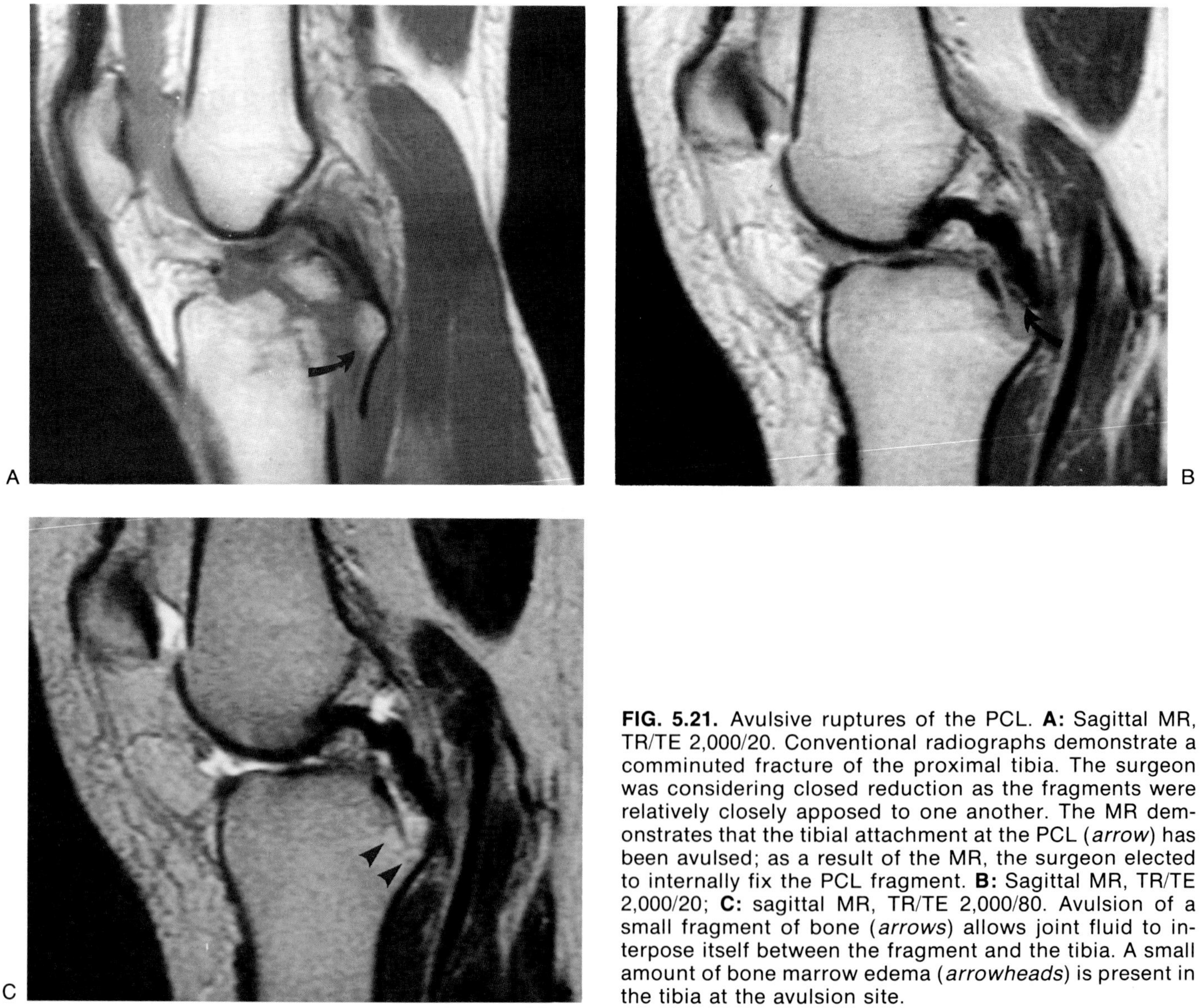

FIG. 5.21. Avulsive ruptures of the PCL. **A:** Sagittal MR, TR/TE 2,000/20. Conventional radiographs demonstrate a comminuted fracture of the proximal tibia. The surgeon was considering closed reduction as the fragments were relatively closely apposed to one another. The MR demonstrates that the tibial attachment at the PCL (*arrow*) has been avulsed; as a result of the MR, the surgeon elected to internally fix the PCL fragment. **B:** Sagittal MR, TR/TE 2,000/20; **C:** sagittal MR, TR/TE 2,000/80. Avulsion of a small fragment of bone (*arrows*) allows joint fluid to interpose itself between the fragment and the tibia. A small amount of bone marrow edema (*arrowheads*) is present in the tibia at the avulsion site.

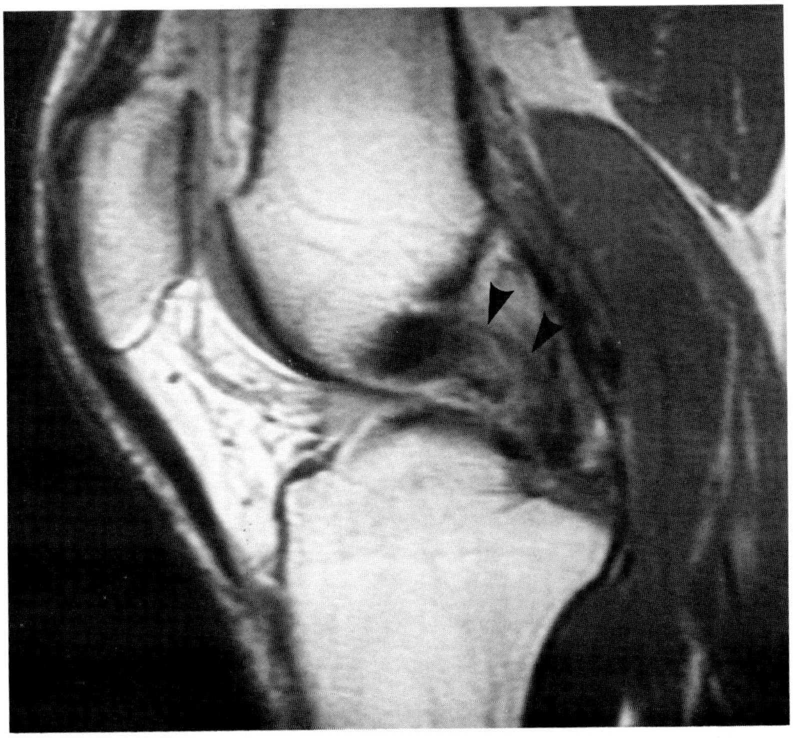

A

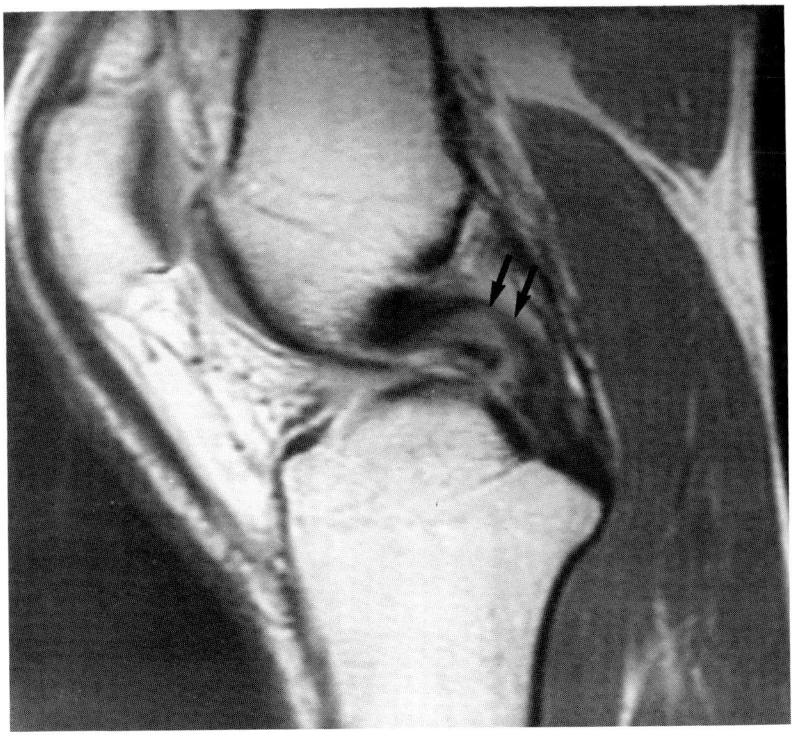

B

FIG. 5.22. Healing of a torn PCL. **A,B:** Sagittal MR, TR/TE 2,000/20. In (A), made the day following a hyperextension injury, the PCL is diffusely widened (*arrowheads*) and demonstrates marked increase in signal throughout its substance. The patient had a stable knee at this time. In (B), made 2 months following (A), the marked widening has resolved. A thin, dark structure (*arrows*) not present in (A) now bridges the gap between the proximal and distal ends of the PCL. On both examinations, the PCL is in continuity.

and failure of the PCL to become more taut during stress (flexion) maneuvers should alert the examiner to the correct diagnosis of chronic PCL disruption. Most often, the chronically torn PCL also demonstrates a mild diffuse increase in ligamentous signal, but some inhomogeneities occur in knees that are thought to be normal and have no arthroscopic correlate (93).

Accuracy

Since PCL ruptures are uncommon, frequently escape clinical (and arthroscopic) detection, and if suspected are often treated conservatively, it is quite difficult to determine the prevalence of such injuries and the efficiency of MR in assessing them. Several recent studies have specifically addressed the ability of MR to assess the integrity of the PCL (43,82,93). One study addressed a very large number of patients (1,014) and encountered eight true positive examinations, 998 true negative, 6 false-positive, and 2 false-negatives (82). There are too few patients with positive MR to statistically validate the examination, although it is obvious that the negative predictive value and specificity are high. False-positives are a more significant occurrence than false-negatives, but given the difficulties encountered in the performance of arthroscopy, it seems more than reasonable that false-negative arthroscopies account for at least some of the false-positive MR examinations.

THE MEDIAL COLLATERAL LIGAMENT

Anatomic Considerations

In order to fully understand the composition of the medial collateral ligament, it is worthwhile to review the anatomy of the medial side of the knee (3,95,110–112). Based on anatomic dissections, three distinct tissue layers, each containing important supporting structures, can be identified along the medial side of the knee. Layer I, first encountered after incision of the skin, is the deep or crural fascia. This fascia provides the point of insertion of the sartorius muscle on the medial side of the knee, and posteriorly it envelopes the gastrocnemius and the structures of the popliteal fossa. Layer I blends with layer II anterior to the medial collateral ligament; the combination serves to form the patellar retinaculum. A layer of fibrofatty tissue lies just deep to layer I; the gracilis and semitendinosus tendons lie in this tissue plane.

Layer II contains a number of important structures, the best known of which is the tibial collateral ligament (TCL) (35,113). The TCL is also known as the superficial layer of the medial collateral ligament complex. The superficial layer, more properly known as the TCL, arises from the medial epicondyle of the femur 5 cm above the joint line and inserts on the tibial metaphysis 6 to 7 cm below the joint line just deep to the pes anserinus tendons. The TCL is approximately 1.5 cm wide. Most of the fibers of the TCL are vertical (the parallel portion), and are oriented in a slightly superoposterior to inferoanterior direction, but one segment of the TCL has a slightly more obliquely posterior course. It extends from the posterior aspect of the TCL, covers the anterior portion of the semimembranosus tendon, and inserts on the joint line. This oblique segment is termed the posterior oblique ligament. Posteromedially, layer II fuses with layer III.

Layer III is the true capsule of the knee joint. Its sites of insertion are the margins of the articular surfaces. Throughout most of its course, the capsule of the joint is a rather thin redundant structure. Just deep to the TCL, the membrane has a focal condensation that is known as the deep medial ligament, the deep layer of the medial collateral ligament, or the medial capsular ligament. The medial capsular ligament is firmly adherent to the medial meniscus. The proximal and distal ends of the medial capsular ligament are known as the meniscofemoral and meniscotibial ligaments (Fig. 5.23).

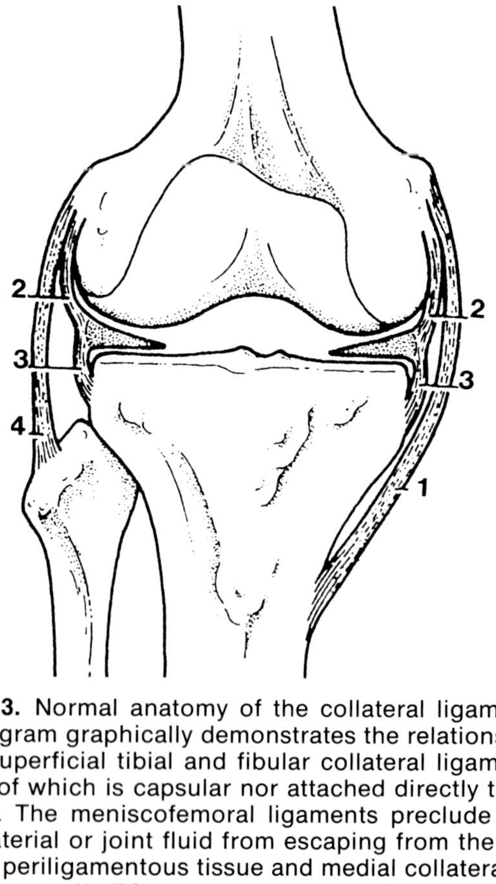

FIG. 5.23. Normal anatomy of the collateral ligaments. This diagram graphically demonstrates the relationships of the superficial tibial and fibular collateral ligaments, neither of which is capsular nor attached directly to the menisci. The meniscofemoral ligaments preclude contrast material or joint fluid from escaping from the joint into the periligamentous tissue and medial collateral ligament bursa. (1, TCL; 2, meniscofemoral ligaments; 3, meniscotibial ligaments; 4, fibular collateral ligament.)

Within the fibrofatty tissue that separates layer II from layer III and therefore separates the TCL from the medial capsular ligament, lies a relatively constant bursa (28,110,112–115). Bursae of variable number and size are located deep to the TCL and are not associated with and do not communicate with any other known bursae about the knee. This intraligamentous bursa may extend either/both above/below the joint line. Clinically, tibial collateral ligament bursitis is manifest by medial joint line pain in a patient without mechanical symptoms or history of acute injury (116).

Clinical Assessment

The function of the TCL reflects its division into parallel and oblique fiber segments (3). The femoral attachment of the parallel fibers is arranged around the axis of flexion of the knee so that its tension remains constant throughout the range of motion; the oblique fibers however, become lax with increasing degrees of knee flexion. The TCL acts as the primary restraint against valgus stress (3,36). If the TCL is sectioned, approximately 4 to 5 mm of opening to valgus stress at the knee can be provoked; if only the medial capsular ligament is cut, there is no significant valgus laxity. The medial collateral ligament also serves to prevent external rotation of the tibia on the femur and to provide a secondary restraint to anterior tibial translation in the ACL-deficient knee.

The extent of a knee injury is determined in part by the nature of the interface between the foot and the surface upon which it is operating. Isolated TCL tears are common in hockey players because the interface between the skate and the ice is quite forgiving; the skate may become trapped in a pure valgus state without the rotary deceleration of the lower limb that commonly occurs at the interface between ski and snow (47). TCL tears account for 61% of all knee injuries that occur during skiing but commonly ACL tears accompany the injury (33). In situations in which a shoe is in contact with turf, medial ligamentous tears are caused by severe valgus stress with contact, the so-called clipping injury, or with valgus stress and external rotation without contract.

If the knee is flexed 90° at the time of injury, the valgus stresses are borne by the medial capsular ligament fibers which tear first; more severe forces tear the TCL and the ACL. Ultimately, the PCL and even the lateral capsular structures may be damaged (in effect, a knee dislocation) (3,52). While the medial meniscus may become detached at the meniscotibial attachment, tears within the substance of the meniscus are uncommon unless a rotary component accompanies the valgus stresses (3). O'Donoghue's "unhappy triad" classically includes a tear of the TCL, the ACL, and the medial meniscus, but in fact it is the lateral meniscus that is more frequently torn. Chronic instability of the knee resulting from a TCL tear will, however, frequently lead to secondary meniscal tears on either side.

Following injury, there is localized swelling and tenderness over the site of injury to the ligament; pain is maximal at 24 to 36 hr. Since the TCL is extracapsular, the presence of a significant joint effusion following a knee injury suggests that intraarticular derangement, often an ACL tear, has also occurred (3,35). Within 1 hr of injury, protective muscle spasm, tenderness, and edema are sufficient to prevent adequate clinical testing for joint laxity. Therefore a rapid physical examination on the playing field or ski slope is indicated, but such examinations of acutely injured knees are painful and challenging (35). Integrity of the TCL and the medial capsule is tested by applying a valgus stress to the knee while it is in hyperextension, in 0° of flexion, and at 30° of flexion. Excessive laxity in hyperextension or full extension nearly always indicates that the TCL and both cruciates are torn, but occasionally the PCL is spared. If the knee is unstable only at 30°, the TCL alone is torn.

The severity of injury of a ligament can be graded according to the amount of ligamentous disruption (3,33,36). A grade 1 sprain is predominantly a microscopic tearing, or stretching, of the ligament and results in no loss of ligament restraining function. In a grade 2 sprain, disruption of a portion of the ligament results in some loss of function. In a grade 3 sprain, or complete rupture, anatomic continuity of the ligament is essentially absent, and function is lost. The degree of injury can be graded clinically according to the amount of medial joint space opening with valgus stress applied at 30° of flexion. In a first degree sprain, there is tenderness at the site of injury and pain, but there is a firm endpoint and there is no identifiable laxity. In second degree injuries, up to 5 mm of excessive laxity is found, but the endpoint remains firm. The middle third of the medial capsule is probably intact. In third degree tears, the endpoint is soft. If 10 mm of laxity is found, the TCL, capsular ligament, and one/both cruciates are undoubtedly torn.

During the first week following ligamentous injury, there is an initial stage of hematoma formation and infiltration of inflammatory cells. Fibroblastic proliferation, collagenization, and organization occur during the subsequent 6 to 12 weeks. Complete mid-substance tears of the TCL that are not surgically repaired heal by scar formation rather than true by ligament regeneration. TCL scar has a reduced strength (40%), and reduced stiffness and stress-relaxation behavior compared to normal TCL (117).

The optimal treatment of injuries to the TCL has been the subject of considerable debate. While it is generally agreed that clinical grade 1 and 2 strains can be treated nonoperatively, grade 3 injuries pose special problems (35,118–121). The issue is clouded by the fact that other significant injuries to the ligaments and menisci commonly complicate complete collateral ligament tears (118). Grade 1 TCL sprains have a 20%, grade 2 a 53%, and grade 3 a 78% likelihood of coincident ACL disruptions (122). Multiple authors have compared the results of surgical versus nonsurgical treatment to patients with isolated grade 3 tears and found that the therapies yielded similar results even though residual laxity after nonoperative therapy may be found (118,119,122). It is critical to the success of nonoperative therapy of an isolated grade 3 TCL tear that one correctly diagnose the lesion, because in complex lesions, the results of nonoperative therapy were judged as unsatisfactory in 79% of patients (3). The TCL-deficient knee is one in which an increase load is borne by the ACL, which is therefore at increased risk for injury (123).

If surgical therapy is necessary, the anterior (tibial) arm of the semitendinosus tendon, the fascia lata, or the adductors can be utilized to reinforce a primary repair of a torn TCL (3).

Rehabilitation treatments for TCL injury vary, but several important principles must be considered. An initial period of immobilization is essential to reduce pain and to permit the repair process to begin. Since normal walking is compressive rather than distractive, walking can be instituted early. Prolonged immobilization leads to muscle atrophy, articular cartilage changes, decreased ligament strength, joint stiffness,

A 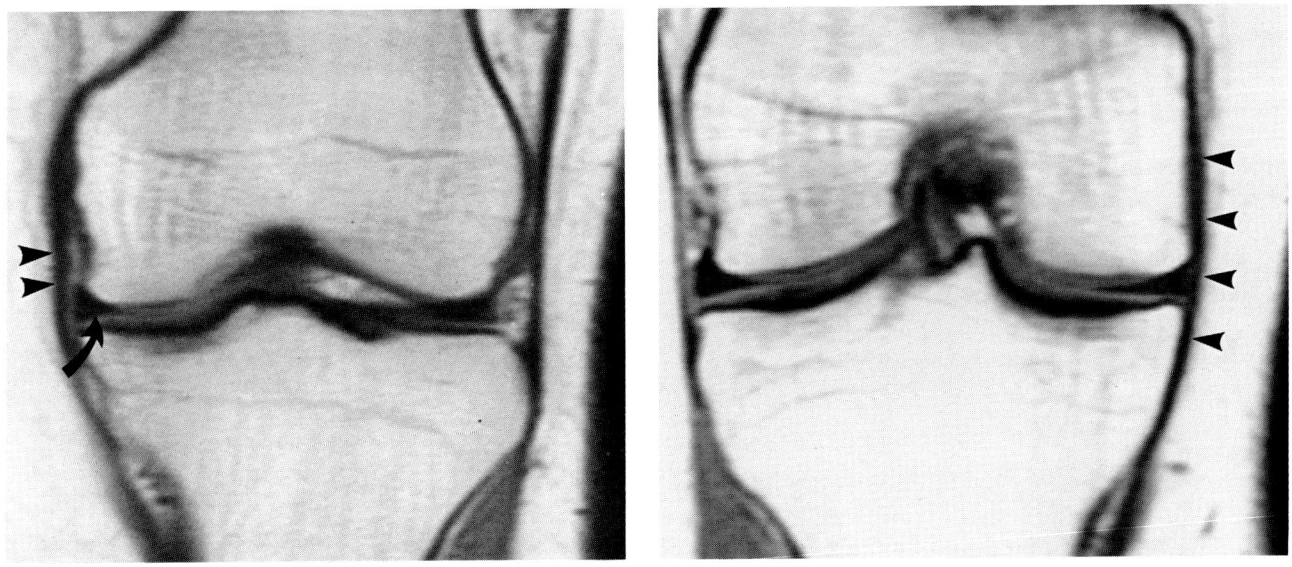 B

C 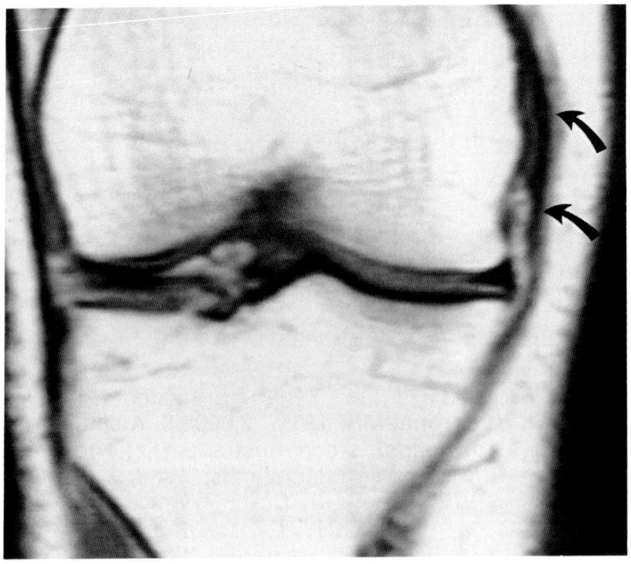

FIG. 5.24. Normal TCL. **A:** Coronal MR, TR/TE 600/20. The TCL (*arrowheads*) is separated from the medial meniscus (*curved arrow*) by a high signal line thought to be a thin layer of fat and a small bursa. This "space" must not be mistaken for a meniscal capsular separation. **B,C:** Coronal MR, TR/TE 800/20. In another patient, the TCL (*arrowheads*) is closely applied to the osseous structures. Ten mm more anteriorly, (C), the TCL blends with the tissues of layer I, forming the patellar retinaculum (*arrows*) which is not as closely related to bone.

decreased strength of the ligament-bone interface, and osteoporosis. Controlled motion in a brace should be instituted early (124). Ultimately, muscle strengthening and agility exercises are necessary to prepare the patient for a return to activity, usually in 4 to 8 weeks (3,12).

Imaging Assessment

The treatment of medial collateral ligament injuries is dependent on: (a) the grade of the TCL tear; (b) the extent of the injury to the medial capsular ligament and the degree, if any, of separation of the meniscus; (c) the integrity of the ACL and the PCL (see above); and (d) the presence of a meniscal tear (see Chapter 4).

The TCL

The structures of the medial collateral ligament complex are most readily reassessed in the coronal plane (Fig. 5.24) (52,54). The TCL is a thin dark stripe (T1 and T2WI) paralleling, and in many cases inseparable from, the dark medial cortices of both the femur and the tibia. Its attachment to the medial femoral epicondyle is usually more easily identified than is the insertion on the tibia 7 cm below the joint line. As one moves further anteriorly, the TCL, which is only 1.5 cm wide, and the structures of layer II blend with layer I and the structures of the patellar retinaculum (111). The retinaculum, also a thin dark stripe, is invariably distinctly separated from the medial aspect of the os-

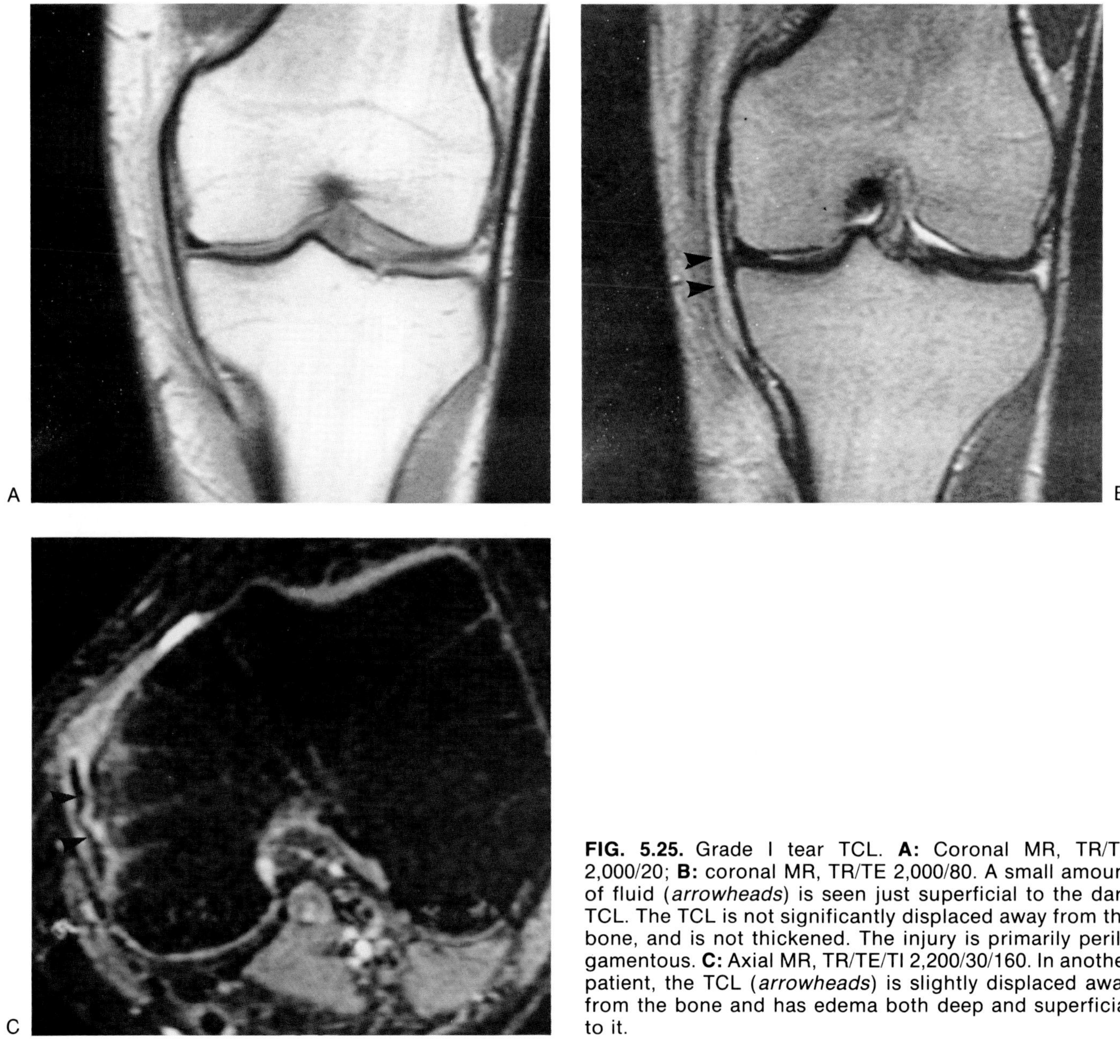

FIG. 5.25. Grade I tear TCL. **A:** Coronal MR, TR/TE 2,000/20; **B:** coronal MR, TR/TE 2,000/80. A small amount of fluid (*arrowheads*) is seen just superficial to the dark TCL. The TCL is not significantly displaced away from the bone, and is not thickened. The injury is primarily periligamentous. **C:** Axial MR, TR/TE/TI 2,200/30/160. In another patient, the TCL (*arrowheads*) is slightly displaced away from the bone and has edema both deep and superficial to it.

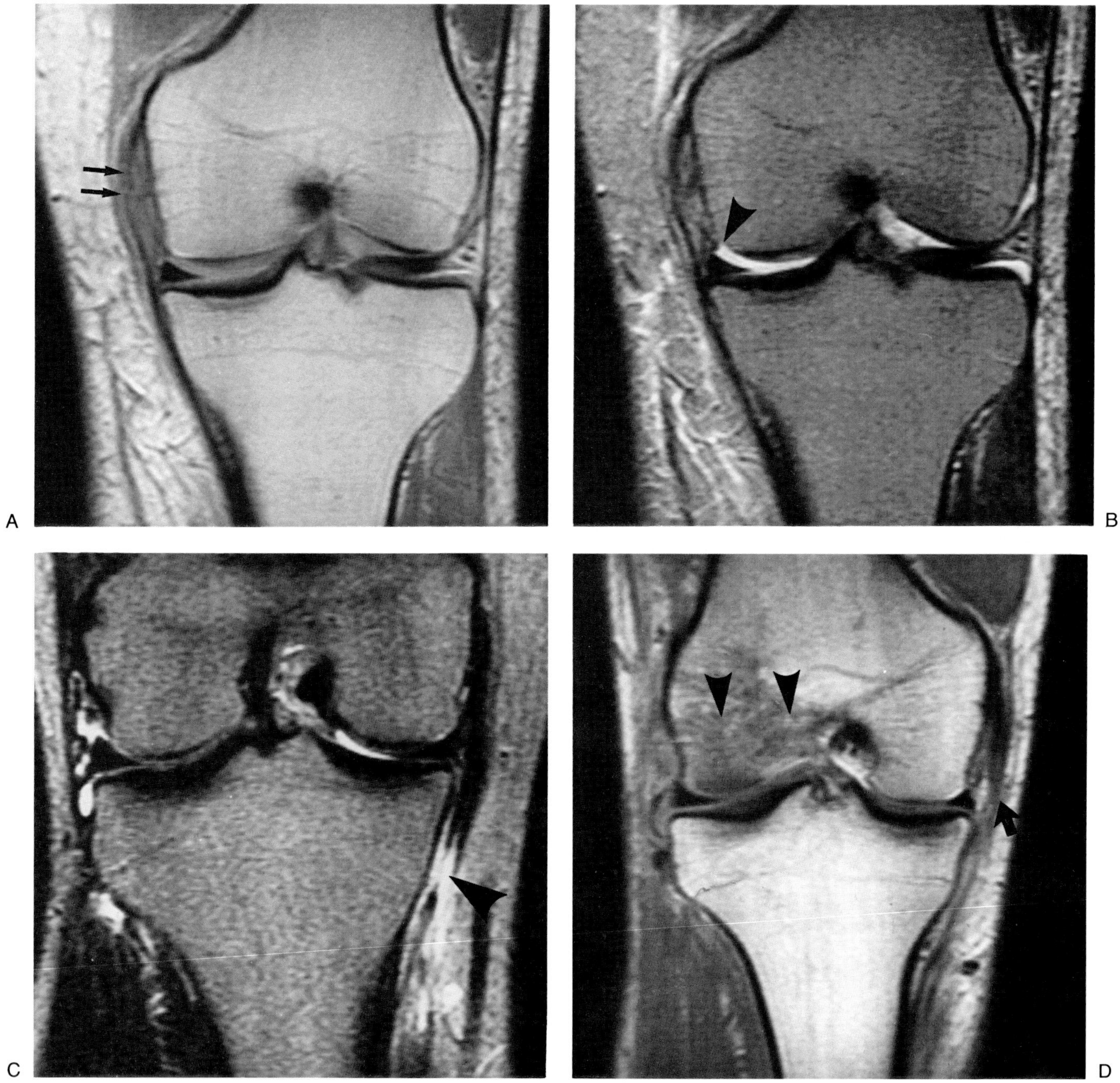

FIG. 5.26. Proximal grade 2/3 TCL tears with an intact medial capsular ligament. **A,B:** Coronal MR, TR/TE 2,000/20 and 2,000/80. The proximal TCL (*arrows*) is displaced away from the femur, and the continuity of the ligament is lost. On the long TR/TE sequence, the mass of blood and torn ligament do not increase in signal intensity. Joint fluid extends over the top of the meniscus (*arrowhead*), but does not extravasate into the subligamentous space. **C:** Distal grade 2/3 TCL tear; coronal MR, TR/TE 2,000/80. The distal TCL is detached from the bone; the site of the injury (*arrowhead*) is characterized by extremely bright fluid. Most commonly, the injury site does not significantly increase in signal on long TR/TE sequences. **D:** Coronal MR, TR/TE 800/20. This TCL is interrupted at and below the joint line (*arrow*) (grade 2/3 tear). The diffuse area of decreased signal in the lateral femoral meta-epiphysis (*arrowheads*) represents a bone bruise, a frequent accompaniment of high grade TCL tears.

seous structures and must not be mistaken for a torn displaced TCL.

Precise anatomic-pathologic correlation of the extent of the TCL tear is lacking, but anecdotal experience with grading of TCL tears exists (47,52,53). In grade 1 sprains, there is only microscopic tearing of the ligament, and the predominant alterations occur in the periligamentous tissue (Fig. 5.25). On MR, the thin dark stripe of the TCL is maintained, but deep subcutaneous and periligamentous edema is present. The soft tissue reaction most commonly increases in signal intensity on long TR/TE sequences, especially if the injury has occurred within 24 hr of the MR examination.

We choose to classify extensive TCL injuries as grade 2/3 sprains since we cannot reliably distinguish between the two grades (Fig. 5.26) (47). In severe injuries, the TCL is thickened and often serpiginous. The

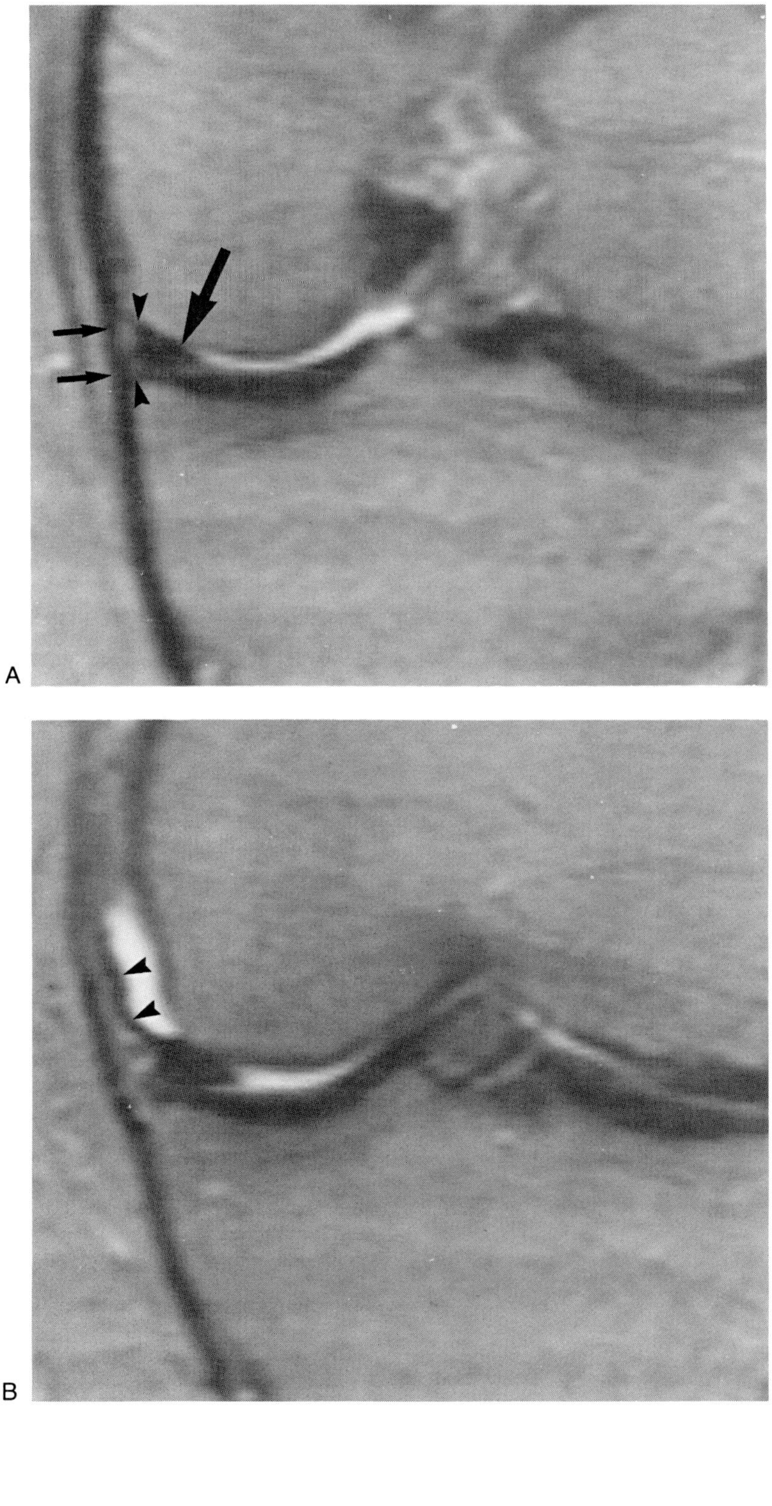

FIG. 5.27. Normal capsular relationships. **A,B:** Coronal MR, TR/TE 2,000/80. (A) is 10 mm posterior to (B). In (A), the TCL (*small arrows*) is clearly seen as a dark band along the medial aspect of the femur and tibia. The medial meniscus (*arrow*) is separated from the TCL by fat. The expected position of the meniscofemoral and meniscotibial ligaments (*arrowheads*) is identified, but in this knee without joint fluid, these small structures are not demonstrated. Further anteriorly in (B), the TCL is no longer seen, and the capsule is more redundant (*arrowheads*), blending with the medial aspect of the suprapatellar bursa. The diagnosis of a meniscofemoral ligament disruption should not be made on sections anterior to the TCL.

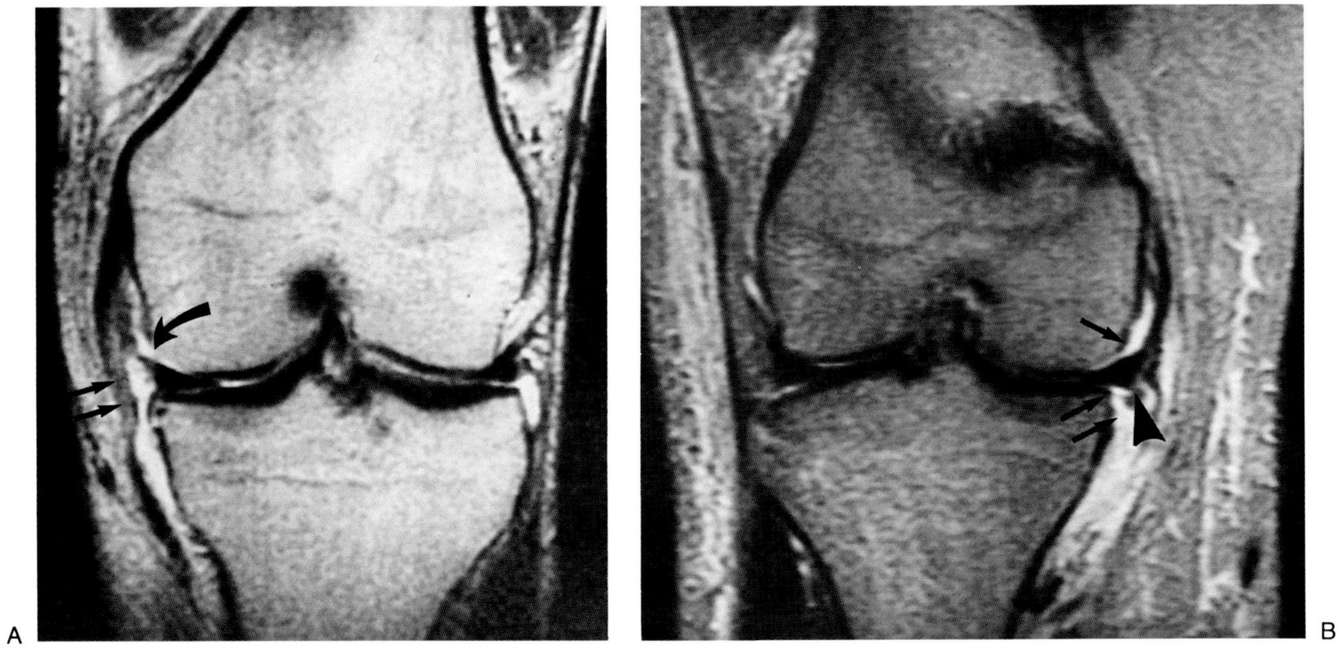

FIG. 5.28. Grade 2/3 TCL tears with rupture of the medial capsular ligament. **A:** Coronal MR, TR/TE 2,000/80. The TCL is torn at and below the joint line (*arrows*). The meniscofemoral ligament (*curved arrow*) is detached from the femur. There is continuity of the joint space and the subligamentous tissues into which joint fluid flows. **B:** Coronal MR, TR/TE 2,000/80. In another patient, both the meniscofemoral (*arrow*) and the meniscotibial (*double arrows*) ligaments are torn, and the medial meniscus (*arrowhead*) is displaced out of the joint. The distal TCL is torn.

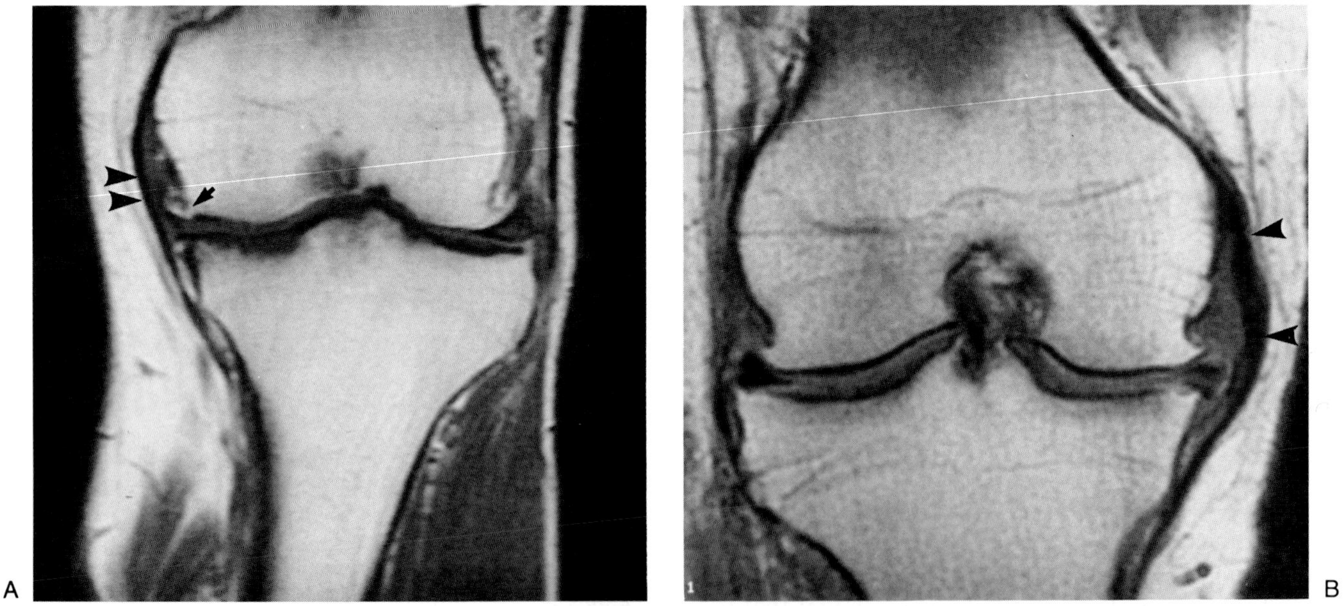

FIG. 5.29. Osteoarthritis of the knee. **A:** Coronal MR, TR/TE 600/20. The TCL (*arrowheads*) is displaced away from the femur and tibia by a combination of osteophytes (*small arrow*), varus deformity of the knee, and adjacent soft tissue thickening. **B:** Coronal MR, TR/TE 800/20. In another patient, the TCL (*arrowheads*) is markedly thickened and redundant. The soft tissues in the position of the collateral ligament bursa are also thickened.

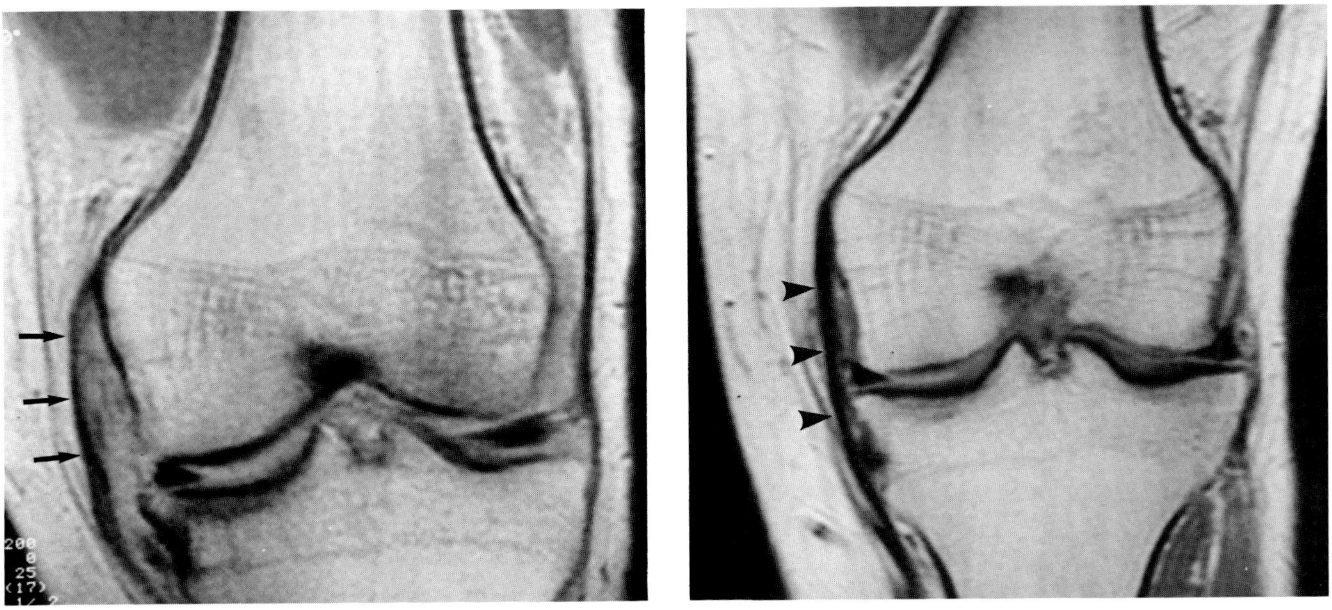

A B

FIG. 5.30. Longitudinal follow-up of a TCL tear. **A,B:** Coronal MR, TR/TE 600/20. The patient suffered a skiing injury resulting in a tear of the TCL and the ACL 36 hr prior to the MR. The proximal TCL (*arrows*) is markedly displaced away from the bone. The follow-up MR (B) made 1 year later demonstrates a thicker but intact and virtually normal appearing TCL (*arrowheads*). The soft tissue reaction has subsided.

distance between the subcutaneous fat and the medullary fat of the osseous structures is increased due to deep soft tissue edema, and edema/hemorrhage in between the TCL and the capsular ligament. The normally dark signal of the TCL is lost; portions of, or occasionally the entire TCL, demonstrate a diffuse increase in signal on T1WI that may increase, but often remain unchanged or even decrease in signal intensity on long TR/TE studies. The combination of hyperacute/acute blood and the distorted dark ligament may produce a low signal abnormality.

Grade 2/3 sprains of the TCL are commonly accompanied by large, poorly defined areas of decreased signal in the lateral femoral or lateral tibial meta-epiphyses (Fig. 5.26). These bone bruises, thought to represent areas of trabecular microfracture, are rarely found in grade 1 sprains.

The TCL is separated from the medial capsular ligament by a bursa and a small amount of fat (Fig. 5.24) (115,116). This stripe of fat is usually resolved on high resolution, T1W coronal images of the knee. The presence of this "gap" must not be mistaken for a meniscocapsular separation (52,53,114). Many radiologists trained in the era of arthrography were taught that the medial meniscus was attached to the medial collateral ligament; in fact, the meniscus is attached only to the capsular ligament, and not the TCL. The bursa that is situated between the layers of the medial collateral lig-

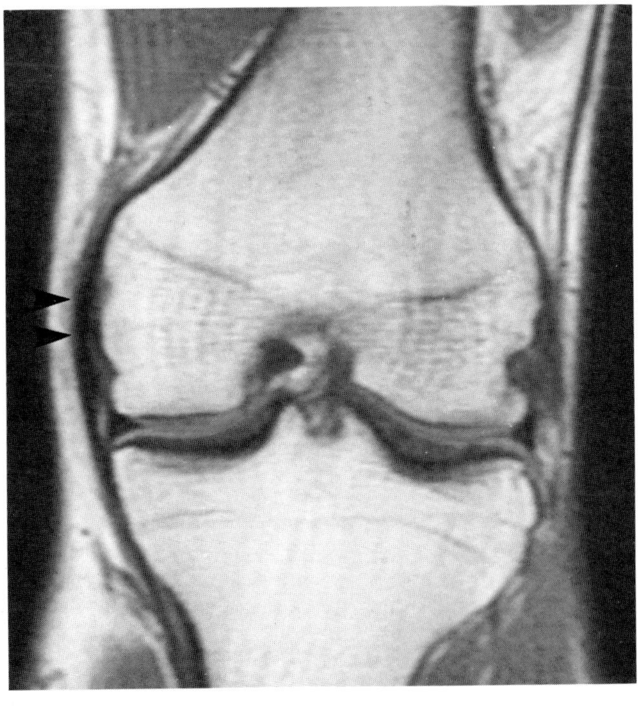

FIG. 5.31. Chronic tear of the TCL. Coronal MR, TR/TE 2,000/20. The patient had suffered a ski injury 2 years earlier, and was treated conservatively. The MR demonstrates thickening of the proximal one-third of the TCL (*arrowheads*) with little surrounding soft tissue reaction. Although this knee was stable, it is not possible to accurately assess the functional integrity of the TCL by MR.

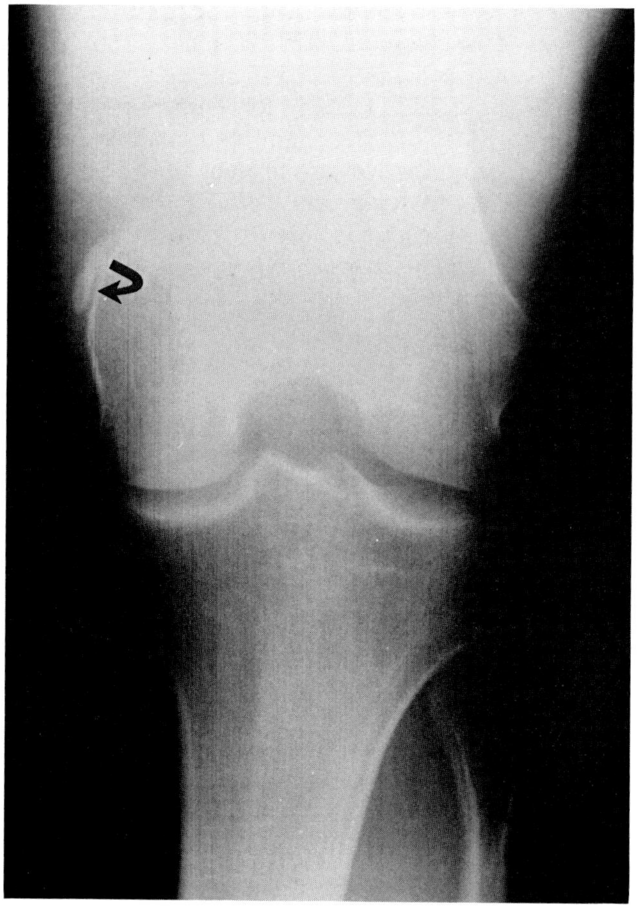

A

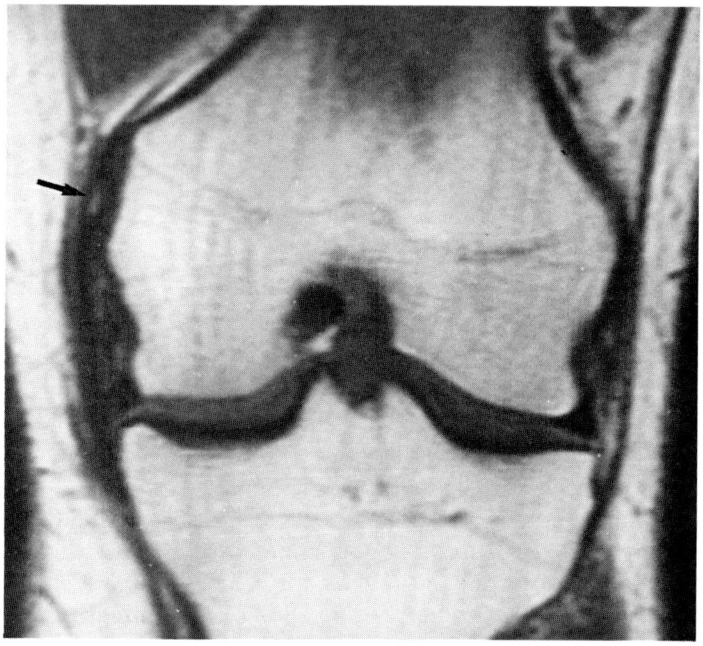

B

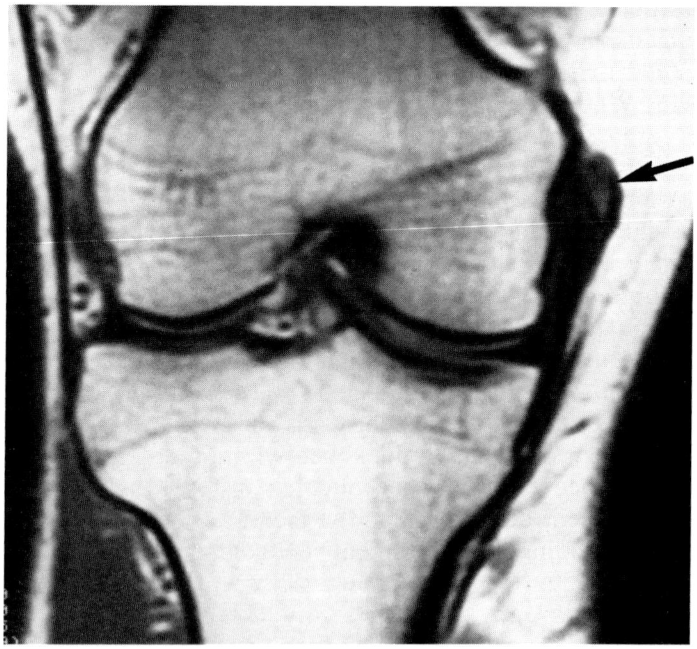

C

FIG. 5.32. Pelligrini-Stieda disease. **A:** The conventional radiograph demonstrates calcification at the femoral attachment of the TCL (*arrow*). **B:** Coronal MR, TR/TE 600/20. The calcification, apparently containing a minimal amount of medullary bone (*arrow*) is evident on the MR. The entire TCL is thickened, displaced from the underlying bone and is diffusely increased in signal intensity. **C:** Coronal MR, TR/TE 800/20. In another patient, the calcification/ossification (*arrow*) of the proximal TCL is larger than in (B). Although the injury to this knee was remote, the ligament is markedly thickened (*arrowheads*).

ament may become inflamed and produce distention of the bursa, at which time it becomes a source of clinical symptoms. Once distended, the bursa may produce a soft tissue mass and displace the TCL, findings suggestive of a TCL injury or a parameniscal cyst associated with a meniscal tear (116). Clinical differential diagnosis is difficult because of the proximity of the bursa to the medial meniscus. TCL bursitis is characterized on MR by the presence of an elongate, occasionally septated fluid collection adjacent to the medial joint line, either spanning the joint line or extending only inferiorly. MR differential diagnosis is a challenge; consideration must be given to the diagnosis of meniscal cyst, ganglion, meniscocapsular separation secondary to trauma, and anserine bursitis.

The Medial Capsular Ligament

The integrity of the attachments of the medial meniscus to the medial capsular ligament is best assessed in coronal plane utilizing long TR/TE or radial T2* sequences (Figs. 5.26 and 5.27) (47,53). In the normal knee, the medial capsular ligament cannot be identified as a distinct structure; it blends imperceptibly with the periphery of the medial meniscus. When the joint is filled with fluid, however, the position and the integrity of the capsular ligaments can be inferred on high resolution T2W coronal images. The intraarticular fluid courses over and under the meniscus to the extreme medial margin of the femoral condyle and tibial plateau. When the meniscofemoral and/or the meniscotibial ligaments are disrupted, one can readily identify joint fluid extravasating from the joint beyond the capsular ligaments into the space normally occupied by the intraligamentous bursa. Displacement of the detached meniscus may also occur (Fig. 5.28).

The diagnosis of acute rupture of the TCL on MR examinations is usually straightforward, but soft tissue changes that occur in osteoarthritis may simulate a ligamentous disruption (Fig. 5.29) (47,52). A patient with medial compartment osteoarthritis (OA) may have a TCL that is abnormally thickened with an increase in the distance between the subcutaneous fat and the medullary fat of the femur. The increased distance is due to buckling of the ligament secondary to the commonly encountered varus deformity seen in OA, soft tissue hypertrophy, joint instability, and displacement of the TCL by osteophytes and the capsule distended with an effusion. While this appearance would not be mistaken for an acute tear, a chronic tear might have a similar appearance.

As the TCL heals, the torn ligament is replaced by dense collagenous scar, a tissue with few mobile protons and therefore one which will produce rather little MR signal. If one longitudinally examines a torn TCL, one will identify progressive "re-ligamentization"; the gray to bright (T2WI) acutely torn ligament will become darker, less distorted, and will achieve an apparent continuity that was not present (Fig. 5.30) (53). The end result is most commonly a TCL that is slightly thicker than normal and perhaps minimally serpiginous. The relatively normal MR appearance of the chronically torn TCL may belie its functional insufficiency (Figs. 5.31 and 5.32).

Calcification/ossification of the hematoma occurring at the site of a proximal TCL tear is known as Pelligrini-Stieda disease. Small amounts of calcium, readily detected on a conventional radiograph, blend imperceptibly with the dark TCL, and generally escape detection by MR. If ossification and medullary bone formation occurs, its characteristic bright signal (T1WI) will be evident (Fig. 5.32).

THE LATERAL COLLATERAL LIGAMENT

Clinical Assessment

The lateral collateral ligamentous anatomy is more complex than the medial and it is arbitrarily divided into anterior, middle, and posterior thirds (94,125). Layer I is composed of the more anterior iliotibial tract and posterior biceps tendon. Layer II is formed by the retinaculum of the quadriceps and the patellofemoral ligaments. Layer II is the deepest of the three and is in fact the lateral part of the joint capsule. It is attached to the edges of the tibia and femur circumferentially at the proximal and distal ends of the knee joint. The fabellofibular ligament and the arcuate ligament are found in layer III. The arcuate ligament is the most important stabilizer of the posterolateral corner of the knee (94). It is a Y-shaped structure with limbs from the femur to both the fibular head and the posterolateral aspect of the tibia.

The major structures comprising the lateral supporting ligaments include the iliotibial band (the distal tendon of the tensor fascia lata), the fibular collateral ligament, the popliteus tendon, the capsule, and biceps femoris tendon (3). The fibular collateral ligament (FCL) is the primary constraint to lateral joint opening, with varus stress and functions to limit hyperextension as well. Injuries to the lateral compartment are more rare than those of the medial and often are unrecognized because signs of injury may be subtle (95), but significant injuries are more disabling because the lateral ligaments are under maximal tension when the leg is extended during walking.

Complete ruptures of the FCL occur either in midsubstance or as an avulsion from the fibular head following a contact or noncontact hyperextension varus stress. Rupture may rarely be associated with a "pop" at the time of injury, but little or no swelling is detectable. Significant effusion indicates the presence of an associated capsular or cruciate ligament injury since the FCL is a completely extraarticular structure. Lateral ligamentous disruptions may occur following a ski injury in which the tips of the ski cross, permitting internal rotation and varus torques to act at the knee. Lateral disruptions are often the result of more severe trauma than normally occurs in athletic injuries, as evidenced by the high incidence of motor vehicle accidents as the mechanism of injury (94). Large varus forces will result in disruption of the iliotibial band (ITB), the popliteus tendon, and the biceps femoris tendon. Complete lateral compartment disruption, essentially a knee dislocation, may be associated with peroneal nerve injury, vascular compromise, and long-term disability, all of which serve to underscore the severity of the injury (94).

Proper treatment of an FCL tear remains controversial. Both operative and nonoperative methods have been advocated, with little evidence to favor one method over the other. Disability and instability following an FCL tear is usually not significant (35), but repair may be warranted in those patients with exces-

A
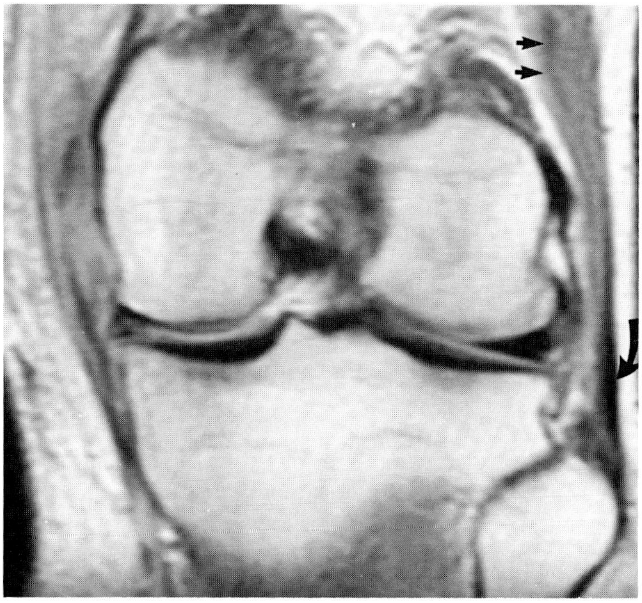

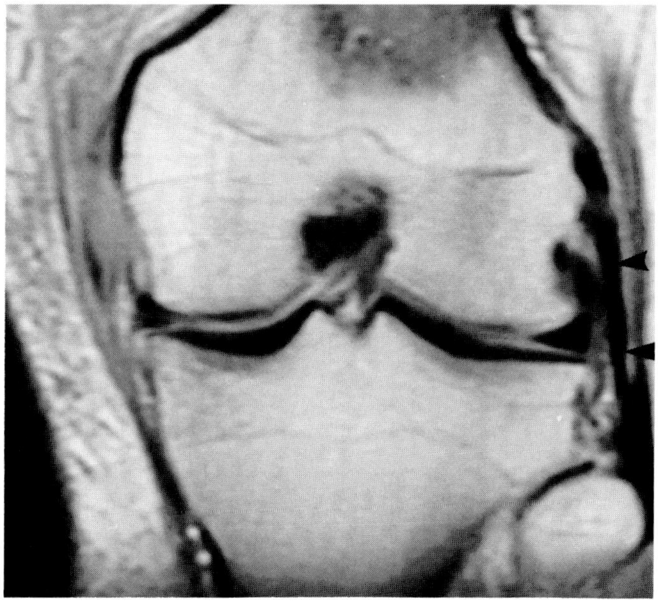
B

C
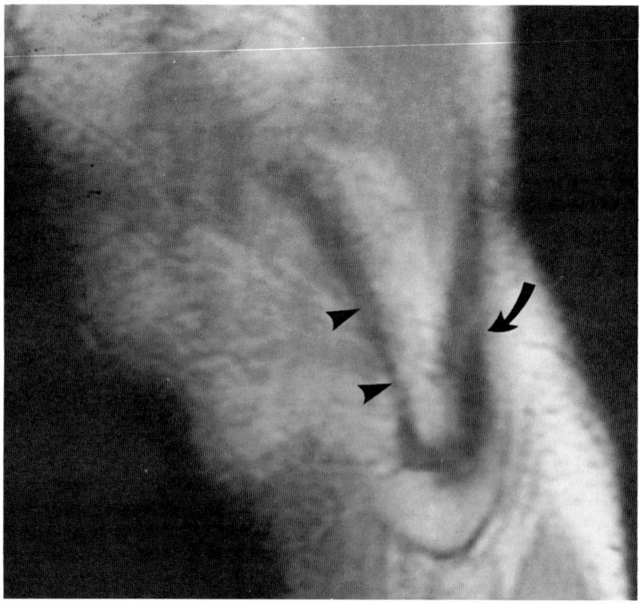

FIG 5.33. Normal FCL and biceps. **A,B:** Coronal MR, TR/TE 2,000/20. (B) is 5 mm anterior to (A). The biceps muscle (*arrows*), tendon (*curved arrow*), and FCL (*arrowheads*) comprise the conjoint tendon, one of the main lateral supports of the knee. **C:** Sagittal MR, TR/TE 2,000/20. Both the FCL (*arrowheads*) and the biceps tendon (curved arrow) are fortuitously identified on this sagittal image.

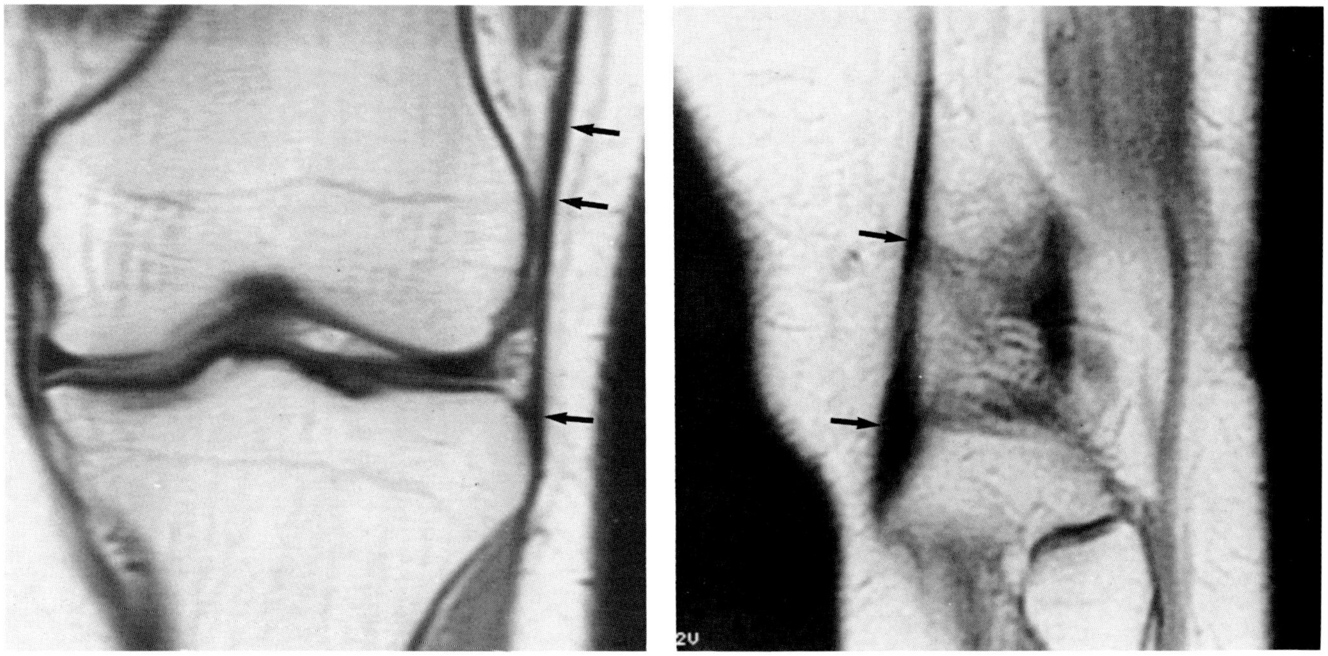

A B

FIG. 5.34. Normal iliotibial band. **A:** Coronal MR, TR/TE 2,000/20. The iliotibial band (*arrows*) is clearly seen inserting onto the anterolateral tibia at the Gerdy tubercle. **B:** Sagittal MR, TR/TE 2,200/20. It is unusual to see the iliotibial band (*arrows*) on a sagittal image. It inserts well anterior to the FCL and the middle third of the capsular ligament.

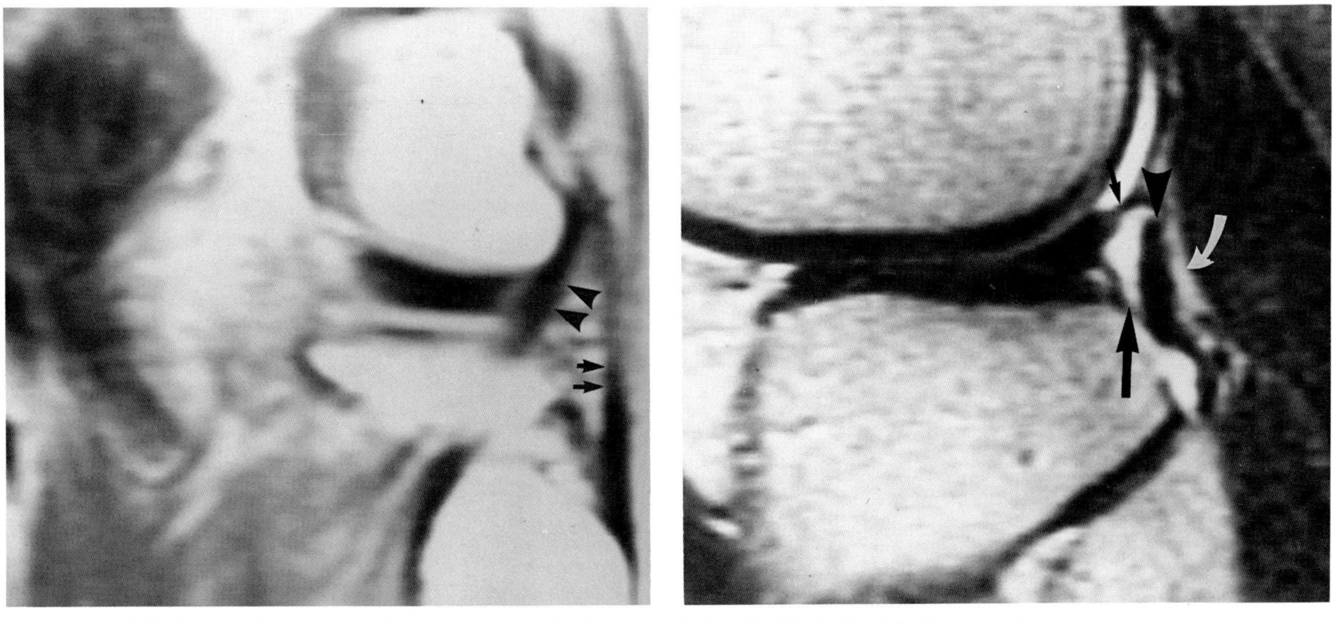

A B

FIG. 5.35. Normal popliteus tendon. **A:** Coronal MR, TR/TE 2,000/20. The popliteus tendon (*arrowheads*) arises from a groove in the lateral femoral epicondyle, courses inferiorly and medially to join its muscle belly on the posterior tibia. The FCL (*arrows*) is just superficial to the popliteus. **B:** Sagittal MR, TR/TE 2,000/80. The popliteus tendon (*arrowhead*) perforates the meniscocapsular junction of the posterior lateral meniscus. As a result, only thin bridges of tissue, the superior (*small arrow*) and inferior fascicles (*large arrow*), provide attachment of the meniscus to the capsule (*curved arrow*, pericapsular fat).

sive physiologic varus, in high performance athletes where varus instability may produce functional impairment, or in patients in whom the rupture has occurred through the mechanism of avulsion of a significant fragment of bone.

Imaging Assessment

Conventional radiographic signs of lateral compartment injury include abnormal widening of the joint space, avulsion of the tibial insertion of the lateral capsular ligament (Segond fracture), avulsion of the Gerdy tubercle, or a fracture of the fibular head (avulsion of the arcuate complex or the biceps femoris tendon) (77,78,94). Arthrographically, only rupture of the middle third of the capsular ligament can be diagnosed since it is the only component of the lateral collateral ligament complex that is capsular; in such cases, extravasation of contrast out of the joint may only be seen within 48 hr of injury.

The normal FCL arises from the lateral epicondyle just above the groove for the popliteus tendon; it courses in a posterior, inferior, and lateral direction

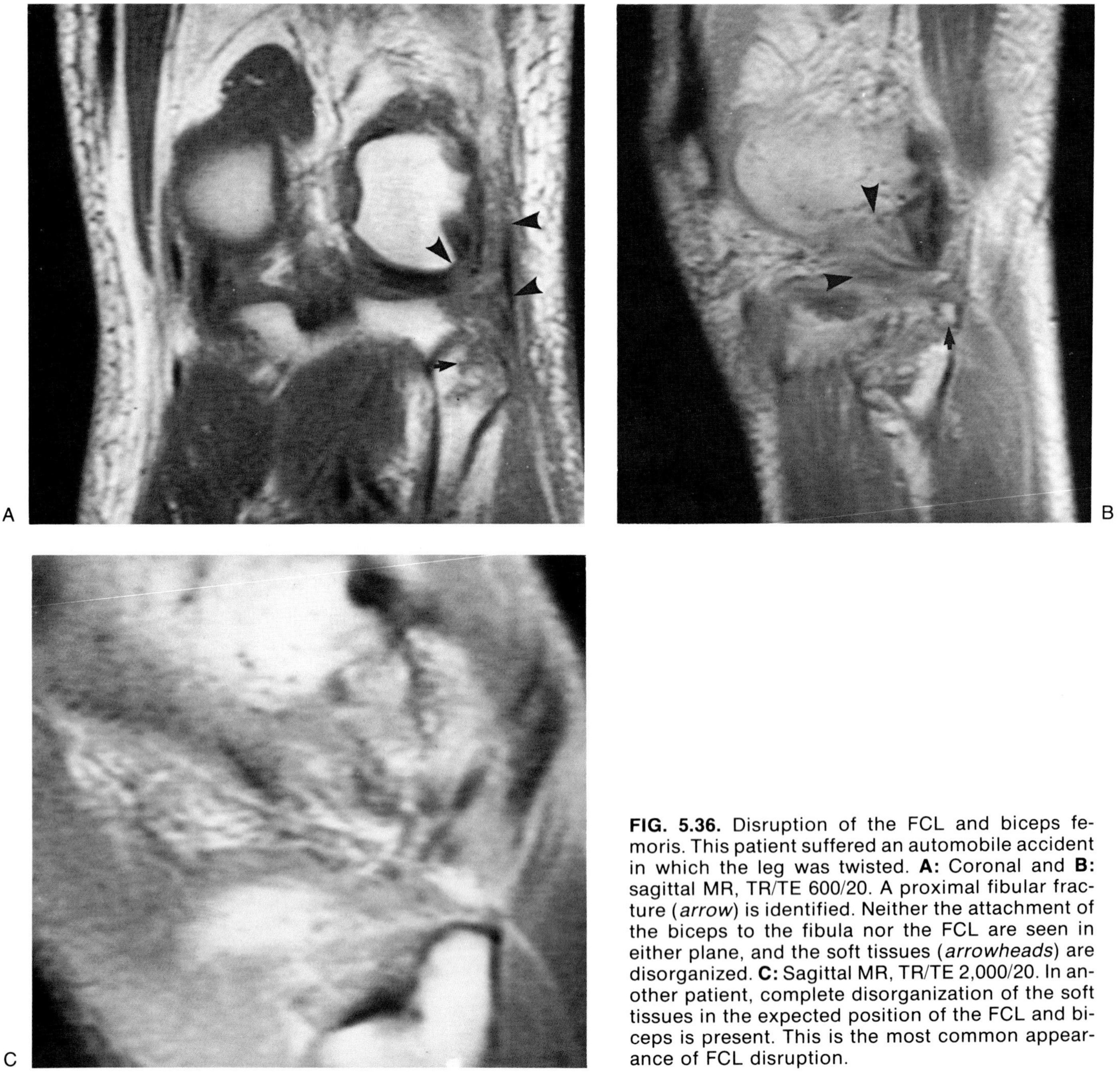

FIG. 5.36. Disruption of the FCL and biceps femoris. This patient suffered an automobile accident in which the leg was twisted. **A:** Coronal and **B:** sagittal MR, TR/TE 600/20. A proximal fibular fracture (*arrow*) is identified. Neither the attachment of the biceps to the fibula nor the FCL are seen in either plane, and the soft tissues (*arrowheads*) are disorganized. **C:** Sagittal MR, TR/TE 2,000/20. In another patient, complete disorganization of the soft tissues in the expected position of the FCL and biceps is present. This is the most common appearance of FCL disruption.

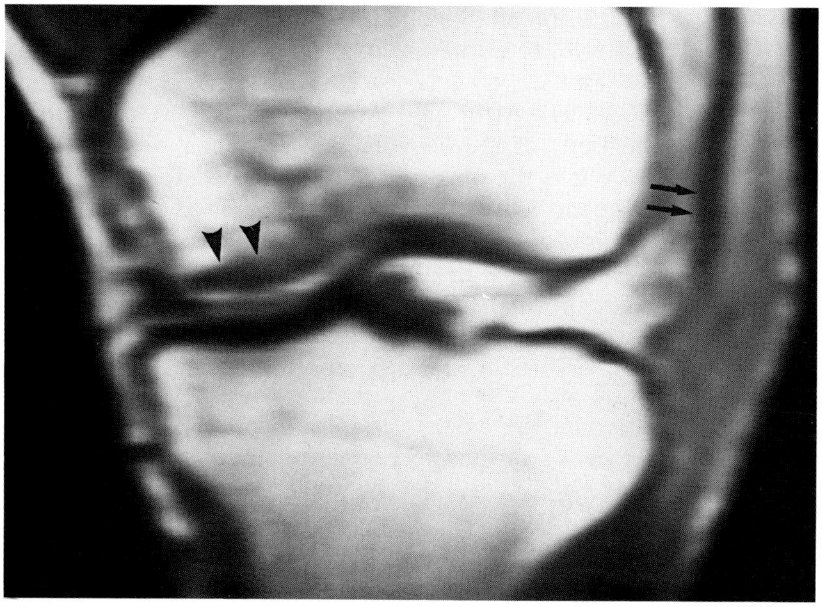

FIG. 5.37. Rupture of the iliotibial band. Coronal MR, TR/TE 800/20. The iliotibial band (*arrows*) is retracted and thickened; it is detached from its tibial insertion adjacent to which there is considerable soft tissue swelling. A subchondral signal abnormality, representing an osteochondral injury (*arrowheads*), is present in the medial femoral condyle.

A 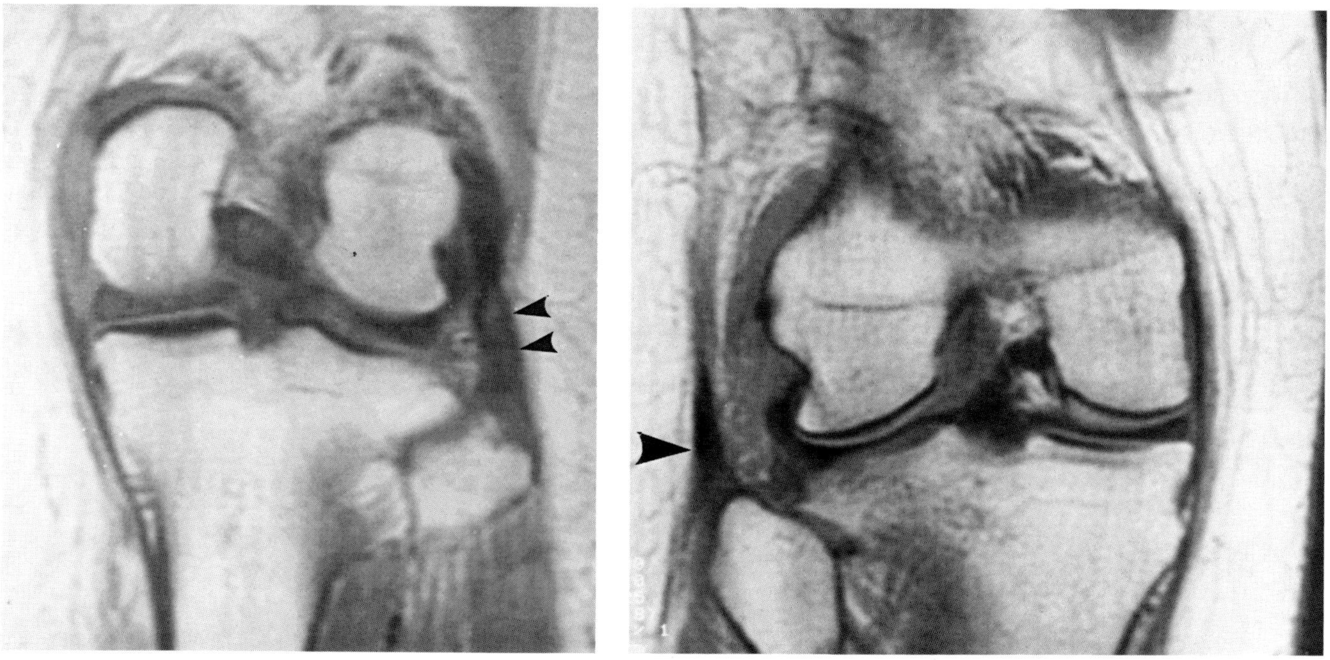 B

FIG. 5.38. Chronic tears of the FCL. **A,B:** Coronal MR, TR/TE 800/20. The FCL (*arrowheads*) is markedly thickened, and demonstrates an increase in signal intensity (A). In another patient, the FCL is completely absent (*arrowhead*, biceps tendon) (B).

on its course to its attachment to the fibular head (Fig. 5.33). Just prior to its insertion, the FCL joins with the tendon of the biceps femoris where they intertwine as a conjoint tendon (47). Both are cord-like, dark structures at all imaging parameters. Because of its slight lateral and posterior-directed course, the FCL is only fortuitously seen on sagittal or coronal images, but is more reliably found on coronal sequences (53). The biceps tendon may be found on either coronal or sagittal images.

The iliotibial band is a thin, dark, vertically oriented stripe arising as a thickening of the fascia lata and coursing distally in the bright subcutaneous fat of the lateral thigh to its insertion on the lateral tibia just below the plateau anteriorly (Gerdy tubercle). It is identified 100% of the time on coronal sequences, but is uncommonly found in the sagittal plane (Fig. 5.34). The popliteus tendon can be found on both the coronal and sagittal sequences (Fig. 5.35). It arises in a groove in the lateral femoral epicondyle, and courses posteriorly, inferiorly, and medially to join its muscle belly on the posterior aspect of the tibia. Along its course, the popliteus tendon perforates the posterior lateral meniscocapsular junction.

Injury to the lateral supporting structures share several characteristics (52). They occur almost exclusively at the distal (tibial and fibular) attachments. The ITB and the FCL may be injured by means of avulsion; identification of their different insertion sites will allow for correct diagnosis (Figs. 5.36 and 5.37). Acute substance tears of any of the lateral structures result in interruption of their contours, a wavy appearance, and localized fluid within/around them, but most commonly, extensive disorganization of the surrounding soft tissues with areas of both increased and decreased signal (T1 and T2WI) are found. In chronic injuries, complete absence, a serpiginous contour, and focal areas of abnormal signal (T1WI) within the ligament are common (Figs. 5.38 and 5.39). Extensive varus stresses may result in medial compartment trabecular microfracture (Fig. 5.37).

It is worthwhile mentioning two disorders of the lateral supporting structures that, while uncommon, are a source of clinical symptoms. Runners may develop pain over the posterolateral aspect of the knee which has been ascribed to popliteus tendonitis or ruptures (Fig. 5.40) (126,127). The popliteus helps prevent forward displacement of the femur on the tibia and helps maintain internal rotation of the tibia. Running downhill produces stress on the tendon, and subsequently pain. The diagnosis of tendonitis or a rupture of the popliteus is quite problematic. MR may offer the ability to accurately diagnose and assess this condition.

The iliotibial band friction syndrome presents with pain over the lateral femoral condyle in patients in training programs that include walking and running (128). Pain is maximal when the knee is flexed; the symptoms are thought to be due to the movement of the band from a position anterior to the condyle in full extension, to posterior in full flexion. Localized tenderness can be elicited in a small area over the lateral femoral condyle when the knee is held in 30° of flexion.

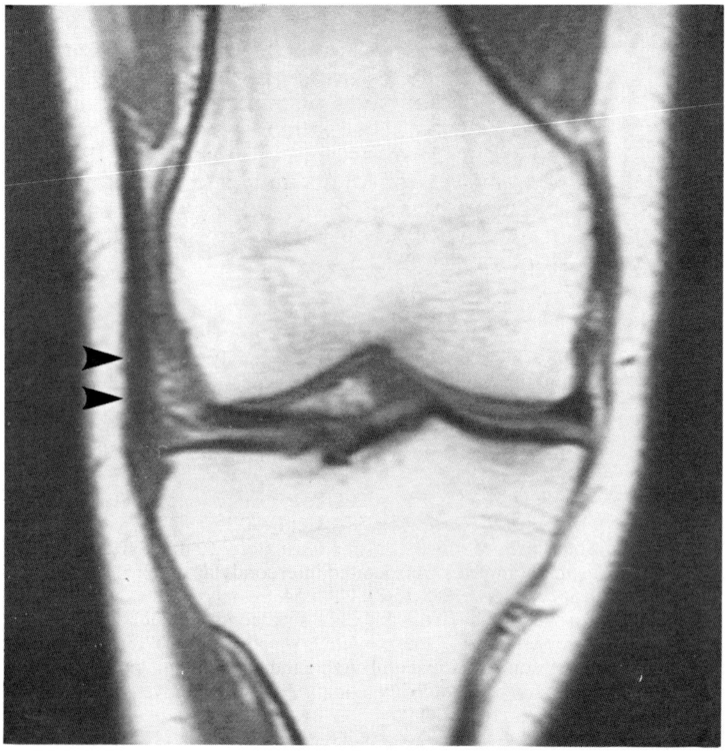

FIG. 5.39. Chronic tear of the iliotibial band. Coronal MR, TR/TE 800/20. The iliotibial band (*arrowheads*) is markedly thickened, but does not demonstrate any signal abnormality.

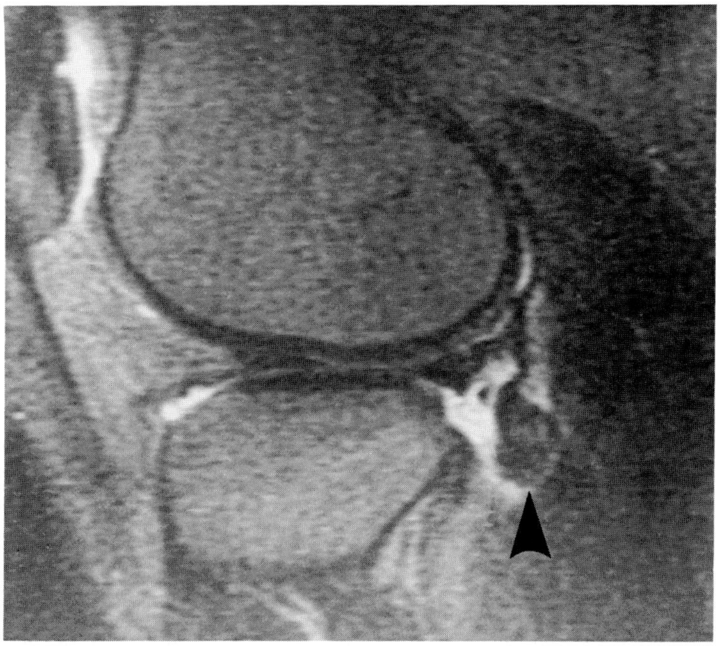
A

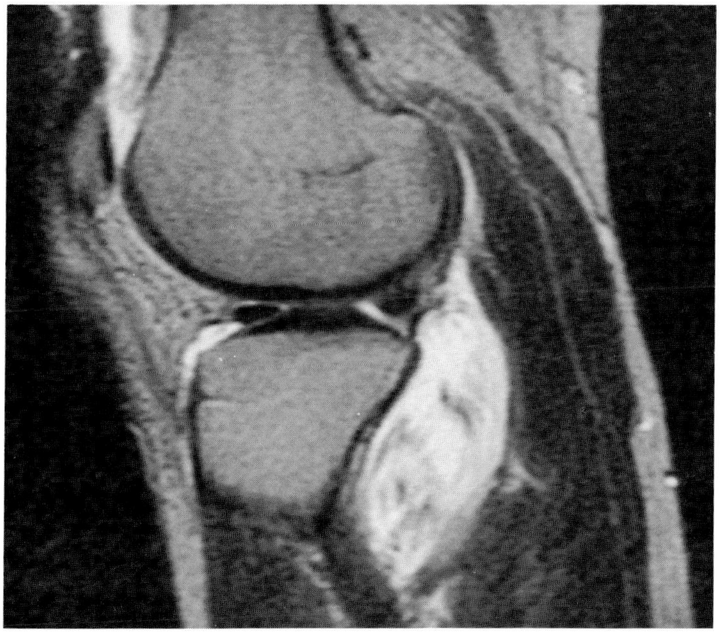
B

FIG. 5.40. Popliteus muscle/tendon repture (presumed). **A,B:** Sagittal MR, TR/TE 2,000/80. While running downhill, these patients experienced pain along the posterolateral aspect of the knee. In (A) there is extravasation of fluid from the joint just in front of the popliteus muscle (*arrowhead*) which is rounded and was discontinuous on multiple images. The patient's symptoms resolved in 4 weeks without surgery. In (B), extremely bright fluid replaces the normal popliteus muscle in this patient with a clinically suspected but unconfirmed popliteus rupture.

REFERENCES

1. Feagin J. *The crucial ligaments*. New York: Churchill Livingstone; 1988.
2. Kennedy JC, Weinberg HW, Wilson AS. The anatomy and function of the anterior cruciate ligament. *J Bone Joint Surg* 1974;56A;223–235.
3. Nicholas J, Hershman E, eds. *The lower extremity and spine in sports medicine*. St Louis: C.V. Mosby Company; 1986.
4. Dehaven K, Diagnosis of acute knee injuries with hemarthrosis. *Am J Sports Med* 1980;8(1):9–14.
5. Vellet AD, Marks P, Fowler P, et al. Accuracy of nonorthogonal magnetic resonance imaging in acute disruption of the anterior cruciate ligament. *Arthroscopy* 1989;5(4):287–293.
6. Odensten M, Lysolm J, Gillquist J. The course of partial anterior cruciate ligament ruptures. *Am J Sports Med* 1985;13:183–186.
7. Noyes FR, Bassett RW, Grood ES, et al. Arthroscopy in acute traumatic hemarthrosis of the knee. *J Bone Joint Surg* 1980;62A:687–695.
8. Feagin JA, Lambert KL, Cunningham RR, et al. Consideration of the anterior cruciate ligament injury in skiing. *Clin Orthop* 187;216:13–18.
9. Feagin JA. The syndrome of the torn anterior cruciate ligament. *Orthop Clin North Am* 1979;10:81–90.
10. Souryal T, Moore H, Evans J. Bilaterality in anterior cruciate ligament injuries: Associated intercondylar notch stenosis. *Am J Sports Med* 1988;15(8):449–454.
11. Norwood LA, Cross MJ. The intercondylar shelf and the anterior cruciate ligament. *Am J Sports Med* 1977;5:171–176.
12. Anderson AF, Lipscomb AB, Liudah KJ, et al. Analysis of the intercondylar notch by computed tomography. *Am J Sports Med* 1987;15:547–552.
13. Houseworth SW, Mauro VJ, Mellon BA, et al. The intercon-

dylar notch in acute tears of the anterior cruciate ligament: A computer graphics study. *Am J Sports Med* 1987;15:221–224.
14. Harilainen A, Myllynen P, Rauste J, et al. Diagnosis of acute knee ligament injuries: The value of stress radiography compared with clinical examination, stability under anaesthesia and arthroscopic or operative findings. *Ann Chir Gynaecol* 1986;75:37–43.
15. Noyes F, De Lucas J, Torvik P. Biomechanics of anterior cruciate ligament failure; an analysis of strain rate sensitivity and mechanisms of failure in primates. *J Bone Joint Surg* 1974;56A:236–253.
16. Daniel D, Akeson W, O'Connor J. Knee ligaments. Structure, function, injury, and repair. New York: Raven Press; 1991.
17. Finsterbush A, Frankl U, Matan Y, et al. Secondary damage to the knee after isolated injury of the anterior cruciate ligament. *Am J Sports Med* 1990;18(5):475–479.
18. Buckley SL, Barrack RL, Alexander H. The natural history of conservatively treated partial anterior cruciate ligament tears. *Am J Sports Med* 1989;17:221–225.
19. Fruensgaard S, Johannsen H. Incomplete ruptures of the anterior cruciate ligament. *J Bone Joint Surg [Br]* 1989;71:526–530.
20. Sandberg R, Balkfors B. Partial ruptures of the anterior cruciate ligament. *Clin Orthop* 1987;220:176–178.
21. Warren RF, Levy IM. Meniscal lesions associated with anterior cruciate ligament injury. *Clin Orthop* 1983;172:32–37.
22. Syndor R, Andrews J. Combined arthroscopy and minireconstruction techniques in the acutely torn anterior cruciate ligament. *Orthop Clin North Am* 1985;16(2):171–179.
23. Eady J, Cardenas C, Sopa D. Avulsion of the femoral attachment of the anterior cruciate ligament in a seven-year-old child. A case report. *J Bone Joint Surg [Am]* 1982;64:1376–1378.
24. McLeod WD. The biomechanics and function of the secondary restraints to the anterior cruciate ligament. *Orthop Clin North Am* 1985;16(2):165–169.
25. Lee J, Yao L, Phelps C, et al. Anterior cruciate ligament tears: MR imaging compared with arthroscopy and clinical tests. *Radiology* 1988;166:861–864.
26. Boeree NR, Watkinson AF, Ackroyd CE, et al. Magnetic resonance imaging of meniscal and cruciate injuries of the knee. *J Bone Joint Surg [Br]* 1991;73:452–457.
27. Butler D, Noyes F Grood E. Ligamentous restraints to anterior-posterior drawer in the human knee. A biomechanical study. *J Bone Joint Surg [Am]* 1980;63:259–270.
28. Casscells SW. *Arthroscopy:* Diagnostic and surgical practice. Philadelphia: Lea & Febiger; 1984.
29. Noyes F, Bassett R, Grood E, et al. Arthroscopy in acute traumatic hemarthrosis of the knee. *J Bone Joint Surg* 1980;62:687–695.
30. Torg J, Conrad W, Kalen V. Clinical diagnosis of anterior cruciate ligament instability in the athlete. *Am J Sports Med* 1976;4:84–93.
31. Braunstein E. Anterior cruciate ligament injuries: A comparison of arthrographic and physical diagnosis. *AJR* 1982; 138(3):423–425.
32. Zarins B, Nemeth VA. Acute knee injuries in athletes. *Orthop Clin North Am* 1985;16(2):285–302.
33. Howe J, Johnson RJ. Knee injuries in skiing. *Orthop Clin North Am* 1985;16(2):303–313.
34. Warren RF, Marshall JL. Injuries of the anterior cruciate and medial collateral ligaments of the knee. *Clin Orthop* 1978;136:198–211.
35. Andrish JT. Ligament injuries of the knee. *Orthop Clin North Am* 1985;16(2):273–284.
36. Baugher WH, White GM. Primary evaluation and management of knee injuries. *Orthop Clin North Am* 1985;16(2):315–327.
37. Noyes FR, Grood ES, Cummings JF, et al. An analysis of the pivot shift phenomenon. The knee motions and subluxation induced by different examiners. *Am J Sports Med* 1991; 19(2):148–155.
38. Jackson D, Drez D, eds. *The anterior cruciate deficient knee.* St. Louis: C.V. Mosby; 1987.
39. Noyes FR, Cummings JF, Grood ES, et al. The diagnosis of knee motion limits, subluxations, and ligament injury. *Am J Sports Med* 1991;19(2):163–171.
40. Harter RA, Osternig LR, Singer KM, et al. Instrumented Lachman tests for the evaluation of anterior laxity after reconstruction of the anterior cruciate ligament. *J Bone Joint Surg [Am]* 1989;71:975–983.
41. Pavlov H, Schneider R. Extrameniscal abnormalities as diagnosed by knee arthrography. *Radiol Clin North Am* 1981;19:287–304.
42. Feagin JA, Curl WW. Isolated tear of the anterior cruciate ligament: 5-year follow-up study. *Am J Sports Med* 1976;4:95–100.
43. Jackson DW, Jennings LD, Maywood RM, et al. Magnetic resonance imaging of the knee. *Am J Sports Med* 1989;16:28–34.
44. Sherman O, Fox J, Snyder, et al. Arthroscopy—"No problem" surgery: An analysis of 2640 cases. *J Bone Joint Surg* 1986;68(A):256–265.
45. Dandy DJ, Flanagan JP, Steenmeyer V. Arthroscopy and the management of the ruptured anterior cruciate ligament. *Clin Orthop* 1982;167:43–49.
46. Bach BR, Jr., Warren RF. "Empty wall" and "vertical strut" signs of ACL insufficiency. *Arthroscopy* 1989;5:137–140.
47. Mink JH, Deutsch AL. *The knee.* IN: MRI of the musculoskeletal system. A teaching file. Mink JH, Deutsch AL, eds. New York: Raven Press, 1990;251–387.
48. Kennedy JC, Hawkins RJ, Willis RB, et al. Tension studies of human knee ligaments. Yield point, ultimate failure, and disruption of the cruciate and tibial collateral ligaments. *J Bone Joint Surg* 1976;58A:350–355.
49. Sherman MF, Warren RF, Marshall JL, et al. A clinical and radiographical analysis of 127 anterior cruciate insufficient knees. *Clin Orthop* 1988;227:229–237.
50. Pavlov H. The radiographic diagnosis of the anterior cruciate ligament deficient knee. *Clin Orthop* 1983;172:57–64.
51. Buttner-Janz K, Schellnack K, Rieder T. Noninvasive diagnosis of cruciate ligament damage with particular reference to computed tomography with arthrography. *Am J Sports Med* 1989;17:501–504.
52. Mink JH. *The ligaments of the knee*, IN: Magnetic resonance imaging of the knee. Mink JH, Reicher MA, Crues JV, eds. New York: Raven Press; 1987;93–111.
53. Stoller DW. *Magnetic resonance imaging in orthopaedics and rheumatology.* Philadelphia: J.B. Lippincott, 1989.
54. Beltran J. *MRI of the musculoskeletal system.* Philadelphia: J.B. Lippincott; 1990.
55. Mink JH, Levy T, Crues JV. Tears of the anterior cruciate ligament and menisci of the knee: MR imaging evaluation. *Radiology* 1988;167:769–775.
56. Barry KP, Mesgarzadeh M, Moyer R, et al. Patterns and accuracy of diagnosis of anterior cruciate ligament tears with MR imaging. *Radiology* 1991;181(P):303.
57. Vahey TN, Broome DR, Kayes KJ, et al. Acute and chronic tears of the anterior cruciate ligament: Differential features at MR imaging. *Radiology* 1991;181:251–253.
58. Glashow J, Katz R, Schneider M, et al. Double-blind assessment of the value of magnetic resonance imaging in the diagnosis of anterior cruciate and meniscal lesions. *J Bone Joint Surg [Am]* 1989;71(1):113–119.
59. Chan WP, Fritz RC, Stoller DW. MR "anterior drawer" sign: A useful sign in the diagnosis of complete anterior cruciate ligament tears. *Radiology* 1991;181(P):178.
60. Bromley JW, Cohen P. Ganglion of the posterior cruciate ligament. Report of a case. *J Bone Joint Surg* 1965;47A(6):1247–1249.
61. Garcia A, Hodler J, Vaughn L, et al. Case report 677: Intra articular ganglion arising from the posterior cruciate ligament. *Skeletal Radiol* 1991;20:373–375.
62. Yasuda K, Majima T. Intra-articular ganglion blocking extension of the knee: Brief report. *J Bone Joint Surg [Br]* 1988;70:837.
63. Burk DL, Dalinka MK, Kanal E, et al. Meniscal and ganglion cysts of the knee: MR evaluation. *AJR* 1988;150:331–336.
64. Lee KR, Cox GG, Neff JR, et al. Cystic masses of the knee: Arthrographic and CT evaluation. *AJR* 1987;148:329–334.
65. Rosen MA, Jackson DW, Berger PE. Occult osseous lesion documented by magnetic resonance imaging associated with

anterior cruciate ligament ruptures. *Arthroscopy* 1991;7(1):45–51.

66. Mink JH, Deutsch AL. Occult cartilage and bone injuries of the knee: Detection, classification, and assessment with MR imaging. *Radiology* 1989;170:823–829.
67. Lynch TCP, Crues JV, Morgan FW, et al. Bone abnormalities of the knee: Prevalence and significance at MR imaging. *Radiology* 1989;171:761–766.
68. Wirth B, Spritzer C. Ancillary findings of anterior cruciate ligament injuries on MR images. *Radiology* 1990;177(P):263.
69. Teresi L, O'Sullivan R, Watanabe A, et al. MR imaging of the knee: Association of occult osseous lesions and medial collateral ligament injury with anterior cruciate ligament tears. *Radiology* 1990;177(P):295.
70. Wrench JD, Spritzer CE, Martinez S. MR imaging in acute knee trauma. *J Magn Reson Imaging* 1991;1(2):175.
71. Yao L, Lee JK. Avulsion of the posteromedial tibial plateau by the semimembranosus tendon: Diagnosis with MR imaging. *Radiology* 1989;172:513–514.
72. Vellet AD, Marks PH, Fowler PJ, et al. Occult posttraumatic osteochondral lesions of the knee: Prevalence, classification, and short-term sequelae evaluated with MR imaging. *Radiology* 1991;178:271–276.
73. Lee JK, Yao L. Stress fractures: MR imaging. *Radiology* 1988;169:217–220.
74. Remer EM, Fitzgerald SW, Friedman H, et al. *Mechanisms of anterior cruciate ligament injury: MR characterization.* IN: 77th Scientific Assembly and Annual Meeting Radiological Society of North America. Chicago; 1991.
75. Goldman A, Pavlov H, Rubenstein D. The Segond fracture of the proximal tibia: A small avulsion that reflects major ligamentous damage. *AJR* 1988;151(6):1163–1167.
76. Weber W, Neumann C, Barakos J, et al. Lateral tibial rim (Segond) fractures: MR imaging characteristics. *Radiology* 1991;180:731–734.
77. Dietz GW, Wilcox DM, Montgomery JB. Segond tibial condyle fracture: Lateal capsular ligament avulsion. *Radiology* 1986;159:467–469.
78. Woods G, Stanley R, Tullos H. Lateral capsular sign: X-ray clue to significant knee instability. *Am J Sports Med* 1979;7:27–33.
79. Deutsch AL, Mink JH. Magnetic resonance imaging of musculoskeletal injuries. *Radiol Clin North Am* 1989;27:983–1002.
80. Yao L, Lee JK. Occult intraosseous fracture: detection with MR imaging. *Radiology* 1988;167:749–751.
81. Reiser M, Rupp N, Pfandner K, et al. Imaging of cruciate ligament injuries using MR tomography. *ROFO* 1986;145(2):193–198.
82. Fischer SP, Fox JM, Del Pizzo W, et al. Accuracy of diagnoses from magnetic resonance imaging of the knee. *J Bone Joint Surg* 1991;73A:1–10.
83. Christel P, Roger B, Witvoet J, et al. Validity and diagnostic value of magnetic resonance imaging in traumatic menisco-ligamentous pathology of the knee. Prospective study of 22 knees. *Rev Chir Orthop* 1988;74:402–412.
84. Jung T, Rodriguez M, Augustiny N, et al. 1.5 T MRI arthrography and arthroscopy in the evaluation of knee lesions. *ROFO* (Germany, West) 1988;148:390–393.
85. Fritz R, Helms C, Genant H, et al. Frequency and significance of equivocal meniscal and anterior cruciate ligament tears in MR imaging of the knee. *Radiology* 1990;177(P):225.
86. Jacobs B, Applegate GR, Recht M, et al. Comparison of frequency-selective fat saturation with spin-echo MR images in the detection of knee pathology. *J Magn Reson Imaging* 1991;1(2):176.
87. Crues J, Mink J, Levy T. Meniscal tears of the knee: Accuracy of MR imaging. *Radiology* 1987;164:445–448.
88. Kelly MA, Flock TJ, Kimmel JA, et al. MR imaging of the knee: Clarification of its role. *Arthroscopy* 1991;7(1):78–85.
89. Shellock F, Foo T, Deutsch, et al. Patellofemoral joint: Evaluation during active flexion with ultrafast spoiled GRASS MR imaging. *Radiology* 1991;180:581–585.
90. Niitsu M, Anno I, Fukubayashi T, et al. Tears of cruciate ligaments and menisci: Evaluation with cine MR imaging. *Radiology* 1991;178:859–864.
91. Van Dommelen B, Fowler P. Anatomy of the posterior cruciate ligament. A review. *Am J Sports Med* 1989;17(1):24–29.
92. Kennedy J. The posterior cruciate ligament. *J Trauma* 1967;7:367–377.
93. Grover JS, Bassett LW, Gross ML, et al. Posterior cruciate ligament: MR imaging. *Radiology* 1990;174:527–530.
94. Delee J, Riley M, Rockwood C. Acute straight lateral instability of the knee. *Am J Sports Med* 1983;11:404–410.
95. Hughston J, Andrews J, Cross M, et al. Classification of knee ligament instabilities. Part I: The medial compartment and cruciate ligaments. *J Bone Joint Surg [Am]* 1976;58:159–172.
96. Torisu T. Isolated avulsion fracture of the tibial attachment of the posterior cruciate ligament. *J Bone Joint Surg [Am]* 1977;59:68–72.
97. Kannas P, Jarvinen M. Bilateral isolated rupture of the posterior cruciate ligament. *Orthopedics* 1989;12(7):995–997.
98. Trickey EL. Injuries to the posterior cruciate ligament: Diagnosis and treatment of early injuries and reconstruction of late instability. *Clin Orthop* 1980;147:76–81.
99. Clendenin MB, DeLee JC, Heckman JD. Interstitial tears of the posterior cruciate ligament of the knee. *Orthopedics* 1980;3(8):764–772.
100. Loos WC, Fox JM, Flazina ME, et al. Acute posterior cruciate ligament injuries. *Am J Sports Med* 1981;9(2):86–92.
101. Hughston JC. The absent posterior drawer test in some acute posterior cruciate ligament tears of the knee. *Am J Sports Med* 1988;16(1):39–43.
102. Fowler PJ, Messieh SS. Isolated posterior cruciate ligament injuries in athletes. *Am J Sports Med* 1987;15(6):553–557.
103. Cross MJ, Fracs MB, Powell JF. Long-term followup of posterior cruciate ligament rupture: A study of 116 cases. *Am J Sports Med* 1984;12(4):292–297.
104. Torg JS, Barton TM, Pavlov H, et al. Natural history of the posterior cruciate ligament-deficient knee. *Clin Orthop* 1989;246:208–216.
105. Eriksson E, Haggmark T, Johnson RJ. Reconstruction of the posterior cruciate ligament. *Orthopedics* 1986;9(2):217–220.
106. Clancy WG, Shelbourne KD, Zoellner GB, et al. Treatment of knee joint instability secondary to rupture of the posterior cruciate ligament. *J Bone Joint Surg [Am]* 1983;65:310–322.
107. Jackson RW, Dandy DJ. Arthroscopy of the knee. New York: Grune & Stratton; 1976.
108. Yerys P. Arthroscopic posterior cruciate ligament repair. *Arthroscopy* 1991;7(1):111–114.
109. Oleaga L, Kressel H. High signal intensity in the posterior cruciate ligament on sagittal gradient-echo images: Normal variant. *Radiology* 1990;177(P):295.
110. Clemente C. *Gray's anatomy*. Philadelphia: Lea & Febiger; 1985.
111. Warren LF, Marshall JL. The supporting structures and layers on the medial side of the knee. An anatomic analysis. *J Bone Joint Surg [Am]* 1979;61:56–62.
112. Brantigan OC, Voshell AF. The tibial collateral ligament: Its function, its bursae, and its relationship to the medial meniscus. *J Bone Joint Surg* 1943;25:121–131.
113. Slocum D, Larson R. Rotary instability of the knee: Its pathogenesis and a clinical test to demonstrate its presence. *J Bone Joint Surg [Am]* 1968;50:211–225.
114. Lee JK, Yao L. Tibial collateral ligament bursa: MR imaging. *Radiology* 1991;178:855–857.
115. Stuttle FL. The no-name and no-fame bursa. *Clin Orthop* 1959;15:197–199.
116. Kerlan RK, Glousman RE. Tibial collateral ligament bursitis. *Am J Sports Med* 1988;16(4)344–346.
117. Frank C, Amiel D, Woo SL-Y, et al. Normal ligament properties and ligament healing. *Clin Orthop* 1985;196:15–25.
118. Jones RE, Henley MB, Francis P, Nonoperative management of isolated grade III collateral ligament injury in high school football players. *Clin Orthop* 1986;213:137–140.
119. Indelicato PA. Non-operative treatment of complete tears of the medial collateral ligament. *J Bone Joint Surg* 1983; 65A:323.
120. Kannus P. Long-term results of conservatively treated medial collateral ligament injuries of the knee. *Clin Orthop* 1988; 226:103–112.

121. Holden DL, Eggert AW, Butler JE. The nonoperative treatment of Grade I and Grade II medial collateral ligament injuries to the knee. *Am J Sports Med* 1983;11(5):340–344.
122. Fetto JF, Marshall JL. Medial collateral ligament injuries of the knee: A rationale for treatment. *Clin Orthop* 1978;132:206–218.
123. Brunner MC, Garrett JC, Stiles RG, et al. A directed approach to MRI of the athletic knee. *Am J Sports Med* 1990;30(2):222–228.
124. Noyes FR. Function properties of knee ligaments and alterations induced by immobilizations. *Clin Orthop* 1977;123:210.
125. Seebacher JR, Inglis AE, Marshall JL, et al. The structures of the postero-lateral aspect of the knee. *J Bone Joint Surg* 1982;64:536–545.
126. Mayfield GW. Popliteus synovitis. *Am J Sports Med* 1977;5(1):31–36.
127. Rose DJ, Parisien JS. Popliteus tendon rupture. *Clin Orthop* 1988;226:113–117.
128. Renne JW. The iliotibial band friction syndrome. *J Bone Joint Surg* 1975;57A(8):1110–1111.

ADDITIONAL READINGS

Graf BK, Lange RH, Fujisaki CK, et al. Anterior cruciate ligament tears in skeletally immature patients: Meniscal pathology at presentation and after attempted conservative treatment. *Arthroscopy* 1992;8:229–233.

Kaplan PA, Walker CW, Kilcoyne RF, et al. Occult fracture patterns of the knee associated with anterior cruciate ligament tears: assessment with MR imaging. *Radiology* 1992;183:835–838.

Murphy BJ, Smith RL, Uribe JW, et al. Bone signal abnormalities in the posterolateral tibia and lateral femoral condyle in complete tears of the anterior cruciate ligament: a specific sign? *Radiology* 1992;182:221–224.

Schils JP, Spindler KP, Piraino DW, et al. Bone bruises and chondral lesions in anterior cruciate ligament tears: A prospective study. *JMRI* 1992;2(P):94.

Shelbourne KD, Porter DA. Anterior cruciate ligament-medial collateral ligament injury: Nonoperative management of medial collateral ligament tears with anterior cruciate ligament reconstruction. *Am J Sports Med* 1992;20:283–286.

CHAPTER 6

The Extensor Mechanism and Patellofemoral Joint

Andrew L. Deutsch and Frank G. Shellock

Complaints attributable to the extensor mechanism and patellofemoral joint represent important reasons for which patients seek orthopedic attention. Acute and chronic injuries of the constituent components of the extensor mechanism are common both on an accidental and sports-related basis. Additionally, the mechanical demands of normal daily activity apply significant compressive forces across the patellofemoral articulation that may contribute to early cartilage degeneration. The capacity to accommodate these forces may be significantly affected by abnormalities of anatomy and alignment that, while the subject of extensive study, remain incompletely understood.

Since the publication of the first edition of this text, there has been a tremendous increase in the application of MR toward depiction of abnormalities involving the extensor mechanism and patellofemoral joint (1–15). The multiplanar capability of MR coupled with its broad range of contrast resolution make this technique ideally suited for evaluation of this anatomic region. The articular cartilage of the patella, which in areas is the thickest in the body (16), has served as a model for articular cartilage investigation with MR (2–21). The muscular and tendinous components of the extensor mechanism are also readily depicted by MR (1,4,6,7,14,15). Rapid technical advances with MR, including ultrafast imaging, have been used for assessment of patellofemoral biomechanics and have already contributed observations that have challenged many longstanding conceptions (3,8–10,22–29). The information provided by this emerging application of MR has already been demonstrated to have important management implications.

In this chapter we review the current status of MR imaging in assessment of the extensor mechanism and patellofemoral joint. Anatomical considerations with an emphasis on functional anatomy and applied biomechanics will be addressed. Abnormalities of the muscular and tendinous components of the extensor mechanism will be reviewed, with an emphasis on accidental and sports-related injuries. In consideration of the place of chondromalacia in any discussion of the patellofemoral joint, this subject will be explored in detail and background information on articular cartilage will be offered to provide the interested reader with a basis for understanding articular cartilage derangements as well as the current MR approaches directed at their depiction. Abnormalities of the patella including fracture and dislocation will be examined. Finally, the emerging role of MR in the dynamic assessment of patellofemoral alignment and motion will be considered.

THE EXTENSOR MECHANISM

Anatomic and Biomechanical Considerations

The extensor mechanism of the knee comprises the quadriceps femoris muscles and tendon, the patella, supporting soft tissue and retinacular structures, and the patella tendon (30–33). The quadriceps is composed of four muscles: the rectus femoris, vastus medialis, vastus intermedius, and vastus lateralis (Fig. 6.1). Distally, component fibers of the three vasti join the rectus femoris tendon to form the quadriceps ten-

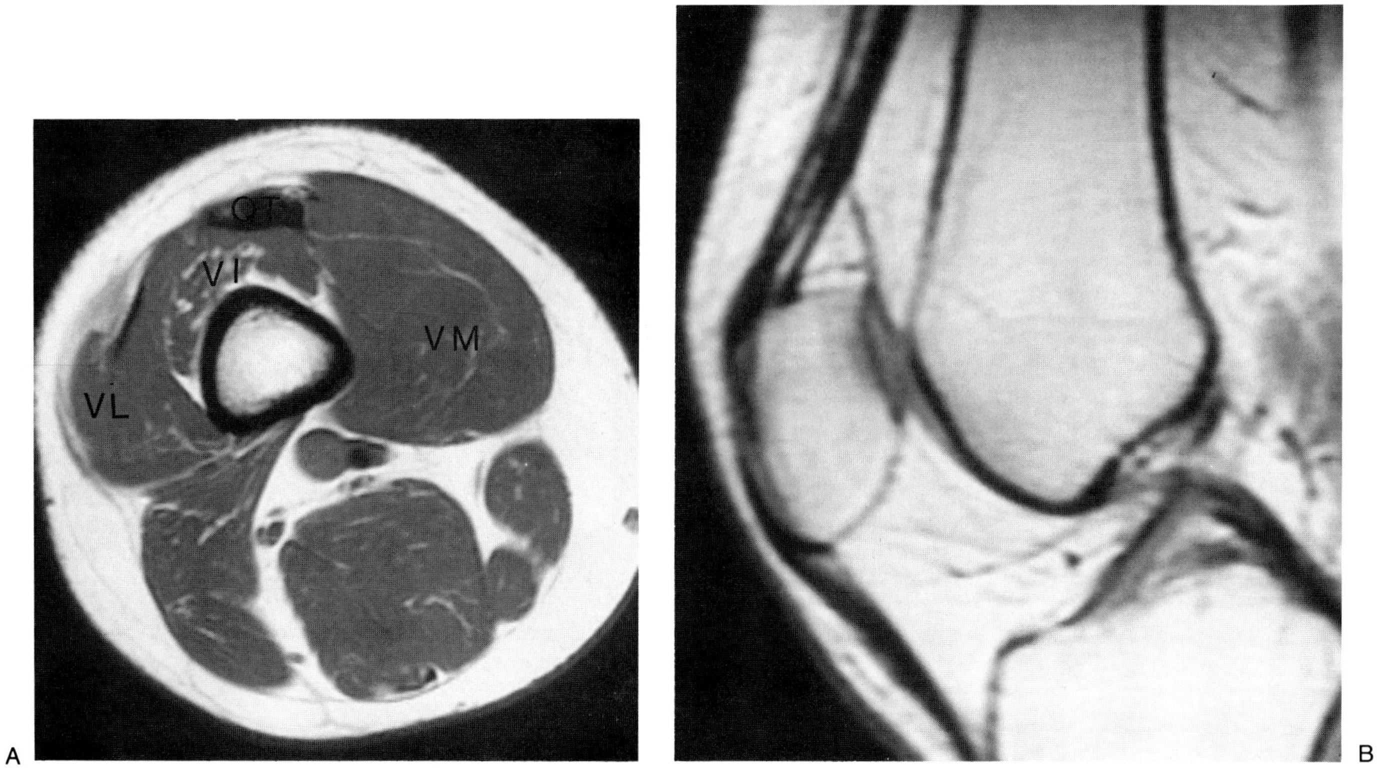

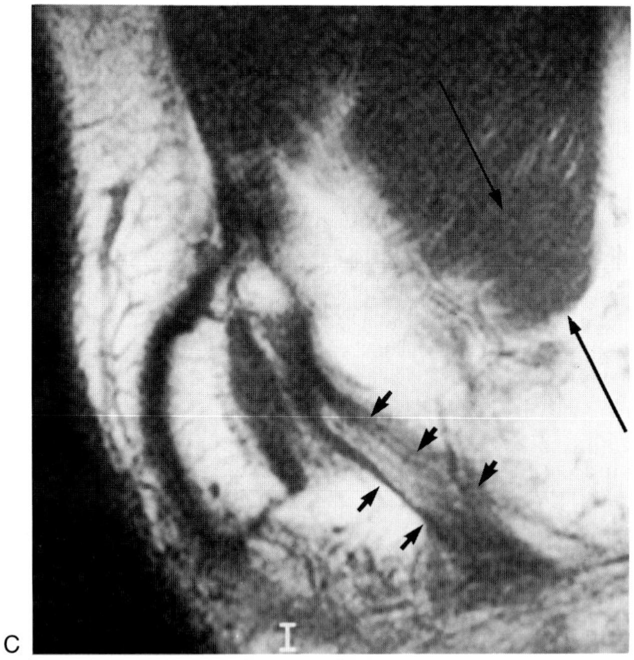

FIG. 6.1. Quadriceps mechanism, normal anatomy. **A:** Axial TR/TE 500/20. The constituent muscular components of the quadriceps are demonstrated in axial section. VM, vastus medialis; VL, vastus lateralis; VI, vastus intermedius; QT, quadriceps tendon. **B:** Sagittal TR/TE 500/20. The distal quadriceps tendon frequently demonstrates a multilaminar appearance reflecting its constituent components. This intratendinous signal should not be mistaken for tendonitis or a partial tear. The patellar tendon is demonstrated as a continuation of the quadriceps from the lower pole of the patellar to its insertion on the tibial tuberosity. The anterior edge should be straight with the patient's knee positioned in extension. **C:** Oblique coronal TR/TE 500/20. The course of the vastus medialis obliquus is demonstrated as it extends to insert upon the proximal medial aspect of the patella (*long arrows*). This component of the quadriceps mechanism is particularly important for stability of the patellofemoral joint. The patellotibial ligament, reflecting a condensation of the medial retinaculum, is seen coursing from patella to tibia (*short arrows*).

don which inserts on the superior border of the patella. A trilaminar appearance to the quadriceps tendon is normally demonstrated on sagittal MR images and correlates with the component fibers of the vasti (Fig. 6.2). The vastus intermedius has a virtually straight insertion into the quadriceps tendon in relation to the long axis of the thigh. The vastus medialis longus fibers make an approximate 50° angle as they insert into the quadriceps tendon (32). The distal third of the vastus medialis, referred to as the vastus medialis obliquus (VMO), has a 65° orientation of its distal fibers which insert not only into the distal quadriceps tendon, but also into the proximal third of the medial border of the patella (Fig. 6.1), (33,34). The vastus lateralis fibers angle 30° to 40° into the quadriceps tendon and terminate in an extremely strong and well-developed ten-

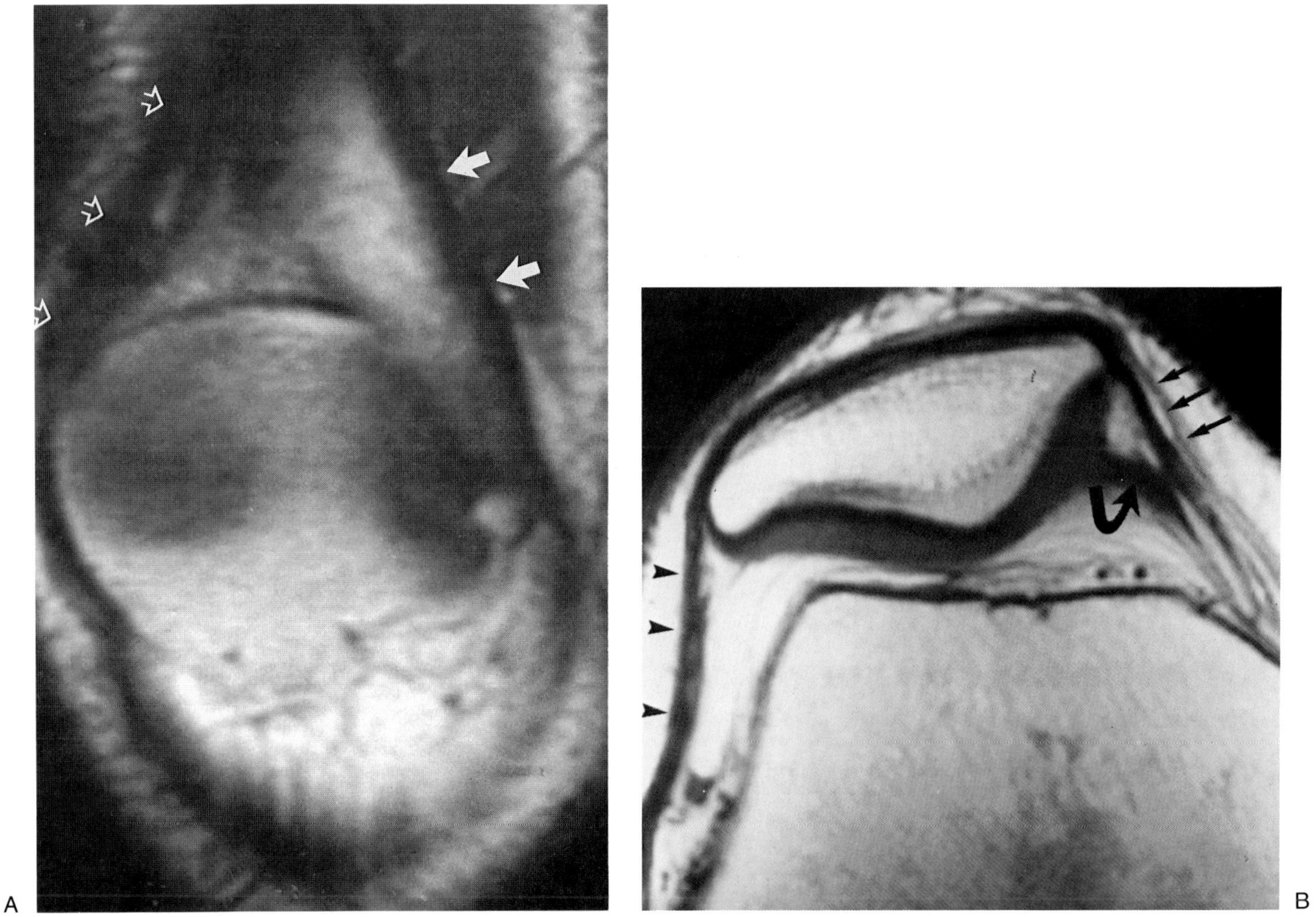

FIG. 6.2. Patella retinaculum, normal appearance. **A:** Coronal TR/TE 500/20. The medial (*open arrows*) and lateral retinaculum are well depicted on coronal sections through the patella and distal quadriceps. **B:** Axial TR/TE 500/20. The supporting retinacular structures of the patellofemoral joint are best depicted in the axial plane. The lateral retinaculum (*arrowheads*) frequently appears slightly thicker than the medial (*arrows*). A medial band-like structure, contributing to an apparent triangular structure (*curved arrow*), most likely represents compressed synovial structures.

dinous attachment at the superolateral border of the patella (34). These angles of insertion are critical for the stability of the patella. A lesser angulation of the VMO would be insufficient to provide dynamic stability to the patella at the moment when the more powerful and longitudinally oriented lateral components produce contraction (33). Insufficiency of the VMO has been associated with patellar subluxation and dislocation (33).

Patellar stability is in large part determined by a balance between dynamic forces provided by the medial and lateral vasti muscles, and static stabilizers formed from fibrous expansions of the vasti and referred to as the medial and lateral retinaculum (Fig. 6.2) (31,35). The static stabilizers preventing passive lateral or medial hypermobility include the medial and lateral patellofemoral and patellotibial ligaments, the iliopatellar band, and the lateral meniscopatellar ligament (31,33,34). The patellofemoral ligaments are transverse condensations of the medial and lateral retinacula at the distal portion of the quadriceps expansions. They are frequently thin as a consequence of stress shielding by the dynamic muscular restraints (31). The medial patellotibial ligament is an oblique condensation of the medial retinaculum which originates inferiorly on the patella and inserts on the anteromedial border of the tibia (Fig. 6.2) (31,33,34). The lateral patellofemoral ligament is the anterior margin of the fascia lata parallel to the patellar tendon and connects the iliotibial tract and tibia distally to the iliopatellar band and patella proximally. Both patellotibial ligaments are obvious well-developed structures with no counterbalancing dynamic restraints (31). It is also important to note that the extensor retinaculum inserts distal to the tibial plateau, leaving some space between the insertion and the articular cartilage of the tibia. This may allow some mechanical advantage through gliding of the retinaculum over the proximal tibia in a manner similar to a bursal structure. The length of the patellar tendon determines the position of the patella in relation to the femur in various degrees of knee flexion. The patella tendon should approximate the greatest vertical dimension of the patella. Variations in the length of the patella have been associated with patellar instability (36,37). The normal patella tendon to patella length ratio is 1.0. Variations of greater than 20% have been considered indicative of either superior displacement (patella alta) or inferior displacement (patella baja) (36,37). This evaluation can be accomplished on lateral radiographs or sagittal MR images. The cross-sectional imaging display of first CT and now MR allows for direct assessment of the position of the patella in relation to the trochlear sulcus. This has led to the concept of "functional" patella alta when the patella is beyond the confines of the bony sulcus. This analysis may not correlate with the more classic description.

Extensor Mechanism Injuries

Extensor mechanism (EM) injuries may be seen in both elderly individuals and highly trained athletes. EM injuries may occur from either direct or, more commonly, indirect (overstretch) trauma (38–44). Disruption may occur at one of multiple sites including (a) tendo-osseous junction (suprapatellar, infrapatellar, tendotubercle), (b) intratendinous, (c) intramuscular (most commonly rectus femoris), and (d) myotendinous (21,42). The extensor mechanism tendons tear uncommonly unless predisposed to rupture by a variety of degenerative, metabolic (hyperparathyroidism, gout, and diabetes), inflammatory, or autoimmune (systemic lupus erythematosus [SLE], rheumatoid arthritis) conditions (45,46). MR is capable of depicting the entire spectrum of extensor mechanism injuries ranging from mild muscular strains to partial and complete tendon disruptions. The focus of this section will be on the latter injuries, with the former considered in detail in Chapter 10.

Imaging Technique

For assessment of the extensor mechanism, we most commonly employ a standard send-receive extremity coil (GE). For evaluation of quadriceps injuries, the position of the surface coil should be modified from that utilized for the standard knee examination to allow inclusion of the distal thigh in the field of view. Only the affected limb is examined to optimize image quality by permitting the use of a small field of view (12–14 cm). For evaluation of the patellar tendon and retinacular structures, standard knee positioning techniques suffice.

Sagittal T1-weighted and STIR sequences are obtained to most optimally depict the full longitudinal extent of the injury. For T1-weighted sequences, a 256 × 256 matrix is commonly utilized with one excitation. Section thickness is 3 mm without interslice gap (interleaved acquisition). Such an imaging protocol yields an in plane spatial resolution of 0.47–0.62 mm. For the STIR sequence, the matrix is reduced to 256 × 128 and again one excitation is performed. Sections are routinely 4 mm with a 2 mm interslice gap. T2-weighted axial images are routinely obtained in cases of suspected quadriceps injuries. This sequence provides a series of images of both good anatomic detail (proton density images) and high tissue contrast (T2-weighted). For standard T2-weighted images, a 256 × 128 matrix is utilized and two excitations are performed. The TR is set at 2,500 to 3,000 msec to allow for the required number of sections, and the slice thickness is maintained at 3 to 4 mm without interslice gap. More recently, the advent of "fast" spin-echo techniques has allowed the successful reduction in the imaging time necessary to complete long TR/TE sequences while taking advantage of higher resolution (256 matrix) imaging. We have replaced our standard T2 sequences detailed above with the following technique. The TR is chosen between 3,000 and 3,500, depending on the number of slices necessary to cover the selected area; the double echo TE combination is typically set at 18 to 21/120 msec. A 256 × 256 matrix size and 3 mm skip with 1 mm spacing with one excitation completes the imaging selections. This sequence can be accomplished in approximately one-half the time of the conventional sequence while achieving a study of higher resolution. The axial sequence permits the best assessment of tendon contour, the extent of intratendinous involvement, and the relationship to surrounding structures, including the muscular constituents of the quadriceps muscle.

Quadriceps Tendon

Background

Quadriceps tendon ruptures occur most often at the tendo-osseous junction. In elderly patients, quadriceps tears commonly occur when the patient is descending stairs. The stress necessary to rupture the mechanism is generated by the opposing forces of quadriceps contraction in conjunction with forced flexion of the knee. In athletes, quadriceps tears often occur with running, with the lead foot on the ground and the individual decelerating, resulting in the full weight of the individual adding to the force of the foot strike (42). During deceleration, the quadriceps is maximally tense, and overstretch injuries may occur secondary to the developed force, which has been estimated to exceed 17 times body weight (42).

The central and anterior segment of the tendon usually tears first at a level approximately 2 cm above the upper pole of the patella (47). Diagnosis may be difficult on physical examination, particularly in the presence of a severe hemarthrosis, and the condition may go unrecognized (44). Of the 17 patients reported by Ramsey and Mueller, 7 (41%) had a delay in diagnosis ranging from 14 days to 1 year after the injury because the condition was not recognized by the examining physician (44). In patients with partial tears and in whom the retinaculum may be intact, active extension may still be possible. Delay in diagnosis may complicate repair or reconstruction, underscoring the need for an accurate means of evaluation (41,48). Conventional radiographic findings are nonspecific and diagnosis may be difficult. Findings include (a) hemarthrosis, (b) blurring of the distal quadriceps tendon, (c) a suprapatellar soft tissue mass, and (d) low patellar position (not invariable) and redundant patellar tendon (42,49). Arthrography has been utilized in quadriceps

A

B

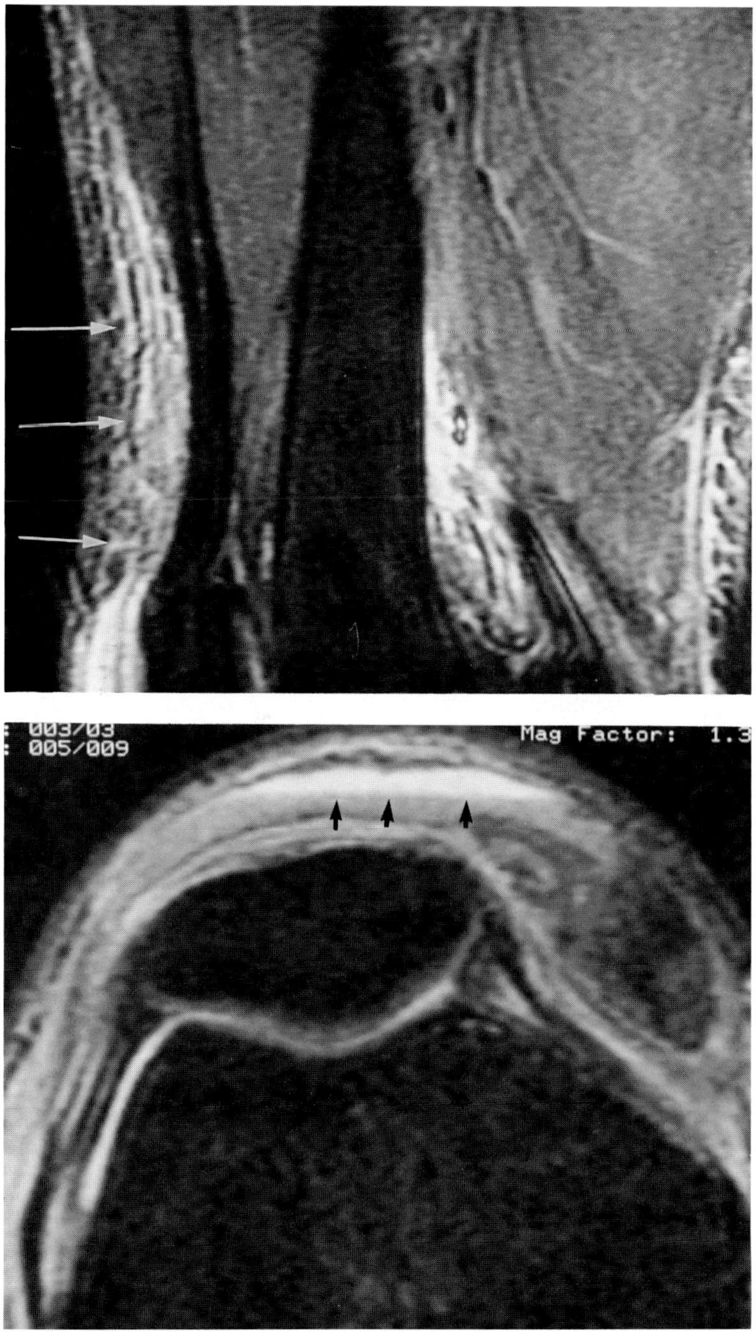

FIG. 6.3. Superficial quadriceps injury. **A:** Sagittal STIR TR 2,200, TE 35, TI 160. The patient was referred for further characterization of a suspected quadriceps tear. He had marked quadriceps weakness and was unable to extend his knee. On this image, extensive high signal edema is seen superficial to an intact quadriceps tendon. **B:** Axial STIR TR 2,200, TE 35, TI 160. Superficial hematoma with a fluid/fluid level is identified. The quadriceps tendon was intact and the findings significantly affected management of this patient who was tentatively scheduled for surgery.

disruption (50,51). Injection of contrast into the suprapatellar bursa may allow detection of full thickness tears with extension of contrast into the soft tissues anteriorly related to associated bursal disruption. Many partial and all intratendinous tears cannot be diagnosed by this technique.

MR Imaging

MR is an exceptional means for evaluating quadriceps injuries (6,7,14). Injuries to superficial structures can mimic quadriceps tendon tears and can be readily distinguished from injuries to the tendon itself (Fig.

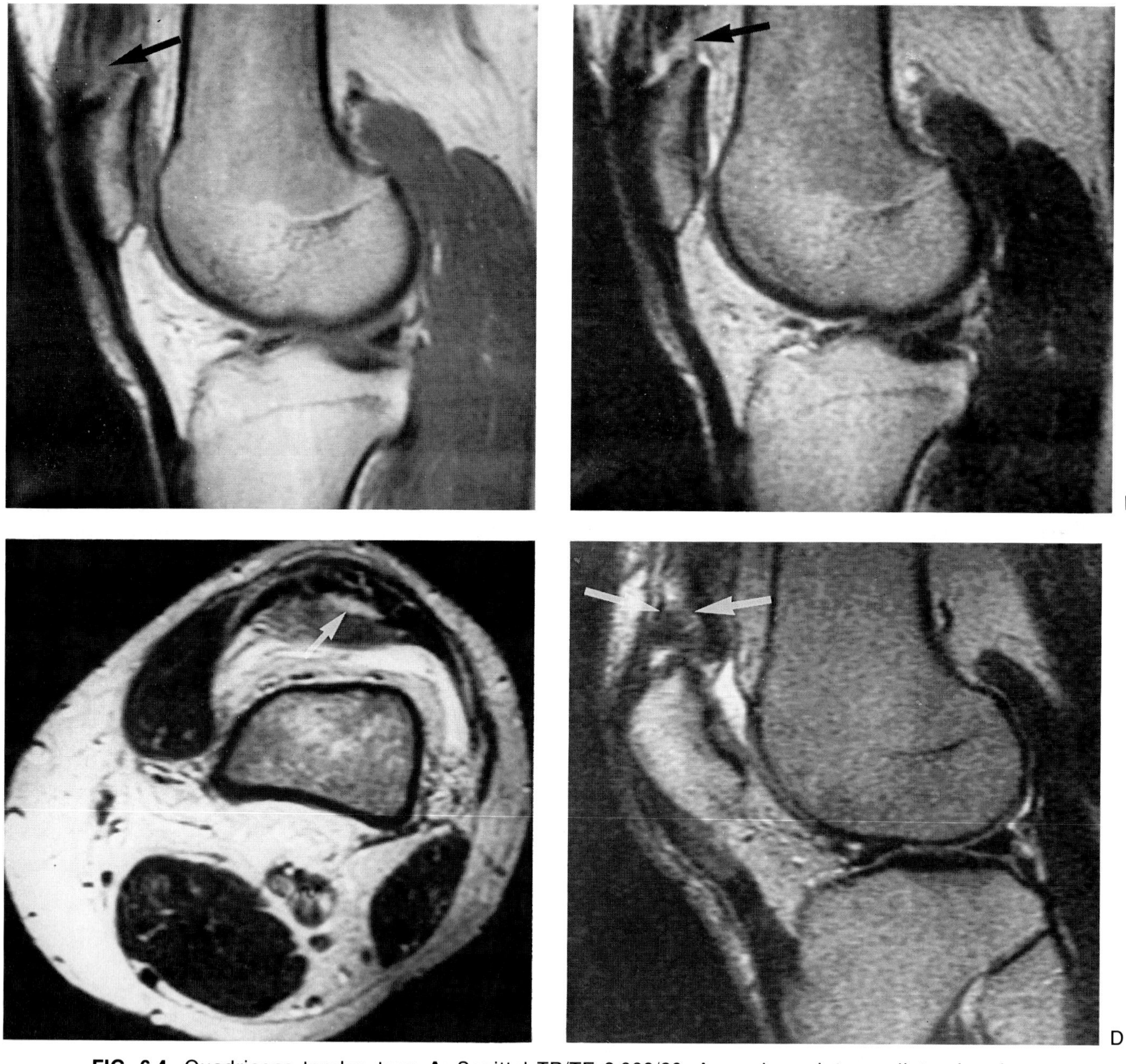

FIG. 6.4. Quadriceps tendon tear. **A:** Sagittal TR/TE 2,000/20. Amorphous intermediate signal intensity largely obscures and replaces the normal trilaminar appearance of the distal quadriceps tendon at the level of its insertion into the upper pole of the patella (*arrow*). **B:** Sagittal TR/TE 2,000/80. On the T2-weighted image of this patient, there is increased signal representing edema and interstitial hemorrhage at the site of tendon disruption (*arrow*). **C:** Axial TR/TE 2,000/80. On this axial section the tendon is thickened and the extent and location of the tear, which predominantly involves the posterior third of the tendon, is well depicted (*arrow*). **D:** Sagittal TR/TE 2,000/80. Complete quadriceps tendon tear demonstrated in a different individual. The tendon is completely obscured by edema and hemorrhage at the site of disruption (*arrows*). Also note the abnormal tilt of the patella, loss of the normal straight contour of the patella tendon, and evidence of increased signal within the proximal patellar tendon reflective of chronic patellar tendonitis.

6.3). Quadriceps tears may be complete or incomplete (42). The precise location, extent, and degree of separation of the tear can be defined and quantitated (Fig. 6.4). With complete disruptions, sagittal images often best depict the degree of distraction of the major components. The continuity of the normally dark tendon is absent. A gap is present and the involved segment of the tendon may be completely filled with signal (Fig. 6.4). The proximal tendon is commonly retracted and the defect filled with edema and hemorrhage. In addition, tilting of the patella in a "front/back" direction and redundancy of the patellar tendon are commonly noted secondary signs of the injury. A "wrinkled" appearance to the anterior margin of the patellar tendon has been emphasized as an important indicator of quadriceps mechanism insufficiency, although in our own experience, a slightly undulating contour to the patellar tendon has frequently been demonstrated in the absence of a demonstrable quadriceps abnormality. Partial tears of the quadriceps may best be characterized utilizing axial plane images (Fig. 6.4). T2-weighted and STIR sequences which depict fluid and edema with high signal intensity are preferred for detection, while T1-weighted and proton density sequences provide enhanced anatomic depiction (52). In cases of quadriceps tendon disruption, the extreme sagittal and axial images may reveal the tear extending into the retinaculum (Fig. 6.5).

The presence of hemorrhage and edema both within and surrounding the acutely injured tendon and muscle can be demonstrated. The appearance of blood on MR will be influenced by whether it is interstitial or contained within a true hematoma (see Chapter 10) (53). The MR appearance of interstitial hemorrhage within soft tissue generally parallels that of and is often indistinguishable in appearance from that of nonhemorrhagic edema (21,53). It has been suggested that this is due to the separation of serum from the cellular components of blood in the soft tissues (54). Hematomas in the acute stage may appear isointense compared with surrounding soft tissues on T1-weighted sequences and may on T2-weighted sequences be isointense (oxyhemoglobin in intact red blood cells) or of decreased intensity (magnetic susceptibility effect of deoxyhemoglobin contained within intact red cells) (54). Subacute hematomas demonstrate increased intensity on T-1 weighted images as a result of the paramagnetic effect of methemoglobin. Older hematomas often demonstrate a low signal intensity border secondary to hemosiderin deposition.

Patellar Tendon

Tendonitis

Tendonitis refers to an injury or symptomatic degeneration of a tendon with associated secondary inflammation of the investing tissues (55). If symptoms have been present for less than 2 weeks, the inflammatory condition is referred to as acute tendonitis; if the situation has persisted for longer than 6 weeks, chronic tendonitis is said to be present. Tendons may be invested with a sheath lined by synovium or alter-

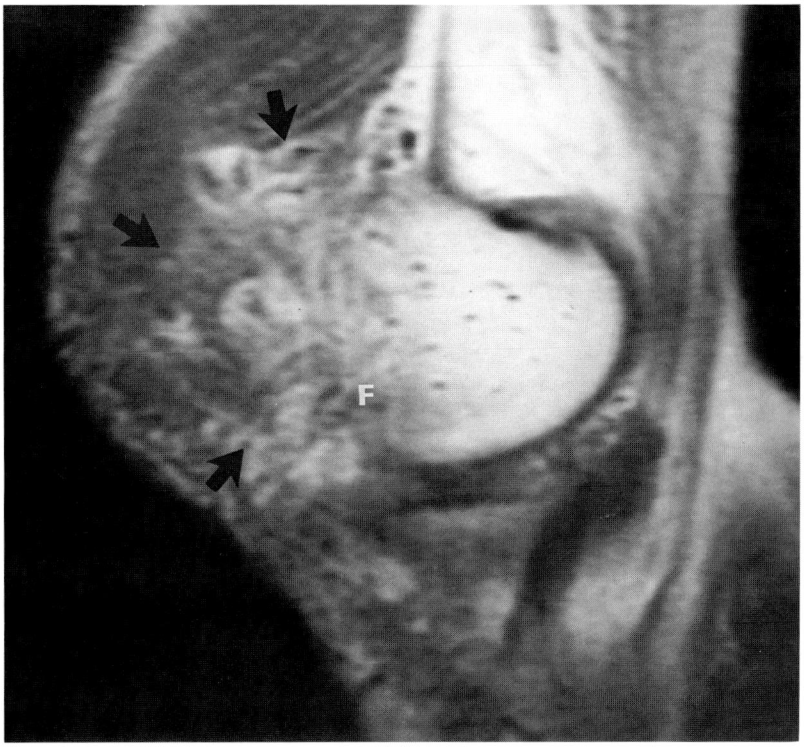

FIG. 6.5. Retinacular tear. Sagittal TR/TE 800/20. The patient had sustained a complete quadriceps tear which on this extreme sagittal image can be seen to extend into the medial retinaculum. There is marked disorganization (*arrows*) of the soft tissues anterior to the distal femur (F). (From ref. 55, with permission.)

natively may be surrounded by only loose connective tissue. Tenosynovitis refers to inflammation of the investing soft tissue of the tendon, whether that tissue is lined by synovium or not.

Acute patellar tendonitis represents a painful condition around the tendon that may result from one of a number of increased demand situations, such as sudden increase in athletic training frequency (55). Patients with suspected acute patellar tendonitis rarely undergo diagnostic imaging, and the authors have not had the opportunity to image a suspected case. In experience with documented acute Achilles tendonitis, however, the tendon itself has been noted to remain normal in size and to be without significant intratendinous signal. MR detectable changes in this setting have predominated in the peritenon (55).

Chronic tendonitis is associated with the development of longitudinal splits within the tendon, and the

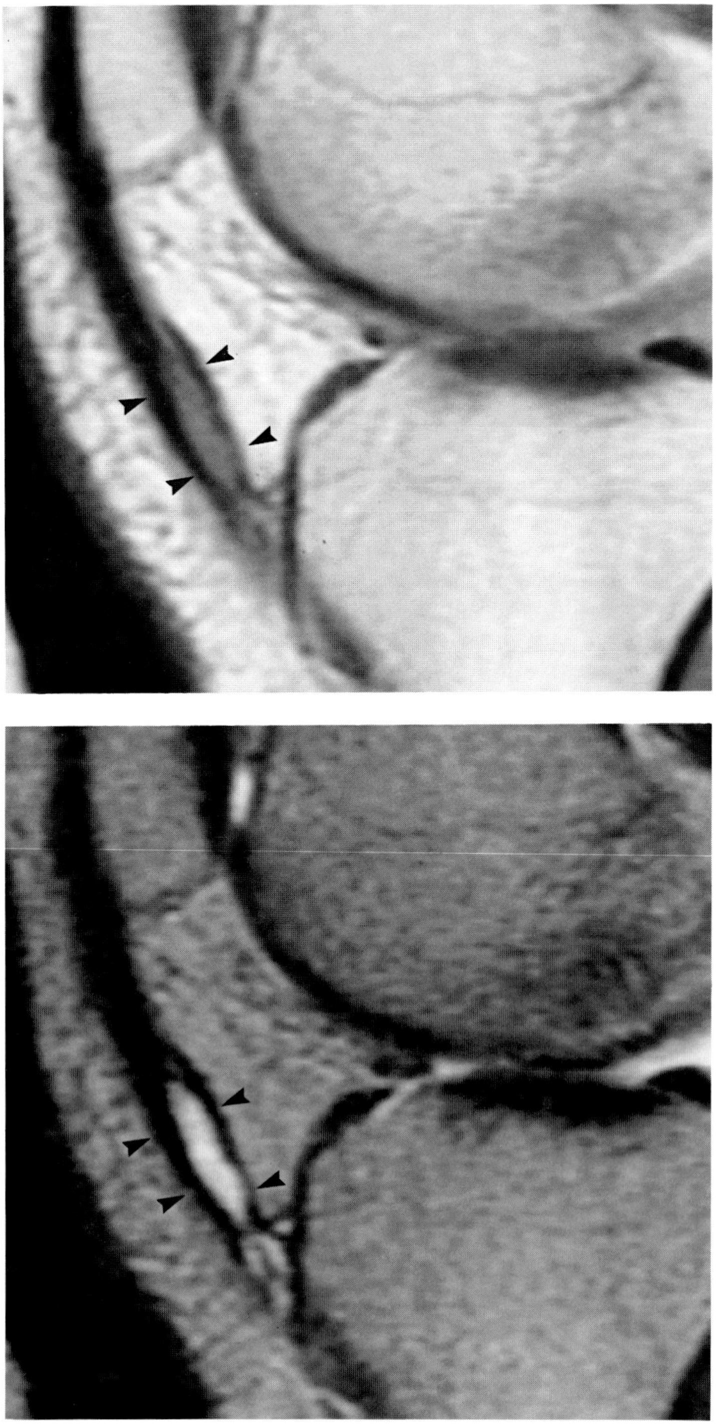

FIG. 6.6. Chronic complicated patellar tendonitis. **A:** Sagittal TR/TE 2,000/20. There is a linear to ovoid focus of intermediate signal intensity with the distal third of the patellar tendon (*arrowheads*). **B:** Sagittal TR/TE 2,000/80. The signal becomes markedly hyperintense on the T2-weighted image (*arrowheads*). There is no superficial edema to suggest an acute tear. The increased signal intensity on this sequence is known to correlate pathologically with areas of partial disruption and zones of inflammation and necrosis in patients with chronic tendonitis.

development of small foci of mucinous degeneration. On MR, chronic patellar tendonitis is manifest as slight to moderate widening of the tendon contour which is well demonstrated in both the sagittal and axial planes (1,55). Foci of signal alteration may be demonstrated on T1-weighted and proton density images which, in the case of uncomplicated chronic tendonitis, may decrease in intensity on T2-weighted images (1).

With uncomplicated chronic tendonitis, the patellar tendon becomes enlarged, but the tendon maintains low signal on both long and short TR imaging sequences. The development of chronic tears with inflammation and necrosis produces areas of high signal on the T2-weighted sequence (Fig. 6.6) (1). In this setting, the appearance of chronic tendonitis on MR may be quite similar to that of acute partial intrasubstance tears, and the difference between the two may be based solely on history. Acute patellar tears, however, are commonly associated with subcutaneous edema. The term "jumper's knee" refers to one form of chronic patellar tendon stress that is commonly encountered in individuals participating in activities requiring frequent jumping such as basketball and volleyball (Fig. 6.7) (56). The condition is manifest clinically by pain

A

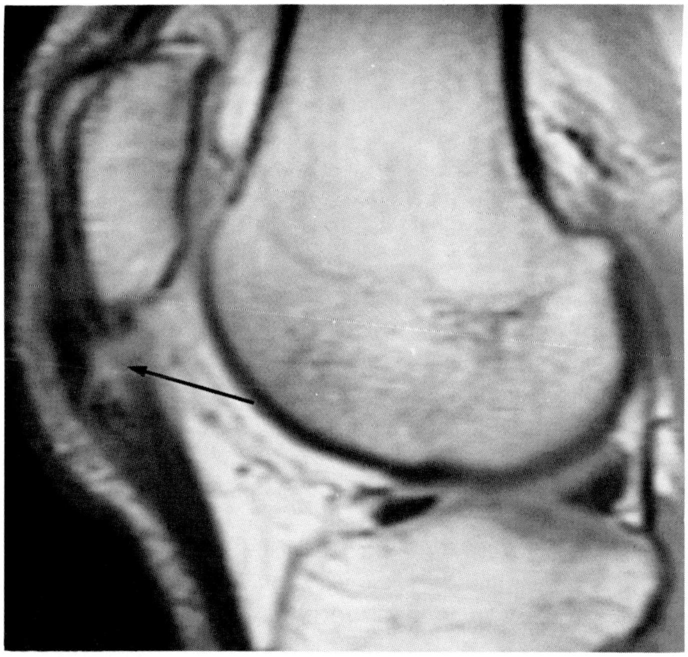

B

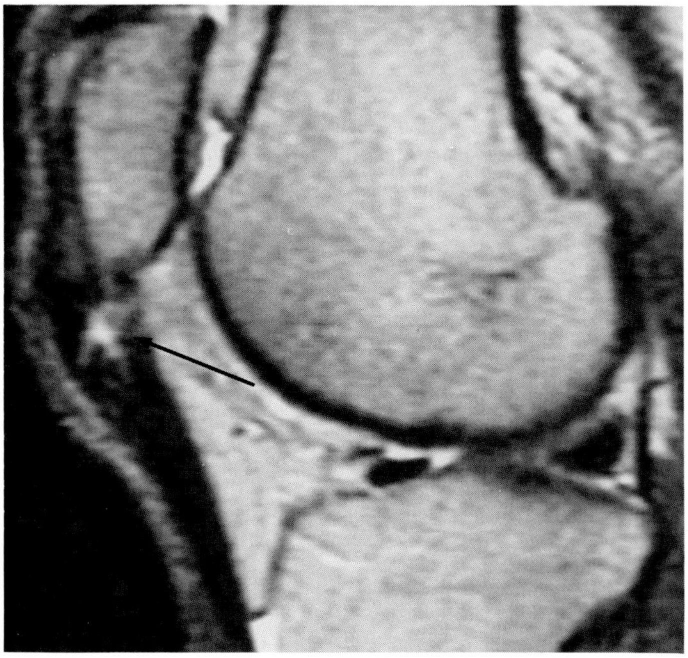

FIG. 6.7. Patellar tendonitis (Jumper's Knee). **A:** Sagittal TR/TE 2,000/20. A stellate focus of increased signal intensity is seen within a widened proximal infrapatellar tendon of a professional basketball player (*arrow*). **B:** Sagittal TR/TE 2,000/80. On the more heavily T2-weighted image, the signal becomes hyperintense, suggesting at least partial disruption of the tendon seen in conjunction with chronic tendonitis (*arrow*).

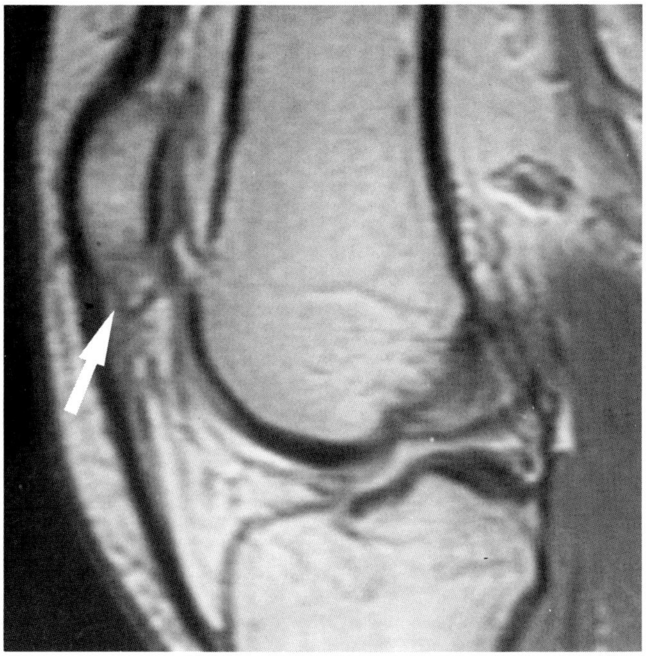

FIG. 6.8. Sinding-Larsen-Johansson disease. Sagittal TR/TE 2,000/20. An ossific fragment is seen relating to the lower pole of the patella (*arrow*) along with subtle signal changes along the proximal patellar tendon. The patient had no history of trauma, although there was a several year history of complaints relating to the lower pole of the patella.

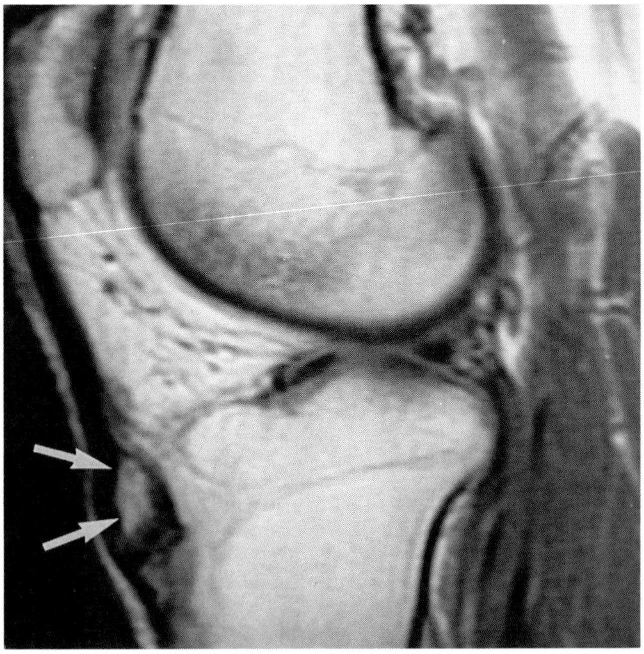

FIG. 6.9. Osgood-Schlatter disease. Sagittal TR/TE 2,000/20. A well-defined ossific fragment is seen within a widened insertion of the patellar tendon (*arrows*). There is slight fragmentation of the tibial tuberosity. The patient had a previous history of Osgood-Schlatter disease.

and on examination by tenderness to palpation. The repeated stress creates micro-tearing, the tissue response to which results in ingrowth of bone, synovium, or cartilage within the tear. The findings most typically involve the proximal patellar tendon.

Sinding-Larsen-Johansson disease represents a syndrome most commonly seen in adolescents consisting of localized tenderness over the lower pole of the patella accompanied by radiographic evidence of osseous fragmentation. It is currently believed that the condition results from a traction phenomenon in which contusion or tendonitis in the proximal attachment of the patellar tendon can be followed by calcification and ossification (57). During the active phase of the disease, the presence of soft tissue swelling allows differentiation of this condition from normally occurring ossification variants in this location (57). On MR, the typical fragmentation of the lower pole of the patella is well demonstrated, as are corresponding changes within the proximal patellar tendon (Fig. 6.8).

Osgood-Schlatter disease represents another type of chronic stress related condition of the patellar tendon that is common in adolescents (57). This condition results from an avulsive force upon the tibial apophysis transmitted via the patellar tendon secondary to quadriceps contraction. A traumatically induced partial disruption of the patellar tendon from the developing tibial tuberosity results in avulsion of fragments of cartilage and bone. Three classic radiographic signs are essential to the diagnosis of this condition: (a) soft tissue swelling anterior to the tibial tubercle, (b) thickening and irregularity of the patellar tendon, and (c) fragmentation and irregularity of the tibial tubercle (21, 58). On MR, signal and morphological changes within the distal tendon are seen in association with osseous changes in the region of the tibial tubercle (Fig. 6.9). These changes must be differentiated from multiple ossification centers of the tibial tuberosity that may be seen as variations in ossification (57). Additionally, a bursa may form between the residual ununited ossicles and the underlying tibial tuberosity which can give rise to symptoms of bursitis in some individuals.

Rupture

Ruptures of the patellar tendon are the least common cause of extensor mechanism disruption (42). In contrast to patients with quadriceps rupture, patients with patellar tendon disruption are typically younger, frequently male, and often involved in sports or active recreational activities (42,43,46,59–61). The most common site of rupture is at the junction of the tendon with the lower pole of the patella (Fig. 6.10). Mid-substance patellar tendon tears are less common and may be seen as a result of a single episode of severe trauma

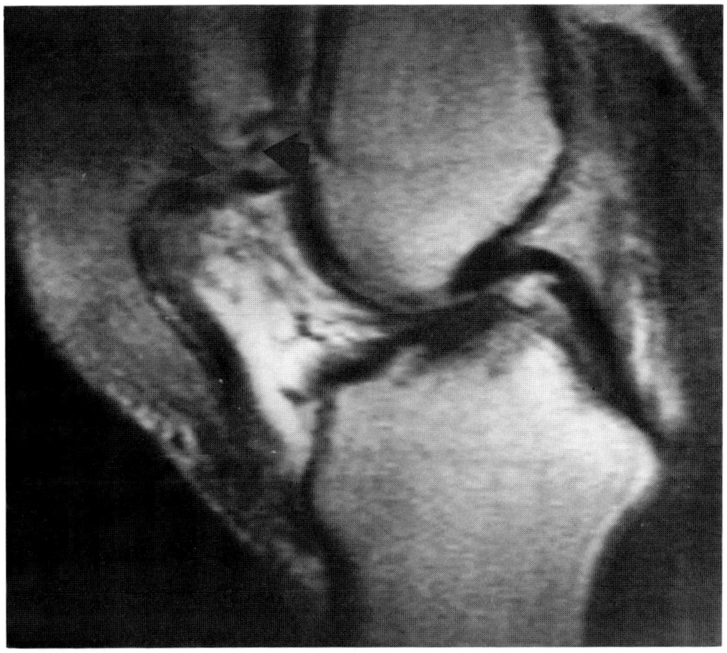

A

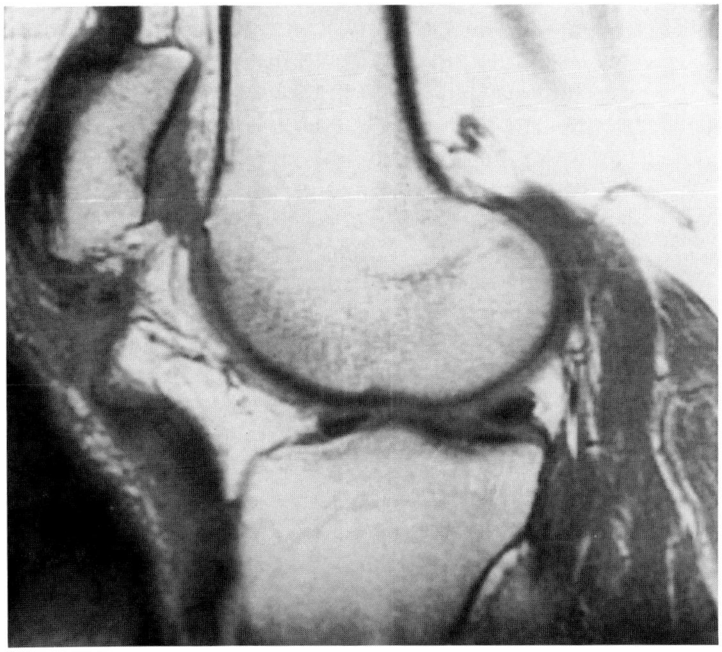

B

FIG. 6.10. Patellar tendon disruption. **A:** Sagittal TR/TE 500/20. A weekend basketball player experienced the sudden onset of pain after coming down with a rebound. The patellar tendon is avulsed from the inferior pole of the patella. (From ref. 55, with permission.) **B:** Sagittal TR/TE 800/20. Complete disruption of the patellar tendon is demonstrated in another individual. The site of disruption is well identified and again is near the proximal insertion. (Courtesy of Scott Kingston, M.D. Los Angeles, California.)

likely related to forced knee flexion against a contracted quadriceps mechanism (61,62). Tears at the level of the lower pole of the patellar predominate in younger individuals. As with all extensor mechanism tears, patellar tendon tears are associated with coexistent conditions which may predispose the tendon to rupture (46,60).

On conventional radiographs, elevation of the patella, indistinctness of the tendon, and at times an avulsion fracture at the inferior pole may be present (42,58). On MR, the continuity of the normally dark tendon is absent. The tendon may be seen to be completely detached from the lower pole of the patella (Fig. 6.10). With complete tears the patella may be displaced superiorly. On T2-weighted and STIR sequences, the site of disruption is filled with high signal reflecting edema and hemorrhage. An unusual variation of a complete patellar tendon tear occurs with a diffuse interstitial tear; the tendon is grossly "intact," but it assumes a serpiginous contour with increased signal intensity (55). Acute avulsive bony injuries are associated with less severe degrees of internal tendon abnormality, although physical impairment is marked.

The etiology of tendinous disruption is incompletely

understood (61). Two major theories have been proposed based on studies of the more commonly disrupted Achilles tendon (61). The first suggests that insufficient healing related to micro-trauma leads to progressive loss in tendon strength and eventual rupture. This theory would likely best explain the etiology of ruptures that occur in patients with systemic diseases (e.g., SLE) or on corticosteroids, and in athletes with repetitive submaximal trauma. Decreased vascularity at the rupture site is the most frequently cited physiological cause for tendon degeneration leading to rupture (61). Tears of the Achilles tendon, which derives its vascularity from its musculotendinous and osteotendinous junctions, occur most typically at a vascular watershed area 2 to 3 cm from the calcaneus (61). The patella tendon differs from the Achilles tendon in several ways. It does not have a paratenon, and the vascular supply originates directly from the infrapatellar fat pad superiorly and the anterior soft tissues inferiorly, providing a rich vascular supply throughout the length of the tendon (61). This may explain the uncommon occurrence of intrasubstance tears of the patella tendon in contrast to the situation observed with the Achilles tendon. The attachment of the patellar tendon to bone, however, is avascular, and inability to repair injury in the avascular area may explain the propensity for infrapatellar tendon ruptures to occur at the osteotendinous junction.

The second theory of tendon rupture proposes that excessive loads can be sufficient to rupture otherwise normal tendons. This theory would best account for ruptures seen following motor vehicle accidents or falls with large hyperflexion forces against the contracting quadriceps. In studying a weight lifter whose infrapatellar tendon ruptured while undergoing kinetic analysis, Zernicke and Jobe found that the tendon ruptured after stress greater than 17.5 times body weight (63). In studies of the strength of the patellar tendon as an anterior cruciate ligament substitute, Noyes et al. reported the strength of the central third as 175% of that of an intact ACL (64). By extrapolation, the strength of the intact infrapatellar tendon would be 525% of the average ACL strength. Given the strength of the tendon, it is understandable why transverse patellar fractures are seen more commonly as a result of hyperflexion injuries than patellar tendon disruptions (61).

Bursa

There are 12 bursa commonly described around the knee, four of which relate to the extensor mechanism. The largest of the bursa, the suprapatellar bursa, may for practical purposes be considered an extension of the joint proper. On MR, the suprapatellar bursa has an inverted U-shaped appearance, with the medial and lateral reflections actually being more dependent than the antefemoral component. Three bursae relate to the patellar tendon and can be demonstrated on MR. The subcutaneous prepatellar bursa is located between the

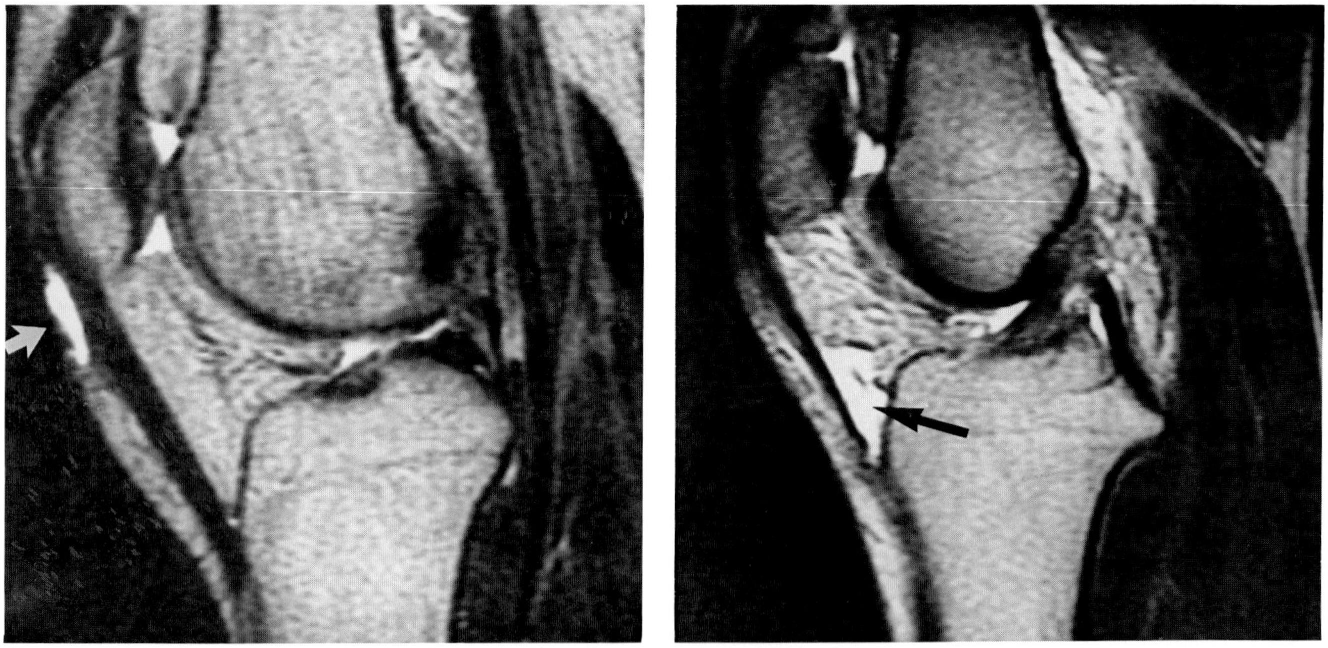

A B

FIG. 6.11. Prepatellar and infrapatellar bursitis. **A:** Sagittal TR/TE 2,000/80. An elongated high signal intensity fluid collection (*arrow*) is seen corresponding to the subcutaneous infrapatellar bursa. **B:** Sagittal TR/TE 2,000/80. High signal intensity fluid is seen distending the deep infrapatellar bursa (*arrow*). Normally no or only a minimal amount of fluid is present in this location.

skin and outer surface of the patella and commonly extends from the mid-plane of the patella to cover the upper half of the patellar tendon. Its superficial location subjects the bursa to repeated trauma (e.g., housemaids' knee) (Fig. 6.11). The subcutaneous infrapatellar bursa lies between the lower part of the tuberosity of the tibia and the skin. The deep infrapatellar bursa lies between the patellar tendon and the upper part of the tuberosity of the tibia (Fig. 6.11). It is small and separated from the synovial membrane of the knee by the infrapatellar fat pad. It represents another functionally important mechanical arrangement of the extensor mechanism by allowing "gliding" of the tibia around the femoral condyles. Any destruction or significant alteration of this structure effectively fixes the patellar tendon insertion more proximally and exerts a different force on the tibiofemoral joint.

Synovial Plicae

Synovial plicae are remnants of synovial tissue that in early development originally divided the joint into three separate compartments, and which may be found normally in the adult knee (51,65,66). While often of no clinical significance, these structures may become pathologically thickened and produce symptoms that mimic other causes of internal derangement, such as meniscal tears. In addition, persistence of these structures in their embryonic form as complete septa may

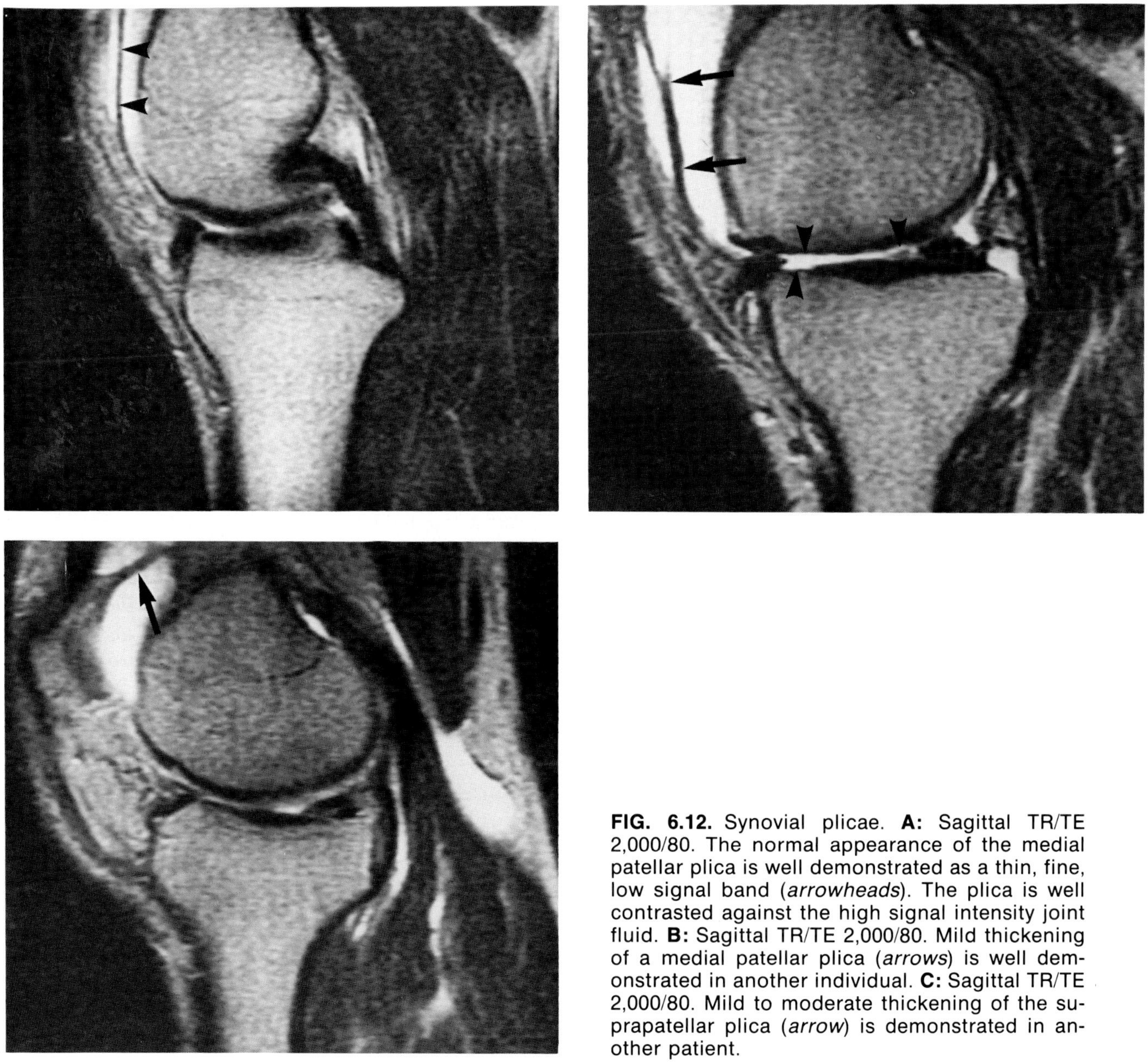

A B C

FIG. 6.12. Synovial plicae. **A:** Sagittal TR/TE 2,000/80. The normal appearance of the medial patellar plica is well demonstrated as a thin, fine, low signal band (*arrowheads*). The plica is well contrasted against the high signal intensity joint fluid. **B:** Sagittal TR/TE 2,000/80. Mild thickening of a medial patellar plica (*arrows*) is well demonstrated in another individual. **C:** Sagittal TR/TE 2,000/80. Mild to moderate thickening of the suprapatellar plica (*arrow*) is demonstrated in another patient.

create a variety of intraarticular compartment syndromes (51,66).

The three most common plicae are classified according to the partitions from which they took origin: suprapatellar, medial patellar, and infrapatellar (66). Of these, the infrapatellar is most frequent and the medial patella the least commonly encountered. The suprapatellar plicae represent the embryonic remnant of the septum that divided the suprapatellar cavity from the medial and lateral joint compartments (66). It may vary widely in appearance, and most commonly takes one of three forms: (a) variably sized, crescent-shaped fold originating cephalad to the patella along the medial aspect of the undersurface of the quadriceps tendon and extending inferiorly along the medial aspect of the joint, (b) an intact septum completely dividing the suprapatellar pouch from the remainder of the knee, or (c) intact septum except for a variably sized central opening referred to as the porta (51,66). The medial patella plica has been variably referred to as a wedge, band, or shelf. This synovial remnant has its origin on the medial wall of the knee joint near the suprapatellar plica, and courses obliquely downward relative to the patella, to insert into the synovium which covers the infrapatellar fat pad. The infrapatellar plica, or ligamentum mucosum, represents a vestige of the partition that once separated the medial and lateral compartments (66). Its course parallels that of the anterior cruciate ligament and it demonstrates a fan-shaped configuration, narrow at the level of the intercondylar notch and progressively widening as it descends to cover the infrapatellar fat pad.

The medial patella plica is the most common of these remnants to become pathologically thickened and symptomatic. This may occur secondary to a number of postulated initiating events. The resulting less elastic and thickened plica no longer glides normally, but rather snaps against the underlying femoral condyle. Repeated irritation and abrasion result in erosive changes of the articular cartilage of the condyle or even the patella. A dull, aching pain medial to the patella aggravated by flexion, or a clicking sensation without locking or giving way are common clinical complaints (51). A palpable or audible snap may occur on knee movement, and a symptomatic plica may be palpated as a tender, band-like structure parallel to the medial border of the patella. A pathological plicae has been reported to simulate a monoarticular arthritis.

Synovial plicae may be identified on conventional arthrography, CT-arthrography, and MR imaging (51). The decided advantage of MR is the lack of need for an intraarticular contrast agent, direct multiplanar image acquistion, and direct depiction of the articular cartilage. On MR, the normal plica is seen as a thin, fine band of low signal intensity, best identified contrasted against high signal intensity joint fluid on T2- and T2*-weighted sequences (Fig. 6.12). Thickening of plicae is well demonstrated on MR, particularly if sufficient joint fluid is present to provide a natural arthrographic effect (Fig. 6.12).

PATELLOFEMORAL JOINT

Anatomic Considerations

The patella is a sesamoid bone contained within the quadriceps tendon. It functions both to protect the femoral articular surface and to increase the efficiency of the quadriceps mechanism by virtue of the fulcrum effect (33,35,67–70). Its width (51–57 mm) and height (47–58 mm) are remarkably constant, although its thickness measured at the midequitorial plane is quite variable (31,71). The posterior surface can be divided into two parts with the inferior portion, which is nonarticular, representing nearly 25% of the total height (31). The upper three-fourths of the posterior surface is covered by hyaline cartilage which in the central portion of the patella is thicker (5–6 mm) than in any other articulation (31). The thickness of the articular cartilage of the patella varies with age, reaching its maximum thickness by age 25 (72).

The articular surface of the patella is roughly oval in shape and is divided into medial and lateral facets by a vertical ridge that is oriented in the long axis of the patella (31). The facets may be equal in size, although predominance of the lateral facet is the most common situation. The medial facet demonstrates considerable anatomical variation and is typically shorter and more steeply angled than the lateral. It is subdivided by a small vertical ridge (secondary ridge) into the medial facet proper, and a smaller "odd" facet along its medial border (31). The medial facet is usually flat or convex, with considerable variation related to the thickness of the articular cartilage. The lateral facet is both longer and wider than the medial and is concave in both vertical and transverse planes (31). Horizontal ridges that divide the articular surface into thirds may be present and often predominate laterally. These latter divisions become significant as the inferior, middle, and superior thirds of the patella are successively brought into contact with the femoral articular surface during flexion (72,73).

In his classic paper, Wiberg proposed a three-part classification to encompass the majority of patellar facet configurations (74). The system was based on the configuration of the subchondral bone of the facets as depicted on tangential conventional radiographs. In the Type I configuration, both facets are gently concave, symmetrical, and nearly equal in size (Fig. 6.13). This is the least commonly encountered form (75). In Type II the medial facet is smaller than the lateral. The

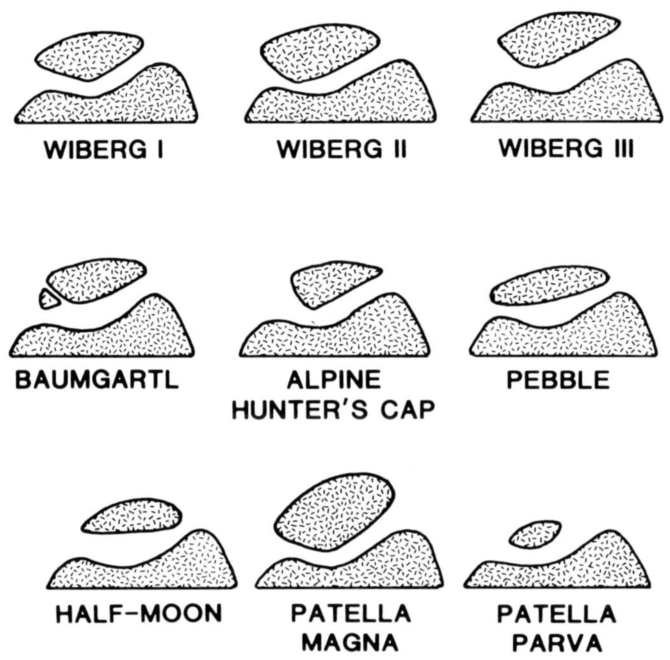

FIG. 6.13. Patellar configurations. Schematic diagram demonstrating the more common variations in patella configuration. (Modified from ref. 31.)

lateral facet remains concave; the medial is flat or convex. This is the most common form accounting for up to 65% of patellae (75). The changes from Type I to II represent a continuum and may be subtle. In the Type III configuration, the medial facet is distinctly smaller with marked lateral predominance. This configuration accounts for approximately 25% of patellae (75). Type III was considered dysplastic by Wiberg although he was unable to demonstrate a higher incidence of chondromalacia than with the other forms (31,74). This lack of association of Type III with chondromalacia has also been noted by other investigators (31). Studies utilizing the tomographic capabilities of CT and MR have demonstrated a change in apparent patella configuration from one section to another (76), contributing to difficulty in categorization of patella configuration. The system proposed by Wiberg, however, remains in common clinical use (23,77,78).

Ficat proposed an alternate classification based on the angle subtended by the two major facets as depicted on conventional radiographs (Fig. 6.13) (79,80). The so-called pebble-shaped patella is characterized by an angle greater than 140°. An angle of 90° to 100° most closely corresponds to the Wiberg Type III. The Alpine hunter's cap deformity is characterized by a patella approaching a single articular facet with the angle 90°. This hemipatella configuration with one articular facet is commonly observed with lateral instability and is associated with hypoplasia of the vastus medialis and decreased depth of the trochlear sulcus (31). The half-moon patella is characterized by an acute angle with a single articular facet. It is associated with permanent subluxation/dislocation of the patella (31).

The trochlear surface of the femur is divided into medial and lateral facets. The contour of the trochlear parallels that of the patella. The lateral trochlear predominates, having a greater radius and articulating surface. The lateral trochlear provides a bony buttress that prevents lateral patellar displacement. Lateral femoral condylar hypoplasia results in an abnormally shallow trochlear sulcus and is associated with patellar instability (see below) (36,73,81).

Patellar Articular Cartilage

The articular cartilage of the patella in areas is the thickest in the body, and has served as a model for the study of the normal composition and structure of articular cartilage as well as for investigating cartilage degeneration, osteochondral injuries, and cartilage biomechanics (31). Disorders of the articular surface of the patella and that of the corresponding trochlear articular surface of the distal femur (e.g., chondromalacia, osteochondral fractures) assume a predominant role in any consideration of the patellofemoral joint. Given the unique opportunity that the relatively thick and readily demonstrable patellar articular cartilage has provided for studying cartilage injury and degeneration, the basic composition, organization, and biomechanics of articular cartilage will be briefly reviewed to provide the interested reader with the requisite background for understanding articular cartilage pathology as well as the present imaging approaches directed at its depiction.

Basic Composition, Organization, and Biomechanical Considerations

Articular cartilage consists of cells (e.g., chondrocytes) embedded in an abundant extracellular matrix (82). In contrast to other parenchymal tissues, the cellular component of cartilage contributes relatively little to the total volume (10% or less) (83). The important biomechanical properties of cartilage, including loadbearing, resilience, and durability, depend principally on the chemical nature and complex spatial arrangement of the constituent components of the extracellular matrix (82–84).

Water constitutes the largest component of the extracellular matrix with tissue fluid accounting for 60% to 80% of the wet weight of cartilage (82–86). This high water content distinguishes articular cartilage from most other connective tissues (82). Approximately 90% of the water is extracellular. The water content of articular cartilage is generally highest next to the articular surface, although its variation in successive layers is not large (82,85). Structural macromolecules including collagens, proteoglycans, and noncollagenous proteins or glycoproteins comprise 20% to 40% of the weight of cartilage and function to maintain and organize the tissue fluid and impede its flow through the matrix, accounting for the biomechanical properties of cartilage (17,84).

Collagens form the fibrillar meshwork that gives cartilage its tensile strength and form (82–86). Proteoglycans, which form the major macromolecules of cartilage ground substance, bind to the collagen meshwork or become mechanically trapped within it (82,85). The proteoglycans are highly viscous, hydrophilic macromolecules capable of entraining an amount of water 50 times their weight in free solution (84). When immobilized as in cartilage by a meshwork of strong collagen fibrils, they can resist very large compressive loads (84). Proteoglycans are composed of a protein core with attached chains of sulfated polysaccharides that radiate out from the core (Fig. 6.14) (82,83). Within the cartilage matrix, the proteoglycans form aggregates which help to immobilize them within the collagen fibrillar meshwork (Fig. 6.14) (84). These aggregates comprise 100 to 200 proteoglycans monomers attached to a single backbone of hyaluronic acid (82,84). Link proteins stabilize the association between proteoglycan monomers and the hyaluronic acid backbone (82,83).

The physical and chemical properties of the proteoglycans are important to the resiliency of the articular cartilage; their mechanical properties are governed by both the composition of the tissue and the movement of interstitial fluid within the tissue (17,82,84). The proteoglycans, which are compressed to only a fraction of their natural aqueous volume by the surrounding meshwork of collagen fibrils, contribute to the turgor of articular cartilage and its ability to resist compressive loads (84). This resistance to compression in large part relates to electrostatic properties of the proteoglycans (82,84). The glycosaminoglycan chains which comprise up to 95% of the proteoglycan molecules (82) are characterized by a large number of negative charges. As a consequence, they exert an enormous electronegative effect, causing the monomers to remain extended in space and to resist compression into a smaller volume (82,84). With the application of a compressive force, the negatively charged glycosaminoglycan chains are driven closer together (Fig. 6.15). Interstitial fluid is extruded and flows through the interstices of the matrix along the pressure gradient created by the load (17). The compressed negatively charged proteoglycan macromolecules exert a resistive force because of electrostatic charge repulsion, ultimately resulting in a state of equilibrium whereby the matrix-resistive force caused by tissue consolidation balances the applied compressive forces (17,82,84). The movement of fluid through the tissue, which is dependent on the frictional interactions between the fluid and the "branches" of the proteoglycan macromolecules, governs both the rate and absolute amount of deformation (82). With release of the load, the molecules are allowed to expand again (82).

In compression and fluid expression experiments, a higher permeability for fluid flow has been found in the superficial cartilaginous zone than in the deep zone, a finding which likely reflects the lower concen-

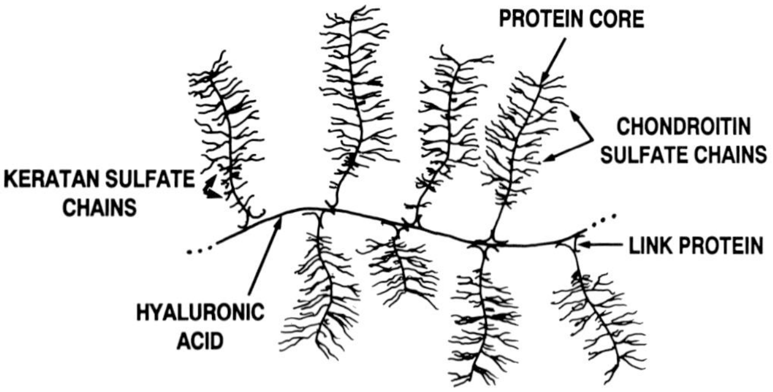

FIG. 6.14. Proteoglycan aggregate. Schematic drawing of a proteoglycan aggregate. Multiple proteoglycan monomers noncovalently associate with a central hyaluronic acid filament via link proteins to form an aggregate. The monomers consist of a central protein core with multiple covalently bound chondroitin and keratin sulfate chains. (Adapted from Buckwalter JA. From ref. 82, with permission.)

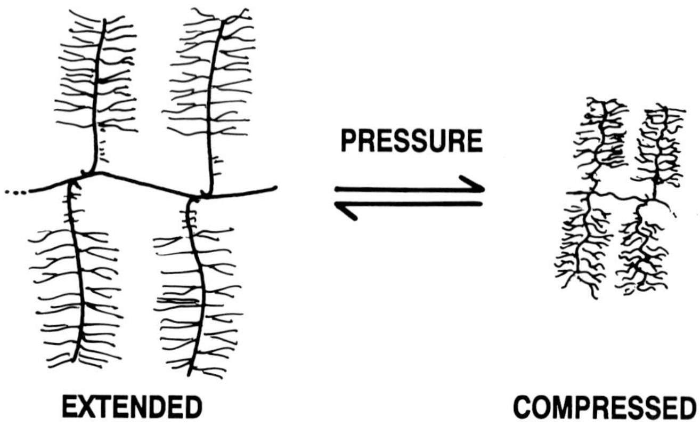

FIG. 6.15. Diagrammatic representation of the reversible compressibility of proteoglycan monomers in solution. The negative charges of the glycosaminoglycan chain repel each other, expanding the domain of the proteoglycans. There is decreased charge density and decreased density of chondroitin sulfate chains in the expanded state (*left*). With the application of compression, the negatively charged chains are driven closer together. With release of compression, the molecules are again allowed to expand. In tissue, as opposed to in solution, the collagen framework limits the expansion of the proteoglycan monomers. (Adapted from Buckwalter JA. From ref. 82, with permission.)

tration of proteoglycans in the former (17). Thus, as a consequence of its higher hydraulic permeability and fluid flow to nonweightbearing areas, the superficial layer appears to act as a cushion, initially absorbing and distributing the impact of a compressive force (17). Only with higher or prolonged strain is the compressive force also translated to the deeper cartilage zones. In contrast, surface fibrillated cartilage demonstrates far greater deformation as a consequence of permeability increases secondary to cartilage disruption, with resulting decreased resistance to fluid motion (17).

In tissue, the collagen meshwork limits the expansion of proteoglycans which could, if fully extended, fill a volume many times larger than the tissue that contains them (82). If the collagen fibril meshwork is disrupted, as may occur with chondral injuries, the matrix swells as proteoglycans expand, increasing the concentration of water and decreasing the proteoglycan concentration (82,84). The ability of MR to detect regional differences in bulk water may form the basis for noninvasive imaging of such structural alterations. Indeed, MR has demonstrated signal changes in cartilage reflecting the effects of compressive loading in experimental studies (17).

One of the unique features of articular cartilage, in distinction to other hyaline cartilages, is the highly ordered structure that changes from joint surface to subchondral bone (82–85). Cell shape and size, collagen fibril diameter, collagen fibril orientation relative to the articular surface, and matrix water content and proteoglycan content all change with increasing depth from the articular surface (82–85). The arrangement of the collagen fibril ultrastructure, in particular, reflects the functional requirements of articular cartilage (84). Four distinct layers or zones have been described and are referred to as the superficial or tangential zone, the middle or transitional zone, the deep or radial zone, and the zone of calcified cartilage (Fig. 6.16) (82–84).

The lamina splendens (LS) represents a thin, cell-free layer of matrix that grossly corresponds to a fine, clear film that can be mechanically stripped from the articular cartilage (82,83,85). It consists of a tightly packed, tangentially arrayed layer of fine fibrils and may function in the boundary lubrication mode of synovial joints (82,83). Deep to the lamina splendens, the elongated chondrocytes arrange themselves so that their long axis lies parallel to the articular surface (82,85,86) This zone is referred to as the superficial or transitional zone (TAN). Within this zone the collagen fibers are arranged in tight bundles oriented parallel to the articular surface (Fig. 6.16) (82,84,85). This orientation of the collagen fibrils reflects the functional requirements of the tissue (84). The principal mechanical properties of the collagen fibrils are their tensile stiffness and strength. As such, they make their greatest contribution when oriented in line with the axis of tensile load (84). The spatial arrangement of the collagen in this zone reflects the large tensile stresses that develop parallel to the articular surface with compressive loading.

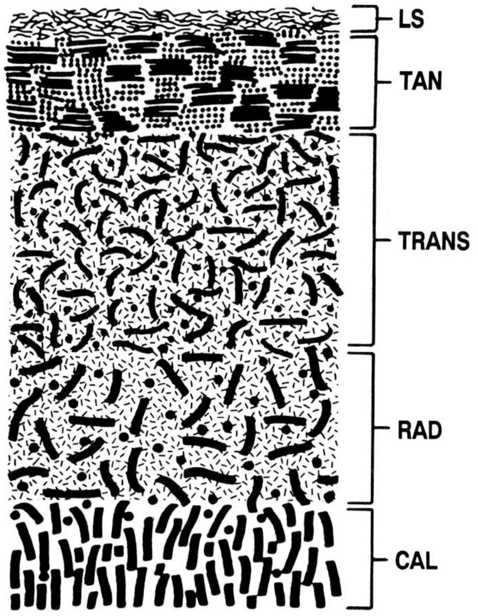

FIG. 6.16. Schematic representation of the organization of articular cartilage. The lamina splendens (LS) is a layer of fine fibrils several micra thick that covers the articular surface. The tangential zone (TAN) consists of closely arranged bundles of collagen fibers arranged parallel to the articular surface. The collagen fibers in the transitional zone (TRANS) are randomly arranged. Collagen fibers of the radial zone (RAD) are larger in diameter but also randomly arranged. The collagen fibers of the calcified zone (CAL) are arranged perpendicular to the articular surface. (Adapted from Lane JM, Weiss C. From ref. 82, with permission.)

The transitional zone (TRANS) lies directly beneath the superficial zone and comprises approximately 40% to 60% of the tissue's thickness (Fig. 6.16) (82,84,86). The collagen fibers are larger than in the superficial zone, loosely packed, randomly oriented, and homogeneously distributed (82,86). The deep or radial zone (RAD) usually forms the largest component of articular cartilage. Collagen fibers of the radial zone are also randomly arranged, but are of larger diameter. This zone contains the largest collagen fibrils of articular cartilage and the highest proteoglycan content (82). The water content, which progressively decreases from the superficial to deep zones, is lowest in this layer (87). This distribution of water in part reflects a counterbalance between the osmotic effect of the proteoglycans and the tension effect of the collagen fiber scaffold (82,84).

In the calcified zone and in the deepest area of the radial zone, the orientation of the collagen fibers is perpendicular to the subchondral bone (82). Collagen fibrils of the radial zone penetrate directly into the calcified cartilage. The "tidemark" represents a histologically demonstrable line dividing the radial and calcified zones and may function as an anchor for the collagen fibrils (82,84). No structural continuity between hyaline cartilage and subchondral bone is present and the collagen fibers of the calcified zone are not anchored to bone. The attachment of hyaline cartilage to subchondral bone has been likened to that of a jigsaw puzzle whereby the interlocking of the irregular surfaces of the calcified zone and underlying bone provides the necessary attachment (85).

MR Imaging of Articular Cartilage

The reported appearance of articular cartilage, and the ability of MR to depict chondral defects and biomechanical changes, has varied among the various approaches adopted by different investigators. These methods have included T1-weighted (short TR/TE), T2-weighted (long TR/TE), various gradient-echo, fat suppression, and inversion recovery pulse sequences, as well as T1-weighted sequences in association with gadolinium-enhanced MR arthrographic techniques (2,17–20,88–93). No method to date has proven clearly superior. The challenge has been to depict both the early changes contained within the cartilage, and the later changes manifest at the articular surface and at the underlying subchondral bone interface. The patella, which has the thickest articular cartilage in the body, has served as the model for many of the investigations of articular cartilage.

Utilizing conventional T1-weighted sequences, normal articular cartilage most commonly demonstrates a homogeneous intermediate signal intensity (Fig. 6.17). A subtle bilaminar appearance, with the signal intensity from the basal two-thirds of the cartilage being lower than that of the superficial level, has been reported (19), although this has been more apparent in vitro (disarticulated patella cadaver specimens) than in vivo and has not been substantiated in other studies (17,19). No zonal differences in the appearance of articular cartilage have been routinely evident in our experience on T1-weighted sequences in high resolution clinical imaging of the patella. On proton density and T2-weighted sequences, articular cartilage has also most consistently demonstrated a homogeneous appearance without reflection of the zonal architecture and differential water distribution of the cartilage (2). In an experimental study utilizing bovine patella, however, a bilaminar appearance to the cartilage was described using long TR/TE and heavily T1-weighted inversion recovery sequences (17). The bilaminar appearance consisted of a superficial zone, characterized by longer T1 and T2 values, correlating with the tangential and transitional zones of normal articular cartilage, and a second MR imaging zone, with shorter T1 and T2 values, corresponding to the deeper layers of the articular cartilage. This zonal signal intensity pattern was inverted on heavily T2-weighted spin-echo sequences and the findings correlated with differences between the better hydrated superficial zone and lower water content of the deeper zone (17). This zonal pattern could not be demonstrated with short TR/TE and T2*-weighted sequences in this study (17).

In a more recent high resolution MR study of human articular cartilage, a trilaminar appearance to the cartilage was identified utilizing T1- and T2-weighted spin-echo images (94). The major difference between this and previous studies related to the depiction in this study of a superficial low signal intensity zone which corresponded in location, although not exactly in thickness, to the superficial zone in correlative histological sections. The depiction of this superficial zone appears dependent on high spatial resolution, and was more consistently depicted on experimental imaging of core cylinders of cartilage in which partial volume effects were minimized (94). A second zone of intermediate to high signal intensity corresponded approximately to the transitional zone in histological sections. A third zone, which was depicted as a broad band of low signal intensity, corresponded approximately to the deep radiate and calcified cartilage zones and the adjacent subchondral bone. The apparent thickness of this latter zone is directly affected by chemical shift effects and thus the direction of the frequency-encoding gradient. The determinants of the signal intensity in the different zones remain incompletely understood. The authors suggest that factors other than differences in water concentration (which they believe are insufficient) must be considered in accounting for the MR appearance of articular cartilage, including the possibility that differences in orientation of collagen fibers may affect signal intensity by means of their effect on magnetic susceptibility (94).

With gradient-echo techniques, image contrast is dependent not only on tissue relaxation times (T1 and T2) and timing parameters (TR and TE), but is additionally affected by pulse flip angle, the presence or absence of steady state conditions, and the mode of acquisition (sequential vs. interleaved). A complete discussion of contrast phenomenology for gradient-echo imaging is well beyond the scope of this text and the reader is referred to many excellent sources (95–98). For purposes or articular cartilage and joint assessment, gradient-echo techniques emphasizing T2* contrast which provide an arthrogram effect are of greatest interest. Under steady state conditions, in which the persistence of the transverse magnetization determines the degree of signal intensity, an arthrogram effect (high signal from long T2 joint fluid) can

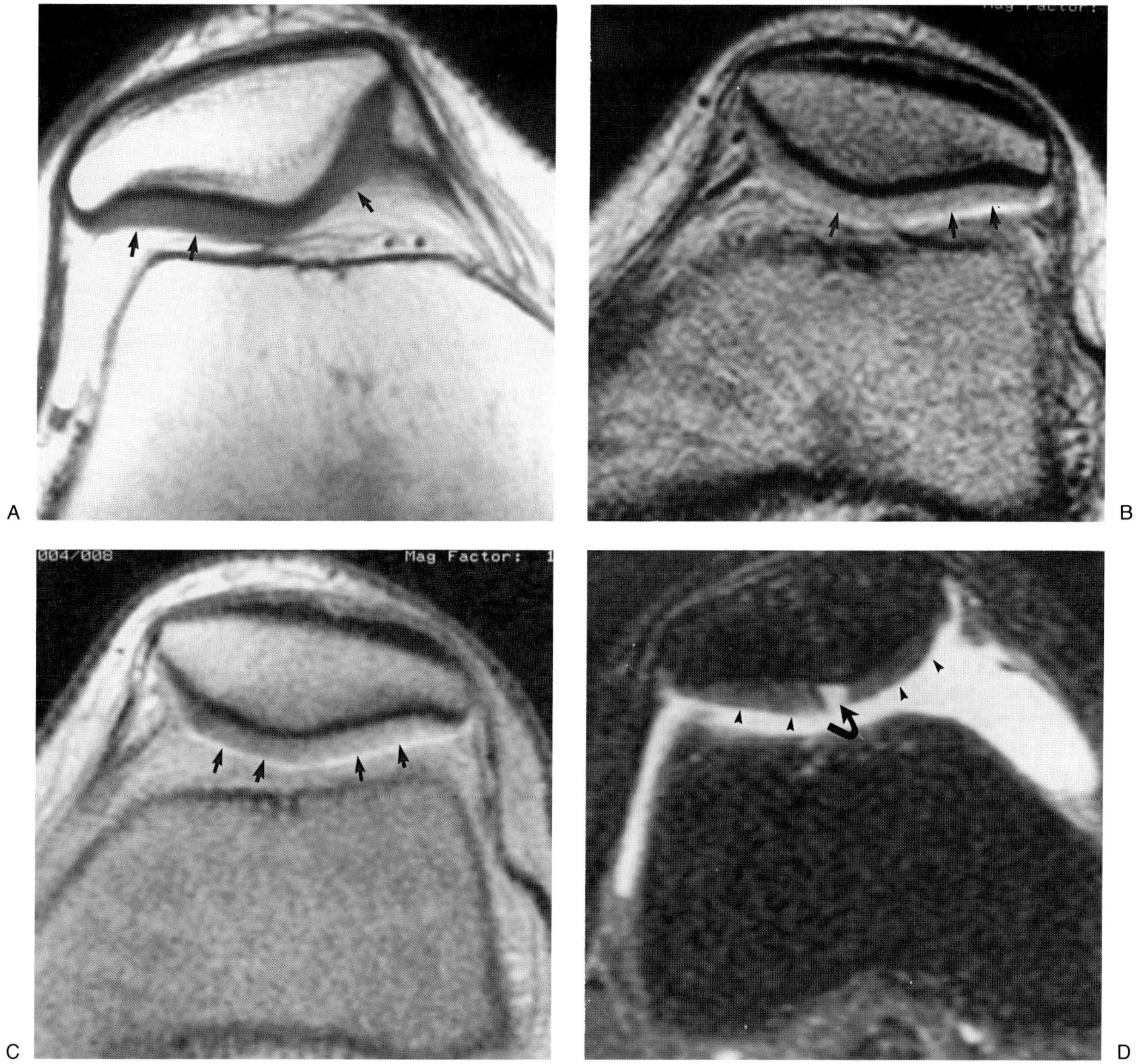

FIG. 6.17. Normal patellar articular cartilage. **A:** Axial TR/TE 500/20. On this T1-weighted magnified view of the patella, the articular cartilage is well depicted and demonstrates an homogeneous intermediate signal intensity (*arrows*). **B:** Axial TR 33 TE 6 Flip angle 45°. Single acquisition gradient recalled sequence obtained under steady state conditions (GRASS). The articular cartilage demonstrates uniform signal intensity (*arrows*) and is well contrasted against high signal intensity joint fluid. **C:** Axial TR/TE 600/200. Fat suppression image. A frequency selective presaturation pulse was applied to fat immediately prior to the excitation pulse, effectively destroying the longitudinal magnetization of fat. With this technique the articular cartilage is of uniform, slightly increased signal intensity (*arrowheads*). **D:** On this STIR sequence, the cartilage is of uniform slightly decreased signal intensity (*small arrowheads*), and it is well contrasted against low signal intensity subchondral bone and high signal intensity joint fluid. A localized full thickness cartilaginous defect is well depicted (*curved arrow*).

be produced with short TR times (20–50 msec) and relatively high flip angles. High quality T2* images can also be produced in the absence of steady state conditions with relatively longer TR (>> 200 msec) and TE times and small flip angles. In general, regardless of the flip angle chosen, T2* weighting increases with an increase in TE. As previously mentioned, the acquisition mode will also affect image contrast with gradient-echo techniques. With sequential acquisition modes that preserve the steady state (e.g., GRASS), joint fluid will demonstrate an arthrogram effect utilizing short TRs (TR less than T2) and moderate to high flip angles. With slice interleaved (multislice) GRASS (MPGR), the longer TR times required for the multislice mode contribute to spoil the steady state condition. With TR times in excess of 200 msec, MPGR sequences with short TE times and small flip angles produce density weighted images, with large flip angles, T1-weighted images result. Konig et al., utilizing a low flip angle (12°) gradient-echo sequence (FLASH), reported this sequence superior to spin-echo images in depicting early cartilage degeneration (90). The signal changes seen on MR (areas of decreased signal within the otherwise high signal intensity cartilage) correlated with a localized reduction of proteoglycan content on histological evaluation of the specimens.

Chemical shift fat suppression techniques have also been applied to depiction and characterization of articular cartilage (88). Two principal approaches, *phase contrast* and *selective presaturation*, have been most extensively investigated (99). The most widely known phase contrast technique is that described by Dixon, which incorporates two separate imaging sequences. Two identical spin-echo sequences are performed with the 180° refocusing pulse of the second sequence applied at a short time interval earlier than that of the first sequence, producing an out of phase, or opposed, echo (99,100). Adding the initial images (in phase) and opposed images produces an image of the predominant spectral component, while subtracting the opposed from the in-phase image produces an image of the minor component (99,100). As a consequence, the Dixon method image achieved is based on spectral components, but it does not produce a fat or water image per se (phase ambiguity) (99).

Modifications of the Dixon method incorporating phase correction have been devised in an attempt to overcome some of the limitations mentioned (99,101). These methods are capable of producing true fat and water images and are accomplished by applying image processing techniques to the phase information in the opposed image (99,101). Chopper averaging, a technique employed for combining data from multiple acquisitions, can also be utilized to enhance the Dixon method and eliminate misregistration artifact between acquisitions (99,102). Phase information is preserved as the data from the opposed and in-phase acquisitions are combined prior to reconstruction. True fat and water images are generated for each acquisition.

The other principal method for accomplishing chemical shift imaging is *selective presaturation* (99). With this technique, a frequency selective presaturation pulse is applied to water or fat immediately before the excitation pulse. This presaturation pulse effectively destroys the longitudinal magnetization of the water or fat, and when the excitation pulse is applied, there will have been insufficient time for longitudinal recovery (99). As such the "presaturated" component does not contribute to the final image. With presaturation techniques (CHESS, ChemSat), section coverage is decreased although imaging times are not necessarily increased. While ideal spectral selectivity may not be accomplished, separate fat and water images are available with this method (99). With presaturation techniques, the articular cartilage of the patella demonstrates uniform, slightly increased signal intensity that is well contrasted against high signal intensity joint fluid (Fig. 6.17).

Totterman (88) utilized chopper fat suppression (CFS), which is a phase-sensitive implementation of the Dixon fat suppression technique, to accomplish significant improvement in visualization of the hyaline cartilage of the knee in comparison to conventional spin-echo sequences (88). Utilizing a long TR/TE (TR 1,500–2,000, TE 60–80) CFS sequence, hyaline cartilage appeared as an intermediate signal intensity structure and was well contrasted against high signal intensity joint fluid. Konig et al. compared the use of standard spin-echo (SE) images with "water" images obtained utilizing a chemical shift selective sequence (CHESS) for visualization of hyaline articular cartilage (90). With CHESS, hyaline cartilage demonstrated high signal intensity and was well contrasted against low signal intensity subchondral bone, synovial tissues, and ligamentous structures, although contrast between joint fluid and hyaline cartilage was better accomplished in this study with conventional long T2-weighted images. CHESS proved superior to SE sequences in depicting early cartilage degeneration manifest on MR as localized decreases in cartilage signal intensity (90). These signal changes correlated histologically with localized reductions of proteoglycan content within the cartilage (90).

The authors have have considerable experience utilizing short tau inversion recovery (STIR) sequences for purposes of accomplishing fat suppression and evaluating articular cartilage. With this pulse sequence, "water" protons are made to dominate the image by utilizing a short inversion time (TI) recovery pulse sequence and adjusting TI to the null crossing point of the longitudinal magnetization of fat (88). With

STIR imaging, the effects of T1 and T2 relaxation are additive, contributing to increased lesion conspicuity. This pulse sequence allows sensitive depiction of the subchondral bone, articular cartilage, and the articular surface, which is particularly well delineated when contrasted against high signal intensity synovial fluid. Normal articular cartilage demonstrates a relatively homogeneous intermediate signal intensity appearance utilizing this pulse sequence (Fig. 6.17). This technique has been quite effective, in the authors' experience, for assessment of the relatively thick articular cartilage of the patella. We routinely obtain an axial STIR sequence as a complement to the basic sagittal and coronal spin-echo sequences on all studies of the knee (see above).

Experimental Chondral Lesions

The ability of MR to depict experimentally produced cartilage abnormalities has been evaluated in a number of investigations (17,18,20). In an early study utilizing relatively limited in-plane resolution (0.95 mm/pixel) Wojtys and colleagues were able to depict 15 of 16 experimentally produced chondral lesions of cadaver patellae that were 4 mm or greater in width and all those 3 mm and greater in depth. In addition, seven of nine lesions 1 mm in depth were identified in this group (20). Using a cadaveric model, Handelberg et al. detected all experimentally created lesions which ranged from 0.8 to 5 mm in diameter and 1 to 2 mm in depth (18). The images were obtained on a high field strength system (1.5 T) utilizing a T2-weighted spin-echo sequence, and the superficial defects were sharply contrasted against high signal intensity joint fluid. Double contrast CT-arthrography, performed after the MR examination, missed 50% of the 1.5 and 2 mm lesions and detected none of the 0.8 mm holes. In contrast to this favorable experience, Gylys-Morin reported significant limitations for detecting experimentally created defects in the femoral condyles utilizing standard spin-echo imaging and advocated the use of intraarticular gadolinium, in the presence of which all lesions 2 mm or greater could be detected (89). While the results of these investigations are quite impressive with regard to articular cartilage lesion detectability, they may not directly correlate with the ability to detect naturally occurring chondral lesions, as the sharp margins produced by the drill holes of experimentally created chondral defects may enhance their detection as compared to naturally occuring cartilage abnormalities.

MR has also been utilized to study the repair characteristics of articular cartilage in experimental models of degenerative arthritis (103). Braunstein et al. demonstrated relative increased thickness of articular cartilage in dogs 3 years following anterior cruciate ligament (ACL) transection (103). This hypertrophic response of articular cartilage is consistent with biochemical observations of increased matrix production manifest by net increased proteoglycan synthesis and concentration in cartilage matrix following ACL transection (103). Using T2*-weighted images, the signal intensity of the hypertrophied cartilage remained similar to that of the contralateral control knee, an observation the investigators attributed to increased proteoglycan concentration, rather than merely to an increase in bulk water content contributing to the increased thickness of the repaired cartilage (103).

Chondromalacia

Background Considerations

In the strictest sense, the term chondromalacia refers to "softening" of the articular cartilage of the patella (31,79,105–113). However, the term has become poorly defined and has unfortunately been applied in a more generic manner to encompass a wide spectrum of disorders associated with anterior knee pain (31,114,115). As such, the voluminous literature on the subject of patellar chondromalacia is replete with contradictory observations on clinical presentations, incidence, location, extent, and foremost, etiology and relationships of chondromalacia to other patellofemoral disorders (114–116). As a clinically defined syndrome, no constant correlation between histological retropatellar articular cartilage changes and clinical symptomatology has been observed (117,118).

Outerbridge described the earliest finding in chondromalacia as the change in appearance of the articular cartilage from its normal bluish-white, smooth, glistening surface to a dull and even slightly yellowish-white appearance (108,109,119). Upon probing, Outerbridge described the cartilage to be soft and readily depressed and the condition to be most commonly encountered along the medial facet (Fig. 6.18). Ficat and Hungerford referred to this initial stage as "closed" chondromalacia and described its possible progression to a localized surface swelling or "blister" with the overlying cartilage remaining grossly intact (Fig. 6.18) (79,80). While corroborating the gross morphology described by Outerbridge, these authors described the most common location as along the lateral facet. Insall et al. also described the initial stages of chondromalacia as swelling and softening ("closed"), but described the most common location based on the observation of 105 patellae as being the mid-point of the median ridge with extension equally onto the medial and lateral patellar facets (107). Over 70% of the lesions were contained within an ellipse located transversely across the central area of the patella, with the

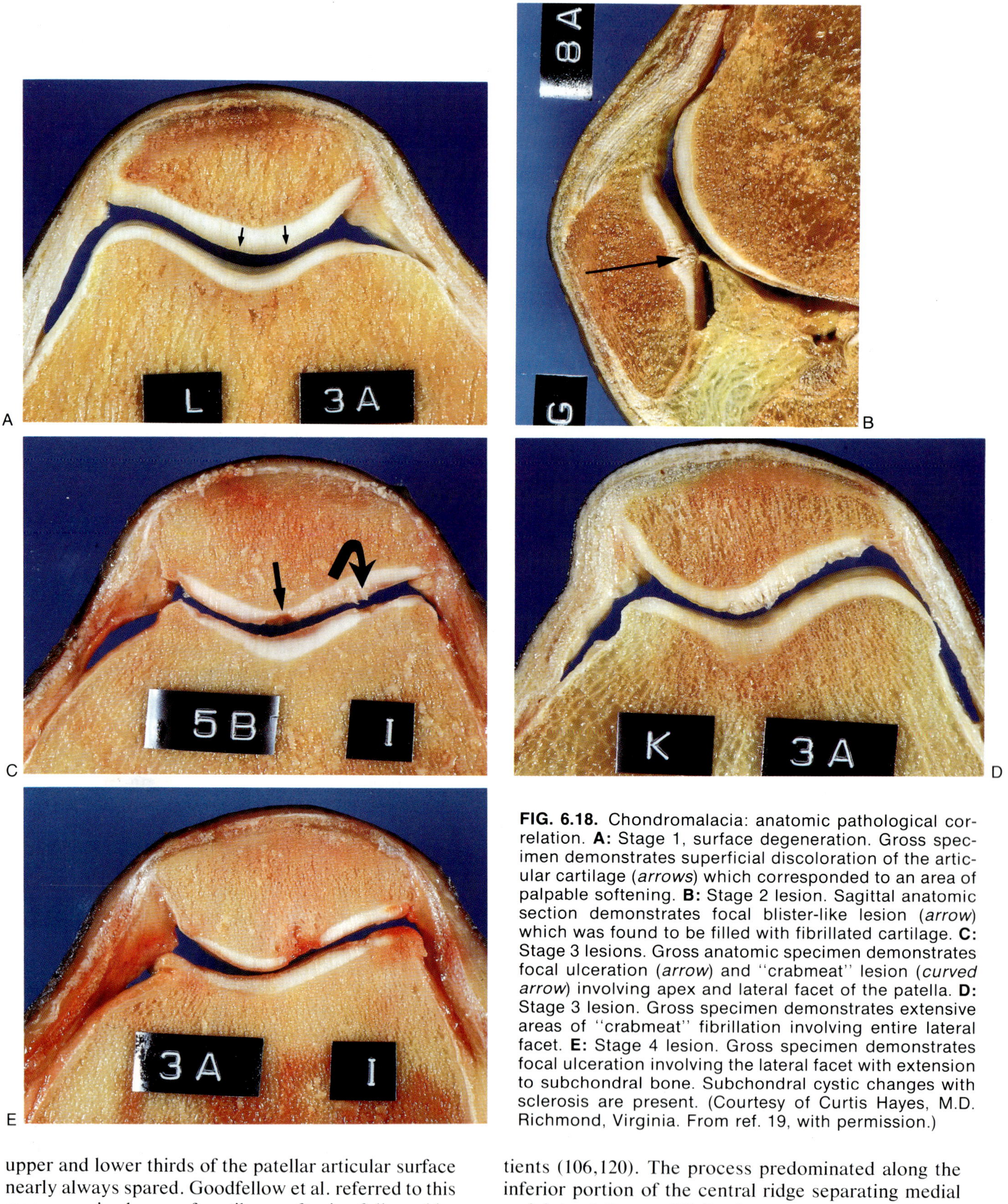

FIG. 6.18. Chondromalacia: anatomic pathological correlation. **A:** Stage 1, surface degeneration. Gross specimen demonstrates superficial discoloration of the articular cartilage (*arrows*) which corresponded to an area of palpable softening. **B:** Stage 2 lesion. Sagittal anatomic section demonstrates focal blister-like lesion (*arrow*) which was found to be filled with fibrillated cartilage. **C:** Stage 3 lesions. Gross anatomic specimen demonstrates focal ulceration (*arrow*) and "crabmeat" lesion (*curved arrow*) involving apex and lateral facet of the patella. **D:** Stage 3 lesion. Gross specimen demonstrates extensive areas of "crabmeat" fibrillation involving entire lateral facet. **E:** Stage 4 lesion. Gross specimen demonstrates focal ulceration involving the lateral facet with extension to subchondral bone. Subchondral cystic changes with sclerosis are present. (Courtesy of Curtis Hayes, M.D. Richmond, Virginia. From ref. 19, with permission.)

upper and lower thirds of the patellar articular surface nearly always spared. Goodfellow et al. referred to this macroscopic change of cartilage softening followed by blister formation as basal degeneration and considered this to represent a localized posttraumatic form of cartilage injury which usually occurred in younger patients (106,120). The process predominated along the inferior portion of the central ridge separating medial and lateral facets. While the consensus prevails regarding the description of the initial stages (grades 1–2) of the condition, considerable disagreement about the most common location of involvement exists. As previously discussed, the relationship of these changes to clinical symptomatology is at best inconsistent.

The initial stages of "closed" or surface intact chondromalacia may progress toward the loss of surface integrity with the development of cracks, fissures, fibrillation, and eventually gross ulceration extending into subchondral bone (Fig. 6.18) (31,106,121). Various investigators have employed different terminology and grading systems to describe this sequence of events. Ficat and Hungerford distinguished between "closed" or surface intact chondromalacia (grade 1) and "open" (grade 2) chondromalacia in which fissures and ulcerations were present (79,80). Insall et al. distinguished between fissuring (grade 2) and fibrillation (grade 3), although the criteria were qualitative (107). Goodfellow et al. distinguished between the previously described basilar degeneration and superficial degeneration, a second "type" of chondromalacia most commonly encountered along the odd facet of the patella and thought related to decreased contact (106,120). In superficial degeneration, initial changes were manifest at the surface with flaking progressing to fibrillation and fissure formation. While the early stages of these two "types" of chondromalacia differed according to Goodfellow, the more advanced stages were quite similar at gross examination, with ulceration and gross fibrillation (grade 3) and crater formation and subchondral eburnation (grade 4) predominating. Shahriaree incorporated these concepts into an arthroscopic grading system (121,122). In this four-part system, grade 1 correlates with the initial stage of softening and "swelling" of the articular cartilage (Fig. 6.18A). Grade 2 represents progression of surface deformity with "blister" formation (Fig. 6.18B). In grade 3, open surface changes are apparent with surface irregularity and cartilage fibrillation (Fig. 6.18C,D). In grade 4, frank ulceration progresses to subchondral bone (Fig. 18E). While no grading system has received universal adoption and significant shortcomings of each have been described, it is conceptually important and clinically valuable in interpreting MR images to appreciate the sequential stages through which the condition progresses.

Histological and ultrastructural studies have demonstrated that initial changes in chondromalacia patellae occur within the matrix below the articular surface (111). This contrasts with the pathogenesis of osteoarthritis in which the earliest changes are observed at the cartilage surface, where collagen fiber bundles fray in the superficial zone (104). In the earliest stages of closed chondromalacia (e.g., softening), the initial findings are seen in the transitional zone (31,111). The surface tangential network of collagen fibers is spared at this stage. The most conspicuous matrix changes are seen in the form of matrix streaks, which arise from the chondrocyte lacunae and represent local reorientation of the collagen fibers from the normal three-dimensional distribution of fibers into tightly packed (collapsed) segments (111). There is an apparent overall decrease in proteoglycan from the streaks (111). The soft, edematous quality of this cartilage may be explained by the loss of an intact collagen network that would allow proteoglycan aggregates to expand into a larger molecular domain. The ultrastructural changes observed support a repeated stress- or load-induced etiology for the cartilage changes (111). The pericellular matrix appears particularly vulnerable to stress-induced change, either because it is inherently weaker, or perhaps because the lacunae act as mechanical stress-risers.

MR Imaging

The appearance of chondromalacia and the ability of MR to depict and characterize the extent of the condition appear to vary with the techniques employed to depict the changes. With standard T1- and T2-weighted spin-echo techniques, MR has reliably demonstrated the more advanced stages of chondromalacia, but difficulty with regard to depiction of the earlier grades has been reported by many investigators (2,19). In a study of naturally occurring chondral lesions in 14 cadavers, as well as in a prospective clinical series, Hayes et al. (123) used high resolution T1-weighted sequences but were unable to correlate MR signal changes with areas of cartilage softening (grade 1 chondromalacia). Areas of subtle hypointensity within the cartilage were commonly identified in both symptomatic and asymptomatic individuals in this series. However, these areas did not correlate with arthroscopic findings and their significance remains uncertain. This difficulty in detecting the initial stages of chondromalacia has been corroborated by other investigators. Yulish compared the MR and arthroscopic findings in 19 patients (12 with MR prior to arthroscopy and 7 with MR following arthroscopy) (2). There was agreement between MR and arthroscopy in 17 of the 19 cases. MR was unable to demonstrate changes in one patient in whom cartilage softening was noted on arthroscopy, and was false-positive for an apparent cartilage fracture in another patient. The authors described areas of "hypointensity with swelling" on relatively T1-weighted images which they speculated might represent the MR correlates of early basal degeneration as described by Goodfellow. These could not be corroborated at arthroscopy, thus their significance at this time remains undetermined.

In a prospective clinical series including 54 knees examined by MR prior to arthroscopy, Handelberg reported an overall MR accuracy of 81.5%, with a sensitivity of 100% but a specificity of only 50% due to

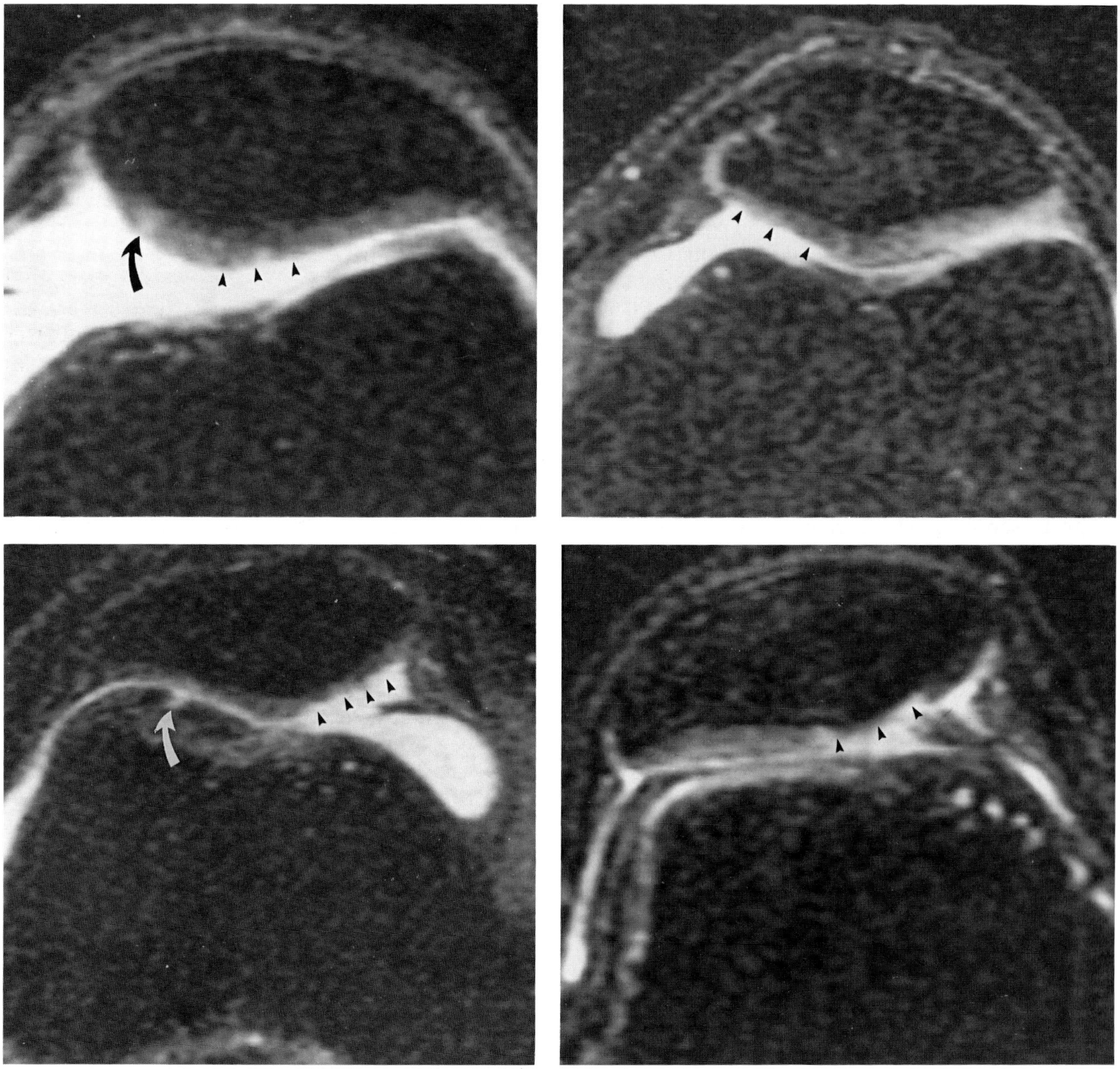

A B C D

FIG. 6.19. Chondromalacia. **A:** Axial STIR TR 2200 TE 35 TI 160. A faint globular focus of signal is seen within the articular cartilage of the medial facet (*curved arrow*). This reportedly correlated with an area of "softening" at arthroscopic probing. At the median ridge and extending along the lateral facet, superficial signal changes are identified and correlated with minimal softening and fibrillation (*arrowheads*). **B:** Axial STIR TR 2200 TE 35 TI 160. There is mild thinning of the articular cartilage overlying the medial facet of the patella. Gentle undulations are seen on the articular surface; these are well contrasted against the fluid within the suprapatellar bursa (*arrowheads*). The findings reflect mild chondromalacia. **C:** Axial STIR TR 2200 TE 35 TI 160. Along the medial facet there is more extensive irregularity of the articular surface with a roughly serrated appearance (*arrowheads*). This correlates with early "crabmeat" fibrillation at arthroscopy. Mild increased signal is also seen within the cartilage of the lateral facet. Cartilage loss with erosive changes manifest as areas of increased signal intensity extend into the subchondral bone along the lateral trochlear surface (*curved arrow*). **D:** Axial STIR TR 2200 TE 35 TI 160. There is extensive full thickness articular cartilage loss centered along the medial ridge (*arrowheads*). The findings correlated with extensive ulceration of the articular cartilage with findings extending to the level of subchondral bone.

false-positive examinations (18). Again the problem related to early stage disease, with the false-positive MR diagnoses related to the depiction of linear areas of decreased signal intensity within the deep zones of the articular cartilage. The authors suggested that these might represent the correlates of very early histological aberrations. Arthroscopically, however, these findings did not correlate with cartilage softening or other visual evidence of grade 1 chondromalacia. Histologically, as previously discussed, the initial changes in chondromalacia are seen in the transitional zone. Clarification of the significance of these low signal intensity foci awaits further MR/histological correlation.

While difficulties have been encountered in correlating MR signal changes with the initial stage of chondromalacia, MR has been reasonably good at depicting the more advanced and arguably more clinically significant lesions (2,18,19). On high resolution T1-weighted sequences, localized "blisters," correlating with arthroscopic grade 2 changes, have been described as low signal intensity areas extending to the cartilage surface. These areas focally bulge the affected articular surface, the margins of which, however, remain sharp (19). Cartilage fibrillation at the articular surface leading to a "crabmeat" appearance (grade 3) can consistently be demonstrated on MR utilizing a variety of approaches (Fig. 6.19) (2,18,19). On short TR/TE sequences, these changes are depicted as areas of decreased signal intensity extending to the surface, accompanied by a conspicuous loss of the normal sharp interface at the cartilage surface (19). On T2- and T2*-weighted sequences, the changes of cartilage fibrillation are clearly seen highlighted against the arthrographic effect of high signal intensity synovial fluid. These findings correlate with the arthroscopic "brush border" sign and are representative of grade 3 changes. Focal erosions may also be evident in grade 3 lesions. Grade 4 chondromalacic changes are manifest as full thickness cartilage defects extending to the level of subchondral bone accompanied by bony signal changes (19). These changes are seen most commonly as decreased intensity within the subchondral and cancellous bone underlying the affected region on T1-weighted sequences, which variably increases on more heavily T2-weighted sequences. The presence of a patellar subchondral signal abnormality invariably reflects advanced grade 3–4 chondromalacia (19).

For the past 3 years, the authors have utilized STIR sequences for evaluating the patellar articular cartilage and they prefer it to both T1- and T2-weighted spin-echo imaging as well as to gradient-echo techniques. In a comparative clinical study, this sequence was found to be far more sensitive than a T-1 weighted sequence for depiction of chondromalacia and patellar chondral defects (124). As previously discussed, normal articular cartilage demonstrates a relatively homogenous intermediate to low signal intensity appearance utilizing this sequence. Early changes of closed chondromalacia have occasionally been depicted as areas of increased signal within the articular cartilage (Fig. 6.19). Such changes may correlate with the histologically known localized increase in water content present during this stage. While the more extensive of these changes have variably correlated with areas of softening at arthroscopic probing, no histological correlation has been obtained.

It is emphasized that our experience with arthroscopically confirmed cases of early (grade 1) chondromalacia remains limited and the ability of MR to reliably depict these changes remains questioned. Early surface fibrillation, however, is well depicted on high resolution MR images, particularly in the presence of sufficient synovial fluid to provide an arthrogram effect (Fig. 6.19). The more advanced "crabmeat" changes and areas of fissuring and ulceration (grade 3) are consistently evident on this pulse sequence (Fig. 6.19). Cartilage ulceration extending into subchondral bone (grade 4) is also readily demonstrated and can be precisely mapped (Fig. 6.19). The signal changes within the bone underlying the affected facet are clearly seen as areas of increased signal intensity.

In our clinical experience, the addition of a routine axial STIR sequence through the patellofemoral joint has been invaluable for diagnosis. The technique reliably detects areas of "open" or surface-evident chondromalacia, including relatively mild fibrillation. The extent of more advanced chondromalacic change of the articular surface with fissuring, ulceration, and any involvement of subchondral bone can be accurately characterized. The configuration of the patella and the corresponding trochlear sulcus and their relationship to each other (congruence) can be assessed (see below). Abnormalities of the trochlear articular surface that may significantly impact function of the patellofemoral joint are readily seen (Fig. 6.20). The supporting retinacular structures can be directly depicted, and in cases of patellar trauma ranging from contusion to fracture and dislocation, this sequence allows sensitive detection of any and all osseous, cartilaginous, and ligamentous injuries (see below). Additionally, the axial projection is useful for both the detection and characterization of juxtaarticular processes.

Osteochondritis Dissecans

The term osteochondritis dissecans (OCD) refers to the separation of a segment of articular cartilage with or without a corresponding segment of subchondral bone, from the remainder of the underlying bone. While believed by most authorities to be posttraumatic, there is rarely a history of a single inciting event

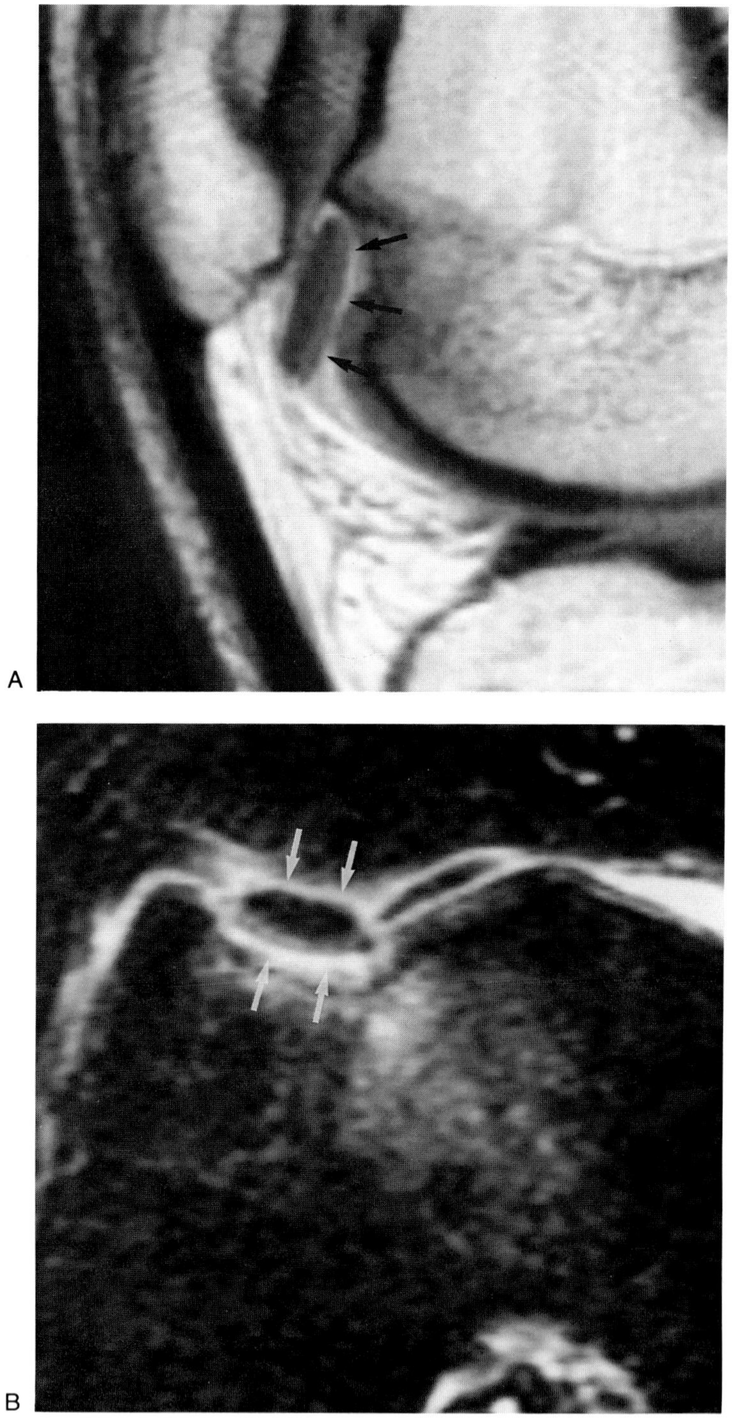

FIG. 6.20. Patellofemoral joint, large osteochondral fragment. **A:** Sagittal TR/TE 500/20. A 22-year-old woman had experienced prior direct trauma to the patellofemoral joint. A large, elliptical-shaped osteochondral fragment related to the anterolateral aspect of the lateral femoral condyle is readily depicted. The fragment has rotated 180° and the articular surface now faces the bed of origin (*arrows*). **B:** Axial STIR TR 2,200, TE 35, TI 160. The fragment is again well depicted and is surrounded by high signal intensity joint fluid (*arrows*). The site of origin from the trochlear articular surface is well defined.

(58). This is in contrast to acute osteochondral injuries as may be seen with patella dislocation. The fate of the resulting osteochondral fragment is variable and it may remain in place (*in situ*), become partially detached, or become completely detached and free within the joint (58,125,126). With completely displaced fragments, the site of origin of the lesion can be identified (Fig. 6.21) and the detached chondral or osteochondral fragment can be sought elsewhere in the joint. Osteochondritis dissecans of the patella is rare in comparison to involvement of the femoral condyles (58). Clinically, patients typically present with arthralgia and often a joint effusion. Men are more commonly affected than women and presentation is typically in the second decade (58,125). The medial facet of the patella is most commonly involved, with lateral facet lesions accounting for approximately 30% of cases (58). The upper pole is typically spared, with involvement of the middle and lower portion of the bone being most common (58).

The principal differential diagnostic considerations include chondromalacia, dorsal defect of the patella (DDP), and osteochondral fractures related to direct impaction or lateral dislocation (58). The localized nature of the abnormality and its involvement of subchondral bone distinguish osteochondritic lesions of the patella from chondromalacia. DDP is readily distinguished by its classic superolateral location. Osteochondral injuries of the patella are typically noted along the extreme medial facet and often are associated with a corresponding "kissing lesion" involving the lateral margin of the lateral femoral condyle (see below). In the acute setting, associated abnormalities of the medial retinaculum will also be present.

Dorsal Defect of the Patella

Dorsal defect of the patella (DDP) is a benign lesion most commonly encountered incidentally on conventional radiographic examinations (127). Located along the superolateral aspect of the articular surface of the patella, DDP presents radiographically as a round, well-defined radiolucency with sclerotic margins (127).

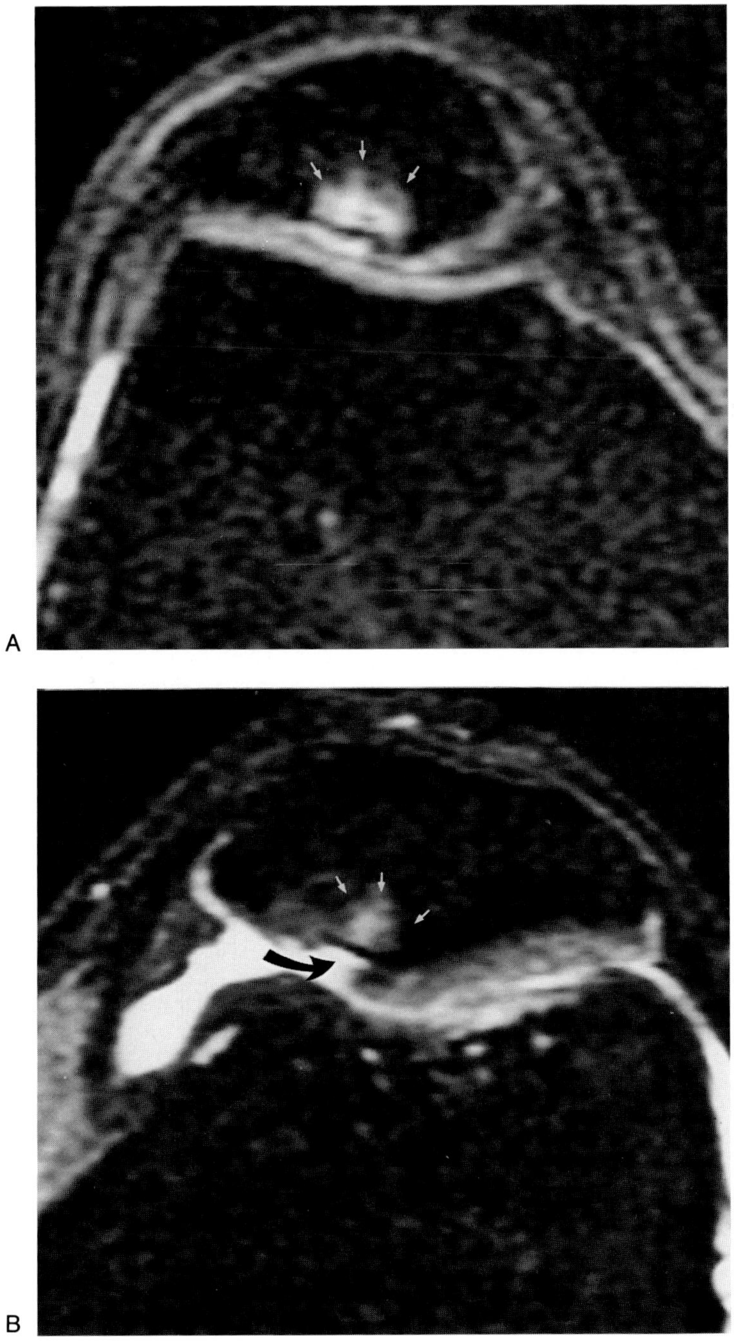

FIG. 6.21. Osteochondritis dissecans. **A:** Axial STIR TR 2200 TE 35 TI 160. The patient is a 14-year-old boy with a multiple year history of patella osteochondritis dissecans. A large osteochondral defect is at the border of the median ridge and lateral facet. There is full thickness loss of articular cartilage (*arrowheads*). The bed of the displaced fragment is filled with intermediate signal intensity granulation tissue, with signal changes extending well into subchondral bone. **B:** Axial STIR TR 2,200, TE 35, TI 160. On a slightly more cephalad section, there is a large, hyperintense geode extending well into subchondral bone (*small arrows*). The full thickness articular cartilage defect representing the displaced fragment's site of origin is well depicted (*curved arrow*).

The overlying articular cartilage is invariably intact. Histologically, DDP consists of nonspecific fibrous tissue. The lesion most commonly presents during the second decade, although the reported age range of presentation is wide. DDP is reported to frequently heal spontaneously by sclerosis. Its major importance is in recognition of its innocuous nature and in distinguishing it from differential considerations including osteochondritis dissecans, neoplasm (most commonly chondroblastoma), and Brodie's abscess (128).

On MR with T1-weighted sequences, DDP has been reported to demonstrate mixed signal intensity in relation to the remainder of the articular cartilage (129). The lesion is predominantly decreased in signal intensity with reported central areas of increased intensity. Radiographically lytic lesions have been reported to demonstrate signal increased intensity relative to cartilage on gradient-echo images and sclerotic lesions low signal intensity utilizing these sequences (129).

Patella Fracture

In the adult, the patella itself is the weakest link in the quadriceps mechanism. Injury to the patella may

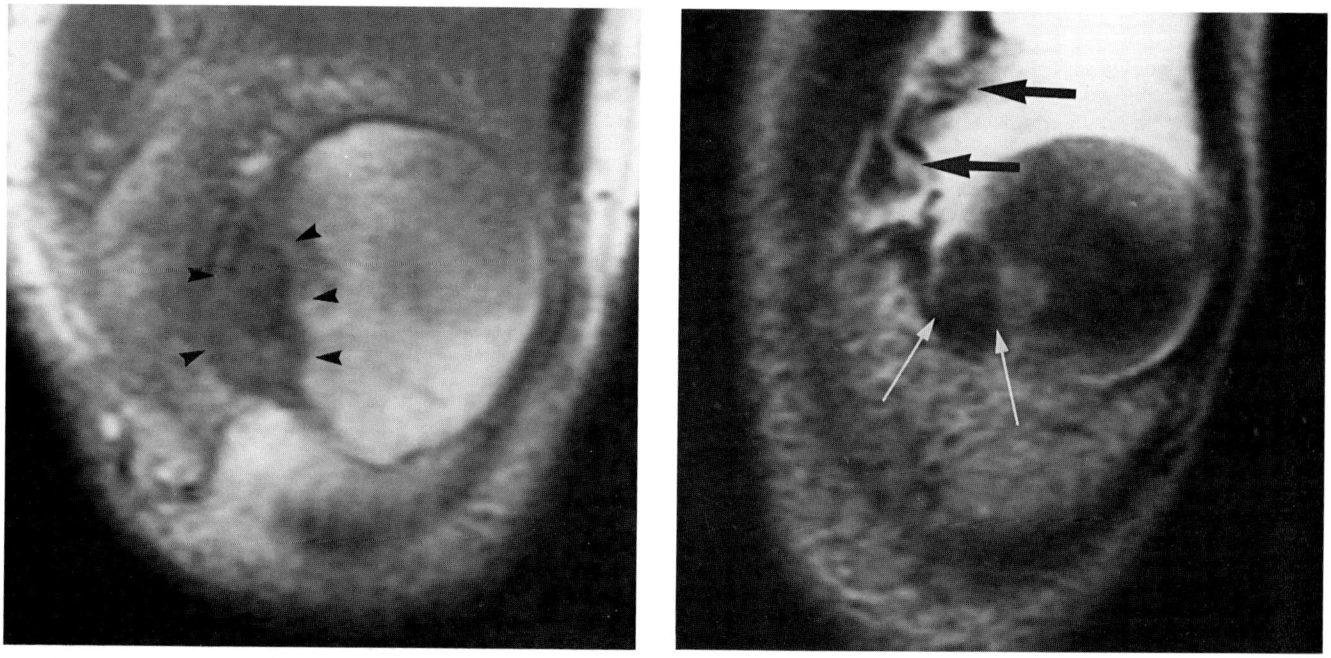

A B

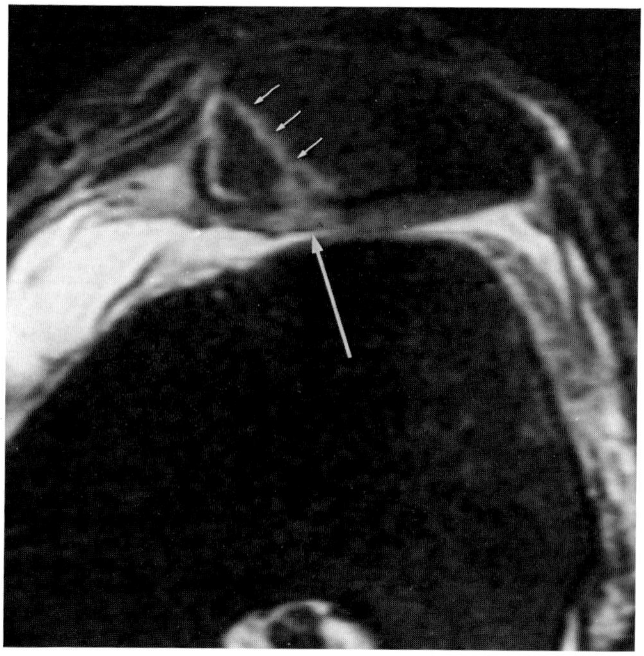

C

FIG. 6.22. Patellar fracture. **A:** Coronal TR/TE 2,000/20. On this far anterior coronal section from an acute knee MR, a fracture involving the medial facet of the patella is identified. There is marked low signal intensity, possibly reflecting edema or hemorrhage within the fragment (*arrowheads*). **B:** Coronal TR/TE 2,000/80. On this more heavily T2-weighted sequence, high signal intensity fluid within the suprapatellar bursa is more apparent. The medial retinaculum demonstrates a markedly irregular and serpiginous appearance consistent with extensive disruption (*black arrows*). The displaced fracture involving the inferomedial aspect of the patella is well demonstrated (*white arrows*). **C:** Axial STIR TR 2200 TE 35 TI 160. An oblique fracture is seen transversing the medial facet of the patellar in another individual (*small arrows*). There is associated injury involving the articular cartilage overlying the medial facet, manifest as a focus of increased signal (*long arrow*) compared with the normal signal of cartilage of the lateral facet. The ability of MR to depict both the subchondral injury and the involvement of the overlying cartilage represents a significant advantage for this diagnostic method.

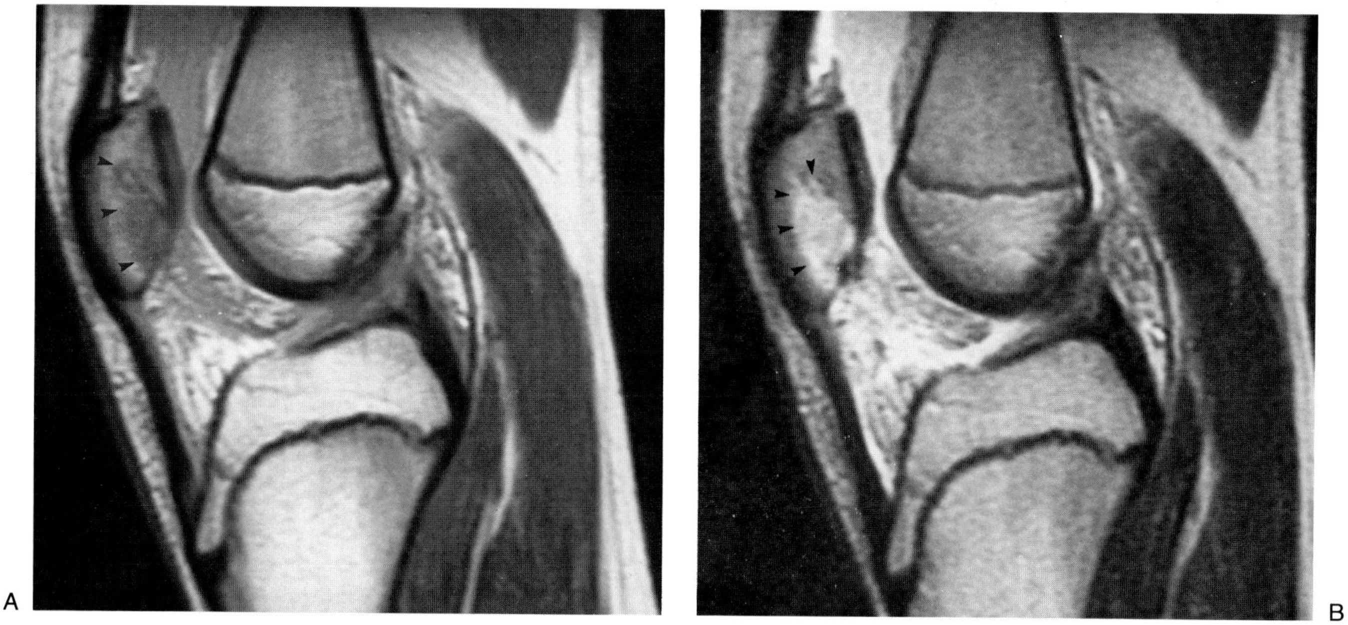

FIG. 6.23. Patellar bruise. **A:** Sagittal TR/TE 2,000/20. A contusional injury of the patella in this adolescent resulted from direct impaction upon the dashboard of an automobile. Diffuse low signal intensity, reflective of marrow edema, is seen within the patella (*arrowheads*). **B:** Sagittal TR/TE 2,000/80. On the more heavily T2-weighted sequence, the area of injury demonstrates significantly increased signal intensity (*arrowheads*).

be secondary to direct or indirect trauma. Transverse fractures predominate (50–80%) and usually result from indirect force (57). Longitudinal, stellate, and comminuted fractures are less common and usually occur as a result of direct impaction. With indirect trauma, the degree of separation of the fragments relates to the extent of associated disruption of the medial and lateral retinaculum. With patellar fracture secondary to direct trauma without retinacular disruption, the fragments are not widely displaced. Conventional radiographs generally suffice for diagnosis of patella fractures. MR may be valuable in the evaluation of the supporting retinacular structures, as well as in the assessment of the status of the articular surface (Fig. 6.22). Additionally, MR is increasingly demonstrating injuries that likely have been previously occult to plain radiographic detection. Bone contusions (bruises) as well as frank radiographically occult fractures have been demonstrated with MR (Fig. 6.23). While the patella is a relatively uncommon site for a stress fracture, such injuries have been described there as a complication of endoprosthesis and cruciate ligament reconstruction (see Chapter 7) as well as on an exercise-related basis (130).

Patella fractures need to be distinguished from bipartite and multipartite patellae and this distinction can usually be readily made with plain radiographs. While multipartite patellae have long been considered an in-

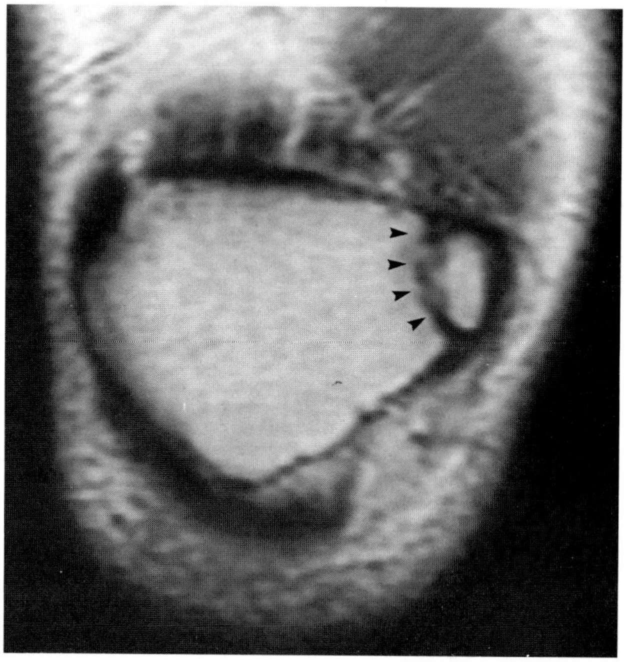

FIG. 6.24. Bipartite patella. Coronal TR/TE 800/20. A well circumscribed ossific fragment is seen in the characteristic location along the superolateral aspect of the patella (*arrowheads*). There is no surrounding edema to suggest an acute abnormality.

nocuous variant of ossification, investigators recently have emphasized local pain accompanying this entity, and histological evaluation has suggested that the basis for a bipartite patella may relate to chronic chondroosseous tensile failure of the bone in the immature skeleton. On MR, the bipartite patella is well demonstrated and the interface is of low signal intensity. In distinction to acute fractures, no surrounding marrow or soft tissue edema is encountered and MR can be of value in making the distinction from acute fracture in cases where this presents a diagnostic problem (Fig. 6.24).

Patellar Dislocation

Patellar dislocation refers to the complete displacement of the patella from the confines of the bony trochlear sulcus of the distal articular surface of the femur (131–134). The typical mechanism of injury is a twisting action with the femur rotating internally on a fixed foot (132–134). Less commonly, patellar dislocation may result from a blow to the outside of the knee producing a typical valgus mechanism. In the latter case, the true nature of the injury may be initially missed and confused clinically with medial ligamentous injury (135).

The natural history of most acute dislocations is for the patella to relocate prior to the patient's clinical presentation (133,134). Physical findings vary widely and the diagnosis may not be initially considered. Many patients present with the patella relocated and a large, tense hemarthrosis. In this setting, patellar dislocation may be confused clinically with anterior cruciate ligament disruption. Indeed, in a clinical series of patients undergoing surgery for suspected ACL disruption, 10% of cases were ultimately determined to represent patellar dislocations (136). In some patients, tenderness or a defect in the VMO may be demonstrable or a medial retinacular hematoma may be present (Fig. 6.25) (133). Other patients may have only mild swelling which is predominantly periarticular. In less extensive injuries no palpable medial defect may be present. Lateral patellar hypermobility accompanied by apprehension and pain can usually be demonstrated.

Lateral patellar dislocation results in demonstrable injury to both soft tissue and bony structures (98). The medial capsule and retinaculum along with the medial patelloepicondylar and patellotibial ligaments may be ruptured. In addition to the static supporting structures, the VMO, which contributes dynamic medial patellar support, may be ruptured. VMO rupture may occur at one of several sites: (a) at the level of insertion into the patella, (b) interstitial rupture in its mid-portion, or (c) tear from its origin at the adductor tubercle

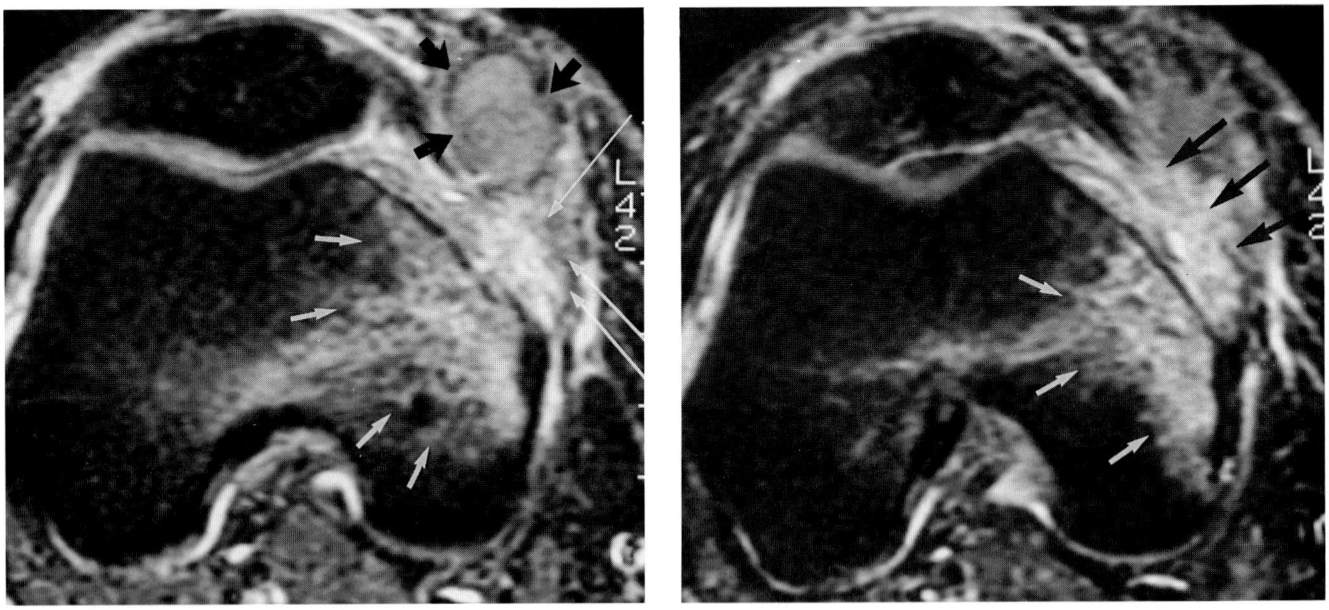

A B

FIG. 6.25. Patellar dislocation; complete retinacular disruption and retinacular hematoma. **A:** Axial STIR TR 2,200, TE 35, TI 160. A 16-year-old girl sustained an unusual lateral dislocation secondary to a medially directed blow. There is extensive edema seen throughout the medial femoral condyle (*small white arrows*) related to direct bony contusion. An intermediate signal intensity mass representing an acute hematoma is identified along the superficial retinaculum (*black arrows*). The retinaculum itself is completely disrupted (*long white arrows*). **B:** Axial STIR TR 2,200, TE 35, TI 160. On a contiguous section the extent of retinacular disruption is well demonstrated (*black arrows*). The contusional injury of the medial femoral condyle is again seen (*white arrows*).

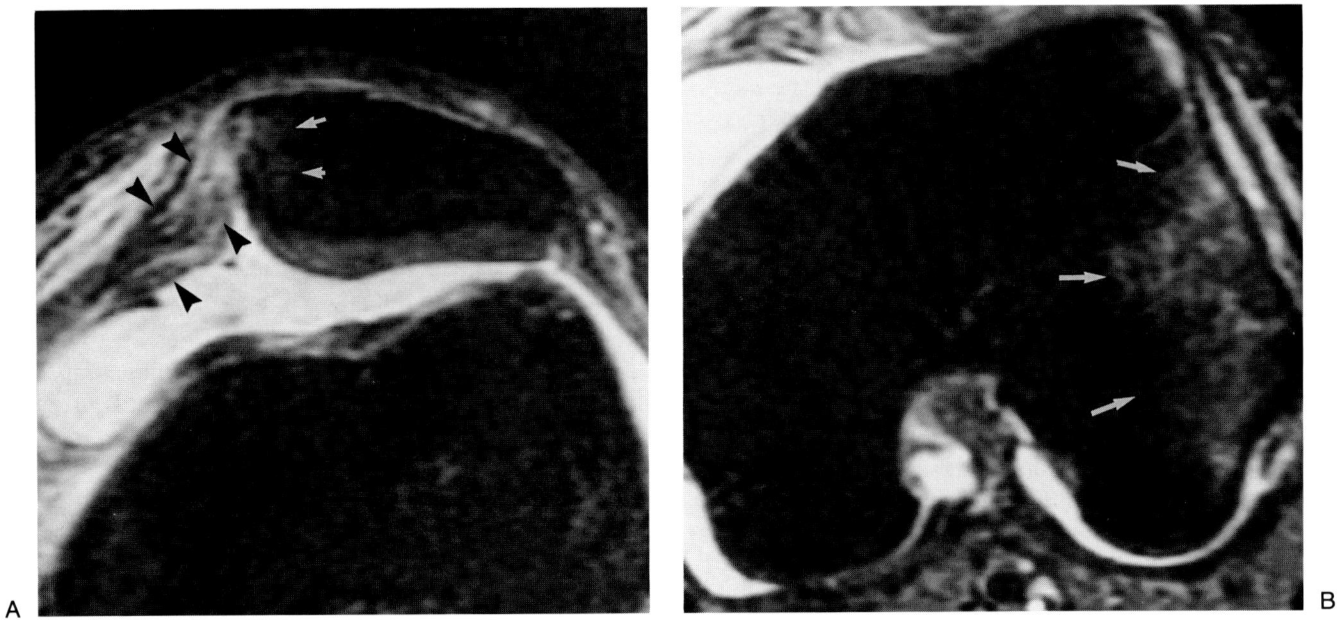

FIG. 6.26. Patellar dislocation. Retinacular disruption and typical bony contusional injuries. **A:** Axial STIR TR 2,200, TE 35, TI 160. This case demonstrates the more typical findings associated with lateral patellar dislocation. High signal intensity reflecting edema and or hemorrhage is seen within the subchondral bone of the medial facet of the patella (*small arrows*), and there is extensive disruption of the medial retinaculum (*arrowheads*). **B:** Axial STIR TR 2,200, TE 35, TI 160. A corresponding site of edema is seen within the subchondral and cancellous bone of the lateral femoral condyle at the site of impaction upon relocation of the patella (*arrowheads*). This combination of findings should strongly alert the interpreting radiologist to the likelihood of a recent patellar dislocation.

(135). Articular cartilage injury is common and occurs typically along the medial edge of the patella and lateral femoral trochlea as the patella impacts at the time of spontaneous reduction. Loose purely chondral or osteochondral fragments may result.

Conventional radiographs may allow demonstration of an osteocartilaginous fragment, although the limits of both physical and roentgenographic evaluation have been emphasized by many investigators (58). In contrast, however, MR is particularly well suited to eval-

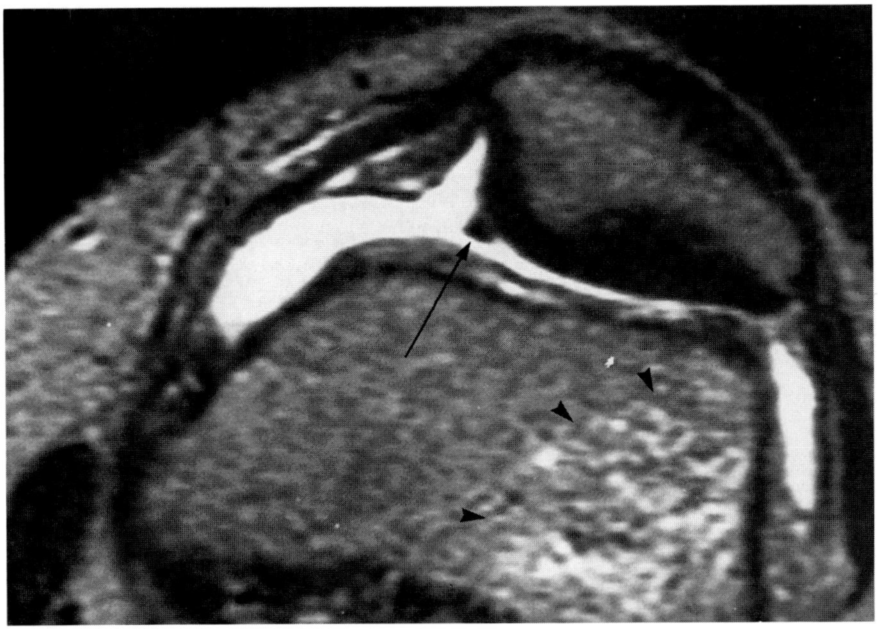

FIG. 6.27. Patellar dislocation; cartilage flap. Axial STIR TR 2,200, TE 35, TI 160. A partial thickness flap of articular cartilage is seen along the medial patellar facet well contrasted against high signal intensity joint fluid (*long arrow*). High signal intensity edema is seen within a typical contusional injury within the lateral femoral condyle (*arrowheads*).

uation of all the components of the extensor mechanism that might be potentially injured in patellar dislocation and can be easily obtained in the acute setting. In addition, MR can precisely delineate and characterize anatomical features of the patellofemoral joint that might underlie and predispose to recurrent subluxation and dislocation (see below).

In a recent study, Lance et al. reported on 22 patients with a final diagnosis of acute patellar dislocation who underwent MR examination (137). In 10 of the 22 patients (46%), the diagnosis was unsuspected by the referring orthopedist at the time of referral for the MR examination and the diagnosis was established retrospectively based on the MR findings. MR demonstrated hemarthrosis in 21/22 patients (96%), and disruption of the medial retinaculum in 18/22 (82%) (Figs. 6.25 and 6.26). Abnormal subchondral and cancellous bone signal was detected within the lateral femoral condyle in 18/22 (82%) patients, and within the medial facet of the patella in 9/22 (44%) patients (Fig. 6.26). This signal presumably reflects trabecular microdisruption, with edema and interstitial hemorrhage occurring secondary to direct impaction. A purely chondral flap involving the medial facet of the patella was identified in one patient (Fig. 6.27). Loose osteochondral fragments were detected in six cases, and confirmed at surgery in three patients. Lateral patellar displacement was detected in nearly half the cases. No cruciate ligament abnormalities were detected, although meniscal tears were detected in two patients and confirmed at surgery in both.

PATELLAR TRACKING DISORDERS

The high incidence of anterior knee pain and poor correlation of chondromalacic changes with clinical symptomatology have led investigators to attempt to move precisely determine the potential pathophysiologic mechanisms underlying the clinical complaints (114,138). In particular, attention has been directed toward both normal patellofemoral alignment, as well as possible mechanical aberrations in alignment of the patella in relation to the trochlear sulcus; the latter as a potential cause for secondary chondromalacia and clinical symptomatology. Recent studies have supported the contention that shearing stresses secondary to abnormal excursion of the patella on the femoral condyle are related to the development of lesions and degeneration of the articular cartilage (117).

Conventional radiography, primarily in the form of tangential or axial (e.g., "sunrise") views, has been the most commonly employed diagnostic method for assessment of patellofemoral relationships. Since initially introduced by Settegest in 1921, multiple variations in technique have been proposed by different investigators, and from these, innumerable indices have been developed in an attempt to define normal patellofemoral relationships and identify subtle derangements in alignment (31,114,139–141). Of these multiple methods and indices, those popularized by Merchant and Laurin have been the most commonly used in the United States (138–140). From a tangential radiograph produced with the knee flexed 45°, Merchant introduced the concept of the congruence angle for detection of early patellar malalignment (Fig. 6.28) (140). In recognition of the need to study patellofemoral relationships in earlier degrees of flexion, Laurin described three patellofemoral indices based on a 20° axial radiograph (Fig. 6.29) (139).

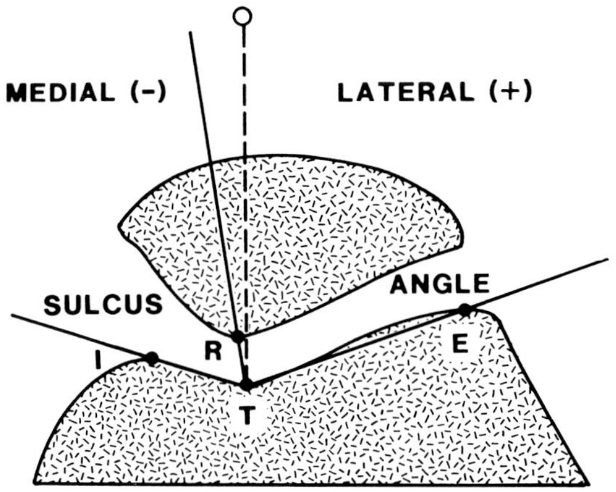

FIG. 6.28. Congruence angle. The sulcus angle ETI is bisected by a neutral reference line (TO). The apex of the median patellar ridge is connected to the lowest point of the trochlear sulcus. When this line (RT) is medial to the neutral reference line, the angle is given a negative value; when lateral, a positive value. (Modified from ref. 140.)

Despite the wide and continuing application of conventional radiographic techniques for evaluating patellofemoral alignment, significant inadequacies in the information provided by these methods has become increasingly apparent. The limitations of conventional tangential radiographs relate primarily to technical constraints which do not allow projections to be obtained at less than 20° and more commonly at 30° to 40° of knee flexion. Biochemical cadaveric studies and the introduction of axial plane imaging utilizing CT and MR have allowed assessment of patellofemoral relationships in extension and in the initial degrees of flexion. The results of these studies have challenged many longstanding conceptions regarding normal patellofemoral dynamics and have pioneered the way for a more sophisticated approach to the determination of normal patellofemoral dynamics and assessment of patellofemoral disorders. Both CT and MR have been utilized clinically to document disorders of patellar alignment

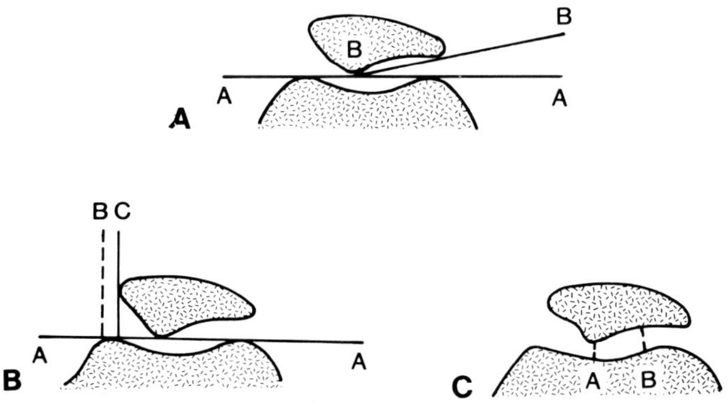

FIG. 6.29. Patella relationships after Laurin. **A:** Lateral patellofemoral angle. Two lines, one connecting the apices of the femoral condyles (AA) and the other along the lateral facet of the patella (BB), are defined. The angle of intersection usually opens laterally; less commonly the lines are parallel. (Modified from ref. 139.) **B:** Lateral patellar displacement. A line is drawn connecting the highest points of the medial and lateral femoral condyles (AA). A perpendicular to that line at the medial edge of the medial femoral condyle (B) normally lies 1 mm or less medial to the patella (line C). (Modified from ref. 139.) **C:** Lateral patellofemoral index. The index represents the ratio between the narrowest medial patellofemoral distance measured at the lateral edge of the medial facet (a) and the narrowest lateral patellofemoral distance (b). The ratio is normally one or less. (Modified from ref. 139.)

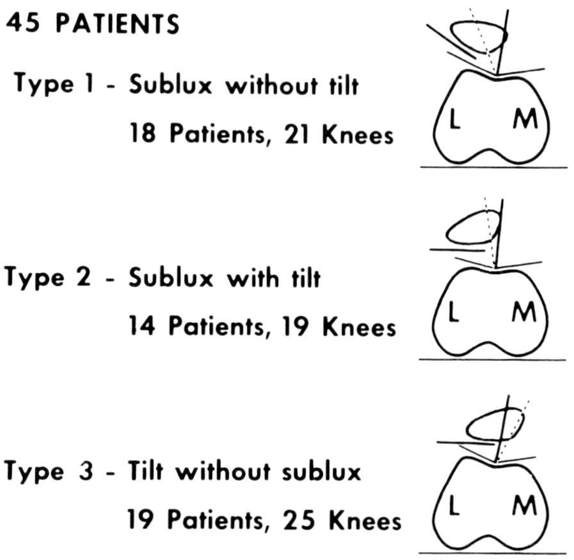

FIG. 6.30. Malalignment patterns determined by CT. Utilizing this classification, tilt may occur with or without patellar subluxation. (From ref. 144, reproduced from ref. 31, with permission.)

during the initial degrees of knee flexion, that previously would have gone undetected by conventional radiographic methods imaging the patellofemoral at 30° to 40° of flexion.

The introduction of CT to the evaluation of the patellofemoral joint by Delgado-Martins in 1979 allowed the first axial plane imaging assessment of patellofemoral relationships in extension and in the initial degrees of flexion. In this study, the normal position of the patella in extension was recognized as being slightly lateral in relation to the trochlear groove (142). Boven et al., in reporting on 71 patients undergoing CT and conventional radiography, further substantiated the ability of CT to detect patellar subluxation and tilt compared with plain radiography (143). Schutzer et al. described three malalignment patterns based on the patellar tilt and congruence angles obtained utilizing CT: type 1, subluxated but nontilted patella; type 2, subluxated and tilted; and type 3, patellar tilt without subluxation (Fig. 6.30) (144,145). Of particular note was the observation that nearly 40% of the patients with a type 1 or type 2 malalignment pattern were completely reduced at a knee flexion angle of 30°; a finding which further underscores the limitations of conventional axial radiographs (rarely technically obtainable at less than 30° flexion) in assessing patellofemoral alignment. The type 3 pattern, according to the authors, correlates with the findings expected in the excessive lateral pressure syndrome initially described by Ficat. In patients with this malalignment pattern, the lateral patellar facet is forced into the femoral trochlear, causing excessive lateral loads that are potentially damaging to patellar cartilage.

Patellofemoral Motion Studies

Computed Tomography

Recent technical advances have allowed patellofemoral motion to be evaluated by ultrafast CT methods (146). This technique offers the opportunity for objective observations of the dynamic influences of muscle contraction on the patellofemoral joint as the knee is actively moved through a range of motion from extension to 90° of flexion. Stanford et al., obtained ten images at each of eight levels encompassing an 8 cm length of the knee in 7 sec as the knee was moved from 90° flexion to full extension and back to 90° flexion (146). The processed images were played back in a closed loop movie format allowing for visualization of real time axial images of patellofemoral motion during muscle contraction and leg movement. In their quantitative data analysis, Stanford et al., corroborated the initial CT observations of Delgado-Martins that the patella often lies laterally and incongruently in contact with the lateral femoral condyle in full extension. They observed lateral subluxation to be most apparent during the initial 20° of knee flexion, again underscoring the value of CT as compared to conventional radiography for this analysis.

MR

The authors of this chapter have described a method of analysis of patellofemoral joint dynamics utilizing MR imaging that has been widely applied to the evaluation of patellar alignment and tracking (25,29,147). In a manner similar to that described for "cine-CT" (146), T1-weighted spin-echo axial images are obtained at selected locations through the patellofemoral joint, and subsequently displayed both as static images and simulated real-time images (3,8–10). Termed kinematic MRI, the technique utilizes a special patient-activated positioning device designed to allow simultaneous axial plane imaging of both knees from extension to 30° of flexion at 5° increments. The patient is placed in a prone position on the device, with special care taken to position the lower extremities to ensure that the individual's baseline alignment is maintained. This positioning scheme is unique because it does not inhibit rotational movements of the lower extremities, which may be partially responsible for abnormal patellar tracking to occur during flexion of the patellofemoral joint (9,74,120,134,148). Prone positioning of the patient does not alter the kinematics of the patellofemoral joint during flexion when compared with the supine position (9).

Three to four different slice locations through the femoral trochlear groove (depending on the height of the patella) are chosen which depict the path of excursion of the patella during flexion. In deference to concerns about the normal position of the patella with the knee extended (i.e., lateralized or centered) (23,144,149), only the images obtained with the patellofemoral joint flexed from 5° to 30° are analyzed for the kinematic study. While static images are obtained (150,151), we have considered it more appropriate and practical to use qualitative rather than quantitative criteria to describe the patella in relation to the femoral

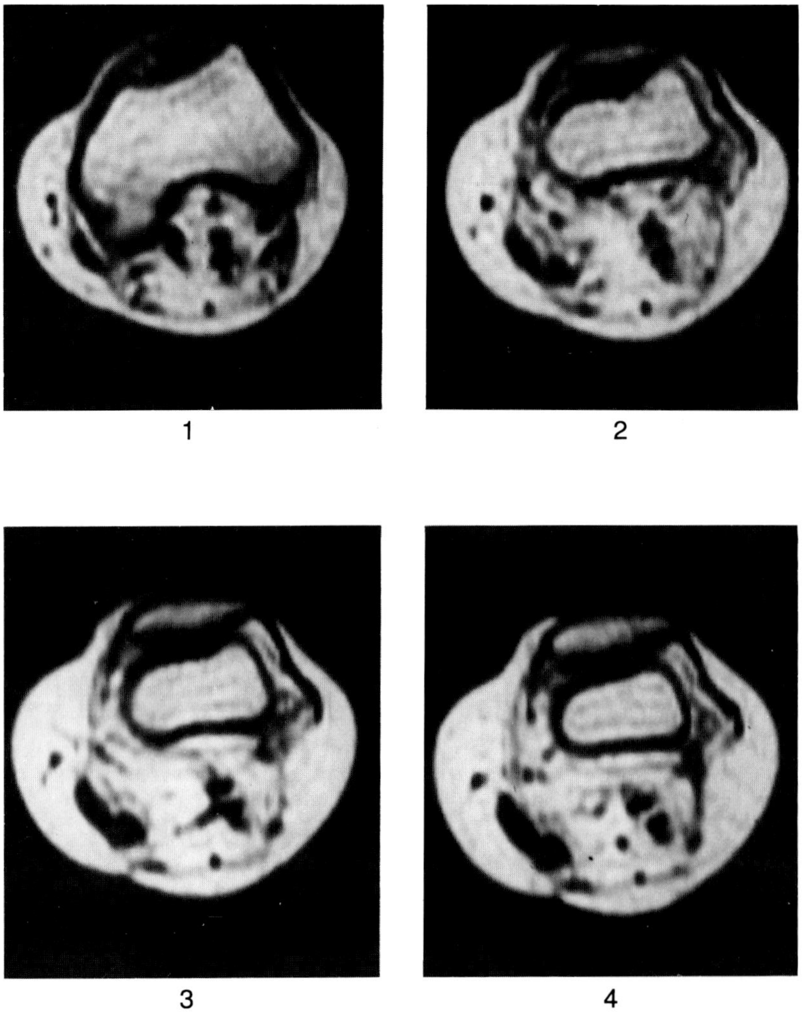

FIG. 6.31. Patellar alta. Sequential axial plane images TR/TE 400/20. The inferior pole of the patella (31-1) is positioned well above the femoral trochlear groove with the knee in extension. In addition, the patella is poorly formed (31-2,31-3,31-4) and the femoral trochlear groove has a hypoplastic medial aspect.

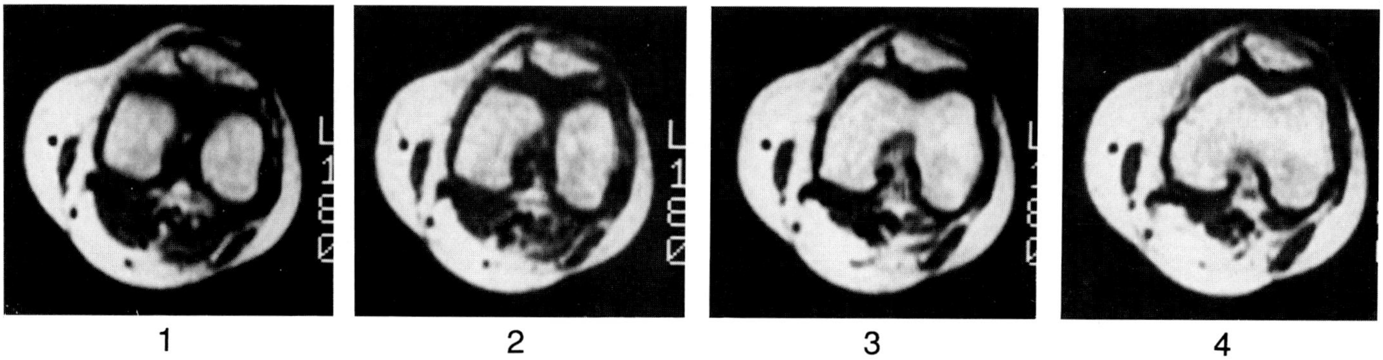

FIG. 6.32. Patella baja. Sequential axial plane images TR/TE 400/20. The inferior pole of the patella (32-1) is positioned below the femoral trochlear groove with the patient's knee in extension. The patella and femoral trochlear grooves have normal shapes.

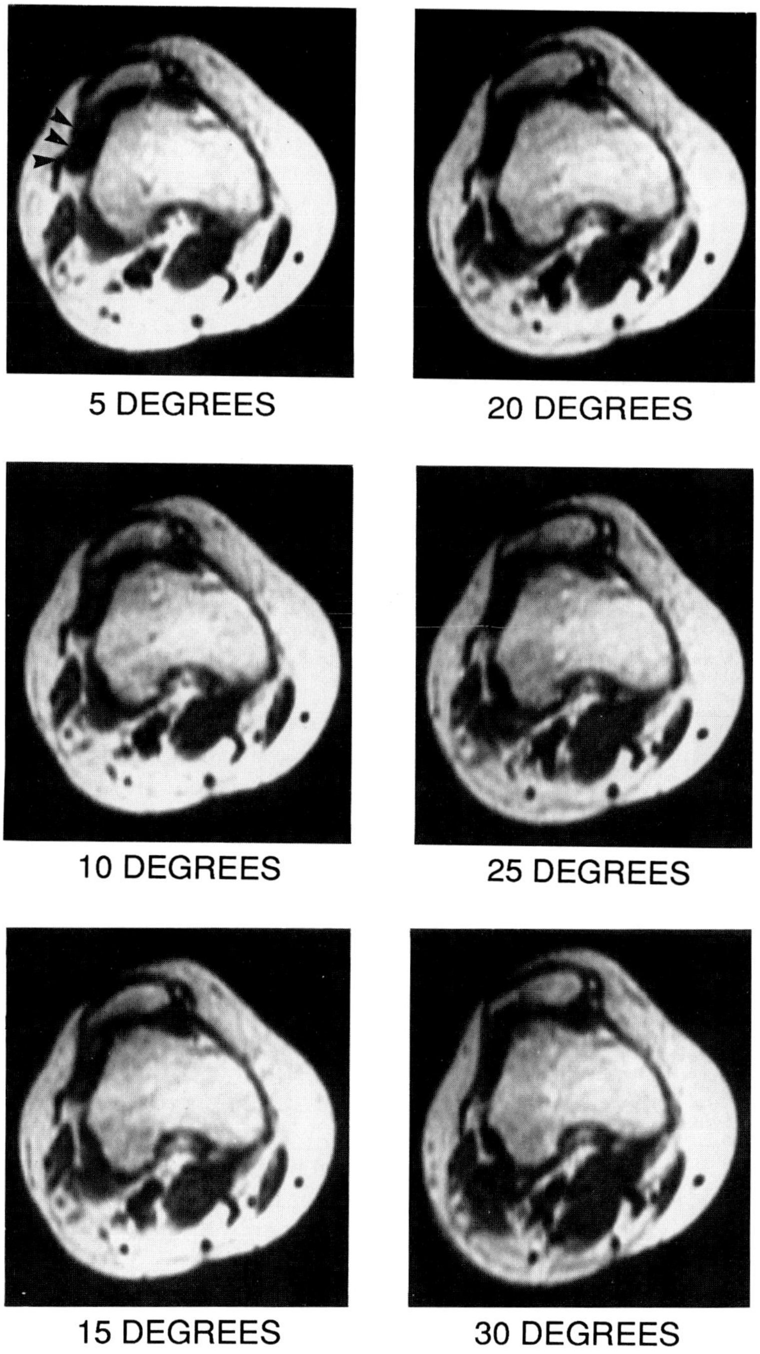

FIG. 6.33. Persistent lateral subluxation following lateral retinacular release. Axial TR/TE 400/20. This patient had persistent symptoms following a lateral retinacular release (note the thickened lateral retinaculum, *arrowheads*) performed to treat anterior knee pain and suspected lateral subluxation. The kinematic MRI of the patellofemoral joint shows striking persistent lateral subluxation of the patella that is evident throughout the range of motion studied.

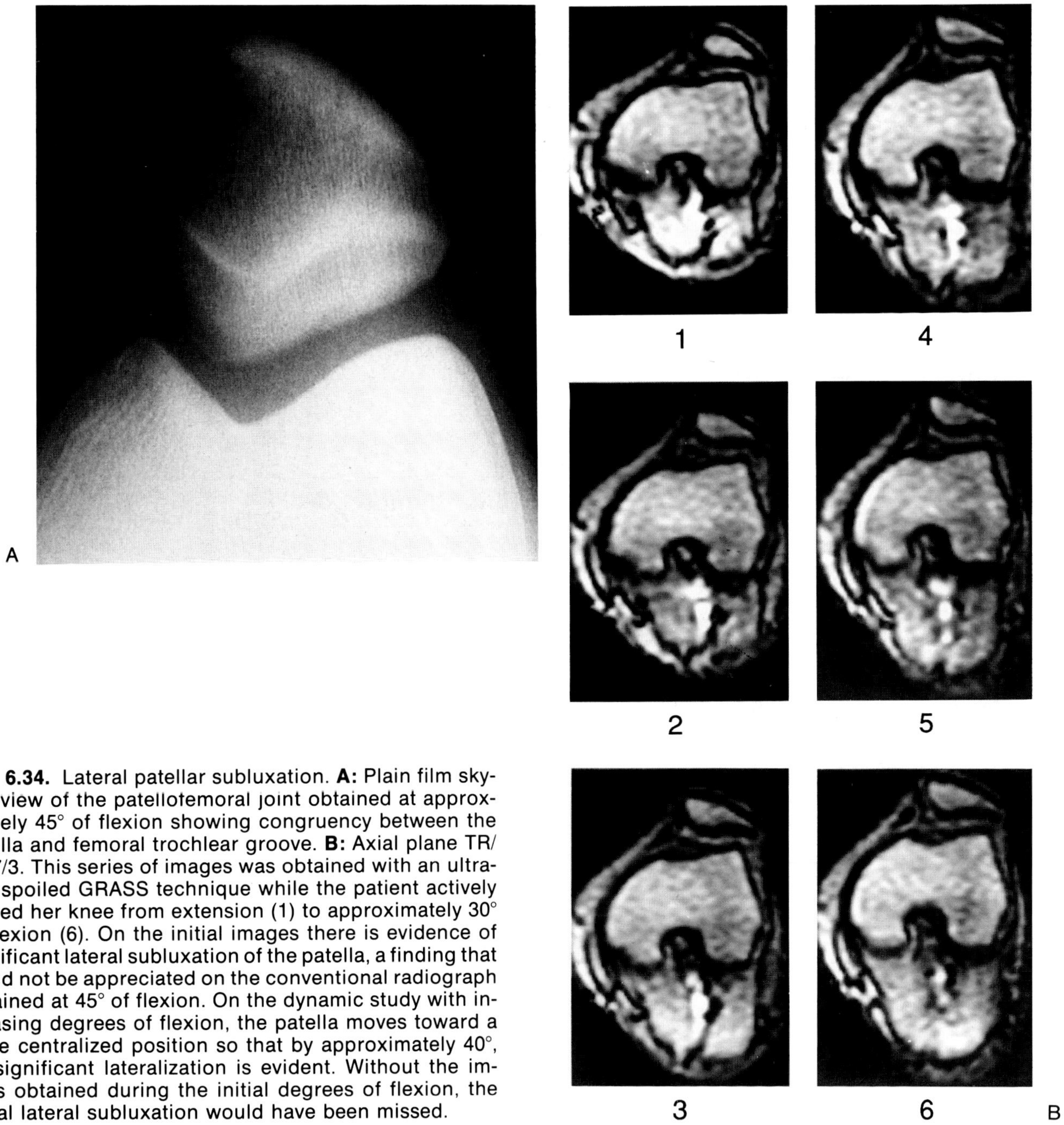

FIG. 6.34. Lateral patellar subluxation. **A:** Plain film skyline view of the patellofemoral joint obtained at approximately 45° of flexion showing congruency between the patella and femoral trochlear groove. **B:** Axial plane TR/TE 7/3. This series of images was obtained with an ultrafast spoiled GRASS technique while the patient actively moved her knee from extension (1) to approximately 30° of flexion (6). On the initial images there is evidence of significant lateral subluxation of the patella, a finding that could not be appreciated on the conventional radiograph obtained at 45° of flexion. On the dynamic study with increasing degrees of flexion, the patella moves toward a more centralized position so that by approximately 40°, no significant lateralization is evident. Without the images obtained during the initial degrees of flexion, the initial lateral subluxation would have been missed.

trochlear groove during flexion of the patellofemoral joint (3,8–11,148,150,151). This preference for dynamic image display has also been advocated by Stanford utilizing ultrafast CT (146), and by Kujala utilizing MR (149,152). This may in part be accounted for by the observation that, in contrast to static conventional radiographs, the changes in anatomic configuration of the patella and trochlear seen through the tomographic capabilities of MR and CT preclude simple application of the previously described parameters developed for conventional radiographs (3,8–11,148,150,151).

The configurations of the patella and femoral trochlear groove are well demonstrated on the kinematic MRI study by the sequential axial plane images obtained through the patellofemoral joint during extension. Abnormalities of patella alignment related to dysplasia of either the patella and or trochlear sulcus, as well as to patella height, can be readily demonstrated. Normal patellar height is considered present on the kinematic MR examination when the inferior pole of the patellar is positioned in the superior aspect of the femoral trochlear groove with the knee in extension

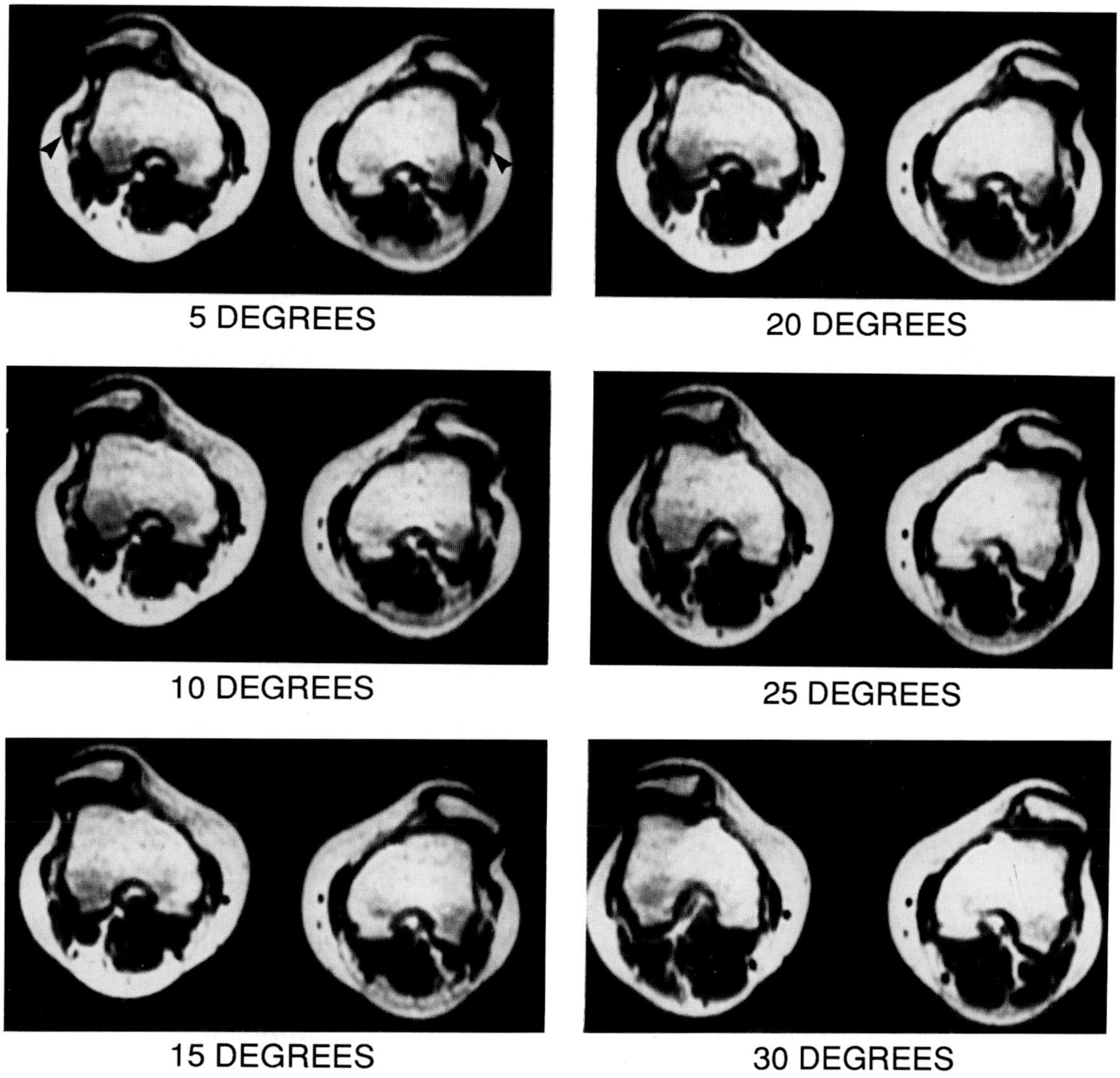

FIG. 6.35. Bilateral subluxation of the patellae with redundant lateral retinacula. Axial plane TR/TE 400/20. Bilateral lateral subluxation of the patellae with redundant lateral retinacula noted at the initial degrees of flexion (*arrowheads*), 5° images. The presence of a redundant lateral retinaculum suggests that the lateral patellar subluxation is not caused by excessive force produced by exorbitant stretching of this particular soft tissue structure.

(70,74,120). Patella alta (Fig. 6.31) is present when the inferior pole of the patella is positioned above the superior aspect of the femoral trochlear groove with the joint extended; patella baja (Fig. 6.32), an abnormally low position of the patella, is present when the entire patella is positioned in or below the femoral trochlear groove with the joint extended. This determination often differs from that accomplished with the conventional criteria developed for axial radiographs (74,114, 120,134,139,140,153), and introduces the concept of "functional" patella alta or baja relating patella position to the trochlea sulcus directly through the use of axial tomographic techniques. The latter criteria is considered to provide a more precise means of determining patellar height.

Normal alignment and tracking of the patella is considered to be present when the median ridge of the patella is positioned in the femoral trochlear groove as it travels in a vertical plane during flexion of the knee, without transverse displacement of the medial or lateral facets of the patella (3,8,9). This orientation causes the patella to appear "centered" in the femoral trochlear groove. Any deviation of this normal pattern of patellar alignment or tracking seen on one or more slice locations at 5° of flexion or greater on the kinematic MRI study is regarded as an abnormality (1–3,11–15).

Four different patterns of patella malalignment seen with MR have been described (3,8–11,26,29,148, 150,151).

Lateral Subluxation of the Patella

On the kinematic MRI study, lateral subluxation of the patella is diagnosed when the median ridge of the patella is laterally displaced relative to the femoral trochlear groove and the lateral facet of the patella overlaps the lateral aspect of the femoral trochlear (3,8–11,147,148,150,151). This is the most common form of patellar malalignment and occurs with varying degrees of severity (Figs. 6.33–6.36) (74,79). MR may demonstrate abnormalities of articular cartilage at the impact site between the patella and femoral trochlear groove as a result of chronic hyperpressure that develops from contact stress due to lateral subluxation of the patella (29,74,80,106,154,155). This is best detected on higher resolution sequences optimized for assessment of articular cartilage. One of the unique observations afforded by the high soft tissue contrast

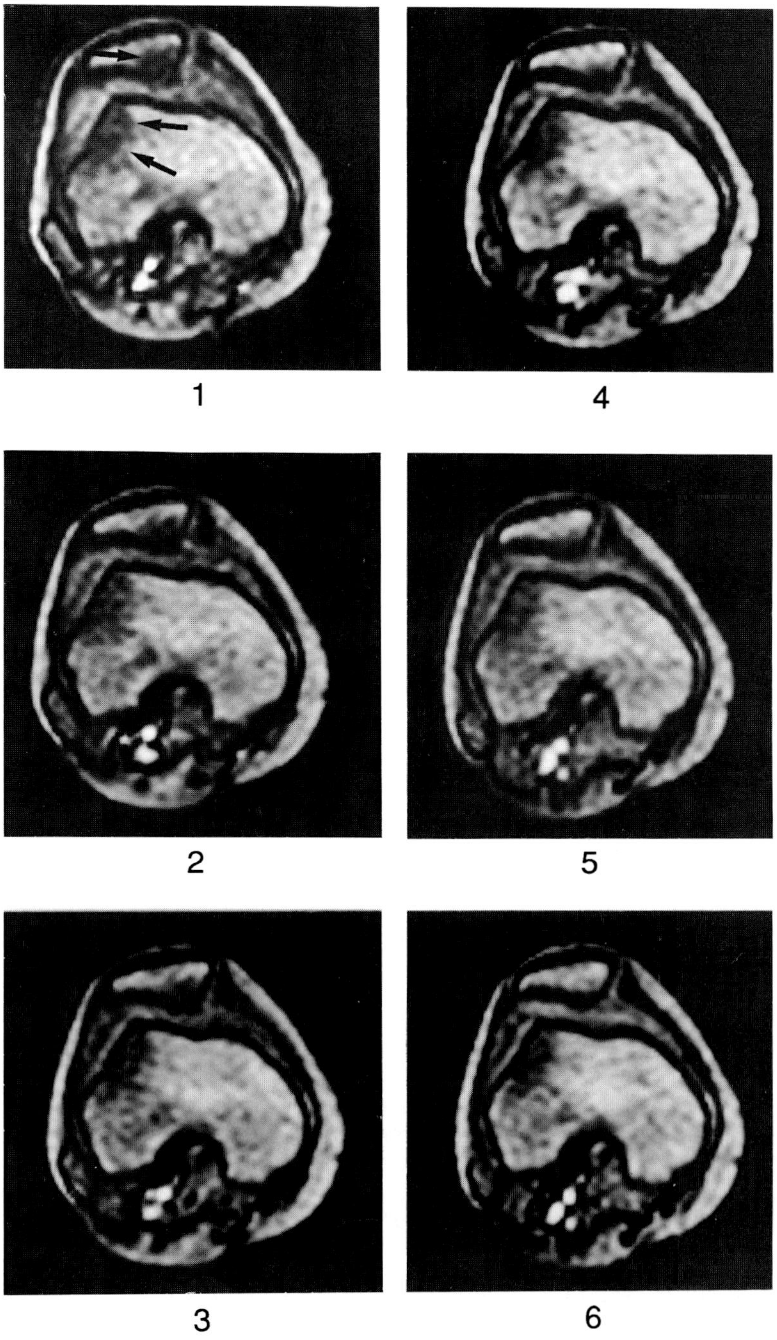

FIG. 6.36. Dynamic examination following traumatic dislocation. Axial TR/TE 7/3. Ultrafast spoiled GRASS technique performed while the patient actively moved knee from extension (1) to approximately 45° of flexion (6). Lateral subluxation of the patella seen in a patient who recently sustained a traumatic dislocation and subsequent reduction of the patella. Note the evidence of contusional injuries along the median ridge of the patella and lateral femoral condyle (*arrows*).

resolution of kinematic MRI has been the presence of a redundant lateral retinaculum associated with lateral subluxation of the patella (Fig. 6.35). This may have important implications regarding the consideration of surgical release of the lateral retinaculum (11,150), which is presently the most commonly performed procedure for treatment of suspected lateral subluxation of the patella and consequent patellofemoral pain. Redundancy of the lateral retinaculum suggests insufficiency of the medial stabilizing structures (e.g., VMO) as opposed to excessive "tightness" of the lateral retinaculum as the underlying etiology of the patellar malalignment.

Excessive Lateral Pressure Syndrome

Lateral patellar tilt or excessive lateral pressure syndrome (ELPS) was first described by Ficat and Hungerford (80). This form of patellar malalignment is a clinical-radiologic entity, clinically characterized by anterior knee pain and radiologically by tilting of the patella with functional patellar lateralization, usually onto a dominant lateral facet. A small amount of lateral displacement of the patella may or may not be present during joint flexion, as increasing tension from one or more overly taut soft-tissue structure(s) tilt the patella in a lateral fashion (Fig. 6.37).

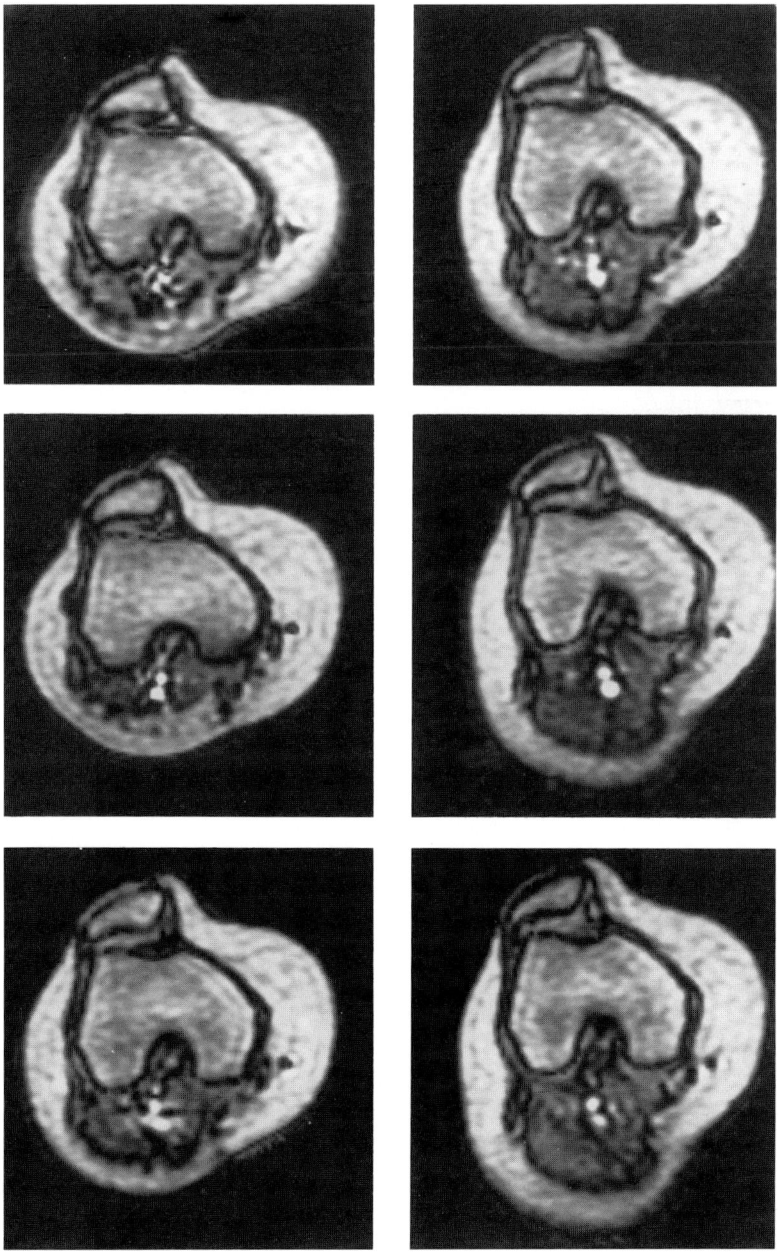

FIG. 6.37. Excessive lateral pressure syndrome. Axial TR/TE 7/3. Ultrafast spoiled GRASS technique performed while the patient actively moved knee from extension (1) to approximately 45° of flexion (6). Excessive lateral pressure syndrome (ELPS) and slight lateral subluxation of the right patella is evident. Note that the lateral tilting of the patella increases at the higher increments of flexion.

There are several reasons to believe that the main pathologic component of this abnormality is the lateral retinaculum. Because the patellar tilting that occurs with ELPS may be either transient (i.e., centralization or correlation of the patellar malalignment occurs during flexion during joint flexion) or progressive during increasing degrees of joint flexion, kinematic MR imaging of the patellofemoral joint is particularly useful for identifying and characterizing this abnormality.

The underlying hyperpressure found in ELPS can produce significant destruction of the articular cartilage. Lateral joint line narrowing is often present as a result of a decrease in cartilage thickness or due to gross cartilage degeneration along the median ridge and lateral facet of the patella that is produced by the excessive lateral pressure. A dysplastic patella and/or femoral trochlea are typically modified by use, and it is therefore likely to be involved in modifying the associated bone anatomy. ELPS is usually treated effectively by surgical release of the lateral retinaculum.

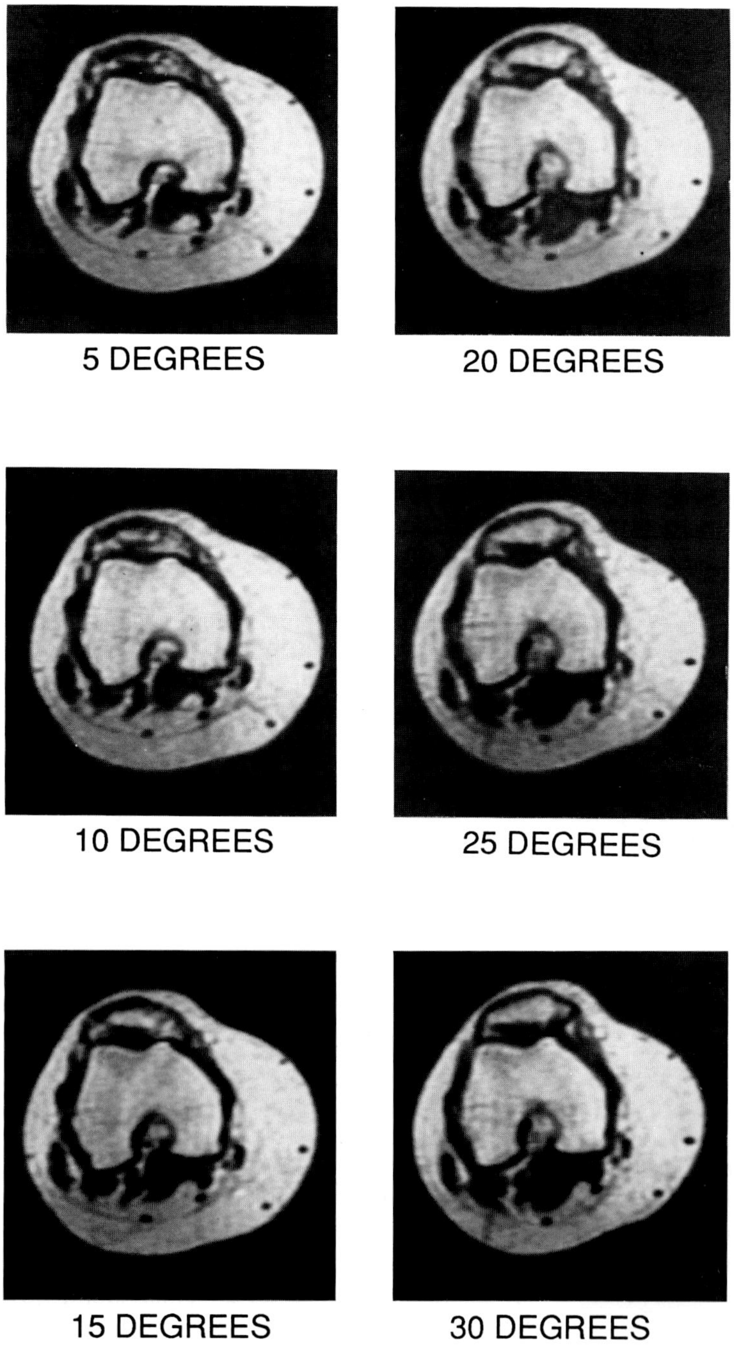

FIG. 6.38. Medial subluxation following previous lateral retinacular release. Axial TR/TE 400/20. Medial subluxation of the patella in a patient with prior lateral retinacular release. The femoral trochlear groove is relatively flattened, so that it does not provide a buttressing structure for the patella.

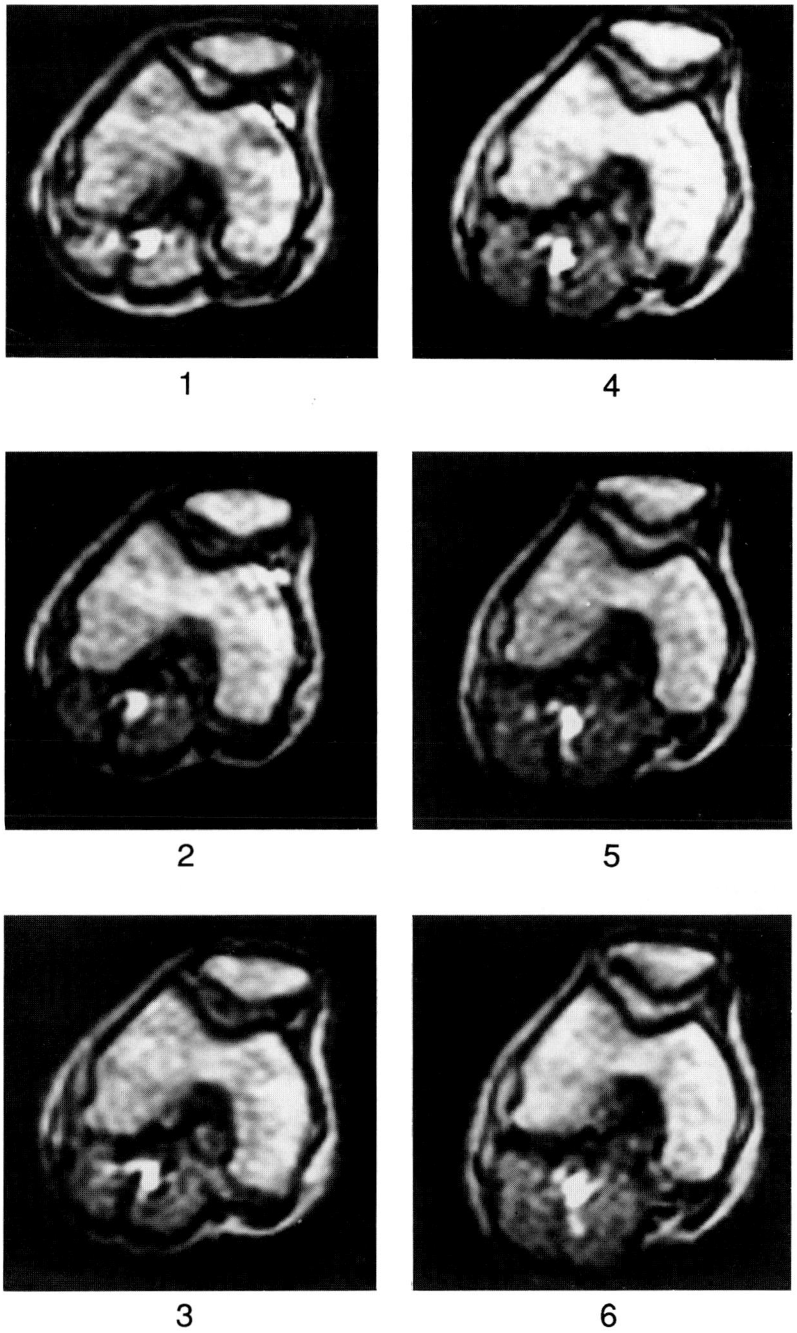

FIG. 6.39. Medial subluxation. Axial TR/TE 7/3. This series of images was obtained with an ultrafast spoiled GRASS technique while the patient actively moved knee from extension (1) to approximately 45° of flexion (6). There is evidence of medial subluxation of the patella in a patient without prior surgery. Note that there is centralization of the patella by the higher increments of flexion (5, 6). The findings of medial subluxation were not evident on a conventional axial radiograph of the patella obtained at approximately 45°.

Medial Subluxation of the Patella (Patella Adentro)

Medial subluxation of the patella, or patella adentro, is distinguished by a medial displacement of the median patellar ridge relative to the femoral trochlear groove (Figs. 6.38–6.41) (8–10,26,147). Medial subluxation has been reported as a complication of prior surgical realignment procedures for presumed lateral subluxation (Fig. 6.38) and also has been reported in patients with no prior surgery (Fig. 6.39). Similar to other forms of abnormal patellar alignment, medial subluxation of the patella may cause localized hyperpressure and, therefore, there may be associated with MR demonstrable cartilage abnormalities along the medial patellar facet (8–10,26).

A variety of causative mechanisms are suspected to be responsible for medial subluxation of the patella that exist either individually or in combination, including (a) an insufficient lateral retinaculum, (b) overtaut medial retinaculum, (c) abnormal patellofe-

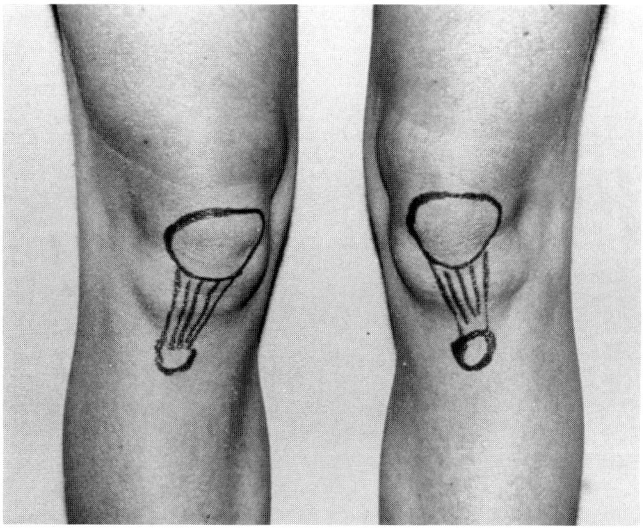

FIG. 6.40. Internal rotation of lower extremities. Bilateral internal rotation of the lower extremities. This condition is commonly found in association with medial subluxation of the patella. (Courtesy of James Fox, M.D. Van Nuys, California.)

moral anatomy, and (d) unbalanced quadriceps muscles (8,26). In addition, excessive internal rotation of the lower extremities is a common clinical finding with medial subluxation of the patella (Fig. 6.40). The presence of medial subluxation of the patella is especially important to identify and to distinguish from lateral or tilting forms of patellar subluxation because various surgical stabilization techniques (e.g., medial transposition of the extensor mechanism, lateral retinacular release, etc.) can further increase medial subluxation of the patella and not only lead to failure of the procedure, but can also exacerbate the patient's symptoms (156). Of additional importance is that most physical rehabilitation techniques, including the use of patellofemoral braces (Fig. 6.41) and McConnel taping, are usually implemented to correct lateral subluxation of the patella (113) and need to be modified when treating patients with medial subluxation of the patella.

Lateral-to-Medial Subluxation of the Patella

Lateral-to-medial subluxation of the patella is a pattern of abnormal patella tracking whereby the patella is in a slightly laterally subluxated position during the early increments of knee flexion, moves across the

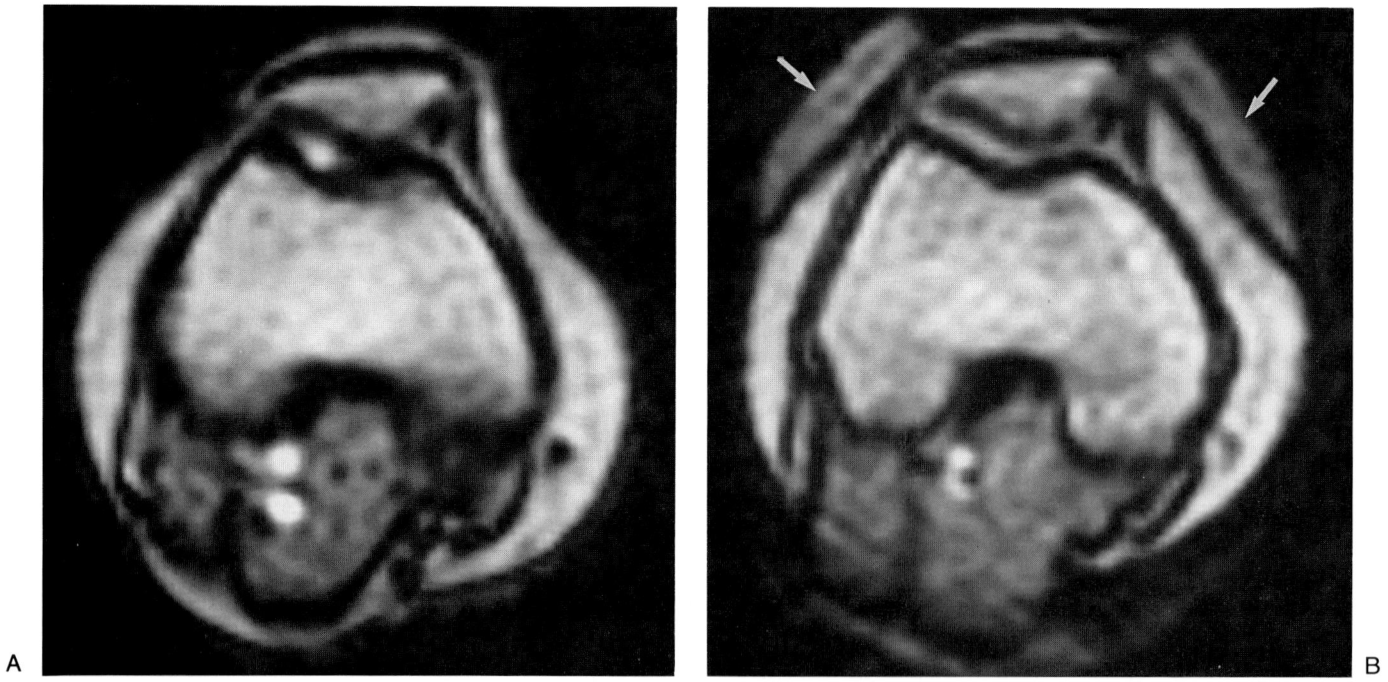

FIG. 6.41. Dynamic examination pre- and postbracing. **A:** Axial TR/TE 7/3. Selected image obtained at approximately 30° of flexion with ultrafast spoiled GRASS technique while patient flexed the knee. Medial subluxation of the patella is identified. **B:** Axial TR/TE 7/3. Selected image obtained at approximately 30° of flexion with ultrafast spoiled GRASS technique while patient actively flexed the knee. A specialized patellar alignment brace (*arrows*) was applied to this patient's knee which served to centralize and improve the tracking of the patella. Such MR-compatible devices allow direct monitoring of the potential effects of different therapeutic regimens.

femoral trochlear groove or femoral trochlea as flexion continues, and displaces medially at the higher increments of flexion (Fig. 6.42). This type of abnormal patellar tracking is commonly exhibited by the patellofemoral joint that tends to have patella alta, a poorly developed femoral trochlear groove, and/or misshapen patella (8,10). Again, there is no mechanical restraint provided by the bony anatomy, and patellofemoral instability results when these structural abnormalities exist. The actual mechanisms responsible for lateral-to-medial subluxation of the patella are complicated and not well-understood. Multiple disordered or uncoordinated forces apparently act on the patella to produce this form of abnormal patellar movement (147). As with other patellar malalignment syndromes, kinematic MRI is particularly suited to identifying this unusual type of patellar tracking because it shows the various positions and movements of the patella at several different increments of joint flexion.

Ultrafast MRI of the Patellofemoral Joint

Recently, new MRI pulse sequences have been developed that allow rapid imaging with sufficient image

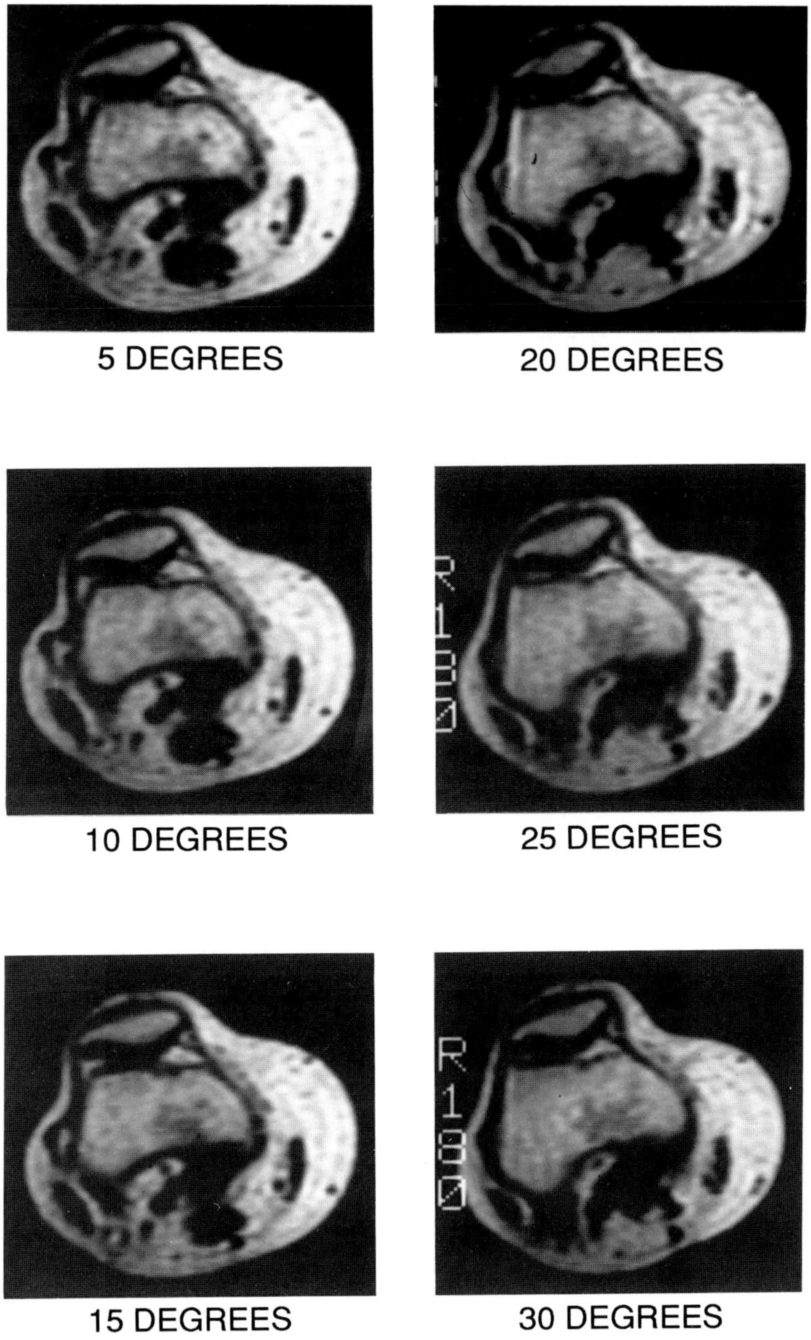

FIG. 6.42. Lateral to medial subluxation. Axial TR/TE 400/20. There is a slight lateral subluxation of the patella at 5–15° of flexion. At 20° and 25° of flexion, the patella centralizes. At 30° of flexion, there is a slight medial subluxation of the patella. This kinematic pattern is indicative to lateral-to-medial subluxation of the patella. Characteristic of these patients is a markedly dysplastic (flattened) trochlear groove contributing little support to the patella. The morphology of the trochlear groove is well demonstrated on the MR study.

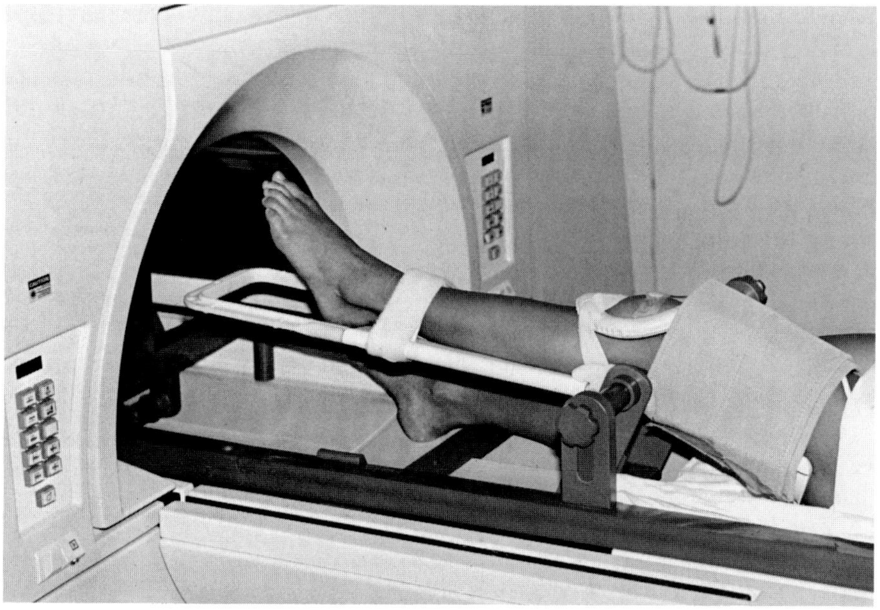

A

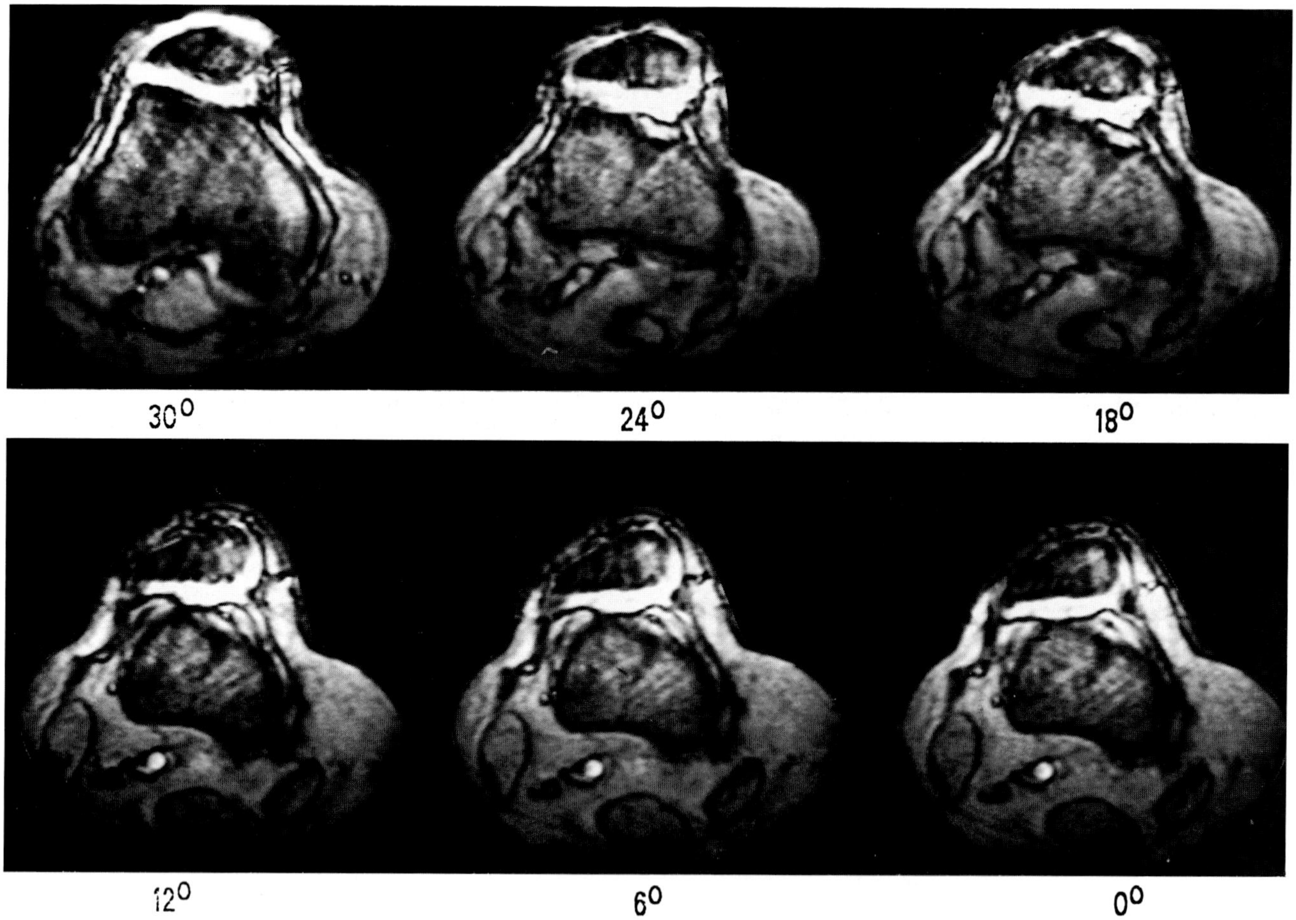

B

FIG. 6.43. Motion-triggered cine MRI. **A:** Positioning device for motion-triggered cine MRI of active patellofemoral joint studies. **B:** Axial TR/TE 41/22 (gradient echo). This study was obtained with the motion-triggered cine MRI technique of studying the patellofemoral joint. Lateral subluxation of the patella is evident (note that images are showing mid-slice location of the patella and where it articulates with different parts of the femoral trochlea). (Courtesy of J. Brossman, M.D. and C. Muhle, M.D. Kiel, Germany.)

quality to permit the evaluation of physiologic motion (25,27–29,97,157,158). Similar to ultrafast CT (146), Shellock et al., have demonstrated that ultrafast MRI may be used to obtain multiple images at a temporal resolution (i.e., images obtained at the rate of approximately one per second) suitable for examining the patellofemoral joint during active movement (12,13, 25,29). This provides a more physiologic and realistic assessment of the kinematic aspects of this joint because of the contribution of the activated muscles and soft tissue structures (Figs. 6.34, 6.36, 6.39, and 6.41) (12,13,25).

Another more elaborate MRI technique has also been developed to image the patellofemoral joint: motion triggered cine MR imaging of active joint movement. This procedure also examines the joint during active motion but requires approximately 5 to 6 mm for the entire study, the use of a special positioning device, and electronic triggering hardware in order to accomplish the examination (Fig. 6.43). Preliminary results using this MRI method have suggested that it may be helpful for studying patellofemoral joint kinematics.

SUMMARY

The extensor mechanism, and patellofemoral joint in particular, have been extensively studied but remain incompletely understood by those charged with caring for patients with complaints attributable to this important biomechanical apparatus. Compounding the difficulty for the orthopedic surgeon is the fact that the components of the extensor mechanism, with the exception of the articular surface of the patella, are not accessible to conventional radiographic or direct arthroscopic assessment. These considerations underscore the contribution that noninvasive assessment of the extensor apparatus with MR can make. All constituent components comprising the extensor mechanism are well delineated on MR and the multiplanar capacity of MR further facilitates diagnostic evaluation. Emerging technology, including the development of new pulse sequences and ultrafast imaging, has brought us to the precipice of real-time imaging, with the promise of further enhancing our understanding of patellofemoral dynamics in particular. This chapter has attempted to summarize the current use of MR with regard to depiction of abnormalities of the extensor mechanism and to provide the interested reader with sufficient background material to understand many of the complexities inherent in evaluation of this region. It is anticipated that the application of MR assessment of the extensor mechanism and the patellofemoral joint will continue to increase.

REFERENCES

1. Bodne D, Quinn SF, Murray WT, et al. Magnetic resonance images of chronic patellar tendinitis. *Skeletal Radiol,* 1988;17(1):24–28.
2. Yulish BS, Montanez J, Goodfellow DB, et al. Chondromalacia patella: Assessment with MR imaging. *Radiology* 1987;164:763–766.
3. Shellock FG, Mink JH, Fox JM. Patellofemoral joint: Kinematic MR imaging to assess tracking abnormalities. *Radiology* 1988;168(2):551–553.
4. Gould ES, Taylor S, Naidich JB, et al. MR appearance of bilateral, spontaneous patellar tendon rupture in systemic lupus erythematosus. *J Comput Assist Tomgr* 1987;11(6):1096–1097.
5. Frocrain L, Rochcongar P, Gouriff Y, et al. Role of MRI in the study of Achilles tendon and patellar tendon. *J Radiol* 1988;69(12):731–734.
6. Daffner RH, Riemer BL, Lupetin AR, et al. Magnetic resonance imaging in acute tendon ruptures. *Skeletal Radiol* 1986;15(8):619–621.
7. Beltran J, Noto AM, Herman LJ, et al. Tendons: High-field-strength surface coil MR imaging. *Radiology* 1987;162:735–740.
8. Shellock FG, Mandelbaum B. Kinematic MRI of the joints. IN: Mink JH, Deutsch AD, eds. *MRI of the musculoskeletal system: A teaching file.* New York: Raven Press; 1990.
9. Shellock FG, Mink JH, Deutsch AD, et al. Evaluation of patellar tracking abnormalities using kinematic MR imaging. Clinical experience in 130 patients. *Radiology* 1989;172:799–804.
10. Shellock FG, Mink JH, Deutsch AD. Kinematic MRI of the joints. Magn Reson Q 1991 (in press).
11. Shellock FG, Mink JH, Deutsch AD, et al. Evaluation in patients with persistent symptoms after lateral retinacular release by kinematic magnetic resonance imaging of the patellofemoral joint. *Arthroscopy* 1990;6:226–234.
12. Shellock FG, Cohen MS, Brady T, et al. Evaluation of patellar alignment and tracking: Comparison between kinematic MRI and true dynamic imaging by hyperscan MR. *JMRI* 1991 (in press).
13. Shellock FG, Foo TK, Duetsch AD, et al. Evaluation of the patellofemoral joint during active flexion by ultra-fast spoiled GRASS MR imaging. *Radiology* 1991 (submitted).
14. Barasch E, Lombardi L, Arena L, et al. MRI visualization of bilateral quadriceps tendons rupture in a patient with secondary hyperparathyroidism: Implications for diagnosis and therapy. *Comput Med Imaging Graph* 1989;5:407.
15. Berlin RC, Levinsohn EM, Chrisman H. The wrinkled patellar tendon: An indication of abnormality in the extensor mechanism of the knee. *Skeletal Radiol* 1991;20:181–185.
16. Hall FM, Wyshak G. Thickness of articular cartilage in the normal knee. *J Bone Joint Surg* 1980;62A:408–413.
17. Lehner KB, Rechl HP, Gmeinwieser JK, et al. Structure, function, and degeneration of bovine hyaline cartilage: Assessment with MR imaging in vitro. *Radiology* 1989;170:495.
18. Handelberg F, Shahabpour M, Casteleyn PP. Chondral lesions of the patella evaluated with computed tomography, magnetic resonance imaging, and arthroscopy. *J Arthrsc* 1990;6(1):24–29.
19. Hayes CW, Sawyer RW, Conway WF. Patellar cartilage lesions: In vitro detection and staging with ME imaging and pathologic correlation. *Radiology* 1990;176:479–483.
20. Wojtys E, Wilson M, Buchwalter K, et al. Magnetic resonance imaging of knee hyaline cartilage and intraarticular pathology. *Am J Sports Med* 1987;15(5):455–463.
21. Mink JH, Deutsch A. The knee. IN: Mink J, Deutsch A, eds. *MRI of the musculoskeletal system: A teaching file.* New York: Raven Press; 1990.
22. Kujala UM, Osterman K, Kormano M, et al. Patellofemoral relationship in recurrent patellar dislocation. *Radiology* 1990;175:886.
23. Kujala UM, Osterman K, Kormano M, et al. Patellar motion analyzed by magnetic resonance imaging. *Acta Orthop Scand* 1989;60(1):13–16.
24. Kujala UM, Osterman K, Kormano M, et al. Patellofemoral

relationships in recurrent patellar dislocation. *J Bone Joint Surg* 1989;71B:788–792.
25. Shellock FG, Foo T, Mink JH. Dynamic imaging of the joints by ultrafast MRI techniques. IN: Proceedings of the European Congress of Radiology, 1991. Vienna.
26. Shellock FG, Fox JM, Deutsch AD, et al. Medial subluxation of the patella: Radiologic and physical findings. *Radiology*, 1991.
27. Muhle C, Brossman J, Schroeder C, et al. MRI-movies of the knee. A comparison of motion triggered active movements and pseudo cine studies. *Book of Abstracts, Society of Magnetic Resonance in Medicine*. 1991;1:60.
28. Shellock FG, Cohen MS, et al. Evaluation of patellar alignment and tracking: Comparison between kinematic MRI and "true" dynamic imaging by hyperscan MRI. *J Magn Reson Imaging* 1991;1:148.
29. Shellock FG, Foo TKF, Deutsch AD, et al. Patellofemoral joint: Evaluation during active flexion with ultrafast spoiled GRASS MR imaging. *Radiology* 1991;180:581–585.
30. Resnick D, Niwayama G. Anatomy of individual joints. IN: Resnick D, Niwayama G, eds. *Diagnosis of bone and joint disorders*. Philadelphia: WB Saunders; 1988:647–755.
31. Fulkerson JP, Hungerford DS. *Disorders of the patellofemoral joint*, 2nd ed. Baltimore: Williams & Wilkins; 1990.
32. Kaplan E. Some aspects of functional anatomy of the human knee joint. *Clin Orthop* 1962;23:18.
33. Fox JM, Sherman OH, Pevsner D. Patellofemoral problems and malalignment. IN: John MD, McGinty B, eds. *Arthroscopic surgery update*. Rockville: Aspen; 1985:31–57.
34. Hallisey M, Doherty N, Bennett W, et al. Anatomy of the junction of the vastus lateralis tendon and the patella. *J Bone Joint Surg* 1987;69A:545.
35. Hehne HJ. Biomechanics of the patellofemoral joint and its clinical relevance. *Clin Orthop* 1990;258:73–85.
36. Insall J, Salvati E. Patella position in the normal knee joint. *Radiology* 1971;101:101–104.
37. Insall J, Goldberg V, Salvati E. Recurrent dislocation and the high-riding patella. *Clin Orthop* 1972;88:67–69.
38. Jelaso DV, Morris GA. Rupture of the quadriceps tendons: Diagnosis by arthroscopy. *Radiology* 1975;116:621–622.
39. King JB, Perry DJ, Kumar SJ. Lesions of the patellar ligament. *Radiology* 1990;177:294.
40. Siwek CW, Rao JP. Ruptures of the extensor mechanism of the knee joint. *J Bone Joint Surg [Am]* 1981;63A:932–937.
41. Larsen E, Lung PM. Ruptures of the extensor mechanism of the knee joint. Clinical results and patellofemoral articulation. *Clin Orthop [Am]* 1986;213:150–153.
42. Nance EPJ, Kaye JJ. Injuries of the quadriceps mechanism. *Radiology* 1982;142:301–307.
43. Kricun R, Kricun ME, Arangio GA, et al. Patellar tendon rupture with underlying systemic disease. *AJR* 1980;135(4).
44. Ramsey RH, Mueller GE. Quadriceps tendon rupture: A diagnostic trap. *Clin Orthop* 1970;70:161.
45. Stephens BO, Anderson GVJ. Simultaneous bilateral quadriceps tendon rupture: A case report and subject review. *J Emerg Med [Am]* 1987;5(6):481–485.
46. Sherlock DA, Hughes A. Bilateral spontaneous concurrent rupture of the patellar tendon in the absence of associated local or systemic disease. *Clin Orthop [Am]* 1988;237:179–183.
47. Weissman BN, Sledge CB. The knee. IN: Weissman BN, Sledge CB, eds. *Orthopedic Radiology*. Philadelphia: W.B. Saunders; 1986;497–587.
48. Levy M. Goldstein J, Rosner M. A method of repair for quadriceps tendon or patellar ligament (tendon) ruptures without cast immobilization. Preliminary report. *Clin Orthop* 1987;218:297–301.
49. Newburg A, Wales L. Radiographic diagnosis of the quadriceps tendon rupture. *Radiology* 1977;125:367.
50. Aprin H, Broukhim B. Early diagnosis of acute rupture of the quadriceps tendon by arthrography. *Clin Orthop* 1985;195:185–190.
51. Resnick D. Arthrography, tenography, and bursography. IN: Resnick D, Niwayama G, eds. *Diagnosis of bone and joint disorders*. Philadelphia: W.B. Saunders; 1988.
52. Fleckenstein JL, Shellock FG. Exertional muscle injuries: magnetic resonance imaging evaluation. *Top Magn Reson Imaging* 1991;3:50–70.
53. Mitchell D, et al. The biophysical basis of tissue contrast in extracranial MR imaging. *AJR* 1987;149:831.
54. Lynch TCP, Crues JV, Morgan FW, et al. Bone abnormalities of the knee: Prevalence and significance at MR imaging. *Radiology* 1989;171:761.
55. Mink JH, Deutsch AL, Kerr R. Tendon injuries of the lower extremity: Magnetic resonance assessment. *Top Magn Reson Imaging* 1991;3:23–38.
56. Blazina M, Kerlan R, Jobe F, et al. Jumper's knee. *Orthop Clin North Am* 1973;4:665.
57. Resnick D, Niwayama G. Diagnosis of bone and joint disorders, 2nd ed. Philadelphia: W.B. Saunders; 1988.
58. Resnick D. Physical injury. IN: Resnick D, Niwayama G. eds. *Diagnosis of bone joint disorders*, 2nd ed. Philadelphia: W.B. Saunders; 1988;2756–3008.
59. Donati RB, Cox S, Echo BS, et al. Bilateral simultaneous patellar tendon rupture in a female collegiate gymnast. A case report. *Am J Sports Med* 1986;14(3):237–239.
60. Peiro A, Ferrandis R, Garcia L, et al. Simultaneous and spontaneous bilateral rupture of the patellar tendon in rheumatoid arthritis: A case report. *Acta Orthop Scand* 1975;46:700.
61. Burgess RC, Guise ER. Infrapatellar tendon ruptures. *Orthopedics* 1985;8:362–364.
62. Daffner R, Reimer B, Lupetin A, et al. Magnetic resonance imaging in acute tendon ruptures. *Skeletal Radiol* 1986;15:619–621.
63. Zernicke RF, Jobe FW. Human patellar tendon rupture, a kinetic analysis. *J Bone Joint Surg* 1977;59A:179–183.
64. Noyes RF, Butler DL, Paulos LE, et al. Intra-articular cruciate reconstruction. 1: Perspectives on graft strength, vascularization, and immediate motion after replacement. *Clin Orthop* 1983;172:71–77.
65. Apple JS, Martinez S, Hardaker WT, et al. Synovial plicae of the knee. *Skeletal Radiol* 1982;7:251–254.
66. Deutsch AL, Resnick D, Dalinka MK, et al. Synovial plicae of the knee. *Radiology* 1981;141:627–634.
67. Hungerford D, Barry M. Biomechanics of the patellofemoral joint. *Clin Orthop* 1979;144:9.
68. Kaufer H. Patellar biomechanics. *Clin Orthop* 1979;144:51–54.
69. Townsend PR, Rose RM. The biomechanics of the human patella and its implications for chondromalacia. *J Biomechan* 1977;10:403–407.
70. Nordin M, Frankel VH. *Basic biomechanics of the musculoskeletal system,* 2nd ed. Philadelphia: Lea and Febiger; 1989.
71. Terry GC. The anatomy of the extensor mechanism. *Clin Sports Med* 1989;8(2):163–177.
72. Reider B, Marshall JL, Koslin B, et al. The anterior aspect of the knee joint: An anatomical study. *J Bone Joint Surg (Am)* 1981;63:351–356.
73. Carson WG, James SL, Larson RL, et al. Patellofemoral disorders: Physical and radiographic evaluation. Part 2: radiographic evaluation. *Clin Orthop* 1984;185:178–186.
74. Wiberg G. Roentgenographic and anatomic studies on the femoro-patellar joint. With special reference to chondromalacia patellae. *Acta Orthop Scand* 1941;12:319–410.
75. Hennsge J. Arthorsis deformans des patella gleitweges. *Zentrabl Chir* 1962;32:1381–1387.
76. Boven F, Bellemans M, Geurts J, et al. A comparative study of the patello-femoral joint on axial roentgenogram, axial arthogram and computed tomography following arthrography. *Skeletal Radiol* 1982;8:179–181.
77. Inque M, Shino K, Hirose H, et al. Subluxation of the patella: Computed tomography analysis of patellofemoral congruence. *J Bone Joint Surg* 1988;70A:1331.
78. Martinez S, Korobkin M, Fondren FB, et al. A device for computed tomography of the patellofemoral joint. *AJR* 1983;140:400–401.
79. Ficat RP, Phillipe J, Hungerford DS. Chondromalacia patellae: A system of classification. *Clin Orthop* 1979;144:55–62.
80. Ficat RP, Hungerford DS. Disorders of the patellofemoral joint. Baltimore: Williams & Wilkins; 1977.

81. Malghem J, Baudouin M. Depth insufficiency of the proximal trochlear groove on lateral radiographs of the knee: Relation to patellar dislocation. *Radiology* 1989;170:507–510.
82. Buckwalter JA, Rosenberg LC, Hunziker EB. Articular cartilage: Composition, structure, response to injury, and methods of facilitating repairs. IN: Ewing JW, ed. *Articular cartilage and knee joint function: Basic science and arthroscopy*. New York: Raven Press; 1990:19–54.
83. Buckwalter JA. Articular cartilage. IN: *American Academy of Orthopaedic Surgeons instructural course lectures*. St. Louis: C.V. Mosby; 1983:349–370.
84. Mow V, Rosenwasser M. Articular cartilage: Biomechanics. IN: Woo SL, Buckwalter JA, eds. *Injury and repair of the musculoskeletal soft tissue*. Park Ridge, IL: American Academy of Orthopaedic Surgeons; 1988:427–463.
85. Arnoczky SP. Articular cartilage injury and repair. IN: Twelfth international seminar on operative arthroscopy. Kauai, Hawaii: 1990.
86. Buckwalter JA, Hunziker E, Rosenberg LC, et al. Articular cartilage: Composition and structure. IN: Woo SL, Buckwalter JA, eds. Injury and repair of the musculoskeletal soft tissues. Park Ridge, IL: American Academy of *Orthopaedic Surgeons:* 1988:405–425.
87. Reiser MF, Bongartz G, Erlemann R, et al. Magnetic resonance in cartilaginous lesions of knee joint with three-dimensional gradient-echo imaging. *Skeletal Radiol* 1988;17(7):465–471.
88. Totterman S, Weiss SL, Szumowski J, et al. MR fat suppression technique in the evaluation of normal structures of the knee. *JCAT* 1989;13:473.
89. Gylys-Morin VM, Hajek PC, Sartoris DJ, et al. Articular cartilege defects: Detectability in cadaver knees with MR. *AJR* 1987;148:1153–1157.
90. Konig H, Sauter R, Deimling M, et al. Cartilage disorders: Comparison of spin-echo, CHESS, and FLASH sequence MR images. *Radiology* 1987;164:753–758.
91. Yulish BS, Mulopulos GP, Goodfellow DB, et al. MR imaging of osteochondral lesion of talus. *J Comput Assist Tomogr* 1987;11(2):296–301.
92. De Smet AA, Fisher DR. Value of MR imaging in staging osteochondral lesions of the talus (osteochondritis dissecans): Results in 14 patients. *AJR* 1990;154:555.
93. Nelson DW, DiPaola J, Colville M, et al. Osteochondritis dissecans of the talus and knee: Prospective comparison of MR and arthroscopic classifications. *J Comput Assist Tomogr* 1990;14(5):804–808.
94. Modl JM, Sether LA, Haughton VM, et al. Articular cartilage: Correlation of histologic zones with signal intensity at MR imaging. *Radiology* 1991;181:853–855.
95. Haacke E. Image behavior: Resolution, signal-to-noise, contrast and artifacts. IN: Magnetic resonance imaging of the spine, 1989. Chicago: Year Book; 1989:1–34.
96. Weherli FW. *Principles of magnetic resonance*. IN: Stark DD, Bradley WG, eds. *Magnetic resonance imaging*. St Louis: C.V. Mosby; 1987:3–23.
97. Haake EM, Tkach JA. Fast MR imaging: Techniques and clinical applications. *AJR* 1990;155:951–964.
98. Hammerle CP, Jacob RP. Chondral and osteochondral fractures after luxation of the patella and their treatment. *Arch Orthop Trauma Surg*, 1980;97:207.
99. Yao L, Seeger LL. MR effective at detecting pathology of marrow space. *Diagn Imaging* 1991:116.
100. Dixon WT. Simple proton spectroscopic imaging. *Radiology* 1984;153:189–194.
101. Borrello JA, Chenevert TL, Meyer CR. Chemical shift-based true water and fat images: Regional phase correction of modified spin-echo MR images. *Radiology* 1987;164:531–537.
102. Szumowwski J, Plewes DB. Separation of lipid and water MR imaging signals by chopper averaging in the time domain. *Radiology* 1987;(165):274–250.
103. Braunstein EM, Brandt KD, Albrecht M. MRI demonstration of hypertrophic articular cartilage repair in osteoarthritis. *Skeletal Radiol* 1990;19:335–339.
104. Mankin HJ. The response of articular carilage to mechanical injury. *J Bone Joint Surg* 1982;64A:460.
105. Hirsch C. Contribution to the pathogenesis of chondromalacia of the patella: A physical, histologic and chemical study. *Acta Chir Scand* 1944;90(Suppl 83).
106. Goodfellow J, Hungerford DS, Woods C. Patello-femoral joint mechanics and pathology. II Chondromalacia patella. *J Bone Joint Surg* 1976;58[B]:291–299.
107. Insall J, Falvo KA, Wise DW. Chondromalacia patellae. A prospective study. *J Bone Joint Surg* 1976;58A:1–8.
108. Outerbridge RE. The etiology of chondromalacia patellae. *J Bone Joint Surg* 1961;43B:752–757.
109. Outerbridge RE. Further studies on the etiology of chondromalacia patellae. *J Bone Joint Surg* 1964;46B:179–190.
110. Bentley G, Dowd G, Orth MC. Current concepts of etiology and treatment of chondromalacia patellae. *Clin Orthop* 1984;189:209–228.
111. Ojno O, Naito J, Iguchi T, et al. An electron microscopic study of early pathology in chondromalacia of the patella [see comments]. *J Bone Joint Surg* 1988;70A(6):883–899.
112. Kempson GE. The mechanical properties of articular cartilage. IN: Sokoloff L, ed. *The joints and synovial fluid*. New York: Academic Press; 1980:177–283.
113. Dehaven KE, Solan WA, Mayer PJ. Chondromalacia patellae in athletes. Clinical presentation and conservative management. *J Sports Med* 1979;7:5–11.
114. Minkoff J, Fein L. The role of radiography in the evaluation and treatment of common anarthrotic disorders of the patellofemoral joint. *Clin Sports Med* 1989;8:203.
115. Merchant AC. Classification of patellofemoral disorders. *Arthroscopy* 1988;4(4):235–240.
116. Kaufer H, Michigan AA. Mechanical function of the patella. *J Bone Joint Surg* 1971;53A(8):1551–1560.
117. Mori Y, Kuroki Y, Yamamoto R, et al. Clinical and histological study of patellar chondropathy in adolescents. *Arthroscopy* 1991;7:182–197.
118. Royle SG, Noble J, Davies DRA, et al. The significance of chondromalacic changes on the patella. *Arthroscopy* 1991;7(2):158–160.
119. Outerbridge RE. The problem of chondromalacia patellae. 1975:177–196.
120. Goodfellow JW, Hungerford DS, Zindel M. Patello-femoral joint mechanics and pathology: I. Functional anatomy of the patellofemoral joint. *J Bone Joint Surg* 1976;58B:287.
121. Shahriaree H. Chondromalacia patellae. IN: Lowe EP, ed. *O'Connor's textbook of arthroscopic surgery*. Philadelphia: Lippincott; 1984;237–262.
122. Shahriaree H. Chondromalacia. *Contemp Orthop* 1985;11:27–39.
123. Chrisman OD. The role of articular cartilage in patellofemoral pain. *Orthop Clin* 1986;xx:xxx–xxx.
124. Deutsch AL, Fox J, Mink JH. MR imaging of chondromalacia patella. 1991 (in preparation).
125. Edwards DH, Bentley G. Osteochondritis dissecans patellae. *J Bone Joint Surg [Br]* 1977;59:58.
126. Herzberger M, Schuler P, Rossak K, Osteochondrosis dissecans patellae. *Z Orthop* 1982;120:268.
127. Goergen TG, Resnick D, Greenway G, et al. Dorsal defect of the patella (DDT): A characteristic radiographic lesion. *Radiology* 1979;130:333.
128. Kujala UM, Osterman K, Aalto T, et al. L'incongruenza patellofemorale e' spesso bilaterale. *J Sports Traumatol Rel Res* 1990;12(1):25–31.
129. Ho VB, Kransdorf MJ, Jelinek JS, et al. Dorsal defect of the patella: MR features. *J Comput Assist Tomogr* 1991;15(3):474–476.
130. Jerosch JG, Castro WHM, Jantea C. Stress fracture of the patella. *Am J Sports Med* 1989;17:579–580.
131. Heywood AWB. Recurrent dislocation of the patella. *J Bone Joint Surg* 1961;43B:507–517.
132. Cash JD, Hughston JC. Treatment of acute patellar dislocation. *Am J Sports Med* 1988;16:244–249.
133. Hawkins RJ, Bell RH, Anisette G. Acute patellar dislocations. *Am J Sports Med* 1986;14(2):117–120.
134. Larson RL. Subluxation-dislocation of the patella. IN: Ken-

nedy JC, ed. *The injured adolescent knee*. Baltimore: Williams & Wilkins; 1979;161–204.
135. Hughston J. Subluxation of the patella. *J Bone Joint Surg* 1968;50A:1003–1026.
136. Feagin JA Jr. Subluxation of the patella versus ACL injury. IN: Feagin JA Jr, ed. *The crucial ligaments*. New York: Churchill Livingstone; 1988.
137. Lance E, Deutsch AL, Mink JH. MR imging of patella dislocation 1991 (submitted).
138. Merchant AC. Patellofemoral malalignment and instabilities. IN: Ewing JW, ed. *Articular cartilage and knee joint function: Basic science and arthroscopy*. New York: Raven Press: 1990;79–91.
139. Laurin CA, Dussault R, Levesque HP. The tangential X-ray investigation of the patellofemoral joint: X-ray technique, diagnostic criteria and their interpretation. *Clin Orthop* 1979;144:16–26.
140. Merchant AC, Mercer RL, Jacobsen RH, et al. Roentgenographic analysis of patellofemoral congruence. *J Bone Joint Surg* 1974;56A(7):1391–1396.
141. Aglietti P, Insall JN. Patellar pain and incongruence. *Clin Orthop* 1983;176:217–224.
142. Delgado-Martins H. A study of the position of the patella using computerized tomography. *J Bone Joint Surg* 1979;61:443–444.
143. Boven F, Belleman MA, Geurts J, et al. The value of computed tomography scanning in chondromalacia patella. *Skeletal Radiol* 1982;8:183–185.
144. Schutzer SF, Ramsby GR, Fulkerson JP. Computed tomographic classification of patellofemoral pain patients. *Orthop Clin North Am* 1986;17(2):235–248.
145. Schutzer SF, Ramsby GR, Fulkerson JP. The evaluation of patellofemoral pain using computerized tomography. A preliminary study. *Clin Orthop* 1986;204:286–293.
146. Stanford W, Phelan J, Kathol MH, et al. Patellofemoral joint motion: Evaluation by ultrafast computed tomography. *Skeletal Radiol* 1988;17:487–492.
147. Shellock FG. Patellofemoral joint abnormalities in athletes: Evaluation by kinematic magnetic resonance imaging. *Top Magn Reson Imaging* 1991;3:71–95.
148. Shellock FG, Mink JH, Deutsch AD, et al. Kinematic magnetic resonance imaging for evalutaion of patellar tracking. *Physician Sports Med* 1989;17:99–108.
149. Kujala U. Knee injuries in athletes. *Sports Med* 1986;3:447–460.
150. Shellock FG, Mink JH, Fox J, et al. Kinematic MRI evaluation of symptomatic patients following two or three patellar realignment surgeries. *JMRI* 1991;148:1.
151. Shellock FG, Deutsch AD, Mink JH, et al. Identification of medial subluxation of the patella in a dancer using kinematic MRI of the patellofemoral joint: A case report. *Kinesiol Med Dance* 1991;13:1–9;1991.
152. Kujala U. Knee exertion injuries in adolescents and young adults. A study with special reference to anatomic predisposition. Publications of the Social Insurance Institution, 1986;64.
153. Carson WGJ, James SL, Larson RL, et al. Patellofemoral disorders: Physical and radiographic evaluation. Part II. Radiographic examination. *Clin Orthop* 1984;185:178–186.
154. Moller BN, Krebs B, Jurik AG. Patellofemoral incongruence in chondromalacia and instability of the patella. *Acta Orthop Scand* 1986;57:232–234.
155. Insall J, Falvo KA, Wise DW. Patellar pain and incongruence. II: Clinical application. *Clin Orthop* 1983;176:225–323.
156. Hughston J, Deese M. Medial subluxation of the patella as a complication of lateral retinacular release. *Am J Sports Med* 1988;16:383–388.
157. Foo TKF, Bernstein MA, Holsinger AE, et al. Ultra-fast spoiled gradient recalled (SPGR) imaging aquisition. *Book of Abstracts, Society of Magnetic Resonance in Medicine* 1990;1: 1308.
158. Shellock FG, Mink JH, Foo TF, et al. MRI of the patellofemoral joint: Comparison between passive positioning and active movement imaging techniques. *Radiology* 1992 (in press).
159. Deutsch AL. *Osteochondral lesions of the talar dome*. IN: Deutsch AL, Mink JH, Kerr R, eds. MR imaging of the foot and ankle. New York: Raven Press, 1991.

CHAPTER 7

The Postoperative Knee

Andrew L. Deutsch and Jerrold H. Mink

> No knee is so bad that it cannot be made worse by surgery.
>
> J. Hughston, M.D.

POSTOPERATIVE MENISCUS

The appearance of the postoperative meniscus has received little attention in the magnetic resonance (MR) literature (1–5), despite the fact that the potential role for MR in the evaluation of patients with persistent or recurrent symptoms following knee surgery may well rival or arguably potentially exceed its role in assessment of the virgin knee. Evaluation of the postoperative meniscus with MR often poses a difficult diagnostic challenge. The appearance of the postoperative meniscus is highly variable, reflecting the multiple treatment options available to the arthroscopist. Evaluation of these menisci requires an understanding of the nature and natural history of different meniscal lesions, their consequences, alternative treatment options, and complications of this treatment. As a consequence of the increasing recognition of the deleterious effects of meniscectomy, the current emphasis in orthopedic practice has increasingly been toward meniscal preservation. This may take the form of conservative care, partial meniscectomy, or most recently, meniscal repair. In this section we will review salient aspects of meniscal therapy in an effort to provide the reader with a basis for understanding the appearance of the postoperative meniscus. The expected MR characteristics of the normal postoperative meniscus as well as the criteria for diagnosis of recurrent meniscal injury will be described.

Biomechanics of the Postoperative Knee

Once considered a functionless anatomic vestige and regularly excised with impunity, the important biomechanical properties of the menisci have been convincingly substantiated in multiple scientific investigations (6). The menisci are composed of two principal arrangements of collagen fiber bundles; randomly oriented surface fibers radiating from the periphery to the inner margin, and peripheral longitudinally oriented fibers anchored at the anterior and posterior horns (6). Biomechanically, the circumferential arrangement of the collagen within the periphery is the most important (6). The menisci contribute to joint stability and congruency and absorb force applied across the joint through the process of elastic deformation. They increase the surface area threefold per condyle and thereby increase the contact area on the tibia. Krause et al. demonstrated that menisci deform during loading and subsequently generate hoop stresses within their substance (7). The menisci have been demonstrated to absorb 40% to 70% of static loads applied across the joint (8). This function may be of even greater importance at higher loads (7,8).

Experimental studies have demonstrated that stresses are higher across the knee following meniscectomy. Hargraves and Seedhom have shown that even partial meniscectomy significantly reduces the capacity of the meniscus to assist in load transmission (9). Partial medial meniscectomy reduces the load transmission by 60%, whereas partial lateral meniscectomy has been reported to decrease load transmission by up to 33%. Thus, even partial meniscectomy may fail to avert the long-term degenerative changes that are commonly observed following total meniscectomy.

Shrive et al. demonstrated a loss of weightbearing function by placing a cadaveric knee in a compressive device and measuring joint space narrowing prior to and following meniscectomy (10). Following meniscectomy, the degree of joint space narrowing increased twofold. The investigators attributed this narrowing to both loss of the spacer effect of the menisci and a reduction of the contact area. Upon removal of the cartilage, the contact area decreases to 40% of normal and results in increased contact stress and loss of the normal protective effect of the menisci on articular cartilage (10). Utilizing pressure-sensitive films to measure the contact stresses of the knee, Fukubayashi and Kurosawa demonstrated that the actual average contact stress increases two to three times following meniscectomy (11).

In a study utilizing human knees, Baratz et al. investigated the contact area and peak local contact stresses in the normal knee, the knee having undergone partial meniscectomy, and the knee having undergone total meniscectomy (12). The partial meniscectomy knee showed a decrease contact area of 10% and an increase in peak local contact stresses of 65%. When a complete meniscectomy was performed, the contact area decreased by 75% and the peak local contact stresses increased by 235%. Baratz et al. also investigated the biomechanical effects of partial meniscal tears (13). In knees with nondisplaced peripheral tears, no change in contact area and only slight (4%) increase peak contact stresses were noted when compared to the normal knee. When an experimentally created longitudinal tear underwent open suture repair, there was no change in contact area and peak local stresses were normal. These findings support the rationale of meniscal preservation and repair of these peripheral tears. When a segmental partial meniscectomy which extended to the rim was performed, the contact area decreased by 66%, and the peak local contact stresses increased by 110%. With total meniscectomy, the contact area also decreased by 66%, and the peak local contact stresses increased by 163%. The nearly equivalent effect of *segmental* meniscectomy and *total* meniscectomy, in loadbearing terms, has been attributed to the loss of the hoop tension effect provided by the longitudinal circumferential collagen fibers within the periphery of the meniscus (13). As previously discussed, these important fibers normally function to counteract outward displacing forces and result in the development of critical "hoop forces."

Shrive et al. reported that the hoop tension was lost when a single radial cut was made out to the capsular margin, and that a single radial cut through the meniscus was equivalent to meniscectomy in loadbearing terms (10). When a radial cut is made which disrupts the peripheral circumferential fibers, the meniscus expands outward due to the loss of the hoop tension forces, allowing compressive forces to be transmitted through direct cartilage contact. From these considerations it is clear that in performing a subtotal meniscectomy, *preservation of the meniscal rim* becomes very important in preventing articular cartilage pressure.

Confirmation of these findings is provided by the work of Ahmed and Burke who studied the contact stresses for a bucket-handle tear where the peripheral rim is left intact (14). Contour maps for a nondisplaced bucket-handle tear showed that the meniscus with such a tear functions like an intact meniscus as long as the loose fragment remains in an anatomical position. The contour map of the meniscus following removal of the bucket-handle tear demonstrates an increase in contact stress which is not as great as the increase after a total meniscectomy.

In biomechanical considerations, meniscal function depends on the integrity of the peripheral circumferential fibers providing "hoop tension" as well as the extraarticular tissues which anchor the menisci. Intact menisci increase the contact area and reduce the contact stresses applied across articular surfaces. Partial *segmental* meniscectomy which disrupts the rim also destroys the hoop tension, thus reducing the ability of the meniscus to perform their load sharing function. In contrast, a *partial* meniscectomy which *preserves* a rim of tissue in place maintains significant stress protection for the articular cartilage.

Natural History of Meniscal Pathology

Given the recognized deleterious effects of meniscectomy on articular cartilage, it is appropriate to briefly consider both the prevalence and natural history of meniscal pathology with regard to their implications for therapy. Multiple cadaver studies have been performed and have helped to elucidate both the prevalence and morphology of meniscal tears, many of which may well have been asymptomatic during life (8,15–17). Casscells demonstrated that a third of cadaveric knees had either meniscal or chondral pathology (15). In this series, 52% had degenerative disease and normal menisci, and 8% had meniscal pathology without articular cartilage damage. Casscells concluded that a simple direct relationship between meniscal tears and degeneration of articular cartilage could not be made (15). Noble and Hamblen, in a series of 100 randomly obtained cadaver knees from individuals with an average age of 65 years, found that 73% had at least one abnormal meniscus (17). The most common lesion was a horizontal cleavage tear of the meniscus. The authors speculated that the majority of the individuals were likely asymptomatic. In a series comprising younger cadaver knees, Fahmy et al. detected meniscal pathology in 57% with the horizontal cleavage tear again predominating (16). It is notable that in those specimens demonstrating degenerative joint disease, the best articular cartilage was demonstrated immediately beneath the menisci.

A review of these studies suggests that from 50% to 75% of cadaver knees can be expected to demonstrate meniscal tears, many if not most of which may have been asymptomatic in life (8). The most common meniscal lesion in these series, the horizontal cleavage tear, represents a degenerative lesion that may reflect a normal manifestation of aging (8). These menisci may still continue to perform their biomechanical functions with regard to load distribution, shock absorption, and knee stability. Their removal may deprive the underlying articular cartilage of the protection normally afforded by the meniscus, and may indeed result in a deterioration of the patient's condition and be a factor responsible for "failed" meniscectomy (8).

In recognition of these considerations, a more conservative approach toward meniscal resection, and in particular for degenerative tears, has been advocated by several investigators (6,18–20). It has been suggested in a study by Fahmy et al. that even badly damaged menisci can still protect articular surfaces (16). Lotke and associates advocated removal of degenerative meniscal tears only when an acute tear or mechanical symptoms were present (21). Out of concern for the possible development of degenerative disease, Jones et al. recommended that degenerative tears not be removed in patients older than 40 years (22). In a study of unsatisfactory results following arthroscopic partial meniscectomy, Ferkel et al. identified a high percentage of poor results in horizontal cleavage tears and questioned whether all such tears needed to be removed (20).

Effects of Meniscectomy

Since the time of Fairbank's (23) initial report of the roentgenographic changes demonstrable post meniscectomy, multiple investigators have corroborated the findings of radiographically demonstrable changes of degenerative joint disease following meniscectomy (7,8,18,19,21–33). In his classic work, Fairbank followed 107 post meniscectomy cases at intervals ranging from 3 months to 14 years post surgery (23). He described radiographically evident joint space narrowing, ridge formation, and/or flattening of the femoral condyle in over 60% of the cases, and suggested that these changes were due to the loss of weightbearing function of the meniscus. Gear in 1967 described radiographic changes consisting of joint space narrowing and marginal lipping with osteophyte formation in 62% of knees followed for a minimum of 10 years post meniscectomy (30). Jackson, in a radiographic study utilizing the patient's opposite knee as a control, found a significant increase in degenerative changes in 577 knees following meniscectomy (31). In a long-term follow-up study of 99 patients, Johnson et al. found a 39% incidence of degenerative changes evident radiographically in postmeniscectomy knees (34). This compared to the incidence of degenerative changes of only 6% identified utilizing the uninjured contralateral knee as a control. In a prospective long-term (14 year) follow-up of open meniscectomy in athletes, Jorgensen et al. described progressive radiographic changes in 89% of cases (35).

The deleterious effects of meniscectomy on articular cartilage have also been described in arthroscopic and investigational series (33,36,37). Dandy and Jackson reported on a series of 174 knees with persistent symptoms following meniscectomy and demonstrated chondromalacic changes of the femoral condyle that they believed responsible for the patient's symptoms in 40% of cases (33). Friedman et al. reported on a group of 44 patients with failed partial meniscectomies who subsequently underwent repeat arthroscopy (36). Twenty-nine patients had medial compartment chondromalacia, of which approximately two-thirds was considered grade 3–4, and 16 patients had lateral compartment chondromalacia, again approximately two-thirds of which was grade 3–4. Cox et al. found that meniscectomies in canine knees lead to gross and microscopic degenerative changes (37). These investigators also noted that partial meniscectomies lead to less severe degenerative changes, and additionally suggested that there was a direct relationship between the degree of degenerative change and the amount of meniscus removed. Studies in humans as well as in animal models have demonstrated that the degree of radiographic degenerative change following meniscectomy is directly related to the amount of meniscus removed (8).

The histologic and biochemical changes that occur within articular cartilage following meniscectomy have been studied in animal models as well as in human knees (38). These changes relate to both the cellular constituents and extracellular matrix of articular cartilage. Fundamental to an understanding of these changes is knowledge of the normal structure of articular cartilage, and the interested reader is referred to a more detailed discussion of this subject (see Chapter 3).

The early stages of degenerative arthritis are characterized by a loss of proteoglycan aggregates from the extracellular matrix, resulting in swelling of the cartilage (38). This does not appear to be the result of decreased proteoglycan synthesis by the chondrocytes (indeed, proteoglycan synthesis is increased), but rather may result from insufficient synthesis of link protein required for stabilization of proteoglycan aggregates. Another and possibly more significant factor responsible for the increase in nonaggregating proteoglycans observed in postmeniscectomy cartilage is the presence of proteolytic enzymes that have substrates in the extracellular matrix (38). These enzymes, classified as metalloproteases, are capable of cleaving the core protein and link proteins of proteoglycan aggregates. While present in normal articular cartilage and secreted by chondrocytes, experimental studies have demonstrated a significant increase in the levels of metalloproteases in the cartilage of degenerative arthritis. An inhibitor of the metalloproteases [tissue inhibitor of metalloproteases (TIMP)] is normally present in the matrix and synovial fluid. In degenerated postmeniscectomy cartilage, the level of TIMP remains stable while protease secretion is increased, thereby creating an imbalance contributing to cartilage destruction (38).

One of the earliest detectable changes in degenerative articular cartilage is swelling of the cartilage with increased water content. This is believed to be secondary to mechanical failure of the collagen network. Electron microscopic studies have revealed separation of the radial fibers and cobblestone-like wrinkling of the articular surface in addition to swelling in specimens of tibial and femoral articular cartilage 20 days postmeniscectomy (38). Rapid turnover with an increased synthesis of type II collagen can be demonstrated within hours of surgery and remains elevated even when the joint appears to be in end-stage disease (38).

Current Concepts of Partial Meniscectomy

The concerns with regard to the possible deleterious effects of meniscectomy on articular cartilage have led to the common practice of preserving as much stable meniscus as possible when partial meniscectomy is undertaken. With partial meniscectomy, resection of the unstable component of the meniscus is accomplished utilizing a knife, meniscutome, or basket forceps, and the remaining meniscus is contoured most commonly utilizing a shaver. As much peripheral rim as possible is left intact (20,39). The meniscus may "remodel" to a variable degree post meniscectomy. Arnoczky et al. studied remodeling of the meniscus following partial meniscectomy in 15 mongrel dogs (40). In ten of the subjects some evidence of remodeling was present, although this was not uniform either between menisci or within an individual meniscus. The remodeling of the meniscal surface appears to occur through the organization and maturation of a fibrin clot which adheres to the cut edge of the meniscus. Changes may be seen as early as 2 weeks post operation with the development of a layer of translucent tissue capping the cut edge of the meniscus. Over time this material demonstrates a more fibrous nature.

Meniscal Repair

Increasing recognition of the deleterious effects of both total and partial meniscectomy has given impetus to efforts directed at meniscal salvage and alternate methods of therapy (39,41–50). Conservative treatment of certain meniscal tears and efforts directed at surgical repair of other tears reflect this trend toward meniscal preservation. The scientific basis for this approach lies in the ability of certain meniscal lesions, specifically those within or directly contiguous with the well vascularized outer rim of the meniscus, to demonstrate healing (46,51).

In multiple elegant investigations, Arnoczky et al. have demonstrated the blood supply to the meniscus that provides the basis for meniscal healing and underlies the rationale for attempts at meniscal repair (44,46,51). Knowledge of this vascular anatomy is fundamental to understanding which meniscal lesions are amenable to possible healing and surgical repair. Branches from the superior, inferior, and lateral geniculate arteries form a premeniscal capillary plexus that supplies the peripheral border of the meniscus throughout its attachment to the joint capsule. Vascular injection studies have demonstrated that the degree of vascular penetration is 10% to 30% of the width of the medial meniscus and 10% to 25% of the width of the lateral meniscus (Fig. 7.1) (46).

The peripheral meniscal blood supply is capable of supporting a reparative response similar to that observed in other connective tissue (46,51). Following injury in the peripheral vascularized zone, a fibrin clot rich in inflammatory cells forms and provides a "scaffold" through which the capillary plexus proliferates (44). This sequence eventuates in a cellular fibrovascular scar that bridges the tear and appears continuous with adjacent normal meniscal fibrocartilage. Experimental studies have shown that radial lesions of the meniscus that extend to the synovium are completely healed with fibrovascular scarring by 10 weeks (46). Progressive remodeling of this scar into more normal appearing fibrocartilage occurs over a period of months, although histologic differences can continue to be demonstrated between this reparative scar and native fibrocartilage.

The observation that lesions connected to the peripheral vascular supply of the meniscus can heal has led to the investigation of techniques to "extend" the level of the repair to avascular meniscal lesions (44). One technique involves creating vascular channels through the outer third of the meniscus to allow connection of a tear within the avascular zone with the peripheral blood supply. While experimentally this has been demonstrated to be a viable technique, concern arises that creation of a large enough access channel may disrupt the normal peripheral architecture of the

FIG. 7.1. Meniscal vascularity. Frontal section of the medial meniscus following vascular perfusion with India ink. Branching radial vessels from the perimeniscal capillary plexus (PCP) can be seen penetrating the peripheral border of the meniscus. (Courtesy of Dr. Steven Arnoczky New York, New York. From ref. 175.)

meniscus and ultimately contribute to its destabilization (46,51). Other methods of vascular ingrowth have been proposed and include creation of vascular tunnels, pedicle grafts of synovium, and synovial abrasion. Recently, the placement of an exogenous fibrin clot placed in a stable lesion in the avascular portion of the meniscus has been demonstrated to support a reparative response similar to that seen in the vascular area (42,51). This technique may allow repair of avascular lesions anywhere in the meniscus or optimize repair of lesions in areas of marginal vascularity, although further investigation and clinical experience with this method will be required.

When describing meniscal tears with regard to their potential for healing and repair, it is useful to classify the lesions according to their proximity to the vascular supply and the vascular appearance of the surfaces of the tear (44). Utilizing this system, peripheral meniscocapsular separations are referred to as *"red-red"* tears and have a functional blood supply on both the capsular and meniscal sides of the tear. These lesions lie within 3 mm of the meniscosynovial junction and have the best prognosis for healing (44). A *"red-white"* tear refers to a meniscal rim tear through the peripheral vascular zone. These lesions have a blood supply to the peripheral rim, with the central surface of the lesion devoid of functioning vessels. In theory, these lesions should have sufficient vascularity to heal by fibrovascular proliferation (44). *"White-white"* tears refer to meniscal lesions completely within the avascular zone (greater than 5 mm from the meniscosynovial junction) and are thus without a blood supply. Theoretically, these tears cannot heal unless they are provided with a means of healing potential through vascular access channels, synovial abrasion, or possibly by use of the previously described exogenous fibrin clot.

In general, full thickness longitudinal tears contained within the periphery of the meniscus are the best candidates for repair (42–44,48). The tears should be longer than 5 mm to 10 mm in length and the central component of good quality. Short vertical tears, stable partial thickness tears involving less than 50% of the vertical height of the meniscus, and shallow radial tears of 3 mm depth or less are not considered candidates for repair. These tears can be freshened with a shaver or rasper to facilitate healing and can be managed without placement of sutures. Meniscal repair can be combined with partial meniscectomy. More complex tears including double flaps (single longitudinal with broken handle), double longitudinal, and radial tears can be approached with specialized techniques including fascial sheaths (42).

Meniscal repair can be accomplished with open surgical methods as well as by means of closed arthroscopic techniques. Advocates of open repair suggest that meniscal rim and capsular preparation is better and placement of vertically oriented sutures to reattach the capsular bed to the entire height of the meniscal rim is easier (43,48). With the development of arthroscopic techniques, meniscal repair could be extended beyond the periphery to the more central red-white tears. There are two principal techniques for the arthroscopic repair of meniscal tears: the "inside-out" technique and the "outside-in" technique (44). With the inside-out technique, long flexible needles are passed through a curved cannula and placed across the meniscal tear, allowing placement of a horizontal mattress suture which is tied subcutaneously. Particularly with this arthroscopic technique, special attention must be directed toward the neurovascular bundle which may be damaged by blind passage of the needles (44,52,53). With medial meniscus repair the saphenous nerve is at risk; with lateral repairs near the popliteal hiatus, the peroneal nerve may be injured. The outside-in technique was developed to facilitate repair and avoid injury to the neurovascular structures (44). Concern about neurovascular complications has prompted many surgeons to utilize a limited open approach in which an open incision permits visualization of the vital structures during needle passage and suture ligation, while still allowing arthroscopic guidance of suture placement within the meniscus to be accomplished (44).

The reported success rate for meniscal repair has been impressive (42–44,49,54). These results have been based on second look arthroscopy, arthrography, and clinical follow-up. One caveat to using clinical follow-up as a criteria for success is the possibility of including cases with only partial healing and/or asymptomatic failures in the results (42). These limitations should be considered when evaluating reported success rates. In a detailed study of 80 open peripheral repairs with an average of 5-year follow-up, DeHaven and Sebastianelli reported re-tears in nine patients (11%), with only three of these tears at the original repair site (43). Additionally, six of the nine tears occurred in anterior cruciate ligament (ACL)-deficient knees that were not reconstructed at the time of meniscal repair. Using arthroscopic techniques, reported healing rates have ranged from 78% to 100% (44). More recent reports using second look arthroscopy have attempted to classify the healing into three groups: (a) completely healed, (b) partially healed (at least 50% thickness), and (c) failure to heal (44). Using these criteria, the literature would suggest that 75% of meniscal repairs will heal completely, 15% will partially heal, and 5% to 10% will fail to heal (44). Analysis of reports on meniscal repair suggest that factors contributing to successful results include (a) acute tears, (b) peripheral tears (within 5 mm of the meniscosynovial junction), (c) stable knees (intact ACL or ACL reconstruction), (d) patient, age (younger is better), and (e) use of synovial abrasion or exogenous blood clot (44).

MR IMAGING

Meniscectomy and Partial Meniscectomy

The MR appearance of the postoperative meniscus has received little attention in the imaging literature (1–5). As can be appreciated by the aforementioned considerations, the appearance of the meniscus following surgery is variable, reflecting the multiple options available to the arthroscopist at the time of surgery. Principal among the factors determining the appearance of the postoperative meniscus will be the type, location, and size of the initial tear, and the techniques utilized and approaches adopted by the surgeon in treating the lesion. The extent of the resection performed for any given tear may vary considerably among surgeons. Knowledge of the prior tear pattern and surgical procedure performed is often critical to accurate evaluation. This information, however, is often not available at the time of interpretation of the MR examination, which contributes to diagnostic difficulty. The principal role of MR in the patient with either persistent or recurrent symptoms following meniscal surgery is in assessment of the stability of the remaining meniscal tissue, and differentiating meniscal from extrameniscal causes of pain. This assessment includes evaluation of the meniscal remnant proper and overlying articular cartilage, detection of any loose osteocartilagenous bodies, evaluation of ligament stability, detection of synovitis, and exclusion of unusual complications such as the development of osteonecrosis.

In a preliminary study, the authors of this chapter reviewed the appearance of 39 menisci status post partial meniscectomy, all of which subsequently underwent repeat arthroscopy (4). It was found helpful for purposes of analysis to divide the menisci into three groups based on extent of prior resection. Our subsequent experience, and that of others (1) has supported the concept of postoperative meniscal grouping based on size of the meniscal remnant. Menisci in which only minimal resection had been performed and in which no more than 25% of the meniscus had been removed were classified as group 1 menisci. Menisci in which between 25% and 75% of the meniscus had been resected were grouped together (group 2). Menisci in which extensive resection had been accomplished (>75%) were classified as group 3.

Group 1

In the author's initial study, which remains reflective of our continuing experience, group 1 menisci, in which relatively minimal resection had been accomplished (Fig. 7.2), could be accurately evaluated utilizing slight modifications of conventionally accepted criteria. Of 11 group 1 menisci evaluated in this study, nine were re-torn and two were stable. There was 100% agreement between MR and arthroscopy in evaluation of this group. Eight of nine re-torn menisci were diagnosed on the basis of the classic criteria of grade 3 signal within the meniscus. The two stable menisci were characterized by mildly blunted tips without other abnormality (Fig. 7.3). In a separate study, Smith and Totty reported on the MR appearance of 23 postoperative menisci with arthroscopic correlation and also found it useful to group the menisci by size for assessment of stability (1). In patients in whom the

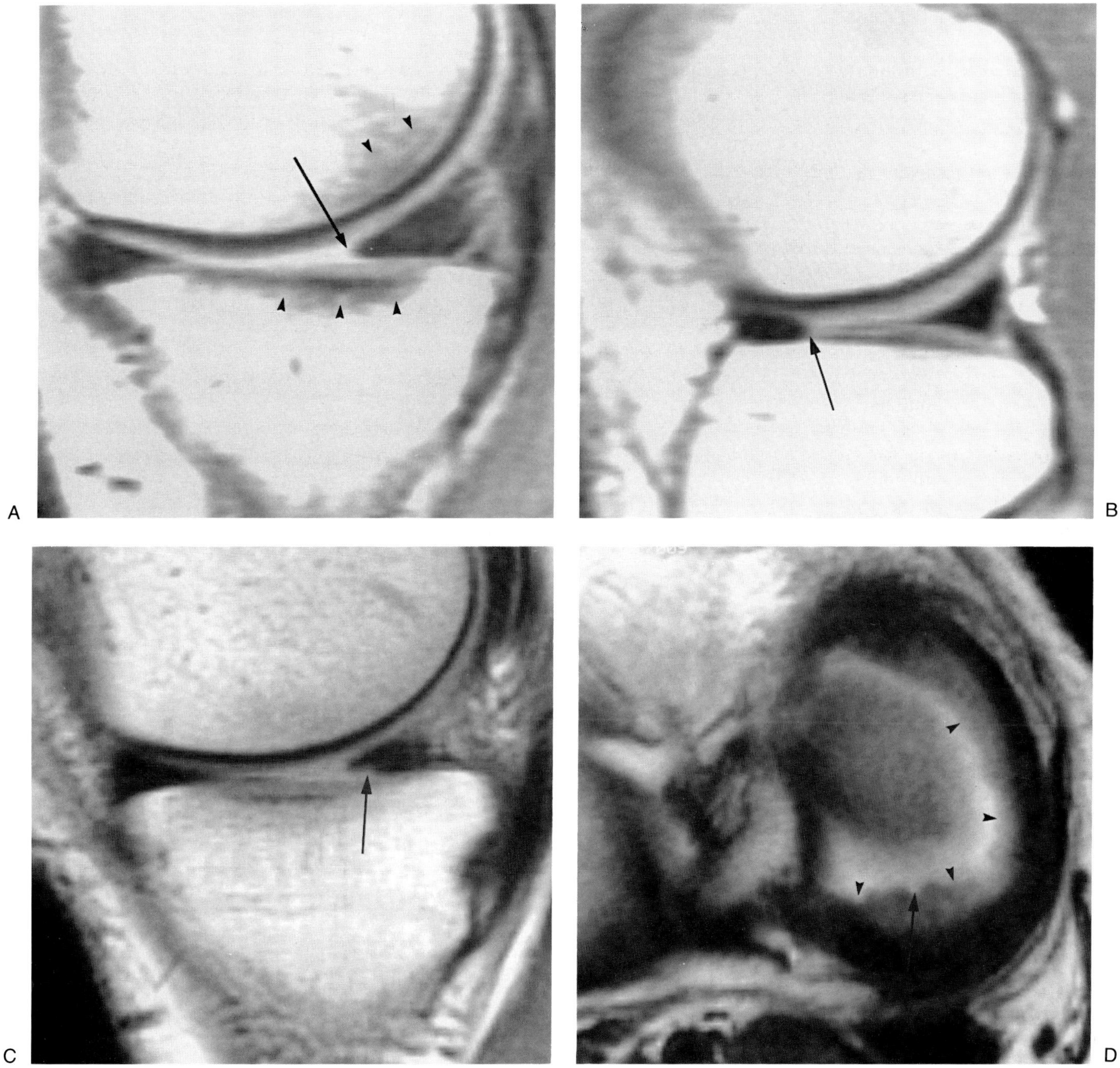

FIG. 7.2. Group 1, stable. **A:** Sagittal MR TR/TE 2,000/20, group 1. Magnified high contrast meniscal "windows." The patient is status post partial medial meniscectomy. There is mild blunting of the apex of the posterior horn, a normal finding following partial meniscectomy (*arrow*). The meniscus is near "normal" in size. Note areas of subchondral decreased signal intensity (better seen on routine windows) within the subchondral bone of the femoral condyle and tibial plateau (*arrowheads*). These findings, presumably reflective of reactive stress changes within cancellous and subchondral bone, are common following meniscal surgery. **B:** Sagittal MR, TR/TE 2,000/20. Sagittal section through the lateral meniscus demonstrates a nearly normal in size meniscus with slight blunting of the apex of the anterior horn (*arrow*). No intrameniscal signal abnormality is present. Such blunting is acceptable in postoperative menisci, although it is critical to determine that a prior partial meniscectomy has been accomplished. In the absence of prior meniscal surgery at this site, the findings reflect a small, displaced tear. **C:** Sagittal MR, TR/TE 2,000/20. There is mild foreshortening and blunting of the posterior horn of the medial meniscus in this patient status post partial meniscectomy. A small focus of signal is seen along the undersurface of the newly contoured free edge (*arrow*). **D:** Axial MR, TR/TE 2,000/80. On this axial section through the meniscus, the small focus of signal seen in Fig. 7.2C is seen to represent a small notch in the newly contoured free edge related to the prior surgery (*arrow*). The free edge of the meniscus is highlighted by high signal intensity joint fluid (*arrowheads*). Minor contour irregularities are commonly seen in stable menisci following surgery, although diagnostic difficulty with regard to meniscal re-tear occurs not uncommonly.

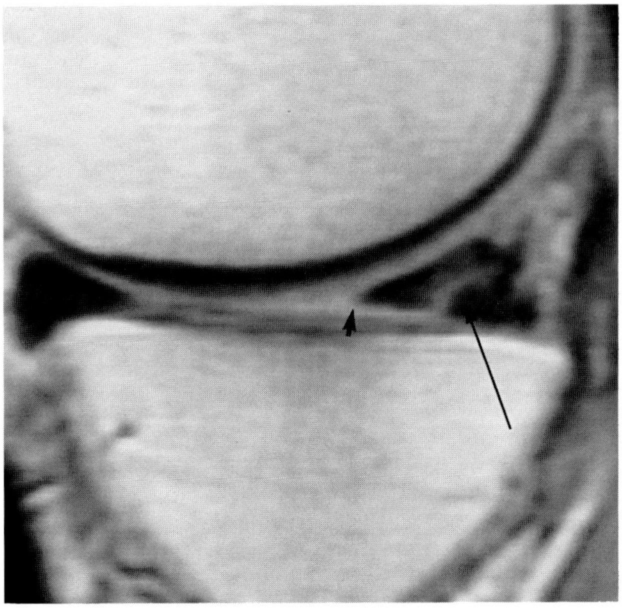

A

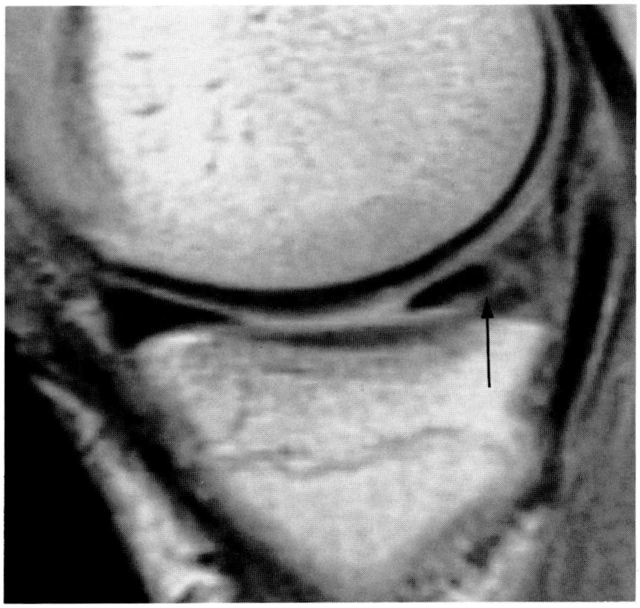

B

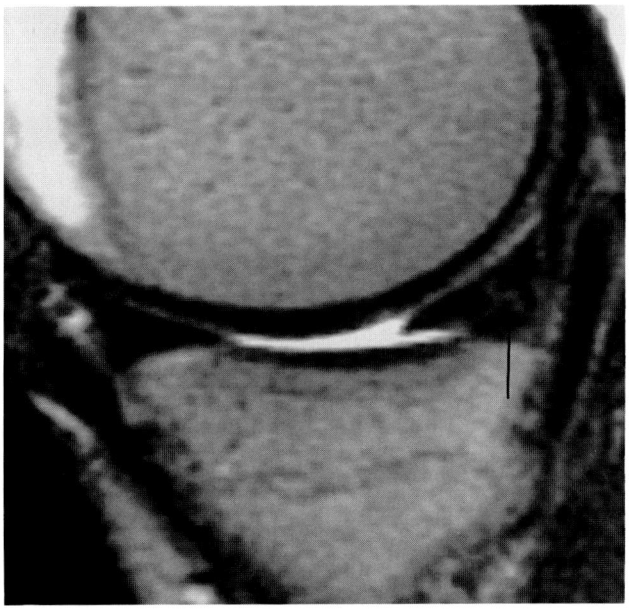

C

FIG. 7.3. Meniscal re-tear (group 1). **A:** Sagittal MR, TR/TE 2,000/20. The posterior horn of the medial meniscus is minimally decreased in size with a slightly blunted tip consistent with limited prior meniscal resection (*short arrow*). A curvilinear line of grade 3 signal is seen traversing the posterior horn (*long arrow*). The criteria of grade 3 signal can be applied to postoperative menisci such as this one with nearly the same confidence as for the virgin meniscus. **B:** Sagittal MR, TR/TE 2,000/20. The posterior horn of the medial meniscus of a different patient is foreshortened and the apex is mildly blunted, findings typical of prior partial meniscectomy. A line of grade 3 signal is seen along the undersurface of the posterior horn remnant (*arrow*) apparently remote from the site of prior meniscal surgery. **C:** Sagittal MR, TR/TE 2,000/80. The line of signal is persistent but not hyperintense (*arrow*). A meniscal re-tear was demonstrated on follow-up arthroscopy.

residual meniscal size was near normal (greater than two-thirds of the meniscal length remaining), MR was highly accurate in detecting meniscal re-tears using conventional criteria. Blunting of the meniscal apex and signal inhomogeneity were common in stable menisci in this group. Based on this reported experience, it appears that for postoperative menisci which are only minimally altered in size and morphology, the presence of grade 3 signal can be employed to diagnose a meniscal tear with an accuracy comparable to that for the virgin meniscus. Blunting of the meniscal tip is an acceptable postoperative appearance for the stable Group 1 meniscus. An important caveat in this regard, however, is to determine that prior partial meniscectomy has been performed. Prior arthroscopy, in and of itself, cannot be equated with meniscal surgery. Meniscal blunting in the absence of prior surgery represents evidence of a meniscal tear. Additionally, while mild blunting of the postoperative meniscal apex is generally acceptable, other significant alterations in meniscal shape and/or size can be utilized to confidently detect meniscal re-tears in this group (Fig. 7.4).

Group 2

Menisci comprising group 2, in which more extensive resection has been performed but in which a significantly sized remnant remains, are among the most problematic for assessment of meniscal stability (Fig. 7.5). In this group, significant shortcomings are encountered in the application of both conventional cri-

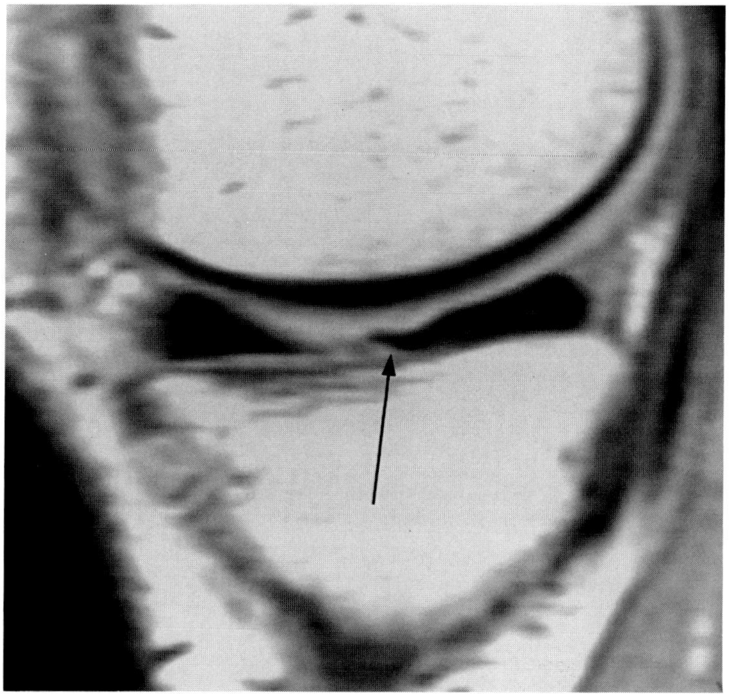

FIG. 7.4. Meniscal re-tear (group 1) abnormal morphology, sagittal MR, TR/TE 2,000/20. The patient is status post medial meniscectomy and now presents for reevaluation following a new injury. Instead of a smoothly rounded tip as may normally be depicted postoperatively, the apex of the meniscus demonstrates a clearly abnormal contour (*arrow*). A flap tear was identified at subsequent arthroscopy. Morphologic criteria can also be utilized in assessment of group 1 menisci.

teria for meniscal tears, the presence of grade 3 signal, and the presence of irregular and misshapen meniscal morphology. Of 14 group 2 menisci evaluated in the author's study, eight were re-torn and the remainder were stable on follow-up arthroscopy. There was agreement between MR and arthroscopy in only nine of 14 cases. In four of the cases falsely positive for meniscal re-tear, apparent grade 3 signal was present

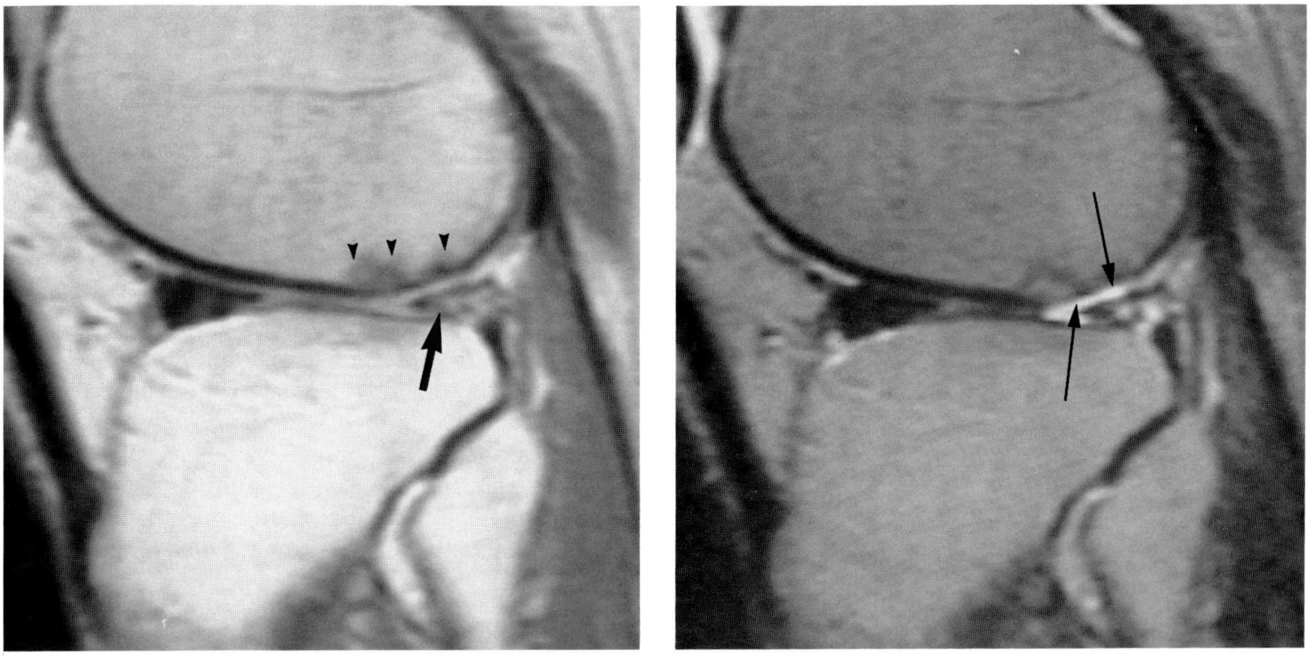

A B

FIG. 7.5. Group 2 menisci. **A:** Sagittal MR, TR/TE 2,000/20. The patient is status post partial lateral meniscectomy involving the posterior horn and midzone. The anterior horn is normal. The posterior horn is small but has been recontoured to nearly a triangular appearance (*arrow*). There is loss of articular cartilage and changes extending into subchondral bone (*arrowheads*), findings which are typical following partial meniscectomy. **B:** Sagittal MR, TR/TE 2,000/80. The synovial fluid provides an arthrogram effect and enhances depiction of meniscal morphology. There is full thickness loss of articular cartilage in this patient three years status post surgery (*arrows*). The remnant is degenerated but stable.

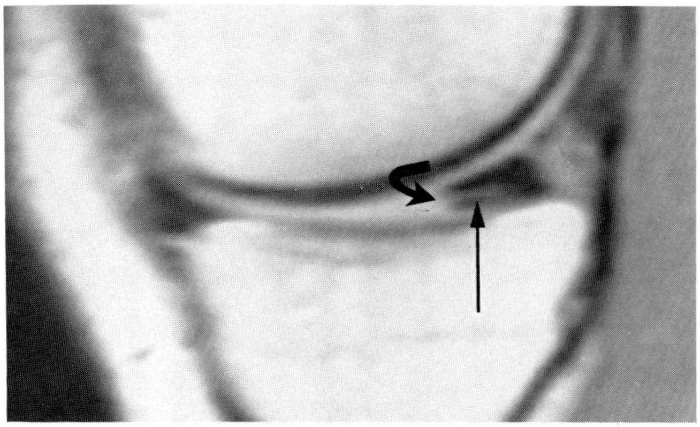

FIG. 7.6. False-positive meniscal re-tear. Sagittal MR, TR/TE 2,000/80. A line of linear signal traverses the posterior horn remnant (*arrow*) and extends to the recontoured meniscal free edge (*curved arrow*). The findings suggest grade 3 signal and thus a re-tear of the posterior horn remnant. The meniscus, however, was considered stable at repeat arthroscopy (see Fig. 7.8).

in the recontoured menisci which were subsequently considered stable at arthroscopy (Fig. 7.6 and see below). An additional factor greatly complicating interpretation of these menisci is the fact that the morphology of the newly contoured meniscal free edge may be highly variable and is some cases clearly irregular even in stable menisci (Fig. 7.7). Difficulty in assessment of menisci having undergone more exten-

A

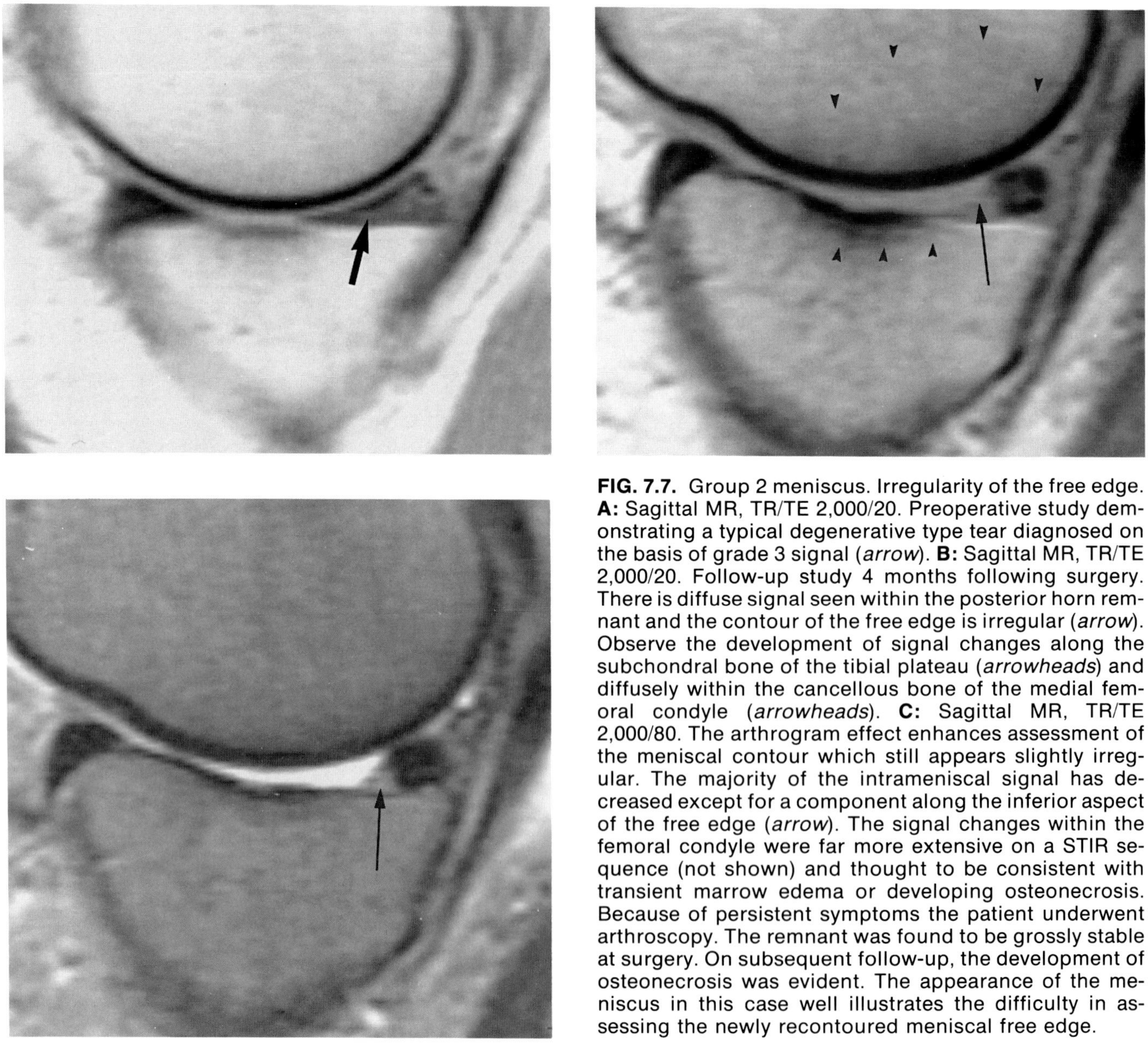

B

C

FIG. 7.7. Group 2 meniscus. Irregularity of the free edge. **A:** Sagittal MR, TR/TE 2,000/20. Preoperative study demonstrating a typical degenerative type tear diagnosed on the basis of grade 3 signal (*arrow*). **B:** Sagittal MR, TR/TE 2,000/20. Follow-up study 4 months following surgery. There is diffuse signal seen within the posterior horn remnant and the contour of the free edge is irregular (*arrow*). Observe the development of signal changes along the subchondral bone of the tibial plateau (*arrowheads*) and diffusely within the cancellous bone of the medial femoral condyle (*arrowheads*). **C:** Sagittal MR, TR/TE 2,000/80. The arthrogram effect enhances assessment of the meniscal contour which still appears slightly irregular. The majority of the intrameniscal signal has decreased except for a component along the inferior aspect of the free edge (*arrow*). The signal changes within the femoral condyle were far more extensive on a STIR sequence (not shown) and thought to be consistent with transient marrow edema or developing osteonecrosis. Because of persistent symptoms the patient underwent arthroscopy. The remnant was found to be grossly stable at surgery. On subsequent follow-up, the development of osteonecrosis was evident. The appearance of the meniscus in this case well illustrates the difficulty in assessing the newly recontoured meniscal free edge.

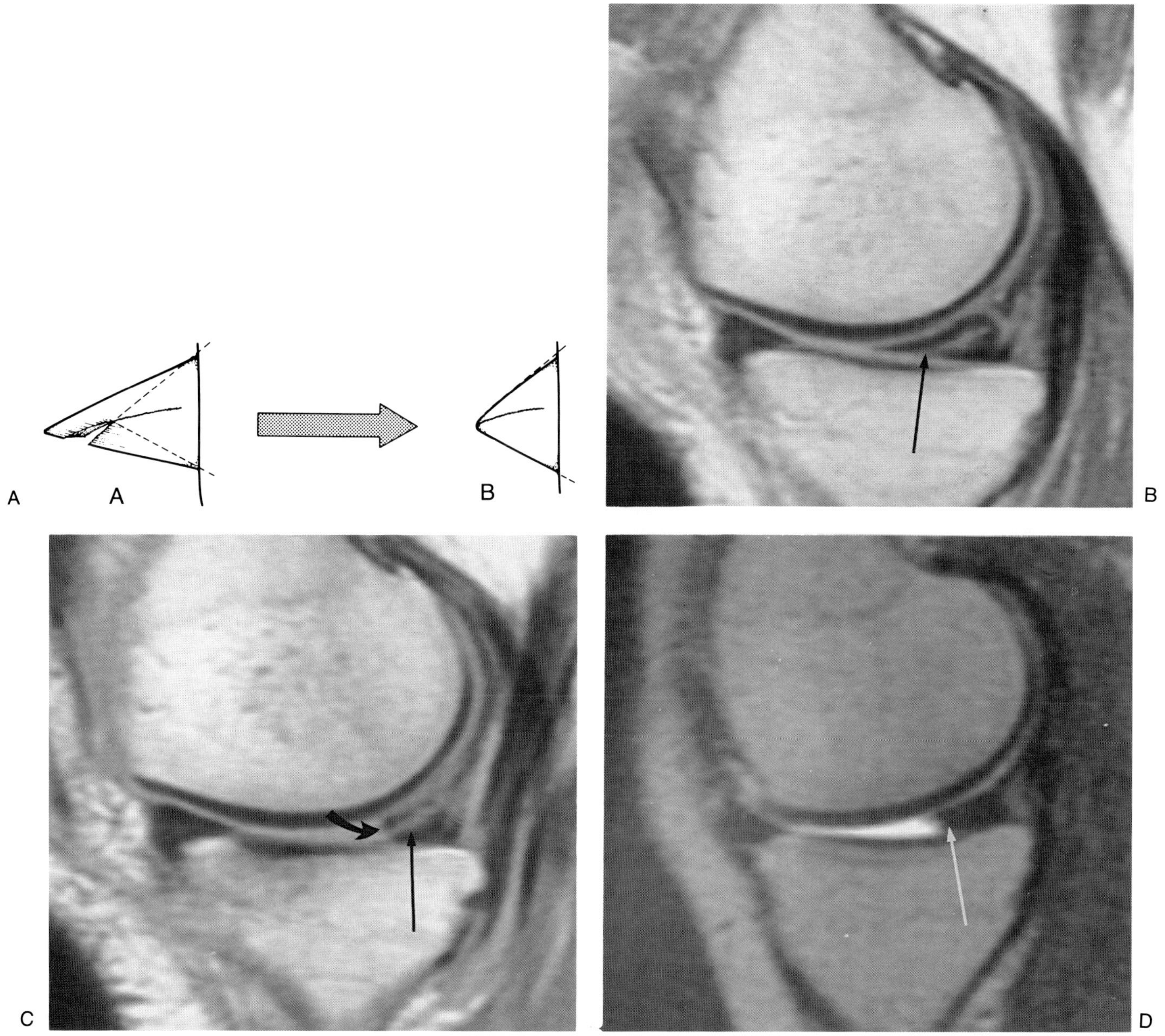

FIG. 7.8. Intrameniscal signal conversion. **A:** Schematic diagram demonstrating the potential conversion of intrasubstance grade 2 signal to mimic a grade 3 tear. The extent of the arthroscopically demonstrable meniscal tear is less than the degree of abnormal signal seen on the MR (A). Following partial meniscal resection (*dotted lines*), the postoperative meniscus (B), which is clinically stable, could contain intrasubstance signal that now appears to extend toward the articular surface and could thus be confused with grade 3 signal indicative of a meniscal tear. (From ref. 176, with permission.) **B:** Sagittal MR, TR/TE 2,000/20. Preoperative MR demonstrating nondisplaced horizontal tear of the posterior horn of the medial meniscus (arrow). **C:** Sagittal MR, TR/TE 2,000/20. Postoperative study (group 2 meniscus) demonstrating blunting of the meniscal tip (*curved arrow*) and intrameniscal signal extending to the recontoured meniscal margin (*long arrow*) in a manner directly analogous to that illustrated in the schematic diagram. The meniscus was proven stable at follow-up arthroscopy. **D:** Sagittal MR, TR/TE 2,000/80. The existing synovial fluid provides a natural contrast medium and highlights the meniscal contour. No high signal intensity fluid is seen extending into the previously identified line to suggest fluid tracking within a meniscal tear (*arrow*). Indeed, the intrameniscal signal is not demonstrable on this sequence. Such findings can be of assistance in excluding re-tear of the meniscal remnant with increased confidence.

sive resection (one-third to two-thirds of the meniscus resected) was also reported in the series of Smith and Totty (1). In their series, marked contour irregularity simulating fragmentation was commonly encountered (nearly 30% of meniscal segments), significantly complicated interpretation, and contributed to the lack of recognition of nondisplaced tears in some cases (1).

Several factors which contribute to the diagnostic difficulty observed particularly in this group need to be considered. As previously discussed, a basic tenent

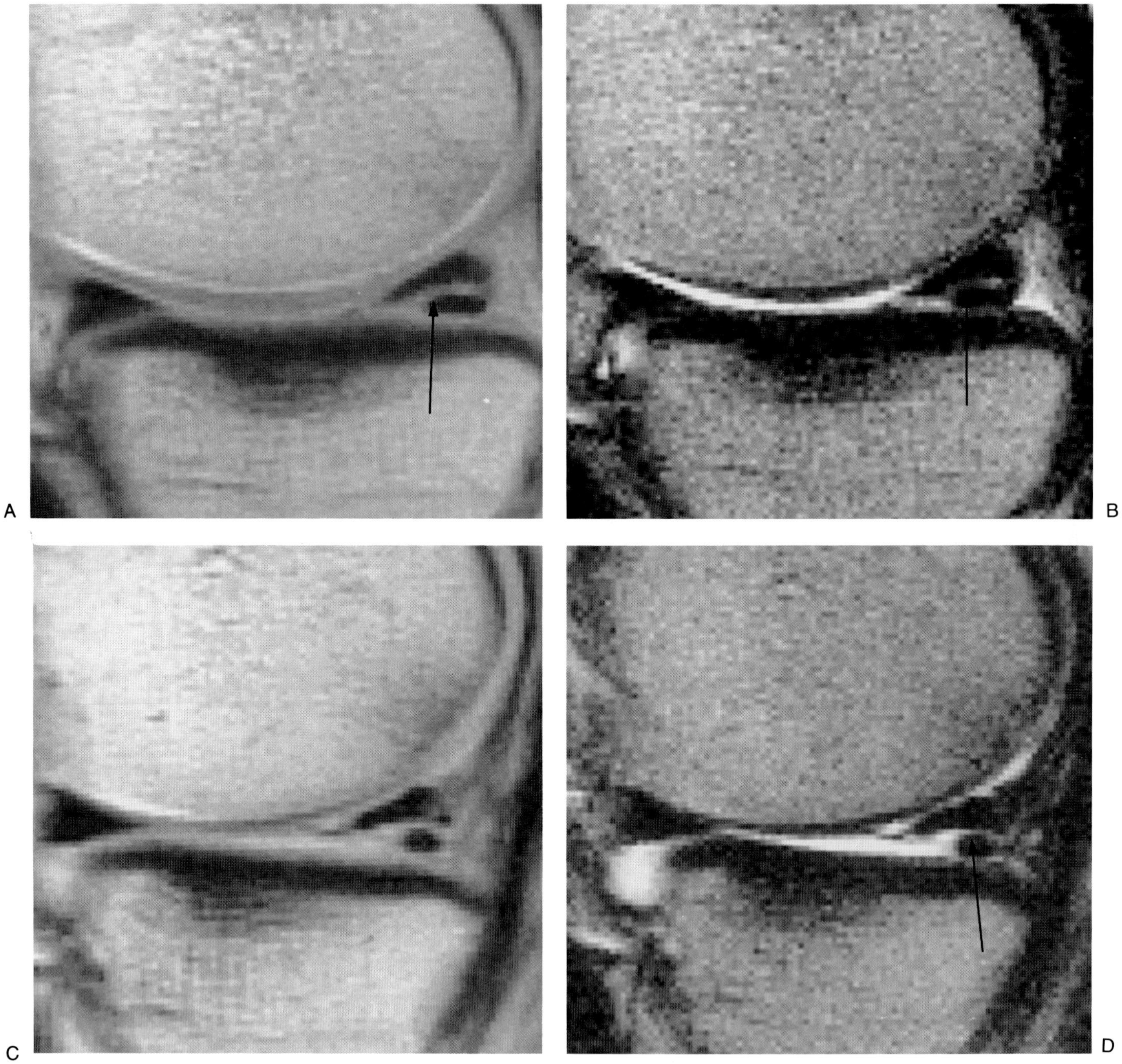

FIG. 7.9. Meniscal re-tear with high signal intensity. **A:** Sagittal MR, TR/TE 2,500/25. Preoperative study demonstrating classic grade 3 signal within the posterior horn of the medial meniscus (*arrow*). **B:** Sagittal MR, TR/TE 2,500/90. Preoperative study demonstrating high signal intensity within the area of altered signal on the T2-weighted image (*arrow*). The findings of a meniscal tear were confirmed at surgery at which time the patient underwent a partial meniscectomy. **C:** Sagittal MR, TR/TE 2,500/25. Postoperative study. The patient returned for a repeat examination following a new injury. A well-defined focus of grade 3 signal is seen involving the undersurface of the posterior horn remnant (*arrow*). **D:** Sagittal MR, TR/TE 2,500/90. Postoperative study. The focus of grade 3 signal demonstrates striking increased intensity (*arrow*), presumably reflecting the extension of synovial fluid into the tear. A re-tear of the posterior horn remnant was identified at surgery. (Courtesy of William Totty, M.D. St. Louis, Missouri.)

of present arthroscopic surgery is restraint in meniscal resection (39). The challenge for the orthopedic surgeon is in balancing the need for debridement of unstable components of the tear with preservation of as much tissue as possible to evert the deleterious consequences upon the articular cartilage commonly seen following meniscectomy. In degenerative horizontal cleavage tears, this may involve resection of only one of two components of the tear with recontouring of the remaining meniscus. Alternatively, out of concern for leaving a potentially unstable remnant that might be subject to re-tear, some arthroscopists may elect to more extensively resect the same meniscal lesion. These considerations underscore the difficulty inherent in both morphologic analysis of these menisci and assessment of residual intrameniscal signal. If the surgeon, in an effort to preserve the peripheral rim, elects to preserve the "nonsignificant" horizontal component of a complex tear, it may contribute to a false-positive diagnosis of re-tear on MR. In patients with horizontal cleavage or "fish mouth" tears, the line of abnormal signal depicted on MR frequently extends from the articular surface of the meniscus to near the capsular margin. The recognized extent of the tear at arthroscopy, however, is almost invariably less than the extent of the signal abnormality depicted on MR. When the meniscus is cut and contoured to the base of the tear, it becomes theoretically possible to create from the remaining intrasubstance (grade 2) signal an appearance that could mimic a grade 3 tear (Fig. 7.8). We have determined this occurrence signal "conversion," and it is important to recognize this possible occurrence and not to necessarily interpret such potentially stable menisci as being "re-torn."

The presence of increased signal intensity within the focus of grade 3 signal on T2-weighted images may represent a highly specific finding for a meniscal tear and may allow far greater confidence in suggesting the presence of a significant meniscal re-tear on the basis of intrameniscal signal grading (Fig. 7.9) (5,55,59a).

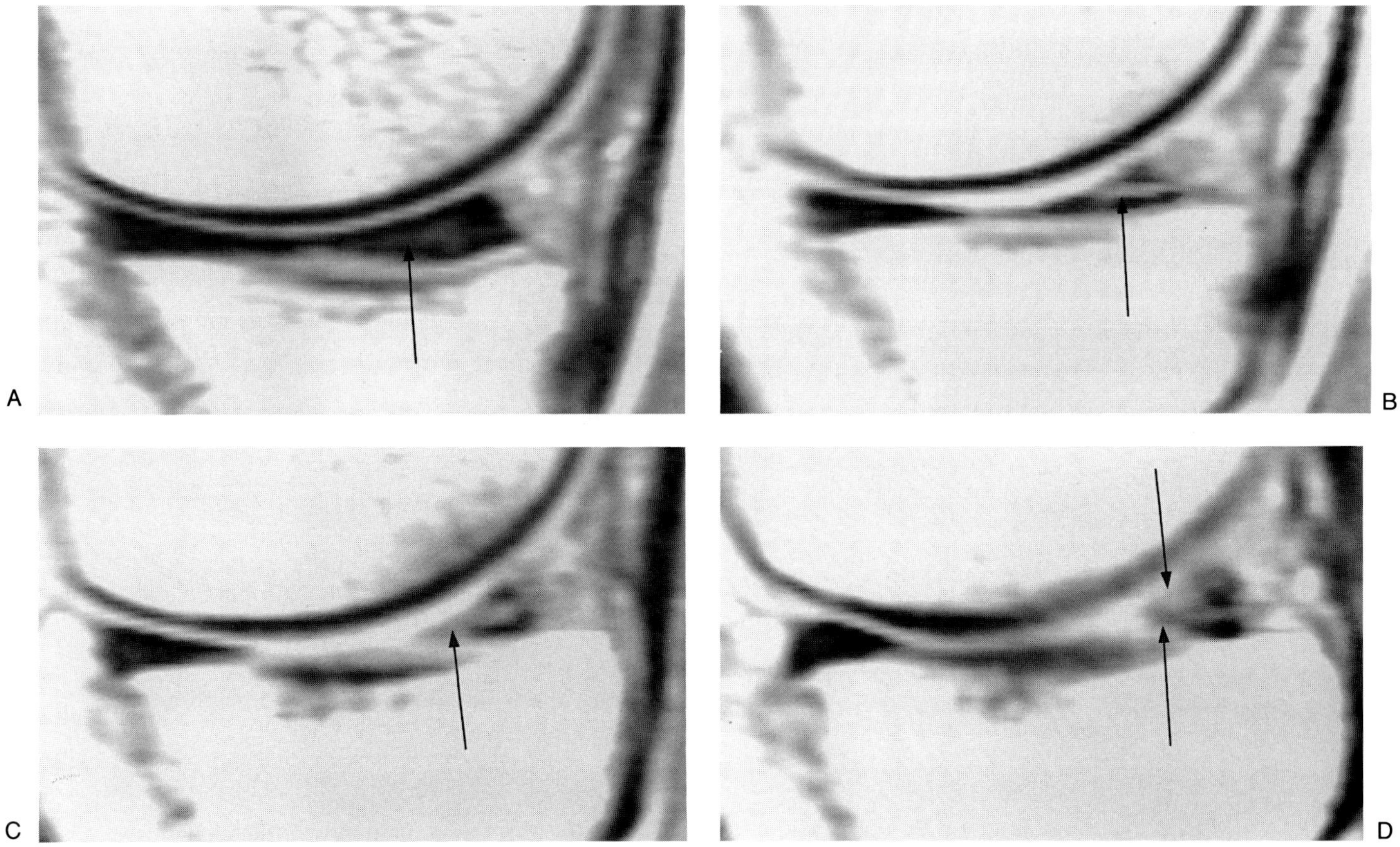

FIG. 7.10. Group 2, meniscal re-tear, abrupt change in meniscal morphology. **A:** Sagittal MR, TR/TE 2,000/20. Preoperative study demonstrating a horizontal cleavage tear within the posterior horn of the medial meniscus (*arrow*). **B:** Sagittal MR, TR/TE 2,000/20. Postoperative study demonstrating signal equivocally extending to the articular surface (*arrow*) and quite possibly reflecting "signal conversion." **C:** Sagittal MR, TR/TE 2,000/20. On the adjacent section diffuse signal is seen within the tip of the remnant. The contour remains grossly normal. **D:** Sagittal MR, TR/TE 2,000/20. An abrupt change in morphology is noted with the meniscus appearing increased in height. The contour is grossly irregular (*arrows*). A flap tear folded upon itself was detected at subsequent arthroscopy.

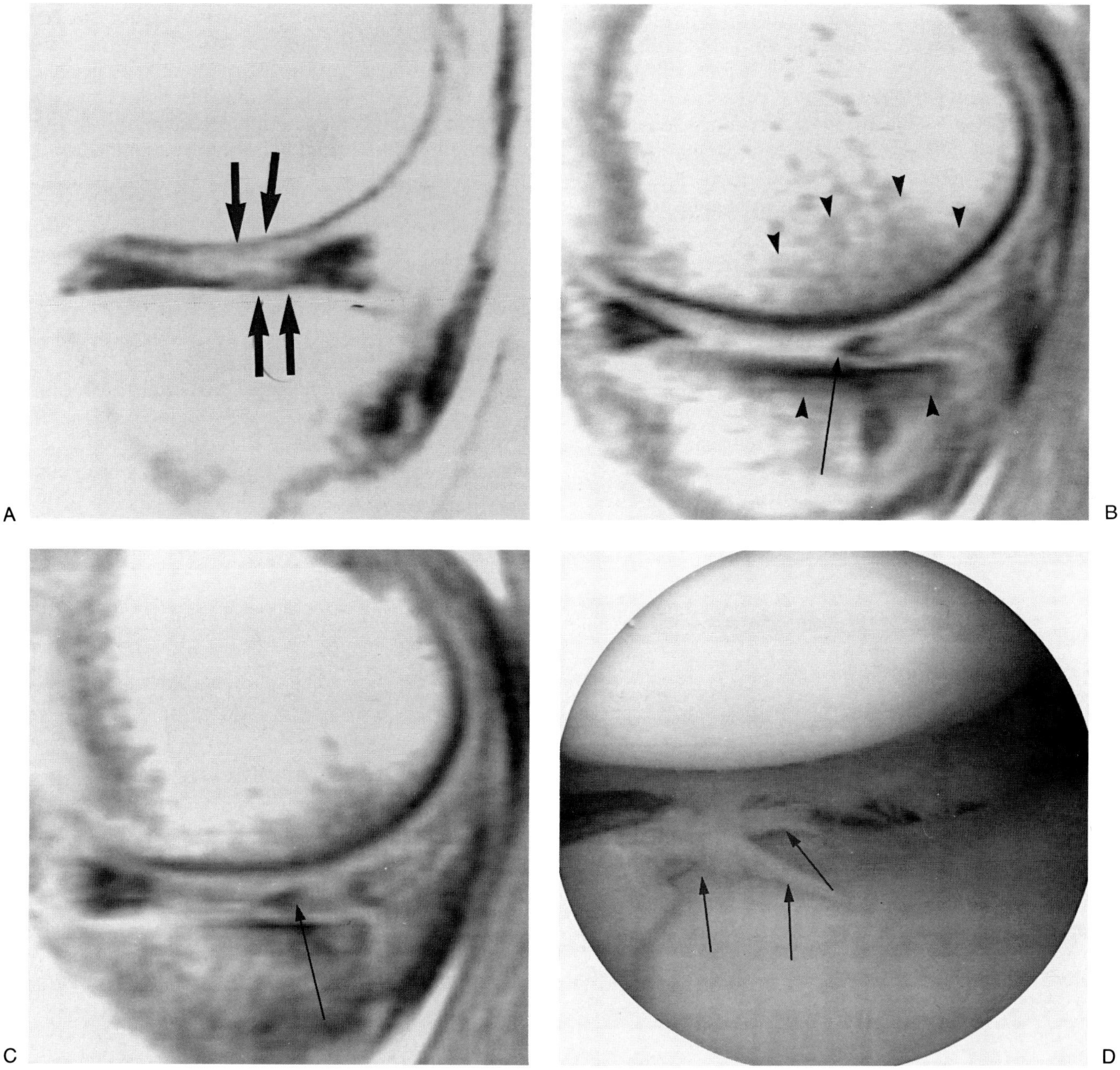

FIG. 7.11. Group 2, displaced meniscal re-tear. **A:** Sagittal MR, TR/TE 2,000/20. Preoperative study demonstrates an extensive tear involving the mid zone and junction with the posterior horn of the medial meniscus (*arrows*). The patient underwent partial meniscectomy in association with an anterior cruciate ligament reconstruction. **B:** Sagittal MR, TR/TE 2,000/20. Fourteen months after surgery, the patient experienced a reinjury and a repeat MR study was performed. On this section the posterior horn of the medial meniscus is blunted (*arrow*). Extensive changes have developed within the subchondral and cancellous bone of the medial femoral condyle and proximal tibial plateau (*arrowheads*). **C:** Sagittal MR, TR/TE 2,000/20. On a more central section, a complex tear with grossly abnormal morphology and centrally displaced fragment is evident (*arrow*). Such morphologic findings are among the most specific findings for significant meniscal re-tears and are generally far more reliable than intrameniscal signal changes in these menisci. **D:** Intraoperative photograph from the time of the patients repeat arthroscopy demonstrates the markedly fragmented appearance of the meniscus (*arrows*).

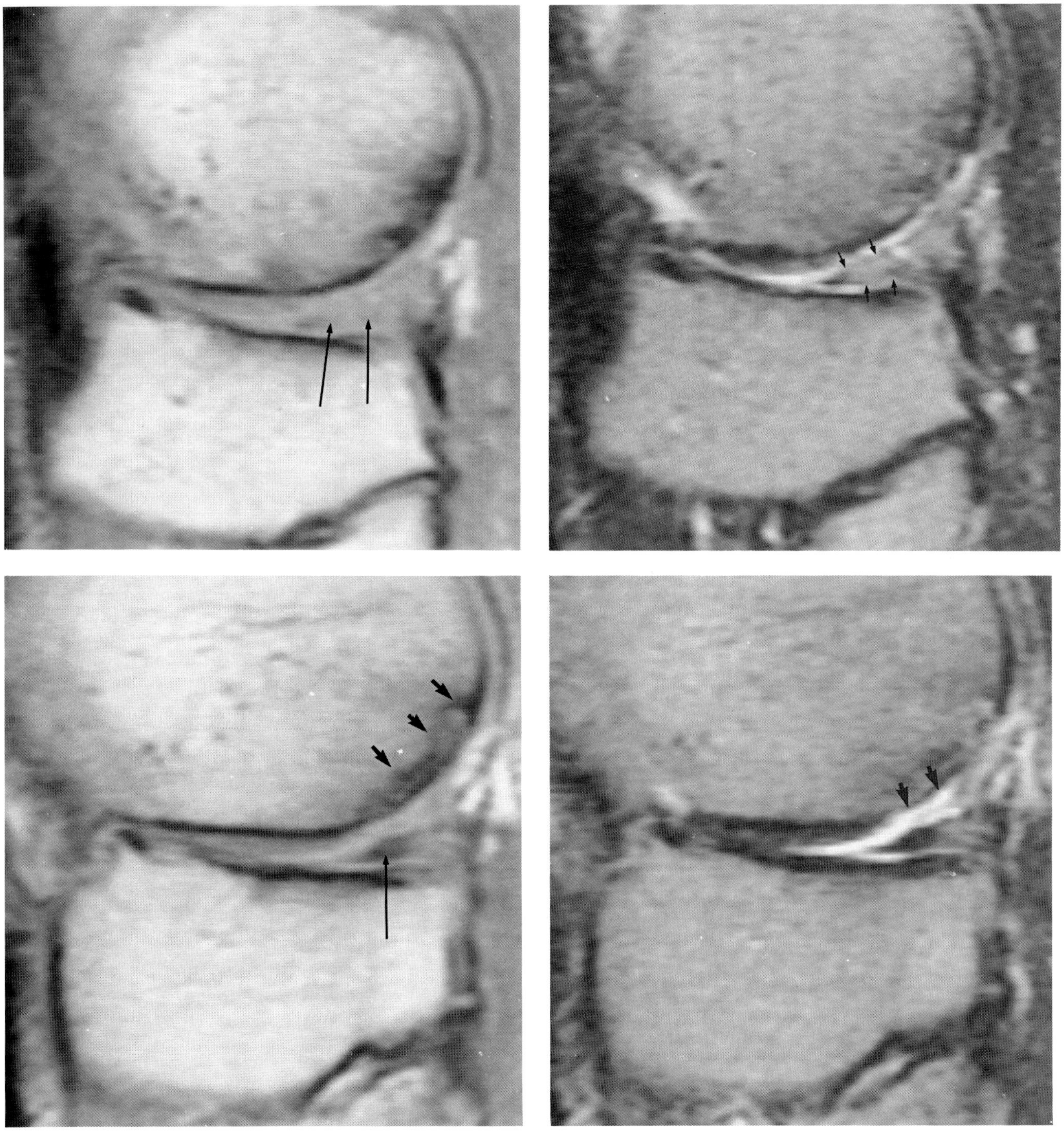

A B C D

FIG. 7.12. Group 3, degenerated but stable remnant. **A:** Sagittal MR, TR/TE 2,000/20. Professional football player status post two previous procedures on this knee. There is a small posterior horn remnant with mild diffuse increased signal (*long arrow*). Grade 4 chondromalacic changes are noted, most extensively involving the posterior-lateral aspect of the lateral femoral condyle. **B:** Sagittal MR, TR/TE 2,000/80. On the T2-weighted image, the extent of the full thickness cartilage loss is better appreciated. The meniscal contour is better defined and the intrameniscal signal has decreased. **C:** Sagittal MR, TR/TE 2,000/20. On an adjacent section, diffuse increased signal is seen throughout the meniscus which can be distinguished only with difficulty from surrounding synovial fluid (*arrows*). Such diffuse intrameniscal signal changes are commonly seen in small postoperative menisci. **D:** Sagittal MR, TR/TE 2,000/80. The arthrogram effect provided by the T2-weighted sequence allows significantly enhanced depiction of the gross contour of the meniscal remnant (*arrows*). The intrameniscal signal is increased in intensity and diffusely distributed throughout the meniscus. At surgery, the meniscus was described as degenerated and mildly fibrillated but without evidence of an unstable fragment.

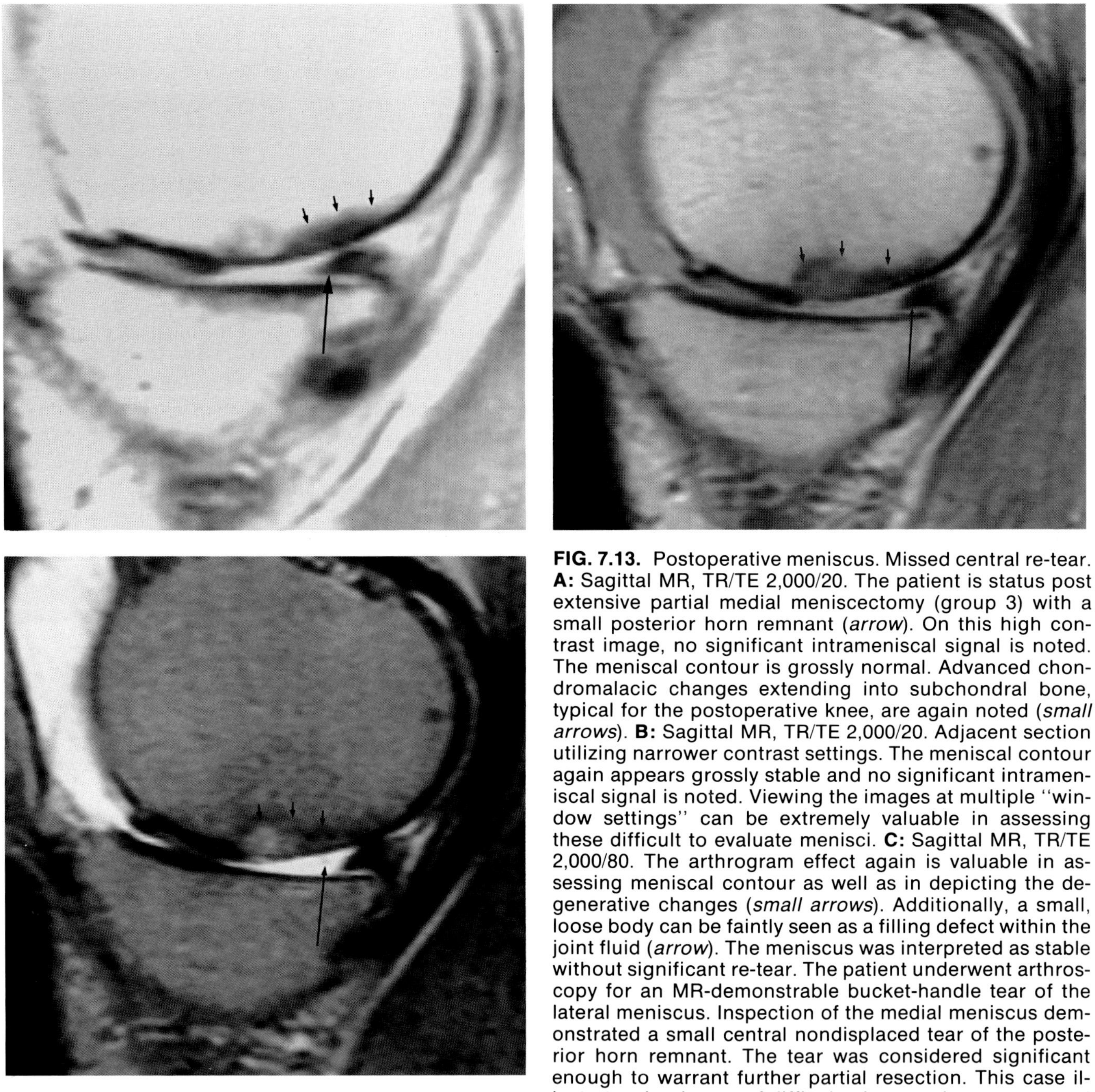

FIG. 7.13. Postoperative meniscus. Missed central re-tear. **A:** Sagittal MR, TR/TE 2,000/20. The patient is status post extensive partial medial meniscectomy (group 3) with a small posterior horn remnant (*arrow*). On this high contrast image, no significant intrameniscal signal is noted. The meniscal contour is grossly normal. Advanced chondromalacic changes extending into subchondral bone, typical for the postoperative knee, are again noted (*small arrows*). **B:** Sagittal MR, TR/TE 2,000/20. Adjacent section utilizing narrower contrast settings. The meniscal contour again appears grossly stable and no significant intrameniscal signal is noted. Viewing the images at multiple "window settings" can be extremely valuable in assessing these difficult to evaluate menisci. **C:** Sagittal MR, TR/TE 2,000/80. The arthrogram effect again is valuable in assessing meniscal contour as well as in depicting the degenerative changes (*small arrows*). Additionally, a small, loose body can be faintly seen as a filling defect within the joint fluid (*arrow*). The meniscus was interpreted as stable without significant re-tear. The patient underwent arthroscopy for an MR-demonstrable bucket-handle tear of the lateral meniscus. Inspection of the medial meniscus demonstrated a small central nondisplaced tear of the posterior horn remnant. The tear was considered significant enough to warrant further partial resection. This case illustrates the degree of difficulty that may be encountered in distinguishing re-torn and stable menisci.

This observation of the high predictive value of hyperintensity along the course of grade 3 signal on T2-weighted images was first made in regard to tears of virgin menisci, and as such is applicable to both operated and nonoperated menisci (56,57). While apparently highly specific for meniscal tear, this finding is not highly sensitive (401 in Tolin's series) (59a). The presence of high signal intensity throughout the course of grade 3 signal, however, should allow a highly confident diagnosis of meniscal re-tear to be offered.

As a consequence of wide variation in the extent of meniscal resection, assessment of the shape of the residual meniscus is fraught with difficulty, as has been discussed. While even irregularity of the recontoured meniscal free edge may not allow consistently correct prediction of meniscal re-tear, the presence of an abrupt change in meniscal morphology (Fig. 7.10) can be highly suggestive of a re-tear, and the detection of a displaced fragment is a highly reliable finding which should be sought closely on every study (Fig. 7.11).

Unfortunately, it is our experience that such findings will be present in only a minority of cases, but when present they allow a confident diagnosis to be established in this otherwise difficult to evaluate group of menisci. Additionally, the detection of a signal or morphologic abnormality at a site remote from the original lesion represents another useful criteria for establishing the presence of a new meniscal tear. This requires knowledge of the location and, ideally, morphology of the earlier tear but this information is often not immediately or easily obtainable. Comparison with any preoperative MR studies, operative reports (particularly with accompanying schematic diagrams), or videotapes of the previous arthroscopy can be of particular value (see below).

Group 3

The third group of postoperative menisci are those in whom extensive prior meniscal resection has been performed (greater than 75%) and in whom only a small peripheral rim remains present (Fig. 7.12). In the author's preliminary series of 14 such menisci, 3 were considered re-torn and 11 were stable at follow-up arthroscopy. There was agreement between MR and arthroscopy of 12 of the 14 patients (85%). Meniscal remnants in this group may typically demonstrate diffuse increased signal intensity throughout the remnant, and wide "meniscal" windows are typically of less value in evaluating these menisci. Because diffuse intrameniscal signal is often of little value in distinguishing between stable and re-torn menisci in this group, we have found it useful to utilize more extensively the T2-weighted images (Fig. 7.12). In many instances the morphology of these menisci is better appreciated on these images, particularly if sufficient joint fluid is present to provide for a natural arthrogram effect. As discussed, meniscal assessment on the basis of signal grading may be extremely difficult in this group, and it may not be possible to recognize small, nondisplaced tears within these remnants (Fig. 7.13). Gross morphologic alterations, and evidence of a displaced fragment in particular, represent the most specific criteria for an unstable remnant (Fig. 7.14). A potential pitfall in this regard, however, is the juxtaposition of loose intraarticular fragments in the immediate vicinity of the meniscal remnant. Such fragments, which are particularly common in this group of knees, can closely mimic a displaced meniscal tear (Fig. 7.15).

Loose osteochondral bodies and degenerative changes of the articular cartilage are common following partial meniscectomy and are nearly universal in patients long term following extensive meniscal resections (group 3). An additional advantage of T2- and T2*-weighted sequences (if sufficient joint fluid is present), or of MR-arthrography (see below), is in the characterization of the status of the articular cartilage. Localized contour defects within the cartilage can be readily seen highlighted by the contrast between cartilage and high signal intensity joint fluid (Figs. 7.12,

A

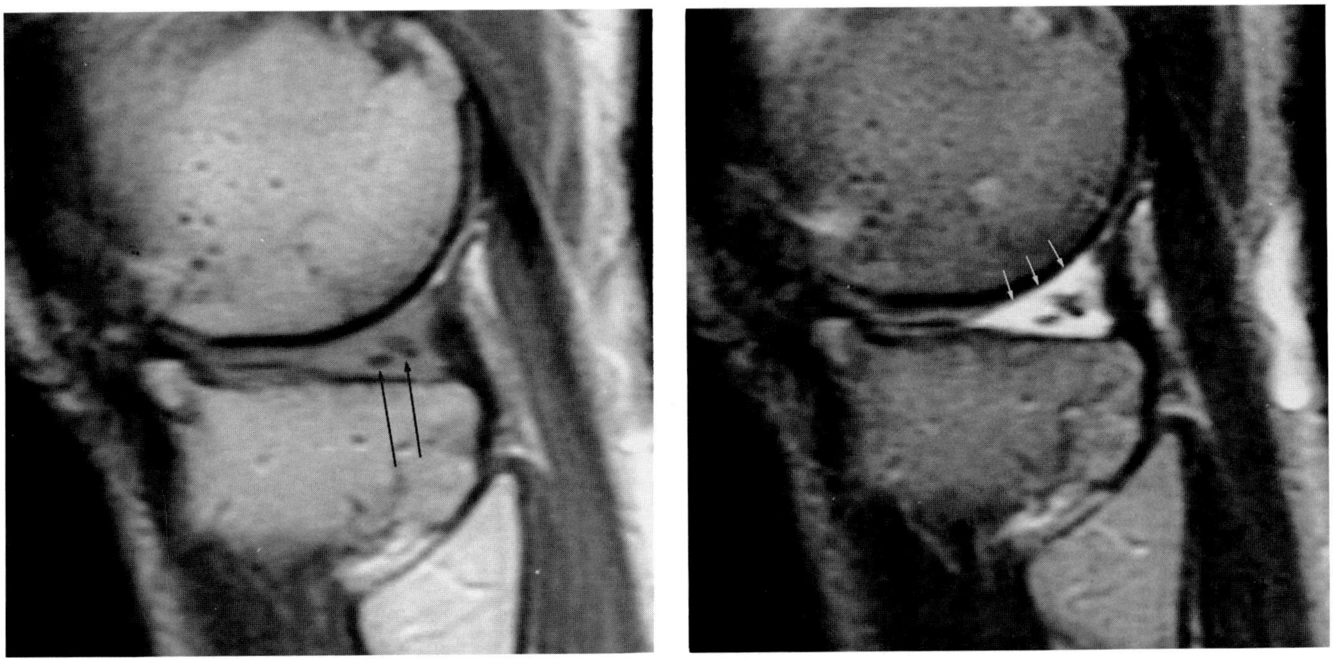

B

FIG. 7.14. Meniscal re-tear (group 3), abnormal morphology. **A:** Sagittal MR, TR/TE 2,000/20. There is a small posterior horn remnant (group 3) with markedly abnormal morphology consistent with fragmentation of the remnant (*arrows*). **B:** Sagittal MR, TR/TE 2,000/80. The meniscal fragments are easily seen contrasted against the high signal intensity joint fluid. There is smooth, full thickness loss of articular cartilage (*small arrows*).

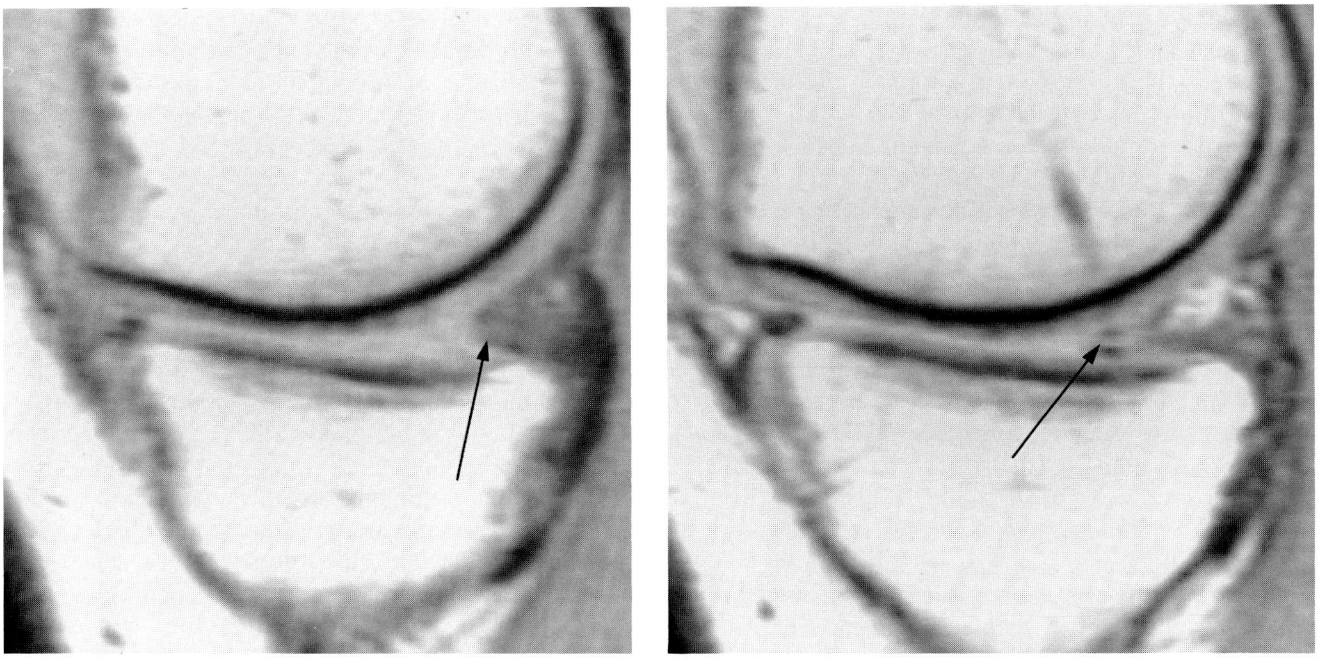

A B

FIG. 7.15. Loose body simulating meniscal re-tear (group 3). **A:** Sagittal MR, TR/TE 2,000/20. There is a small posterior horn remnant with mildly diffuse intrameniscal signal (*arrow*). **B:** Sagittal MR, TR/TE 2,000/20. On a contiguous section, two small, low signal intensity structures are seen juxtaposed to the meniscal rim simulating a fragmented meniscus (*arrow*) but in actuality representing small loose osteocartilagenous bodies. At surgery, the meniscal rim was mildly frayed and multiple small, loose bodies were identified within the joint. Differentiating a "fragmented" rim from a degenerated rim with contiguous loose bodies may be extremely difficult.

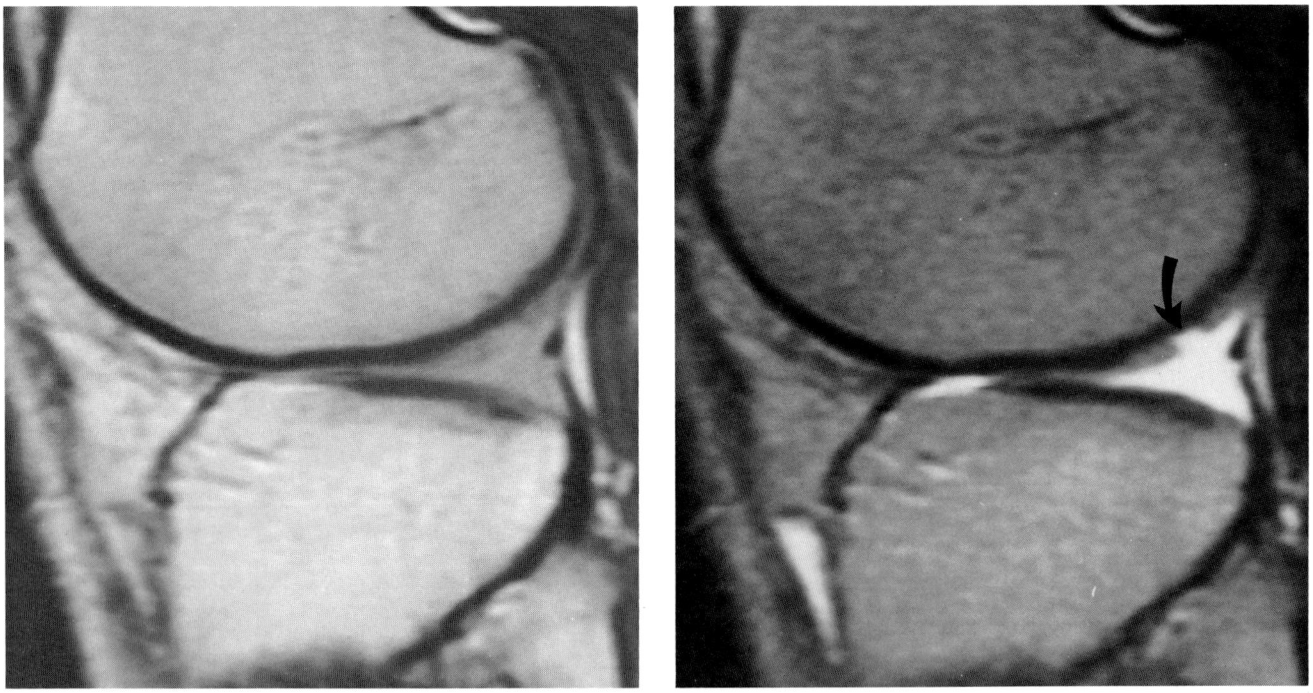

A B

FIG. 7.16. Articular cartilage defect, value of long TR/TE imaging. **A:** Sagittal MR, TR/TE 2,000/20. The patient is status post near subtotal meniscectomy. On this intermediate pulse sequence, there is little contrast between synovial fluid and articular cartilage. **B:** Sagittal MR, TR/TE 2,000/80. Utilizing the natural contrast provided synovial fluid, intermediate signal intensity articular cartilage is readily distinguished from high signal intensity synovial fluid. A large focal articular cartilage defect is well demonstrated by the bright joint fluid extending into the defect (*curved arrow*).

7.16, and 7.17). Such findings may be subtle and must be actively sought at the time of interpretation of MR scans on postoperative patients. Corresponding changes within the subchondral bone are often manifest as areas of lower signal underlying the articular cartilage. With advanced malacic changes, the likelihood of detached areas of cartilage and or subchondral bone increases. Such loose chondral and/or osteochondral bodies within the joint may demonstrate a varied appearance, and their detection with MR is often difficult. When a lesion is ossified, it has medullary bone in its center and has a dark cortical rim on at least one side. Such loose bodies are generally of high signal intensity on T1-weighted sequences and become of lower signal intensity on longer TR/TE sequences. Some loose bodies are heavily calcified but do not have a medullary center; they are of low signal intensity on all pulse sequences and their identification is facilitated when they are contrasted against high signal intensity joint fluid. Purely chondral fragments may also be difficult to detect. On T1-weighted sequences, they have an intensity that is often similar to joint fluid, and again are often best demonstrated highlighted against high signal intensity joint fluid on T2- and T2*-weighted sequences (Fig. 7.17). In the author's experience, most loose bodies have had mixed characteristics, and in cases of multiple loose bodies, each may have different components. Nodular synovitis, seen in some postoperative patients as well as in those with inflammatory arthritis, may present an appearance similar to loose bodies. Synovial nodules, however, may be distinguished by their typical nondependent position.

MR-Arthrography

The difficulties presented by signal conversion and other equivocal signal changes in postoperative menisci can potentially be approached by using MR as an arthrotomographic technique. If sufficient joint fluid is present, the morphology of the postoperative meniscus is often best appreciated on T2- or T2*-weighted images. This is particularly apparent when extensive intrameniscal signal is present. In the absence of sufficient synovial fluid to act as a natural contrast agent, saline or a diluted solution of saline and gadopentetate dimeglumine (Gd-DTPA) can be utilized. For purposes of MR-arthrography, Gd-DTPA can be diluted at a ratio of 1:100 with saline, and 30 cc to 40 cc of this contrast agent injected into the knee joint. A conventional examination is performed prior to the administration of the contrast material and a T1-weighted sagittal sequence is obtained immediately following the arthrocentesis. If iodinated contrast material is admin-

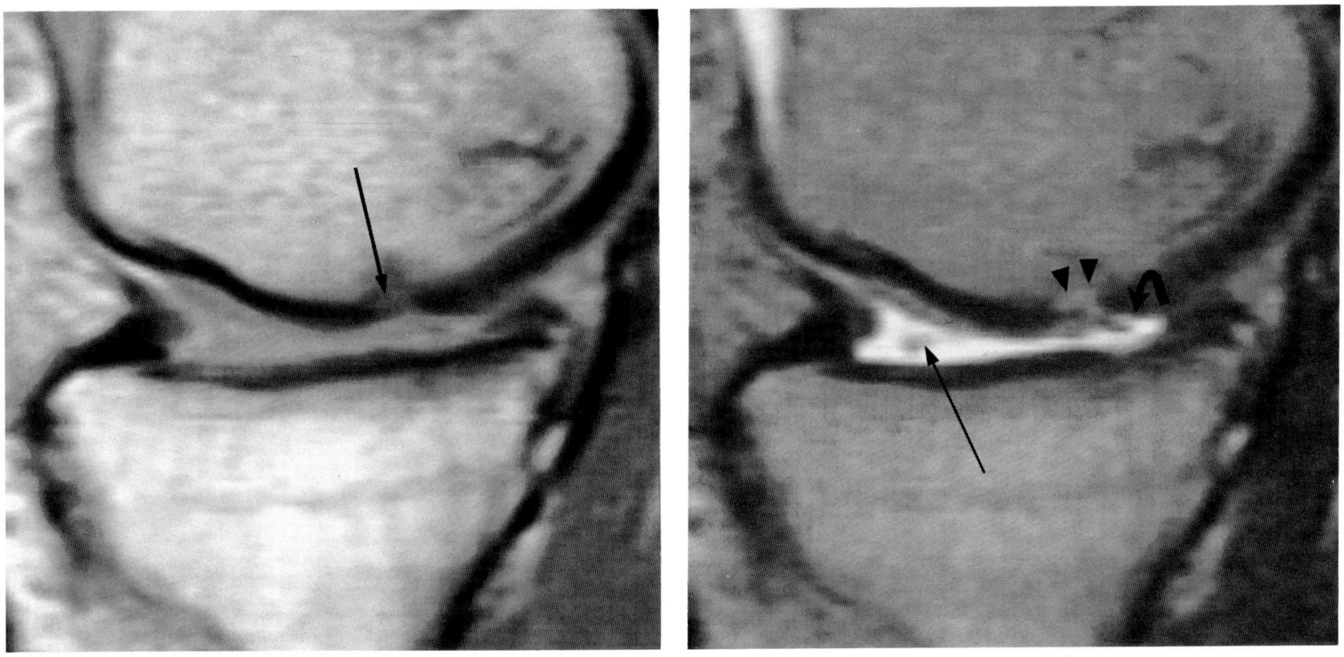

A B

FIG. 7.17. Articular cartilage changes following meniscectomy. **A:** Sagittal MR, TR/TE 2,000/20. The patient is status post extensive partial meniscectomy. There is focal interruption of the subchondral bone plate (*arrow*) with changes extending into the cancellous bone of the femoral condyle. **B:** Sagittal MR, TR/TE 2,000/80. High signal intensity joint fluid extends into ulcers within the articular surface of the femoral condyle (*arrowheads*) and there is slight increased signal within the bony changes of the femoral condyle (*arrows*). A loose intraarticular body is appreciated only on the long TR/TE sequence as a filling defect within the joint fluid (*arrow*). The findings are typical of the postmeniscectomy knee.

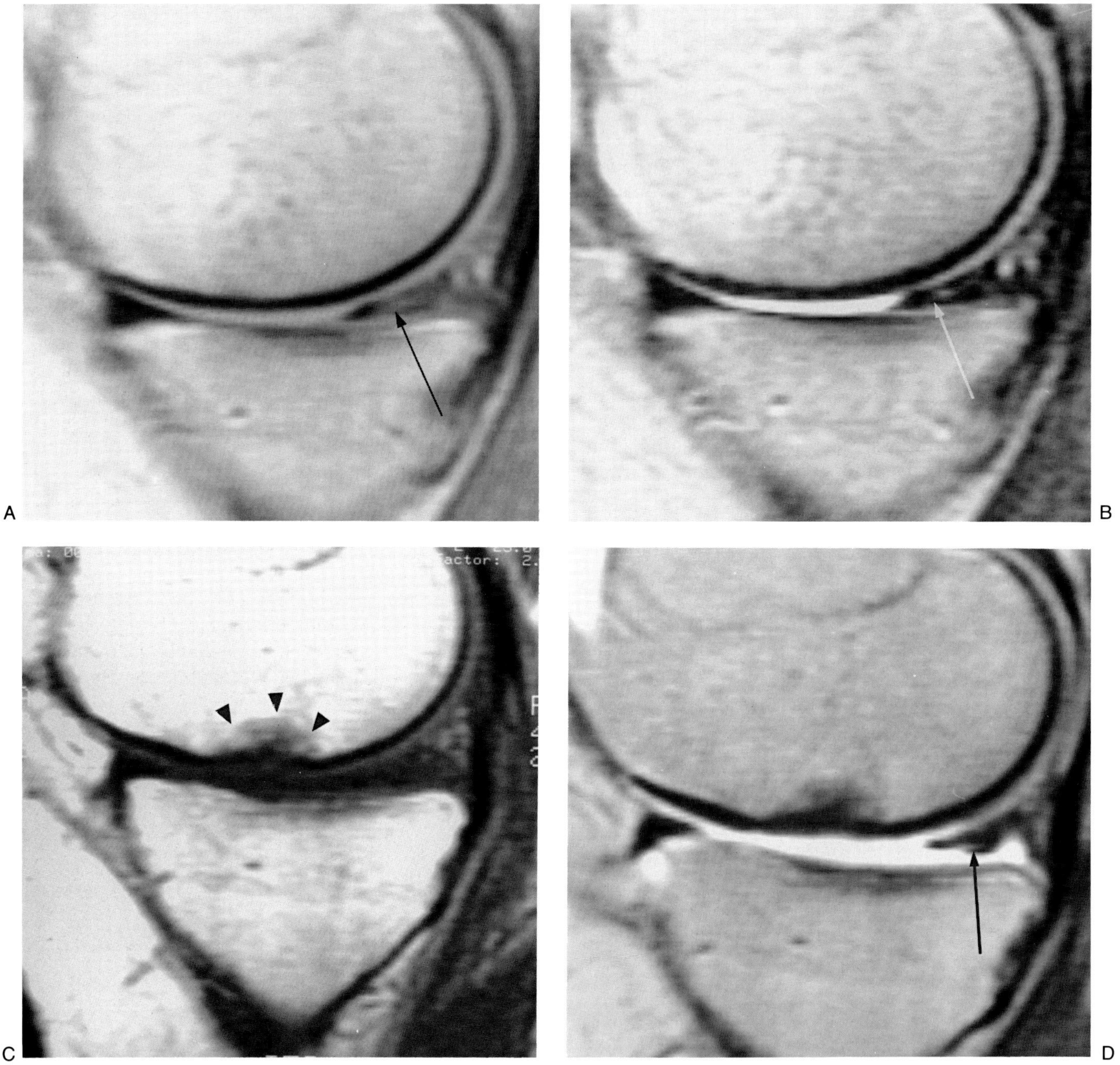

FIG. 7.18. MR arthrography, meniscal re-tear. **A:** Sagittal MR, TR/TE 2,000/20. The patient is status post partial medial meniscectomy. A line of signal equivocally extends to the articular surface (*arrow*). The meniscal tip is slightly deformed. **B:** Sagittal MR, TR/TE 2,000/80. On the T2-weighted image, a segment of the signal demonstrates increased intensity (*arrow*), a finding suggesting a meniscal re-tear. **C:** Sagittal MR, TR/TE 500/20. On a more central section obtained prior to the intraarticular administration of contrast, there is increased signal intensity seen diffusely within the meniscus, and the meniscal contour is not well defined. A localized articular cartilage and subchondral defect is noted (*arrowheads*). **D:** Sagittal MR, TR/TE 500/20. Section corresponding to Fig. 7.18C obtained following the intraarticular administration of contrast. A thin line of gadolinium-DTPA extends into the undersurface of the posterior horn of the medial meniscus (*arrow*). The contour of the meniscus is better defined following contrast administration.

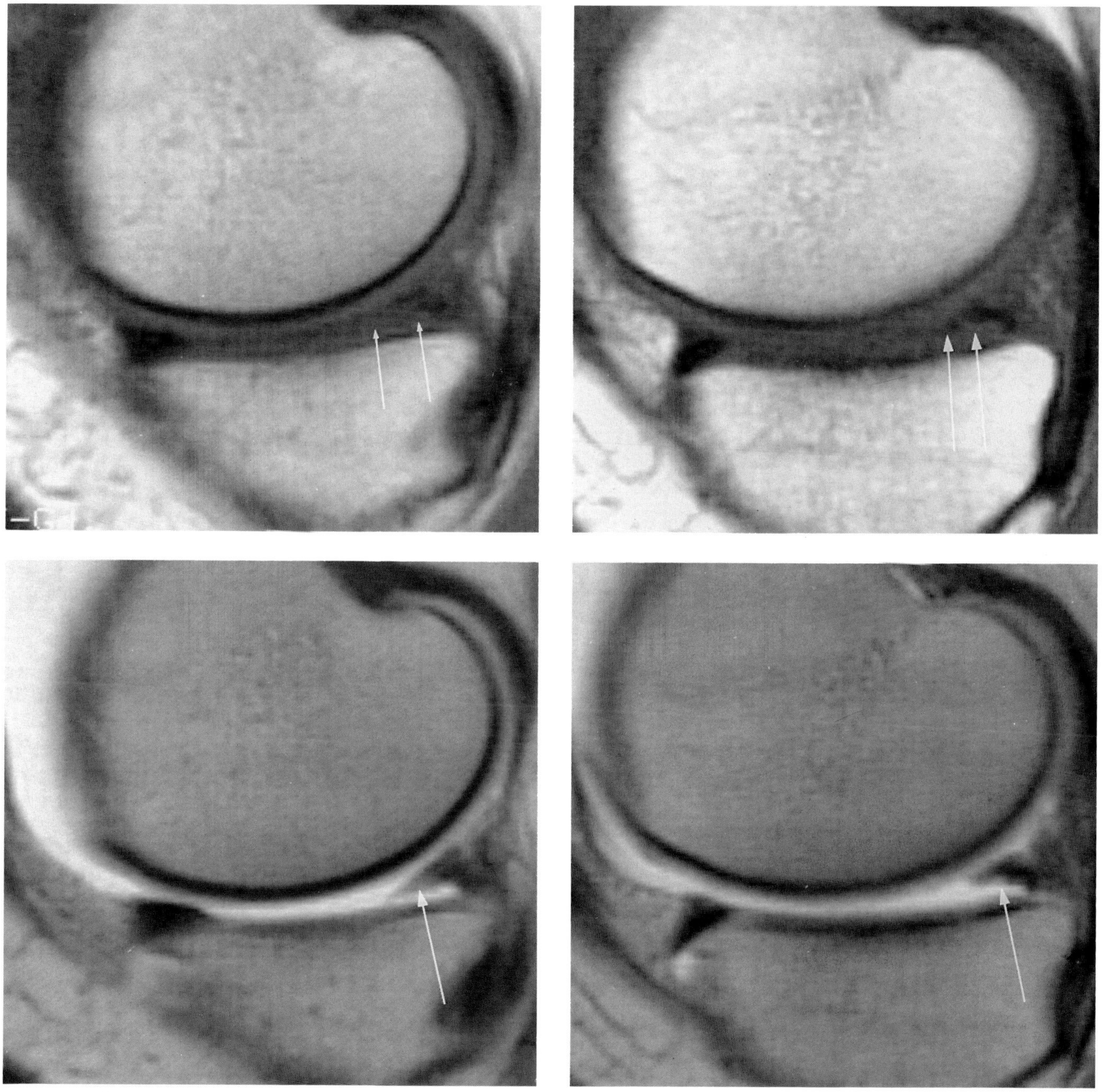

A B C D

FIG. 7.19. MR arthrography utilizing intraarticular gadolinium-DTPA. **A:** Sagittal MR, TR/TE 2,000/20. Magnified view of the medial meniscus of a patient status post partial meniscectomy who presents with recurrent symptoms. There is poorly defined increased signal throughout the posterior horn remnant (*arrows*), a not uncommon finding in the postoperative meniscus. **B:** Sagittal MR, TR/TE 2,000/20. On a slightly more central section through the same meniscus, the morphology of the meniscus is difficult to evaluate as a consequence of the diffuse increased signal (*arrows*). The anterior horn is stable. **C:** Sagittal MR, TR/TE 500/20. Magnified view of the medial meniscus obtained following the intraarticular administration of 30 cc of a diluted solution of Gd-DTPA and saline. There is marked improvement in the delineation of the meniscal morphology in this group 2 meniscus. There is slight imbibition of contrast within the apex of the posterior horn (*arrow*). The articular cartilage is without significant malacic change. **D:** Sagittal MR, TR/TE 500/20. On a section through the more central aspect of the meniscus, contrast minimally extends into the undersurface of the posterior horn (*arrow*). Minimal fraying of the meniscus with a small, incomplete undersurface tear was present at repeat arthroscopy.

istered within the joint prior to the administration of gadolinium to ensure an intraarticular location, it is important that it be limited to 1 cc to 2 cc. This relates to a concern regarding the potential of the iodinated contrast to change the binding capacity of gadolinium to the DTPA moiety, allowing free gadolinium in the joint to become absorbed into articular cartilage with potential deleterious effects (58).

Present experience with MR-arthrography remains quite preliminary. The principal diagnostic criteria of contrast imbibition and assessment of altered morphology are similar to those established for conventional radiographic arthrography. (59). The extension of contrast material into the substance of the meniscus is considered diagnostic of a meniscal re-tear (Fig. 7.18). The presence of contrast material has significantly improved the depiction of meniscal morphology, particularly in the presence of diffuse increased intra-meniscal signal (Fig. 7.19). This improved anatomical depiction on the arthrographic studies has contributed to the diagnosis of mensical re-tears and the important depiction of mechanically unstable meniscal flaps (Fig. 7.20).

In our experience and that reported by Tolin et al., (59a) MR-arthrography is of greatest value in the evaluation of Group 2 and 3 menisci and can significantly improve the diagnostic accuracy of MR in evaluation of these post-operative menisci. In Group 1 menisci, the addition of contrast does not appear to afford a significant advantage over conventional MR and as such need not be performed in this group. (59a). Potential advantages of Gd-arthrography include the lower viscosity of the solution compared to synovial fluid which may provide an advantage with regard to contrast imbition, as well as the ability to utilize T1 weighted images with their inherent increased signal to noise properties. While further studies will be required to determine whether the apparent advantages afforded by MR-arthrography will outweigh the inconvenience, invasive nature, and increased cost of the procedure, the authors are presently encouraged by their early experience.

Postmeniscectomy Osteonecrosis

Osteonecrosis of the femoral condyle following meniscectomy is not among the frequently reported complications of this procedure. Brahme et al., however, reported on seven patients with persistent pain following arthroscopic partial meniscectomy who developed marked MR changes of bone marrow edema, a finding which the authors considered presumptive evidence of osteonecrosis (60). All of the reported patients demonstrated no signal abnormalities of subchondral or medullary bone on preoperative studies. At surgery, meniscal tears were documented in all individuals and partial meniscectomies were accomplished. Additionally, all individuals demonstrated moderate to severe

A 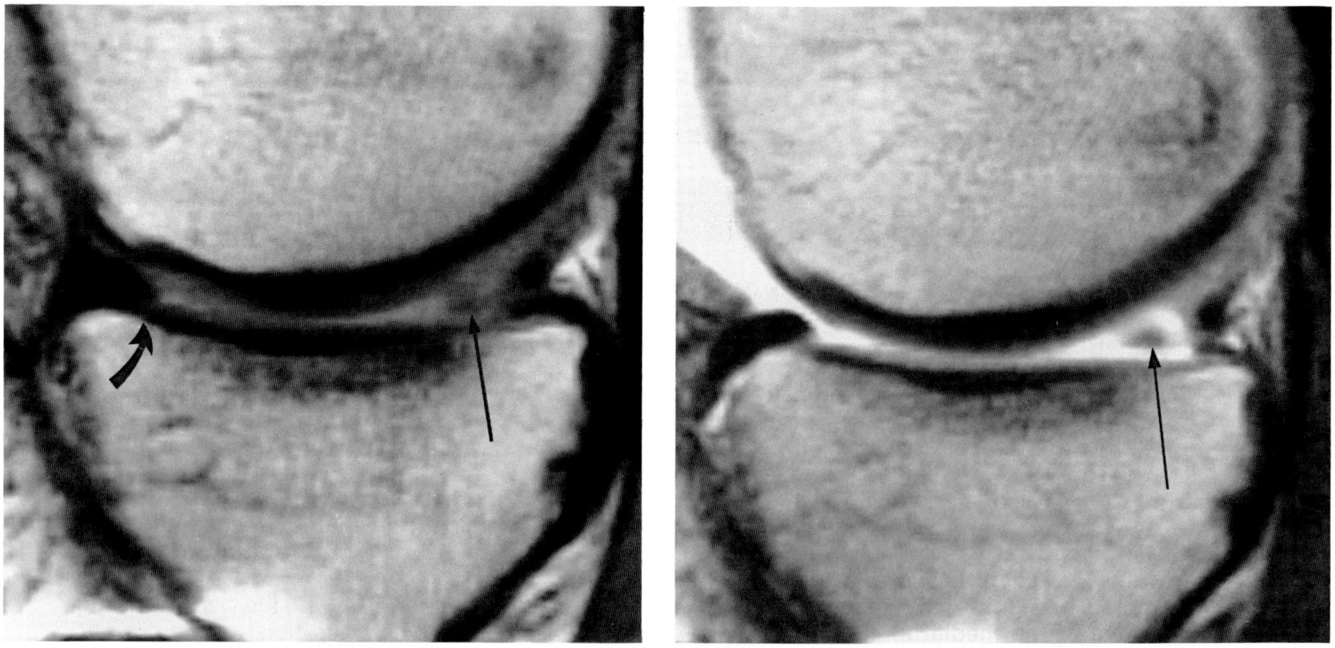 B

FIG. 7.20. MR arthrography, meniscal re-tear. **A:** Sagittal MR, TR/TE 500/20. The patient is status post partial meniscectomy. The tip of the anterior horn is mildly blunted consistent with prior surgery (*curved arrow*). The posterior horn remnant is irregular with the suggestion of a small, centrally displaced flap (*arrow*). **B:** Sagittal MR, TR/TE 500/20. Following the intraarticular administration of Gd-DTPA, the depiction of the centrally displaced fragment is markedly improved (*arrow*).

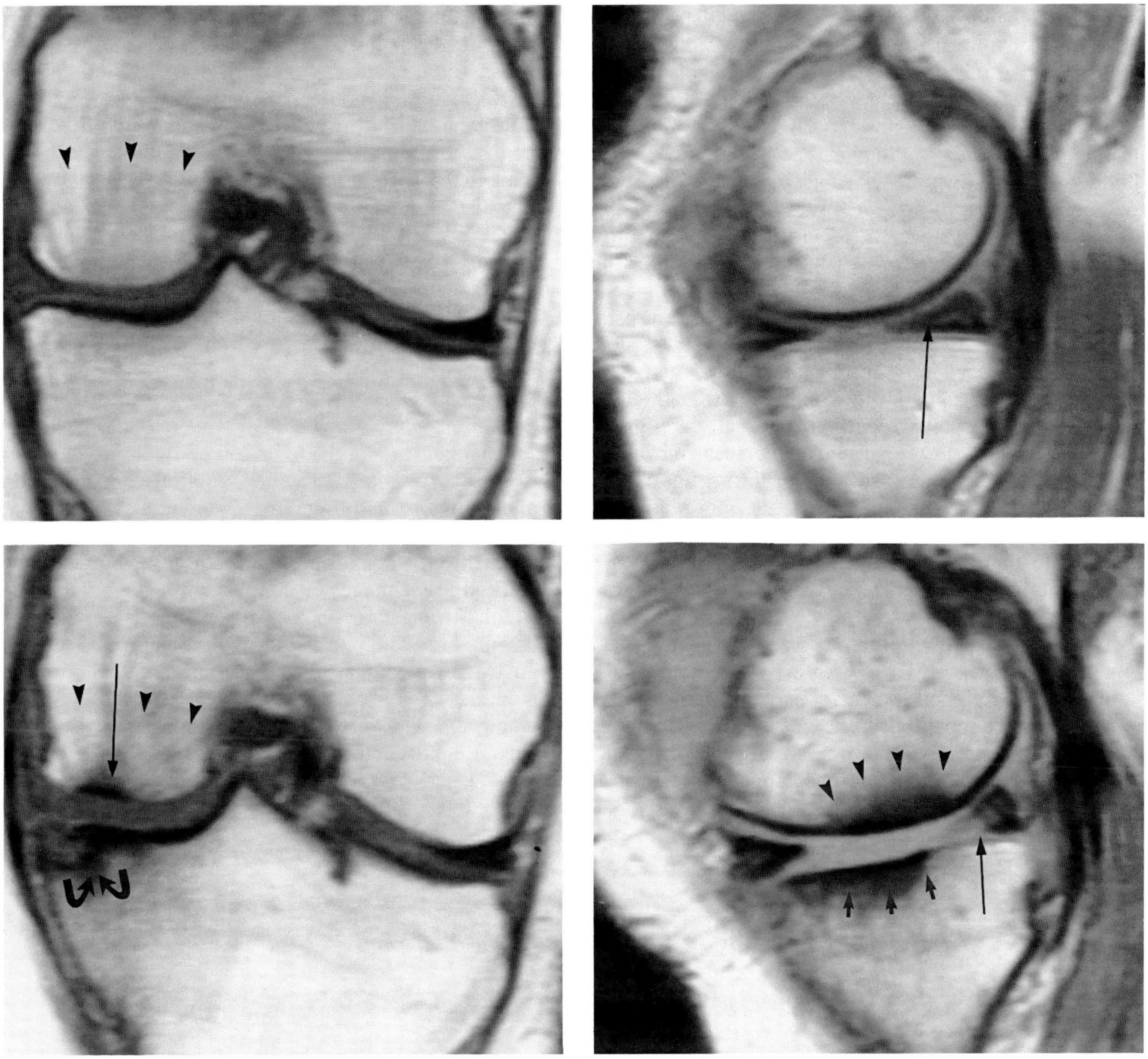

A B C D

FIG. 7.21. Presumed spontaneous osteonecrosis following arthroscopic partial meniscectomy. **A:** Coronal MR, TR/TE 500/30. Preoperative examination demonstrating normal signal intensity within the medial femoral condyle (*arrowheads*). **B:** Sagittal MR, TR/TE 2,000/20. Preoperative examination. Magnified view of the medial meniscus demonstrates a horizontal clevage tear (*arrow*) and normal signal intensity within the medial femoral condyle. **C:** Coronal MR, TR/TE 500/30. The patient returned for evaluation 2 months after surgery complaining of increasing pain in the medial aspect of his left knee. There is poorly defined, slightly decreased signal intensity within the medial femoral condyle (*arrowheads*) and a localized linear to crescent-shaped focus of markedly low signal paralleling the subchondral bone plate of the weightbearing surface of the medial femoral condyle (*long arrow*). A prominent low signal intensity focus is also seen within the proximal medial tibial plateau (*curved arrows*). **D:** Sagittal MR, TR/TE 2,000/20. Postoperative examination. The meniscal remnant is stable (*arrow*). There is striking decreased signal within the osteonecrotic segment of subchondral bone along the weightbearing aspect of the medial femoral condyle. Changes are also noted within the tibial plateau (*small arrows*).

chondromalacia of the subsequently affected femoral condyle and underwent cartilage shaving. On MR examinations performed 2 to 14 months postoperatively, signal changes demonstrating focal low signal intensity located subcortically were seen on T1-weighted images of the femoral condyle adjacent to the site of the previous meniscal tear. On T2-weighted images, central high signal intensity areas were seen within the low signal area in five of seven patients, but no distinct "reactive interface," characteristic of osteonecrosis, was demonstrated. Histologic correlation was not available. We have also had experience with extensive subchondral and medullary signal abnormalities occurring in patients with persistent symptoms following arthroscopic partial meniscectomy (Figs. 7.21–7.23). While the precise histologic correlates of these findings remain unknown, subsequent radiographic manifestations have further supported the likelihood of osteonecrosis in at least some of the cases. The contribution of the arthroscopic procedure to the development of these findings has not been established. All of the reported patients underwent both partial meniscectomy and cartilage shaving for advanced chondromalacia. It is possible that microtrauma associated with the procedure precipitated the development of the changes within the underlying bone (60). Alternatively, the re-

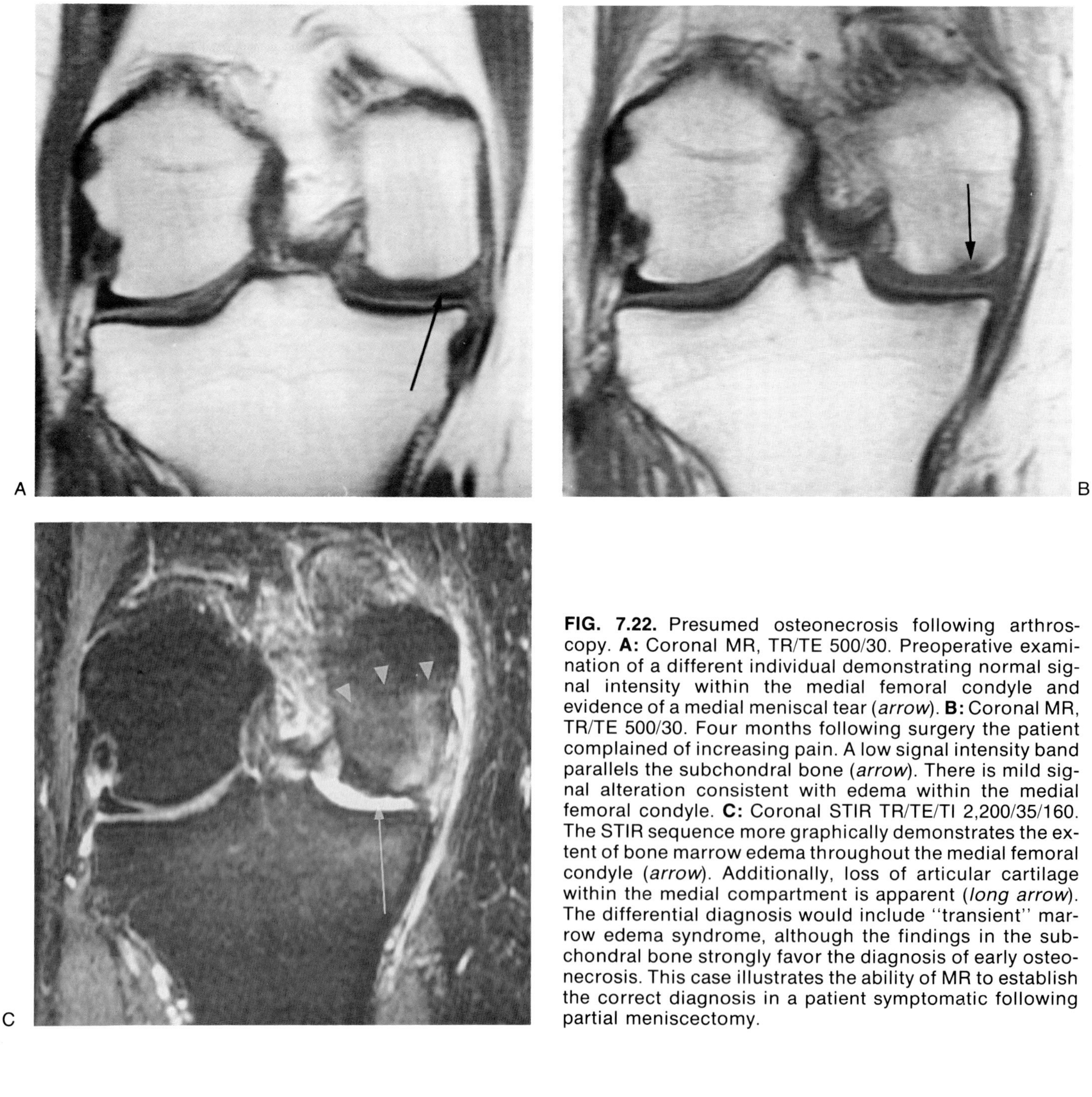

FIG. 7.22. Presumed osteonecrosis following arthroscopy. **A:** Coronal MR, TR/TE 500/30. Preoperative examination of a different individual demonstrating normal signal intensity within the medial femoral condyle and evidence of a medial meniscal tear (*arrow*). **B:** Coronal MR, TR/TE 500/30. Four months following surgery the patient complained of increasing pain. A low signal intensity band parallels the subchondral bone (*arrow*). There is mild signal alteration consistent with edema within the medial femoral condyle. **C:** Coronal STIR TR/TE/TI 2,200/35/160. The STIR sequence more graphically demonstrates the extent of bone marrow edema throughout the medial femoral condyle (*arrow*). Additionally, loss of articular cartilage within the medial compartment is apparent (*long arrow*). The differential diagnosis would include "transient" marrow edema syndrome, although the findings in the subchondral bone strongly favor the diagnosis of early osteonecrosis. This case illustrates the ability of MR to establish the correct diagnosis in a patient symptomatic following partial meniscectomy.

A
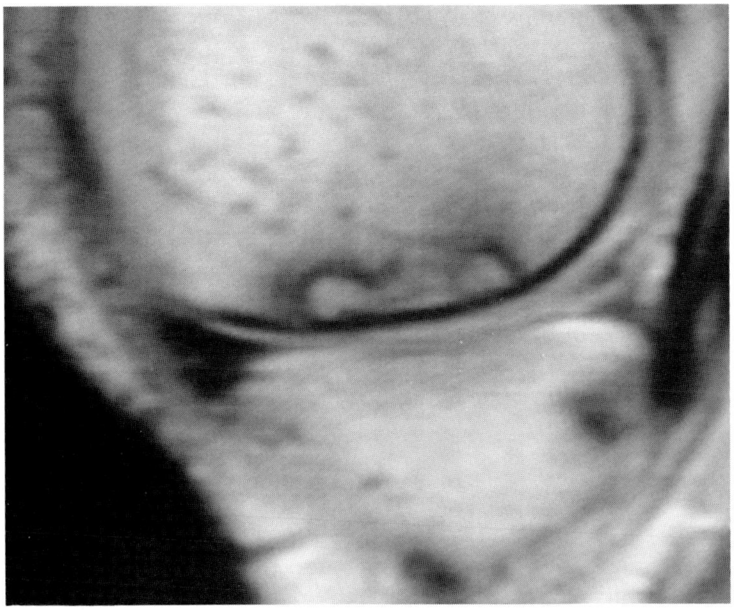

B
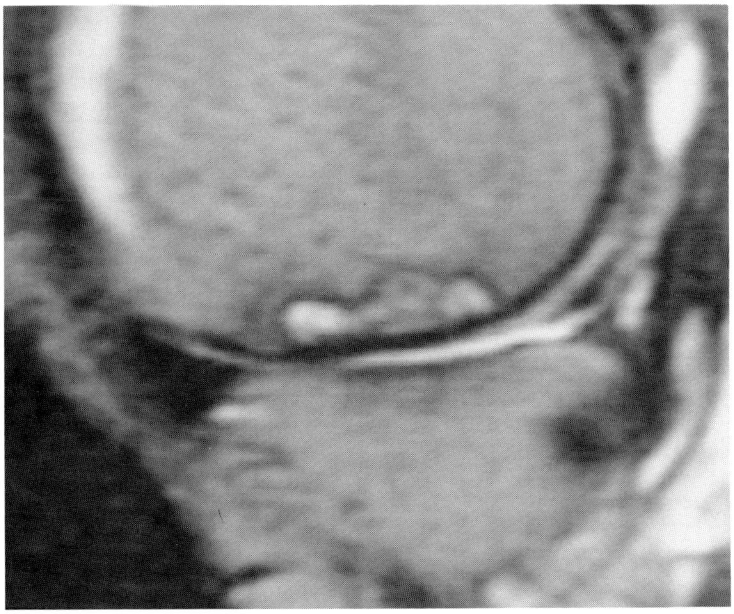

FIG. 7.23. Presumed osteonecrosis following partial meniscectomy. **A:** Sagittal MR, TR/TE 2,000/20. The patient is status post partial medial meniscectomy. There is a circumscribed focus of altered signal intensity paralleling the subchondral bone of the weightbearing aspect of the medial femoral condyle (*arrowheads*). **B:** Sagittal MR, TR/TE 2,000/20. On the T2-weighted sequence, the presumed osteonecrotic segment becomes relatively hyperintense (*arrowheads*). There is diffuse loss of articular cartilage.

moval of meniscal fibrocartilage in a patient with already compromised articular cartilage may have critically increased bone to bone contact resulting in microfractures, one of the factors associated with the development of osteonecrosis (60).

Conservative Management and Meniscal Repair

The MR appearance of menisci with peripheral tears that were either conservatively managed or underwent suture repair has been the subject of several recent investigations (2,5,61). These studies have demonstrated striking persistence of signal changes within menisci of patients who have become asymptomatic and presumably have healed meniscal lesions. The persistence of these signal changes presents a significant challenge for interpretation of MR scans in patients with continued symptoms or re-injury following conservative management and operative repair.

Deutsch et al. reported on six menisci with arthroscopically proven small peripheral tears that were treated conservatively with brace immobilization (2). All patients in this group were considered to be clinically stable with presumably healed meniscal tears at the time of repeat MR examination. On follow-up MR studies performed with a minimum follow-up interval of 3 months and a maximum of 13 months, no significant change in appearance of the tear was observed in any patient in whom the tear was diagnosed originally on the basis of grade 3 signal (Fig. 7.24). In our subsequent experience, we have had longer term follow-up on conservatively treated peripheral tears, and while abnormal signal has persisted, in at least one case it could be documented to have significantly decreased (Fig. 7.25). It is certainly possible that with increasing time, the signal within some ''healed'' tears may markedly diminish or potentially disappear. The important point at present, however, is that signal may persist in some tears that are apparently stable and clinically healed with the clear potential for misinterpretation of the findings as indicative of re-tears.

Twelve menisci in the author's study underwent arthroscopically guided suture repair. Preoperative MR images of nine of these menisci were available and demonstrated the tears in each case as a line of grade 3 signal within the peripheral third of the meniscus. All menisci were considered clinically stable at the time of follow-up. On follow-up MR studies obtained 3 to 27 months after repair (mean, 8 months), persistent signal intensity, judged to be not significantly changed from that in the preoperative study, was evident (Fig. 7.26). Additionally, three patients with a total of four repaired menisci were evaluated by second look arthroscopy. The tears were considered healed by inspection and probing in all four cases. In the one patient in whom both pre- and postoperative studies were available for review, no significant change in the appearance of the meniscus was identified, and persistent grade 3 signal was present on the MR examination. In two other menisci, both considered to be healed at second look arthroscopy, grade 3 signal was evident on the follow-up study (Fig. 7.27). In the other meniscus, grade 2 signal was present on the follow-up study. As no preoperative study was available for review in this patient, it cannot be determined whether any significant change occurred as a consequence of the repair.

The finding of persistent grade 3 signal in menisci following meniscal repair has been corroborated in other investigations. Kent et al. reported on the longitudinal follow-up appearance of menisci that had undergone repair (61). Follow-up studies were performed in 17 patients at varying intervals following meniscal repair. Detectable signal was seen in 15 of 15 patients imaged at 7 weeks post repair and in 6 of 6 patients imaged at 6 months. Farley et al. also reported on the MR appearance of meniscal tears following surgical repair and corroborated the difficulty of using the presence of persistent grade 3 signal as the basis for

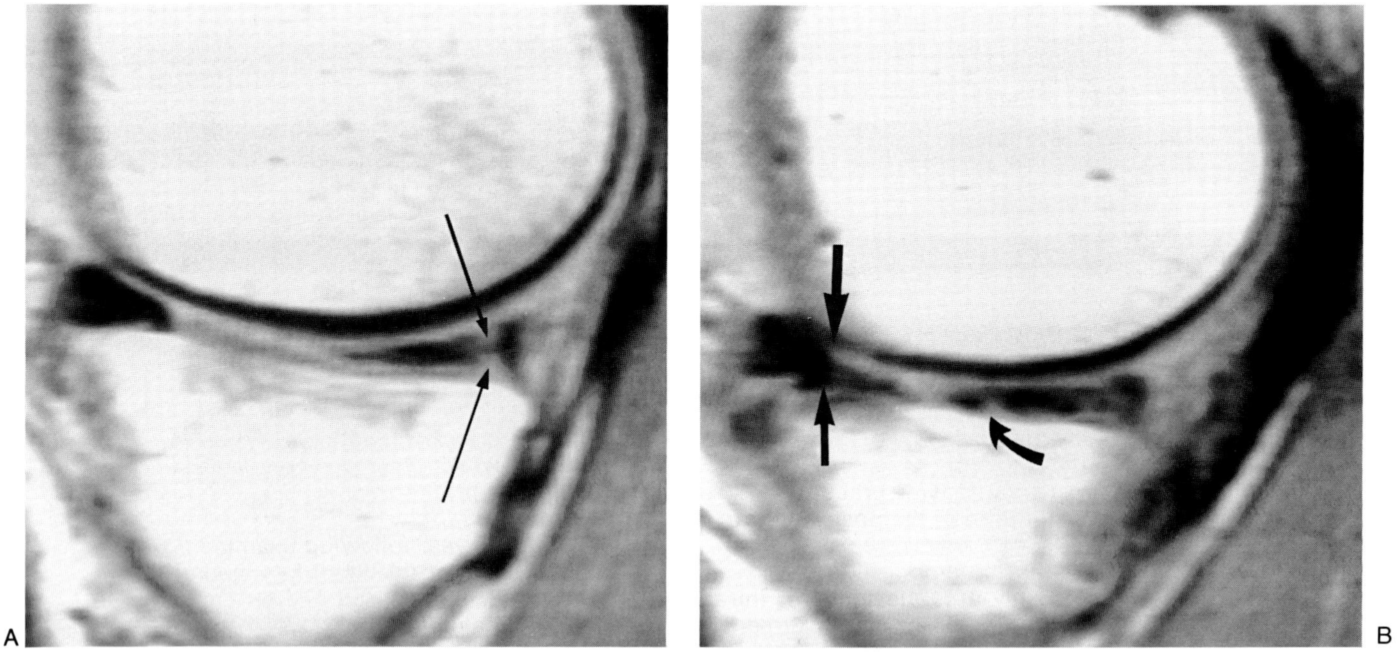

FIG. 7.24. Peripheral meniscal tear: Conservative therapy and subsequent reinjury. **A:** Sagittal MR, TR/TE 2,000/20. The patient had undergone arthroscopy 5 years earlier and had a peripheral tear documented within the posterior horn of the medial meniscus. The tear was treated conservatively without suture repair and the patient subsequently became asymptomatic. The patient returned following an acute knee injury and underwent MR evaluation. Within the peripheral third of the posterior horn a well-defined, vertically oriented line of signal is seen (*arrows*). This correlated with the site of the previous tear according to surgical records. At subsequent arthroscopy this site was entirely stable and healed. **B:** Sagittal MR, TR/TE 2,000/20. Magnified view of the same meniscus demonstrates a flap tear involving the central third of the posterior horn (*curved arrow*) and truncation of the anterior horn (*arrows*). The findings were confirmed at arthroscopy.

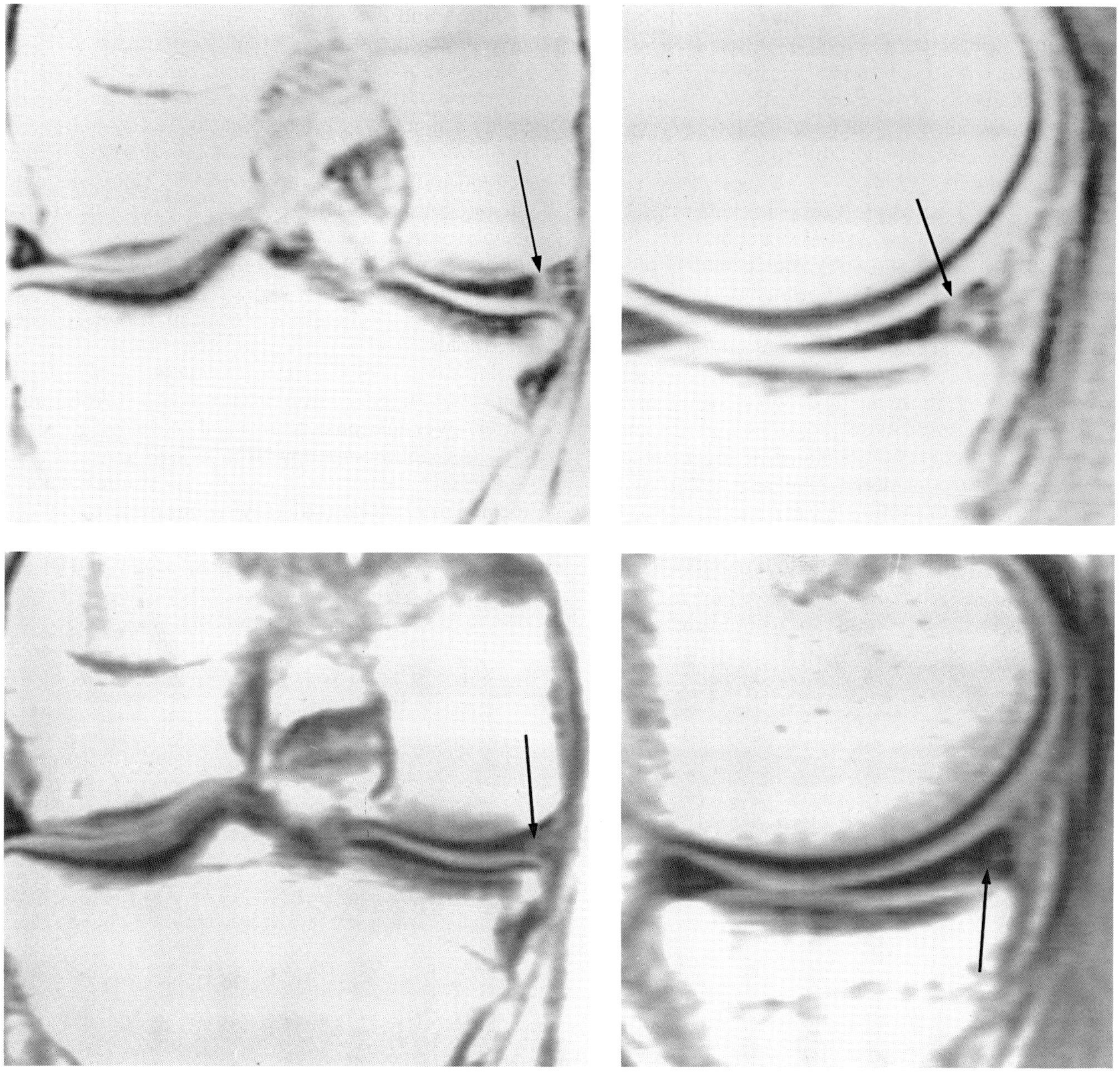

A B C D

FIG. 7.25. Peripheral meniscal tear: Conservative treatment. **A:** Coronal MR, TR/TE 800/20. A magnified high contrast image of the meniscus demonstrates the classic appearance of a vertical tear through the red-white junction of the medial meniscus (*arrow*). The tear was nondisplaced. The patient was treated with brace immobilization. **B:** Sagittal MR, TR/TE 2,000/20. The tear involving the outer third of the posterior horn is again well demonstrated on this magnified high contrast sagittal view (*arrow*). **C:** Coronal MR, TR/TE 800/20. Three year follow-up examination. The patient was asymptomatic at this time. The previously demonstrated signal has markedly diminished with only a subtle focus of intrameniscal present (*arrow*). **D:** Sagittal MR, TR/TE 2,000/20. On the follow-up exam performed at a time when the patient was asymptomatic and clinically stable, the signal focus has markedly diminished, although not entirely resolved (*arrow*).

A
B

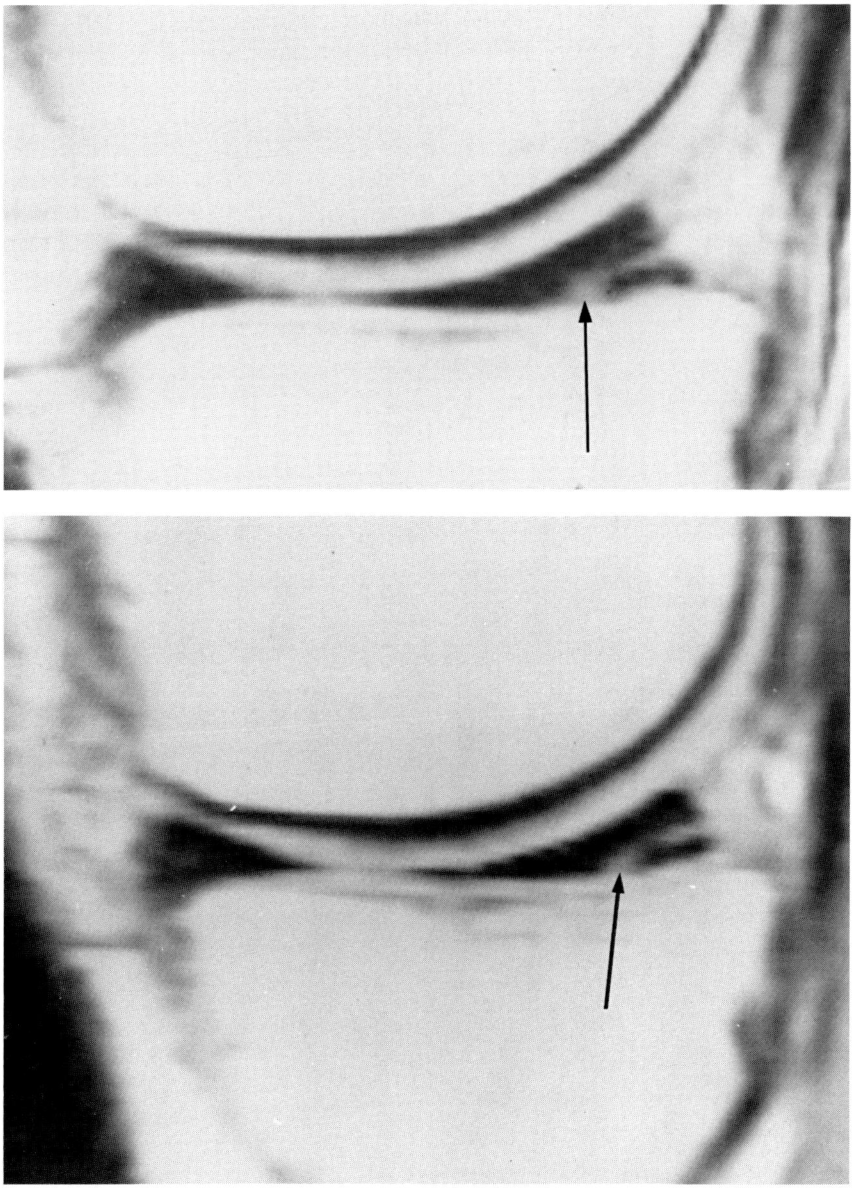

FIG. 7.26. Meniscal repair: Persistent signal. **A:** Sagittal MR, TR/TE 800/20. An oblique line of high signal intensity traverses the outer third of the posterior horn of the medial meniscus (*arrow*). The patient underwent arthroscopically directed repair of this meniscal tear. **B:** Sagittal MR, TR/TE 800/20. Follow-up MR performed 27 months after surgery at a time when the patient was asymptomatic and presumably healed. The signal from the tear persists without dramatic change. (From ref. 2, with permission.)

the diagnosis of nonhealing or re-tear (5). In studying a group of 30 menisci (29 repaired, one conservatively managed), grade 3 signal on T1- or intermediate-weighted images was present in 27 of the 30 cases. Of five patients with preoperative MR examinations available for comparison, the signal on the postoperative study was unchanged in four (all repaired) and decreased in one (conservatively managed).

The MR findings were compared to the arthrographic appearance of the menisci in 23 cases reported by Farley et al. (5). In seven of nine cases considered healed by accepted arthrographic criteria (no contrast extension into the repair site) and in three of four cases considered partially healed by arthrography, MR images demonstrated grade 3 signal extending to both articular surfaces on T1- or intermediate-weighted images. In ten cases, the MR findings of grade 3 signal correlated with arthrographic findings of meniscal re-tear. Closer correlation between MR and arthrography was obtained if only the T2-weighted images were utilized for diagnosis and the criteria for re-tear was unequivocal T2 increase in signal intensity through a full thickness defect extending to two articular surfaces. None of the menisci deemed healed by arthrography demonstrated such signal changes on T2-weighted images. Unfortunately, this MR criteria was present in only six of ten menisci considered re-torn by arthrographic criteria. Only two of the menisci in this series were subjected to repeat arthroscopy and both of these patients were symptomatic. Both MR and arthrography correctly demonstrated re-tears in each of these menisci. As a consequence, the stability of the menisci in the cases of disagreement between MR and arthrography, and thus determining which study was more "correct," remains unknown. While it is recognized that grade 3 signal on MR may be present in stable menisci, arthrography, conversely, may potentially be falsely negative. This is a consequence of the need for a lack of watertight apposition of the tear margins for the exam to be positive.

The findings of these studies emphasize that the diagnosis of re-tear of the meniscus on the basis of grade 3 signal on T1- or intermediate-weighted images in patients following meniscal repair should be made with caution. Hyperintensity of signal on T2-weighted sequences allows added confidence in diagnosis, but this finding, while apparently more specific, is not highly sensitive.

Arthrography, if positive, would confirm the presence of re-tear in these cases, although if negative, it does not entirely exclude it. The potential role for intraarticular paramagnetic contrast MR (MR arthrog-

A

B

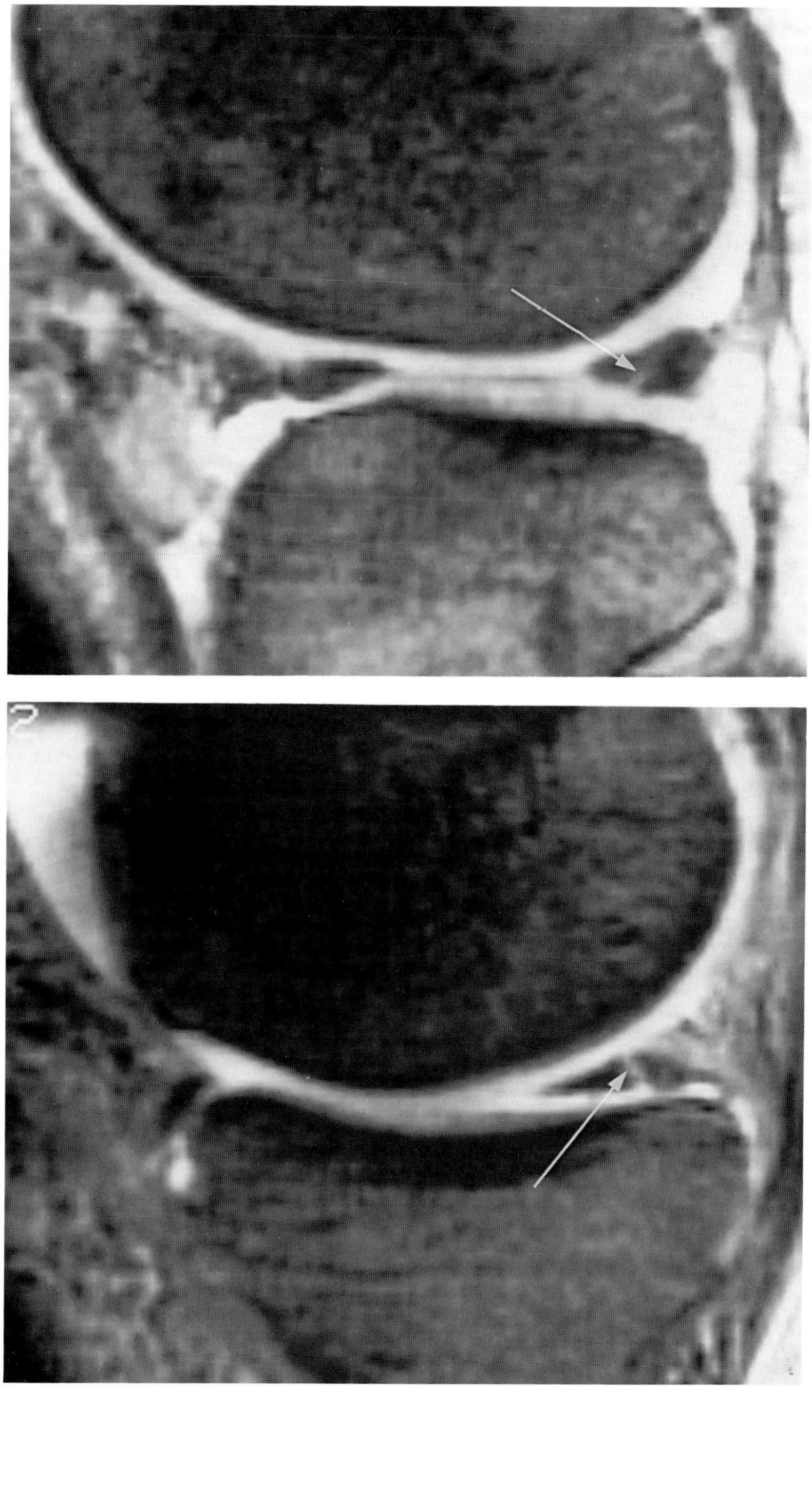

FIG. 7.27. Meniscal repair, persistent signal. **A:** Sagittal MPGR/TR/TE 20/600/15. A faint line of increased signal intensity is seen within the posterior horn of the lateral meniscus of this patient status post meniscal repair (*arrow*). No preoperative study was obtained. At surgery, the meniscus was considered healed without evidence of a tear. **B:** Sagittal MPGR/TR/TE 20/600/15. A prominent line of increased signal is seen extending vertically through the peripheral third of the posterior horn of the lateral meniscus of this patient status post meniscal repair (*arrow*). At second look arthroscopy performed 6 months following the initial surgery, the meniscus was healed to inspection and probing. (From ref. 2, with permission.)

raphy) has not been the subject of critical study in this patient group, but may potentially contribute in selected cases.

Persistent signal within previously repaired menisci may be accounted for by histologic differences between the reparative tissue and native meniscal fibrocartilage (2). The significance of the persistent signal with regard to the structural integrity of the reparative tissue and the potential risk of re-tear remains unanswered (2). The extent to which signal may eventually modulate in healed meniscal tears has also not been established. We have had subsequent experience with at least two cases in which the signal from a previous tear became markedly diminished on follow-up (Fig. 7.25). The extent and time frame for this to occur awaits further study. It is clear from present experience, however, that persistent signal at the site of the previous meniscal tear is a common occurrence following meniscal repair, and certainly cannot be utilized as the sole criterion for the diagnosis of re-tear in these patients.

In addition to the detection of grade 3 signal as the basis for the diagnosis of meniscal re-tear, two other observations may be used to assess these postoperative menisci (5). The first of these is the ability of MR to demonstrate the presence of a new tear at a site

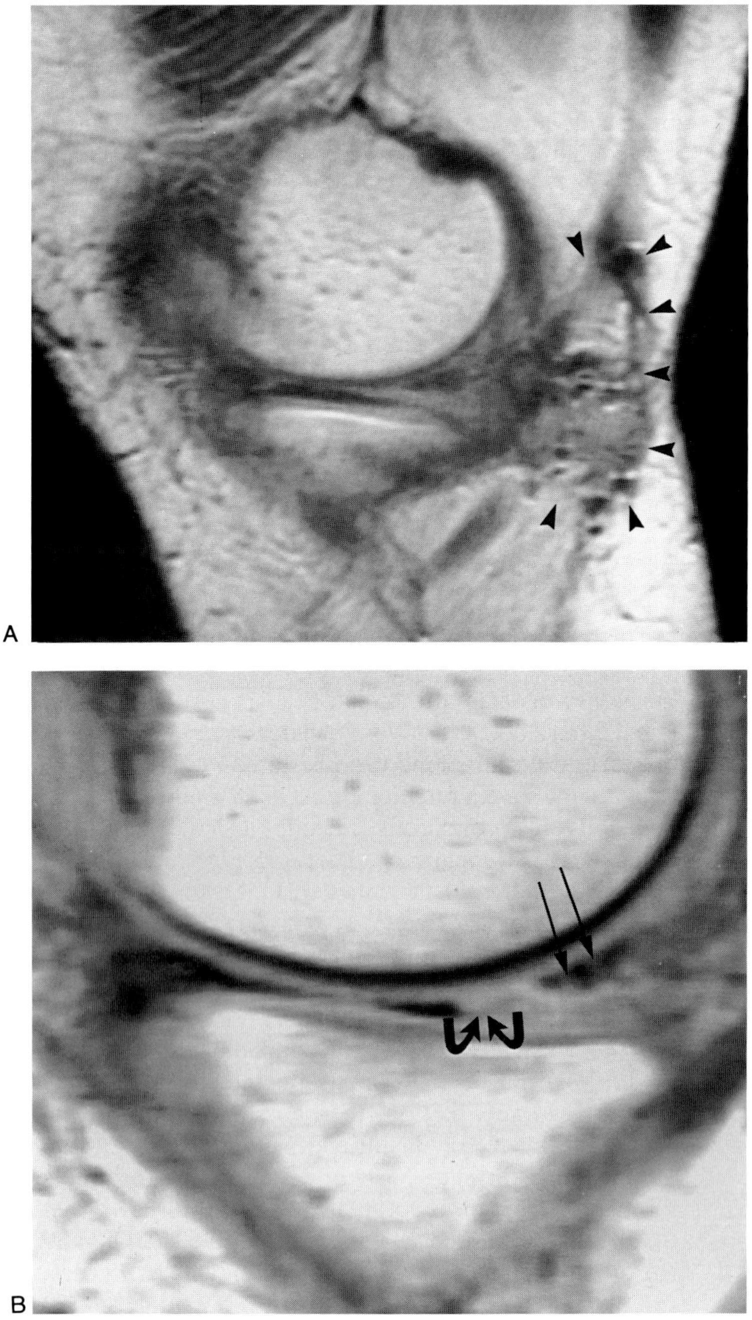

FIG. 7.28. Failed meniscal repair. **A:** Sagittal MR, TR/TE 2,000/20. Extensive micrometallic artifact is seen in and around the posteromedial corner of the knee (*arrowheads*). Prominent scar is also noted. The patient had undergone prior meniscal repair and had sustained damage to the saphenous nerve. **B:** Sagittal MR, TR/TE 2,000/20. Magnified view of the same patient demonstrates wide displacement of the patient's repaired meniscus (*curved arrows*). The sutures are seen coursing through the peripheral fragment (*long arrows*). MR can be valuable in assessing the integrity of the repaired meniscus and any possible complications of the procedure.

remote from the previous meniscal repair or conservatively treated tear (Fig. 7.24) (5). Knowledge of the location and morphology of the previous tear can aid considerably in diagnosis, particularly in distinguishing a new tear from potential persistent changes related to the prior injury. Information regarding the site of the previous tear, while often not immediately available, should be specifically sought. In this regard, operative reports themselves, if unaccompanied by schematic diagrams, are often of limited value. Correlation with the videotape of the prior arthroscopy represents another potential means to obtain the required information, but this is often not available or is at best impractical in most settings. Prior MR examinations, if available, generally provide the best means for comparison and establishment of the location and morphology of the previous tear. In addition to the detection of high signal intensity within the focus of grade 3 signal and the establishment of a new injury at a site remote from the previous tear, the detection of a displaced fragment on MR represents another firm criteria for detecting a re-tear of a previously repaired or partially resected meniscus (Figs. 7.24 and 7.28).

In addition to evaluation of meniscal healing, MR may also be applied to the assessment of complications of meniscal repair. While the rate of complications has decreased with increasing surgical experience, injuries to the neurovascular structures remain one of the primary areas of concern. Injury to the saphenous nerve and vein may occur with medial repairs, and injury to the peroneal nerve may occur with lateral repairs near the popliteal hiatus. Extensive metallic particles and localized hematomas have been demonstrated with MR in patients with neurovascular injuries following repair (Fig. 7.28).

Summary

Evaluation of the previously injured meniscus is complex and requires an understanding of the nature and natural history of such lesions, their consequences, alternative treatment options, and complications of this treatment. Considerable attention has been directed to the potentially deleterious effects of meniscectomy, and the current emphasis in orthopedic practice has increasingly been on meniscal preservation. This may take the form of conservative care, partial meniscectomy, or most recently meniscal repair. MR can be an extremely useful technique for evaluation of patients with persistent or recurrent symptoms following prior knee surgery. It is the only diagnostic method capable of noninvasively differentiating extraarticular from potentially correctable intraarticular causes of pain. With regard to the meniscus, considerable difficulty has been experienced in assessment of meniscal stability in the postoperative setting. In only limited situations can the criteria described for the virgin meniscus be applied to the postoperative situation with the same high accuracy and predictive value. The mere presence of grade 3 signal within menisci having undergone partial resection or meniscal repair need not indicate evidence of a re-tear. High signal intensity on T2-weighted images within full thickness lines of grade 3 signal appears to represent a highly specific, albeit not highly sensitive, criteria for meniscal re-tear. The detection of displaced fragments as well as the establishment of signal or morphologic alterations at sites remote from the original tear also represent highly valuable criteria for the confident establishment of meniscal re-tears. In addition to assessment of meniscal stability, the deleterious effects on articular cartilage of prior injury and surgery can be documented utilizing MR, but they must often be specifically sought and require meticulous attention to technique in order to document and completely characterize their presence and extent.

ASSESSMENT OF THE ANTERIOR CRUCIATE LIGAMENT RECONSTRUCTION

Clinical and Anatomic Considerations

Tears of the anterior cruciate ligament (ACL) have been well recognized since the turn of the century, and the first attempt at reconstruction was reported in 1917. Nonetheless, more has been written on the torn ACL in the past 5 years than in the preceding 80 years (62).

The need for reestablishment of the stabilizing function of the torn ACL has become obvious because of the rather dismal natural history of the ACL-deficient knee. Many authors have consistently noted the high proportion of poor long-term results in those patients who do not have a functioning ACL (63–68). The long-term adverse conditions include articular cartilage damage which occurs in 58% to 96% of knees, osteoarthritis, and secondary meniscal tears (69). Meniscal tears are known to occur concurrently with tears of the ACL, but secondary tears due to wear damage occur in 16% to 22% of patients (68–73). A knee that has suffered an ACL tear is a knee at risk; it frequently follows a relentless downhill course of deterioration which may not become apparent until the secondary restraints to subluxation (the capsular ligaments) become stretched and insufficient (44,65,67,70,73–78). Kennedy performed a longitudinal study of patients who had sustained a complete tear of the ACL; one-half of the group underwent conservative therapy and the other half had ACL reconstruction (74). At 7 years, the untreated group had deteriorated far more significantly than had the treated group. In a group of high school and college athletes who had suffered an ACL tear, less than half were able to return to their sports activities by 6 months because of pain and swelling, and only 11% were able to play strenuous sports 5 years later (63). It is apparent that the syndrome of progressive deterioration of the knee occurs in the athletically active individual who continues that lifestyle (63). Not all patients who suffer an ACL tear will require reconstruction. Patients who develop (a) functional instability that cannot be controlled with behavior modification and bracing, (b) symptomatic instability during activities of daily living, and (c) a significantly positive pivot shift test should be considered for operation (79,80).

Primary Repairs

Two types of complete ACL ruptures occur, and each requires a different surgical philosophy and approach. In the first, less common type, the rupture occurs via bony avulsion, usually at the tibial intercondylar eminence (73,74). Reattachment of significant avulsive lesions is universally advocated since fixation is by means of bone to bone, and the likelihood of a successful repair is high.

The vast majority of ACL ruptures occur within the substance of the ligament. The first primary repair of an intrasubstance tear was attempted in the early 1900s. Early animal studies suggested that primary repairs of a torn ligament were feasible but clinical studies in humans have been mixed (81,82). Feagin et al. reviewed the results of primary repairs at 5 years post injury. Instability occurred in 94%, and fewer than 10% had an asymptomatic stable knee 5 years after repair (83). Marshall et al. and Kaplan et al. reviewed the records of 70 patients who underwent primary repair of the ACL (82,84). They originally recommended repair in the athlete, but when they reviewed the long-term results at an average of nearly 7 years from repair, they concluded that primary repair is an unpredictable operative procedure (84). There is now a general opinion that primary repair of midsubstance ACL tears is frequently unsuccessful regardless of technique, and that the repairs deteriorate with time (79,82–87).

Reconstruction: Surgical Considerations

The first reported surgical reconstruction of the ACL was done by Hey Grove in 1917 when he utilized the iliotibial band (62). Since that time, 60 different reconstructive ACL procedures have been advocated, each with varying degrees of success (76). The following sections are included to provide the reader with a limited review of the surgical procedures available, an understanding of the terminology employed in knee reconstruction, and the biomechanical principles which guide reconstruction therapy.

Reconstructions of the ACL have been divided into those that are either intraarticular or extraarticular, and those that combine some features of each. Since neither intra- nor extraarticular stabilizations provide complete control of both abnormal rotation and subluxation, they may therefore be combined (76,88–93). A potential advantage of the combined procedure is the ability of the extraarticular graft to protect the intraarticular component during the early postoperative period, when the intraarticular graft may be maximally weak (94,95). Significant long-term gain from combining procedures is unlikely, but the issue continues to be debated (89,96).

The goal of *extraarticular* procedures (examples include the MacIntosh, iliotibial band tenodesis, Losee, James, Andrews, Hughston, Ellison, and biceps transfer operations) is to prevent the anterior subluxation of the tibia in relation to the lateral femoral condyle by providing a static check rein effect on the lateral tibial plateau as the knee approaches terminal extension (6,90). Most lateral procedures utilize some section of the iliotibial band (ITB) to keep the tibia in a position behind the transverse center of rotation of the femur (90,97). In the MacIntosh extraarticular reconstructive procedure, a 16 cm long strip of the distal ITB is freed up proximally while its distal attachment is left in place. The proximal end of the distal segment of the ITB is routed beneath the fibular collateral ligament, through a tunnel in the distal intermuscular septum, back under the fibular collateral ligament, where it is reattached to the Gerdy tubercle.

Noyes et al. described six parameters of an ideal *intraarticular* ACL substitute (98):

1. The substitute must have properties similar to those of the native ACL. Length of available material, stiffness, and graft geometry must be considered in selection of an appropriate material.
2. A ligament substitute should have intact vascular and neural attachments.
3. The graft substitute must be readily available and must be an expendable structure if it is autogenous.
4. The graft should have bone at each end, allowing for rigid fixation, early motion, and advanced rehabilitation.
5. The operation must be technically feasible.
6. The operation must have an acceptable morbidity rate.

A wide variety of natural and synthetic materials have been adopted for intraarticular reconstruction of the ACL. *Autograft* substitutes include the gracilis, semimembranosus, patellar, or semitendinosus tendons; the fascia lata; and the meniscus (41,99–110). *Allo-*

grafts, in the form of fresh frozen or freeze dried patellar or Achilles tendons, or bovine tendon xenografts, also have been used. Rejection is not a significant problem, but recently considerable concern has been expressed about the ability of standard procedures to ensure viral sterilization (100,111–113). When considering the use of synthetics, problems of safety and biocompatibility, strength and yield, and adequate fixation must be addressed. The materials most commonly used include polytetrafluoroethylene (Gore-tex), carbon fiber, teflon, Dexon, proplast, high strength Dacron, and polypropylene braid (LAD), but to date, only the Gore-tex, LAD, and Dacron materials have been approved for human use (6,80,99,101, 104,114–116). The Gore-tex prosthesis was designed as a permanent replacement for the ACL and it enjoys the most popularity; over 15,000 replacements have been performed (101–103,115). It is the strongest of the available synthetics, and it is easily fixed to bone and allows for fibrous ingrowth into its porous microstructure. Its use is restricted to those patients who have failed reconstruction with an autogenous substance.

Multiple combinations of graft material and operative techniques have been advocated as ideal methods of reconstruction (73,79,86,89,91,106,110,118–123), but today a graft derived from either the middle or the medial third of the patellar tendon is the reconstruction material of choice (77,79,85,92,96,112,124–130). The patellar tendon meets all of Noyes' criteria of an ideal intraarticular reconstruction material. Of the autogenous materials available, the patellar tendon is the strongest under uniaxial tension tests. At the time of harvesting the patellar tendon is approximately 150% as strong as the native ACL, while the hamstring grafts (semitendinosus or gracilis) are only 50% as strong (98,127,128). The patellar tendon blood supply can be preserved, although the need for this property has been disputed recently (111,131). The utilization of bone to bone fixation at both ends allows for rapid and strong fixation of the graft, earlier joint motion, and a shorter rehabilitation period. Finally, removal of a portion of the patellar tendon is associated with rather little morbidity (see below). Recent studies have found that in patients treated with a bone-patellar tendon-bone autograft (BPTB) who were followed for a minimum of 2 years, 95% no longer experienced clinical instability, and their pivot shift test was negative in 84% (130).

During harvesting, one-third of the patellar tendon is dissected free. The insertion sites of the patellar tendon into the patella and the tibial tubercle are removed in continuity with the graft (79,96,132) (Fig. 7.29). This

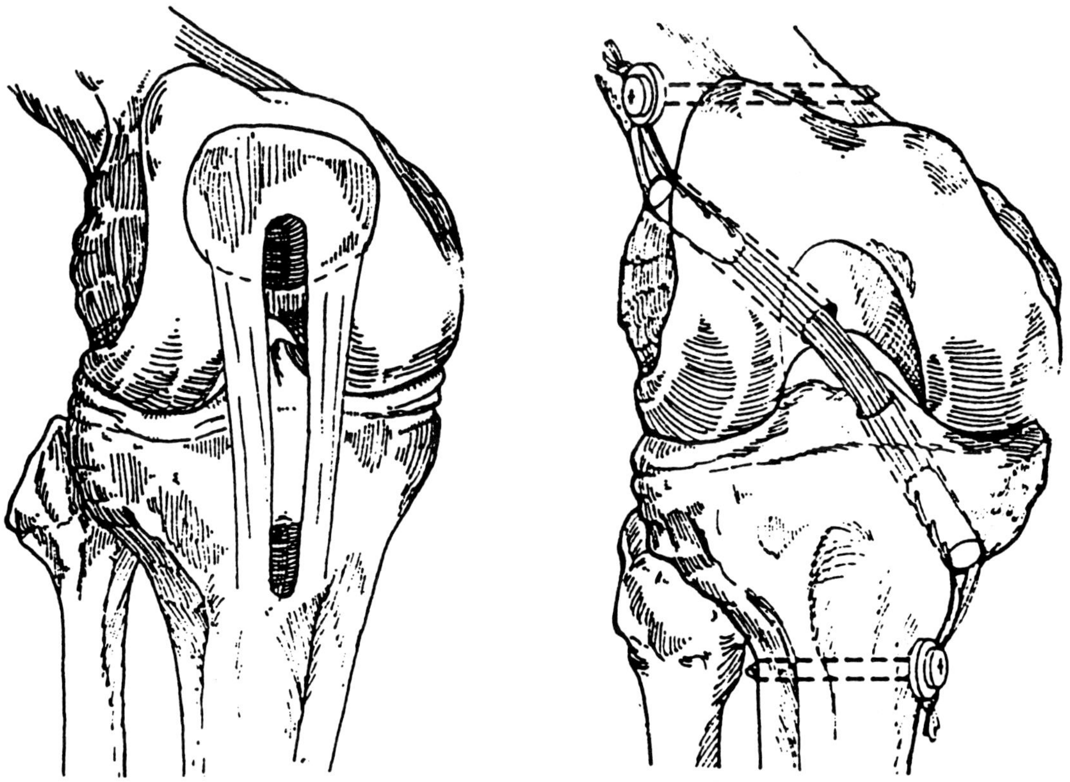

A

B

FIG. 7.29. A,B: Harvesting a patellar tendon autograft. These diagrams demonstrate the method in which a graft is harvested and fixed. The central one-third of the patellar tendon is dissected and removed along with small plugs of bone from both the patella and the tibial tubercle. The plugs are pushed into the femoral and tibial tunnels, and in this case are secured by means of bicortical screws.

procedure yields a tendinous graft that is 10 mm wide and 3 to 4 mm thick. An alternative procedure involves the use of the hamstring tendons, especially in young women with preexisting patellofemoral pain syndromes. The semitendinosus and gracilis tendons are harvested and transferred as free double looped grafts without a bony attachment. Regardless of which autogenous material is used, the graft is placed into a 8 mm to 10 mm bone tunnel drilled just anterior and medial to the anatomic center of the native ACL insertion on the tibia (91,115). On the femoral side of the joint, the drill hole is superior and posterior to the anatomic center of the origin of the ACL. The proper position of each tunnel is determined by intraoperative tension testing. The graft is twisted upon itself at the time of final placement; experiments have shown that this maneuver increases graft strength (115). The plugs of bone at each end are rigidly fixed by one of several methods. An interference fit screw may be placed within the tunnel to push the bone plug against the side wall (Fig. 7.30). Alternatively, staples or a bicortical screw may be used to secure suture which has been attached directly to the bone plugs or soft tissue components of the graft (79,133–135). Finally, chamfering (radiusing) of the openings of the bony tunnels eliminates sharp edges which could damage the graft; a notchplasty removes acquired osteophytes and enlarges the aperture of the notch to at least 23 mm (6,80,136). Drilling tunnels across an open physis in adolescents who require ACL reconstruction is contraindicated; the trauma will interfere with growth. An alternative procedure involves pulling the proximal end of the semimembranosus/gracilis tendons through the joint under the anterior horn of the medial meniscus, to the back of the joint where they are attached to the lateral femoral condyle (122).

Adherence to the surgical principles of isometry and impingement-free movement are of the utmost importance in assuring stability and longevity of any type of graft. Isometry is the ability of a graft to maintain a constant length and tension during flexion and extension of the knee (69,76,80,115,129,137,138). Lack of length and tension change is critical since multiple studies have shown that the native ACL has a normal length change of 2 mm (5%) beyond which ligamentous damage begins to occur (115,137,139). Overloading the graft will result in limitation of motion and graft failure due to stretching of the neoligament. Underloading of the graft will not restore knee stability, which was the original goal of the procedure (76). While the achievement of isometric tracking is dependent on the interrelationship of the positions of the femoral and the tibial sites, moderate variations in the *tibial* attachment site of the graft only minimally affect isometry (6,76,133,139).

Impingement refers to interference with the smooth, frictionless motion of the graft. While occasionally impingement is secondary to fibrous or fibrocartilaginous scar tissue (140), most cases are due to contact of the graft with a rough or protuberant bony surface. Impingement may not be associated with restricted joint motion, pain, or instability, but abnormal contact of the graft may lead to excessive wear on the graft, and ultimately to its failure (102). Impingement most commonly occurs within and at the exit point of the grafts from the bony tunnels, the side wall, and the roof of the intercondylar notch (96,132). Impingement is most evident in the chronically unstable knee where osteophyte proliferation and intercondylar notch encroachment are marked (66,102,115). The role of a small intercondylar notch in the pathogenesis of ACL tears, and the need for a capacious notch to allow for unimpinged action of the reconstructed ACL, has been well documented (66,102,115,136,141–144).

The most difficult site at which to identify impingement at the time of surgery is at the anterior inferior edge of the intercondylar roof (roof impingement). Roof impingement is essentially a radiographic diagnosis since it occurs in the terminal 5° to 10° of knee extension where the surgeon may not have an optimal view of the bone-graft interface. Roof impingement is said to be present when the tibial tunnel and its contained graft lie at least in part anterior to the projected inferior slope of the intercondylar roof. The originally described "ideal" position of the tibial tunnel is now considered to be too far anterior, and it frequently results in roof impingement (96,136,145). The clinical diagnosis of roof impingement may also be extremely difficult since impinged grafts are generally stable and capable of restraining abnormal movement of the tibia (the pivot shift test is normal) (145). Other clinical parameters of strength and stability—the single leg hop test, the Lysholm scoring scale, the manual maximum KT-1000 laxity test—may not demonstrate significant differences between grafts that are and those that are not impinged (136,145). The incidence of pain and persistent effusion is more frequent in patients with roof impingement than in those without, and impinged knees may lack several degrees of extension (142).

Biology of the Graft

In 1982, Arnoczky et al. investigated the natural history of the patellar tendon graft following implantation (146). He found that harvested grafts were actually avascular, but by 6 weeks after implantation, they become completely enveloped in a vascular synovial sheath arising from the vessels of the infrapatellar fat pad and the adjacent synovium. Ischemic necrosis oc-

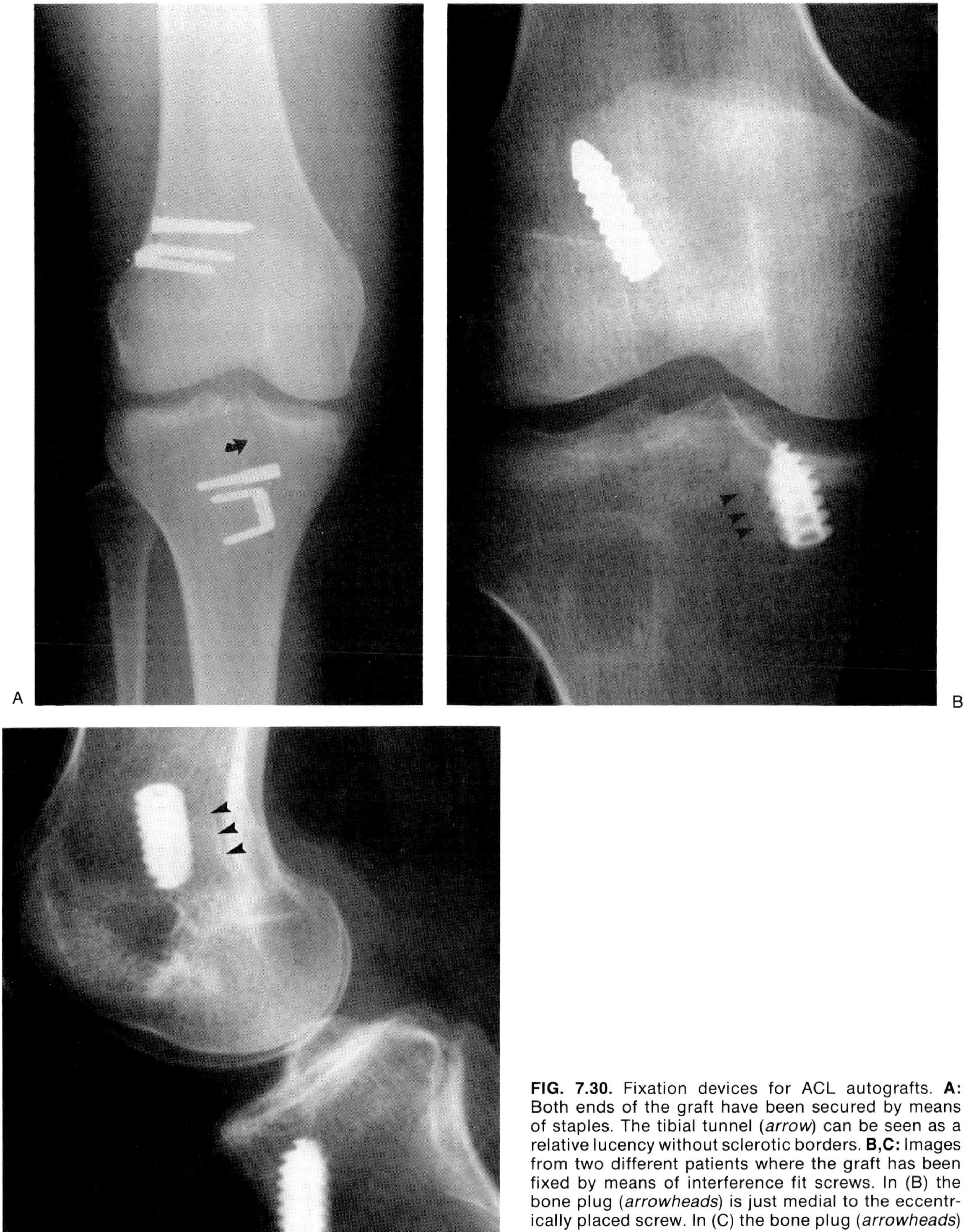

FIG. 7.30. Fixation devices for ACL autografts. **A:** Both ends of the graft have been secured by means of staples. The tibial tunnel (*arrow*) can be seen as a relative lucency without sclerotic borders. **B,C:** Images from two different patients where the graft has been fixed by means of interference fit screws. In (B) the bone plug (*arrowheads*) is just medial to the eccentrically placed screw. In (C) the bone plug (*arrowheads*) can be identified just posterior to the femoral screw.

curring in the central core of the graft is followed by revascularization progressing from the endosteal vessels of the bony tunnels and from the fat pad. Revascularization is complete by 20 weeks, after which the vascularity on the surface of the graft diminishes. Over a 6 to 12 month period the graft undergoes a remarkable metamorphosis, maturing into well-arranged collagen bundles, spindle-shaped nuclei, and a normal degree of cellularity. It is transformed into a tissue virtually identical to a normal ACL morphologically, histologically, and biomechanically through a process termed ligamentization (85,115,124,146–150).

In assessing the relationship of vascularity and water content to tensile strength, McFarland et al. discovered that grafts became more vascular, more hydrated, and less strong 4 weeks postimplantation (127). By 16 weeks the grafts regain some strength, but remain only 40% as strong as controls. The data suggest that increasing vascularity of a graft correlates with inflammation due to cell death, and that reorganization of the graft in the early postoperative period leaves the tissue with significantly less strength. Given the fact that patellar tendons are initially stronger than the native ACL in uniaxial tension tests, it is possible that the ligamentized patellar tendon at 1 year is equal in strength to the normal ACL, although extrapolation of this data from monkeys to humans may not be valid (85,115).

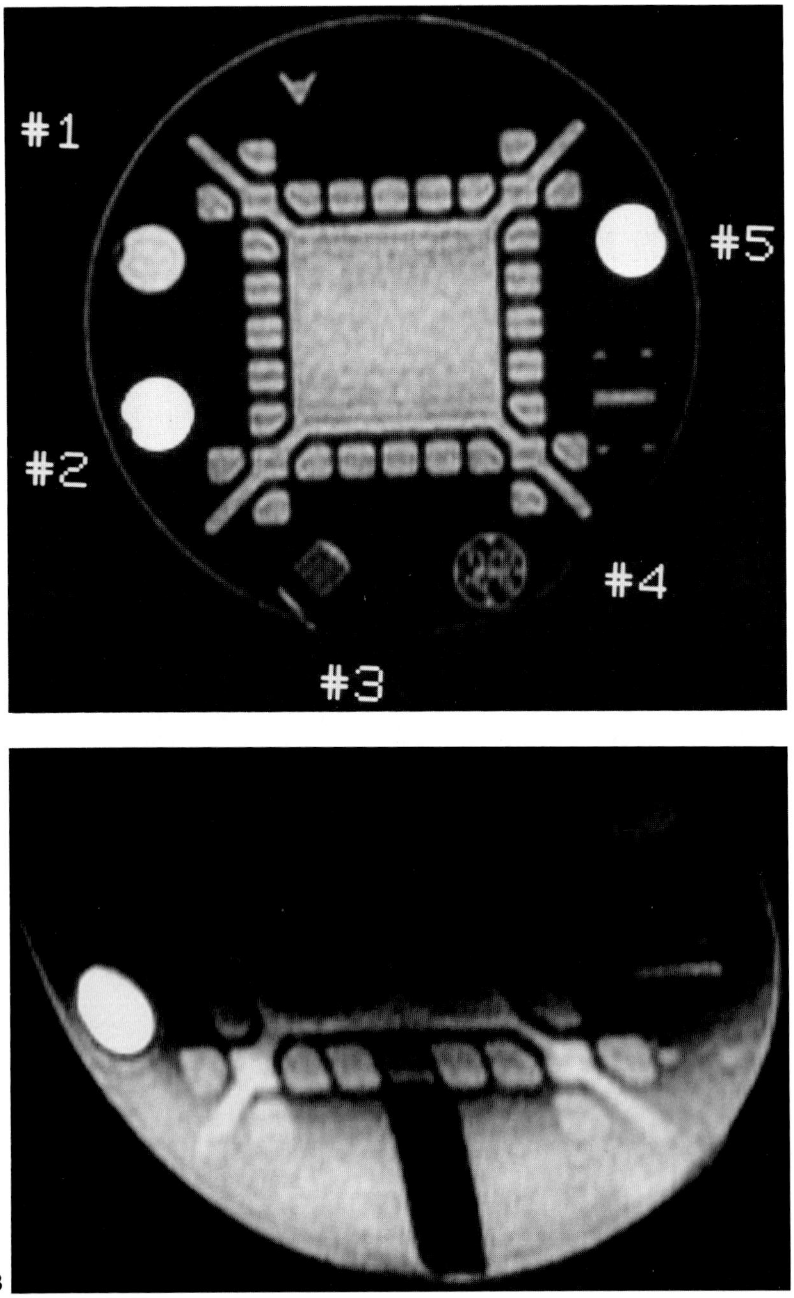

A

B

FIG. 7.31. Ferromagnetic properties of orthopedic hardware. **A:** A series of screws, staples, and washers often used for ACL graft fixation have been positioned around the phantom with only the most minimal degree of image distortion at the sites of devices #3 and #4 (1, cortical screw; 2, cortical screw; 3, staple plate; 4, fixation staple; 5, Acufex interference fit screw) (135). **B:** An interference fit screw (Perfix interference fit screw, Instrument Makar, Okemos, MI) has been placed on top of the MR phantom. Because of the high ferromagnetic property of the screw, there is gross image distortion.

Four arthroscopic phases of the maturation process have been correlated with histologic changes (117,150). In the first 3 months postoperatively, a thick layer of vascularized synovial tissue originating from the fat pad surrounds the graft (type 1 reaction). The grafts contain many fibroblasts with spindle-shaped nuclei within their substance. The synovial reaction becomes thinner and the graft becomes thicker due to fibrous tissue ingrowth and commencement of transformation into ligament-like tissue over the first year (types 2 and 3 reaction). At the end of the first year the synovial membrane is very thin and rather poorly vascularized, and the graft has a gross appearance nearly identical to that of a normal ACL (type 4 reaction).

Surgical Complications

All of the intraarticular procedures used to reconstruct the ACL may be associated with superficial and deep wound infections, hemorrhage, popliteal artery injury, skin slough, flexion contracture, persistence effusions, thrombophlebitis, and intraarticular fibrosis (69,114,151–153). Removing one-third of the patellar tendon may alter the biomechanics of the extensor mechanism. Loss of quadriceps strength, loss of range of motion, medial and lateral patellar subluxation, and patella baja all have been reported (69,96,115,128,151, 152,154–156). More dramatically, fracture of the inferior pole of the patella and rupture of the patellar tendon have been reported to occur in 1% to 2% of patients (123,128,133,157,158). The latter complication may result from interference with the vascular supply, or loss of strength of the tendon secondary to decreased mass following resection (157).

Synthetic replacements are subject to abrasion by bony impingement with the release of small wear particles which can result in recurrent/persistent effusions, infection, and synovitis (94,114,159). More importantly, however, it has been postulated that, through activation of synoviocytes and chondrocytes, wear particles may have the potential to induce osteoarthritis (160). Rupture of a Gore-tex graft is not uncommon (103).

Imaging Assessment

Technical Considerations

Although MR assessment of the reconstructed ACL is still a relatively new technique which continues to evolve, MR has already demonstrated several advantages over second look arthroscopy in studying the condition of the graft. Arthroscopy provides only a limited view of the anterior surface of the graft (117); MR, on the other hand, can provide assessment of the entire intraarticular volume of the graft as well as the intraosseous portions. It is highly impractical to perform routine second look arthroscopy because of the expense and morbidity; MR allows one to noninvasively evaluate the graft in a less expensive manner. Finally, the natural history of the graft core, including necrosis and subsequent revascularization, may be studied by MR using intravascular contrast agents (gadolinium); the arthroscopist is only able to visualize the surface of the graft, and is not able to assess its viability and vascularity without biopsy.

MR imaging of the reconstructed ACL is complicated by the presence of postoperative edema and tissue distortion, and fixation hardware. The extent of artifact formation from the hardware is highly dependent on the composition of the metal in the fixation devices, the quantity of metal used, and the proximity of the devices to the joint line. Stainless steel may be highly ferromagnetic and may generate significant metallic artifact (Fig. 7.31) (135). Titanium hardware has been shown to distort the image less than some varieties of stainless steel; therefore even if titanium screws are relatively close to the joint line, image distortion may not be significant (142). When designing an imaging protocol for the knee that has had an ACL reconstruction, one must not forget that patients are not referred for MR exclusively for assessment of the graft. Patients may incur secondary meniscal tears and articular cartilage damage. In order to properly evaluate the knee for all possible causes of joint

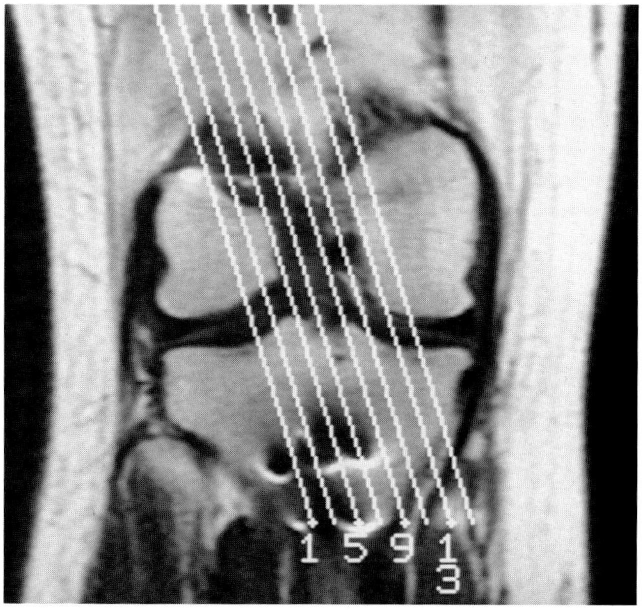

FIG. 7.32. Oblique scanning plane for the ACL reconstruction. Coronal MR, TR/TE 2,000/20. The coronal image is examined to identify the course of the reconstruction, and the sagittal plane is prescribed obliquely along the course of the neoligament.

pain/dysfunction postreconstruction, we have established the following imaging protocol. We begin with a standard coronal and sagittal knee MR examination. The coronal images through the midportion of the intercondylar notch are specifically examined for the reconstructed ligament and a customized oblique sagittal scanning plane parallel to the superolateral to inferomedial trajectory of the graft is selected (Fig. 7.32) (136,142). We have found this plane preferable to a standard true sagittal plane for graft assessment, although in cases in which the graft has been perfectly placed, the entire ACL is often well seen on the standard sagittal images. It is not optimal to utilize the oblique sagittal plane *in lieu of* the orthogonal plane because some grafts may be oriented in a rather steep degree of obliquity, and the oblique sagittal plane may produce difficult to read images of the menisci and other intraarticular structures.

Selection of an imaging sequence must balance the need for relatively high signal, high soft tissue contrast, and reasonable imaging times. We have found both very short TR and very long TR sequences unacceptable, and now choose T1 and/or proton density sequences (TR/TE 1,200/40) as being optimal for assessment of anatomy and internal signal of grafts (134,136, 142,161). The normal BPTB graft is typically 10 mm wide but only 3 mm to 4 mm thick, and it is impossible to predict the precise orientation of any portion of the graft along its course in the intercondylar notch. We therefore choose a 3 mm slice thickness without an interslice gap in an effort to maximize graft visualization.

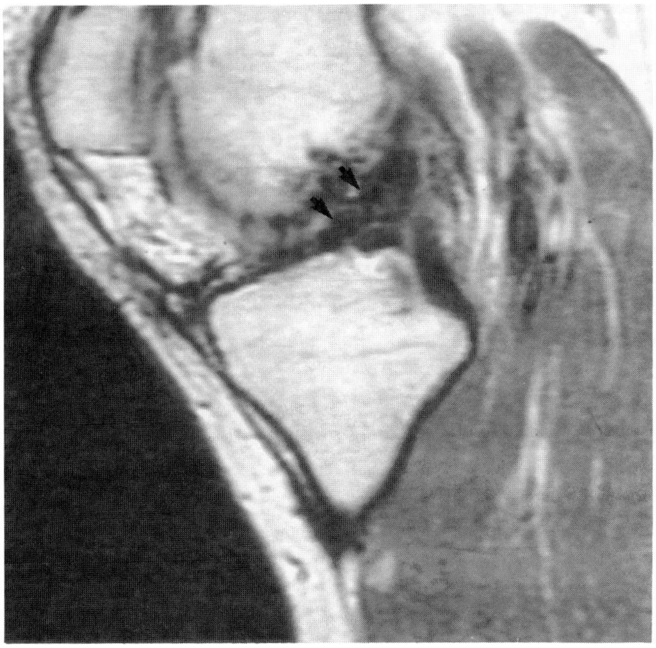

FIG. 7.33. Primary repair of a torn ACL. Sagittal MR, TR/TE 1,200/40. This patient is known to have had a torn ACL, and underwent a primary repair. A linear, dark structure (*arrows*) is in the expected course of the anterior cruciate ligament. The somewhat nodular character of the repaired ligament is due to the presence of suture material. At second look arthroscopy, the repair was intact and the knee was stable.

Primary Repair

A number of conventional radiographic signs may indicate failure of an attempted primary *repair* of an ACL tear: (a) peaking of the intercondylar tubercles; (b) spurring of the intercondylar eminence; (c) osteophytosis of the inferior pole of the patella; (d) stenosis of the notch; and (e) osteoarthritic changes of the knee joint itself (83).

Successful primary repairs of the ACL can be identified on MR (Fig. 7.33). The sutured ligament is a single dark band similar to a normal ACL, but most commonly focal "nodules" along the expected course of the ACL, representing suture material, can be identified, and a blooming effect from micrometallic debris is often seen in the intercondylar notch.

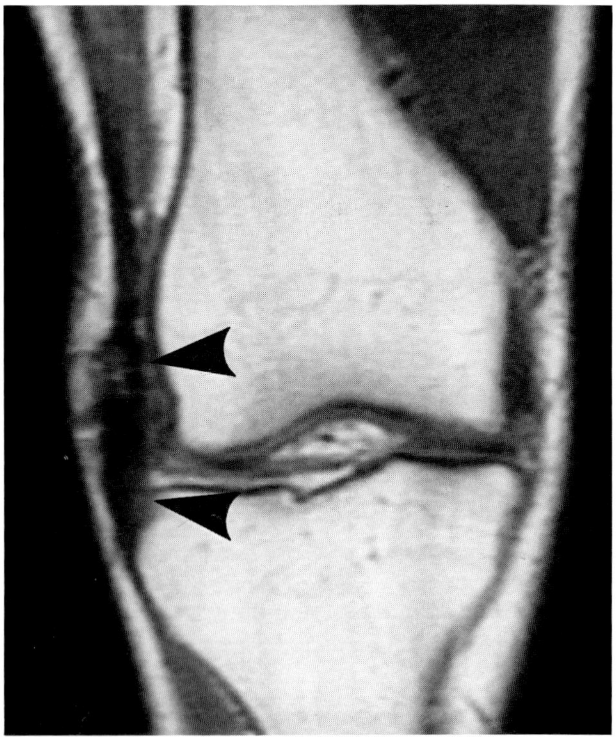

FIG. 7.34. Extraarticular reconstruction. Coronal MR, TR/TE 2,000/20. Following removal of a portion of the ITB for reconstruction, the ITB appears thickened due to micrometallic artifact and scar (*arrowheads*).

Extraarticular Reconstructions

As noted above, the vast majority of extraarticular reconstructions utilize a segment of the ITB to prevent abnormal rotation of the tibia. The MR findings in patients who have undergone extraarticular reconstructions of the ACL are confined to the area of the ITB (Fig. 7.34). Harvesting results in the deposition of micrometallic artifact in the bed of the graft; healing of the graft donor site occurs by scar formation. The metal and the scar both have very low signal on both T1- and T2-weighted images, and the end result is a thickening of the naturally dark ITB. The intricate anatomic alterations occurring around the fibular collateral ligament and intermuscular septum are not resolved by MR.

Intraarticular Reconstructions

Patellar Tendon/Hamstring Grafts

Tunnels to accommodate the graft are drilled into the femur and the tibia at sites that have been identified as being isometric (133). These tunnels may be seen on conventional radiographs by visualization of their thin, sclerotic edges which often are visible by 4 months post implantation (Figs. 7.35 and 7.36) (136). Since MR is a tomographic technique, it has enhanced capabilities to optimally assess tunnel position. As noted above, the femoral tunnel site is largely responsible for the isometry of the graft. The isometric position of the femoral tunnel can be visualized on the sagittal MR (or the lateral radiograph) as the intersection of two lines; the posterior femoral cortex and the posterior edge of the intercondylar roof (the posterosuperior edge of the intercondylar roof corresponds nearly precisely to the posterior edge of the physeal scar) (76,115,133). The femoral tunnel opening on the coronal MR is at either the 11 o'clock or 1 o'clock position, depending on whether one is dealing with the right or left knee (Figs. 7.35, 7.36, and 7.37).

On the sagittal MR, the impingement-free tibial tunnel opens distally below the tibial tubercle and courses posterosuperiorly to exit the tibial articular surface immediately anterior to the anterior tibial spine (Figs. 7.35, 7.36, and 7.38). The location of the center of a well-placed tibial tunnel is 42% of the sagittal distance from the front of the tibial plateau; those grafts which

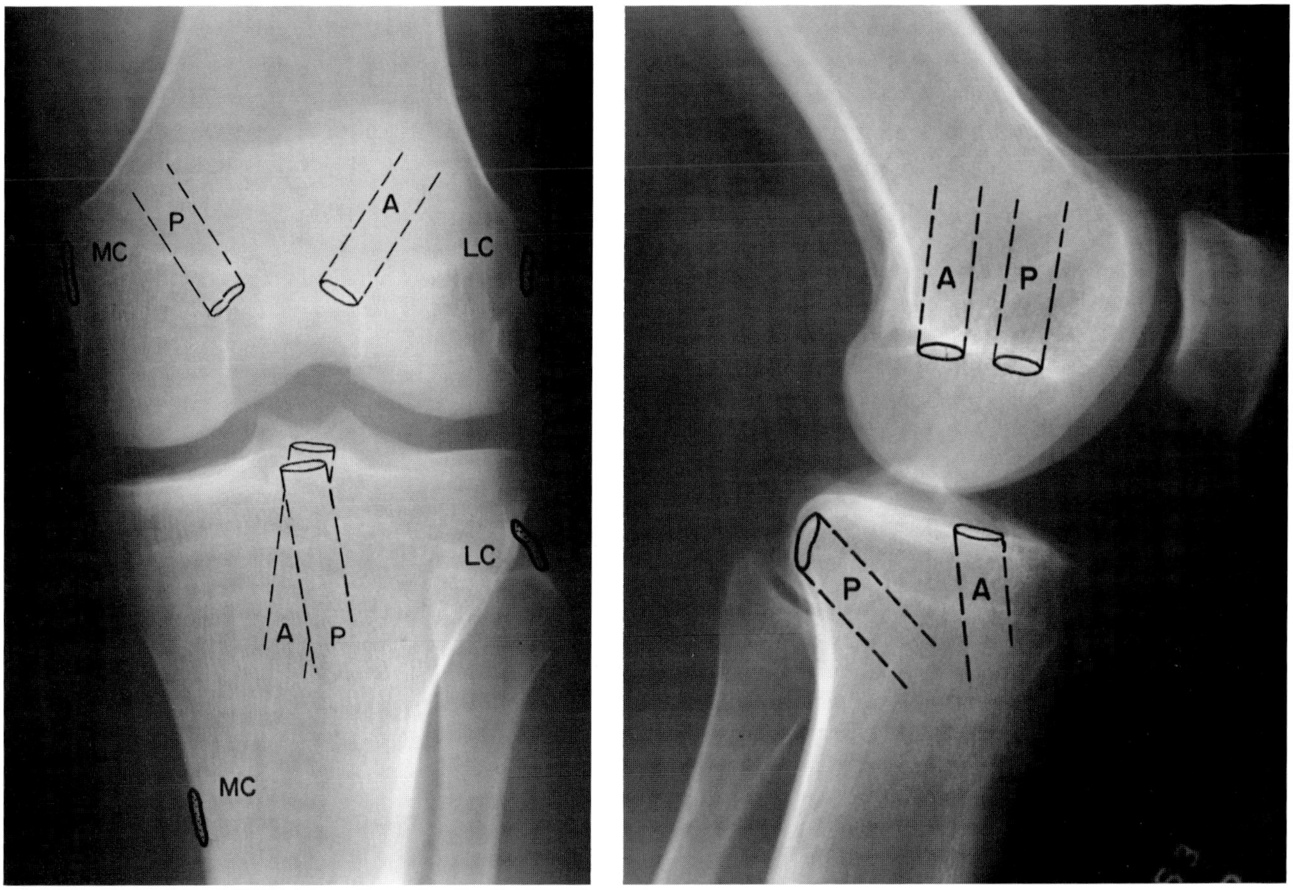

FIG. 7.35. Normal tunnels. The position of the anterior (A) and posterior (P) cruciate ligament tunnels are depicted on these radiographs. On the anteroposterior radiograph, the attachment sites of both the medial collateral ligament (MC) and lateral collateral ligament (LC) are identified. (Case courtesy of B.J. Manaster, M.D. From ref. 133, with permission.)

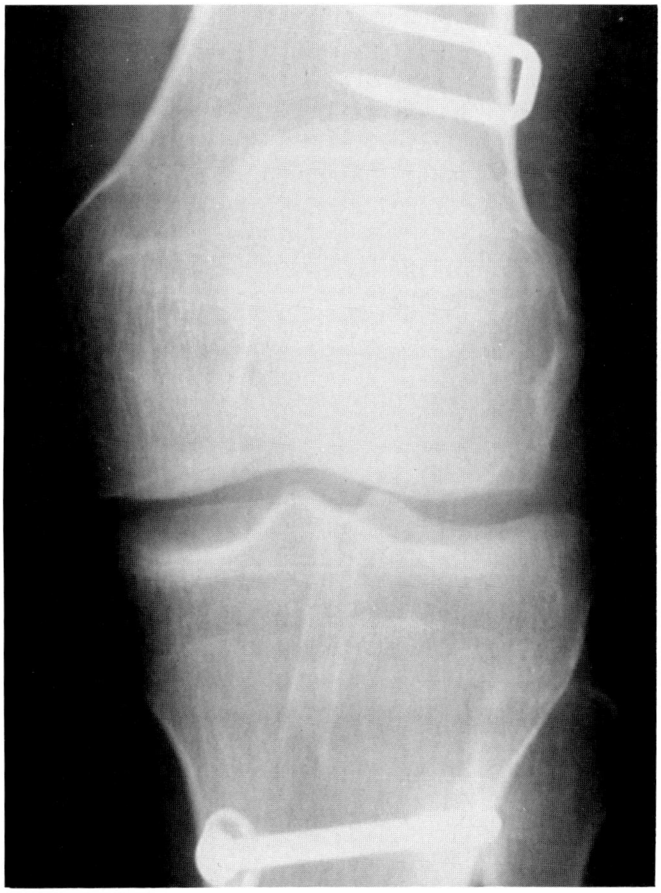
A

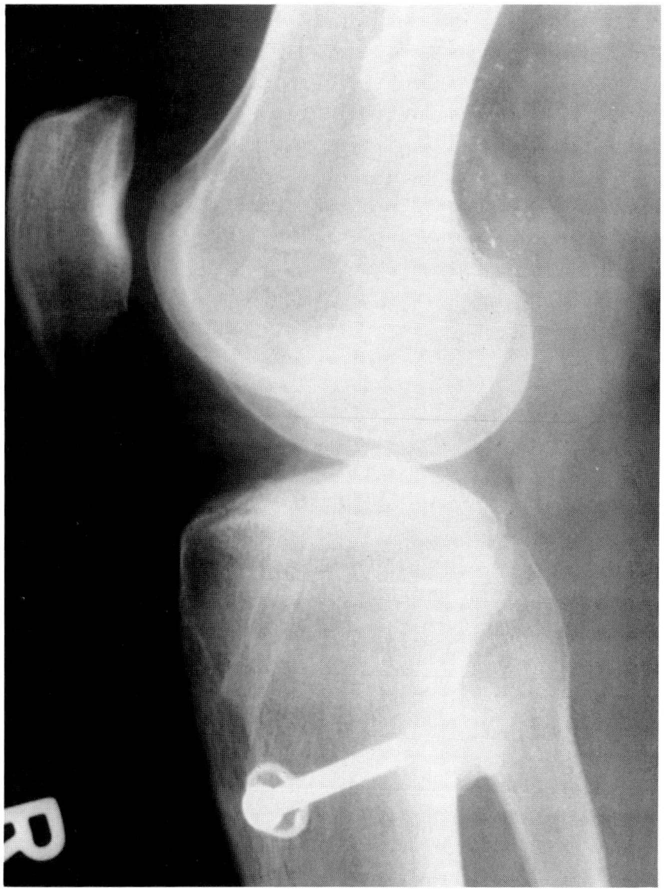
B

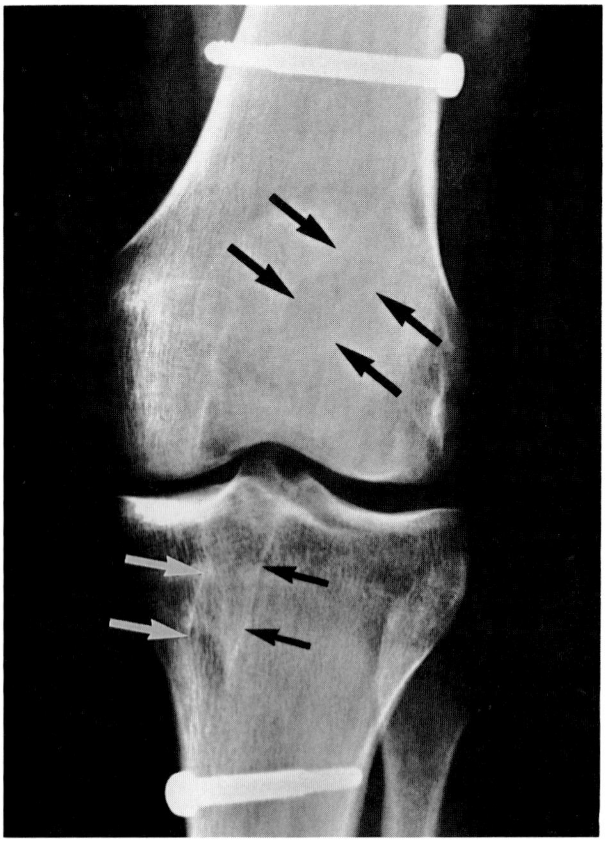
C

FIG. 7.36. Graft tunnels. **A,B:** The tibial tunnel can be identified by virtue of its sclerotic margins. The normal tibial tunnel opens in front of the intercondylar eminence. **C:** Both femoral and tibial tunnels (*arrows*) have sclerotic margins.

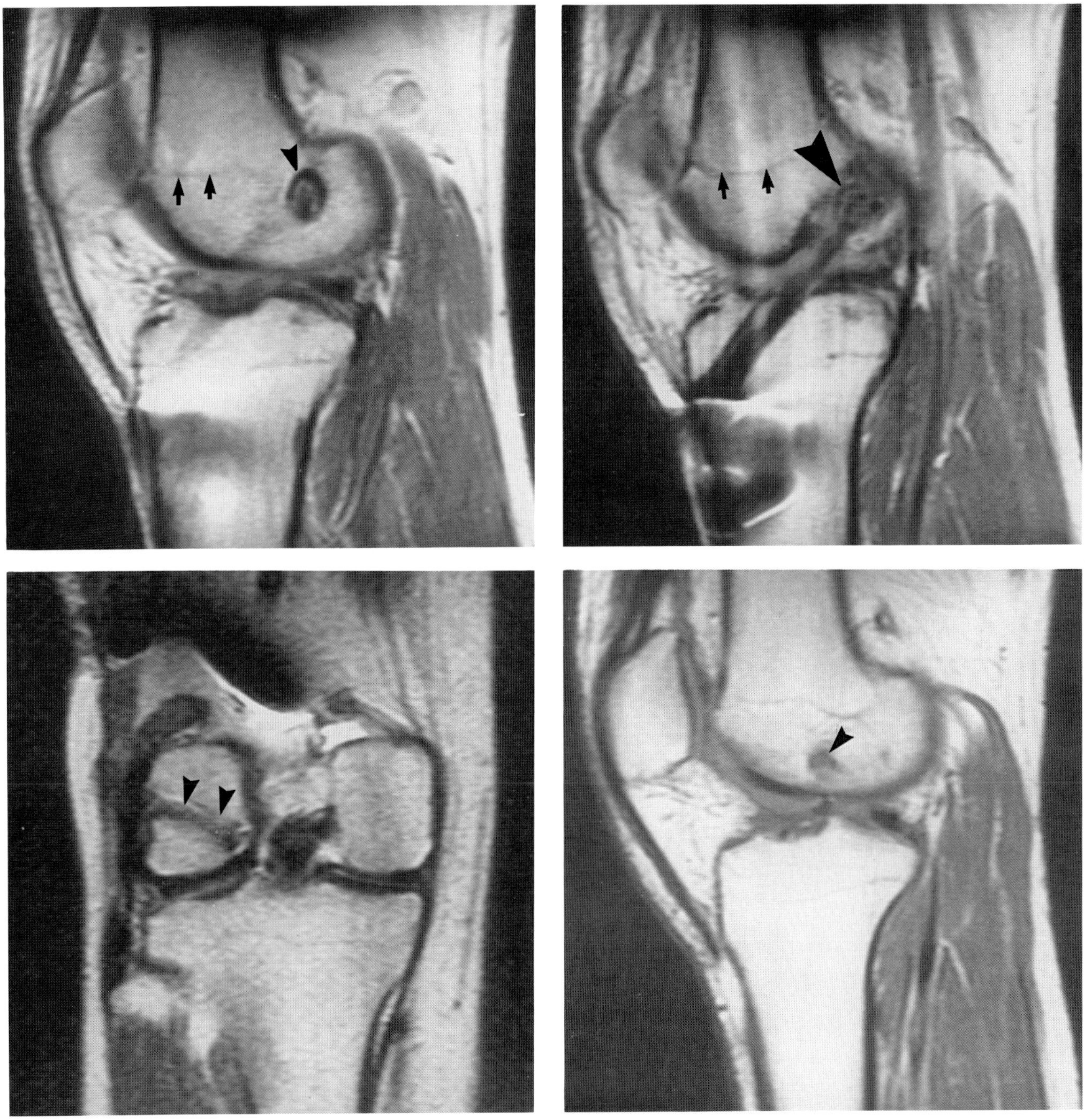

A B C D

FIG. 7.37. Normal and abnormal femoral tunnel positions. **A,B:** Adjacent 3 mm sagittal MR, TR/TE 800/20. The isometric femoral tunnel (*arrowheads*) opens at the intersection of the line described by the posterior femoral cortex and the physeal scar (*arrows*). **C:** Coronal MR, TR/TE 2,000/80. **D:** Sagittal MR, TR/TE 800/20. In another patient, the femoral tunnel was placed quite low in position (*arrowheads*). The normal tunnel opens at the 11 o'clock position when viewed coronally, although in this knee the tunnel opens at 9 o'clock. Additionally, the tunnel opens too far anteriorly.

demonstrate roof impingement have tibial tunnels that are only 30% of the distance from the anterior to the posterior edges of the tibial plateau (145). On the coronal MR images, the tibial tunnel opening into the joint should be centered on the intercondylar eminence. The position of the tibial tunnel in large part determines whether the graft will suffer from impingement. Since roof impingement is difficult to diagnose either clinically or at the time of surgery, the criteria for diagnosis are radiographic. Roof impingement is said to be present when a significant portion of the tibial tunnel lies anterior to the inferiorly projected slope of the inter-

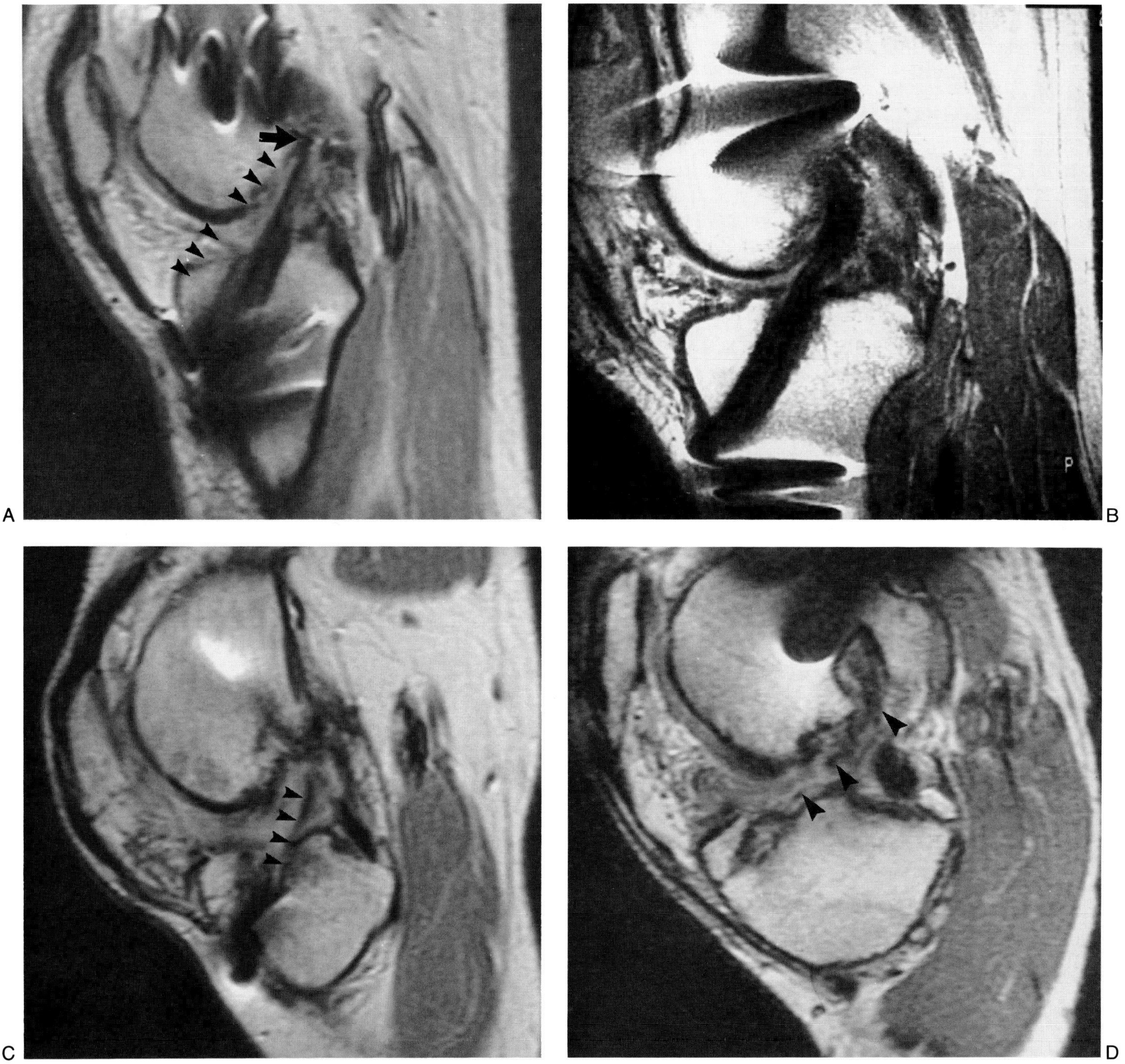

FIG. 7.38. Stable ACL autografts. **A:** Sagittal MR, TR/TE 1,200/40. This patient had a hamstring graft placed 4 months prior. The image demonstrates two principles of ACL reconstructive surgery: (a) the high, posterior origin of the graft in the joint (*arrow*) indicates that the graft has been placed isometrically and (b) the entire tibial tunnel lies posterior to the projected slope of the intercondylar roof (*arrowheads*), ensuring impingement free motion. The metallic artifact from the securing hardware does not degrade the image of the intraarticular segment of the graft. **B:** Sagittal MR, TR/TE 1,200/40. This patient had a double looped hamstring graft positioned 8 weeks earlier. The thick, dark graft is seen to have a uniform low signal intensity. (Case courtesy of Stephen Howell, M.D.) **C:** Sagittal MR, TR/TE, 2,000/20. This graft was placed 6 months prior to this MR. In spite of the fact that the knee is stable, there is no evidence of impingement, and the patient is asymptomatic, the graft itself (*arrowheads*) is very thin throughout its course. Is this the MR representation of the process of ligamentization? **D:** Sagittal MR, TR/TE 2,000/20. This college football player suffered a tear of the ACL requiring reconstruction using the patellar tendon 6 years earlier. The patient presently has no complaints and a normal clinical examination. Isolated segments of the graft (*arrowheads*) are seen, with apparent discontinuities in its structure. In all four of these patients, there is no evidence of impingement by the intercondylar roof, the graft positions are isometric, and the knees are stable.

condylar roof (Figs. 7.38 and 7.39). It is critical to recognize that *the position of the bone plug, the securing staples, or the interference fit screws is not the site of exit/entrance of the graft into the bony tunnels* (Fig. 7.30).

Although many different autologous and allogenous materials have been utilized for ACL reconstruction, the BPTB, the hamstring tendon grafts, and Achilles tendon allografts are currently the most popular. Preliminary evidence suggests that the natural history of these grafts is identical, and in our experience, the MR appearance differs only in the surgical changes occurring at the donor site.

Multiple studies have focused on the MR appearance of the normal BPTB and hamstring graft (134,136, 142,161–167). The immediate postinsertion signal of a graft is identical to the signal of the posterior cruciate ligament; the harvesting, preparing, sizing, and insertion of the graft do not result in an appreciable signal increase (136,142,161). The portions of the graft that lie within the tunnels also demonstrate low signal. Ninety percent of BPTB grafts in clinically stable knees during the first year post implantation appear as solid, low signal bands traversing the joint (Fig. 7.38) (161). If clinical stability is assumed to represent the "gold standard," MR achieves a sensitivity of 90%, a specificity of 100%, and an accuracy of 90% in determining the status of the neoligament (134,161). When

A

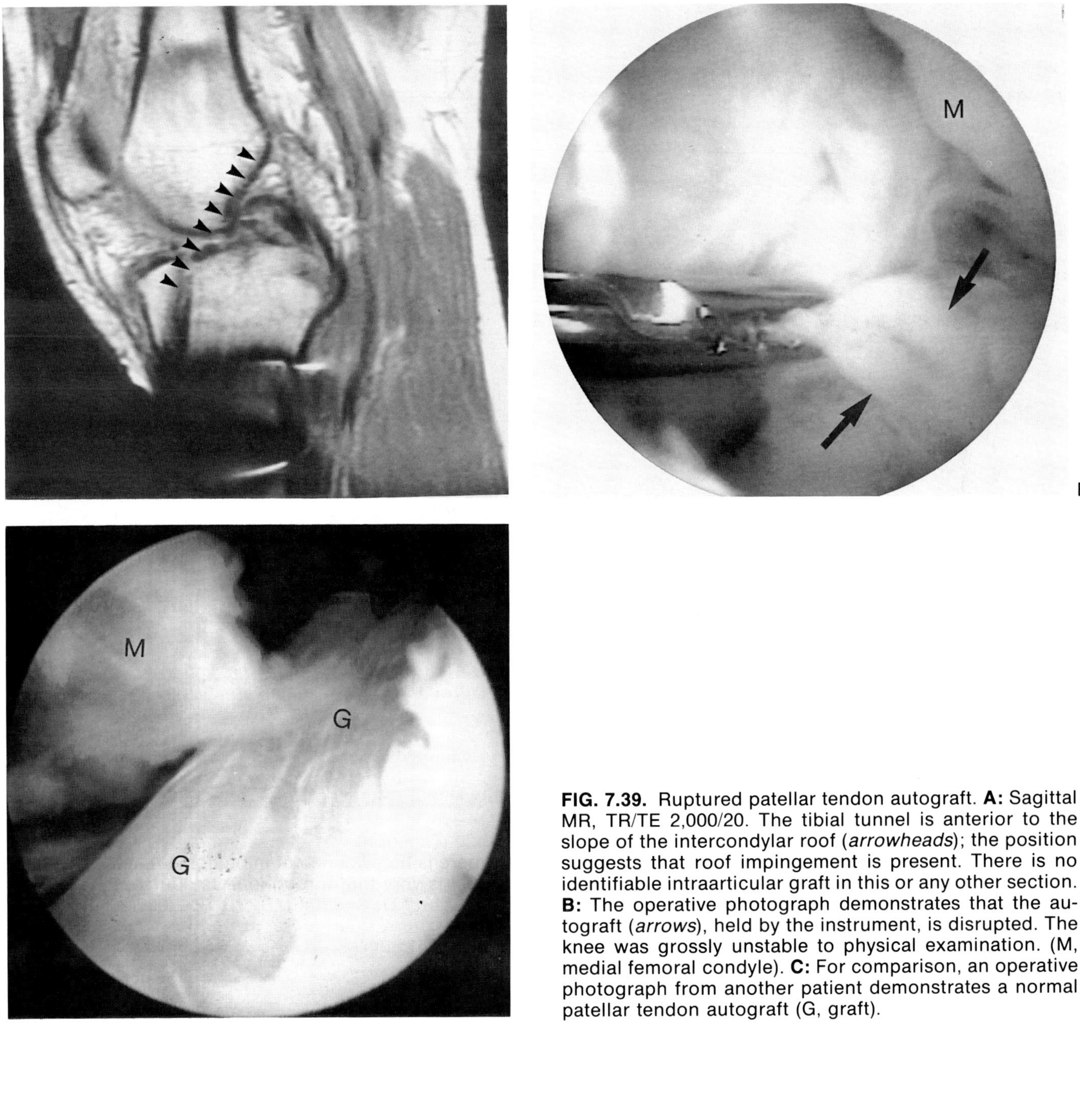

B

C

FIG. 7.39. Ruptured patellar tendon autograft. **A:** Sagittal MR, TR/TE 2,000/20. The tibial tunnel is anterior to the slope of the intercondylar roof (*arrowheads*); the position suggests that roof impingement is present. There is no identifiable intraarticular graft in this or any other section. **B:** The operative photograph demonstrates that the autograft (*arrows*), held by the instrument, is disrupted. The knee was grossly unstable to physical examination. (M, medial femoral condyle). **C:** For comparison, an operative photograph from another patient demonstrates a normal patellar tendon autograft (G, graft).

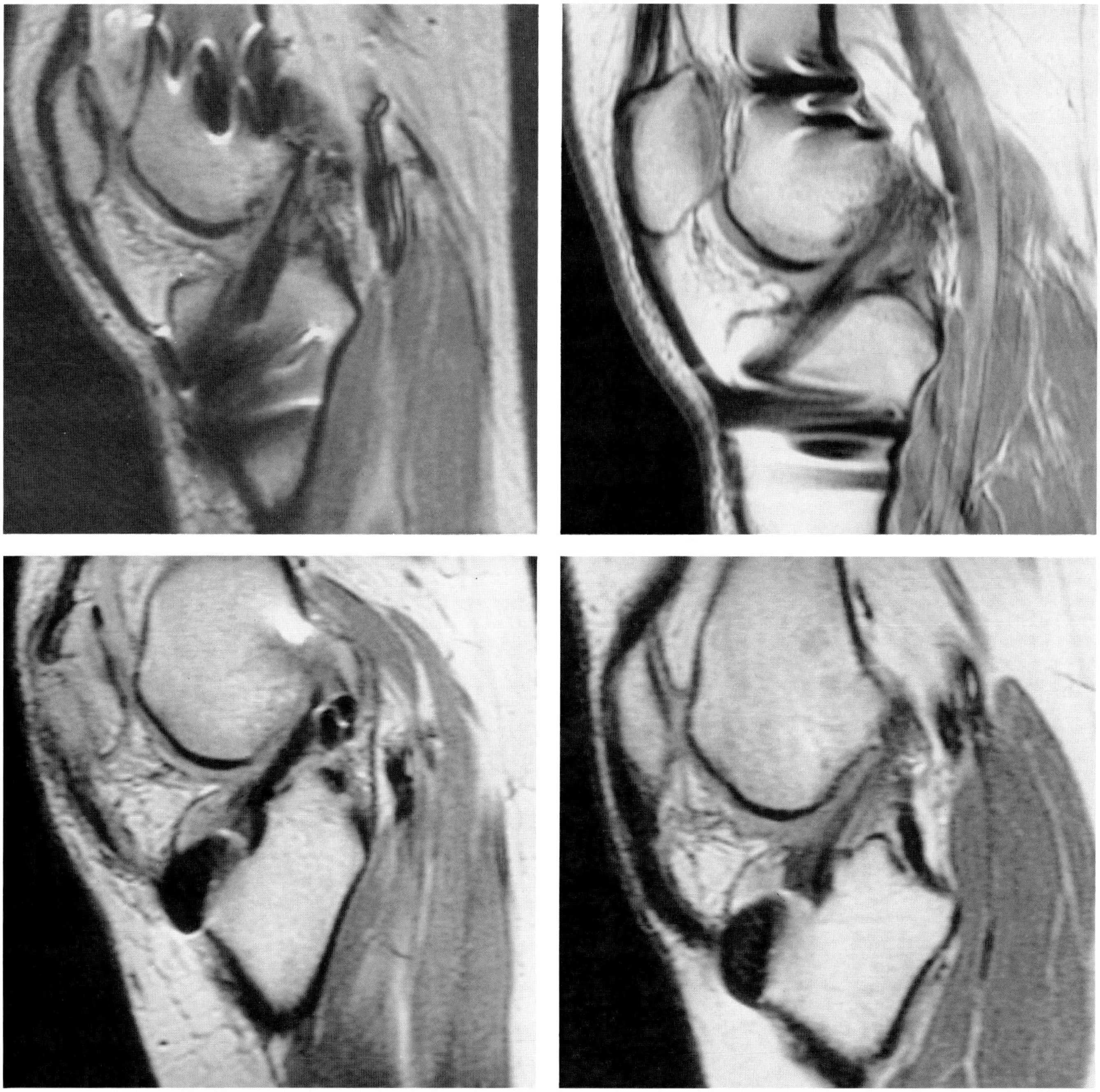

A B C D

FIG. 7.40. Normal temporal changes in autografts. **A,B:** Sagittal MR, TR/TE 1,200/40. (A) was made 5 months prior to (B). In (A), the hamstring graft is a solid, dark band with only minimal intratendinous signal inhomogeneities. In (B) the graft appears wider, with a somewhat linear signal increase within the graft. The patient is totally asymptomatic and has a stable knee to examination. **C,D:** Sagittal MR, TR/TE 2,000/20. (C) is 3 months postoperative; (D) is 6 months postoperative. In another patient, the entire graft has acquired a mild but definite increase in signal over time. The graft appears somewhat thinner than it did in (C).

comparing MR to arthroscopy in the limited number of cases for which these data are available, MR accurately identified 13 intact and 5 torn reconstructions.

Ten percent of normal grafts *as judged by clinical stability alone* (without reference to imaging assessment) demonstrate an intermediate signal intensity (Fig. 7.38) (161,162). Occasionally, the increase in signal intensity is so significant that the graft appears discontinuous (134,136,145,162). This finding has led some authors to conclude that MR is an unreliable method of evaluating the integrity of an ACL graft (134,162). In our experience and that of others, an increase in signal, even to the extent that the graft is very poorly delimited, is *not necessarily* indicative of a rerupture (145). Rerupture without a secondary traumatic event is unusual. In six of our own cases in which the graft was not visible, and in which second look arthroscopy was undertaken, only two grafts were reruptured (Fig. 7.39). Both patients had a tibial tunnel that was definitely anteriorly positioned, both had an unstable knee and no identifiable structure in the lateral intercondylar notch, and both suffered a reinjury. Because significant signal increase may normally occur in an ACL graft, the MR diagnosis of rerupture may be quite difficult.

There are two potential explanations for an increase in graft signal in clinically stable knees in which second look arthroscopy reveals an intact graft (134,145): (a) normal temporal changes due to physiologic ligamentization; (b) graft impingement. In patients in whom sequential MR studies are performed within 4 months of reconstruction, between 4 and 8 months, and again after 9 months postoperatively, MR demonstrates that there is a tendency for some grafts to acquire an increase in signal intensity of a moderate to a marked degree by the second MR examination, and a trend to diminish this newly acquired signal by the time of the last examination (Figs. 7.38 and 7.40) (163,166). It is

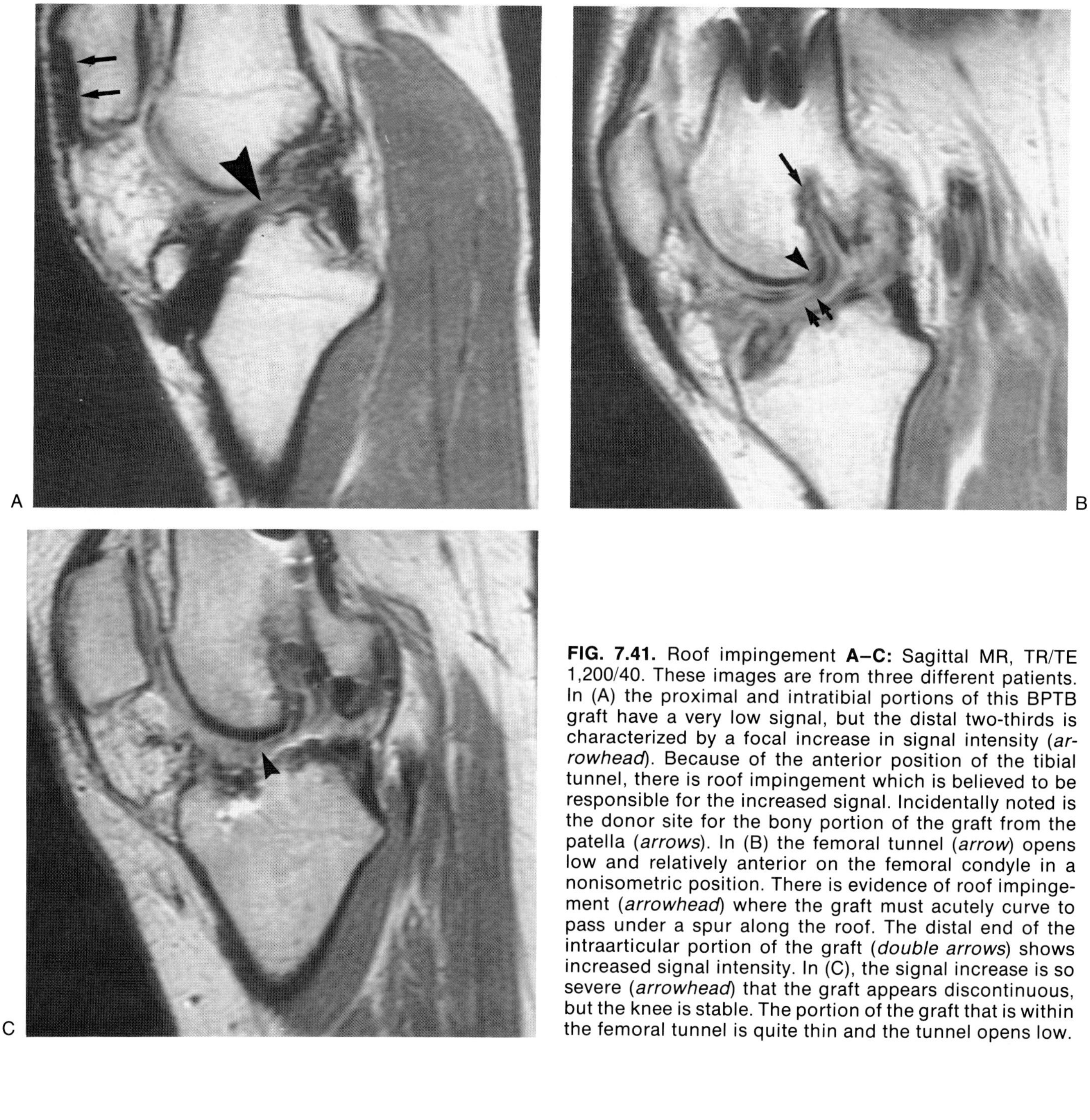

FIG. 7.41. Roof impingement **A–C:** Sagittal MR, TR/TE 1,200/40. These images are from three different patients. In (A) the proximal and intratibial portions of this BPTB graft have a very low signal, but the distal two-thirds is characterized by a focal increase in signal intensity (*arrowhead*). Because of the anterior position of the tibial tunnel, there is roof impingement which is believed to be responsible for the increased signal. Incidentally noted is the donor site for the bony portion of the graft from the patella (*arrows*). In (B) the femoral tunnel (*arrow*) opens low and relatively anterior on the femoral condyle in a nonisometric position. There is evidence of roof impingement (*arrowhead*) where the graft must acutely curve to pass under a spur along the roof. The distal end of the intraarticular portion of the graft (*double arrows*) shows increased signal intensity. In (C), the signal increase is so severe (*arrowhead*) that the graft appears discontinuous, but the knee is stable. The portion of the graft that is within the femoral tunnel is quite thin and the tunnel opens low.

possible that this appearance can be explained by the process of ligamentization. The synovial membrane, which engulfs the ACL reconstruction, has an intermediate signal character on T1 and proton density sequences, and can potentially hamper graft visualization. The central ischemic necrosis and peripheral revascularization may sufficiently alter the relaxation characteristics of the graft to cause it to appear isointense with the intraarticular soft tissues. Because of the partial volume effect of the intermediate signal synovial sheath with the dark ligament, the graft might appear smaller than it actually is. If these processes of synovial reaction, graft edema, and revascularization are extensive, it is possible for a totally normal graft to be difficult to depict, thereby accounting for the reported inability of some examiners to identify the neoligament in a clinically stable knee (134). We have examined multiple stable grafts over the first year post implantation. At least one-half of the grafts demonstrated a mild diffuse increase in signal intensity (T1WI) that developed over the first 3 months. The signal increase is diffuse, imparting an overall gray or occasionally a striped appearance to the graft.

Clinically unrecognized roof impingement has been identified as a cause for a regionalized increase in graft signal (136,142). Howell has described the MR changes in the graft which result from occult roof impingement. In these cases, the tibial tunnel is placed too far anterior, and the anteroinferior edge of the roof impacts the superior surface of the graft during full extension (136,145). Two groups of patients were followed by means of sequential MR studies; one group had and the other group did not have imaging evidence of roof impingement. Those without roof impingement had grafts which demonstrated low signal intensity throughout the period of assessment (up to 1 year). On the other hand, those with radiographic evidence of impingement developed regionalized signal increases in the distal two-thirds of the intraarticular segment of the graft (Fig. 7.41). This localized increase in signal is thought to be secondary to entrapment of the graft in extension between the tibial plateau below and the inferior lip of the intercondylar roof above (137,145).

Sequential studies on patients with the presumptive diagnosis of roof impingement [made by the finding of increased signal (T1WI) in the distal graft] suggest that these changes become well established during the first 3 months following surgery. It is during this time period that the patient attempts full extension, the position that would most likely result in graft impingement (136). The regionalized signal increase tends to remain constant over at least 3 years (145). While frequently the signal increases are mild, they are occasionally so severe as to suggest that the graft is severely attenuated or in fact discontinuous. Resection of the offending anteroinferior lip of the roof or replacement of the tibial tunnel in proper position has been shown to relieve the impingement, and the graft in such cases returns to a normal, low signal state within 12 weeks after delayed roofplasty (136,142). Therefore, increased signal within ACL grafts may *not* be a normal part of the ligamentization process, but rather a result of clinically occult roof impingement. Since the clinical diagnosis of impingement may be difficult, MR may be the first indicator that a problem exists. The potential long-term consequences (e.g., rerupture, progressive instability) of untreated roof impingement are not known at this time. Preliminary data in animals suggest that an increase in signal intensity of the graft corresponds to an increase in the water concentration and a reduction in strength (127).

Sidewall impingement is more readily visible to the arthroscopist than is roof impingement, and it is more readily diagnosed at the time of arthroscopy. We have encountered a single case in which a BPTB graft was seen to be draped over the medial aspect of the lateral femoral condyle (Fig. 7.42).

One of the reported advantages of using the patellar tendon for reconstruction is the ability to achieve bone to bone fixation by permitting the plugs to incorporate into the femoral and tibial shafts. Radiographic evidence of incorporation may be quite variable, however. In some patients, the plug cannot be identified on conventional radiographs and assessment of incorporation may not be possible. In other patients, con-

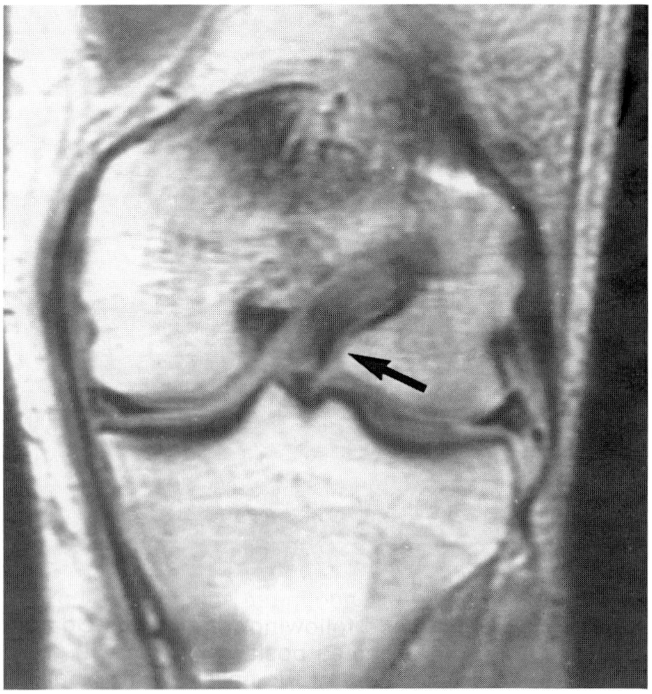

FIG. 7.42. Sidewall impingement. Coronal MR, TR/TE 2,000/20. The opening of the femoral tunnel is somewhat low in the notch. The graft curves over the lateral femoral condyle (*arrow*).

ventional radiographic evidence for graft-to-bone union may be seen as early as 3 weeks after operation. Of the patients followed for 6 weeks, 79% showed conventional radiographic signs of plug incorporation; in those followed for 5 months or more, 95% of the plugs were judged to be incorporated (133). MR, by virtue of its tomographic capabilities, has a similar ability to make this judgment. The intraosseous portion of the graft in the tibia demonstrates MR evidence of bone ingrowth by 6 months in 20% of patients; by 1 year, 35% of the knees have bony incorporation (136). It should be noted, however, that in those grafts which

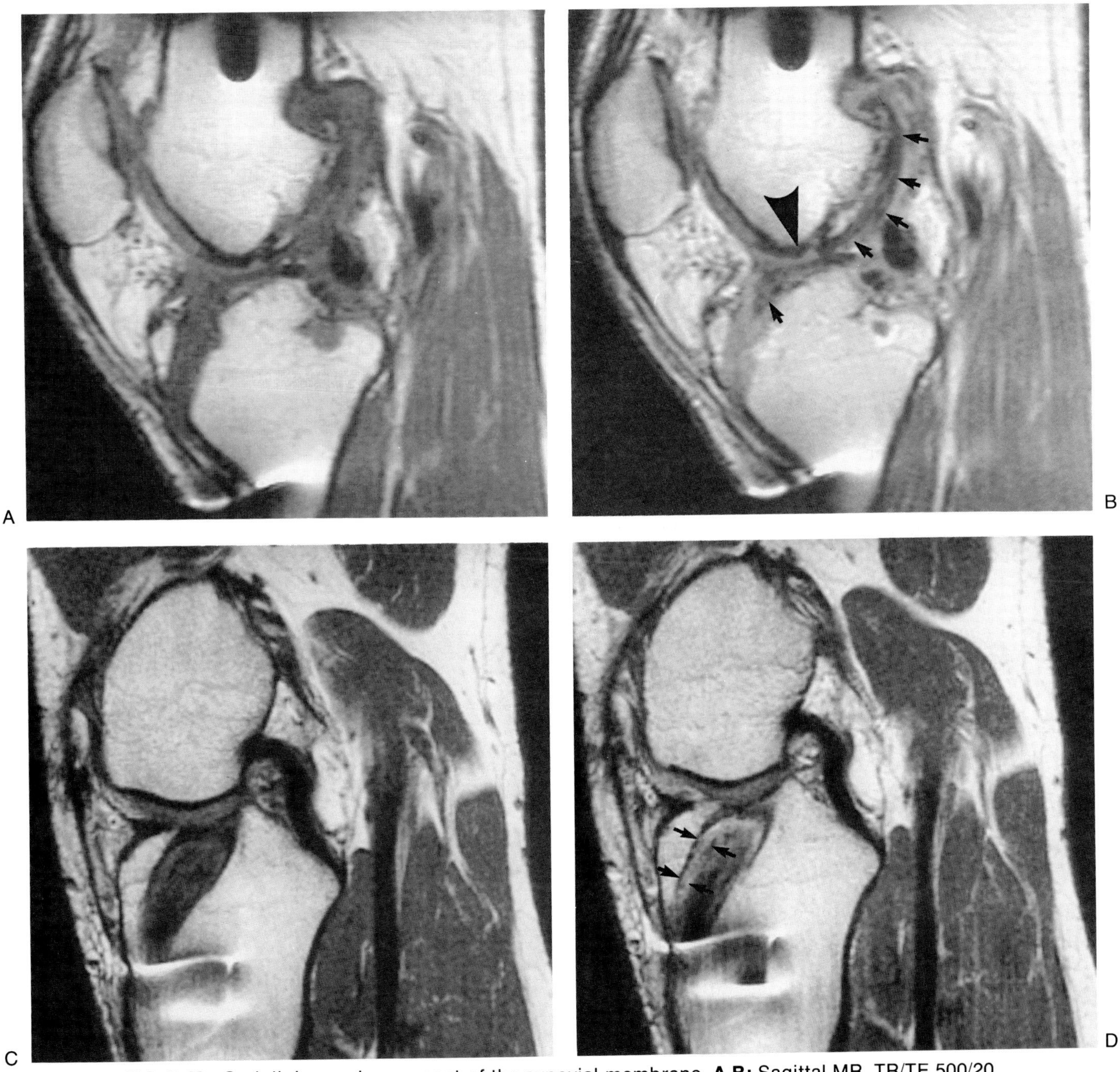

FIG. 7.43. Gadolinium enhancement of the synovial membrane. **A,B:** Sagittal MR, TR/TE 500/20. (B) is made following intravenous Gd administration. In (A), a tubular soft tissue mass is seen in the expected position of the ACL graft. The dark graft itself is not identified. Following Gd enhancement of the envelope, the graft (*arrows*) is readily seen to be intact, although roof impingement is present (*arrowhead*) and the graft appears focally discontinuous. **C,D:** Gadolinium enhancement in the tibial tunnel; sagittal MR, TR/TE 500/20. (C) is before and (D) is following intravenous Gd administration. In (C), the tunnel has a uniformly intermediate signal; the edge of the tunnel increases in signal (*arrows*) after Gd. The uptake is in the vascular membrane that lines the tunnel and covers the surface of the graft.

have been secured by means of an interference fit screw, the local metal artifact may totally preclude assessment of bony ingrowth.

Uses of Gadolinium

Gadolinium-enhanced MR gives the radiologist the opportunity to noninvasively study the physiologic events occurring around the graft (167,169,147,169). As noted above, the graft becomes cloaked by a vascular envelope 10 to 20 weeks after implantation. It is this period of revascularization by means of vessels from synovial folds, endosteal vessels, and remnants of the native torn ACL that ultimately reestablishes the strength of the graft. Gadolinium-DTPA (Gd) is a paramagnetic intravenous contrast agent which has been shown to enhance neovascularized tissues on MR images. Gd enhancement occurs most commonly within the synovial membrane around the graft (Fig. 7.43) (136,166,168,169). Only minor degrees of signal enhancement characteristically occur in the synovial envelope during the first 4 months following implantation; by 7 months, however, approximately one-half of the grafts demonstrate significant synovial enhancement (169). Rarely does the synovial membrane demonstrate moderate signal enhancement after 7 months. In those patients who underwent Gd-enhanced studies both early and late in the first postoperative year, the degree of membrane enhancement decreased in 45%, remained unchanged (minimal) in 45%, and increased in 10%. This temporal pattern of perigraft enhancement correlates extremely well with the known physiologic events occurring in the knee, e.g., synovial proliferation, vascular ingrowth, reestablishment of vascularization, and eventual subsidence of the inflammatory reaction of the host. Graphic demonstration of these physiologic events by MR might possibly be valuable in designing and timing of rehabilitation programs.

The pattern of Gd enhancement of the femoral and tibial tunnels has also been examined. Since there is an endosteal blood supply to the graft, it is not surprising that the bony tunnels demonstrate Gd enhancement of varying degrees, although the level of intensity tends to decrease with time (Fig. 7.43) (85,148,169). The edges of the tunnel and the superficial layers of the graft may demonstrate signal increase following Gd injection as a result of the increased surface vascularity.

While Gd enhancement of the synovial envelope is physiologic, focal intraligamentous enhancement is abnormal and must suggest the diagnosis of unrecognized roof impingement (Fig. 7.44). The abnormal signal decreases somewhat over 2 to 3 years. The specificity or sensitivity of intraligamentous signal increase is not known, but we have not identified such enhancement in otherwise normal grafts.

A 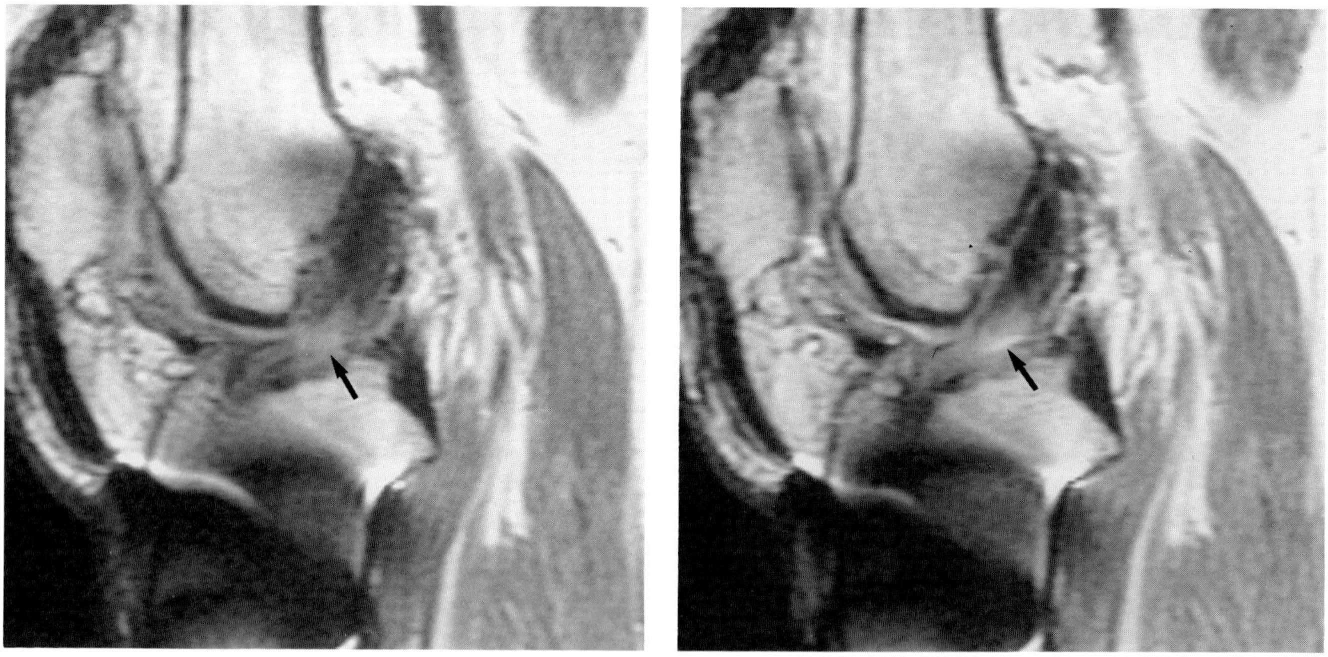 B

FIG. 7.44. Graft impingement assessed with Gd. **A,B:** Sagittal MR, TR/TE 500/20. (B) is made following intravenous Gd administration. In both images, the distal two-thirds of the graft demonstrates a focal increase in signal (*arrow*), but the signal alteration is more evident following Gd. In (B) the entire graft is more easily identified since the synovial membrane has enhanced. The tibial tunnel and distal graft lie anterior to the projected slope of the intercondylar roof.

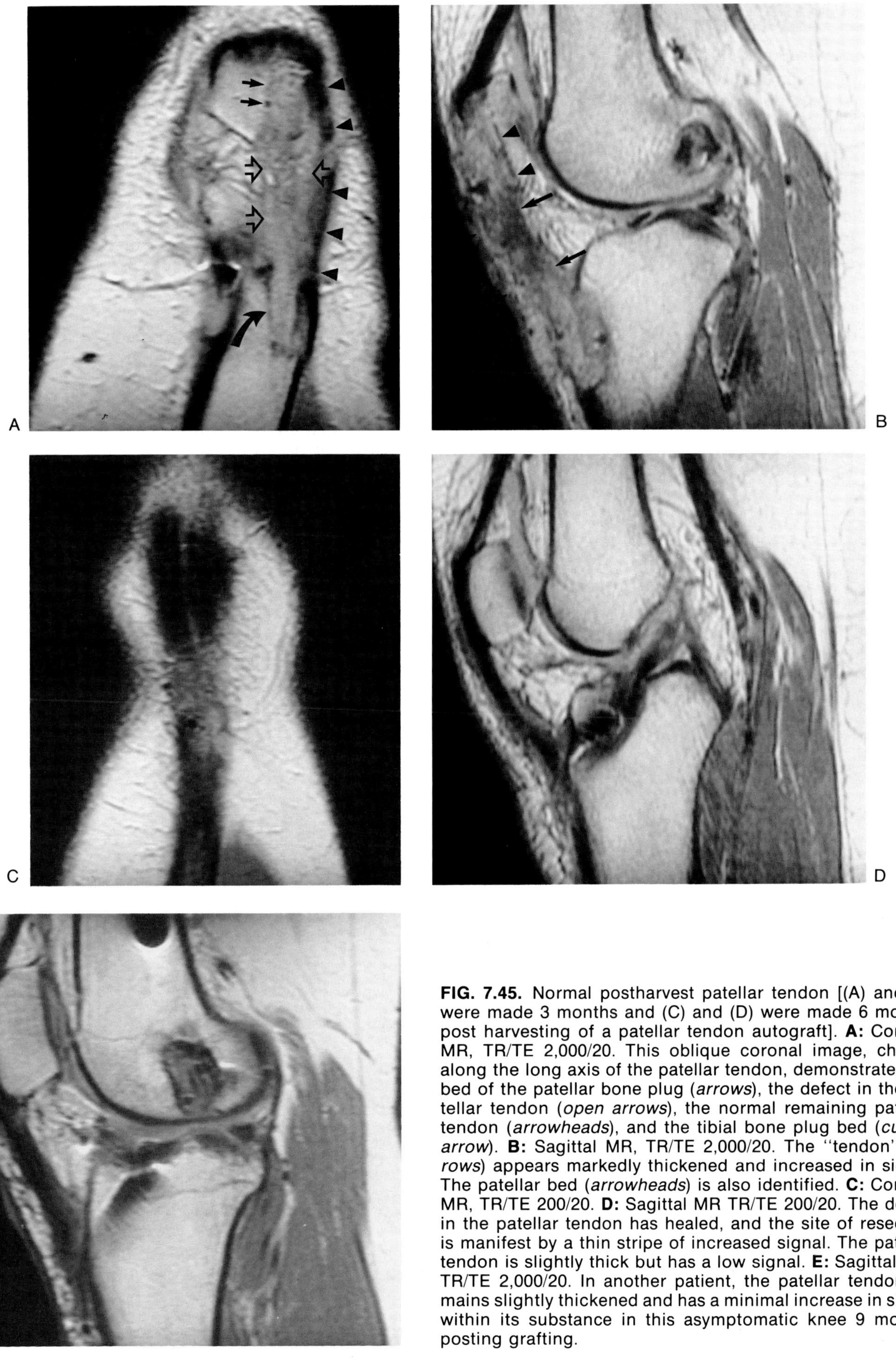

FIG. 7.45. Normal postharvest patellar tendon [(A) and (B) were made 3 months and (C) and (D) were made 6 months post harvesting of a patellar tendon autograft]. **A:** Coronal MR, TR/TE 2,000/20. This oblique coronal image, chosen along the long axis of the patellar tendon, demonstrates the bed of the patellar bone plug (*arrows*), the defect in the patellar tendon (*open arrows*), the normal remaining patellar tendon (*arrowheads*), and the tibial bone plug bed (*curved arrow*). **B:** Sagittal MR, TR/TE 2,000/20. The "tendon" (*arrows*) appears markedly thickened and increased in signal. The patellar bed (*arrowheads*) is also identified. **C:** Coronal MR, TR/TE 200/20. **D:** Sagittal MR TR/TE 200/20. The defect in the patellar tendon has healed, and the site of resection is manifest by a thin stripe of increased signal. The patellar tendon is slightly thick but has a low signal. **E:** Sagittal MR, TR/TE 2,000/20. In another patient, the patellar tendon remains slightly thickened and has a minimal increase in signal within its substance in this asymptomatic knee 9 months posting grafting.

Alterations in the Patellar Tendon

The portion of the patellar tendon that remains after harvesting demonstrates both signal and size alterations that change over time (Fig. 7.45) (161,167). As expected, the immediate postharvesting MR appearance of the remaining patellar tendon is quite abnormal; 87% of patellar tendons reveal a diffuse increase in signal on T1 and T2 WI that commonly persists for 6 months. This increase in signal is thought to be a function of edema from the harvesting and repair processes occurring at the donor site. By 12 months, asymptomatic patellar tendons return to their normal low signal appearance, or demonstrate only a linear increase in signal at the site of the graft harvest.

Thickening of the remaining patellar tendon is seen immediately after harvesting and persists even in asymptomatic patients. At 12 months symptomatic tendons always measure greater than 10 mm in anteroposterior dimension (normal 7–8 mm). Persistent signal at the patellar tendon donor site and thickening of the patellar tendon after 3 months correlates with residual clinical symptoms (161,167). In the first 6 months postoperatively, patella baja (measured by the degree of patellofemoral overlap) is common, but by 12 months the patella returns to its normal position (170). Patella baja has been attributed to quadriceps weakness which generally returns to baseline values by 1 year postoperatively, but adhesions between the fat pad and the tendon, and arthrofibrosis may contribute to patellar tendon shortening.

Many of the complications of patellar tendon autografting including patellar fractures, patellar tracking disorders, patellar tendonitis, and rupture can be assessed by MR (Fig. 7.46) (161,171). The two patellar fractures that we have encountered were both transverse, nondisplaced lesions that were manifest by a line of decreased signal on T1WI associated with a

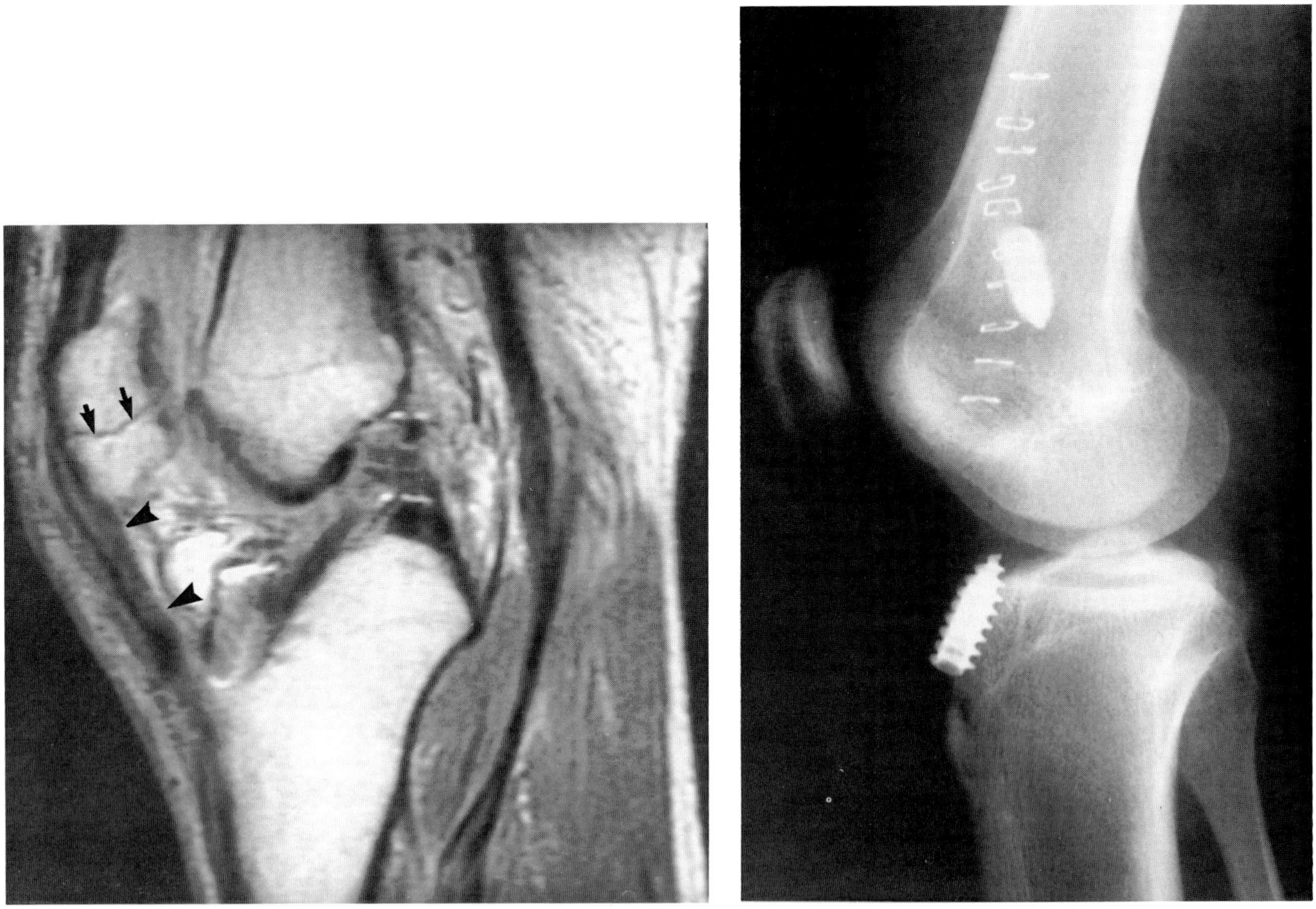

A B

FIG. 7.46. Operative complications. **A:** Postoperative patellar fracture; sagittal MR, TR/TE 2,000/20. This patient had a BPTB graft harvested 2 months earlier. There is a transverse, nondisplaced fracture in the patella (*arrows*). The remaining patellar tendon (*arrowheads*) is thickened and demonstrates an increase in signal intensity. The upper portion of the intraarticular segment of the graft demonstrates a diffuse increase in signal and is never optimally seen. **B:** The lateral radiograph in another patient reveals that the interference screw has pulled out of the tibia.

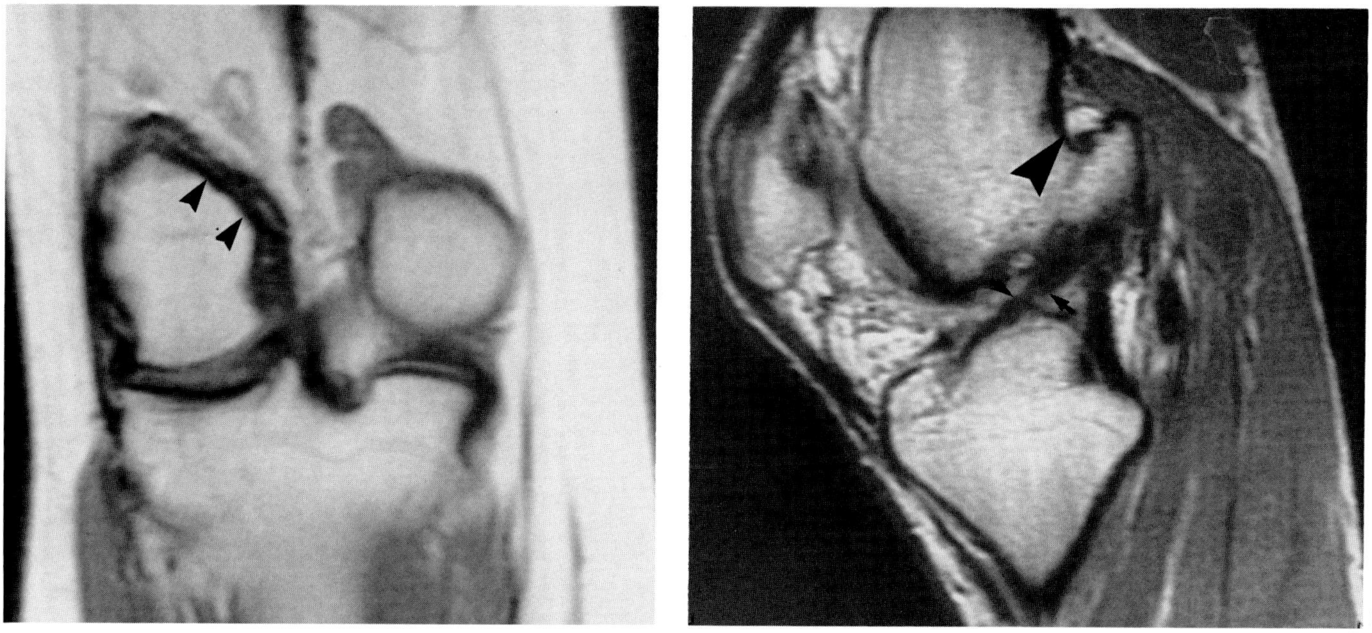

A B

FIG. 7.47. MacIntosh over-the-top intraarticular reconstruction. **A:** Coronal MR, TR/TE 800/20. **B:** Sagittal MR, TR/TE 800/20. During this procedure, the ITB is detached proximally and brought over the top of the lateral femoral condyle into the joint and eventually into the tibia. The bed of the graft along the posterior lateral femoral condyle is deepened by means of a surgically created groove (*arrowheads*) which can be identified on both coronal and sagittal images. A portion of the graft itself (*arrows*) can be seen on this nonoblique image.

surrounding zone of edema. The diagnosis of patellar tendon rupture following harvesting of its central one-third may be clinically difficult since there is no dramatic event at the time of occurrence, and the anticipated postoperative extensor lag further obscures the correct diagnosis. The concept of stress radiography of the patellar ligament has been advocated as a means to diagnose this condition (172); MR offers the additional advantage of allowing direct visualization of the status of the tendon and the condition of the patella itself.

Other Autogenic Grafts

It is not yet clearly established whether other autologous, allogenic, or synthetic material might have a different natural history and different MR appearance from that described for autogenic patellar tendon or hamstring grafts. Moeser et al. performed MR on 17 individuals who had undergone a MacIntosh lateral substitution over the top repair utilizing a rolled fascia lata graft; all had stable knees (Fig. 7.47) (172). Nearly one-half had no identifiable graft, and more than one-

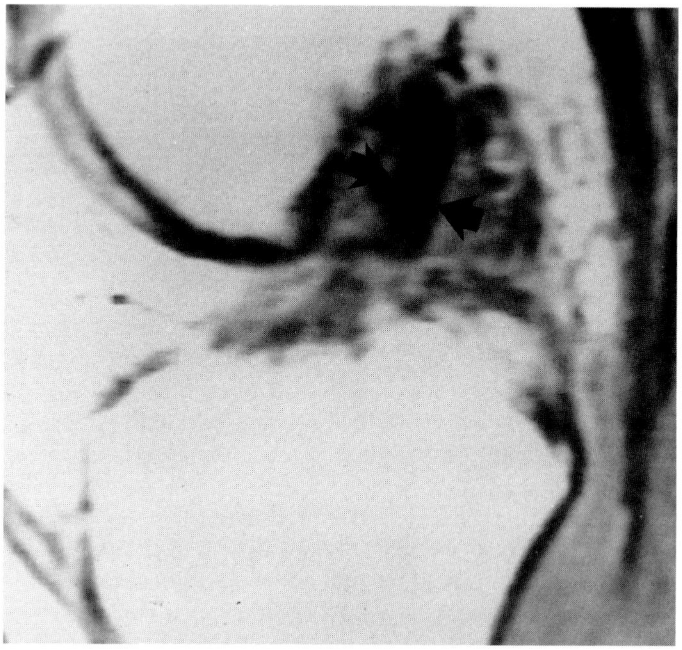

FIG. 7.48. Meniscal autograft. Sagittal MR, TR/TE 2,000/20. The medial meniscus was used to replace this patient's torn ACL. The proximal two-thirds of the graft (*arrows*) is well seen, but no attachment to the tibia can be identified. The graft has a meniscoid appearance.

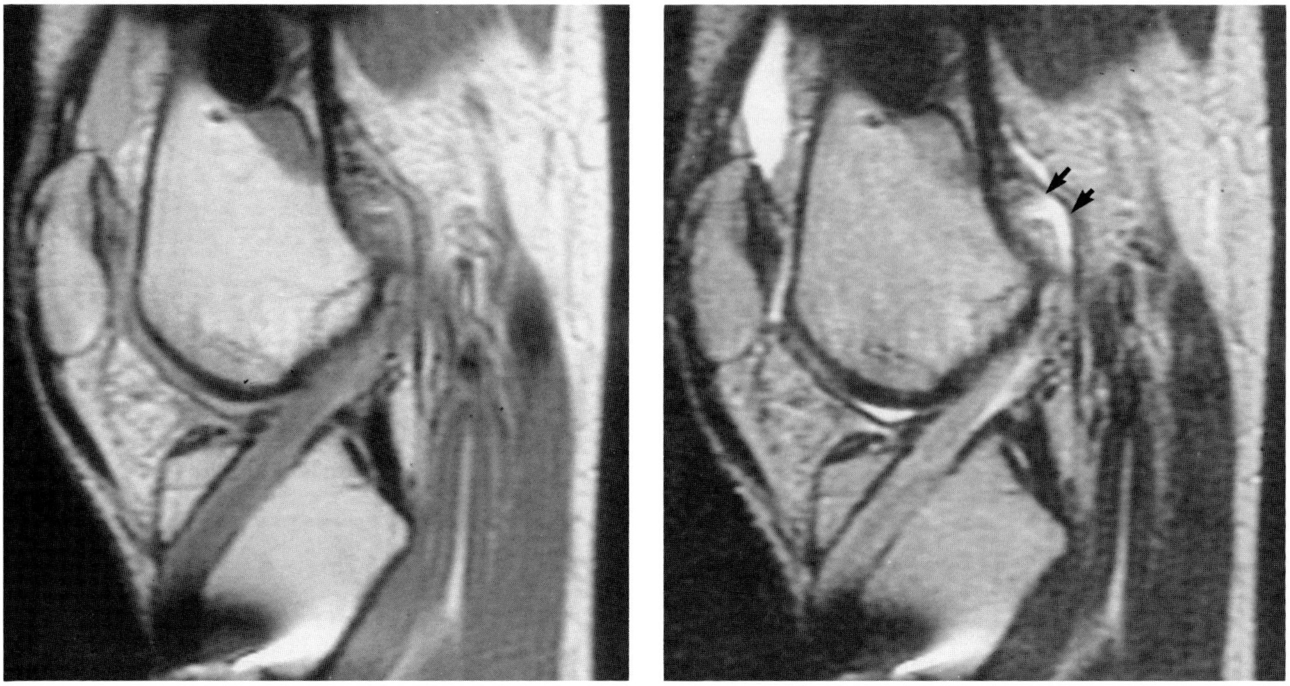

FIG. 7.49. Normal Gore-tex graft. **A:** Sagittal MR, TR/TE 2,000/20. **B:** Sagittal MR, TR/TE 2,000/80. Placement of a Gore-tex graft is achieved via an over-the-top position. On proton density sequences (A), the normal Gore-tex graft has an intermediate signal that is significantly higher than tendinous autografts. On the long TR sequence (B), there is a slight increase in signal intensity. The small amount of fluid (*arrows*) surrounding the upper portion of the graft is normal.

third had a poorly defined ligament. He postulated that soft tissues in the intercondylar notch obscured the definition of the graft. The fascia lata graft is considerably thinner than the BPTB graft, a fact which may help explain why the graft was not visible in apparently normal knees. Other less common reconstructions may be used to reestablish the ACL. The ITB is left attached to the Gerdy tubercle but its proximal end is severed and is brought "over the top" of the lateral femoral condyle, pulled through the joint along the course of the native ACL, and placed within a standard tibial tunnel. The bed of the graft over the posterior aspect of the lateral femoral condyle is surgically deepened in an attempt to achieve a more isometric reconstruction (Fig. 7.47).

The larger medial meniscus is utilized for replacing the torn ACL only in those knees in which the meniscus had to be removed for mechanical reasons; normal menisci are never harvested for this purpose. Interestingly, the meniscus may maintain its comma-shaped configuration on MR images and, as expected, has a low signal intensity at all imaging parameters (Fig. 7.48).

Synthetic Reconstructions

We have examined eight patients with Gore-tex grafts that have had clinically stable knees and no imaging evidence of impingement (Fig. 7.49) (173). The Gore-tex graft is placed in an over-the-top position. All grafts have demonstrated intermediate signal intensity on proton density images, yielding a definitely higher signal than is present in a BPTB graft; it is not known, however, if the signal of Gore-tex grafts is time-dependent. This signal increase is evident throughout the length of the graft. The entire graft becomes slightly but definitely brighter on long TR sequences. We have had the opportunity to review five cases of torn Gore-tex grafts (Fig. 7.50). In one case, the MR revealed a focal fluid collection high in the intercondylar notch adjacent to the site at which the neoligament entered the femoral tunnel. In two other cases, the synthetic graft was retracted toward the femoral attachment and was surrounded by a well-defined fluid collection. In all cases in which significant fluid collections were associated with a Gore-tex graft, surgery revealed graft disruption (174).

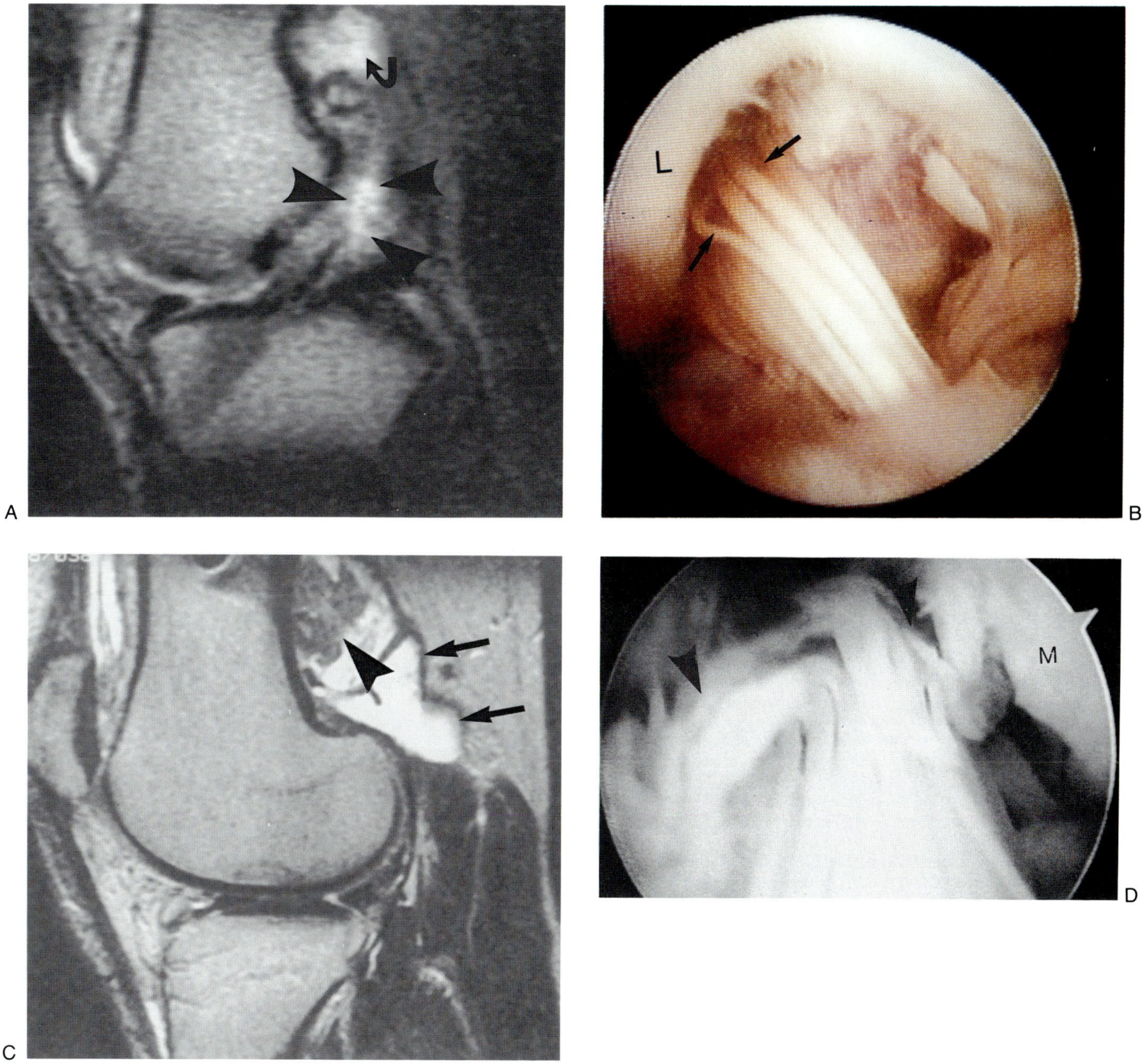

FIG. 7.50. Ruptured Gore-tex grafts. **A:** Sagittal MR, TR/TE 2,000/80. Two fluid collections are seen in association with this graft. One is high in the intercondylar notch (*arrowheads*), and the other is along the posterior femur where it is associated with the fixation hardware (*curved arrow*). **B:** This arthroscopic image demonstrates dehiscence of the proximal fibers of the Gore-tex graft (*arrows*). (L, lateral femoral condyle.) **C:** Sagittal MR, TR/TE 2,000/80. [(C) and (D) are from another patient]. This image is made just lateral to the notch. A fluid-filled mass (*arrows*) associated with the proximal hardware is identified, and within it is solid appearing tissue (*arrowhead*). **D:** At surgery, a completely ruptured Gore-tex graft was found retracted up to the hardware. The intraoperative photograph demonstrates the frayed distal segment of the graft (*arrowheads*). (M, medial femoral condyle.) [(A) and (B) from ref. 176, with permission.]

REFERENCES

1. Smith DK, Totty WG. The knee after partial meniscectomy: MR imaging features. *Radiology* 1990;176:141–144.
2. Deutsch AL, Mink JH, Fox JM, et al. Peripheral meniscal tears: MR findings after conservative treatment or arthroscopic repair. *Radiology* 1990;176:485–488.
3. Katz LD, DePuy J, Pope CF, et al. *MR imaging of meniscal injury and healing in a rabbit model.*
4. Deutsch AL, Mink JH. *MR imaging of the postoperative meniscus.*
5. Farley TE, Howell SM, Love KF, et al. Meniscal tears: MR and arthrographic findings after arthroscopic repair. *Radiology* 1991;180:517–522.
6. Renstrom P, Johnson RJ. Anatomy and biomechanics of the menisci. *Clin Sports Med* 1990;9(3):523–537.
7. Krause WE, Pope MD, Johnson RJ, et al. Mechanical changes in the knee after meniscectomy. *J Bone Joint Surg* [*Am*] 1976; 58:599.
8. Kitziger KJ, DeLee JC. Failed partial meniscectomy. *Clin Sports Med* 1990;9(3):641–660.
9. Hargraves D, Seedhom B. On the bucket handle tears: partial or total meniscectomy? A quantitative study. *J Bone Joint Surg* [*Br*] 1979;61:381.
10. Shrive N, O'Conner J, Goodfellow. J. Loadbearing in the knee joint. *Clin Orthop* 1978;131:279.
11. Fukubayashi T, Kurosawa H. The contact area and pressure distribution pattern of the knee: A study of normal and osteoarthritic knee joints. *Acta Orthop Scand* 1980;51:871–879.
12. Baratz M, Fu F, Mengat R. Meniscal tears: The effect of meniscectomy and of repair in the intra-articular contact areas and stress in the human knee. *Am J Sports Med* 1986;14:270–275.
13. Baratz M, Fu F, Rehak D, et al. Peripheral tears of the meniscus. The effect of open versus arthroscopic repair on intraarticular contact stresses in the human knee. *Am J Sports Med* 1988;16:1–6.
14. Ahmed A, Burke DL. In vitro measurements of static pressure distribution in synovial joints. Part I. Tibial surface of the knee. *J Biomech Eng* 1983;105:216–225.
15. Casscells SW. The torn or degenerated meniscus and its relationship to degeneration of the weight-bearing areas of the femur and tibia. *Clin Orthop* 1978;132:196–200.
16. Fahmy NRM, Williams EA, Noble J. Meniscal pathology and osteoarthritis of the knee. *J Bone Joint Surg* [*Br*] 1983;65:24–28.
17. Noble J, Hamlen DL. The pathology of the degenerative meniscus lesion. *J Bone Joint Surg* [*Br*] 1975;57:180–186.
18. Noble J, Erat K. In defense of the meniscus. A prospective study of 200 meniscectomy patients. *J Bone Joint Surg* [*Br*] 1980;62:7.
19. Northmore-Ball M, Dandy DJ. Long-term results of arthroscopic partial meniscectomy. *Clin Orthop* 1982;167:34–42.
20. Ferkel RD, Davis JR, Friedman MJ, et al. Arthroscopic partial medial meniscectomy: An analysis of unsatisfactory results. *Arthroscopy* 1985;1:44–52.
21. Lotke PA, Lefkoe RT, Ecker ML. Late results following medial meniscectomy in an older population. *J Bone Joint Surg* [*Am*] 1981;63:115–119.
22. Jones RE, Smith EC, Reisch JS. Effects of medial meniscectomy in patients older than forty years. *J Bone Joint Surg* [*Am*] 1978;60:783–786.
23. Fairbank TJ. Knee joint changes after meniscectomy. *J Bone Joint Surg* [*Br*] 1948;30:664–670.
24. Allen PR, Denham RA, Swan AV. Late degenerative changes after meniscectomy. Factors effecting the knee after operation. *J Bone Joint Surg* [*Br*] 1984;66:666–671.
25. Tapper EM, Hoover NW. Late results after meniscectomy. *J Bone Joint Surg* [*Am*] 1969;51:517.
26. Veth RPH. Clinical significance of knee joint changes after meniscectomy. *Clin Orthop* 1985;198:56.
27. Orthmore-Ball M, Dandy DJ. Long-term results of arthroscopic partial meniscectomy. *Clin Orthop* 1982;167:34–42.
28. Lysholm J, Gilquist J, Lijedahl S-O. Long-term results after early treatment of knee injuries. *Acta Orthop Scand* 1982; 53:109–118.
29. McGinty JB, Geuss LF, Martin RA. Partial or total meniscectomy—A comparative analysis. *J Bone Joint Surg* [*Am*] 1977; 59:763–766.
30. Gear MWL. The late result of meniscectomy. *Br J Surg* 1967; 54:270–272.
31. Jackson JP. Degenerative changes following knee meniscectomy. *Br Med J* 1968;2:525.
32. Dandy DJ, Jackson RW. The diagnosis of problems after meniscectomy. *J Bone Joint Surg* [*Br*] 1975;57:349.
33. Dandy DJ, Jackson RW. Meniscectomy and chondromalacia of the femoral condyle. *J Bone Joint Surg* [*Am*] 1975;57:1116–1119.
34. Johnson RJ, Kettelkamp DB, Clark W, et al. Factors affecting late results after meniscectomy. *J Bone Joint Surg* [*Am*] 1974; 56:719.
35. Jorgensen U, Sonne-Holm S, Lauridsen F, et al. Long term followup of meniscectomy in athletes—A perspective longitudinal study. *J Bone Joint Surg* [*Br*] 1987;69:80.
36. Friedman MJ, Brna JA, Gallick GS, et al. Failed arthroscopic meniscectomy: Prognosis factors for repeat arthroscopic examination. *Arthroscopy* 1987;3:99–105.
37. Cox JS, Nye CE, Schaefer WW, et al. The degenerative effects of partial and total resection of the medial meniscus in dogs' knees. *Clin Orthop* 1975;109:178–182.
38. Lanzer WL, Komenda G. Changes in articular cartilage after meniscectomy. *Clin Orthop* 1990;252:41–48.
39. DeHaven K. Decision-making factors in the treatment of meniscus lesions. *Clin Orthop* 1990;252:49–54.
40. Arnoczky SP, Warren RF, Kaplan N. Meniscal remodeling following partial meniscectomy—An experimental study in the dog. *Arthroscopy* 1985;1:247–252.
41. Garrett JC, Stevensen RN. Meniscal transplantation in the human knee: A preliminary report. *Arthroscopy* 1991;7(1):57–62.
42. Henning CE. Current status of meniscus salvage. *Clin Sports Med* 1990;9:567–576.
43. DeHaven KE, Sebastianelli WJ. Open meniscus repair. Indications, technique, and results. *Clin Sports Med* 1990;9:577–587.
44. Cooper DE, Arnoczky SP, Warren RF. Arthroscopic meniscal repair. *Clin Sports Med* 1990;9:589–607.
45. King D. The healing of the semilunar cartilages. *J Bone Joint Surg* 1936;18:333–342.
46. Arnoczky SP, Warren RF. The microvasculature of the meniscus and its response to injury. An experimental study in the dog. *Am J Sports Med* 1983;11:131–141.
47. Hanning CE. Arthroscopic repair of meniscus tears. *Orthopaedics* 1983;6:1130.
48. DeHaven KE. Peripheral meniscus repair. An alternative to meniscectomy. *Orthop Trans* 1981;5:399–400.
49. Warren RF. Arthroscopic meniscal repair. *Arthroscopy* 1985; 1:170–172.
50. Cassidy RE, Shaffer AJ. Repair of peripheral meniscal tears: A preliminary report. *Am J Sports Med* 1981;9:209.
51. Arnoczky SP, McDevitt RF, Warren J, et al. Meniscal repair using an exogenous fibrin clot—An experimental study in the dog. *Orthop Trans* 1986;10:327.
52. Small NC. Complications in arthroscopic meniscal surgery. *Clin Sports Med* 1990;9:609–617.
53. Sprague NF. *Complications in arthroscopy*. New York: Raven Press, 1989.
54. DeHaven KE, Black KP, Griffiths JH. Open meniscus repair. Techniques and two to nine year results. *Am J Sports Med* 1989;17:788–795.
55. Totty W. *MR imaging of the post operative knee.* In: Annual Meeting International Skeletal Society. San Diego, California, 1991.
56. Crues JV, Mink J, Levy TL, et al. Meniscal tears of the knee: Accuracy of MR imaging. *Radiology* 1987;164:445–448.
57. Mink JH, Levy T, Crues JV. Tear of the anterior cruciate ligament and menisci of the knee: MR imaging evaluation. *Radiology* 1988;167:769–774.

58. Brahme S. *MR arthrography of the shoulder.* IN: Annual Meeting International Skeletal Society San Diego, California, 1991.
59. Freiberger RH, Killoran PJ, Cardona G. Arthrography of the knee by double contrast method. *AJR* 1966;97:736–747.
59a. Tolin BS, Fox JM, Del Pizzo W, Applegate, GR, Flannigan BD. MR arthrography of the post op meniscus in the knee. American Orthopaedic Society for Sports Medicine, 18th annual meeting, July 1992 (*in preparation*).
60. Brahme SK, Fox JM, Ferkel RD, et al. Osteonecrosis of the knee after arthroscopic surgery: Diagnosis with MR imaging. *Radiology* 1991;178:851–853.
61. Kent R, Pope CF, Dillon EH, et al. *MR imaging of the repaired meniscus: 12-month follow-up.*
62. Burnett GM, Fowler PJ. Reconstruction of the anterior cruciate ligament: Historic overview. *Orthop Clin North Am* 1985;16: 143–158.
63. Noyes FR, Mooar P, Matthews D, et al. The symptomatic anterior cruciate deficient knee. Part I: The long term functional disability in athletically active individuals. *J Bone Joint Surg [Am]* 1983;65A:154–162.
64. Fetto JF, Marshall JL. The natural history and diagnosis of anterior cruciate ligament insufficiency. *Clin Orthop* 1980;147: 29–38.
65. Kannus P, Jarvinen M, Finland T. Conservatively treated tears of the anterior cruciate ligaments. *J Bone Joint Surg* 1987;69A: 1007–1012.
66. Marshall JL, Warren RF, Wickiewicz TL, et al. The anterior cruciate ligament: A technique of repair and reconstruction. *Clin Orthop* 1979;143:97–106.
67. Feagin JA. The syndrome of the torn anterior cruciate ligament. *Orthop Clin North Am* 1979;10:81–90.
68. Hawkins RJ, Misamore GW, Merritt TR. Follow-up of the acute nonoperated isolated anterior cruciate ligament tears. *Am J Sports Med* 1986;14:205–210.
69. Johnson RJ, Eriksson E, Haggmark T, et al. Five- to ten-year follow-up evaluation after reconstruction of the anterior cruciate ligament. *Clin Orthop* 1984;183:122–140.
70. Pattee GA, Fox JM, Del Pizzo W, et al. Four to ten year follow-up of unreconstructed anterior cruciate ligament tears. *Am J Sports Med* 1989;17:430–435.
71. Finsterbush A, Frankl U, Matan Y, et al. Secondary damage to the knee after isolated injury of the anterior cruciate ligament. *Am J Sports Med* 1990;18:475–479.
72. Bray RC, Dandy DJ. Meniscal lesions and chronic anterior cruciate ligament deficiency. Meniscal tears occurring before and after reconstruction. *J Bone Joint Surg* 1989;71: 128–130.
73. Syndor RW, Andrews JR. Combined arthroscopy and "mini-reconstruction" techniques in the acutely torn anterior cruciate ligament. *Orthop Clin* 1985;16:171–179.
74. Kennedy JC, Weinberg HW, Wilson AS, et al. The anatomy and function of the anterior cruciate ligament as determined by clinical and morphological studies. *J Bone Joint Surg* 1974;56: 223–235.
75. Sherman MF, Warren RF, Marshell JL, Savatsky GJ. A clinical and radiographical analysis of 127 anterior cruciate insufficient knees. *Clin Orthop* 1988;227:229–237.
76. Melhorn JM, Henning CE. The relationship of the femoral attachment site to the isometric tracking of the anterior cruciate graft. *Am J Sports Med* 1987;15:539–542.
77. Jones KG. Reconstruction of the anterior cruciate ligament. A technique using the central one-third of the patellar ligament. *J Bone Joint Surg [Am]* 1963;45:925–932.
78. Ballock T, Woo SL-Y, Lyon RM, et al. Use of patellar tendon autograft for anterior cruciate ligament reconstruction in the rabbit: A long-term histologic and biomechanical study. *J Orthop Res* 1989;7:474–485.
79. Clancy WG. Intra-articular reconstruction of the anterior cruciate ligament. *Orthop Clin North Am* 1985;16:181–189.
80. Friedman MJ, Ferkel RD. *Prosthetic ligament reconstruction of the knee.* Philadelphia: WB Saunders, 1988.
81. Higgins RW, Steadman JR. Anterior cruciate ligament repairs in world class skiers. *Am J Sports Med* 1987;15:439–447.
82. Marshall JL, Warren RF, Wickiewicz TL. Primary surgical treatment of anterior cruciate ligament lesions. *Am J Sports Med* 1982;10:103–106.
83. Feagin JA, Cabaud H, Curl W. The anterior cruciate ligament: Radiographic and clinical signs of successful and unsuccessful repairs. *Clin Orthop* 1982;164:54–58.
84. Kaplan N, Wickiewicz TL, Warren RF. Primary surgical treatment of anterior cruciate ligament ruptures. A long-term follow-up study. *Am J Sports Med* 1990;18:354–358.
85. Clancy WG, Narechania RG, Rodenberg TD, et al. Anterior and posterior cruciate ligament reconstruction in rhesus monkeys. *J Bone Joint Surg* 1981;63:1270–1284.
86. Engebretsen L, Benum P, Fasting O, et al. A prospective, randomized study of three surgical techniques for treatment of acute ruptures of the anterior cruciate ligament. *Am J Sports Med* 1990;18:585–590.
87. Andersson C, Odensten M, Gillquist J. Early arthroscopic evaluation of acute repair of the anterior cruciate ligament. *Arthroscopy* 1989;5:331–335.
88. Zarins B. Combined intra-articular and extra-articular reconstruction for anterior tibial subluxation. *Orthop Clin North Am* 1985;16:223.
89. Strum GM, Fox JM, Ferkel RD, et al. Intraarticular versus intraarticular and extraarticular reconstruction for chronic anterior cruciate ligament instability. *Clin Orthop* 1989;245:188–198.
90. Carson WG. Extra-articular reconstruction of the anterior cruciate ligament: Lateral procedures. *Orthop Clin* 1985;16:191–211.
91. Clancy WG, Nelson DA, Reider B, et al. Anterior cruciate ligament reconstruction using one-third of the patellar ligament, augmented by extra-articular tendon transfer. *J Bone Joint Surg* 1982;64:352.
92. Paterson FWN, Trickey EL. Anterior cruciate ligament reconstruction using part of the patellar tendon as a free graft. *J Bone Joint Surg* 1986;68B:453–457.
93. Rackemann S, Robinson A, Dandy DJ. Reconstruction of the anterior cruciate ligament with an intra-articular patellar tendon graft and an extra-articular tenodesis. Results after six years. *J Bone Joint Surg [Br]* 1991;73:368–373.
94. McCarthy JA, Steadman JR, Dunlap J, et al. A non parallel nonisometric synthetic graft augmentation of a patellar tendon anterior cruciate ligament reconstruction. A model for assessment of stress shielding. *Am J Sports Med* 1990;1;43–49.
95. Noyes FR, Barber SD. The effect of an extra-articular procedure on allograft reconstructions for chronic ruptures of the anterior cruciate ligament. *J Bone Joint Surg* 1991;73A:882–891.
96. Daniel D, Akeson W, O'Connor J. *Knee ligaments: Structure, function, injury, and repair.* New York: Raven Press; 1990.
97. Arnold JA. A lateral extra-articular tenodesis for anterior cruciate ligament deficiency of the knee. *Orthop Clin* 1985; 16:213.
98. Noyes FR, Butler DL, Paulos LE, et al. Intra-articular cruciate reconstruction: I: Perspectives on graft strength, vascularization, and immediate motion after replacement. *Clin Orthop* 1983;172:71–77.
99. Arnoczky S, Warren R, Mine J. Replacement of the anterior cruciate ligament using a synthetic prosthesis: An evaluation of graft biology in the dog. *Am J Sports Med* 1986;14:1–6.
100. Indelicato P, Bittar ES, Prevot TJ, et al. Clinical comparison of freeze-dried and fresh frozen patellar tendon allografts for anterior cruciate ligament reconstruction of the knee. *Am J Sports Med* 1990;18:335–342.
101. Woods GA, Indelicato PA, Prevot TJ. The Gore-tex anterior cruciate ligament prosthesis. Two versus three year results. *Am J Sports Med* 1991;19:48–55.
102. Ferkel RD, Fox JM, Wood D, et al. Arthroscopic "second look" at the Gore-tex ligament. *Am J Sports Med* 1989;17:147–153.
103. Indelicato PA, Pascale MS, Huegel MO. Early experience with the GORE-TEX polytetrafluoroethylene anterior cruciate ligament prosthesis. *Am J Sports Med* 1989;17:55–62.
104. Lukianov AV, Richmond JC, Barrett GR, et al. A multicenter study on the results of anterior cruciate ligament reconstruction

using a Dacron ligament prosthesis in "salvage." *Am J Sports Med* 1989;17:380–386.

105. Linn RM, Fischer DA, Smith JP. *Allograft replacement of the anterior cruciate ligament.* IN: *American Academy of Orthopedic Surgeons*, Anaheim, California, 1991.
106. Friedman MJ, Sherman OH, Fox JM, et al. Autogenic anterior cruciate ligament reconstruction of the knee. A review. *Clin Orthop* 1985;196:9–14.
107. Ivey FM, Blazina ME, Fox JM, et al. Intraarticular substitution for anterior cruciate insufficiency. A clinical comparison between patellar tendon and meniscus. *Am J Sports Med* 1980; 8:405–410.
108. Collins HR, Hughston JC, Dehaven KE, et al. The meniscus as a cruciate ligament substitute. *J Sports Med* 1974;2:11–21.
109. Macnicol MF, Penny ID, Sheppard L, Early results of the Leeds-Keio anterior cruciate ligament replacement. *J Bone Joint Surg [Br]* 1991;73:377–380.
110. Ferkel RD, Fox JM, Del Pizzo W, et al. Reconstruction of the anterior cruciate ligament using a torn meniscus. *J Bone Joint Surg [Am]* 1988;70:715–723.
111. Arnoczky S, Warren R, Ashlock M. Replacement of the anterior cruciate ligament using a patellar tendon allograft: An experimental study. *J Bone Joint Surg* 1986;68:376–385.
112. Roberts TS, Drez D, McCarthy W, et al. Anterior cruciate ligament reconstruction using freeze-dried, ethylene oxide-sterilized, bone-patellar tendon-bone allografts. Two year results in thirty six patients. *Am J Sports Med* 1991;19:35–41.
113. Jackson DW, Grood ES, Cohn BT, et al. The effects of in situ freezing on the anterior cruciate ligament. *J Bone Joint Surg* 1991;73A:201–213.
114. Woods GW. Synthetics in anterior cruciate ligament reconstruction: A review. *Orthop Clin North Am* 1985;16:227–235.
115. Jackson DW, Drez D. *The anterior cruciate deficient knee. New concepts in ligament repair.* St. Louis: CV Mosby, 1987.
116. Pinar H, Gillquist J. Dacron augmentation of a free patellar tendon graft: A biomechanical study. *Arthroscopy* 1989;5:328–330.
117. Yasuda K, Tomiyama Y, Ohkoshi Y, et al. Arthroscopic observations of autogeneic quadriceps and patellar tendon grafts after anterior cruciate ligament reconstruction of the knee. *Clin Orthop* 1989;246:217–224.
118. Kurosaka M, Yoshiya S, Andrish JT. A biomechanical comparison of different surgical techniques of graft fixation in anterior cruciate ligament reconstruction. *Am J Sports Med* 1987; 15:225–229.
119. Lipscomb AB, Johnston RK, Synder RB, et al. Secondary reconstruction of anterior cruciate ligament in athletes by using the semitendinosus tendon. Preliminary report of 78 cases. *Am J Sports Med* 1979;7:81–84.
120. Hardin GT, Shelbourne KD, Rettig AC, et al. *Comparison of open versus arthroscopically-assisted anterior cruciate ligament reconstruction with autogenous patellar tendon.* IN: *American Academy of Orthopedic Surgeons*. Anaheim, California, 1991.
121. Paulos LE, Butler DL, Noyes FR, et al. Intra-articular reconstruction. II: Replacement with vascularized patellar tendon. *Clin Orthop [Am]* 1983;;172:78–84.
122. Brief LP. Anterior cruciate ligament reconstruction without drill holes. *Arthroscopic Surg* 1991;7:350–357.
123. Carson WGJ. A new technique of harvesting patellar tendon autografts for anterior cruciate ligament reconstruction. *Arthroscopic Surg* 1991;7:368–374.
124. Amiel D, Kliner JB, Akeson WH. The natural history of the anterior cruciate ligament autograft of patellar tendon origin. *Am J Sports Med* 1986;14:449.
125. Kohn D. Arthroscopic evaluation of anterior cruciate ligament reconstruction using a free patellar tendon autograft. *Clin Orthop* 1990;254:220–224.
126. Shelbourne KD, Whitaker J, McCarroll JR, et al. Anterior cruciate ligament injury: Evaluation of intra-articular reconstruction of acute tears without repair. Two to seven year follow-up of 155 athletes. *Am J Sports Med* 1990;18:484–489.
127. McFarland EG, Morrey BF, An KN, et al. The relationship of vascularity and water content to tensile strength in a patellar tendon replacement of the anterior cruciate in dogs. *Am J Sports Med* 1986;14:436–448.
128. Noyes FR, Butler DL, Grood ES, et al. Biomechanical analysis of human ligament grafts used in knee-ligament repairs and reconstruction. *J Bone Joint Surg* 1984;66A:344–352.
129. Clancy WG. Anterior cruciate ligament functional instability. A static intra-articular and dynamic extra-articular procedure. *Clin Orthop* 1983;172:102–106.
130. Moran DJ, Andrews JR. *Arthroscopically-assisted ACL reconstruction using patellar tendon autograft: Results after two years.* IN: *American Academy of Orthopedic Surgeons*. Anaheim, California, 1991.
131. Paulos LE, Butler DL, Noyes FR, et al. Intra-articular cruciate reconstruction. II: Replacement with vascularized patellar tendon. *Clin Orthop* 1983;172:78–84.
132. Feagin J. *The crucial ligaments.* New York: Churchill Livingstone, 1988.
133. Manaster BJ, Remley K, Newman AP, et al. Knee ligament reconstruction: Plain film analysis. *AJR* 1988;150:337–342.
134. Rak KM, Gillogly SD, Schaefer RA, et al. Anterior cruciate ligament reconstruction: Evaluation with MR imaging. *Radiology* 1991;178:553–556.
135. Shellock FG, Mink JH, Curtin S, Friedman MJ. MRI and orthopedic implants used for anterior cruciate ligament reconstruction: Assessment of ferromagnetism and artifacts. *Journal of Magnetic Resonance Imaging* 1992;2:247–248.
136. Howell SM. Serial magnetic resonance imaging of hamstring anterior cruciate ligament autografts during the first year of implantation. *Am J Sports Med* 1991;19:42–47.
137. Penner DA, Daniel DM, Wood P, et al. An in vitro study of anterior cruciate ligament graft placement and isometry. *Am J Sports Med* 1988;16:238–243.
138. Hewson GF, Jr. Drill guides for improving accuracy in anterior cruciate ligament repair and reconstruction. *Clin Orthop* 1983; 172:119.
139. Grood ES, Hefzy MS, Butler DL, et al. On the placement and initial tension of anterior cruciate ligament substitutes. *Trans Orthop Res Soc* 1983;8:92.
140. Cannon WDJ, Vittori JM. The role of arthroscopic debridement after anterior cruciate ligament reconstruction. *Arthroscop Surg* 1991;7:344–349.
141. Houseworth SW, Mauro VJ, Mellon BA, et al. The intercondylar notch in acute tears of the anterior cruciate ligament: A computer graphic study. *Am J Sports Med* 1987;15:221–224.
142. Howell SM. Pixel signal intensity measurements of unimpinged and impinged anterior cruciate ligament grafts: A prospective and reconstructive MR study. *Radiology* 1990;177:295.
143. Norwood LA, Cross MJ. The intercondylar shelf and the anterior cruciate ligament. *Am J Sports Med* 1977;5:171–176.
144. Souryal TO. Letter to the editor. *Arthroscopy* 1991;7:126.
145. Howell SM, Berns GS, Farley TE. Unimpinged and impinged anterior cruciate ligament grafts: MR signal intensity measurements. *Radiology* 1991;179:639–643.
146. Arnoczky SP, Tarvin GB, Marshall JL. Anterior cruciate ligament replacement using patellar tendon. *J Bone Joint Surg* 1982;64A:217–224.
147. Shino K, Inque M, Horibe S. Maturation of allograft tendons transplanted into the knee. *J Bone Joint Surg* 1988;70B:556–560
148. Alm A, Stromberg B. Transposed medial third of patellar ligament in reconstruction of the anterior cruciate ligament. A surgical and morphologic study in dogs. *Acta Chir Scand* 1974; 445(Suppl):37–49.
149. Alm A, Gillquist J, Stromberg B. The medial third of the patellar ligament in reconstruction of the anterior cruciate ligament. A clinical and histologic study by means of arthroscopy or arthrotomy. *Acta Chir Scand* 1974;445(Suppl):5–14.
150. Abe S, Hirohata K, Iguchi T, et al. *Remodeling and maturation of the autogenous graft for anterior cruciate ligament reconstruction: Second look arthroscopic and histologic study.* IN: *American Academy of Orthopedic Surgeons*. Anaheim, California, 1991.

151. Hughston JC. Complications of anterior cruciate ligament surgery. *Orthop Clin* 1985;16:237–240.
152. O'Brien S, Warren RF, Pavlov H, et al. Reconstruction of the chronically insufficient anterior cruciate ligament with the central third of the patellar ligament. *J Bone Joint Surg* 1991;73A: 278–285.
153. Strum GM, Friedman MJ, Fox JM, et al. Acute anterior cruciate ligament reconstruction. Analysis of complications. *Clin Orthop [Am]* 1990;253:184–189.
154. Sachs RA, Daniel DM, Stone ML, et al. Patellofemoral problems after anterior cruciate ligament reconstruction. *Am J Sports Med* 1989;17:760–765.
155. Harner CD, Paul JJ, Fu FH, et al. *Loss of knee motion following arthroscopic anterior cruciate ligament reconstruction.* IN: *American Academy of Orthopedic Surgeons.* Anaheim, California, 1991.
156. Ott JW, Graf BK, Keene JS, et al. *Risk factors for restricted motion after anterior cruciate ligament reconstruction: A retrospective study of 373 patients.* IN: *American Academy of Orthopedic Surgeons*, Anaheim, California, 1991.
157. Bonano JJ, Krinick RM, Sporn AA. Rupture of the patellar ligament after use of its central third for anterior cruciate ligament reconstruction. *J Bone Joint Surg* 1984;66A:1294–1297.
158. Burnstein MI, Ellis BI, Teitge RA, et al. Radiographic features of anterior cruciate ligament reconstruction. *Henry Ford Hosp J* 1986;43:270–274.
159. Greis PE, Evans CH, Fu FH, et al. *Immunohistochemical evaluation of prosthetic anterior cruciate ligament wear debris effects on synovial cells.* IN: *American Academy of Orthopedic Surgeons*, 1991, Anaheim, California.
160. Olson EJ, Kang JD, Fu FH, et al. The biomechanical and histological effects of artificial ligament wear particles: In vitro and in vivo studies. *Orthop Trans* 1989;13:56–57.
161. Gillogly SD, Schaefer RA, Rak KM, et al. *The accuracy of magnetic resonance imaging in assessment of patellar tendon autograft anterior cruciate ligament reconstruction.* IN: *American Academy of Orthopedic Surgeons.* Anaheim, California, 1991.
162. Schultz TK, Black KP, Kneeland BJ, et al. *Magnetic resonance imaging evaluation of an autogenous patellar tendon graft following anterior cruciate ligament reconstruction.* IN: *American Academy of Orthopedic Surgeons.* Anaheim, California, 1991.
163. Bachmann G, Sens MA, Rauber K. *MR imaging in follow-up of ligament reconstruction of the knee.* IN: 77th Scientific Assembly and Annual Meeting, Radiological Society of North America. Chicago, 1991.
164. Rak KM, Aurura CO, Gillogly SD, et al. *MR evaluation of the knee following anterior cruciate ligament reconstruction.* IN: 77th Scientific Assembly and Annual Meeting, Radiological Society of North America. Chicago, 1991.
165. Smith A, Pope CF, Dillon E, et al. *Serial MR images of anterior cruciate ligament autografts during the initial 2 years after surgery.* IN: 77th Scientific Assembly and Annual Meeting Radiological Society of North America. Chicago, 1991.
166. Tosch US, Sander B, Lais ES, et al. *MR follow-up of Gd-DTPA-enhanced autologous patellar ligament transplants for anterior crucial ligament repair.* IN: 77th Scientific Assembly and Annual Meeting Radiological Society of North America. Chicago, 1991.
167. Totterman SM, Adams MJ, Langhans MJ, et al. *MR imaging of knee after reconstruction of anterior cruciate ligament.* IN: 77th Scientific Assembly and Annual Meeting Radiological Society of North America. Chicago, 1991.
168. Schaefer RS, Gillogly SD, Rak KM, et al. *Evaluation of patellar tendon autograft revascularization in anterior cruciate ligament reconstruction using Gadolinium-DTPA enhanced magnetic resonance imaging.* IN: *A.A.O.S.* Anaheim, California, 1991.
169. Schaefer RS, Gillogly SD, Rak KM, et al. *Evaluation of patellar tendon autograft revascularization using Gadolinium-DTPA enhanced magnetic resonance imaging.* (in press).
170. Smith A, Pope CF, Jokl P, et al. *The MRI appearance of the healing patellar tendon: 12 month follow-up.* IN: American Roentgen Ray Society. Boston, MA, 1991.
171. Rijke AM, Goitz HT, McCue FC, et al. *Stress radiography of the patellar ligament following patellar tendon-bone reconstruction of the anterior cruciate ligament.* IN: American Roentgen Ray Society. Boston, MA, 1991.
172. Moeser P, Bechtold RE, Clark T, et al. MR imaging of anterior cruciate ligament repair. *J Comput Assist Tomogr* 1989;13:105–109.
173. Kolin ES, Mink JH, Deutsch AL, et al. *MR imaging of Goretex ACL reconstructions* (in press).
174. Arnoczky SP, Warren RF. Microvasculature of the human meniscus. *Am J Sports Med* 1982;10:90–95.
175. Mink JH, Deutsch AL. The knee. IN: Mink JH, Deutsch AL, eds. *MRI of the Musculoskeletal system: A teaching file.* New York: Raven Press; 1990.
176. Mink JH, Reicher M, Crues JV III. *MRI of the knee.* New York: Raven Press; 1987.

ADDITIONAL READINGS

Autz G, Goodwin C, Singson RD. Magnetic resonance evaluation of anterior cruciate ligament repair using the patellar tendon double bone block technique. *Skeletal Radiol* 1991;20:585–588.

Cheung Y, Magee TH, Rosenberg ZS, et al. MRI of anterior cruciate ligament reconstruction. *J Comput Assist Tomogr* 1992;16:134–137.

Coupens SD, Yates CK, Sheldon C, et al. Magnetic resonance imaging evaluation of the patellar tendon after use of its central one-third for anterior cruciate ligament reconstruction. *Am J Sports Med* 1992;20:332–335.

Covall DJ, Wasilewski SA. Roentgenographic changes after arthroscopic meniscectomy: Five-year follow-up in patients more than 45 years old. *Arthroscopy* 1992;8:242–246.

Dagher AP, Magid D, Johnson CA, et al. Aneurysmal bone cyst developing after anterior cruciate ligament tear and repair. *AJR* 1992;158:1289–1291.

Marans HJ, Hendrix MR, Paterson RS. A new femoral drill guide for arthroscopically assisted anterior cruciate ligament replacement. *Arthroscopy* 1992;8:234–238.

Meyers JF, Caspari RB, Cash JD, et al. Arthroscopic evaluation of allograft anterior cruciate ligament reconstruction. *Arthroscopy* 1992;8:157–161.

Paulos LE, Rosenberg TD, Grewe SR, et al. The Gore-Tex anterior cruciate ligament prosthesis a long-term followup. *Am J Sports Med* 1992;20:246–252.

Vail TP, Malone TR, Bassett FH. Long-term functional results in patients with anterolateral rotatory instability treated by iliotibial band transfer. *Am J Sports Med* 1992;20:274–282.

CHAPTER 8

Osseous and Cartilaginous Trauma

Gary M. Blum, Phillip F. J. Tirman, and John V. Crues III

At the inception of clinical magnetic resonance (MR) imaging, many in clinical medicine presumed that the new modality would not be effective in characterizing the skeletal elements due to the lack of signal arising from cortical bone. This concept has faded in light of the advances that have occurred over the better part of the last decade. In fact, MR has allowed more sensitive detection and more precise characterization of bony abnormalities than conventional imaging modalities, and has even supplemented histologic study to extend the understanding of normal and pathophysiologic mechanisms of bone function, repair, and healing. An appreciation of these processes can help reinforce the rationale underlying the appropriate utilization of musculoskeletal MR imaging, as well as allow the maximum in clinical information to be derived from the studies undertaken. In particular, knowledge of the characteristics and function of the subchondral trabecular bone and its intimate relationships to the overlying cartilage are essential for an understanding of the pathogenic mechanisms in articular trauma. Consideration of the significance of the imaging characteristics sheds light, in turn, on the etiologic and pathophysiologic nature of these entities.

Maintenance of an intact subchondral plate is of critical importance in avoiding debilitating posttraumatic sequelae. Without the smooth support of the intact subchondral plate, the articular cartilage breaks down, leading to degenerative joint disease. Forces of weightbearing through the knee are transmitted through the subchondral plate which is supported by the trabecular meshwork in the metaphysis and epiphysis. Microfractures of the supporting bony trabeculae are common in both normal bone and traumatized bone. If many trabecular microfractures coalesce in a local region of the subchondral bone, then the overlying subchondral plate may not be adequately supported and subchondral collapse may occur. Attention will be directed to the MR characteristics of these and other osseous lesions about the knee, including osteonecrosis, osteochondritis dissecans, stress fracture, complex fractures, and cartilage injury. Imaging pitfalls and normal variants will also be described. Standard fractures which can be diagnosed by plain x-ray will not be stressed in this chapter because emphasis will be placed on bony lesions for which MRI adds significant unique information contributing to patient management or the understanding of the pathogenesis of bony disease.

FUNCTIONAL AND PATHOPHYSIOLOGIC BACKGROUND

Trabecular Bone and the Subchondral Plate

Nearing the broad metaphyses and epiphyses of the distal femur and proximal tibia, the lamellar bone of the diaphyseal cortex thins, and the trabecular meshwork assumes a greater proportion of the weight and volume of the bony mass. Teleologically, this architecture appears to have arisen to support the increased surface area present at the loadbearing (articulating) ends of bone, as compared to the smaller midshaft cross-section. Indeed, the distribution and density of the trabecular network has been likened to a map of the stresses acting upon bone (1). This concept is exemplified by the familiar appearance of the orientation of the trabecular pattern seen through the femoral neck region on a frontal projection radiograph (Fig. 8.1).

The cancellous or trabecular component of bone has dynamic properties that subserve its homeostatic func-

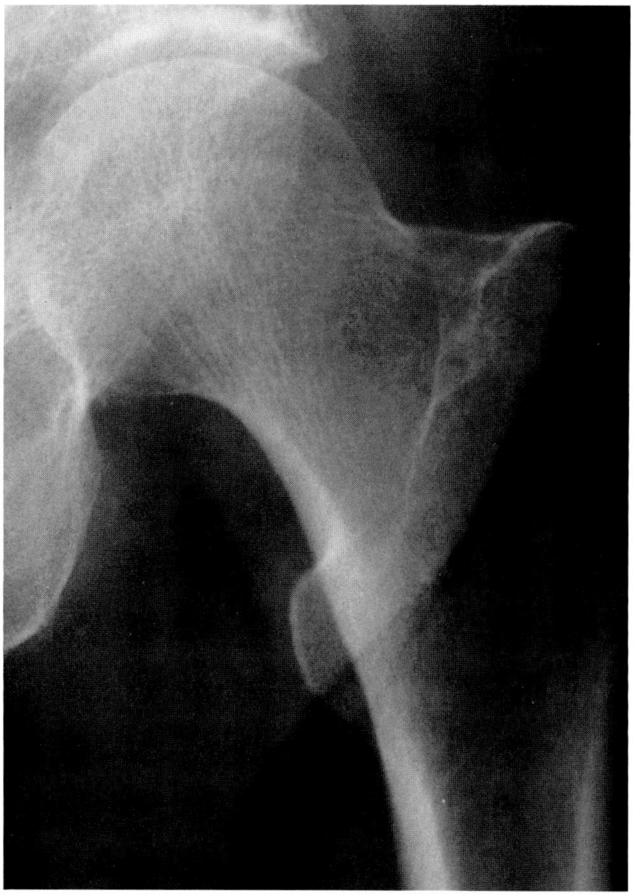

FIG. 8.1. Trabecular pattern, normal femoral neck. The more prominent trabeculae seen medially along the neck and lesser trochanter illustrate the lines of stress produced by ordinary weightbearing activities. This appearance results from active remodeling of the trabecular bony meshwork, mediated through small numbers of physiologic trabecular fractures (Wolff's Law, see text).

tion. Isolated fractures of individual trabeculae are seen on histologic examination of normal individuals (1,2), particularly when specimens are embedded in acrylic and viewed under reflected light (3). These are presumably a result of fatigue, occurring with application of forces well within the physiologic range (i.e., walking) (3). In fact, it has been estimated that the average fatigue life of a single human trabecula—in this case, in the femoral head—is somewhere between 1,000 and 10,000 strides (4). A normally high rate of osteoblastic repair prevents these ordinary stresses from resulting in an eventual net increase in the number of fractured trabeculae. This process of continual remodeling appears to underlie Wolff's Law and explain the stress-map appearance of the normal trabecular pattern (3).

The sponge-like configuration of trabecular bone imparts important local load-attenuating compliance to the tibial plateaus and femoral condyles (although coordinated, opposing contraction of large skeletal muscles provides the highest amplitude shock absorbing system in the neurologically intact patient) (5). The normal compliance of subchondral bone shields the articular cartilage from large, instantaneous pressure loads, thus protecting the articular cartilage from injury. The loss of such compliance appears to result largely from the presence of callus involved in healing supernormal numbers of trabecular fractures (6). Investigation of the possible significance of this subchondral bone stiffening in the development of early osteoarthrosis has spawned many biomechanical and histologic studies, some of which have clarified the reparative responses that are observed by MR in traumatized joints (5,7–9). There is a large humoral component to the early response phase of injury, including development of new and larger vascular channels within the marrow space (10). Osteoid is present in large amounts and healing callus becomes rapidly calcified. Interestingly, localized islands of osteonecrosis are observed as early as 7 days posttrauma (11). The commonality of these microscopic findings with those of primary (so-called nontraumatic or spontaneous) osteonecrosis suggests a possibly similar etiology, as will be discussed in a following section.

Epiphyseal trabecular microfractures, when they occur in greater than what can be considered physiologic numbers, lead to alteration in the loadbearing capacity of the subchondral plate. During the subsequent period of accelerated osteoblastic repair, the supporting trabecular meshwork is weakened (12); each remaining intact trabecula is subjected to increased stresses, and the risk is thereby elevated so that massive numbers of additional fractures will occur with relatively mild trauma. This process can lead to coalescence of neighboring microfractures producing macroscopically visible fractures and eventual collapse (impaction fracture) of the subchondral plate itself (not surprising, since this cortical bony plate is only as thick as a single trabecula) (4). Such a complication can be likened to the collapse of a building's roof following removal of its structural supporting beams.

An interesting and possibly related finding commonly encountered in degenerative or osteoarthrosis change is the subchondral cyst or geode. Resnick and Niwayama (13) point out that neither designation is entirely appropriate; this may reflect the uncertainty as to its nature and origin. These round or pyriform cavities in the end of bone are sharply marginated by sclerotic, thickened trabeculae. They can contain either proteinaceous fluid and debris, or a combination of myxoid, fibrous, and adipose elements, often with foci of metaplastic cartilage. Commonly recognized radiographically and upon histologic examination, geodes can also be seen in association with osteonecrosis; similar appearing pannus-filled cavities occur in rheumatoid arthritis.

Differentiation is usually possible upon examination of associated articular findings. Communication or intrusion of synovial fluid into the subchondral bone was originally the presumed etiology of the geode (13). However, discovery of these cysts in absence of communication to the articular surface and the observed similarity in histologic appearance of cyst contents and surroundings to that of healing trabecular fracture, has led to an alternative explanation: bony contusion. Following focal coalescence of trabecular fractures from repetitive trauma, reparative elements create cavities of hyperemic granulation tissue with resorption of the dead injured bone. Local remodeling of the trabecular meshwork ensues in an attempt to restore loadbearing function (10,14,15). The trabeculae at the margins of the cyst are bolstered with additional mineralized osteoid, producing the sclerotic rim appearance seen radiographically. Cyst/geode contents can be differentiated by appearance on T2-weighted MR sequences: fibrous and myxoid material generally appears intermediate in signal intensity, while fluid constituents are hyperintense. This appearance closely resembles healing traumatic lesions of the trabecular bone, including those of posttraumatic osteonecrosis (*vida infra*). These MR observations lend support to the bony contusion mechanism proposed to explain the origin of subchondral geodes.

Articular Cartilage

Subchondral plate fracture results in alteration in contour of the overlying articulating cartilage. The cartilage is resilient, however, and its surface frequently remains intact initially (16) and appears normal to arthroscopic inspection (17). Cartilage function is heavily dependent on its viscoelasticity and low coefficient of friction—properties which can be degraded by such an injury, more markedly so once the articulating surface becomes disrupted (18). Articular cartilage is poorly tolerant of the increased shear forces that follow loss of joint congruity (as can occur in radiographically and arthroscopically occult bony injury), and it becomes subject to rapidly progressive fissuring (19). The injured articular cartilage is further degraded by enzymes released into the synovial fluid (20–22) and cannot maintain its shock-absorptive function, which it normally accomplishes through transient deformation by shifts in water content (23). The loss of function of the articular cartilage leads to further cartilage and bone trauma from increased joint friction and vibrations. Thus, a vicious cycle is established which must be broken early in the cycle if irreversible articular cartilage injury is to be prevented.

The regenerative and repair capacities of articular cartilage are extremely limited, mainly due to its avascular status. Unless a cartilage injury extends across the subchondral plate and allows influx of blood products, the stages of tissue response seen in most injured tissues, i.e., necrosis, inflammation, and repair, will not occur. Instead, superficial lacerative injury tends to be filled in by a modified form of hyaline cartilage, but only for a matter of weeks to months, with the defect eventually reappearing in its initial form (18). In response to blunt impact there is early death of chondrocytes and release of sterptolysin, metalloproteinase, and other enzymes, with disruption of the supporting collagen network (20–22,24). This leads to abnormal cartilage function and progression to changes of osteoarthrosis, both in the cartilage and in the underlying subchondral bone (9,18,19). One of the earliest changes observed (both on histologic study and by MRI) in cartilage subjected to recurrent mild blunt trauma is an increase in water content in the superficial layers (tangential and transitional zones) (23,25). This appears to result from damage to the collagen fibrils, which permits additional water to become associated with the internal matrix proteoglycans. The resultant relative decrease in concentration of the structural proteoglycans appears to further weaken the restraining collagen framework. The synergism of these insults leads to the ongoing structural derangement that we identify in its clinical "final common pathway" of chondromalacia.

Clinical Implications

Even in the absence of overt cortical or end-plate fracture, recognition of fracture of the trabecular bone meshwork is important. As will be discussed in a subsequent section, continued activity can result in a more severe delayed injury involving collapse of the articulating surface. Animal work has demonstrated a parallel phenomenon. Radin et al. (6,16) exposed rabbit knees to daily dynamic loading that exceeded the stress of normal activity but was insufficient to cause overt fracture. By 3 weeks following onset of this repetitive and relatively mild trauma, histologic study revealed numerous trabecular microfractures and accompanying callus formation. These peaked in number by the sixth week, at which time the mechanical integrity of the subchondral plate was maximally compromised. Bone in each contralateral, unstressed leg remained normal. These changes diminished the compliance of trabecular bone and allowed greater peak forces to be transmitted to the overlying subchondral plate, placing it at greater risk for collapse. Similar changes were confirmed in human cadaver knees (26). These studies suggest that patients with pain on weightbearing and intact subchondral plates with x-ray and arthroscopically occult trabecular bony injuries may be at risk of subchondral plate collapse if they continue aggressive weightbearing. It may be prudent to place such patients on a regimen of weightbearing to tolerance and remove them from sports activities for 4 to 6 weeks to allow early healing of the trabecular injuries. With significant trabecular healing, compliance of the cancellous bone should improve and the patient's risk of the subchondral plate collapse should normalize.

In addition, drugs are being developed which specifically block the actions of degrading enzymes and have significant protective effects on injured articular cartilage in animals (22,24). Detection of shifts in free water concentration between the superficial and deep layers of the articular cartilage—the earliest known changes in cartilage injury and degeneration—could allow identification of patients at risk for progressive posttraumatic degenerative sequelae and may target a cohort of patients who could benefit from such drug therapy. Consequently, early diagnosis of trabecular

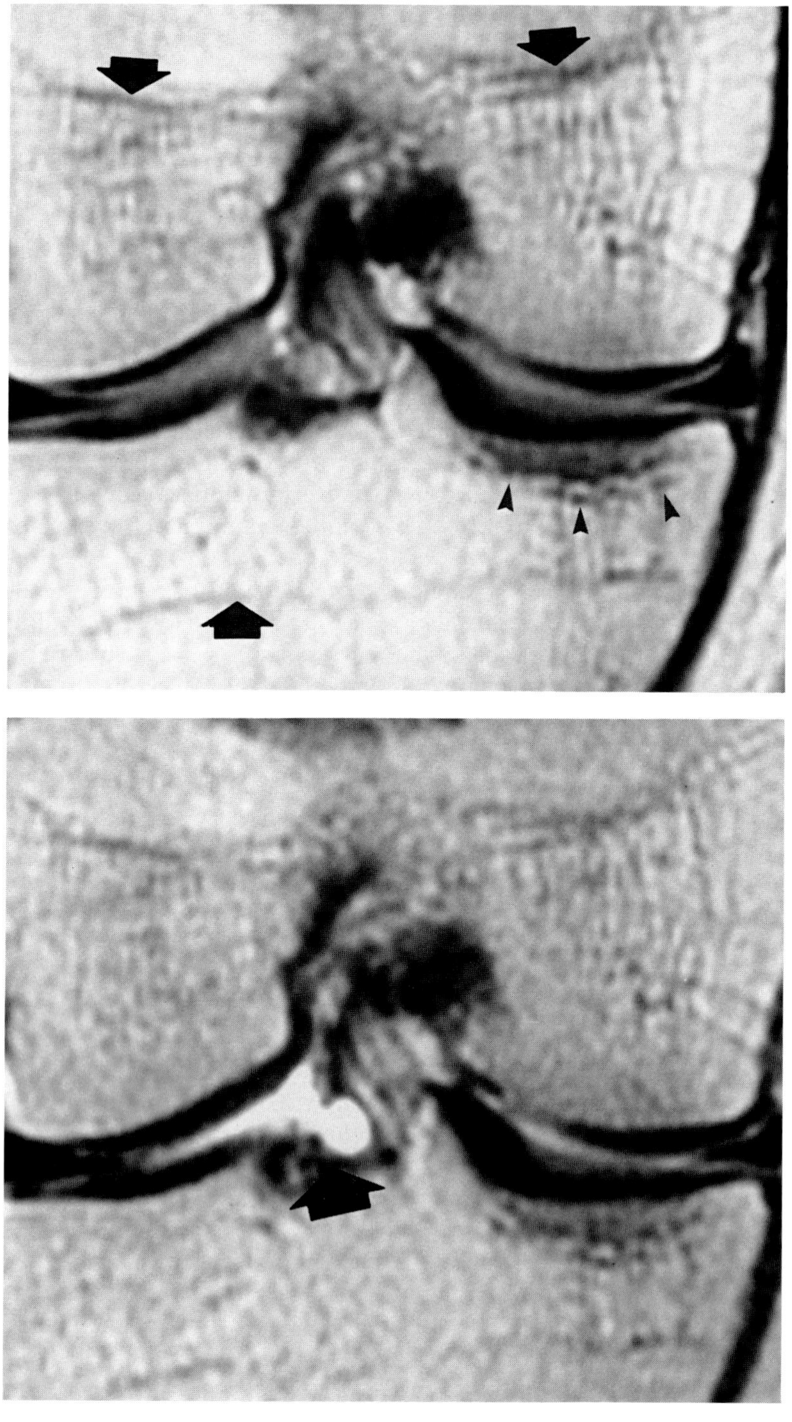

A

B

FIG. 8.2. Normal trabecular pattern; coronal intermediate. **A:** TR/TE 2,000/20. **B:** T2-weighted (TR/TE 2,000/70) images of a normal 24-year-old volunteer show the trabecular struts as a reticulated low signal meshwork against a higher signal background of the largely fatty marrow space. Note the femoral and tibial physeal scars [*arrows* in (A)], appearing as delicate linear bands of diminished signal. There is a normal condensation of trabeculae [*arrowheads* in (A)] beneath the subchondral plate of the medial tibial plateau. The small amount of joint fluid seen as bright signal in (B) (*arrow*) is physiologic.

injury by MRI may help prevent further trabecular fractures, collapse of the subchondral bony plate, and irreparable damage to the articular cartilage.

IMAGING CHARACTERISTICS

The tissue contrast inherent in appropriate spin-echo sequences has been combined with high resolution, small field-of-view, surface coil techniques to provide unprecedented characterization of the cancellous bone (Fig. 8.2), as well as identification of disrupted cortical and subchondral surfaces. Techniques continue to be refined to allow optimal visualization of the articular cartilage itself.

In this section the MR appearance of trabecular bony injury resulting from acute trauma, as well as from more chronic repetitive trauma, will be described. In many cases histologic confirmation has not been obtained. However, through an understanding of the biomechanical principles, the pathophysiology previously outlined, and correlation with the clinical history and mechanism of injury, reliable postulates regarding the significance of imaging findings can be made.

Most clinical investigators have utilized double echo spin-echo sequences, both at high field and lower field strengths (27–32). The double echo spin-echo technique allows simultaneous acquisition of two sets of images with different contrast (see Chapter 1). The short TE images give contrast similar to T1-weighted images with bright fat and excellent anatomic delineation. The long TE images are T2-weighted with bright fluid and high sensitivity to many disease processes (see Chapter 1). Newly developed ultra high resolution spin-echo images may be capable of depicting individual trabeculae (Fig. 8.3). The uses of ultra high resolution imaging for joint pathology are in their earliest stages of clinical evaluation, but will probably be important in future imaging evaluation.

Two different inversion recovery (IR) sequences may be useful in evaluating bony trauma and articular cartilage disease (see Chapter 1). Heavily T1-weighted IR sequences are useful in delineating the superficial layer of articular cartilage (low in signal on T1-weighted images because of increased free water content) from the deeper layers (higher in signal intensity because of lower free water content) (23). Short TI inversion recovery techniques (STIR techniques, see Chapter 1) depict water as bright signal and are sensitive to trabecular bony injury. STIR is also sensitive in detecting articular cartilage injury.

Gradient-echo sequences may be advantageous in evaluating bony and cartilaginous injuries, especially when large flip angles are used (23,33,34). Gradient-echo images may be obtained with short TR acquisitions allowing for the acquisition of very high resolution three-dimensional Fourier-transform images (see Chapter 1). High resolution three-dimensional images may be helpful in detecting articular cartilage injuries (especially on the curved surfaces of the femoral condyles) (33) and can visualize individual trabeculae of the subchondral meshwork (35), although they are more susceptible to patient motion artifact (36). Gradient-echo sequences can be helpful in specific instances (quantification of trabecular bone struts within the cancellous space, for example) (37). The routine use of gradient-echo techniques, however, has signif-

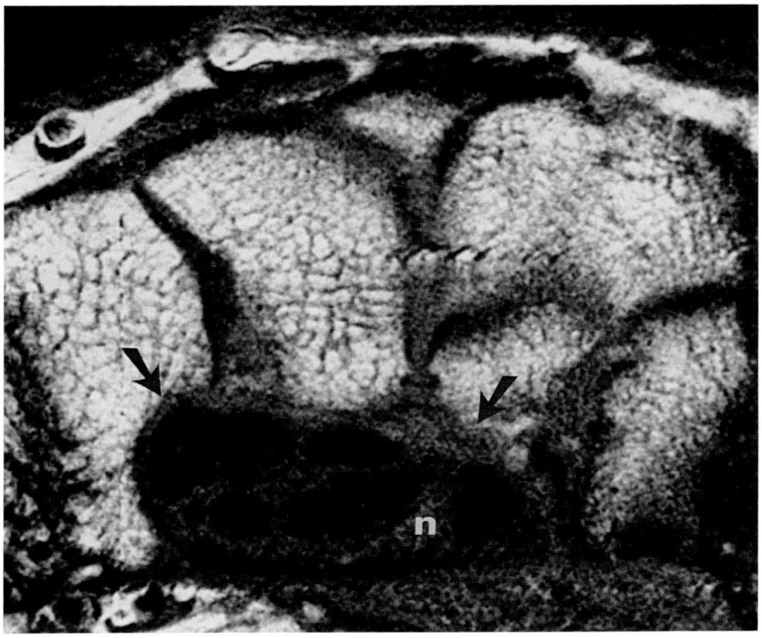

FIG. 8.3. Ultra-high resolution image. Axial image of the wrist in a normal volunteer (TR/TE 600/23) shows detailed trabecular architecture of the proximal carpal bones. The median nerve (n) and grouped flexor tendons of the hand (*arrows*) are also seen to advantage.

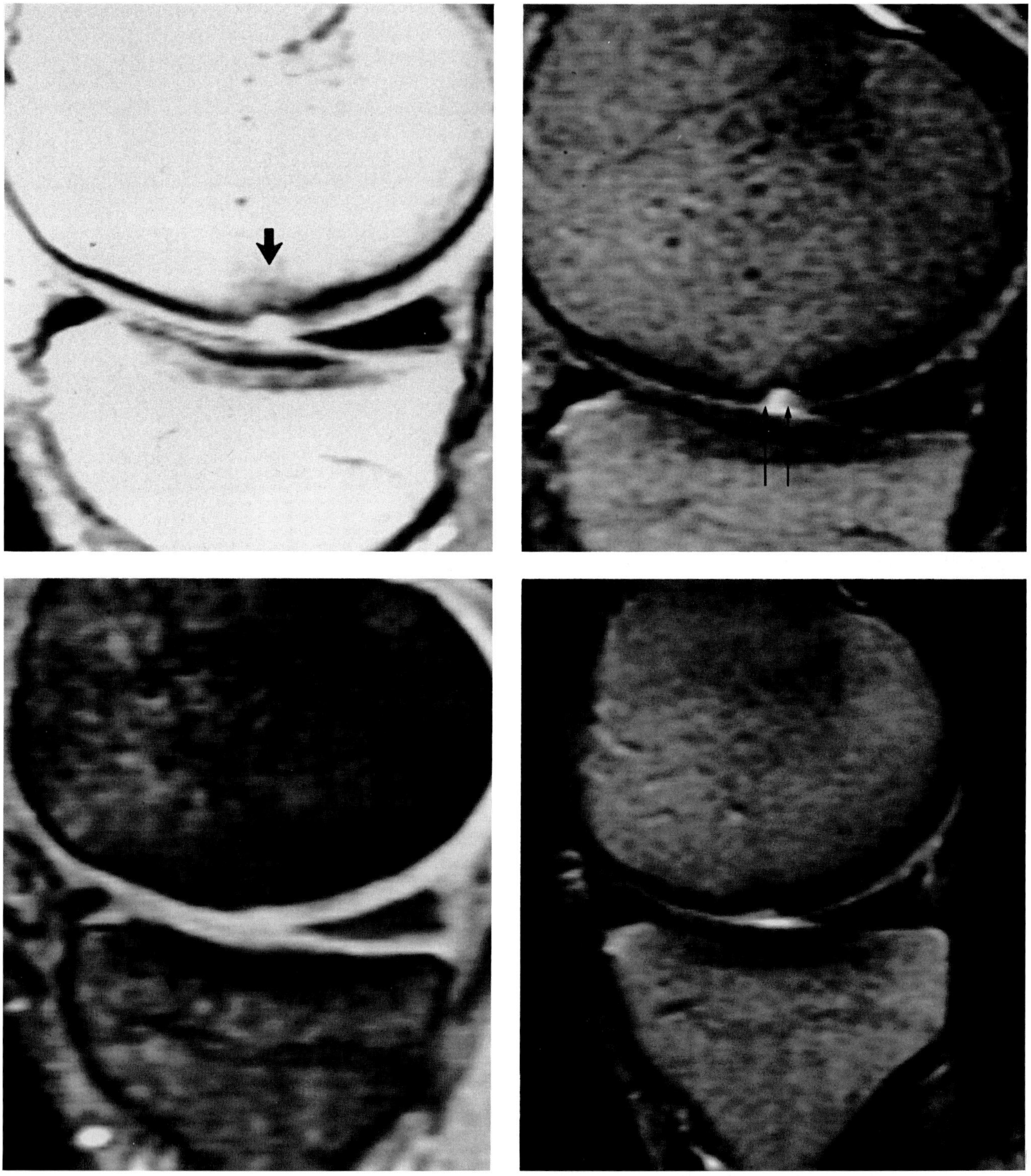

A B C D

FIG. 8.4. Delayed impaction fracture (type 2 bony injury). Professional basketball player was kneed while driving for a lay-up. He continued to play with mild knee pain until approximately 4 weeks later when he suffered acute onset of severe knee pain without trauma, leading to the initial MR scan. **A:** Sagittal MR TR/TE 2,000/30, shows focal impaction of the subchondral plate of the medial femoral condyle, with adjacent signal loss in the trabecular bone (*arrow*). **B;** With T2-weighting, bright joint fluid highlights the associated defect in the articular cartilage (*arrows*) (TR/TE 2,000/80). **C:** Gradient-echo image (TR/TE 100/15/30°) fails to depict the cartilage injury. **D:** Follow-up image (TR/TE 2,000/80) 5 months later shows that the margins of the impaction have smoothed.

icant shortcomings when not accompanied by a long TE spin-echo sequence. Since most gradient-echo pulse sequences depict articular cartilage as high in signal intensity, it is often indistinguishable from joint fluid and significant cartilage disease can be hidden (Fig. 8.4). Images obtained with large flip angles (greater than 50°) can image joint fluid with lower signal intensity than articular cartilage and may circumvent this potential pitfall of gradient-echo imaging (23). Because of the variability of gradient-echo image contrast depending on the choice of flip angle, we have found these sequences to be suboptimal for primary evaluation of bony and soft tissue abnormalities.

Considerations of expense and patient comfort versus the incremental gain in using a combined approach—gradient-echo plus T2-weighted spin-echo—has led most clinical centers to concentrate largely on spin-echo scanning, with gradient-echo techniques reserved for problem-solving use in individual cases. Technical considerations are further discussed in Chapter 1, as well in several other excellent musculoskeletal texts (28,38–41).

A

B

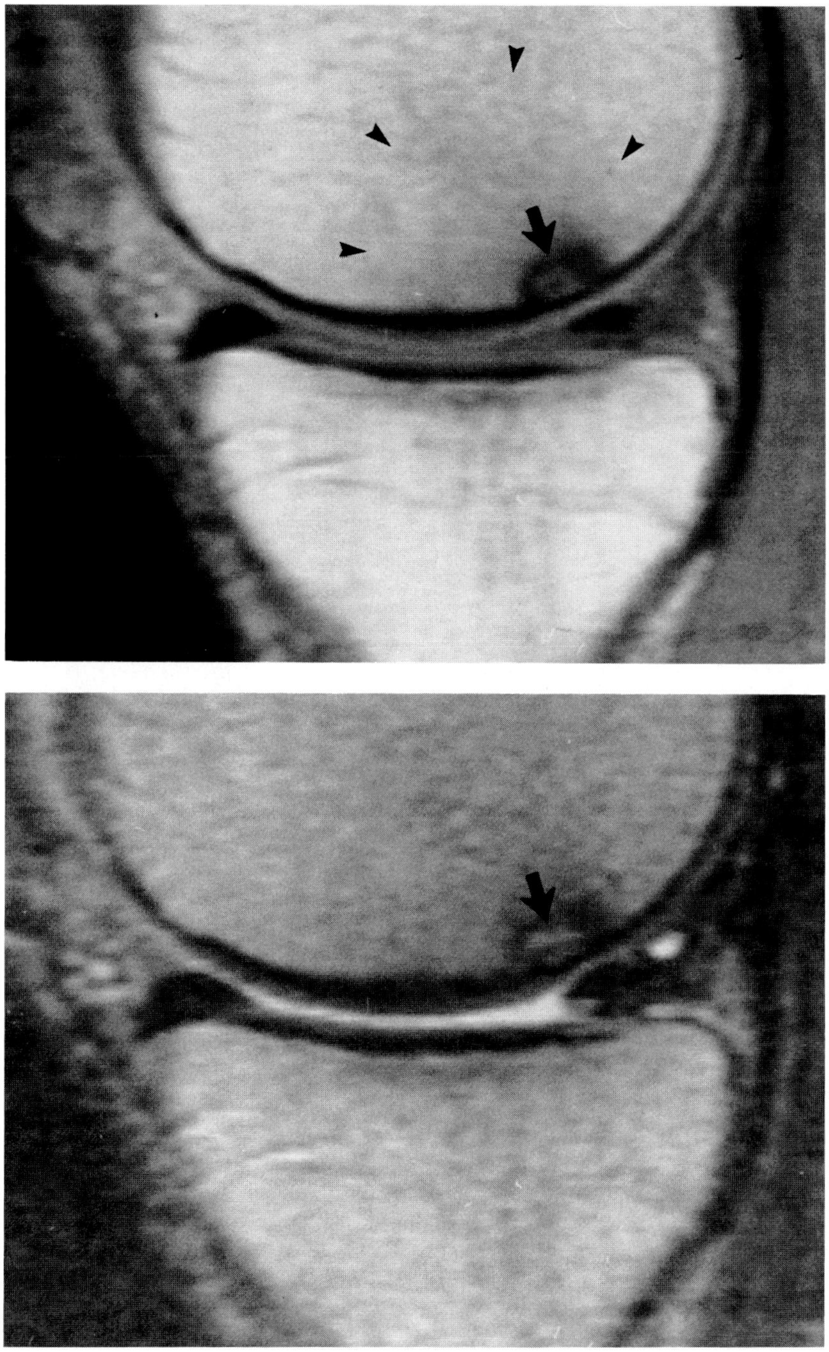

FIG. 8.5. Type 1 trabecular injury. Sagittal images demonstrate a crescentic area of signal loss in the trabecular bone (*straight arrows*) on both **A:** intermediate (TR/TE 2,000/20) and **B:** T2-weighted (TR/TE 2,000/80) images, which corresponds to a geographic lesion as described by Vellet (see text). There is less marked signal loss extending well into the epiphysis (*arrowheads*), as well as mild signal increase on T2 within the center of the lesion (B). The black line representing the subchondral plate is intact, although the overlying cartilage appears focally thinned. The patient had fallen down a flight of stairs 7 weeks prior to imaging. These injuries to supporting trabeculae immediately beneath the articulating surface are often associated with delayed collapse of the subchondral plate (see Figs. 8.4 and 8.15).

Trabecular Bone Injury

Recognition of the efficacy of MRI in detecting and characterizing bony lesions about the knee began with a case report of an overt, complex tibial plateau fracture (42). This was followed by descriptions of signal changes in diaphyseal stress fracture (43,44). Subsequently, the serendipitous identification of MR signal changes in radiographically normal periarticular trabecular bone provided the basis for subsequent refinements in applications. Several series have correlated MR characteristics in posttraumatic trabecular bony lesions with clinical history, findings of other imaging studies, arthroscopic examination, and follow-up MR imaging (17,45,46), thereby providing information regarding the clinical significance and prognosis of such injuries. It is helpful to review several classification schemes that have been developed so that a reasonable analysis of clinical data can be undertaken.

Yao and Lee first presented a series of patients with radiographically occult trabecular injury about the knee (32). They described eight patients, each of whom had sustained an initial injury within days to weeks of MR imaging, all with normal plain radiographs. MR findings were similar to those seen in the shaft of the bone in Stafford's stress fracture patients: irregular foci of increased signal in long TE (T2-weighted images) with corresponding signal loss on short TE (both T1 and intermediate-weighted) images. Two patients with follow-up scans after periods of restricted activity showed resolution of the signal abnormalities. The authors proposed that these epiphyseal lesions might represent "a new category of incomplete fracture that spares the cortex." Considering the numerous studies verifying the common occurrence of trabecular fracture in animals and man (2,3,6,14,15,26,47,48), it is more likely that these lesions were not truly new, but rather newly imaged. This is supported by recognition, in this study and others (17,49), that patients who

A

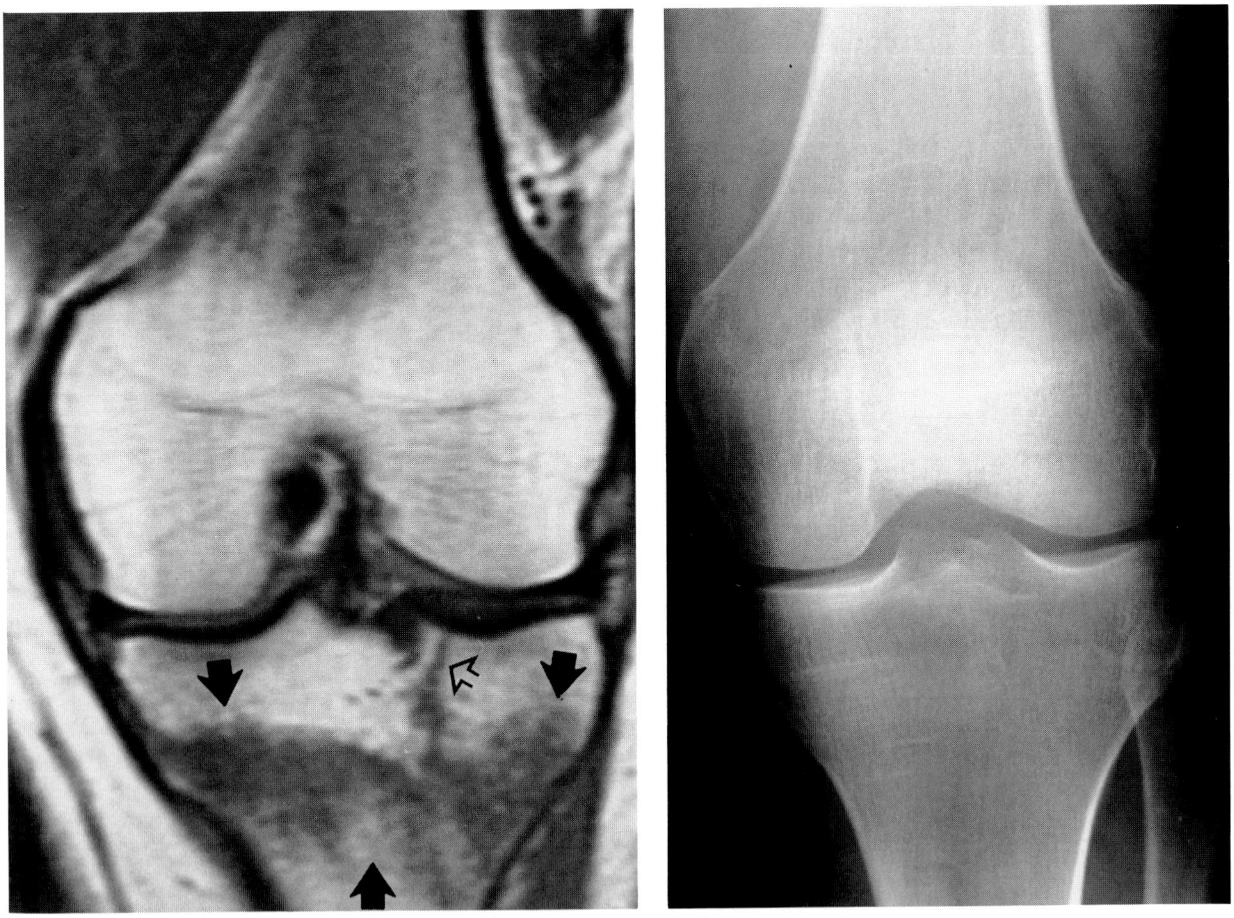

B

FIG. 8.6. Type 1 trabecular injury in a 23-year-old with pain following a fall. **A:** Coronal image (TR/TE 2,000/20) reveals a large region of trabecular signal loss in the tibial epiphysis and metaphysis (*closed arrows*). There is extension of the injury (*open arrow*) to the subchondral plate of the lateral tibial plateau, which remains intact (as confirmed on the T2-weighted image, not shown). Apparent signal loss in the femoral shaft and metaphysis results from partial volume averaging. **B:** Plain radiograph is unremarkable. Delicate transverse densities represent so-called bone bars or growth arrest lines.

undergo concomitant radionuclide evaluation show areas of increased tracer activity which correspond to the MR abnormalities. We suspect that prior to their depiction by MRI, these trabecular microfractures had been insensitively detected (by plain radiography and computed tomography), and imprecisely localized and characterized (by extremely sensitive modalities such as nuclear bone scan).

Additional clinical perspective was provided through the work of Lynch, Crues, and co-workers (45). It was found that trabecular lesions were quite common: 20% of 434 consecutive patients who underwent knee MR evaluation had at least one region of signal abnormality within the medullary space about the knee. These lesions resembled those described by Yao and Lee (*vida supra*) (32). They were categorized according to their imaging characteristics and correlated with ligamentous injury of the knee. Accompanying plain radiographs had all been interpreted prospectively as negative for acute traumatic changes. Abnormalities of intramedullary MR signal were morphologically classified into three types. All involved loss of the normal high signal from the medullary fat on short TE images in the epiphyses, often with ex-

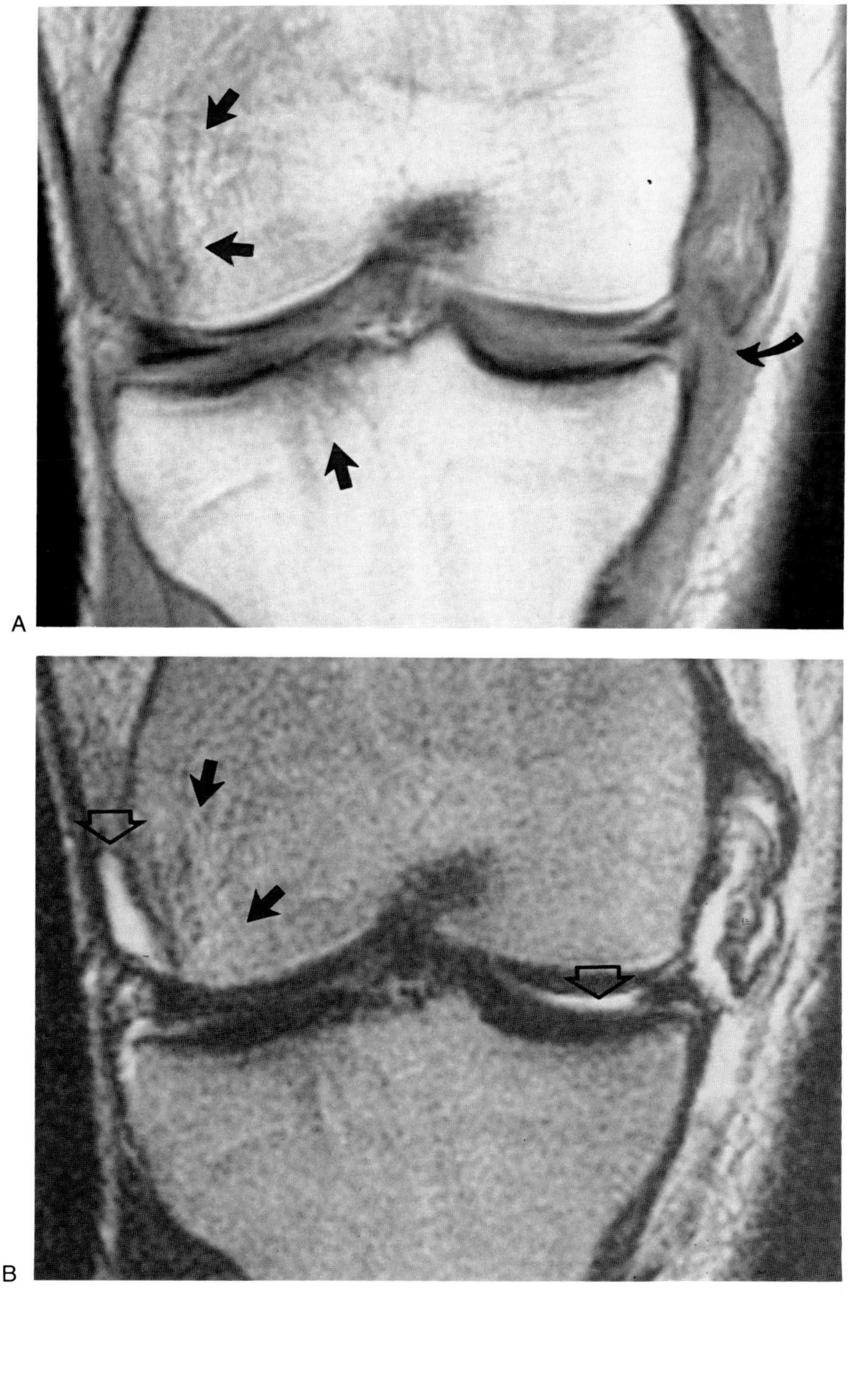

FIG. 8.7. Type 1 trabecular injury in a 20-year-old with trauma resulting in ACL and PCL tears. **A:** Coronal intermediate-weighted image (TR/TE 2,000/20) showed reticulated and coalescent areas of signal loss (*straight arrows*) within the trabecular bone of the lateral femoral condyle and lateral tibial plateau. Preservation of the overlying black line of the subchondral bone plate typifies the type 1 injury. The valgus compressive forces that caused these bony injuries also resulted in complete rupture of the medial collateral ligament (*curved arrow*). **B:** Corresponding T2-weighted image (TR/TE 2,000/80) readily depicts joint fluid (*open arrows*) and reveals subtle increased signal in the areas of trabecular injury (*straight arrows*), most likely representing blood products and the fluid components of the early inflammatory and reparative response.

A

B

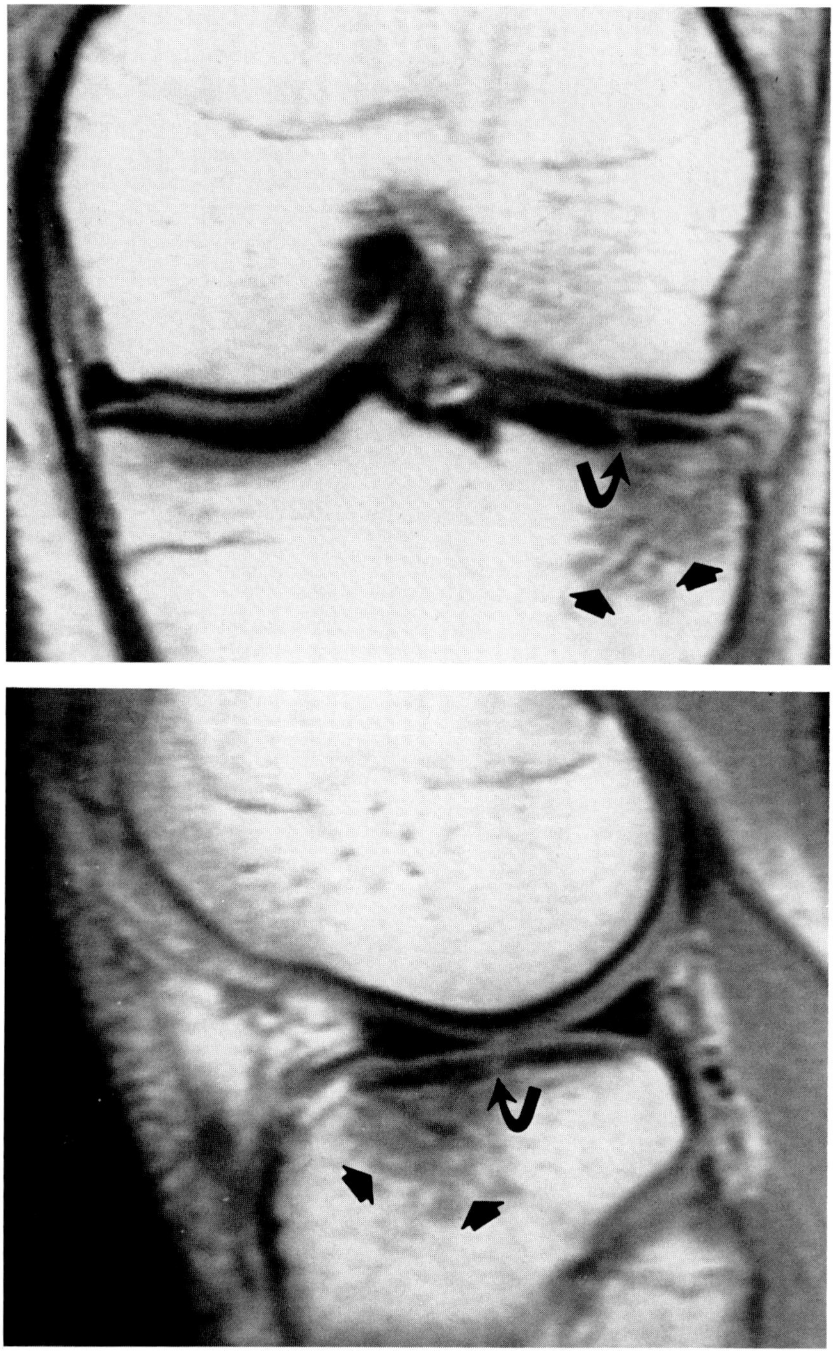

FIG. 8.8. Type 2 bony injury in an 18-year-old following a skateboarding accident. **A:** Coronal intermediate-weighted image (TR/TE 2,000/20) demonstrates amorphous low signal with reticulated periphery (*short arrows*) against the higher signal marrow fat in the lateral tibial plateau. High signal that crosses the subchondral plate (*curved arrow*) characterizes this as a type 2 lesion. **B:** Sagittal intermediate-weighted image (TR/TE 2,000/20) through the same area shows the subchondral plate injury (*curved arrow*) to be quite localized. Trabecular component of the injury (*short arrows*) has a similar appearance to that in (A).

tension into the metaphysis. The *type 1* lesion appeared as an area of reticulated, inhomogeneous signal loss within the trabecular bone on short TE images, often with associated increased signal on T2-weighted images (Figs. 8.5–8.7). The *type 2* abnormalities were similar, but also included an interruption in the smooth black line representing the surface of the weightbearing bone, the subchondral plate (Figs. 8.8–8.10). This interruption was either a "breach" of high signal, or merely an abrupt contour change. Signal changes designated as *type 3* were confined to the immediate juxtacortical region, and consisted solely of profound signal loss on short TE images; increased signal on long TE (T2-weighted images) was not seen. This group largely represented patients who had not suffered acute trauma, and whose radiographs showed changes typical of degenerative bony sclerosis or eburnation (Figs. 8.11 and 8.12). Arthroscopy revealed chondromalacia of the overlying cartilage in 70% of these patients, as opposed to only 28% in patients with types

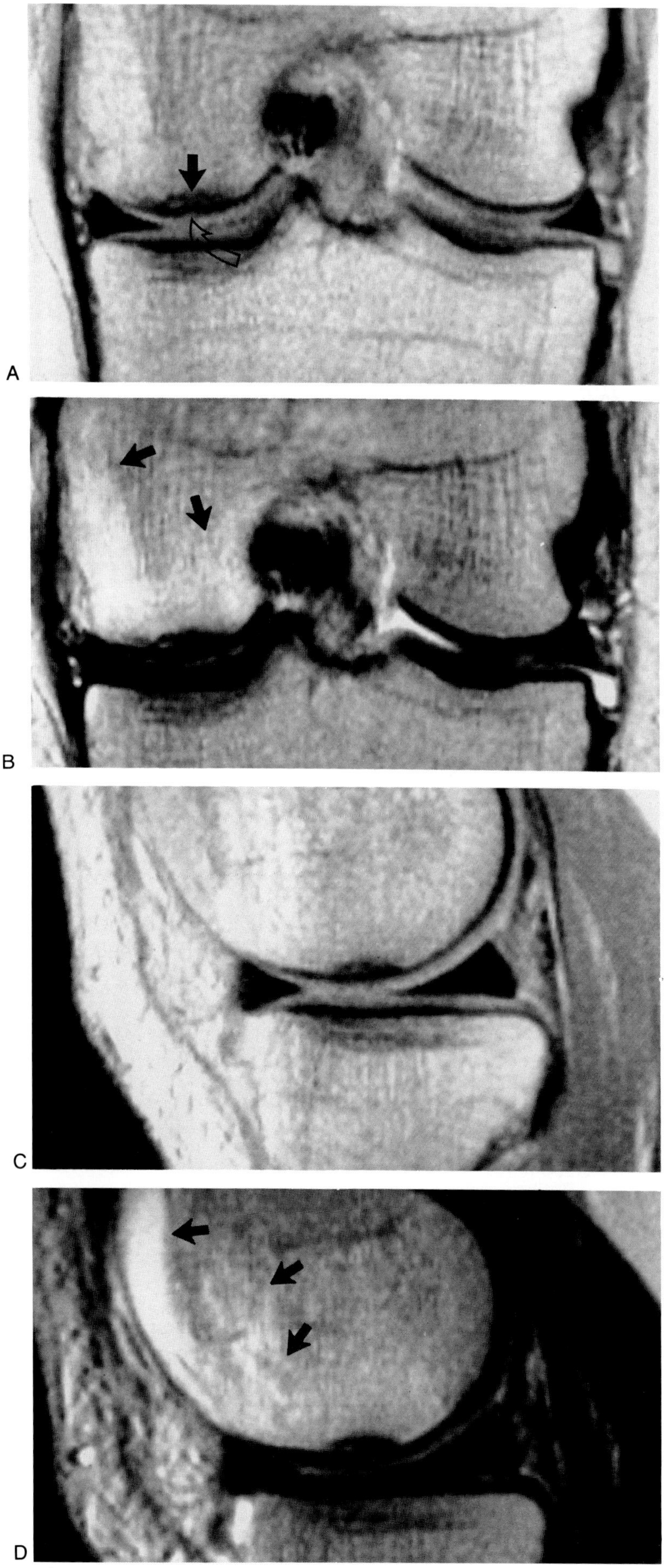

FIG. 8.9. Type 2 bony injury with images obtained on a 0.4 T scanner. **A:** Coronal intermediate-weighted (TR/TE 2,000/20) image shows a crescentic small focus of diminished signal in the trabecular bone close to the subchondral plate of the medial femoral condyle (*straight arrow*). The overlying subchondral plate appears slightly depressed (*curved arrow*), making this a type 2 injury. **B:** Corresponding T2-weighted image (TR/TE 2,000/80) shows in addition a large amorphous region of bright signal in the trabecular bone (*short arrows*). (Subchondral plate abnormality is obscured by increased magnetic susceptibility effect on T2.) Note the absence of diminished signal within this large injury on the intermediate weighted image in (A). At mid or low field strengths this is common, making inspection of the longer TE images especially critical in detection of trabecular injury. **C,D:** Sagittal intermediate- and T2-weighted (TR/TE 2,000/20,80) images just medial to the subchondral plate defect, show the typical appearance of trabecular injury with T2 weighting [(D), *short arrows*], but again note the lack of diminished signal with shorter TE (C), due to reduced T1 effect at lower field strength, in comparison to 1.5 T scanning.

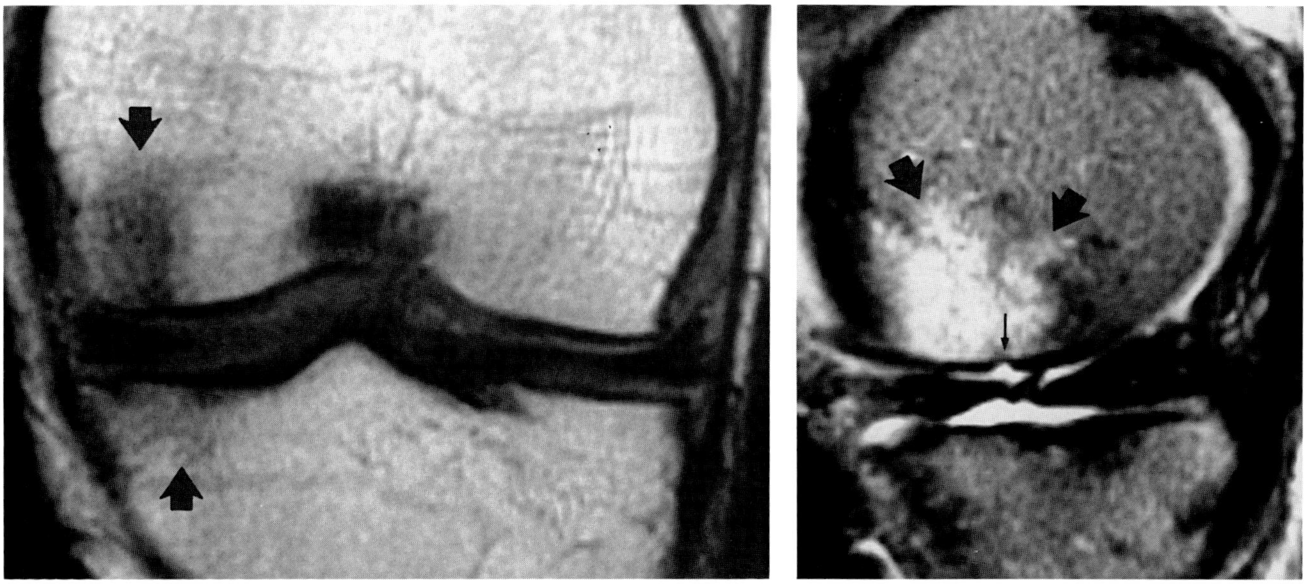

A B

FIG. 8.10. Type 2 bony injury. **A:** Coronal T1-weighted (TR/TE 800/20) image of the right knee in a 66-year-old jogger with medial pain demonstrates amorphous signal loss in the cancellous bone of the medial femoral condyle and medial tibial plateau (*short arrows*). There is a focal contour defect in the femoral subchondral plate overlying the trabecular signal abnormality. **B:** Sagittal T2-weighted (TR/TE 2,117/80) image through this location reveals the circumscribed nature of the end-plate injury (*long arrow*), and also displays the abnormal trabecular region as increased signal (*short arrows*). Due to the arthrogram effect, the severe thinning of articulating cartilage on both sides of the joint is conspicuous with T2 weighting (B).

1 and 2 injury. As expected, type 3 lesions are also found in the patellofemoral compartment, where degenerative bony changes are common (Fig. 8.13).

Fifty of the 73 patients with types 1 and 2 lesions had concomitant ligament rupture, attesting to the magnitude and mechanism of injury. The most common location for the types 1 and 2 injuries was the posterior aspect of the lateral tibial plateau near the proximal tibiofibular joint. Injuries at the anterior (weightbearing) portion of the lateral femoral condyle were nearly as frequent, as were combined injuries. Blows to the anterolateral tibia initially deflect the knee medially and posteriorly, placing compressive forces on the lateral bony compartment. The tibia is also forced posteriorly, placing tension on the anterior cruciate ligament (ACL). Blows from directly lateral result in distractive forces which stretch the medial collateral ligament (MCL). If forces are of sufficient magnitude, ligament rupture occurs which relieves the compressive forces on the lateral compartment (Figs. 8.7 and 8.14). Most fractures occur to the lateral compartment by this mechanism; medial compartment injury (from varus forces) is comparatively rare since the opposite knee provides protection. The compressive mechanism involving the bones of the lateral compartment is easily visualized when viewing examples (Fig. 8.7). Medial compartment injuries can result from direct impact such as falls onto the extended leg or onto the flexed knee, with transmission of forces via the patella.

Those patients who underwent follow-up MR imaging at 3 months demonstrated resolution of the cancellous bone abnormalities. Subchondral plate irregularities (as seen in some type 2 injuries) persisted. Since the initial study, there have been several cases at the authors' facility, mostly in competitive athletes, in which isolated trabecular microfracture (type 1 or type 2 injury) progressed to a more extensive lesion involving the overlying subchondral plate, including collapse and impaction fracture (Figs. 8.15 and 8.16). Recovery from this type of injury is limited, emphasizing the importance of early detection and restriction of weightbearing (*vida infra*).

Mink and Deutsch (31) presented a series of patients with trabecular bone lesions of the distal femur/proximal tibia that were radiographically unsuspected (62 of 66 plain film examinations negative, four equivocal). They confirmed the relationship to acute trauma by noting strong association with ligamentous injury and by correlating lesion location with mechanism of injury. These findings were remarkably similar to those of Lynch and Crues above. The classification system used by Mink and Deutsch included *bone bruise*, which corresponds to the type 1 lesion previously described, as well as *amorphous stress fracture* (identical to bone bruise, but distinguished by a clinical history without acute trauma), and *impacted osteochondral fracture*, roughly comparable to the type 2 lesion of Lynch and Crues. Additionally included in their report were linear stress fracture, displaced osteochondral fracture, and frank fractures of the plateau and/or condyle (this last category ac-

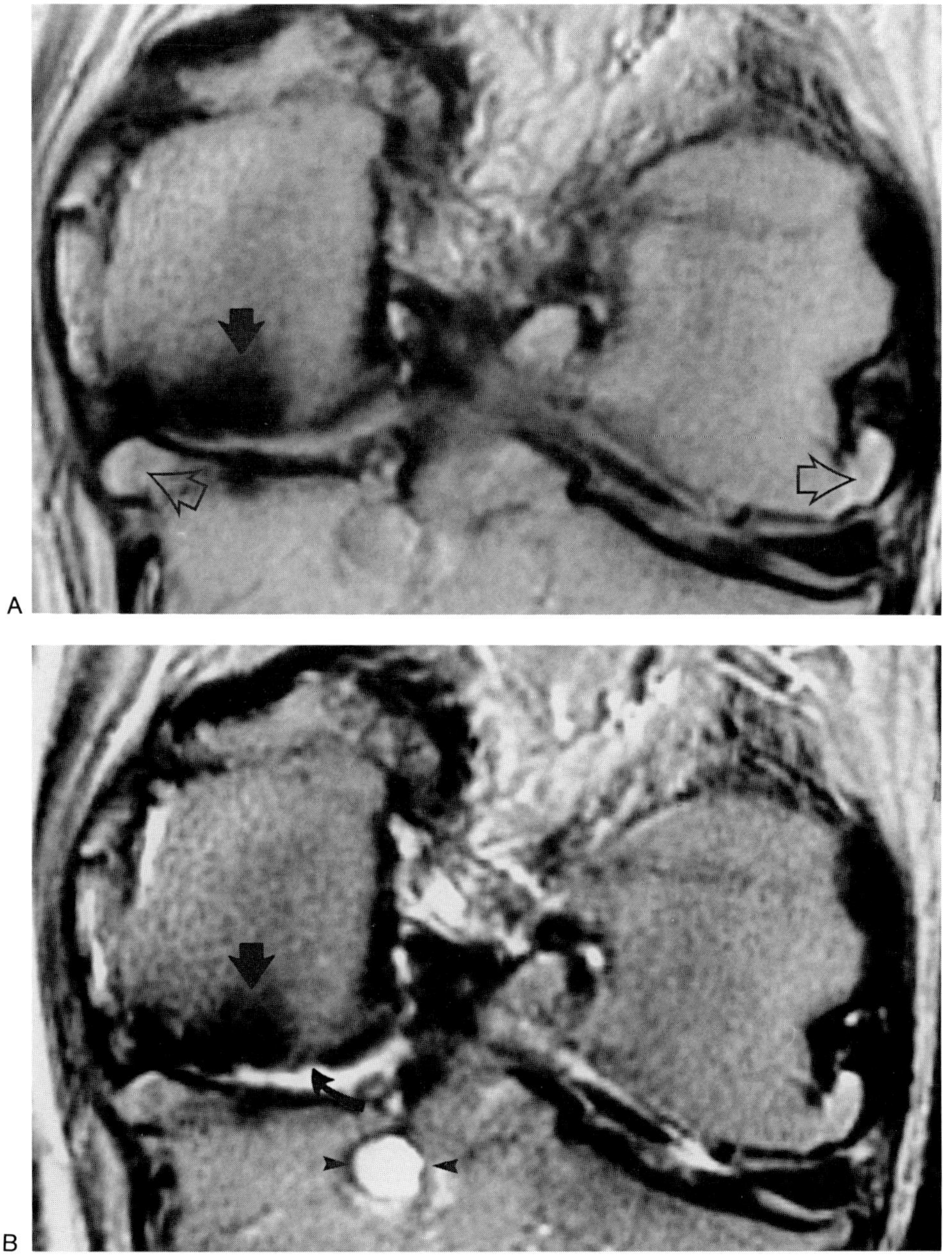

FIG. 8.11. Type 3 bony injury. **A,B:** Coronal (TR/TE 2,000/20,80) images through the posterior portion of the joint show extensive degenerative changes in the left knee of this 77-year-old. There is profound signal loss on both sequences in the juxtaarticular trabecular bone (*short arrows*) of the medial compartment, representing eburnation typical of type 3 injury. Nearly complete denudation of the articular cartilage, particularly in the medial compartment, and irregularity of the subchondral bony surface (*curved arrow*) is depicted. Note osteophytic spurs (*open arrows*) composed of cortical and cancellous bone. A subchondral cyst or geode is seen medially in the tibia (*arrowheads*) as well-circumscribed high signal on the T2-weighted image (B).

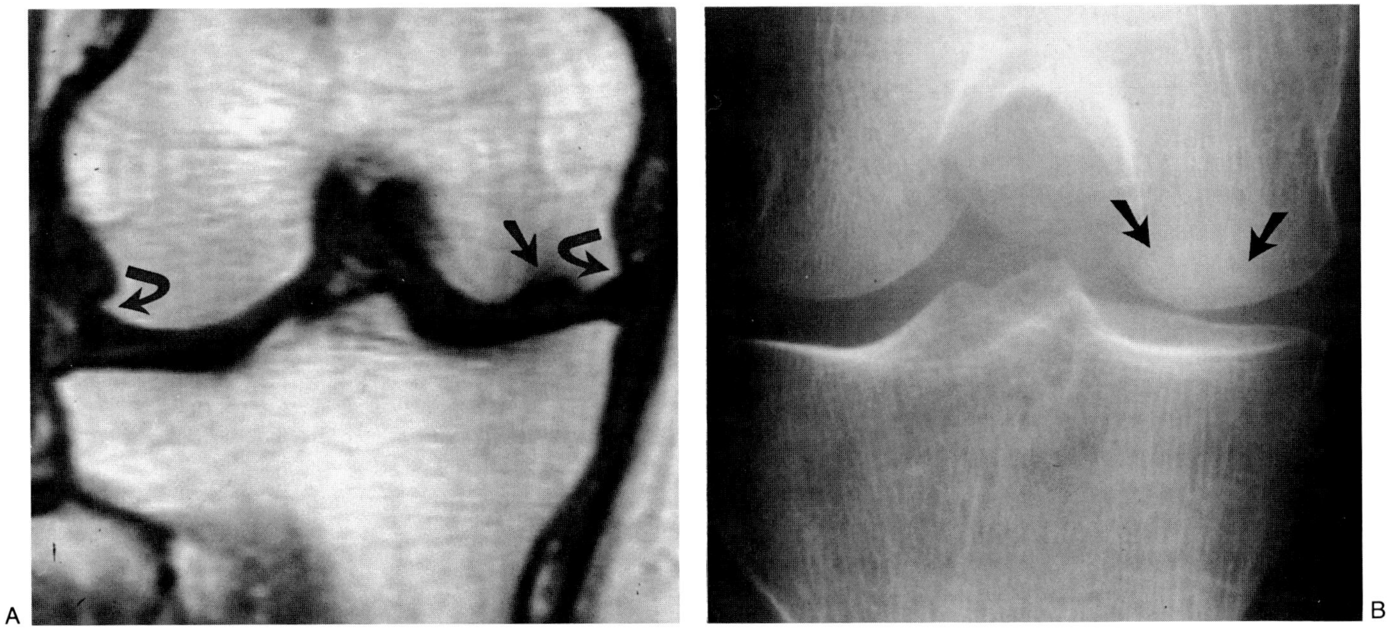

FIG. 8.12. Type 3 trabecular injury in a 73-year-old with chronic right knee pain. **A:** Coronal T1-weighted (TR/TE 800/20) image reveals profound signal loss in the trabecular bone immediately deep to the subchondral plate of the medial femoral condyle (*straight arrow*). Signal from cancellous bone is otherwise normal. There was no increased signal in the abnormal area on corresponding T2-weighted image (not shown). Note the small osteophytes at the margins of the condyles (*curved arrows*). **B:** Plain radiograph shows sclerosis of the juxtaarticular bone of the medial femoral condyle (*arrows*).

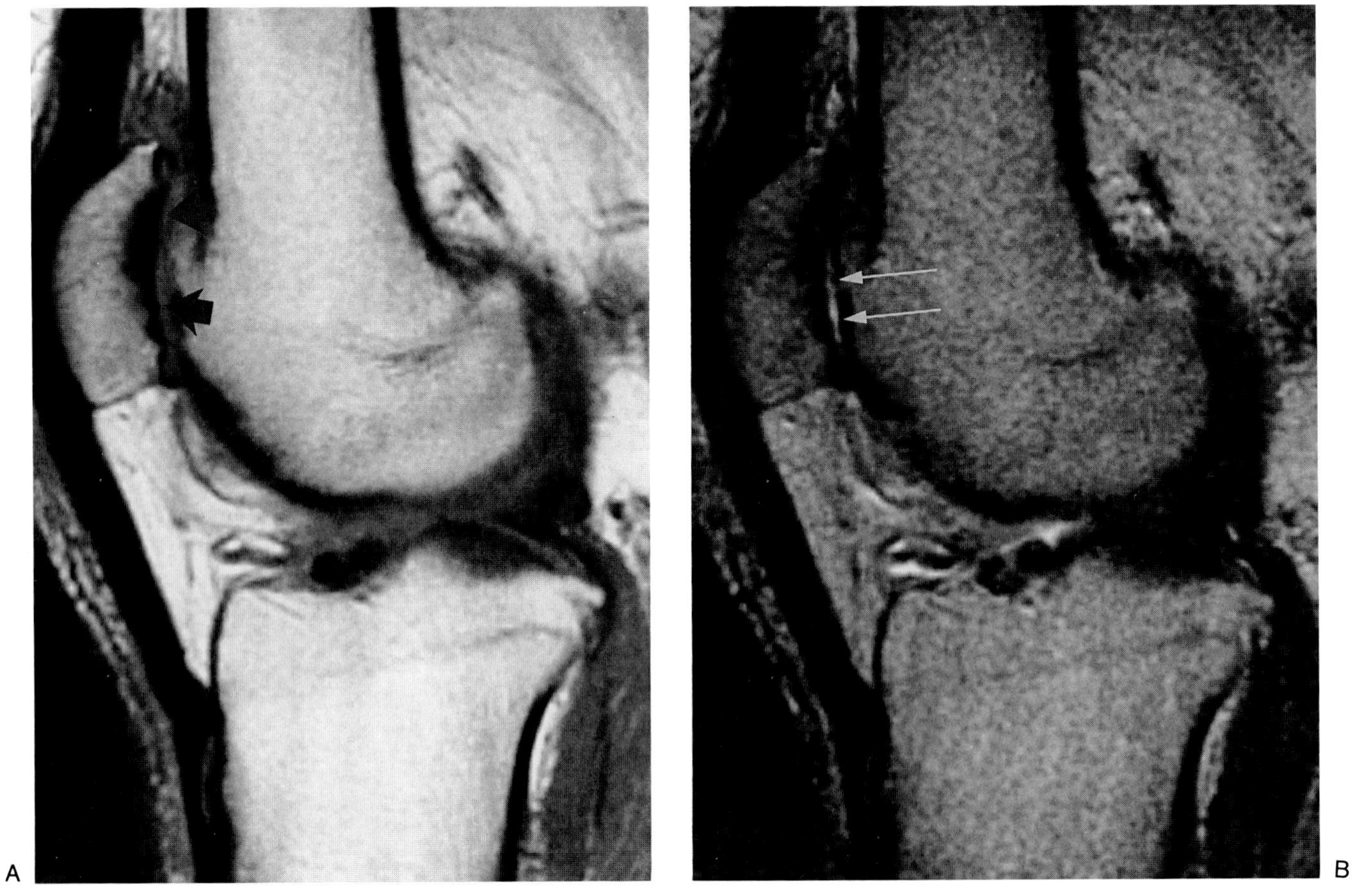

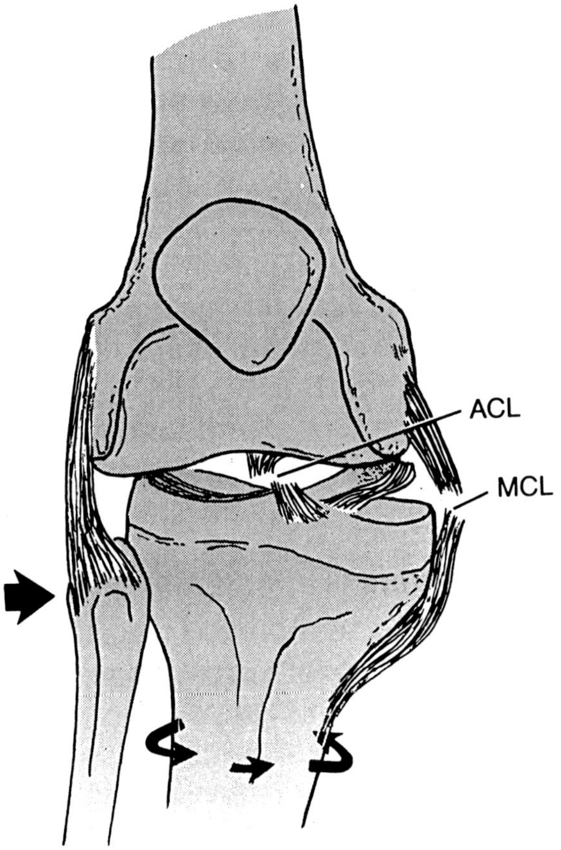

FIG. 8.14. Mechanism of injury, lateral trauma. Valgus forces result in anterior and medial deflection of the tibia, compressing the bones of the lateral compartment. The ACL and MCL can rupture if force is great enough. Medial collateral ligament rupture relieves compression upon the lateral compartment. (Reproduced with permission, ref. 102.)

counted for two of the patients with suspicious plain radiographs). Not surprisingly, abnormalities confined to the trabecular bone were not detected at arthroscopy. Linear stress fractures extending to the metaphyseal cortex, as well as a high percentage of impacted osteochondral fractures, were also arthroscopically occult. Resolution of trabecular lesions was noted in the small number of patients with bone bruise or amorphous stress fracture that underwent follow-up MR imaging between 3 and 8 weeks following initial evaluation.

The work of Vellet, Marks, and colleagues (17) included 120 patients with posttraumatic hemarthrosis who underwent MR imaging and arthroscopy. This series also showed that occult bony injury was common (occurring in 72% of patients), and was additionally important in including correlation with arthroscopy and follow-up MR scanning. The authors described what they considered to be several different types of occult fractures, all having diminished signal intensity on T1-weighted image and increased signal intensity on T2-weighted images. The *reticular* and *geographic I* lesions consisted of globular regions of signal change, more coalescent centrally, distinguished by separation from the subchondral plate (reticular), or contiguous with the subchondral plate (geographic I). By description, then, these lesions—especially the reticular type—would equate to the type 1 changes of Lynch and Crues. The *geographic II* and *linear* lesions included less extensive marrow changes surrounding crescentic or linear bands of diminished T1 signal, progressing on a spectrum of imaging changes to nondisplaced osteochondral fracture. Continuing on this same spectrum, *osteochondral* fractures could be identified that were either radiographically occult or overt. Lastly, *impaction* fractures were always seen in association with reticular or geographic lesions. Interestingly, radionuclide scans were abandoned early in this series when correspondence to MR findings was noted in each case.

Despite the number of subtypes described, the vast majority of lesions in Vellet's series were of the reticular and geographic categories. There was a frequent association, particularly in the follow-up group, between geographic lateral femoral condyle lesions and reticular lesions in the corresponding tibial plateau. At initial arthroscopy, osteochondral pathology corresponding to these MR findings was not identified in any patient, yet at follow-up imaging, two-thirds showed significant sequelae, including cartilage thinning or defects, osteosclerosis, osteochondral fractures, and impaction fractures, which correlated with persistence of clinical symptoms in most cases. In only a small fraction (4 of 15 patients having plain films) were any abnormalities detectable on follow-up radiographs.

A recent series (46) is pertinent in that it confirms the prevalence of trabecular fracture in patients with rupture of the ACL. Bony injuries were seen by MR in 85% of 75 adults. There was no histologic evidence of trabecular fracture in the one patient from whom material was available. However, this is not surprising considering the elusive nature of trabecular fractures by routine light microscopic observation; special tech-

←

FIG. 8.13. Type 3 injury with chondromalacia. In cases of advanced chondromalacia, the cartilage is markedly thinned to essentially absent (TR/TE 2,000/20). **A:** Notice the decrease in joint space width and the presence of irregular decreased signal intensity adjacent to the bony end plate within the patella corresponding to sclerosis (*arrows*). **B:** A thin line of joint fluid (*arrows*) outlines the markedly thinned patellofemoral joint space (TR/TE 2,000/80). This patient had grade 4 chondromalacia.

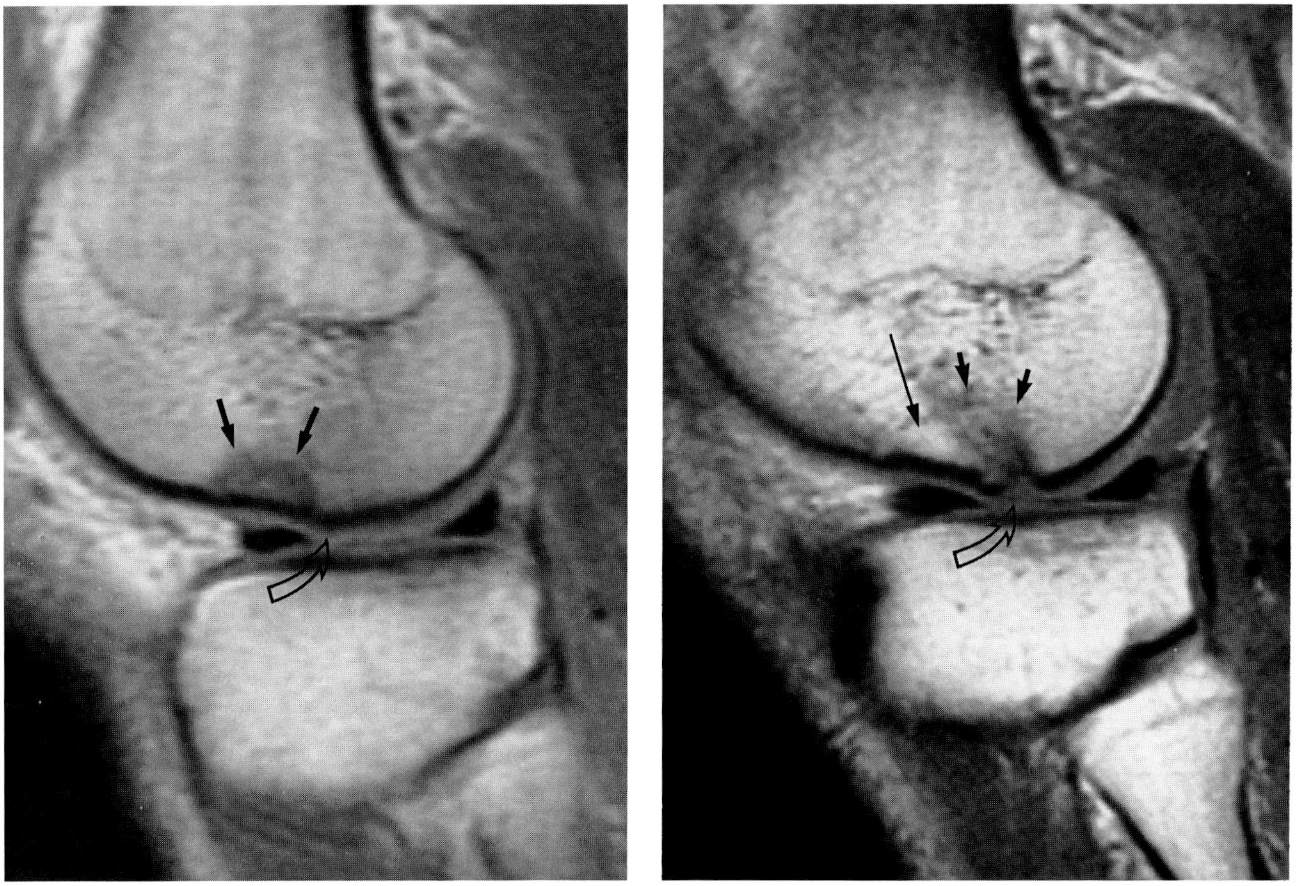

FIG. 8.15. Progression of type 2 bony injury. **A:** Sagittal intermediate-weighted (TR/TE 2,000/20) image following trauma shows juxtaarticular signal loss within the trabecular bone of the lateral femoral condyle (*straight arrows*). The subchondral plate maintains its characteristic low signal, but the abrupt contour change (*curved arrow*) indicates injury. **B:** Persistent pain led to follow-up image 2 months later (TR/TE 2,000/20) which demonstrates return to normal of the focus of low marrow signal (*long arrow*) in the type 2 injury shown in (A). Abnormal intermediate signal now crosses the subchondral plate at the site corresponding to the original contour change (*curved arrow*). There is now trabecular signal loss adjacent to the end-plate injury (*short arrows*), posterior to the original lesion. The bony lesion extended as persistent weightbearing on a deformed end plate with underlying injured trabeculae increased stress on neighboring trabeculae, causing further fracture.

special techniques, infrequently used in clinical laboratories, must be employed (11).

In assessing the significance of the type 1 and 2 cancellous bone abnormalities, it is useful to extract the features common to the several studies summarized. First, these trabecular bone injuries are extremely common. A traumatic origin seems highly likely considering the available clinical history, the frequent association with hemarthrosis and significant ligamentous injury, the location of the lesions (typically beneath the weightbearing surfaces of the lateral condyle and plateau in valgus stress injuries) and the absence of such abnormalities in knees not subjected to trauma. The loss of T1 signal has been interpreted as representing compaction of the trabecular struts, possibly combined with the presence of increased fluid elements (probably a combination of hemorrhage, osteoid, and edema) and cellular components within the marrow space concerned with repair of the cancellous bone (8,50,51). This cellular and humoral response likely accounts for the T2 prolongation observed in the lesions most closely correlated with trauma. The absence of high signal on long TE images, as in the type 3 injury, represents an inactive, sclerotic, more chronic lesion which does not contain the high free water content of the more acute type 1 and 2 lesions. The trabecular bone lesions seem to represent a heterogeneous class of injury which can be identified by MRI before radiographically and/or arthroscopically detectable sequelae become apparent.

The benefit of MR scanning of the posttraumatic knee can be maximized through identification of imaging features which frequently antedate significant osteochondral sequelae. Type 1 lesions overlying the

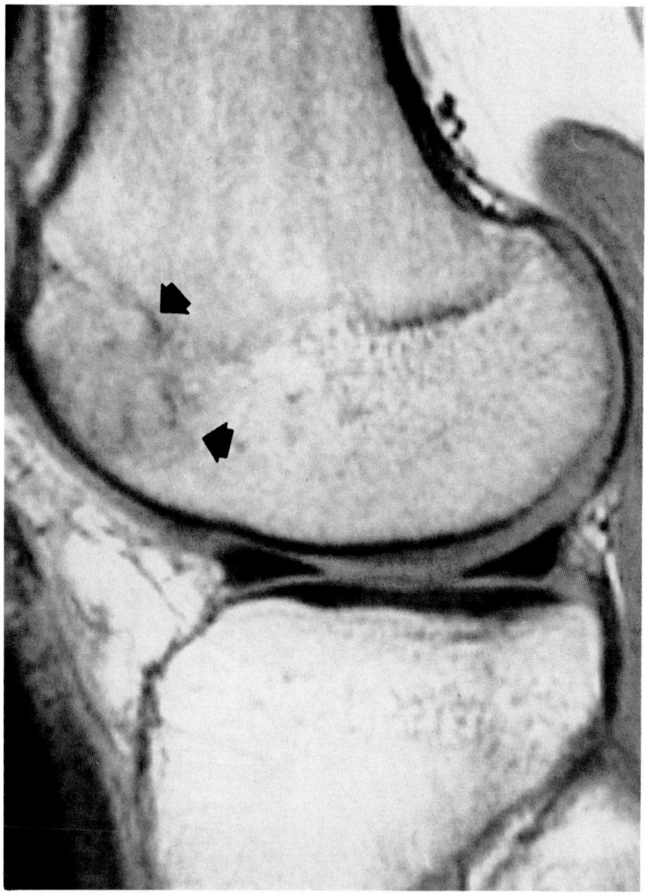
A

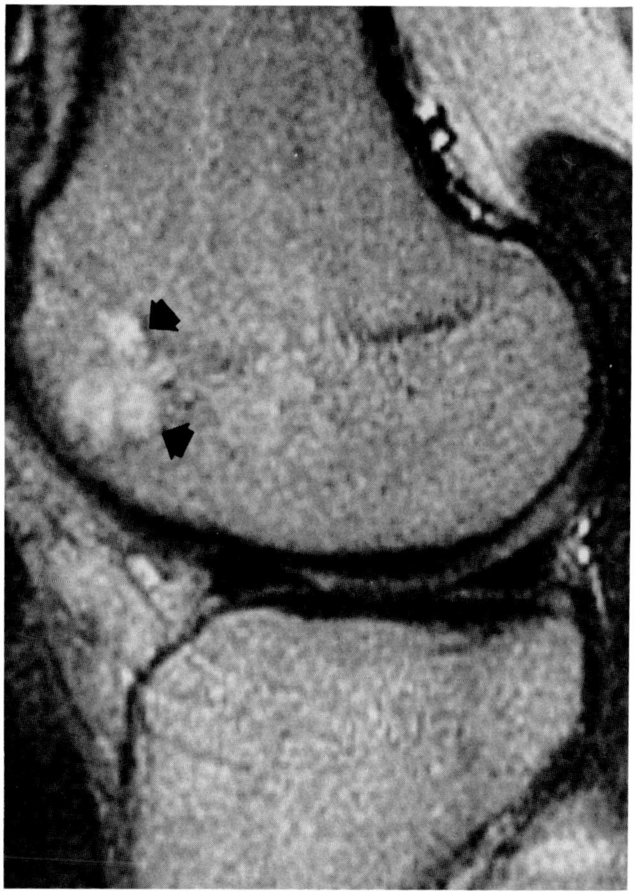
B

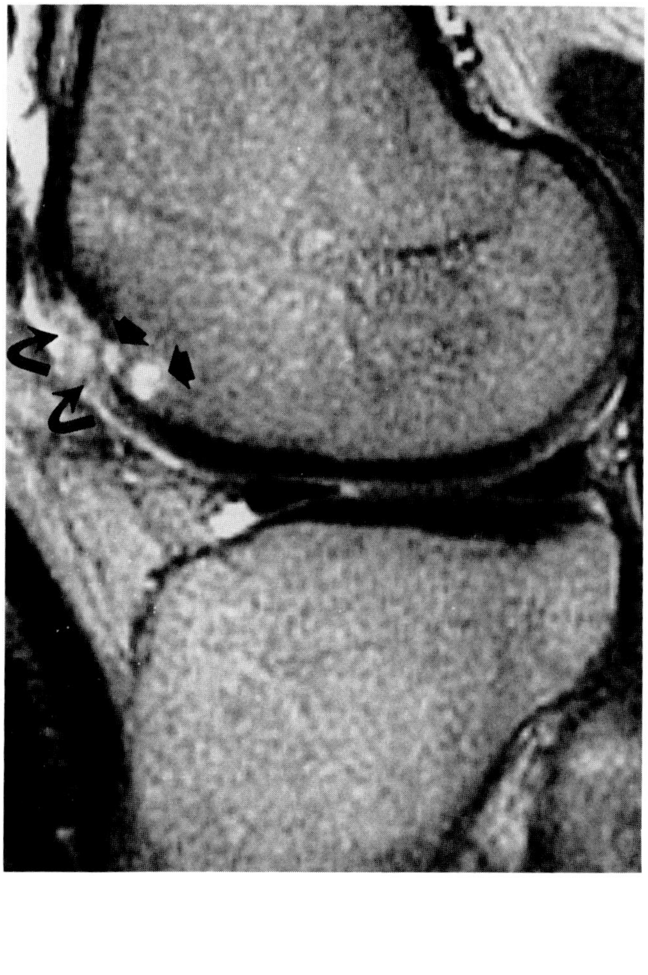
C

FIG. 8.16. Trabecular injury with progression to subchondral collapse in a 28-year-old professional basketball player who sustained a fall, landing on the flexed knee. **A:** Sagittal image (TR/TE 2,100/20) obtained the first week following trauma demonstrates amorphous trabecular signal loss in the anterior lateral femoral condyle (*short arrows*) with an intact subchondral plate. **B:** Corresponding foci of increased signal are seen (*short arrows*) (TR/TE 2,100/80). Note that signal changes abut the subchondral plate, although the appearance is that of type 1 injury. **C:** The patient had persistent, worsening pain. Sagittal image (TR/TE 2,117/80) from follow-up scan 6 months after the initial injury reveals extensive interruption of the subchondral plate (*curved arrows*), as well as more coalescent high signal in the trabecular bone (*short arrows*), some of which may result from communication of joint fluid across the subchondral impaction fracture. This injury is now type 2.

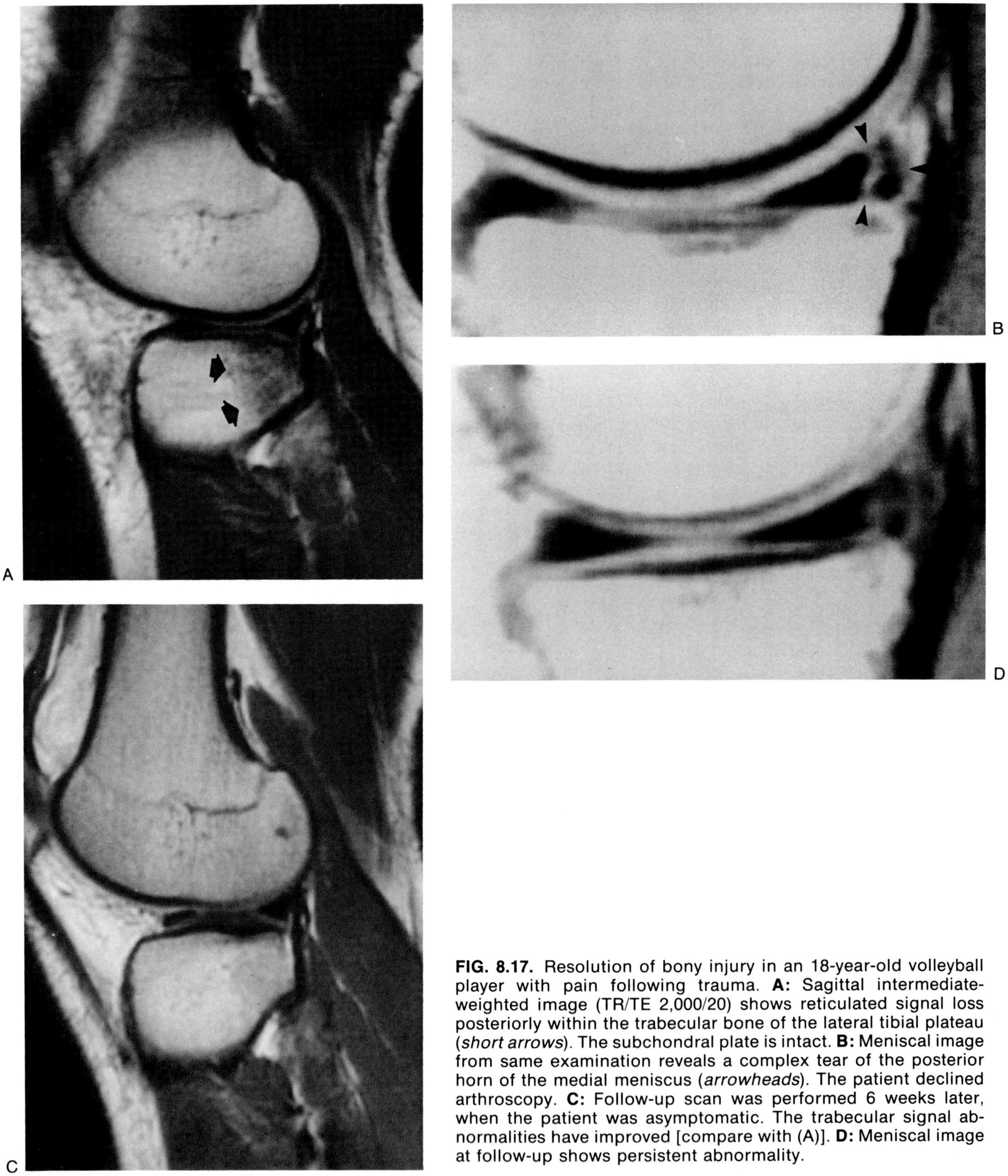

FIG. 8.17. Resolution of bony injury in an 18-year-old volleyball player with pain following trauma. **A:** Sagittal intermediate-weighted image (TR/TE 2,000/20) shows reticulated signal loss posteriorly within the trabecular bone of the lateral tibial plateau (*short arrows*). The subchondral plate is intact. **B:** Meniscal image from same examination reveals a complex tear of the posterior horn of the medial meniscus (*arrowheads*). The patient declined arthroscopy. **C:** Follow-up scan was performed 6 weeks later, when the patient was asymptomatic. The trabecular signal abnormalities have improved [compare with (A)]. **D:** Meniscal image at follow-up shows persistent abnormality.

weightbearing subchondral plate in athletes with pain on standing may be an indication to restrict weightbearing until early healing of the injured trabeculae has occurred. MR findings associated with poorer prognoses include an interrupted subchondral plate in type 2 lesions, especially if the cortical abnormality is associated with juxtacortical trabecular signal abnormality, as in the Vellet and Marks geographic lesions. Other types of lesions such as amorphous signal change confined to the metaphyseal and/or epiphyseal marrow space without pain on weightbearing (Lynch and Crues type 1, Mink and Deutsch bone bruise, Vellet and Marks reticular) tend to resolve without sequelae. Profound signal loss adjacent to the subchondral bone (type 3) usually represents chronic subchondral sclerosis and is stable or slowly progressive. The pathophysiology of trabecular bone injury as discussed in the previous section is helpful in clarifying these empirical observations and determining their clinical significance.

Zones of MR signal abnormality were initially presumed to represent "bone bruises" which uniformly resolved without consequence (32). We now believe that this applies only to a subgroup of these lesions. It is likely that injuries such as the Lynch and Crues type 1 and the Vellet and Marks reticular represent trabecular microfractures without involvement of the subchondral plate or detectable cartilage defect. Frequently, they are found in association with ligamentous injuries which themselves dictate periods of activity limitation after which the trabecular injuries have been noted to heal (31,45,46). In other cases there are associated osteochondral, impaction, or complex frank fractures elsewhere about the knee which mandate a period of restricted weightbearing for the patient. There are other cases of isolated trabecular fracture in which progression to collapse of the subchondral plate has been observed (17,29,52). This type of complication would be anticipated by understanding the loadbearing function of the intact trabecular meshwork, described in the previous section.

Even when uncomplicated by collapse or by cartilage damage, trabecular injury is important to recognize because it can be the source of the patient's presenting symptoms. There is growing recognition of the prevalence of MR signal abnormalities of the menisci in asymptomatic individuals (53–55), and at the authors' facility, several posttraumatic knees have demonstrated coexisting meniscal tears and trabecular injuries; symptoms resolved as the bony injuries healed without repair of the menisci (Fig. 8.17).

While additional long-term clinical studies are necessary to fully understand the relationships between MR findings and prognosis in patients with bony trauma, the series presented thus far seem to indicate that contiguity of trabecular signal abnormalities with the subchondral plate or fracture of the plate itself (even if radiographically occult) places the patient at increased risk for later compromise of the joint surfaces. At the present time it would seem prudent to place patients who manifest a trabecular injury to a weightbearing region with pain on standing on a program of limited weightbearing for 4 to 6 weeks to allow initiation of trabecular healing. [This temporal endpoint is chosen in light of animal data (6,11) that place bone at its most weakened state approximately 4 weeks following injury.]

OSTEOCHONDRAL DEFECTS (OSTEOCHONDRITIS DISSECANS)

Osteochondral lesions [osteochondritis dissecans (OCD), transchondral fractures] refer to chondral and/or transchondral fragmentation of a portion of the articular surface (13,56,57). While first described by Pagct in 1870, Konig coined the term osteochondritis dissecans in 1887 and claimed that it was one of three mechanisms leading to loose bodies of the knee joints (58). The term osteochondritis dissecans is no longer the preferred term for this condition because it is not descriptive of the etiology and is consequently confusing (59). Osteochondral lesions of the femoral condyles are relatively common articular lesions in adolescents, with peak incidence occurring around age 15 and most patients being diagnosed before the age of 18 (60,61). Boys are affected with greater frequency than girls (62), and the right knee is involved slightly more than the left with up to 30% of cases being bilateral (57,63). In Aichroth's series the medial condyle was involved in 85% of cases and the lateral condyle was involved in 15% of cases (61). Three common areas of involvement of the medial condyle include (a) *classic*, affecting the lateral nonweightbearing portion of the medial femoral condyle; (b) *extended classic*, involving additionally a portion of the weightbearing surface; or (c) *infracentral*, solely involving the articular weightbearing surface (61).

Lateral condylar lesions typically involve the weightbearing anterior portion of the condyle (61). The medial facet of the patella and the tibial plateau may rarely be involved (13,64,65). Patients may be asymptomatic or present with pain, limitation of motion, clicking, locking, and/or swelling (13).

Several theories have been put forth to explain the pathogenesis, but it is generally accepted that it is traumatic in origin (13,57,59,63). Aichroth demonstrated significant injury to the knee in 46% of cases, and of these, 60% participated actively in sports (61). It is not always possible to accurately elicit a history of trauma, and some have put forth alternative etiologies including avascular necrosis (57,61,66). Cases of familial osteochondral lesions occurring in dysplastic syndromes have been reported (13,67,68). Osteochondral lesions are the end result of a spectrum of shearing, rotary, or impaction forces occurring at the susceptible locations, especially when superimposed on a predisposing condition which weakens the bones' ability to resist trauma (13,63,69).

Until recently, plain radiographs, arthrography/arthrotomography, computed tomography (CT), and arthroscopy were used to characterize and follow osteochondral lesions. Plain radiographs are sensitive in detecting most osteochondral lesions, but are not accurate in staging the lesions for therapeutic purposes (70). Arthrography and arthroscopy are helpful in staging, but are not as desirable for follow-up evaluations because of their invasive nature. CT may detect the osseous lesion, but has not been proven adequate in therapeutic staging. Bone scintigraphy is useful in staging, but is nonspecific as a diagnostic tool (70). In addition to detecting the bony lesion, MRI can detect the presence of fluid or bone marrow edema adjacent to the osteochondral fragment thus providing a clue to the stability of the lesion which is essential in selecting surgical versus conservative management. MRI may also be used for following the course of the disease over time since it is noninvasive, does not utilize ionizing radiation, demonstrates superior soft tissue contrast resolution, and allows direct visualization of cartilaginous surfaces.

Classically, OCD is a plain film diagnosis. However, those lesions detected on plain films probably represent the end of a spectrum of osteochondral lesions. MRI can accurately detect the less severe lesions which may be radiographically occult (71). By detect-

A

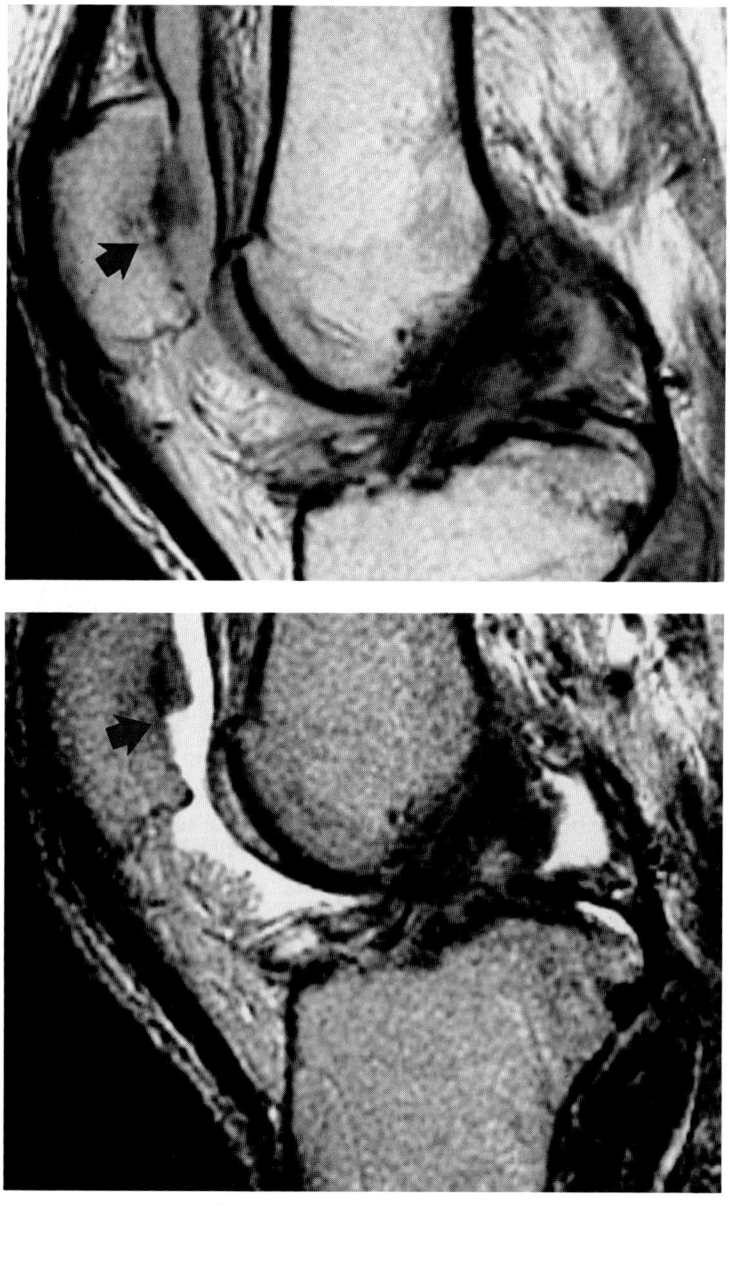

B

FIG. 8.18. Cartilage fracture. **A:** An abrupt signal change in the region of the patellar articular cartilage is seen at the site of the cartilage fracture (*arrow*) (TR/TE 2,000/20). **B:** The conspicuity of the cartilaginous fracture is increased through the "arthrogram effect" of adjacent high intensity joint fluid (*arrow*) (TR/TE 2,000/80). This fracture involves the hyaline cartilage of the medial articular facet, sparing the underlying bone.

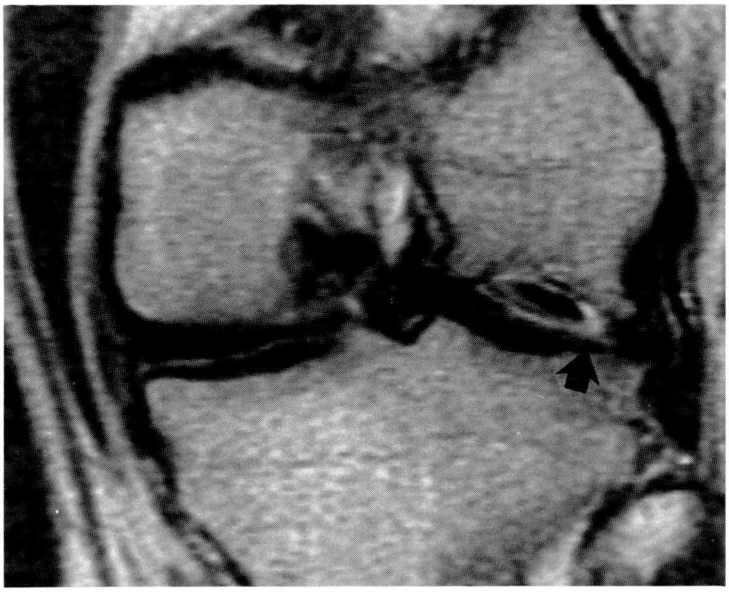

FIG. 8.19. Osteochondral defect. Note the transchondral osseous defect demonstrating increased signal intensity between the osseous fragment and medullary bone as well as through the overlying articular cartilage (*arrow*) (TR/TE 2,000/80).

ing bony injury at an earlier stage, restrictions of weightbearing may allow earlier bony healing before a complete fracture of the subchondral plate has occurred. Maintaining the integrity of the support of the subchondral plate may protect the overlying cartilage from injury. It is the irreversible injury to the articular cartilage which leads to degenerative joint disease and chronic disability. In later stages of the disease, MRI is also helpful in the evaluation of cartilaginous integrity and osteochondral fragment stability in those lesions that are visible on plain radiographs (70). Most osteochondral lesions may be the end stage bony reaction to chronic recurrent trauma; however, some lesions may be the result of a single traumatic episode. In such situations MR is useful in the evaluation of associated traumatic injuries of the knee joint, such as ligamentous and meniscal injuries.

The MR appearance is varied and can range from a chondral fracture alone (Fig. 8.18), a bony lesion with intact overlying cartilage, or injury to both the cartilage and the bone (Fig. 8.19). In classic OCD a bony fragment is present which may be stable and nondisplaced, loose *in situ*, or displaced. Both nondisplaced lesions and loose intraarticular osteochondral fragments may demonstrate either diffuse decreased signal intensity or increased signal intensity centrally surrounded by a black rim on T1-weighted images. The increased signal intensity represents marrow fat, whereas the low signal centrally is from fibrosis, edema, and loss of fatty marrow elements. The low intensity rim seen in both instances represents fibrosis, granulation tissue and ossification. Histologically, some resected lesions demonstrate viable hyaline articular cartilage overlying variable amounts of acellular necrotic bone (72).

The fate of the osteochondral fracture is varied. The fragment may remain in place and heal, or *in situ* lesions may become loose but nondisplaced with the fragment being ballottable arthroscopically (70). In these cases, the articular cartilage is often intact (70). In more advanced cases, fragmentation (Fig. 8.20) or loose body formation may occur.

Determination of lesion stability has important treatment implications. MRI findings associated with stability include a very small sized lesion and absence of an increased signal intensity rim surrounding the fragment adjacent to the donor site on T2-weighted images. Signs of instability include fragment size greater than 1 cm, affinity for bone-seeking radionuclides, and high

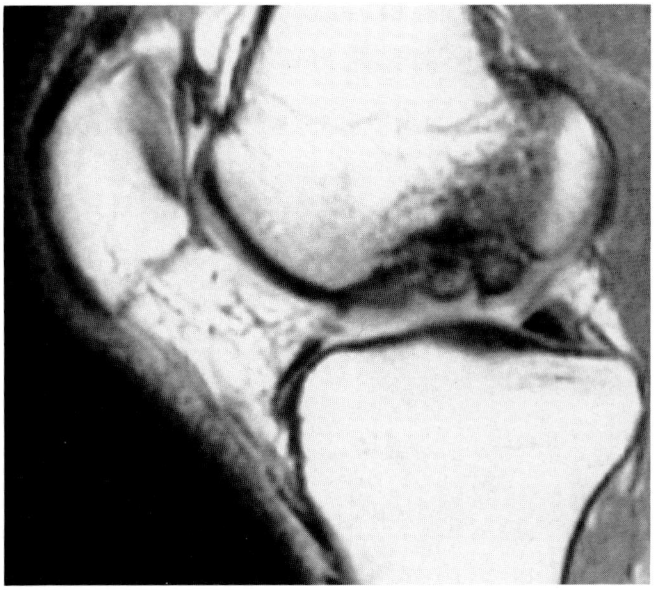

FIG. 8.20. Fragmented osteochondral defect. In this advanced case, the affected osteochondral segment has become fragmented (TR/TE 2,000/20).

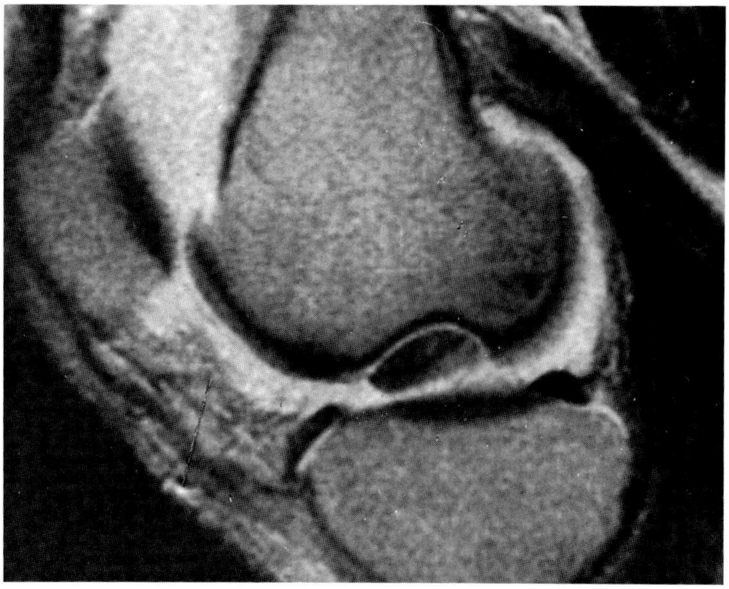

FIG. 8.21. Osteochondral defect. Linear increased signal intensity between the fragment and the donor site is shown on this T2-weighted sagittal image (TR/TE 2,000/80). This finding is associated with a high probability of instability.

signal intensity on T2-weighted images deep to the articular surface adjacent to the donor site (59,70). This increased signal intensity may be linear and discrete (Fig. 8.21) or more diffuse (Fig. 8.22) and has been associated with either synovial fluid insinuated between the fragment and the donor site and/or granulation tissue (59). If the discrete cortical fragment is nondisplaced, it is important to evaluate not only those signs of stability already mentioned but also the integrity of the overlying articular cartilage (71). Disruption of the overlying cartilage is associated with instability (70). Overall, MRI is 92% sensitive and 90% specific in the limited case studies where signs of stability were evaluated (59,70).

The donor site is often irregular in appearance and may demonstrate an amorphous band of decreased signal intensity on all pulse sequences corresponding to the sclerosis often seen on plain films in advanced cases (Fig. 8.23). Premature osteoarthritis is the end result in almost all cases where irreversible damage to the articular cartilage has occurred (13,56,57). The prognosis however, may be dependent on the age of the patient. Young children with thick articular cartilage tend to heal the articular cartilage injuries and remodel the bony defects. Consequently, the prognosis worsens with increasing age of the patient.

The high degree of specificity and sensitivity of MRI in staging bony fragment stability and cartilage integ-

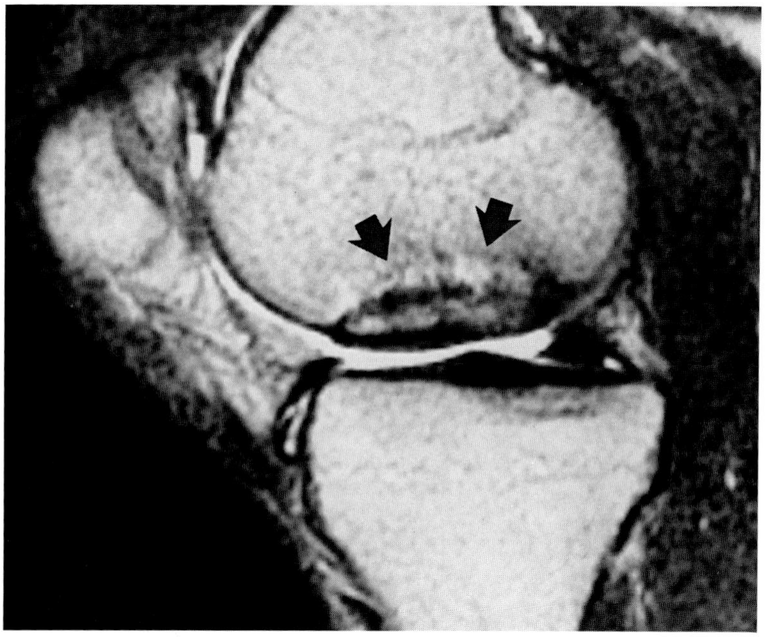

FIG. 8.22. Osteochondral lesion. Diffuse increased signal intensity in the medullary bone adjacent to the bony fragment can be seen on T2-weighted images (TR/TE 2,000/20) with unstable lesions (*arrows*).

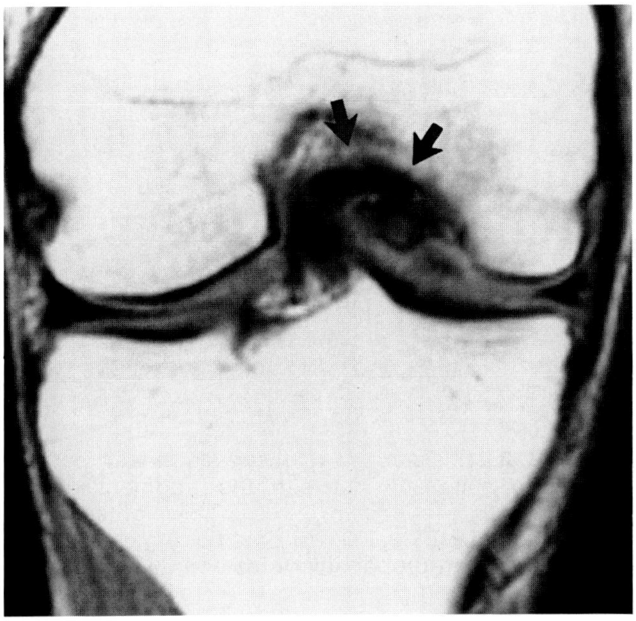

FIG. 8.23. Sclerotic osteochondral lesion. Band of decreased signal intensity is seen involving the donor site adjacent to the osseous fragment (TR/TE 2,000/20). The dark band (*arrows*) corresponds to sclerosis seen on plain films (not shown).

rity helps define the role of MR in the treatment planning of these patients. Stable lesions are usually treated conservatively by joint motion limitation and avoidance of weight bearing for 6 to 12 weeks (70,73). Treatment is generally surgical in those cases where displaced or unstable nondisplaced fragments are present. Surgical removal, internal fixation, or drilling are options available to the surgeon depending on the status of the fragment and the clinical circumstances unique to each patient (56,61,70,74). Free fragments and *in situ* loose fragments with interrupted overlying cartilage are frequently removed.

Some authors suggest that MRI should be reserved for those cases where plain radiography and bone scintigraphy are indeterminate (57,70). However, due to the fact that some lesions are not well seen on plain films, bone scintigraphy lacks diagnostic specificity, and the integrity of other intraarticular structures must be evaluated frequently, MRI is often the diagnostic imaging study of choice in the evaluation of suspected osteochondral lesions. Accurate assessment of the status of the bony lesion, the overlying articular cartilage, and associated internal derangement by a noninvasive method allows informed treatment decisions to be made.

OSTEONECROSIS

Osteonecrosis is defined as the presence of osteocyte death. Empty lacunae are the hallmark on histological examination. Multiple causes are known, including corticosteroid excess, organ transplantation, pancreatitis/alcoholism, caisson disease (hyperbaric states), increased pressure within the marrow space (e.g., leukemia, hemoglobinopathy, glycogen storage disease), radiation, and avascular states, including damage to small blood vessels as in collagen vascular disorders. *Idiopathic osteonecrosis* includes those cases of appropriate clinical, imaging, and histologic findings of osteocyte death without a definable inciting cause. *Avascular necrosis* is a more specific term for death of osteocytes due to primary loss of the nutrient vascular supply to the bone. This may occur from venous hypertension and secondary decreased perfusion as is thought to occur in some forms of osteonecrosis of the hip (75,76) or in true bone infarcts. *Idiopathic* and *avascular osteonecrosis* are discussed in Chapter 9. *Spontaneous osteonecrosis* and *posttraumatic osteonecrosis* are probably bony responses to trauma and will be discussed below.

Spontaneous osteonecrosis was originally described and is most commonly seen in older oseopenic women (77), but has also been recognized frequently in middle-aged men (78,79). Spontaneous osteonecrosis is suspected in patients who have acute onset of knee pain, usually medial in location, in the absence of significant trauma. The medial femoral condyle is the most common site of involvement, although patients with predisposing conditions for osteonecrosis in general (corticosteroid therapy, organ transplantation, lupus erythematosus) frequently have lateral condylar or bicompartment disease (80,81). Patients typically experience continuous pain, frequently with nocturnal exacerbation and only slight relief at rest. In many patients pain resolves within weeks to months. A minority progress to bony collapse and severe degenerative changes, and this discussion will focus on that subset of individuals.

Initially, plain radiographic findings may be normal, but MRI may show characteristic abnormalities. With persistent symptoms, subtle condylar flattening and sclerosis occur later (weeks to months) in the course of the disease (77); appearance of these findings indicates significant progression and correlates with a poor prognosis. Later, a subchondral lucency can be seen, size measurements of which have been used to determine the nature of surgical intervention (77,79,80,82). Finally, progression to collapse of the subchondral plate can be observed. Particularly in the earlier stages, scintigraphy sensitively confirms location of the lesion, which avidly adsorbs pyrophosphate (79,81,83).

The disorder has been likened to osteonecrosis of the hip, which was believed also to occur spontaneously. A vascular etiology has been proposed by many investigators (75–77,80). Histologic study of knee pathologic material in patients with symptoms characteristic of spontaneous osteonecrosis by Ahlback et al. (77) revealed abundant reparative healing response with sparse evidence of dead bone. These findings contrast with avascular necrosis of the hip where necrosis is readily demonstrated histologically. As described at the beginning of this chapter, histologic studies which established the existence of trabecular microfractures frequently revealed bone necrosis in association with both early and late posttraumatic reparative response. Furthermore, the elusive nature of trabecular fractures by light microscopy leads to the intriguing possibility that the microscopic findings originally described by Ahlback may represent late-stage reparative changes following an insult which included trabecular fracture as a primary or subsequent feature. Ahlback initially discounted trauma as an etiology primarily because of the absence of acute trauma in the clinical history and the lack of abnormality on plain radiographs (77).

It is commonly recognized that osteoporosis can weaken the loadbearing capacity of bone such that fracture can result from normal activity (insufficiency fracture); these are typically radiographically occult (*vida infra*). Insufficiency fractures have an avidity for phosphate tracer similar to the lesions of spontaneous osteonecrosis. As will be outlined below, the MR characteristics of spontaneous osteonecrosis closely resemble those of traumatic trabecular fractures. Histologic support for a traumatic etiology for spontaneous osteonecrosis is in fact found in Ahlback's original description (77), in which the abnormal regions of cancellous bone did not exhibit the empty lacunae of necrosis, but "lively synthesis of bone [as] a consistent finding." We suggest that these observations may have been a function of the late stage of the disease during which patients had abnormal radiographs and presented for surgical treatment. Considering the commonality of imaging and histologic findings with those of traumatic conditions, and the frequent occurrence of this entity in osteoporotic individuals, we suggest that minor, repetitive, unrecognized trauma could result in trabecular insufficiency fracture and thus to the symptoms and sequelae of spontaneous osteonecrosis (71). Others have suggested that the location of the typical medial condylar lesion, overlying the medial meniscus, and the frequent sudden onset of symptoms (unusual in spontaneous osteonecrosis of the femoral head) points to a traumatic origin (82,84).

Spontaneous osteonecrosis of the knee may exhibit MR signal changes of bone marrow edema that are also seen in osteonecrosis of the femoral head (including so-called transient osteoporosis or transient bone marrow edema of the hip). Many of its features, however, resemble those seen in traumatic injury to the trabecular bone. These include focal loss of medullary signal on short TE images, usually abutting the subchondral plate. The abnormality is most frequently localized to the weightbearing portion of the medial femoral condyle, and can involve the medial tibial plateau. Lateral compartment involvement is least common, probably because most forces in weightbearing are transmitted through the medial compartment (see Chapter 4). Trabecular signal loss resembles most closely the type 3 bony injury in those patients with late disease (Figs. 8.11 and 8.12). Other lesions of spontaneous osteonecrosis, however, are associated with surrounding areas of reticular or amorphous increased signal on the corresponding T2-weighted images, making them essentially indistinguishable from the types 1 or 2 injuries (*vida supra*). These probably represent early lesions. Some patients exhibit a "double line sign" (Fig. 8.24) on T2-weighted images [although this feature is less frequently seen in osteonecrosis of the knee than hip (81)]. Histologic correlation indicates that the curvilinear high signal band represents the advancing edge of hyperemic granulation tissue against the regressing front of devitalized trabeculae, fibrosis, and cellular debris, which is seen as the low signal companion line (81,85). Presence of the "double line sign" in the knee may be relatively specific for avascular necrosis and, when seen outside of the femoral head, suggests an underlying predisposing systemic etiology (steroid therapy, collagen vascular disease, etc.). In the hip this may be seen in the absence of systemic disease because the venous drainage of the femoral head must traverse the length of the medullary space of the femoral neck. This predisposes the head to vascular compromise (86–88).

Bjorkengren found that prognosis could best be predicted on T2-weighted images because the presence of a low-signal subchondral focus correlated with a poor clinical outcome in their series of 16 patients (83). It should be noted that their scans were performed at 0.3 T. They demonstrated that short TE images were best at depicting the extent of the lesions of spontaneous osteonecrosis, as compared with plain radiographs. This comparison was made because the percentage of cross-sectional area involvement of the articular surface has been used as the major criterion for surgical intervention (79,80). Although this and other size criteria continue to be used, there is evidence that once the lesions of spontaneous osteonecrosis have progressed such that surgical management is indicated, prosthetic joint replacement is the only effective option (75). The previously favored drilling procedures have not been as beneficial as in osteonecrosis of the hip (75,80,83). If a traumatic etiology is accepted, this is not a surprising observation.

Aglietti et al. showed that early detection of osteonecrosis is important in clinical outcome (80). Patients

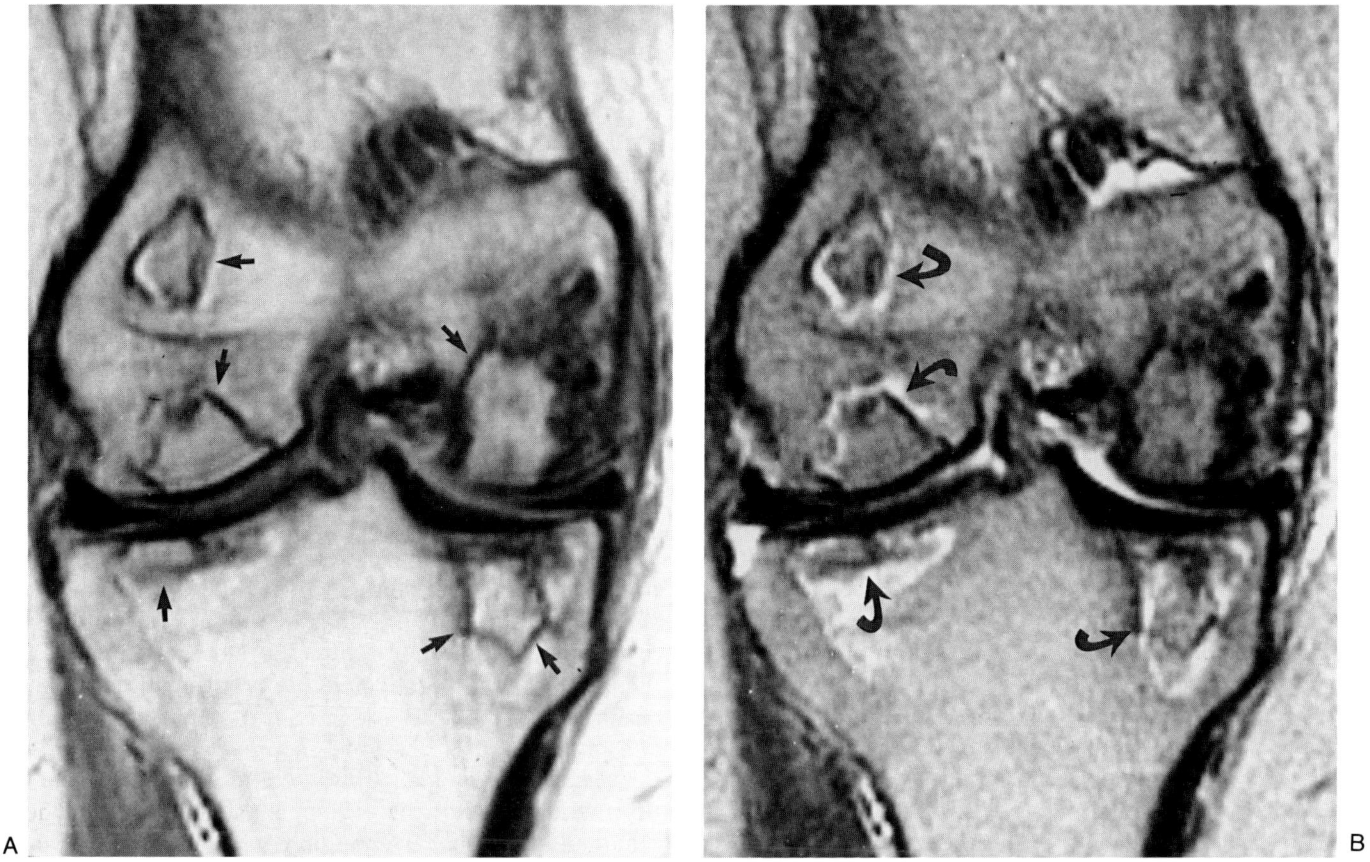

FIG. 8.24. Osteonecrosis. **A:** Coronal intermediate-weighted image (TR/TE 2,000/20) demonstrates numerous well-circumscribed regions of trabecular bone bounded by diminished signal (*arrows*). **B:** The T2-weighted image (TR/TE 2,000/80) shows regions bounded by a double line of high and low signal (*curved arrows*). This "double line sign" is a specific sign of avascular necrosis and suggests a systemic etiology. The patient is a 25-year-old ethanol abuser who had pain for 7 weeks following minor trauma.

with lesions that were discovered early received conservative therapy and 80% had at least a satisfactory long-term result. Lesions discovered later were treated with arthrotomy and curettage, tibial osteotomy to unload the medial compartment, or joint replacement. Since MRI is more sensitive than plain radiographs to early changes of spontaneous osteonecrosis, it should be valuable in permitting a higher proportion of patients to be managed conservatively. Pollack and collaborators showed MR to be additionally useful in revealing unexpected bicompartment or bilateral involvement (81). Insall et al. (76) summarized the differentiation from osteochondritis dissecans, since both entities commonly involve the medial femoral condyle. Most obvious is patient age, which is usually adolescent or young adult in OCD. OCD is typically located toward the notch of the knee, rather than on the central weightbearing surface of the condyle. OCD is more frequently associated with a traumatic history and the onset of pain is (somewhat paradoxically) more insidious. Loose body formation is far more common in OCD, while joint space narrowing and rapid progression of degenerative changes at the joint surface favors osteonecrosis.

While the term *posttraumatic osteonecrosis* has been used to describe instances of osteonecrosis following discrete trauma, it has most commonly been described in the femoral head, talus, and carpal scaphoid. We know of no accounts of this entity in the knee. It may, however, be appropriate to apply this terminology in those cases involving the development of severe knee pain and imaging findings compatible with osteonecrosis weeks to months following trauma. Rather than as sequela to frank fracture, as the typical scenario in other joints, we have seen this constellation of findings in several competitive athletes who continued to engage in sports despite persistent, low-grade knee pain following trauma (Figs. 8.4 and 8.16). This group of patients has been incompletely studied, but it is intriguing to postulate that ongoing aggressive weightbearing in the absence of normal trabecular support may lead, as in osteoporotic elderly patients, to an insufficiency fracture of the trabecular bone with possible subsequent impaction of the articular surface.

It is also possible that continued activity in the presence of the volume-occupying humoral and cellular components of active healing may lead to diminished perfusion of the marrow space, thereby compromising osteocyte nutrition and fostering development of osteonecrosis.

A review of the commonality of histologic and MR imaging features in healing trauma and osteonecrosis indicates that spontaneous osteonecrosis represents an insufficiency fracture in osteopenic bone. Histologic studies performed on patients with suspected spontaneous osteonecrosis of the knee (SONK) have shown prominent reparative response as the primary finding (*vida supra*) (76,77,82,83). This concept has important implications regarding both the use of MRI in detection and evaluation of osteonecrosis, and in choice of therapy. If the marrow signal changes represent reparative response related to trabecular fractures, these patients would then appear to be at increased risk for collapse of the overlying subchondral plate, as occurs in animals (6) and in patients (most frequently competitive athletes) with trabecular bone injury who persist in stressing the joint (29). A conservative initial therapeutic approach, including a period of restricted weightbearing, would seem reasonable considering the lowered support threshold of the trabecular network. Allowing trabecular healing may favor preservation of the overlying subchondral plate and decrease the patient's risk for sequelae of degenerative joint disease. Aggressive surgical management, such as drilling procedures designed to relieve venous hypertension or improve vascularity, may have a limited role to play in posttraumatic osteonecrosis of the knee, because it is probably unlikely that surgical drilling would significantly stimulate repair or vascularity beyond that naturally present.

STRESS FRACTURE

Stress fracture results from repetitive activity, rather than from a single traumatic event. Cortical or cancellous bone, or both, can be involved. Some of the trabecular injuries described previously could be considered stress fractures, although most type 1 and 2 injuries are from acute trauma. The two major types of stress fracture are fatigue fracture and insufficiency fracture. Fatigue fractures occur in normal bone by the application of supranormal stresses. Insufficiency fractures result from ordinary, physiologic forces applied to bone which has deficient elastic resistance. Fatigue fractures in diaphyses of healthy individuals frequently have such a characteristic presentation that MR imaging is not employed in their detection. Metatarsal "march" (actually, foot-stamping) fractures in military recruits and proximal tibial stress fractures in long-distance runners are examples. Alterations in loadbearing patterns as a result of deformity or surgery can also predispose to fatigue fracture (e.g., symphysis pubis following hip arthroplasty). The clinical utility of MR imaging of stress fractures of the knee is probably greatest in cases of insufficiency fractures, which may be unexpected. Multiple factors can predispose to insufficiency fracture, including osteoporosis, osteomalacia/rickets, irradiation, Paget's disease, hyperparathyroidism, metastasis, fibrous dysplasia, and osteoporosis. MR can often characterize the more diffuse underlying process in addition to detecting the presenting lesion. Additionally, the soft tissue assessment provided by MR allows differentiation of fatigue fracture of the medial tibial plateau from pes anserinus bursitis, which can have a similar insidious clinical presentation.

Stafford and co-workers (43) first presented cases in which diametaphyseal lesions showed intracortical fracture lines as increased signal on T2-weighted images, with adjacent signal loss within the marrow space (trabecular bone) on T1-weighted images. Etiology in these cases was confirmed by surgical biopsy (two cases) and by resolution of the imaging abnormality and patient symptoms (one case). Histologic correlation, seldom obtained in more recent studies, was helpful in showing that the trabecular signal changes represented areas of active healing and repair as opposed to neoplasm, a concern raised by history and findings on plain radiographs.

In stress fracture (fatigue or insufficiency), MR images typically show a band-like area, roughly perpendicular to the cortex, of marked intramedullary signal loss on short TE images. There is frequently corresponding high signal on T2-weighted sequences, sometimes with additional subperiosteal regions of bright signal, probably representing confined hemorrhage and/or edema (Fig. 8.25). Changes on long TE images correlate with more recent injury, declining in prominence with time. These injuries frequently have a characteristic diametaphyseal location.

The MR findings have been shown to predate plain radiographic abnormality (44). Radionuclide bone scanning appears in our experience to have higher sensitivity in the early phases of stress fracture of the diaphysis than MRI, possibly because weightbearing in the diaphysis is primarily transmitted through the cortex, whereas weightbearing in the metaphysis and epiphysis is primarily through the trabecular bony struts. Fracture of trabecular bone produces hemorrhage into the surrounding marrow which is easily detected with MRI. Stress injury to the cortex may not produce as extensive hemorrhage into adjacent marrow and may be less easily detected by MRI, especially when chronic. Stress fractures of the cortex do induce a reparative response which avidly takes up bone scanning agents. The superior spatial resolution and tissue con-

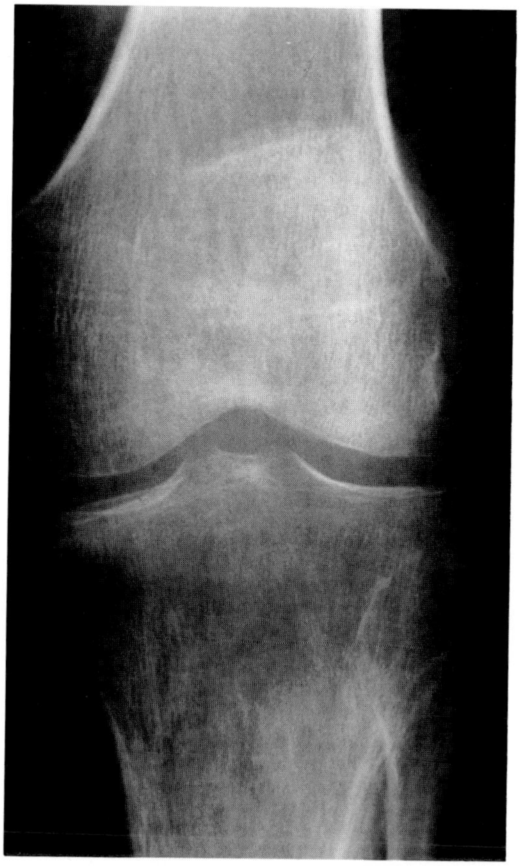

A

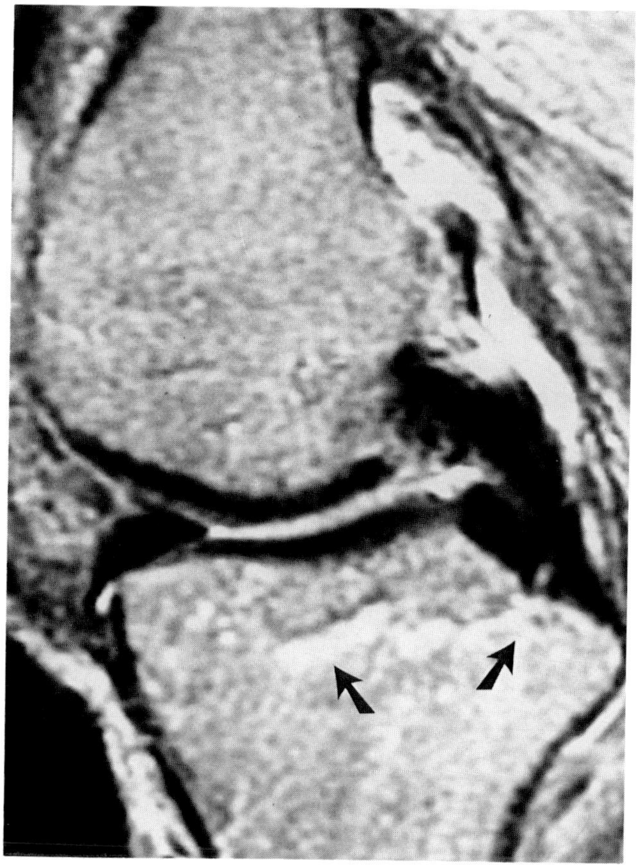

C

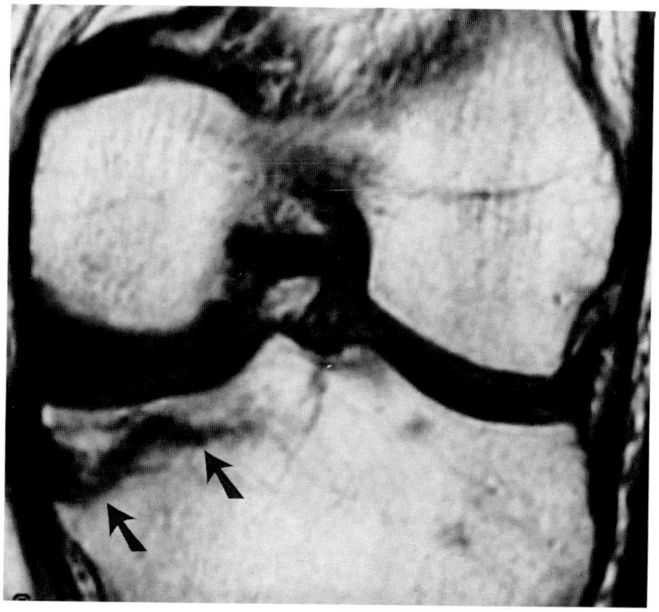

B

FIG. 8.25. Stress fracture in an 84-year-old man with prostate carcinoma who had generalized left knee pain for several weeks. **A:** Plain radiograph demonstrates demineralization, most pronounced in the proximal tibia. **B:** Coronal image (TR/TE 800/20) reveals a thick undulating band of diminished signal (*arrows*) perpendicular to the medial tibial cortex. **C:** With T2-weighting (TR/TE 2,000/80) the lesion is less conspicuous, and is seen largely as increased signal (*arrows*), probably representing presence of humoral constituents of the reparative response. These findings represent a nonpathologic insufficiency fracture, since no underlying focal bone lesion (such as metastasis) is seen.

trast of MRI may provide increased specificity, particularly in cases where the clinical presentation results in a low index of suspicion for such an injury.

COMPLEX FRACTURES

The multiplanar imaging capability of MR can be used to advantage in the evaluation of complex periarticular fractures, including those that are radiographically overt (89,90). Most complex overt fractures of the knee involve the lateral tibial plateau (91). In addition to detecting occult fractures, MR can provide detailed anatomic characterization through multiple orthogonal views with resolution superior to CT reconstructions. Tissue contrast allows simultaneous evaluation of extraosseous structures. This assessment can aid in choice of therapy because fractures with limited depression and displacement are usually treated nonoperatively (91–93), although this remains controversial. Some types of split fractures, and most combined split-depressed fractures, are treated with open reduction. Depiction of displaced small fragments is helpful in surgical planning (Fig. 8.26). Fractures which exceed approximately 5 mm of depression are also usually treated surgically (93). The presence of small fractures at the articulating surface not immediately visualized can be inferred from the presence of intraarticular marrow fat droplets (Fig. 8.27) (49).

Imaging during the recovery period following tibial plateau fracture can provide prognostic information and guide ongoing therapy. Jensen and colleagues (90) found that an overlying intact cartilaginous surface correlated with an excellent functional result. Patients whose defects filled in with fibrocartilage or synovium had a better outcome than those with persistent clefts at the articulating surface. The availability of extremely high signal-to-noise surface coils and accompanying high spatial resolution software may be of additional benefit in providing information regarding joint surface congruity and the status of fracture union. These techniques are presently under investigation (94).

MR has been useful in differentiating fracture from "congenital bipartite patella" (71), often a difficult distinction on plain radiographs. On axial images, the patellar cartilage is smoothly intact across the bony defect with bipartite patella. The articular cartilage may also be intact shortly after patellar fracture, but signs of edema within the bony lesion can be seen. In old fracture the overlying articular cartilage may be disrupted. If the overlying articular cartilage is intact, the lesion probably represents a "leave me alone lesion", whether or not the bony lesion is congenital or due to old trauma.

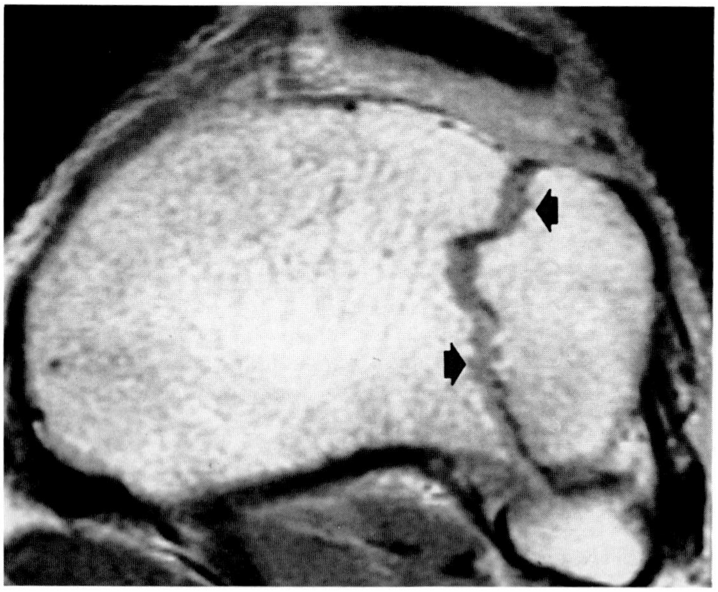
A

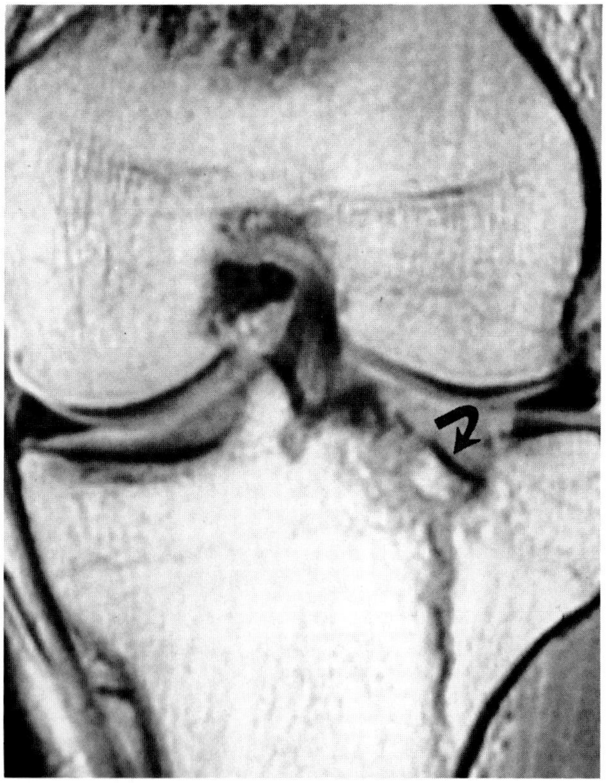
B

FIG. 8.26. Tibial plateau fracture. **A:** An axial image (TR/TE 2,000/20) depicts a displaced split type fracture of the lateral tibial plateau (*arrows*) in this 32-year-old with a skiing injury. The fracture extends to the proximal tibiofibular joint. **B:** A coronal image (TR/TE 2,000/20) reveals the comminuted, impacted nature of the fracture. The significant depression of portions of the articular surface (*curved arrow*) necessitated open reduction.

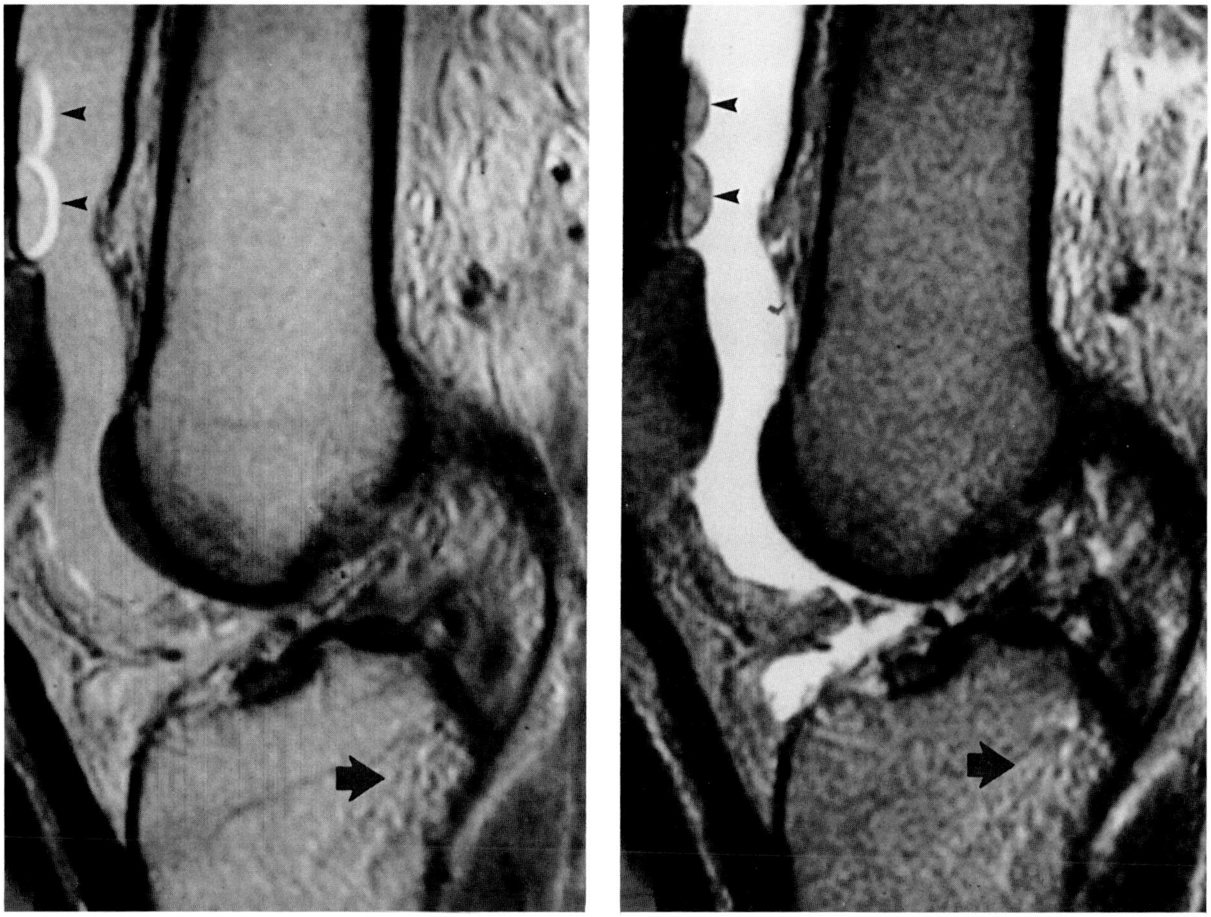

A B

FIG. 8.27. Intraarticular fat. **A,B:** This 29-year-old had traumatic injuries with breach of the subchondral plate (type 2 injury) of the medial tibial plateau. The periphery of a type 1 injury of the lateral tibial plateau is shown (*long arrows*) in (A) (TR/TE 2,000/20) and (B) (TR/TE 2,000/80). Note the moderately large joint effusion/hemarthrosis. In this supine patient, marrow fat (*arrowheads*) lies along the nondependent surface of the suprapatellar recess. These droplets are isointense with marrow signal on both pulse sequences. On the short TE image (A), misregistration artifact outlines the fat droplets posteriorly with high signal as a result of chemical shift.

MRI APPEARANCE OF CARTILAGE INJURY

There is a relative paucity of information in the literature regarding the MRI appearance of acute cartilaginous injury. Plain radiographic evaluation of joint space narrowing, arthrographic depiction of cartilaginous surfaces, and more recently CT arthrotomography have been used to indirectly evaluate cartilaginous contours. MRI with its improved contrast resolution shows promise not only in detecting surface contour abnormality, but also by revealing signal alterations shown to represent traumatic and degenerative processes. With careful attention to technique and heightened observer awareness of the appearance of normal and abnormal cartilage, MR sensitivity can approach, and may ultimately surpass, that of arthroscopic evaluation (71,95).

Utilizing double echo spin-echo technique, short TE sequences display articular hyaline cartilage as relatively homogeneous intermediate or medium gray signal intensity adjacent to the dark low signal subchondral bony end-plate (Fig. 8.28A). On T2-weighted images the articular cartilage/subchondral plate interface is generally lost owing to the loss of signal in hyaline cartilage with longer TE (Fig. 8.28B). In the presence of a joint effusion, the articular surface is well demonstrated due to the contrast between low signal intensity cartilage and high signal intensity fluid. To recreate this ''arthrographic effect'' in the absence of joint fluid, some authors favor the use of an intraarticular contrast agent for evaluating cartilage injuries (96,97).

Gradient-echo sequences provide a uniform high signal intensity appearance to articular cartilage (Fig.

8.28C). This technique can be less sensitive than spin echo in revealing surface abnormalities and in demonstrating alterations in cartilage thickness due to relative isointensity with joint fluid unless a large flip angle of greater than 50° is used (71,96). This large flip angle improves contrast between joint fluid and articular cartilage by increasing the T1-weighting of the images and decreasing the signal from joint fluid. Some authors have found gradient-echo sequences to be superior in demonstrating anatomical detail in the evaluation of the articular cartilages of the curved femoral condyles, especially when high resolution three-dimensional Fourier-transform techniques are used (33,71,96).

As discussed previously, the healing response of articular cartilage to superficial lacerations is limited because of its avascular nature and the consequent absence of an inflammatory response (18). With the spatial resolution obtainable in clinical MR studies, small surface tears may go undetected. However, laceration of the articular cartilage may be seen as focally

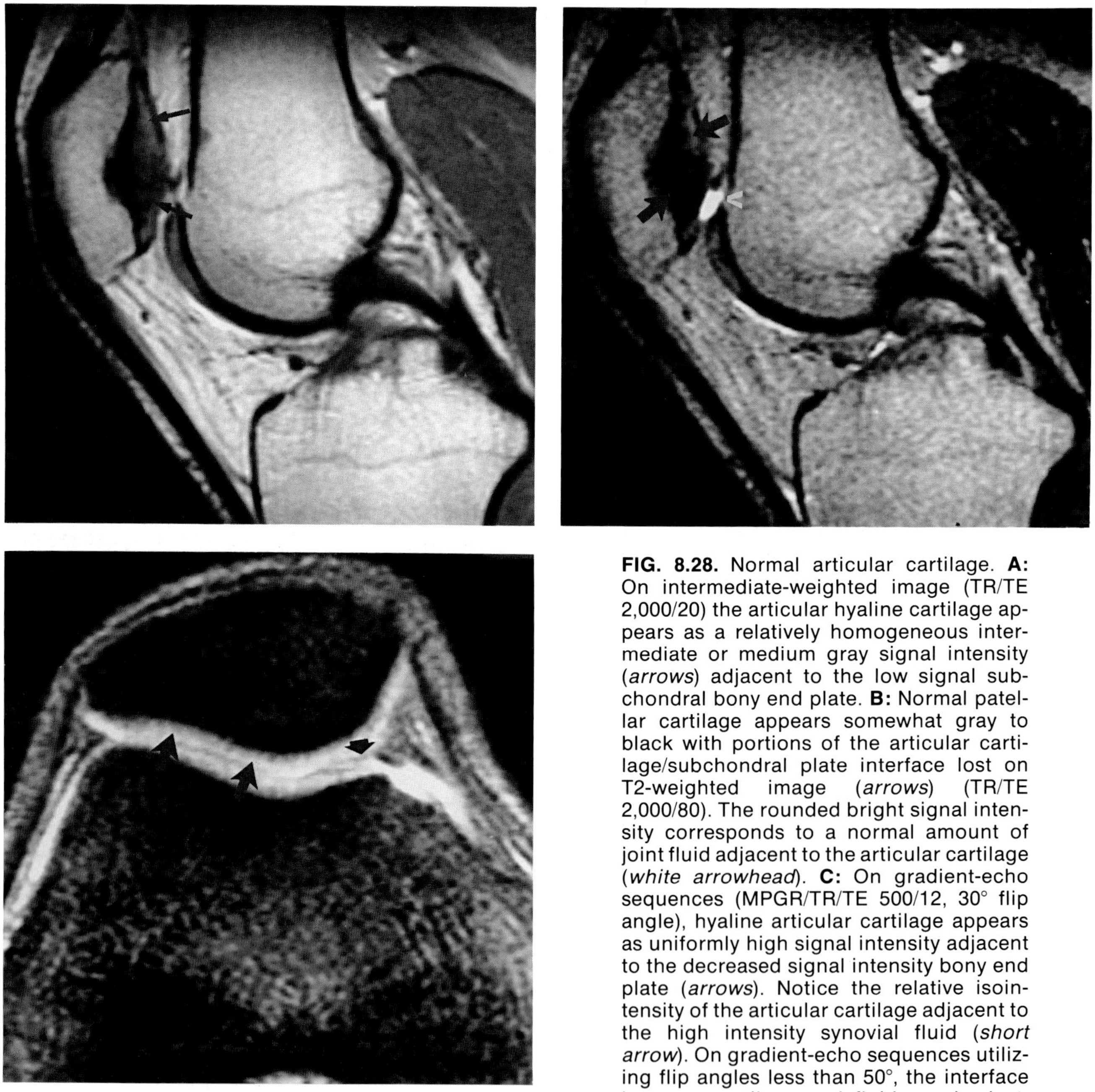

A B C

FIG. 8.28. Normal articular cartilage. **A:** On intermediate-weighted image (TR/TE 2,000/20) the articular hyaline cartilage appears as a relatively homogeneous intermediate or medium gray signal intensity (*arrows*) adjacent to the low signal subchondral bony end plate. **B:** Normal patellar cartilage appears somewhat gray to black with portions of the articular cartilage/subchondral plate interface lost on T2-weighted image (*arrows*) (TR/TE 2,000/80). The rounded bright signal intensity corresponds to a normal amount of joint fluid adjacent to the articular cartilage (*white arrowhead*). **C:** On gradient-echo sequences (MPGR/TR/TE 500/12, 30° flip angle), hyaline articular cartilage appears as uniformly high signal intensity adjacent to the decreased signal intensity bony end plate (*arrows*). Notice the relative isointensity of the articular cartilage adjacent to the high intensity synovial fluid (*short arrow*). On gradient-echo sequences utilizing flip angles less than 50°, the interface between cartilage and fluid can be lost, thus making surface alterations less conspicuous.

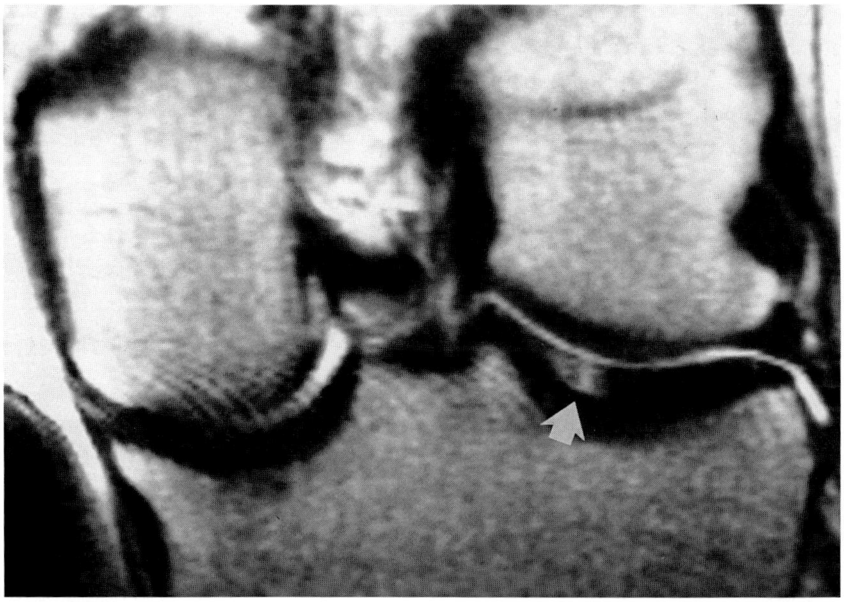

FIG. 8.29. Articular cartilage injury. High signal intensity within the lateral tibial plateau articular cartilage (*arrow*) is visualized in this young man who sustained a fall (TR/TE 2,000/80, 0.4 T). (See text for explanation.) There were no other associated injuries and arthroscopy was cancelled as a result of the MR findings.

increased signal intensity entering or crossing the cartilage stripe, with or without thickening. This appearance probably results from synovial fluid within the tear. Trauma resulting in damage to the full thickness of cartilage can additionally disrupt the subchondral bony plate, allowing a communication with the vascular space and a full inflammatory response (18). This results in accumulation of blood products and edema in the immediate area of the osteochondral fracture. Frequently changes can be observed in the underlying cancellous bone which indicate bony trabecular injury (see sections above). Such influx of inflammatory fluid leads to T2 prolongation, with variably increased signal intensity in the articular cartilage on short and long TE sequences.

Detection of a chondral defect in the absence of disruption of the subchondral plate or bony trabecular injury is enhanced by the presence of a joint effusion which increases the conspicuity of the lesion through the arthrographic effect mentioned above (Fig. 8.4). High impact injury of cartilage, with or without subchondral plate fracture, results in fissuring of the articular surface (Fig. 8.29) (18,19). Increased signal is visible within the injured cartilage when a TE is chosen such that considerable phase coherence persists among mobile protons. This likely represents signal arising from free water molecules, and is analogous to the increase in signal intensity seen in chronically degenerated collagen because of increases in unbound water content (98). Recent studies indicate that arthroscopic evaluation is more sensitive than MR in detecting these superficial lesions (96). When more extensive, these injuries can allow invagination of the synovial membrane and/or insinuation of synovial fluid into the fissures. The injury may then be visible as an increase in signal intensity on T2-weighted sequences (Fig. 8.29).

More forceful impact can lead to actual cartilage shearing and fracture exposing the "bare" bony end plate (Fig. 8.30). The denuded area can be visualized as an interruption in the normal smooth contour of articular cartilage (Fig. 8.18). These severe injuries can lead to cartilaginous loose body formation which may ultimately ossify and be detected as intraarticular marrow signal in unexpected locations.

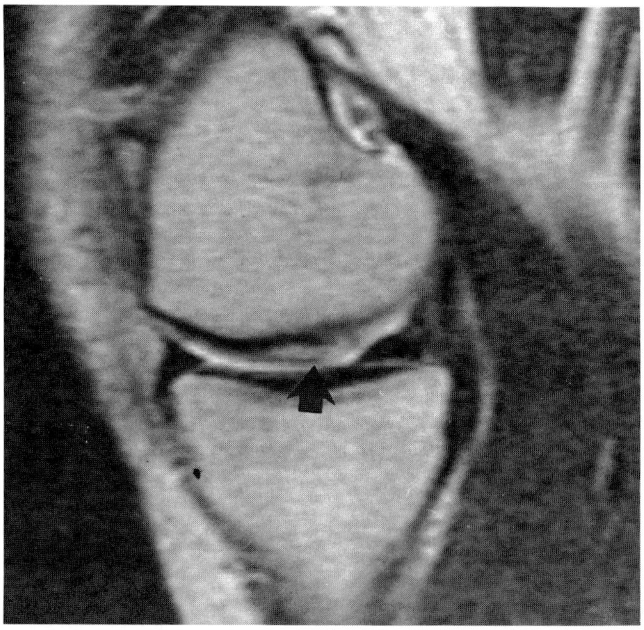

FIG. 8.30. Laceration of the hyaline articular cartilage. The increased intensity stripe in a region of focal cartilaginous thickening (*arrow*) is consistent with cartilaginous laceration on T2-weighted images (TR/TE 2,000/80).

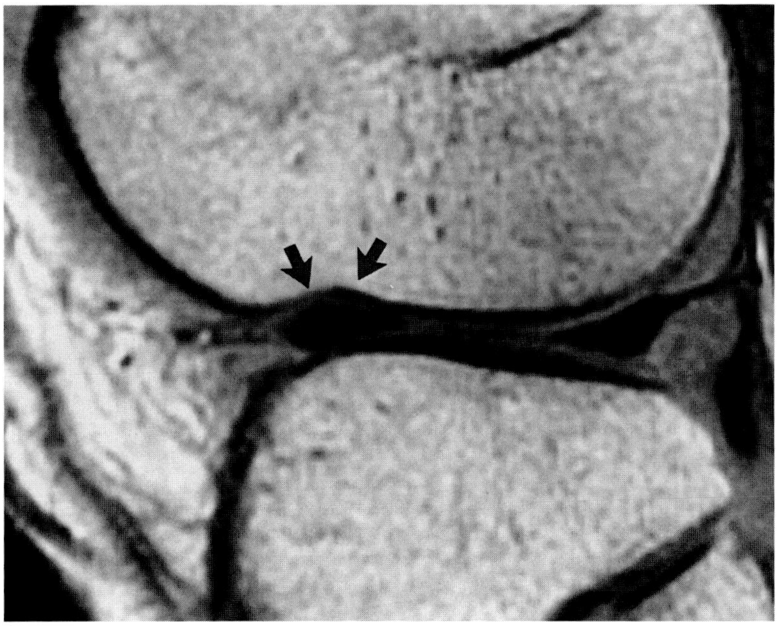

FIG. 8.31. Lateral femoral condyle anterior indentation. Notice the smooth undulation of the lateral femoral condylar articular surface just above the anterior horn of the lateral meniscus (*arrows*) (TR/TE 2,000/20). While this is a common region for impaction fractures (Fig. 8.15A), absence of bone marrow and trabecular changes adjacent to the indentation and appropriate history help one to recognize this normal variation.

Thinning of the cartilage, biochemical changes, and irregularity on a degenerative basis (chondromalacia) can be detected using routine imaging protocols (Figs. 8.11 and 8.13) (95). Lehner et al. (23) showed that heavily T1- and T2-weighted images were able to discriminate the 11% difference in hydration between the superficial and deep layers of bovine articular cartilage. Since increased water content of the (superficial) tangential and transitional layers is the earliest change in chondromalacia, MR has potential to detect degenerative phenomena even before arthroscopically detectable fissuring has occurred. In clinical work, decreased signal intensity within the articular cartilage in early (so-called grade 1) chondromalacia on T1-weighted and gradient-echo acquisitions corresponds to visualized discoloration and cartilage softening at arthroscopy (71,95). It is now recognized that specific enzymes of cartilage injury which may accelerate cartilage breakdown may be released at this stage (20–22,24). Since specific biochemical blockers of these destructive enzymes have been manufactured and appear to protect cartilage from accelerated injury, the importance of detecting early cartilage injury before marked morphologic changes have occurred may be important in selecting patients for preventive therapy when these drugs become available for human use. More advanced stages of chondromalacia (grades 3 and 4) usually result in irregular cartilaginous thinning (Figs. 8.11 and 8.13) and actual dehydration, leading to a decrease in signal intensity on T2-weighted sequences (23). Additional studies are needed to help define the role of MRI in the determination of early changes of chondromalacia. At present, there is controversy concerning the significance of signal alterations on MRI in patients with early chondromalacia (71,95).

PITFALLS IN THE INTERPRETATION OF BONY TRAUMA

Normal cancellous bone in the adult is characterized on both short and long TE images by homogeneous

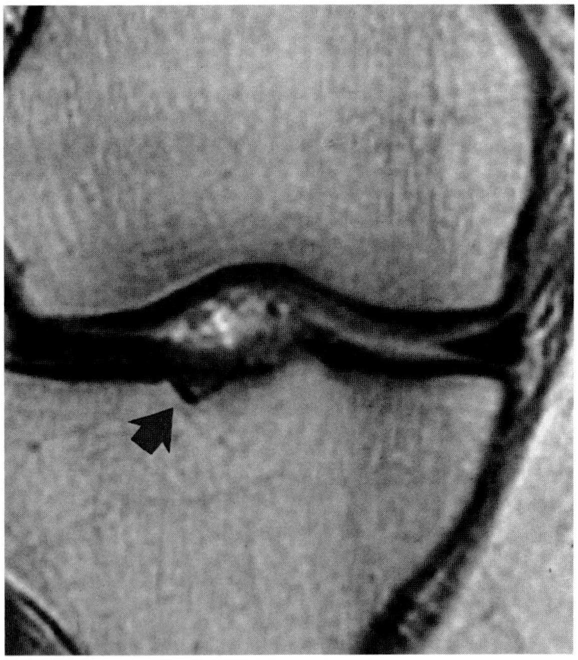

FIG. 8.32. Anterior/lateral tibial indentation. A cortical invagination is frequently seen anteriorly adjacent to the lateral tibial spine of variable depth and irregularity (*arrow*) (TR/TE 2,000/20). This is a normal variant.

high signal intensity of the marrow fat on a background of reticulated lower signal representing the bony trabeculae (Fig. 8.2). Relative contributions of each vary from patient to patient, depending on composition of the marrow, degree of bone mineralization, and individual variability. The normal subchondral plate appears as a smooth, homogeneously low signal stripe. In a few specific areas, subchondral end-plate invaginations and variations can be troublesome and resemble impaction fracture. These normal features generally demonstrate smooth or gently angulated margins and are unassociated with adjacent trabecular signal abnormality, allowing distinction from fracture.

As viewed on the sagittal series, the normal smooth indentation of the articular surface of the lateral femoral condyle overlying the anterior lateral meniscus can be problematic, as this is a frequent site of impaction fracture. However, if this region has a gently undulating contour, low signal is maintained in the subchondral plate, and there are no accompanying marrow changes, it should not be mistaken for fracture (Figs. 8.15A and 8.31). A similar but more subtle contour variance can be seen on sagittal images of the medial femoral condyle near the center of the weightbearing surface.

A cortical invagination with more acutely angulated contours is frequently conspicuous adjacent to the lateral tibial spine, usually on the most anterior images of the coronal series (Fig. 8.32). Again, without surrounding traumatic changes, this is a normal finding. An identical tibial indentation can be seen posteriorly on coronal images medially near the location of the insertion of the posterior cruciate ligament (Fig. 8.33).

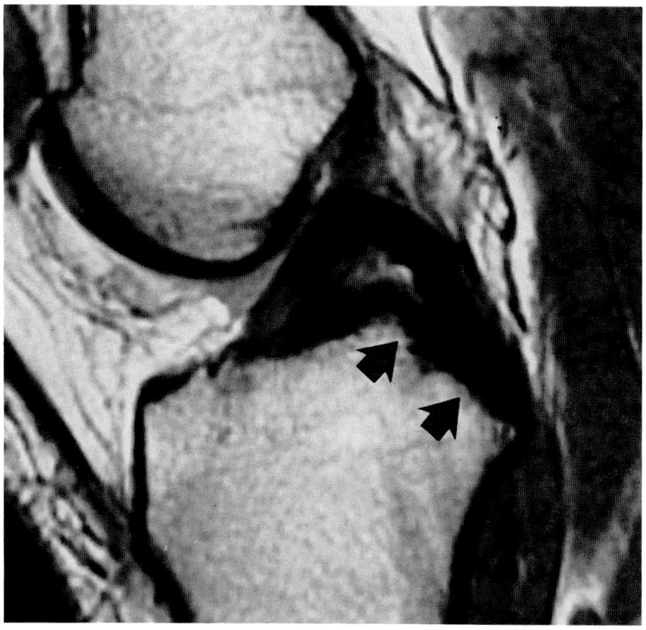

FIG. 8.34. Insertion of posterior cruciate ligament. Insertion of the posterior cruciate ligament on the tibia is normally slanted on mid-sagittal images and should not be mistaken for an impaction fracture (*arrows*) (TR/TE 2,000/20).

The insertion of the posterior cruciate ligament on the tibia is normally slanted inferiorly on sagittal images (Fig. 8.34) and should not be interpreted as an impaction fracture.

Asymmetric distribution of cellular hematopoietic

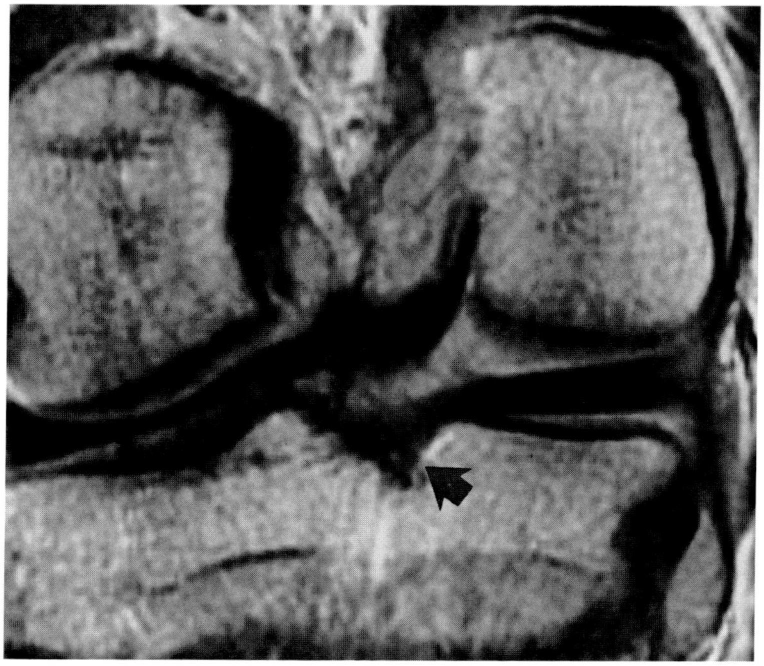

FIG. 8.33. Posterior tibial indentation. Similar to the tibial indentation seen anteriorly, an indentation is present in the majority of cases posteromedially near the insertion of the posterior cruciate ligament on coronal images (*arrow*) (TR/TE 2,000/20). This should not be mistaken for an impaction fracture.

marrow and fatty marrow is a common finding in children and adolescents (Fig. 8.35) (99–101). Occasionally red marrow persists in the appendicular skeleton, and can be seen into adulthood. Islands of fatty marrow adjacent to cellular marrow will demonstrate high signal intensity on a background of lower signal intensity, mimicking bony trabecular injury (Fig. 8.35) or tumor infiltration. This appearance may be at least in part due to chemical shift artifact and can be troublesome. Careful correlation with all pulse sequences as well as with clinical history is frequently necessary to avoid overinterpretation. If necessary, fat saturation imaging or bone scintigraphy may be of benefit in situations where the clinical setting raises concern for real injury or hematologic pathology.

The appearance of normal metaphyseal vessels can be confusing and raise concern for focal trabecular lesions. Typically these are seen anteriorly and are most conspicuous on coronal images. A mirror-image increased/decreased intensity dot appearance (Fig. 8.36A) is typical and is most likely due to flow-related enhancement and misregistration artifact. These may at times appear linear (Fig. 8.36B) or ovoid, depending on vessel course and plane of imaging section, and can mimic a linear fracture or, in the case of small rounded signal voids, bony trabecular injury.

The normal physeal scar in the adult should appear as an undulating line, frequently interrupted, of diminished signal intensity on all pulse sequences. A subacute or chronic stress fracture can have a similar appearance, but usually does not extend the entire transverse diameter of the bone; it is more irregular and thicker in appearance and is often associated with surrounding bone marrow edema or hemorrhage. These adjacent marrow changes are often lacking in the chronic setting (Figs. 8.2 and 8.25). The physis in children, while thicker in appearance, has a predictable appearance, symmetric with other growth plates in the vicinity, and should not pose a diagnostic problem (Fig. 8.37).

A

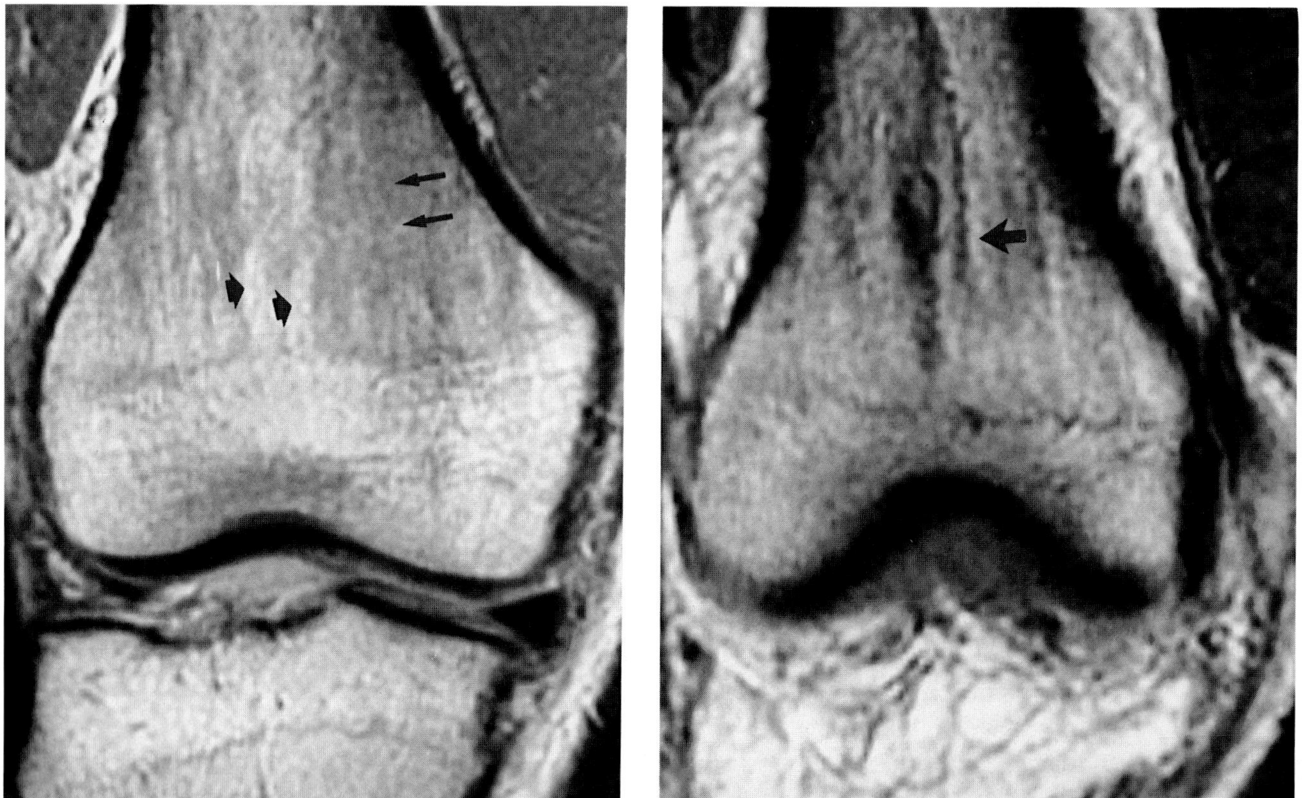

B

FIG. 8.35. Hematopoietic marrow. **A:** Cellular hematopoietic marrow (*thin arrows*) adjacent to fatty marrow (*thick arrows*) in the metaphysis and diaphysis of a young adult (TR/TE 2,000/20). Persistence of cellular hematopoietic marrow is a common finding in children and adolescents of both sexes and in adult women. This is a normal finding. **B:** Cellular marrow with islands of fatty marrow mimicking bony trabecular injury. Occasionally islands of fatty marrow (*arrow*) contained within the cellular hematopoietic marrow can mimic bony trabecular injury (TR/TE 2,000/80). This may be due to the high signal of fatty marrow on short and long TE sequences and the appearance may in part be explained by chemical shift artifact.

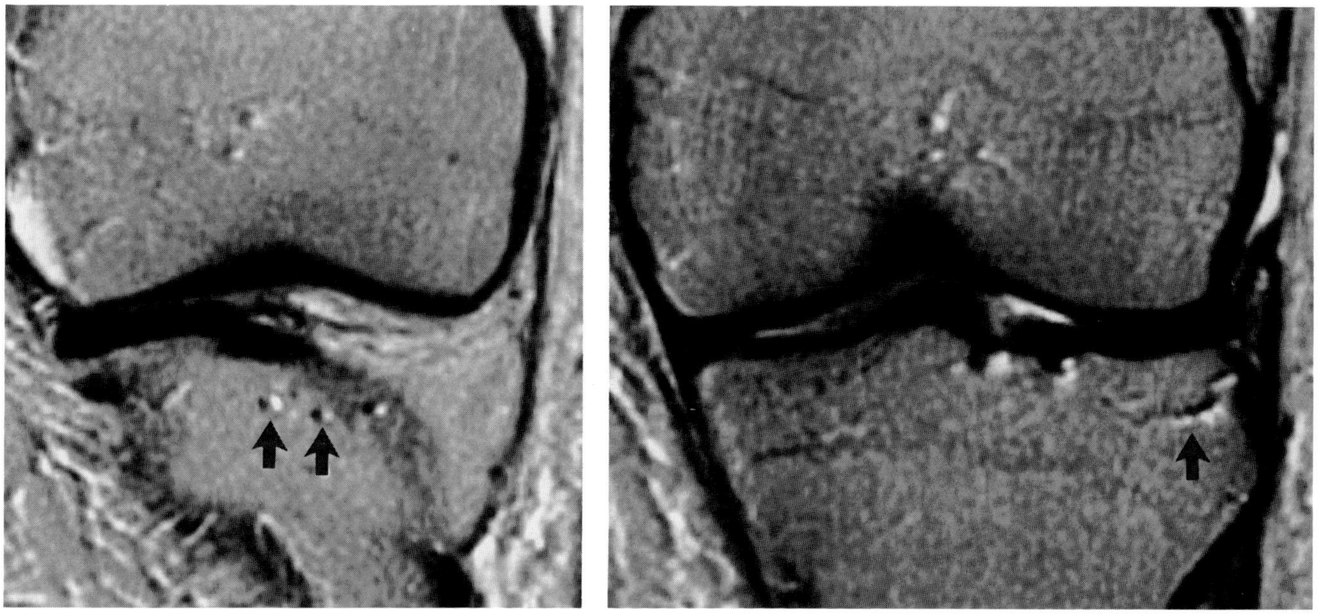

FIG. 8.36. Normal metaphyseal vessels. **A:** A mirror image increased/decreased intensity dot appearance is typical of normal metaphyseal vessels and is felt to be due to flow-related enhancement and misregistration artifact (*arrows*) (TR/TE 2,000/80). These should not be mistaken for regions of bony trabecular injury. **B:** Metaphyseal vessels may at times appear linear (*arrow*) (TR/TE 2,000/80). Lack of traumatic history in addition to the MR appearance which is a result of misregistration artifact are clues to the true nature of the finding.

CONCLUSIONS

The utility of MR in evaluation of several traumatic lesions of the bony and cartilaginous structures of the knee has been outlined in this chapter. Additionally, observations made thus far in the relatively early stages of investigation have led to improved understanding and new concepts regarding pathologic processes affecting the articulating cartilage and weight-bearing bones of the knee. These may directly contribute to clinical success in patient management.

Isolated trabecular microfractures are homeopathic events. Trabecular fractures in massive numbers occur commonly in traumatic injury to the knee that does not

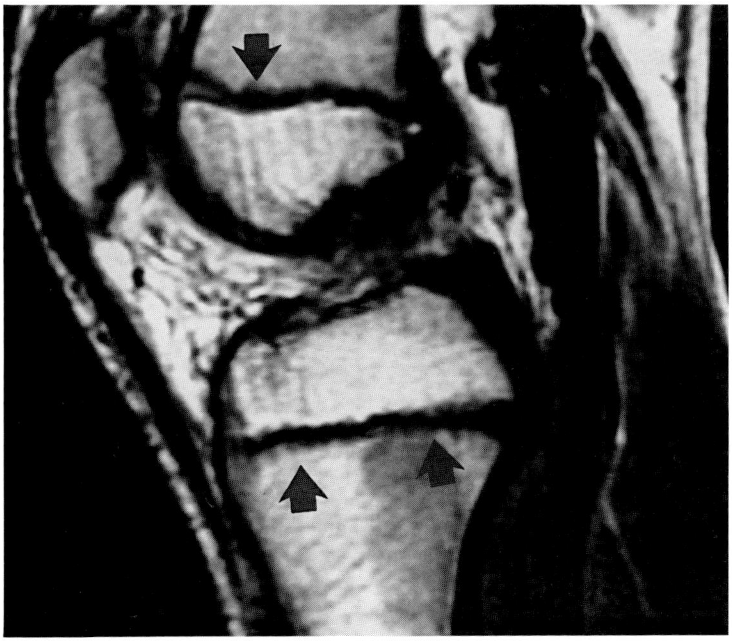

FIG. 8.37. Normal open physeal plate. The normal open physeal plate in children has a characteristic appearance on short TE images (TR/TE 2,000/20) and should not be mistaken as an abnormality (*arrows*).

result in cortical or subchondral plate fracture. Such injury to the cancellous bone is occult to conventional imaging modalities but well depicted on MR because of associated hemorrhage into the surrounding marrow space. In some cases—particularly when they abut the subchondral plate—these injuries are not trivial, and can lead to collapse of the subchondral bone plate and significant degenerative sequelae. Contiguity of MR signal abnormalities with the subchondral bone may be a predictor of vulnerability to impending collapse. Patients with such injuries and persistent pain may benefit from a period of restricted weightbearing while bony healing progresses. This conservative approach should allow patients to achieve an optimal result while minimizing unnecessary surgical intervention. Patients with concomitant meniscal tear and bony injury have had resolution of symptoms upon healing of the bony lesion without specific therapy for the meniscus.

Osteochondral defects are best evaluated by MR, which can determine the location and extent of disease, and can predict stability of osseous lesions with greater accuracy than arthroscopy. This assessment facilitates noninvasive treatment planning and avoids instrumentation in patients in whom surgical therapy is not indicated. Follow-up imaging can monitor disease progression. Osteonecrosis shares many common features with traumatic lesions of the trabecular bone, both on MR images and histopathologically. Many cases considered spontaneous may indeed have a traumatic origin. Again, limited weightbearing in the post-injury period appears preferable to early surgical intervention. Stress fracture about the knee is sensitively detected by MR. Similarly, in cases of complex fractures, usually involving the tibial plateaus, MR provides multiplanar assessment of injury leading to accurate presurgical planning or to appropriate nonoperative management.

While not yet as sensitive as arthroscopy in detection of small lacerations of the articular cartilage surface, MR is superior in demonstrating nonlacerative cartilage injury following blunt trauma. Moderately sized and large tears can be demonstrated reliably. Awareness of these injuries can allow the prescription of appropriate activity restriction and may be beneficial in selecting appropriate patients for drug therapy if blockers of enzymes important in pathogenesis of cartilage injury become available. It is hoped that these advances will allow maximum preservation of the cartilaginous surface.

REFERENCES

1. Fazzalari N, Vernon-Roberts B, Darracott J. Osteoarthritis of the hip. Possible protective and causative roles of trabecular microfractures in the head of the femur. *Clin Orthop* 1987; 216:224–233.
2. Todd R, Freeman M, Pirie C. Isolated trabecular fatigue fractures in the femoral head. *J Bone Joint Surg (Brit)* 1972;54:723–728.
3. Pugh J, Rose R, Radin E. A possible mechanism of Wolff's law: Trabecular microfracture. *Arch Int Physiol Biochim* 1973; 81:27–40.
4. Pugh J, Rose R, Radin E. A structural model for the mechanical behavior of trabecular bone. *J Biomech* 1973;6(6):657–670.
5. Radin E., et al. *The mechanics of joints as it relates to their degeneration.* In: *Symposium on Osteoarthritis*, 1974;34–43.
6. Radin E, et al. Response of joints to impact loading—III. Relationship between trabecular microfractures and cartilage degeneration. *J Biomech* 1973;6:51–57.
7. Pugh J, Rose R, Radin E. Elastic and viscoelastic properties of trabecular bone: Dependence on structure. *J Biomech* 1973; 6:475–485.
8. Pugh J, Radin E, Rose R. Quantitative studies of human subchondral cancellous bone. Its relationship to the state of its overlying cartilage. *J Bone Joint Surg (Am)* 1974;56(2):313–321.
9. Radin E, Paul I, Lowy M. A comparison of the dynamic force transmitting properties of subchondral bone and articular cartilage. *J Bone Joint Surg [US]* 1970;52:444–456.
10. Farkas T, et al. Early vascular changes in rabbit subchondral bone after repetitive impulsive loading. *Clin Orthop [US]* 1987; 219:259–267.
11. Milgram J. Injury to articular cartilage joint surfaces: II. Displaced fractures of underlying bone. A histopathologic study of human tissue specimens. *Clin Orthop* 1986;206:236–247.
12. Radin E. Elastic and viscoelastic properties of trabecular bone: Dependence on structure. *J Biomech* 1973;6(5):475–485.
13. Resnick D, Niwayama M, eds. *Diagnosis of bone and joint disorders*, 2nd ed. Philadelphia: Saunders, 1988:4174ff.
14. Watson M. Microfractures in the head of the femur. *J Bone Joint Surg* [Am] 1975;57A:696–698.
15. Milgram J. Reparative cartilaginous callus in subarticular osteonecrosis of bone. A histopathological study of the femoral head. *Clin Orthop* 1984;186:272–283.
16. Radin E, Rose R. Role of subchondral bone in the initiation and progression of cartilage damage. *Clin Orthop [US]* 1986; 213:34–40.
17. Vellet A, et al. Occult posttraumatic osteochondral lesions of the knee: Prevalence, classification, and short-term sequelae evaluated with MR imaging. *Radiology* 1991;178:271–276.
18. Mankin H. The response of articular cartilage to mechanical injury. *J Bone Joint Surg [US]* 1982;64A:460–466.
19. Repo R, Finlay J. Survival of articular cartilage after controlled impact. *J Bone Joint Surg [US]* 1977;59A:1068–1076.
20. McCachren SS. Expression of metalloproteinases and metalloproteinase inhibitor in human arthritic synovium. *Arthritis Rheum* 1991;34:1085–1093.
21. Kuettner KE, et al. *Articular cartilage and arthritis*, 1st ed. New York: Raven Press, 1992:757.
22. Firestein GS, Paine MM, Littman BH. Gene expression (collagenase, tissue inhibitor of metalloproteinases, complement, and HLA-DR) in rheumatoid arthritis and osteoarthritis synovium. *Arthritis Rheum* 1991;34(9):1094–1105.
23. Lehner K, et al. Structure, function, and degeneration of bovine hyaline cartilage: Assessment with MR imaging in vitro. *Radiology* 1989;170(2):495–499.
24. Gravallese EM, et al. In situ hybridization studies of stromelysin and collagenase messenger RNA expression in rheumatoid synovium. *Arthritis Rheum* 1991;34(9):1076–1084.
25. Maroudas A, et al. *Physiochemical properties and functional behavior of normal and osteoarthritis human cartilage.* In: Kuettner K, et al. eds. *Articular cartilage biochemistry*, Raven Press; 1986:311–329.
26. Radin E. Subchondral bone changes in patients with early degenerative joint disease. *Arthritis Rheum* 1970;13:400–405.
27. Beltran J, et al. The knee: Surface-coil MR imaging at 1.5T. *Radiology* 1986;159:747–751.
28. Berquist T. *MRI of the musculoskeletal system*, 2nd ed. New York: Raven Press, 1990:545.
29. Crues J, Ryu R. MRI of the knee: Part II. *Surg Rounds Orthop* 1989;(May):27–33.

30. Lee J, Yao L. Occult intraosseous fracture: Magnetic resonance appearance versus age of injury. *Am J Sports Med* 1989; 17(5):620–623.
31. Mink J, Deutsch A. Occult cartilage and bone injuries of the knee: Detection, classification, and assessment with MR imaging. *Radiology* 1989;170:823–829.
32. Yao L, Lee J. Occult intraosseous fracture: Detection with MR imaging. *Radiology* 1988;167:749–751.
33. Spritzer C, et al. MR imaging of the knee: Preliminary results with a 3DFT GRASS pulse sequence. *Am J Radiol* 1988; 150:597–603.
34. Ehman R. MR imaging of medullary bone. *Radiology* 1988; 167:867–868.
35. Wehrli FW, et al. Trabecular structure: Preliminary application of MR interferometry. *Radiology* 1991;179(3):615–621.
36. Gallimore G, Harms S. Knee injuries: High-resolution MR imaging. *Radiology* 1986;160:457–461.
37. Sebag G, Moore S. Effect of trabecular bone on the appearance of marrow in gradient-echo imaging of the appendicular skeleton. *Radiology* 1990;174:855–859.
38. Beltran J. *MRI—Musculoskeletal system*. Philadelphia: JB Lippincott, 1990.
39. Mink J, Deutsch A. *MRI of the musculoskeletal system—A teaching file*. New York: Raven Press, 1990:607.
40. Reicher M, Kellerhouse L. *MRI of the wrist and hand*. New York: Raven Press, 1990:203.
41. Zlatkin M. *MRI of the shoulder*. New York: Raven Press, 1991:174.
42. Kean D, et al. NMR imaging of the knee: Examples of normal anatomy and pathology. *Br J Radiol* 1983;56:355–364.
43. Stafford SA, et al. MRI in stress fracture. *AJR* 1986;147:553–556.
44. Lee J, Yao L. Stress fractures: MR imaging. *Radiology* 1988; 169:217–220.
45. Lynch T, et al. Bone abnormalities of the knee: Prevalence and significance at MR imaging. *Radiology* 1989;171:761–766.
46. Rosen M, Jackson D, Berger P. Occult osseous lesions documented by magnetic resonance imaging associated with anterior cruciate ligament ruptures. *J Arthro Rel Surg* 1991; 7(1):45–51.
47. Milgram J. Morphologic alterations of the subchondral bone in advanced degenerative arthritis. *Clin Orthop* 1983;173:293–312.
48. Radin E. Response of joints to impact loading. II: In vivo behavior of subchondral bone. *J Biomech* 1972;5(3):267–272.
49. Stoller D, Crues JI, Mink J. *Detection of stress fractures in the knee using MRI*. In: *ARRS Annual Convention*. 1987.
50. Cameron H, Fornasier V. Trabecular stress fractures. *Clin Orthop Rel Res* 1975;111:266–268.
51. Radin E. A possible mechanism of Wolff's law: Trabecular microfractures. *Arch Int Physiol Biochim* 1973;81:27–40.
52. Crues J, ed. *MRI of the musculoskeletal system*. In: Lufkin R, Bradley W, Brant-Zawadzki M, eds. The MRI teaching file. New York: Raven Press; 1991:247.
53. Kornick J, et al. Meniscal abnormalities in the asymptomatic population at MR imaging. *Radiology* 1990;177:463–465.
54. Stoller D, et al. Meniscal tears: Pathologic correlation with MR imaging. *Radiology* 1987;163:731–735.
55. Smillie IS. *Surgical pathology of the menisci*. In: *Injuries of the knee joint, 4th ed*. London: E & S Livingstone; 1971:39–69.
56. Pappas A. Osteochondrosis Dissecans. *Clin Orthop Rel Res* 1981;158:59–69.
57. Mandell G, Harcke H, Kumar S. Avascular necrosis and related conditions. *Topics Magn Reson Imaging* 1991;4(1):38–44.
58. Barrie H. Osteochondritis dissecans 1887–1987. A centennial look at Konig's memorable phrase. *J Bone Joint Surg [Br]* 1987; 69(5):693–695.
59. De Smet A, et al. Osteochondritis dissecans of the knee: Value of MR imaging in determining lesion stability and the presence of articular cartilage defects. *AJR* 1990;155:549–553.
60. Mubarak SJ, Carroll NC. Juvenile osteochondritis dissecans of the knee: Etiology. *Clin Orthop Rel Res* 1981;157:200–211.
61. Aichroth P. Osteochondritis dissecans of the knee. A clinical survey. *J Bone Joint Surg* 1971;53B(3):440–447.
62. Linden B. The incidence of osteochondritis dissecans in the condyles of the femur. *Acta Arthop Scand* 1976;48:664–667.
63. Munk P, Helms C. *MRI of the knee*, 1st ed. In: Kricun M, ed. Clinical Diagnosis Imaging Series. Gaitherburg, MD: Aspen; 1992:198.
64. Pantazopoulos T, Exarchou E. Osteochondritis dissecans of the patella. *J Bone Joint Surg* 1971;53A(6):1205–1207.
65. Towbin J, Towbin R, Crawford A. Osteochondritis dissecans of the tibial plateau. *J Bone Joint Surg* 1982;64A(5):783–784.
66. Canale S, Belding R. Osteochondral lesions of the talus. *J Bone Joint Surg* 1980;62A:97–102.
67. Phillips HO, Grubb SA. Familial multiple osteochondritis dissecans. Report of a kindred. *J Bone Joint Surg [AM]* 1985; 67(1):155–156.
68. Andrew T, Spivey J, Lindebaum R. Familial osteochondritis dissecans and dwarfism. *Acta Orthop Scand* 1981;52(5):519–523.
69. Milgram JW, Rogers LF, Miller JW. Osteochondral fractures: Mechanisms of injury and fate of fractures. *Am J Radiol* 1978; 130(4):651–658.
70. Mesgarzadeh M, et al. Osteochondritis dissecans: Analysis of mechanical stability with radiography, scintigraphy, and MR imaging. *Radiology* 1987;165:775–780.
71. Crues J, Ryu RKN. *The knee*. In: Stark D, Bradley W, eds. *Magnetic resonance imaging*. St. Louis, MO: CV Mosby; 1992: 2355–2423.
72. Linden B, Telhaag H. Osteochondritis dissecans: A histologic and autoradiographic study in man. *Acta Orthop Scand* 1977; 48(682):682–686.
73. Hughston JC, Hergenroeder PT, Courtney BG. Osteochondritis dissecans of femoral condyle. *J Bone Joint Surg [AM]* 1984;66:1340–1348.
74. Thompson NL. Osteochondritis dissecans and osteochondral fragments managed by Herbert compression screw fixation. *Clin Orthop* 1987;224:71–78.
75. Rozing P, Insall J, Bohne W. Spontaneous osteonecrosis of the knee. *J Bone Joint Surg [AM]* 1980;62(1):2–7.
76. Insall J, Aglietti P, Bullough P. *Osteonecrosis*. In: Insall J, ed. *Surgery of the knee*. London: Churchill; 1984:527–549.
77. Ahlback S, Bauer G, Bohne W. Spontaneous osteonecrosis of the knee. *Arthritis Rheum* 1968;11(6):726.
78. Williams J, Cliff M, Bonakdarpour A. Spontaneous osteonecrosis of the knee. *Radiology* 1973;107:15–19.
79. Lotke PA, Ecker ML. Osteonecrosis of the knee. *J Bone Joint Surg [AM]* 1988;70(3):470–473.
80. Aglietti P, et al. Idiopathic osteonecrosis of the knee. Aetiology, prognosis and treatment. *J Bone Joint Surg [Br]* 1983; 65(5):588–597.
81. Pollack M, et al. Magnetic resonance imaging in the evaluation of suspected osteonecrosis of the knee. *Skeletal Radiol* 1987; 16(2):121–127.
82. Lotke P, Ecker M, Alavi A. Painful knees in older patients. *J Bone Joint Surg [AM]* 1977;59:617.
83. Bjorkengren A, et al. Spontaneous osteonecrosis of the knee: Value of MR imaging in determining prognosis. *AJR* 1990; 154:331–336.
84. Norman A, Baker N. Spontaneous osteonecrosis of the knee and medial meniscal tears. *Radiology* 1978;129:653–656.
85. Lang P, et al. Avascular necrosis of the femoral head: High-field-strength MR imaging with histologic correlation. *Radiology* 1988;169:517–524.
86. Mitchell DG. *MR of the normal and ischemic hip*. In: Kressel HY, ed. *Magnetic Resonance Annual 1988*, New York: Raven Press; 1988:37–69.
87. Mitchell D. Using MR imaging to probe the pathophysiology of osteonecrosis. *Radiology* 1989;171:25–26.
88. Turner D, et al. Femoral capital osteonecrosis: MR finding of diffuse abnormalities without focal lesions. *Radiology* 1989; 171:135–140.
89. Berger PE, et al. MRI demonstration of radiographically occult fractures: What have we been missing? *Radiographics* 1989; 9(89):407–436.
90. Jensen D, et al. Magnetic resonance imaging in the evaluation of sequelae after tibial plateau fractures. *Skeletal Radiol* 1990; 19:127–129.

91. Newberg M, Greenstein R. Radiographic evaluation of tibial plateau fractures. *Radiology* 1978;126(Feb 78):319–323.
92. Deutsch A, Mink J, Shellock F. Magnetic resonance imaging of injuries to bone and articular cartilage: Emphasis on radiographically occult abnormalities. *Orthop Rev* 1990;19:66–75.
93. Rafii M, et al. Computed tomography of tibial plateau fractures. *Am J Roentgenology* 1984;142:1181–1186.
94. Foo TKF, et al. High resolution, short TE and TR gradient echo imaging using partial echo acquisition in the wrist and eye. *Book of Abstracts, Society of Magnetic Resonance in Medicine*. 1991;1:313.
95. Handelberg F, Shahabpour M, Casteleyn PP. Chondral lesions of the patella evaluated with computed tomography, magnetic resonance imaging, and arthroscopy. *Arthroscopy* 1990; 6(1):24–29.
96. Speer K, et al. Magnetic resonance imaging of traumatic knee articular cartilage injuries. *Am J Sports Med* 1991;19(4):396–402.
97. Gylys-Morin V, et al. Articular cartilage defects: Detectability in cadaver knees with MR. *Am J Radiol* 1987;148:1153–1157.
98. Crues J, Fareed D. Magnetic resonance imaging of shoulder impingement. *Topics Magn Reson Imaging* 1991;3(4):39–49.
99. Moore SG, Dawson KL. Red and yellow marrow in the femur: Age-related changes in appearance at MR imaging. *Radiology* 1990;175:219–223.
100. Ricci C, et al. Normal age-related patterns of cellular and fatty bone marrow distribution in the axial skeleton: MR imaging study. *Radiology* 1990;177(1):83–88.
101. Weinreb JC. MR imaging of bone marrow: A map could help. *Radiology* 1990;177(1):23–24.
102. Browner BD, Jupiter JB, Levine AM, Trafton PG. *Skeletal trauma: fractures, dislocations, and ligamentous injuries*. W. B. Saunders Co. 1992, p. 1719.

ADDITIONAL READINGS

Murphy BJ, Smith RL, Uribe JW, et al. Bone signal abnormalities in the posterolateral tibia and lateal femoral condyle in complete tears of the anterior cruciate ligament: a specific sign? *Radiology*, 1992;182:221–224.

Schils JP, Spindler KP, Piraino DW, et al. Bone bruises and chondral lesions in anterior cruciate ligament tears: A prospective study. *JMRI*, 1992;2(P):p. 94.

CHAPTER 9

The Spectrum of Knee Joint Disorders

Murray A. Reicher

This chapter addresses magnetic resonance imaging (MRI) of a diverse group of conditions, including articular cartilage abnormalities, disorders of subchondral bone, osteochondral loose bodies, synovial disease, synovial fluid collections, synovial cysts, osteonecrosis, bone marrow infiltration, tumors, infection, and vascular disease. MRI has become the primary diagnostic technique for evaluating most of these abnormalities. The goal of this chapter is to review the pathophysiology and anatomy of these disorders and to point out the proven and potention roles for MRI.

DISORDERS OF SUBCHRONDRAL BONE AND ARTICULAR CARTILAGE

Normal Anatomy and Physiology

The knee is a movable, diarthrodial joint, its component bony parts covered by hyaline cartilage, and its joint cavity lined by a synovial membrane. Tightly adherent to the bony end plate is the hyaline articular cartilage, a weightbearing and gliding surface which provides for the almost frictionless mobility essential to function (101). Beneath the articular end plate is the cancellous bone, often containing red (hematopoietic) marrow. The knee is a rounded, condylar, cam-shaped, four-bar linkage that allows not only flexion and extension, but also a surprising amount of rotation (101).

The articular cartilage and synovial fluid are responsible for the almost frictionless movement of the articulating surfaces (101). Articular cartilage is dense and white on gross examination, but tends to yellow with age. Its water content is estimated to be over 70%, even in adults, but nevertheless articular cartilage feels semisolid to palpation. Although on gross examination the surface feels smooth, in fact detailed histologic examination reveals gentle undulations and irregular depressions that correspond to the location and shape of cells lying just beneath the surface (101).

Articular cartilage, which is aneural, avascular, and alymphatic, derives its nutrition from synovial fluid. Nutrients first diffuse across the synovial membrane into the synovial fluid and then through the dense matrix of cartilage to reach the chondrocyte (101).

Hyaline articular cartilage may be subdivided into four more or less distinct histologic zones (101,111). The superficial zone, also known as the tangential or gliding zone, occupies approximately 5% of the cartilage thickness and consists of elongated cells oriented parallel to the articular surface. The transitional zone occupies an average of 42% of the cartilage thickness and consists primarily of rounded, randomly distributed cells. The deep radial and calcified zones combine to occupy an average of 53% of the cartilage thickness. In the radial zone, the cells are arranged in short, irregular columns, and in the calcified zone, the cells and matrix are heavily encrusted with hydroxyapatite (101,111). On high resolution spin-echo MR images, three layers of hyaline articular cartilage can be distinguished; a low intensity zone near the articular surface, a zone of higher signal intensity immediately deeper, and a second zone of low signal intensity even deeper (111).

Water content in articular cartilage ranges from 70% to 80%. Collagen is the most prevalent organic constituent, accounting for over 50% of the remaining material (101). Water within articular cartilage may freely exchange with synovial fluid. Extrusion of water from cartilage may aid in joint lubrication. The frictional resistance of animal joints lubricated with synovial fluid has been estimated to be approximately one-tenth of that of an ice skate on ice (101). The knee joint usually functions under mechanical loads approximately two to ten times that of body weight. Under these conditions, fluid from hydrated articular cartilage may be "squeezed" from its proteoglycan-collagen gel. Because this water cannot flow through the subchondral plate or substance of the cartilage in adults, the water is preferentially expressed onto the surface of the cartilage, enhancing joint lubrication. Under physiologic conditions, articular cartilage can be compressed by as much as 80%.

Since articular cartilage depends on diffusion for nutrition, its maximum thickness is limited. The patellar articular cartilage is the thickest in the human body, measuring no more than 6 mm.

Osteoarthritis

The knee is the most common site of osteoarthritis (68,141). Patients generally present with symptoms of pain, tenderness, diminished range of motion, warmth, effusions, or synovial cysts. As osteoarthritis progresses, varus deformity, soft tissue atrophy, and knee joint instability frequently develop. Osteoarthritis begins with articular cartilage degeneration and erosion. Typical pathologic changes include cartilage fibrillation, denudation of osseous surfaces, subchondral cystic lesions, and osteophytosis. Because osteoarthritis is not truly an inflammatory disease, some authors advocate the use of the term osteoarthropathy. The three compartments of the knee (the medial and lateral femorotibial compartments and the patellofemoral compartment) are usually involved unevenly; i.e., the medial femorotibial and patellofemoral compartments are involved most frequently, and often both are involved. Involvement of the patellofemoral compartment usually predominates along the lateral facet of the patella. Isolated patellofemoral disease occurs secondary to calcium pyrophosphate dihydrate deposition disease or hyperparathyroidism. Lateral compartment osteoarthritis is less frequent and is rarely associated with disease of the patellofemoral compartment.

Early disease of articular cartilage is not visible on standard radiographs but may be seen on arthrograms. Arthrography in early osteoarthritis depicts articular cartilage erosion, fragmentation, and imbibition of contrast material (9,55,141). Associated findings include meniscal degeneration, subchondral cysts, and popliteal cysts. Advanced disease is manifested on standard radiographs as compartmental narrowing, marginal osteophytes, juxtaarticular sclerosis, and subchondral cysts. Sclerosis usually occurs in the tibia or in both the femur and the tibia, but rarely in the femur alone (141). Subchondral cysts usually are small and predominate in the tibia. Single photon emission computed tomography (SPECT) has been shown to be more sensitive than either standard radiography or planar scintigraphy for detecting osteoarthropathy of the knee (40).

MRI offers a unique capacity for evaluating osteoarthritis because it is the only technique that allows visualization of the full width of articular cartilage. A wide variety of techniques has been advocated for optimally evaluating articular cartilage, including spin-echo imaging, gradient-echo imaging, fat-suppressed imaging, and MR arthrography. Both two-dimensional and three-dimensional techniques have been advocated. As yet, no single technique has emerged as the clearly most effective. As a general strategy in prescribing MR techniques to evaluate articular cartilage, contrast between joint fluid and articular cartilage should be optimized and spatial resolution maximized.

On T1-weighted or intermediate-weighted spin-echo images, normal articular cartilage is seen as a thin rim of moderate signal intensity covering the articulating ends of the femur, tibia, and patella (see Chapter 3). On T2-weighted spin-echo images, articular cartilage decreases slightly in signal intensity, and since neighboring synovial fluid increases in signal intensity, the contrast between joint fluid and the articular cartilage surface is increased. On high resolution MR images, articular cartilage can be further subdivided into three layers of differing signal intensity.

There are now numerous reports of knee joint articular cartilage thinning and erosion seen on MR images (3,15,27,33,44,59,66,81,91,102,105,108,111,135,139,159,173,182). MRI has been found to be accurate in delineating chondral defects as shallow as 1 mm in cadaver knees and patients (181). Since the signal yielded by articular cartilage is dependent on interaction of water with negatively charged chondroitin sulfate and keratan sulfate polysaccharide chains, alterations in these proteoglycans may be associated with changes in signal (27,80). This may allow detection of histologic changes within articular cartilage associated with aging and degeneration even prior to surface erosion.

Figures 9.1 through 9.4 reveal MR findings typical of osteoarthritis on intermediate- and T2-weighted scans, including joint space narrowing, cartilage erosion, osteophyte formation, subchondral degenerative cyst formation, and subchondral sclerosis. In evaluating older patients, MRI is clinically most useful in distinguishing the relative severity of osteoarthritis versus meniscal derangement, helping the orthopedic surgeon discriminate between the need for arthroscopic debridement and joint replacement.

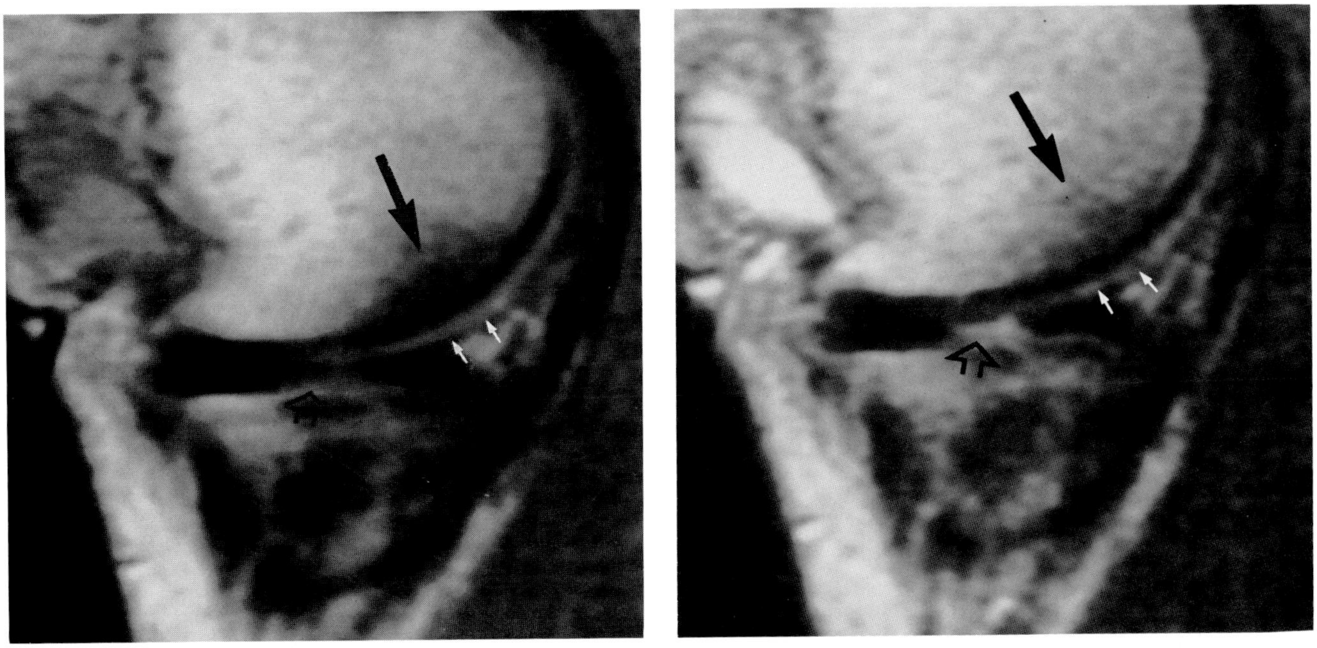

A B

FIG. 9.1. Osteoarthritis. **A:** Intermediate-weighted and **B:** T2-weighted spin-echo images depict subchondral decreased signal intensity, indicating subchondral sclerosis (*black arrow*). Normal thickness articular cartilage is present (*small white arrows*). Incidentally seen is a tear of the middle third of the medial meniscus (*open arrows*).

Osteochondral Fractures and Chondral Fractures

Excluding patients with osteochondritis dissecans, the most common sites of osteochondral fractures are the medial patellar facet and the lateral femoral condyle (9,125). The mechanism of injury is impaction of the patella against the lateral femoral condyle, sometimes associated with patellar dislocation and relocation. In contrast, pure chondral fractures of the knee occur most often in the medial femoral condyle in young adults following a rotational injury or direct blow to the knee (58). With osteochondral fractures,

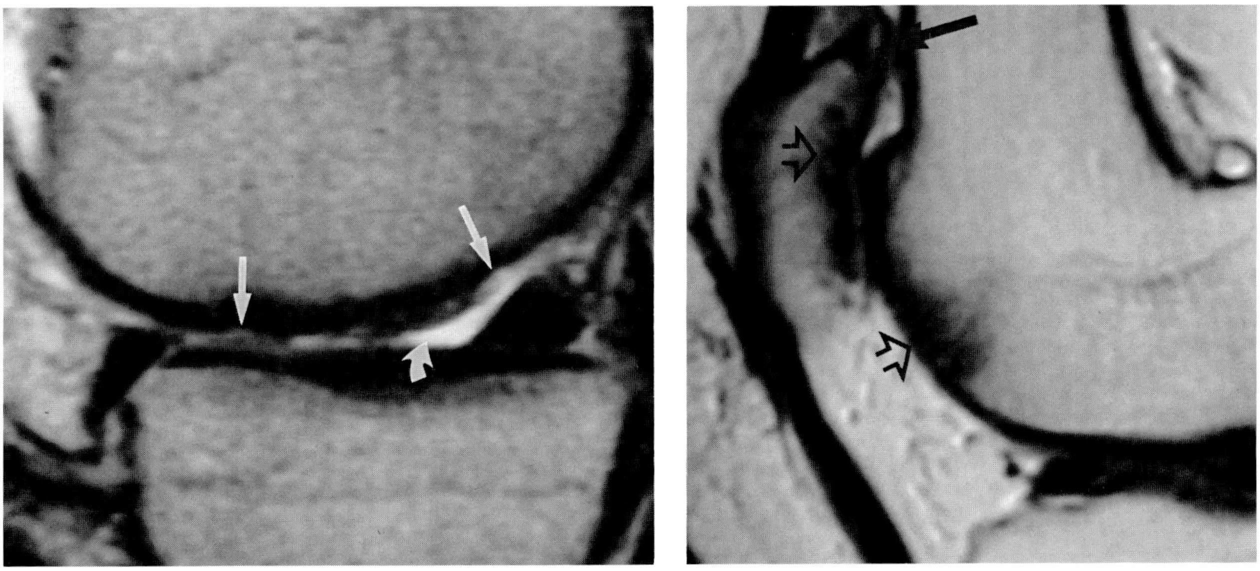

A B

FIG. 9.2. Osteoarthritis. **A:** Sagittal T2-weighted spin-echo image of medial femoral condyle depicts erosions of articular cartilage (*arrows*) outlined by high signal intensity synovial fluid (*curved arrow*). **B:** Patellofemoral osteoarthritis is manifested by erosion of articular cartilage of the patella and trochlear groove as well as osteophyte formation (*straight arrow*) and subchondral erosions (*open arrows*).

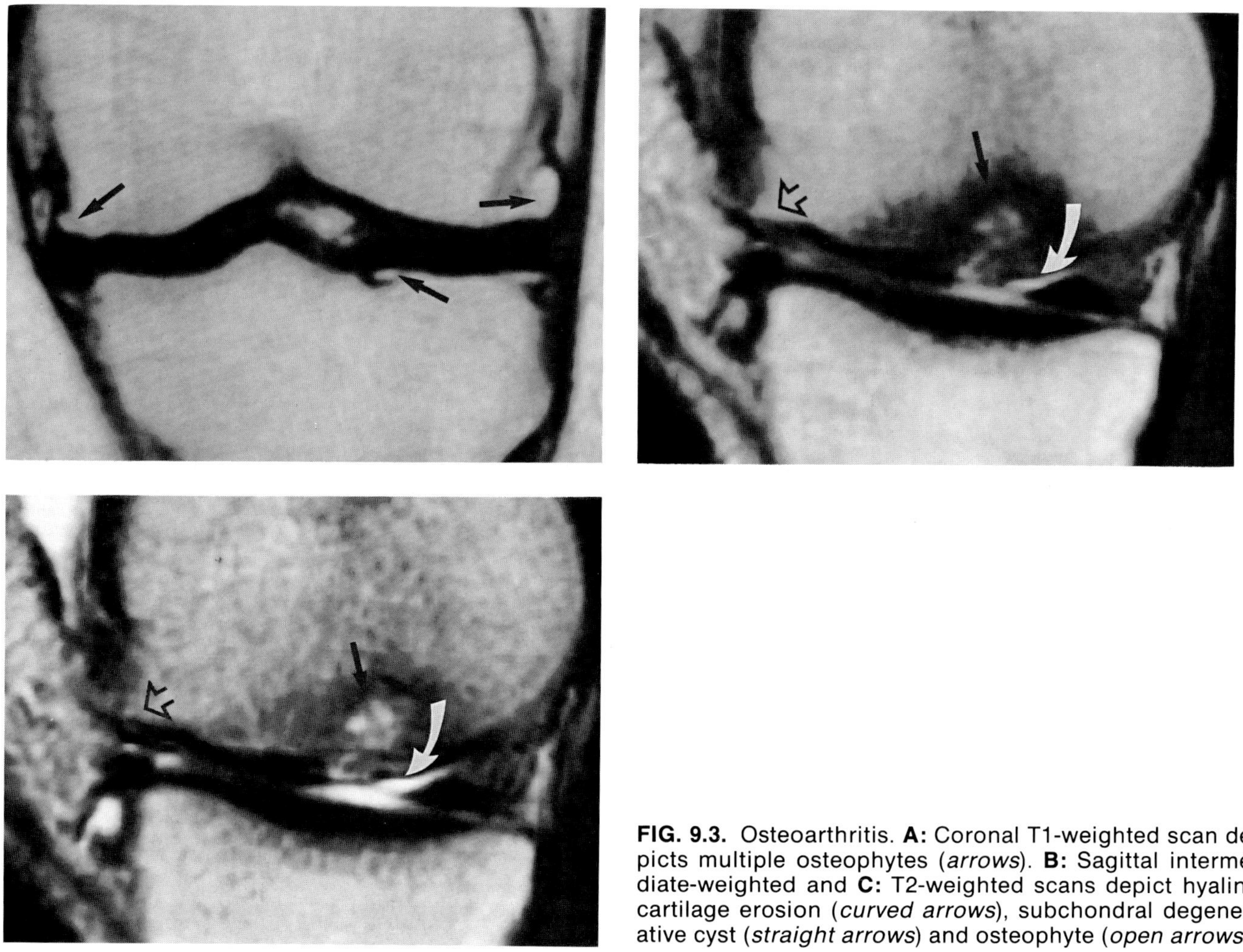

A B C

FIG. 9.3. Osteoarthritis. **A:** Coronal T1-weighted scan depicts multiple osteophytes (*arrows*). **B:** Sagittal intermediate-weighted and **C:** T2-weighted scans depict hyaline cartilage erosion (*curved arrows*), subchondral degenerative cyst (*straight arrows*) and osteophyte (*open arrows*).

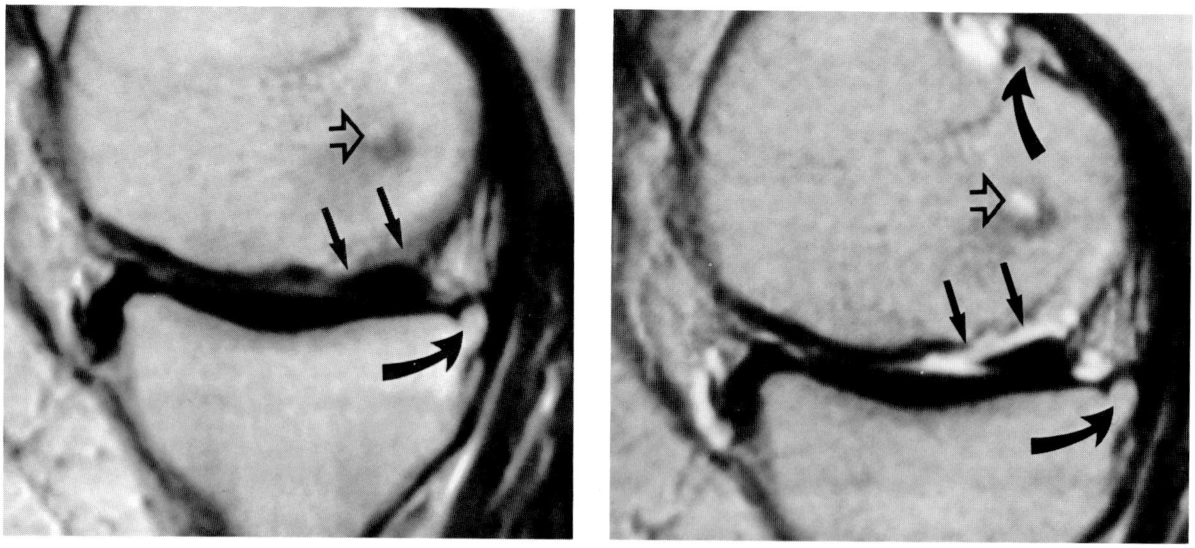

A B

FIG. 9.4. Osteoarthritis. Sagittal **A:** intermediate-weighted spin-echo image and **B:** T2-weighted image depict erosion of hyaline articular cartilage (*straight arrows*), osteophytes (*curved arrows*), and degenerative cysts (*open arrows*). Erosion of hyaline articular cartilage is more easily appreciated on T2-weighted image.

bone fragments frequently are identified on routine radiographs, whereas chondral fractures are generally radiographically occult except for the presence of an effusion. Loose bodies, discussed later in this chapter, are often associated with chondral or osteochondral fractures. Approximately 40% of patients with chondral fractures have associated meniscal tears (58).

There are now numerous reports of MRI findings in chondral and osteochondral fractures, a topic discussed more extensively in Chapter 8. These are depicted as small defects or "dents" in the tibia, femoral condyles, or patella and may be associated with lipohemarthrosis. MRI frequently depicts radiographically occult fractures (3,14,24,33,45,47,99,108,134, 135,165,174,180). MRI has replaced arthrography for detecting chondral and osteochondral fractures, and has proved useful for both establishing the presence of these injuries and determining prognosis. The radiologist performing MRI must recognize these fractures, because often they are serendipitously found in patients being scanned to evaluate clinically suspected meniscal tears. Aggressive surgical therapy, including curettage of the affected cartilage and drilling of subchondral bone, has been previously suggested (58). More recently, MR imaging has revealed that most osteochondral injuries are radiographically occult and heal without surgical intervention.

In the author's experience, T2-weighted spin-echo images seem to be most effective in detecting pure chondral fractures, although no controlled clinical study has firmly established the superiority of any single MR technique. On MR images, pure chondral injuries may have a variety of appearances: linear fissures within the articular cartilage, joint fluid undermining articular cartilage, or frank defects in the articular cartilage (Fig. 9.5). Recently, vacuum phenomenon within the knee joint has been reported to surreptitiously simulate a pure chondral injury (159). Subchondral bony injuries are often present without interruption of the overlying articular cartilage (Fig. 9.6; see also Chapter 8).

The radiologist must be aware that as the normal epiphysis ossifies, the subchondral bone along the posterior margins of the femoral condyles may be irregular

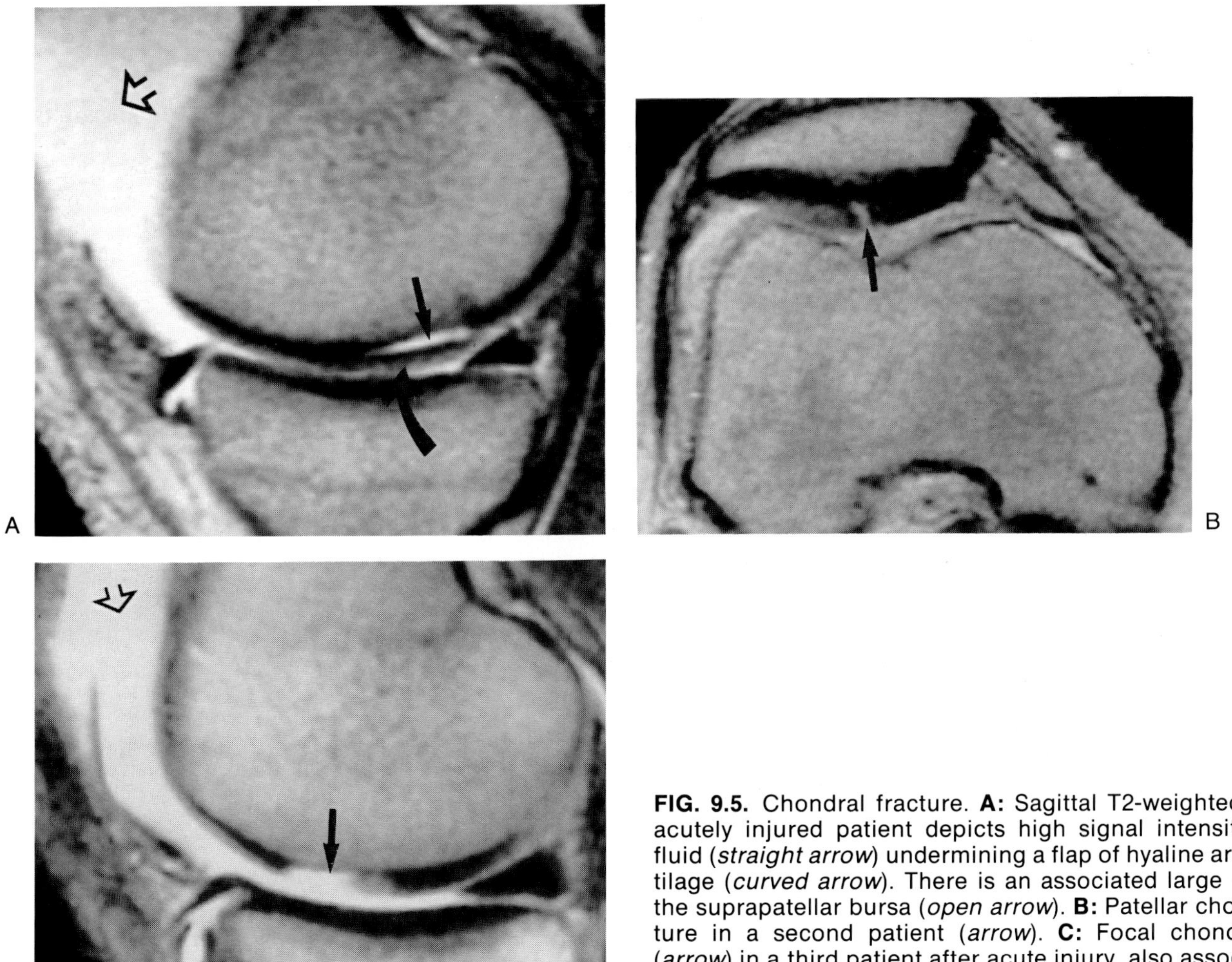

FIG. 9.5. Chondral fracture. **A:** Sagittal T2-weighted image in acutely injured patient depicts high signal intensity synovial fluid (*straight arrow*) undermining a flap of hyaline articular cartilage (*curved arrow*). There is an associated large effusion in the suprapatellar bursa (*open arrow*). **B:** Patellar chondral fracture in a second patient (*arrow*). **C:** Focal chondral defect (*arrow*) in a third patient after acute injury, also associated with acute effusion (*open arrow*).

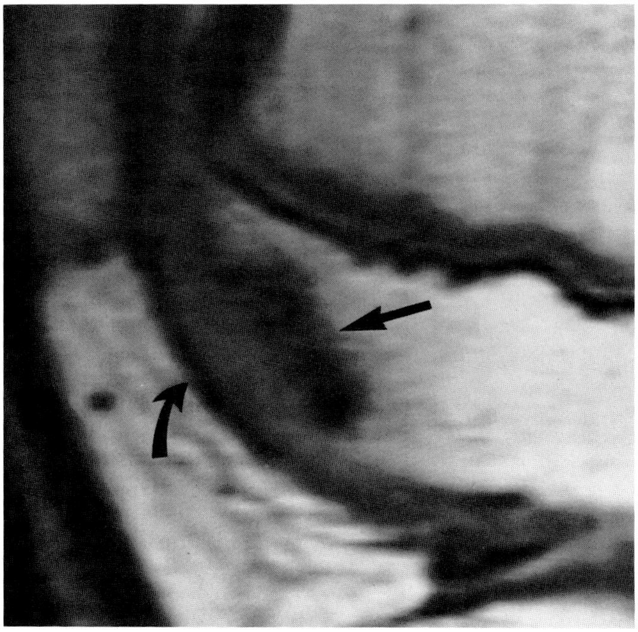

FIG. 9.6. Subchondral bony injury (*arrow*) in adolescent patient, with intact overlying articular cartilage (*curved arrow*). A low signal intensity line demarcates the margin in the intact articular cartilage (*curved arrow*).

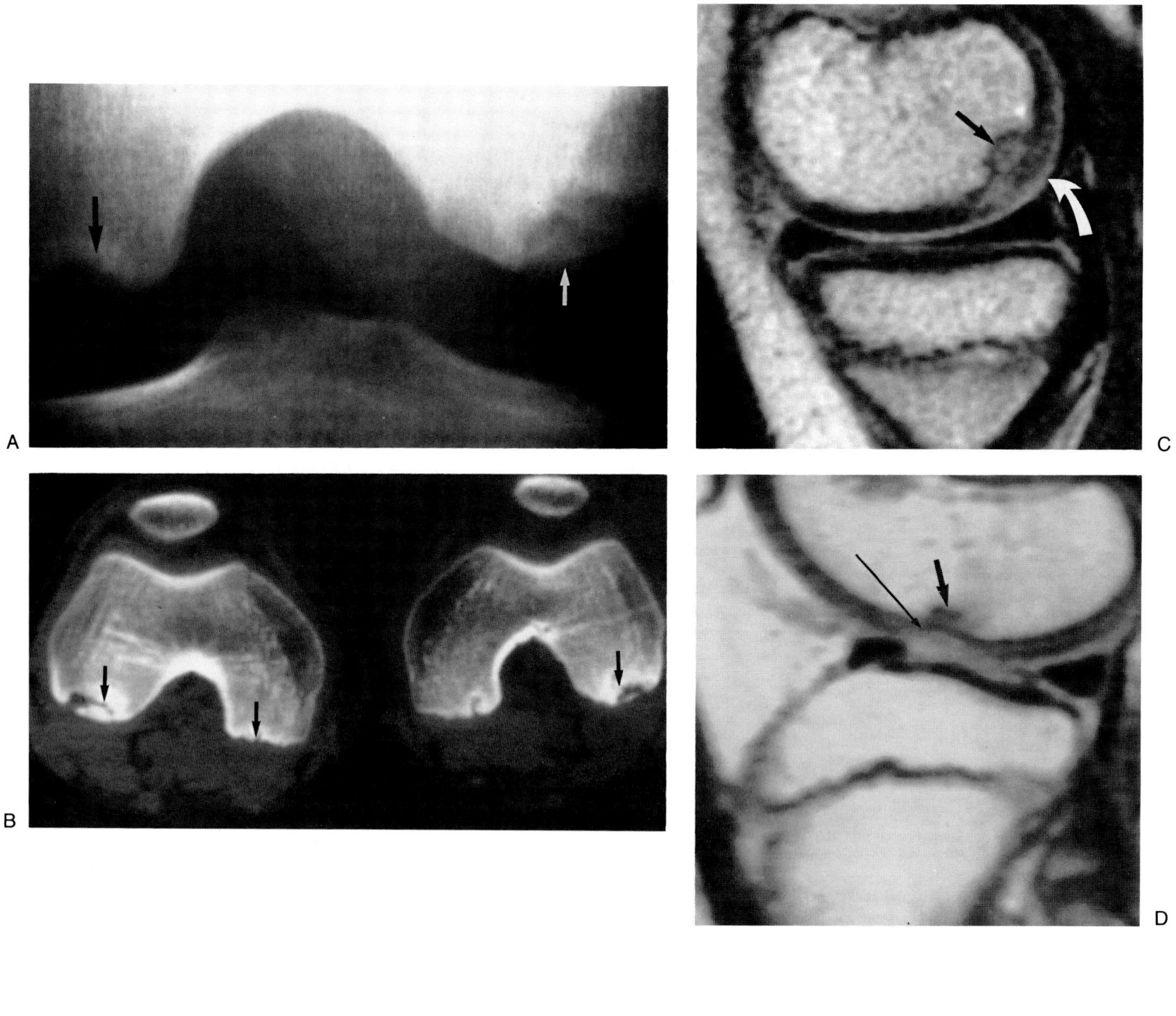

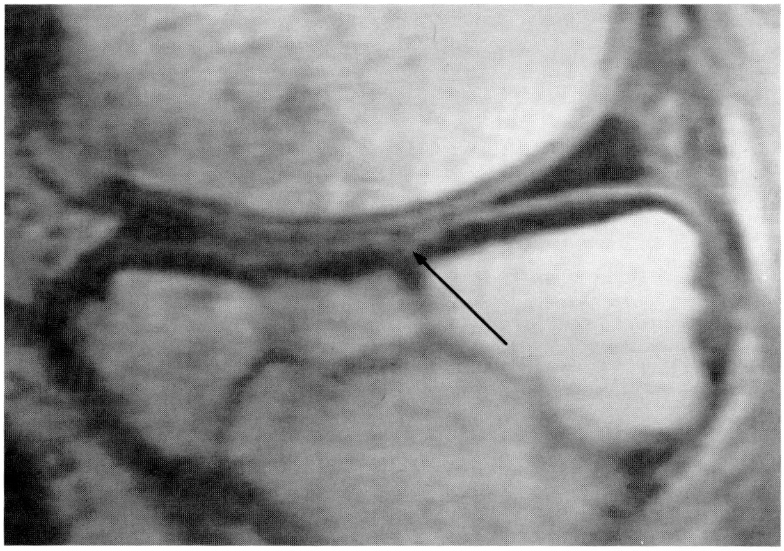

FIG. 9.8. Posttraumatic hyaline cartilage defect. Sagittal image in patient with old healed tibial plateau fracture depicts even subtle old injury to hyaline articular cartilage (*arrow*).

in contour and contain multiple small calcifications or ossifications (Fig. 9.7). This normal variant should not be mistaken for an osteochondral injury (Fig. 9.7).

MRI is also useful in delineating cartilaginous fractures associated with overt bony fractures, such as fractures of the tibial plateau. Thorough assessment of the degree of comminution and displacement of fragments is required to properly plan therapy for tibial plateau fractures. Most fractures involve the lateral tibial plateau and result from an abduction or valgus strain combined with an axial or compression force (131). The anterior portion of the lateral femoral condyle is thus driven into the lateral tibial plateau. Most authors agree that open reduction and internal fixation are indicated when there is associated instability of the anterior cruciate ligament or medial collateral ligament (8,28,131). In addition to delineating fracture fragments and associated soft tissue injuries, MRI may disclose even subtle displacements or fractures of articular cartilage, either acutely or after fracture healing (Fig. 9.8).

Osteochondritis Dissecans

Osteochondral injuries are more extensively discussed in Chapter 8, but osteochondritis dissecans is reviewed here since it is a distinct clinical entity. Osteochondritis dissecans is a painful, usually unilateral disease of children and young adults in which a segment of articular cartilage and an attached fragment of subchondral bond become partially or completely separated from the underlying bone (9,55,70,93,141). The most commonly involved site is the lateral aspect of the medial femoral condyle, although the lateral condyle may be affected in up to 15% to 20% of cases (29,55,70), and the patella may be rarely involved (55,141). Osteochondritis dissecans of the patella usually occurs in the medial facet and involves the middle and lower portions of the patella. An osteochondritis dissecans-like syndrome has also been reported to occur on the anterior articular surface of the lateral femoral condyle, at the level of the superior pole of the patella (31).

Proposed causes of osteochondritis dissecans include (a) patellar impingement on the medial femoral condyle with full flexion of the knee, (b) impingement of the anterior tibial spine on the medial femoral condyle with full knee extension, (c) secondary local ischemia occurring because of repeated trauma, and (d) avascular necrosis secondary to multiple emboli of fat globules or clumped red blood cells (31,70). Most authors favor an acute or chronic traumatic cause. Once separated, the subchondral bone fragment is devascularized and often is covered by fibrous tissue that may prevent revascularization. If the fragment remains partially attached to the underlying bone, revascularization and healing can occur. Alternatively, the frag-

FIG. 9.7. Normal variant vs. osteochrondral injury. **A:** Standard radiograph, **B:** axial CT scan, and **C:** sagittal MR image all demonstrate normal variation in epiphyseal ossification (*arrows*). This finding is typically seen along the posterior convexities of the medial and lateral femoral condyles in children. MRI (C) reveals intact overlying articular cartilage (*curved arrow*). In contrast, osteochondral injury (**D**) occurs in a different location (*arrow*) and is associated with buckling of overlying articular cartilage (*long arrow*).

ment may completely detach and become an intraarticular loose body that can grow, ossify, or calcify. The site of origin may eventually become remodeled and flattened.

Patients with osteochondritis dissecans typically complain of knee pain, effusions, or locking due to loose bodies. Standard radiography may reveal fragmentation of the subchondral bone, but is insensitive early in the course of this disease and cannot be used to assess articular cartilage or the presence of nonradiopaque loose bodies. For these reasons, arthrography and scintigraphy were previously often used to evaluate osteochondritis dissecans (9,55,125). MRI has essentially replaced arthrography and scintigraphy in the evaluation of osteochondritis dissecans because of its superior assessment of articular cartilage and subchondral bony fragments. Precise anatomic assessment of patients with osteochondritis dissecans is clinically important because small fragments that remain attached often heal with conservative therapy, whereas patients with loose fragments are best treated surgically (drilling or pinning large fragments or removing small fragments) (31,70). The prognosis is poor if the lesion is large or involves a weightbearing surface. MR imaging is ideal for evaluating osteochondritis dissecans because of the combined capability of detecting cartilaginous defects, identifying cartilaginous loose bodies, sizing and delineating the precise location of the lesion, and predicting lesion stability (45,66,106,160). The literature thus far suggests that MRI is approximately 90% accurate in determining whether osteochondral fragments are stable or unstable. Dissection of high signal intensity synovial fluid or granulation tissue between the osteochondral lesion and surrounding cancellous bone on T2-weighted images is an indicator of lesion instability. MRI is useful in differentiating the single unfragmented normal signal intensity osteochondral lesion (Fig. 9.9), which may be surgically reattached, from the more common cases of subchondral fragmentation.

Osteochondral Loose Bodies

Intraarticular loose bodies occur when osseous or cartilaginous fragments become detached within a joint. Fragments of meniscal or articular cartilage are nourished by synovial fluid and may grow, calcify, or ossify. Causes include osteochondritis dissecans, chondral fractures, osteoarthritis, chondromalacia, meniscal tears, and synovial chondromatosis. loose bodies may cause recurrent effusions, locking, and restricted range of motion. Although most loose bodies calcify and are therefore visible on standard radiographs, conventional x-rays often do not allow differentiation between a loose body and a meniscal ossicle or calcification of the cruciate ligaments, menisci, or joint capsule.

Arthrography may be employed to detect noncalcified loose bodies, determine the precise locations of loose bodies, and determine if a small calcification

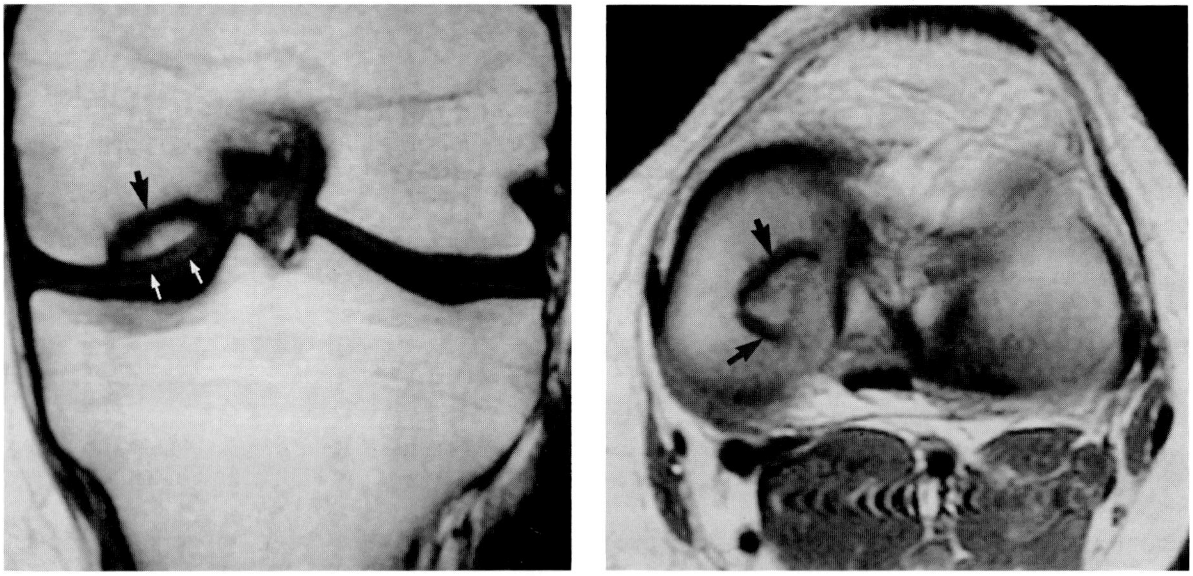

A B

FIG. 9.9. Osteochondritis dissecans. **A:** Coronal and **B:** axial T1-weighted spin-echo images depict classic osteochondritis diseccans lesion (*arrows*). Lesion size, precise location, status of overlying articular cartilage (*white arrows*), and degree of fragmentation are all assessable.

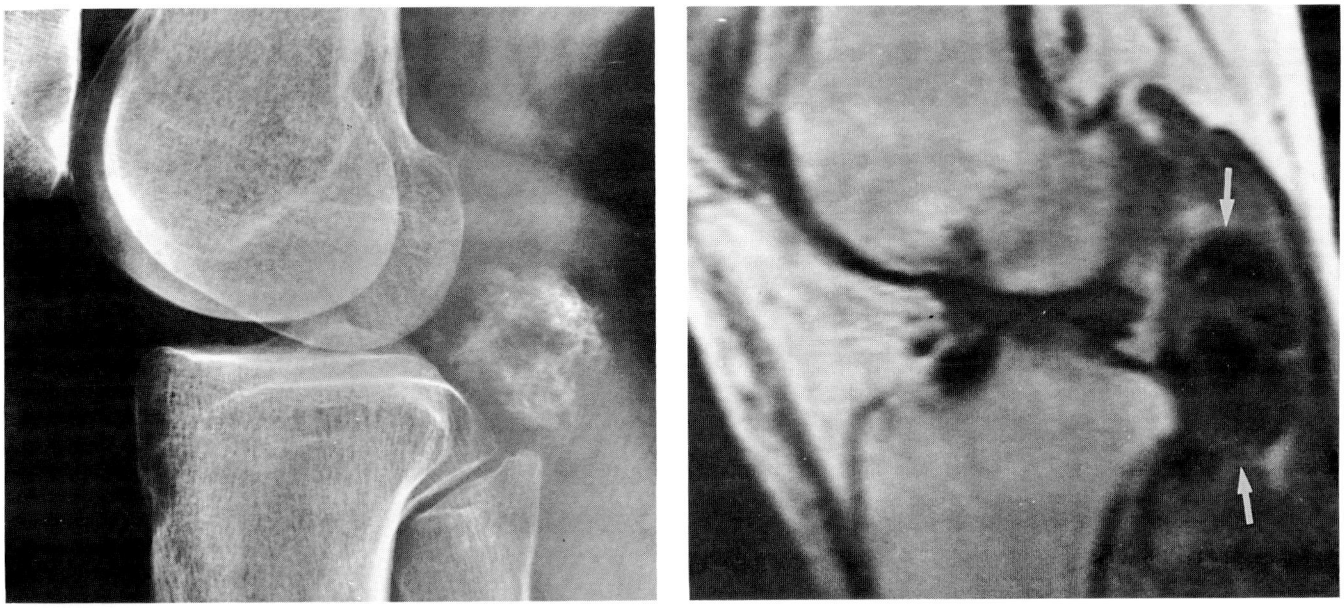

A B

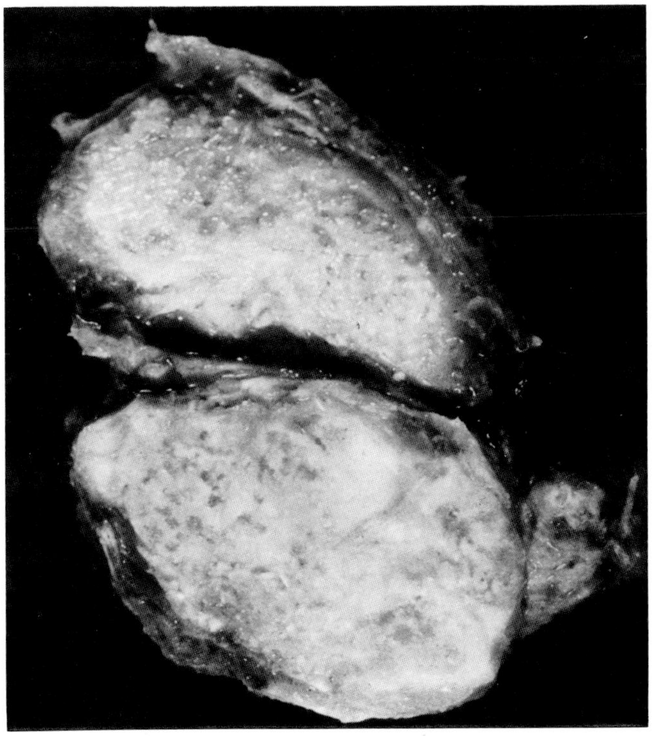

C

FIG. 9.10. Osteochondral loose bodies. **A:** Standard radiograph reveals fluffy calcification in posterior aspect of the knee. The differential diagnosis included the possibilities of a loose body and synovial chondrosarcoma. **B:** Sagittal T1-weighted MRI depicts two well-circumscribed, faceted, low intensity osteochondral bodies (*white arrows*). **C:** Surgical specimen provides precise correlation.

seen on standard radiographs has a large cartilaginous component (9,55,125). Pitfalls in arthrography include confusion between air bubbles and loose bodies when a double contrast technique is used, or obscuration of calcified bodies when a single contrast technique is employed. To avoid these pitfalls, air arthrography has been advocated (125), but air may not adequately cover all joint surfaces (150). CT-arthrography may overcome these problems, but it is costly and time-consuming.

MRI can detect intraarticular loose bodies of the knee joint (66,107,108,134,135,136). Densely calcified cartilage bodies or meniscal fragments usually yield low signal intensity on T1-weighted images (Figs. 9.10

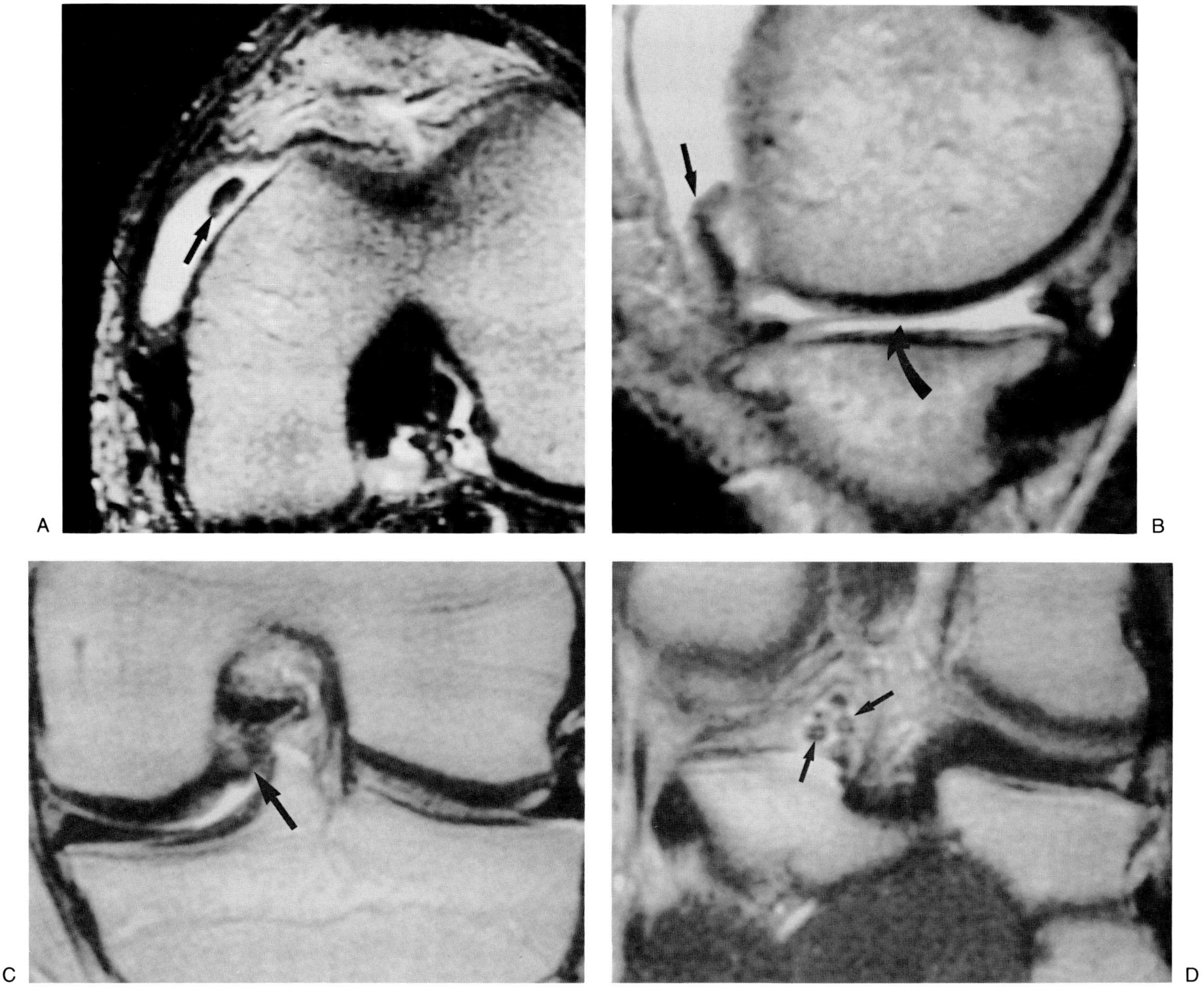

FIG. 9.11. Osteochondral loose bodies. **A:** Axial T2-weighted MRI depicts low signal intensity fibrocartilaginous loose body (*arrow*) surrounded by high signal intensity synovial fluid. **B:** Sagittal T2-weighted MRI depicts same loose body (*straight arrow*) and further reveals that the loose body is, in fact, a displaced meniscal fragment (*curved arrow*). **C,D:** T2-weighted spin-echo images in different patients demonstrating typical low signal intensity loose bodies (*arrows*) are seen.

and 9.11). Ossified fragments may yield high signal on either T1- or T2-weighted images because of the inclusion of marrow fat (Fig. 9.12). Loose bodies most frequently occur in the intercondylar notch, suprapatellar bursa, medial and lateral gutters, or adjacent to the anterior horn of the lateral meniscus. Unusual locations of osteochondral loose bodies include the popliteal tendon sheath bursa (Fig. 9.12) and the infrapatellar fat pad (Fig. 9.13). The radiologist and surgeon must both be aware that osteochondral bodies partially embedded within the infrapatellar fat pad may not be arthroscopically visible unless aggressively pursued (Fig. 9.13).

In the author's opinion, osteochondral loose bodies are optimally detected only when properly performed standard radiographs are carefully compared to MR images. The radiologist must also be cognizant of all normal knee structures which may simulate loose bod-

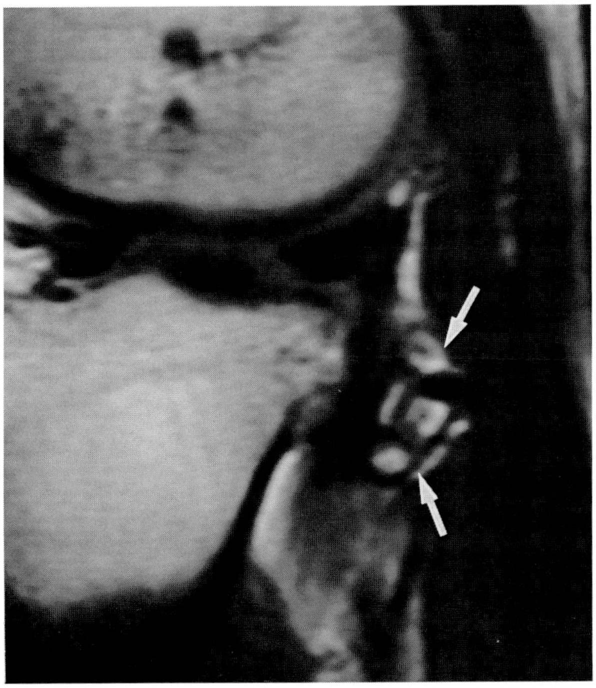

FIG. 9.12. Osteochondral loose bodies. Sagittal T1-weighted MRI depicts multiple high signal intensity osseous bodies within the popliteal tendon sheath bursa (*arrows*).

ies, including the meniscofemoral ligaments of Humphrey and Wrisberg, the transverse geniculate ligament, blood vessels perforating the infrapatellar fat pad, and fat globules or air bubbles within synovial fluid.

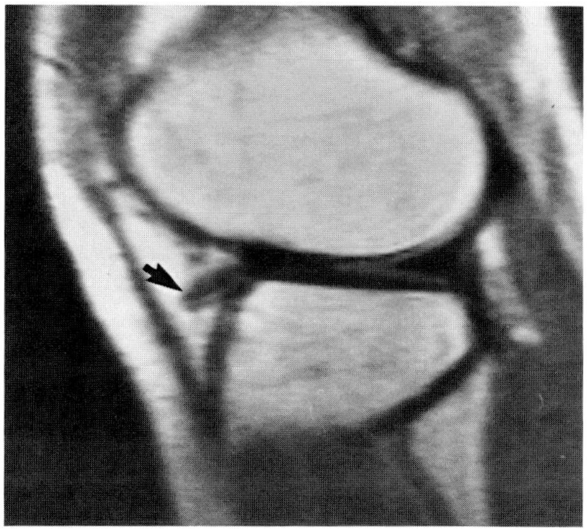

FIG. 9.13. Osteochondral loose body. Sagittal T1-weighted MR image reveals oval low intensity fragment embedded in infrapatellar fat pad. The loose body was not detected at prior arthroscopy, but after MRI, localized incisional arthrotomy was performed and the symptomatic loose body was excised.

OSTEOCHONDROSES

The osteochondroses constitute a heterogeneous group of disorders that tend to involve epiphyses and apophyses of the immature skeleton, leading to fragmentation of subchondral bone. The two disorders involving the knee that are usually classified in this group are Blount disease and Osgood-Schlatter disease.

Blount Disease

This disorder, also known as tibia vara or osteochondrosis deformans tibiae, has two distinct clinical presentations (62,141). The infantile form presents with a painless varus deformity of the knee, is associated with obesity, and affects both knees in 60% of cases. Typically a worsening of physiologic tibial bowing is noted when the child begins to ambulate. It is more common among those of black African heritage and other populations in which children begin ambulating at an early age. Possible causes include trauma, growth arrest, or dysplasia. The adolescent form appears in children 8 to 15 years old. It may be painful, usually is unilateral (90%), and may result in moderate to severe limb shortening (141).

Radiographs in Blount disease demonstrate irregularity and a beak-like projection of the medial tibial metaphysis that points medially and distally. There is hypertrophy of the articular cartilage of the medial tibial plateau that initially maintains a flat articular surface but later becomes angulated along with the underlying bone. MRI can be employed to directly depict the hypertrophied hyaline articular cartilage overlying the medial tibial plateau.

Osgood-Schlatter Disease

This trauma-induced avulsion of the growing tibial tuberosity occurs in adolescents and is bilateral in 25% of cases (154). Boys are affected three times more often than girls. Patients are typically athletically active and often have undergone a growth spurt prior to onset of symptoms. Symptoms include pain and tenderness of the tuberosity and are exacerbated by physical activity. On physical examination, local soft tissue swelling or a firm mass may be palpable. This disease affects not only the tuberosity but also the patellar tendon and infrapatellar fat pad, which may be thickened and edematous. Standard radiographs may demonstrate fragmentation of the tibial tuberosity, but interpretation may be difficult, because boys age 9 to 14 and girls age 8 to 12 normally have multiple ossification centers in the cartilaginous portion of the tuberosity (141). Soft tissue changes such as loss of the sharp inferior angle

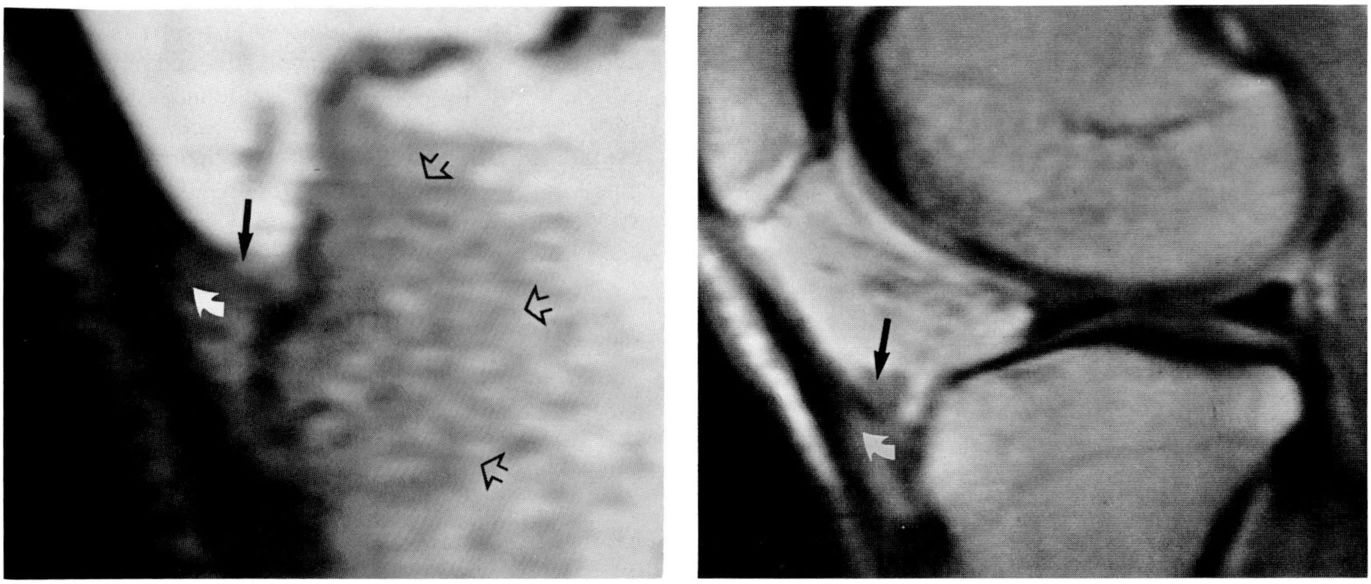

FIG. 9.14. Early Osgood-Schlatter disease. **A,B:** Sagittal T1-weighted scans in two different patients depict effusions within the infrapatellar bursa (*straight arrows*) as well as low signal intensity edema within the bone marrow of the tibial tuberosity (*open arrows*). In each case, tendon fibers are partially disrupted (*curved arrows*). Intratendinous ossifications are also common in Osgood-Schlatter disease.

of the infrapatellar fat pad, soft tissue swelling anterior to the tubercle, and thickening of the patellar tendon are more reliable signs. For this reason, low voltage radiography or xeroradiography was previously advocated in evaluating Osgood-Schlatter disease (141,154). Although data are limited, MRI has thus far been found to be effective for evaluating patellar tendon injuries, the infrapatellar fat pad, and the infrapatellar bursa (66,134,135).

On MR images, Osgood-Schlatter disease is characterized by effusion within the infrapatellar bursa, partial disruption of the inferior margin of the patellar tendon, and often edema within the bone marrow adjacent to the tibial tubercle (Fig. 9.14). Small intratendinous calcifications may not be detectable with MRI. T2-weighted images are useful to differentiate areas of tendinitis from frank tendon tears. In most cases, Osgood-Schlatter disease is accurately clinically diagnosed without the need for MR examination, but in rare cases, particularly when surgical intervention is considered, MR may be useful to stage the severity of tendon disruption and precisely localize the site of disease.

SYNOVIAL DISEASES AND MASSES

Numerous conditions may cause synovial inflammation or thickening, including rheumatoid arthritis, gout, calcium pyrophosphate dihydrate deposition disease, pigmented villonodular synovitis, infection (tuberculous or fungal), hemophilic arthropathy, amyloid arthropathy, synovial chondromatosis, synovial hemangioma, and lipoma arborescens (9,30,37,55,65, 103,107,125,141,143,157,166). Both standard radiography and arthrography are limited for evaluation of these disorders because the primary site of disease, the synovium, cannot be visualized directly using either method. MRI, which yields physiologic and anatomic information, offers a tool for both early diagnosis and assessment of the response to therapy.

Numerous investigators have now reported the use of MRI for evaluation of synovial diseases and intraarticular masses (2,23,27,36,66,73,82,86,88,100,140,157, 166,167,181). As a general rule, multiple techniques are usually necessary to completely evaluate synovial disorders. The following is a brief review of the conditions that have been evaluated with MRI.

Rheumatoid Arthritis

Adult-onset rheumatoid arthritis is characterized by synovial thickening, articular erosions, subluxations, subchondral cysts, effusions, popliteal cysts, and osteoporosis. In contrast to the adult-onset form, in which articular erosion occurs early, joint space narrowing in juvenile rheumatoid arthritis (JRA) is a late finding. Growth disturbances, such as metadiaphyseal constriction and ballooning of distal femoral and proximal tibial epiphyses, are common features of JRA. Other findings common in JRA include osteoporosis,

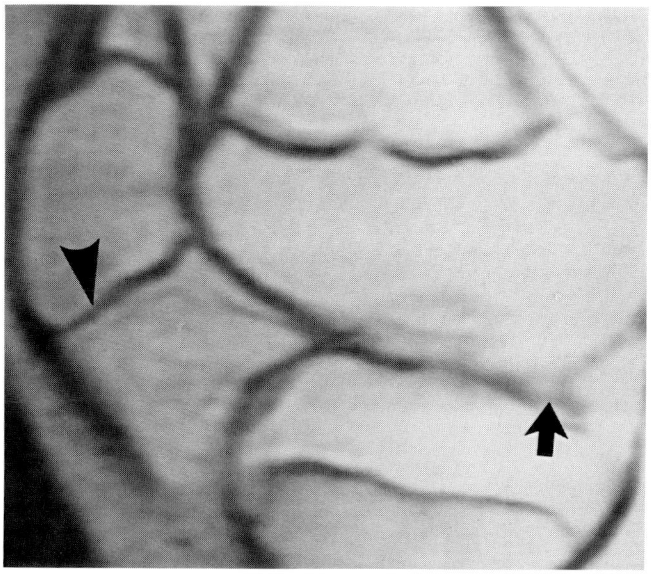

FIG. 9.15. Juvenile rheumatoid arthritis. Sagittal intermediate-weighted MRI depicts typically enlarged epiphysis, meniscal hypoplasia (*arrow*), and hypoplasia or "squaring" of the patellar inferior pole (*arrowhead*).

flattening of the femoral condyles, widening of the intercondylar notch, "squaring" of the inferior margin of the patella, and marginal or central osseous erosions (30,62,103,141). Hypoplasia of menisci has also been reported in JRA, with the menisci ranging in size from slightly small to virtually absent (157).

Prior to MRI, arthrography was occasionally employed to evaluate rheumatoid arthritis. Arthrographic findings, however, are nonspecific and include enlargement of the joint cavity, nodular irregularity or corrugation of the synovial membrane, subchondral cysts, intraarticular filling defects, and lymphatic filling. As a single examination, MRI painlessly evaluates synovial hypertrophy, cartilaginous erosions, subchondral cysts, noncommunicating synovial cysts, meniscal hypoplasia, bony growth disturbances, and associated bone marrow infarctions (Figs. 9.15–9.17). Fibrous pannus typically yields low signal intensity on T2-weighted images (Fig. 9.17). On standard MR images, inflamed synovium, which yields high signal intensity on T2-weighted images, may appear indistinguishable from joint effusion (Fig. 9.18).

Several authors have advocated the use of intravenous gadolinium compounds to improve detection of inflamed synovium (2,19,23,82,88). Proliferative inflamed synovial tissue is generally hypervascular and thus preferentially increases in signal intensity on T1-weighted images performed after intravenous gadolinium infusion. Gradient-echo techniques may also be employed to perform a rapid series of T1-weighted images during which dynamic uptake of intravenous gadolinium by inflamed synovium may be documented (82). Gadolinium enhancement helps differentiate synovitis from joint effusion, and subchondral pannus from subchondral sclerosis. Gadolinium enhancement of inflamed synovium is not specific for rheumatoid arthritis, and may be seen with other inflammatory arthropathies such as psoriatic arthritis (Fig. 9.19).

A

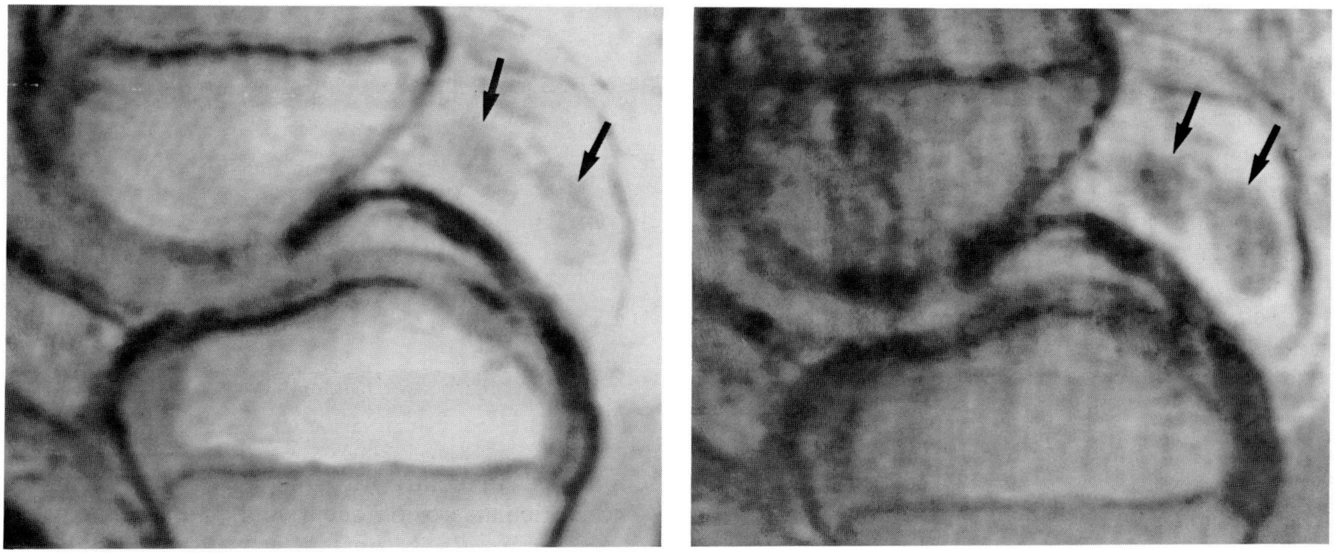

B

FIG. 9.16. Juvenile rheumatoid arthritis. **A:** Sagittal intermediate-weighted and **B:** T2-weighted images reveal effusion and low signal intensity intraarticular nodules (*arrows*) which are masses of fibrous pannus. Progressively decreasing signal intensity with T2 weighting implies fibrosis. Fibrous pannus is most clearly demarcated on T2-weighted images, in contrast to high signal intensity synovial fluid.

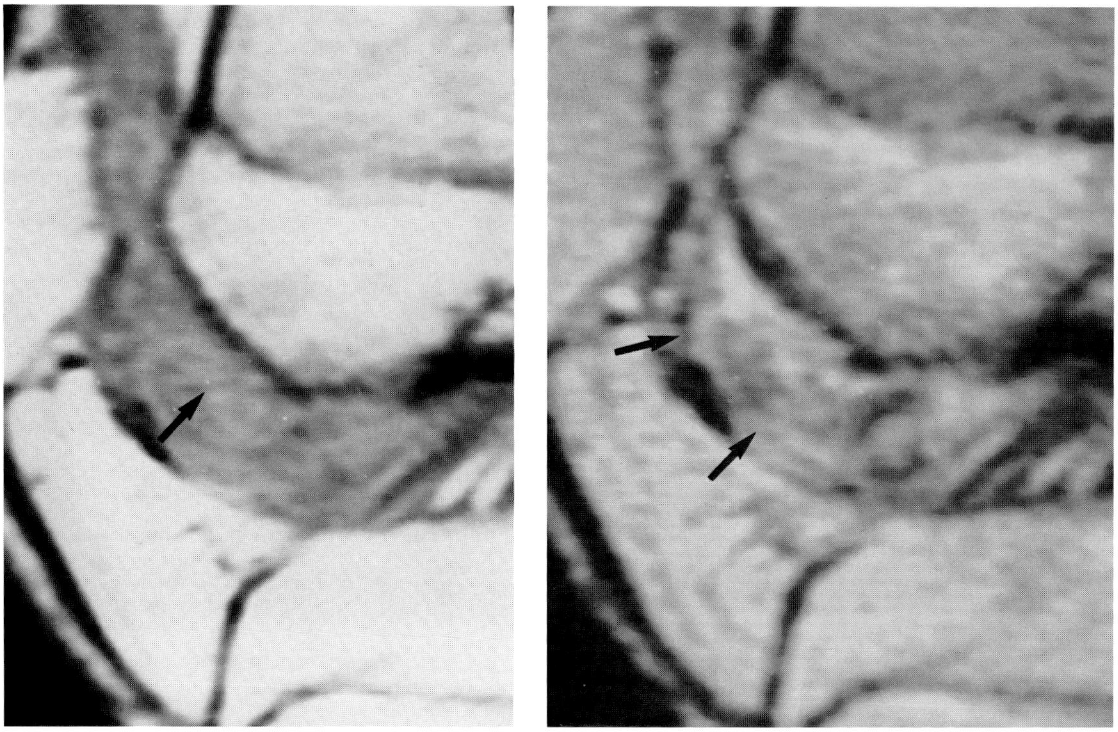

A B

FIG. 9.17. Juvenile rheumatoid arthritis, hypertrophied synovium with fibrous scarring. **A:** Sagittal intermediate-weighted MRI reveals moderate signal intensity collection (*arrow*) which could easily be mistaken for synovial fluid. **B:** On T2-weighted section, signal intensity declines, indicating fibrous pannus. Patellar articular cartilage and articular cartilage of the trochlear groove are severely eroded.

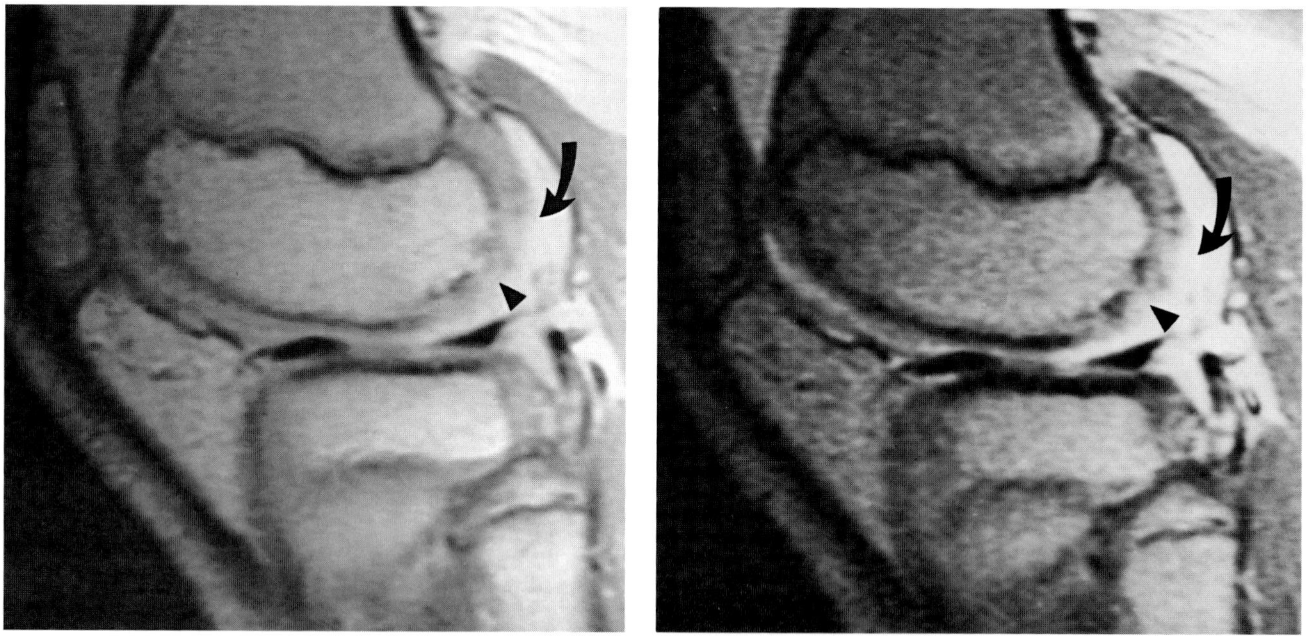

A B

FIG. 9.18. Juvenile rheumatoid arthritis. **A:** Sagittal intermediate-weighted and **B:** T2-weighted MR images depict effusion and inflamed thickened synovium (*curved arrows*) as well as cartilaginous erosion (*arrowheads*). The lateral meniscus is hypoplastic. Inflamed synovium and synovial fluid cannot be reliably distinguished on spin-echo images without intravenous contrast enhancement.

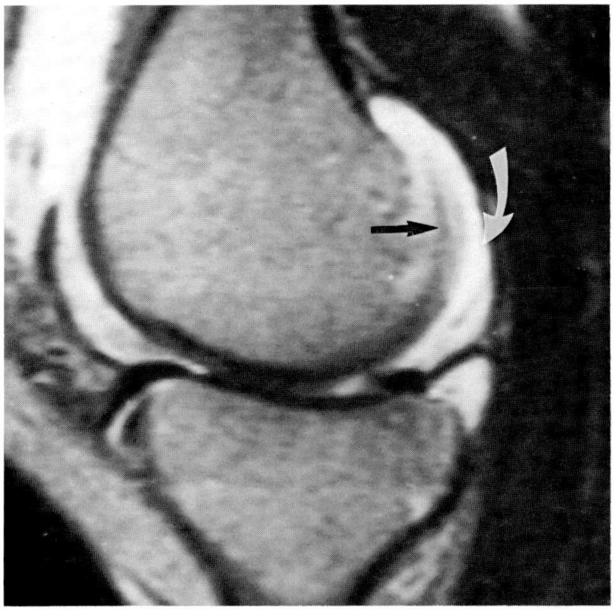
A

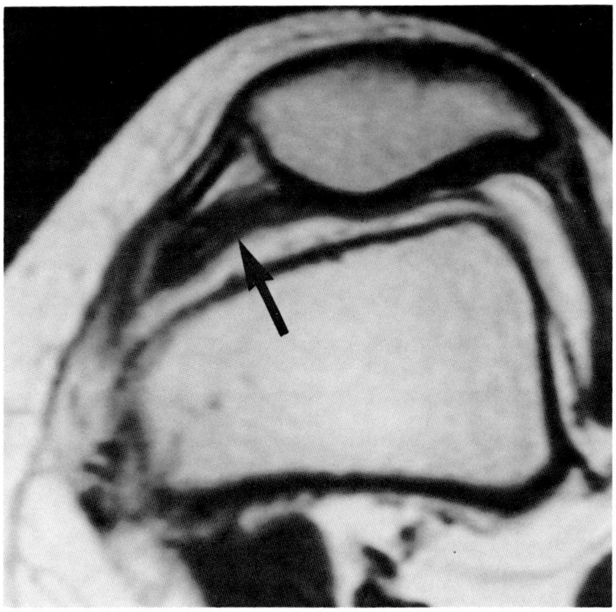
B

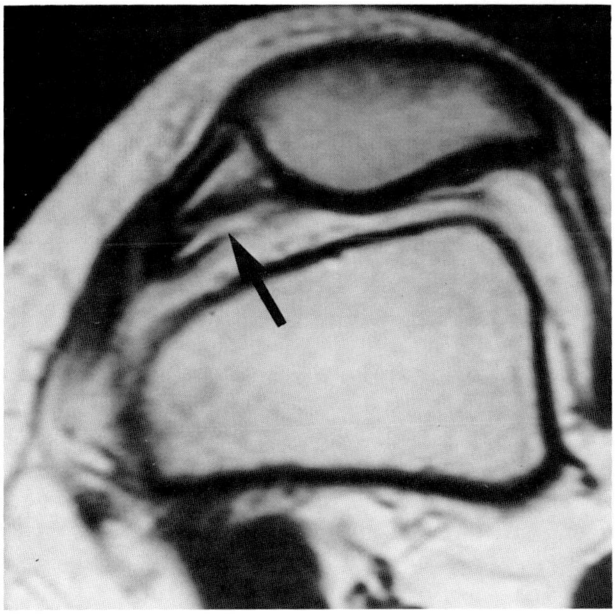
C

FIG. 9.19. Psoriatic arthritis. **A:** Sagittal T2-weighted spin-echo image depicts large effusion and hypertrophied synovium (*curved arrow*) which cannot be reliably distinguished. Articular cartilage along the posterior convexity of the medial femoral condyle is thinned (*straight arrow*). **B:** Axial T1-weighted MRI in the same patient depicts an apparent effusion within the medial recess of the suprapatellar bursa (*arrow*). **C:** After intravenous infusion of gadolinium compound, enhancement of thickened synovium becomes apparent (*arrow*).

Pigmented Villonodular Synovitis

Pigmented villonodular synovitis (PVNS) is a monoarticular synovial disease that affects either the joints or tendon sheaths in young adults (37,49,55,60, 73,97,100,125,141,179). The knee is involved in 80% of articular disease. Other joints that may be involved, in decreasing order of frequency, include the hip, ankle, hand, foot, and elbow. Articular disease has two forms: diffuse and nodular. In the diffuse form, patients typically present with pain, decreased range of motion, joint swelling, warmth, and tenderness. The patient's erythrocyte sedimentation rate is usually normal. Rarely are two or more joints affected (37). In the diffuse form, the joint fluid is usually sanguineous or dark brown, not viscous, and does not usually clot (37). The affected joint typically gradually swells due to effusion and synovial proliferation. Focal masses may be palpable in or surrounding the joint. The principal symptoms are pain and swelling.

The nodular form is four times less common than the diffuse form of PVNS (74,119). The nodular form may cause knee locking, but is usually less symptomatic than the diffuse form, and may be clinically mistaken for a meniscal tear (60,97,179). Patients generally are symptomatic for months or even years before the diagnosis is established. The synovial fluid is less likely to be sanguineous in the nodular form. Radiographs are usually normal (74,119).

Pathologic examination reveals fibrous synovial pro-

A 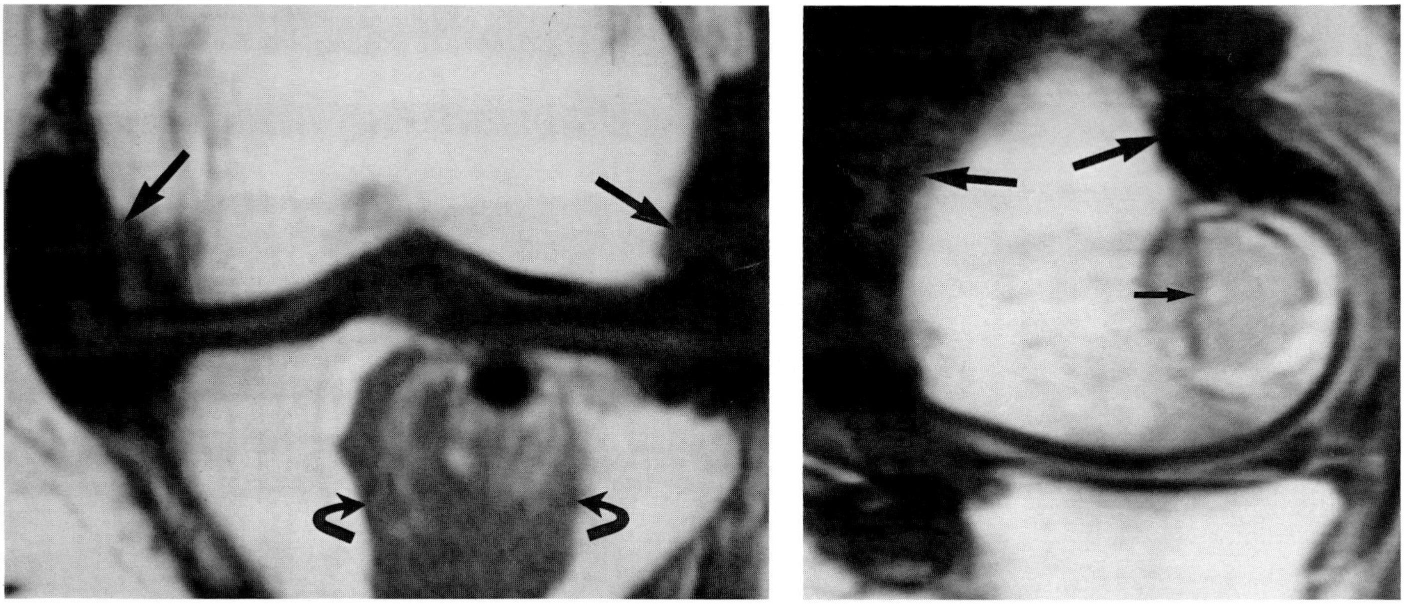 B

FIG. 9.20. Pigmented villonodular synovitis. **A:** Coronal T1-weighted spin-echo image depicts massively hypertrophied low signal intensity synovium (*arrows*) resulting in osseous erosions. A hemorrhagic erosive bony cyst is present in the proximal tibia (*curved arrows*). **B:** Sagittal intermediate-weighted image again depicts massive hypertrophy of low signal intensity synovium (*large arrows*). A hemorrhagic bony cyst is present, containing a serum/sediment level (*small arrow*).

liferation and infiltration by histiocytes, multinucleated giant cells, and hemosiderin-laden macophages (49,60,125). The cause is unknown, but PVNS is postulated to be a benign inflammatory reaction to some unidentified agent (179). The high frequency of bony erosions and the common recurrence after synovectomy in the diffuse form of PVNS support the theory that PVNS is a neoplasm. Histologic criteria of malignancy are lacking, however, and no metastases have ever been reported.

Typical radiographic findings in PVNS include a noncalcified synovial soft tissue mass, subchondral and juxtaarticular erosions and cysts, normal bone density, normal joint space, and lack of hypertrophic bone formation (49,141). Arthrography has been reported in both the diffuse and nodular forms (49,55,60, 97,141). In the diffuse form, arthrograms disclose an enlarged joint cavity and irregular nodular synovial masses with villous projections, an appearance that may be mimicked by many synovial processes. In the

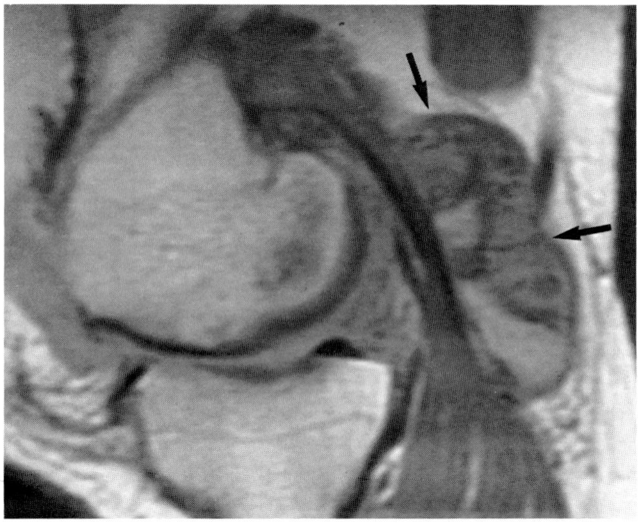

FIG. 9.21. Pigmented villonodular synovitis. Sagittal intermediate-weighted image reveals enlarged gastrocnemius-semimembranosus bursa, packed with low signal intensity PVNS (*arrows*).

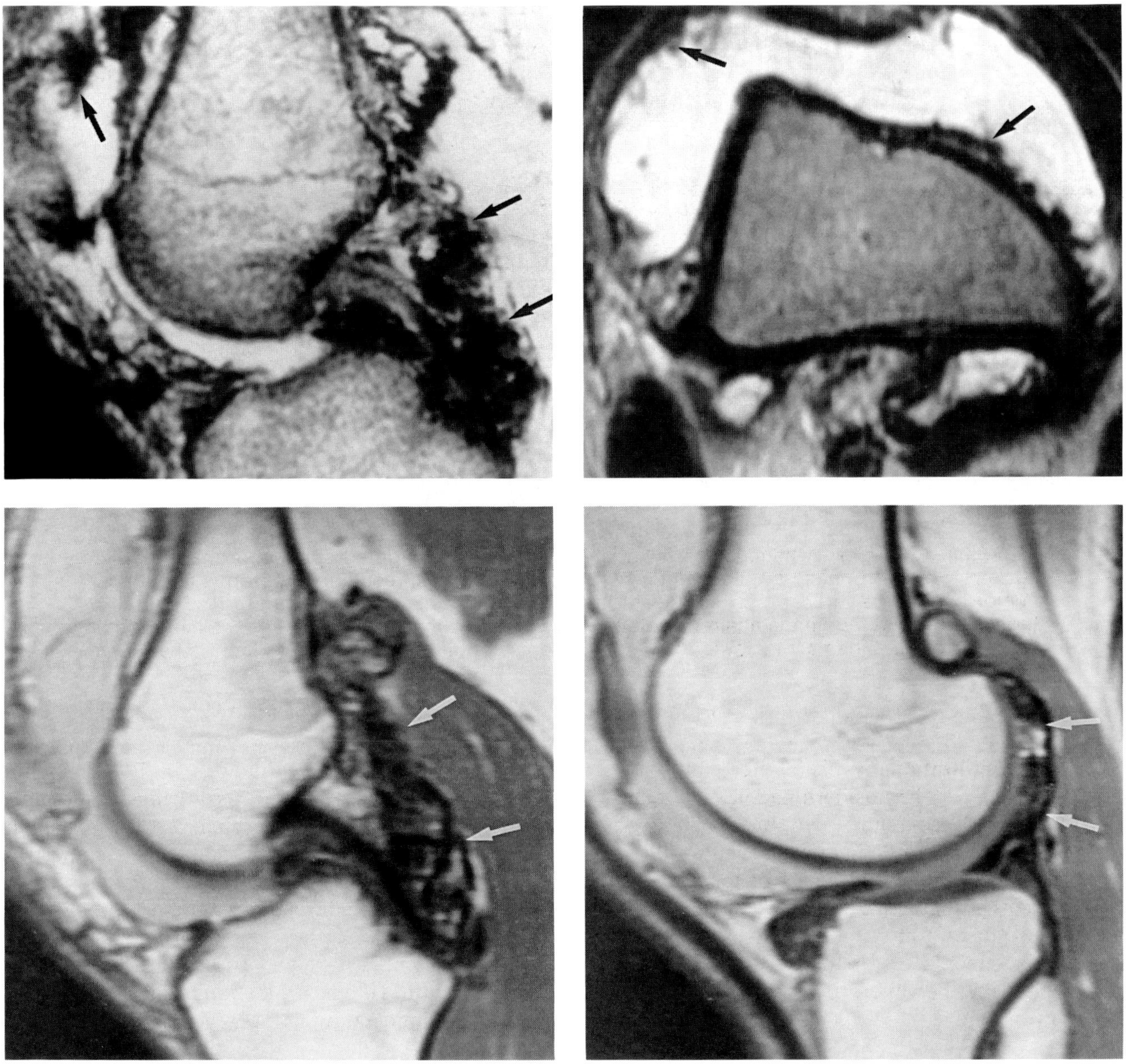

A B C D

FIG. 9.22. Pigmented villonodular synovitis. **A,B:** Sagittal gradient-echo image and axial T2-weighted spin-echo image depict low signal intensity hypertrophied synovium within both anterior and posterior aspects of the knee joint (*arrows*). **C,D:** Sagittal intermediate-weighted scans in another patient reveal extensive disease remaining in the posterior aspect of the knee joint after synovectomy performed via anterior approach.

nodular form, a circumscribed mass may be seen in the infrapatellar region, or arthrography may be normal. Numerous cases of MRI of PVNS have now been reported (73,100,107). Because of the presence of hemosiderin, PVNS typically yields low signal intensity on both T1- and T2-weighted images. Large bony erosions are often seen, and serum sediment levels may be detected within hemorrhagic bony cysts (Fig. 9.20). In the author's experience, PVNS typically extends into the semimembranosus-gastrocnemius bursa, which may be markedly distended (Fig. 9.21). Extension of PVNS along the posterior capsular margin of the knee joint may be clinically and arthroscopically overlooked. This may explain the reported 50% incidence of recurrence after arthroscopic synovectomy (Fig. 9.22).

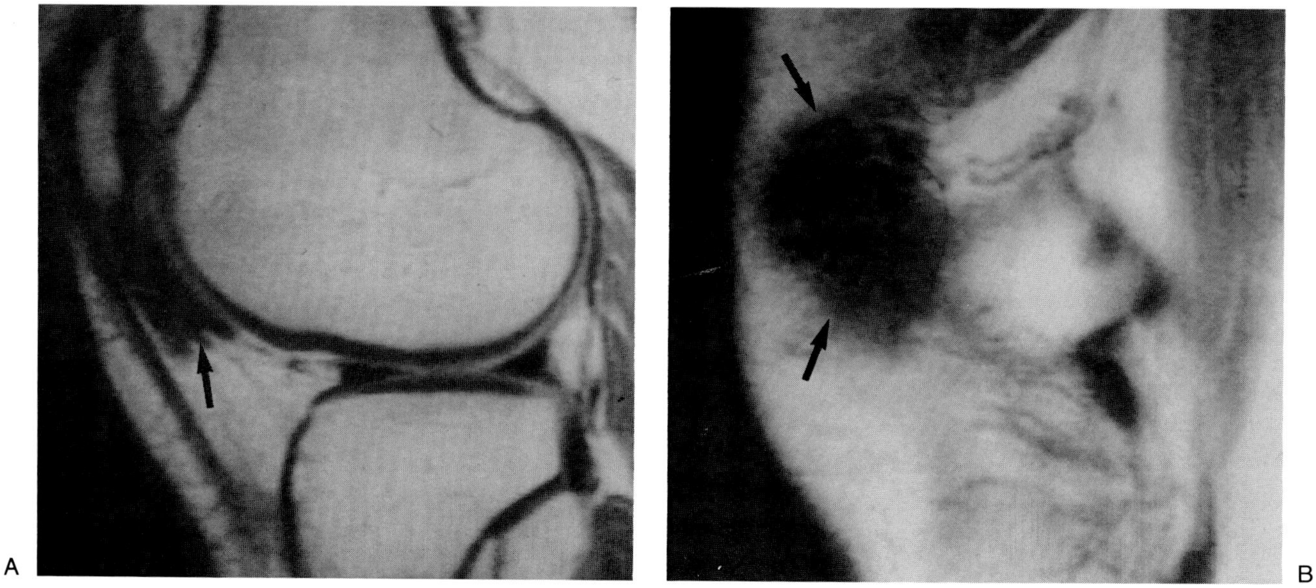

FIG. 9.23. Focal nodular pigmented villonodular synovitis. **A,B:** Sagittal T1-weighted scans were performed in a patient with nonspecific knee pain. A focal low signal intensity mass (*arrows*) was revealed in the infrapatellar fat pad, extending to the lateral peripatellar fat. As preoperatively predicted by the examining radiologist, this mass was *not* arthroscopically visible. Excisional arthrotomy confirmed the diagnosis of focal nodular PVNS.

MR images in the focal nodular form of PVNS typically reveal a focal low signal intensity mass, usually within or adjacent to the infrapatellar bursa. This area is not directly accessible to arthroscopic inspection and, in fact, in the two cases of focal nodular PVNS in this author's personal experience, neither mass was arthroscopically visible (Fig. 9.23).

Successful treatment of PVNS is dependent on local excision, which is more easily accomplished with the nodular form. By delineating posterior capsular and popliteal cyst involvement, MRI may facilitate more complete resection of the diffuse form, which usually requires synovectomy via combined anterior and posterior approaches.

Hemophilia

Knee joint arthropathy occurs frequently in patients with hemophilia secondary to repeated massive and often subclinical hemarthroses. The radiographic findings, which may be virtually indistinguishable from those of JRA, are known to underestimate the extent of disease (163). MRI may identify hemarthroses, subchondral cysts, synovial inflammation, and periarticular fibrosis in patients with hemophilia (86,181). Chronic periarticular and subchondral fibrosis yields low signal intensity on T1- and T2-weighted images. Subchondral cysts and areas of synovial inflammation yield low signal intensity on T1-weighted images, but high signal intensity on T2-weighted images. In chronic hemophilic arthropathy, the synovium is scarred and thickened, resulting in low signal surrounding the knee.

Other rarer causes of low signal intensity tissue infiltrating within the knee joint include amyloid arthropathy and fibrosis of the infrapatellar fat pad (36,107). Dialysis-related amyloidosis is a recognized complication of long-term hemodyalysis, caused by the deposition of amyloid derived from circulating beta 2 microglobulin. In the single case of amyloid arthropathy thus far reported, knee MRI showed a moderate-sized joint effusion, large nodular low signal intensity synovial masses, and extensive osseous erosions (36). Although unproven, one would expect the MR depiction of amyloid deposition in synovium to precede the radiographic appearance of bony erosions and cystic lesions.

Chronic scarring and inflammation within the infrapatellar fat pad, termed Hoffa disease, may be manifested by serpiginous low signal intensity in the infrapatellar region on MR images (107). This clinical condition is characterized by painful enlargement of the fat pad, often accompanied by an effusion. The syndrome typically affects younger patients following trauma and is associated with hypertrophy of the infrapatellar fat pad, which may become entrapped between the femur and the tibia when the knee is suddenly extended, mimicking other causes of internal derangement. Fibrous scarring of the infrapatellar fat pad due to multiple arthroscopic surgeries may also result in an infiltrative pattern of low signal intensity in this region. Although there are some similarities in appearance of hemophilic arthropathy, amyloid arthropathy, PVNS, Hoffa disease, and postsurgical intraarticular scarring, these entities are generally easily distinguishable on the basis of clinical history.

Synovial Chondromatosis

Synovial chondromatosis is caused by focal metaplasia of subsynovial tissue, producing nodules of cartilage within the joint, adjacent bursa, tendon sheath, or paraarticular connective tissue (37). These nodules often become pedunculated, detached from the synovium, and grow as intraarticular loose bodies. The cartilage nodules may calcify or ossify. This disorder typically affects young or middle-aged adults and is most often monoarticular. The knee is the most frequently involved joint. Patients complain of pain, swelling, and symptoms of intraarticular loose bodies. Standard radiographs are generally diagnostic, disclosing multiple intraarticular calcifications in 90% of patients (37).

MRI is generally not necessary to establish the diagnosis of synovial chondromatosis, since in most cases standard radiographs are diagnostic. On rare occasions, however, synovial chondromatosis may be serendipitously discovered, particularly in patients without calcified lesions (Fig. 9.24). Numerous foci of low signal intensity are seen within hypertrophied synovium.

Treatment of synovial chrondromatosis is directed at removing loose bodies and resecting hypertrophied synovium, via either arthroscopy or arthrotomy.

Synovial Hemangioma

Synovial hemangioma most frequently involves the knee joint (37,65,92,107,114,143). Hemangiomas occur most often in adolescents or in young adults, with girls affected more frequently than boys. Since many patients are, in retrospect, symptomatic since childhood, the lesion may represent a congenital vascular malformation. The most common knee symptoms are periodic pain and swelling associated with recurrent hemarthrosis. Hemangiomas may be evident in the skin overlying the affected joint. Similar to PVNS, localized disease is often in the form of a dark grape-like mass, commonly originated from the infrapatellar fat pad. Circumscribed focal lesions are usually capillary or cavernous hemangiomas, whereas in the diffuse form the vascular channels are predominantly venous (37).

The diffuse form is often associated with visceral or cutaneous vascular lesions (found in 40% of patients). Both the Klippel-Trenaunay syndrome (varicose veins, soft tissue and bony hypertrophy, and cutaneous hemangioma) and the Kasabach-Merritt syndrome (capillary hemangiomas and purpura) have been associated with similar arthropathy.

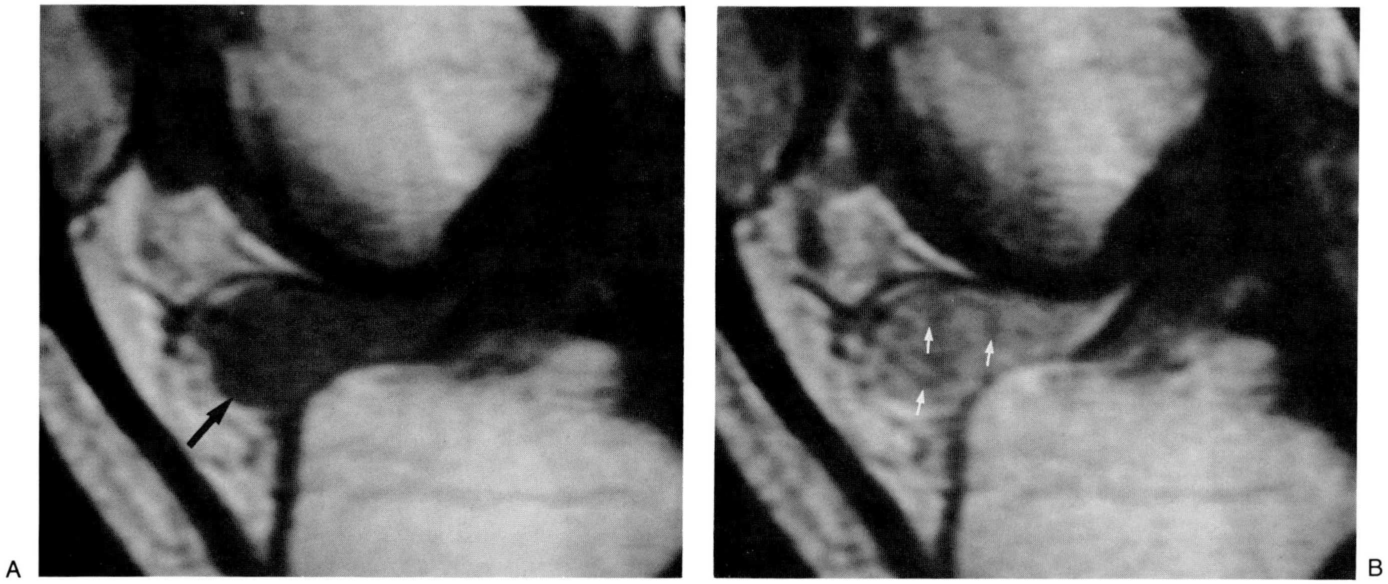

FIG. 9.24. Synovial chondromatosis. **A:** Sagittal intermediate-weighted image depicts moderate signal intensity synovial mass (*arrow*). This could easily be mistaken for simple effusion. **B:** On T2-weighted sequence, this mass did not increase in signal intensity as much as typical synovial fluid. Stippled low signal intensity foci are demonstrated within this synovial mass (*small arrows*) which pathologically corresponded to regions of chondroid metaplasia. This patient did not have radiographically visible intraarticular calcifications.

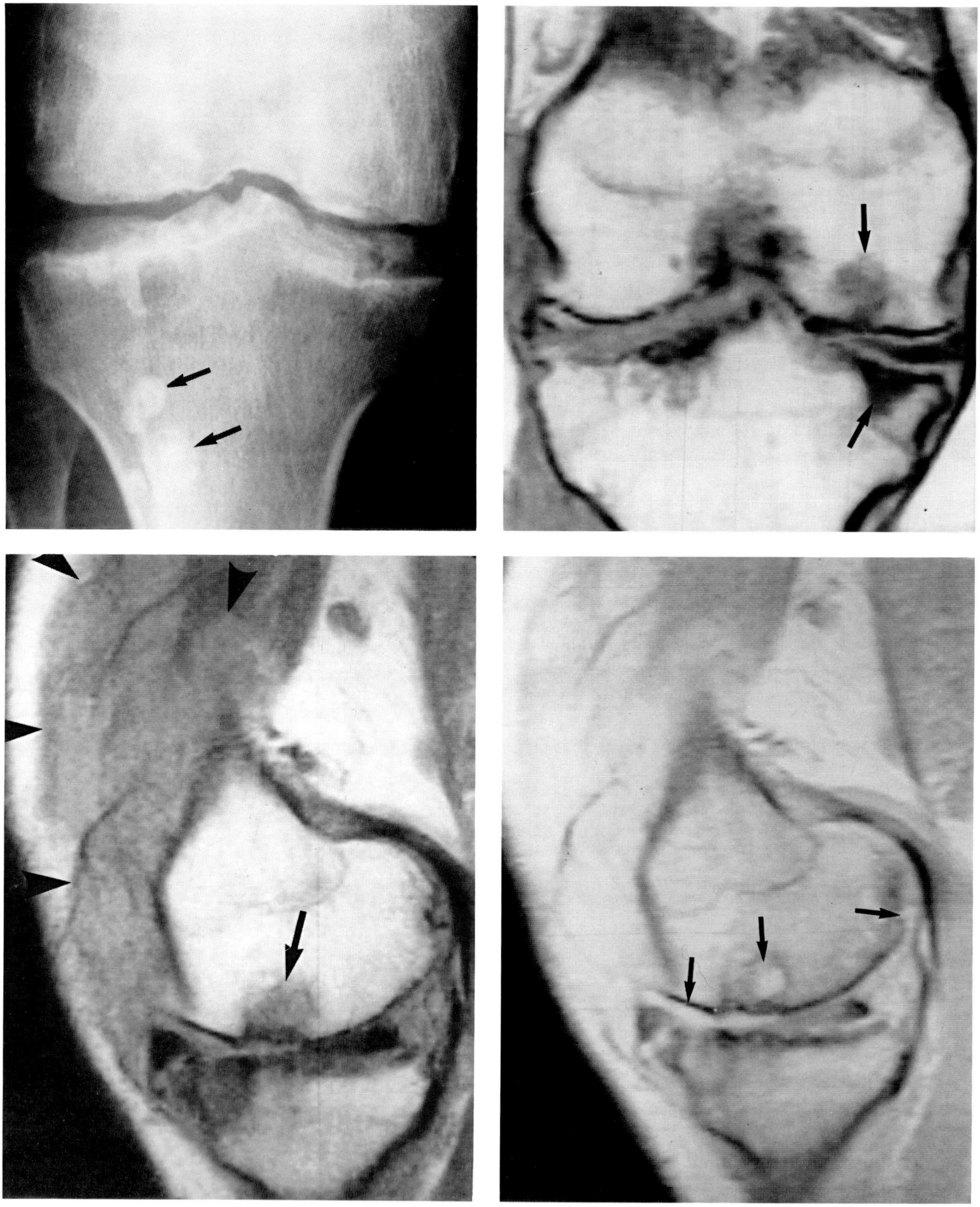

A B C D

FIG. 9.25. Synovial hemangioma. **A:** Standard radiograph depicts multiple phleboliths (*arrows*) as well as extensive osteocartilageinous erosions. **B:** Coronal T1-weighted image, **C:** sagittal T1-weighted image, and **D:** sagittal intermediate-weighted image depict extensive intraarticular mass (*arrowheads*) as well as numerous osseous erosions and subchondral cysts (*arrows*).

As with PVNS, surgical excision of the focal form of the disease is generally simple and effective. Resection of diffuse synovial hemangioma is difficult, and recurrence is common.

Radiographic findings in synovial hemangioma include phleoboliths, soft tissue mass, osteoporosis, epiphyseal overgrowth, squaring of the patella, widening of the intercondylar notch, and irregular cartilaginous erosions. MRI is valuable in delineating the extent of disease, detecting cartilaginous erosions, differentiating the focal from the diffuse form, and thereby facilitating surgical planning (Fig. 9.25).

Intraarticular Lipomatous Masses

Intraarticular lipomas are rare tumors which have been found almost exclusively in the knee joint (37,130,141). Lipomas may also be found in the tendon sheaths of the hand, wrist, foot, or ankle. Focal lipomas have been found adjacent to the patellar tendon or along the anterior surface of the femur (37), but may be seen elsewhere within the knee joint. Patients generally present with symptoms of a soft tissue mass, but may have symptoms of mechanical derangement (Fig. 9.26).

In contrast, lipoma arborescens is characterized by villous or polypoid synovial proliferation, arising *de novo* or occasionally associated with degenerative joint disease, chronic rheumatoid arthritis, or prior trauma (141). Recurrent effusion is common. Standard radiographs may demonstrate radiolucent fat within the suprapatellar bursa. The author is aware of only two cases of lipoma arborescens diagnosed preoperatively with MRI. In both cases, the MR findings consisted of synovial proliferation with tissue identical to fat in signal intensity, interspersed with synovial fluid (Figs. 9.27 and 9.28). Erosions were far less severe than in PVNS.

JOINT FLUID AND CYSTS

The knee joint is lubricated by synovial fluid that contains hyaluronate, a large glycoprotein molecule whose average molecular weight is in excess of 1,000,000 (101). This giant molecule is responsible for the "thixotropic" flow characteristics of synovial fluid (greater viscosity with slower flow) (101). The knee joint is further lubricated by fluid "squeezed out" of articular cartilage during mechanical loading. The normal knee joint, which is the largest joint in the body, contains no more than 4 ml of synovial fluid. Normal joint fluid is clear and colorless; it does not clot. The synovial membrane, which is a vascular connective tissue lining the inner surface of the knee capsule, is responsible for the transfer of blood constituents into the synovial cavity and the formation of synovial fluid. The synoviocytes synthesize and secrete hyaluronate. Synovial fluid provides nutrition for the articular cartilage and the menisci (101,104).

Normal and acutely hemorrhagic synovial fluid yields moderate signal intensity on T1-weighted MR images and high signal intensity on T2-weighted images (10,13,15,56,66,86,134,166). Hemorrhagic effusions may yield every variety of signal intensity, including

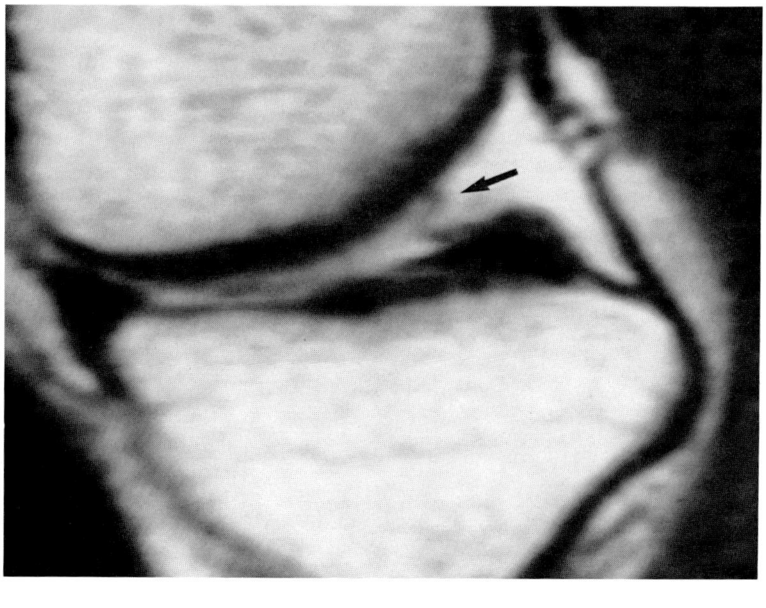

FIG. 9.26. Intraarticular lipoma. Sagittal T1-weighted scan depicts abnormal mass of septated lipomatous tissue (*arrow*) invaginated between the posterior horn of the medial meniscus and the posterior convexity of the medial femoral condyle. The patient's symptoms were those of mechanical derangement. Arthroscopy confirmed the presence of an intraarticular lipoma and symptoms improved after resection.

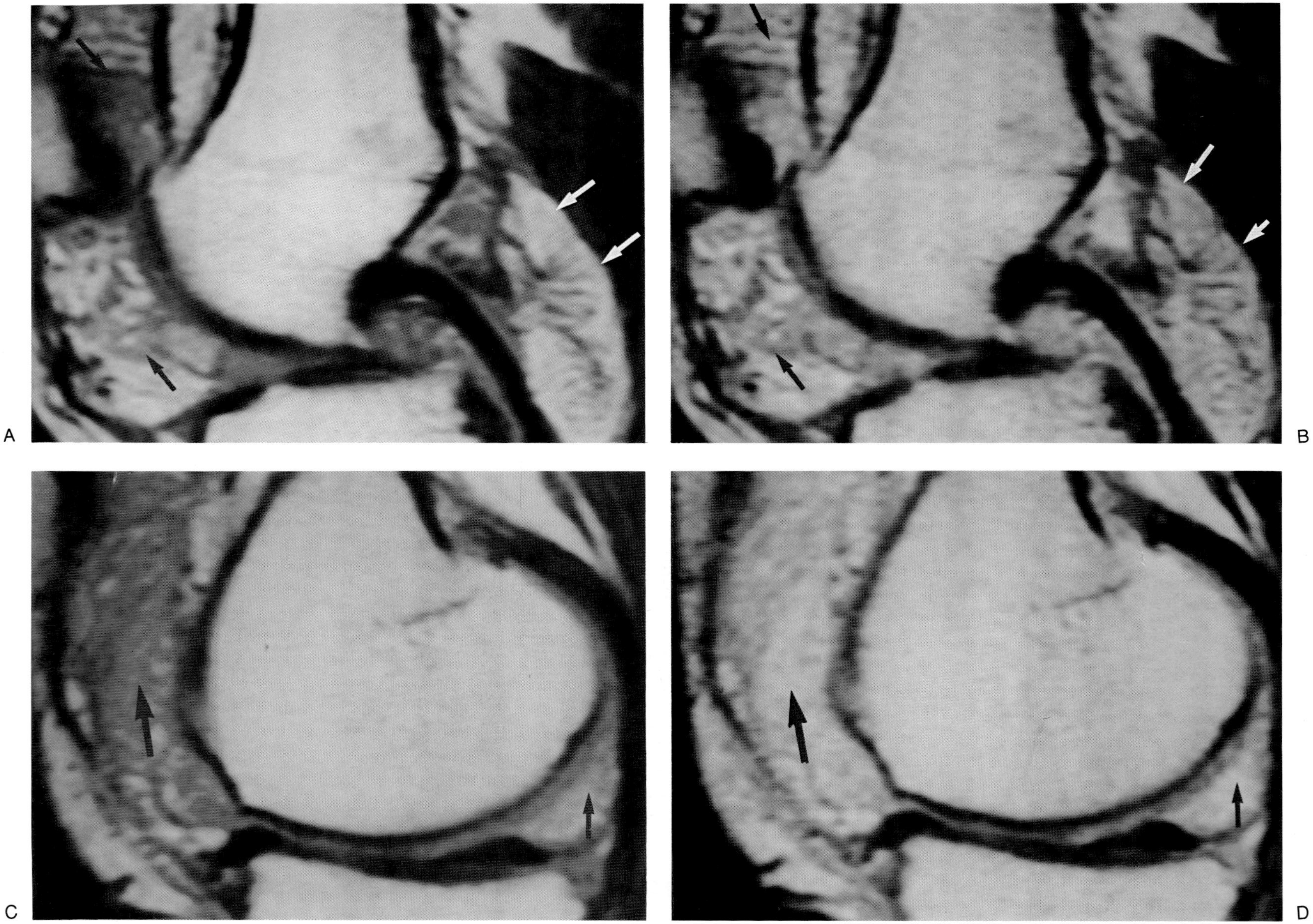

FIG. 9.27. Lipoma arborescens. **A,C:** Sagittal intermediate-weighted scans and **B,D:** T2-weighted scans depict intraarticular mass of lipomatous tissue (*arrows*) interspersed with synovial fluid. Only subtle cartilaginous erosions are present. Lipoma arborescens was accurately preoperatively diagnosed with MRI.

A 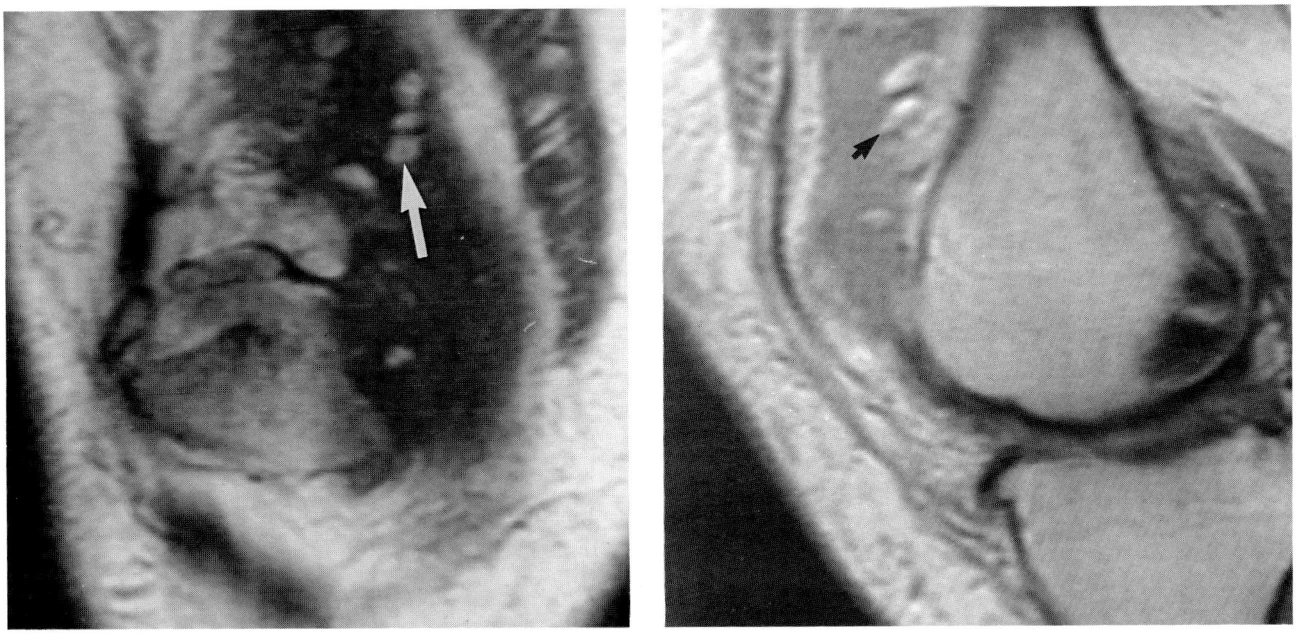 B

C 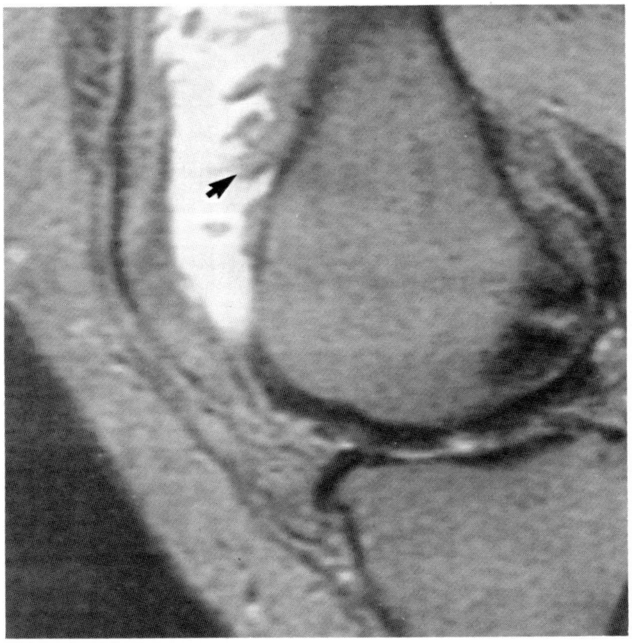

FIG. 9.28. Lipoma arborescens. **A:** Coronal T1-weighted section through the suprapatellar bursa, **B:** sagittal intermediate-weighted scan, and **C:** sagittal T2-weighted scan all depict villus-like projections of hypertrophied synovium, equivalent in signal intensity to fat (*arrows*).

high signal intensity on T1- and T2-weighted images and low signal intensity on T1- and T2-weighted images. Serum-sediment levels are often present in subacute hemorrhagic effusions (Fig. 9.29). Low signal intensity clots may also be seen in acutely hemorrhagic effusions, simulating intraarticular loose bodies (Fig. 9.30). Fat globules or even air bubbles can be seen in effusions (Fig. 9.31). Because MR images are generally acquired with the patient supine, fat droplets or air bubbles tend to rise to the ventral aspect of the suprapatellar bursa.

Although standard radiography may be up to 90% accurate in determining the presence of a tappable fluid collection in the suprapatellar bursa (161), MRI offers the additional advantages of detecting effusions in other sites, characterizing the composition of fluid, and determining the underlying cause. Furthermore, small knee effusions that do not distend the suprapatellar bursa may be missed on physical examination or standard radiography, but detected using MRI (66).

Popliteal Cysts

Popliteal cysts, or Baker cysts, are bursal collections of synovial fluid that usually occur in the semimem-

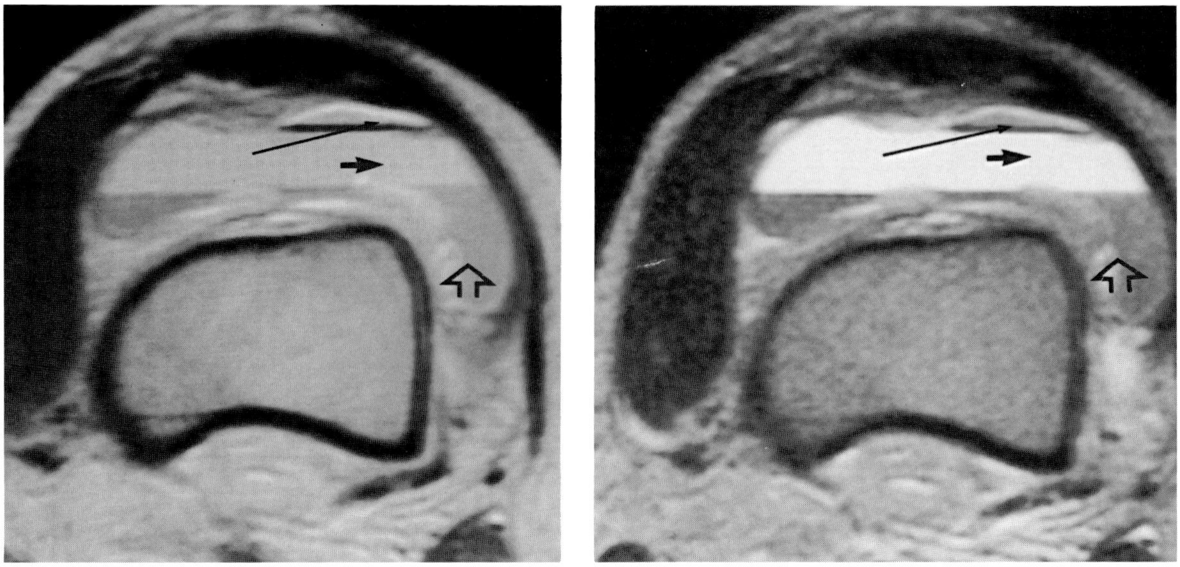

A B

FIG. 9.29. Lipohemarthrosis. **A:** Axial intermediate-weighted image and **B:** T2-weighted image reveal tri-layered synovial fluid collection consisting of fat (*long, thin arrow*), serum (*short arrow*), and sediment (*open arrow*).

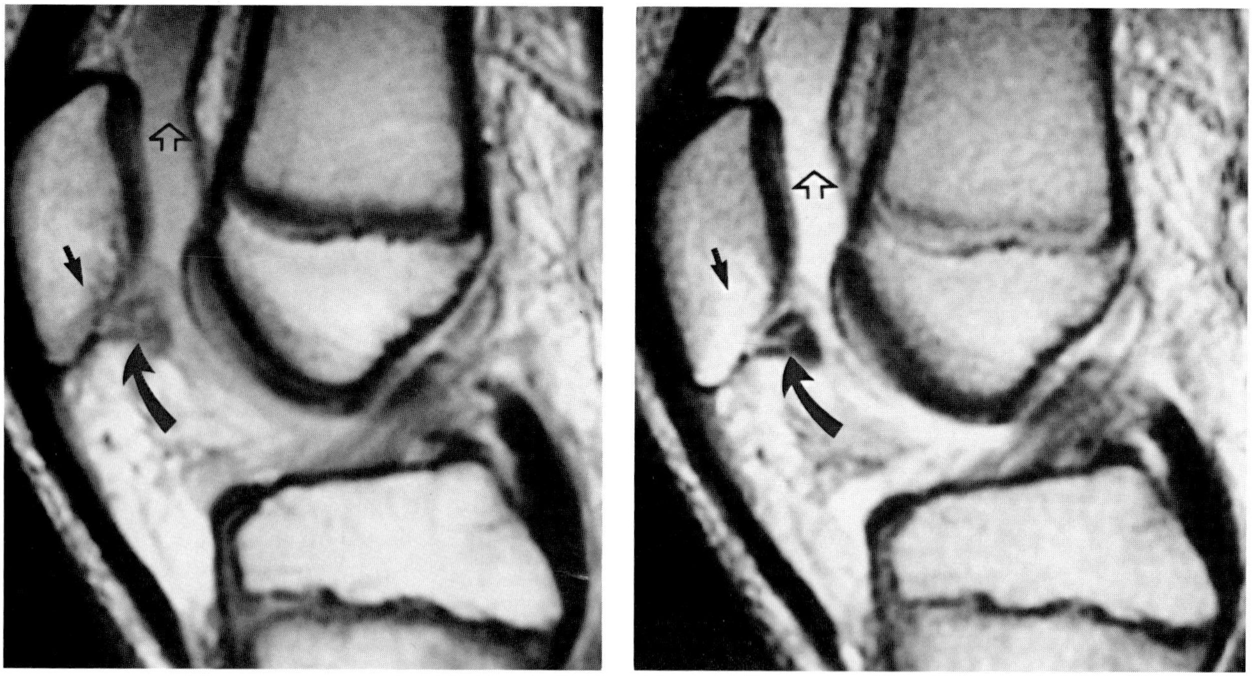

A B

FIG. 9.30. Subacute hemorrhagic effusions. **A:** Sagittal intermediate-weighted image and **B:** T2-weighted image depict subchondral contusion of patella (*straight arrows*), effusion within the suprapatellar bursa (*open arrows*), and low signal intensity thrombus (*curved arrows*).

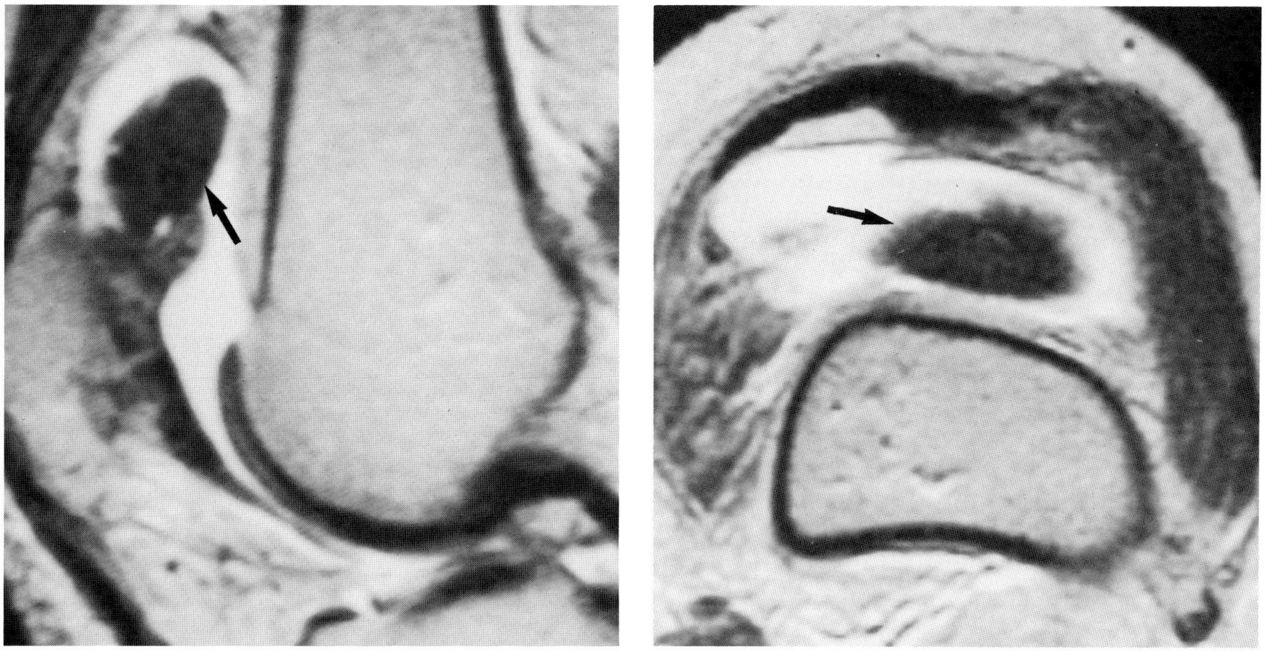

C D

FIG. 9.30. (*Continued*) In another patient, low signal intensity thrombus (*arrow*) is again seen outlined by higher signal intensity synovial fluid on **C:** sagittal and **D:** axial T2-weighted sections through the suprapatellar bursa. Thrombus could be easily mistaken for chondral loose body or even PVNS. Whenever a low signal intensity mass is seen within an effusion, the possibility of an acute thrombus should be strongly considered, particularly if no anatomic source of a loose body is demonstrated.

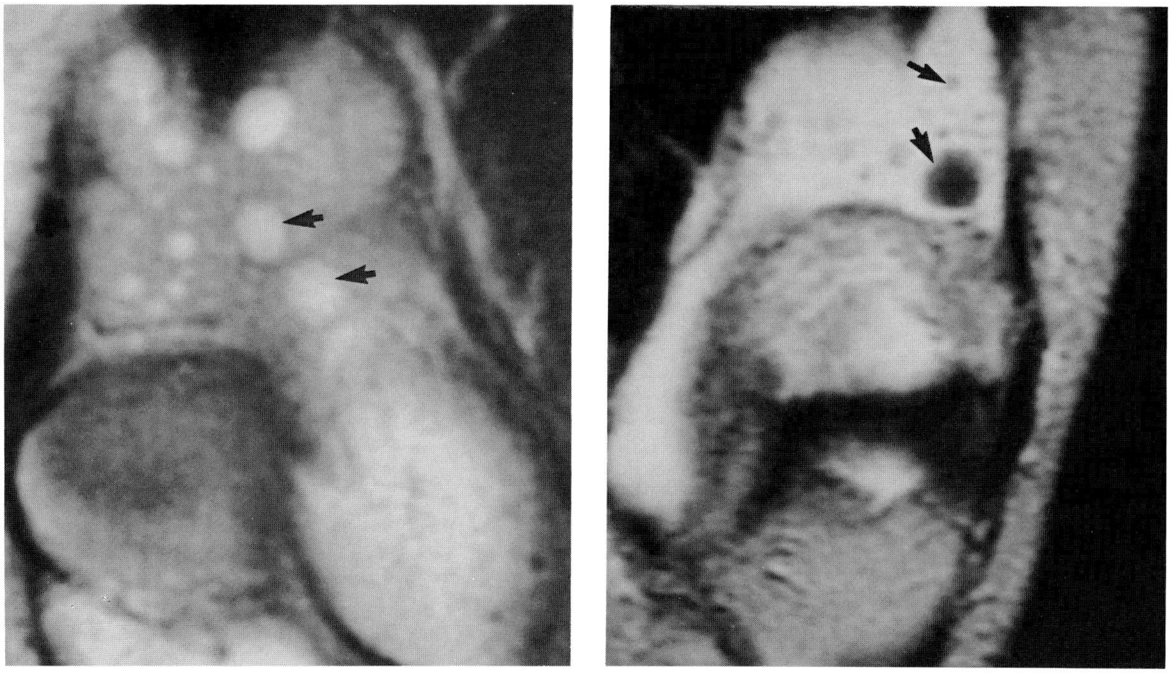

A B

FIG. 9.31. Joint effusions. **A:** Coronal intermediate-weighted scan through the suprapatellar bursa depicts fat globules (*arrows*) floating within the suprapatellar bursa. **B:** In another patient, low signal intensity foci seen within the suprapatellar bursa on coronal T2-weighted image were air bubbles (*arrows*) due to penetrating injury.

A 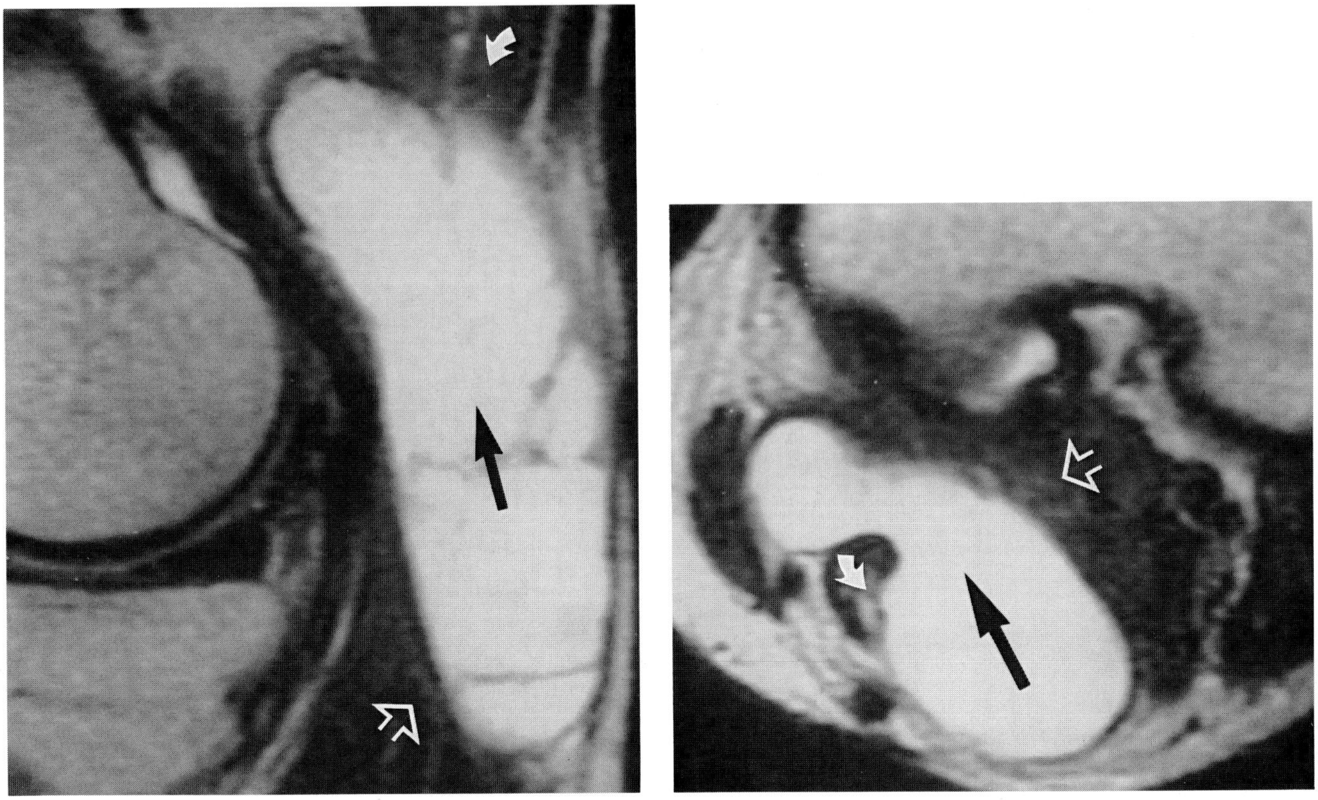 B

FIG. 9.32. Popliteal cyst. **A:** Sagittal T2-weighted image and **B:** axial T2-weighted image reveal most common location of popliteal cyst (*black arrows*) interposed between the semimembranosus (*curved arrows*) and gastrocnemius (*open arrows*) muscles.

branosus-gastrocnemius bursa, but can occur in other locations including the bursa beneath the popliteal tendon, the bursa between the lateral head of the gastrocnemius and the distal biceps femoris tendon, or the tibiofibular joint cavity (9,55,64,94,141). Although 30% to 50% of adults have been found to have communications between the knee joint and the semimembranosus-gastrocnemius bursa, these communications are seldom seen in younger people (94), suggesting that they are acquired through trauma or degeneration of the posterior joint capsule. Cysts, which may or may not be clinically palpable, are typically associated with chronic knee joint effusions. In adults they usually are sequelae of other knee joint abnormalities. Common causes include internal derangement (meniscal or cruciate ligament tear or loose body), osteoarthritis, rheumatoid arthritis, or JRA. Less common causes include other chronic arthritides, chondromalacia, granulomatous synovitis, osteochondritis dissecans, PVNS, and septic arthritis. Idiopathic popliteal cysts occur in children. Occasionally popliteal cysts may rupture, causing synovial fluid to dissect between the soleus and gastrocnemius muscles. Local pain and inflammation result, mimicking thrombophlebitis. Further complicating the clinical assessment in these cases is the fact that cysts may directly compress the popliteal vein and cause deep venous obstruction (169). Other conditions that may clinically mimic popliteal cysts are benign lipomas, popliteal artery aneurysms, malignant tumors, and chronic hematomas (153).

Traditional methods for imaging popliteal cysts include ultrasonography and arthrography (9,55,67,98, 169,178). Sonography is a safe, painless, and inexpensive method, but it can miss small cysts and cannot disclose the underlying intraarticular cause (67). Arthrography is more sensitive than ultrasound (67), but it may miss cysts that do not readily communicate with the joint and has been reported to have a sensitivity of under 50% (178). More recently, CT has been advocated and found to be superior to arthrography, especially for detecting cysts in unusual locations (90,153).

In most clinical practices, MRI of the knee has essentially replaced these other modalities for evaluating cysts in the popliteal region. Simple cysts are well circumscribed and yield low to moderate signal intensity on T1-weighted images and high signal intensity on T2-weighted images. MRI offers several advantages in evaluating popliteal cysts. It is as painless and as safe as ultrasound, yet is more capable of disclosing underlying intraarticular disorders than is arthrography. MRI is extremely sensitive in detecting fluid collections along the posterior capsule of the knee joint and can certainly detect fluid collections within the semi-

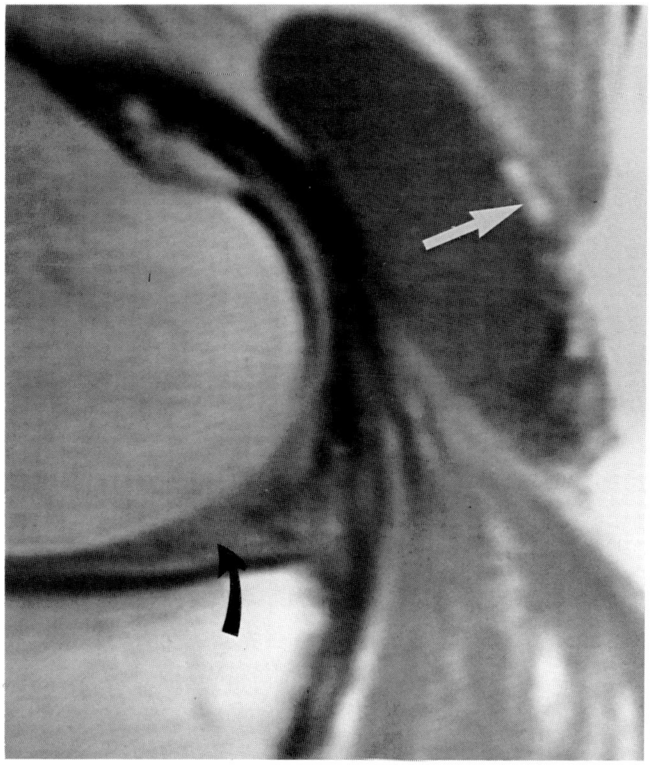

FIG. 9.34. Popliteal cyst. Sagittal T1-weighted scan demonstrates popliteal cyst containing osteocartilaginous debris (*straight arrow*) as well as associated disruption of the posterior horn of the medial meniscus (*curved arrow*).

membranosus-gastrocnemius bursa as small as 1 cc to 2 cc. An understanding of knee joint anatomy and imaging characteristics of synovial fluid allows the diagnosis of a popliteal cyst to be made with complete specificity with MRI (Fig. 9.32). Popliteal cysts may be differentiated from popliteal lymph nodes and can be easily discriminated from venous thrombosis. The site of communication with the posterior capsule of the knee joint is often specifically delineated (Fig. 9.33). Popliteal cysts containing loose bodies, debris, or synovial proliferative disease can be distinguished from simple popliteal cysts (Fig. 9.34; see also Fig. 9.21). Associated knee joint derangement can be simultaneously disclosed. Dissecting popliteal cysts are readily detected (Fig. 9.35), and easily differentiated from deep venous thrombosis.

Imaging in the axial plane is particularly useful for delineating the precise course of popliteal cysts, assessing adjacent vasculature, and discriminating popliteal cysts from other popliteal masses.

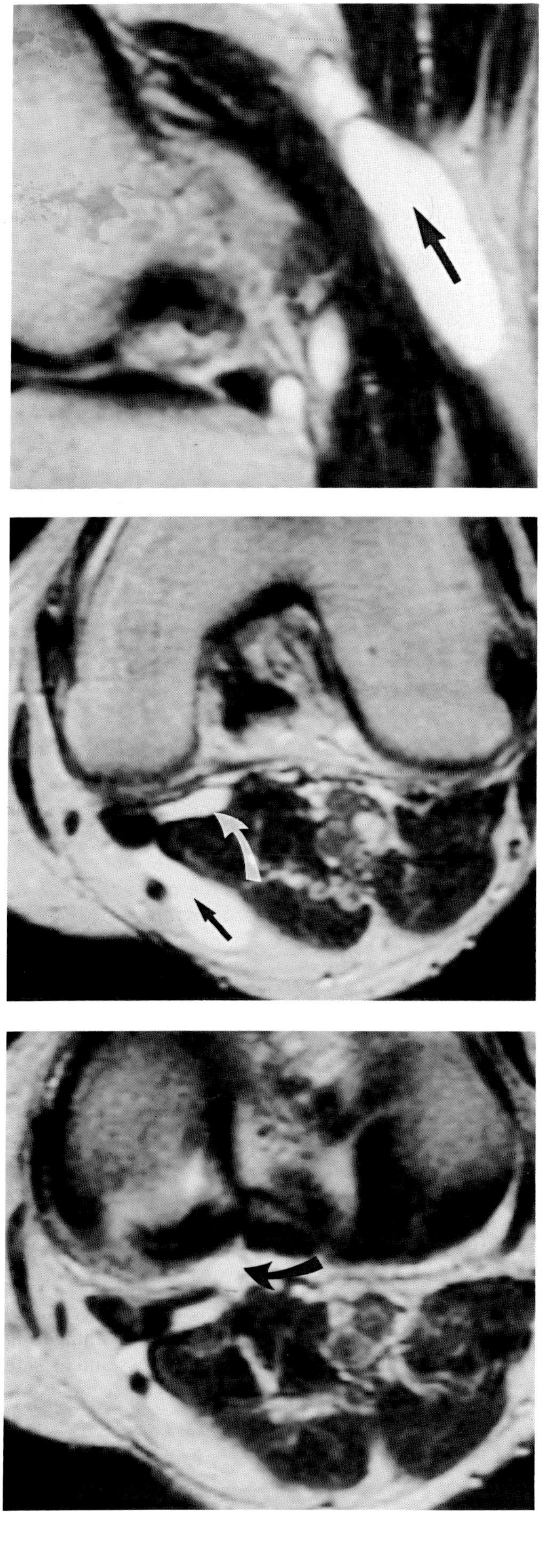

A

B

C

FIG. 9.33. Popliteal cyst. **A:** Sagittal T2-weighted image depicts typical cystic dilatation of the semimembranosus-gastrocnemius bursa (*arrow*). **B,C:** Sequential axial T2-weighted images depict a cyst (*straight arrow*) as well as the narrow neck extending through the posteromesial joint capsule (*curved arrows*).

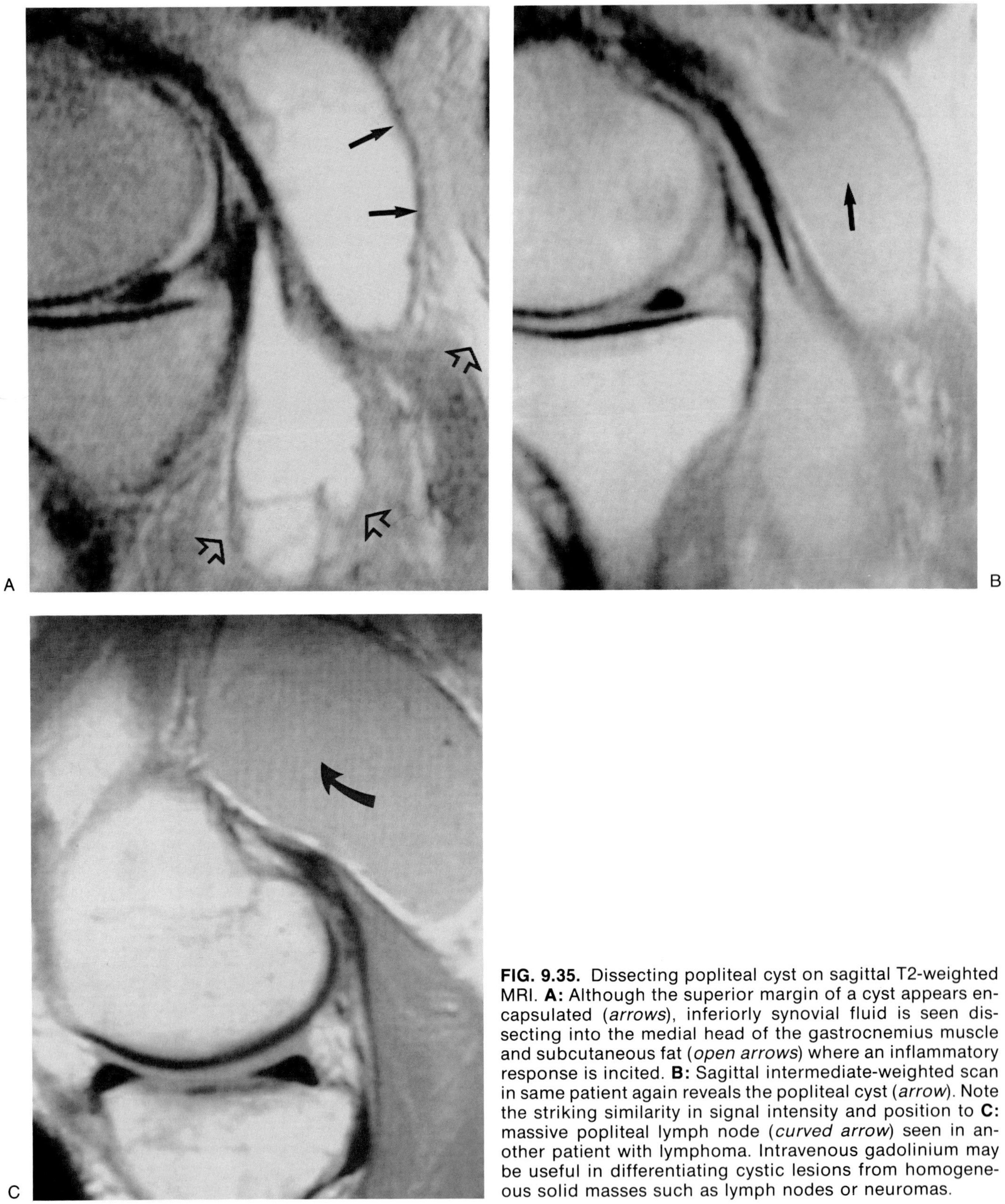

FIG. 9.35. Dissecting popliteal cyst on sagittal T2-weighted MRI. **A:** Although the superior margin of a cyst appears encapsulated (*arrows*), inferiorly synovial fluid is seen dissecting into the medial head of the gastrocnemius muscle and subcutaneous fat (*open arrows*) where an inflammatory response is incited. **B:** Sagittal intermediate-weighted scan in same patient again reveals the popliteal cyst (*arrow*). Note the striking similarity in signal intensity and position to **C:** massive popliteal lymph node (*curved arrow*) seen in another patient with lymphoma. Intravenous gadolinium may be useful in differentiating cystic lesions from homogeneous solid masses such as lymph nodes or neuromas.

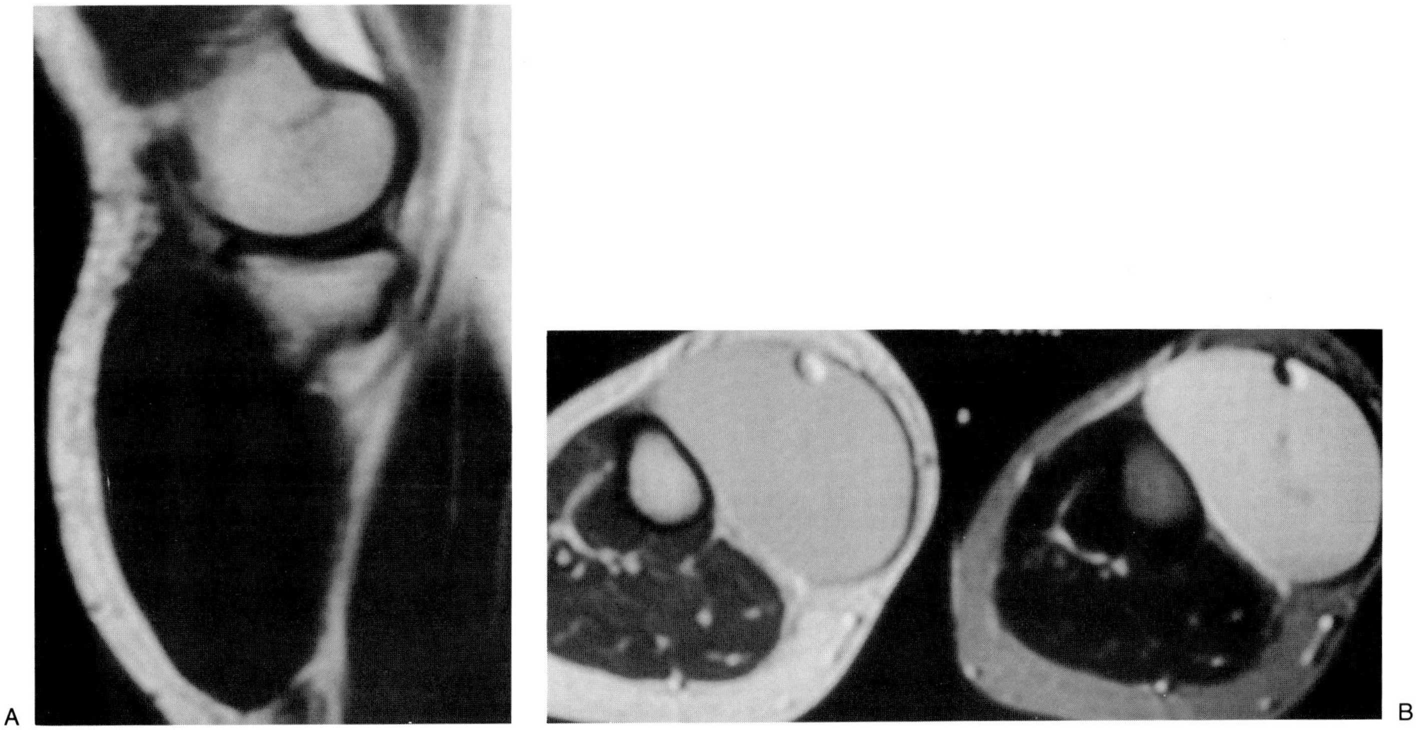

FIG. 9.36. Giant ganglion. A: Sagittal T1-weighted image, B: axial intermediate, and T2-weighted scans all reveal huge cystic mass extending inferiorly from the level of the knee joint. No intraarticular effusion was present, and no patient joint communication was found at MRI or subsequent surgery.

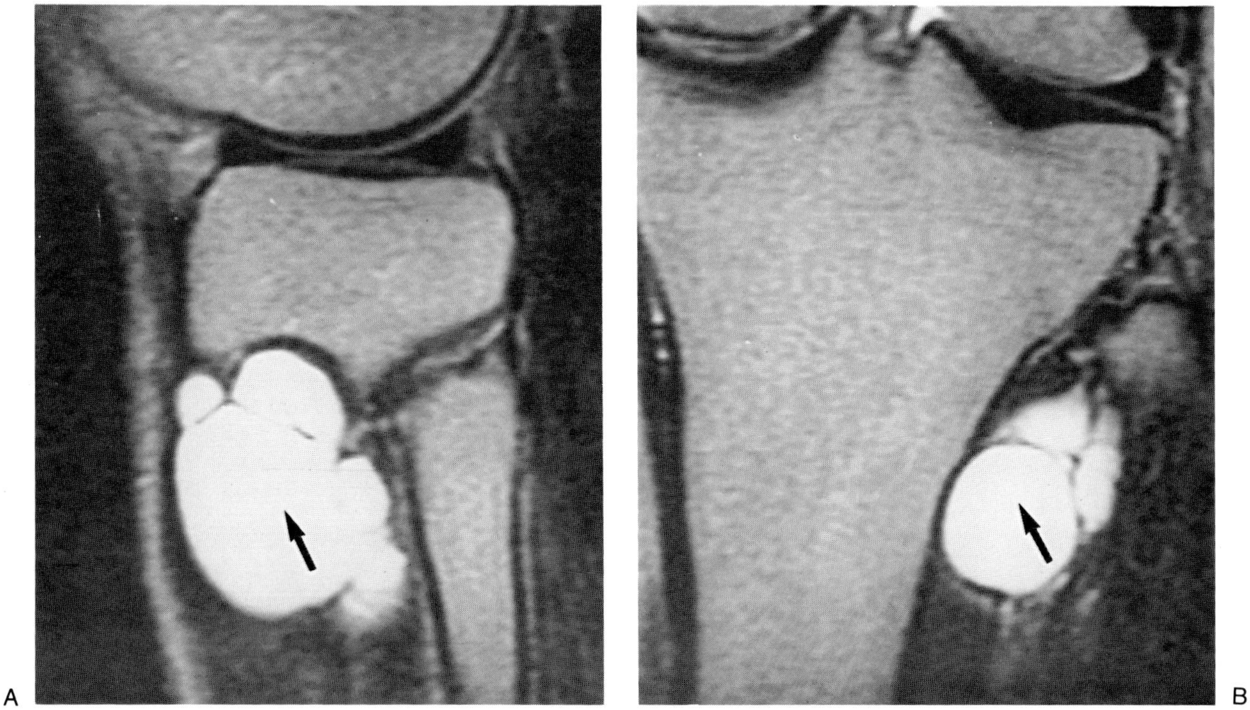

FIG. 9.37. Ganglion of the tibiofibular joint. A: Sagittal and B: coronal T2-weighted images disclose a large septated high signal intensity mass extending adjacent to the tibiofibular articulation (*arrows*).

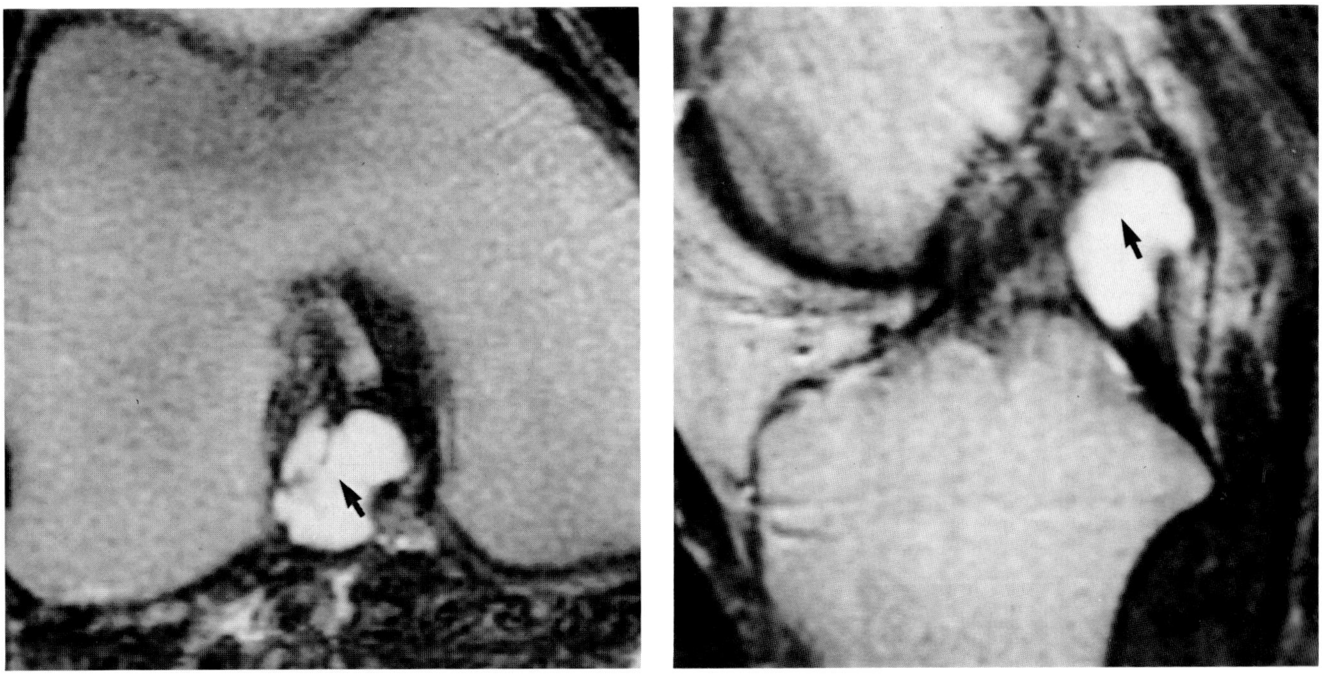

FIG. 9.38. Intraarticular ganglion. **A:** Axial and **B:** sagittal T2-weighted images demonstrate a focal high signal intensity mass (*arrows*) interposed between the anterior and posterior cruciate ligaments.

Ganglia

A ganglion cyst is a soft tissue mass filled with a viscous fluid rich in hyaluronic acid and other mucopolysaccharides, and surrounded by a dense connective tissue capsule. Ganglion cysts can be found in any location, including adjacent to the joint capsule, the pesanserinus tendons, or within the knee joint (26,53,76,77,107,162). Ganglion cysts invariably interconnect to the joint, although the communication may be a fibrous stalk rather than a patent channel (107). Ganglia are often septated, with smaller capsular cysts budding from the main cyst and dissecting along fascial planes. Meniscal cysts, which are invariably associ-

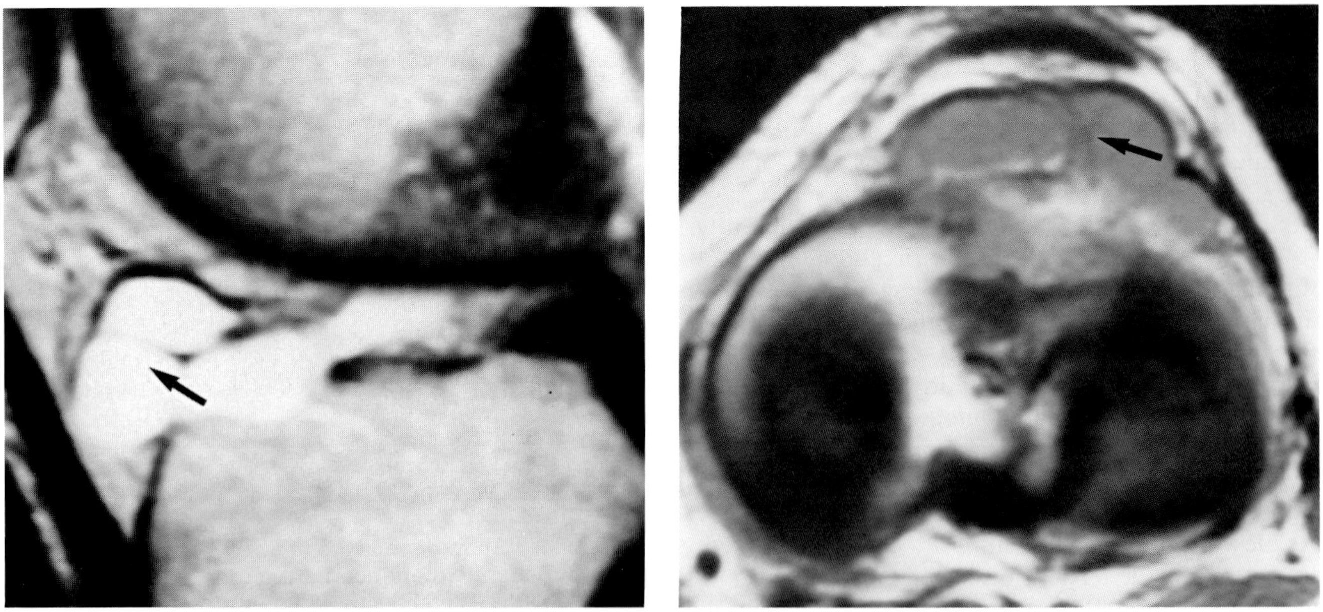

FIG. 9.39. Intraarticular ganglion. **A:** Sagittal T2-weighted image and **B:** axial intermediate-weighted image depict septated intraarticular mass (*arrows*) invaginated into the infrapatellar bursa.

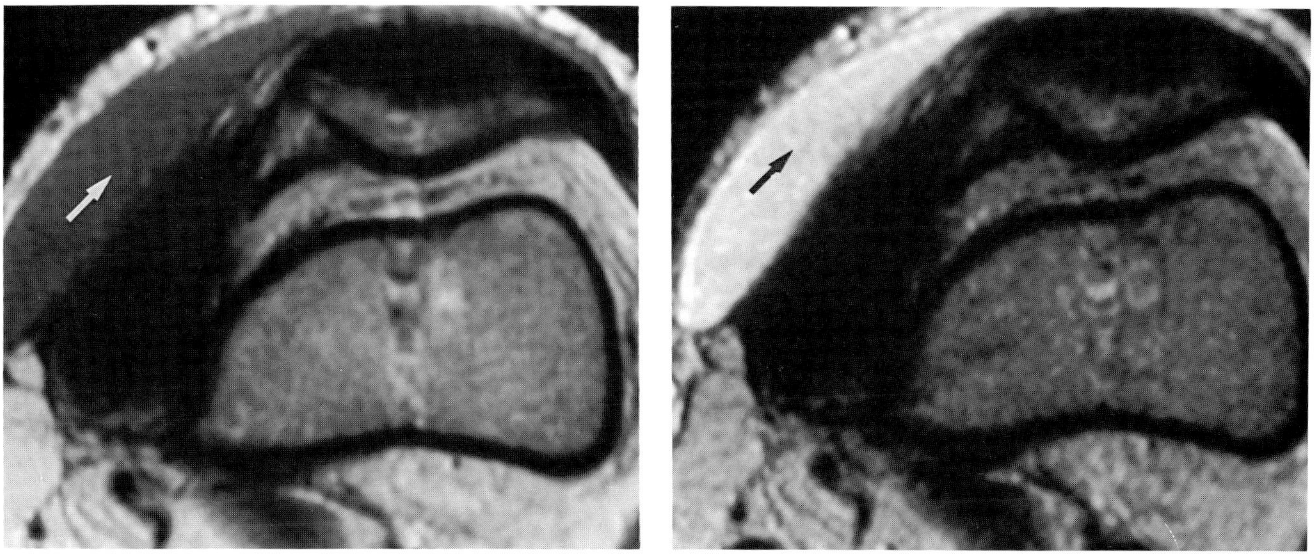

A B

FIG. 9.40. Periarticular hematoma. **A:** Axial intermediate- and **B:** T2-weighted images depict well-circumscribed subcutaneous mass (*arrows*) superficial to the vastus medialis muscle. Although hematomas often contain thrombus or fluid levels, they may be entirely homogeneous.

ated with meniscal derangement, are herein distinguished from ganglia.

Patients with ganglia generally complain of the presence of a mass, although in rare cases internal derangement and neural compression have been reported (76,168).

On MR images, knee ganglia exhibit signal intensity characteristics virtually identical to those of synovial fluid. Both intra- and extraarticular ganglia can be disclosed with MRI (Figs. 9.36–9.39), typically yielding moderately low signal intensity on T1-weighted images and very high signal intensity on T2-weighted images.

Hematomas

Focal hematomas often occur adjacent to the knee joint due to trauma. Occasionally a posttraumatic or postoperative hematoma may persist for months or even years. When the inciting trauma is temporally remote, the patient may present with a nontender soft tissue mass.

In patients with acute trauma, the principal advantage of MRI is in distinguishing extraarticular or superficial hemorrhage from acute internal derangement or periarticular musculotendinous tear (Fig. 9.40). Most hematomas yield moderate signal intensity on T1-weighted images and high signal intensity on T2-weighted images, particularly when chronic, but every possible variation in signal intensity can be observed. Serum-sediment levels are often seen within subacute hematomas.

In one rare case, acute spontaneous hemorrhage in the infrapatellar bursa clinically simulated internal derangement and simulated an infrapatellar mass on MRI. Both high and moderate signal intensity were yielded within a distended infrapatellar bursa on T1-weighted images (Fig. 9.41). Percutaneous aspiration yielded fresh blood and subsequent follow-up confirmed complete clinical and imaging resolution of the mass.

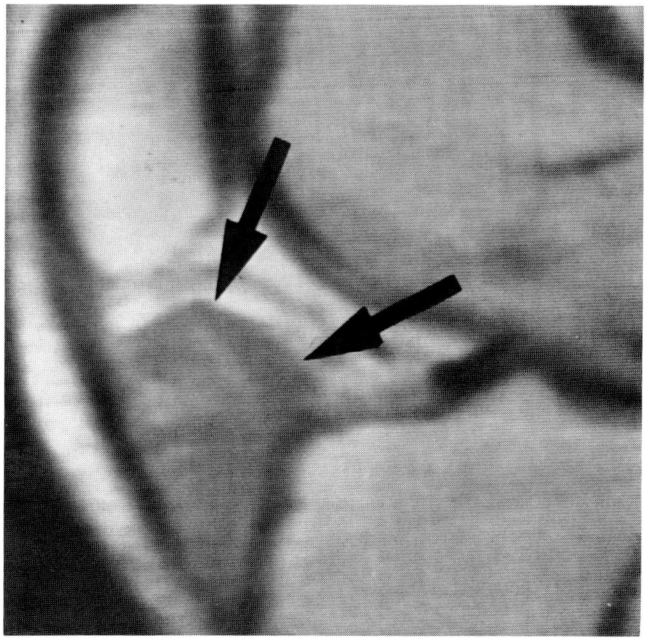

FIG. 9.41. Spontaneous hematoma of the infrapatellar bursa. Patient experienced an acute "pop" in the knee and subsequently developed an infrapatellar mass. Sagittal T1-weighted MRI revealed an inhomogeneous mass in the infrapatellar bursa (*arrows*). Aspiration yielded blood. The mass and the patient's symptoms subsequently resolved without treatment.

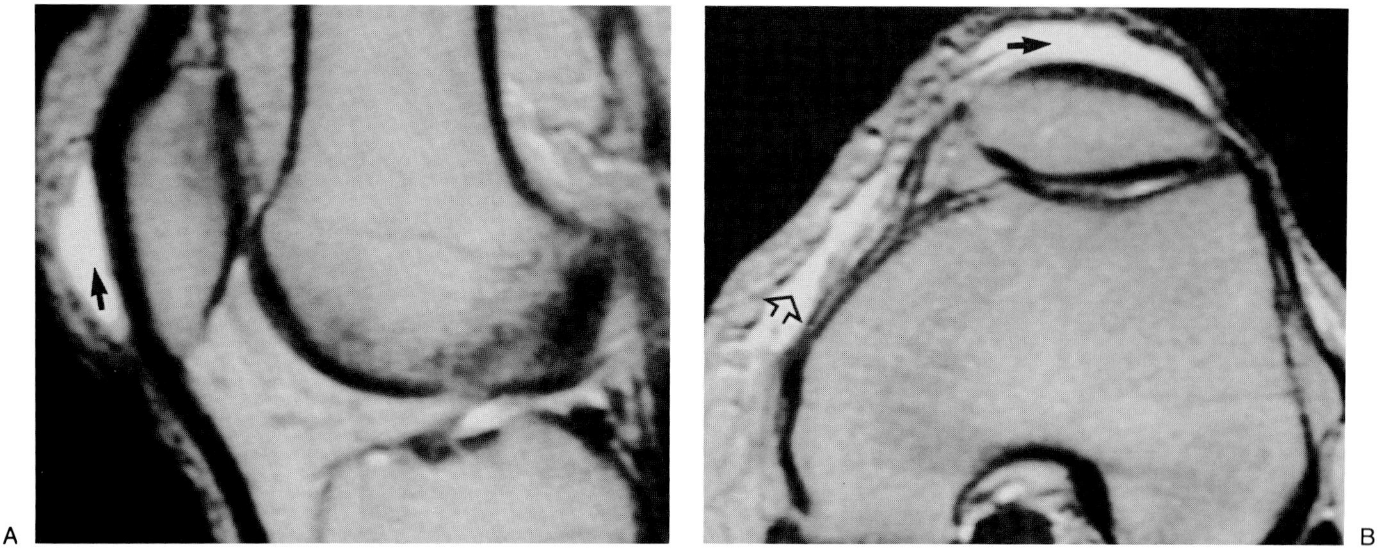

FIG. 9.42. Prepatellar bursitis. **A:** Sagittal and **B:** axial T2-weighted images depict effusion within the prepatellar bursa (*arrows*) and edema in adjacent subcutaneous fat (*open arrow*) due to sterile posttraumatic bursitis.

Bursitis

There are numerous bursae in the knee region, including the prepatellar bursa, the pes anserinus bursa, and the tibial collateral ligament bursa (89). Infection or inflammation of the prepatellar bursa, known as prepatellar bursitis or "housemaid's knee," is the most common affliction of the bursa in the knee region. Infection is most commonly caused by *Staphylococcus aureus*, but many other organisms may cause bursitis. Sterile bursitis is most commonly caused by repetitive trauma. Septic bursitis is most commonly caused by direct trauma with bursal puncture.

On MR images, bursitis is characterized by an effusion in the affected bursa, typically yielding high signal intensity on T2-weighted images (Figs. 9.42–9.45). An inhomogeneous bursal collection or edema extending into the adjacent subcutaneous fat suggests the possibility of superimposed infection, but infectious bursitis cannot be accurately differentiated from sterile bursitis based on imaging criteria alone.

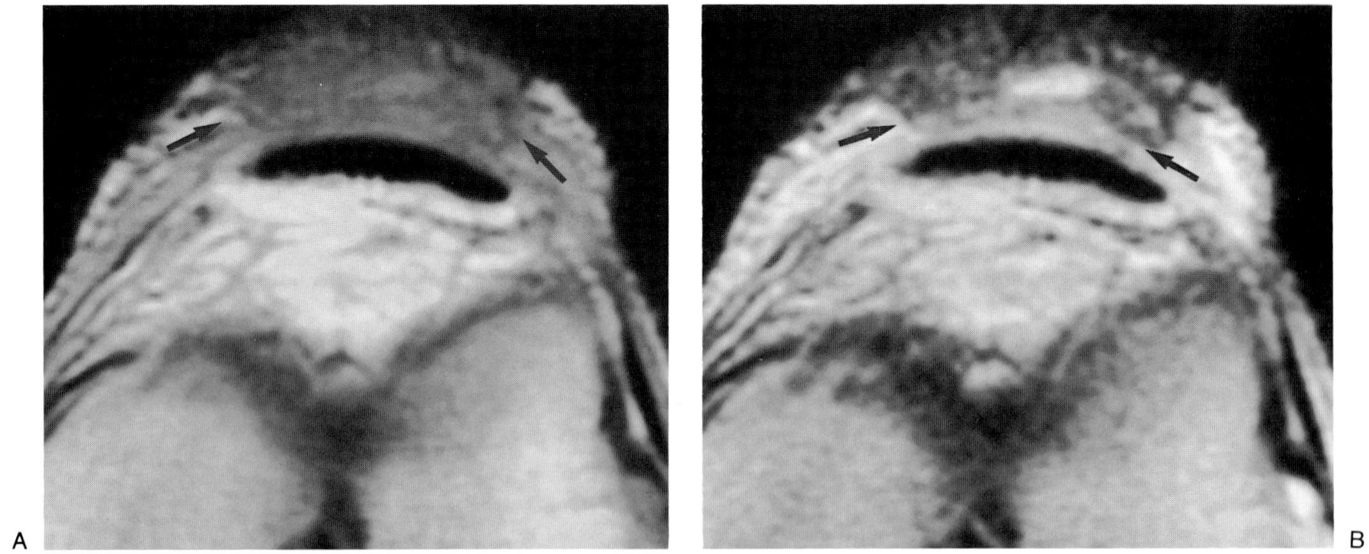

FIG. 9.43. Prepatellar bursitis. **A:** intermediate- and **B:** T2-weighted images depict inhomogeneous fluid collection within the prepatellar bursa (*arrows*). Inhomogeneity suggests the possibility of infection, but nevertheless, septic fluid collections cannot be definitively distinguished from sterile collections with MRI.

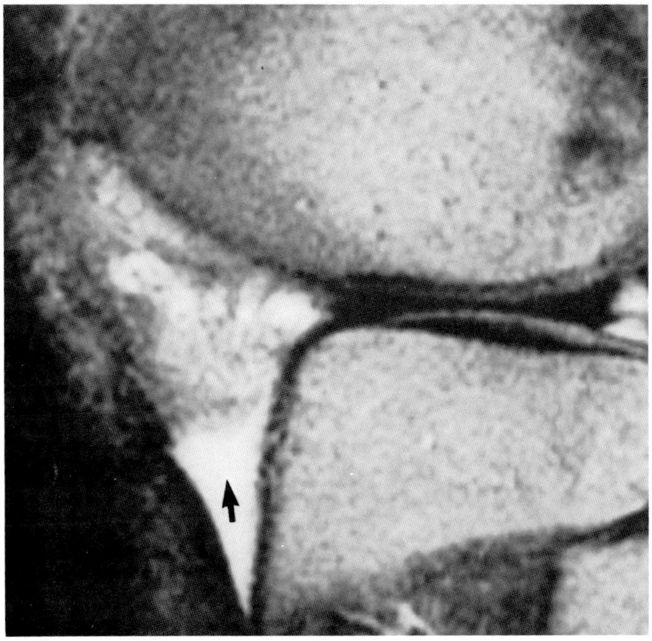

FIG. 9.44. Infrapatellar bursitis. Sagittal T2-weighted scan depicts an isolated fluid collection within the infrapatellar bursa (*arrow*).

Tibial collateral ligament bursitis may cause isolated medial joint line pain, mimicking meniscal derangement (89). MRI has been employed to differentiate tibial collateral ligament bursitis from meniscal disease (89).

AVASCULAR NECROSIS AND BONE MARROW EDEMA

Normal Bone Marrow

Fetal bone marrow is entirely hematopoietic, but soon after birth replacement with fatty marrow begins at the distal ends of the extremities. By young adulthood, hematopoietic marrow is confined to the axial skeleton, proximal humeri, femoral necks, and intertrochanteric regions (85,107,109). Hematopoietic marrow is composed of both cellular elements and varying degrees of fat, but the percentage of fat within red marrow progressively increases with age. On spin-echo images, hematopoietic marrow yields lower signal intensity than fatty marrow on most T1- and T2-weighted spin-echo images (78,107,129). Nonuniform magnetic susceptibility caused by intracellular iron in hematopoietic marrow may contribute to T2 shortening, and thus to the decreased signal intensity of hematopoietic marrow compared with fatty marrow (109). Fatty marrow is characterized by high signal intensity on T1-weighted spin-echo images, with slightly decreased signal intensity on T2-weighted images. On gradient-echo images, T2* shortening may occur in bone marrow contacting trabecular bone, so that fatty marrow in regions of dense trabecular bone may also yield low signal intensity, simulating hematopoietic marrow on scans performed with gradient-echo technique (155,174). Both inversion recovery and chemical sat-

A 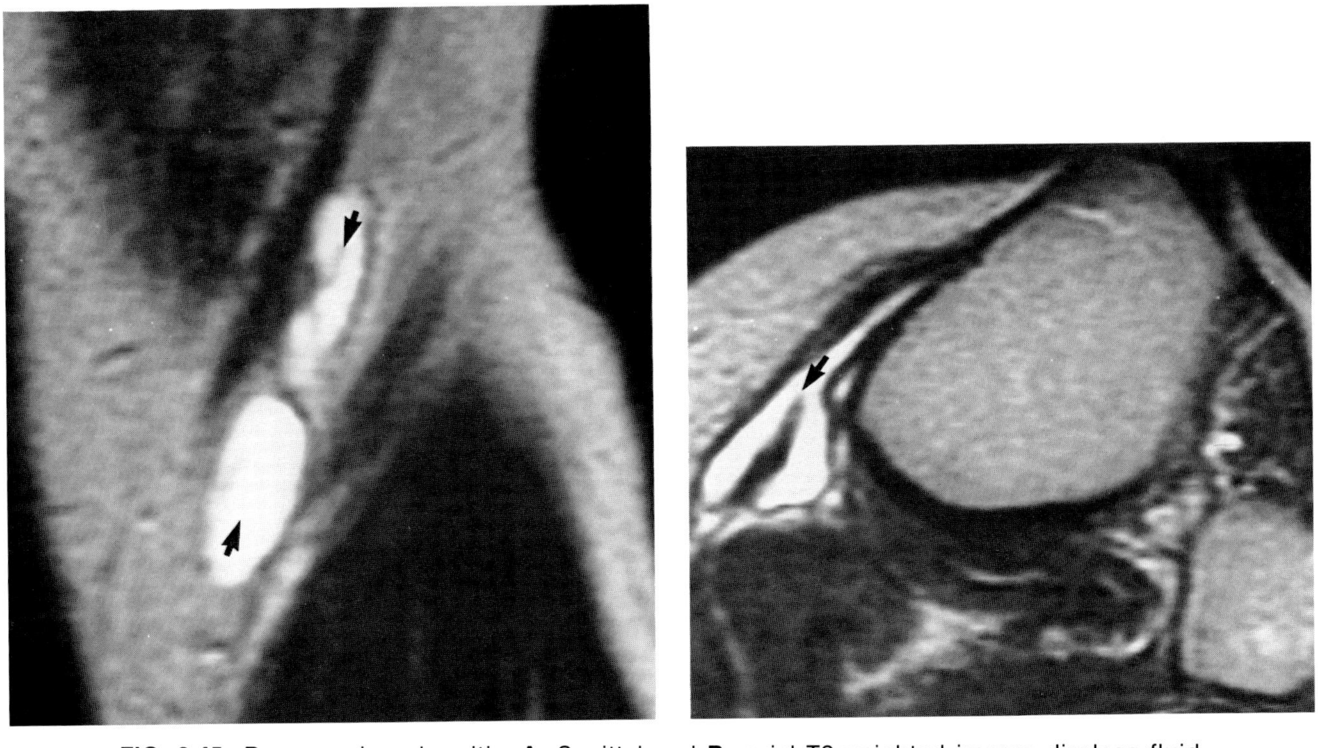 B

FIG. 9.45. Pes anserinus bursitis. **A:** Sagittal and **B:** axial T2-weighted images disclose fluid-distended bursa (*arrows*) tracking along the pes anserinus tendons.

uration techniques can be employed to suppress signal intensity detection from fatty marrow.

In adults, conditions of stress may stimulate hematopoietic hyperplasia, effectively reexpanding hematopoietic elements in areas previously occupied by fatty marrow. This phenomenon is commonly observed in chronic anemias and in a variety of infiltrative bone marrow disorders such as Gaucher disease, myelofibrosis, myeloma, and metastatic disease. This fatty marrow "reconversion" begins proximally and extends distally, the reverse of physiologic marrow conversion (85,127).

In practice, MR images of the knee joint display a great normal variation in the appearance of marrow signal intensity. On T1-weighted spin-echo images, the metaphyses tend to be homogeneously low in signal intensity in early childhood, gradually increasing in signal intensity through adolescence until the adult pattern is reached at ages 21 to 24 (115). Nevertheless, inhomogeneity of bone marrow signal intensity is commonly seen in asymptomatic adults (Fig. 9.46). Bone marrow biopsies have rarely been performed in patients with such regions of low signal intensity within bone marrow on MR images, and have typically shown

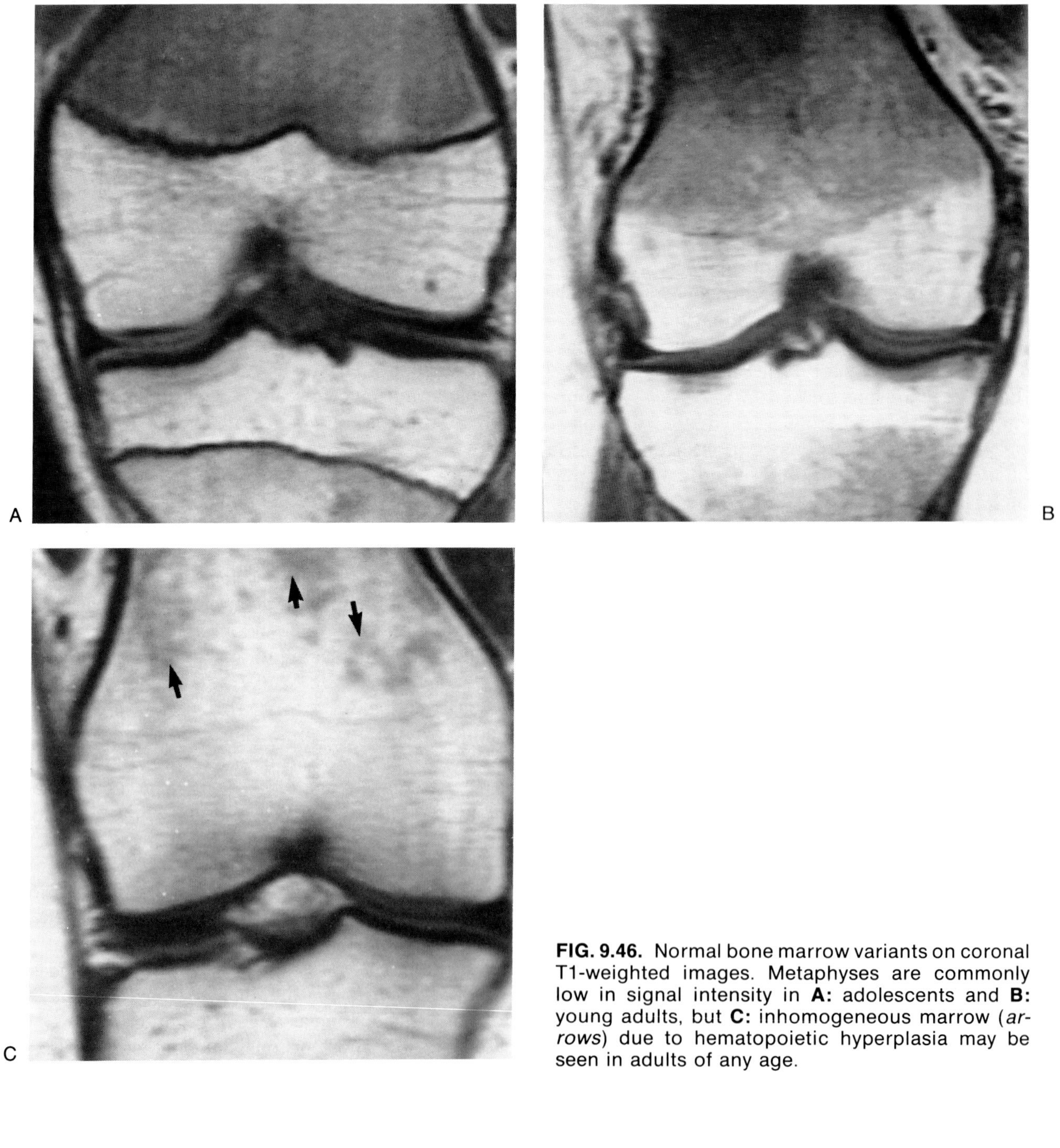

FIG. 9.46. Normal bone marrow variants on coronal T1-weighted images. Metaphyses are commonly low in signal intensity in **A:** adolescents and **B:** young adults, but **C:** inhomogeneous marrow (*arrows*) due to hematopoietic hyperplasia may be seen in adults of any age.

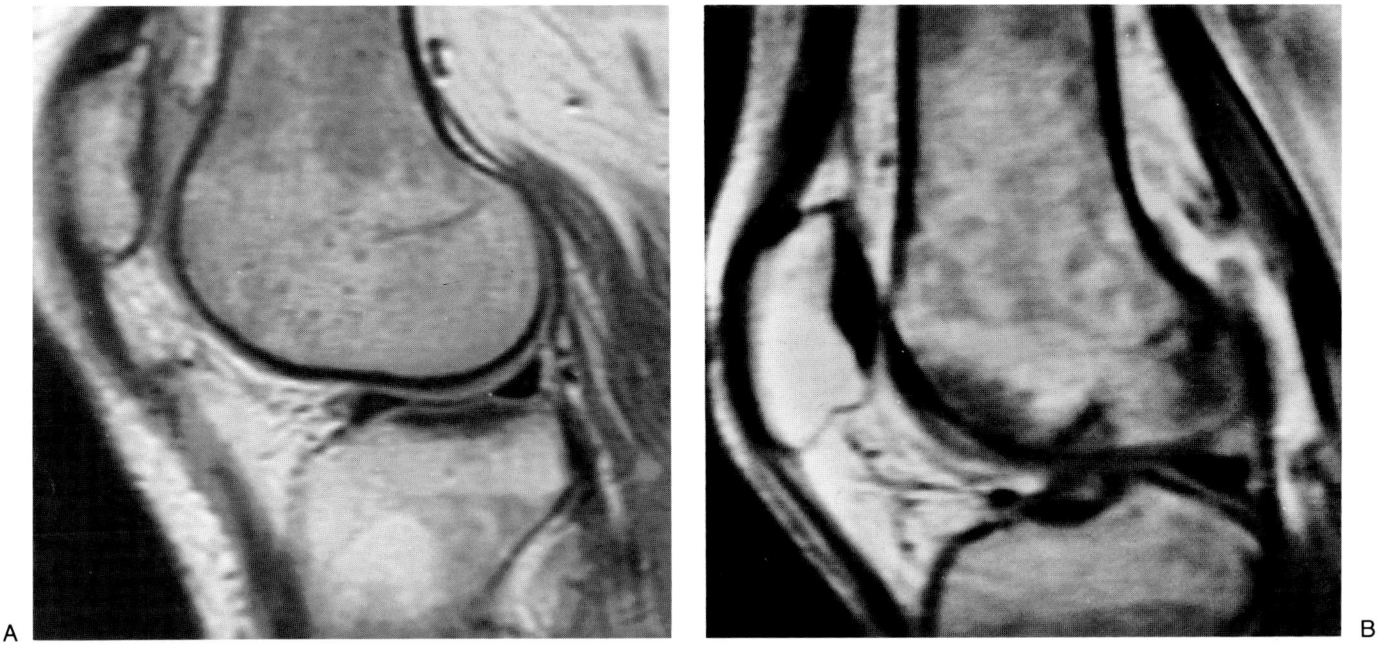

FIG. 9.47. Bone marrow infiltration vs. normal variant on sagittal T1-weighted images. **A:** Normal hematopoietic hyperplasia may be virtually indistinguishable from **B:** bone marrow infiltration (hairy-cell leukemia).

hypercellularity but otherwise normal-appearing hematopoietic marrow (46). In the absence of cortical destruction or soft tissue masses, neoplastic bone marrow infiltration may be entirely indistinguishable from normal hematopoietic hyperplasia on MR images (Fig. 9.47). Early research performed in rats suggests that ultrasmall superparamagnetic iron oxide (USPIO) may be employed as an intravenous contrast agent to help differentiate islands of hypointense red marrow from tumor deposits (158). These ultrasmall supraparamagnetic particles are taken up by phagocytosis, which is typically active in normal red marrow but absent or significantly reduced in bone tumors (142,158).

At the present time, differentiation of normal inhomogeneous marrow due to hematopoietic hyperplasia from bone marrow infiltration should be based on careful clinical history and peripheral blood analysis. In the vast majority of cases, invasive investigation procedures are generally not warranted.

Osteonecrosis

Osteonecrosis, like infarction in other organs, results from a significant reduction or elimination of the blood supply to the infected area. Lesions involving the femoral condyles are next in frequency to those involving the femoral and humeral heads (42). The proximal tibia is also susceptible to osteonecrosis (95). Ischemic necrosis occurs predominantly within areas of fatty marrow because these are the sites of slow sinusoidal blood flow. The epiphyseal ends of long bones are particularly susceptible because arterial access and venous outflow are limited. The femoral condyles depend entirely on the condylar arteries for their blood supply. These vessels fan out toward the joint surface and have few anastomoses, making the subchondral bone vulnerable to ischemia (151). The cells in medullary bone are not equally sensitive to hypoxia. Hematopoietic cells, which are most sensitive to ischemia, undergo necrosis in 6 to 12 hr, osteocytes necrose after 12 to 24 hr, and marrow fat necroses after 2 to 5 days (170).

Causes of osteonecrosis include traumatic disruption of blood supply, hemoglobinopathies, exogenous or endogenous steroids, alcoholism, Gaucher disease, dysbaric disorders, radiation, neuromuscular disease, collagen vascular diseases (systemic lupus erythematosus, rheumatoid arthritis), pancreatitis, and ''idiopathic causes'' (meaning cause unknown) (35,141).

Infarcts have been divided into four histologic zones: a central zone of necrosis, and successive surrounding zones of ischemia, hyperemia, and normal bone. The initial phase is cell death and has no accompanying radiologic changes. Scintigraphy may show decreased isotope uptake due to interruption of the region's blood supply. On T1-weighted spin-echo images, the high MR signal yielded by normal medullary bone (primarily due to the short T1 and short T2 of marrow fat) is diminished by edema (having a longer T1) occurring as a result of infarction (133,170). Brody et al. reported consistent detection of experi-

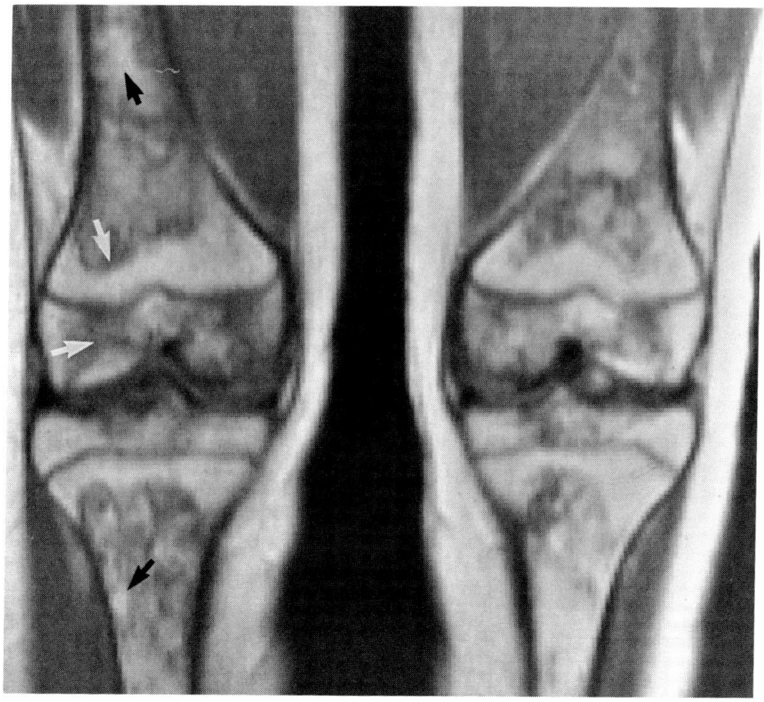

FIG. 9.48. Chronic bone infarction. Coronal T1-weighted spin-echo image demonstrates regions of increased signal attributable to isolated deposits of fat (*black arrows*) and decreased signal due to fibrous reactive tissues and dystrophic calcification (*white arrows*).

mentally induced early avascular necrosis in adult beagle dogs within 7 days of vascular interruption, employing moderately T1-weighted images (22). Cova et al. studied dynamic gadolinium enhancement within bone marrow using dynamic fast MR imaging in a dog model and were able to detect signal intensity differences within bone marrow immediately after devascularization by embolization (41).

Edema due to early infarction in humans may yield high signal intensity on T2-weighted spin-echo images (133). Healing begins with a hyperemic response of the tissues surrounding the ischemic area (75,141). Breakdown products of damaged cells stimulate a response of vascular dilatation, transudation of fluid, fibroprecipitation, and local infiltration of inflammatory cells (75,141). In this phase, which marks the beginning of

A

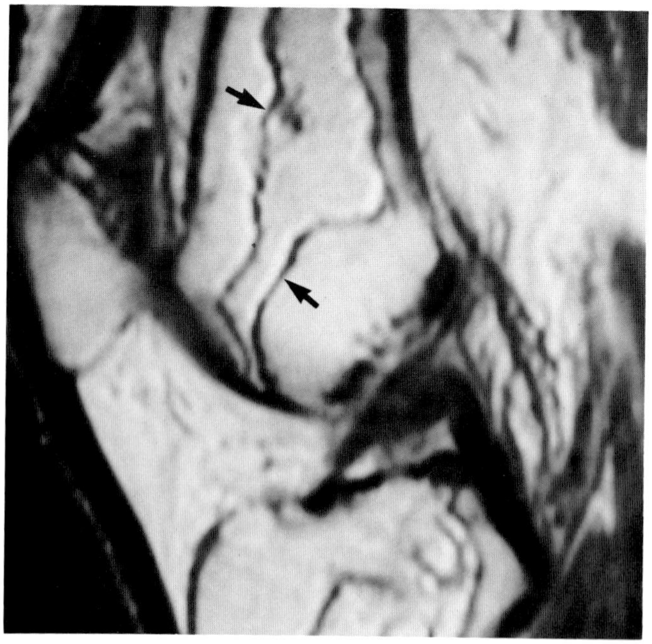

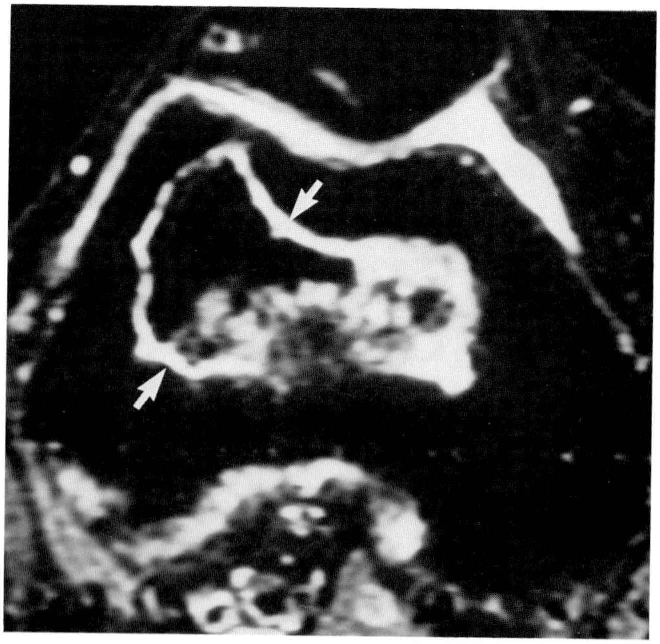

B

FIG. 9.49. Chronic bone infarction. **A:** Sagittal T1-weighted image and **B:** axial STIR image depict bone infarcts with perimeters of fibroblastic reactive tissue (*arrows*) involved in "creeping substitution."

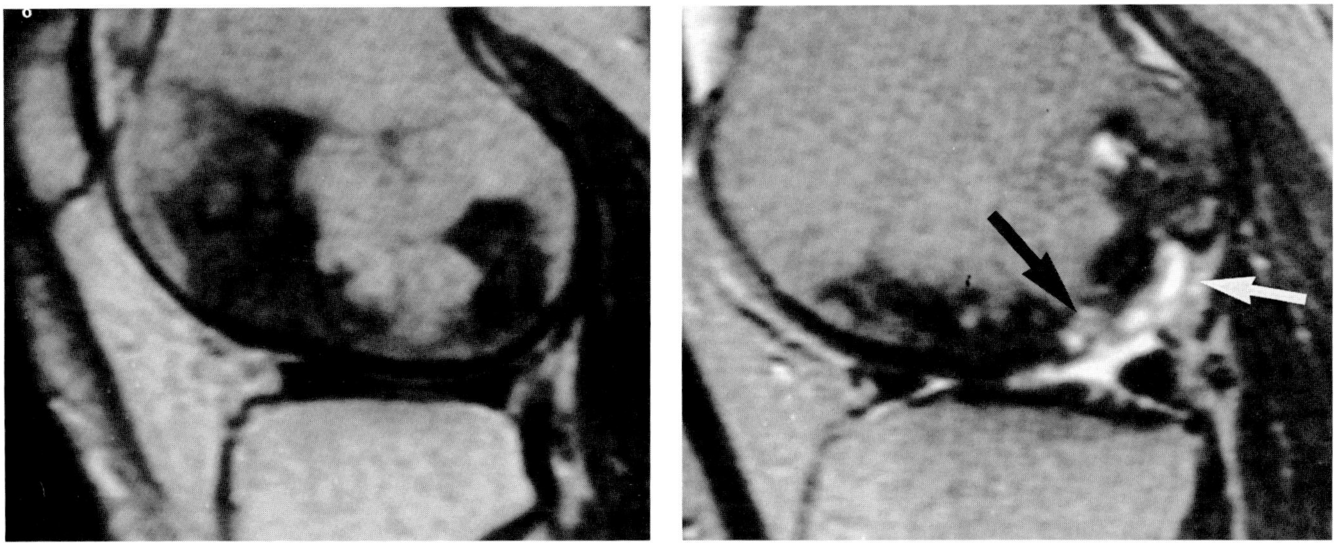

A B

FIG. 9.50. Epiphyseal infarction. **A:** Sagittal T1-weighted image demonstrates low signal intensity bone infarction extending to subchondral bone. **B:** Adjacent sagittal T2-weighted section depicts weakening and collapse of the subchondral plate (*arrows*).

the reestablishment of blood flow, standard radiographs may show relative osteopenia in the hyperemic region surrounding the infarct. As healing continues, a reactive interface of increased vascularity and inflammatory cells is seen surrounding the infarct, and new bone is deposited to reinforce the adjacent viable bone. Standard radiographs may demonstrate mixed density, and scintigraphy may reveal deposition of isotope at the margins of the infarct. Because both the changes accompanying the inflammatory response and the deposition of trabecular bone replace marrow fat, MRI continues to be characterized by loss of signal on T1-weighted images (Fig. 9.48). Further healing is characterized by remodeling of the reactive interface. The margins of the fibroblastic reactive tissue slowly expand inward into the infarct as new bone is formed along the outer circumference, a process sometimes referred to as "creeping substitution." Whereas metaphyseal infarcts are completely surrounded by the reactive interface (Fig. 9.49), those infarcts occurring within epiphyses differ in that one surface is surrounded by subchondral bone and articular cartilage. The subchondral plate may therefore be weakened or reabsorbed wherever it intersects the reactive interface. Weakening of the trabecular bone may cause microfractures and collapse of the articular surface (Fig. 9.50). The articular cartilage may buckle and collapse or may become loose within the joint. Radiographs may reveal articular flattening, subchondral lucencies, and mixed sclerosis. Scintigraphic findings remain positive as long as active bone remodeling continues. When infarction involves the subchondral bone of the femur or tibia, MRI reveals diminished signal from the subchondral region but, in addition, delineates the contour, thickness, and continuity of the overlying articular cartilage (Fig. 9.50). Subacute or old infarctions of the metaphyseal regions may appear as doughnuts with central high signal intensity and surrounding regions of low signal (Fig. 9.51). The central high signal

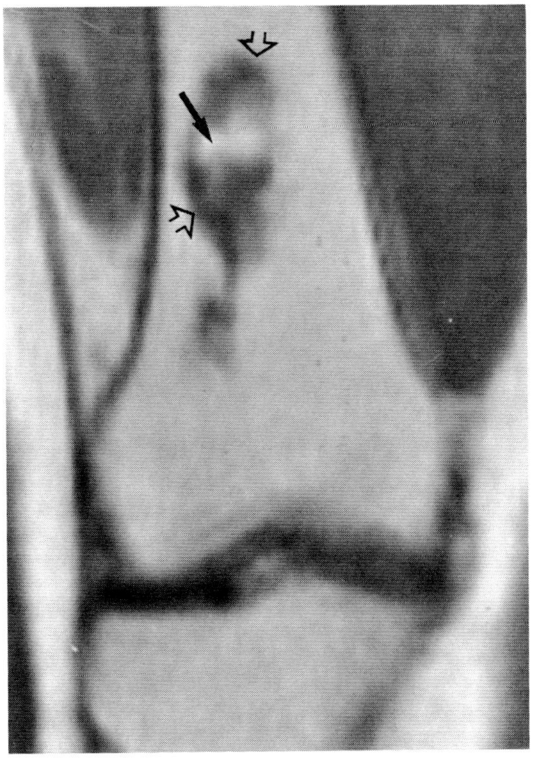

FIG. 9.51. Chronic infarction. Coronal T1-weighted image depicts doughnut pattern with central region of high signal intensity (*black arrow*) and peripheral rim of low signal intensity (*open arrows*).

has been found to correlate pathologically with regions of isolated marrow fat, and the margin of low signal corresponds to regions of fibrosis and dystrophic calcification (12).

Methods for diagnosing osteonecrosis include standard radiographs, tomography, scintigraphy, intraosseous venography, and intraosseous pressure measurements. Because there is high morbidity associated with articular collapse, early detection and preventive therapy are imperative. MRI has been compared with scintigraphy for detecting osteonecrosis and results indicate that MRI is the more sensitive technique (12,39,66,110,134,170). Furthermore, MRI is superior to scintigraphy in determining the extent of disease and, unlike scintigraphy, is often able to distinguish between acute and chronic infarctions. MRI is more specific in diagnosing infarction than is scintigraphy or standard radiography (117). The high spatial resolution and contrast resolution of MRI relative to scintigraphy often allow early specific diagnosis of infarction, whereas any process stimulating bone remodeling may cause increased isotope uptake. There have been anecdotal reports of cases in which MRI was found to be more accurate than even intraosseous venography and intraosseous pressure measurements (12). With regard to MR technique, fat suppression methods such as short TI inversion recovery and chemical fat suppression now appear to be more sensitive than standard spin-echo imaging in detecting bone marrow infarction and other bone marrow pathology.

Spontaneous osteonecrosis of the medial femoral condyle and spontaneous osteonecrosis of the proximal tibia are similar clinical entities affecting the knee that will be addressed separately below.

Spontaneous Osteonecrosis of the Medial Femoral Condyle

This well-recognized clinical entity occurs in middle-aged and elderly individuals and involves the weight-bearing surfaces of the distal femoral condyles, particularly the medial (5,18,63,66,87,95,96,107,121,122, 134,141,147,176). Women are affected three times more often than men. Involvement of both knees in a given patient is unusual. Patients most often report abrupt onset of knee pain that usually persists even at rest. Physical examination may reveal local tenderness in the medial femoral condyle, restricted range of motion, and effusion. Progression of disease is accompanied by subchondral collapse, fragmentation of the articular cartilage, and progressive osteoarthritis. In fact, it is postulated that many cases of medial tibiofemoral osteoarthritis are preceded by spontaneous osteonecrosis. Spontaneous healing can occur if the affected area is small and the stress- or weightbearing surface is not involved (4,5,95,96,176). The prognosis is unfavorable if the lesion is larger than 5 cm or involves more than 40% of the width of the condyle (4).

The cause of this disorder is unknown and there is no association with alcohol or steroid use. Proposed mechanisms include an alteration in the microvascular supply to the affected condyle or necrosis secondary to repeated trauma (4,121,122,141). In one study, a high association with medial meniscal tears was found

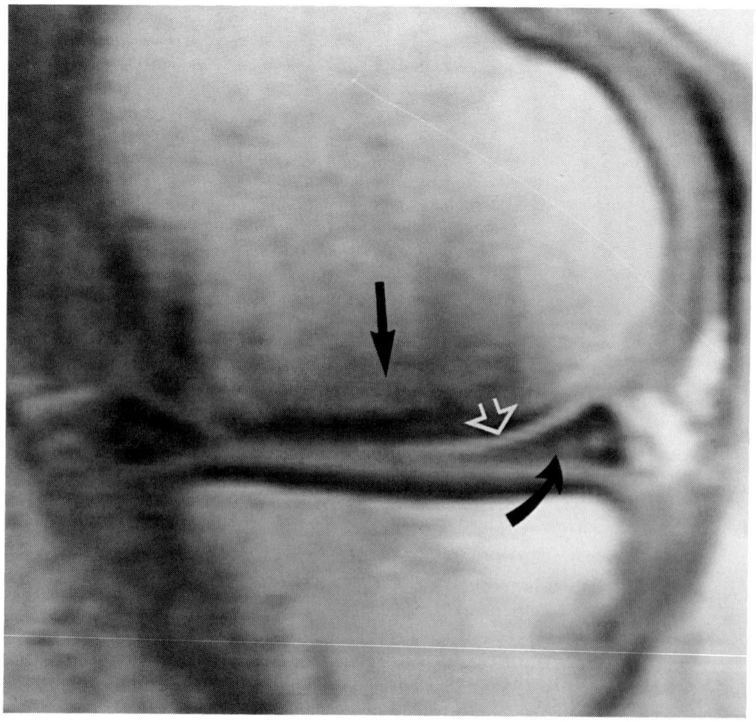

FIG. 9.52. Spontaneous osteonecrosis of the medial femoral condyle. Sagittal T1-weighted spin-echo image depicts subchondral diminished signal intensity along the central weight-bearing surface of the medial condyle (*arrow*), intact overlying articular cartilage (*open arrow*), and associated medial meniscal tear (*curved arrow*).

(121,122), although this association was not reported in another series (95). One investigator has documented seven patients who underwent arthroscopic treatment of meniscal tears in whom preoperative MR scans disclosed no evidence of avascular necrosis, but postoperative MR examinations performed 2 to 14 months after surgery revealed findings typical of spontaneous osteonecrosis (87).

Before MRI, scintigraphy was most commonly employed to diagnose spontaneous osteonecrosis prior to the appearance of overt radiographic abnormalities (63,96,147,176). Unfortunately, bone scan findings are nonspecific because spontaneous osteonecrosis may be mimicked by osteoarthritis, stress fractures, osteonecrosis of other causes, and meniscal injuries. When they appear, radiographic findings in order of occurrence include flattening of the articular surface of the medial femoral condyle, and development of an initially poorly defined subchondral lucency that progresses to a well-defined oval bony defect with a sharp sclerotic margin (5).

There have now been several reports concerning MRI of this disorder (18,27,66,87,128,134). The hallmark is loss of signal from the subchondral bone on T1-weighted spin-echo images (Fig. 9.52) and increased signal in the same region on short TI inversion recovery images. Instability of the osteonecrotic subchondral bone and interruption of the articular cartilage is implied when high signal intensity synovial fluid or granulation tissue is seen dissecting around the margin of necrotic subchondral fragment (Fig. 9.53). Advantages of MRI for assessing spontaneous osteonecrosis of the medial femoral condyle include the ability to assess the condition of the articular cartilage and menisci as well as the capacity to determine the presence and precise location of loose bodies.

Conditions that may be clinically confused with spontaneous osteonecrosis include osteochondritis di-

A B C

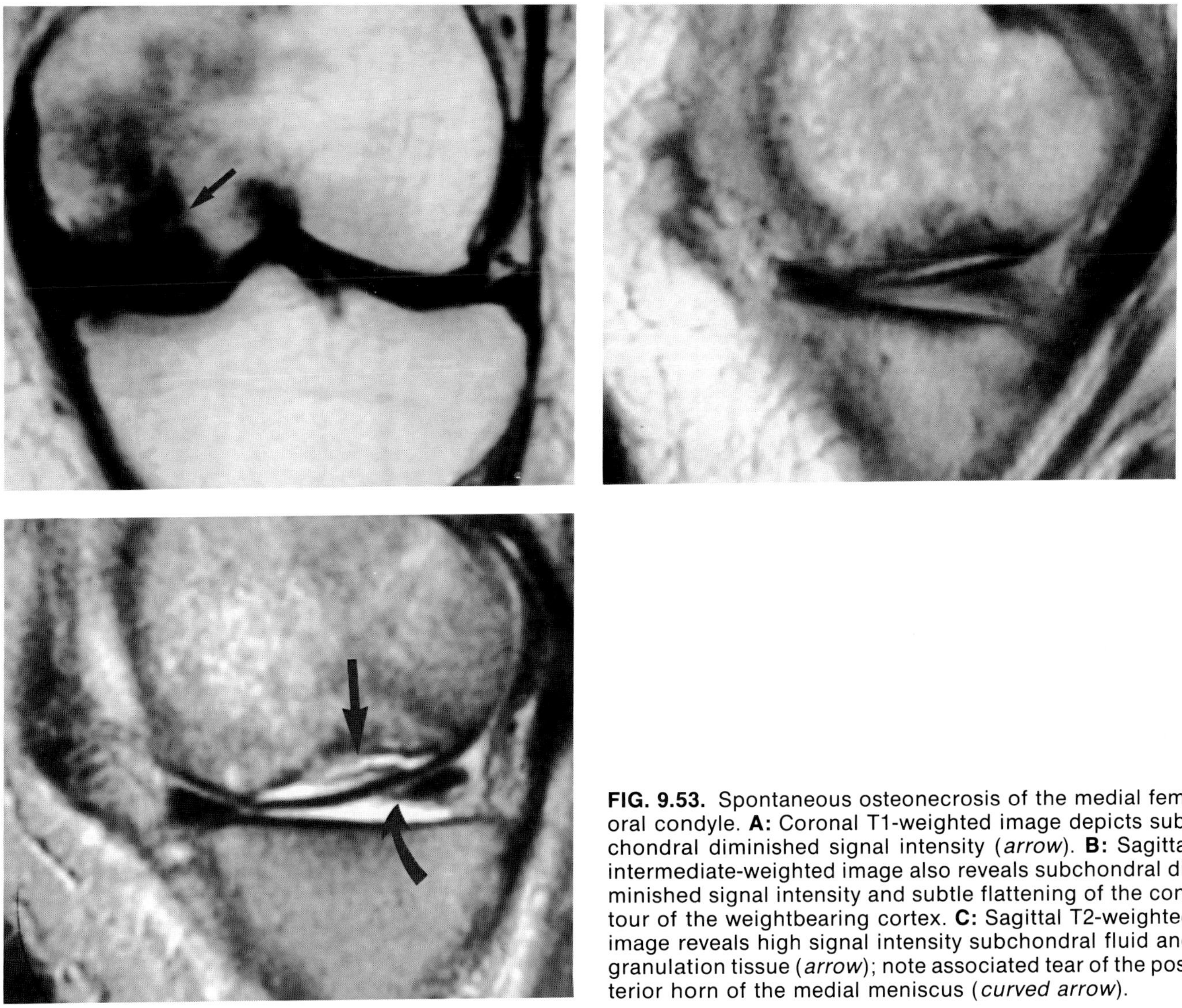

FIG. 9.53. Spontaneous osteonecrosis of the medial femoral condyle. **A:** Coronal T1-weighted image depicts subchondral diminished signal intensity (*arrow*). **B:** Sagittal intermediate-weighted image also reveals subchondral diminished signal intensity and subtle flattening of the contour of the weightbearing cortex. **C:** Sagittal T2-weighted image reveals high signal intensity subchondral fluid and granulation tissue (*arrow*); note associated tear of the posterior horn of the medial meniscus (*curved arrow*).

seccans, osteonecrosis of other cause, osteoarthritis, pes anserinus bursitis, and medial meniscal injury. Osteochondritis diseccans differs in that pain is insidious in onset and young adults and adolescents are primarily affected (5,147). The lesion in osteochondritis dissecans usually occurs on the lateral aspect of the medial condyle, whereas in spontaneous osteonecrosis the lesion involves the weightbearing surface. Loose bodies are common in osteochondritis dissecans but uncommon in spontaneous osteonecrosis. Osteoarthritis is characterized by insidious onset of pain that is relieved by rest, whereas spontaneous osteonecrosis frequently results in abrupt onset of pain that persists at rest. Osteoarthritis usually begins on the tibial side of the joint, whereas spontaneous osteonecrosis begins on the femoral side. In osteoarthritis, cartilaginous erosions occur early; in spontaneous osteonecrosis, erosions occur late.

Osteonecrosis secondary to systemic causes, such as steroid use, differs from spontaneous osteonecrosis in that the former is frequently bilateral, frequently involves the lateral condyle, and usually is accompanied by multiple joint involvement (5). MRI has thus far proven more useful than scintigraphy for differentiating the conditions that may be clinically mistaken for spontaneous osteonecrosis.

Treatment of spontaneous osteonecrosis is dependent on early and accurate diagnosis. Avoidance of weightbearing early in the course of this disease frequently results in complete resolution of symptoms (96). Precise localization of the lesion may facilitate surgical therapy, which can include tibial osteotomy and arthroplasty.

Osteonecrosis of the Medial Tibial Plateau

A syndrome presenting in the same age group and with the same clinical history as spontaneous osteonecrosis of the medial femoral condyle has been described (95). Instead of the femoral condyle, however, the site of osteonecrosis is the tibial plateau, almost always the medial aspect. As with spontaneous osteonecrosis of the medial femoral condyle, radiographs initially show normal findings. Scintigraphy may show deposition of isotope at the site of the lesion. Most patients recover after 9 to 12 months of conservative therapy. In others, the disorder may be a precursor to osteoarthritis. It is interesting that in their original description of the radiographic appearance of spontaneous osteonecrosis of the medial femoral condyle, Ahlback et al. reported sclerosis or defects in the tibial plateau in 20% of patients (5). Osteonecrosis of the tibia has been postulated to be caused by stress microfractures of the plateau, resulting in elevated intraosseous pressure due to either intrusion of synovial fluid or lack of sufficiently rigid subchondral bone to support the load (5). MRI is extremely valuable in diagnosing osteonecrosis of the medial tibial plateau (Fig. 9.54). MRI is employed to detect the lesion, evaluate the overlying articular cartilage, and exclude other clinical diagnoses such as a medial meniscal tear,

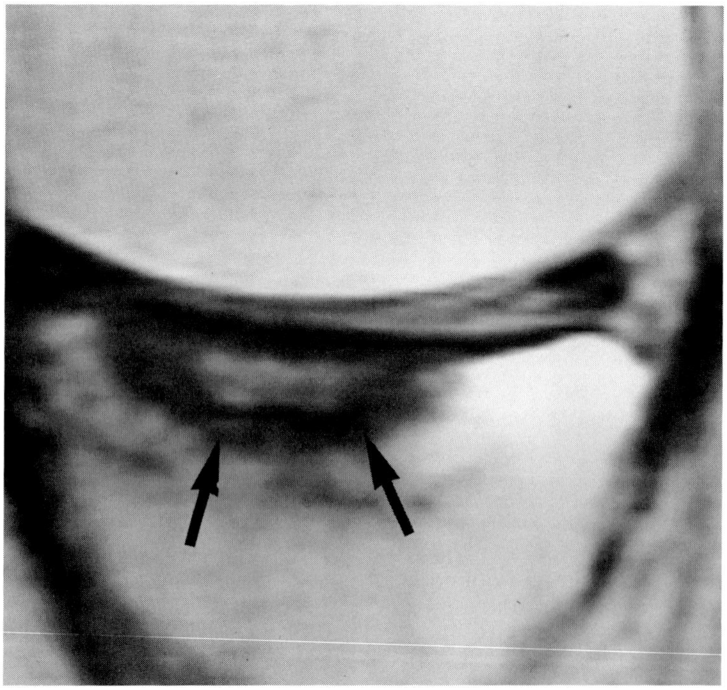

FIG. 9.54. Osteonecrosis of the medial tibial plateau. Sagittal T1-weighted image reveals diminished signal intensity of subchondral bone of the medial tibial plateau (*arrows*).

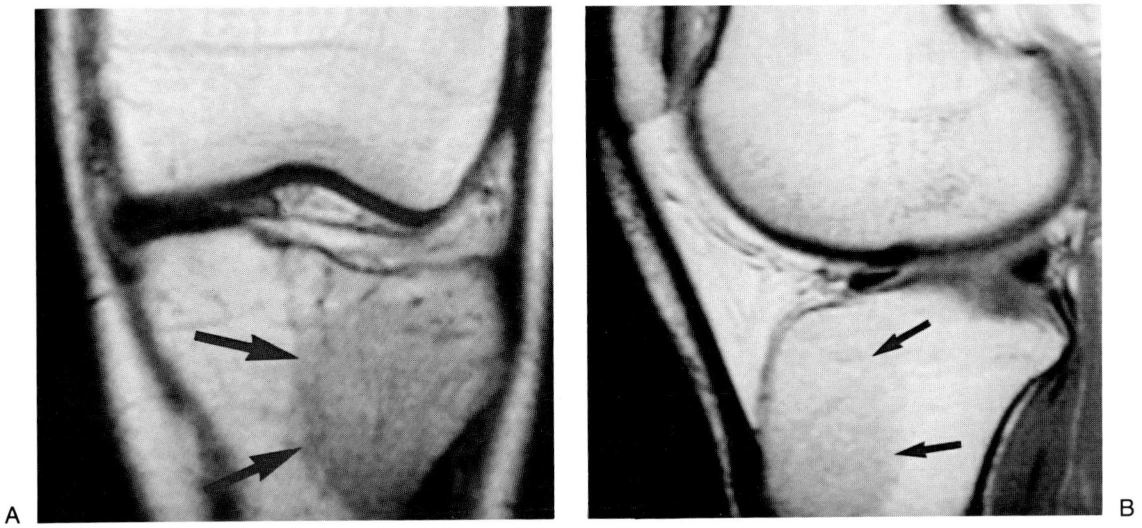

FIG. 9.55. Regional migrating osteoporosis. **A:** Coronal and **B:** sagittal T1-weighted spin-echo images reveal a zonal region of diminished signal intensity within bone marrow (*arrows*). The patient was followed conservatively, but MR appearance worsened after 2 months prompting biopsy, which revealed only osteoporosis. Lesion subsequently completely resolved, both clinically and on MR images.

pes anserinus bursitis, or spontaneous osteonecrosis of the medial femoral condyle.

Regional Migratory Osteoporosis

Regional migratory osteoporosis is a form of transient regional osteoporosis characterized by rapidly developing osteoporosis affecting periarticular bone, absence of a clear-cut inciting event, and a self-limited, reversible course (62,141). The knee, ankle, and foot are most commonly involved (141). This disorder usually appears in the fourth or fifth decade of life, and men are affected three times more commonly than women (62).

Typically pain and swelling develop rapidly, last up to 9 months, and then disappear. Recurrences are common, often involving the articulation nearest the formerly diseased one. Each episode clinically resembles reflex sympathetic dystrophy. The duration of disability averages 3 to 4 months (62). Regional migrating osteoporosis may manifest a zonal pattern, with involvement of the medial or lateral condyle of the femur. Histologic evaluation reveals osteoporotic bone with increased numbers of osteoblasts and osteoclasts (141). The bone marrow is otherwise normal.

On MR images, regional migratory osteoporosis is characterized by a broad, poorly demarcated, periarticular region of bone marrow edema, manifested by low signal intensity on T1-weighted images and relatively high signal intensity on fat-suppressed images. This entity may be impossible to differentiate from early bone marrow infarction, bone marrow infiltration, or edema secondary to trauma (Figs. 9.55 and 9.56). Serial MR examinations may be necessary to follow the resolution of this disorder in order to rule out neoplastic marrow infiltration. When the pattern of bone marrow edema resolves but then recurs in an adjacent region, the diagnosis of regional migratory osteoporosis can be made with certainty.

Ischemic necrosis may develop after biopsy (63). The author has encountered one case of presumed regional migrating osteoporosis which progressed to subchondral avascular necrosis over a period of 5 months, although no biopsy was performed (Fig. 9.57).

Osteoporosis

MR imaging has recently demonstrated clinical potential as a technique for quantifying trabecular bone (54,155,174). Marrow in contact with trabecular bone exhibits a shortened effective transverse relaxation time (T2*) which results in signal loss on gradient-echo images because of local field inhomogeneities. This T2* shortening is increased in regions with more trabecular bone (54,145,155,174). Therefore, in a normal dense trabecular network, T2* shortening should be more pronounced than in rarefied osteoporotic trabeculae (54). Bone marrow in patients with osteoporosis has significantly prolonged T2* values, possibly due to an increase in the intertrabecular space (55).

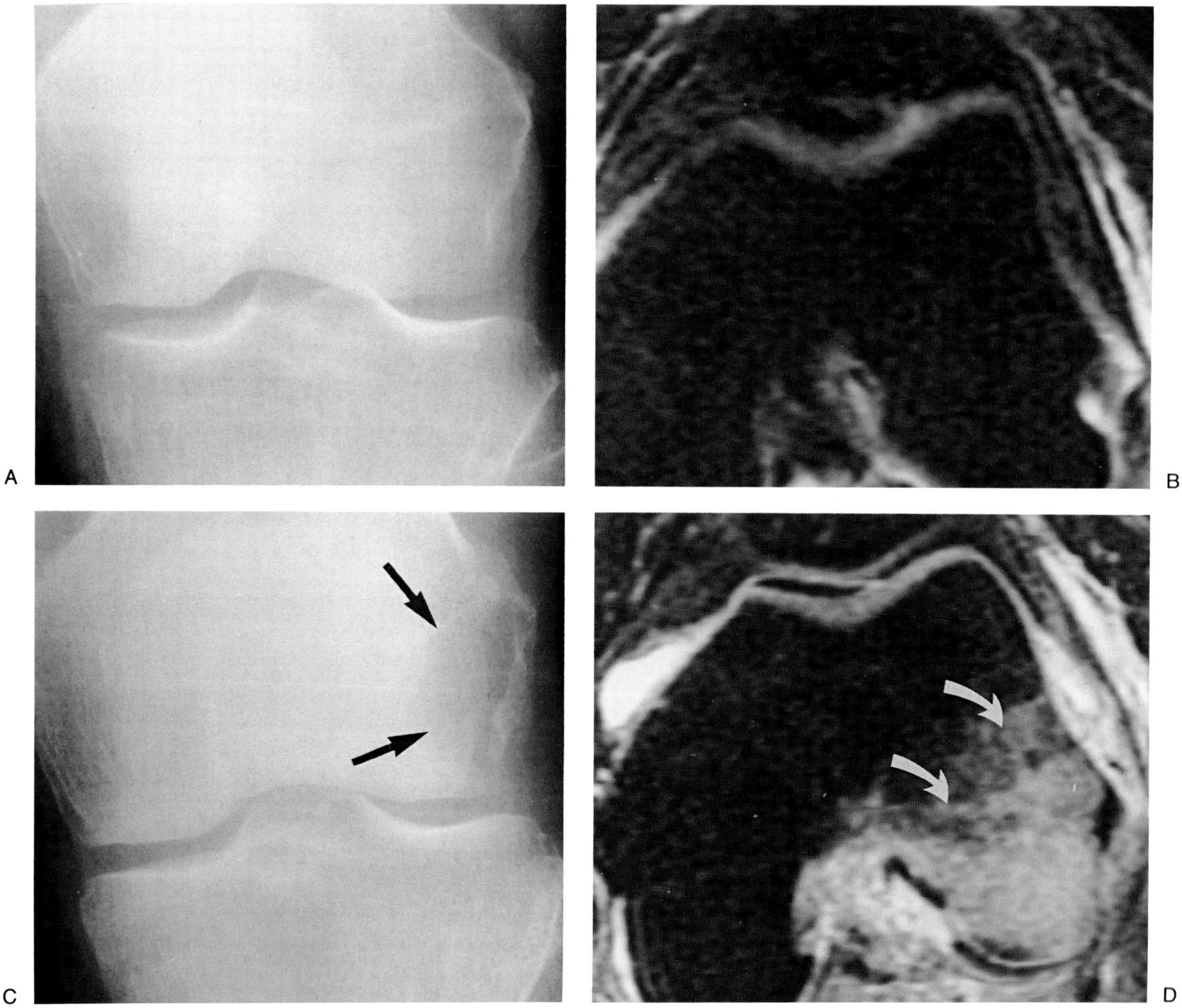

FIG. 9.56. Regional migrating osteoporosis—zonal type. **A:** Initial standard radiograph and **B:** axial STIR MR image are normal. Patient remained symptomatic, and **C:** subsequent standard radiograph revealed subtle focal osteoporosis (*arrows*), with bone marrow abnormality (*curved arrows*) much more convincingly demonstrated on **D:** axial STIR MRI.

Based on these observations, MR techniques are under development to positively identify sites of osteoporosis. This may prove useful in the systemic evaluation of osteoporotic fracture risk. In evaluating knee joint disorders, specific early diagnosis of entities such as regional migratory osteoporosis or reflex sympathetic dystrophy may be facilitated by these future developments.

TUMORS AND TUMOR-LIKE MASSES

Numerous benign and malignant tumors occur within or adjacent to the knee joint, including osteosarcomas, fibrosarcomas, chondrosarcomas, malignant fibrous histiocytomas, synovial sarcomas, giant cell tumors, chondromyxoid fibromas, pigmented vill-

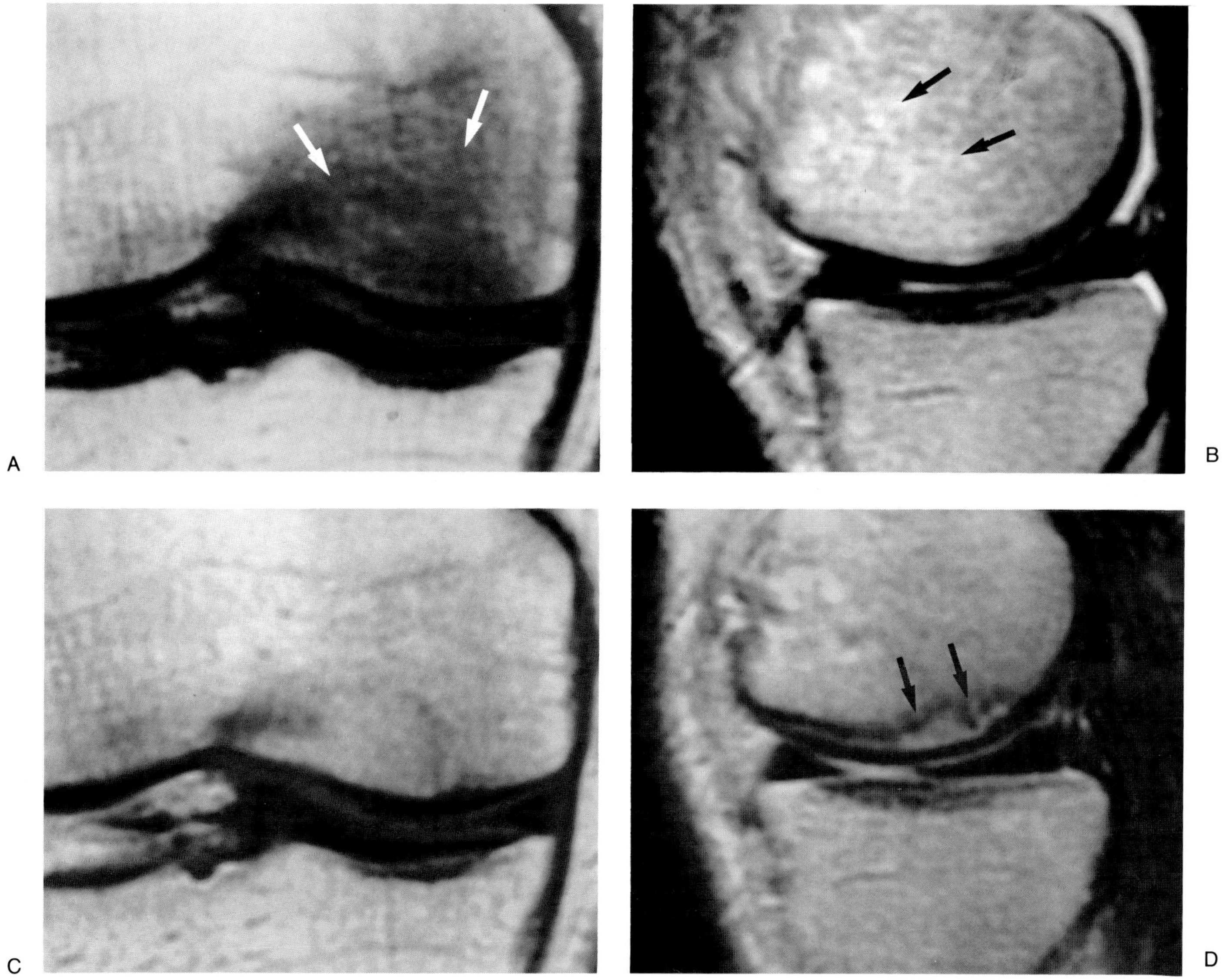

FIG. 9.57. Presumed regional migrating osteoporosis. On initial examination, **A:** coronal T1-weighted MRI demonstrated zonal region of bone marrow edema. **B:** Sagittal T2-weighted MRI also disclosed bone marrow edema with subtle increased signal intensity within the medial femoral condyle (*arrows*). **C:** Several months later, coronal T1-weighted image revealed marked improvement in bone marrow edema, but **D:** sagittal T2-weighted scan disclosed presumed subchondral osteonecrosis (*arrow*). Clinical characteristics were most compatible with regional migrating osteoporosis, but no biopsy was performed.

onodular synovitis, lipomas, osteoid osteomas, and osteochondromas. In addition, metastatic and infiltrating malignancies of medullary bone may certainly involve the knee joint. There are now numerous published reports of MR imaging of bone and soft tissue tumors or bone marrow infiltration (6,11,17,20,21,38,43,49, 52,57,66,69,79,83,84,107,112,113,116,118,124,126,132, 137,138,144,146,148,156,177,183). MRI is now the examination of choice for evaluating clinically indeterminate soft tissue masses. For bone tumors, the plain radiograph is the least expensive test and usually still the most reliable predictor of a lesion's histologic nature. Therefore, for bone tumors, the imaging workup should begin with standard radiographs followed by MRI for precise localization and staging. CT is of additional use only when standard radiographs or MRI fail to fully characterize cortical interruption or associated calcifications.

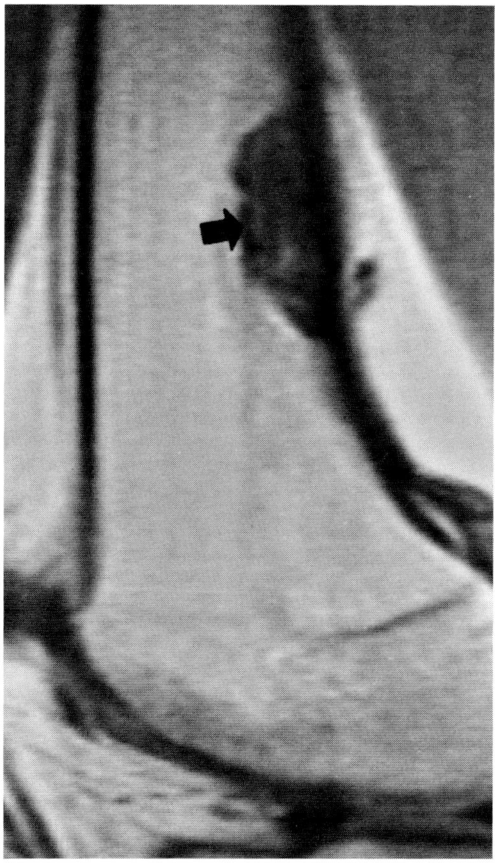

FIG. 9.58. Nonossifying fibroma. Sagittal T1-weighted scan depicts classic low signal intensity, eccentric, juxtacortical lesion, typical of nonossifying fibroma (*arrows*).

Although one report indicates approximately 90% accuracy in differentiating benign from malignant soft tissue lesions using MRI (17), the remaining literature strongly indicates that MRI is not reliable in distinguishing benign from malignant neoplasms of bone or soft tissue. Although not histologically specific, MRI is exquisitely accurate in anatomically delineating both benign and malignant soft tissue lesions. MR imaging should precede biopsy of bone and soft tissue tumors and is often essential in preoperative planning.

In addition to depicting the extent of disease, MRI is also useful in specifically solving a variety of clinical dilemmas:

1. Certain benign osseous and soft tissue entities have a sufficiently specific MR appearance to obviate the need for further diagnostic tests. Examples include nonossifying fibromas (Fig. 9.58), enchondromas (Fig. 9.59), simple lipomas (Fig. 9.60), and osteoid osteoma (Fig. 9.61).
2. Pathologic fluid collections such as popliteal cysts, ganglia, and simple hematomas can be accurately discriminated from soft tissue masses.
3. Often, stress fractures (Fig. 9.62) and avascular necrosis have a sufficiently specific appearance to allow these entities to be distinguished from other neoplastic causes of bone pain.
4. Exostoses or osteochondromas, which commonly occur in the knee region, can be precisely characterized. The presence and thickness of the overlying cartilaginous cap can be easily assessed (Figs. 9.63 and 9.64) and bursae overlying exostoses can be delineated (Fig. 9.65).

(*Text continues on page 381*)

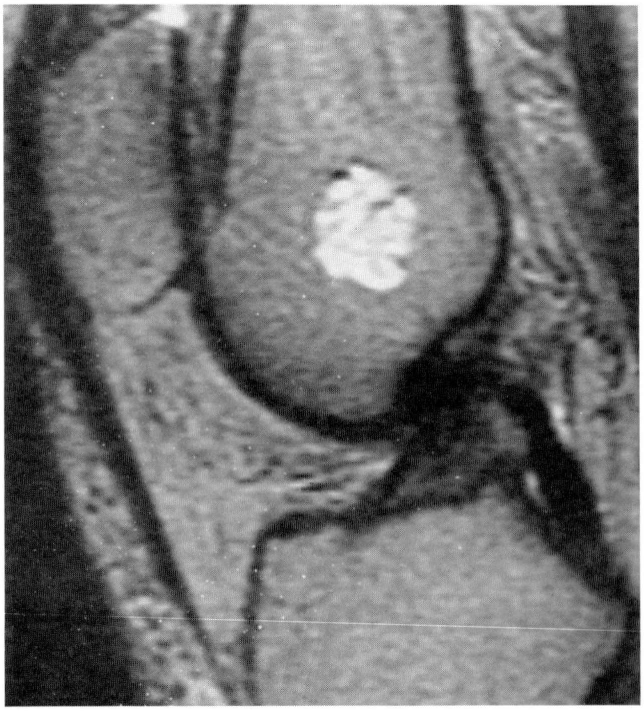

FIG. 9.59. Enchondroma. Sagittal T2-weighted scan depicts septated, well-circumscribed, high signal intensity lesion, typical of nonaggressive enchondroma.

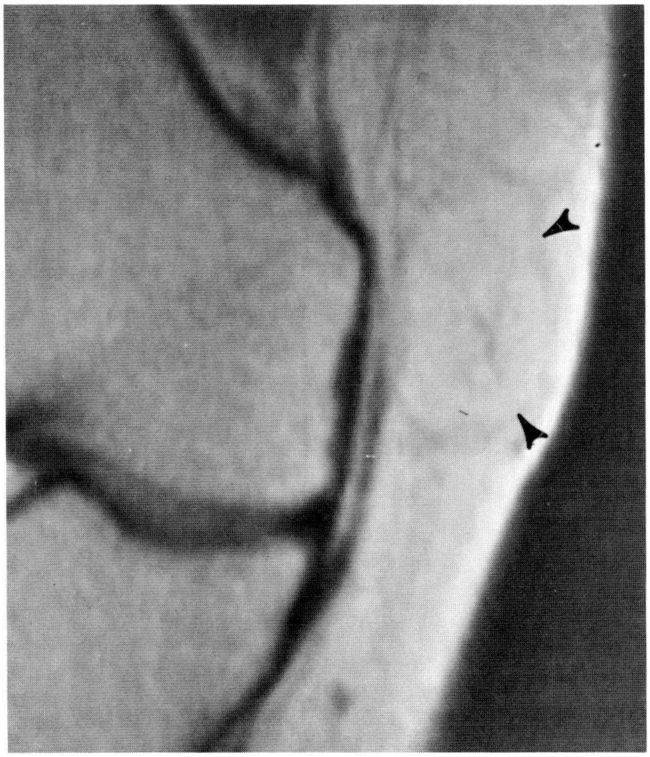

FIG. 9.60. Subcutaneous lipoma. Oval mass (*arrowheads*) identical in signal intensity to adjacent subcutaneous fat, with very thin capsule typical of subcutaneous lipoma. Lipomas may be indistinguishable from adjacent subcutaneous fat on MR images. Therefore, whenever a subcutaneous mass is clinically palpable but not visible on MRI, the diagnosis of a lipoma should be suggested.

A

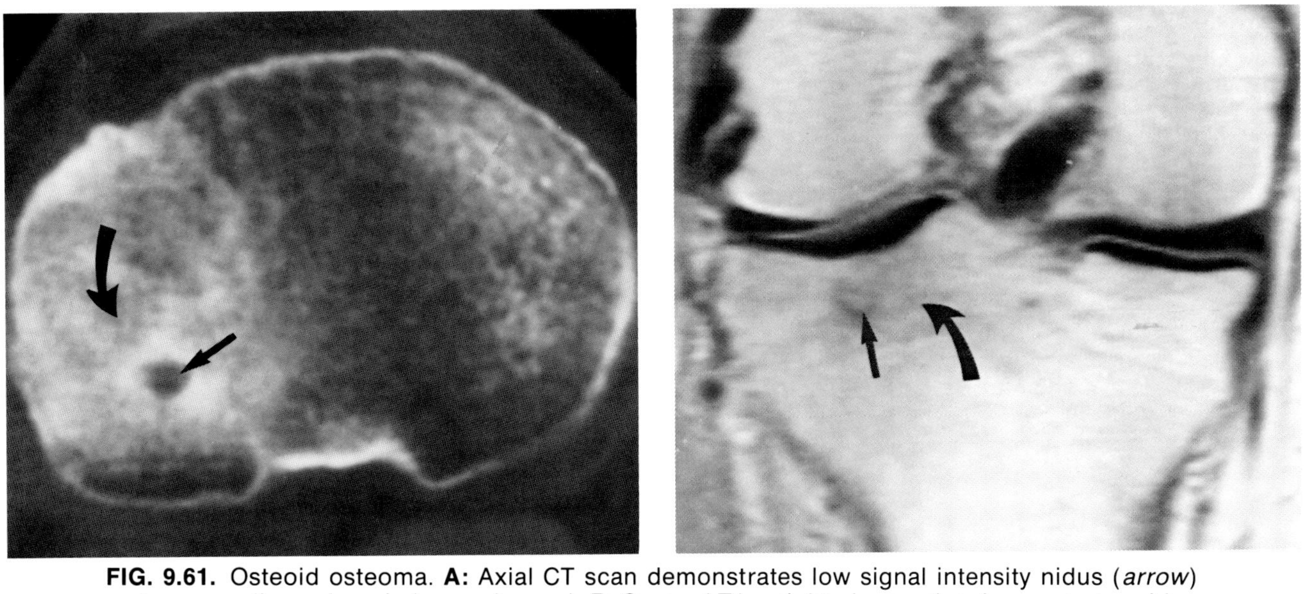

B

FIG. 9.61. Osteoid osteoma. **A:** Axial CT scan demonstrates low signal intensity nidus (*arrow*) and surrounding sclerosis (*curved arrow*). **B:** Coronal T1-weighted scan also demonstrates nidus (*straight arrow*) and adjacent bone marrow reaction (*curved arrow*).

A

B

C

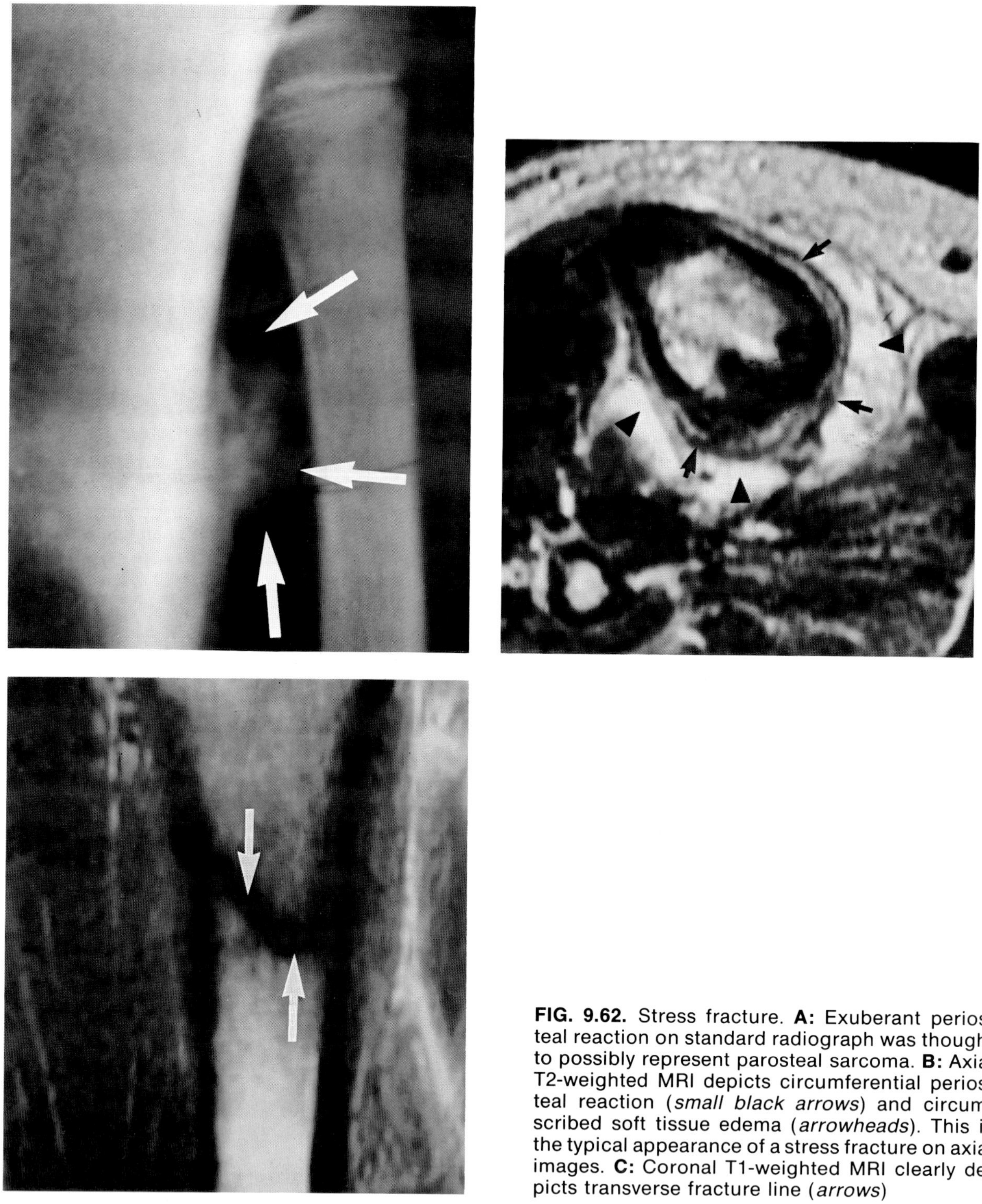

FIG. 9.62. Stress fracture. **A:** Exuberant periosteal reaction on standard radiograph was thought to possibly represent parosteal sarcoma. **B:** Axial T2-weighted MRI depicts circumferential periosteal reaction (*small black arrows*) and circumscribed soft tissue edema (*arrowheads*). This is the typical appearance of a stress fracture on axial images. **C:** Coronal T1-weighted MRI clearly depicts transverse fracture line (*arrows*)

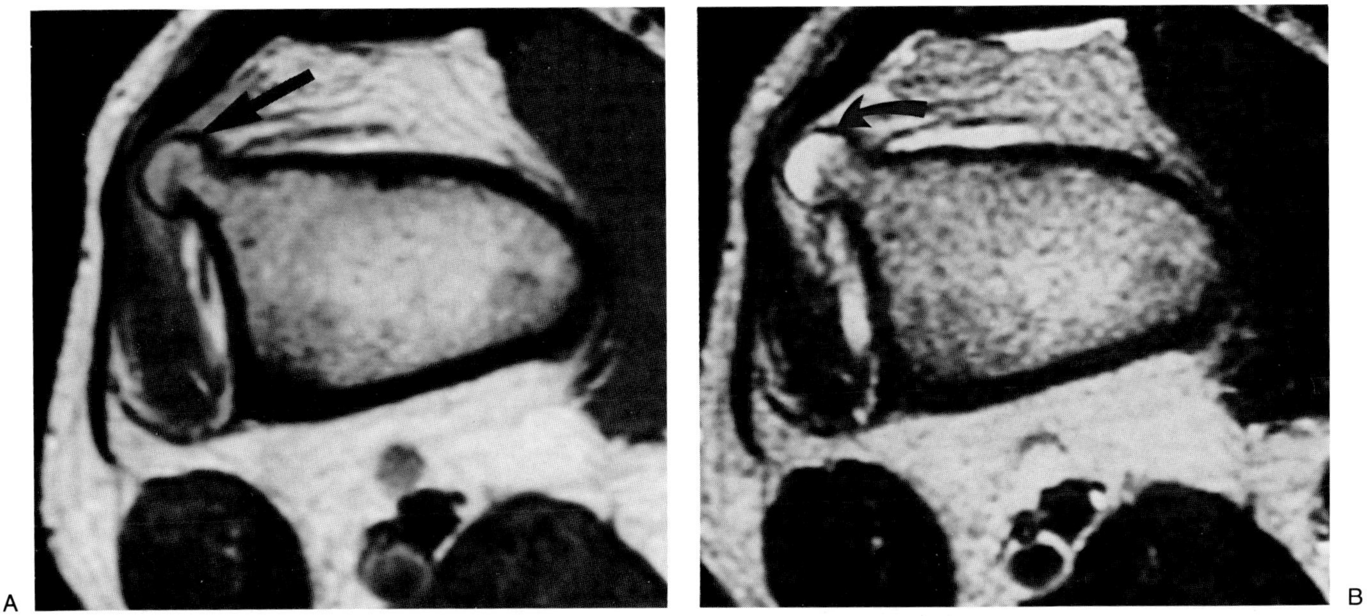

FIG. 9.63. Distal femoral osteochondroma. **A:** Axial intermediate-weighted image depicts exostosis (*arrow*), while **B:** axial T2-weighted scan clearly discloses high signal intensity cartilaginous cap (*curved arrow*).

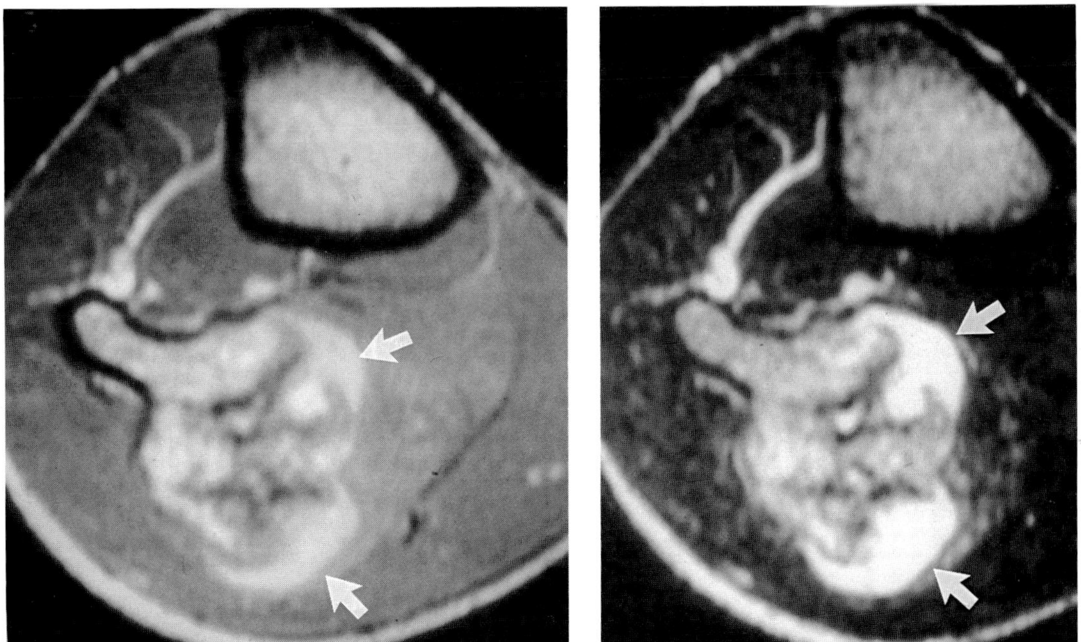

FIG. 9.64. Aggressive-appearing osteochondroma. Both the axial intermediate-weighted image (*left*) and axial T2-weighted image (*right*) depict large proximal fibular osteochondroma with thick, high signal intensity cartilaginous cap (*arrows*).

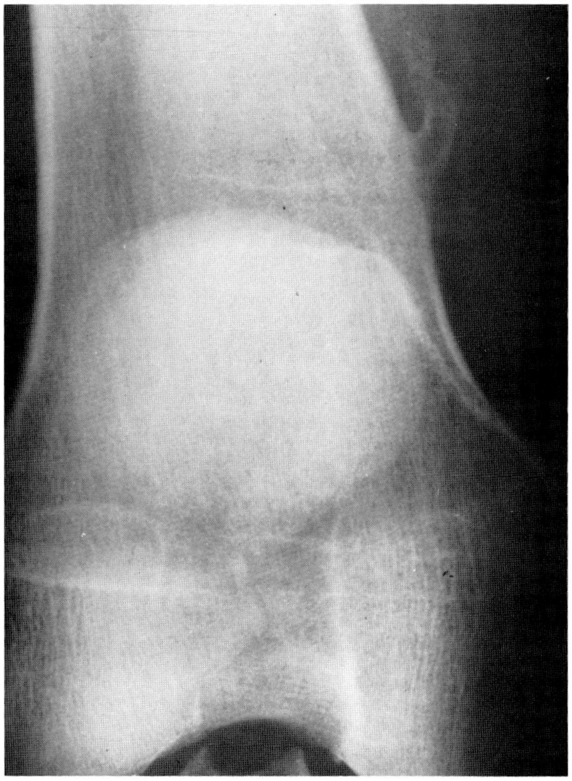

A

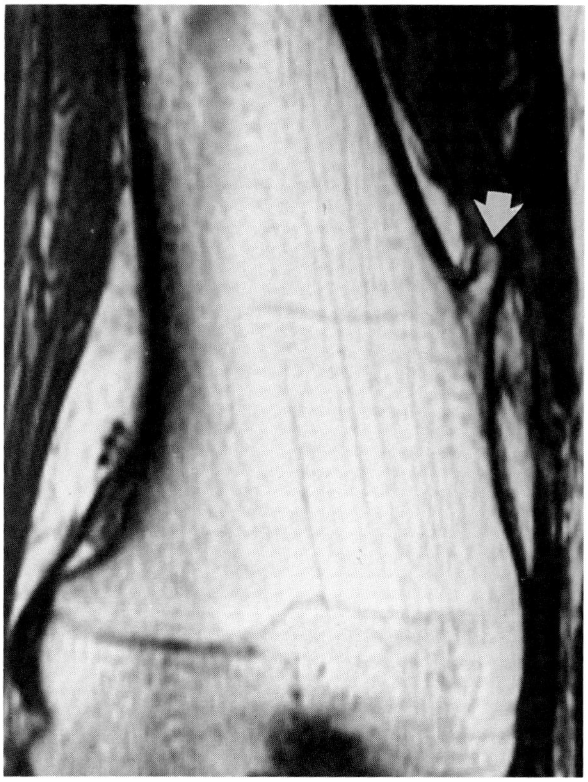

B

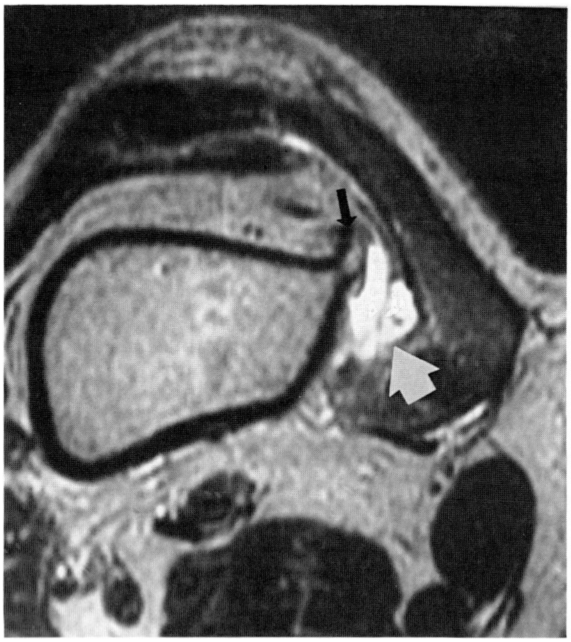

C

FIG. 9.65. Osteochondroma. **A:** Standard radiograph reveals typical osteochondroma pointing away from the distal femoral physis. **B:** T1-weighted MRI depicts osteochondroma and absence of cartilaginous cap (*white arrowhead*). **C:** Axial T2-weighted MRI again depicts osteochondroma (*black arrow*) and adjacent high signal intensity bursa (*white arrowhead*).

Evaluation of bone and soft tissue tumors with MRI is, however, limited in several respects:

1. MRI is not histologically accurate. Although bone tumors associated with frank cortical destruction and infiltration can be accurately classified as malignant (Figs. 9.66–9.68), an aggressive appearance does not always indicate malignancy (Figs. 9.69 and 9.70). Furthermore, a benign, sharply demarcated appearance certainly does not exclude the possibility of a malignant bony lesion (Fig. 9.71).
2. With regard to soft tissue masses in the knee region, although a malignant nature is clearly suggested by an invasive pattern of growth (Fig. 9.72), malignant lesions may also be deceptively well circumscribed (Figs. 9.73 and 9.74).
3. A wide range of benign conditions can mimic the appearance of neoplasm. Benign soft tissue conditions which may mimic neoplasm include infection, complex hematoma (Fig. 9.75), abscess, and even rare conditions such as crystalline masses or melorheostosis (Figs. 9.76 and 9.77). Benign conditions which may simulate the appearance of bone neoplasm include hematopoietic hyperplasia, regional migrating osteoporosis, infection, early bone infarction, and even bone chips related to prior bone grafting procedure (72).

Regarding MR imaging techniques for evaluating bone and soft tissues, certain generalizations can be made:

1. T1-weighted images and/or fat-suppressed images (chemical fat suppression or short TI inversion recovery) usually maximize contrast between neoplasms and bone marrow or subcutaneous fat.
2. T2-weighted and/or short TI inversion recovery images usually display maximal contrast between tumors, normal muscle, and neurovascular structures.
3. Intravenous infusion of gadolinium is generally not useful for evaluating bone or soft tissue tumors. On rare occasions, intravenous gadolinium is useful to differentiate a benign cyst or ganglion from a well-circumscribed tumor such as a neuroma, since only the latter enhances with intravenous gadolinium.

(*Text continues on page 389*)

A

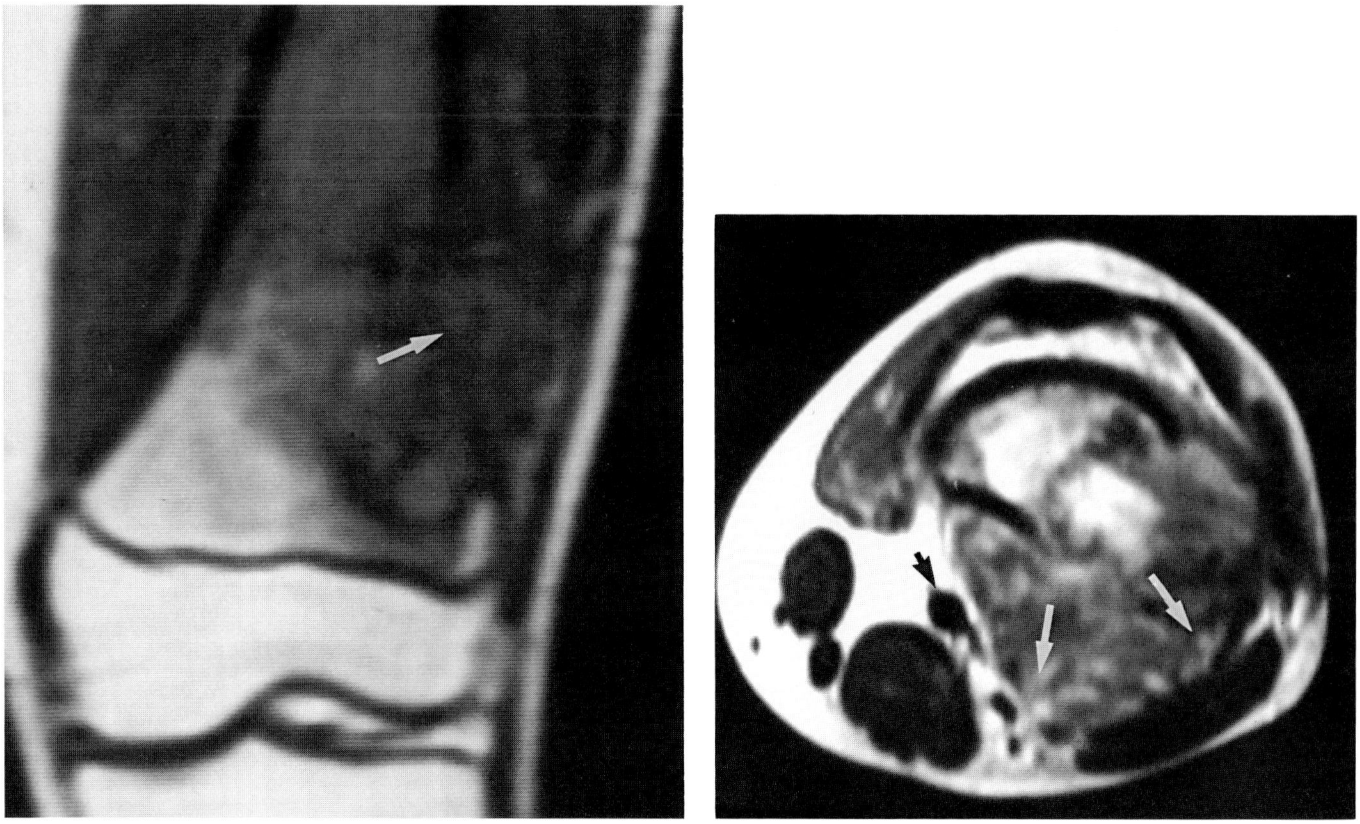

B

FIG. 9.66. Osteogenic sarcoma. **A:** T1-weighted coronal scan depicts low signal intensity mass infiltrating femoral metadiaphysis, breaching lateral cortex (*arrow*). **B:** Axial image reveals extension of mass (*arrows*) into adjacent superior popliteal space. The superior aspect of the popliteal artery (*black arrow*) is demonstrated to be separated from the lesion by a thin fascial plane.

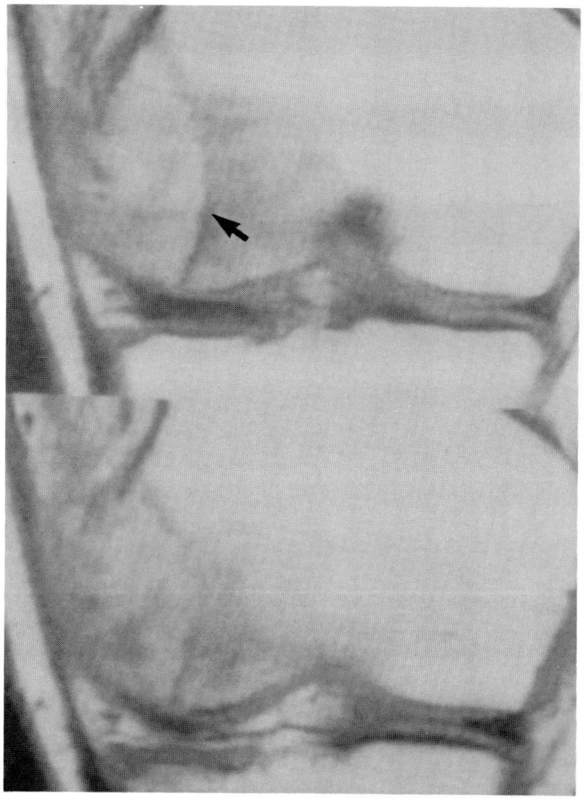

A

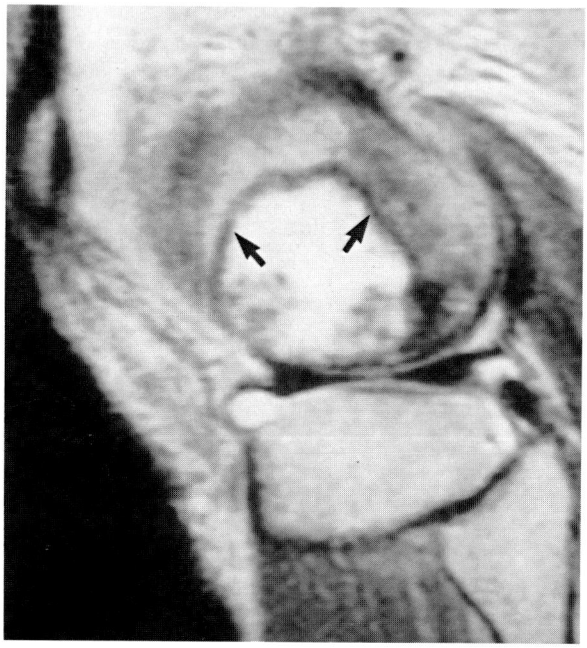

B

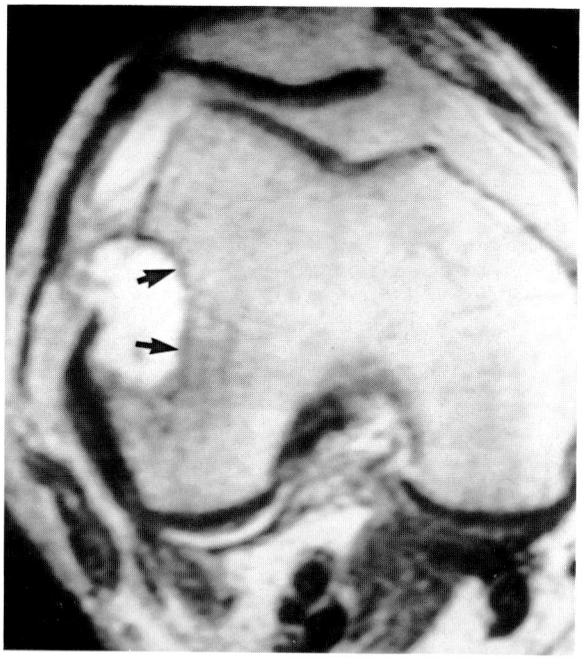

C

FIG. 9.67. Metastatic melanoma. Focally destructive metastatic neoplasm is depicted on **A:** coronal T1-weighted scan, **B:** sagittal T2-weighted image, and **C:** axial T2-weighted image. The lesion appears to be partially encapsulated (*small black arrows*), although the femoral cortex is clearly breached. (Courtesy of Anthony Stauffer, M.D. and Robert Turner, M.D.)

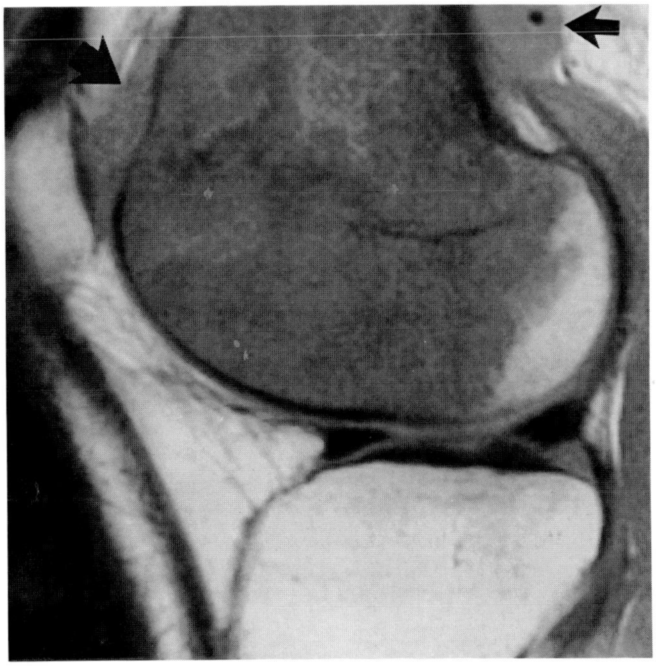

FIG. 9.68. Metastatic lymphoma. Sagittal T1-weighted scan depicts extensive low signal intensity bone marrow infiltration as well as extension into periosseous soft tissues (*arrows*).

A

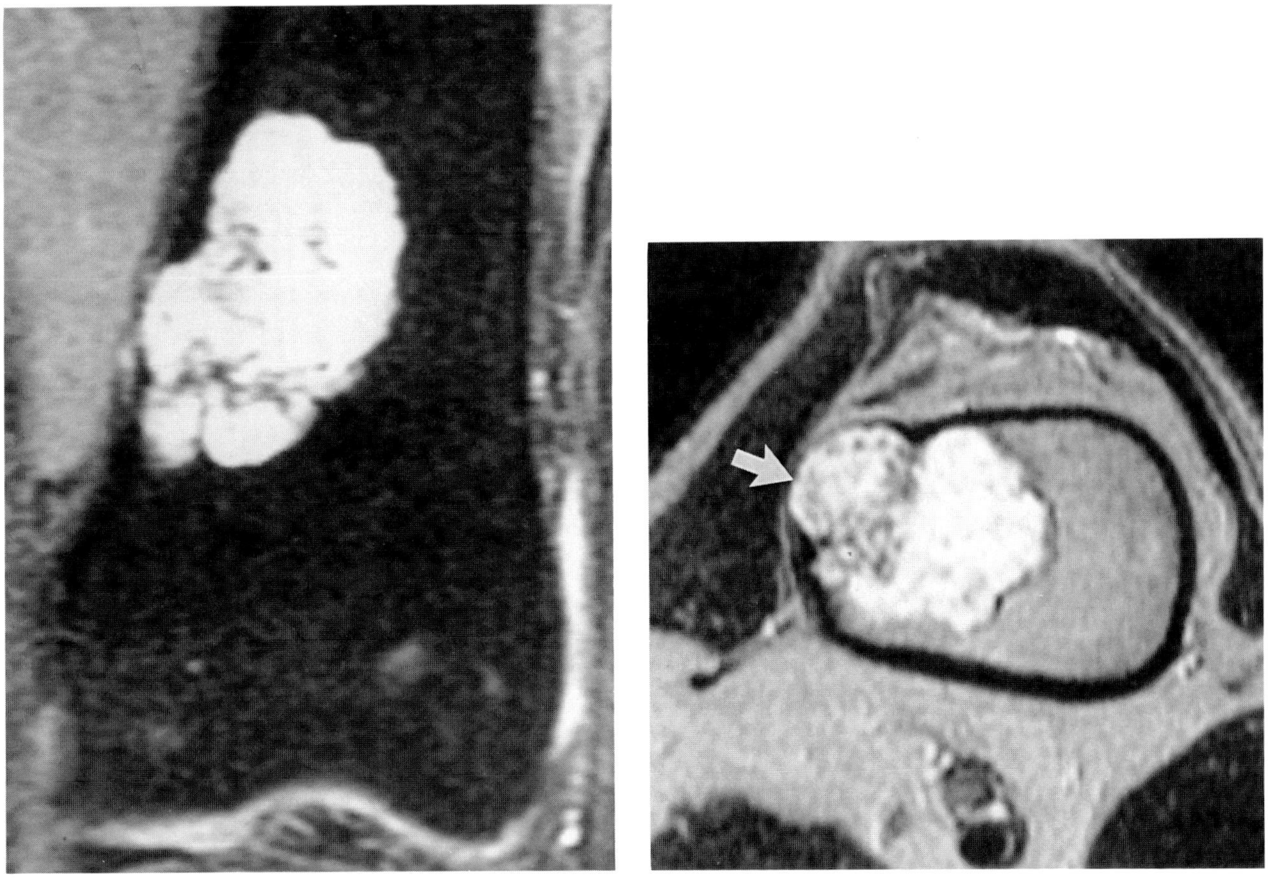

B

FIG. 9.69. Aggressive but benign enchondroma. **A:** Coronal STIR image reveals encapsulated, septated, high signal intensity mass. **B:** Axial T2-weighted MRI discloses cortical expansion (*arrow*) without interruption. On the basis of MRI, one cannot accurately distinguish whether this is an enchondroma or a chondrosarcoma.

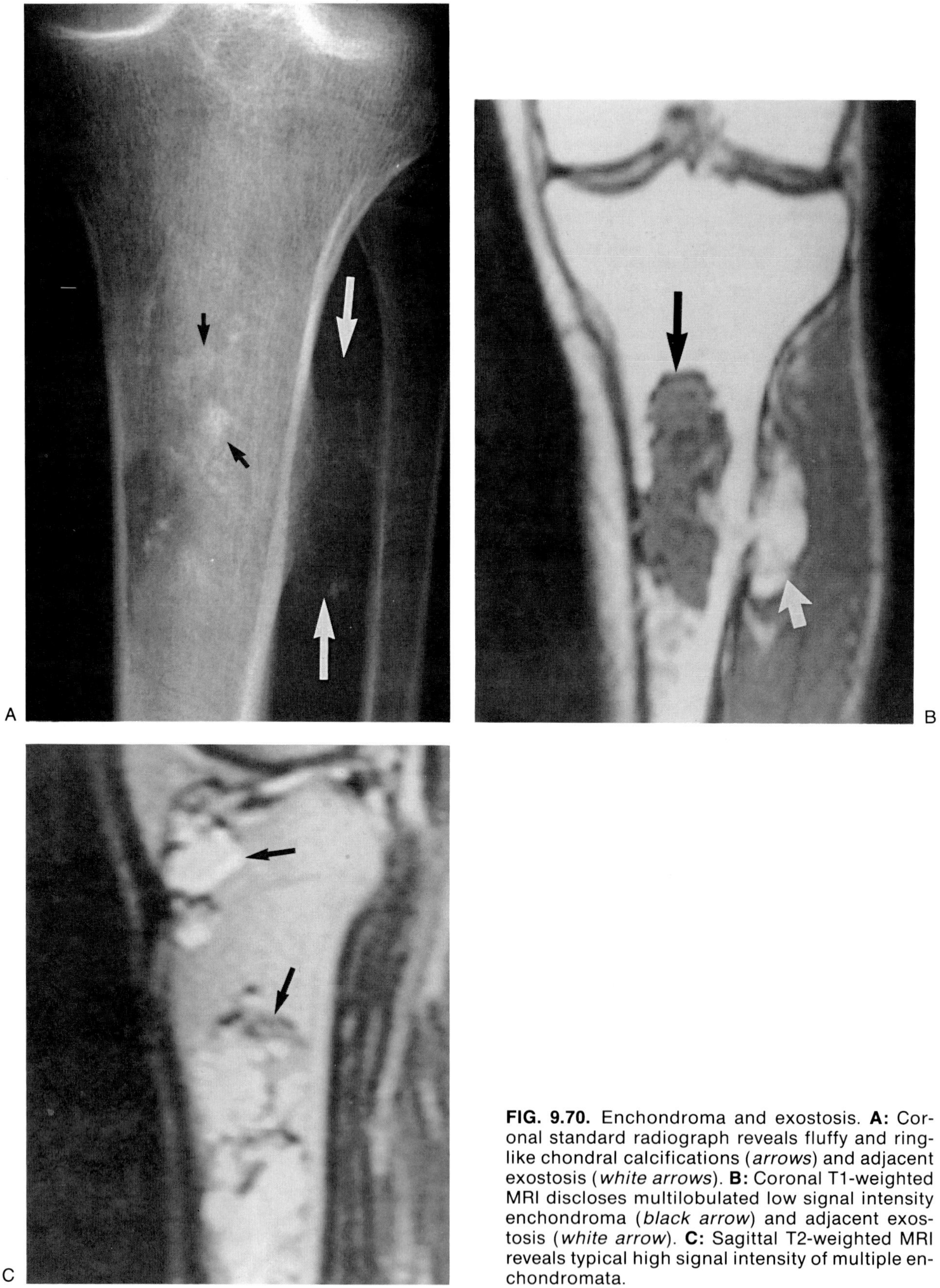

FIG. 9.70. Enchondroma and exostosis. **A:** Coronal standard radiograph reveals fluffy and ringlike chondral calcifications (*arrows*) and adjacent exostosis (*white arrows*). **B:** Coronal T1-weighted MRI discloses multilobulated low signal intensity enchondroma (*black arrow*) and adjacent exostosis (*white arrow*). **C:** Sagittal T2-weighted MRI reveals typical high signal intensity of multiple enchondromata.

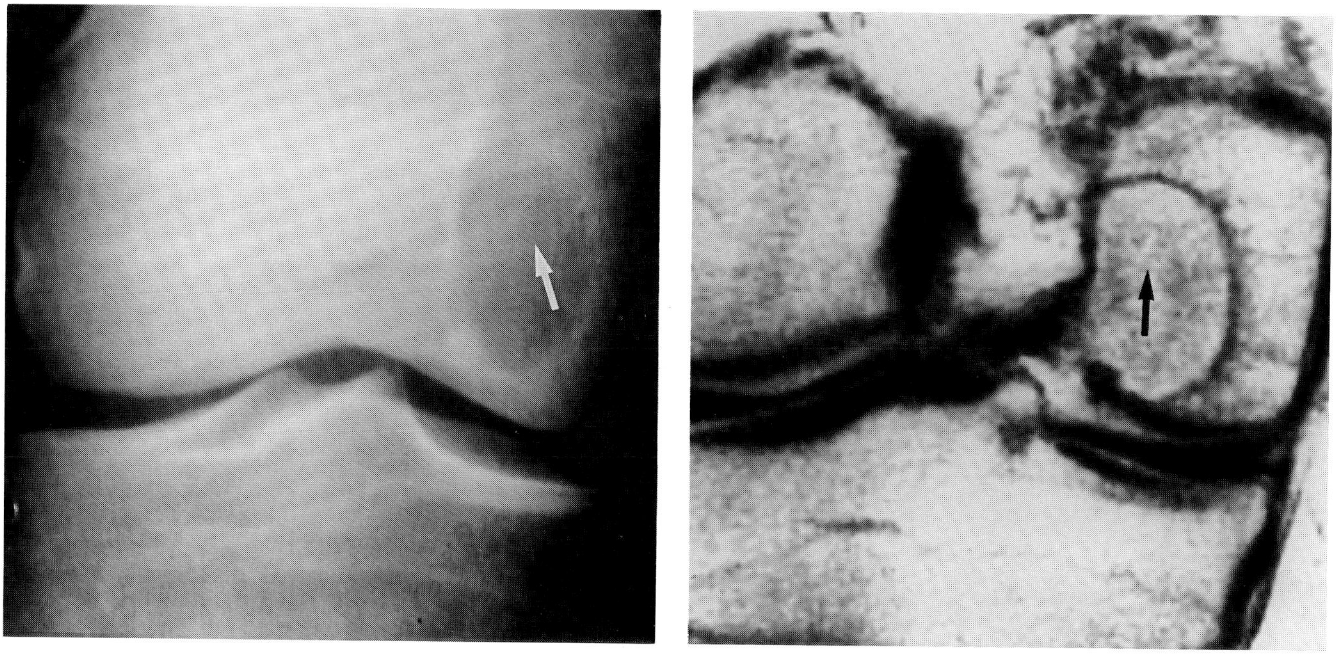

A B

FIG. 9.71. Breast carcinoma metastasis. **A:** Standard radiograph discloses sharply demarcated lytic lesion in the medial femoral condyle with radiographically benign characteristics (*arrow*). **B:** Coronal T2-weighted MRI again suggests benign etiology because of the sharply circumscribed, encapsulated appearance of the lesion (*arrow*). Moral: Imaging does not provide histologic diagnosis.

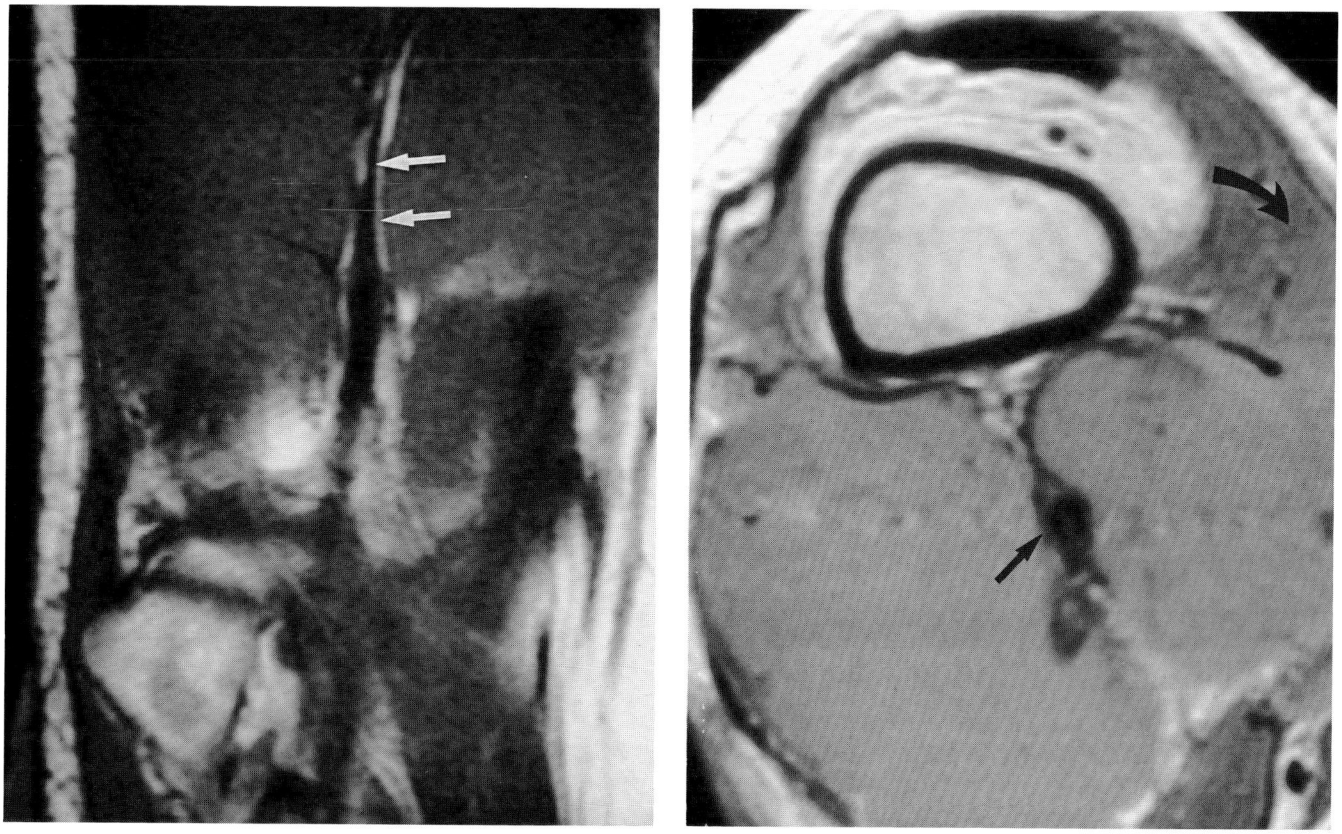

A B

FIG. 9.72. Metastatic lymphoma. **A:** Coronal T1-weighted scan and **B:** axial intermediate-weighted image both depict massive popliteal adenopathy with encasement of popliteal artery (*arrows*) and invasion of vastus medialis muscle (*curved arrow*).

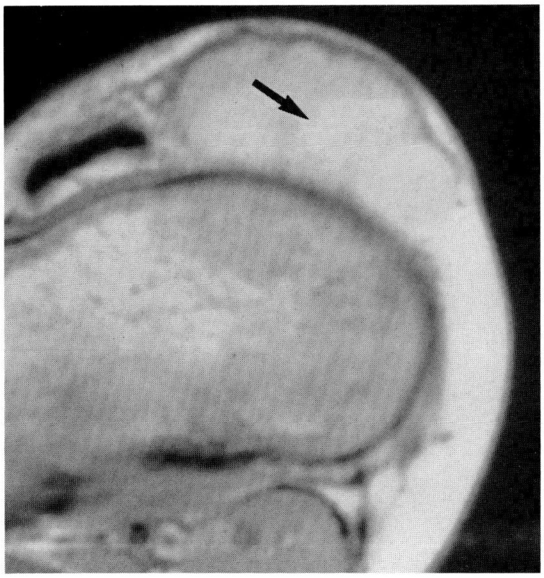

FIG. 9.73. Malignant fibrous histiocytoma. Painless pretibial mass in adolescent patient was clinically felt to be hemangioma after aspiration yielded only blood and physical examination disclosed compressibility. Axial intermediate-weighted MRI disclosed benign-appearing encapsulated subcutaneous mass (*arrow*), but open excisional biopsy was nevertheless recommended by a radiologist wary of "benign-appearing" soft tissue masses. Histologic evaluation revealed malignant fibrous histiocytoma.

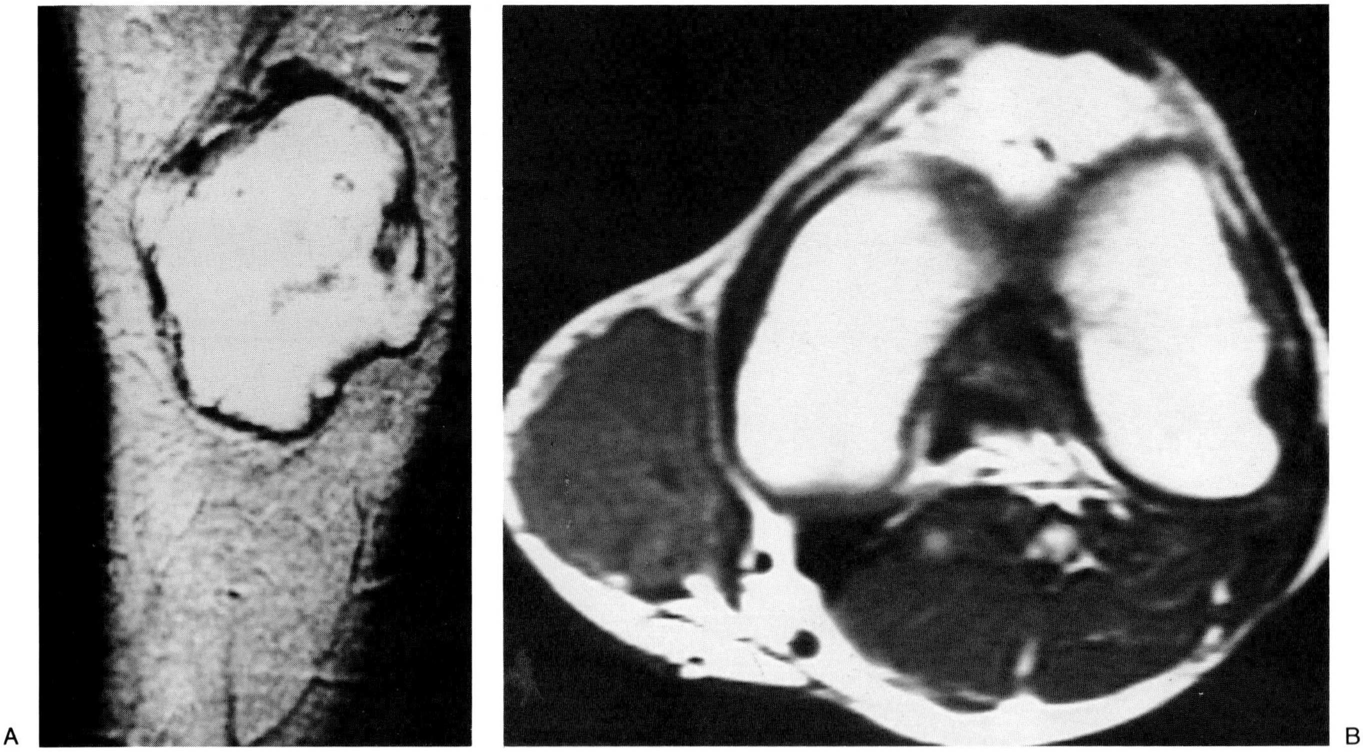

FIG. 9.74. Fibrosarcoma. **A:** Sagittal T2-weighted scan and **B:** axial T1-weighted image depict encapsulated benign-appearing subcutaneous mass without invasive growth pattern. Despite benign appearance, the histologic diagnosis was fibrosarcoma.

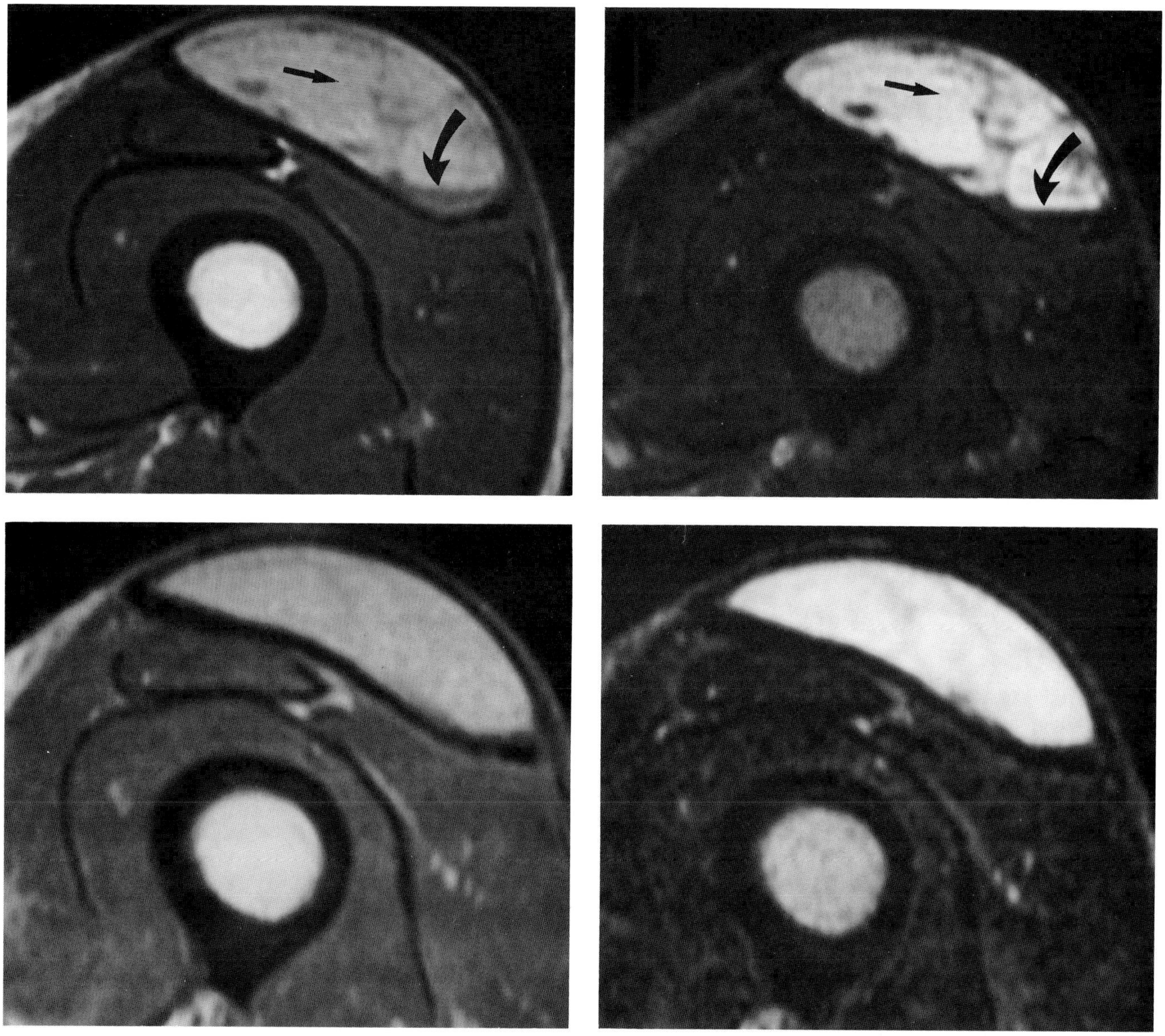

A B C D

FIG. 9.75. Postoperative hematoma. **A:** Axial intermediate- and **B:** T2-weighted images obtained several months after restriction of liposarcoma depict inhomogeneous superficial quadriceps region mass (*arrows*). A shallow fluid level (*curved arrows*) is the only imaging clue that this is a benign postoperative hematoma. Six months later, **C:** axial intermediate- and **D:** T2-weighted images demonstrate that hematoma has become homogeneous.

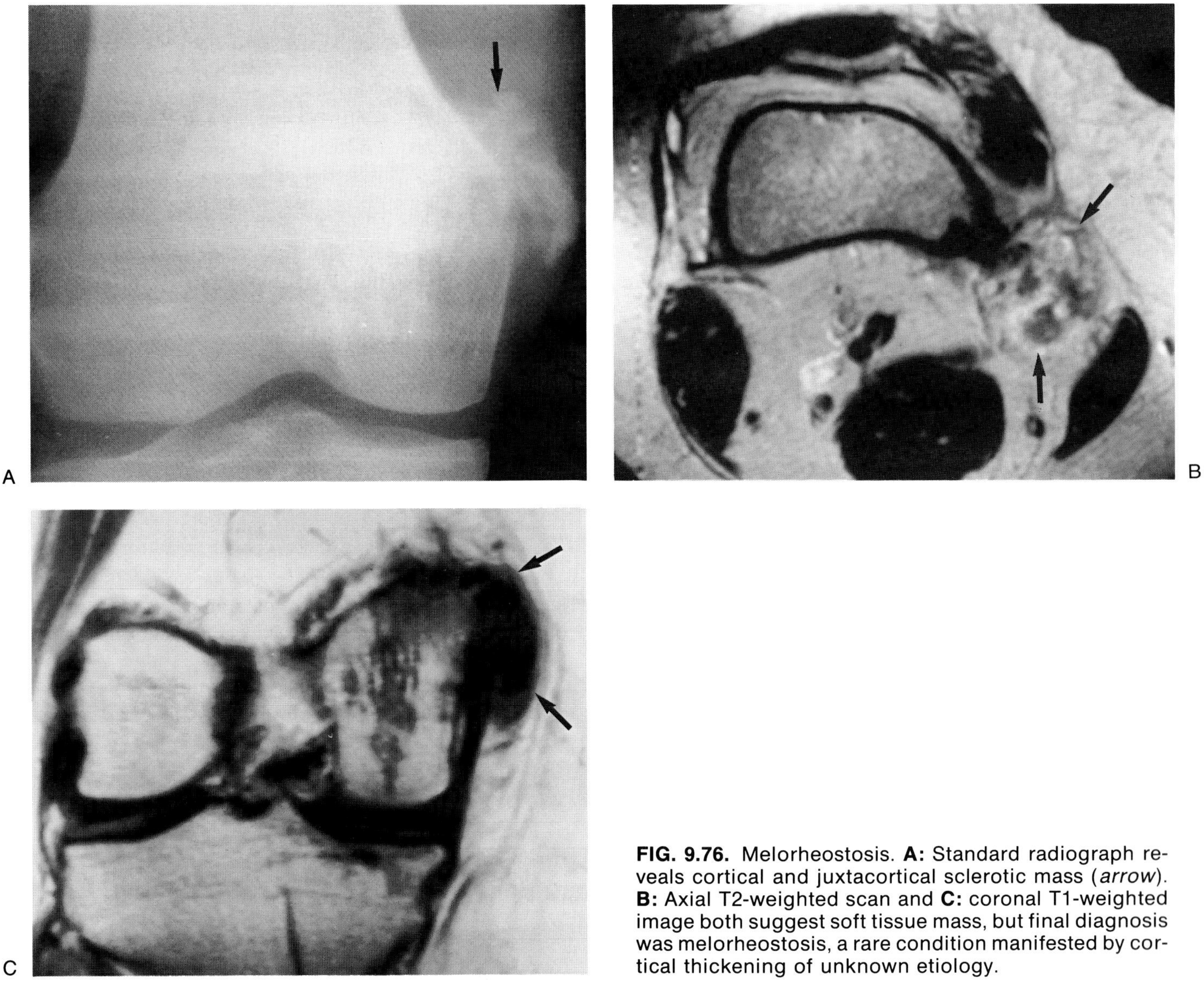

FIG. 9.76. Melorheostosis. **A:** Standard radiograph reveals cortical and juxtacortical sclerotic mass (*arrow*). **B:** Axial T2-weighted scan and **C:** coronal T1-weighted image both suggest soft tissue mass, but final diagnosis was melorheostosis, a rare condition manifested by cortical thickening of unknown etiology.

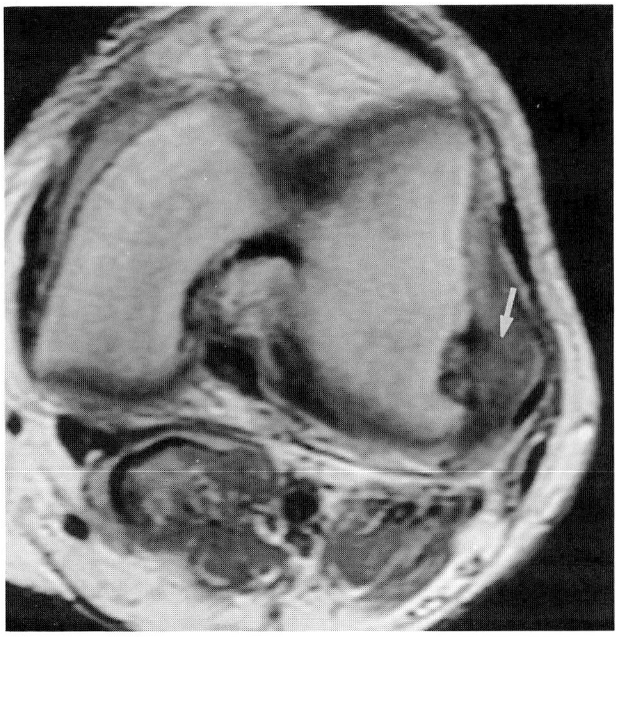

FIG. 9.77. Crystalline mass. Axial intermediate-weighted MRI reveals inhomogeneous mass deep to the fibular collateral ligament and associated with cortical erosion (*arrow*), found to be a mass of calcium hydroxyapatite crystals.

INFECTION

Early detection of osteomyelitis and soft tissue infection is essential to ensure appropriate therapy and optimal clinical outcome. Osteomyelitis may be due to hematogenous or contiguous infection. Hematogenous osteomyelitis most often involves rapidly growing bone, characteristically the metaphyses of the long bones in children such as the distal femur or proximal tibia. In children under age 2 *Haemophilus influenzae* are the most frequent organisms. In older children, *Staphylococcus aureus* and other gram-positive cocci are most often found (152). Specific clinical settings may suggest a likely organism. For example, intravenous drug use is often associated with *Pseudomonas auriginosa* or *Serratia osteomyelitis, salmonella* species are prominent in sickle cell disease, and group E *Streptococci* are common in neonates. Osteomyelitis is typically diagnosed from a combination of clinical examination, positive blood cultures, and typical radiographic findings.

MRI now appears to be the most specific and sensitive examination for diagnosing osteomyelitis, with overall accuracy of diagnosis reported to be as high as 94% (172). MRI is useful in defining the extent of the infectious process as well as in distinguishing osteomyelitis from cellulitis (32,50,61,172). Compared to radionuclide bone scanning, MRI is at least as sensitive and more anatomically specific in evaluating osteomyelitis and soft tissue infection.

When medullary bone is infected, increased intramedullary water resulting from inflammatory exudate, hyperemia, and edema results in diminished signal intensity on T1-weighted spin-echo images. Fat-suppressed or T2-weighted spin-echo images typically reveal high signal intensity collections within the medullary canal, occasionally accompanied by high signal intensity subperiosteal abscesses (Fig. 9.78).

In chronic osteomyelitis, the margins of the intramedullary abscess tend to become more sharply defined and a low signal intensity sclerotic rim may encase the intramedulary infection (Figs. 9.79 and 9.80). Low signal intensity foci representing osteonecrotic sequestra may be seen within the medullary canal. Whenever fluid is seen extending into the metaphyseal region of the medullary canal, particularly if there is associated juxtacortical soft tissue fluid, the diagnosis of osteomyelitis should be strongly considered.

The appearance of stress fractures on MR images can mimic the appearance of osteomyelitis. More specifically, longitudinal stress fractures of the tibia may result in linear cortical disruption, bone marrow edema, and circumferential surrounding soft tissue swelling (7), all features which may also occur sec-

A

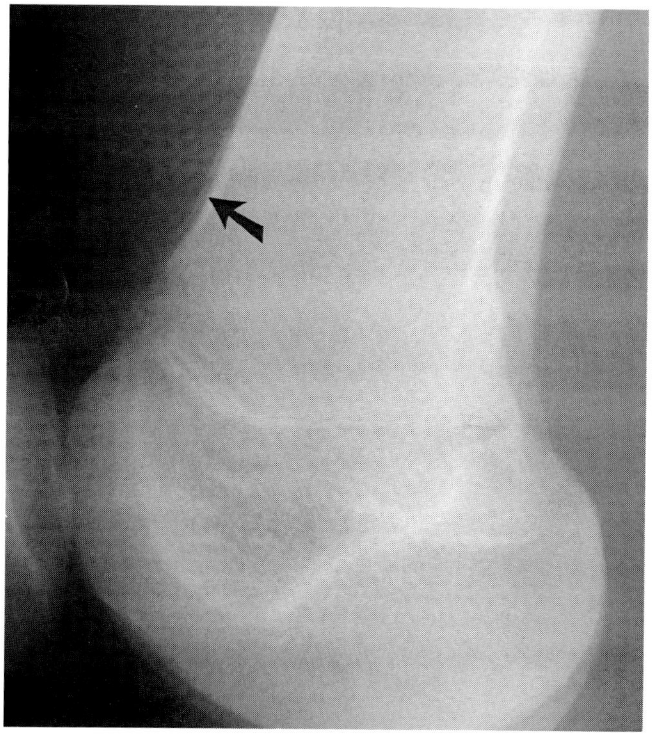

B

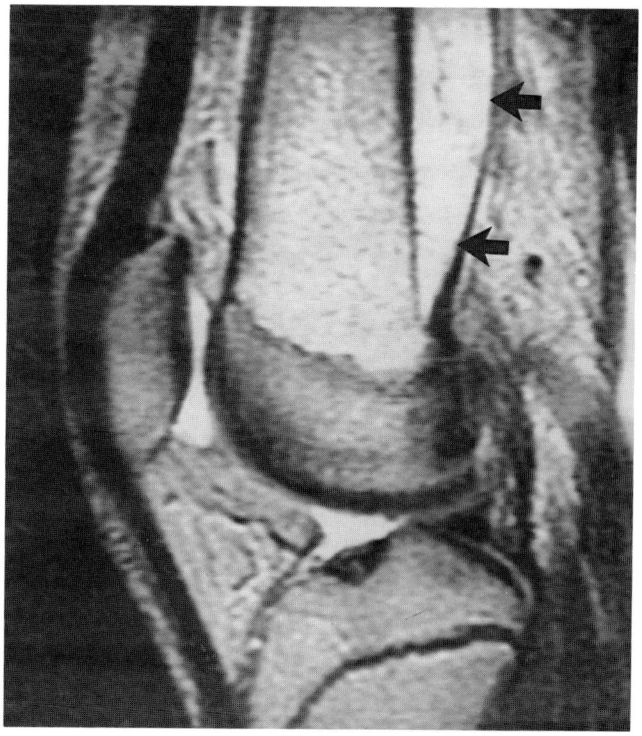

FIG. 9.78. Osteomyelitis. **A:** Standard radiograph is normal except for very tiny area of periosteal elevation (*arrow*). **B:** Sagittal T2-weighted MRI demonstrates that the entire medullary canal is abnormally high in signal intensity and there is an associated large subperiosteal abscess (*arrows*).

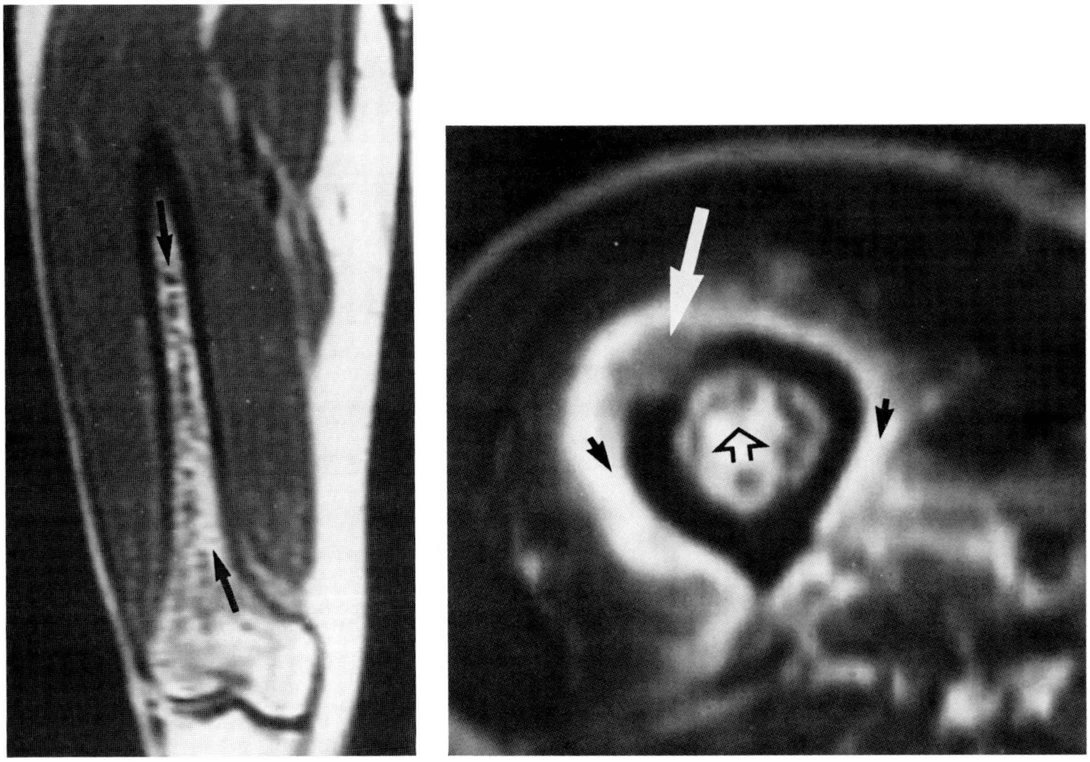

A B

FIG. 9.79. Osteomyelitis. **A:** Coronal T1-weighted MRI discloses long segment of abnormally diminished signal intensity within femoral medullary canal. **B:** Axial T2-weighted MRI discloses focal cortical destruction (*white arrow*), circumferential high signal intensity edema (*small black arrows*), and high signal intensity intraosseous abscess (*arrowhead*).

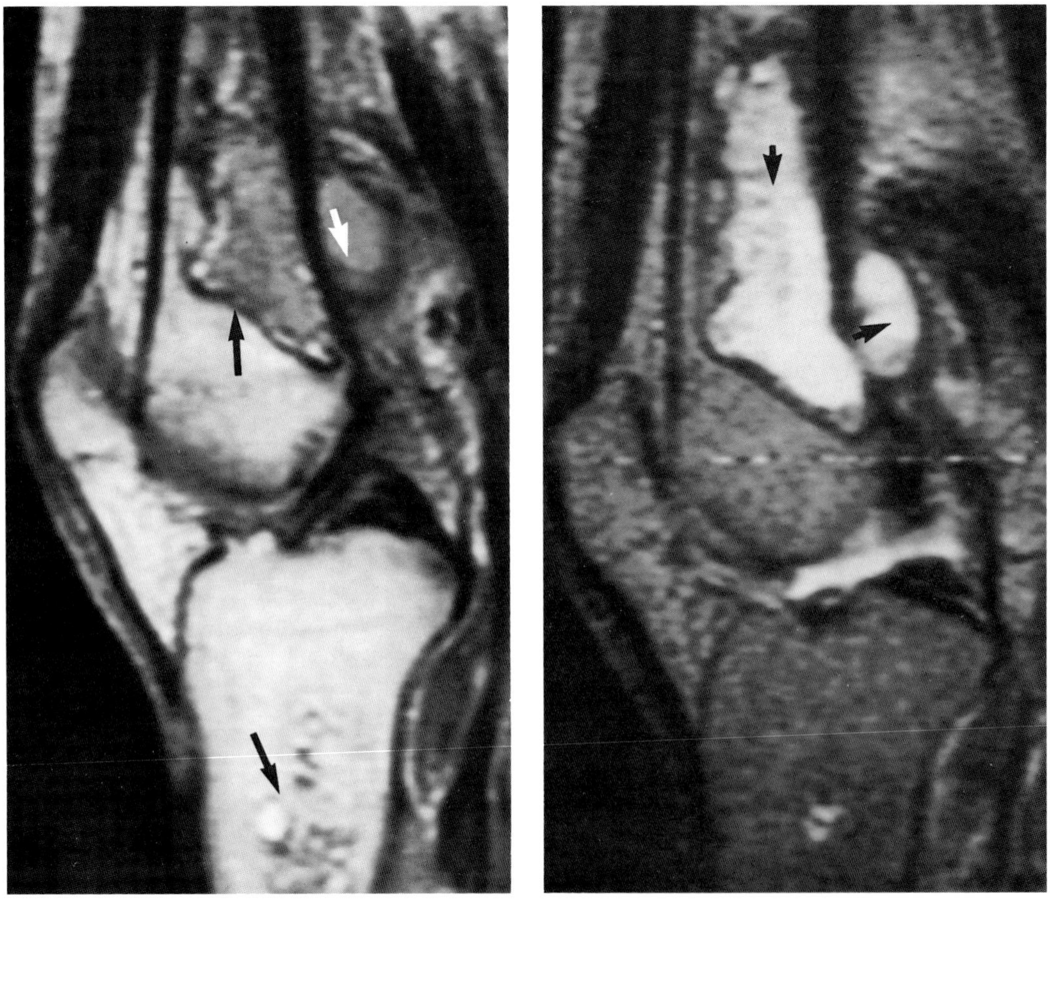

A B

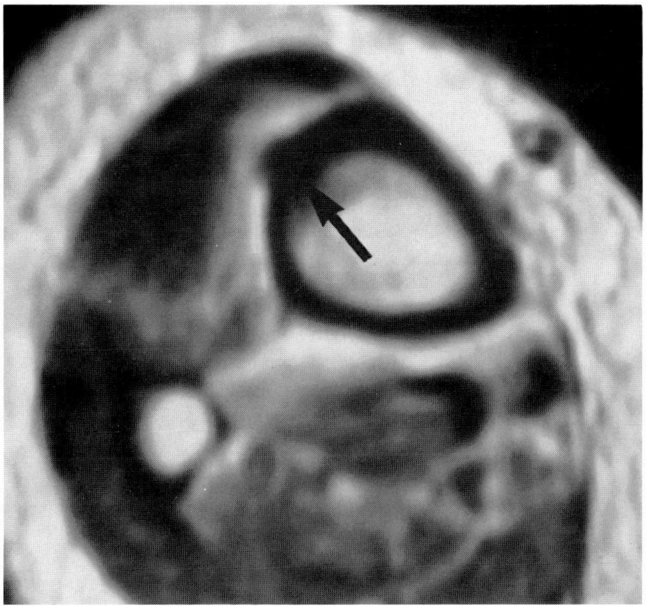

FIG. 9.81. Longitudinal stress fracture. Axial T2-weighted image closely mimics osteomyelitis. Cortical interruption (*arrow*) was due to longitudinal stress fracture and seen on numerous adjacent axial scans following a longitudinal linear course.

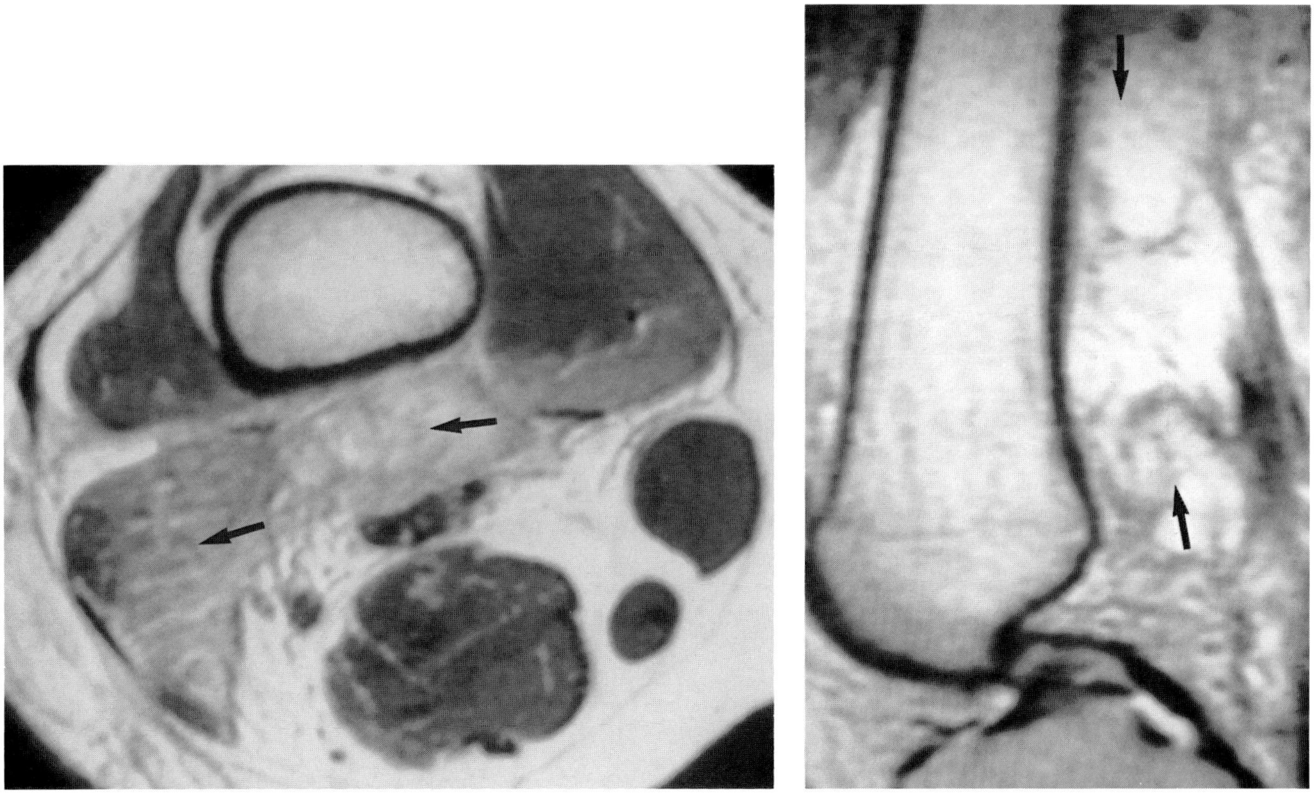

A B

FIG. 9.82. Popliteal space abscess/phlegmon. **A:** Axial intermediate-weighted scan and sagittal **B:** T2-weighted image demonstrate phlegmonous fluid collection in the popliteal space (*arrows*) infiltrating adjacent musculature and even subcutaneous fat without respect for fascial planes.

FIG. 9.80. Infected infarction. **A:** Sagittal intermediate-weighted scan discloses infarcts in the femur and tibia, but an abnormal adjacent soft tissue fluid collection is also seen (*white arrow*). **B:** Sagittal T2-weighted image discloses periosseous soft tissue fluid collection as well as abnormally high signal intensity within the medullary canal (*arrows*) due to infected infarction with soft tissue and intramedullary abscess.

ondary to osteomyelitis (Fig. 9.81). The long linear cortical defect in a longitudinal stress fracture of the tibia is a relatively distinctive feature, however.

Cellulitis and deep soft tissue infections are characterized on MR images by edema within deep or subcutaneous fascial planes. In the acute phase, bacterial infection is extensively permeative and infiltrative, with virtually no respect for fascial planes (Fig. 9.82). Only the most aggressive permeative tumors, such as rhabdomyosarcomas, simulate soft tissue infection on MR images.

One cannot differentiate sterile from infected joint fluid by MR imaging characteristics. When pyarthrosis of the knee is clinically suspected, MR imaging should never result in delay of prompt joint fluid aspiration.

Infection following joint replacement surgery is an uncommon but serious complication. Most infections occur during the early postoperative months (85% within 1 year), but some infections may appear several years after surgery (152). Persistent joint pain may be the only clinical symptom. *S. aureus* or *Staphylococcus epidermidis* are the most common causative organisms; gram-negative bacilli and anaerobic bacteria are less common. MR imaging is not useful for evaluating infected joint prostheses. Palestro et al. recently reported 95% accuracy in detecting infected knee prostheses using the combination of indium 111-labeled leukocyte scintigraphy and sulfur colloid marrow imaging (123). Using this method, infection is diagnosed if activity is observed on labeled leukocyte images without corresponding activity on sulfur colloid images.

MRI can be used to detect infection of synthetic cruciate ligament grafts. Loculated fluid collections within the knee joint, particularly those surrounding a prosthetic cruciate ligament, suggest the presence of infection. Joint aspiration is always necessary to confirm the presence of infection and determine the causative organism.

VASCULAR DISORDERS

Popliteal Artery Aneurysms

The most common site of a peripheral aneurysm is the popliteal artery. Lesions in this site account for 70% of peripheral aneurysms and are bilateral in up to 63% of cases (1,120). Atherosclerosis is the most common cause; other causes include trauma, mycotic infection, and complications of synthetic grafts. Patients usually present with a nonpulsatile popliteal mass. There is a high incidence of complications including thrombosis, ulceration, peripheral ischemia, and distal embolization. In addition, 10% to 15% of popliteal artery aneurysms cause popliteal vein obstruction, an important complication because pulmonary embolism occurs in up to two-thirds of patients with popliteal vein thrombosis. Because of the high incidence of complications, aggressive surgical intervention is generally warranted.

The calcified rim of a popliteal artery aneurysm frequently is visible on standard radiographs. Ultrasound is a safe, inexpensive, and painless method for confirming the presence of an aneurysm and evaluating mural thrombosis (120,149). However, definitive diagnosis of a popliteal aneurysm with ultrasound depends on demonstration of continuity of the lesion with the native vessel, which may be difficult when the popliteal artery is tortuous. The recent addition of color flow Doppler imaging has facilitated ultrasound evaluation of tortuous popliteal arteries and, in addition, has improved the sonographic capacity for diagnosing popliteal vein obstruction. Angiography may be employed to depict the lumen of the aneurysm, but surrounding thrombus is not directly imaged.

MRI is certainly capable of depicting the normal popliteal artery. Newer techniques, such as MR angiography, can depict blood flow within the popliteal artery and trifurcation vessels (Fig. 9.83). The normal popliteal artery is 0.8 $\pm$ 0.2 cm in diameter (120).

Popliteal artery aneurysms can be delineated with MR angiography or on spin-echo MR images (Fig. 9.84). Thrombus, mural calcification, and the relationship of the aneurysm to the native vessel are all well demonstrated. In evaluating popliteal aneurysms, MRI as a single test is potentially capable of answering clinical questions regarding differential diagnosis, arterial and venous patency, mural thrombosis, and aneurysm size and configuration.

Deep Venous Thrombosis

A wide array of new MR techniques for evaluating arterial and venous flow is currently undergoing rapid development. Detection of arterial and venous thrombosis and flow is certainly within the realm of MRI (34,51,164,175). Axial gradient-echo MR imaging can be employed to evaluate deep venous thrombosis at the popliteal level (164). Spritzer et al. have reported sensitivity of 100% and specificity of 92.9% in detecting lower extremity deep venous thrombosis with gradient-echo MR imaging (164). MRI offers several advantages in evaluating deep venous thrombosis, particularly in the knee region: (a) the examination is noninvasive and has high patient acceptance; (b) both lower extremities can be studied simultaneously; and (c) the deep veins of the thigh and pelvis which are not routinely visualized by venography may be simultaneously evaluated. Furthermore, in the knee region, popliteal vein compression by adjacent masses such as

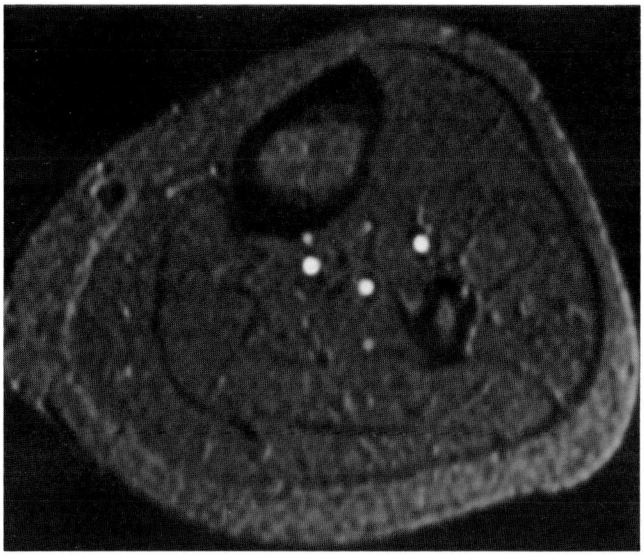

A

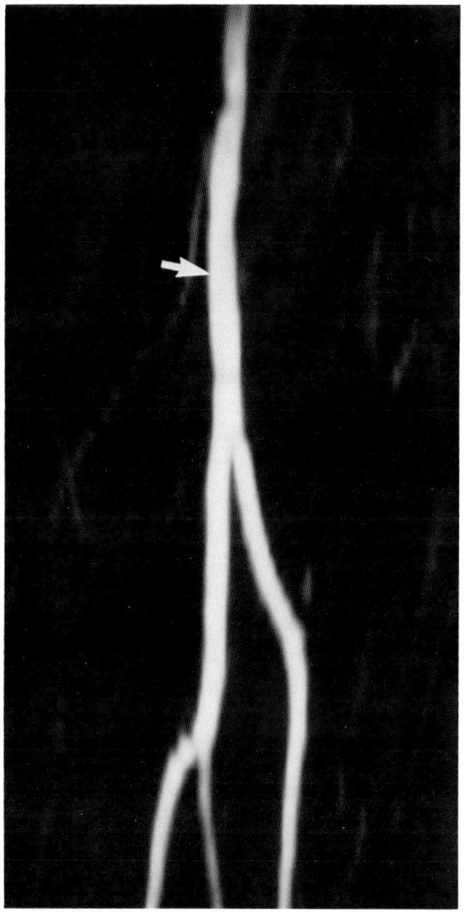

B

FIG. 9.83. MR angiography. **A:** Axial MR angiogram utilizing two-dimensional time-of-flight technique depicts the anterior tibial, posterior tibial, and peroneal arteries in high contrast to adjacent low signal intensity tissues. **B:** Reconstructed image utilizing maximum intensity pixel display method clearly demonstrates the popliteal artery (*arrow*) and trifurcation vessels.

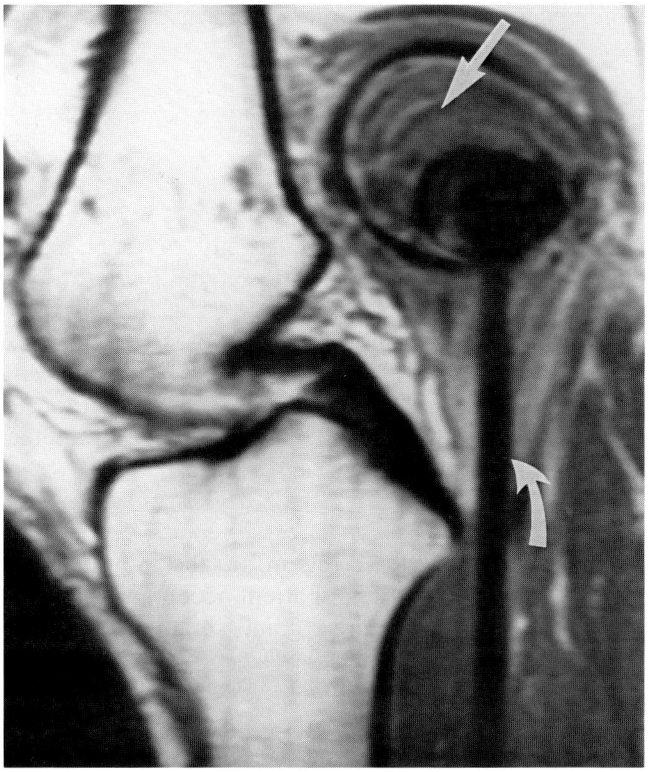

FIG. 9.84. Popliteal artery aneurysm. Large aneurysm (*arrow*) is seen contiguous with more distal popliteal artery (*curved arrow*).

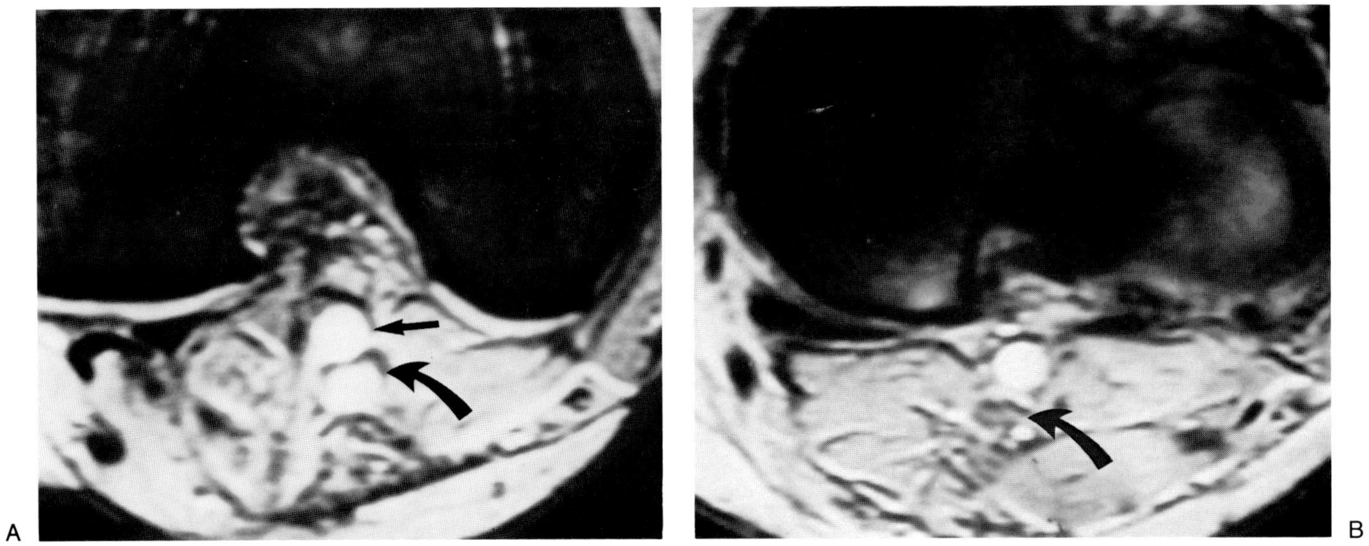

FIG. 9.85. Popliteal vein thrombosis. **A:** Normal axial gradient-echo image depicts high signal intensity within popliteal artery (*straight arrow*) and popliteal vein (*curved arrow*). **B:** Low signal intensity is seen within thrombosed popliteal vein (*curved arrow*).

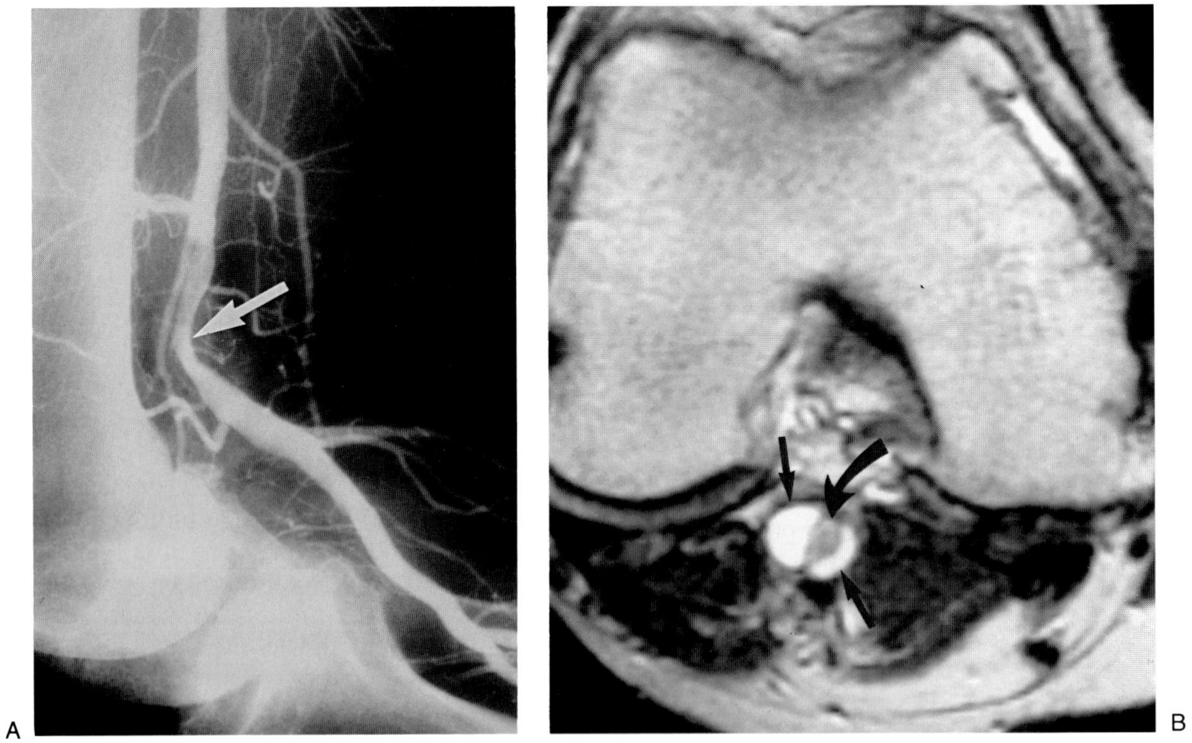

FIG. 9.86. Cystic adventitial disease of the popliteal artery. **A:** Conventional angiogram discloses focal narrowing of the high popliteal artery (*arrow*). **B:** Axial T2-weighted MRI discloses high signal intensity intramural cystic collections (*straight arrows*) and associated popliteal luminal narrowing (*curved arrow*).

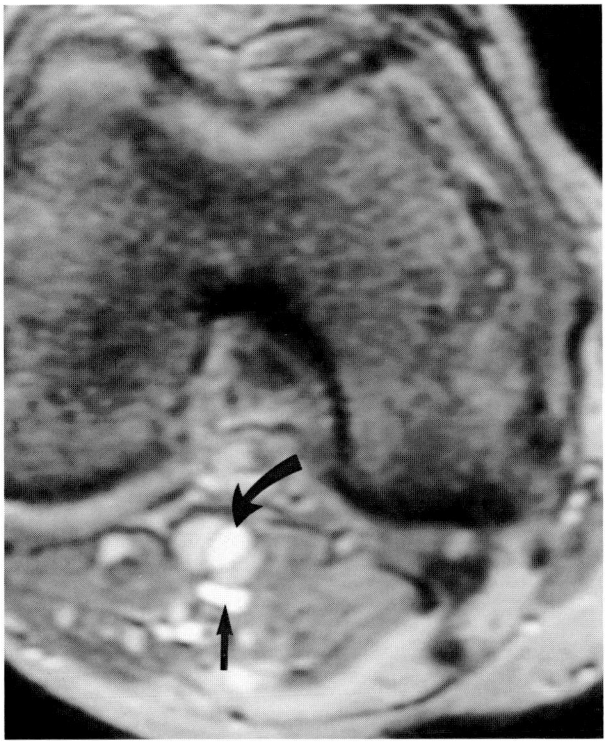

C

FIG. 9.86. (*Continued*) **C:** Axial gradient-echo image again demonstrates cystic adventitial disease, with high signal intensity signifying flow in the narrowed popliteal artery (*curved arrow*) and vein (*straight arrow*). (Courtesy of Mark S. Schechter, M.D.)

aberrant musculature, popliteal cysts, or popliteal aneurysm may be assessed. As previously discussed, popliteal vein thrombosis (Fig. 9.85) can be easily distinguished from dissecting popliteal cyst on MR images.

Adventitial Cystic Disease and Popliteal Artery Entrapment

Adventitial cystic disease of the popliteal artery and entrapment of the popliteal artery are causes of claudication in young, physically active patients which can result in debilitating symptoms and irreversible injury (71). Adventitial cystic disease is predominantly found in men under age 30 and is usually unilateral (16,25,71). Popliteal artery entrapment similarly occurs most commonly in young athletes, but is bilateral in up to 25%. In adventitial cystic disease peripheral pulses are generally lost with knee flexion, whereas popliteal artery entrapment often causes loss of peripheral pulses with knee hyperextension. Adventitial cystic disease is characterized by a thickened cystic region of the arterial adventitia filled with mucoproteins and mucopolysaccharides. The etiology is uncertain.

The diagnostic value of arteriography is limited for evaluation of both of these disorders since arteriography discloses only the vessel lumen and not the cause of luminal narrowing. For this reason, CT has been advocated for evaluation of both cystic adventitial disease and popliteal artery entrapment (71). The author has encountered a single case of cystic adventitial disease on knee MRI, documented by angiography and subsequent surgery (Fig. 9.86). The MR findings consisted of a moderately high signal intensity cyst-like collection within the popliteal artery wall, associated with luminal narrowing.

SUMMARY

An ever-broadening spectrum of pathologic conditions in the knee region can now be accurately assessed with magnetic resonance imaging. Increased experience in the past decade has resulted in greater specificity of diagnosis. Development of new techniques such as fat suppression imaging, refined gradient-echo techniques, and MR angiography has further heightened the diagnostic capacity of MRI. Further improvements in spatial resolution and development of MR spectroscopy promise to further improve the diagnostic capacity of MR imaging.

ACKNOWLEDGMENTS

The authors thank C. Christie Nute, CMT, medical transcriptionist, for all of her outstanding help in preparing the manuscript. I also thank the entire professional and technical staff at Mercy Magnetic Imaging Center, my colleagues at Radiology Medical Group, and the orthopedic surgical staff at Mercy Hospital and Medical Center, San Diego, California, for their invaluable assistance. Deepest gratitude is extended to Dr. Jerrold Mink for his friendship and assistance, as well as for providing many of the images present in this chapter.

REFERENCES

1. Abrams HL. *Angiography: Vascular and interventional radiology*, 3rd ed. Boston: Little, Brown, 1983.
2. Adam G, Dammer M, Bohndorf K, Christoph R, Fenke F, Gunther RW. Rheumatoid arthritis of the knee: Value of gadopentetate dimeglumine-enhanced MR imaging. *AJR* 1991;156:125–129.
3. Adam G, Nolte-Ernsting C, Prescher A, et al. Experimental hyaline cartilage lesions: Two-dimensional spin-echo versus three-dimensional gradient-echo MR imaging. *JMRI* 1991;1:665–672.
4. Aglietti P, Insall JN, Buzzi R, Deschamps G. Idiopathic osteonecrosis of the knee—etiology, prognosis, and treatment. *J Bone Joint Surg* 1983;65:588–597.
5. Ahlback S, Bauer GCH, Bohne WH. Spontaneous osteonecrosis of the knee. *Arthritis Rheum* 1968;11:705–733.
6. Aisen AM, Martel W, Braunstein EM, McMillan KI, Phillips WA, Kling TF. MRI and CT evaluation of primary bone and soft tissue tumors. *AJR* 1987;146:749–756.
7. Allen GJ. Longitudinal stress fractures of the tibia: Diagnosis with CT. *Radiology* 1988;167:799–801.
8. Anderson PW, Harley JD, Maslin PU. Arthrographic evaluation of problems with united tibial plateau fractures. *Radiology* 1976;119:75–78.
9. Arndt RD, Horns JW, Gold RH. *Clinical arthrography*, 2nd ed. Baltimore: Williams & Wilkins, 1985.
10. Baker DG, Schumacher HR, Wolf GL. Nuclear magnetic resonance evaluation of synovial fluid and articular tissue. *J Rheumatol* 1985;12:1062–1065.
11. Balsara ZN, Stainken BF, Martinez AJ. MR image of localized giant cell tumor of the tendon sheath involving the knee. *J Comput Assist Tomogr* 1989;13(1):159–162.
12. Bassett LW, Gold RH, Reicher MA, Bennett LR, Tooke SM. Magnetic resonance imaging in the early diagnosis of ischemic necrosis of the femoral head. *Clin Orthop* 1987;214:237–248.
13. Beltran J, Noto AM, Herman LJ, Mosure JC, Burk JM, Christoforidis AJ. Joint effusions: MR imaging. *Radiology* 1985;158:133–137.
14. Beltran J, Noto AM, Mosure JC, Shamam OM, Weiss KL, Zeulzer WA. The ankle: Surface-coil MR imaging at 1.5 T. *Radiology* 1986;161:203–209.
15. Beltran J, Noto AM, Mosure JC, Weiss KL, Zeulzer W, Christoforidis AJ. The knee: Surface-coil MR imaging at 1.5 T. *Radiology* 1986;159:747–751.
16. Bergan JJ. *Adventitial cystic disease of the popliteal artery*. In: Rutherford RB, ed. *Vascular surgery*. Philadelphia: W Saunders, 1977:569–576.
17. Berquist TH, Ehman RL, King BF, Hodgman CG, Ilstrup DM. Value of MR imaging in differentiating benign from malignant soft-tissue masses: Study of 95 lesions. *AJR* 1990;155:1251–1255.
18. Bjorkengren AG, AlRowaih A, Lindstrand A, Wingstrand H, Thorngren K-G, Pettersson H. Spontaneous osteonecrosis of the knee: Value of MR imaging in determining prognosis. *AJR* 1990;154:331–336.
19. Bjorkengren AG, Geborek P, Rydholm U, Holtas S, Pettersson H. MR imaging of the knee in acute rheumatoid arthritis: Synovial uptake of gadolinium-DOTA. *AJR* 1990;155:329–332.
20. Brady TJ, Rosen BR, Pykett IL, McGuire MH, Munkin HJ, Rosenthal DI. NMR imaging of leg tumors. *Radiology* 1983;149:181–187.
21. Brasch RC, Wesby GE, Gooding CE, Koerper MA. Magnetic resonance imaging of transfusional hemosiderosis complicating thalassemia major. *Radiology* 1984;150:767–771.
22. Brody AS, Strong M, Babikian G, Sweet DE, Seidel FG, Kuhn JP. Avascular necrosis: Early MR imaging and histologic findings in a canine model. *AJR* 1991;157:341–345.
23. Brown DG, Edwards NL, Greer JM, Longley S, Gillespy T III, Panush RS. Magnetic resonance imaging in patients with inflammatory arthritis of the knee. *Clin Rheum* 1990;9(1):73–83.
24. Buckwalter KA, Braunstein EM, Wilson MR, Wojtys EM, Martel W. Evaluation of hyaline cartilage of the knee with MR imaging. *Radiology* 1986;161(P):239.
25. Bunker SR, Lauten GJ, Jutton JE Jr. Cystic adventitial disease of the popliteal artery. *AJR* 1981;136:1209–1212.
26. Burk DL, Dalinka MK, Kanal E, et al. Meniscal and ganglion cysts of the knee: MR evaluation. *AJR* 1988;150:331–336.
27. Burk DL, Kanal E, Brunberg JA, Johnstone GF, Swensen HE, Wolf GL. 1.5T surface-coil MRI of the knee. *AJR* 1986;147:293–300.
28. Burri C, Bartzke G, Coldewey J, Muggler E. Fractures of the tibial plateau. *Clin Orthop* 1979;138:84–93.
29. Cahil B. Treatment of juvenile osteochondritis dissecans and osteochondritis dissecans of the knee. *Clin Sports Med* 1985;4(2):367–384.
30. Calabro JJ. Juvenile rheumatoid arthritis. In: McCarty DJ, ed. *Arthritis and allied conditions: A textbook of rheumatology, 11th ed.* Philadelphia: Lea & Febiger, 1989:913–925.
31. Cayeu PD, Pavlov H, Sherman MF, Goldman AG. Lucent articular lesion in the lateral femoral condyle: Source of patellar femoral pain in the athletic adolescent. *AJR* 1981;137:1145–1149.
32. Chandnani VP, Beltran J, Morris CS, et al. Acute experimental osteomyelitis and asbcesses: Detection with MR imaging versus CT. *Radiology* 1990;174:233–236.
33. Chandnani VP, Ho C, Chu P, Trudell D, Resnick D. Knee hyaline cartilage evaluated with MR imaging: A cadaveric study involving multiple imaging sequences and intraarticular injection of gadolinium and saline solution. *Radiology* 1991;178:557–561.
34. Choyke PL, Kressel HY, Axel L, et al. Vascular occlusions detected by magnetic resonance imaging. *Magn Reson Med* 1985;2:540–554.
35. Citron ND, Patterson FWN, Jacson AM. Neuropathic osteonecrosis of the lateral femoral condyle in childhood. *J Bone Joint Surg* 1986;68:96–99.
36. Cobby MJ, Adler RS, Swartz R, Martel W. Dialysis-related amyloid arthropathy: MR findings in four patients. *AJR* 1991;157:1023–1027.
37. Cohen AS, Canoso JJ. *Tumors of joints and related structures*. In: McCarty, DJ, ed. *Arthritis and allied conditions: A textbook of rheumatology*, 11th ed. Philadelphia: Lea & Febiger, 1989:1492–1508.
38. Cohen MD, Klatte EC, Baehner R, et al. Magnetic resonance imaging of bone marrow disease in children. *Radiology* 1984;151:715–718.
39. Coleman BG, Kressel HY, Dalinka MK, Scheibler ML, Burk DL, Cohen EK. Radiographically negative avascular necrosis: Detection with MR imaging. *Radiology* 1988;168:525–528.
40. Collier BD, Johnson RP, Carrera GF, et al. Chronic knee pain assessed by SPECT: Comparison with other modalities. *Radiology* 1985;157:795–802.
41. Cova M, Kang YS, Tsukamoto H, et al. Bone marrow perfusion evaluated with gadolinium-enhanced dynamic fast MR imaging in a dog model. *Radiology* 1991;179:535–539.
42. Cruess RL. Osteonecrosis of bone, current concepts as to etiology and pathogenesis. *Clin Orthop Rel Res* 1986;208:30–39.
43. Daffner RH, Lupetin AR, Dash N, DeeB ZL, Sefczek RJ, Shapiro RL. MRI in the detection of malignant infiltration of bone marrow. *AJR* 1986;146:353–358.
44. De Smet AA, Fisher DR, Graf BK, Lange RH. Osteochondritis dissecans of the knee: Value of MR imaging in determining lesion stability and the presence of articular cartilage defects. *AJR* 1990;155(3):549–553.
45. De Smet AA, Fisher DRm, Burnstein MI, Graf BK, Lange RH.

Value of MR imaging in staging osteochondral lesions of the talus (osteochondritis dissecans): Results in 14 patients. *AJR* 1990;154:555–558.
46. Deutsch AL, Mink JH, Rosenfelt FP, Waxman AD. Incidental detection of hematopoietic hyperplasia on routine knee MR imaging. *AJR* 1989;152:333–336.
47. Deutsch AL, Mink JH, Waxman AD. Occult fractures of the proximal femur: MR imaging. *Radiology* 1989;170:113–116.
48. Dewhirst MW, Sostman HD, Leopold KA, et al. Soft-tissue sarcomas: MR imaging and MR spectroscopy for prognosis and therapy monitoring. *Radiology* 1990;174:847–853.
49. Dorwart RH, Genant HK, Johnston WH, Morris JM. Pigmented villonodular synovitis of synovial joints—clinical, pathologic, and radiologic features. *AJR* 1984;143:877–885.
50. Erdman WA, Tamburro F, Jayson HT, Weatherall PT, Ferry KB, Peshock RM. Osteomyelitis: Characteristics and pitfalls of diagnosis with MR imaging. *Radiology* 1991;180:533–539.
51. Erdman WA, Weinreb JC, Cohen JM, Buja LM, Chaney C, Peshock RM. Venous thrombosis: Clinical and experimental MR imaging. *Radiology* 1986;161:233–238.
52. Erlemann R, Vassallo P, Bongartz G, et al. Musculoskeletal neoplasms: Fast low-angle shot MR imaging with and without Gd-DTPA. *Radiology* 1990;176:489–495.
53. Fairbank HAT, Lloyd EI. Cysts of the external cartilage of the knee with erosion of the head of the tibia. *Br J Surg* 1934;22:115–118.
54. Faulkner KG, Gluer C-C, Majumdar S, Lang P, Engelke K, Genant HK. Noninvasive measurements of bone mass, structure, and strength: Current methods and experimental techniques. *AJR* 1991;157:1229–1237.
55. Freiberger RH, Kaye JJ. *Arthrography*. New York: Appleton-Century-Crofts, 1979.
56. Gallimore GW, Harms SE. Knee injuries: High resolution MR imaging. *Radiology* 1986;160:457–461.
57. Ghormley RK, Dockerty MB. Cystic myxomatous tumors about the knee: Their relation to cysts of the menisci. *J Bone Joint Surg* 1943;25:306–318.
58. Gilley JS, Gelman MI, Edson M, Metcalf RW. Chondral fractures of the knee. *Radiology* 1981;138:51–54.
59. Glylys-Morin VM, Hajek PS, Sartoris DJ, Resnick D. Articular cartilage defects: Detectability in cadaver knees with MR. *AJR* 1987;148:1153–1157.
60. Goergen TG, Resnick D, Niwayama G. Localized nodular synovitis of the knee: A report of two cases with abnormal arthrograms. *AJR* 1976;126:647–650.
61. Gold RH, Hawkins RA, Katz RD. Bacterial osteomyelitis: Findings on plain radiography, CT, MR, and scintigraphy. *AJR* 1991;157:365–370.
62. Greenfield GB. *Radiology of bone diseases*, 3rd ed. Philadelphia: JB Lippincott, 1980.
63. Greyson ND, Lotem MM, Gross AE, Houpt JB. Radionuclide evaluation of spontaneous femoral osteonecrosis. *Radiology* 1982;142:729–735.
64. Guerra J, Newell JD, Resnick D, Danzig LA. Gastrocnemiosemimembranosus bursal region of the knee. *AJR* 1981; 136:593–596.
65. Halborg A, Hansen H, Sneppen HO. Hemangioma of the knee joint. *Acta Orthop Scand* 1968:39:209–216.
66. Hartzman S, Reicher MA, Bassett LW, Duckwiler GR, Mandelbaum B, Gold RH. MR imaging of the knee. Part II: Chronic disorders. *Radiology* 1987;162:553–557.
67. Herman G, Yeh HC, Lehr-Jannus C, Berson BL. Diagnosis of popliteal cyst: Double contrast arthrography and sonography. *AJR* 1981:137:369–372.
68. Hough AJ Jr, Sokoloff L. Pathology of osteoarthritis. In: McCarty DJ, ed. *Arthritis and allied conditions: a textbook of rheumatology*, 11th ed. Philadelphia: Lea & Febiger, 1989:1571–1594.
69. Hudson TM, Hamlin DJ, Enneking WF, Petterson H. Magnetic resonance imaging of bone and soft tissue tumors: Early experience in 31 patients compared with computed tomography. *Skeletal Radiol* 1985;13:134–146.
70. Hughston JC, Hergenroed PT, Courtenay BG. Osteochondritis dissecans of the femoral condyles. *J Bone Joint Surg [Am]* 1984;66A(9):1340–1348.
71. Jasinski RW, Masselink BA, Partridge RW, Deckinga BG, Bradford PF. Adventitial cystic disease of the popliteal artery. *Radiology* 1987;163:153–155.
72. Jelinek JS, Kransdorf MJ, Moser RP, Temple HT, Lenhart MK, Berrey BH. MR imaging findings in patients with bone-chip allografts. *AJR* 1990;155:1257–1260.
73. Jelinek JS, Kransdorf MJ, Utz JA, et al. Imaging of pigmented villonodular synovitis with emphasis on MR imaging. *AJR* 1989;152:337–342.
74. Johansson JE, et al. Pigmented villonodular synovitis of joints. *Clin Orthop* 1982;163:159–166.
75. Jones JP Jr. Osteonecrosis. In: McCarty DJ, ed. *Arthritis and allied conditions: A textbook of rheumatology*, 11th ed. Philadelphia: Lea & Febiger, 1989:1545–1562.
76. Kaempffe F, D'Amato C. An unusual intra-articular ganglion of the knee with interosseous extension: A case report. *J Bone Joint Surg* 1989;71-A:773–775.
77. Kambolis C, Bullough PG, Jaffe HL. Ganglion cystic defects of bone. *J Bone Joint Surg* 1973:55-A:496–505.
78. Kaplan PA, Asleson RJ, Klassen LW, Duggan MJ. Bone marrow patterns in aplastic anemia: Observations with 1.5-T MR imaging. *Radiology* 1987;164:441–444.
79. Kean DM. Worthington BS, Preston BJ, et al. Nuclear magnetic resonance imaging of the knee: Examples of normal anatomy and pathology. *Br J Radiol* 1983;56:355–364.
80. Koenig SH, Brown RD. The importance of the motion of water for magnetic resonance imaging. *Invest Radiol* 1985;20:297–304.
81. Konig H, Sauter R, Deimling M, Vogt M. Cartilage disorders: Comparison of spin-echo, CHESS, and FLASH sequence MR images. *Radiology* 1987;164:753–758.
82. Konig H, Sieper J, Wolf K-J. Rheumatoid arthritis: Evaluation of hypervascular and fibrous pannus with dynamic MR imaging enhanced with Gd-DTPA. *Radiology* 1990;176:473–477.
83. Kransdorf MJ, Jelinek JS, Moser RP, et al. Soft-tissue masses: Diagnosis using MR imaging. *AJR* 1989;153:541–547.
84. Kransdorf MJ, Meis JM, Jelinek JS. Myositis ossificans: MR appearance with radiologic-pathologic correlation. *AJR* 1991;157:1243–1248.
85. Kricun ME. Red-yellow marrow conversion: Its effect on the location of some solitary bone lesions. *Skeletal Radiol* 1985;14:10–19.
86. Kulkarni MV, Drolshagen LF, Kaye JJ, et al. MR imaging of hemophiliac arthropathy. *J Comput Assist Tomogr* 1986;10:445–449.
87. Kursunoglu-Brahme S, Fox JM, Ferkel RD, Friedman MJ, Flannigan BD, Resnick DL. Osteonecrosis of the knee after arthroscopic surgery: Diagnosis with MR imaging. *Radiology* 1991;178:851–853.
88. Kursunoglu-Brahme S, Riccio T. Weisman MH, et al. Rheumatoid knee: Role of gadopentetate-enhanced MR imaging. *Radiology* 1990;176:831–835.
89. Lee JK, Yao L. Tibial collateral ligament bursa: MR imaging. *Radiology* 1991;178(3):855–857.
90. Lee KR, Tines SC, Price HI, Desmet AA, Neff JR. The computed tomographic findings of popliteal cysts. *Skeletal Radiol* 1983;10:26–29.
91. Lehner KB, Rechl HP, Gmeinwieser JK, Heuck AF, Lukas HP, Kohn HP. Structure, function, and degeneration of bovine hyaline cartilage: Assessment with MR imaging in vitro. *Radiology* 1989;170:495–499.
92. Lewis RC, Coventry MB, Soule EH. Hemangioma of the synovial membrane. *J Bone Joint Surg* 1959;41A:264–270.
93. Linden B. The incidence of osteochondritis dissecans in the condyles of the femur. *Acta Orthop Scand* 1976;47:664–667.
94. Lindgreen PG, Willen R. Gastrocnemius-semimembranosus bursa and its relation to the knee joint. Anatomy and histology. *Acta Radiol (Diagn)* 1977;18:497–512.
95. Lotke PA, Ecker ML. Osteonecrosis-like syndrome of the medial tibial plateau. *Clin Orthop Rel Res* 1983;176:148–153.
96. Lotke PA, Ecker ML, Alauc A. Painful knees in older patients: Radionuclide diagnosis of possible osteonecrosis with spontaneous resolution. *J Bone Joint Surg* 1977;59:617.
97. Lowenstein MB, Smith JRV, Cole S. Infrapatellar pigmented

villonodular synovitis—arthrographic detection. *AJR* 1980;135:279–282.
98. Lukes PJ, Herberts P, Zachrisson BE. Ultrasound in the diagnosis of popliteal cysts. *Acta Radiol (Diagn) (Stockh)* 1980;21:663–665.
99. Lynch TCP, Crues JV III, Morgan FW, Sheehan WE, Harter LP, Ryu R. Bone abnormalities of the knee: Prevalence and significance at MR imaging. *Radiology* 1989;171(3):761–766.
100. Mandelbaum BR, Grant TT, Hartzman S, et al. The use of MRI to assist in diagnosis of pigmented villonodular synovitis of the knee joint. *Clin Orthop Rel Res* 1988;231:135–139.
101. Mankin HJ, Radin E. Structure and function of joints. In: McCarty DJ, ed. *Arthritis and allied conditions: A textbook of rheumatology*, 11th ed. Philadelphia: Lea & Febiger, 1989:189–200.
102. McAlindon TE, Watt I. McCrae F, Goddard P, Dieppe PA. Magnetic resonance imaging in osteoarthritis of the knee: Correlation with radiographic and scintigraphic findings. *Ann Rheum Dis* 1991;50(1):14–19.
103. McCarty DJ, ed. *Arthritis and allied conditions: A textbook of rheumatology*, 11th ed. Philadelphia: Lea & Febiger, 1989.
104. McCarty DJ. Synovial fluid. In: McCarty DJ, ed. *Arthritis and allied conditions: A textbook of rheumatology*, 11th ed. Philadelphia: Lea & Febiger, 1989:69–90.
105. McCauley TR, Kier R, Lynch KJ, Joki P. Chondromalacia patellae: Diagnosis with MR imaging. *AJR* 1992;158:101–105.
106. Mesgarzadeh M, Sapega AA, Bonakdarpour A, et al. Osteochondritis dissecans: Analysis of mechanical stability with radiography, scintigraphy, and MR imaging. *Radiology* 1987;18:386–390.
107. Mink JH, Deutsch AL, eds. *MRI of the musculoskeletal system*. New York: Raven Press, 1990.
108. Mink JH, Deutsch AL. Occult cartilage and bone injuries of the knee: Detection, classification, and assessment with MR imaging. *Radiology* 1989;170:823–829.
109. Mitchell DG, Rao VW, Dalinka W, et al. Hematopoietic and fatty bone marrow distribution in the normal and ischemic hip: New observations with 1.5-T MR imaging. *Radiology* 1986;161:199–202.
110. Mitchell MD, Kundel HL, Steinberg ME, Kressel HY, Alavi A, Axel L. Avascular necrosis of the hip: Comparison of MR, CT, and scintigraphy. *AJR* 1986;147:67–71.
111. Modl JM, Sether LA, Haughton VM, Kneeland JB. Articular cartilage: Correlation of histologic zones with signal intensity at MR imaging. *Radiology* 1991;181:853–855.
112. Moon KL, Genant HK, Davis PL, et al. Nuclear magnetic resonance imaging in orthopedics: Principles and applications. *J Orthop Res* 1983;1:101–114.
113. Moon KL, Genant HK, Helms CA, Chatetz NI, Crooks LE, Kaufman L. Musculoskeletal applications of nuclear magnetic resonance. *Radiology* 1983;147:161–171.
114. Moon NF. Synovial hemangioma of the knee joint. *Clin Orthop* 1973;90:183.
115. Moore SG, Dawson KL. Red and yellow marrow in the femur: Age-related changes in appearance at MR imaging. *Radiology* 1990;175:219–223.
116. Morton MJ, Berquist TH, McLeod RA, Unni KK, Sim FH. MR imaging of synovial sarcoma. *AJR* 1991;156:337–340.
117. Munk PL, Helms CA, Holt RG. Immature bone infarcts: Findings on plain radiographs and MR scans. *AJR* 1989;152:547–549.
118. Murphy WD, Hurst GC, Duerk JL, Feiglin DH, Christopher M, Bellon EM. Atypical appearance of lipomatous tumors on MR images: High signal intensity with fat-suppression STIR sequences. *JMRI* 1991;1:477–480.
119. Myers BW, Masi AT, Feigenbaum SL. Pigmented villonodular synovitis and tenosynovitis: A clinical epidemiology study of 166 cases and literature review. *Medicine* 1980;59:233–238.
120. Neiman HL, Yao JST, Silver TM. Grey-scale ultrasound diagnosis of peripheral arterial aneurysms. *Radiology* 1979;130:413–416.
121. Norman A, Baker ND. Spontaneous osteonecrosis of the knee and medial meniscal tears. *Radiology* 1978;129:653–656.
122. Norman A, Baker ND. Osteonecrosis of the knee and medial meniscal tears [letter to the editor]. *Radiology* 1979;133:828.
123. Palestro CJ, Swyer AJ, Kim CK, Goldsmith SJ. Infected knee prosthesis: Diagnosis with In-111 leukocyte, Tc-99m sulfur colloid, and Tc-99m MDP imaging. *Radiology* 1991;179:645–648.
124. Pan G, Raymond AK, Carrasco CH, et al. Osteosarcoma: MR imaging after preoperative chemotherapy. *Radiology* 1990;174:517–526.
125. Pavlov H, Schneider R. Extrameniscal abnormalities as diagnosed by knee arthrography. *Radiol Clin North Am* 1981;19:287–304.
126. Petasnick JP, Turner DA, Charles JR, Gitelis S, Zacharlas CE. Soft-tissue masses of the locomotor system: Comparison of MR imaging with CT. *Radiology* 1986;160:125–133.
127. Piney A. The anatomy of the bone marrow. *Br Med J* 1922;2:792–799.
128. Pollack MS, Dalinka MK, Kressel HY, et al. Magnetic resonance imaging in the evaluation of suspected osteonecrosis of the knee. *Skeletal Radiol* 1987;16:121–127.
129. Porter EA, Shields AF, Olson DO. Magnetic resonance imaging of bone marrow disorders. *Radiol Clin North Am* 1986;24(2):269–289.
130. Pudlowski RM, Gilula LA, Kyriakos M. Intraarticular lipoma with osseous metaplasia: Radiologic-pathologic correlation. *AJR* 1979;132:471–473.
131. Raffi M, Firooznia H, Golimbu C, Bonamo J. Computed tomography of tibial plateau fractures. *AJR* 1984;142:1181–1186.
132. Rao VM, Dalinka MK, Mitchell DG, et al. Osteopetrosis: MR characteristics of 1.5T. *Radiology* 1986;161:217–220.
133. Rao VM, Fichman M, Mitchell DG, Steiner RM, Ballas SK, Axel L. Painful sickle cell crisis: Bone marrow patterns observed with MR imaging. *Radiology* 1986;161:211–215.
134. Reicher MA, Basset LW, Gold RH. High-resolution magnetic resonance imaging of the knee joint: Pathologic correlations. *AJR* 1985;145:903–909.
135. Reicher MA, Hartzman S, Bassett LW, Mandelbaum B, Duckwiler G, Gold RH. MR imaging of the knee. I. Traumatic disorders. *Radiology* 1987;162:547–551.
136. Reicher MA, Hartzman S, Duckwiler GK, Bassett LW, Anderson LJ, Gold RH. Meniscal injuries: Detection using MR imaging. *Radiology* 1986;159:753–757.
137. Reis ND, Ben-Arieh Y. Localization of a small tumor in the leg by magnetic resonance imaging. *J Bone Joint Surg* 1986;68:929–931.
138. Reis ND, Lanir A, Benmair J, Hadar H. Magnetic resonance imaging in orthopaedic surgery—a glimpse into the future. *J Bone Joint Surg* 1985;67:659–664.
139. Reiser MF, Bongartz G, Erlemann G, et al. Magnetic resonance in cartilaginous lesions of the knee joint with three-dimensional gradient-echo imaging. *Skeletal Radiol* 1988;17:465–471.
140. Rennebohm RM, Towbin RB, Crowe WE, Levinson JE. Popliteal cysts in juvenile rheumatoid arthritis. *AJR* 1983;140:123–125.
141. Resnick D, Niwayama G. *Diagnosis of bone and joint disorders*. Philadelphia; WB Saunders, 1981.
142. Roessner A, Zwaldo G, Vollmer E, et al. Biologic characterization of human bone marrow. IX. Occurrence of macrophages. *Pathol Res Pract* 1987;182:336–343.
143. Rosales Wynne-Roberts C, Anderson C, Turano AM, Baron M. Synovial haemangioma of the knee: Light and electron microscopic findings. *J Pathol* 1977;123:247–254.
144. Rosenthal DI, Scott JA, Barranger J, et al. Evaluation of Gaucher disease using magnetic resonance imaging. *J Bone Joint Surg* 1986;68:802–808.
145. Rosenthal H, Thulborn KR, Rosenthal DI, Rosen BR. Magnetic susceptibility effects of trabecular bone on magnetic resonance bone marrow imaging. *Invest Radiol* 1990;25:173–178.
146. Rosenthal RE, Wozney P. Diagnostic value of gadopentetate dimeglumine for 1.5-T MR imaging of musculoskeletal masses: Comparison with unenhanced T1- and T2-weighted imaging. *JMRI* 1991;1:547–551.
147. Rozing PM, Insall J, Bohne W. Spontaneous osteonecrosis of the knee. *J Bone Joint Surg* 1980;62:2–7.
148. Sanchez RB, Quinn SF, Walling A, Estrada J, Greenberg H. Musculoskeletal neoplasms after intraarterial chemotherapy: Correlation of MR images with pathologic specimens. *Radiology* 1990;174:237–240.

149. Sarti DA, Louie JS, Lindstrom RR, Nies K, London J. Ultrasonic diagnosis of a popliteal artery aneurysm. *Radiology* 1979;121:707–708.
150. Sartoris DJ, Kursunoglu S, Pineda C, Kerr R, Pute D, Resnick D. Detection if intra-articular osteochrondral bodies in the knee using computed arthrotomography. *Radiology* 1985;155:447–450.
151. Scapinelli R. Studies on the vasculature of the human knee joint. *Acta Anat* 1968;70:305–311.
152. Schmid FR. Principles of diagnosis and treatment of bone and joint infections. In: McCarty DJ, ed. *Arthritis and allied conditions: A textbook of rheumatology*, 11th ed. Philadelphia: Lea & Febiger, 1989:1863–1891.
153. Schwimmer M, Edelstein G, Heiken JP, Gilula LA. Synovial cysts of the knee: CT evaluation. *Radiology* 1985;154:175–177.
154. Scotti DM, Sadhu VK, Heimberg F, O'Hara AE. Osgood-Schlatter's disease, an emphasis on soft tissue changes in roentgen diagnosis. *Skeletal Radiol* 1979;4:21–25.
155. Sebag GH, Moore SG. Effect of trabecular bone on the appearance of marrow in gradient-echo imaging of the appendicular skeleton. *Radiology* 1990;174:855–859.
156. Seeger LL, Basset LW, Bertino F, Hanafee W. Magnetic resonance imaging for demonstrating undetected bone metastasis. Presented at the 16th International Congress of Radiology. Honolulu, July 8–12, 1985.
157. Senac MO Jr, Deutsch D, Bernstein BH, et al. MR imaging in juvenile rheumatoid arthritis. *AJR* 1988;150:873–878.
158. Seneterre E, Weissleder R, Jaramillo D, et al. Bone marrow: Ultrasmall superparamagnetic iron oxide for MR imaging. *Radiology* 1991;179:529–533.
159. Shogry MEC, Pope TL Jr. Vacuum phenomenon simulating meniscal or cartilaginous injury of the knee at MR imaging. *Radiology* 1991;180:513–515.
160. Sims RE, Genant HK. Magnetic resonance imaging of joint disease. *Radiol Clin North Am* 1986;56:118–125.
161. Singer AM, Naimark A, Felson D, Shapiro JH. Comparison of overhead and cross-table lateral views for detection of knee-joint effusions. *AJR* 1985;144:973–975.
162. Soren A. Pathogenesis and treatment of ganglion. *Clin Orthop* 1966;48:173–179.
163. Speer DP. Early pathogenesis of hemophiliac arthropathy. Evolution of the subchondral cyst. *Clin Orthop* 1984;185:250–265.
164. Spritzer CE, Sostman HD, Wilkes DC, Coleman RE. Deep venous thrombosis: Experience with gradient-echo MR imaging in 66 patients. *Radiology* 1990;177:235–241.
165. Stoller DW, Crues JV, Mink JH. Detection of stress fractions in the knee using MRI. In: *Proceedings of the American Roentgen Ray Society, 1987 Annual Meeting, April 26 to May 1.*
166. Sundaram M, McGuire MH, Fletcher J, Wolverson MK, Heiberg E, Shields JB. Magnetic resonance imaging of lesions of synovial origin. *Skeletal Radiol* 1986;15:110–116.
167. Sundaram M, McLeod RA. MR imaging of tumor and tumorlike lesions of bone and soft tissue. *AJR* 1990;155:817–824.
168. Sutro CT. Intraneural ganglion of the common peroneal nerve. Bull Hosp Joint Dis 1960;21:330–331.
169. Swett HA, Jaffe RB, McIff EB. Popliteal cysts: Presentation as thrombophlebitis. *Radiology* 1975;115:613.
170. Totty WG, Murphy WA, Ganz WI, Kumar B, Daum WJ, Siegel BA. Magnetic resonance imaging of the normal and ischemic femoral head. *AJR* 1984;143:1273–1280.
171. Trigaux J-P, Van Beers B, De Wispelaere J-F. Anatomic relationship between the popliteal artery and vein: A guide to accurate angiographic puncture. *AJR* 1991;157:1259–1262.
172. Unger E, Moldofsky P, Gatenby R, Hartz W, Broder G. Diagnosis of osteomyelitis by MR imaging. *AJR* 1988;150:605–610.
173. Vellet AD, Marks PH, Fowler PJ, Munro TG. Occult post-traumatic osteochondral lesions of the knee: Prevalence, classification, and short-term sequelae evaluated with MR imaging. *Radiology* 1991;178:271–276.
174. Wehrli FW, Ford JC, Attie M, Kressel HY, Kaplan FS. Trabecular structure: Preliminary application of MR interferometry. *Radiology* 1991;179:615–621.
175. White EM, Edelman RR, Wedeen VJ, Brady TJ. Intravascular signal in MR imaging: Use of phase display for differentiation of blood-flow signal from intraluminal disease. *Radiology* 1986;161:245–249.
176. Williams JL, Cliff MM, Bonakdarpour A. Spontaneous osteonecrosis of the knee. *Radiology* 1973;107:15–19.
177. Wilson AJ, Kyriakos M, Ackerman LV. Chondromyxoid fibroma: Radiographic appearance in 38 cases and in a review of the literature. *Radiology* 1991;179:513–518.
178. Wolfe RD, Colloff B. Popliteal cysts: An arthrographic study and review of the literature. *J Bone Joint Surg* 1972;54:1057.
179. Wolfe RD, Giuliano VJ. Double-contrast arthrography in the diagnosis of pigmented villonodular synovitis of the knee. *AJR* 1970;110:793–799.
180. Yao L, Lee JK. Occult intraosseous fracture: Detection with MR imaging. *Radiology* 1988;167:749–751.
181. Yulish BS, Lieberman J, Mulopulus GP, et al. MR imaging of arthropathies of juvenile arthritis and hemophilia. *Radiology* 1986;161(P):258.
182. Yulish BS, Montanez J, Goodfellow DB, Bryan PJ, Mulopulos GP, Modic MT. Chondromalacia patellae: Assessment with MR imaging. *Radiology* 1987;164:763–766.
183. Zimmer WD, Berquist TH, McLeod RA, et al. Bone tumors: Magnetic resonance imaging versus computed tomography. *Radiology* 1985;155:709–718.

ADDITIONAL READINGS

Chan WP, Lang P, Stevens MP, et al. Osteoarthritis of the knee: comparison of radiography, CT, and MR imaging to assess extent and severity *AJR*, 1991;157:799–806.

Mirowitz SA, Totty WG, Lee JKT. Characterization of musculoskeletal masses using dynamic Gd-DTPA enhanced spin-echo MRI. *J Comput Assist Tomogr*, 1992;16:120–125.

CHAPTER 10

Muscle Injuries[1]

Jerrold H. Mink

While great progress has been made in the understanding of the function, pathology, and repair processes of cartilage, bone, and ligamentous tissue, there is rather scant literature regarding similar properties of skeletal muscle. This is surprising in view of the fact that muscle injuries account for half of all injuries that occur in some sports, and that they are among the most common, lingering, and disabling conditions treated by orthopedic surgeons (1). Most muscle injuries, commonly known as muscle pulls or strains, result from rapid, violent contraction of the affected muscle during an excessively forceful stretch, and they frequently occur when exercise is intense, unaccustomed, and of long duration. Although injuries to muscles and tendons have been separated into different chapters in this and other books, one should be aware that muscles and tendons are so intimately related to one another that it is best to refer to them as a musculotendinous unit (2,3). The unit is composed of a contractile element, the muscle, and a passive element, the tendon. In fact, most injuries to muscle occur at the musculotendinous junction.

Individual muscle cells are known as muscle fibers, structures that are elongated multinucleated cells spanning up to 30 cm in length. These fibers originate and terminate in an attachment to a tendon or other connective tissue. The cytoplasm of a muscle cell is known as sarcoplasm, which is divided into longitudinal threads called myofibrils. They are crossed by striations known as A, I, H, and Z bands that differ in their chemical and optical properties.

[1] This chapter is being reprinted from *MRI of the Foot and Ankle* (A. L. Deutsch, J. H. Mink, and R. Kerr, eds), Raven Press: New York, 1992; 281–311.

Myofilaments are substructures of the myofibrils. The thicker myofilaments are called myosin; the thinner ones are known as actin. Both types of myofilaments are crossed by Z bands creating serially repeating regions known as sarcomeres. During muscle contraction, the actin filaments are seen to slide toward the center of the sarcomere in relation to the myosin. This action brings the Z bands closer together and produces shortening of the entire muscular unit. By actively shortening, muscle is able to produce force; by contracting and resisting lengthening, muscles serve to regulate movement or decelerate motion (4).

Two types of muscle fibers have been identified (1,4,5). These types, I and II, were initially distinguished from one another on the basis of color since type I fibers are red and type II are white. These fiber types are currently separable on the basis of histochemical differences. Type I fibers are characterized by the presence of low myofibrillar adenosine triphosphatase (ATPase) activity, whereas type II fibers have high ATPase activity. Histologic analysis reveals that type I fibers have abundant mitochondria and have bulky sarcomeres. Type II fibers have electron dense granules, composed of stores of glycogen throughout their sarcoplasm. Their sarcomeres are more streamlined and their membrane system is better developed for rapid transmission of nerve impulses into the fiber. Types I and II fibers show distinctly different physiologic properties reflecting their differing structure. Type I fibers are more resistant to fatigue than type II, and have a slower contraction time. Additionally, the force per unit in a type I fiber is much less than type II. Type I fibers are therefore best adapted for prolonged activity of relatively low intensity, those involving repetitive action, postural activity, or endurance-type activity. In contrast, type II fibers have a well-developed anaerobic energy system. They are best adapted for intense activities demanding speed and strength of short duration. They rapidly generate peak tension levels that fall back to baseline more quickly than type I fibers. Because of these properties, type I fibers have been designated as slow twitch, and type II as fast twitch fibers. Fast twitch fibers dominate in those human muscles that cross two joints, such as the hamstring muscles, rectus femoris, triceps, and biceps.

Since exercise is specific for muscle fiber type, and since individuals differ in their fiber type body composition, it is possible that people may be better suited for one type of activity rather than another (5). Biopsy studies of athletes have shown a significantly higher proportion of type I fibers in long distance runners and a higher proportion of type II fibers in sprinters; the muscle fiber composition of strength-trained athletes were mixed. It does not appear, however, that exercise, endurance, or strength training programs are capable of producing any significant change in the relative percentage of type I and II fibers in the adult population.

IMAGING OF MUSCLES

Before the advent of magnetic resonance imaging (MRI), nuclear medicine techniques, ultrasound, and computed tomography (CT) were used to assess muscle disorders (6–12). Technetium-99m is successful in localizing area of muscle damage by virtue of the deposition of the nuclide into areas of injured cell membranes and accumulation of intracellular calcium. The relative degree of deposition of the nuclear agent is proportional to the extent of muscle damage and the serum creatine phosphokinase. Scintigrams are abnormal within several hours of exercise-induced muscle injury. The pattern is most intense in 24 to 48 hr and within 1 week returns to normal (7,8,9). CT has been used extensively in assessment of soft tissue tumors, tumorlike conditions, and metabolic disorders such as the muscular dystrophies (10–15). There is, however, scant evidence of its use in evaluation of acute and chronic muscle injuries. Acute strains of the hamstring group are manifest on CT as focal areas of low attenuation, most commonly occurring in the long head of the biceps femoris. These areas are thought to represent edema and inflammation rather than blood, since hematomas are hyperdense on CT (12,16,17).

While still in a relative degree of infancy, MRI has rapidly established itself as the premier imaging modality for assessment of the full spectrum of muscle injuries, tumors, and infections (12,16–38).

Normal skeletal muscle is characterized by an intermediate to long T1 relaxation time and by a short T2 relative to other soft tissues. Muscle injuries are characterized by a long T2 relaxation time, but T1 values are quite similar to normal muscle; injuries are therefore most conspicuous on T2 and even more dramatically on inversion recovery (STIR) sequences, both of which optimize contrast between normal muscle and edema/hemorrhage (17,29,34,37). T2-weighted images (T2WI), however, may fail to distinguish edema from fatty infiltration of a muscle in those cases in which the degree of edema infiltration of the muscle is only moderate in degree (39). Normally, edema becomes brighter than fat on long TR sequences, but when the edema is "mingled" with the lower signal muscle fibers, the distinction between fat and edema may not be very evident. The T2 of fat is about 60 ms, which is about twice that of muscle (30 ms). When the degree of edema within a muscle is sufficient to lengthen the muscle T2 $>$ 60 ms, edematous muscle is easily distinguished from fat. However, when there is mild or moderate edema so that the T2 of the edematous muscle is between 30 and 60 ms, it may be impossible to distinguish fatty infiltration from edema. Inversion recovery sequences by virtue of their high contrast, high conspicuity properties do not suffer from this difficulty (39). STIR sequences have the advantage that edema-associated increases in muscle T1 and T2 times are generally additive to signal intensity, as is spin density. Unfortunately, STIR sequences are characterized by decreased signal-to-noise ratios and heightened susceptibility to flow artifacts, and they provide only a limited number of slice locations/unit time (34). They must therefore must be performed in conjunction with T1 and/or proton density sequence images for optimal anatomic delineation.

We choose to examine a suspected muscle injury by means of an extremity coil, although the body coil is used to accommodate a larger body part (e.g., the upper thigh) and examine a larger field of view (when comparison with the opposite side is necessary). An oil marker is usually placed on the skin at the site of the "mass" or the point of maximal tenderness. The first sequence performed is a short repetition time (TR) study in the coronal plane. This sequence allows one to be certain of proper centering and to localize the superficial skin marker. Short TR sequences, although they are insensitive for detection of acute muscle injury and masses, are helpful for demonstrating atrophy, for detection of related osseous lesions, for quantification of muscle volumes, for detection of subacute hemorrhage, and for assessment of abnormal muscle architecture (29). An axial T2-weighted sequence is usually performed next, employing a TR = 2500–3500, echo time (TE) = 20/120. The axial plane is ideal for determination of (a) the transverse extent of a lesion, and (b) the relationship of the lesion to the surrounding muscle groups and the neurovascular bundle. Occasionally, a STIR axial sequence is substituted for the long TR sequence. Finally, a sagittal or coronal STIR sequence (depending on the location of the lesion) is performed to describe most graphically the longitudinal extent of the lesion (29,34).

HEMATOMA AND INTERSTITIAL HEMORRHAGE

In order to understand better the MR appearance of muscle injuries, it is important to distinguish between a hematoma and interstitial hemorrhage (2,16,37,40). A hematoma refers to a confluent collection of blood within a relatively restricted area; the lesion is well defined and has a distinctly masslike character. Interstitial hemorrhage refers to bleeding that insinuates itself between the fibers of the damaged connective tissue. The girth of the affected limb and the affected muscle is increased but there is no discrete mass.

The MR appearance of a hematoma is affected by both the field strength of the MR imager as well as the stage of evolution of the mass (37,41) (Fig. 10.1). On T1WI, a hematoma less than 48 hr old has a signal intensity similar to that of muscle; on STIR studies, it is usually dark. As the lesion ages, the water content decreases and the protein content increases. Both events tend to decrease the T1 and the T2 relaxation times moderately. Therefore, a subacute or chronic hematoma (7–300 days) usually has a signal intensity quite close to that of fat on a T1-weighted image. While the tissues are affected by the presence of oxyhemoglobin, deoxyhemoglobin, and intact red cells, the observed effect of T1 shortening is due predominantly to oxydative denaturation of hemoglobin, resulting in methemoglobin production (37,42,43). Methemoglobin, containing iron in a ferric form, is highly paramagnetic. The local concentration of methemoglobin increases over the 80 to 90 hr following injury; the T1 shortening of the hematoma parallels the time course of methemoglobin production.

While subacute hematomas characteristically demonstrate high signal on T1WI, especially along their rims, the appearance is not specific. A similar finding has been identified in some sarcomas and abscesses (33,44,45) (see Miscellaneous Muscle Disorders, below). On T2WI and occasionally T1WI, a central low intensity focus is commonly seen within a hematoma on images obtained at high field strengths. Its presence relates to preferential T2 shortening secondary to deoxyhemoglobin within intact red cells. This effect is 100 times stronger at high field strengths, and is no longer evident following lysis of the erythrocytes. Finally, the wall of a subacute/chronic hematoma is usually very dark, reflecting the presence of fibrosis and/or hemosiderin deposition (43). The residual of a longstanding hematoma whose liquid component has resorbed may be only a masslike region of decreased signal. In our experience, most hematomas do not demonstrate a "pure" appearance that precisely fits into one of the descriptions above. More commonly, considerable inhomogeneity of the lesion with a wide variety of signal intensities on both T1WI and T2WI is present, and is due to presumed repeated/continuing hemorrhage (Fig. 10.1).

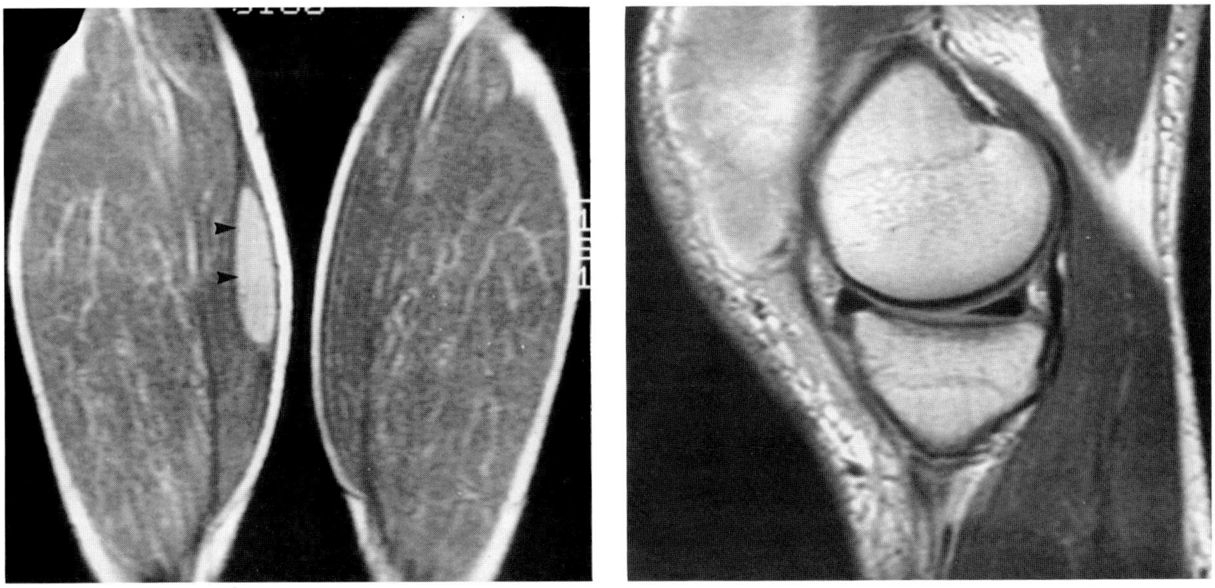

A B

FIG. 10.1. Hematoma. **A:** Coronal MR, TR/TE 300/15. This classic ovoid hematoma (*arrowheads*) in the medial head of the gastrocnemius is bright on T1WI, reflecting the characteristic short T1 of subacute lesions. **B** and **C** are from another patient. **B:** Sagittal MR, TR/TE 2000/20.

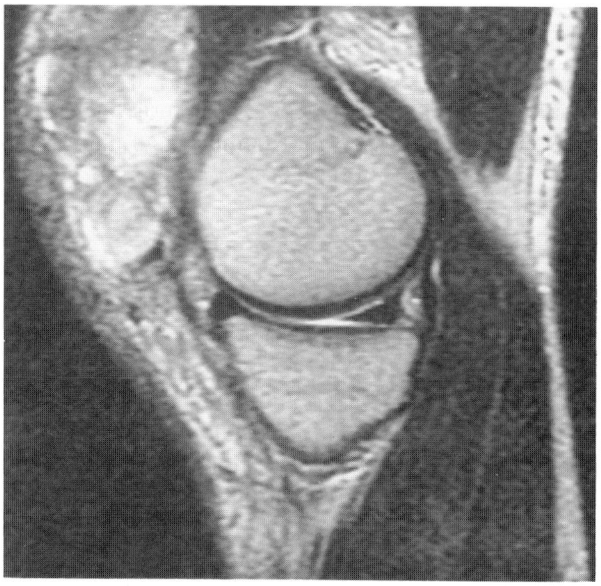

C

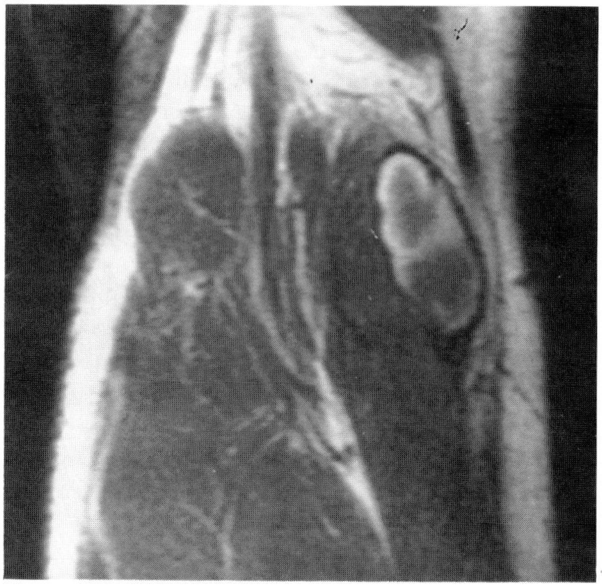

D

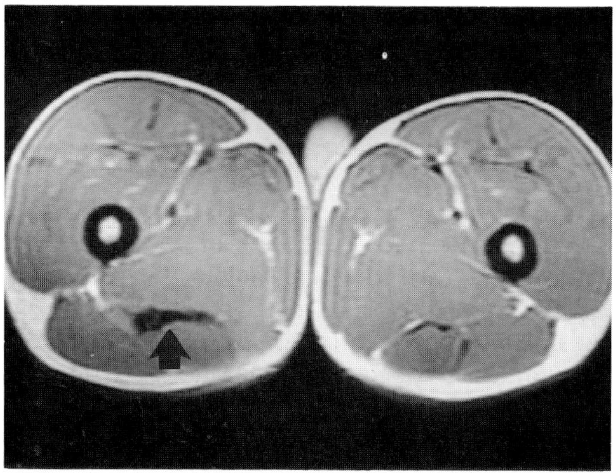

E

FIG. 10.1. (*Continued*) **C:** Sagittal MR, TR/TE 2000/80. These images are from another patient who suffered direct trauma to the leg 3 weeks prior. This hematoma has areas of high and low signal on both T1 and T2WI. The marked inhomogeneity, due to recurrent bleeding, is quite typical of a subacute hematoma. **D:** Axial MR, TR/TE 2500/80. In a different patient, another appearance of a subacute hematoma is identified. The mass in the gastrocnemius has a very dark rim due to deposition of hemosiderin. The lesion is predominantly bright, reflecting the long T2; the intermediate mass in the center represents intact red cells. **E:** Axial MR, TR/TE 2000/20. "Collapsed" hematoma following a muscle strain. The patient had suffered a muscle injury 6 months prior. In this case, the hematoma has largely resolved and the dark, hemosiderin-laden wall and fibrosis is all that remains.

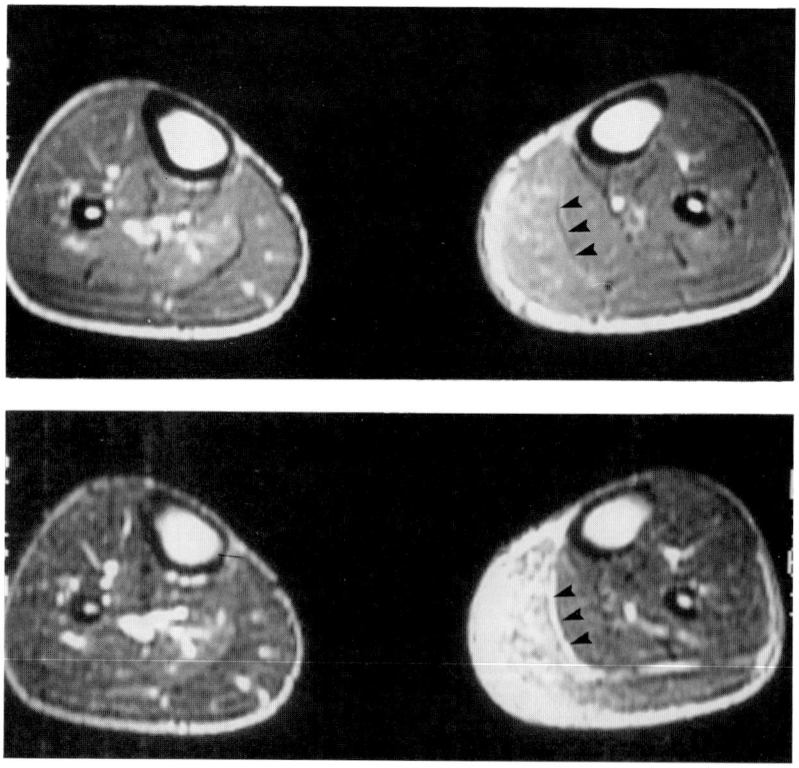

A

B

FIG. 10.2. Acute interstitial hemorrhage. **A:** Axial MR, TR/TE 2000/20. **B:** Axial MR TR/TE 2000/80. This professional football player was struck in the calf with a helmet incurring a painful, swollen calf. MR examination demonstrates abnormal signal in the medial head of the gastrocnemius (*arrowheads*). The intermediate signal on the proton density and the high signal on T2WI are consistent with edema fluid, yet the mechanism of injury and the marked discoloration of the skin that was present overwhelmingly suggests that significant blood must be present.

The time-dependent effects of hemoglobin breakdown products on relaxation times, so important in determining the MR signal and image character in a hematoma, are much less significant in interstitial hemorrhage (16,37,40) (Fig. 10.2). In the latter, the dominant signal characteristic is one of edema rather than blood.

On a short TR sequence, interstitial hemorrhage has a signal character quite similar to that of muscle. In general, the T2 of oxygenated liquid blood is relatively long (100–200 ms) because of the high free water content relative to soft tissue. The precise reason why edema, rather than blood, dictates the signal character is unknown, but several postulates may be offered: (a) parenchymal blood incites a significant inflammatory response in adjacent tissues, quite different from a hematoma. The inflammatory reaction causes additional prolongation of T2 by further increasing the free water content (16,37). It is the inflammatory response and its attendant water increase that determines the signal, and (b) the existing cellular framework of the soft tissue alters the breakdown process and removal of blood products (37,40). MR signal in interstitial hemorrhage is affected more by the fluid than by the protein component. In summary, T1WI of both acute hematoma and acute interstitial/intraparenchymal hemorrhage demonstrates a signal similar to that of muscle. As a hematoma ages, its decreasing T1 value results in increasing signal on T1WI, whereas the T1 value for interstitial hemorrhage does not change. On T2WI, hematomas develop progressively decreasing T2 values, but the T2 remains sufficiently long that the lesion remains bright on T2WI. The signal of hemorrhage is not significantly affected by time; hemorrhage has an initially and persistently long T2, and remains bright on long TR sequences (17,40).

DIRECT MUSCLE INJURY

Laceration

Muscle lacerations result from penetrating injury. Although the wound itself heals rapidly, the site of the laceration is permanently characterized by a dense connective tissue scar. There is little true muscle regeneration across the scar site, and the muscle distal to the laceration shows little evidence of reinnervation (4,46). Therefore although the repair process results in reestablishment of structural continuity, restoration of function is incomplete. Only one-half of the ability of a lacerated muscle to generate tension, and two-thirds of its ability to shorten, are recovered (46). The MR appearance of a lacerated muscle is dictated by the extent of the injury (Fig. 10.3). There is typically a transverse defect in the continuity of the muscle, and the gap between the torn fibers usually fills with a combination of both blood and edema, creating a mass with mixed signal characteristics on both T1WI and T2WI. If the laceration has occurred near one end of the muscle, the tendon may also be lacerated, and in such cases the functional defect may be profound.

A

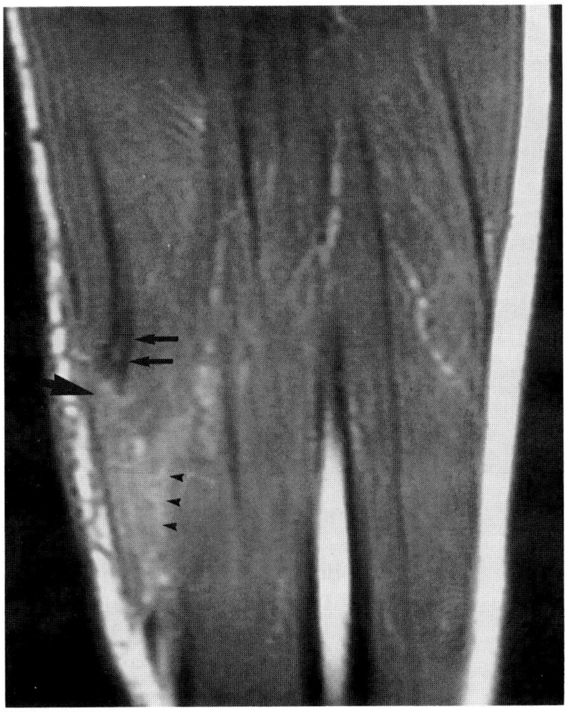

FIG. 10.3. Laceration of the anterior tibial muscle/tendon. **A:** Sagittal MR, TR/TE 600/20. There is a focal alteration in signal (*arrowheads*) of the tibialis anterior muscle. The tendon (*arrows*) is focally widened and is interrupted at the site of the laceration (*large arrow*).

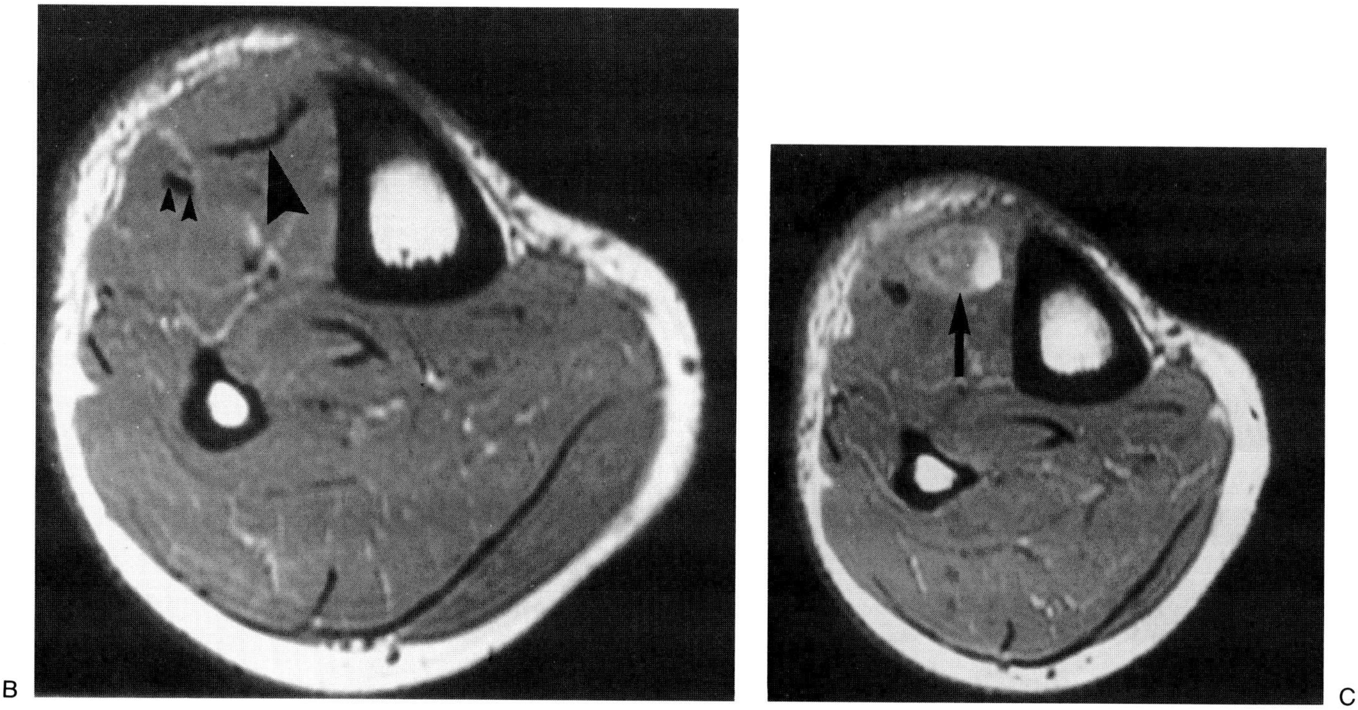

FIG. 10.3. (*Continued*) **B,C:** Axial MR, TR/TE 800/20; B is 2 cm above C. In B, the tibialis anterior tendon (*arrowhead*) is forming deep within the muscle on this image made above the level of the laceration. The extensor hallucis tendon (*arrowheads*) is also easily identified. In C, a mass of mixed signal (*arrow*) is seen replacing the normal tibialis anterior muscle/tendon; the mass is a mixture of blood and edema filling the site of the tear.

Contusion

A contusion of muscle is defined as a compressive- or concussive-type injury resulting from direct trauma by a blunt object. The term "charley horse" is synonymous with contusion. This type of injury is quite common in contact sports wherein the impact of one body part, often the knee, produces the injury in a large muscle, often the quadriceps. The blow results in capillary rupture and interstitial hemorrhage followed by edema and an inflammatory reaction. A contusion is differentiated from a muscular rupture by virtue of the uninterrupted ability of the muscle to function, and a contusion is differentiated from a strain by the mechanism of injury. Immediately following the injury, the affected individual may well continue to participate in the activity, but by the following day, muscle spasm sets in, leading to pain, swelling, disability, and stiffness. Contusions are graded in severity by the restriction in the range of motion of the adjacent joint (47). A mild contusion is one in which active or passive range of motion is limited by less than a third of normal. Affected patients have an average of 6 days of disability. A moderate contusion is one in which muscle spasm limits active motion to one-third to two-thirds of normal; these patients suffer disability for an average of 56 days. A severe contusion results in greater than two-thirds of loss of motion (3). Initial therapy should be directed toward limitation of the extent of the injury. Rest, ice, compression, and elevation are the mainstays of early therapy. Oral anti-inflammatory agents, proteolytic agents, steroids, aspiration of the hematoma (if one has formed), ultrasound, short wave diathermy, and exercise have all been used in therapy with varying degrees of success, but the disability period does not appear to be significantly shortened in any case (47). Extended immobilization is not necessarily beneficial, but the affected limb should be protected until a full range of motion is achieved. If unprotected, the individual may sustain a reinjury and further hemorrhage, and may potentially incur the development of myositis ossificans (3,46).

The MR appearance of a muscular contusion is a reflection of the composition of the inflammatory response which is edema, although surely blood, as evidenced by skin discoloration, is frequent (Fig. 10.2). It has been shown that the early posttraumatic change in muscle, caused by direct or stretch injury, is fiber swelling; the free water associated with this swelling causes the early diffuse increase in T2 (37). CT studies carried out in the hyperacute situation have confirmed this impression. The attenuation values of injured tissue are low relative to muscle, indicating that inflammation and/or edema is the major component of the muscle strain injury (12,37).

A contused muscle has four MRI characteristics: (a)

the affected muscle is usually slightly increased in girth; (b) there is an increase is signal intensity within the muscle on a STIR or a long TR sequence. On short TR studies, the edema appears isointense or hypointense to muscle, and may be very difficult to distinguish from normal muscle; (c) because the inflammatory liquid is dispersed within and between the muscle fibers, the edema manifests a feathery, interstitial pattern; and (d) the integrity of the muscle is not violated; there are no significant areas of fiber disruption.

Myositis ossificans circumscripta (MO) is the localized formation of nonneoplastic heterotopic bone and cartilage in the soft tissues in proximity to bone (48). Histologically, there is rapidly proliferating mesenchymal tissue that may occasionally be mistaken for a sarcoma. Instead of orderly scar formation, osteogenesis occurs in the soft tissues. The classically described pattern of the ''zoning'' pattern of maturation of the cellular elements is the most important diagnostic feature. Ossification actually occurs in the interstitial blood rather than primarily in the muscle (2). The reasons proposed for the development of heterotopic ossification include unresolved blood clot, infection of the hematoma, tearing of the periosteum with seeding of cells into the hematoma, or a change in local pH. Symptoms of MO include pain, warmth, tenderness, loss of function out of proportion to the severity of the muscle contusion, and a tendency for symptoms to increase during the period in which one would expect them to decrease. Patients affected with MO present

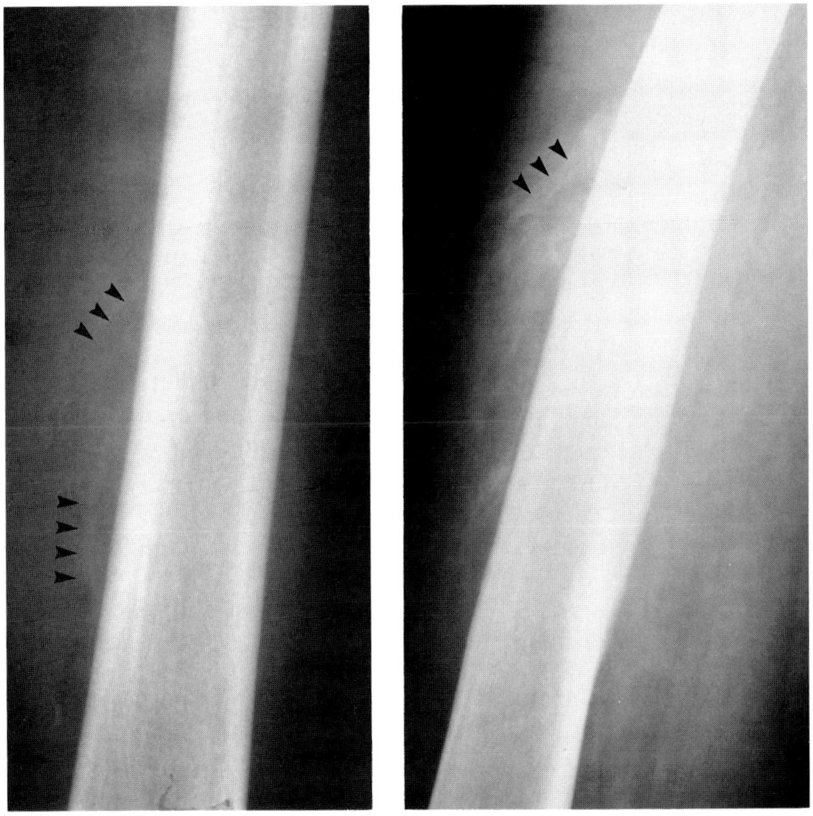

A,B

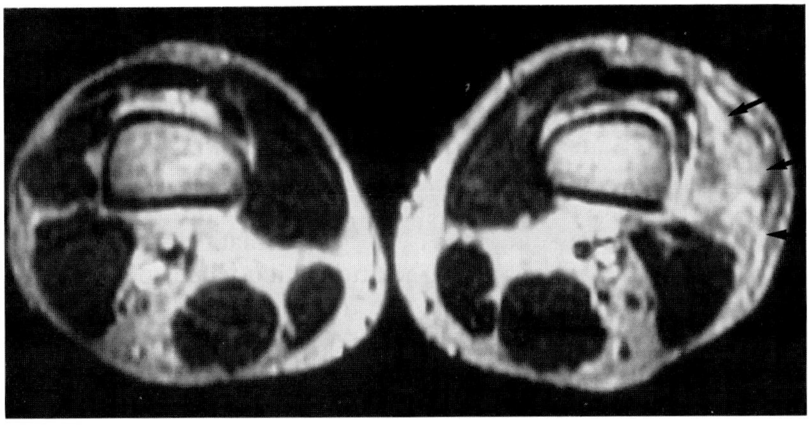

C

FIG. 10.4. Myositis ossificans. The patient suffered a deep thigh bruise in a football game. **A** was made 1 month prior to **B.** Although the views are not precisely comparable, it is readily apparent that there has been a marked increase in the calcifications/ossification (*arrowheads*) occurring in the soft tissues during this rather short interval. **C:** Axial MR, TR/TE 2000/80. This MR was obtained in another patient. The vastus lateralis muscle (*arrows*) is diffusely filled with high signal, which is mixed in character. The MR appearance of myositis ossificans is nonspecific; this biopsy-proved case of myositis ossificans has a similar appearance to that of other muscle injuries or even neoplasm. The conventional radiographs in this case showed calcifications at the time of the MR.

with a large firm tender mass within the contused muscle and restriction of motion of the adjacent joint. Flocculent calcification typical of MO can first be identified on conventional radiographs by 3 weeks postinjury, and by 6 weeks, structural organization is evident (Fig. 10.4). This calcification is predominantly along the periphery of the lesion. By 4 to 6 months, the bony mass will become mature, joint motion will be regained, and the mass of bone will begin to resorb (2,3). Rarely, resection of a symptomatic bony mass is necessary. Treatment consists of rest and immobilization to prevent further ossification. Proteolytic enzymes, intramuscular hyaluronidase, and radiotherapy are of little value.

The MR appearance of a muscle afflicted with MO is quite nonspecific, and has imaging characteristics quite similar to that of other types of muscle injury (Fig. 10.4). The affected muscle is often slightly enlarged in circumference, and demonstrates a diffuse increase in signal intensity on T2WI. There may be several areas of fluid confluence, suggesting hematoma formation. The conventional radiographically evident calcifications that are diagnostic of MO are generally not visible on MR.

INDIRECT MUSCLE INJURY: EXERCISE-INDUCED MUSCLE ALTERATIONS

Muscle pain is a frequent result of excessive exercise, especially following periods of relative inactivity. Several clinical syndromes characterized by muscle pain after exercise are differentiated by the time of onset of pain relative to the inciting event. Pain may begin during exercise (strain or muscle tear) or it may be delayed for 1 to 2 days (delayed onset muscle soreness).

In order to understand better exercise-related muscle injury, it is necessary to define the different types of muscle contractions. Isometric contraction produces tension without a change in length. Isotonic contraction is one in which the load is constant, and the muscle length changes during contraction. Two types of isotonic action are recognized. If the muscle shortens during contraction, a concentric action is said to have occurred; if the muscle lengthens during contraction, an eccentric action has occurred. An example of both types of isotonic muscle action can be found in the weightlifter performing a biceps-related exercise (Fig. 10.5). If a weight is placed in the hand while the arm is fully extended, and the hand is then brought up to the chest by flexing the arm at the elbow, the biceps muscle has actively shortened (concentric action). If the weight in the hand is then slowly lowered to the side, the biceps muscle is actively lengthening against

FIG. 10.5. Concentric and eccentric muscle action. This diagram depicts the two essential types of muscle action. In the right hand, the figure pulls the weight to his chest, and in doing so, shortens the biceps muscle. The action is termed concentric. In the left hand, the biceps lengthens while performing work, the essence of eccentric motion.

resistance (eccentric action). Muscles produce greater tension when they stretch than when they shorten. Eccentric contractions are therefore capable of producing greater force than concentric actions, and are considered the primary cause of exertion-related sports injury (1,46,49). Additionally, eccentric actions, as opposed to concentric actions, are associated with reduced oxygen consumption and less lactate production for the same power output. Eccentric muscle contractions require less energy than concentric actions for a given work load (50).

Injury prone muscles have three common characteristics: (a) they commonly perform eccentric actions. Eccentric actions of muscles are frequently those that limit, restrict, or regulate motion; for instance, some of the action of running is eccentric. Hamstring strains most frequently occur during intense bursts of speed when the hamstring is decelerating the rapidly extending knee. The muscle is therefore contracting while being forcefully lengthened; (b) they cross two joints, and they are subject to stretch at more than one site. The biceps brachii, the gastrocnemius, and the hamstring group are examples; and (c) strains are most likely to occur in those muscles used in sports or in activities requiring sudden increases or decreases in speed. These muscles commonly have a greater number of fast twitch type II fibers (49,51). While the hamstrings are the most frequently strained muscle group, the medial head of the gastrocnemius muscle, the rectus femoris, the hip adductors, and the hip flexors are commonly injured (6,46,51–59).

MRI of Physiologic Alterations in Muscle

The simplest type of exercise-induced change in muscle is found in a muscle that undergoes mild, nonexhaustive repetition (Fig. 10.6). Exercise is known to alter both the amount and distribution of water and electrolytes in muscle. Submaximal exercise transiently increases total muscle water content, and, subsequently, transient changes in tissue relaxation times can be detected by MRI in a forearm muscle that has

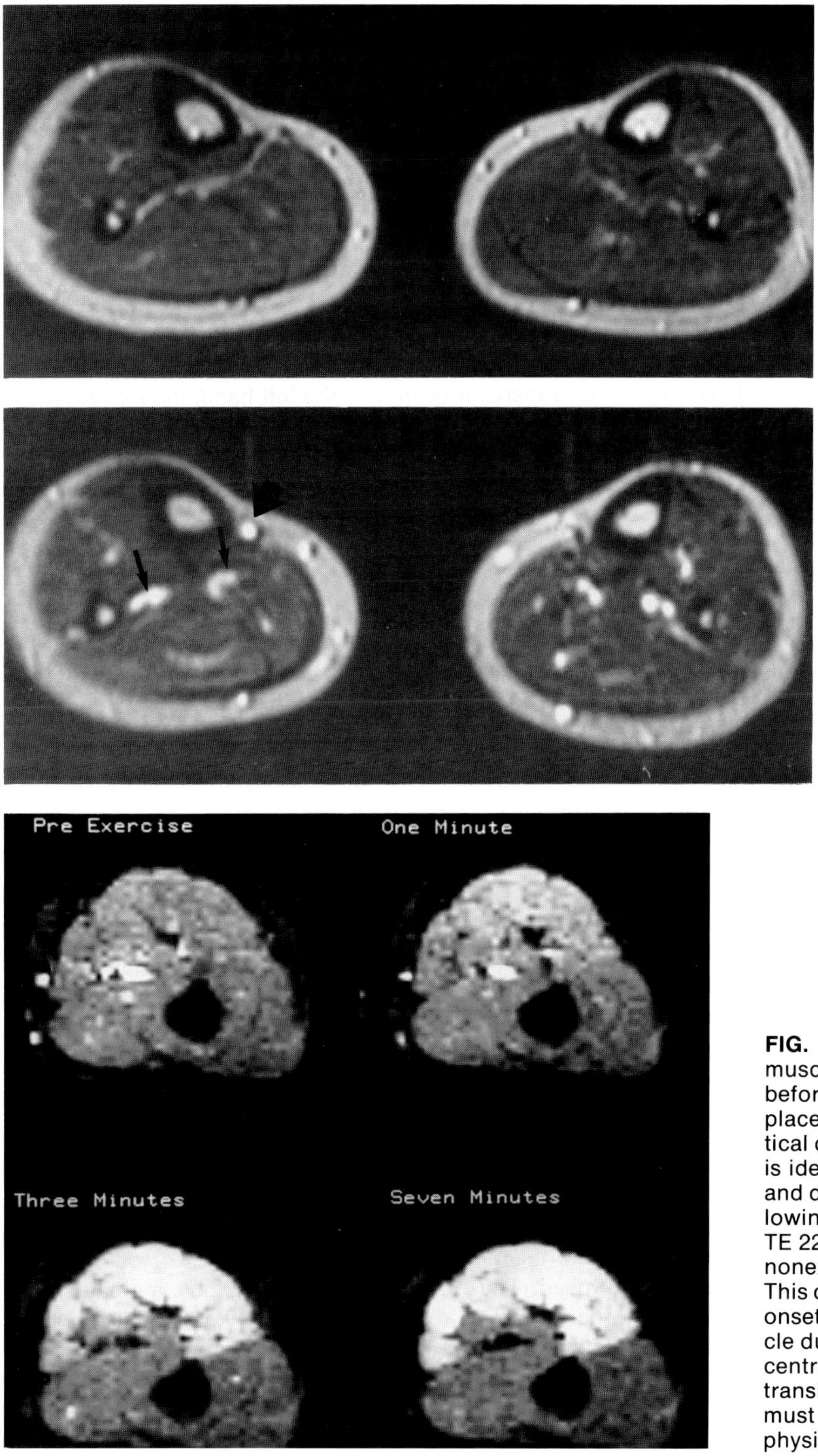

FIG. 10.6. Transient effects of exercise on muscle. **A, B:** Axial MR, TR/TE 2000/80. A is before, and B is made 5 min after jogging in place. The signal of the musculature is identical on A and B, although a "venogram effect" is identified; both the superficial (*large arrow*) and deep (*arrows*) veins become distended following exercise as seen in B. **C:** Axial MR TR/TE 2200/35/160. This patient performed a mild, nonexhaustive exercise of the biceps muscle. This case demonstrates the rapid (within 1 min) onset of signal increase (T2WI) within the muscle due to benign increases in local water concentrations. Intramuscular signal increases are transient and resolve quite quickly; imaging must be performed rapidly to detect these physiological changes.

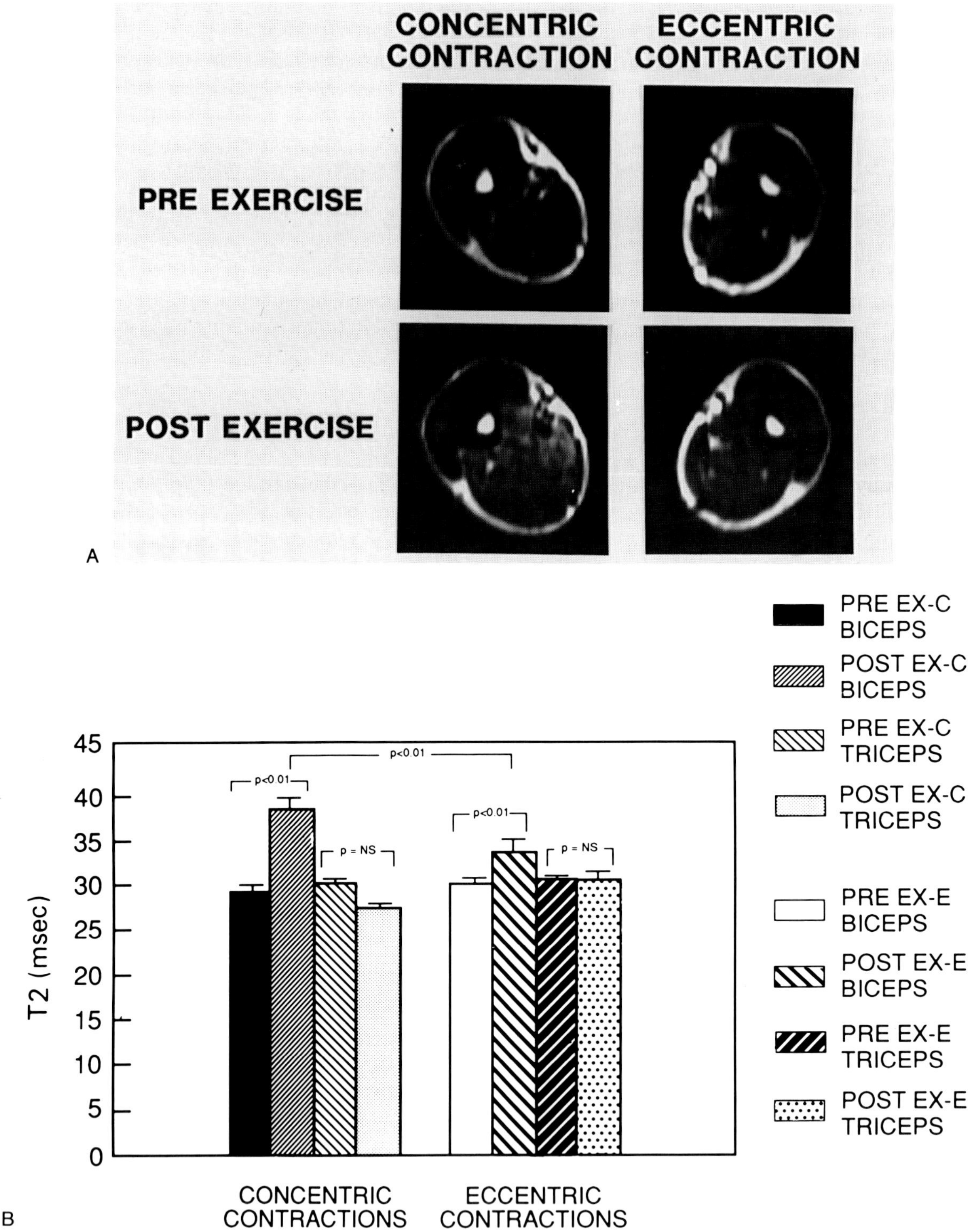

FIG. 10.7. Physiological increases in T2 following exercise. **A:** Axial MR, TR/TE 2000/80. These images are of the upper arms of a patient obtained before (*top row*) and immediately after (*bottom row*) the patient performed concentric (*left column*) and eccentric (*right column*) actions. The active muscles, the biceps and brachialis, that performed the concentric actions are differentiated easily on postexercise images by an increase in signal intensity. By comparison, little or no change occurs in signal intensity of active muscles that performed eccentric action. **B:** The graph shows T2 values of biceps and triceps measured before and after extremities performed concentric and eccentric actions (EX-C, concentric; EX-E, eccentric). (Courtesy of F. Shellock, Ph.D., Los Angeles, CA.) (From ref. 50, with permission.)

performed as few as eight contractions (60). This initial increase in water content occurs primarily in the extracellular space, while more severe degrees of activity leads to increases in intracellular water and sodium (18,50,61). The water content in muscle increases as a result of the movement of water across the capillary wall, increased local tissue osmolality, and increased local blood flow.

Immediate postexercise T2 relaxation times increase regardless of the type of exercise (concentric vs. eccentric) that the individual performs, but T2 values for muscles that perform concentric actions are statistically significantly higher than those for muscles that perform eccentric actions (50,62). The changes in MR signal are readily detectable on images immediately following concentric, but not eccentric, muscle actions (50) (Fig. 10.7). These immediate postexercise T2 values and corresponding images of them are a reflection of the work done by muscle; MR signal increases on long TR/TE or STIR sequences correlate weakly with the level of exertion, but strongly with the mean force during exercise (18,61). In summary, the MR changes in muscle signal immediately following nonexhaustive exercise is a reflection of the work done by the muscle and the alterations in intra- and extracellular water content. The effect of local blood flow is not critical to the appearance of these changes; vascular occlusion does not prevent their appearance (18). If the exercise is submaximal and muscle damage does not occur, these transient physiologic changes, including venous dilatation, increase in muscle signal on contrast-oriented examinations and the increase in T2 values, reverts to normal within minutes of cessation of activity.

Strains

An acute strain of a muscle is defined as a painful stretch-induced injury resulting from a single applied violent force (2). An example of such an injury might be found in the sprinter who rapidly and forcefully contracts his quadriceps muscles as he pushes out from the blocks to start a race. Contusions and lacerations are a direct injury; strains have an indirect mechanism of production.

Muscle strains are typically divided into three grades that are distinguished from one another by the degree of disruption that occurs, but one should be clearly aware that it may be difficult to make this distinction reliably (2,3,29,53). A grade 1 strain is one in which there is a minor degree of tearing of the muscle fibers so that there is less than 5% of loss of motion distal to the injury; there is no permanent defect. Pathologically, first degree strains are characterized by edema and low grade inflammation without significant muscle fiber disruption. Rest, ice, compression, and elevation are the mainstays of local therapy. A grade 2 strain is a more significant but still incomplete disruption of the muscle and is best classified as a partial tear. The clinical feature that distinguishes a first from a second degree strain is the loss of strength. Protection from further injury and proper therapy will nearly always result in rapid recovery with resolution by 10 to 14 days. A third degree strain, or complete rupture, is one in which there is extensive interruption of the muscle with a mop-end appearance of the torn edges. It is associated with a nearly complete loss of function, and therapy must be aimed at prompt restoration of the integrity of the musculotendinous unit. Spasm, retraction, and shortening occur immediately, and rapid contracture and necrosis make late repair unfeasible (2).

The direct mechanically induced injury effect of a strain on muscle is relatively minor (3,63,64). Rather, it is the subsequent biochemical response of the body to the injury that produces the histologic and clinical syndrome that is recognized as a strain. Following the injury, the affected cells self-destruct, releasing proteolytic enzymes; the exposed collagen fibers attract platelets that release further enzymes resulting in rapid breakdown of damaged muscle fibers by lysosomal proteases, free radicals, and other substances produced by macrophages. These events reach a peak several days following the exercise. Elevated temperatures, lowered pH, and other metabolic conditions further alter the situation. Eventually phagocytic inflammation, granulation tissue formation, revascularization, and complete restoration of functional myofibers occurs.

While contusions produce injury at the point of impact, strains produce a lesion at the histologic musculotendinous junction where injury to elastic, noncontractile tissue appears to be responsible for symptoms (1,46,65,66). Pain receptors are known to be most prevalent in the region of the tendon and related connective tissues; additionally, the muscle fibers nearest the tendon junction do not stretch exactly as the midportion of the muscle fiber does. It should be noted that when making reference to the musculotendinous junction, one is not referring to the macroscopic transition. In the biceps femoris, the proximal tendon and the muscle/tendon junction extend approximately 60% of the total length of the muscle; the distal tendon and muscle/tendon junction occupy 66% of the muscle length. Therefore, the anatomic and functional muscle tendon junction extend the entire length of the hamstring group, and injury can occur anywhere along the musculotendinous junction (12).

Following muscle strain, a predictable course of functional alterations occur (49,67,68) (Fig. 10.8). Strength decreases immediately following injury and the strained muscle is capable of generating only 70% of its normal force. At 24 hr, muscle function is at its

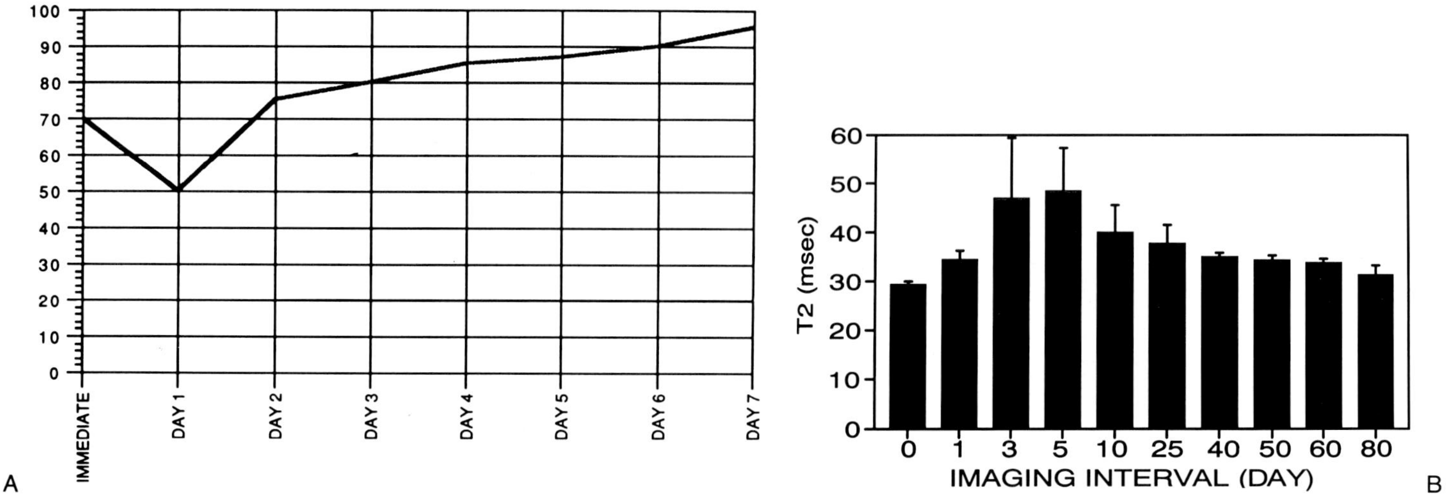

FIG. 10.8. Muscle strength and T2 after muscle strain. **A:** Muscle strain reduces muscle strength, which falls to its nadir on the first day. By the second week, function has returned to normal. **B:** The T2 of muscle following strain varies with the curve of the strength graph.

lowest level (50% of control); recovery begins by 48 hr. The muscle is capable of generating 75% of its control contractile force by 2 days, and by 7 days, the muscle returns to near normal (90%) strength. Measurement of relaxation times of affected muscles in patients with acute strains reveal that both T1 and T2 values are prolonged to approximately two times normal. T2 values of injured muscles increase with increasing degrees of damage, perhaps related to a more extensive inflammatory/repair process and accompanying edema. The observed decrease in function is due to both an absolute decrease in the ability of the muscle to generate force and a voluntary patient component secondary to pain.

The ideal method of treating muscle strains is to prevent them, and a number of activities have been shown to decrease the incidence of such events. Proper warm-up and stretching exercises have been advocated as one method to reduce injury. Stretching increases flexibility in the lower extremity and warming up (isometric preconditioning) is of benefit in preventing muscular injury, increasing muscle contraction efficiency, and increasing the elasticity of the muscle tendon unit (46,69–71).

In contradistinction to the physiological pattern of the MR signal that accompanies exercise, patterns of injury to muscle make their appearance 24 to 48 hr after the event. The appearance of a muscle strain is dependent upon the degree of the injury and the time from injury that the MR is acquired. The MR appearance of a grade 1 strain of a muscle is similar to that of a contusion (Figs. 10.9, 10.10, 10.11). The slightly enlarged muscle has a feathery, interstitial increase in signal (long TR/TE or STIR sequence), due primarily to edema, and there is no interruption of muscle continuity (37,72). Perifascial fluid collections are common. A grade 2 strain has an MR appearance that reflects the greater degree of injury (37) (Figs. 10.12, 10.13, 10.14). Physiologically, the primary difference between the grade 1 and 2 strains is the fact that in the

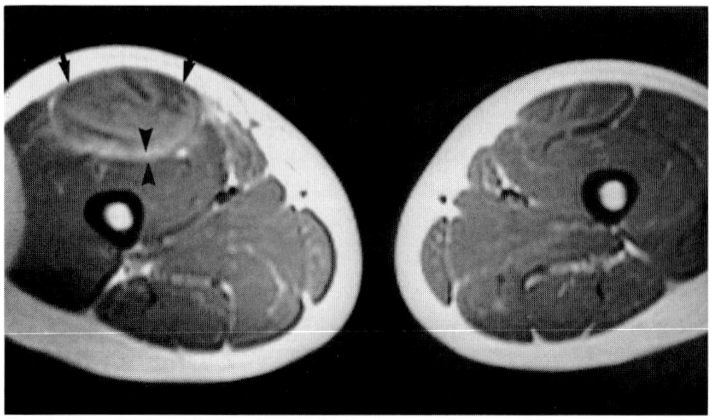

FIG. 10.9. Grade 1 strain of the rectus femoris. Axial MR, TR/TE 2000/20. This patient experienced pain in the right thigh after beginning a foot race, and underwent an MR the following morning. The rectus muscle (*arrows*) is enlarged and demonstrates a slight increase in signal. A bright perifascial rim (*arrowheads*) defines the boundary of the muscle.

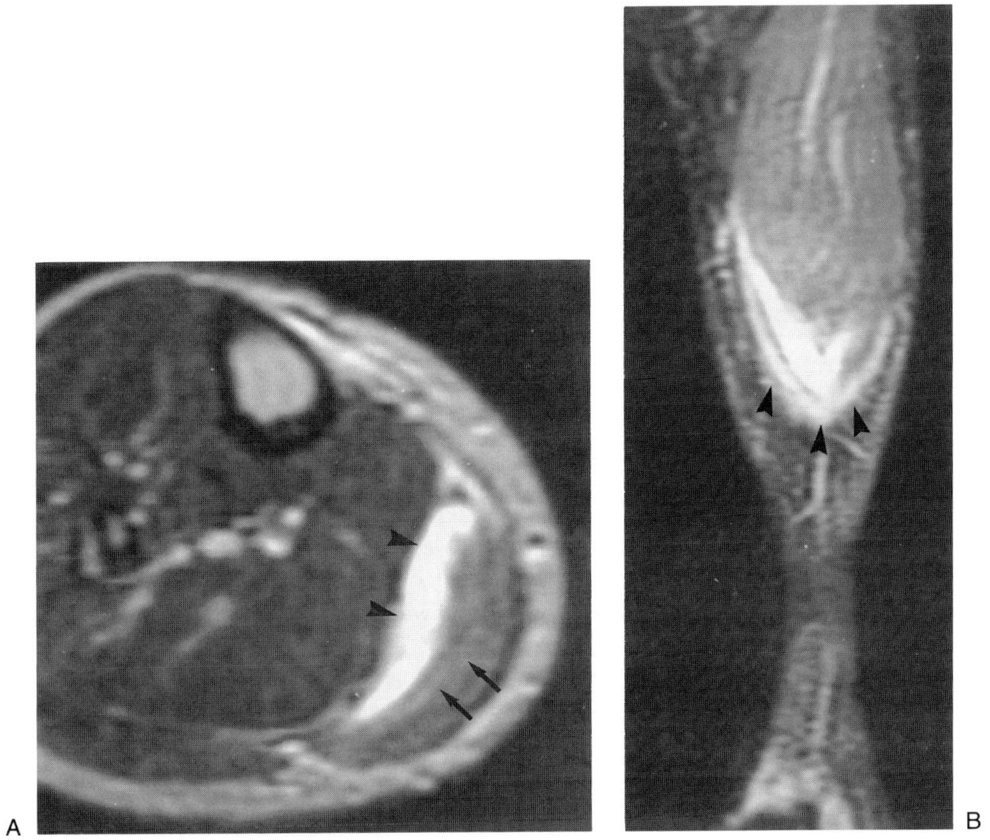

FIG. 10.10. Grade 1 muscle strain. **A:** Axial MR, TR/TE 2000/80. The gastrocnemius muscle demonstrates a mild increase in signal, but a perifascial fluid collection (*arrowheads*) is the most dramatic finding on the image. **B:** Coronal MR, TR/TE/TI 2200/35/160. *Arrowheads* define the fluid collection just deep to and distal to the gastrocnemius.

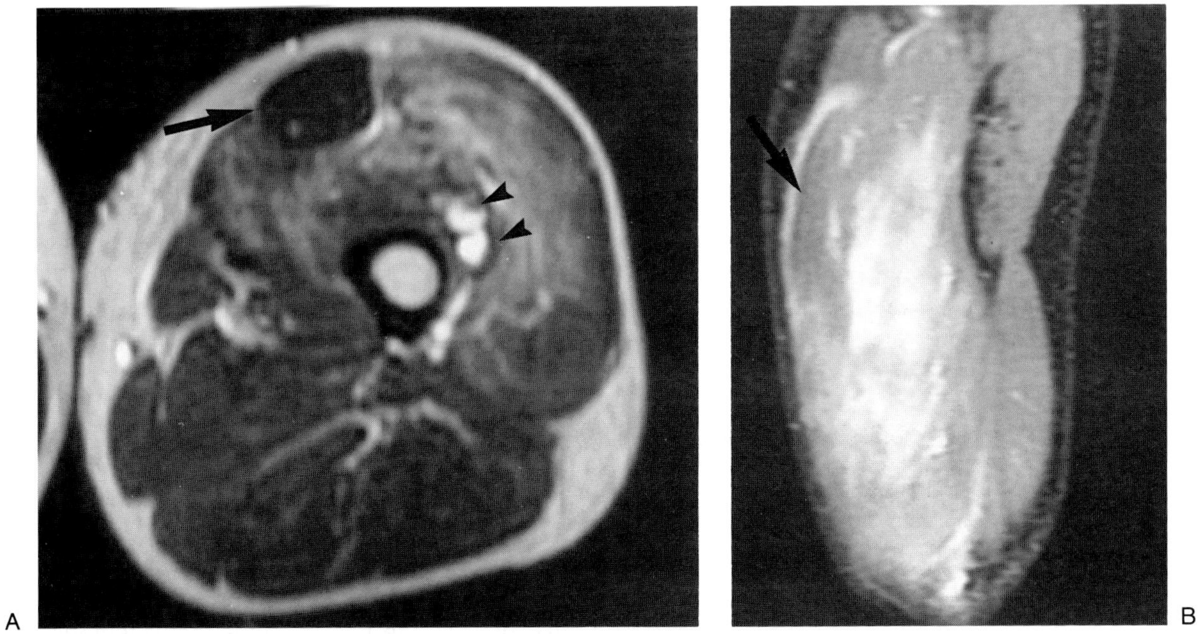

FIG. 10.11. Grade 1 muscle strain with small hematomas. **A:** Axial MR, TR/TE 2000/80. **B:** Sagittal STIR MR, TR/TE/TI 2200/35/160. This individual suffered repeated stretch injuries to the thigh. The entire vastus group is diffusely edematous. The rectus femoris (*large arrow*) is spared. Within the vastus intermedius is one, and perhaps several small focal fluid collections (*arrowheads*). These collections were bright on T1WI and are thought to represent hematomas from previous injuries. The STIR sequence is more dramatic than T2WI in demonstrating edema.

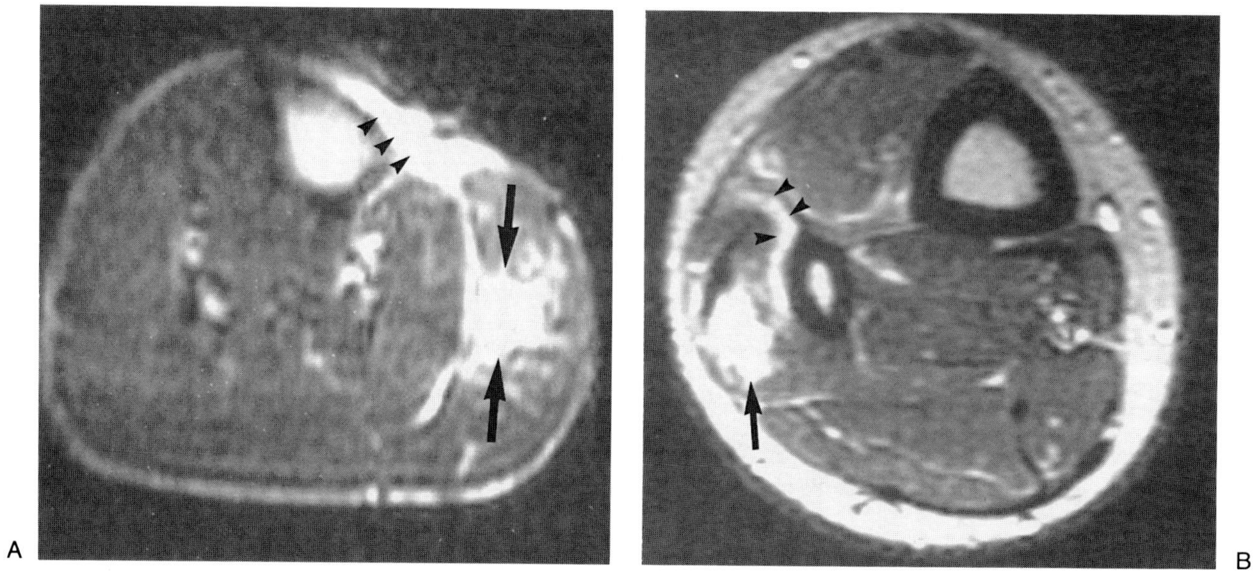

FIG. 10.12. Grade 2 strain of the gastrocnemius. **A:** This 26-year-old man suffered an acute muscle injury during a foot race. Axial MR, TR/TE 3000/80. The image demonstrates a focal stellate interruption (*arrows*) of the gastrocnemius. Focal interruption of the muscle fibers is characteristic of a grade 2 strain. A perifascial fluid collection is present just anterior to the muscle injury (*arrowheads*). **B:** Axial MR, TR/TE 2000/80. Another professional football player suffered a grade 2 strain (*arrows*) in the peroneal muscles, and there is an adjacent perifascial fluid collection (*arrowheads*).

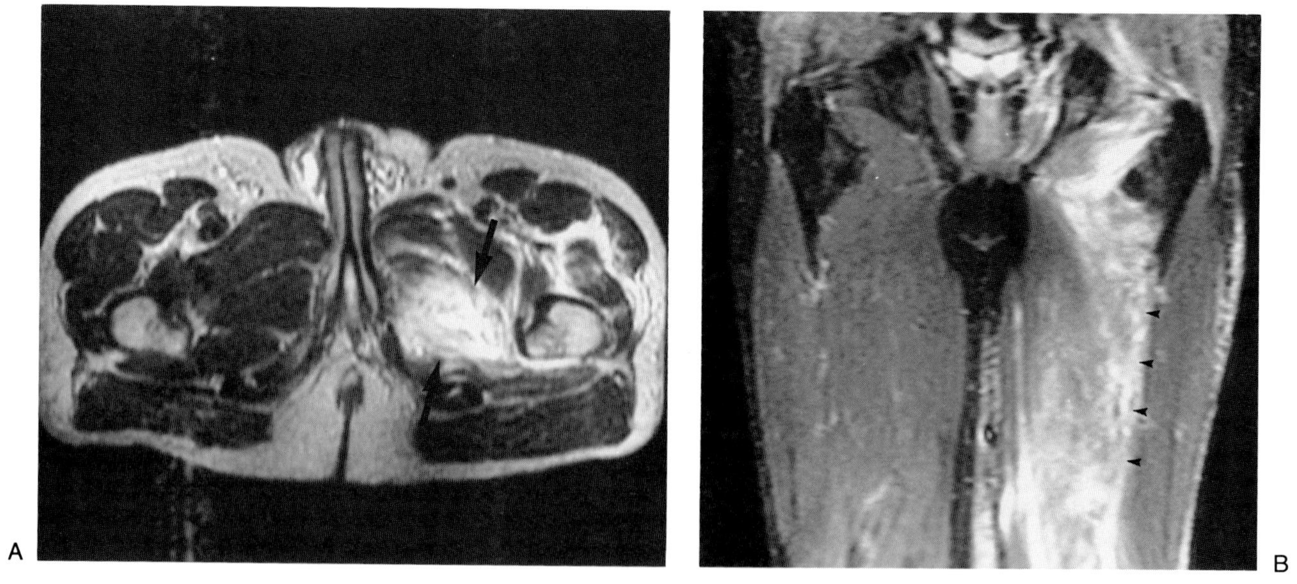

FIG. 10.13. Grade 2 strain of the adductors. **A:** Axial MR, TR/TE 2500/80. **B:** Coronal MR, TR/TE/TI 2200/35/160. This skier suffered abduction injury with the sudden onset of thigh pain. The obturator externus and the quadratus femoris (*arrows*) demonstrate diffuse edema within their substance with focal areas of apparent muscular discontinuity. The coronal sequence (B) demonstrates the full extent of the injury to the adductors (*arrowheads*).

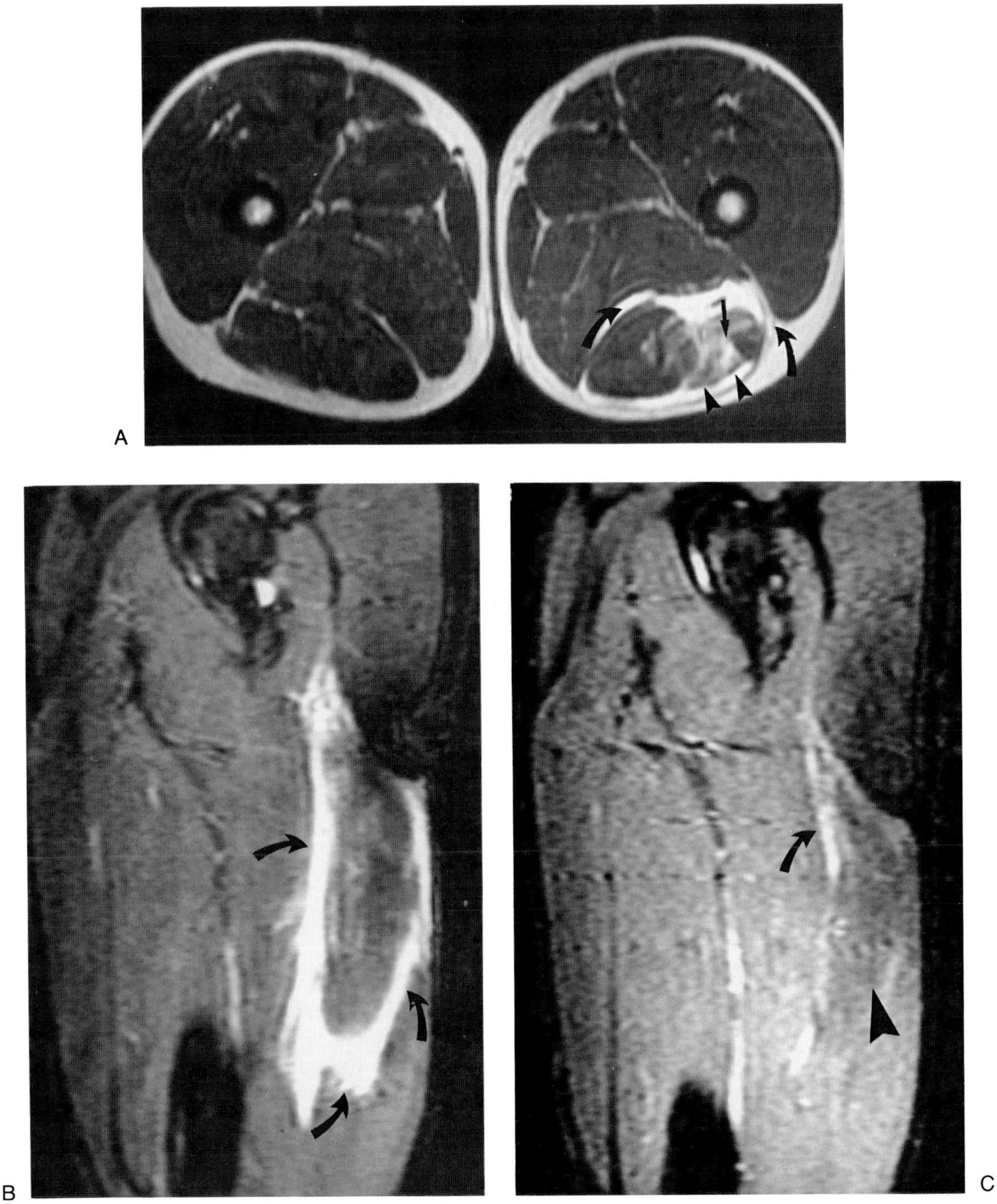

FIG. 10.14. Repetitive grade 2 hamstring strain. This professional football player suffered an acute hamstring strain during push-off from the line of scrimmage; the initial injury is depicted in **A** (axial MR TR/TE 2000/80) and **B** (sagittal MR, TR/TE/TI 2200/35/160). The biceps femoris muscle demonstrates edema throughout its substance (*arrowheads*) with several small focal areas perhaps representing limited muscle fiber interruption (*small arrow*) (grade 2 strain). Most striking however, is the ring of perifascial edema (*curved arrows*). **C:** Sagittal MR, TR/TE/TI 2200/35/160. Three weeks later, following conservative therapy and nearly complete rest, the ring of edema is virtually gone (*curved arrows*) and the patient was asymptomatic. The biceps muscle is slightly darker than the others, perhaps secondary to diffuse hemosiderin deposition.

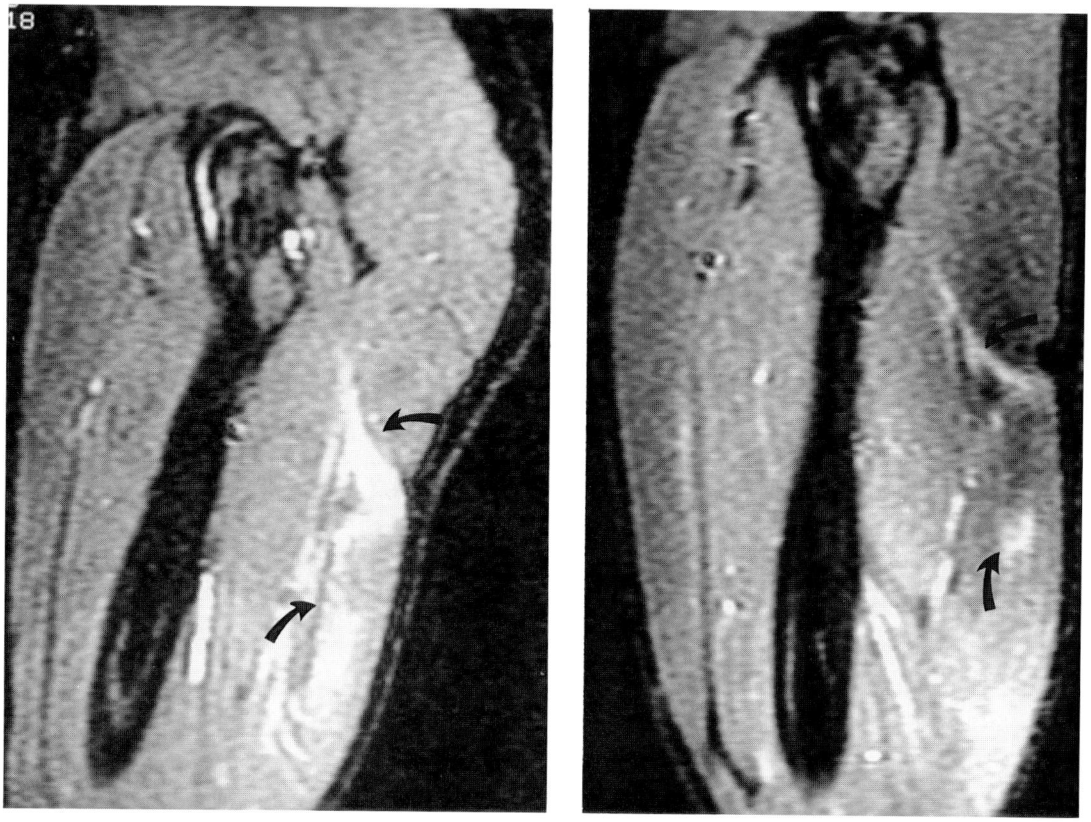

D E

FIG. 10.14. (*Continued*) **D:** Sagittal MR, TR/TE 2200/35/160. Two weeks after C, the patient resumed working out, and again experienced the sudden onset of pain. The MR image again demonstrates intramuscular edema and perifascial fluid (*curved arrows*). **E:** Sagittal MR, TR/TE/TI 2200/35/160. The patient was again placed at rest, and 2 weeks later, a repeat MR demonstrates virtual resolution of the intramuscular and perifascial edema (*curved arrows*). While perifascial fluid collections may be rapidly mobilized if the part is protected, continued activity may lead to hematoma formation.

latter, damage to the musculotendinous unit results in loss of function. The feathery MR appearance of interstitial blood/edema typical of a grade 1 strain is present in the background, but additionally, a focal mass-like lesion and/or a stellate defect can inevitably be found within the injured muscle. While the MR signal character of this lesion is dictated primarily by the presence of water/edema, signal increases on short TR sequences, typical of blood, may rarely be seen (72). Depending on the age of the tear, there may either be a focal increase in the mass of the muscle, or if sufficient posttraumatic atrophy has occurred, the bulk may be decreased. Just as in the clinical situation, the imaging differentiation of grade 1 and 2 strains is often quite difficult.

Longitudinal MR examination of patients suffering first and second degree acute muscle strains reveals a predictable sequence of signal alterations (66,73). Soon after injury, there is a diffuse increase in signal within multiple muscles in the same compartment with the muscle that is in fact injured, but rapidly the pattern of injury becomes localized to those muscles that are actually injured. MR reveals that the signal increase is initially most evident within the center of the affected muscle. The signal abnormality spreads centrifugally to involve the entire muscle more homogeneously and becomes maximal by day 3 to 5. By 36 hr, a thin rim of increased signal (T2WI) appears in the perifascial space and intermuscular septa, closely applied to the injured muscle. Within the first week following injury, linear, streaky low signal alterations become evident in the fat adjacent to the injured muscle. The precise nature of these streaks is unknown, although similar lesions detected by CT have been attributed to myoglobin. If this supposition is true, the presence of these lines would indicate that the muscle injury is relatively acute. By 2 weeks, the abnormal signal within a strained muscle is often nearly resolved. However, the lesion will persist well beyond the time of clinical recovery (defined as pain-free motion). The clinical examination may therefore overestimate the degree of muscle recovery, an observation that could potentially have implications for rehabilitation programs.

The MR appearance of grade 3 strains is very striking. A definite interruption in the continuity of the entire muscle is evident. The torn edges of the muscle retract and assume a lobulated or serpiginous contour. The gap between the torn edges may fill in with blood in sufficient quantity that a true hematoma forms. Gross functional impairment necessitating surgical reconstruction usually makes MRI superfluous.

Delayed Onset Muscle Soreness

Delayed onset muscle soreness (DOMS) is defined as pain in skeletal muscle that occurs some time following unaccustomed muscular exertion (74). The clinical feature that distinguishes DOMS from strain is the time of onset of pain and dysfunction. The soreness progressively increases in intensity over the first 24 to 48 hr following exercise, peaks at 2 to 3 days, and usually resolves by 7 days (66,75). The most predictable method of producing DOMS is to have a muscle perform eccentric contractions. Although patients experience acute fatigue sooner after performing concentric rather than eccentric work, they will not develop DOMS (65,76,77).

DOMS is associated with a transient decrease in muscle performance (Fig. 10.15). This reduction is due both to reduced voluntary effort secondary to pain and a lowered inherent capacity of the muscle to produce force (68,74). There is no evidence that even repeated episodes of DOMS are associated with long-term loss of function or permanent muscle damage.

Although DOMS is quite prevalent and has been experienced by most adults on countless occasions, rather little is known about the basic cellular and biochemical mechanisms involved in its production. Possible explanations for the observed muscle soreness include accumulation of metabolic waste, structural damage, and elevated muscle temperatures (74). Muscle-based enzymes such as creatine phosphokinase (CPK) and lactic dehydrogenase (LDH) are routinely elevated in patients with DOMS, and the peak elevation of the enzymes occurs from 18 to 30 hr postexercise, the same time at which DOMS is maximal (75). Sequential muscle biopsies taken up to 1 week postexercise from the soleus muscle of individuals suffering from DOMS reveal no evidence of gross morphological fiber disruption or evidence of ischemic tissue injury. At the ultrastructural level, however, frequent myofibrillar disturbances and Z band streaming (a sign of muscle injury) is observed. Histologically, the pattern of injury and decrease in muscle function is evident immediately after exercise, when there is no soreness, but tissue damage increases over 24 to 48 hr (19,74). It remains unknown whether these findings reflect direct mechanical Z band disruption or whether they are a result of secondary activation of lysosomal enzymes producing inflammation and secondary injury, but it is clear that overloaded muscles do have their contractile machinery partially distorted. Biochemically, the injury manifests itself as an elevation of serum enzymes. Serum LDH, transaminases, and CPK activities increase in patients with DOMS, with the maximum values appearing at 8 to 24 hr postexercise (75,76). Finally, there does not appear to be any treatment for DOMS. Antiinflammatory medications, linaments, methanol, and camphor have little value in actually reducing soreness. Remarkably, performance of precisely the same motor task that caused the muscle soreness may temporarily result in alleviation of discomfort, but unfortunately, with cessation of exercise, the soreness returns.

Exertional rhabdomyolysis may be thought of as an extreme degree of DOMS that results in release of cellular contents into the general circulation (14,35,77). Pathologically, myocyte swelling, inflammation, and hyaline degeneration are present (35). Crush injuries, burns, prolonged muscular compression, toxin expo-

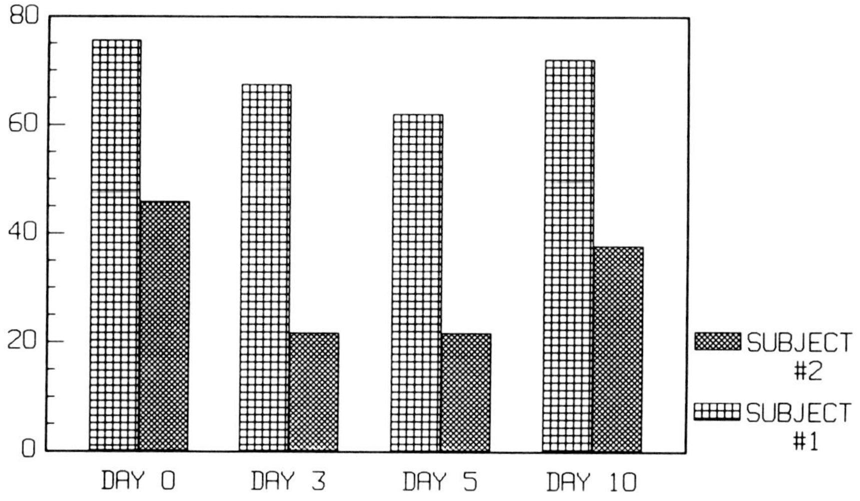

FIG. 10.15. Muscle strength during DOMS. The graph depicts the change in the ability of a muscle to produce force following DOMS. Both patients demonstrate maximal loss of muscle strength on days 3–5, and show a return to baseline strength by the second week.

sure, hypoxia, abnormalities of energy production, and extremely intense exercise, especially in hot climates, can be associated with rhabdomyolysis and even frank muscle necrosis. The diagnosis is confirmed by finding myoglobin in the urine. Patients with rhabdomyolysis experience intense pain and weakness following exercise. Acute renal failure, tetany, compartment syndromes, and disseminated intravascular coagulation may complicate rhabdomyolysis; while these complications may prove life-threatening, they fortunately are rare (74).

The MR findings of postexercise soreness are similar to those described for grade 1 strains (Fig. 10.16). Edema within muscle bundles and perifascial fluid collections dominate the MRI. The similar appearance of muscle strains and DOMS makes it difficult to distinguish reliably between the two clinical syndromes on the basis of MR changes alone (73). Within 24 hr of exhaustive exercise, there is a predictable but transient increase in T1 and T2 relaxation times that can be visualized on STIR and long TR/TE sequences as an increase in signal intensity (29,66,73). This increase in signal is not initially limited to the muscles performing work, but rather extends over a more broad region. Patients do not generally experience soreness at this time. Within 48 to 72 hr, the signal changes become clearly demarcated, involving only those muscles that are sore. The MR signal intensity is proportional to the extent of ultrastructural myofibrillar alterations, and the volume of abnormal muscle on MR (32). The relative increase in T2 values correlates with the degree of functional impairment and pain, all of which become maximally severe by day 3 (29,68). T2 relaxation times remain statistically significantly increased for prolonged periods of time following the resolution of soreness. Shellock et al. longitudinally studied a group of patients that exercised to failure and ultimately developed soreness. While soreness, pain, and joint stiffness were present only in the first week following eccentric muscle actions, MRI showed subclinical signal abnormalities in the biceps muscle that lasted as long as 75 days after the disappearance of symptoms (66) (Fig. 10.16).

Correlation of serum creatine kinase (CK) and signal alterations seen on MRI demonstrate that whereas there was a positive correlation between the relaxation times, CK values, and pain, the MR signal abnormalities preceded and persisted longer than other commonly used indicators of injury (73,75). This finding may have an effect on the timing and the intensity of rehabilitation of athletes.

In patients with rhabdomyolysis, MRI reveals abnormal signal in the affected muscle. Both T1 and T2 values increase in muscles undergoing rhabdomyolysis; subsequently, there is an increase in signal on T2 and STIR sequences. It has been postulated that the signal alterations are secondary to the ubiquitous edema and inflammation that is present; the contribution of the macromolecular changes in damaged muscle is difficult to evaluate (14). MRI is extremely valuable in patients whose clinical syndrome suggests that decompressive fasciotomy is indicated; MRI can preoperatively localize the affected muscles. MRI cannot, at this time, be used to determine whether or not muscle is irreversibly damaged; preliminary evidence suggests that MR spectroscopy may be valuable in this regard (35).

The exquisite sensitivity of MRI to the detection of pathologic findings far exceeds its specificity. Changes on MRI due to exertional exercise are quite similar to those due to radiation, denervation, and infectious and inflammatory disorders such as polymyositis (29).

Compartment Syndromes

Compartment syndrome is a pathologic condition of skeletal muscle in which increased interstitial pressure within an anatomically confined space interferes with the function of the muscle and neurovascular components (4,37,78). The critical signs and symptoms of compartment syndrome are: (a) pain out of proportion to the injury; (b) weakness and pain on passive stretch; (c) hypoesthesia in the distribution of the nerves traversing the affected compartment; and (d) tenseness of the boundaries of the compartment (79). In patients with compartment syndrome, nuclear medicine flow studies reveal a relative decrease in tracer activity in the affected compartment secondary to the decrease in capillary perfusion, although blood flow images do not demonstrate ischemia (36). The anterior and lateral compartments of the leg and the anterior (volar) compartment of the arm are the most frequently affected compartments. Common etiologies of acute compartment syndrome include hemorrhage, direct muscular trauma, nondisplaced tibial shaft fractures, intense muscular activity, and muscle rupture (4,78,80). The diagnosis is confirmed when intracompartmental pressures are 15 to 20 mm Hg at rest (normal = 0–4), rising to 75 mm Hg during exercise (normal = 50) (7). These elevated pressures compromise capillary perfusion but are not sufficient to occlude major arteries. Therefore, the pulse distal to the affected compartment is present until the very late stages of the syndrome. If sufficiently severe and untreated, an acute compartment syndrome may lead to muscle necrosis and fibrous contracture with permanent neurological damage (e.g., Volkmann's contracture). Therapy includes immediate decompression by means of fasciotomy and debridement of necrotic tissue.

Chronic exertional compartment syndromes (CCS) are more common than the acute variety. CCS is defined as a condition in which exercise-related pain in-

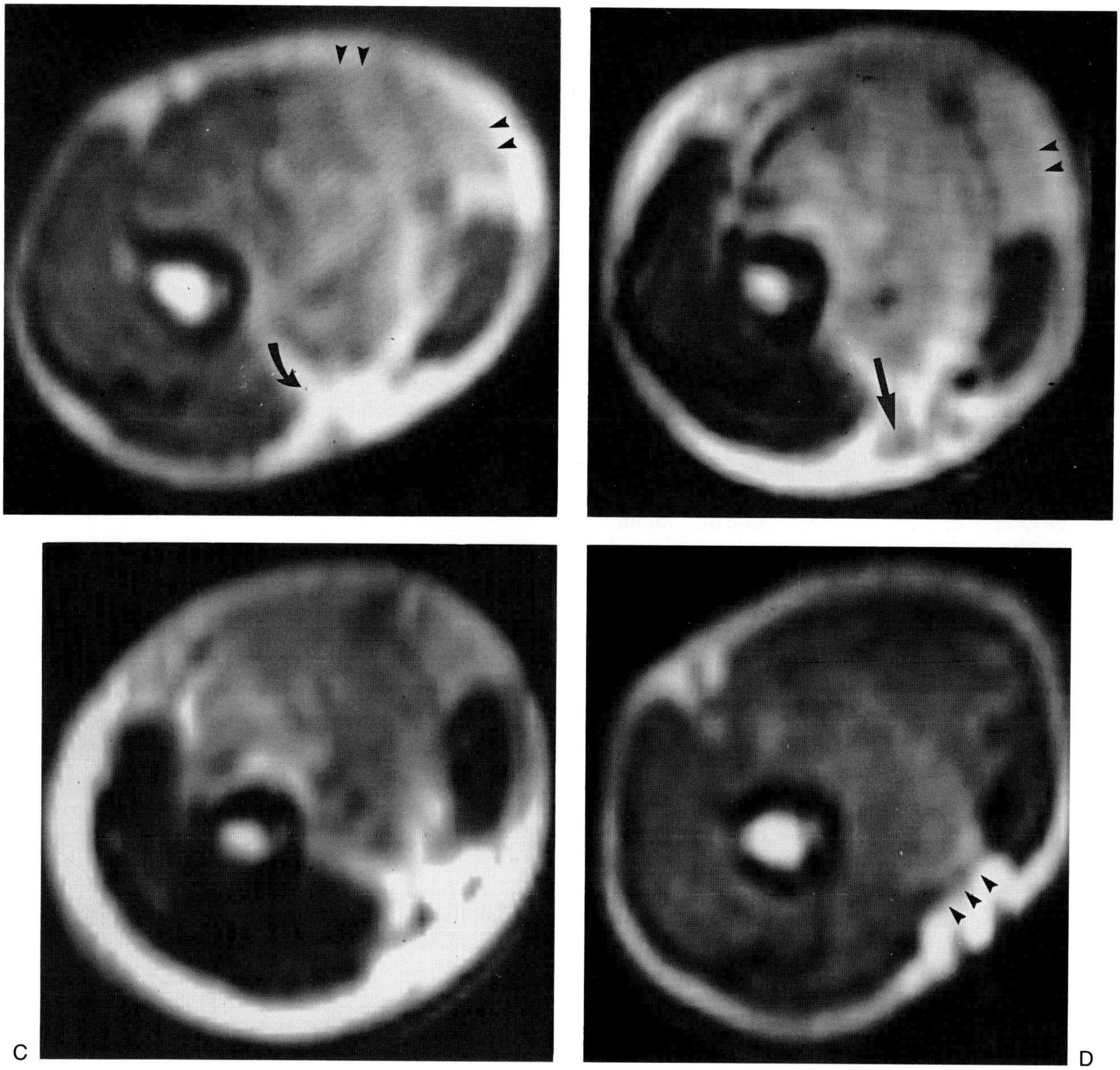

A B C D

FIG. 10.16. DOMS. **A–D:** Axial MR, TR/TE 2000/80. This patient exercised to exhaustion by performing eccentric muscle contractions. Images of the biceps muscle were obtained on day 3 (A), 5 (B), 10 (C), and 25 (D) following exercise. On day 3, all of the muscles have a slight increase in signal intensity, with the biceps (*arrowheads*) revealing the most evident abnormality. By day 5, the signal alteration is confined to the symptomatic injured biceps muscle. There is a small fluid collection present (*curved arrow*) that is not optimally seen because of the windowing necessary to demonstrate the muscle abnormality best. Although the patient was asymptomatic by day 7, the abnormality can be identified on studies carried out on day 25. Several streaky low signal abnormalities (*large arrow*) are present in the subcutaneous tissue that are thought to represent myoglobin.

terferes with the performance of recreational athletics (7,36). It is possible that either abnormal blood flow or buildup of noxious metabolites is responsible for symptoms. Patients present with deep pain over the anterior aspect of the leg which occurs after prolonged exercise; the pain is severe enough to limit the patient's activity. In CCS, the postexercise pressures remain abnormally elevated. At 1 min postexercise, intracompartmental pressures are 30 mm Hg and at 5 min postexercise, the value is at least 20 mm Hg (7,81). Treatment includes fasciotomy if activity alteration fails to relieve symptoms (79,82).

The MR findings in patients with acute compartment syndrome include swelling of the affected limb, re-

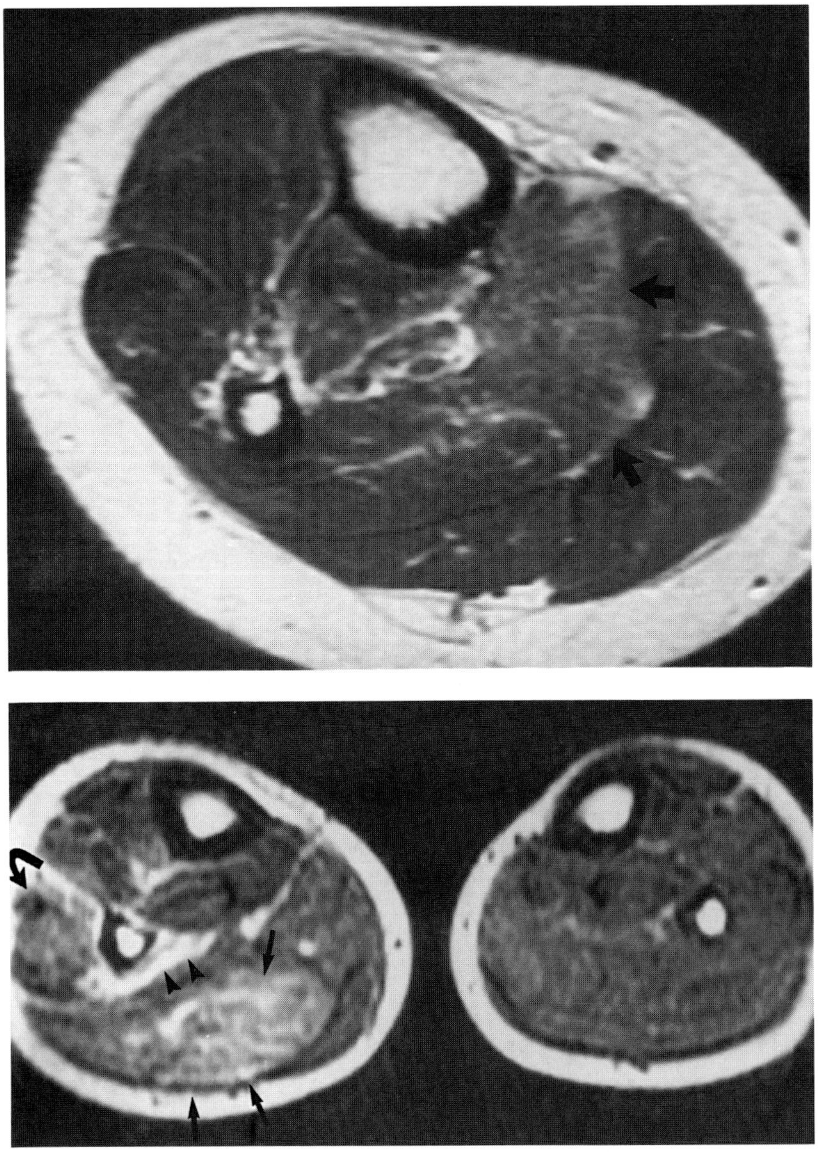

A

B

FIG. 10.17. Acute compartment syndrome. Both patients suffered injuries to the leg, and presented with severe leg pain and intact pulses. **A:** Axial MR, TR/TE 2000/80. There is a slight increase in signal in the deep posterior compartment. **B:** Axial MR TR/TE 2000/80. In this patient, there is increase in signal in the deep posterior compartment (*arrows*) and a slight increase in signal in the lateral compartment (*curved arrow*). A small perifascial fluid collection (*arrowheads*) is present. Involvement of multiple compartments often goes undiagnosed by interventional techniques.

sulting in increased girth, and an increase in signal intensity of muscle on contrast-oriented examinations (T2 or STIR sequences) (Fig. 10.17). Interestingly, the signal alteration may not always be limited to the one fascial compartment in which the pressures are elevated (35,37). In patients with symptoms suggestive of chronic compartment syndrome, transient increases in signal may be found within the muscle of one compartment postexercise. These foci often rapidly disappear concurrent with the resolution of symptoms. The signal alterations that are imaged in patients with CCS can be quantified by measurement of the T1 and T2 relaxation times. Both T1 and T2 values achieve a peak immediately postexercise, and regression analysis of percent change in relaxation times versus time indicates that there is a relationship with recovery (36). The curves of T1 and T2 values versus time are similar to those curves generated by direct intracompartmental pressure measurements. It has been suggested that qualitative assessment of blood flow to injured muscle in patients with compartment syndrome may be evaluated using gadolinium (39).

Potential advantages of MRI over pressure measurements include the noninvasive nature of MRI and its ability to examine multiple compartments simultaneously. MRI has shown abnormalities in more than one compartment when the clinical situation suggested only a single compartment was affected; in those cases, surgical decompression was in fact necessary in more than one compartment (36).

MISCELLANEOUS MUSCLE DISORDERS

Atrophy

Muscle atrophy may result from either a lack of use (disuse atrophy) or from denervation (denervation atrophy) (41,43). Histologically, replacement of normal muscle by fat cells is identified (43). Following immobilization, muscle strength rapidly declines. This decrease is due to both a diminution in muscle size and a decrease in tension per unit of muscle cross-sectional area. The loss of mass is greatest soon after immobilization; atrophy begins after a lag of 1 to 3 days and is greatest during the next 5 days (83). After as little as 1 to 2 months, muscles may atrophy to one-half of their normal size, but disuse atrophy can be reversed if periods of immobilization are short. However, the longer the state of disuse atrophy is allowed to exist, the longer it takes to reverse the process. In fact, after 4 months, muscle atrophy from disuse may no longer be reversible.

Disuse atrophy with fatty replacement results in a diffuse marbling of increased signal intensity within the striated muscle (T1WI) (17) (Fig. 10.18). Most commonly there is a loss of muscle bulk. Muscle denervation results in detectable increases of both T1 and T2 values (41,43,84). The decrease in the size of the muscle fibers occurs concurrently with a relative shrinking of the myoplasm and a compensatory increase in the size of the extracellular fluid space; both of these events lead to an increase in the T2 values of the denervated tissues.

MRI has been used to assess and to characterize the nature and extent of benign and malignant soft tissue tumors (12,18–28,38). MRI is generally unable to distinguish benign from malignant lesions reliably, either by measurement of relaxation times or by imaging characteristics (17). While the overwhelming majority of neoplasms have decreased signal on T1WI, hemangiomas are one notable exception (24,85) (Fig. 10.19). Vascular malformations are among the most common congenital abnormalities occurring in man. They are considered hamartomas rather than true neoplasms because of the frequent finding of vascular, adipose, and fibrous tissue, smooth muscle, and rarely bone within the lesion (28). While a variety of names have been suggested for these lesions (capillary hemangioma, cavernous hemangioma, venous malformation, angiolipoma, etc.), hemangioma is the single best term for these benign growths characterized by an increased number of vascular channels and overgrowth of the endothelial lining cells. Skeletal muscle hemangiomas occur deep within striated muscle. Benign intramuscular hemangiomas have a fatty component within the tumor vascular matrix, a finding that may contribute to the high signal on T1WI. It is this property that may suggest the diagnosis of this benign lesion, and perhaps spare the patient biopsy; however, hemorrhage within a malignant lesion may also result in high signal on T1WI. Additionally, it is common to identify focal muscle atrophy adjacent to a hemangioma (24). This common tumor has a serpiginous configuration with high signal intensity on long TR/TE sequences due to the presence of vascular-lined channels, a finding that is not found in soft tissue sarcomas.

High signal on T2WI may be found in otherwise normal skeletal muscle adjacent to malignant tumors of either bone or soft tissue (17,25). If the obvious tumor mass is well demarcated from the surrounding muscle, the increased signal intensity is most likely secondary to edema. If the distinction between muscle and tumor is not clear, abnormal signal in the adjacent muscle probably reflects local tumor infiltration. Extensive

A

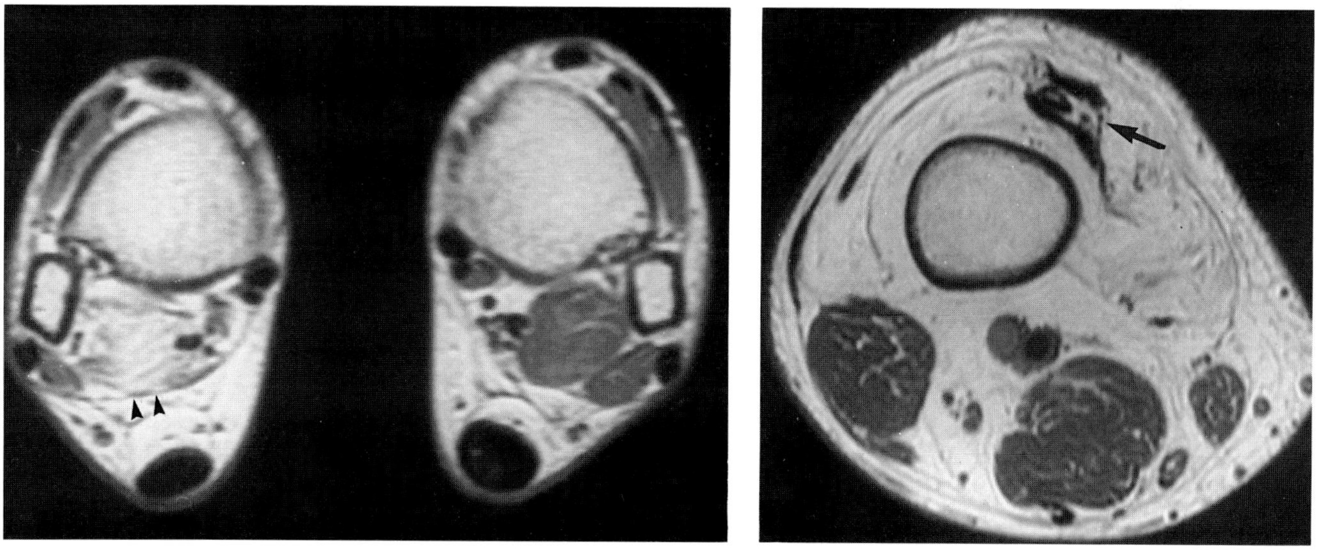

B

FIG. 10.18. Muscle atrophy. **A:** Atrophy of FHL. Axial MR, TR/TE 500/20. The FHL muscle (*arrowheads*) is diffusely infiltrated with fat. The extensor digitorum and extensor hallucis muscles and the peroneus brevis are also somewhat atrophic. Incidentally noted is bilateral Achilles tendonitis. **B:** Severe quadriceps atrophy. Axial MR, TR/TE 300/20. The quadriceps muscles are totally atrophic and blend nearly imperceptibly with the subcutaneous fat. The quadriceps tendon (*arrow*) is all that is visible in the extensor mechanism.

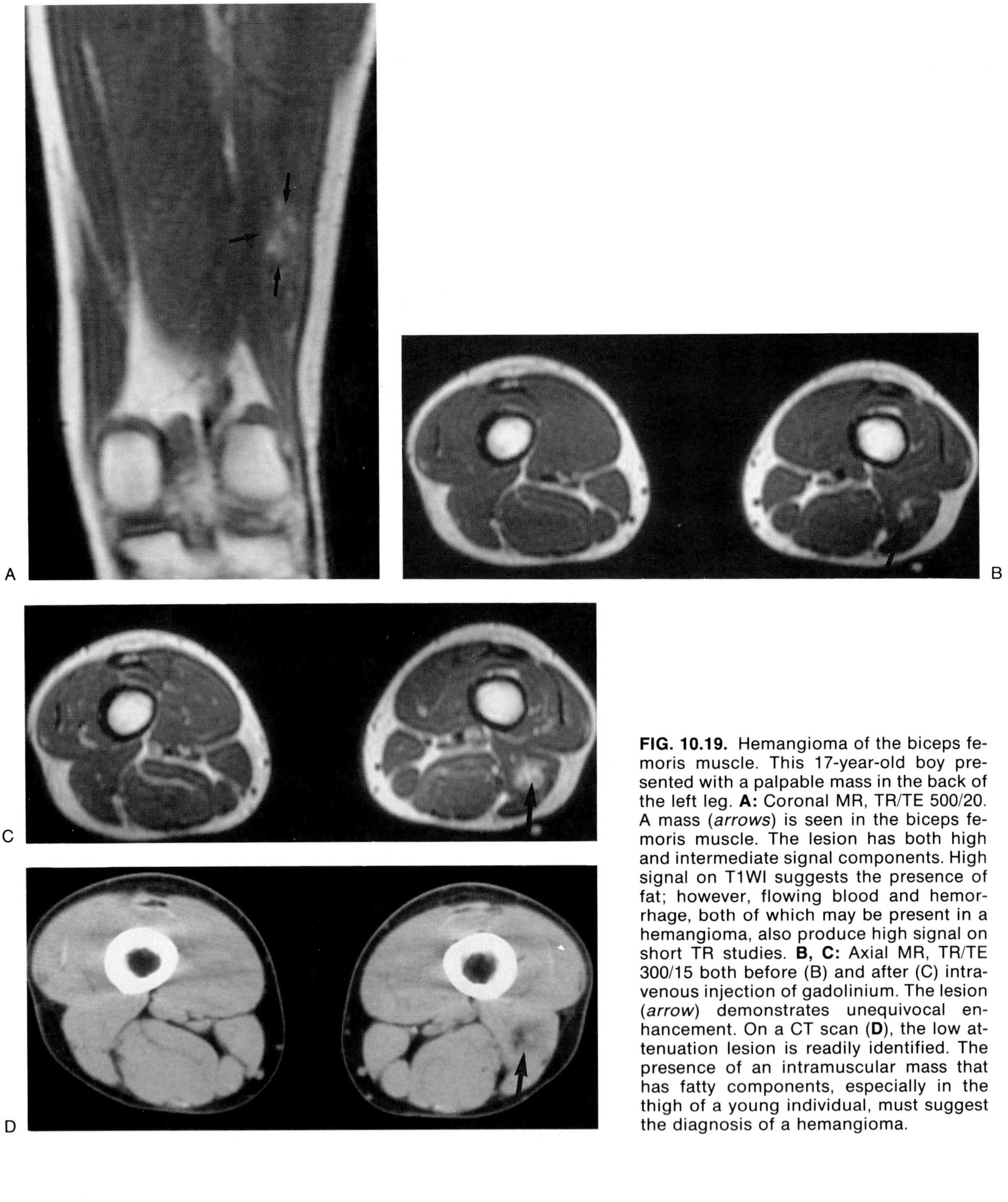

FIG. 10.19. Hemangioma of the biceps femoris muscle. This 17-year-old boy presented with a palpable mass in the back of the left leg. **A:** Coronal MR, TR/TE 500/20. A mass (*arrows*) is seen in the biceps femoris muscle. The lesion has both high and intermediate signal components. High signal on T1WI suggests the presence of fat; however, flowing blood and hemorrhage, both of which may be present in a hemangioma, also produce high signal on short TR studies. **B, C:** Axial MR, TR/TE 300/15 both before (B) and after (C) intravenous injection of gadolinium. The lesion (*arrow*) demonstrates unequivocal enhancement. On a CT scan (**D**), the low attenuation lesion is readily identified. The presence of an intramuscular mass that has fatty components, especially in the thigh of a young individual, must suggest the diagnosis of a hemangioma.

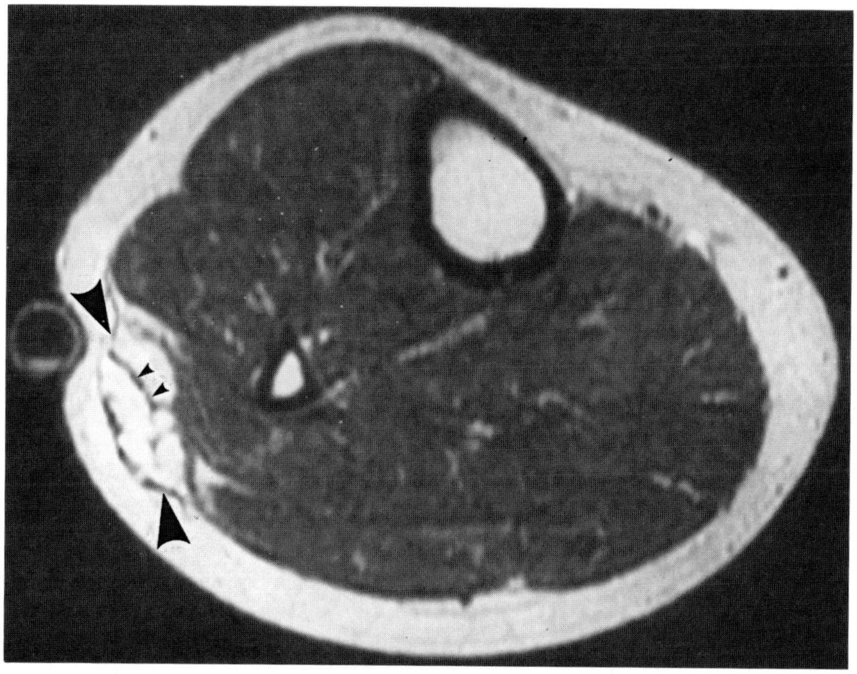

E

FIG. 10.19. (*Continued*) **E:** Axial MR, TR/TE 2000/80. In another patient with a biopsy-proven hemangioma, the MR demonstrates the classic serpiginous character of the lesion (*arrowheads*); multiple fine septae (*tiny arrowheads*) are typical.

muscle edema associated with malignant musculoskeletal tumors is frequently an ominous finding (38).

Bacterial pyomyositis is an uncommon pyogenic disorder that is usually caused by *S. aureus* (45). While patients infected with the human immunodeficiency virus (HIV) suffer from this disease, myositis occurs most frequently in normal men (86). The muscle of the thighs, calves, and buttocks are most commonly affected; multifocal disease is frequent. Low grade fever, weakness, and leukocytosis may be the only positive findings, but the affected skeletal muscle may be rapidly destroyed and replaced by pus. Prompt diagnosis is critical as bacterial myositis is one of the more readily treatable complications of HIV. Nuclear medicine techniques, sonography, and CT scanning have all been used to diagnose pyomyositis, but MRI appears to be more sensitive and specific than these other modalities (Fig. 10.20). T1WI images of some patients with bacterial myositis demonstrate a rim of high signal intensity separating a central region that is isointense with the surrounding normal muscle (33,45). This finding is peculiar since it has not been recognized in abscesses in other locations. Potential etiologies for this high signal region include methemoglobin production

A

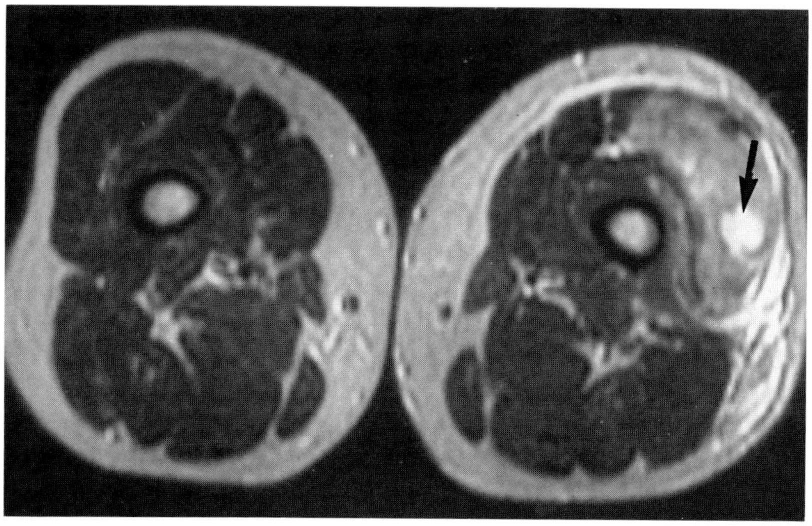

FIG. 10.20. Bacterial pyomyositis. **A:** Axial MR, TR/TE 2500/80. This immunologically normal woman developed subacute pain in the left thigh without a known injury. The MR demonstrates a diffuse increase in signal throughout the vastus lateralis with a focal area of fluid within the muscle (*arrow*). At surgery, an *S. aureus* abscess was drained.

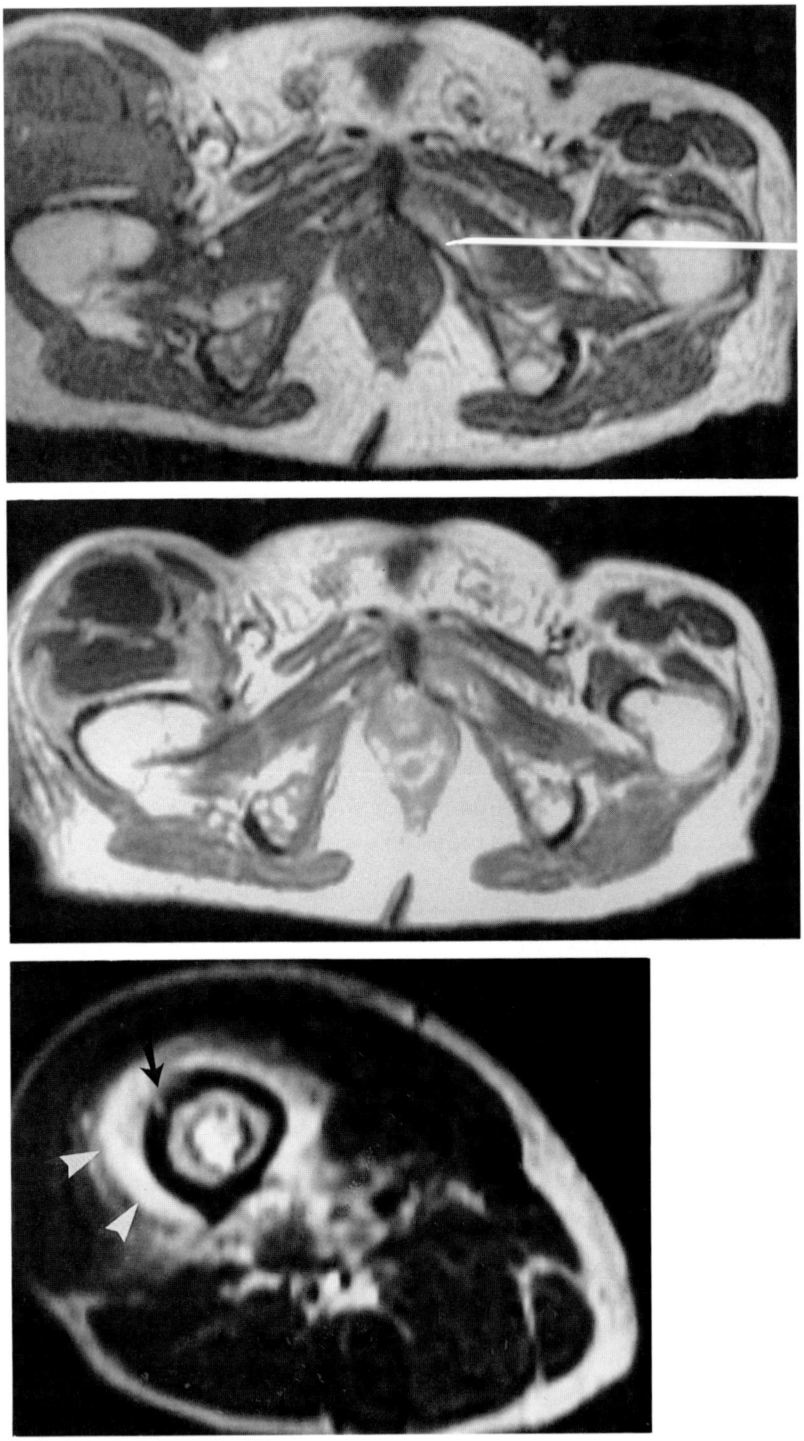

B

C

D

FIG. 10.20. (*Continued*) **B, C:** Pyomyositis in a patient with HIV. Axial MR, TR/TE 300/15 made before (B) and after (C) the intravenous administration of gadolinium. A large soft tissue mass is present anterior to the right hip. Following intravenous gadolinium injection, the rim is seen to markedly enhance (*arrows*). An *S. aureus* abscess was drained. **D:** Osteomyelitis. Axial MR, TR/TE 2200/80. In another patient, biopsy of the femur revealed an *S. aureus* infection in the medullary canal. The infection had spread to the adjacent musculature that was also (but secondarily) infected. The circumferential ring of edema/pus, the cortical violation, and the medullary abscess distinguish this osteomyelitis from a primary pyomyositis.

as a result of subacute bleeding, or bacterial or macrophage sequestration of iron. In some patients with bacterial myositis, the abscess does not have this high intensity rim, although the entire mass has a slightly increased signal intensity when compared with normal skeletal muscle. On T2WI and STIR, all patients with bacterial myositis show large, well-defined central regions of markedly increased signal, surrounded by a broad zone of diffuse muscle edema. MRI is extremely valuable in helping to assess those patients presenting with fever, muscle pain, and a swollen limb. Differential considerations would include osteomyelitis, thrombophlebitis, septic arthritis, cellulitis, and pyomyositis. Osteomyelitis is distinguished by the significant and early involvement of the bone marrow. Thrombophlebitis shows marked signal alterations in

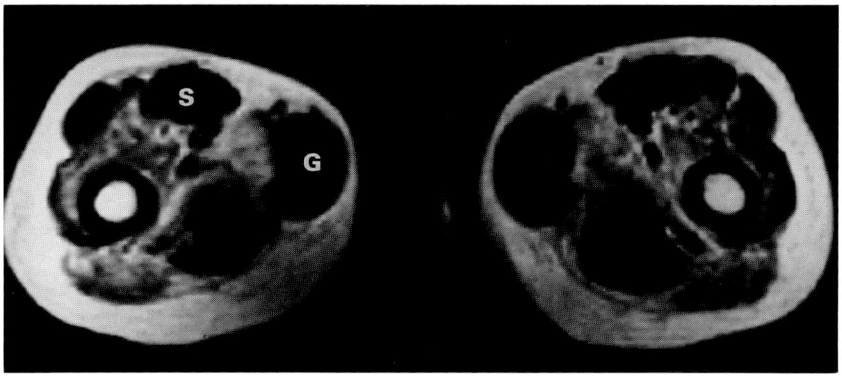

FIG. 10.21. Muscle atrophy and hypertrophy in Duchenne's muscular dystrophy. Axial MR, TR/TE 500/30. Symmetric proximal diminution in muscle volume is evident. The gracilis (*G*) and sartorius (*S*) muscles are spared. (Case courtesy of J. Fleckenstein, M.D., Dallas, TX) (From ref. 39, with permission.)

the intermuscular fascia caused by obstructive edema with relative sparing of the muscles. Cellulitis should have rather little effect on the muscles themselves.

Rapid fatigueability, myalgia, and exercise-induced muscle contracture are characteristic of myophosphorylase deficiency (MPD) and phosphofructokinase deficiency (PFKD). Studies have shown a high prevalence of focal muscle abnormalities in patients with glycolytic myopathies and the ability of MRI to detect them (87). The coexistence of muscle edema, atrophy, and fatty infiltration should suggest a myopathic or neurogenic disorder (29). MRI has been successful in assessing the changes in muscles found in patients with Duchenne's muscular dystrophy, fascioscapulohumeral muscular dystrophy, and limb girdle muscular dystrophy (88) (Fig. 10.21). In the former, a pattern of pseudohypertrophy may be present, but all patients have decreased thigh muscle mass. Interestingly, the sartorius and gracilis muscles are relatively spared. In patients with limb girdle dystrophy, all have fatty replacement of the thigh. Disease duration does not correlate with severity as depicted by MRI (29).

Polymyositis, an idiopathic inflammatory myopathy, is a noninfectious disorder of muscle that may be

A

B

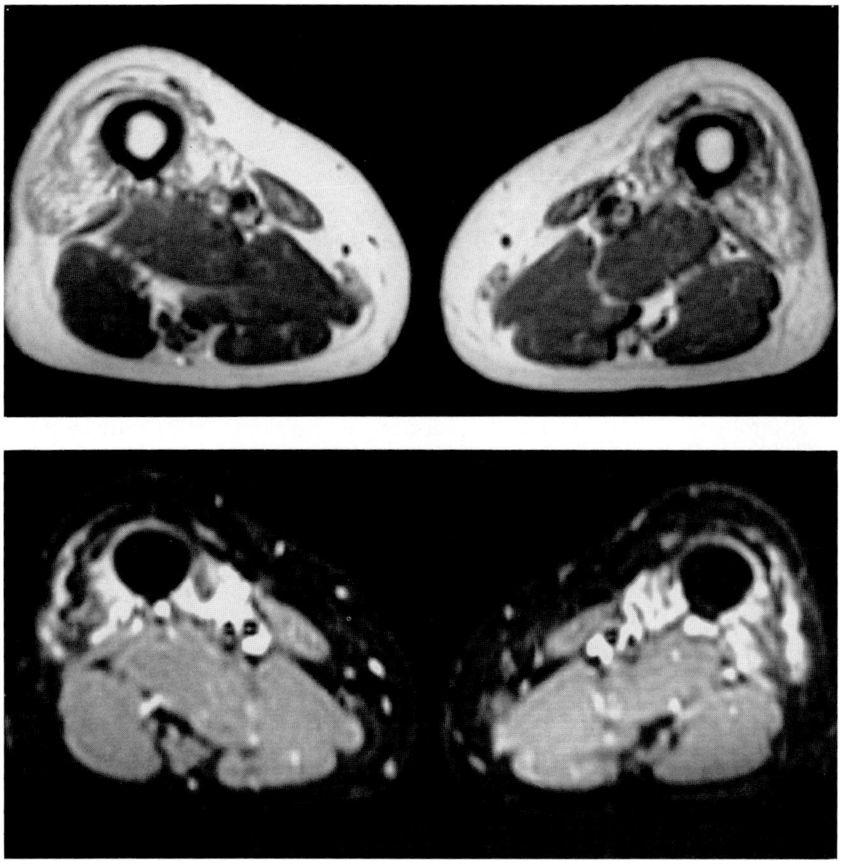

FIG. 10.22. Inclusion body myositis. **A:** Axial MR, TR/TE 700/20. **B:** Axial MR, TR/TE/TI 2200/35/160. The T1WI reveals the marked degree of atrophy of the quadriceps muscles that has occurred bilaterally; the STIR sequence most graphically depicts areas of active muscle inflammation that are intermingled with the atrophic zones. The anterior compartment predominance of the disease should suggest the specific diagnosis of inclusion body myositis.

associated with collagen vascular disease (89). Profound muscular weakness that is out of proportion to the degree of wasting is common; symmetric weakness of the limb girdle musculature and neck flexors, muscle tenderness, and ultimately atrophy characterize the disease.

Patients with chronic polymyositis have proximal and symmetric atrophy but highly focal involvement has been reported. The sartorius and gracilis are frequently spared. Although compensatory hypertrophy is characteristically found in patients with neurogenic muscle disorders, the same is not true for polymyositis, a fact that may be specific for this disease (39). Biopsy reveals necrosis of skeletal muscle fibers. MRI detects the foci of muscle inflammation as mottled or infiltrative increased signal intensity of T2WI (Fig. 10.22). The abnormalities are scattered and noncontiguous, a finding in direct contradistinction to the findings in neoplasms. MRI may be useful in following response to steroid administration (41).

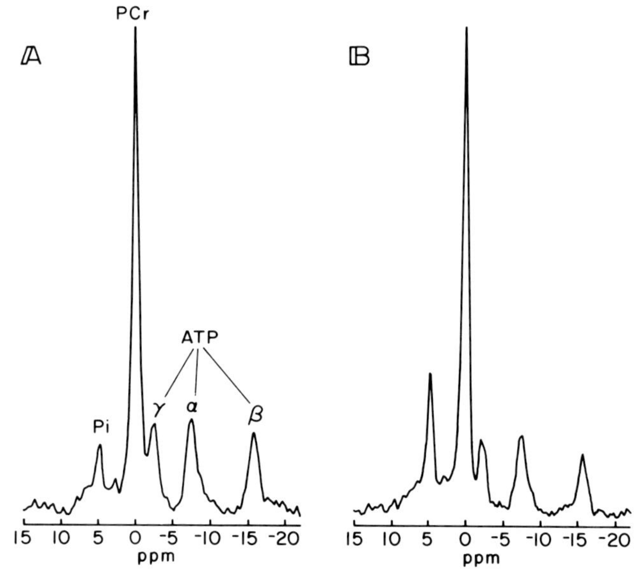

FIG. 10.23. 31-P MRS spectra from wrist flexor muscle of one patient before (**A**) and 24 hr after (**B**) 60 lengthening contractions. Each spectrum consists of 60 scans collected over 5 min. The five major peaks are (from *left* to *right*) inorganic phosphate (*Pi*), phosphocreatine (*PCr*), and the three peaks from ATP (*gamma, alpha,* and *beta*). For comparative purposes, the height of each spectrum was normalized to the same value of PCr. Compared with PCr, Pi is elevated and ATP is reduced following the eccentric action. (Case courtesy of K. McCully, Ph.D., Philadelphia, PA.) (From ref. 124, with permission.)

MUSCLE SPECTROSCOPY

Magnetic resonance spectroscopy (MRS) is an important tool for the elucidation of the structure and function of macromolecules. More recently, attempts have been made to apply the same principles to biological systems.

An MRS spectrum is a representation of the intensity of energy released in the resonance process versus applied frequency (Fig. 10.23). The spectrum is usually depicted as a graph of resonant energy on the ordinate and frequency of resonance on the abscissa. Peaks on this graph represent frequencies where many nuclei within the sample resonate; troughs represent frequencies with a paucity of resonating nuclei. By observing the spectroscopic graph, detailed information about the structure of molecules within the sample can be determined.

In the absence of an applied magnetic field gradient, the hydrogen nuclei should resonate at a frequency specified by the strength of the main magnetic field, Bo, times the gyromagnetic ratio, g. However, in the presence of large molecules with many surrounding electrons, the precise magnetic field inside of the molecule's electron cloud is not identical to the main applied magnetic field. The electrons "screen" a portion of the magnetic field, so that the field in the midst of the electron cloud is lower than Bo. Hydrogen nuclei imbedded in a molecule's electron cloud will then resonate at a lower frequency than nuclei in an isolated small molecule, such as free water. This shift to lower frequencies is a sensitive and often specific indicator of the chemical environment surrounding the nucleus and can reveal a wealth of chemical information. Today, most investigations into the physiology of muscle by means of MRS is being done by examination of phosphorus in its high energy forms, adenosine mono- (AMP), di- (ADP) and tri- (ATP) phosphate.

MRS is a noninvasive technique that can obtain detailed chemical information while cells are functioning. MRS has become preeminently important in the evaluation of muscle chemistry and metabolism (88,90–96). MRS's ability to determine chemical concentrations within living tissues has been instrumental in evaluating a variety of important chemical parameters. These include electrolyte shifts (97–100), intracellular pH (101,102), body composition (103,104), energy metabolism (88,90–96), and inborn errors of metabolism (105–107).

Energy metabolism is critically important to the proper function of muscle. MRS allows *in vivo* monitoring of relative concentrations of high energy phosphate metabolites, such as phosphocreatine, ATP, ADP, and AMP. Quantitative analysis of 31-P spectra follows the bioenergetics of exercise response by using the fact that the concentration of phosphocreatine is relatively high in resting muscle but decreases rapidly with exercise (39). These metabolic changes are accompanied by a complementary increase in inorganic phosphate and a reduction in pH. Within minutes, all of these parameters return to their resting values. Changes in these and other high energy metabolites have proved helpful in evaluating normal muscle function (108), athletic conditioning (96,109–122), normal response to exercise (123), and muscle response to injury (124–126). The mechanisms of fatigue are complex and in addition to examination of high energy metabolites, changes in muscle response to electrical events may be important (123). Muscle metabolism is regulated by a variety of factors. The effects of hormonal influences are important and can be monitored with MRS (127). The detrimental effects of sepsis on muscle function can be quantitated and hopefully better understood with MRS (128). Nutritional effects can also be detected with spectroscopic methods (129).

Although MRS has proved useful in the evaluation of a broad spectrum of muscle function, chemistry, and disease, perhaps its greatest potential use is in the work-up of specific metabolic muscle diseases (130–132). Significant metabolic alterations have been detected in diabetes (133,134), fibromyalgia (135), and malignant hyperthermia (136,137). Molecular changes from denervation from multiple causes including trauma, congenital (21), amyotrophic lateral sclerosis (88,94), and poliomyelitis (94) have characteristic MRS signatures. In-born errors of muscle metabolism and other specific biomolecular abnormalities must be clearly differentiated and defined before specific therapy can be developed. Unfortunately, many of these conditions overlap in clinical presentations. MRS's specific ability to pinpoint abnormal biochemistry will be important to further advances in this field. MRS has been used in the evaluation of many of these diseases including paramyotonia congenita (105), mitochondrial cytopathy (106), McArdle's syndrome (107), Becker's dystrophy (138), rickets (139), PFKD (140,141), phosphoglycerate mutase deficiency (142), Duchenne's muscular dystrophy (143), limb-girdle muscular dystrophy (88), fascioscapulohumeral dystrophy (88), Kearn-Sayre mitochondrial muscle disease (88), mitochondrial myopathies (144), myotonic dystrophy (94,103), and Werdnig-Hoffmann disease (94,103). This list will undoubtedly expand as activity in this field increases.

Muscle ischemia is a common clinical problem, especially in patients with peripheral vascular disease. There is currently tremendous interest in the use of MRS as a clinical tool in assessing this disorder; abnormal levels of phosphocreatine and inorganic phosphate have been associated with regions of ischemia (145–152). MRS may prove to be helpful in determining the significance of angiographically proven vascular lesions and may be helpful in determining location, timing, and extent of vascular interventions. The clinical response to vascular intervention may also be best determined by quantitating the metabolic response of ischemic muscle to attempted reperfusion. Compartment syndromes may involve similar pathophysiology and MRS has been used for evaluating this condition (153).

A recent area of research interest in MRS is the evaluation of inflammatory myopathies (30). The use of MRS in clinical decision making in this condition requires further evaluation (31,154).

MRI has revolutionized the detection and staging of soft tissue and muscle neoplasia (15,22,23,26,38, 155,156). MRI is highly sensitive in the detection of soft tissue and muscle neoplasia, but its specificity is usually not adequate for treatment planning. It is hoped that the ability of MRS for biochemical analysis will lead to accurate, noninvasive diagnostic criteria for clinical planning. MRI and MRS already show great promise in following neoplasm response to therapy (157,158).

Diagnosis and treatment planning would benefit by having the ability to monitor changes in metabolic parameters using a noninvasive method. The combination of MRI and multivoxel localized spectroscopy can be used to obtain images of the variations in distribution of both P-31 and proton metabolites. In this technique, spatial localization is provided by the same phase-encoding gradients as in MRI, thereby allowing direct comparison with the anatomic images. Because of the low signal to noise ratio of 31-P MRS, spatial resolution is limited. From a clinical point of view, it is essential that accurate methods of quantifying spectral data are available in a form readily interpreted by the clinician. The success of this technique remains to be documented.

Much of the initial enthusiasm for MRI in medicine in the 1970s and early 1980s was fueled by the prospects of using the rich biochemical information generated by MRS for medical diagnosis. MRS was heralded as the "noninvasive biopsy." While enthusiasm for MRS has waned in comparison with the revolution in medical diagnosis provided by MRI, the fundamental information obtainable by MRS continues to be profound. As experience is gained in this complicated field, it is likely that MRS will eventually play a more significant role in medicine than imaging.

REFERENCES

1. Garrett WE, Safran MR, Seaber AV, et al. Biomechanical comparison of stimulated and nonstimulated skeletal muscle pulled to failure. *Am J Sports Med* 1987;15(5):448–454.
2. O'Donoghue DH. *Treatment of injuries to athletes*, 4th ed. Philadelphia: WB Saunders; 1984.
3. Zarins B, Ciullo JV. Acute muscle and tendon injuries in athletes. *Clin Sports Med* 1983;2(1):167–182.
4. Garrett WEJ. Basic science of musculotendinous injuries, IN: *The lower extremity and spine in sports medicine*. JA Nicholas, EB Hershman, ed. St Louis. CV Mosby: 1986;42–58.

5. Garrett WE, Mumma M, and Lucaveche C. Ultrastructural differences in human skeletal muscle fiber types. *Orthop Clin* 1983;14(2):413–425.
6. Sutro C, Sutro W. The medial head of the gastrocnemius. A review of the basis for partial rupture and for intermittent claudication. *Bull Hosp Jt Dis Orthop Inst* 1985;45(2):150–157.
7. Matin P. Basic principles of nuclear medicine techniques for detection and evaluation of trauma and sports medicine injuries. *Semin Nucl Med* 1988;18(2):90–112.
8. Valk P. Muscle localization of Tc-99m MDP after exertion. *Clin Nucl Med* 1984;9(9):493–494.
9. Matin P, Lang G, Carretta R, et al. Scintigraphic evaluation of muscle damage following extreme exercise: Concise communication. *J Nucl Med* 1983;24(4):308–311.
10. Shirkhoda A, Mauro MA, Staab EV, et al. Soft-tissue hemorrhage in hemophiliac patients. Computed tomography and ultrasound study. *Radiology* 1983;147:811–814.
11. Termote J, Baert A, Crolla D, et al. Computed tomography of the normal and pathologic muscular system. *Radiology* 1980; 137:439–444.
12. Garrett WE, Rich FR, Nikolaou PK, et al. Computed tomography of hamstring muscle strains. *Med Sci Sports Exerc* 1989; 21(5):506–514.
13. Vukanovic S, Wettstein P. CT localization of myonecrosis for surgical decompression. *AJR* 1980;135:1298–1299.
14. Lamminen AE, Hekali PE, Tiula E, et al. Acute rhabdomyolysis: Evaluation with magnetic resonance imaging compared with computed tomography and ultrasonography. *Radiology (Br)* 1989;62:326–331.
15. Aisen AM, Martel W, Braunstein EM, et al. MRI and CT evaluation of primary bone and soft tissue tumors. *AJR* 1986; 146:749–756.
16. Dooms GC, Fisher MF, Hricak H, et al. MR imaging of intramuscular hemorrhage. *J Comput Assist Tomogr* 1985;9:908–913.
17. Fisher MR, Dooms GC, Hricak H, et al. Magnetic resonance imaging of the normal and pathologic muscular system. *Magn Reson Imaging* 1986;4:491–496.
18. Fleckenstein JL, Canby RC, Parkey RW, et al. Acute effects of exercise on MR imaging of skeletal muscle in normal volunteers. *AJR* 1988;151:231–237.
19. Newham DJ, McPhail G, Mills KR, et al. Ultrastructural changes after concentric and eccentric contractions of human muscle. *J Neurol Sci* 1983;61:109–122.
20. Milgram JE. Muscle ruptures and avulsions with particular reference to the lower extermities. A.A.O.S. Instructional Course Lectures 1953;10:233–243.
21. Zochodne DW, Thompson RT, Driedger AA, et al. Metabolic changes in human muscle denervation: Topical 31P NMR spectroscopy studies. *Magn Reson Med* 1988;7:373–383.
22. Frank JA, Ling A, Patronas NJ, et al. Detection of malignant bone tumors: MR imaging versus scintigraphy. *AJR* 1990; 155:1043–1048.
23. Wetzel LH and Levine E. Soft-tissue tumors of the foot: Value of MR imaging for specific diagnosis. *AJR* 1990;155:1025–1030.
24. Yuh WTC, Kathol MH, Sein MA, et al. Hemangiomas of skeletal muscle: MR findings in five patients. *AJR* 1987;149:765–768.
25. Beltran J, Simon DC, Katz W, et al. Increased MR signal intensity in skeletal muscle adjacent to malignant tumors: Pathologic correlation and clinical relevance. *Radiology* 1987; 162:251–255.
26. Sundaram M, McLeod RA. Review article. MR imaging of tumor and tumorlike lesions of bone and soft tissue. *AJR* 1990; 155:817–824.
27. Hanna SL, Fletcher BD, Parham DM, et al. Muscle edema associated with musculoskeletal tumors: MR imaging findings and clinical significance. *J Magn Reson Imaging* 1991;1(2):194.
28. Kaplan PA and Williams SM. Mucocutaneous and peripheral soft-tissue hemangiomas: MR imaging. *Radiology* 1987; 163:163–166.
29. Fleckenstein JL and Shellock FG. Exertional muscle injuries: MRI evaluation. *Topics Magn Reson Imaging* 1991;3:50–70.
30. Park JH, Vansant JP, Kumar NG, et al. Dermatomyositis: Correlative MR imaging and P-31 MR spectroscopy for quantitative characterization of inflammatory disease. *Radiology* 1990; 177(2):473–479.
31. Fraser DD, Frank JA, and Dalakas MC. Inflammatory myopathies: MR imaging and spectroscopy. *Radiology* 1991; 179(2):341–342.
32. Giddings CJ, Nurenberg P, Stray-Gundersen J, et al. Muscle injury assessed with magnetic resonance imaging and electron microscopy following downhill running in humans. *Med Sci Sports Exerc* 1991;23 Suppl(4).
33. Yuh WTC, Schreiber AE, Montgomery WJ, et al. Magnetic resonance imaging of pyomyositis. Skeletal *Radiol* 1988; 17:190–193.
34. Greco A, McNamara M, Escher R, et al. Spin-echo and STIR MR imaging of sports-related muscle injuries at 1.5T. *Radiology* 1990;177 (P):306–312.
35. Zagoria RJ, Karstaedt N, Koubek TD. MR imaging of rhabdomyolysis. *J Comput Assist Tomogr* 1986;10:268–270.
36. Amendola A, Rorabeck CH, Vellett D, et al. The use of magnetic resonance imaging in exertional compartment syndromes. *Am J Sports Med* 1990;18(1):29–34.
37. Ehman RL, Berquist TH, Magnetic resonance imaging of musculoskeletal trauma. *Radiol Clin North Am* 1986;24(2):291–319.
38. Hanna S, Fletcher B, Parham D. Muscle edema in musculoskeletal tumors: MR imaging characteristics and clinical significance. *J Magn Reson Imaging* 1991;1:441–449.
39. Fleckenstein JL, Weatherall PT, Bertocci LA, et al. Locomotor system assessment by muscle magnetic resonance imaging. *Magn Reson Q* 1991;7:79–103.
40. Swensen SJ, Keller PL, Berquist TH, et al. Magnetic resonance imaging of hemorrhage. *AJR* 1985;145:921–927.
41. MRI of the musculoskeletal system: A teaching file. JH Mink, AL Deutsch, eds. Raven Press; New York: 1990.
42. Unger EC, Glazer HS, Lee JKT, et al. MRI of extracranial hematomas: Preliminary observations. *AJR* 1986;146:403–407.
43. Deutsch AL, Mink JH. Magnetic resonance imaging of musculoskeletal injuries. *Radiol Clin North Am* 1989;27(5):983–1002.
44. Sundaram M, McGuire MH, Herbold DR, et al. High signal intensity soft tissue masses on T1 weighted pulsing sequences. *Skeletal Radiol* 1987;16:30–36.
45. Fleckenstein JL, Burns DK, Murphy FK, et al. Differential diagnosis of bacterial myositis in AIDS: Evaluation with MR imaging. *Radiology* 1991;179:653–658.
46. Garrett WEJ, Injuries to the muscle-tendon unit. *Instr Course Lect* 1988;37:275–282.
47. Jackson DW and Feagin, JA. Quadriceps contusions in young adults. Relationship of severity of injury to treatment and prognosis. *J Bone Joint Surg (Am)* 1973;55:95–105.
48. Amendola M, Glazer GM, Agha Z, et al. Myositis ossificans circumscripta: Computed tomographic diagnosis. *Radiology* 1983;149:775–779.
49. Garrett WE. Muscle strain injuries: Clinical and basic aspects. *Med Sci Sports Exerc* 1990;22(4):436–443.
50. Shellock FG, Fukunaga T, Mink JH, et al. Acute effects of exercise on MR imaging of skeletal muscle: Concentric vs eccentric actions. *AJR* 1991;156:765–768.
51. Garrett WE, Califf JC, Bassett FH. Histochemical correlates of hamstring injuries. *Am J Sports Med* 1984;12(2):98–103.
52. Anzel SH, Covey KW, Weiner AD, et al. Disruption of muscles and tendons an analysis of 1,014 cases. *Surgery* 1959;45(3):406–414.
53. Baker BE. Current concepts in the diagnosis and treatment of musculotendinous injuries. *Med Sci Sports Exerc* 1984; 16(4):323–327.
54. Burkett LN. Causative factors in hamstring strains. *Med Sci Sports Exerc* 1970;2(1):39–42.
55. Arner O, Lindholm A. What is tennis leg? Acta Chir Scand 1958;116:73–75.
56. Miller WA. Rupture of the musculotendinous juncture of the medial head of the gastrocnemius muscle. *Am J Sports Med* 1977;5(5):191–193.
57. Miller AP. Strains of the posterior calf musculature ("tennis leg"). *Am J Sports Med* 1979;7(3):172–174.

58. Burkett LN. Investigation into hamstring strains: The case of the hybrid muscle. *J Sports Med* 1976;3(5):228–231.
59. Symeonides PP. Isolated traumatic rupture of the adductor longus muscle of the thigh. *Clin Orthop* 1973;88:64–66.
60. Peshock RM, Fleckenstein JL, Malloy CR, et al. Echoplanar imaging in the evaluation of skeletal muscle exercise. *J Magn Reson Imaging* 1991;1(2):149.
61. Fisher MJ, Meyer RA, Adams GR, et al. Direct relationship between proton T2 and exercise intensity in skeletal muscle MR images. *Invest Radiol* 1989;25:480–485.
62. Cohen MS, Shellock FG, Nadeau KA, et al. Acute muscle T2 changes during exercise. SMRM Book of Abstracts 1991;107.
63. Fisher BD, Baracos VE, Shnika TK, et al. Ultrastructural events following acute muscle trauma. *Med Sci Sports Exerc* 1990;22:185–193.
64. Armstrong RB. Initial events in exercise-induced muscular injury. *Med Sci Sports Exerc* 1990;22(4):429–435.
65. Newham DJ, Mills KR, Quigley BM, et al. Pain and fatigue after concentric and eccentric muscle contractions. *Clin Sci* 1983;64:55–62.
66. Shellock FG, Fukunaga T, Mink JH, et al. Exertional muscle injury: Evaluation of concentric versus eccentric actions with serial MR imaging. *Radiology* 1991;179:659–664.
67. Nikolaou PK, MacDonald BL, Glisson RR, et al. Biomechanical and histological evaluation of muscle after controlled strain injury. *Am J Sports Med* 1987;15(1):9–14.
68. Shellock FG, Fukunaga T, Day K, et al. Serial MRI and cybex testing evaluations of exertional muscle injury: Concentric vs eccentric actions. *Med Sci Sports Exerc* 1991;23:S110.
69. Wikorsson-Moller M, Oberg B, Ekstrand J, et al. Effects of warming up, massage, and stretching on range of motion and muscle strength in the lower extremity. *Am J Sports Med* 1983; 11(4):249–252.
70. Safran MR, Garrett WE, Seaber V, et al. The role of warmup in muscular injury prevention. *Am J Sports Med* 1988; 16(2):123–128.
71. Taylor DC, Dalton JC, Seaber AV, et al. Viscoelastic properties of muscle-tendon units. The biomechanical effects of stretching. *Am J Sports Med* 1990;18(3):300–309.
72. De Smet AA, Heiner JP, et al. Magnetic resonance imaging of muscle tears. *Skeletal Radiol* 1990;19:283–286.
73. Fleckenstein JL, Weatherall PT, Parkey RW, et al. Sports-related muscle injuries: Evaluation with MR imaging. *Radiology* 1989;172:793–798.
74. Armstrong RB. Mechanisms of exercise-induced delayed onset muscular soreness: A brief review. *Med Sci Sports Exerc* 1984; 16(6):529–538.
75. Tiidus PM, Ianuzzo CD. Effects of intensity and duration of muscular exercise on delayed soreness and serum enzyme activities. *Med Sci Sports Exerc* 1983;15:461–465.
76. Schwane JA, Johnson SR, Vandenakker CB, et al. Delayed-onset muscular soreness and plasma CPK and LDH activities after downhill running. *Med Sci Sports Exerc* 1983;15:51–56.
77. Zabetakis P. Muscle soreness and rhabdomyolysis, IN: *The lower extremity and spine in sports medicine*. JA Nicholas, EB Hershman, eds. CV Mosby: St Louis. 1986:59–71.
78. Anouchi YL, Parker RD, Seitz WH. Posterior compartment syndrome of the calf resulting from misdiagnosis of a rupture of the medial head of the gastrocnemius. *J Trauma* 1987; 27(6):678–680.
79. Matsen FA, Winquist RA, Krugmire RB, et al. Diagnosis and management of compartmental syndromes. *J Bone Joint Surg (Am)* 1980;62:286–291.
80. Arciero RA, Shishido NS, Parr TJ. Acute anterolateral compartment syndrome secondary to rupture of the peroneus longus muscle. *Am J Sports Med* 1984;12(5):366–367.
81. Pedowitz RA, Hargens AR, Mubarak SJ, et al. Modified criteria for the objective diagnosis of chronic compartment syndrome of the leg. *Am J Sports Med* 1990;18(1):35–40.
82. Rorabeck CH. The diagnosis and management of chronic compartment syndrome. *Instr Course Lect* 1989;38:466–472.
83. Booth FW. Physiologic and biomechanical effects of immobilization on muscle. *Clin Orthop* 1987;219:15–20.
84. Polak JF, Jolesz FA, Adams DF. Magnetic resonance imaging of the skeletal muscle prolongation of T1 and T2 subsequent to denervation. *Invest Radiol* 1988;23:365–369.
85. Cohen EK, Kressel HY, Perosio T, et al. MR imaging of soft-tissue hemangiomas: Correlation with pathologic findings. *AJR* 1988;150:1079–1081.
86. Scott JA, Palmer EL, Fischman AJ. HIV-associated myositis detected by radionuclide bone scanning. *J Nucl Med* 1989; 30(4):556–558.
87. Fleckenstein JL, Peshock RM, Lewis SF, et al. Magnetic resonance imaging of muscle injury and atrophy in glycolytic myopathies. *Muscle Nerve* 1989;12:849–855.
88. Murphy WA, Totty WG. Carroll JE. MRI of normal and pathologic skeletal muscle. *AJR* 1986;146:565–574.
89. Plotz PH, Dalakas M, Leff RL, et al. Current concepts in the idiopathic inflammatory myopathies: Polymyositis, dermatomyositis, and related disorders. *Ann Intern Med* 1989;111:143–157.
90. Achten E, Van CM, Willem R, et al. 31P-NMR spectroscopy and the metabolic properties of different muscle fibers. *J Appl Physiol* 1990;68(2);644–649.
91. McCully KK, Kent JA, Chance B. Muscle injury and exercise stress measured with 31-P magnetic resonance spectroscopy. *Prog Clin Biol Res* 1989;315(1):197–207.
92. Chance B, Younkin DP, Kelley R, et al. Magnetic resonance spectroscopy of normal and diseased muscles. *Am J Med Genet* 1986;25(4):659–679.
93. Vock P, Hoppeler H, Hartl W, et al. Combined use of magnetic resonance imaging (MRI) and spectroscopy (MRS) by whole body magnets in studying skeletal muscle morphology and metabolism. *Invest Radiol* 1985;20(5):486–491.
94. Barany M, Siegel IM, Venkatasubramanian PN, et al. Human leg neuromuscular diseases: P-31 MR spectroscopy. *Radiology* 1989;172(2):503–508.
95. Barany M, Langer GB, Glick RP, et al. In vivo H-1 spectroscopy in humans at 1.5 T. *Radiology* 1988;167(3):839–844.
96. Edwards RH. New techniques for studying human muscle function, metabolism, and fatigue. *Muscle Nerve* 1984;7(8):599–609.
97. Naritomi H, Kanashiro M, Sasaki M, et al. In vivo measurements of intra- and extracellular Na^+ and water in the brain and muscle by nuclear magnetic resonance spectroscopy with shift reagent. *Biophys J* 1987;52(4):611–616.
98. Syme PD, Dixon RM, Allis JL, et al. A non-invasive method of measuring concentrations of rubidium in rat skeletal muscle in vivo by 87Rb nuclear magnetic resonance spectroscopy: Implications for the measurement of cation transport activity in vivo. *Clin Sci*.1990;78(3):303–309.
99. Higgins RM, Richardson AJ, Endre ZH, et al. Hypophosphataemia after renal transplantation: Relationship to immunosuppressive drug therapy and effects on muscle detected by 31P nuclear magnetic resonance spectroscopy. *Nephrol Dial Transplant* 1990;5(1):62–68.
100. Le Rumeur E, Le Moyec L, Yvin JC, et al. Assessment of intracellular magnesium depletion in rat striated muscle by in vivo 31P NMR. *Magn Reson Med* 1990;13(3):504–506.
101. Wray S. Regulation of intracellular pH in rat uterine smooth muscle, studied by 31P NMR spectroscopy. *Biochim Biophys Acta* 1988;972(3):299–301.
102. Victor RG, Bertocci LA, Pryor SL, et al. Sympathetic nerve discharge is coupled to muscle cell pH during exercise in humans. *J Clin Invest* 1988;82(4):1301–1305.
103. Barany M, Venkatasubramanian PN, Mok E, et al. Quantitative and qualitative fat analysis in human leg muscle of neuromuscular diseases by 1H MR spectroscopy in vivo. *Magn Reson Med* 1989;10(2):210–226.
104. Leroy WA, Duboc D, Bittoun J, et al. Spectroscopic MRI: A tool for the evaluation of systemic lipid storage disease. *Magn Reson Imag* 1990;8(4):511–515.
105. Lehmann HF, Hopfel D, Rudel R, et al. In vivo P-NMR spectroscopy: Muscle energy exchange in paramyotonia patients. *Muscle Nerve* 1985;8(7):606–610.
106. Narayana PA, Slopis JM, Jackson EF, et al. In vivo muscle magnetic resonance spectroscopy in a family with mitochon-

drial cytopathy: A defect in fat metabolism. *Magn Reson Imaging* 1989;7(2):133–139.
107. Argov Z, Bank WJ, Maris J, et al. Muscle energy metabolism in McArdle's syndrome by in vivo phosphorus magnetic resonance spectroscopy. *Neurology* 1987;37(11):1720–1724.
108. Dawson MJ. The relation between muscle contraction and metabolism: Studies by 31P nuclear magnetic resonance spectroscopy. *Adv Exp Med Biol* 1988;226(1):433–448.
109. Fleckenstein JL, Bertocci LA, Nunnally RL, et al. Exercise-enhanced MR imaging of variations in forearm muscle anatomy and use: Importance in MR spectroscopy. *AJR* 1989; 153(4):693–698.
110. Shenton DJ, Heppenstall RB, Chance B, et al. Electrical stimulation of human muscle studied using 31P-nuclear magnetic resonance spectroscopy. *J Orthop Res* 1986;4(2):204–211.
111. Park JH, Brown RL, Park CR, et al. Energy metabolism of the untrained muscle of elite runners as observed by 31P magnetic resonance spectroscopy: Evidence suggesting a genetic endowment for endurance exercise. *Proc Natl Acad Sci USA* 1988;85(23):8780–8784.
112. McCully KK, Kent JA, Chance B. Application of 31P magnetic resonance spectroscopy to the study of athletic performance. *Sports Med* 1988;5(5):312–321.
113. McCully KK, Boden BP, Tuchler M, et al. Wrist flexor muscles of elite rowers measured with magnetic resonance spectroscopy. *J Appl Physiol* 1989;67(3):926–932.
114. Boicelli CA, Baldassarri AM, Borsetto C, et al. An approach to noninvasive fiber type determination by NMR. *Int J Sports Med* 1989;10(1):53–54.
115. Park JH, Brown RL, Park CR, et al. Functional pools of oxidative and glycolytic fibers in human muscle observed by 31P magnetic resonance spectroscopy during exercise. *Proc Natl Acad Sci USA* 1987;84(24):8976–8980.
116. Sapega AA, Sokolow DP, Graham TJ, et al. Phosphorus nuclear magnetic resonance: A non-invasive technique for the study of muscle bioenergetics during exercise. *Med Sci Sports Exerc* 1987;19(4):410–420.
117. Inch WR, Serebrin B, Taylor AW, et al. Exercise muscle metabolism measured by magnetic resonance spectroscopy. *Can J Appl Sport Sci* 1986;11(2):60–65.
118. Taylor DJ, Styles P, Matthews PM, et al. Energetics of human muscle: Exercise-induced ATP depletion. *Magn Reson Med* 1986;3(1):44–54.
119. Kuno S, Akisada M, Katsuta S, et al. Evaluation of exercise muscle energetics by NMR. *Ann Physiol Anthropol* 1990; 9(2):235–239.
120. Lopata M, Onal E, Aronson R, et al. Effects of inspiratory loading on 31 phosphorus magnetic resonance (PMR) spectroscopy of the inspiratory intercostal muscles in normal humans. *Chest* 1990;3(1):97.
121. Helpern JA, Kao W, Gross B, et al. Interleaved 31P NMR with transcutaneous nerve stimulation (TNS): A method of monitoring compliance-independent skeletal muscle metabolic response to exercise. *Magn Reson Med* 1989;10(1):50–56.
122. Clark BJ, Acker MA, McCully K, et al. In vivo 31P-NMR spectroscopy of chronically stimulated canine skeletal muscle. *Am J Physiol* 1988;254(2):258–266.
123. Le Rumeur E, Le Moyec L, Toulouse P, et al. Muscle fatigue unrelated to phosphocreatine and pH: An " in vivo" 31-P NMR spectroscopy study. *Muscle Nerve* 1990;13(5):438–444.
124. McCully K, Argov Z, Boden B, et al. Detection of muscle injury in humans with 31-P magnetic resonance spectroscopy. *Muscle Nerve* 1988;11(3):212–216.
125. Kariya Y, Itoh M, Nakamura T, et al. Magnetic resonance imaging and spectroscopy of thigh muscles in cruciate ligament insufficiency. *Acta Orthop Scand* 1989;60(3):322–325.
126. Taylor DJ, Brosnan MJ, Arnold DL, et al. Ca^{2+}-ATPase deficiency in a patient with an exertional muscle pain syndrome. *J Neurol Neurosurg Psychiatry* 1988;51(11):1425–1433.
127. Argov Z, Renshaw PF, Boden B, et al. Effects of thyroid hormones on skeletal muscle bioenergetics. In vivo phosphorus-31 magnetic resonance spectroscopy study of humans and rats. *J Clin Invest* 1988;81(6):1695–1701.
128. Jacobs DO, Maris J, Fried R, et al. In vivo phosphorus 31 magnetic resonance spectroscopy of rat hind limb skeletal muscle during sepsis. *Arch Surg* 1988;123(11):1425–1428.
129. Pichard C, Vaughan C, Struk R, et al. Effect of dietary manipulations (fasting, hypocaloric feeding, and subsequent refeeding) on rat muscle energetics as assessed by nuclear magnetic resonance spectroscopy. *J Clin Invest* 1988;82(3):895–901.
130. Sugie H, Tsurui S, Sugie Y, et al. Study of metabolic myopathies using 1H NMR spectroscopy—analysis of muscle metabolites and muscle autolytic change. *Rinsho Shinkeigaku* 1990;30(3):320–323.
131. Challiss RA, Blackledge MJ, Radda GK. Spatial heterogeneity of metabolism in skeletal muscle in vivo studied by 31P-NMR spectroscopy. *Am J Physiol* 1988;3(1):1–11.
132. Meyer RA, Kuchmerick MJ, Brown TR. Application of 31P-NMR spectroscopy to the study of striated muscle metabolism. *Am J Physiol* 1982;242(1).
133. Challiss RA, Vranic M, Radda GK. Bioenergetic changes during contraction and recovery in diabetic rat skeletal muscle. *Am J Physiol* 1989;1(1):1:29–37.
134. Challiss RA, Blackledge MJ, Radda GK. Spatially resolved changes in diabetic rat skeletal muscle metabolism in vivo studied by 31P-n.m.r. spectroscopy. *Biochem J* 1990;268(1):111–115.
135. Kushmerick MJ. Muscle energy metabolism, nuclear magnetic resonance spectroscopy and their potential in the study of fibromyalgia. *J Rheumatol Suppl* 1989;19(1):40–46.
136. Kozak RG, Gascard JP, Redouane BK. Detection of peranesthetic malignant hyperthermia by muscle contracture tests and NMR spectroscopy. *Ann Fr Anesth Reanim* 1986;5(6):584–589.
137. Foster PS, Hopkinson K, Denborough MA. 31P-NMR spectroscopy: The metabolic profile of malignant hyperpyrexic porcine skeletal muscle. *Muscle Nerve* 1989;12(5):390–396.
138. Newman RJ. An in vivo study of muscle phosphate metabolism in Becker's dystrophy by 31P NMR spectroscopy. *Metabolism* 1985;34(8):737–740.
139. Mize CE, Corbett RJ, Uauy R, et al. Hypotonia of rickets: A sequential study by P-31 magnetic resonance spectroscopy. *Pediatr Res* 1988;24(6):713–716.
140. Argov Z, Bank WJ, Maris J, et al. Muscle energy metabolism in human phosphofructokinase deficiency as recorded by 31P nuclear magnetic resonance spectroscopy. *Ann Neurol* 1987; 22(1):46–51.
141. Giger U, Argov Z, Schnall M, et al. Metabolic myopathy in canine muscle-type phosphofructokinase deficiency. *Muscle Nerve* 1988;11(12):1260–1265.
142. Argov Z, Bank WJ, Boden B, et al. Phosphorus magnetic resonance spectroscopy of partially blocked muscle glycolysis. An in vivo study of phosphoglycerate mutase deficiency. *Arch Neurol* 1987;44(6):614–617.
143. Newman RJ, Bore PJ, Chan L, et al. Nuclear magnetic resonance studies of forearm muscle in Duchenne dystrophy. *Br Med J [Clin Res]* 1982;284(6322):1072–1074.
144. Arnold DL, Taylor DJ, Radda GK. Investigation of human mitochondrial myopathies by phosphorus magnetic resonance spectroscopy. *Ann Neurol* 1985;18(2):189–196.
145. Authier B, Rossi A, Albrand JP, et al. Effects of acute arterial occlusion on muscle energy metabolism. An experimental model using phosphorus NMR spectroscopy in the rat. *J Mal Vasc* 1987;12(4):323–328.
146. Williams DM, Fencil L, Chenevert TL. Peripheral arterial occlusive disease: P-31 MR spectroscopy of calf muscle. *Radiology* 1990;175(2):381–385.
147. Hands LJ, Bore PJ, Galloway G, et al. Muscle metabolism in patients with peripheral vascular disease investigated by 31P nuclear magnetic resonance spectroscopy. *Clin Sci* 1986; 71(3):283–290.
148. Heppenstall RB, Scott R, Sapega A, et al. A comparative study of the tolerance of skeletal muscle to ischemia. Tourniquet application compared with acute compartment syndrome. *J Bone Joint Surg (Am)* 1986;68(6):820–828.
149. Keller U, Oberhansli R, Huber P, et al. Phosphocreatine content and intracellular pH of calf muscle measured by phospho-

rus NMR spectroscopy in occlusive arterial disease of the legs. *Eur J Clin Invest* 1985;15(6):382–388.
150. Balschi JA, Bittl JA, Springer CJ, et al. 31P and 23Na NMR spectroscopy of normal and ischemic rat skeletal muscle. Use of a shift reagent in vivo. *Nmr Biomed* 1990;3(2):47–58.
151. Rexroth W, Semmler W, Guckel F, et al. Assessment of muscular metabolism in peripheral arterial occlusive disease using 31P nuclear magnetic resonance spectroscopy. Comparison with metabolite concentrations in femoral blood. *Klin Wochenschr* 1989;67(16):804–812.
152. Blum H, Schnall MD, Chance B, et al. Intracellular sodium flux and high-energy phosphorus metabolites in ischemic skeletal muscle. *Am J Physiol* 1988;3(1):1–8.
153. Heppenstall RB, Sapega AA, Scott R, et al. The compartment syndrome. An experimental and clinical study of muscular energy metabolism using phosphorus nuclear magnetic resonance spectroscopy. *Clin Orthop* 1988;226:138–155.
154. Park JH, Gibbs SJ, Price RR, et al. Inflammatory myopathies: MR imaging and spectroscopy. *Radiology* 1991;179(2):343–344.
155. Berquist TH, Ehman RL, King BF, et al. Value of MR imaging in differentiating benign from malignant soft-tissue masses: Study of 95 lesions. *AJR* 1990;155:1251–1255.
156. Pettersson H, Eliasson J, Egund N, et al. Gadolinium-DTPA enhancement of soft tissue tumors in magnetic resonance imaging—Preliminary clinical experience in five patients. *Skeletal Radiol* 1988;17:319–323.
157. Karczmar GS, Meyerhoff DJ, Boska MD, et al. P-31 spectroscopy study of response of superficial human tumors to therapy. *Radiology* 1991;179(1):149–153.
158. Lemmi, MA, Fletcher, BD, Marina, NM, et al. Use of MR imaging to assess results of chemotherapy for Ewing sarcoma. *AJR* 1990;155:343–346.

ADDITIONAL READINGS

de Kerviler E, Leroy-Willig A, Jehenson P, et al. Exercise-induced muscle modifications: study of healthy subjects and patients with metabolic myopathies with MR imaging and P-31 spectroscopy. *Radiology*, 1991;181:259–264.

Duewell S, Hagspiel KD, Zuber J, et al. Swollen lower extremity: role of MR imaging. *Radiology*, 1992;184:227–231.

Hernandez RJ, Keim DR, Chenevert TL, et al. Fat-suppressed MR imaging of myositis. *Radiology*, 1992;182:217–219.

Huber DJ, Sumers E, Klein M. Soft tissue pseudotumor following intramuscular injection of "DPT": a pitfall in magnetic resonance imaging. *Skeletal Radiol*, 1987;16:469–473.

Lamminen AE, Hekali PE, Tiula E, et al. Acute rhabdomyolysis: evaluation with magnetic resonance imaging compared with computed tomography and ultrasonography. *Br J Radiol*, 1989;62:326–330.

Resendes M, Helms CA, Fritz RC, et al. MR appearance of intramuscular injections. *AJR*, 1992;158:1293–1294.

CHAPTER 11

Pitfalls in Interpretation

Jerrold H. Mink

Magnetic resonance (MR) is a highly accurate modality for evaluation of suspected meniscal and ligamentous abnormalities of the knee joint (1,2), but difficulties in interpretation occur (3–7). Errors in interpretation of knee magnetic resonance examinations occur because of (a) the unique physical principles that govern MR imaging, (b) the ability of MR to delineate structures not appreciated by other imaging modalities, (c) the tomographic nature of MR, and (d) representation of structures in unfamiliar planes. Recurring interpretive problems warrant special attention.

MENISCI

Posterior Horn of the Lateral Meniscus

There are multiple causes for false-positive MR findings of a meniscal tear in the posterior horn of the lateral meniscus. The posterior horn of the lateral meniscus has been the cause of interpretive difficulty for arthrographers because of the peculiar and unique relationship of the popliteus tendon and its sheath to the capsule and meniscus (8–14), and subsequently, the arthrographic accuracy of lateral meniscal evaluation is less than that of the medial meniscus (15–17). The popliteus tendon causes similar problems for the "MR arthrographer" (6,7). An oblique line of intermediate-to-bright signal on T1 or proton density sequences is often seen to course in an anterosuperior to posteroinferior direction across the posterior horn of the lateral meniscus (Fig. 11.1). The "lesion" should be classified as a grade 3 signal because it extends to the superior surface and therefore carries an extremely high probability that it is an arthroscopically detectable tear. The "tear," however, is the popliteus tendon sheath and the small amount of fat that lies against it.

It is worthwhile to review the complex and variant anatomy of this region by means of sequential MR images made from the extreme lateral aspect of the tibial plateau toward the central attachment (6). The popliteus tendon arises from a groove in the lateral femoral condyle just under the fibular collateral ligament and turns downward posteriorly and medially where it eventually joins its muscle belly on the posterior tibia. The tendon is covered anteromedially by a synovial-lined sheath that communicates with the knee joint. The lateral meniscus is firmly attached to the capsule along its cephalocaudal extent throughout the anterior horn and the midzone of the lateral meniscus, but in the posterior horn, the popliteus tendon violates the meniscocapsular junction superiorly. On the first or occasionally the second image from the periphery, the only capsular attachment of the meniscus is the inferior fascicle (Fig. 11.1). As the imaging plane moves further medially, the superior attachment is reconstituted, and the tendon lies in a "tunnel": the superior fascicle is above, the meniscus and the popliteus bursa are anterior, and the inferior fascicle is below. Even further medially, the popliteus tendon interposes itself between the capsule and inferior fascicle so that the superior fascicle is the only peripheral attachment. This intricate anatomy is most readily appreciated on T2-weighted images in patients with large joint effusions.

The pseudotear of the posterior horn of the lateral meniscus can be found in 27% of routine MR examinations (6,18). Several features distinguish the pseudotear from a true tear. The pseudotear invariably courses in a 30° to 45° anterosuperior to posteroinferior direction, and the "peripheral fragment," which is in fact the popliteus tendon, can be followed from the popliteal fossa above the joint line to its attachment on the posterior tibia below the joint line. The width

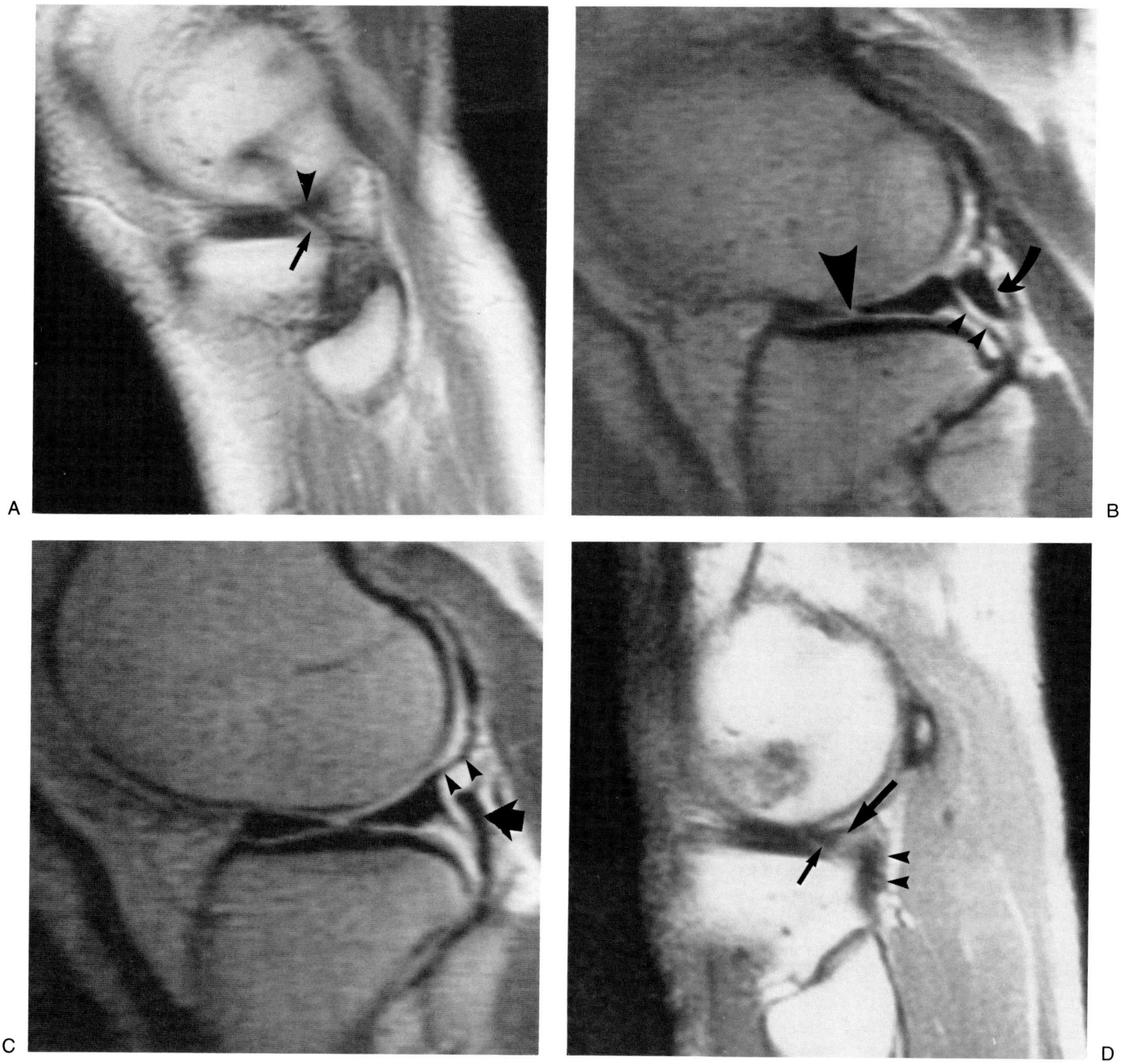

FIG. 11.1. Pseudotear of the posterolateral meniscus. **A:** Sagittal MR, TR/TE 2,000/20. An oblique line of high signal (*arrow*) extends across the posterior horn of the lateral meniscus at approximately 30° angle to the vertical. The location and appearance is overwhelmingly suggestive of a pseudotear caused by the popliteus tendon sheath **B,C:** Sagittal MR, TR/TE 2,000/80. These are adjacent images of the knee in another patient with a tear (*large arrowhead*) of the midzone of the lateral meniscus. In (B), the popliteus tendon (*arrow*) is violating the superior capsular margin, and the superior fascicle is absent. The inferior fascicle (*arrowheads*) is the only peripheral attachment. In (C), the superior fascicle has reestablished itself (*arrowheads*) and the inferior fascicle appears torn as the popliteus tendon interposes itself between the meniscus and capsule. In all images, the popliteus bursa, filled with fluid, is clearly identified just anterior to the tendon. **D:** True tear of the posterior horn of the lateral meniscus. Sagittal MR, TR/TE 800/20. The oblique line of abnormal signal in the posterior horn of the lateral meniscus (*small arrow*) suggests the presence of an oblique pseudotear. However, this section is not the most peripheral section through the meniscus where one expects to see the pseudotear; in fact, the actual position of the popliteus tendon sheath (*large arrow*) is posterior to the surgically proven tear. The popliteus tendon (*arrowheads*) has traversed the meniscus and lies below the tibial plateau.

of the pseudotear is only 1 mm, and is seen on the most peripheral image through the "bow-tie"; rarely, it is seen on two images. True tears have a much less constant appearance.

Any structure except the meniscus or the cruciate ligaments that lies in the joint and projects between the condyles is abnormal; detached meniscal fragments and osteochondral loose bodies are the most common offenders. The diagnosis of a displaced meniscal tear is made by finding a portion of the meniscus in an abnormal position, most commonly in the intercondylar notch. Occasionally, however, onc may find a normal fragment of lateral meniscus posteriorly in the notch on high resolution coronal images (Fig. 11.2).

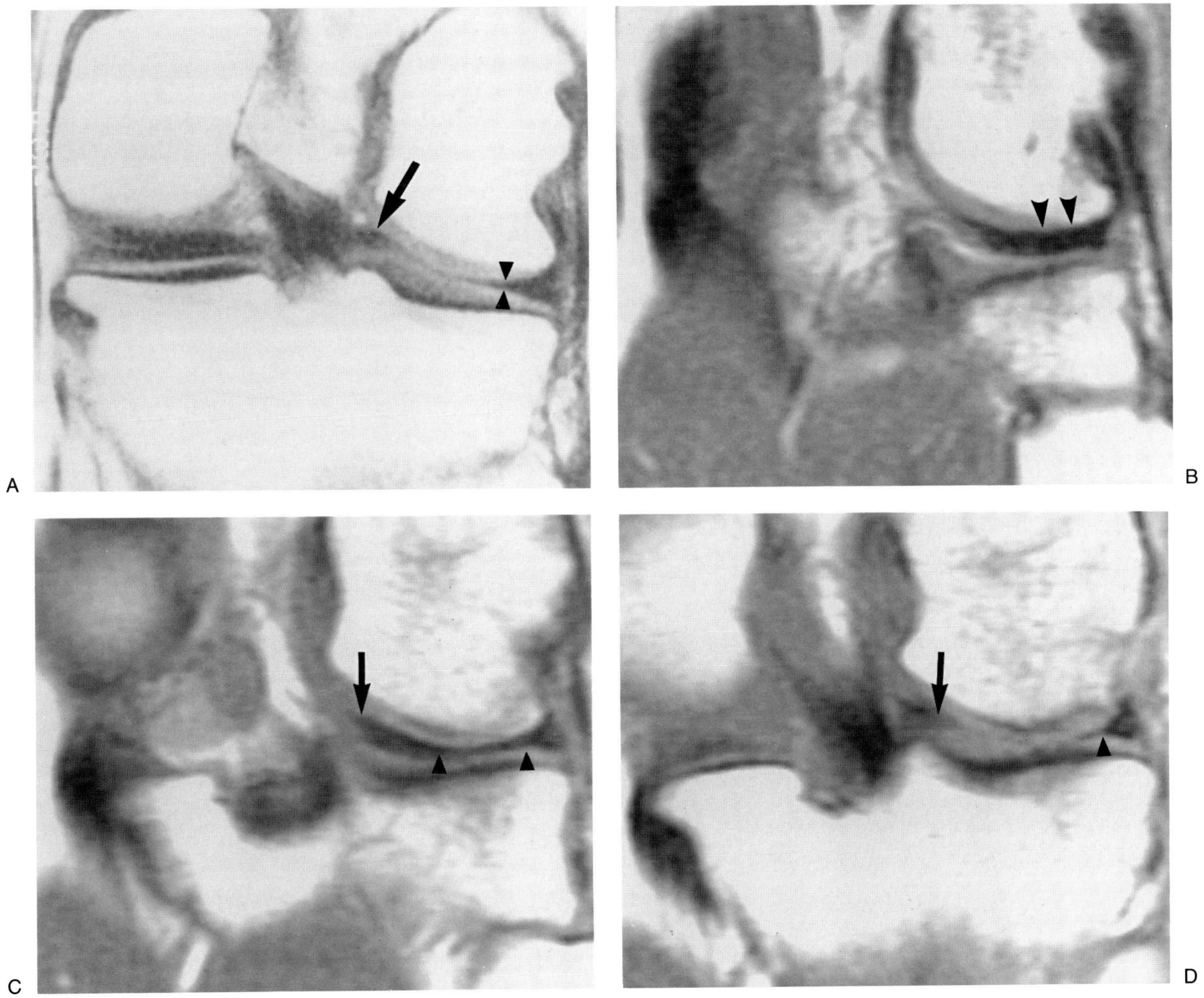

FIG. 11.2. Pseudo and true flap tears of the posterior horn of the lateral meniscus. **A:** Coronal MR, TR/TE 2,000/20. A central pseudofragment (*arrow*) of the lateral meniscus is the attachment of the meniscus to the intercondylar region. When such a pseudofragment is present, the "peripheral fragment" of the lateral meniscus (*arrows*) has a sharp triangular configuration. **B–D:** Normal coronal anatomy of the posterior horn of the lateral meniscus. The three images are 5 mm apart; all are coronal MR, TR/TE 2,000/20. The most posterior section (B) intersects the extreme posterior margin of the lateral meniscus; the meniscal tissue has a bow-tie appearance quite analogous to the image depicted on sagittal sequences through the midzone. In (C), the image plane has moved anteriorly into the midsubstance of the posterior horn, and consequently the width of meniscus visualized is greater. Further, anteriorly (D), the imaging plane has "crossed" the free edge, and the imaging plane now intersects two noncontiguous portions of the posterior horn. It is on this last image that confusion with a free meniscal fragment may occur. (*Figure continues on next page*)

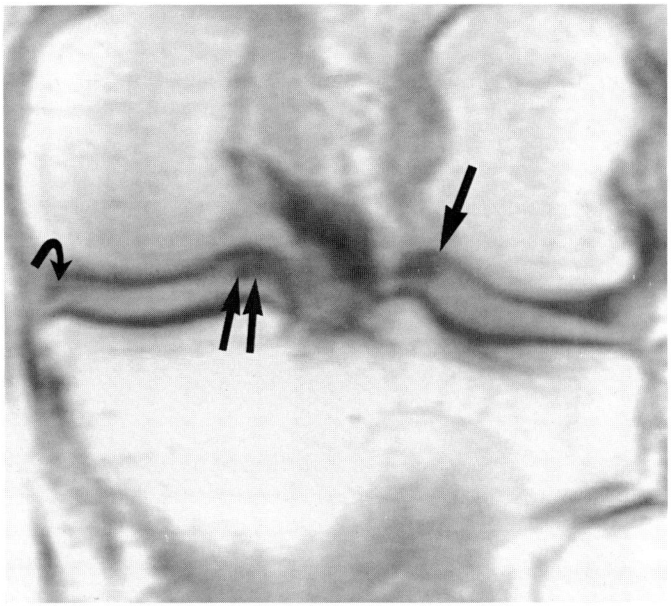

E

FIG. 11.2. (*Continued*) **E:** Coronal MR, TR/TE 2,000/20. In another patient, central "fragments" of both menisci look alike (*double arrows*, medial; *large arrow*, lateral meniscus). The peripheral portion of the *lateral* meniscus has a sharp triangular configuration as it should; the peripheral segment of the *medial* meniscus (*curved arrow*) is markedly deformed. At surgery, a displaced flap tear of the medial meniscus was resected; the lateral meniscus was normal.

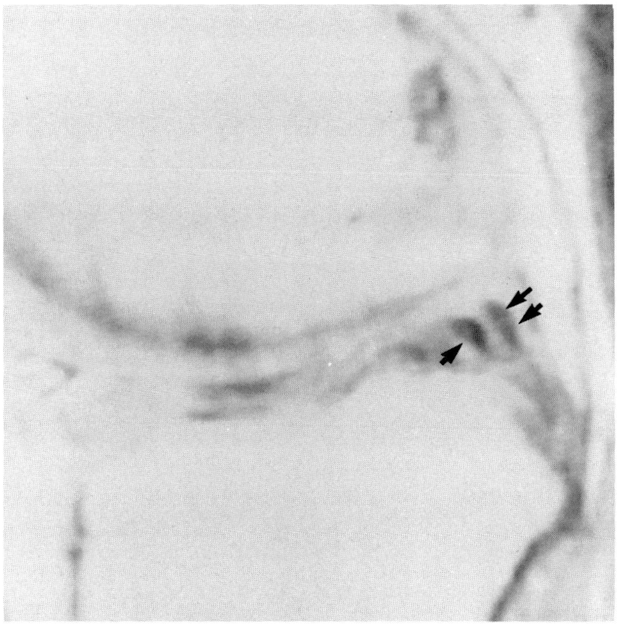

A

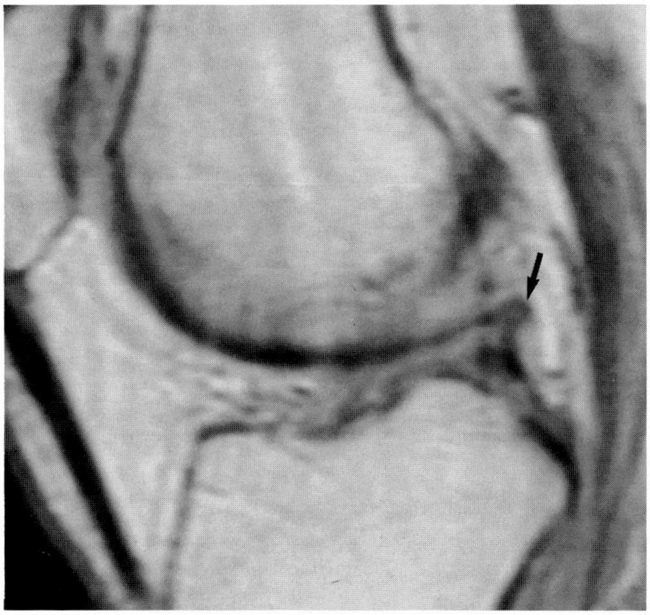

B

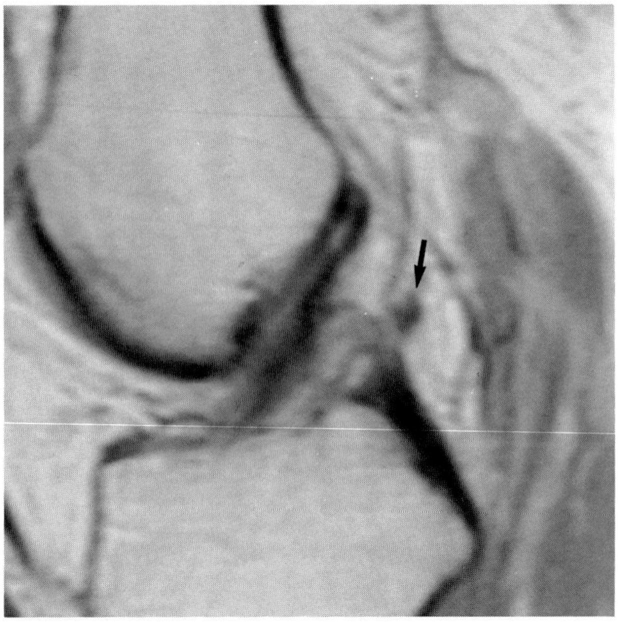

C

FIG. 11.3. Pseudotear of the lateral meniscus; effect of the MFL. **A:** Sagittal MR, TR/TE 2,000/20. The two "lesions" in the posterolateral compartment of the knee represent the central attachment of the lateral meniscus (*arrow*) and the more posteriorly positioned origin of the ligament of Wrisberg (*double arrows*). The appearance simulates a tear of the meniscus or intraarticular loose bodies. **B–E:** The normal MFL. Sagittal MR, TR/TE 2,000/20; the images are 5 mm apart. In another patient, the sequence of images depicts the path of the ligament of Wrisberg (*arrow*), from its origin from the lateral meniscus to its attachment to the posterior medial femoral condyle. **F:** Sagittal MR, TR/TE 2,000/20. The ligaments of Humphrey (*arrow*) and Wrisberg (*arrowhead*) are fortuitously both identified on either side of the posterior cruciate ligament. They must not be mistaken for meniscal fragments or osteochondral loose bodies. **G:** Sagittal MR TR/TE 800/20. The two dark structures in the posterior knee joint resemble portions of a torn meniscus. The origin of the ligament of Wrisberg (*arrowhead*) from the meniscal attachment (*arrow*) accounts for the appearance.

The appearance results because the imaging plane, which is a straight line, intersects two nonadjacent points on the semicircular meniscus. The more lateral fragment represents the portion of the meniscus at the junction of the posterior horn with the midzone; the more central fragment is the extreme central portion of the meniscus and/or the meniscal attachment to the intercondylar notch and the posterior cruciate ligament. This central pseudofragment is seen on one or occasionally two images if the scan is performed at a 3 mm or 4 mm section thickness. This effect of mistaking the partially sectioned posterior horn/central attachment for a free meniscal fragment is much less common on the medial side of the knee because the medial meniscus is more comma-shaped than the lateral; the likelihood of acquiring an image that obliquely sections two noncontiguous segments of the medial posterior horn is much less likely. Excessive external rotation of the knee will increase the probability of depicting a lateral pseudotear, and perhaps the number of sections on which the pseudofragment is seen. This pitfall can be avoided by assessing the degree of internal or external rotation of the knee, by carefully examining the sagittal sections to ascertain that the posterior horn has a totally normal configuration, and by finding the central fragment on only one or occasionally two successive images.

A third pseudotear of the posterior horn of the lateral meniscus can be encountered on sagittal images quite near the intercondylar notch (Fig. 11.3). In order to

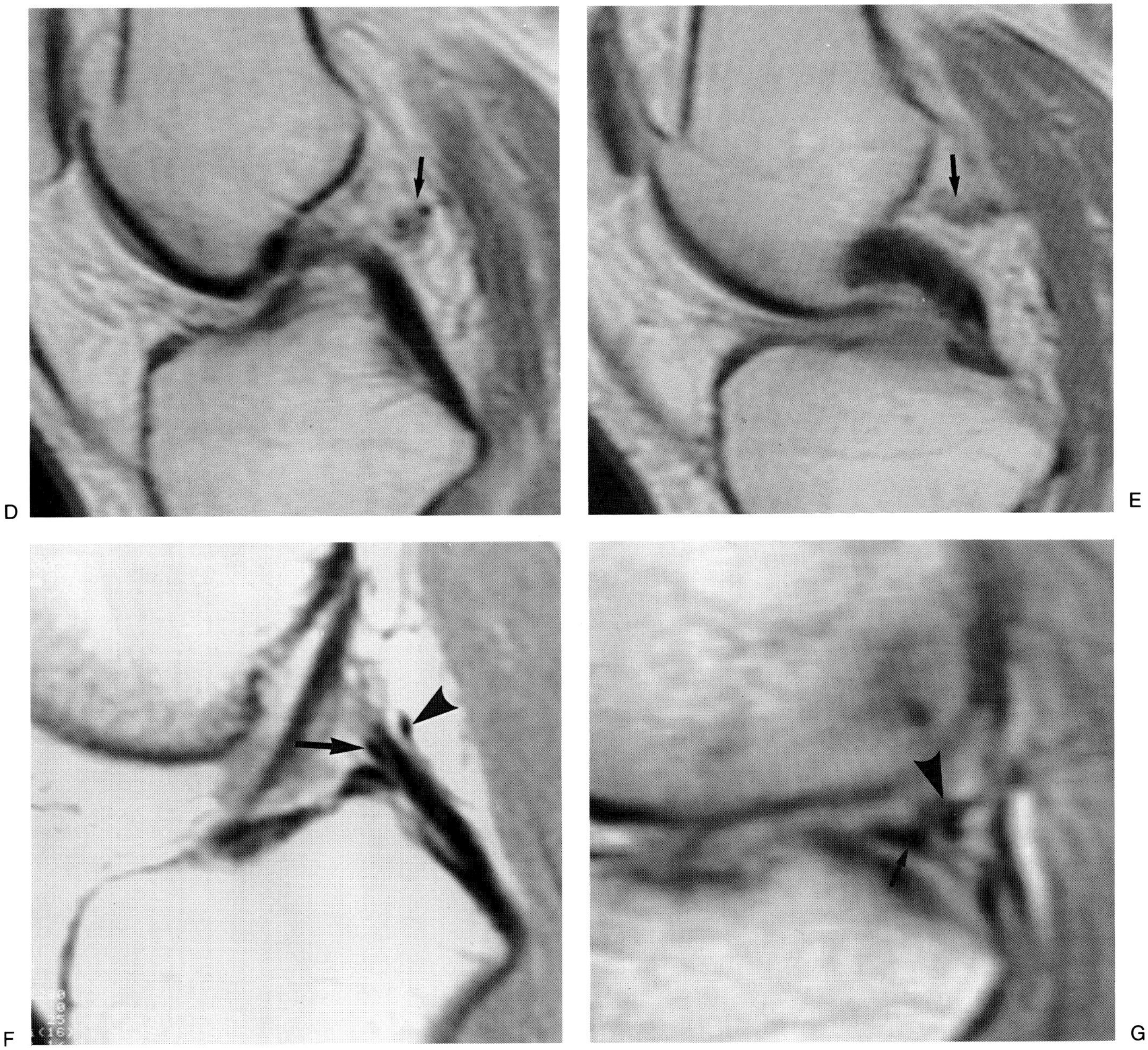

FIG. 11.3. (*Continued*)

fully understand this potential pitfall, one must understand the anatomy of the meniscofemoral ligament (MFL), the cord-like attachment of the lateral meniscus to the medial femoral condyle (19,20). The MFL was first described by Poirier and Charpy in 1892; they coined the term "the third cruciate ligament" because of the relationship of this ligament to the origin of the posterior cruciate ligament (PCL). Anatomically, two distinct branches can be identified. The more anterior is known as the ligament of Humphrey. It is directed in an oblique orientation from the posterior lateral meniscus, anterior to the PCL toward its medial femoral condylar insertion site. It is usually one-third of the PCL diameter. The posterior branch of the MFL, also known as the ligament of Wrisberg, or Roberts ligament, has a similar orientation but runs immediately posterior to the PCL. It is the larger of the two branches, and may be as large as one-half the diameter of the PCL. Most commonly, the MFL attaches directly to the lateral meniscus; occasionally it attaches indirectly to the meniscus via the posterior capsule and/or the popliteus tendon (20). The MFL functions to pull the posterior horn of the lateral meniscus anteriorly and medially during flexion, balancing the action of the popliteus muscle whose fibers attach to the posterior and lateral aspects of the meniscus.

While both branches of the MFL may exist, there are considerable variations in their frequency and size. Various anatomic dissections have found that the branches of the meniscofemoral ligaments may be found in 80% to 100% of knees; the anterior is found in 34% and the posterior in 60% of the specimens (21). In a recent MR study, the ligament of Humphrey was visualized 33% of the time as a small ovoid low signal focus just anterior to the PCL, or more commonly in our experience, as a discrete low signal bulge inseparable from the concave surface of the PCL (Fig. 11.3). The ligament of Wrisberg, visualized in 33% of MR cases, is virtually always identified as a discrete structure behind the PCL. Both ligaments are visualized in only 3% of MR studies. Either of these two low signal foci may be mistaken for free meniscal or osteochondral fragments (18,22,23). Conversely, true loose bodies must not be mistaken for the normal MFL (Fig. 11.4). Errors of interpretation will be avoided if one carefully follows the "lesion" on consecutive sagittal images.

When the origin of the MFL parallels and closely approximates the posterior margin of the lateral meniscus, a pseudotear is produced by the presence of fat interposed between the meniscal attachment and the meniscofemoral ligament. As the meniscus blends imperceptibly with its attachment, it loses its normal triangular configuration, becomes more rhomboid, and/or may have a blunted tip. It is at this point that the meniscus may demonstrate a line of increased sig-

A

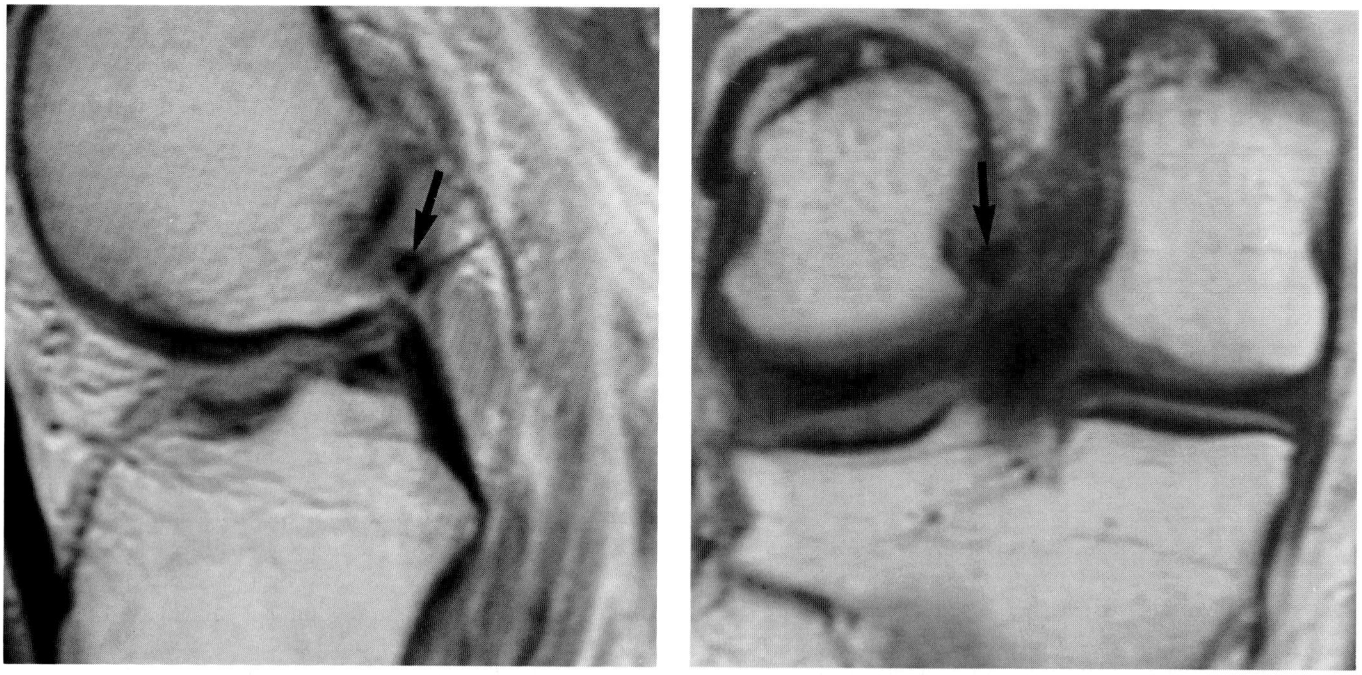

B

FIG. 11.4. Pseudoligament of Wrisberg. **A:** Sagittal and **B:** coronal MR, TR/TE 2,000/20. A rounded lesion (*arrow*) in the posterior capsule, immediately posterior to the posterior cruciate ligament, simulates the ligament of Wrisberg. However in (B), the "ligament" is seen to represent a round osteochondral fragment. In the coronal plane, the ligament of Wrisberg would be expected to have a horizontal course.

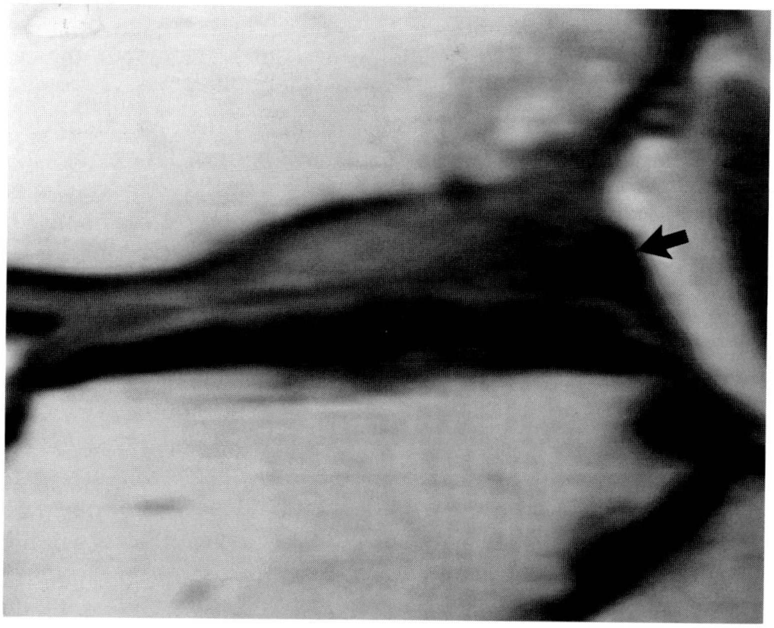

FIG. 11.5. Pseudo-retained posterior horn of the lateral meniscus. Sagittal MR, TR/TE 800/20. In this patient who has had a complete lateral meniscectomy, the popliteus tendon (*arrow*) simulates a retained posterior horn. The correct diagnosis is made by examining the full series of sagittal images.

nal running from its superior to its inferior surface; the lesion might well be classified as a grade 3 signal because of its extent to the articular surfaces (Fig. 11.3). This line is either nearly vertical in orientation (16%) or more commonly is slightly oblique (84%), extending from superoanteriorly to inferoposteriorly (19). While the abnormal morphology of the meniscus at this point is often sufficient evidence that the plane of section is quite central and that one is no longer imaging the fibrocartilaginous meniscus (the image plane actually intersects the more fibrous meniscal attachment), one occasionally encounters a nearly triangular meniscal attachment.

If the lateral meniscus has been completely resected, the popliteus tendon can be mistaken for a retained posterior horn (Fig. 11.5). Errors will be avoided if the suspected meniscal fragment can be followed both above and below the joint line.

Anterior Horn, Lateral Meniscus

A meniscal pseudotear may also be found in the anterior horn of the lateral and occasionally the medial meniscus (7,18,23,24). A line of intermediate-to-bright signal, seen on T1- or occasionally T2-weighted images, courses obliquely through the anterosuperior corner of the anterior horn (Fig. 11.6). This line extends from the superior margin of the meniscus anteriorly and courses obliquely posteriorly and distally. This might be inadvertently classified as a tear given its extension to an articular surface. The complex may be mistaken for a loose body. The central attachments of the anterior horn of the lateral meniscus account for this variant appearance which is present in 22% to 30% of MR studies (6,18,25,26). The meniscus is secured (a) directly to the tibial intercondylar eminence and anterior cruciate ligament by central attachments and (b) to the medial meniscus via the intermeniscal, or transverse meniscal ligament (TML). The TML arises as a rounded band of tissue from the anterosuperior corner of the lateral meniscus immediately above and anterior to the rhomboid central attachment, courses between the Hoffa fat pad and the tibial attachment of the anterior cruciate ligament, and ultimately inserts onto the anterosuperior corner of the anterior horn of the medial meniscus. The TML is variable in size, and is often completely absent. The entire medial to lateral course of the transverse meniscal ligament is found in only 15% of MR examinations (6,18,24). Diagnostic errors will be avoided by examining the sequence of images and demonstrating that the structure in question extends completely across the joint. Additionally, isolated tears of the anterior horn of the lateral meniscus are uncommon (27). Loose bodies or displaced flaps of meniscus in the anterior aspect of the joint may simulate a normal TML (Fig. 11.7).

Herman and Beltran have offered another explanation for the pseudotear of the anterior horn of the lateral meniscus (24). He noted that the lateral inferior genicular artery (LIGA) may be relatively closely applied to the meniscus and that the fat between the artery and the meniscus may simulate a tear. The LIGA arises from the popliteal artery at the level of the proximal tibiofibular joint and courses lateral to the anterior aspect of the knee where it joins with other arteries to form the genicular anastomosis. The approximation to the joint line and subsequent confusion with a meniscal

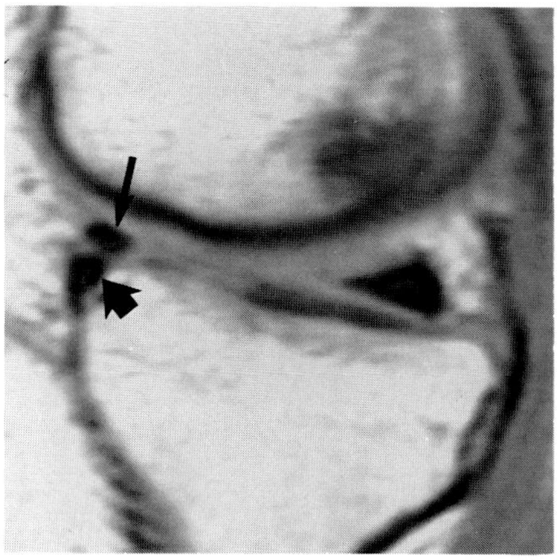

A

FIG. 11.6. Pseudotear of the anteromedial meniscus; effect of the TML. **A:** Sagittal MR, TR/TE 2,000/20. Two distinct "fragments" are seen in the expected position of the anterior horn of the medial meniscus. The upper "lesion" (*long arrow*) is the TML, while the lower (*short arrow*) is the meniscal attachment to the tibia. **B:** This series of sagittal MRs demonstrates the course of the transverse meniscal ligament, and its role in producing a pseudotear of the anterolateral, and occasionally the anteromedial meniscus. The superior and posterior position of the transverse meniscal ligament (*arrow*) relative to the medial meniscal attachment (*arrowhead*) can be identified. The transverse meniscal ligament crosses the joint anterior to the ACL, and inserts on the anterosuperior corner of the central attachment of the anterior horn of the lateral meniscus (*double arrow*) where it again may simulate a tear.

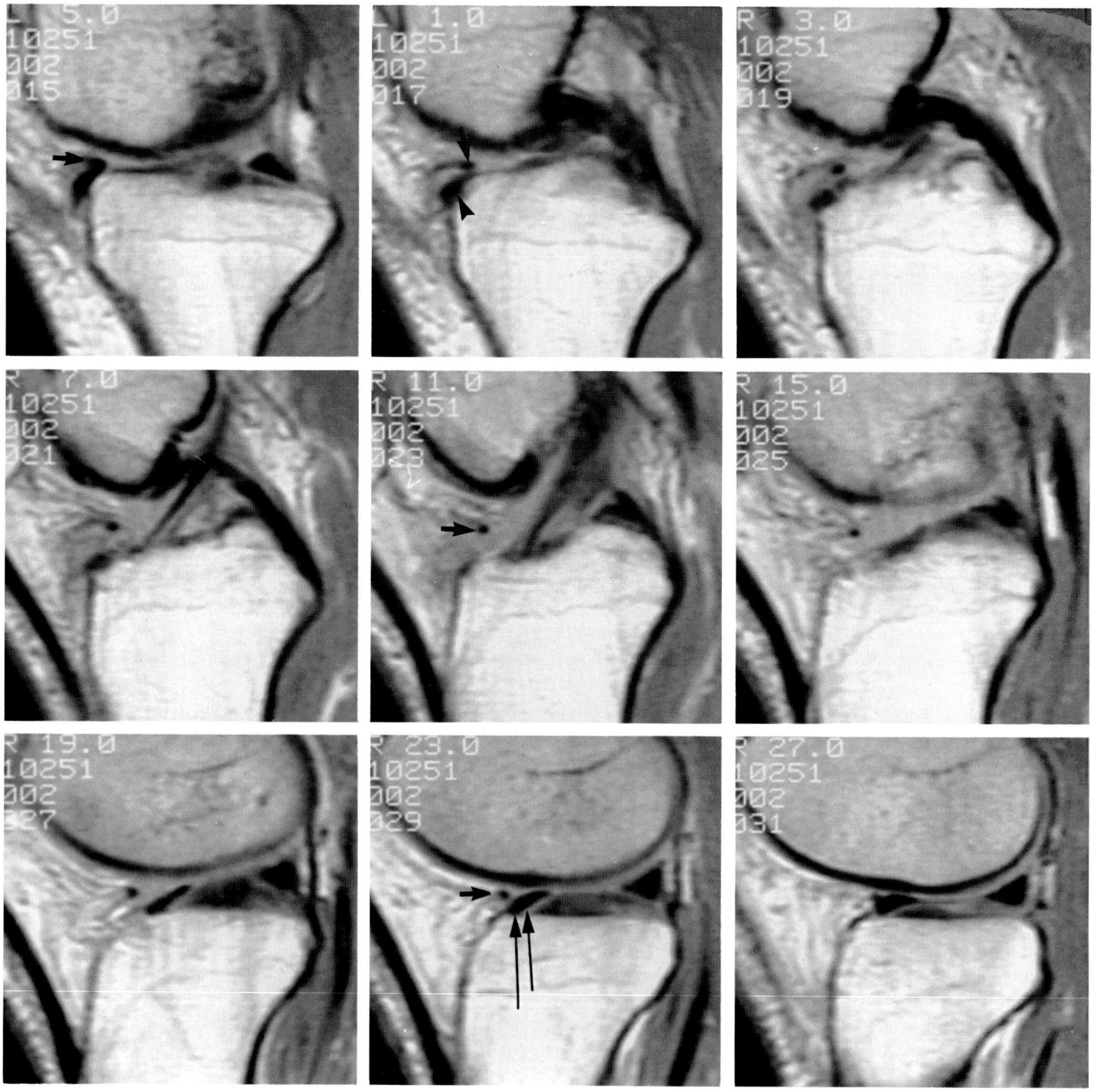

B

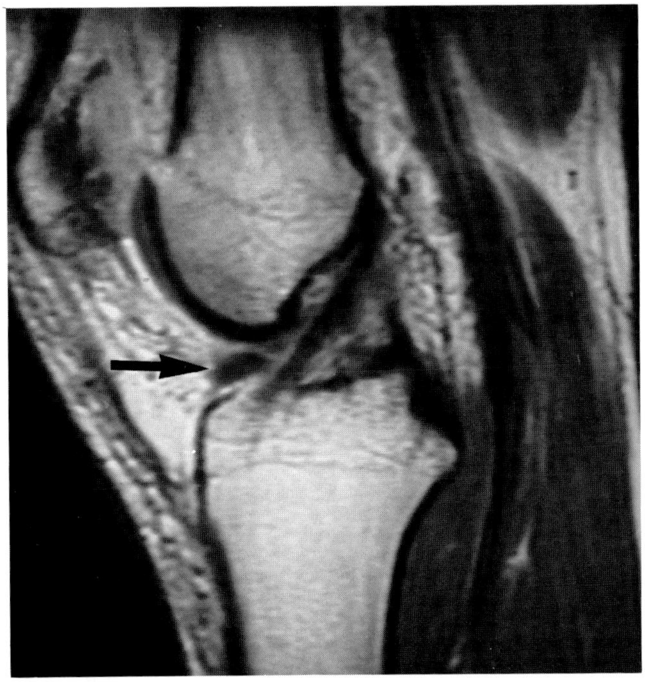

FIG. 11.7. Loose body simulating the TML. Sagittal MR, TR/TE 2,000/20. A surgically proven osteochondral loose body (*arrow*) resides in approximately the expected position of the TML.

tear is only seen laterally since the *medial* inferior genicular artery curves around the tibia well below the joint line. On any given image, it may be impossible to distinguish between the LIGA and the TML. Both structures are dark on standard imaging sequences, the LIGA because of flowing blood within and the transverse ligament because it contains type 1 collagen and few excitable protons. If the structure in question can be followed uninterruptedly from the medial to the lateral meniscus, it probably represents the TML; if it branches, it is likely to be a vascular structure. In our experience, mistaking the LIGA for a tear is exceedingly less common than mistaking the TML for a tear. It is reasonable to conclude that on the lateral side of the knee, either the TML or the LIGA can produce a pseudotear effect, but on the medial side, only the TML has this ability. Most importantly, regardless of which anatomic variation accounts for the confusion, pseudotears in the anterior horn commonly occur.

Mistaking the ligamentous attachments of the menisci for the fibrocartilages themselves can result in false-positive diagnoses because the attachments typically have abnormal signal within them and are rhomboid instead of triangular (Fig. 11.8). This appearance is most confusing in the anterior horn of the lateral meniscus, where the attachment is typically seen on one or two images. The meniscal horns are roughly equilateral triangles with their apices pointing transversely across the plateau, whereas the anterior ligamentous attachment is nearly rhomboid and points obliquely upward. The transition point between the meniscus and ligamentous attachment approximates the site of origin of the transverse ligament; tissue central to the transverse ligament origin is not the fibrocartilage. True central detachments are exceedingly uncommon but the posterior horn of the medial meniscus may be detached during avulsion of the posterior or anterior cruciate ligaments (28).

Peripheral Meniscal Concavity

Pseudotears of the menisci may occur at their periphery where a horizontal linear focus of increased signal may simulate a tear on sagittal MR images. Within the vertical center of the periphery of the meniscus, one may encounter a horizontal line of increased signal (T1WI). This linear signal simulates a grade 2, or in very rare instances a grade 3, intrameniscal signal (Fig. 11.9). The linear high signal focus may be due to partial volume effect of the pericapsular fat adjacent to the peripheral meniscal concavity with the otherwise dark meniscus. The appearance may occur in as many as 29% of medial and 6% of lateral menisci (24). The effect of the normal concavity of the peripheral margin of the meniscus simulating intrameniscal signal has been examined by means of a Plexiglass model immersed in an oil bath (24). Sections made through the outer margin of the Plexiglass "meniscus" demonstrate a bright horizontal line running the length of the disk; on the next section toward the center of

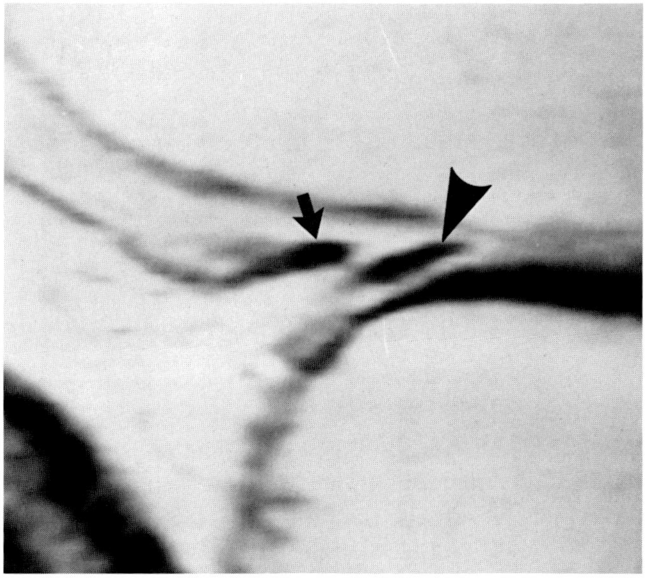

FIG. 11.8. Central attachment of the anterior horn of the lateral meniscus. Sagittal MR, TR/TE 2,000/20. The central attachment of the meniscus (*arrowhead*), which is normally rhomboid, may be mistaken for a deformed, torn meniscus (*arrow*, TML).

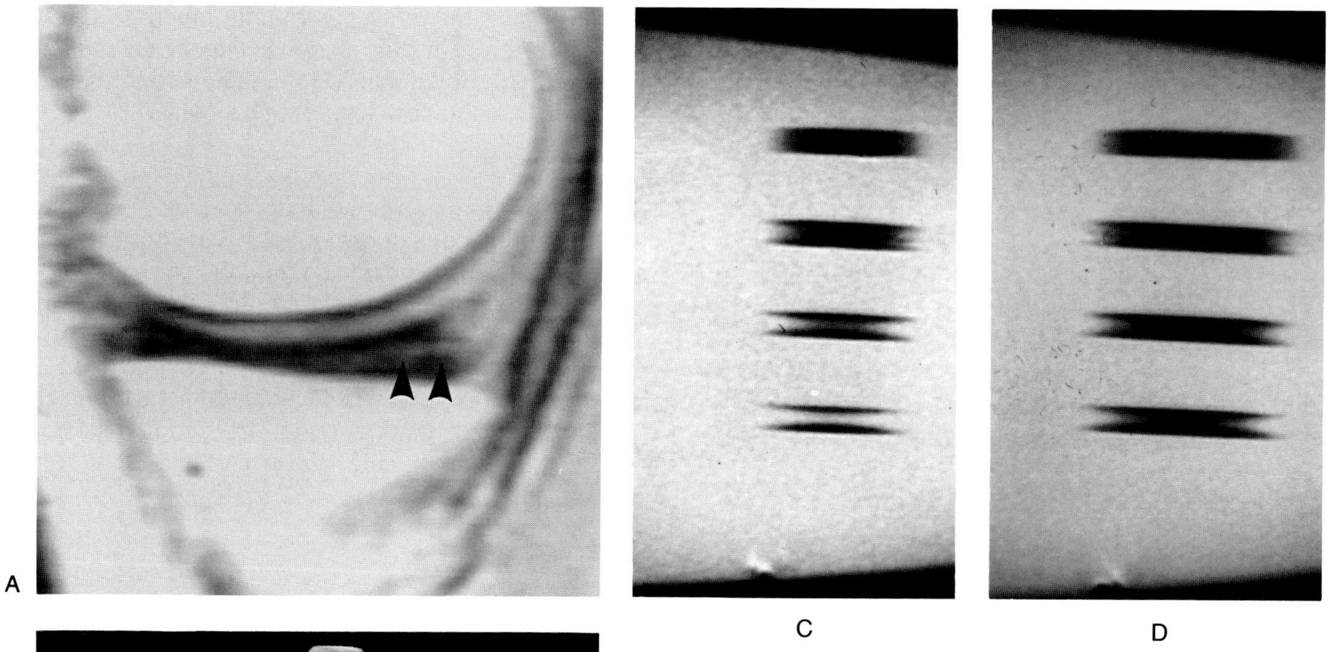

FIG. 11.9. Peripheral meniscal concavity artifact. **A:** Sagittal MR (TR/TE 1,500/20, 3 mm slice thickness). The image through the medial meniscus demonstrates a linear horizontal line (*arrowheads*) caused by volume averaging of fat in the concavity of the meniscal edge with the dark meniscus. **B:** Plexiglass model used to simulate the meniscus. The edges of the uppermost disc are straight. The edges of lower three discs are beveled 1 mm, 2 mm, and 3 mm, respectively. **C:** Sagittal scans obtained through the edge of the Plexiglass discs (TR/TE 1,500/20, 3 mm slice thickness). The images demonstrate a bright horizontal line caused by volume averaging of the low signal intensity Plexiglass disc and the high signal intensity of the oil bath. This artifact is most apparent in the discs with the deepest bevel (*lowest image*). The images in (**D**) are 5 mm medial to (C). The artifact is now seen best at the anterior and posterior margins of each of the discs. (Courtesy of J. Beltran, M.D.) (From ref. 51, with permission.)

the model, only the extreme anterior and posterior segments of the line remain visible. Because the meniscus, just like the Plexiglass model, is curved, sections closer to the center of the joint or model include a portion of the outer margin of the meniscus anteriorly and posteriorly, both/either of which may encompass short linear artifacts within them. It is this high signal focus in the anterior and posterior meniscal horns that may be mistaken for a tear. This pseudotear effect is less evident on images obtained with a 5 mm rather than with a 3 mm section thickness, but is more dramatic in those menisci with deeper peripheral concavities. Confusion and error can be eliminated merely by examining high resolution coronal images. In cases of a pseudotear, the concave margin of the meniscus will be appreciated; failure to identify any abnormal signal within the meniscus on coronal images in a patient in whom the sagittal images depict grade 2 or 3 signal is suggestive of the pseudotear effect of the peripheral meniscal concavity.

Free Edge Evaluation

The grading system for evaluation of meniscal signal is critically dependent on whether intrameniscal signal does (grade 3) or does not (grade 1 or grade 2) extend to an articular surface. It should not be forgotten, however, that the free central edge is also an articular surface. The meniscus is a triangle with a wide base that is lying on its side. The base of the triangle is the periphery, and the apex of the triangle is the central free edge. The base, or periphery, attains a maximal height of 4 to 5 mm; the meniscus rapidly tapers to a wafer thin central tip. On MR, an image may be encountered in which the triangular anterior and posterior horns are joined by a thin, low signal bridge representing the free edge. Signal within this thin tissue bridge that "joins" the anterior and posterior horns may indicate the presence of either a radial or flap tear. When viewing MR images, one does not know whether or not the apparent signal is true intrameniscal signal or a partial volume effect of the meniscus with the opposing tibial and femoral articular cartilage surfaces (Fig. 11.10). The composition of the image is dependent on the exact relationship of the free meniscal edge to the center of the 5 mm thick section. In a similar manner, *coronal* images may suffer from partial volume effect

A

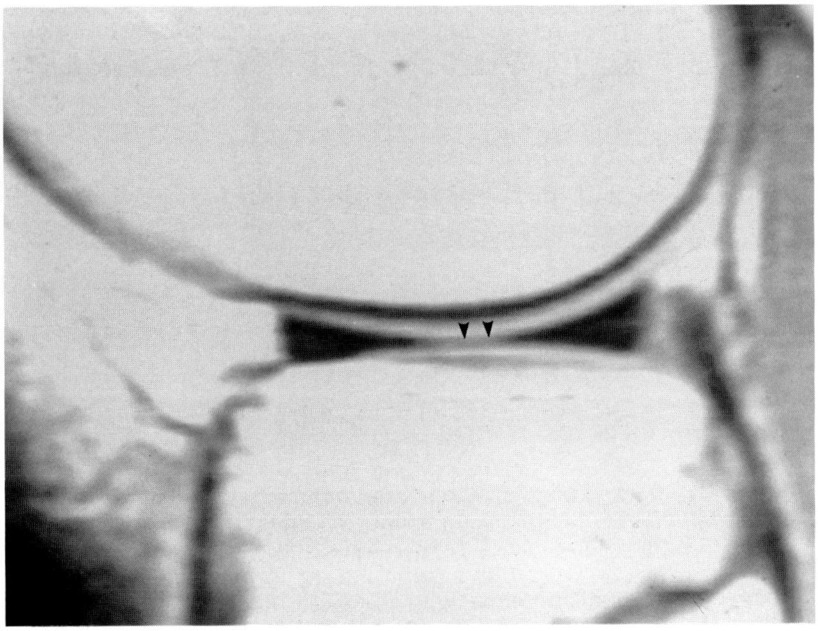

B

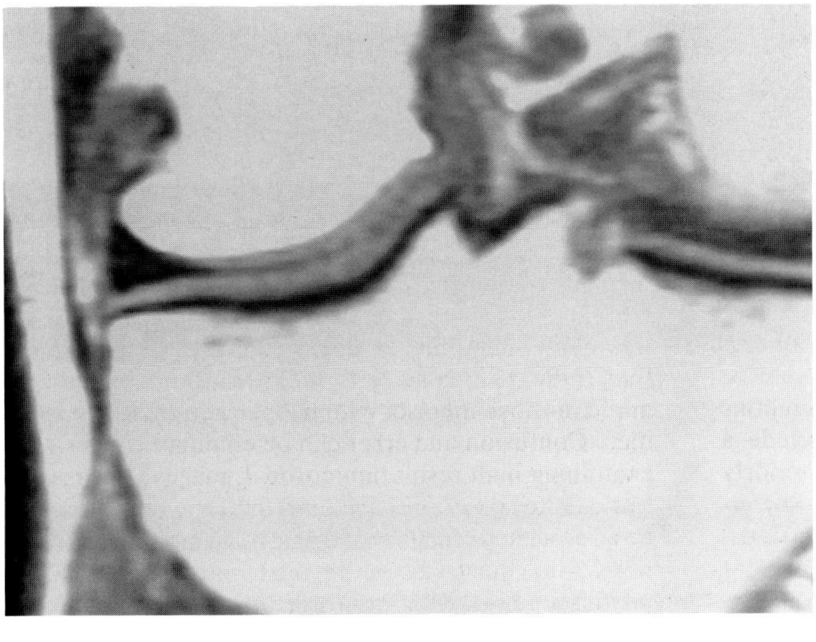

FIG. 11.10. Evaluation of the free edge. **A:** Sagittal MR, TR/TE 2,000/20. **B:** Coronal MR, TR/TE 2,000/20. In (A), a faint signal (*arrowheads*) is seen between the tips of the anterior and posterior horns of this lateral meniscus. It is uncertain whether the signal represents partial volume effect with the co-apted articular cartilages of the femur and tibia, or a tear at the free edge of the lateral meniscus. In (B), however, a sharp central tip of the midzone of the meniscus is identified, confirming that no tear is present.

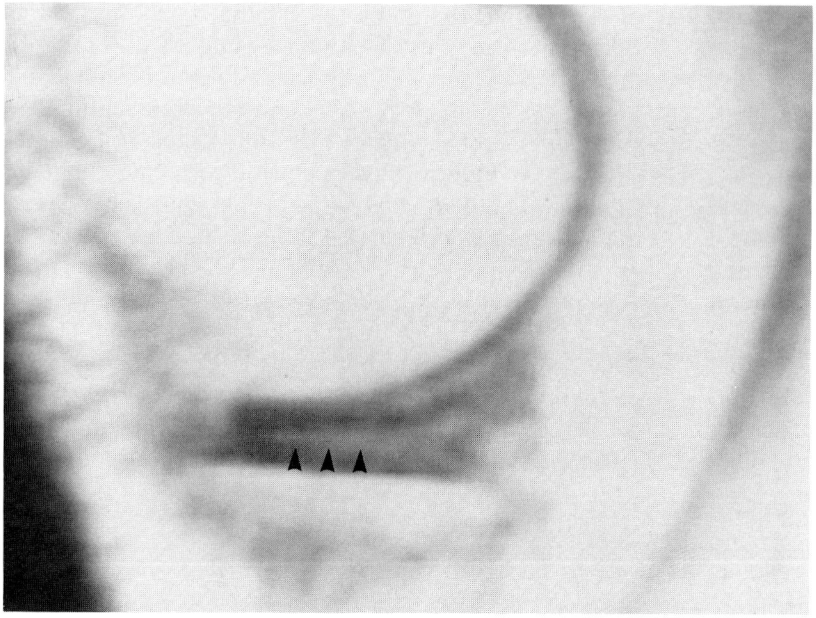

C

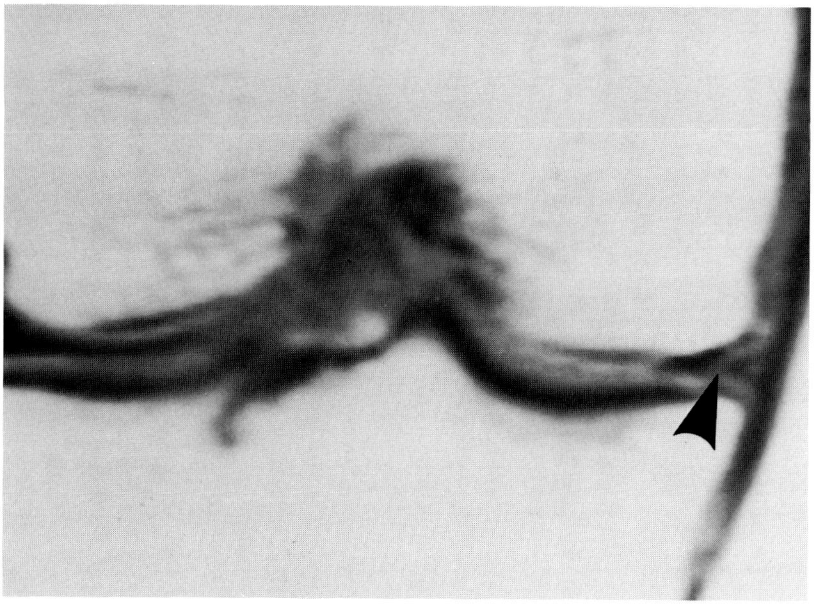

D

FIG. 11.10. (*Continued*) **C,D:** Imaging of a tear in only one plane. Sagittal (C) and coronal (D) MR, TR/TE 2,000/20. In (C) there is faint inhomogeneous signal within the posterior horn/midzone of this medial meniscus, but there is no extension to the articular surface. The coronal image (D) demonstrates a nearly vertical tear (*arrowhead*) reaching the inferior surface. Localized, vertically oriented tears are best seen on coronal images.

along the free meniscal edge of the posterior horn of either meniscus. Examination of high resolution images in both coronal and sagittal planes will reduce the likelihood of partial volume effect along the free edge causing confusion with a true meniscal tear.

The logic applied to assessment of the free edge should not be interpreted to mean that all tears will be visible on images made in both coronal and sagittal imaging planes. Unequivocal grade 3 signal seen on one image in one plane indicates a high likelihood that a tear will be present at arthroscopy. Steeply oriented, nearly vertical tears will appear in one plane as a definite tear, but in the 90° orthogonal plane may only appear as diffuse signal increase. Typically, vertically oriented tears are best seen in the coronal plane (Fig. 11.10).

Menisci of Abnormal Size

Menisci that are too big, too small, or asymmetric in configuration must be regarded as abnormal. Both menisci are normally 3 to 5 mm in height peripherally (26,29). The average breadth of the anterior horn of the medial meniscus is 6 mm, and the posterior horn is 12 mm. The C-shaped lateral meniscus is 10 to 12 mm in all three horns. Because the midzone of each meniscus is 10 to 12 mm (average, 11.6 mm), there are two 5 mm thick MR images through it. The more central of the two images typically falls at, or quite near, the free edge. Central to the free edge are two or sometimes three images depicting the triangular anterior and posterior horns pointing toward one another.

Discoid lateral menisci are present in 1% to 2% of arthrographic series and 2% to 5% of surgical series, although the true incidence is undoubtedly higher since only those discoid menisci that become symptomatic come to diagnosis (30). The incidence of discoid lateral meniscus is 15 times that of discoid medial meniscus (31). Discoid meniscus is unlikely to be a failure of embryogenesis, as suggested by Smillie, but rather is a result of meniscal hypertrophy (32). The lack of a normal meniscotibial attachment, a constant finding in patients with discoid menisci, results in impaired mobility; the meniscus fails to move completely forward during extension and is pulled laterally by strong meniscofemoral attachments during flexion. This abnormal anteroposterior and mediolateral movement predisposes to enlargement (33,34). While this hypothesis is attractive and is held by many investigators, it does not account for alterations in the fibular head and fibular musculature, the lateral malleolus, the tibial spines, the inferior genicular arteries, and the femoral condyles that have been noted in patients with discoid menisci (35).

The MR diagnosis of discoid meniscus is made by the finding of an abnormally sized meniscus (Fig. 11.11). Silverman et al. defined a discoid meniscus on MR as when three or more 5 mm thick contiguous sag-

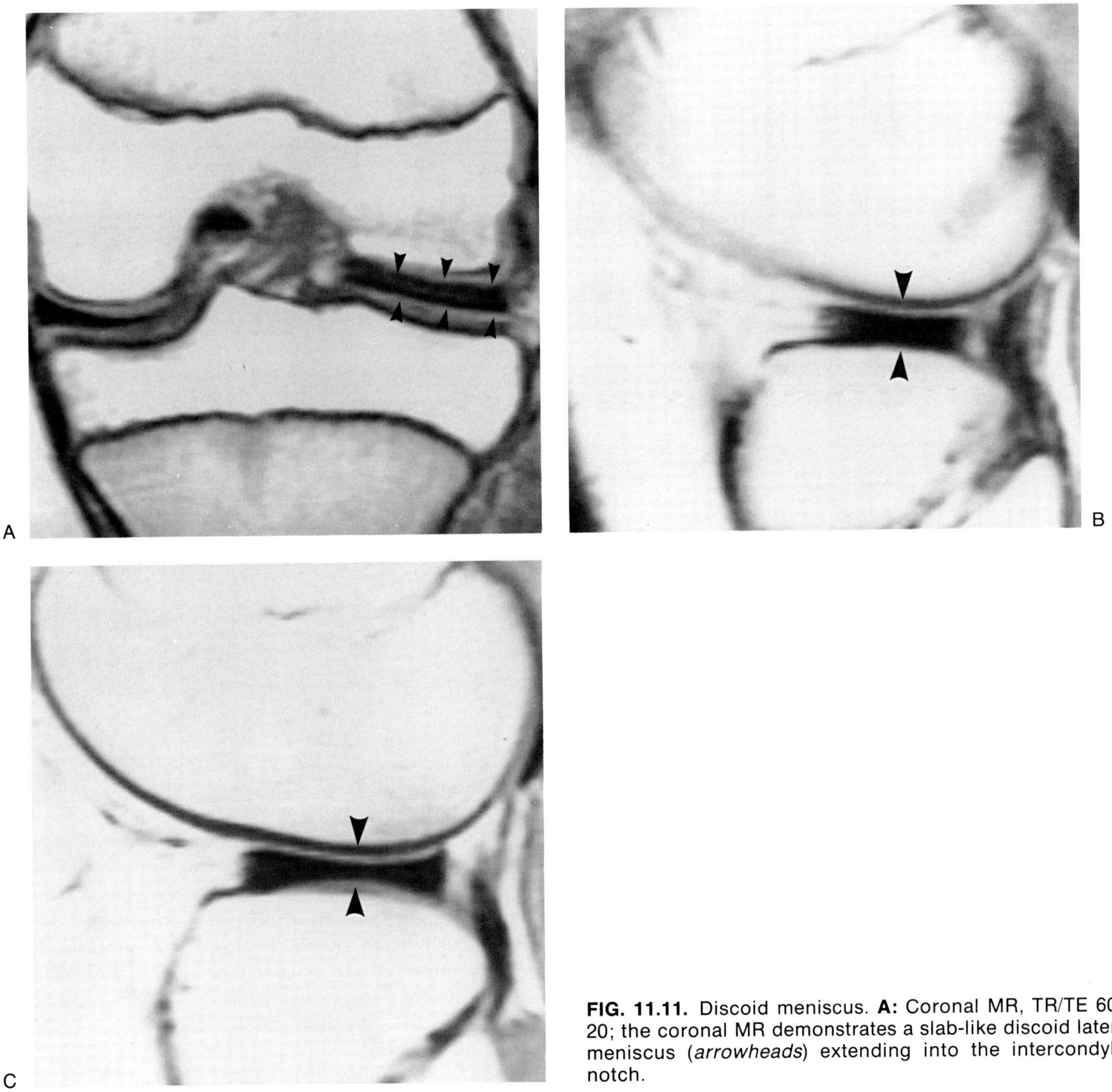

FIG. 11.11. Discoid meniscus. **A:** Coronal MR, TR/TE 600/20; the coronal MR demonstrates a slab-like discoid lateral meniscus (*arrowheads*) extending into the intercondylar notch.

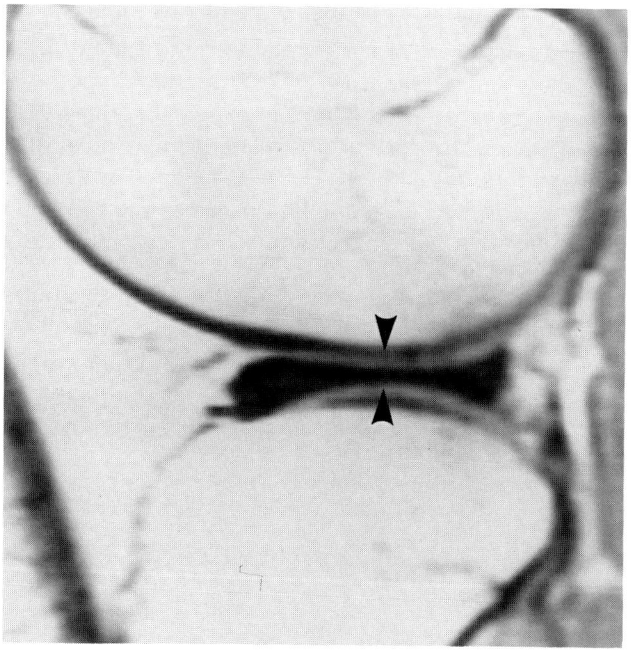

D

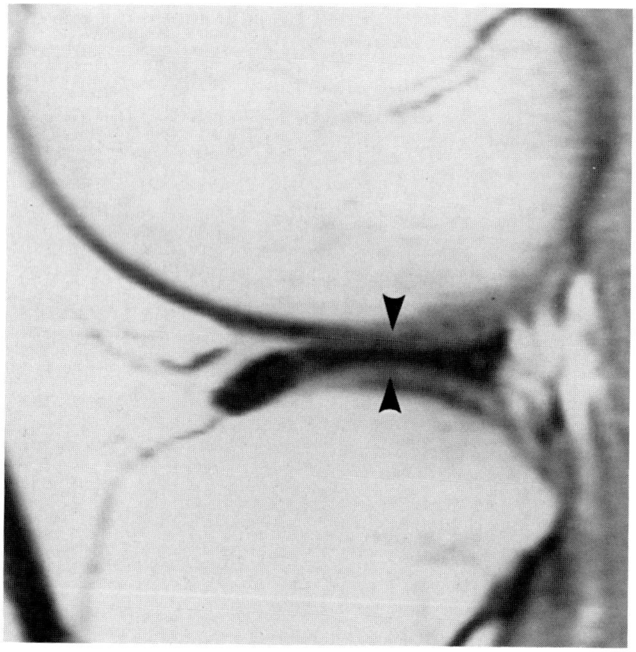

E

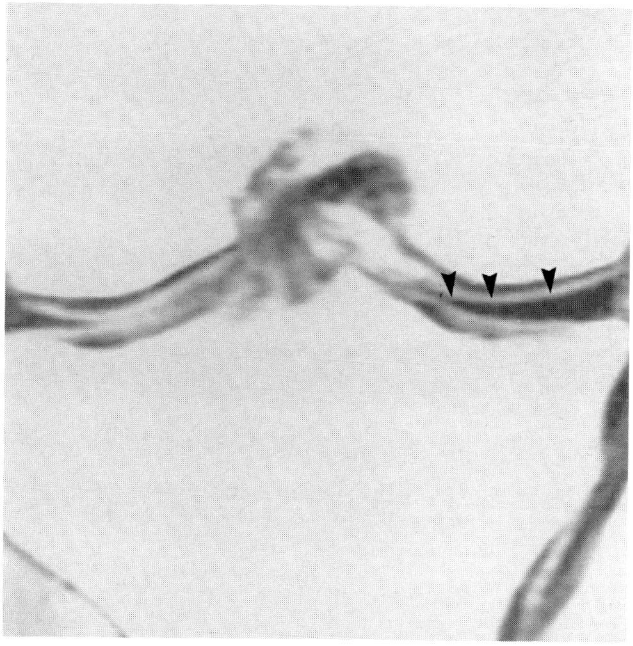

F

FIG. 11.11. (*Continued*) **B–E:** Sagittal MR, TR/TE 2,000/20. These images from another patient are 5, 10, 15, and 20 mm from the lateral edge of the tibia, and demonstrate continuation of meniscal tissue (*arrowheads*) between anterior and posterior horns, findings diagnostic of a discoid lateral meniscus. **F:** Discoid medial meniscus. Coronal MR, TR/TE 2,000/20. An extremely large but centrally tapering medial meniscus (*arrowheads*) is identified.

ittal images demonstrate continuity of the meniscal tissue between the anterior and posterior horns (35). While variations in discoid menisci occur, the slab variety does not demonstrate the normal tapering of height as one moves from the periphery of the joint to the center so that the midzone (bow-tie) height remains constant. Additionally, discoid menisci are often characterized by a 2 mm difference in height of the periphery of the affected meniscus when compared to the opposite meniscus. The routine use of high resolution coronal images most graphically depicts the abnormally shaped meniscus which often covers the entire tibial plateau and extends into the intercondylar notch.

Menisci may be focally or diffusely too small. *Focal* alterations of meniscal size, especially those in which the posterior horn of the medial meniscus is smaller than the anterior horn, frequently indicate the presence of a large flap tear or bucket-handle tear (Fig. 11.12). A meniscus that is *diffusely* "too small" poses a more complex diagnostic problem. The dimensions given earlier are merely guidelines and it is essential that suspect menisci be compared with their contralateral mates to help exclude normal variations in size. "Hypoplastic" menisci, that is, menisci that are idiopathically significantly smaller than normal, have been reported, and in our experience and that of others have

A

B

C

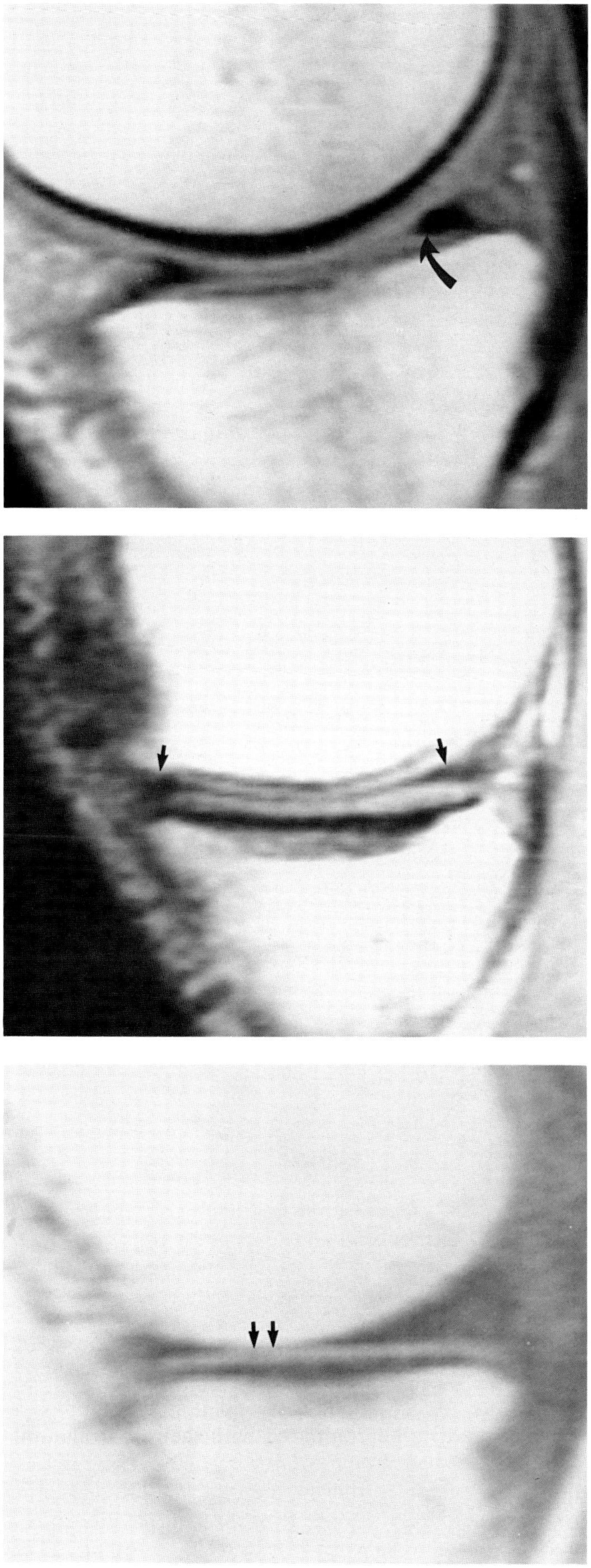

FIG. 11.12. Small menisci. **A:** Abnormal size ratio of meniscal horns. Sagittal MR, TR/TE 800/20. The posterior horn of the medial meniscus is approximately equal in size to the anterior. This change in the ratio of the sizes of the horns to one another should raise the question of a meniscal tear, but in addition, the tip of the posterior horn is blunted (*arrow*). A bucket-handle tear was found at surgery. **B:** Diffusely small meniscus, status post "bucket-handle-ectomy." Sagittal MR, TR/TE 800/20. Only tiny portions of the anterior and posterior horns of the medial meniscus (*arrows*) are identified in this patient who had an extremely large bucket-handle fragment removed. **C:** Hypoplastic medial meniscus. Sagittal MR, TR/TE 800/20. A very small body of the medial meniscus (*arrows*), and an even small anterior horn are present. An MR of the opposite knee demonstrated similar findings. At surgery the menisci were small but otherwise normal.

been found by arthrography in young women (26). The diagnosis of hypoplastic meniscus must remain one of exclusion.

A distinctly small meniscus may represent the peripheral segment of a bucket-handle tear, and before making the diagnosis of a hypoplastic meniscus, one must carefully examine the intercondylar notch on both the coronal and sagittal images (9,26,27). Oblique, longitudinal tears may result in displacement of a fragment of the meniscus into the intercondylar notch, and if the plane of the tear is parallel to the upper or lower margins of the meniscus, the remaining periphery may resemble a small but otherwise normal meniscus. In such instances there will be only one, instead of the normal two, MR images intersecting the bow-tie of the meniscus as seen on a sagittal series. Other clues to the correct diagnosis of an occult bucket-handle tear include a small meniscus with an asymmetric triangular shape and/or an increased apical angle.

A small meniscus may be found in the otherwise normal postoperative patient who has had removal of a previously displaced meniscal tear (Fig. 11.12) (26,36). Most surgeons resect the displaced portion of a torn meniscus and trim the remaining peripheral portion into a triangular shape. In postoperative patients in whom the centrally displaced fragment has been resected, the MR image of the peripheral rim will yield an image virtually identical to that of the *unoperated* patient with a displaced tear, but in the former, no central fragment can be found and postoperative changes are invariably present in the fat pad. Linear or nodular foci of decreased signal can be found in the fat pad which represent both micrometallic artifact from the instrumentation and scar in the healing ar-

A

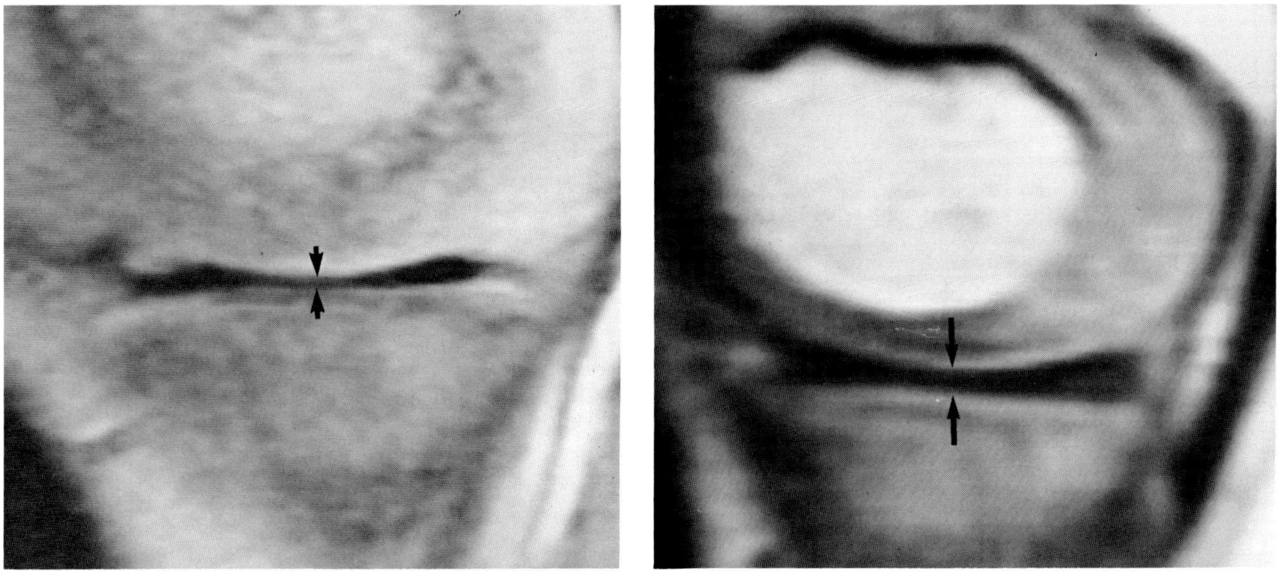

B

C

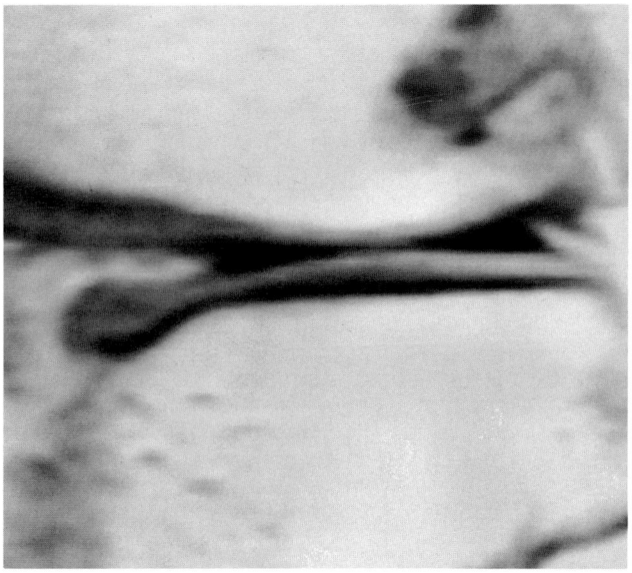

FIG. 11.13. Small menisci secondary to inflammatory arthritis. **A,B:** Sagittal MR, TR/TE 800/20. This 9-year-old boy has JRA (A). The height of the midzone of the medial meniscus (*arrows*) is extremely small, a finding which is more dramatic when compared to the height of the medial meniscus in a normal 8-year-old boy (B). **C:** Sagittal MR, TR/TE 800/20. In another patient, the lateral meniscus is quite small for an adult male. It was enveloped in a thickened synovial membrane at surgery. (From ref. 6, with permission.)

throscopic portals. Such fat pad alterations may provide a clue to the origin of the small meniscus. Finally, following a complete meniscectomy, fibrous tissue may proliferate from the highly vascularized meniscocapsular region and grow into the area of the resected meniscus. The regenerated tissue can resemble a meniscus in shape, but is only 2 to 4 mm in breadth (36,37). Critical assessment of the postoperative meniscus can be found in Chapter 7.

Patients that are afflicted with primary inflammatory arthridities commonly have small menisci (Fig. 11.13). (38). Arthroscopy in patients with juvenile rheumatoid arthritis (JRA) demonstrates a thickened, hypertrophied synovial membrane throughout the knee, which envelopes the menisci and causes difficulty in direct visualization. Thirteen of 20 knees afflicted with JRA have demonstrated distinctly abnormal menisci on MR examinations. In six of the knees, the menisci were so severely atrophic that they were barely visible; in the other seven the menisci were extremely small and had lost their normal triangular configuration. Additionally, we have seen abnormally small menisci in three adults. One patient had Reiter disease affecting the knee for years, and arthroscopy revealed a hypertrophied, inflamed synovium and small menisci. One patient had known rheumatoid arthritis, but the other patient did not have a known systemic disorder. The etiology for small menisci in patients with inflammatory disorders is unknown, but several postulates may be offered. The menisci derive their nutrition from the synovial fluid; "starvation" secondary to a markedly thickened synovium might result in atrophy of the menisci. Failure of proper meniscal development in patients with childhood arthritic disorders is another possible cause, but would not account for similar findings in adult patients. Release of proteolytic enzymes by inflamed synovium may lead to degeneration of the menisci.

Diffusely small menisci may be found in patients with longstanding neurologic disorders such as polio (Fig. 11.14).

Buckled Meniscus

Arthrographers have recognized that the menisci may appear transiently abnormal in configuration as a result of their elasticity and the stresses placed upon them during the performance of an arthrogram (39,40). The phenomenon is manifest by distortion of the meniscal shape so severe as to suggest a gross tear with amputation of the central portion of the meniscus. After further manipulation, however, the appearance returns to normal. The term "buckled meniscus" has

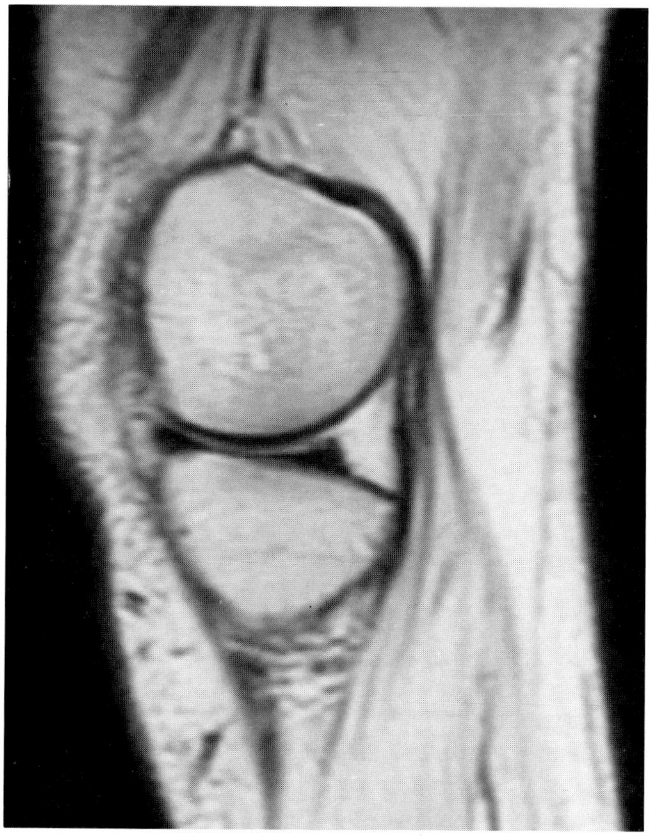

FIG. 11.14. Small meniscus in a patient with polio. Sagittal MR, TR/TE 2,000/20. This 50-year-old patient has had polio for many years. There is virtually no identifiable musculature in the leg. The midzone of the medial meniscus is virtually nonexistent, and the height of the meniscus is approximately 2 ½ mm.

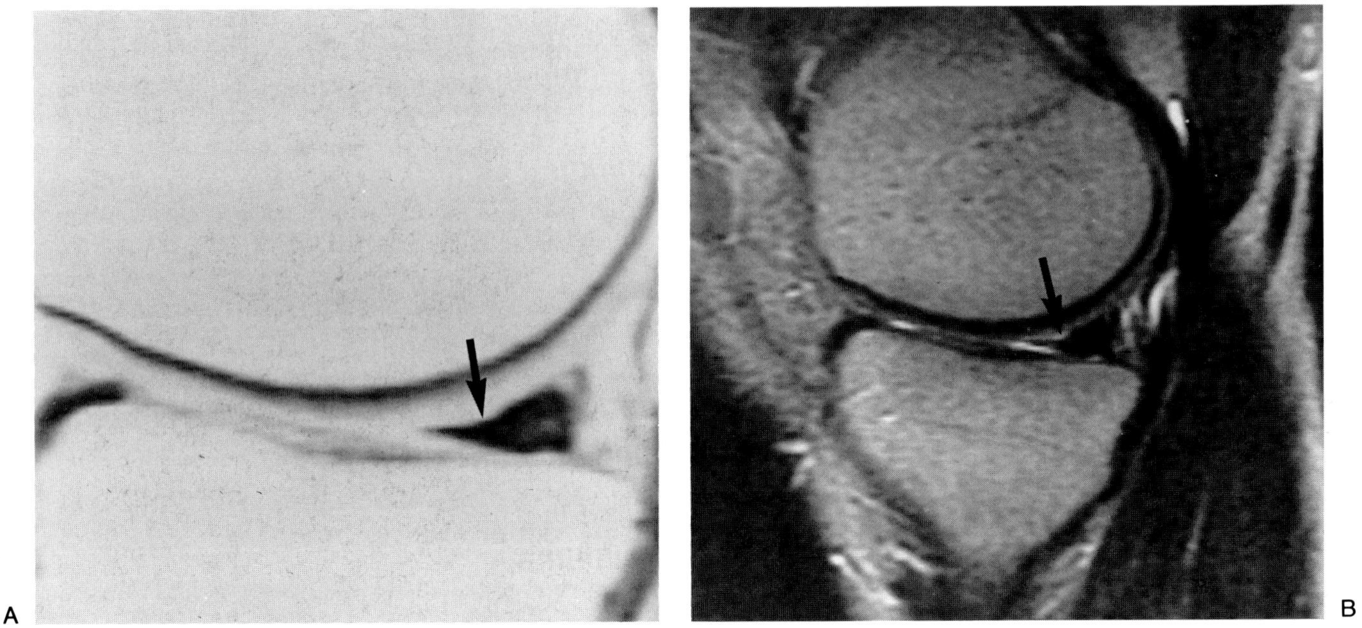

FIG. 11.15. Buckled medial meniscus. **A:** Sagittal MR, TR/TE 2,000/20. **B:** Sagittal MR, TR/TE 2,000/80. Both images made at the same location demonstrate a focal buckling of the superior surface (*arrow*) of the posterior horn of the medial meniscus. At surgery, the meniscus was quite pliable but otherwise normal.

been coined to describe this observation. Arthroscopy in such patients has demonstrated morphologically normal menisci which buckle upon themselves during the stressful maneuvers that are used during surgery to improve visualization.

We have identified a very small number of patients who have undergone MR examinations of the knee and have demonstrated slight buckling of the posterior horn of the medial meniscus; the superior surface demonstrates a wavy character and the meniscus appears foreshortened (Fig. 11.15). Unfortunately, the "MR arthrographer" does not have the ability to reposition the knee immediately and re-image the meniscus to confirm that a pseudotear is present. Re-imaging the patient on a different day may be worthwhile in indeterminate cases.

Meniscal Artifacts

High signal linear intrameniscal foci simulating meniscal tears may be caused by one of several artifacts. Intraarticular gas, either idiopathic (vacuum phenomenon) or iatrogenic (arthrography), may produce a high signal (T1WI) artifact within the meniscus which simulates a meniscal tear (Fig. 11.16) (41). The pitfall is secondary to the production of magnetic susceptibility artifacts. These artifacts occur where adjacent tissues vary greatly and abruptly in magnetic susceptibility; they are caused by field inhomogeneities resulting from intrinsic magnetic field gradients generated in such a region. These local inhomogeneities produce areas of high and low signal intensity similar to, but less predictable than, those seen with chemical shift artifact. Bone and air have less susceptibility, and ferromagnetic metals have more susceptibility than soft tissues, and it is at boundaries between these tissues where artifacts occur. If gas within the knee joint collects in a triangular configuration between the femoral or tibial articular cartilage and the meniscus, it may appear as a signal void separated from the low signal meniscus by a thin, high signal line, an appearance highly suggestive of a meniscal tear. Gradient-recalled images in patients with intraarticular gas distributed in a linear or triangular configuration demonstrate artifactual blooming of the signal void, resulting in dark and bright streaks as a result of misregistration of spatial information; in these cases, the findings suggest either/both meniscal and cartilaginous disease. Many of the patients with this reported artifact have had intraarticular gas present on a conventional radiograph, a finding which may help in differential diagnosis, but it is the blooming effect on gradient sequences that should alert the examiner to this pitfall.

Truncation artifacts produce high signal lines that simulate meniscal tears (Fig. 11.17) (42). These artifacts result from the use of Fourier-transform methods to construct MR images of high contrast boundaries, such as between the spinal cord and cerebrospinal fluid, between cortical bone and medullary fat, or be-

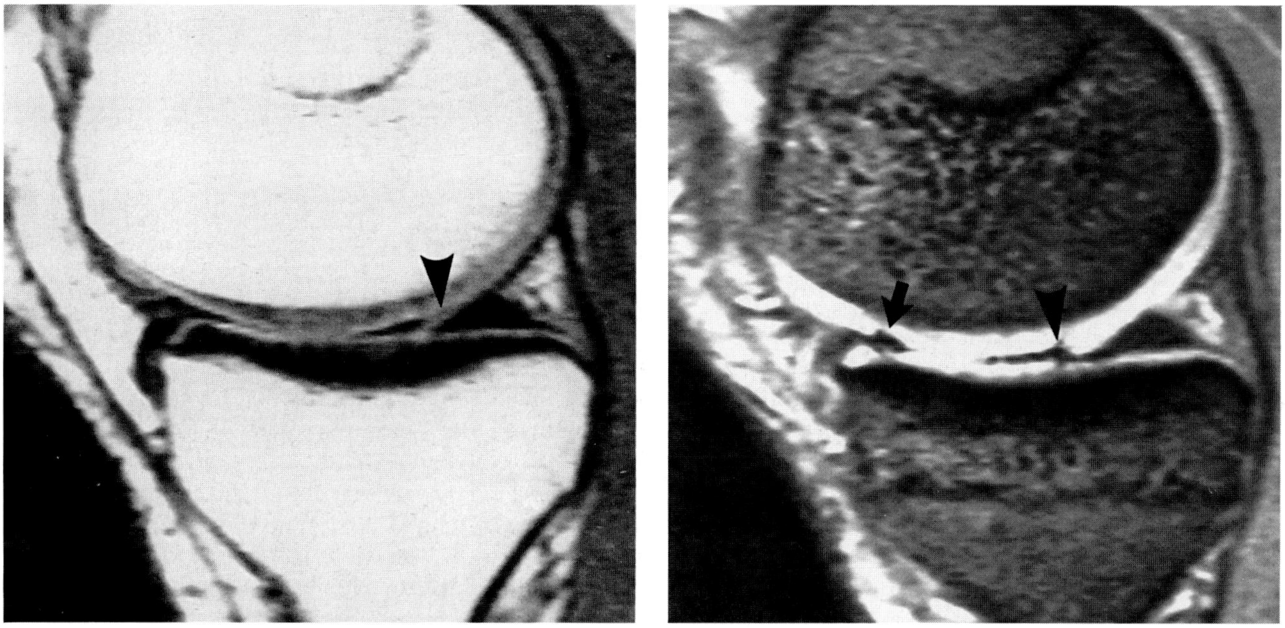

A B

FIG. 11.16. Pseudotear of the medial meniscus secondary to intraarticular gas. **A:** Sagittal MR, TR/TE 800/20. An oblique line of high signal (*arrowhead*) extends obliquely across the posterior horn of the medial meniscus, an appearance overwhelmingly suggestive of a tear. **B:** Sagittal MR, TR/TE 500/13. The oblique line demonstrates a "blooming effect" (*arrowhead*) both in the posterior horn and to a lesser degree adjacent to the anterior horn (*arrow*). Intraarticular gas, confirmed on a conventional radiograph, was responsible for this appearance. (Case courtesy of Mark Shogry, M.D. From ref. 53, with permission.)

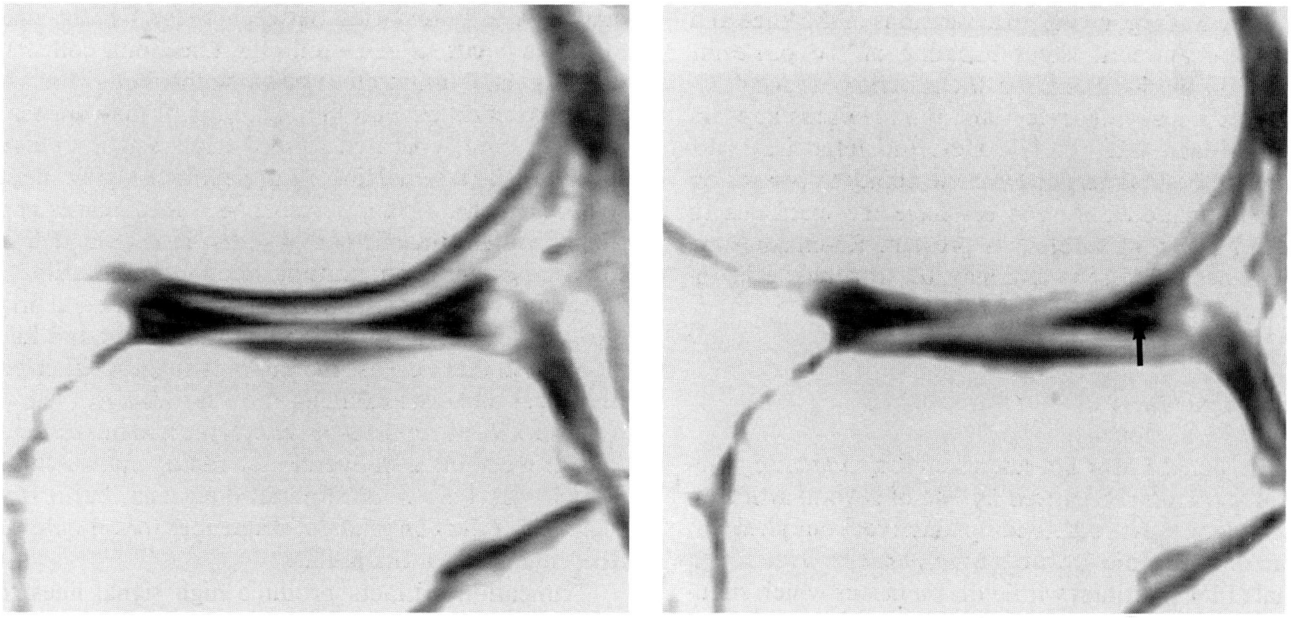

A B

FIG. 11.17. Truncation artifact. **A,B:** Sagittal MR, TR/TE 2,000/20. In (A), the phase-encoded direction is anterior-posterior, and the posterior horn of the lateral meniscus does not demonstrate any significant linear signal. In (B), the phase-encoded direction has been charged to superior-inferior and only 128 phase-encoded steps were utilized. A faint truncation artifact (*arrow*) is now evident.

tween articular cartilage and the menisci. When an attempt is made to describe such boundaries, an inaccurate representation may occur. These truncation artifacts are a series of high and low signal intensity lines adjacent to, and parallel to, a high contrast boundary. When a high intensity artifact is projected over a low signal intensity structure such as the meniscus, a pseudotear of the meniscus may result. The most prominent artifact is most likely to occur at a distance of 2 pixels from the high contrast boundary between the meniscus and the articular cartilage.

Conventional knee MR may be performed with either a 128 or 256 acquisition matrix in the phase-encoded direction which can be either anteroposterior or superoinferior. Truncation artifacts are most evident on those studies in which 128 phase-encoding steps are oriented in a superoinferior direction. If such boundary artifacts occur, they can be minimized by increasing the phase-encoding matrix to 192 or 256, and/or by redirecting a 128 step phase-encoding direction to anteroposterior rather than superoinferior. Recognition of the characteristics of truncation artifacts will help avoid mistaking them for meniscal tears. The artifacts are relatively subtle and have a uniform thickness; they are parallel to, and 2 pixels distant from, the high contrast boundary from which they originate. Careful inspection may reveal that these lines extend beyond the margins of the menisci (they do in fact exist across the image, but are seen only when projected over a low signal structure). In our experience, truncation artifacts have been only rarely confused with meniscal tears.

THE LIGAMENTS

Several variant appearances of the anterior cruciate ligament (ACL) may mimic pathologic states. The ACL is composed of two primary fiber bundles, the larger and stronger anteromedial and the smaller posterolateral band, each of which is composed of individual fascicles. While the normal ACL usually appears as a single solid structure extending from femur to tibia in the lateral aspect in the intercondylar notch, it is occasionally possible to identify the bands or individual fascicles on high resolution MR images (Fig. 11.18). One must not mistake the bundles and the tiny amounts of fat interposed between them for longitudinal tears of the ACL. Rarely, the bundles are so separated as to suggest a "double ACL."

Acute partial and complete ACL tears are frequently associated with a soft tissue mass at the proximal attachment of the ligament to the femur. This mass, which represents hemorrhage, edema, and the torn fi-

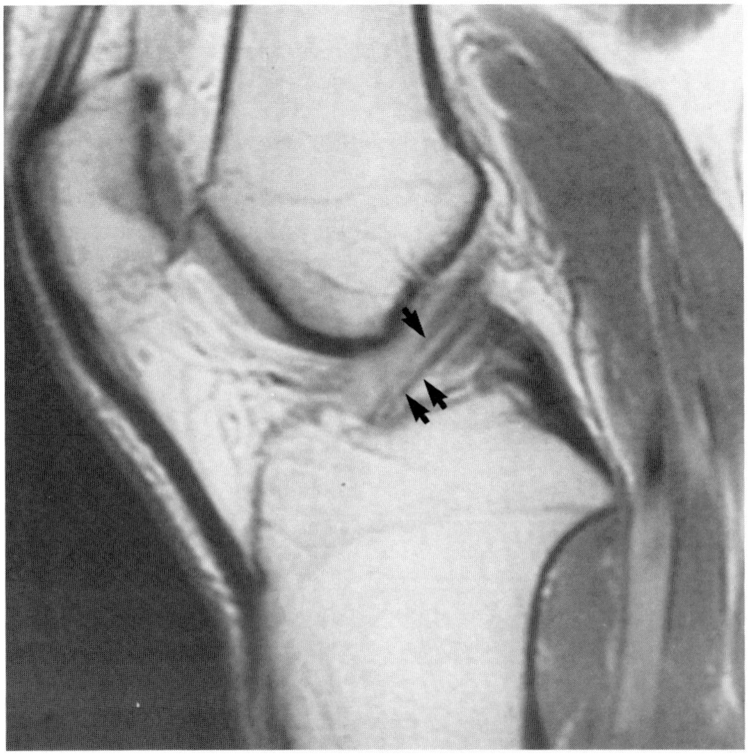

FIG. 11.18. "Double" ACL. Sagittal MR, TR/TE 2,000/20. The individual fiber bundles of the anterior cruciate ligament (*arrows*) are remarkably well distinguished.

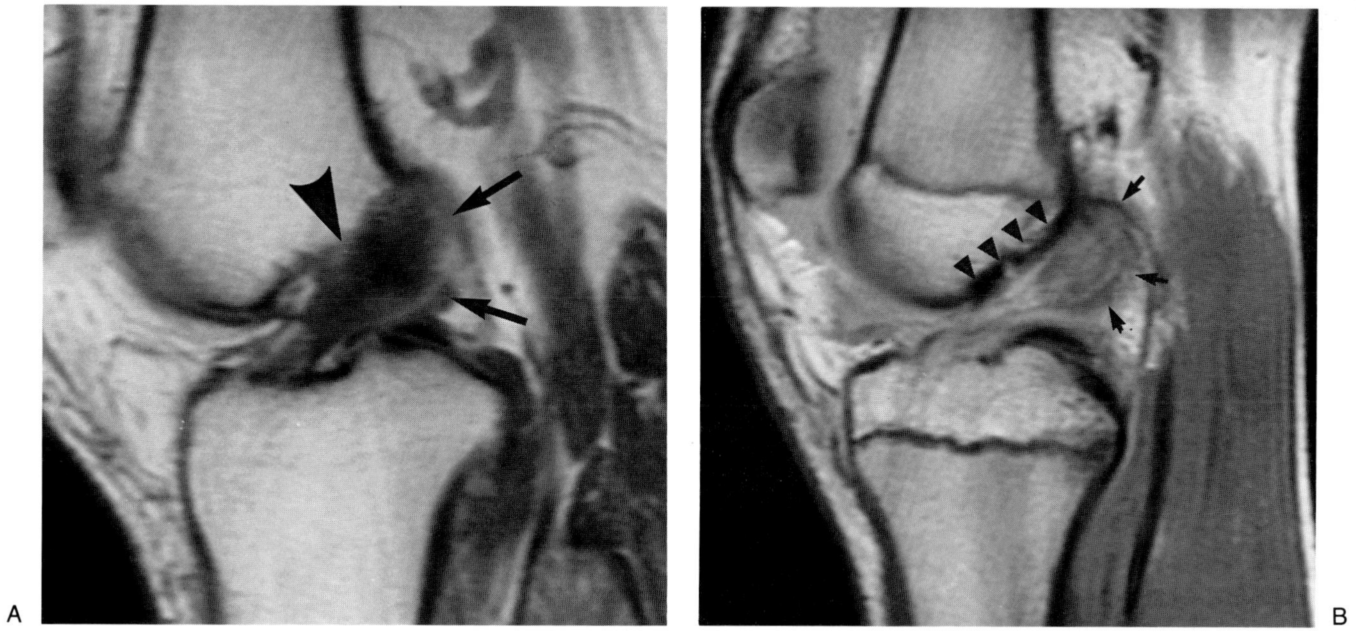

FIG. 11.19. Pseudotear of the ACL. **A:** Sagittal MR, TR/TE 2,000/20. Occasionally, significant partial volume effect of the lateral femoral condyle with the proximal ACL produces a pseudomass (*arrows*). The cortical margin of the intercondylar roof appears to end (*arrowhead*) instead of continuing to the posterior femoral cortex. **B:** True tear of the proximal ACL. Sagittal MR, TR/TE 2,000/20. Similar to (A) above, a soft tissue mass (*arrows*) is seen in the expected position of the proximal ACL. The intercondylar roof (*arrowheads*) is clearly delineated, indicating that there is little partial volume effect and that the soft tissue mass does in fact represent a torn ACL.

bers of the ACL, has an intermediate signal on T1WI, portions of which inhomogeneously increase in signal on the long TR sequences. During conventional knee MR studies, one may encounter a proximal ligamentous "mass" in a patient in whom there is no clinical evidence or history to suggest an ACL tear. This pseudomass represents a partial volume effect of the ACL with the cortical bone of the medial aspect of the lateral femoral condyle (Fig. 11.19) (see Chapter 5). This is most likely to occur if one has selected a 5 mm section thickness, and if the knee is excessively externally rotated. Three clues may be helpful in deciding whether or not one is dealing with a pseudomass. (1) One must ascertain whether or not the plane of the image is truly through the lateral intercondylar notch. Such a plane is characterized by visualization of the dark, uninterrupted line representing the intercondylar roof. If the complete posterosuperior to anteroinferior extent of the roof cannot be depicted, the image plane may suffer from partial volume effect, and the suspected mass possibly represents a pseudomass. (2) Examination of high resolution coronal MR images will demonstrate a normal proximal ligament in cases of a pseudomass. Partial volume effect in the coronal plane is a much less significant problem.

Rarely, the ACL may tear distally, and the remaining ligament may become balled up in the proximal, posterior aspect of the joint, simulating either a double PCL or a large loose body.

A variety of intraarticular normal and pathologic conditions may produce a band-like low signal structure paralleling the normal posterior cruciate ligament (PCL) (Fig. 11.20). This arrangement has given rise to the term "double PCL." In our experience, the most frequent cause of this lesion is a centrally displaced fragment from a bucket-handle tear of the medial meniscus. The fragment orients itself parallel to and in front of the anterior margin of the PCL. The proximal attachment of a large ligament of Wrisberg or Humphrey may simulate this anomaly.

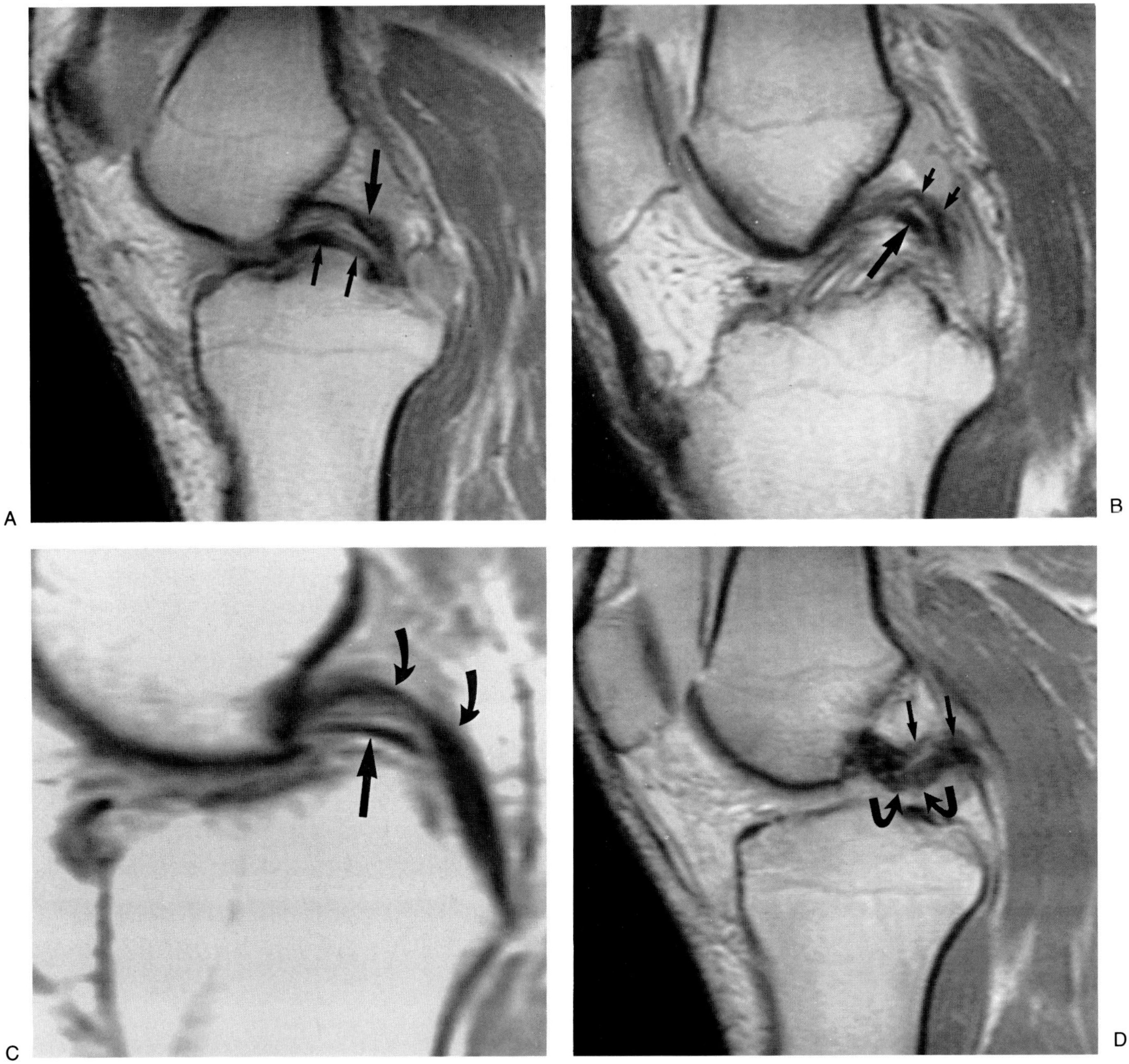

FIG. 11.20. "Double PCL," a spectrum of disorders. **A:** Sagittal MR, TR/TE 2,000/20. Two parallel linear dark structures are seen within the joint. The upper (*arrow*) is the normal PCL, while the lower (*double arrows*) is a centrally displaced fragment of a bucket-handle tear. **B:** Normal ligament of Wrisberg. Sagittal MR, TR/TE 2,000/20. Two parallel dark structures in the joint represent the normal proximal PCL (*arrow*) and a large ligament of Wrisberg (*double arrows*). At surgery, the patient had no detectable PCL abnormality. **C:** Normal ligament of Humphrey. Sagittal MR, TR/TE 2,000/20. The normal PCL (*curved arrows*) is seen just above an unusually long and large segment of the ligament of Humphrey (*arrow*). This image most closely simulates that in (A), and in this case, examination of all available images was necessary to confirm that there was no evidence of a displaced meniscal fragment. **D:** Failed PCL augmentation. Sagittal MR, TR/TE 2,000/20. The image suggests the presence of two PCLs, one on top of the other. This patient had had an attempted reconstruction of a torn PCL using a tendinous autograft. Sequential images suggested that the upper half of the PCL (*arrows*) is the remnant of the native PCL, and the lower half (*curved arrows*) is the failed tendinous reconstruction.

BONE AND CARTILAGE

Contour Anomalies of the Femoral Condyles

Gentle concavities of the anterior weightbearing aspects of either femoral condyle accommodate the anterior horns of the menisci during full knee extension (Fig. 11.21) (25,43,44). On conventional radiographs, these grooves are smooth and have intact cortical bone at their bases. On MR, the articular cartilage that overlies these grooves is directly visualized; it is concave but smooth in outline. The underlying cortical (dark) and medullary (bright) bone have normal signal on T1WI. These normal contour alterations may be accentuated by the occurrence of an osteochondral impaction injury that frequently accompanies ACL ruptures (45–47). The tibia may transiently anteriorly sublux during ACL rupture; the posterior aspect of the lateral tibial plateau will impact the anterolateral femoral condyle precisely within the groove. Alternatively, the knee may hyperextend with the leg internally rotated, producing an isolated tear of the ACL and driving the anterior horn of the lateral meniscus up into the lateral femoral condyle, again in the normal condylar concavity.

On conventional radiographs the groove may appear normal even following tibial subluxation if the degree of impaction is not significant enough to affect the cortical bone. MR, however, demonstrates the full spectrum of the injury by virtue of its ability to directly depict articular cartilage (46). In some cases only the articular cartilage is impacted; in others there is full

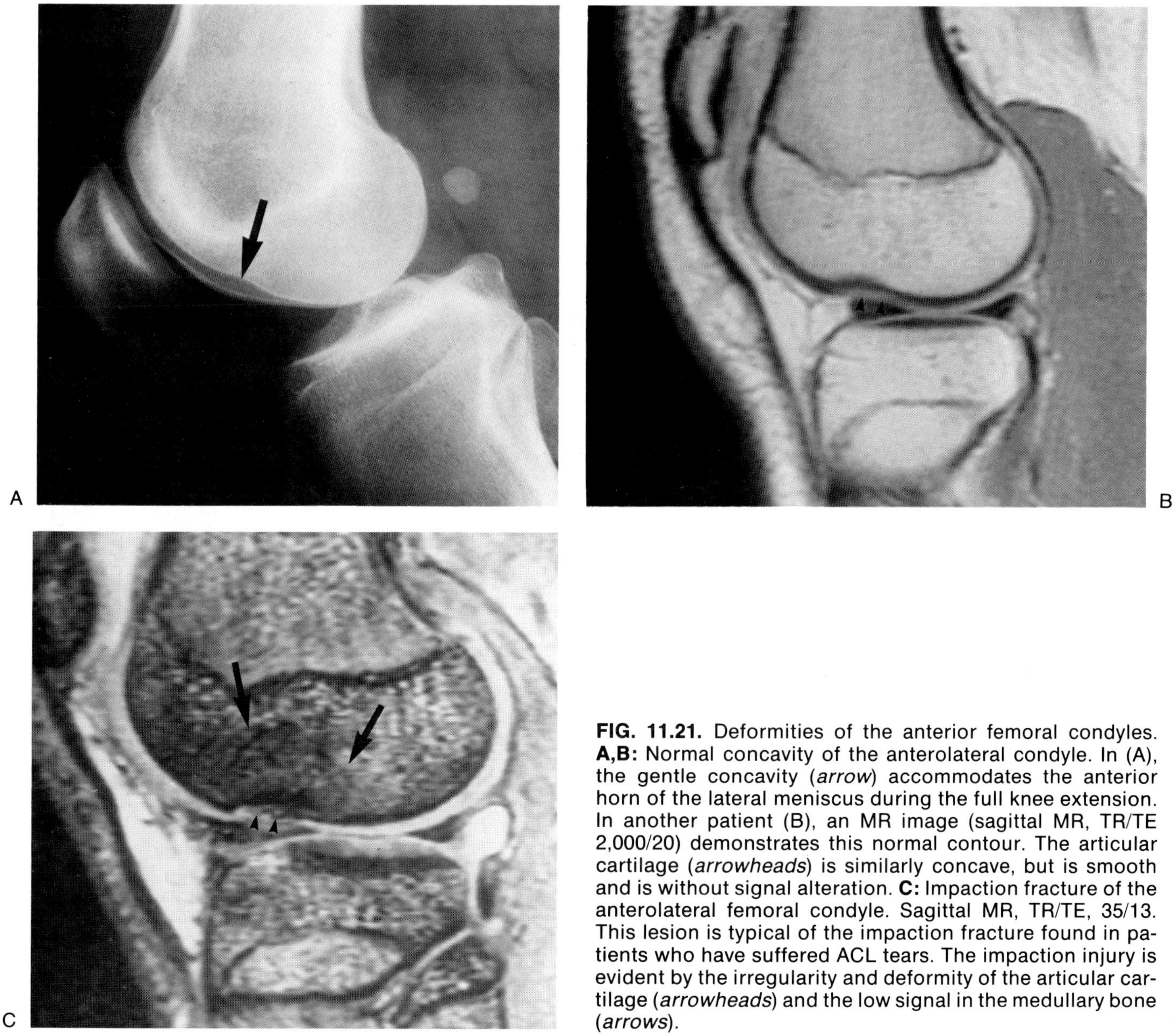

FIG. 11.21. Deformities of the anterior femoral condyles. **A,B:** Normal concavity of the anterolateral condyle. In (A), the gentle concavity (*arrow*) accommodates the anterior horn of the lateral meniscus during the full knee extension. In another patient (B), an MR image (sagittal MR, TR/TE 2,000/20) demonstrates this normal contour. The articular cartilage (*arrowheads*) is similarly concave, but is smooth and is without signal alteration. **C:** Impaction fracture of the anterolateral femoral condyle. Sagittal MR, TR/TE, 35/13. This lesion is typical of the impaction fracture found in patients who have suffered ACL tears. The impaction injury is evident by the irregularity and deformity of the articular cartilage (*arrowheads*) and the low signal in the medullary bone (*arrows*).

thickness fracture of the cartilage with exposure of the underlying bone. Subchondral edema/hemorrhage decreases the normally high signal of the intramedullary fat (T1WI). The overlying articular cartilage may be fibrillated, or the underlying bone may be exposed with a gross osteochondral fracture. In our experience and that of others, however, arthroscopists frequently report that the cartilage is intact and smooth in patients whose MR examination shows marked subchondral abnormalities (46–48). Young cartilage is quite elastic, and following an impaction injury the cartilage is capable of rebounding into normal position. Osteochondral fractures may progress to focal or diffuse areas of articular cartilage damage.

Stress Fractures

Both stress fractures and true traumatic fractures may be detected by MR prior to, or exclusive of, their appearance on conventional radiographs (45,46,48). One must be careful not to mistake the normal physeal

A

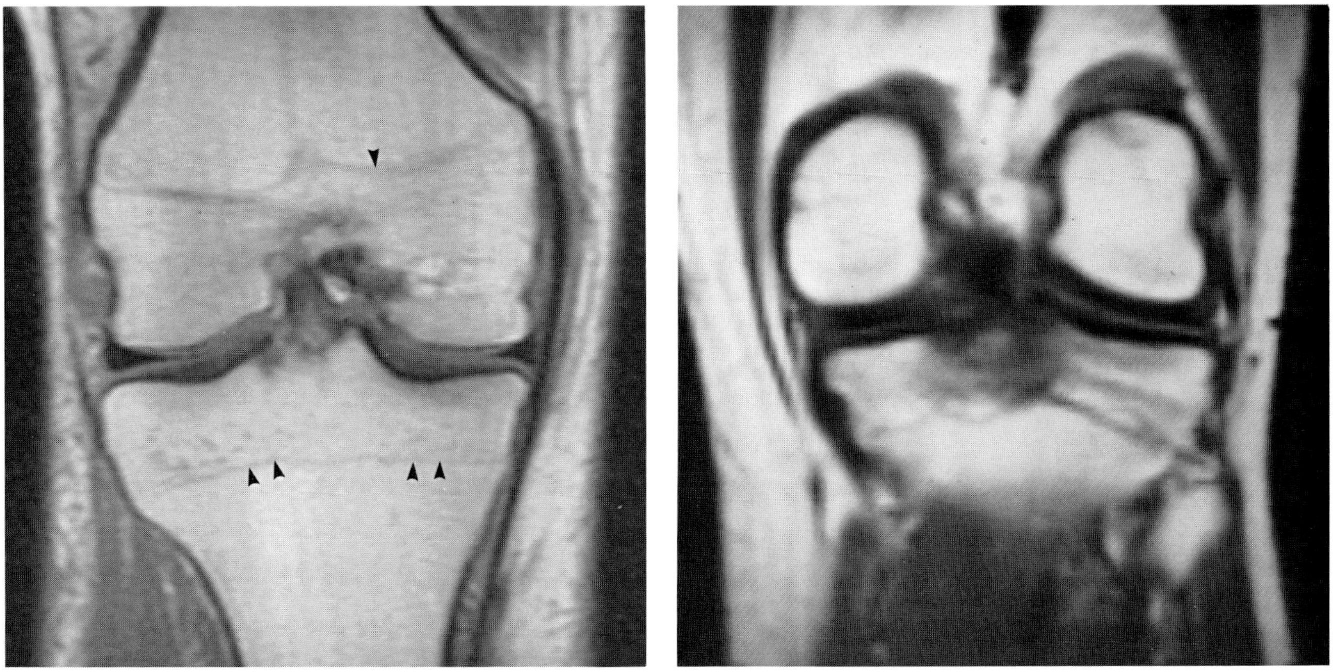

B

C

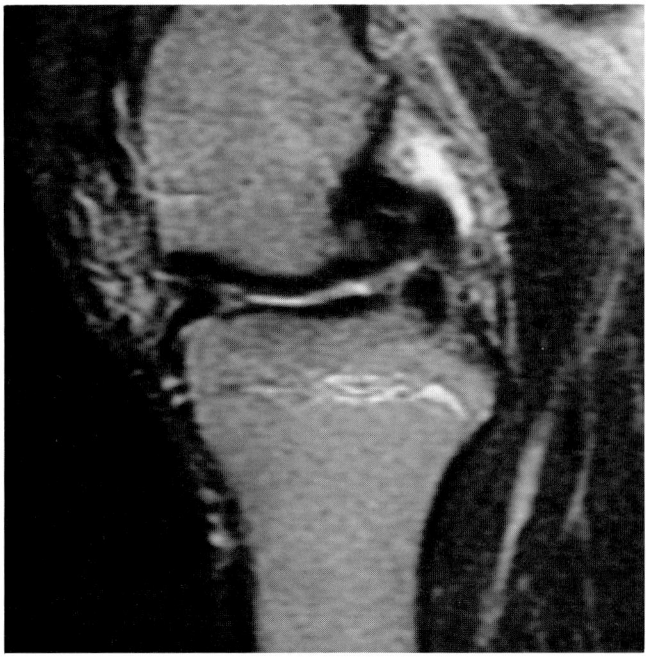

FIG. 11.22. The physis in an adult. **A:** Coronal MR, TR/TE 2,000/20. Portions of the femoral and tibial physeal scars (*arrowheads*) can nearly always be identified in adult patients. **B:** Plateau fracture. Coronal MR, TR/TE 800/20. There are multiple lines in the expected position of the lateral tibial physis; these represent fracture lines in this patient who suffered trauma. **C:** Sagittal MR, TR/TE 2,000/80. A line of increased signal is seen in the expected position of the physis in this adult whose clinical findings suggest a stress fracture. The adult physis is never bright on any MR sequence.

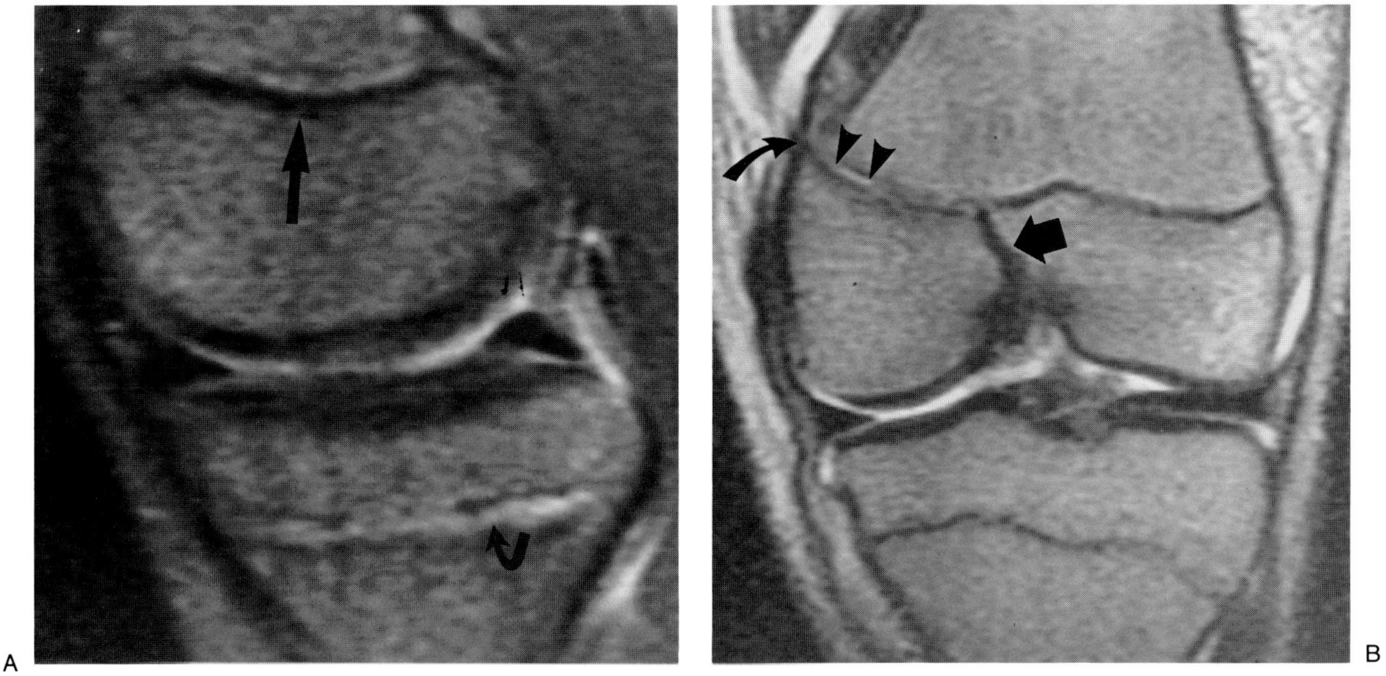

FIG. 11.23. The physis in a child. **A:** Sagittal MR, TR/TE 2,000/80. The tibial physis (*curved arrow*) is bright, while the femoral physis (*arrow*) is dark. This differing appearance is secondary to chemical shift artifact, a situation which can be confirmed by reversing the frequency-encoding direction from inferior/superior to superior/inferior. **B:** Coronal MR, TR/TE 2,000/80. The medial side of the physis is brighter (*arrowheads*) than the lateral. This patient suffered a vertical intercondylar fracture (*arrow*) which extended through the physis and produced a step off (*curved arrow*) medially. (From ref. 6, with permission.)

scar for a fracture (Fig. 11.22) (6). The physeal scar is manifest in adults of any age as a sharp line of decreased signal (T1 and T2WI). The differentiation of a fracture from the normal adult physis is made if (a) the line of decreased signal is seen not at the expected site of the physis, (b) a thick band is present, (c) multiple lines are seen, or (d) the line becomes bright on a T2WI. In adults, the physeal scar never increases in signal intensity, unlike fractures which invariably have sufficient edema that they become bright. In children, however, high signal associated with the physis may be normal (Fig. 11.23). Chemical shift artifact occurs in the frequency-encoded direction; if it is cephalad, the tibial physis in a child may show increased signal intensity on T2WI, simulating a fracture with edema. Confirmation that this appearance is, in fact, artifactual can be made by repeating the examination with reversal of the frequency-encoded direction, at which time the tibial physis will become normal and the femoral physis will become bright. Asymmetry of the signal on one side of the physeal scar when compared with the other must be highly suggestive of a fracture extending through the physis, regardless of the direction of frequency-encoding (Fig. 11.23).

Articular Cartilage Loss

High field strength imagers produce a chemical shift artifact that can alter the apparent thickness of the femoral or tibial cartilage and cortical bone if the frequency-encoded direction is superior/inferior or inferior/superior (Fig. 11.24). If the frequency-encoded direction is inferior/superior, the most peripheral sections through the femoral condylar cartilage and cortical bone can appear severely ''thinned'' because of both the chemical shift effect and the curved contours of the structure involved. It may therefore be ''normal'' not to see any articular cartilage peripherally, and the diagnosis of chondromalacia may be falsely applied. By reversing the frequency-encoded direction, one will reverse the relative cartilage thickness over the tibia and femur.

Loose Bodies

Osteocartilaginous bodies may have a variety of MR signal characteristics, depending on their composition. Those fragments that are composed predominantly of mature marrow fat have a high signal on T1WI; if heav-

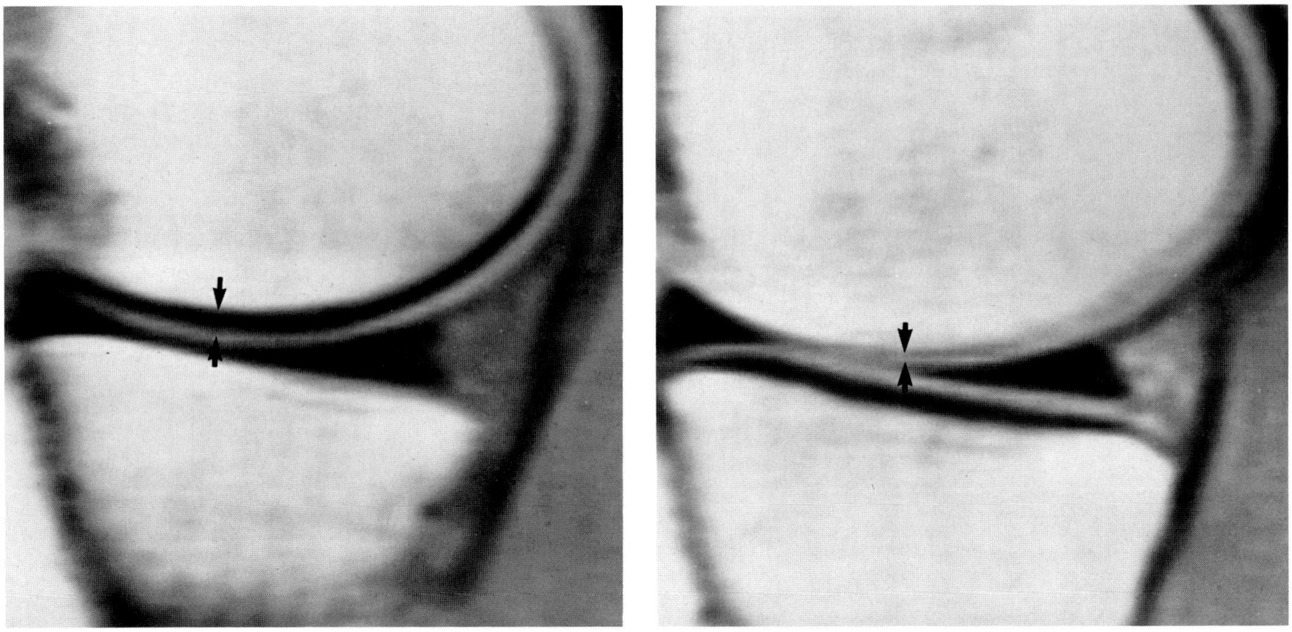

FIG. 11.24. Normal articular cartilage; effect of chemical shift. **A,B:** Sagittal MR, TR/TE 800/20. In (A), the frequency-encoded direction is superior to inferior. The femoral articular cartilage and cortical bone (*arrows*) appear markedly thicker than they do on (B), in which the frequency and coded direction is inferior to superior. In the same manner, the tibial articular cartilage, which appears nonexistent in (A), is dramatically thicker in (B). (From ref. 6, with permission.)

ily calcified, the body is often quite dark at all imaging sequences. Purely chondral loose bodies have an intermediate signal. While loose bodies, by definition, may be found in any portion of the knee joint at any given moment, most can be found in the most capacious segments of the capsule, namely, the suprapatellar bursa and the posterior capsule. It is common to find a large mass in the suprapatellar bursa on the extreme lateral sagittal images just lateral to the patellofemoral articular (Fig. 11.25). This "lesion" is bright on T1Wl, intermediate on proton density and T2W sequences, and very dark on STIR images. When carefully followed on sequential sagittal, or extreme coronal images, this large fatty "loose body" is seen to be continuous with the prefemoral fat. The characteristic location, the continuity with extraarticular fat, and the inability to identify an ossified loose body on conventional radiographs will suggest to the examiner that the abnormality is a pseudolesion. Such pseudo loose bodies resulting from partial volume effect of fat with cortical and medullary bone may be found in other locations about the knee, especially near the intercondylar notch. Hypertrophic nodules of synovium and subsynovial tissue may protrude into the joint and simulate loose bodies.

Joint Effusion

The patient undergoing MR is supine and the leg is fully extended. Fluid is therefore distributed differently than during conventional radiography when the patient is in a lateral decubitus position (49–51). The distribution of fluid during MR is determined by the degree of knee flexion, the degree of external/internal rotation of the knee, and the amount of fluid present.

Fluid normally accumulates in the most dependent portion of the knee joint that is available. When the knee is slightly flexed, the posterior capsule and the gastrocnemii muscles relax and fluid accumulates behind the femoral condyles and the PCL. If, however, the knee is in full extension, the gastrocnemii displace the fluid into the intercondylar region and ultimately into the suprapatellar bursa. The suprapatellar bursa has an inverted U appearance, with the medial and lateral reflections actually being more dependent than the antefemoral component. Fluid will, therefore, first be seen in these recesses rather than in the radiographic suprapatellar bursa. The medial and lateral reflections are well depicted on coronal and axial images. Because the knee is usually externally rotated during MR, small amounts of fluid will preferentially fill the

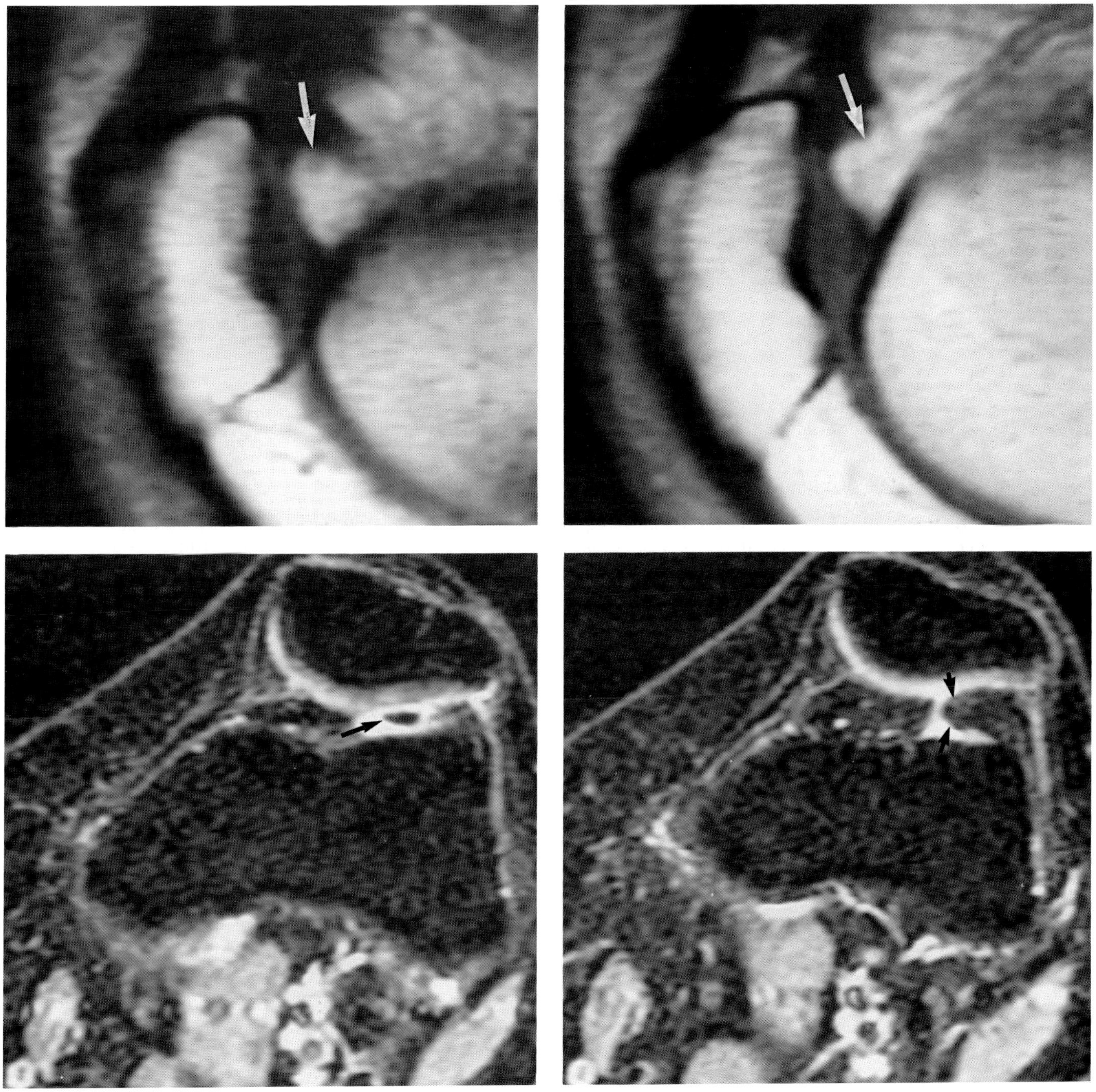

A B C D

FIG. 11.25. Pseudo loose bodies. **A,B:** Sagittal MR, TR/TE 800/20; (A) is 5 mm medial to (B). A fatty "mass" (*arrow*) is seen between the patella and the anterolateral femoral condyle. The lesion appears to be within the suprapatellar bursa. In (B), however, the "lesion" is seen to be continuous with the prefemoral fat. Invagination of the bursa by fat is common in patients with small joint effusions. **C,D:** Axial MR, TR/TE/TI 2,200/35/160. (C) is 5 mm proximal to (D). In (C), an ovoid focus of markedly decreased signal (*arrow*) is seen to be surrounded by joint fluid. This mimics an intraarticular loose body. In (D), however, the "lesion" is seen to represent the tip of the prefemoral fat invaginating the fluid-filled suprapatellar bursa.

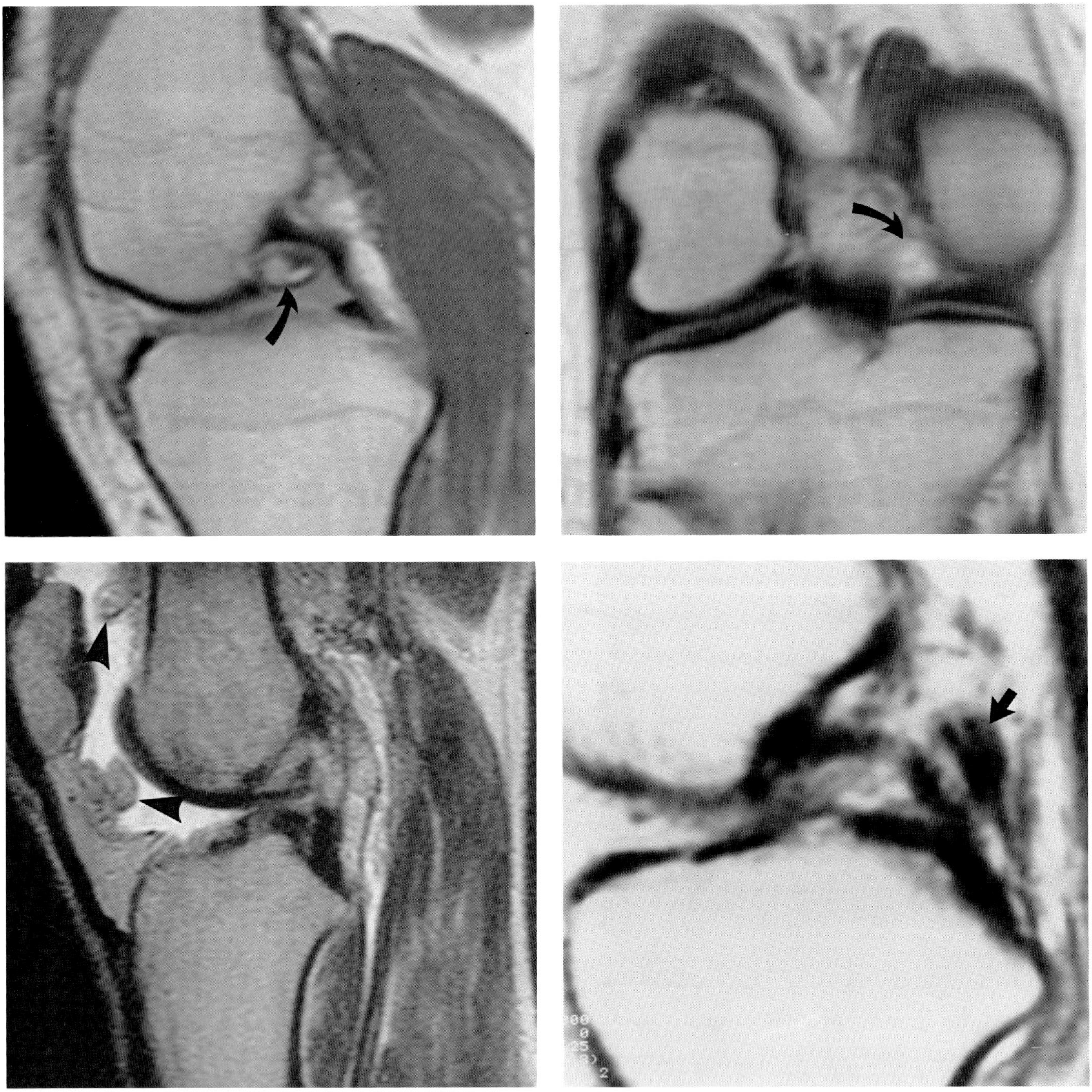

E F G H

FIG. 11.25. (*Continued*) **E:** Sagittal MR, TR/TE 2,000/20. **F:** Coronal MR, TR/TE 800/20. In a third patient, a loose body (*curved arrow*) appears to be present under the PCL on the sagittal image. This appearance is secondary to partial volume effect of this rounded focus of fat (*curved arrow*) with the cortical edge of the medial femoral condyle. **G:** Sagittal MR, TR/TE 2,000/80. Two nodular masses (*arrowheads*) protrude into the joint fluid and simulate loose bodies. At surgery, lobulated hypertrophied synovium was present but no loose bodies were seen. **H:** Sagittal MR, TR/TE 2,000/20. A large ligament of Wrisberg (*arrow*) simulates a loose body. [(A) and (B) from ref. 6, with permission.]

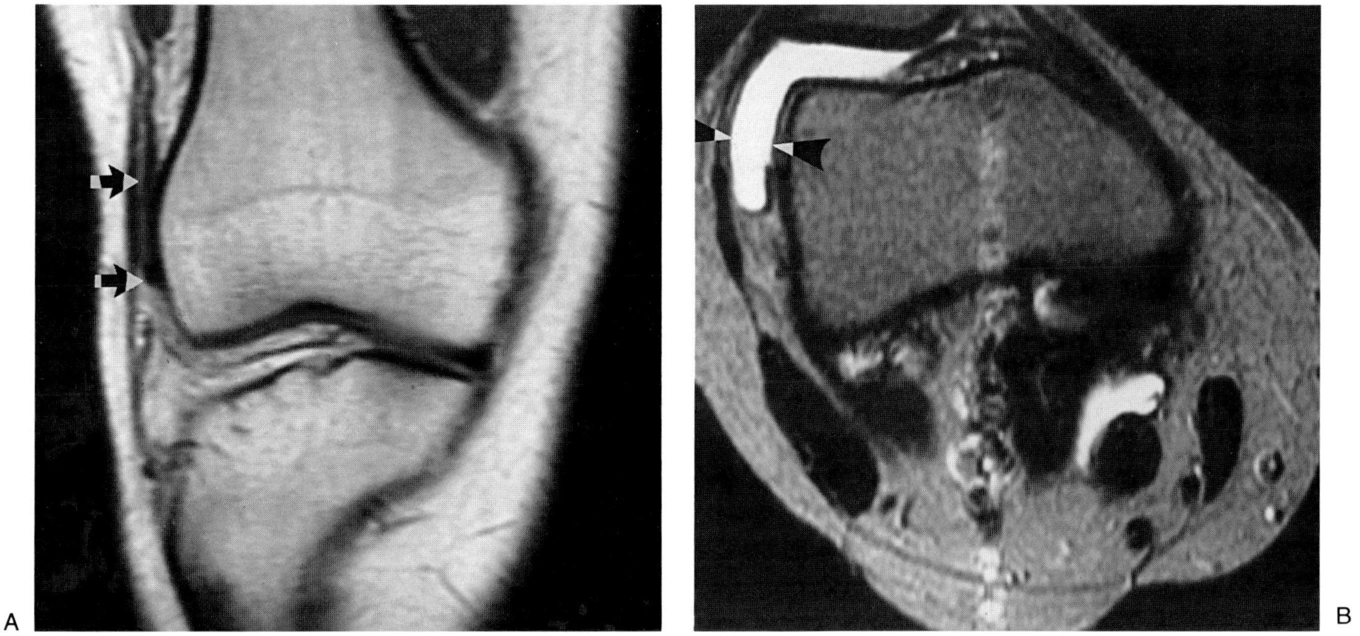

FIG. 11.26. Joint effusion. **A:** Coronal MR, TR/TE 800/20. A collection of fluid (*arrows*) is seen lateral to the lateral femoral condyle, and just medial to the iliotibial band. **B:** Axial MR, TR/TE 2,000/80. Because of the normal external rotation position of the knee, small amounts of joint fluid preferentially fill the lateral "saddlebag" of the suprapatellar bursa.

lateral rather than the medial recess. The distended lateral gutter often appears as a soft tissue mass just lateral to the femur, and just medial to the iliotibial band (Fig. 11.26).

REFERENCES

1. Crues JV, Mink JH, Levy TL, et al. Meniscal tears of the knee: Accuracy of MR imaging. *Radiology* 1987;164:445–448.
2. Mink JH, Levy T, Crues JV. Tears of the anterior cruciate ligament and menisci of the knee: MR imaging evaluation. *Radiology* 1988;167:769–774.
3. Bellon E, Haacke E, Coleman P, et al. MR artifacts: A review. *AJR* 1986;147:1271–1281.
4. Patton J, Kulkarni M, Craig J, et al. Techniques, pitfalls and artifacts in magnetic resonance imaging. *Radiographics* 1987;7:505–519.
5. Richardson ML, Helms CA. Artifacts, normal variants, and imaging pitfalls of musculoskeletal magnetic resonance imaging. *Radiol Clin North Am* 1986;24(2):145–177.
6. Mink J. *Pitfalls in interpretation.* IN: Mink J, Reicher MA, Crues JV III, eds. Magnetic resonance imaging of the knee. New York: Raven Press; 1987.
7. Mink JH, Stoller DW, Martin C. *MR imaging of the knee: Pitfalls in interpretation.* IN: Radiological Society of North America 72nd scientific assembly and annual meeting. Chicago: 1986.
8. Hall F. Pitfalls in knee arthrography. *Radiology* 1976;118:55–62.
9. Hall F, Further pitfalls in knee arthrography. *J Can Assoc Radiol* 1987;29:179–184.
10. Harley J. An anatomic-arthrographic study of the relationships of the lateral meniscus and the popliteus tendon. *AJR* 1977;128:181–187.
11. Jelaso D. The fascicles of the lateral meniscus: An anatomic-arthrographic correlation. *Radiology* 1975;114:335–339.
12. Pavlov H, Goldman A. The popliteus bursa: An indicator of subtle pathology. *AJR* 1980;134:313–321.
13. McIntyre JL. Arthroscopy of the lateral meniscus. *Radiology* 1972;105:531–536.
14. Wickstrom KT, Spitzer RM, Olsson HE. Roentgen anatomy of the posterior horn of the lateral meniscus. *Radiology* 1975;116:617–619.
15. Watt I, Tasker T. Pitfalls in double contrast knee arthrography. *Br J Radiol* 1980;53:754–759.
16. Nicholas JA, Freiberger RH, Killoran PJ. Double contrast arthrography of the knee: Its value in management of 225 knee derangements. *J Bone Joint Surg* 1970;52:203–220.
17. Thijn CJP. Accuracy of double-contrast arthrography and arthroscopy of the knee joint. *Skeletal Radiol* 1982;8:187–192.
18. Watanabe AT, Carter BC, Teitelbaum GP, et al. Common pitfalls in magnetic resonance imaging of the knee. *J Bone Joint Surg* 1989;71A(6):857–862.
19. Vahey TN, Bennett HT, Arrington LE, et al. MR imaging of the knee: Pseudotear of the lateral meniscus caused by the meniscofemoral ligament. *AJR* 1990;154:1237–1239.
20. Van Dommelen B, Fowler P. Anatomy of the posterior cruciate ligament. A review. *Am J Sports Med* 1989;17(1):24–29.
21. Brantigan OC, Voshell AF. Ligaments of knee joints: Relationship of ligament of Humphrey to ligament of Wrisberg. *J Bone Joint Surg* 1946:28:66–67.
22. Carpenter WA. Meniscofemoral ligament simulating a tear of the lateral meniscus: MR features. *J Comput Assist Tomogr* 1990;14(6):1033–1034.
23. Watanabe AT, Carter BC, Teitelbaum GP, et al. Normal variations in MR imaging of the knee: Appearance and frequency. *AJR* 1989;153:341–344.
24. Herman LJ, Beltran J. Pitfalls in MR imaging of the knee. *Radiology* 1988;167:775–781.
25. Clemente LD. *Gray's anatomy.* Philadelphia: Lea & Febiger; 1985.
26. Ricklin P, Ruttiman A, Del Buono MS. *Meniscus lesions: Practical problems of clinical diagnosis, arthrography and therapy.* New York: Grune & Stratton, 1971.
27. DeSmet AA. Meniscal tears on knee arthrography: Patterns of arthrographic abnormalities. *Skeletal Radiol* 1985;14:280–285.
28. Torisu T. Isolated avulsion fracture of the tibial attachment of the posterior cruciate ligament. *J Bone Joint Surg* 1977;59:68–72.

29. Ferrer-Roca O, Vilalta C. Lesions of the meniscus: Macroscopic and histologic findings. *Clin Orthop* 1980;146:289–300.
30. Hall FM. Arthrography of the discoid lateral meniscus. *AJR* 1977;128:993–1002.
31. Dickason JM, Del Pizzo W, Blazina M, et al. A series of ten discoid medial menisci. *Orthopedics* 1982;168:75–79.
32. Smillie IS. The congenital discoid meniscus. *J Bone Joint Surg* 1948;30:671–682.
33. Dalinka MK, Lally JF, Gohel VK. Arthrography of the lateral meniscus. *AJR* 1974;121:79–85.
34. Kaplan EB. Discoid lateral meniscus of the knee joint. *J Bone Joint Surg* 1957;39:77–87.
35. Silverman JM, Mink JH, Deutsch AL. Discoid menisci of the knee; MR imaging appearance. *Radiology* 1989;173:351–354.
36. Debnam JW, Staple TW. Arthrography of the knee after meniscectomy. *Radiology* 1974;113:67–71.
37. Arndt RD, Horns JW, Gold RH. *Arthrography* Baltimore: Williams and Wilkins, 1981.
38. Senac MO, Deutsch D, Bernstein BH, et al. MR imaging in juvenile rheumatoid arthritis. *AJR* 1988;150:873–878.
39. Hall FM. Pitfalls in the assessment of the menisci by knee arthrography. *Radiol Clin North Am* 1981;19(2):305–328.
40. Hall FM. Buckled meniscus. *Radiology* 1978;126:89–90.
41. Shogry MEC, Pope TL. Vacuum phenomenon simulating meniscal or cartilaginous injury of the knee at MR imaging. *Radiology* 1991;180:513–515.
42. Turner DA, Rappaport MI, Erwin WD, et al. Truncation artifact: A potential pitfall in MR imaging of the menisci of the knee. *Radiology* 1991;179:629–633.
43. Harrison H, Wood M, Keats T. The grooves of the distal articular surface of the femur–A normal variant. *AJR* 1976;126:751–754.
44. Noyes FR, Mooar PA, Matthews DS, et al. The symptomatic anterior cruciate-deficient knee. Part I. The long term functional disability in athletically active individuals. *J Bone Joint Surg* [AM] 1983;65:154–162.
45. Deutsch AL, Mink JH. Magnetic resonance imaging of musculoskeletal injuries. *Radiol Clin North Am* 1989;27(5)983–1001.
46. Mink JH, Deutsch AL. Occult cartilage and bone injuries of the knee: Detection, classification, and assessment with MR imaging. *Radiology* 1989;170:823–829.
47. Munk PL, Helms CA. *MRI of the knee*. Gaithersburg, Maryland: Aspen, 1992.
48. Yao L, Lee JK. Occult intraosseous fracture: Detection with MR imaging. *Radiology* 1988;167:749–751.
49. Hall F. Radiographic diagnosis and accuracy in knee joint effusions. *Radiology* 1975;115:49–54.
50. Singer AM, Naimark A, Felson D, et al. Comparison of overhead and cross-table lateral views for detection of knee-joint effusion. *AJR* 1985;144:973–975.
51. Beltran J, Noto A, Herman L, et al. Joint effusions: MR imaging. *Radiology* 1986;158:133–137.

CHAPTER 12

Cost-Effectiveness of Magnetic Resonance Imaging

Patrick A. Ruwe and Shirley McCarthy

The history and physical examination of the knee in patients with knee complaints are known to be nonspecific in determining the cause of internal derangement. In a recent comprehensive study, physical evaluation of the knee was evaluated in 161 patients with knee pain of at least 1 year duration (1). The predictive value of five common clinical tests (McMurray, flexion pinch test, Apley grind test, joint line tenderness, extension block) for the detection of meniscal tears was evaluated in a prospective fashion and compared to arthroscopic findings. The authors found a correlation between joint line tenderness and meniscal pathology in 68 of 80 patients with a proven meniscal tear. Overall the sensitivity was 85.5% and the specificity only 29.4%. Sixty-five patients had pain in forced flexion for a sensitivity of only 50% and a specificity of 68.2%. McMurray test and extension block were insensitive in predicting meniscal tears. In patients with anterior cruciate ligament tears, only the presence of block to extension showed predictive value for meniscal tear. The authors concluded that clinical diagnosis has considerable limitations when used to confirm meniscal lesions of the knee.

In patients in whom the clinical diagnosis is uncertain, physicians must therefore turn to other diagnostic techniques to determine the appropriate treatment. Prior to the advent of magnetic resonance imaging (MRI), plain radiography, arthrography, computed tomographic arthrography, and diagnostic arthroscopy were employed. Plain radiography is clearly insensitive, while arthrography is invasive although considerably more accurate with the most skilled users (1–3). CT arthrography may be accurate (4) but it is expensive, invasive, and exposes the patient to ionizing radiation. These factors, among others, has led to the use of arthroscopy as the diagnostic procedure of choice of orthopaedic surgeons (5).

While diagnostic arthroscopy is expensive and invasive, proponents point to its accuracy and to the surgeon's ability to diagnose and treat pathology with a single procedure (6). Unfortunately, diagnostic arthroscopy will sometimes demonstrate no abnormality; minor, possibly nonpathologic lesions such as plicae or chondromalacia patellae; or a totally different diagnosis all together (7). Consequently, patients may be subjected to an unnecessary operative procedure with its associated risks and potential morbidity.

MRI, a relatively new imaging technique for the knee, has been shown in multiple studies to have excellent sensitivity and specificity, particularly for meniscal pathology (4,8–11). The wide range of reported accuracies can be partly attributed to scanner type, imaging technique, radiologist's experience, and choice of gold standard. For example, accuracy in diagnosing medial meniscus tears increased from 86% to 93%, sensitivity from 84% to 97%, and negative predictive value from 86% to 96% when comparisons were drawn between a mobile .35 Tesla and a stationary 1.5 Tesla machine (12). Variables in this study were minimized: the same technique and the same radiologist were used. Using the same patient for both scanners would be ideal but has yet to be reported. Using a 1.5 T magnet, Mink et al. reported an overall MRI accuracy of 93% (13). They observed that the false-positive MR imaging rates varied from 6% to 16% among 37 general orthopedic surgeons; however, it was only 6% for the knee subspecialist arthroscopist.

Arthroscopy is usually used as the gold standard for

determining MRI accuracy, but in actuality, the accuracy of arthroscopy and MRI are not dissimilar. Each are highly dependent on the surgeon's operative expertise and the radiologist's technical and interpretative expertise, respectively.

Quinn et al. (14) reported that some MRI false-positives as measured against arthroscopic findings may be attributed to inadequate visualization of the meniscus at arthroscopy. They conducted a study of 254 consecutive patients who underwent both knee MR and arthroscopy. All the arthroscopic procedures were videotaped. There were 17 false-positive MR errors as determined by the results of the operative report or the statement of the surgeon. Fifty-three percent (9/17) represented high signal lines that extended to the articular surface of the posterior horn which were incompletely visualized when the arthroscopic videotapes were reviewed. In 11.8% (2/17), postmeniscectomy irregularities were found but were not considered clinically significant by the arthroscopist. Forty-one percent (7/17) were high signal abnormalities involving the free edge or the entire meniscus that did not correlate with clinically significant tears. Of 32 false-negative errors, 46.7% (15/32) were small tears not necessitating meniscectomy. Seventeen of thirty-two (17/32) of the false-negative errors were treated with a partial meniscectomy, and nine of these were radial tears or abnormalities of the inner third of the meniscus. Considering only significant tears missed, the false-negative rate was 3% (17/508).

The authors conclude that false-positive MR studies involving the posterior horn may actually indicate the inability of the arthroscopist to adequately visualize the meniscus. Not all segments of the posterior horn can be easily visualized. False-positive MR scans may also reflect the subjectivity with which postsurgical appearances and free edge irregularities are labeled. False-negative MR scans often represent small tears diagnosed by compression or probing at arthroscopy, most of which are managed nonsurgically and hence are not a retrospective justification for arthroscopy.

MRI is useful to discern between those patients without significant internal derangements and those patients who would benefit from therapeutic arthroscopy (4,9,15). In addition to guiding treatment, MRI further aids the surgeon in terms of preoperative planning. Plans for meniscal repair can be made in appropriate patients based on the MR scan. Isolated ACL injuries, based on instability patterns, can be treated electively while young patients with combined meniscal tears and ACL disruption can be treated with meniscal repair and cruciate ligament reconstruction.

A limitation of many of the studies comparing MRI to arthroscopy is a failure to use the final outcome of the patient as the standard of comparison (16,17). The success of MRI can be judged by two criteria: patient outcome and cost-effectiveness. If MR can preclude an unnecessary arthroscopy, then MR would be judged as beneficial. Furthermore, if a significant number of patients avoid the cost of diagnostic arthroscopy, then MR in all patients may also save money, i.e., be cost-effective. A few recent studies have attempted to determine if MRI in all patients with suspected internal derangement was cost-effective compared to diagnostic arthroscopy in diagnosing internal derangement of the knee. Boden et al. (6) concluded on the basis of a cost analysis formula that diagnostic arthroscopy would be more cost-effective than MRI if 78% of the patients underwent arthroscopy. The study cohort consisted of a select population which excluded patients who were older, had no history of specific injury, had surgery, and/or had knee complaints of greater than 6 months duration. In their study, MRI was not cost-effective because 87% of the patients underwent arthroscopy. This study had several limitations, however (18). First, the final outcome of the patients was not ascertained, i.e., did the patient benefit from arthroscopy? The authors provided no information on the clinical course for the 47 patients who received arthroscopy. Patients who failed to improve following arthroscopy would hardly be a triumph for arthroscopy. Second, the authors failed to define MRI criteria. Third, no criteria for avoiding a diagnostic arthroscopy were stated. It is not clear why or how the surgeon used MRI and if the decision to arthroscope the patient was made before and despite the MRI.

In a recent study attempting to determine if MRI is cost-effective and if it reduces the need for diagnostic knee arthroscopy, only 43% of patients underwent immediate therapeutic arthroscopy based on MRI findings (19). During a 9-month period, 103 consecutive patients with knee complaints sufficient to warrant a diagnostic arthroscopy and compatible with the diagnosis of internal knee derangement underwent a 1.5 T MRI. The criteria for inclusion were persistent knee pain, recurrent knee effusions, repeated episodes of giving-way, a positive McMurray's sign, a positive flexion pinch test, or acute trauma. Patients with a normal MR scan, or MR diagnoses of isolated ligament tears or intrasubstance meniscal changes, did not receive diagnostic arthroscopy. These patients were evaluated at a minimum of 17 months after the MR scan by questionnaire or by phone interview.

Patients were classified as successes if they had not had a subsequent positive diagnostic arthroscopy and were functionally normal without significant activity limitation, had received a subsequent negative arthroscopy, or had received ligament reconstruction for known chronic instability (and had no other treatable lesion in the knee). Patients were classified as failures if they received a subsequent positive arthroscopy or had persistent complaints limiting activity. Cost-effec-

(103 × 3900) −	[(103 × 1000) +	(44 × 3900) +	(6 × 3900)] =	103,700
All patients Surgery	All patients MRI	Surgery Patients	Failures	Savings

FIG. 12.1. Formula used to calculate cost-effectiveness of MRI versus diagnostic arthroscopy at the Yale University Sports Medicine Center.

(103 × 4300) −	[(103 × 1000) +	(44 × 4300) +	(6 × 4300)] =	124,900
All patients Surgery	All patients MRI	Surgery Patients	Failures	Savings

FIG. 12.2. Use of formula for costs in Los Angeles results in a greater savings.

tiveness was based on the estimated costs of $1,000 per knee MR scan for all patients and a savings of $3,900 for each patient not receiving a diagnostic arthroscopy. The cost-effectiveness of all patients with knee complaints receiving MR was calculated as follows: the cost of all patients receiving a MR scan, plus the cost of all patients who received an immediate therapeutic arthroscopy, plus all patients who were classified as failures, was subtracted from the theoretical cost of all patients receiving an immediate diagnostic arthroscopy (Fig. 12.1).

As a result of an MRI, 43% (44/103) of patients received immediate surgery. The remaining 57% (59/103) did not have immediate surgery and 93% (55/59) were available for follow-up at a mean of 22 months (17/28). Forty-nine out of fifty-five (89%) were successes; 40 had normal function and no limitation in activity, six had chronic ACL reconstruction, and three had negative arthroscopies. Eleven percent were failures; four had continued limitation in activity and two had successful arthroscopic treatment of plicae.

All patients in this study would have had a diagnostic arthroscopy prior to the use of MRI. Based on the MR diagnoses, however, 51% of patients avoided a potentially unnecessary diagnostic arthroscopy resulting in a net savings of $103,700 for the 103 patients.

The above findings that many patients with knee complaints will improve with nonoperative therapy is consistent with the results of the study by Hede et al. (20) who found that 14 of 36 (39%) patients decided not to proceed with knee arthroscopy while on the surgical waiting list. Boeree et al. (21) have suggested that the accuracy of MRI should allow more appropriate selection of patients for whom arthroscopy would be beneficial. The authors postulate that this would eliminate the need for arthroscopy in one-third to one-half of those considered on clinical grounds to have meniscal or cruciate derangement.

Finally, in interpreting the results of these and other studies, several limitations should be considered. First, reproduction of good results could only be anticipated in centers with state-of-the-art MR scanners and experienced MR radiologists interpreting the scans. Second, costs described in these studies are specific to the reporting institution and cannot necessarily be extrapolated to other sites. For example, in Los Angeles, the cost savings would be $124,900 (Fig. 12.2). Finally, not all costs are included in cost analysis equations, specifically costs related to lost work time, physical therapy, and arthroscopic complications.

In conclusion, using patient outcome as the gold standard, MRI is cost-effective compared to diagnostic arthroscopy in patients with knee complaints consistent with internal derangement. The results of MR imaging can be used as a guide for the management of knee pain.

REFERENCES

1. Fowler PJ. The predictive value of five clinical signs in the evaluation of meniscal pathology. *Arthroscopy* 1989;5:184–186.
2. Kaye JJ. Knee arthrography today. *Radiology* 1985;157:265–266.
3. Selesnick FH, Noble HB, Bachman DC, Steinberb F. Internal derangement of the knee: Diagnosis by arthrography, arthroscopy and arthrotomy. *Clin Orthop Rel Res* 1985;198:26–30.
4. Manco LG, Kavanaugh JH, Lozman J, Colman ND, Bilfield BS, Fay FJ. Diagnosis of meniscal tears using high-resolution computed tomography. *J Bone Joint Surg* 1987;69A:498–502.
5. Jackson RW. Arthroscopy of the knee. IN: Evarts CW, ed. *Surgery of the musculoskeketal system*. New York: Churchill Livingstone, 1985.
6. Boden SD, Labaopoulos PA, Vailas JL. MR Scanning of the acutely injured knee: Sensitive, but is it cost effective? *Arthroscopy* 1990;6:306–310.
7. Johnson L. Impact of diagnostic arthroscopy on the clinical judgment of an experienced arthroscopist. *Clin Orthop Rel Res* 1982; 167:75–83.
8. Crues JV III, Mink J, Levy T, Lotysch M, Stoller DW. Meniscal tears of the knee: Accuracy of MR imaging. *Radiology* 1987; 164:445–448.
9. Crues JV, Ryu R, Morgan FW. Meniscal pathology: The expanding role of MRI. *Clin Orthop Rel Res* 1990;252:80–87.
10. Reicher MA, Hartzman S, Bassett LW, Duckwiller GR, Mandelbaum B, Gold RH. MRI imaging of the knee. *Radiology* 1987; 162:547–551.
11. Silva I, Silver DM. Tears of the meniscus as revealed by MRI. *J Bone Joint Surg* 1988;70-A:199–202.
12. Fischer SP, Fox JM, Del Pizzo W, Friedman MJ, Snyder SJ,

Ferkel RD. Accuracy of diagnoses from magnetic resonance imaging of the knee. *J Bone Joint Surg* 1991;73A:2–10.
13. Mink JH, Levy T, Crues JV III. Tears of the anterior cruciate ligament and menisci of the knee: MR imaging evaluation. *Radiology* 1988;167:769–774.
14. Quinn S, Brown T. Evaluation of the menisci: How reliable and relevant is arthroscopic correlation? *Radiology* 1991;181;843–847.
15. Hartzman S, Reicher MA, Bassett LW, Duckwiler GR, Mandelbaum B. MRI imaging of the knee. *Radiology* 1987;162:553–557.
16. Fuchs V, Garber A. The new technology assessment. *N Engl J Med* 1990;323:673–677.
17. Rettig R. Technology assessment. An update. *Invest Radiol* 1991;26:165–173.
18. Wright J, McCarthy S, Lynch K, Ruwe P, Randall L, Jokl P. letter to the editor. *Arthroscopy* 1991;7:336–337.
19. Ruwe P, McCarthy S, Wright J, Randall L, Lynch K, Jokl P. Does MR imaging effectively replace diagnostic arthroscopy? *Radiology* 1992;183:335–339.
20. Hede A, Hempel-Poulson S, Jensen JS. Symptoms and level of sports activity in patients awaiting arthroscopy for meniscal lesions of the knee. *J Bone Joint Surg* 1990;72A:550–552.
21. Boeree NR, Watkinson AF, Ackroyd CE, Johnson C. Magnetic resonance imaging of meniscal injuries of the knee. *J Bone Joint Surg* 1991;73B:452–457.

Subject Index